MANY-DIMENSIONAL MODAL LOGICS: THEORY AND APPLICATIONS

STUDIES IN LOGIC

AND

THE FOUNDATIONS OF MATHEMATICS

VOLUME 148

ELSEVIER
NORTH
HOLLAND

AMSTERDAM • BOSTON • HEIDELBERG • LONDON • NEW YORK • OXFORD
PARIS • SAN DIEGO • SAN FRANCISCO • SINGAPORE • SYDNEY • TOKYO

MANY-DIMENSIONAL MODAL LOGICS: THEORY AND APPLICATIONS

D.M. GABBAY
Department of Computer Science, King's College
Strand, London WC2R 2LS
United Kingdom

A. KURUCZ
Department of Computer Science, King's College
Strand, London WC2R 2LS
United Kingdom

F. WOLTER
Department of Computer Science, University of Liverpool
Peach Street, Liverpool L69 7ZF
United Kingdom

M. ZAKHARYASCHEV
Department of Computer Science, King's College
Strand, London WC2R 2LS
United Kingdom

ELSEVIER
NORTH
HOLLAND

AMSTERDAM • BOSTON • HEIDELBERG • LONDON • NEW YORK • OXFORD
PARIS • SAN DIEGO • SAN FRANCISCO • SINGAPORE • SYDNEY • TOKYO

ELSEVIER SCIENCE B.V.
Sara Burgerhartstraat 25
P.O. Box 211, 1000 AE Amsterdam, The Netherlands

First edition 2003

Library of Congress Cataloging in Publication Data
A catalog record from the Library of Congress has been applied for.

British Library Cataloguing in Publication Data
A catalogue record from the British Library has been applied for.

ISBN: 0 444 50826 0
ISSN: 0049-237X

Transferred to digital printing 2006

With love and gratitude to our parents
this book is dedicated

'You've got to learn to think multi-dimensionally...

If you'd like to know, I can tell you that in your universe you move freely in three dimensions that you call space. You move in a straight line in a fourth, which you call time, and stay rooted to one place in a fifth, which is the first fundamental of probability. After that it gets a bit complicated, and there's all sorts of stuff going on in dimensions 13 to 22 that you really wouldn't want to know about.'

Douglas Adams 'Mostly Harmless'

Preface

Modal logic is a discipline of many facets. It was baptized in philosophy, and for a long time it was known as 'the logic of necessity and possibility.' The modal analysis of the 'mathematical necessity'—provability—brought modal logic to the foundations of mathematics. The discovery of topological and algebraic semantics for modal logic connected it with general topology and universal algebra, and the fact that first-order logic can be regarded as a propositional modal logic opened a 'modal perspective' in classical mathematical logic. But the most amazing metamorphosis happened when it turned out that modal logic could provide languages for talking about various relational structures, such as state transition systems for computer programs, semantic networks for knowledge representation, or attribute value structures in linguistics—languages that combined both sufficient expressive power and effectiveness! This opened new rich and rapidly growing application fields in computer science, artificial intelligence, linguistics, as well as in mathematics itself.

A great many systems with various kinds of modal operators have been constructed in the last few decades in order to provide effective formalisms for talking about time, space, knowledge, beliefs, actions, obligations, etc.: temporal, spatial, epistemic, dynamic, deontic, and so forth. However, modern applications often require rather complex formal models and corresponding languages that are capable of reflecting different features of the application domain. For instance, to analyze the behavior of a multi-agent distributed system we may need a formalism containing both epistemic operators for capturing the knowledge of agents and temporal operators for taking care of the development of this knowledge in time. In other words, we should construct a suitable *combination* of epistemic and temporal logics. Borrowing the geometrical terminology, one can call the resulting hybrid a *many-dimensional modal logic* (later on we shall see that this name is not merely a nice metaphor).

The algorithmic properties of such many-dimensional hybrids that appeared more or less independently and with different motivations in computer science, artificial intelligence, algebraic logic, and modal logic, turn out to be quite different from those of their well-known and well-behaved one-dimensional components. In particular, the complexity of decision algorithms may increase dramatically, even up to undecidability; two fairly simple finitely axiomatizable systems may give rise to a hybrid which is not even recursively enumerable, etc.

To study the computational behavior of many-dimensional modal logics is the main aim of this book. More precisely, as suggested by its title, our aim is twofold. On the one hand, we are concerned with providing a solid mathematical foundation for the discipline characterized in (Blackburn *et al.* 2001) as

> ...*a branch of modal logic dealing with special relational structures in which the states, rather than being abstract entities, have some inner structure. More specifically, these states are tuples or sequences over some base set* ... *Furthermore, the accessibility relations between these states are (partly) determined by this inner structure of the states.*

On the other hand, we show that many seemingly different applied many-dimensional systems (e.g., multi-agent systems, description logics with epistemic, temporal and dynamic operators, spatio-temporal logics, (fragments of) first-order temporal or epistemic logics, etc.) fit in perfectly with this theoretical framework, and so their computational behavior can be analyzed using the developed machinery. Thus, we contribute to filling in the gap between the mathematical theory of modal logic and applications in computer science and artificial intelligence, which were developing in parallel, often independently of each other (witness description logics created in the field of knowledge representation and proved to be terminological variants of well-known modal logics). This gap is clearly reflected in the existing literature. Take, for example, two recent books: (Fagin *et al.* 1995) is an excellent exposition of applied epistemic logic, but it avoids difficult proofs, say, the complexity results are only formulated; the first monograph on multi-dimensional modal logics (Marx and Venema 1997), on the contrary, considers mostly mathematical aspects of modal systems originating in algebraic logic.

This book also reflects a new direction in applied logic in general and in modal logic in particular that has become apparent in the last few years: we mean the direction towards constructing and investigating complex combined systems out of relatively simple ones. It has manifested itself in a number of international conferences (e.g., 'Frontiers of Combining Systems' FroCoS'96–FroCoS'02) and subsequent volumes (Baader and Schulz 1996, de Rijke and Gabbay 2000, Kirchner and Ringeissen 2000, Armando 2002), special issues of *Notre Dame Journal of Formal Logic* (de Rijke and Blackburn 1996) and *Studia Logica* (Gabbay and Pirri 1997, Kurucz *et al.* 2002), and the monograph (Gabbay 1999).

Complex logical systems and knowledge representation formalisms present many challenging problems for investigation. In this book we concentrate on three of them that are regarded as fundamental in both mathematical logic and theoretical computer science. These are:

- The *decision problem*, i.e., to find out whether there exists an algorithm that is capable of deciding a given reasoning task (say, satisfiability or entailment) for a logic.
- The *complexity problem*, i.e., to find lower and upper bounds for the computational complexity of a possible algorithmic solution to the decision problem.
- The *axiomatization problem*, i.e., to give an effective (preferably finite) syntactical characterization of a semantically defined logic, or to provide an adequate semantics for a syntactically defined one.

The first two problems are concerned with *effectiveness* (or programmability) and *efficiency* (or fast programmability) of logical systems, which make them fundamental in artificial intelligence and other practical fields of computer science as well. The third problem is connected with the proof-theoretic approach: it can be understood as describing the most essential features of a logic starting from which all others are derivable. Thus, the axiomatization problem underpins possible implementations of decision procedures.

The direct *practical* use of decidability, complexity and axiomatizability results may be not that obvious.

'Negative' results, say undecidability, are clearly useful: they warn us not to waste time with implementing this or that 'decision' algorithm. These results lead to a new research programme: (1) to find semi-decision (i.e., sound but incomplete) procedures, (2) to search for decidable fragments of the logic in question, (3) to modify the logic by making it decidable, etc. Similarly, a result establishing a high computational complexity or non-axiomatizability may force the researcher to devise another language to model her application domain.

On the other hand, a positive decidability result does not yet guarantee that trying to implement a decision procedure is not a waste of time: after all the British Museum algorithm is also a decision procedure. It may seem that the only use of a positive solution to the decision problem is the conclusion that it is not provable (with the existing concepts in recursion and complexity theory) that implementing a decision procedure is hopeless.[1] This interpret-

[1]Even this cautious claim may appear too strong. Here, for example, is a citation from (Johnson 1990): 'Even those unimpressed with the difficulty of problems in 2- or 3-EXPTIME will have to admit that if a problem is decidable but not in ELEMENTARY, it might as well not be decidable at all.' So, according to Johnson, there are concepts in com-

ation of decidability results can hardly justify spending an enormous amount of energy for obtaining them.

Actually, it is not only the result itself, not only the first 'sanity check' of the system, not only an indication of how to implement a decision procedure that motivates the researcher, but also a deep insight into properties of the logical system that can be extracted from decidability and axiomatizability *proofs*.[2] It is the *proof* of a complexity result that provides the researcher with insights into the sources of high computational costs and can be used to guide the search for more efficient languages—see (Gurevich 1990, 1995) for a similar position. To conclude this discussion: we believe that the decidability, complexity, and axiomatizability theorems we present in this book are of interest mainly because their proofs provide a better understanding of the logical formalisms we consider and, at the same time, are often useful for the design of practical procedures (to a certain extent, this is illustrated in Chapter 15 which presents tableau calculi for modal description logics).

The book is organized in the following way. Part I may serve as an easy introduction to modal logic and its applications. Chapters 1 and 2 introduce the basic modal logics we deal with, explain their roots, motivations, syntax, semantics and application fields. At the end of Chapter 2 we show useful 'semantical level' reductions between many of these logics. In Chapter 3, we consider a number of many-dimensional systems constructed in logic, artificial intelligence and computer science, and establish connections between the various formalisms.

Part II is the technical core of the book. Here we develop a mathematical theory for handling a spectrum of 'abstract' combinations of modal logics, ranging from fusions (known also as independent joins or dovetailings), in which the modal operators of different components do not interact at all, to products of modal logics, where such an interaction is rather strong. The ideas, tools and techniques developed in Part II as well as the obtained results will be used in the subsequent two parts to investigate the computational behavior of first-order modal and temporal logics and some combined knowledge representation formalisms.

In Part III we consider first-order modal and temporal logics in the two-

---

plexity theory which allow us to classify certain decidable problems as 'non-implementable' ones. We do not agree with this for the simple reason that a non-elementary problem may be solvable in polynomial time in all *practical cases*. For certain highly complex problems a useful and *complete* decision algorithm may exist; see e.g. (Horrocks *et al.* 1999, Hustadt and Schmidt 2000). Klarlund *et al.* (2002), writing about the prover MONA in monadic second-order logic, observe that 'perhaps surprisingly, this [NONELEMENTARY] complexity also contributes to successful applications, since it is provably linked to the succinctness of the logic.'

[2]That is why decidability results obtained by means of embeddings into extremely expressive formalisms (like monadic second-order logics $S1S$ or $S\omega S$) may appear somewhat disappointing.

dimensional perspective. It has been known since the 1960s that it is extremely hard to deal with these logics (and that they can be very important for applications). But in contrast to classical predicate logic, where the early undecidability results of Turing and Church stimulated research and led to a rich and profound theory concerned with classifying fragments of first-order logic according to their decidability (see, e.g., Börger *et al.* 1997), there were no serious attempts to convert the 'negative' results in first-order modal and temporal logic into a classification problem. Apparently, the extremely weak expressive power of the modal and temporal formulas required to prove undecidability left no hope that any useful decidable fragments located 'between' propositional and first-order modal and temporal logics could ever be found. However, the studies of many-dimensional propositional modal logics have brought a new insight into the first-order case. In Part III we present a number of recent results concerning decidable and axiomatizable fragments of various first-order modal and temporal logics, and try to draw a borderline between 'the decidable' and 'the undecidable.'

Part IV applies the developed techniques and obtained results to analyze the computational behavior of three kinds of knowledge representation formalisms: temporal epistemic logics, description logics with modal and temporal operators, and spatio-temporal logics. In particular, we show how the method of quasimodels, developed for proving decidability, can be used for devising tableau decision procedures for some of these logics.

The genre of the book can be defined as a research monograph. It brings the reader to the front line of current research in the field by showing both recent achievements and directions of future investigations (in particular, multiple open problems). On the other hand, well-known results from modal and first-order logic are formulated without proofs and supplied with references to accessible sources.

The intended audience of this book is primarily those researchers who use logic in computer science and artificial intelligence. More specific areas are, e.g., knowledge representation and reasoning, in particular, terminological, temporal and spatial reasoning, or reasoning about agents. For 'pure' logicians Parts II and III may be of chief interest. Logicians looking for possible applications may find some useful ideas in Parts I and IV. And we also believe that researchers from certain other disciplines, say, temporal and spatial databases or geographical information systems, will benefit from this book as well.

We conclude this preface by putting the subject of the book into a more general perspective of ongoing and future research. As we have said above, many-dimensional modal logics are just one example of 'combined logical systems.' And even within this smaller area there may be different ways of constructing complex logics out of relatively simple ones. Our approach is basically semantical: given two (or more) classes of 'one-dimensional' Kripke frames characterizing logics L_i, we construct a class of two- (or higher-) dimensional Kripke structures reflecting some desirable features of the target combination of the L_i, and then investigate the logic determined by this class. However, logics do not always come equipped with their Kripke semantics. They may be given by some kind of algebraic structures or purely syntactically, as Hilbert-, Gentzen-, tableau-, resolution-, etc. style calculi. Thus, we need a spectrum of *methodologies* providing us with means of combining homogeneously given logics (say, tableau systems) and perhaps meta-methodologies for combining methodologies. These challenging problems are far beyond the scope of this book; many of them are still open for investigation.

Acknowledgments: We are indebted to our friends and colleagues Ian Hodkinson, Robin Hirsch, Carsten Lutz, Roman Kontchakov and Valentin Shehtman who not only helped us with numerous comments, but also supplied us with some material for this book. Thanks are also due to Alessandro Artale, Franz Baader, Brandon Bennett, Alexander Chagrov, Anthony Cohn, Stéphane Demri, Ulle Endriss, Jim Farrugia, Enrico Franconi, David Gabelaia, Maarten Marx, Szabolcs Mikulás, Milenko Mosurovich, Alexander Rabinovich, Mark Reynolds, Aleksey Romanov, Holger Sturm, Yde Venema, and Rostislav Yavorski for stimulating discussions and remarks on the preliminary versions of the book.

In different periods, the work on the book was supported by UK EPSRC grants no. GR/L85978, GR/M36748 GR/N23028/01, GR/R45369/01 and GR/R42474/01, by Hungarian National Foundation for Scientific Research grants no. T30314 and 035192, by the DFG-grants Wo 583/3-1, Wo 583/3-3, and by grant no. 99–01–00968 from the Russian Foundation for Basic Research.

King's College London D.G.
King's College London A.K.
University of Liverpool F.W.
King's College London M.Z.

Leipzig–Liverpool–London
Winter 2002–2003

Contents

I Introduction 1

1 Modal logic basics **3**

1.1 Modal axiomatic systems 3
1.2 Possible world semantics 9
1.3 Classical first-order logic and the standard translation 15
1.4 Multimodal logics 20
1.5 Algebraic semantics 27
1.6 Decision, complexity and axiomatizability problems 31

2 Applied modal logic **41**

2.1 Temporal logic 41
2.2 Interval temporal logic 49
2.3 Epistemic logic 55
2.4 Dynamic logic 61
2.5 Description logic 65
2.6 Spatial logic 78
2.7 Intuitionistic logic 92
2.8 'Model level' reductions between logics 96

3 Many-dimensional modal logics **111**

3.1 Fusions 111
3.2 Spatio-temporal logics 114
3.3 Products 125
3.4 Temporal epistemic logics 134
3.5 Classical first-order logic as a propositional multimodal logic . 141
3.6 First-order modal logics 143
3.7 First-order temporal logics 157
3.8 Description logics with modal operators 164
3.9 **HS** as a two-dimensional logic 179
3.10 Modal transition logics 180
3.11 Intuitionistic modal logics 187

II Fusions and products 195

4 Fusions of modal logics **197**
4.1 Preserving Kripke completeness and the finite model property . 197
4.2 Algebraic preliminaries 200
4.3 Preserving decidability of global consequence 205
4.4 Preserving decidability 208
4.5 Preserving interpolation 216
4.6 On the computational complexity of fusions 218

5 Products of modal logics: introduction **221**
5.1 Axiomatizing products 222
5.2 Proving decidability with quasimodels 235
5.3 The finite model property 244
5.4 Proving undecidability 252
5.5 Proving complexity with tilings 262

6 Decidable products **273**
6.1 Warming up: $\mathbf{K}_n \times \mathbf{K}_m$ 274
6.2 **CPDL** $\times$ $\mathbf{K}_m$ 285
6.3 Products of epistemic logics with $\mathbf{K}_m$ 297
6.4 Products of temporal logics with $\mathbf{K}_m$ 303
6.5 Products with **S5** 319
6.6 Products with multimodal **S5** 334

7 Undecidable products **343**
7.1 Products of linear orders with infinite ascending chains 344
7.2 Products of linear orders with infinite descending chains 354
7.3 Products of Dedekind complete linear orders 361
7.4 Products of finite linear orders 364
7.5 More undecidable products 367

8 Higher-dimensional products **377**
8.1 $\mathbf{S5} \times \mathbf{S5} \times \cdots \times \mathbf{S5}$ 379
8.2 Products between $\mathbf{K4}^n$ and $\mathbf{S5}^n$ 391
8.3 Products with the fmp 394
8.4 Between $\mathbf{K}^n$ and $\mathbf{S5}^n$ 396
8.5 Finitely axiomatizable and decidable products 410

9 Variations on products **417**
9.1 Relativized products 418
9.2 Valuation restrictions 434

10 Intuitionistic modal logics **439**

10.1 Intuitionistic modal logics with $\Box$ 439

10.2 Intuitionistic modal logics with $\Box$ and $\Diamond$ 445

10.3 The finite model property 453

III First-order modal logics 461

11 Fragments of first-order temporal logics **465**

11.1 Undecidable fragments 465

11.2 Monodic formulas, decidable fragments 471

11.3 Embedding into monadic second-order theories 481

11.4 Complexity of decidable fragments of $\mathsf{QLog}_{\mathcal{SU}}(\mathbb{N})$ 488

11.5 Satisfiability in models over $\langle\mathbb{N},<\rangle$ with finite domains 501

11.6 Satisfiability in models over $\langle\mathbb{R},<\rangle$ with finite domains 509

11.7 Axiomatizing monodic fragments 527

11.8 Monodicity and equality 538

12 Fragments of first-order dynamic and epistemic logics **547**

12.1 Decision problems 547

12.2 Axiomatizing monodic fragments 550

IV Applications to knowledge representation 565

13 Temporal epistemic logics **567**

13.1 Synchronous systems 569

13.2 Agents who know the time and neither forget nor learn 575

14 Modal description logics **583**

14.1 Concept satisfiability 585

14.2 General formula satisfiability 591

14.3 Restricted formula satisfiability 605

14.4 Satisfiability in models with finite domains 610

15 Tableaux for modal description logics **615**

15.1 Tableaux for $\mathcal{ALC}$ 616

15.2 Tableaux for $\mathbf{K}_{\mathcal{ALC}}$ with constant domains 623

15.3 Adding expressive power to $\mathbf{K}_{\mathcal{ALC}}$ 640

16 Spatio-temporal logics **647**

16.1 Modal formalisms for spatio-temporal reasoning 648

16.2 Embedding spatio-temporal logics in first-order temporal logic 654

16.3 Complexity of spatio-temporal logics 661

16.4 Models based on Euclidean spaces 672

Epilogue **679**

Bibliography **685**

List of tables **725**

List of languages and logics **727**

Symbol index **733**

Subject index **737**

Part I

Introduction

Chapter 1

Modal logic basics

This chapter can serve as a concise introduction to modal logic. We define a number of basic modal systems, introduce the possible world semantics for propositional multimodal logics, establish connections with classical first-order logic, and discuss the decision, complexity and axiomatization problems which will be investigated later on in the book for much more complex many-dimensional modal systems. As all the results of this chapter are well documented in the accessible literature, we omit the proofs and provide the reader with references to available textbooks. The reader familiar with elements of modal logic can safely skip this introduction and proceed with Chapter 2.

1.1 Modal axiomatic systems

Modal logic originated in philosophy. The creator of the first modal systems, C.I. Lewis (1918, 1932), constructed them as an auxiliary tool in his attempts to solve the paradoxes of 'material' (i.e., Boolean) implication.[1] His idea was to replace the material implication 'if φ then ψ' with the 'strict' implication 'it is necessary that if φ then ψ'. And for this purpose Lewis constructed five axiomatic systems with simple names: **S1**–**S5**.[2] It seems that the only intuition behind them was whether they could help to get rid of the paradoxes. In any case, Lewis never clarified his understanding of the notions of *necessity* and *possibility*. Yet, at least two of his systems, **S4** and **S5**, became celebrities in *modal logic*.

[1]Here is one such paradox: 'If the moon is made of green cheese then $2 \times 2 = 4$.' We have to regard this statement as true in Boolean logic if we agree that $2 \times 2 = 4$. Those who want to learn more about the paradoxes of material implication are referred to (Zeman 1973) and (Anderson and Belnap 1975).

[2]Actually, **S5** was introduced before Lewis by H. McColl (1906).

Approximately at the same time when Lewis formulated **S4** in (Lewis and Langford 1932), the very same logic was also constructed by Orlov (1928) and Gödel (1933). However, their aim was different. Both of them tried to interpret the intuitionistic logic of Brouwer by embedding it into classical logic extended with an operator 'it is provable.'[3]

Unlike Lewis who used the necessity operator implicitly, having hidden its properties in strict implication, Orlov and Gödel added it to classical propositional logic[4] explicitly, thus arriving to the *propositional modal language* which will be denoted in this book by $\mathcal{ML}$.

The *alphabet* of $\mathcal{ML}$ consists of

- a (fixed, countably infinite) list $p_0, p_1, \dots$ of *propositional variables*;
- the *logical constants*: $\top$ ('true') and $\bot$ ('false');
- the *Boolean logical connectives*: $\wedge$ ('and'), $\vee$ ('or'), $\rightarrow$ ('implies'), and $\neg$ ('not');
- the *modal operators*: $\Box$ ('it is necessary') and $\Diamond$ ('it is possible');
- the *punctuation symbols*:) and (.

Propositional variables can be thought of as ranging over arbitrary propositions—sentences in some (say, natural) language whose content can be evaluated as true or not true. Starting from these variables and the logical constants, we construct inductively *well-formed formulas* of $\mathcal{ML}$ (*$\mathcal{ML}$-formulas*, for short, or simply *formulas*, if $\mathcal{ML}$ is understood) intended for representing compound propositions:

- all propositional variables and the constants $\top$ and $\bot$ are *$\mathcal{ML}$-formulas* (these are called *atomic formulas* or simply *atoms*);
- if φ and ψ are $\mathcal{ML}$-formulas then so are $(\varphi \wedge \psi)$, $(\varphi \vee \psi)$, $(\varphi \rightarrow \psi)$, $(\neg\varphi)$, $(\Box\varphi)$, and $(\Diamond\varphi)$.[5]

Sometimes we use $(\varphi \leftrightarrow \psi)$ as an abbreviation for $((\varphi \rightarrow \psi) \wedge (\psi \rightarrow \varphi))$. In our metalanguage, we may denote propositional variables by lower case Roman letters like p, q, r, possibly with subscripts or superscripts; lower case Greek letters like φ, ψ, χ are reserved for formulas, and upper case letters Σ, Δ, etc. for sets of formulas. To simplify notation, we use the following

[3]We discuss intuitionistic logic and this embedding in Section 2.7.

[4]Orlov's system was based on a logic weaker than classical propositional logic; actually, it was the first system of *relevant logic*.

[5]To be absolutely precise, we should add here that no other objects, different from those defined above, can be called $\mathcal{ML}$-formulas. We will never formulate statements like this explicitly, relying upon the reader's common sense.

standard conventions on formula representation: we assume $\neg$, $\Box$ and $\Diamond$ to bind formulas stronger than $\wedge$ and $\vee$, which in turn are stronger than $\rightarrow$ and $\leftrightarrow$, and omit those brackets that can be uniquely recovered according to this priority of connectives. Thus, the $\mathcal{ML}$-formula

$$(\Box((p_0 \wedge p_1) \rightarrow ((\Diamond p_0) \vee p_1)))$$

can be shortened to

$$\Box(p_0 \wedge p_1 \rightarrow \Diamond p_0 \vee p_1).$$

Instead of $(\dots((\varphi_1 \vee \varphi_2) \vee \varphi_3) \vee \dots \vee \varphi_n)$ and $(\dots((\varphi_1 \wedge \varphi_2) \wedge \varphi_3) \wedge \dots \wedge \varphi_n)$ we write, respectively, $\varphi_1 \vee \varphi_2 \vee \dots \vee \varphi_n$ and $\varphi_1 \wedge \varphi_2 \wedge \dots \wedge \varphi_n$, or $\bigvee_{i=1}^n \varphi_i$ and $\bigwedge_{i=1}^n \varphi_i$. By definition, $\bigvee_{i\in\emptyset} \varphi_i$ is $\bot$, while $\bigwedge_{i\in\emptyset} \varphi_i$ is $\top$.

Given a formula φ, we write $\varphi(q_1, \dots, q_n)$ to indicate that all propositional variables occurring in φ are among $q_1, \dots, q_n$; $sub\,\varphi$ denotes the set of all *subformulas* of φ (i.e., the formulas used in the construction of φ according to the definition above, including φ itself). Say, if φ is $\Box(p_0 \wedge p_1 \rightarrow \Diamond p_0 \vee p_1)$ then

$$sub\,\varphi = \{p_0, p_1, p_0 \wedge p_1, \Diamond p_0, \Diamond p_0 \vee p_1, p_0 \wedge p_1 \rightarrow \Diamond p_0 \vee p_1, \varphi\}.$$

A logical system in general, and a modal system in particular, is supposed to single out and describe those formulas that represent certain 'true' propositions no matter what values (propositions) are assigned to their variables. There are two main ways of defining logics: *semantical* and *syntactical*. Usually, the semantical and syntactical definitions complement each other: the former explains the (intended) meaning of the logical constants and connectives, while the latter provides us with a reasoning machinery.

We illustrate the semantical approach by reminding the reader of the *classical semantics* of the sublanguage $\mathcal{L}$ of $\mathcal{ML}$ that results by omitting the modal operators $\Box$, $\Diamond$ and all formulas containing them. There is a very simple interpretation of this language based on the assumption that every proposition is either true or false. Having assigned one of these truth-values T (for true) or F (for false) to each propositional variable, we can then compute the truth-value of an $\mathcal{L}$-formula (under that assignment) using the well-known '*Boolean truth-tables*,' reflecting the above readings of the logical connectives:

ψ	χ	$\psi \wedge \chi$	$\psi \vee \chi$	$\psi \rightarrow \chi$	$\neg\psi$
F	F	F	F	T	T
F	T	F	T	T	T
T	F	F	T	F	F
T	T	T	T	T	F

(of course, the logical constants $\top$ and $\bot$ are always evaluated as T and F, respectively). *Classical propositional logic* **Cl** can be defined then as the set of all those $\mathcal{L}$-formulas that are true under every such assignment.

Now, returning to modal logic, we see that this semantical definition of **Cl** cannot be extended to the modal language in a straightforward way. The apparent reason is that the modal operators are not *truth-functional*: the truth-value of a formula of the form $\Box\varphi$ can depend not only on whether φ is true or false. For example, we most likely agree that the proposition 'it is necessary that $2 \times 2 = 4$' is true, while 'it is necessary that NATO bombs Belgrade' is undoubtedly false, although both propositions '$2 \times 2 = 4$' and 'NATO bombs Belgrade' are true.[6]

Perhaps this is one of the reasons why the first modal logics were defined in another, *syntactical*, way with the help of *inference systems* (*calculi*). In this book we consider mainly *Hilbert-style inference systems*.[7] To define such a system, one has to indicate which formulas are regarded as *axioms* of the system and to specify its *inference rules*. A *derivation of a formula* φ in the system is a finite sequence of formulas ending with φ and such that each formula in the sequence is either an axiom or obtained from earlier formulas in the sequence by applying one of the inference rules. The *logic* of this inference system is defined then as the set of all derivable formulas. To put it another way, the logic defined by the system is the smallest set of formulas which contains the axioms and is closed under the inference rules.

For example, classical propositional logic **Cl** can be defined by the following Hilbert-style calculus:

Axioms:

(A1) $p_0 \to (p_1 \to p_0)$,
(A2) $(p_0 \to (p_1 \to p_2)) \to ((p_0 \to p_1) \to (p_0 \to p_2))$,
(A3) $p_0 \wedge p_1 \to p_0$,
(A4) $p_0 \wedge p_1 \to p_1$,
(A5) $p_0 \to (p_1 \to p_0 \wedge p_1)$,
(A6) $p_0 \to p_0 \vee p_1$,
(A7) $p_1 \to p_0 \vee p_1$,
(A8) $(p_0 \to p_2) \to ((p_1 \to p_2) \to (p_0 \vee p_1 \to p_2))$,
(A9) $\bot \to p_0$,
(A10) $p_0 \vee (p_0 \to \bot)$.

Inference rules:

Modus ponens (MP): given formulas φ and $\varphi \to \psi$, derive ψ;

Substitution (Subst): given a formula $\varphi(p_1, \dots, p_n)$, derive the formula $\varphi\{\psi_1/p_1, \dots, \psi_n/p_n\}$ which is obtained by uniformly substituting formulas $\psi_1, \dots, \psi_n$ instead of the variables $p_1, \dots, p_n$ in φ, respectively.

[6] This was written on 27 April 1999.

[7] For other kinds of modal inference systems the reader is referred to (Fitting 1983, Wansing 1996) and references therein; see also Chapter 15 below.

As the well-known *soundness and completeness theorem* of classical propositional logic says, the logic defined by this calculus coincides with **Cl** (see e.g., Chagrov and Zakharyaschev 1997, Enderton 1972).

The above calculus does not involve the connective $\neg$ and the constant $\top$. We can define them as abbreviations:

$$\neg\varphi = \varphi \to \bot, \qquad \top = \bot \to \bot.$$

(These abbreviations 'agree' with the classical semantics in the sense that the truth-values of the left-hand and the right-hand sides of these equalities are the same under any assignment.) Moreover, in classical logic we can further reduce the number of basic logical connectives, say, to $\wedge$ and $\neg$, or to $\vee$ and $\neg$, by defining

$$\varphi \to \psi = \neg(\varphi \wedge \neg\psi), \qquad \varphi \vee \psi = \neg(\neg\varphi \wedge \neg\psi), \qquad \bot = p_0 \wedge \neg p_0$$

(a corresponding inference system can be found e.g. in (Shoenfield 1967)). Throughout the book we will often use this fact to shorten inductive definitions and proofs.

Let us now return to modal logic. If we agree to accept the reasoning principles of classical propositional logic, then modal calculi can be constructed by adding to the Hilbert-style calculus for **Cl** those axioms and inference rules that reflect our understanding of the modal operators. A set of $\mathcal{ML}$-formulas which contains the axioms (A1)–(A10), the modal axiom

(K) $\quad \Box(p_0 \to p_1) \to (\Box p_0 \to \Box p_1)$,

and is closed under MP, Subst, and the rule of

Necessitation (RN): given φ, derive $\Box\varphi$

is called a *modal logic*.[8] The possibility operator $\Diamond$ can be regarded as an abbreviation for $\neg\Box\neg$ (or, we can add the axiom $\Diamond p_0 \leftrightarrow \neg\Box\neg p_0$). The *minimal modal logic* is denoted by **K**: it is defined by the inference system having (A1)–(A10) and (K) as its axioms and MP, Subst and RN as its inference rules. Every other modal logic L can be obtained by extending this system with a (possibly infinite) set Σ of *extra axioms*. In this case we write

$$L = \mathbf{K} \oplus \Sigma.$$

If Σ can be chosen finite, then we call L *finitely axiomatizable*. In general, given a modal logic L and a set Δ of $\mathcal{ML}$-formulas,

$$L \oplus \Delta$$

[8] Actually, such logics are usually called *normal* modal logics. We omit the epithet 'normal' because no non-normal modal logics are considered in this book.

denotes the smallest modal logic containing $L \cup \Delta$. We write $L \oplus \varphi$ whenever $\Delta = \{\varphi\}$. Using this notation, we can define the Lewis systems **S4** and **S5** as follows:

$$\begin{aligned}\mathbf{S4} &= \mathbf{K} \oplus \Box p_0 \to \Box\Box p_0 \oplus \Box p_0 \to p_0,\\ \mathbf{S5} &= \mathbf{S4} \oplus \Diamond p_0 \to \Box\Diamond p_0.\end{aligned}$$

These axioms and rules of **S4** were first introduced by Orlov (1928) and Gödel (1933) in order to characterize the operator 'it is provable.' For example, $\Box p_0 \to \Box\Box p_0$ means that, given a proof of p_0, we can prove that it is indeed a proof, and $\Box p_0 \to p_0$ says that everything provable is true. Gödel observed, however, that the $\Box$ of **S4** cannot be understood as the formal provability in axiomatic theories like Peano Arithmetic *PA* (the formula $\Box\neg\Box\bot$, provable in **S4**, would mean then that *PA* can prove its own consistency, contrary to Gödel's second incompleteness theorem). This observation gave rise to a new branch of mathematical logic—provability logic—studying the laws of formal provability that are provable in *PA* and other theories (see, e.g., Boolos 1993). One of the most important modal systems constructed in provability logic is known as the *Gödel-Löb logic* **GL**. It can be obtained from **S4** by replacing $\Box p_0 \to p_0$ with the *Löb axiom*

$$\Box(\Box p_0 \to p_0) \to \Box p_0$$

or, using the above notation,

$$\mathbf{GL} = \mathbf{K} \oplus \Box p_0 \to \Box\Box p_0 \oplus \Box(\Box p_0 \to p_0) \to \Box p_0.$$

Solovay (1976) showed that **GL** adequately describes those properties of Gödel's provability predicate $Bew(x)$ which are provable in *PA*. Recently Artemov (see (Artemov 2001) and references therein) has constructed a *logic of proofs* **LP** extending **Cl** with atomic formulas of the form 't is a proof of φ' and showed that by replacing in **LP** all such formulas (and their subformulas) with $\Box\varphi$ we again get **S4**.

The $\Box$ of **S5** can also be read as 'I know' or 'Mr X believes.' By accepting one or more of the axioms of **S5** as properties of knowledge or belief we can obtain new modal systems, like **T** and **K4**:

$$\begin{aligned}\mathbf{T} &= \mathbf{K} \oplus \Box p_0 \to p_0,\\ \mathbf{K4} &= \mathbf{K} \oplus \Box p_0 \to \Box\Box p_0.\end{aligned}$$

A more detailed discussion of these *epistemic logics* can be found in Section 2.3.

The interpretation of $\Box$ as 'it is obligatory' and $\Diamond$ as 'it is permitted' gives another family of modal logics known as *deontic*. It is a natural principle of reasoning about norms (coming from law, moral, etc.) that everything obligatory is permitted. The minimal deontic logic **D** reflecting this principle is defined as

$$\mathbf{D} = \mathbf{K} \oplus \Box p_0 \to \Diamond p_0.$$

We shall see many other modal systems later on in this book. At the moment we have got enough examples to illustrate the semantical side of modal logic.

1.2 Possible world semantics

The provability interpretation of the necessity operator $\Box$ and its relation to intuitionism gave a strong impetus to mathematical studies in modal logic, which resulted, in particular, in establishing connections with algebra and topology by McKinsey and Tarski (1944, 1946, 1948), and finally led to the discovery of relational representations of modal algebras by Jónsson and Tarski (1951); see Section 1.5 for some details. This *relational semantics* was also invented by philosophers: Carnap (1942, 1947), Prior (1957), Kanger (1957a,b), Hintikka (1957, 1961, 1963), and Kripke (1959, 1963a,b) who apparently were not aware of (Jónsson and Tarski 1951).[9] In philosophy, this semantics can be traced back to the Leibnizean understanding of necessity as truth in all *possible worlds*. Let us imagine a system of 'worlds' which can have some alternatives (for instance, as an alternative to our world we can consider another one where NATO does not bomb Belgrade and the coalition forces do not bomb Baghdad). Denoting the *alternativeness relation* by R, we write xRy to say that y is an alternative (or possible) world for x. Every world x 'lives' under the laws of classical logic: an atomic proposition is either true or false in it, and the truth-values of compound non-modal propositions are determined by the Boolean truth-tables. A modal formula $\Box\varphi$ is then regarded to be true in a world x if φ is true in all worlds that are alternative to x; $\Diamond\varphi$ is true in x if φ is true in at least one world y such that xRy. It is not hard to capture this intuitive picture in a precise definition.

Systems of worlds with alternativeness relations can be represented by relational structures $\mathfrak{F} = \langle W, R\rangle$ in which W is a non-empty set and R a binary relation on W. Such structures are known in modal logic as *Kripke frames* or simply *frames*. Elements of W are called *worlds*, *states* or, more neutrally, *points*. If xRy, we say that y is *accessible* from x, or x *sees* y. Other synonyms are: y is a *successor* of x, x is a *predecessor* of y.

A *valuation* in a frame $\mathfrak{F} = \langle W, R\rangle$ is a map $\mathfrak{V}$ associating with each propositional variable p of $\mathcal{ML}$ a set $\mathfrak{V}(p)$ of points in W (which is understood as the set of those worlds where p holds true). A *Kripke model* for $\mathcal{ML}$ is a pair $\mathfrak{M} = \langle \mathfrak{F}, \mathfrak{V}\rangle$, where $\mathfrak{F} = \langle W, R\rangle$ is a frame and $\mathfrak{V}$ a valuation in $\mathfrak{F}$. We

[9] See, however, a footnote at the beginning of Section 2 in (Kripke 1963a).

say that the model $\mathfrak{M}$ is *based on* the frame $\mathfrak{F}$, or that $\mathfrak{F}$ is the *underlying* frame of $\mathfrak{M}$. Let φ be an $\mathcal{ML}$-formula and x a point in W. The *truth-relation* $(\mathfrak{M}, x) \models \varphi$, read as

$$\text{`}\varphi \textit{ is true at } x \textit{ in } \mathfrak{M}\text{,'}$$

is defined by induction on the construction of φ as follows:

$$
\begin{array}{lll}
(\mathfrak{M}, x) \models p & \text{iff} & x \in \mathfrak{V}(p) \ (p \text{ a propositional variable});\\
(\mathfrak{M}, x) \models \top; & & \\
\text{not } (\mathfrak{M}, x) \models \bot; & & \\
(\mathfrak{M}, x) \models \psi \land \chi & \text{iff} & (\mathfrak{M}, x) \models \psi \text{ and } (\mathfrak{M}, x) \models \chi;\\
(\mathfrak{M}, x) \models \psi \lor \chi & \text{iff} & (\mathfrak{M}, x) \models \psi \text{ or } (\mathfrak{M}, x) \models \chi;\\
(\mathfrak{M}, x) \models \psi \to \chi & \text{iff} & (\mathfrak{M}, x) \models \psi \text{ implies } (\mathfrak{M}, x) \models \chi;\\
(\mathfrak{M}, x) \models \neg\psi & \text{iff} & \text{not } (\mathfrak{M}, x) \models \psi;\\
(\mathfrak{M}, x) \models \Box\psi & \text{iff} & (\mathfrak{M}, y) \models \psi \text{ for all } y \in W \text{ such that } xRy;\\
(\mathfrak{M}, x) \models \Diamond\psi & \text{iff} & (\mathfrak{M}, y) \models \psi \text{ for some } y \in W \text{ such that } xRy.
\end{array}
$$

If $(\mathfrak{M}, x) \models \varphi$ does not hold then we write $(\mathfrak{M}, x) \not\models \varphi$ and say that $\mathfrak{M}$ *refutes* φ at x. Instead of $(\mathfrak{M}, x) \models \varphi$ and $(\mathfrak{M}, x) \not\models \varphi$ we write simply $x \models \varphi$ and $x \not\models \varphi$, if $\mathfrak{M}$ is understood. The *truth-set of* φ *in* $\mathfrak{M}$ is defined as $\mathfrak{V}(\varphi) = \{x \in W \mid x \models \varphi\}$.

Let $\mathfrak{M} = \langle \mathfrak{F}, \mathfrak{V} \rangle$ be a model based on the frame $\mathfrak{F} = \langle W, R \rangle$. A formula φ is said to be *true in* $\mathfrak{M}$ (in symbols: $\mathfrak{M} \models \varphi$) if $x \models \varphi$ for all $x \in W$, that is, if $\mathfrak{V}(\varphi) = W$. Dually, φ is *satisfied in* $\mathfrak{M}$ if $\mathfrak{V}(\varphi)$ is not empty. We say that φ is *valid* in the frame $\mathfrak{F}$ (or $\mathfrak{F}$ *validates* φ) and write $\mathfrak{F} \models \varphi$ if $\mathfrak{V}(\varphi) = W$ for every valuation $\mathfrak{V}$ in $\mathfrak{F}$, or equivalently, if φ is true in all models based on $\mathfrak{F}$. And φ is *satisfiable in* $\mathfrak{F}$, if it is satisfied in some model based on $\mathfrak{F}$. It should be clear that φ is valid in $\mathfrak{F}$ iff $\neg\varphi$ is not satisfiable in $\mathfrak{F}$. For a set Γ of $\mathcal{ML}$-formulas, we say that $\mathfrak{F}$ is a *frame for* Γ if all formulas from Γ are valid in $\mathfrak{F}$. In this case we write $\mathfrak{F} \models \Gamma$. A formula φ *is* Γ*-satisfiable* if it is satisfiable in a frame for Γ.

Now we can give a semantical characterization of (at least some) modal logics by establishing a connection between logics and frames. Let $\mathcal{C}$ be an arbitrary class of frames. It is not hard to check that

$$\mathsf{Log}\,\mathcal{C} = \{\varphi \in \mathcal{ML} \mid \forall \mathfrak{F} \in \mathcal{C}\ \mathfrak{F} \models \varphi\}$$

is a modal logic. We call it the *logic of* $\mathcal{C}$.

A modal logic L is said to be *sound with respect to* $\mathcal{C}$ (or $\mathcal{C}$*-sound*) if $\mathfrak{F} \models \varphi$ for all $\varphi \in L$ and all $\mathfrak{F} \in \mathcal{C}$, that is, $L \subseteq \mathsf{Log}\,\mathcal{C}$. L is *complete with respect to* $\mathcal{C}$ (or $\mathcal{C}$*-complete*) if $\varphi \in L$ whenever φ is valid in every frame in $\mathcal{C}$, that is, $\mathsf{Log}\, \mathcal{C} \subseteq L$. We say that L is *determined* (or *characterized*) *by* $\mathcal{C}$ if L is both $\mathcal{C}$-sound and $\mathcal{C}$-complete, that is, $L = \mathsf{Log}\, \mathcal{C}$. If L is determined by some class of frames, we call L *Kripke complete*. It is worth noting that a Kripke complete logic L can be characterized by different classes of frames (we shall see many examples in what follows). If L is Kripke complete then it is clearly determined by the class $\mathsf{Fr}L$ of all frames for L, i.e., $L = \mathsf{Log}\, \mathsf{Fr}L$.

> Although Kripke frames provide us with a rather transparent semantical instrument for dealing with modal languages, this instrument is far from being universal: as was shown by Fine (1974a) and Thomason (1974a), not every modal logic is Kripke complete. Equivalently, there exist two (actually uncountably many; see (Blok 1978) or (Chagrov and Zakharyaschev 1997)) distinct modal logics having precisely the same Kripke frames. It is worth noting, however, that every consistent modal logic L is determined by its *Kripke models* in the sense that $\varphi \notin L$ iff there is a model $\mathfrak{M}$ such that all formulas of L are true in $\mathfrak{M}$, while φ is not. Moreover, L is determined by a single model $\mathfrak{M}_L$ known as the *canonical model* for L: for every formula φ, we have $\varphi \in L$ iff $\mathfrak{M}_L \models \varphi$. For more details and proofs consult (Chagrov and Zakharyaschev 1997, Blackburn *et al.* 2001).

A very attractive feature of the possible world semantics is that many standard modal logics are determined by 'natural' classes of frames. Let us see first what kind of frames correspond to the modal logics introduced in the previous section. First of all, we have:

Theorem 1.1. K *is determined by the class of all frames.*

Before describing frame classes for the other logics, we remind the reader that a binary relation R on a set W is said to be *transitive* if

$$\forall x, y, z \in W(xRy \wedge yRz \rightarrow xRz).$$

R is *reflexive* if

$$\forall x \in W\ xRx.$$

A transitive and reflexive relation on W is called a *quasi-order* on W. We denote by R^* the *reflexive and transitive closure* of a binary relation R on W (in other words, R^* is the smallest quasi-order on W to contain R).

R is *symmetric* if

$$\forall x, y \in W(xRy \rightarrow yRx).$$

A symmetric quasi-order is called an *equivalence relation* on W. If

$$\forall x, y \in W\ xRy,$$

then R is said to be *universal* on W. R is *serial* on W if

$$\forall x \in W \exists y \in W\ xRy.$$

We say that a frame $\mathfrak{F} = \langle W, R \rangle$ is serial, if R is serial on W; $\mathfrak{F}$ is a *quasi-ordered frame* or simply a *quasi-order*, if R is a quasi-order on W, and so forth.

One of the first remarkable results obtained by Kripke (1959, 1963a) was the following completeness theorem (see, e.g., Hughes and Cresswell 1996, Chagrov and Zakharyaschev 1997):

Theorem 1.2. *The logics* **D**, **T**, **K4**, **S4** *and* **S5** *are Kripke complete.*

- Fr**D** *is the class of all serial frames;*
- Fr**T** *is the class of all reflexive frames;*
- Fr**K4** *is the class of all transitive frames;*
- Fr**S4** *is the class of all quasi-ordered frames;*
- Fr**S5** *is the class of all frames with equivalence accessibility relations.*

Note that **S5** is also determined by the class of all universal frames which is a proper subclass of Fr**S5**. The class of serial frames clearly coincides with the class of frames validating the formula $\Diamond\top$; in fact, $\Diamond\top$ is an alternative extra axiom of **D**:

$$\mathbf{D} = \mathbf{K} \oplus \Diamond\top.$$

Frames for **GL** are somewhat more complex. A binary relation R on a set W is said to be *irreflexive* if xRx holds for no $x \in W$. An irreflexive and transitive relation is known as a *strict partial order*. Call a sequence $x_0, x_1, x_2, \dots$ of points in W a *strictly ascending chain* if $x_0 R x_1 R x_2 \dots$ and $x_n \neq x_{n+1}$, for all $n < \omega$. A binary relation R is called *Noetherian* if there is no infinite strictly ascending chain of points in W. The following result is due to Segerberg (1971):

Theorem 1.3. GL *is Kripke complete.* Fr**GL** *is the class of all Noetherian strict partial orders.*

Many other 'mathematically natural' frame classes give rise to 'sensible' modal logics as well. Here are a few examples. The meaning of some of these logics will be explained later on in the book.

$$\begin{aligned}
\mathbf{Alt} &= \mathbf{K} \oplus \Diamond p_0 \to \Box p_0,\\
\mathbf{DAlt} &= \mathbf{Alt} \oplus \Box p_0 \to \Diamond p_0 = \mathbf{D} \oplus \Diamond p_0 \to \Box p_0,\\
\mathbf{KD45} &= \mathbf{K4} \oplus \Box p_0 \to \Diamond p_0 \oplus \Diamond p_0 \to \Box\Diamond p_0,\\
\mathbf{K4.3} &= \mathbf{K4} \oplus \Box(\Box^+ p_0 \to p_1) \vee \Box(\Box^+ p_1 \to p_0),\\
\mathbf{GL.3} &= \mathbf{GL} \oplus \Box(\Box^+ p_0 \to p_1) \vee \Box(\Box^+ p_1 \to p_0),
\end{aligned}$$

$$\begin{aligned}
\mathbf{S4.3} &= \mathbf{S4} \oplus \Box(\Box p_0 \to p_1) \vee \Box(\Box p_1 \to p_0),\\
\mathbf{Grz} &= \mathbf{S4} \oplus \Box(\Box(p_0 \to \Box p_0) \to p_0) \to p_0,\\
\mathbf{Grz.3} &= \mathbf{Grz} \oplus \Box(\Box p_0 \to p_1) \vee \Box(\Box p_1 \to p_0).
\end{aligned}$$

Here, by definition, $\Box^+\varphi = \varphi \wedge \Box\varphi$ and $\Diamond^+\varphi = \varphi \vee \Diamond\varphi$.

A binary relation R on a set W is

- *antisymmetric* if $\forall x, y \in W\ (xRy \wedge yRx \to x = y)$;
- *functional* if $\forall x, y, z \in W\ (xRy \wedge xRz \to y = z)$;
- *Euclidean* if $\forall x, y, z \in W\ (xRy \wedge xRz \to yRz)$;
- *weakly connected* if $\forall x, y, z \in W\ (xRy \wedge xRz \to yRz \vee y = z \vee zRy)$.

A transitive, reflexive and antisymmetric R is called a *partial order*.

Theorem 1.4. *The logics* **Alt**, **DAlt**, **KD45**, **K4.3**, **GL.3**, **S4.3** *and* **Grz** *are Kripke complete.*

$$\begin{aligned}
\mathsf{Fr}\mathbf{Alt} &= \{\mathfrak{F} \mid \mathfrak{F} \textit{ is functional}\};\\
\mathsf{Fr}\mathbf{DAlt} &= \{\mathfrak{F} \mid \mathfrak{F} \textit{ is functional and serial}\};\\
\mathsf{Fr}\mathbf{KD45} &= \{\mathfrak{F} \mid \mathfrak{F} \textit{ is serial, transitive and Euclidean}\};\\
\mathsf{Fr}\mathbf{K4.3} &= \{\mathfrak{F} \mid \mathfrak{F} \textit{ is transitive and weakly connected}\};\\
\mathsf{Fr}\mathbf{GL.3} &= \{\mathfrak{F} \mid \mathfrak{F} \textit{ is a Noetherian weakly connected strict partial order}\};\\
\mathsf{Fr}\mathbf{S4.3} &= \{\mathfrak{F} \mid \mathfrak{F} \textit{ is a weakly connected quasi-order}\};\\
\mathsf{Fr}\mathbf{Grz} &= \{\mathfrak{F} \mid \mathfrak{F} \textit{ is a Noetherian partial order}\};\\
\mathsf{Fr}\mathbf{Grz.3} &= \{\mathfrak{F} \mid \mathfrak{F} \textit{ is a Noetherian weakly connected partial order}\}.
\end{aligned}$$

We defined modal logics as certain *sets* of $\mathcal{ML}$-formulas. It is natural to ask which of the constructed logics is 'stronger' or 'weaker' with respect to the set-theoretic inclusion $\subseteq$. The family of all modal logics together with $\subseteq$ form a structure the algebraists call a *lattice*. **K** is the least (smallest) element of the lattice. The greatest (largest) one is clearly $\mathsf{Log}\,\emptyset$, i.e., the set of all $\mathcal{ML}$-formulas, called the *inconsistent logic* (because it contains both φ and $\neg\varphi$). An interesting observation, due to Makinson (1971), is that there are precisely two maximal (with respect to $\subseteq$) consistent modal logics. These are

$$\mathbf{Verum} = \mathsf{Log}\,\{\bullet\} = \mathbf{K4} \oplus \Box p,$$

$$\mathbf{Triv} = \mathsf{Log}\,\{\circ\} = \mathbf{K4} \oplus \Box p \leftrightarrow p,$$

where $\bullet$ denotes a single irreflexive point (i.e., the frame $\langle\{w\}, \emptyset\rangle$) and $\circ$ a single reflexive point (i.e., the frame $\langle\{w\}, \langle w, w\rangle\rangle$). Thus, every consistent

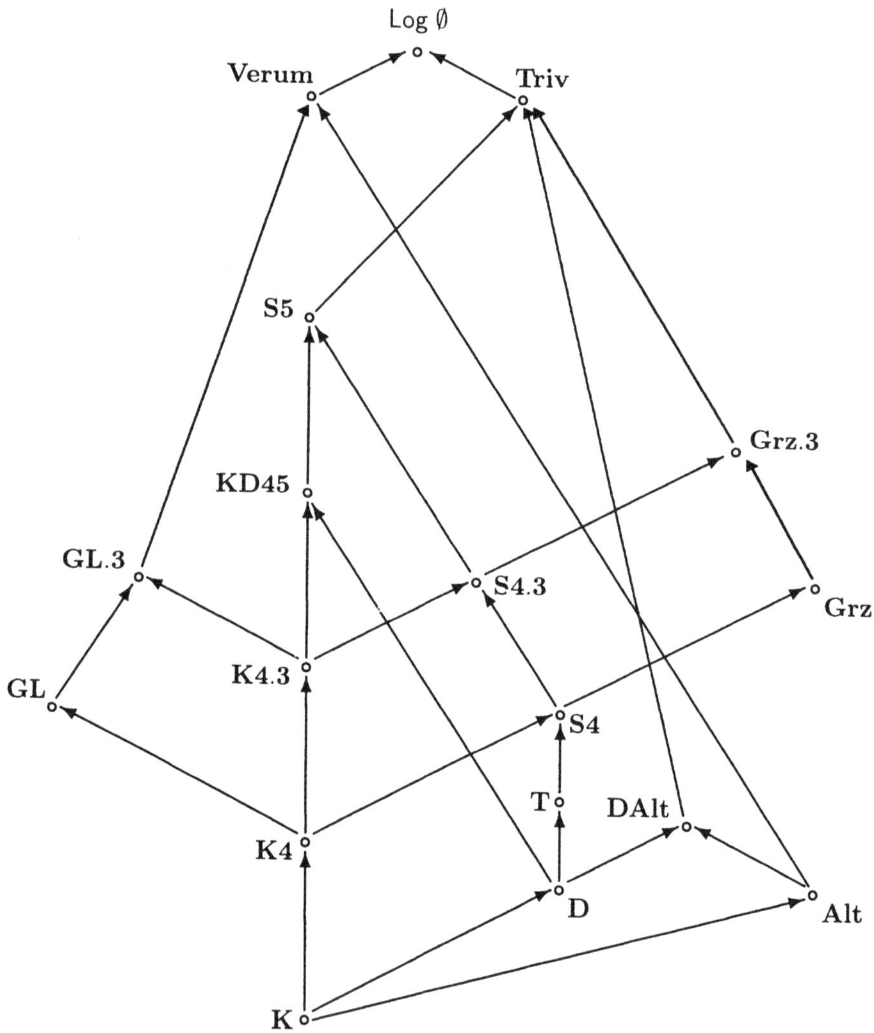

Figure 1.1: Lattice of 'standard' modal logics.

modal logic is contained either in **Verum** or in **Triv** or in both. To put it another way, according to Makinson's theorem, at least one of the frames $\bullet$ or $\circ$ is a frame for every consistent modal logic. As an exercise, the reader can check the correctness of the diagram in Fig. 1.1, where an arrow from L_1 to L_2 means that $L_1 \subseteq L_2$.

1.3 Classical first-order logic and the standard translation

In this book we consider many different logical formalisms. To understand how they are related to each other, to compare their expressive power and thereby to elucidate possible areas of applications are among the main aims of the book.

Perhaps the best known connection of that sort is the *standard translation* which embeds modal languages into the language of *classical first-order* (or *quantified*) *logic*. Although we assume some familiarity with the syntax and semantics of first-order logic,[10] here we give a brief summary of the basic definitions and properties we use later on.

Classical first-order logic

The *first-order* (or *quantified*) *language* $\mathcal{QL}$ we deal with in this book is based on the following alphabet:

- *predicate symbols*: $P_0, P_1, \dots$ (or $P, Q, R, S, \dots$);
- *individual constants*: $c_0, c_1, \dots$ (or $a, b, c, d, \dots$);
- a countably infinite list of *individual variables*: $x_0, x_1, \dots$ (or $x, y, z, \dots$);
- the *logical constants*: $\top$ and $\bot$;
- the *Boolean logical connectives*: $\wedge$, $\vee$, $\rightarrow$ and $\neg$;
- the *universal quantifier* $\forall$;
- the *existential quantifier* $\exists$.

The predicate symbols and the individual constants together form the *signature of* $\mathcal{QL}$. As usual, we assume that each predicate symbol is of some fixed arity ≥ 0, that the signature contains countably infinitely many predicate

[10] Good introductions to first-order logic are, e.g., (Enderton 2001, Shoenfield 1967, Barwise 1977).

symbols of each arity, and that the set of individual constants is also countably infinite. And, of course, we assume that the language $\mathcal{QL}$ is recursive in the sense that we can always effectively recognize its predicate symbols with their arities, individual constants and variables. Sometimes we consider *sublanguages* of $\mathcal{QL}$ with smaller signatures (but the same set of non-signature symbols as in $\mathcal{QL}$).

Note that *equality* $=$ is not a symbol of $\mathcal{QL}$, and that $\mathcal{QL}$ contains no function symbols different from constants. Occasionally we shall use the language $\mathcal{QL}^=$ whose alphabet extends that of $\mathcal{QL}$ with the binary predicate symbol $=$.

Individual variables and constants are also known as *terms*. Formulas of any sublanguage of $\mathcal{QL}$ are defined inductively as follows. If P is an n-ary signature predicate symbol and $\tau_1, \ldots, \tau_n$ are signature terms, then $P(\tau_1, \ldots, \tau_n)$ is an (*atomic*) *formula*. (If P is binary then we sometimes write $\tau_1 P \tau_2$ instead of $P(\tau_1, \tau_2)$.) Logical constants are (atomic) formulas as well.

In the case of $\mathcal{QL}^=$, we also have $\tau_1 = \tau_2$ as atomic formulas for all τ_1 and τ_2.

If φ, ψ are formulas and x an individual variable, then $\varphi \wedge \psi$, $\varphi \vee \psi$, $\varphi \to \psi$, $\neg\varphi$, $\forall x \varphi$ and $\exists x \varphi$ are formulas. The conventions on punctuation and formula representation of Section 1.1 are extended by the following one: $\forall x$ and $\exists x$ have the same priority as $\neg$. An occurrence of a variable x in a formula φ is *bound* if this occurrence lies under the scope of $\forall x$ or $\exists x$; otherwise the occurrence of x is *free*. Formulas without free variables are called *sentences*. If φ is a formula, τ a term, and x a variable, then $\varphi\{\tau/x\}$ denotes the result of the simultaneous substitution of τ for all free occurrences of x in φ. Say that τ *is free for* x *in* φ, if no variable in τ becomes bound in $\varphi\{\tau/x\}$. We write $\varphi(x_0, \ldots, x_n)$ to indicate that all free variables of φ are among $x_0, \ldots, x_n$.

$\mathcal{QL}$ and its sublanguages are interpreted in *first-order structures* (or $\mathcal{QL}$-*structures*, if we want to mention the signature explicitly) of the form

$$I = \left\langle D^I, P_0^I, \ldots, c_0^I, \ldots \right\rangle,$$

where

- D^I is a nonempty set, the *domain* of I;
- for any predicate symbol P_i in the signature, P_i^I is a relation on D^I of the same arity as P_i;
- for any individual constant c_i in the signature, c_i^I is an element of D^I.

An *assignment in* I is a function $\mathfrak{a}$ from the set of individual variables to D^I. The *value* $\mathfrak{a}(\tau)$ *of a term* τ *in* I *under the assignment* $\mathfrak{a}$ is $\mathfrak{a}(x)$ if τ is a variable x, and c^I if τ is a constant c.

The *truth-relation* $I \models^{\mathfrak{a}} \varphi$ (in words: 'φ *is true in* I *under the assignment* $\mathfrak{a}$') is defined by induction on the construction of φ in the following way:

- $I \models^{\mathfrak{a}} P_i(\tau_1, \dots, \tau_n)$ iff $\langle \mathfrak{a}(\tau_1), \dots, \mathfrak{a}(\tau_n) \rangle \in P_i^I$;
- $I \models^{\mathfrak{a}} \top$ and $I \not\models^{\mathfrak{a}} \bot$;
- $I \models^{\mathfrak{a}} \psi \wedge \chi$ iff $I \models^{\mathfrak{a}} \psi$ and $I \models^{\mathfrak{a}} \chi$;
- $I \models^{\mathfrak{a}} \psi \vee \chi$ iff $I \models^{\mathfrak{a}} \psi$ or $I \models^{\mathfrak{a}} \chi$;
- $I \models^{\mathfrak{a}} \psi \to \chi$ iff $I \models^{\mathfrak{a}} \chi$ whenever $I \models^{\mathfrak{a}} \psi$;
- $I \models^{\mathfrak{a}} \neg\psi$ iff $I \not\models^{\mathfrak{a}} \psi$;
- $I \models^{\mathfrak{a}} \forall x\psi$ iff $I \models^{\mathfrak{b}} \psi$ for every assignment $\mathfrak{b}$ in I such that $\mathfrak{a}(y) = \mathfrak{b}(y)$ for all variables y different from x;
- $I \models^{\mathfrak{a}} \exists x\psi$ iff $I \models^{\mathfrak{b}} \psi$ for some assignment $\mathfrak{b}$ in I such that $\mathfrak{a}(y) = \mathfrak{b}(y)$ for all variables y different from x.

In the case of $\mathcal{QL}^=$, we add one more item: $I \models^{\mathfrak{a}} \tau_1 = \tau_2$ iff $\mathfrak{a}(\tau_1) = \mathfrak{a}(\tau_2)$.

It should be clear that the truth of a formula $\varphi(x_1, \dots, x_n)$ in I under an assignment $\mathfrak{a}$ depends only on the values $a_1 = \mathfrak{a}(x_1), \dots, a_n = \mathfrak{a}(x_n)$. So instead of $I \models^{\mathfrak{a}} \varphi$ we sometimes write $I \models \varphi[a_1, \dots, a_n]$.

If $I \models^{\mathfrak{a}} \varphi$ holds for all assignments $\mathfrak{a}$ in I, then we say that φ *is true in* I and write $I \models \varphi$. A *set* Γ *of formulas is true in* I if every formula in Γ is true in I. Γ *is true in a class* $\mathcal{C}$ of first-order structures (in symbols: $\mathcal{C} \models \Gamma$) if $I \models \Gamma$ for all $I \in \mathcal{C}$. The *theory of* $\mathcal{C}$ is the set of sentences that are true in $\mathcal{C}$. We say that a set Γ of sentences *implies* a sentence φ (or φ is a *consequence* of Γ) and write $\Gamma \models \varphi$, if φ is true in I whenever Γ is true in I, for every first-order structure I. Given a sublanguage of $\mathcal{QL}$, we define *classical first-order logic* (of that sublanguage) as the set of formulas that are true in all first-order structures (of the appropriate signature) and denote this logic (slightly abusing notation) by **QCl** ('quantified **Cl**'). Formulas in **QCl** are often called *classically valid*.

Similarly, $\mathbf{QCl}^=$ denotes *classical first-order logic with equality*.

A class $\mathcal{C}$ of first-order structures of some signature $\mathcal{S}$ is said to be *first-order definable* if there is a set Γ of $\mathcal{QL}^=$-sentences of signature $\mathcal{S}$ such that, for every $\mathcal{S}$-structure I,

$$I \in \mathcal{C} \quad \text{iff} \quad I \models \Gamma.$$

In this case we also say that $\mathcal{C}$ is *definable by* Γ. It should be clear that if $\mathcal{C}$ is first-order definable, then it is definable by the theory of $\mathcal{C}$.

Syntactically, **QCl** can be defined by a calculus with the following axiom schemata and inference rules.

Axiom schemata:

- (A1)–(A10) of classical propositional logic **Cl** in Section 1.1 regarded as *axiom schemata* (in the sense that the propositional variables p_i can be replaced by arbitrary formulas of the given sublanguage of $\mathcal{QL}$);
- $\forall x\varphi \to \varphi\{\tau/x\}$, where τ is free for x in φ;
- $\varphi\{\tau/x\} \to \exists x\varphi$, where τ is free for x in φ.

Inference rules:

- modus ponens (MP);
- given $\psi \to \varphi$, derive $\psi \to \forall x\varphi$, whenever x is not free in ψ;
- given $\varphi \to \psi$, derive $\exists x\varphi \to \psi$, whenever x is not free in ψ.

A formula φ is *derivable* from a set Γ of formulas if there is a sequence of formulas ending with φ and such that each member of the sequence is either in Γ, or a substitution instance of an axiom schema, or obtained from some earlier members of the sequence by applying an inference rule. According to Gödel's completeness theorem, for every set Γ of sentences and every sentence φ, we have $\Gamma \models \varphi$ iff φ is derivable from Γ.

The standard translation

Consider the sublanguage of $\mathcal{QL}$ with countably many unary predicate symbols $P_0, P_1, \dots$, and a single binary predicate symbol R. The *standard translation* $\cdot^\star$ of $\mathcal{ML}$-formulas into this first-order language is defined inductively as follows,[11] where x is a fixed individual variable:

$$
\begin{aligned}
p_i^\star &= P_i(x) & & \\
\top^\star &= \top & \bot^\star &= \bot \\
(\varphi \wedge \psi)^\star &= \varphi^\star \wedge \psi^\star & (\varphi \vee \psi)^\star &= \varphi^\star \vee \psi^\star \\
(\varphi \to \psi)^\star &= \varphi^\star \to \psi^\star & (\neg\varphi)^\star &= \neg\varphi^\star \\
(\Box\psi)^\star &= \forall y\,(xRy \to \psi^\star\{y/x\}) & (\Diamond\psi)^\star &= \exists y\,(xRy \wedge \psi^\star\{y/x\}).
\end{aligned}
$$

Here y is a fresh variable not occurring in $\psi^\star$. As $\varphi^\star$ always has at most one free variable, we can define the translation $\cdot^\star$ in such a way that $\varphi^\star$ contains

[11] Here and below we provide the inductive definition of the translation for all logical constants and connectives $\top, \bot, \wedge, \vee, \to, \neg, \Box$, and $\Diamond$, although it would suffice to define it for, say, $\wedge$, $\neg$, and $\Box$. The reason is that later on in this book we shall consider a similar translation for intuitionistic modal logic, where the connectives are not interdefinable.

at most two variables altogether (simply re-use the available second variable, which is not free in $\psi^\star$, in the definitions of $(\Box\psi)^\star$ and $(\Diamond\psi)^\star$).

Every Kripke model $\mathfrak{M} = \langle \mathfrak{F}, \mathfrak{V} \rangle$ can be regarded as a first-order structure

$$I(\mathfrak{M}) = \left\langle D^{I(\mathfrak{M})}, R^{I(\mathfrak{M})}, P_0^{I(\mathfrak{M})}, P_1^{I(\mathfrak{M})}, \dots \right\rangle$$

for the above sublanguage of $\mathcal{QL}$, where $D^{I(\mathfrak{M})}$ is the set of worlds in $\mathfrak{F}$, $P_i^{I(\mathfrak{M})} = \mathfrak{V}(p_i)$, for every i, and $R^{I(\mathfrak{M})}$ is the accessibility relation in $\mathfrak{F}$. It is easily seen that for every $\mathcal{ML}$-formula φ, every Kripke model $\mathfrak{M}$ and every world w in $\mathfrak{M}$, we have

$$(\mathfrak{M}, w) \models \varphi \quad \text{iff} \quad I(\mathfrak{M}) \models \varphi^\star[w].$$

Conversely, every first-order structure of the form $I = \left\langle D^I, R^I, P_0^I, \dots \right\rangle$ can be considered as a Kripke model

$$\mathfrak{M}(I) = \langle \mathfrak{F}(I), \mathfrak{V}(I) \rangle,$$

where $\mathfrak{F}(I) = \left\langle D^I, R^I \right\rangle$ and $\mathfrak{V}(I)(p_i) = P_i^I$, for every i. And then we have

$$I \models \varphi^\star[w] \quad \text{iff} \quad (\mathfrak{M}(I), w) \models \varphi,$$

for all $\mathcal{ML}$-formulas φ, first-order structures I, and $w \in D^I$. Therefore,

$$\varphi \in \mathbf{K} \quad \text{iff} \quad \varphi^\star \in \mathbf{QCl}.$$

Note that starting from a model $\mathfrak{M}$ based on a universal frame $\mathfrak{F}$ (where all points are accessible from each other) we obtain a first-order structure $I(\mathfrak{M})$, where $\varphi^\star$ is equivalent to the formula $\varphi^\dagger$ defined by taking:

$$\begin{aligned} p_i^\dagger &= P_i(x) & & \\ \top^\dagger &= \top & \bot^\dagger &= \bot \\ (\varphi \wedge \psi)^\dagger &= \varphi^\dagger \wedge \psi^\dagger & (\varphi \vee \psi)^\dagger &= \varphi^\dagger \vee \psi^\dagger \\ (\varphi \to \psi)^\dagger &= \varphi^\dagger \to \psi^\dagger & (\neg\varphi)^\dagger &= \neg\varphi^\dagger \\ (\Box\psi)^\dagger &= \forall x \psi^\dagger & (\Diamond\psi)^\dagger &= \exists x \psi^\dagger. \end{aligned}$$

Since **S5** is characterized by universal frames, we then have

$$\varphi \in \mathbf{S5} \quad \text{iff} \quad \varphi^\dagger \in \mathbf{QCl}.$$

Observe that x is the only variable which can occur in $\varphi^\dagger$ and that $\varphi^\dagger$ is a *monadic* formula, that is, a formula having only unary predicate symbols. On the other hand, it should be clear that every $\mathcal{QL}$-formula with one variable is equivalent to a one-variable monadic formula. Thus, modulo equivalence, the translation $\cdot^\dagger$ is one-one and onto the set of all one-variable first-order formulas. In other words, the logic **S5** can be regarded as the one-variable fragment of classical first-order logic (Wajsberg 1933).

1.4 Multimodal logics

The modal language $\mathcal{ML}$ introduced in Section 1.1 contains only one pair of (dual) modal operators. It is not hard, however, to imagine situations when two or even more such pairs are required. For instance, to represent beliefs of n agents developing in time we may need $n+1$ pairs of boxes and diamonds—one pair to talk about time and one pair for each agent to represent its beliefs. We shall see many examples of this kind later on in the book. Here we discuss how to extend the concepts and results of the previous sections to *multimodal logic*.

For each natural number $n > 0$, the *propositional n-modal language* $\mathcal{ML}_n$ is defined in almost the same way as the language $\mathcal{ML}$. The only difference is that now we have n necessity and n possibility operators $\Box_1, \dots, \Box_n$ and $\Diamond_1, \dots, \Diamond_n$, respectively, and that $\Box_i\varphi$ and $\Diamond_i\varphi$ are formulas of $\mathcal{ML}_n$ whenever $1 \le i \le n$ and φ is an $\mathcal{ML}_n$-formula. The *modal depth* $md(\varphi)$ of an $\mathcal{ML}_n$-formula φ is defined inductively as follows:

$$\begin{aligned}
md(\alpha) &= 0, \text{ for atomic } \alpha,\\
md(\psi \odot \chi) &= \max\{md(\psi), md(\chi)\}, \text{ for } \odot \in \{\wedge, \vee, \to\},\\
md(\neg\psi) &= md(\psi),\\
md(\Box_i\psi) &= md(\psi) + 1, \text{ for } 1 \le i \le n,\\
md(\Diamond_i\psi) &= md(\psi) + 1, \text{ for } 1 \le i \le n.
\end{aligned}$$

To introduce n-modal logics syntactically we need the axiom (K) and the necessitation rule formulated for each of the boxes $\Box_1, \dots, \Box_n$. More precisely, for each $i = 1, \dots, n$ let

$(\mathrm{K})_i$ $\quad \Box_i(p_0 \to p_1) \to (\Box_i p_0 \to \Box_i p_1)$,

$(\mathrm{RN})_i$ $\quad$ given φ, derive $\Box_i\varphi$.

A set of $\mathcal{ML}_n$-formulas is called an *n-modal logic* if it contains the axioms (A1)–(A10) and $(\mathrm{K})_i$, for $1 \le i \le n$, and is closed under the rules MP, Subst and $(\mathrm{RN})_i$, for all $i = 1, \dots, n$. (As before, the possibility operators $\Diamond_i$ are regarded as abbreviations for $\neg\Box_i\neg$.) We define $\mathbf{K}_n$ as the *minimal* (i.e., smallest) *n-modal logic*. In general, for a n-modal logic L_0 and a set Γ of $\mathcal{ML}_n$-formulas, we denote by

$$L_0 \oplus \Gamma$$

the smallest n-modal logic containing $L_0 \cup \Gamma$. Logics of the form $\mathbf{K}_n \oplus \Gamma$, for a recursive set Γ are called (*recursively*) *axiomatizable*. If Γ is finite then $\mathbf{K}_n \oplus \Gamma$ is called *finitely axiomatizable*.[12]

[12]There exist other kinds of axiomatic systems for multimodal logics, for instance, those using the so-called *irreflexivity rules* (e.g., given $\neg(p \to \Diamond_i p) \to \varphi$, derive φ, provided that

As examples of multimodal logics we give here the n-modal variants of **K4**, **T**, **S4**, **KD45** and **S5**:

$$\begin{aligned}
\mathbf{K4}_n &= \mathbf{K}_n \oplus \{\Box_i p_0 \to \Box_i\Box_i p_0 \mid 1 \le i \le n\},\\
\mathbf{T}_n &= \mathbf{K}_n \oplus \{\Box_i p_0 \to p_0 \mid 1 \le i \le n\},\\
\mathbf{S4}_n &= \mathbf{K4}_n \oplus \{\Box_i p_0 \to p_0 \mid 1 \le i \le n\},\\
\mathbf{KD45}_n &= \mathbf{K4}_n \oplus \{\Box_i p_0 \to \Diamond_i p_0, \Diamond_i p_0 \to \Box_i\Diamond_i p_0 \mid 1 \le i \le n\},\\
\mathbf{S5}_n &= \mathbf{S4}_n \oplus \{\Diamond_i p_0 \to \Box_i\Diamond_i p_0 \mid 1 \le i \le n\}.
\end{aligned}$$

The axioms of $\mathbf{K4}_n$ require each $\Box_i$ to behave like a **K4**-box. Similarly, $\mathbf{T}_n$, $\mathbf{S4}_n$, $\mathbf{KD45}_n$ and $\mathbf{S5}_n$ are the n-modal logics each box in which behaves like a **T**-, **S4**-, **KD45**- and **S5**-box, respectively. No axiom with two different boxes is postulated. In other words, there is no interaction between different modal operators. These logics are the simplest examples of *fusions* of unimodal logics to be discussed in detail in Section 3.1 and Chapter 4.

Let us now introduce the possible world semantics for n-modal logics. Recall that the operator $\Box$ is interpreted by means of the accessibility relation R between worlds in a Kripke frame $\langle W, R\rangle$. To interpret $\mathcal{ML}_n$-formulas, we need n accessibility relations $R_1, \dots, R_n$, one for each $\Box_i$. Thus we come to the notion of an *n-frame* as a structure of the form $\mathfrak{F} = \langle W, R_1, \dots, R_n\rangle$ consisting of a non-empty set W of worlds and n binary relations $R_1, \dots, R_n$ on W.

As before, a *valuation* in an n-frame $\mathfrak{F}$ is a map $\mathfrak{V}$ associating with each propositional variable p a subset $\mathfrak{V}(p)$ of W. The pair $\mathfrak{M} = \langle \mathfrak{F}, \mathfrak{V}\rangle$ is a *model* for $\mathcal{ML}_n$. The inductive definition of the *truth-relation* $\models$ in $\mathfrak{M}$ is a straightforward generalization of that for the unimodal case: we simply replace the clauses for $\Box$ and $\Diamond$ with

$$\begin{aligned}
(\mathfrak{M}, x) \models \Box_i\psi \quad &\text{iff} \quad (\mathfrak{M}, y) \models \psi \text{ for all } y \in W \text{ such that } xR_iy;\\
(\mathfrak{M}, x) \models \Diamond_i\psi \quad &\text{iff} \quad (\mathfrak{M}, y) \models \psi \text{ for some } y \in W \text{ such that } xR_iy,
\end{aligned}$$

for all $i = 1, \dots, n$. Now, given a class $\mathcal{C}$ of n-frames, we define the *logic of* $\mathcal{C}$ by taking

$$\mathsf{Log}\,\mathcal{C} = \{\varphi \in \mathcal{ML}_n \mid \forall \mathfrak{F} \in \mathcal{C}\ \mathfrak{F} \models \varphi\}.$$

We will not reformulate here the other syntactical and semantical definitions of the previous sections for the language $\mathcal{ML}_n$ relying upon the reader's common sense.

---

p does not occur in φ). We do not consider such axiomatizations in this book and refer the reader to (Gabbay 1981a, Marx and Venema 1997).

Theorem 1.5. *The n-modal logics* $\mathbf{K}_n$, $\mathbf{K4}_n$, $\mathbf{T}_n$, $\mathbf{S4}_n$, $\mathbf{KD45}_n$ *and* $\mathbf{S5}_n$ *are complete with respect to the classes of their n-frames, viz.,*

$$\begin{aligned}
\mathsf{Fr}\mathbf{K}_n &= \{\mathfrak{F} \mid \mathfrak{F} \text{ is an } n\text{-frame}\};\\
\mathsf{Fr}\mathbf{K4}_n &= \{\langle W, R_1, \dots, R_n\rangle \mid R_i \text{ is transitive, } 1 \le i \le n\};\\
\mathsf{Fr}\mathbf{T}_n &= \{\langle W, R_1, \dots, R_n\rangle \mid R_i \text{ is reflexive, } 1 \le i \le n\};\\
\mathsf{Fr}\mathbf{S4}_n &= \{\langle W, R_1, \dots, R_n\rangle \mid R_i \text{ is a quasi-order on } W,\ 1 \le i \le n\};\\
\mathsf{Fr}\mathbf{KD45}_n &= \{\langle W, R_1, \dots, R_n\rangle \mid R_i \text{ is serial, transitive,}\\
&\qquad \text{and Euclidean, } 1 \le i \le n\};\\
\mathsf{Fr}\mathbf{S5}_n &= \{\langle W, R_1, \dots, R_n\rangle \mid R_i \text{ is an equivalence relation on } W,\\
&\qquad 1 \le i \le n\}.
\end{aligned}$$

Note that this theorem is a special case of Theorem 4.1 claiming that Kripke completeness is preserved under the formation of fusions.

The following theorem is an illustration of the use of the standard translation:

Theorem 1.6. *Let L be an n-modal logic such that $L = \mathsf{Log}\,\mathcal{C}$, for some class $\mathcal{C}$ of frames which is first-order definable in the language with n binary predicate symbols and equality. Then L is determined by the class of its countable frames.*

Proof. Let Γ denote the first-order theory defining $\mathcal{C}$. Suppose that $\varphi \notin L$, i.e., $(\mathfrak{M}, w) \not\models \varphi$ for some model $\mathfrak{M}$ based on a frame in $\mathcal{C}$ and some world w in $\mathfrak{M}$. Consider $\mathfrak{M}$ as a first-order structure $I(\mathfrak{M})$ of the language having n binary and countably many unary predicate symbols (see Section 1.3). Then $\Gamma' = \Gamma \cup \{\exists x \neg\varphi^\star(x)\}$ holds in $I(\mathfrak{M})$ (where $\varphi^\star$ is the standard translation of φ). By the downward Löwenheim–Skolem–Tarski theorem, there is a countable first-order structure J such that $J \models \Gamma'$. Consider now J as a modal model $\mathfrak{M}(J)$. It is clearly based on a countable frame for L and refutes φ. ❑

Truth-preserving operations

We conclude this section by introducing three important truth-preserving operations on n-frames and models.

P-morphism. Given two n-frames

$$\mathfrak{F} = \langle W, R_1, \dots, R_n\rangle \quad \text{and} \quad \mathfrak{G} = \langle V, S_1, \dots, S_n\rangle,$$

a map f from W to V is called a *p-morphism from* $\mathfrak{F}$ *to* $\mathfrak{G}$ if it satisfies the following two conditions, for all $x, y \in W$, $z \in V$ and $i = 1, \dots, n$:

- if xR_iy then $f(x)S_if(y)$,
- if $f(x)S_iz$ then there is $y \in W$ such that xR_iy and $f(y) = z$.

A function f satisfying only the former condition is called a *homomorphism from* $\mathfrak{F}$ *to* $\mathfrak{G}$. If a p-morphism f is onto then we say that $\mathfrak{G}$ is a *p-morphic image* of $\mathfrak{F}$, or $\mathfrak{F}$ *maps p-morphically onto* $\mathfrak{G}$. A p-morphism f from $\mathfrak{F}$ onto $\mathfrak{G}$ is called a *p-morphism from a model* $\mathfrak{M} = \langle \mathfrak{F}, \mathfrak{V} \rangle$ *onto a model* $\mathfrak{N} = \langle \mathfrak{G}, \mathfrak{U} \rangle$ if, for every propositional variable p and every point $x \in W$, we have $x \in \mathfrak{V}(p)$ iff $f(x) \in \mathfrak{U}(p)$. It is readily checked by induction that for all $\mathcal{ML}_n$-formulas φ and all $x \in W$,

$$(\mathfrak{M}, x) \models \varphi \quad \text{iff} \quad (\mathfrak{N}, f(x)) \models \varphi. \tag{1.1}$$

It follows, in particular, that if $\mathfrak{F}$ maps p-morphically onto $\mathfrak{G}$ and $\mathfrak{F} \models \varphi$ then $\mathfrak{G} \models \varphi$ as well, for every $\mathcal{ML}_n$-formula φ, or, to put it another way, φ is satisfiable in $\mathfrak{F}$ whenever it is satisfiable in $\mathfrak{G}$.

An n-frame $\mathfrak{F} = \langle W, R_1, \dots, R_n \rangle$ is called *rooted* if there is a $w_0 \in W$ such that $W = \{w \in W \mid w_0 R^* w\}$, where

$$R = \bigcup_{1 \le i \le n} R_i.$$

Such a w_0 is called a *root of* $\mathfrak{F}$. Given a rooted n-frame $\mathfrak{F}$ with root w_0, we can construct another n-frame $\mathfrak{G} = \langle V, S_1, \dots, S_n \rangle$ by taking V to be the set consisting of $\langle w_0 \rangle$ and all the tuples $\langle w_0, R_{i_1}, w_1, \dots, R_{i_k}, w_k \rangle$, $k > 0$, of points in W and accessibility relations $R_{i_j} \in \{R_1, \dots, R_n\}$ such that $w_j R_{i_{j+1}} w_{j+1}$ whenever $j < k$ and, for any two points $\langle w_0, \dots, w_k \rangle$ and x in V,

$$\langle w_0, \dots, w_k \rangle \, S_j x \quad \text{iff} \quad \exists w \in W \; x = \langle w_0, \dots, w_k, R_j, w \rangle .$$

The frame $\mathfrak{G}$ is called the *unraveling* of $\mathfrak{F}$ (see Fig. 1.2 in which the accessibility relation in tuples is omitted). Two properties of $\mathfrak{G}$ make the unraveling construction important in modal logic. First, it is not hard to show (see, e.g., Chagrov and Zakharyaschev 1997, Blackburn *et al.* 2001) that the map $\langle w_0, \dots, w_k \rangle \mapsto w_k$ is a p-morphism from $\mathfrak{G}$ onto $\mathfrak{F}$. And second, $\mathfrak{G}$ has a rather special form known as an *intransitive tree*. A general definition is as follows.

A rooted frame $\mathfrak{F} = \langle W, R_1, \dots, R_n \rangle$ is said to be a *tree* if all the R_i are pairwise disjoint and, for every $x \in W$, the set $W_x = \{y \in W \mid yR^*x\}$ is finite and linearly ordered by the reflexive and transitive closure R^* of the relation $R = \bigcup_{1 \le i \le n} R_i$ (its restriction to W_x, to be more precise). $\mathfrak{F}$ is called *intransitive* if for any R_i, R_j $(1 \le i, j \le n)$ we have

$$\forall x, y, z \in W \; (xR_iy \wedge yR_jz \rightarrow \neg xR_iz \wedge \neg xR_jz).$$

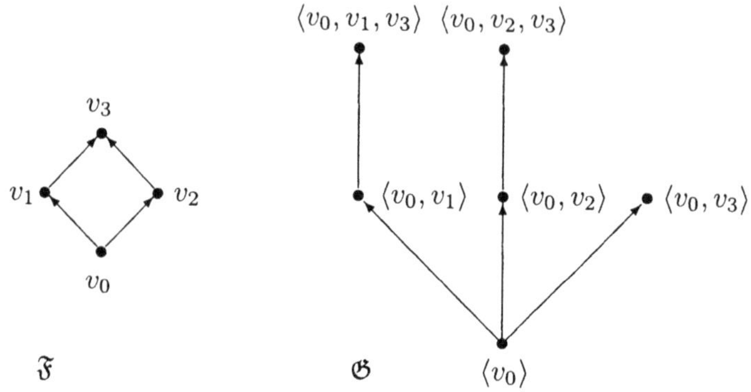

Figure 1.2: $\mathfrak{G}$ is the unraveling of transitive $\mathfrak{F}$.

An intransitive frame is clearly irreflexive Let $\mathfrak{F} = \langle W, R_1, \dots, R_n \rangle$ be an intransitive tree and $x, y \in W$. A *path of length* k from x to y in $\mathfrak{F}$ is a sequence $\langle x_0, \dots, x_k \rangle$ such that $x_0 = x$, $x_k = y$ and $x_i R_j x_{i+1}$, for each $i < k$ and some j, $1 \leq j \leq n$. By the definition, there is a unique path from the root of $\mathfrak{F}$ to x. The length of this path is called the *co-depth* of x and denoted by $cd(x)$. (Thus, the co-depth of the root in $\mathfrak{F}$ is 0.) If the set $\{cd(x) \mid x \in T\}$ is bounded, then the *depth* of $\mathfrak{F}$ is the maximum of $cd(x)$ for $x \in W$. By the *depth* $d(x)$ of x in $\mathfrak{F}$ we understand the depth of the subtree of $\mathfrak{F}$ with root x. (Thus, the depth of a leaf in $\mathfrak{F}$ is 0.)

Now, returning back to the unraveling, we obtain the following remarkable result:

Proposition 1.7. *Every rooted n-frame is a p-morphic image of some intransitive tree.*

An immediate consequence of this proposition is that $\mathbf{K}_n$ is characterized by the class of intransitive trees. Moreover, one can easily strengthen this observation to the following one:

Proposition 1.8. *If an $\mathcal{ML}_n$-formula φ is satisfiable in a frame then it is also satisfiable in a finite intransitive tree of depth $\leq md(\varphi)$.*

Another important transformation of frames is known as *bulldozing*. It operates on transitive frames by 'bulldozing' their 'clusters' into infinite ascending chains of points. For simplicity we define bulldozing for 1-frames.

Let $\mathfrak{G} = \langle V, S \rangle$ be a transitive frame and $x \in V$. The *cluster generated by* x is the set

$$C(x) = \{x\} \cup \{y \in V \mid xSy \text{ and } ySx\}.$$

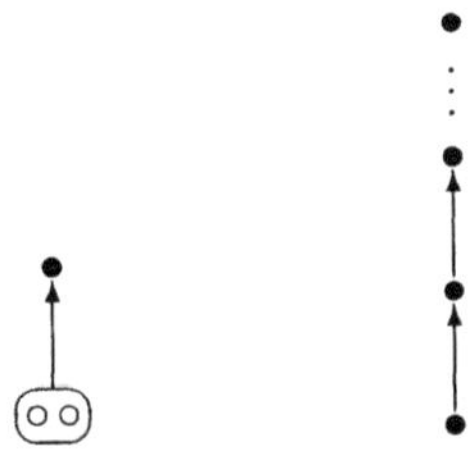

Figure 1.3: Bulldozing $\mathfrak{G}$.

We distinguish between three types of clusters: a *proper cluster* contains at least two points (which see each other), a *simple cluster* consists of a single reflexive point, and a *degenerate cluster* consists of a single irreflexive point. Now, with every $x \in V$ we associate a set x^+ which is $\{\langle x, i\rangle \mid i = 0, 1, \dots\}$ if $C(x)$ is nondegenerate and $\{\langle x, 0\rangle\}$ if $C(x)$ is degenerate. Let W be the union of all x^+. Fix some well-ordering $x_0, x_1, \dots, x_\xi, \dots$ of each cluster C in $\mathfrak{G}$ and define a relation R on W by taking

$$\langle x_\xi, i\rangle \, R \, \langle x_\zeta, j\rangle \text{ iff either } i < j \text{ or } \xi < \zeta \text{ and } i = j, \tag{1.2}$$

when $C(x_\xi) = C(x_\zeta)$ and, for distinct $C(x)$ and $C(y)$,

$$\langle x, i\rangle \, R \, \langle y, j\rangle \quad \text{iff} \quad xSy$$

(see Fig. 1.3). It is easy to see that R is a strict partial order and that the map $f : \langle x, i\rangle \mapsto x$ is a p-morphism from $\mathfrak{F} = \langle W, R\rangle$ onto $\mathfrak{G}$. Note that if $\mathfrak{G}$ is reflexive then we can make $\mathfrak{F}$ a partial order by replacing $<$ in (1.2) with $\leq$. Thus we have:

Proposition 1.9. (i) *Every transitive frame is a p-morphic image of some strict partial order.*

(ii) *Every quasi-ordered frame is a p-morphic image of some partial order.*

As an easy consequence of Proposition 1.9 we obtain, for instance, the following

Theorem 1.10. (i) **K4** *is characterized by the class of strict partial orders.*

(ii) **S4** *is characterized by the class of partial orders.*

Generated subframe. We say that an n-frame $\mathfrak{G} = \langle V, S_1, \dots, S_n\rangle$ is a *subframe* of an n-frame $\mathfrak{F} = \langle W, R_1, \dots, R_n\rangle$ if $V \subseteq W$ and, for all $i = 1, \dots, n$, S_i is the restriction of R_i to V (i.e., $S_i = R_i \cap (S \times S)$). A subframe $\mathfrak{G}$ of $\mathfrak{F}$ is called a *generated subframe* of $\mathfrak{F}$ if for every $y \in W$, we have $y \in V$ whenever

xR_iy for some $x \in V$ and $1 \leq i \leq n$ (in other words, V is *upward closed in* $\mathfrak{F}$). A model $\mathfrak{N} = \langle \mathfrak{G}, \mathfrak{U} \rangle$ is a *generated submodel* of $\mathfrak{M} = \langle \mathfrak{F}, \mathfrak{V} \rangle$ if $\mathfrak{G}$ is a generated subframe of $\mathfrak{F}$ and $\mathfrak{U}$ is the restriction of $\mathfrak{V}$ to V (i.e., $\mathfrak{U}(p) = \mathfrak{V}(p) \cap V$ for all p). It should be clear that in this case we have

$$(\mathfrak{M}, x) \models \varphi \quad \text{iff} \quad (\mathfrak{N}, x) \models \varphi, \tag{1.3}$$

for every $\mathcal{ML}_n$-formula φ and every point $x \in V$. If $\mathfrak{G}$ is a generated subframe of $\mathfrak{F}$ and V is the *upward closure of* some set $X \subseteq W$ (i.e., V is the smallest upward closed set in $\mathfrak{F}$ containing X), then we say that $\mathfrak{G}$ is *generated by* X. A generated submodel $\mathfrak{N} = \langle \mathfrak{G}, \mathfrak{U} \rangle$ of $\mathfrak{M} = \langle \mathfrak{F}, \mathfrak{V} \rangle$ is called in this case a *submodel generated by* X. Note that if $\mathfrak{G}$ is generated by a singleton $\{x\}$ then $\mathfrak{G}$ is rooted, with x being its root. For a class $\mathcal{C}$ of n-frames, denote by $\mathcal{C}^r$ the class of all rooted subframes of frames in $\mathcal{C}$; $\mathsf{Fr}^r L$ is the class of all rooted n-frames for an n-modal logic L. It follows from (1.3) that if $(\mathfrak{M}, x) \models \varphi$ then $(\mathfrak{M}^x, x) \models \varphi$, where $\mathfrak{M}^x$ is the submodel of $\mathfrak{M}$ generated by $\{x\}$. So we have the following:

Proposition 1.11. *If an n-modal logic L is determined by a class $\mathcal{C}$ of n-frames then*

$$L = \mathsf{Log}\, \mathcal{C}^r = \mathsf{Log}\, \mathsf{Fr}^r L,$$

that is, L is determined by the class of its rooted frames.

We say that a (strict) partial order $\langle W, R \rangle$ is a (*strict*) *linear order* if it is *connected*, i.e., for any distinct points $x, y \in W$, either xRy or yRx. It is straightforward to see that a rooted, transitive, weakly connected frame is connected. Therefore we obtain an easy consequence of (1.1), Propositions 1.9 and 1.11, and Theorem 1.4:

Theorem 1.12. (i) **K4.3** *is characterized by the class of strict linear orders.*

(ii) **S4.3** *is characterized by the class of linear orders.*

(iii) **GL.3** *is characterized by the class of all finite strict linear orders as well as by the single frame $\langle \mathbb{N}, > \rangle$ (or by the frame obtained by adding a root to $\langle \mathbb{N}, > \rangle$).*

(iv) **Grz.3** *is characterized by the class of all finite linear orders as well as by the single frame $\langle \mathbb{N}, \geq \rangle$ (or by the frame obtained by adding a root to $\langle \mathbb{N}, \geq \rangle$).*

Disjoint union. Let $\mathfrak{F}_j = \left\langle W_j, R_1^j, \dots, R_n^j \right\rangle$, for $j \in J$, be a family of n-frames with pairwise disjoint sets of worlds, i.e., $W_j \cap W_k = \emptyset$ for all distinct $j, k \in J$. (If this is not the case, we can always take suitable isomorphic copies of the $\mathfrak{F}_j$.) The *disjoint union* of $\mathfrak{F}_j$ is simply the n-frame

$$\sum_{j \in J} \mathfrak{F}_j = \left\langle \bigcup_{j \in J} W_j, \bigcup_{j \in J} R_1^j, \dots, \bigcup_{j \in J} R_n^j \right\rangle.$$

The *disjoint union of models* $\mathfrak{M}_j = \langle \mathfrak{F}_j, \mathfrak{V}_j \rangle$, $j \in J$, is the model

$$\sum_{j\in J} \mathfrak{M}_j = \left\langle \sum_{j\in J} \mathfrak{F}_j, \bigcup_{j\in J} \mathfrak{V}_j \right\rangle.$$

Again, for all $\mathcal{ML}_n$-formulas φ, $j \in J$ and $x \in W_j$, we have

$$(\mathfrak{M}_j, x) \models \varphi \quad \text{iff} \quad \left(\sum_{j\in J} \mathfrak{M}_j, x \right) \models \varphi. \tag{1.4}$$

Summarizing (1.1), (1.3) and (1.4), we can formulate the following:

Theorem 1.13. *For every $\mathcal{ML}_n$-formula φ,*
(i) *if $\mathfrak{G}$ is a p-morphic image of $\mathfrak{F}$, then $\mathfrak{F} \models \varphi$ implies $\mathfrak{G} \models \varphi$;*
(ii) *if $\mathfrak{G}$ is a generated subframe of $\mathfrak{F}$, then $\mathfrak{F} \models \varphi$ implies $\mathfrak{G} \models \varphi$;*
(iii) *if $\mathfrak{F}_j \models \varphi$, for all $j \in J$, then $\sum_{j\in J} \mathfrak{F}_j \models \varphi$.*

In other words, for every n-modal logic L, $\mathsf{Fr}L$ is closed under the formation of p-morphic images, generated subframes and disjoint unions.

1.5 Algebraic semantics

As was said in Section 1.2, Kripke frames were constructed first as relational (or Stone–Jónsson–Tarski) representations of *modal algebras* (see Jónsson and Tarski 1951, Dummett and Lemmon 1959). Unlike Kripke frames, modal algebras can be viewed as a straightforward translation of the language of modal logic into the language of algebra (see, e.g., the construction of Lindenbaum algebras in (Chagrov and Zakharyaschev 1997) or (Goldblatt 1989)), which makes the algebraic semantics adequate for *all* modal logics.

Although not so intuitive and transparent as the possible world semantics, the algebraic semantics brings us to the realm of *universal algebra* and makes its rich and well-developed machinery available for studying modal logics. In this section we give a brief overview of (a very small number of) elementary algebraic concepts we need in what follows. For more detailed expositions see (Burris and Sankappanavar 1981, Chagrov and Zakharyaschev 1997, Goldblatt 1989).

To begin with, we remind the reader that a *Boolean algebra* is a structure of the form

$$\mathfrak{A} = \left\langle A, \wedge^{\mathfrak{A}}, \neg^{\mathfrak{A}}, 0^{\mathfrak{A}}, 1^{\mathfrak{A}} \right\rangle,$$

in which A, the *universe* of $\mathfrak{A}$, is a non-empty set, $\wedge^{\mathfrak{A}}$ is a binary operation on A, $\neg^{\mathfrak{A}}$ a unary one, $0^{\mathfrak{A}}, 1^{\mathfrak{A}} \in A$, and the following conditions hold for all $a, b, c \in A$:

- $a \wedge^{\mathfrak{A}} b = b \wedge^{\mathfrak{A}} a$ and $a \vee^{\mathfrak{A}} b = b \vee^{\mathfrak{A}} a$ (commutativity of $\wedge^{\mathfrak{A}}$ and $\vee^{\mathfrak{A}}$);
- $a \wedge^{\mathfrak{A}} (b \wedge^{\mathfrak{A}} c) = (a \wedge^{\mathfrak{A}} a) \wedge^{\mathfrak{A}} c$ and $a \vee^{\mathfrak{A}} (b \vee^{\mathfrak{A}} c) = (a \vee^{\mathfrak{A}} a) \vee^{\mathfrak{A}} c$ (associativity of $\wedge^{\mathfrak{A}}$ and $\vee^{\mathfrak{A}}$);
- $(a \wedge^{\mathfrak{A}} b) \vee^{\mathfrak{A}} b = b$ and $(a \vee^{\mathfrak{A}} b) \wedge^{\mathfrak{A}} b = b$ (absorption);
- $a \wedge^{\mathfrak{A}} (b \vee^{\mathfrak{A}} c) = (a \wedge^{\mathfrak{A}} b) \vee^{\mathfrak{A}} (a \wedge^{\mathfrak{A}} c)$ and $a \vee^{\mathfrak{A}} (b \wedge^{\mathfrak{A}} c) = (a \vee^{\mathfrak{A}} b) \wedge^{\mathfrak{A}} (a \vee^{\mathfrak{A}} c)$ (distributivity);
- $a \wedge^{\mathfrak{A}} \neg^{\mathfrak{A}} a = 0^{\mathfrak{A}}$ and $a \vee^{\mathfrak{A}} \neg^{\mathfrak{A}} a = 1^{\mathfrak{A}}$,

where, by definition, $a \vee^{\mathfrak{A}} b = \neg^{\mathfrak{A}}(\neg^{\mathfrak{A}} a \wedge^{\mathfrak{A}} \neg^{\mathfrak{A}} b)$. It is readily checked that for every set W, the structure

$$\langle 2^W, \cap, -, \emptyset, W \rangle$$

is a Boolean algebra, where 2^W is the set of all subsets of W, and $\cap$ and $-$ are the usual set-theoretic intersection and complementation in W, respectively. By the *Stone representation theorem*, every Boolean algebra is embeddable into such a set-algebra. For more details on Boolean algebras and their connections with classical propositional logic the reader is referred to (Monk 1988, Rasiowa and Sikorski 1963, Sikorski 1969).

An *n-modal algebra* is a Boolean algebra with extra n operations modeling the n boxes $\Box_i$, viz., a structure of the form

$$\mathfrak{A} = \left\langle A, \wedge^{\mathfrak{A}}, \neg^{\mathfrak{A}}, 0^{\mathfrak{A}}, 1^{\mathfrak{A}}, \Box_1^{\mathfrak{A}}, \ldots, \Box_n^{\mathfrak{A}} \right\rangle,$$

where $\langle A, \wedge^{\mathfrak{A}}, \neg^{\mathfrak{A}}, 0^{\mathfrak{A}}, 1^{\mathfrak{A}} \rangle$ is a Boolean algebra and, for each $i = 1, \ldots, n$, $\Box_i^{\mathfrak{A}}$ is a unary operation on A such that $\Box_i^{\mathfrak{A}} 1^{\mathfrak{A}} = 1^{\mathfrak{A}}$ and, for all $a, b \in A$,

$$\Box_i^{\mathfrak{A}}(a \wedge^{\mathfrak{A}} b) = \Box_i^{\mathfrak{A}} a \wedge^{\mathfrak{A}} \Box_i^{\mathfrak{A}} b.$$

$\mathcal{ML}_n$-formulas are interpreted in $\mathfrak{A}$ by means of *valuations* $\mathfrak{V}$ *in* $\mathfrak{A}$ which map these formulas into A in such a way that, for all $\varphi, \psi \in \mathcal{ML}_n$, and all $i = 1, \ldots, n$, we have

$$\begin{aligned}
\mathfrak{V}(\varphi \wedge \psi) &= \mathfrak{V}(\varphi) \wedge^{\mathfrak{A}} \mathfrak{V}(\psi), \\
\mathfrak{V}(\neg \varphi) &= \neg^{\mathfrak{A}} \mathfrak{V}(\varphi), \\
\mathfrak{V}(\Box_i \varphi) &= \Box_i^{\mathfrak{A}} \mathfrak{V}(\varphi).
\end{aligned}$$

It follows that the *value* $\mathfrak{V}(\varphi)$ *of a formula* φ *under* $\mathfrak{V}$ is uniquely determined by the values $\mathfrak{V}(p)$ of the propositional variables p occurring in φ. The pair $\mathfrak{M} = \langle \mathfrak{A}, \mathfrak{V} \rangle$ is called an *algebraic model for* $\mathcal{ML}_n$ *based on* $\mathfrak{A}$.

A formula φ is said to be *true in* $\mathfrak{M}$, $\mathfrak{M} \models \varphi$ in symbols, if $\mathfrak{V}(\varphi) = 1^{\mathfrak{A}}$; φ is *satisfied in* $\mathfrak{M}$ if $\mathfrak{V}(\varphi) \neq 0^{\mathfrak{A}}$. We say that φ is *valid in* $\mathfrak{A}$ and write $\mathfrak{A} \models \varphi$ if φ is true in all models based on $\mathfrak{A}$.

Given a class $\mathcal{C}$ of n-modal algebras, the set

$$\mathsf{Log}\, \mathcal{C} = \{\varphi \in \mathcal{ML}_n \mid \forall \mathfrak{A} \in \mathcal{C}\ \mathfrak{A} \models \varphi\}$$

is always a n-modal logic. It is called the *logic of* $\mathcal{C}$. If $L = \mathsf{Log}\, \mathcal{C}$, we say that L is *determined* (or *characterized*) by $\mathcal{C}$. An n-modal algebra $\mathfrak{A}$ validating all formulas in some n-modal logic L is called an *algebra for* L; in this case we write $\mathfrak{A} \models L$. The class of all n-modal algebras for L is denoted by $\mathsf{Alg}L$.

In contrast to the possible world semantics, the algebraic one is able to characterize all n-modal logics:

Theorem 1.14. *For every n-modal logic L, we have $L = \mathsf{Log}\, \mathsf{Alg}L$.*

(The proof of this theorem is similar to that of Theorem 4.4.)

There can be a slightly different view on the algebraic semantics. Algebras in general are first-order structures for a language having only function symbols in its signature. In particular, n-modal algebras are first-order structures of the signature having a binary function symbol $\wedge$, unary function symbols $\neg$, $\Box_i$ $(1 \leq i \leq n)$, and individual constants 0 and 1. $\mathcal{ML}_n$-formulas can be regarded then as *terms*, and—if we also have equality—equations of the form $\varphi = 1$ are well-formed formulas of this first-order language. And then for every n-modal logic L and every $\mathcal{ML}_n$-formula φ,

$$\varphi \in L \quad \text{iff} \quad \mathsf{Alg}L \models \varphi = 1.$$

Thus, various problems concerning n-modal logics can be straightforwardly reformulated as problems concerning equational theories of classes of n-modal algebras and vice versa.

Every n-frame $\mathfrak{F} = \langle W, R_1, \ldots, R_n \rangle$ gives rise to the n-modal algebra

$$\mathfrak{F}^+ = \left\langle 2^W, \cap, -, \emptyset, W, \Box_1^{\mathfrak{F}^+}, \ldots, \Box_n^{\mathfrak{F}^+} \right\rangle,$$

where, for all $X \subseteq W$ and $i = 1, \ldots, n$,

$$\Box_i^{\mathfrak{F}^+} X = \{x \in W \mid \forall y \in W\ (xR_iy \rightarrow y \in X)\}.$$

Moreover, an $\mathcal{ML}_n$-formula is valid in $\mathfrak{F}$ iff it is valid in $\mathfrak{F}^+$. Thus, for any class $\mathcal{C}$ of n-frames,

$$\mathsf{Log}\, \mathcal{C} = \mathsf{Log}\, \{\mathfrak{F}^+ \mid \mathfrak{F} \in \mathcal{C}\}.$$

Note, however, that it is not the case that for every n-modal algebra $\mathfrak{A}$ there is an n-frame $\mathfrak{F}$ such that $\mathfrak{A} = \mathfrak{F}^+$ (for instance, there are countable algebras, while $\mathfrak{F}^+$ is either finite or uncountable).

The truth-preserving operations on n-frames considered in Section 1.4 correspond to the well-known algebraic operations of taking subalgebras, homomorphic images, and direct products. Suppose we are given two n-modal algebras

$$\mathfrak{A} = \left\langle A, \wedge^{\mathfrak{A}}, \neg^{\mathfrak{A}}, 0^{\mathfrak{A}}, 1^{\mathfrak{A}}, \Box_1^{\mathfrak{A}}, \dots, \Box_n^{\mathfrak{A}} \right\rangle,$$
$$\mathfrak{B} = \left\langle B, \wedge^{\mathfrak{B}}, \neg^{\mathfrak{B}}, 0^{\mathfrak{B}}, 1^{\mathfrak{B}}, \Box_1^{\mathfrak{B}}, \dots, \Box_n^{\mathfrak{B}} \right\rangle.$$

Then $\mathfrak{A}$ is called a *subalgebra of* $\mathfrak{B}$ if $A \subseteq B$, $0^{\mathfrak{A}} = 0^{\mathfrak{B}}$, $1^{\mathfrak{A}} = 1^{\mathfrak{B}}$, and for all $a, a' \in A$, $1 \le i \le n$,

$$\begin{aligned} a \wedge^{\mathfrak{A}} a' &= a \wedge^{\mathfrak{B}} a', \\ \neg^{\mathfrak{A}} a &= \neg^{\mathfrak{B}} a, \\ \Box_i^{\mathfrak{A}} a &= \Box_i^{\mathfrak{B}} a. \end{aligned}$$

A *homomorphism* from $\mathfrak{A}$ to $\mathfrak{B}$ is a map $h : A \to B$ such that, for all $a, a' \in A$, $1 \le i \le n$,

$$\begin{aligned} h(a \wedge^{\mathfrak{A}} a') &= h(a) \wedge^{\mathfrak{B}} h(a'), \\ h(\neg^{\mathfrak{A}} a) &= \neg^{\mathfrak{B}} h(a), \\ h(\Box_i^{\mathfrak{A}} a) &= \Box^{\mathfrak{B}} h(a). \end{aligned}$$

If h is onto then $\mathfrak{B}$ is called a *homomorphic image of* $\mathfrak{A}$. Now suppose that $\mathfrak{A}_j$, for $j \in J$, is a family of n-modal algebras of the form

$$\mathfrak{A}_j = \left\langle A_j, \wedge^{\mathfrak{A}_j}, \neg^{\mathfrak{A}_j}, 0^{\mathfrak{A}_j}, 1^{\mathfrak{A}_j}, \Box_1^{\mathfrak{A}_j}, \dots, \Box_n^{\mathfrak{A}_j} \right\rangle.$$

The *direct product*

$$\mathfrak{A} = \prod_{j \in J} \mathfrak{A}_j$$

of the $\mathfrak{A}_j$ is defined by taking A to contain all functions g from J into $\bigcup_{j \in J} A_j$ such that $g(j) \in A_j$, for all $j \in J$, and defining the operations $\wedge^{\mathfrak{A}}$, $\neg^{\mathfrak{A}}$, $1^{\mathfrak{A}}$, $0^{\mathfrak{A}}$ and $\Box_i^{\mathfrak{A}}$ $(1 \le i \le n)$ component-wise. For example, $\Box_i^{\mathfrak{A}} g$ is defined by taking

$$(\Box_i^{\mathfrak{A}} g)(j) = \Box_i^{\mathfrak{A}_j} g(j),$$

for $1 \le i \le n$ and $j \in J$. One can show (see, e.g., Chagrov and Zakharyaschev 1997) that, given two n-frames $\mathfrak{F}$ and $\mathfrak{G}$,

- if $\mathfrak{G}$ is a p-morphic image of $\mathfrak{F}$, then $\mathfrak{G}^+$ is (isomorphic to) a subalgebra of $\mathfrak{F}^+$;

- if $\mathfrak{G}$ is a generated subframe of $\mathfrak{F}$, then $\mathfrak{G}^+$ is a homomorphic image of $\mathfrak{F}^+$.

Similarly, given a family $\mathfrak{F}_j$ $(j \in J)$ of n-frames,

- the n-modal algebra $(\sum_{j\in J}\mathfrak{F}_j)^+$ is isomorphic to $\prod_{j\in J}\mathfrak{F}_j{}^+$.

For more information on the duality between frames and modal algebras see (van Benthem 1984, Goldblatt 1989, Chagrov and Zakharyaschev 1997, Blackburn *et al.* 2001).

We conclude this section by formulating the *Birkhoff variety theorem* from universal algebra specialized for n-modal algebras.

Theorem 1.15. *For every n-modal logic L, the class $\mathsf{Alg}L$ is closed under the formation of subalgebras, homomorphic images, and direct products. Moreover, if $L = \mathsf{Log}\,\mathcal{C}$ for some class $\mathcal{C}$ of n-modal algebras, then $\mathsf{Alg}L$ is the closure of $\mathcal{C}$ under taking subalgebras, homomorphic images, and (isomorphic copies of) direct products.*

The proof of this theorem can be found, e.g. in (Burris and Sankappanavar 1981) (in a universal algebraic setting) or in (Chagrov and Zakharyaschev 1997).

1.6 Decision, complexity and axiomatizability problems

In general, neither the syntactical nor the semantical characterizations of a modal logic L provides us with a means to decide, given an arbitrary formula φ, whether $\varphi \in L$. If an *algorithm* (or a *program*) capable of solving this *decision problem* does exist, then L is called *decidable*; otherwise it is *undecidable*. The existence of a decision algorithm for L does not yet guarantee that it can be used in practice: the amount of computational resources it requires may be astronomic. That is why we need to know the optimal computational complexity of the decision problem for L. Problems of this sort are briefly discussed in this section. It is beyond the scope of the book to give a formal treatment of the concepts from *computability theory* such as *algorithm*, *recursive* (or *computable*) *function* and *set*, *recursive enumerability*, etc. The reader can find all these in (Barwise 1977, Enderton 2001, Shoenfield 1967) and other textbooks on mathematical logic and recursion theory.

Let us begin with complexity problems. A standard way of measuring the difficulty of problems like '$\varphi \in L$?' is by the amount of time (number of steps) and/or space (memory) required by the decision algorithm to solve the problem, depending on the size of φ. The size or the *length* $\ell(\varphi)$ of a formula φ is usually defined as the number of symbol occurrences[13] in φ. Here we give

[13]There are other ways of defining $\ell(\varphi)$. For instance, one can understand by $\ell(\varphi)$ the number of subformulas of φ or the size of memory required to store the symbols in φ (thereby taking into account the difference between p_0 and p_{2003}). However, in this book the complexity of the decision algorithms is not affected by the choice we make.

a brief overview of the notions from *complexity theory* we use in this book; for more details consult (Garey and Johnson 1979, Hopcroft and Ullman 1979, Papadimitriou 1994).

Following the standard terminology, we call an algorithm *deterministic* if each step of the algorithm is uniquely determined. On the other hand, a *nondeterministic* algorithm may guess at each step which of a finite number of possible next steps to take. We say that a problem '$x \in X$?' belongs to the *complexity class*

- P if it is solvable by a deterministic algorithm in polynomial time of the size of x;

- EXPTIME if it is solvable by a deterministic algorithm in exponential time of the size $|x|$ of x, i.e., in time $\leq 2^{|x|^k}$, for some $k > 0$;

- 2EXPTIME if it is solvable by a deterministic algorithm in doubly exponential time of the size of x, i.e., in time $\leq 2^{2^{|x|^k}}$;

- ELEM if it is solvable by a deterministic algorithm in time $f(|x|)$ where f is an elementary recursive function of the size of x, i.e., there is a natural number n such that

$$\forall x \; f(|x|) \leq \left.2^{2^{\cdot^{\cdot^{\cdot^{2^{|x|}}}}}}\right\} n.$$

The problem '$x \in X$?' is in

- NP if it is solvable by a nondeterministic algorithm in polynomial time of the size of x, and it is in

- NEXPTIME if it is solvable by a nondeterministic algorithm in exponential time of the size of x.

Finally, we say that the problem is in

- PSPACE if it is solvable by a deterministic algorithm using polynomial space of the size of x;

- EXPSPACE if it is solvable by a deterministic algorithm using exponential space of the size of x.

According to Savitch's theorem (see, e.g., Papadimitriou 1994), nondeterminism does not increase the level of space complexity. So the complexity classes of nondeterministic polynomial space and nondeterministic exponential space coincide with PSPACE and EXPSPACE, respectively. It should be clear that the complexity class of nondeterministic elementary time is the same as ELEM and that

$$\begin{aligned} \mathrm{P} \subseteq \mathrm{NP} \subseteq \mathrm{PSPACE} \subseteq \mathrm{EXPTIME} \subseteq \mathrm{NEXPTIME} \subseteq \\ \subseteq \mathrm{EXPSPACE} \subseteq \mathrm{2EXPTIME} \subseteq \mathrm{N2EXPTIME} \subseteq \cdots \subseteq \mathrm{ELEM}. \end{aligned}$$

It is also known that

$$\begin{array}{ll} \mathrm{P} \neq \mathrm{EXPTIME}, & \mathrm{NP} \neq \mathrm{NEXPTIME}, \\ \mathrm{PSPACE} \neq \mathrm{EXPSPACE}, & \mathrm{EXPTIME} \neq \mathrm{2EXPTIME}, \\ \mathrm{NEXPTIME} \neq \mathrm{N2EXPTIME}, & \mathrm{N2EXPTIME} \neq \mathrm{ELEM}. \end{array}$$

Whether the remaining inclusions are strict or not is one of the most challenging open problems in complexity theory.

Given a problem of the form '$x \in X$?', its *complement* is the problem '$x \notin X$?'. For any complexity class $\mathcal{C}$, the class co$\mathcal{C}$ consists of all problems whose complement is in $\mathcal{C}$. It is not hard to see that for deterministic classes $\mathcal{C} = \mathrm{co}\mathcal{C}$, while for nondeterministic $\mathcal{C}$ it is not known whether this equality holds.

We say that a problem A is $\mathcal{C}$-*hard*, for a complexity class $\mathcal{C}$ (above P), if every problem $B \in \mathcal{C}$ can be *polynomially reduced* to A, i.e., there is a recursive function (program) f which, given a word b (in the language of B), in deterministic polynomial time returns a word $f(b)$ (in the language of A) such that $b \in B$ iff $f(b) \in A$. A problem is called $\mathcal{C}$-*complete*, if it is $\mathcal{C}$-hard and belongs to $\mathcal{C}$. Thus a standard technique for proving $\mathcal{C}$-completeness of a problem A is to show first that A is in $\mathcal{C}$, and then give a polynomial time reduction of some $\mathcal{C}$-complete problem B to A. However, if our aim is to establish that A is undecidable, then any recursive reduction of an undecidable problem B to A will suffice.

For an n-modal logic L, the question if 'there is an algorithm which, given an $\mathcal{ML}_n$-formula φ, decides whether φ belongs to L' is called the *decision problem* or the *validity problem for* L. We say that L is *decidable* (or $\mathcal{C}$-complete) if the decision problem for L is decidable (respectively $\mathcal{C}$-complete). Closely related is the *satisfiability problem for* L: 'given φ, decide whether φ is satisfiable in a frame for L.' It should be clear that, for any Kripke complete logic L, φ is in L iff $\neg\varphi$ is not satisfiable in a frame for L. Thus, validity and satisfiability are complementary problems connected by a very simple reduction $\varphi \rightsquigarrow \neg\varphi$: one is decidable iff the other is decidable; if one is $\mathcal{C}$-complete for some complexity class $\mathcal{C}$, then the other is co$\mathcal{C}$-complete.

Let us consider, for instance, classical propositional logic **Cl**. Given a formula φ, we can nondeterministically assign (guess) truth-values to the propositional variables in φ, and then compute the truth-value of φ in polynomial time. Thus the satisfiability problem for **Cl** is in NP. Moreover, according to Cook's theorem (see e.g. Papadimitriou 1994), it is NP-complete. It follows that the decision problem for **Cl** is coNP-complete. A similar algorithm can be used to show the NP-completeness of the satisfiability problem for a modal logic characterized by a single *finite* frame—such logics are called *tabular*. However, most of the modal logics considered in this book are not tabular.

If a modal logic L is recursively axiomatizable then we can recursively enumerate all formulas in L by systematically constructing all possible derivations. So we would have a decision algorithm for L if we could enumerate those formulas that are not in L. Obviously, this can be done if

- L has the *finite model property* (*fmp*, for short), i.e., L is characterized by the class of its *finite frames*[14] and
- the class of finite frames for L is recursively enumerable (up to isomorphism), which is clearly the case if L is finitely axiomatizable (for then we can even decide whether a given finite frame is a frame for L).

However, the fmp itself says nothing about the complexity of the decision algorithm. We can get more information about complexity by establishing a stronger property which is sometimes called the *effective* (or *bounded*) *finite model property* (*efmp*): L has the efmp if there is a recursive function f such that, for any formula φ, $\varphi \notin L$ iff there is a frame $\mathfrak{F}$ for L such that $\mathfrak{F} \not\models \varphi$ and $\mathfrak{F}$ contains at most $f(\ell(\varphi))$ points. If f is a polynomial or exponential function then we say that L has the *polynomial* or, respectively, *exponential* fmp. Suppose now that for L the problem whether a frame $\mathfrak{F}$ belongs to $\mathsf{Fr}L$ can be decided in polynomial time in the size of $\mathfrak{F}$. (Obviously, this is the case for all modal logic introduced so far in this book.) Then the polynomial and exponential fmp provide satisfiability checking algorithms that are in NP and NEXPTIME, respectively: given a formula φ, we guess a model $\mathfrak{M} = \langle \mathfrak{F}, \mathfrak{V} \rangle$ of size polynomial or exponential in φ, and check whether $\mathfrak{F} \in \mathsf{Fr}L$ and $(\mathfrak{M}, x) \models \varphi$ for some x in $\mathfrak{F}$. While the NP upper bound obtained in this way is always optimal (because **CL** is already NP-hard), the NEXPTIME-upper bound can often be improved by more fine-tuned arguments. Here are some examples; for proofs consult (Ladner 1977, Ono and Nakamura 1980, Chagrov and Zakharyaschev 1997).

Theorem 1.16. *All the logics* **K**, **S4**, **S5**, **KD45**, **K4.3**, **S4.3**, **GL**, **GL.3**, **K4**, **T** *and* **D** *are decidable. Moreover,*

(i) **S5**, **KD45**, **K4.3**, **GL.3** *and* **S4.3** *have the polynomial fmp and are* coNP*-complete.*

---

[14] An equivalent formulation (which has actually given the name to the property): L has the fmp if, for every $\varphi \notin L$, there is a finite model $\mathfrak{M}$ such that $\mathfrak{M} \models L$ and $\mathfrak{M} \not\models \varphi$. For a proof that the two formulations are equivalent see e.g. (Chagrov and Zakharyaschev 1997).

(ii) **K**, **S4**, **GL**, **K4**, **T** *and* **D** *have the exponential fmp and are* PSPACE-*complete.*

An example of a coNEXPTIME-complete finitely axiomatizable logic with the exponential fmp can be found in Section 5.5; examples of EXPTIME-complete finitely axiomatizable logics with the exponential fmp are provided by Theorem 1.26.

In Chapter 4 we shall see that many properties (such as the fmp and decidability) of unimodal logics are preserved under joining them into multimodal ones without postulating any interactions between their modal operators. However, this does not always apply to the complexity of the decision algorithms, as the following theorem suggests:

Theorem 1.17. (i) $\mathbf{K}_n$, $\mathbf{T}_n$, $\mathbf{K4}_n$ *and* $\mathbf{S4}_n$ *are* PSPACE-*complete, for all* $n > 0$.

(ii) *If* $n > 1$ *then* $\mathbf{S5}_n$ *and* $\mathbf{KD45}_n$ *are* PSPACE-*complete as well.*

The proof can be found in (Halpern and Moses 1992). Note that the addition of interaction axioms involving different boxes (say, commutativity) can drastically change the character of these rather 'harmless' logics; see Chapter 8.

From a purely logical point of view, the most important reasoning task for a logic L is to recognize, given two arbitrary formulas ψ and φ, whether φ is a *logical consequence* of ψ in L. The notion of logical consequence may be different depending on applications.

Given an n-modal logic L, we say that an $\mathcal{ML}_n$-formula φ is a *global consequence of* ψ *in* L and write $\psi \vdash_L^* \varphi$, if φ belongs to the smallest set of $\mathcal{ML}_n$-formulas which contains $L \cup \{\psi\}$ and is closed under the inference rules MP and RN_i ($1 \leq i \leq n$).

A formula φ is said to be a *local consequence of* ψ *in* L (in symbols: $\psi \vdash_L \varphi$) if φ belongs to the smallest set of $\mathcal{ML}_n$-formulas which contains $L \cup \{\psi\}$ and is closed under MP only. The consequence relation $\vdash_L$ can be easily reduced to validity in L via the following equivalence known as the *deduction theorem*. For all $\mathcal{ML}_n$-formulas φ and ψ,

$$\psi \vdash_L \varphi \quad \text{iff} \quad \psi \to \varphi \in L.$$

It follows, in particular, that L is decidable if and only if the local consequence relation $\vdash_L$ is decidable, and that the decision problems for L and $\vdash_L$ always have the same complexity.

For many n-modal logics L there is a very natural semantical interpretation of $\vdash_L^*$ and $\vdash_L$.

Theorem 1.18. (i) *For any Kripke complete n-modal logic L, we have $\psi \vdash_L \varphi$ iff $(\mathfrak{M}, x) \models \varphi$ whenever $(\mathfrak{M}, x) \models \psi$, for every model $\mathfrak{M}$ based on a frame for L and every point x in $\mathfrak{M}$.*

(ii) *For all n-modal logics L defined above, we have $\psi \vdash_L^* \varphi$ iff $\mathfrak{M} \models \varphi$ whenever $\mathfrak{M} \models \psi$, for every model $\mathfrak{M}$ based on a frame for L.*

Logics L such that, for all formulas φ and ψ, $\psi \vdash_L^* \varphi$ iff $\mathfrak{M} \models \psi$ implies $\mathfrak{M} \models \varphi$, for every model $\mathfrak{M}$ based on a frame for L, are called *globally Kripke complete*. Clearly, every globally Kripke complete logic is Kripke complete; however, the converse does not hold (Kracht 1999). On the other hand, we have the following general completeness result:

Theorem 1.19. *Suppose $L = \mathsf{Log}\, \mathcal{C}$ for some first-order definable class $\mathcal{C}$ of n-frames. Then L is globally Kripke complete.*

(This result does not seem to be stated explicitly in the literature. It follows from the Fine–van Benthem Theorem, according to which any logic $\mathsf{Log}\, \mathcal{C}$ with first-order definable $\mathcal{C}$ is canonical, and from the fact that any canonical logic is globally Kripke complete. For details consult (Zakharyaschev *et al.* 2001, Chagrov and Zakharyaschev 1997).)

If $\vdash_L^*$ can be characterized by models $\mathfrak{M}$ based on countable (or finite) frames, then we say that $\vdash_L^*$ is *determined by countable (finite) frames*. The following theorem can be proved similarly to Theorem 1.6:

Theorem 1.20. *Suppose $L = \mathsf{Log}\, \mathcal{C}$ for some first-order definable class $\mathcal{C}$ of n-frames. Then $\vdash_L^*$ is determined by countable frames.*

We also have:

Theorem 1.21. *For all the logics* **K**, **S4**, **S5**, **KD45**, **K4.3**, **S4.3**, **GL**, **K4**, **T** *and* **D**, *the global consequence relation is determined by finite frames.*

For more details and further references consult (Goranko and Passy 1992, Zakharyaschev *et al.* 2001).

It is not so simple to reduce the global consequence relation $\vdash_L^*$ to validity in L. The deduction theorem for $\vdash_L^*$ is not constructive in general. To formulate it, we require the following notation. For a formula φ, put $\boldsymbol{M}^0_{(n)}\varphi = \boldsymbol{M}^{\leq 0}_{(n)}\varphi = \varphi$ and, for $k \geq 0$,

$$\begin{aligned} \boldsymbol{M}^{k+1}_{(n)}\varphi &= \textstyle\bigwedge_{i=1}^{n} \Box_i \boldsymbol{M}^k_{(n)}\varphi, \\ \boldsymbol{M}^{\leq k+1}_{(n)}\varphi &= \textstyle\bigwedge_{j=0}^{k+1} \boldsymbol{M}^j_{(n)}\varphi. \end{aligned}$$

In the unimodal case, let $\Box^0\varphi = \Box^{\leq 0}\varphi = \varphi$ and, for $k \geq 0$,

$$\begin{aligned} \Box^{k+1}\varphi &= \Box\Box^k\varphi, \\ \Box^{\leq k+1}\varphi &= \textstyle\bigwedge_{j=0}^{k+1} \Box^j\varphi. \end{aligned}$$

Theorem 1.22. *For every n-modal logic L and all $\mathcal{ML}_n$-formulas φ and ψ,*

$$\psi \vdash_L^* \varphi \quad \textit{iff} \quad \exists m \geq 0 \left(\boldsymbol{M}_{(n)}^{\leq m} \psi \to \varphi \in L \right).$$

In particular, in the unimodal case we have

$$\psi \vdash_L^* \varphi \quad \textit{iff} \quad \exists m \geq 0 \left(\Box^{\leq m} \psi \to \varphi \in L \right).$$

In general, the parameter m above is not a computable function of $\ell(\varphi)$ and $\ell(\psi)$, even under the condition that L is decidable. In fact, we shall see later on in the book a number of natural decidable n-modal logics L for which $\vdash_L^*$ is undecidable. Note, however, that for every unimodal logic L containing the axiom $\Box p_0 \to \Box\Box p_0$ (saying that frames for L are transitive) we have

$$\psi \vdash_L^* \varphi \quad \text{iff} \quad \psi \wedge \Box\psi \to \varphi \in L. \tag{1.5}$$

In particular, there is no difference between the complexity of the decision problems for L and $\vdash_L^*$.

For some non-transitive unimodal logics there also exists a computable bound on m; say, for **K**, **D** and **T**, $m \leq 2^{|sub\psi \cup sub\varphi|}$ (see e.g. Chagrov and Zakharyaschev 1997). This upper bound cannot be reduced substantially. As follows from (Spaan 1993) and (Ladner 1977), the global consequence relations for these logics are computationally more complex than the local ones:

Theorem 1.23. *For $L = \mathbf{K}_n, \mathbf{D}_n, \mathbf{T}_n$, the problem of whether $\psi \vdash_L^* \varphi$ holds is* EXPTIME-*complete.*

The global consequence relation $\vdash_L^*$ can be reduced to the decision problem for a logic closely related to L. Recall that semantically $\psi \vdash_L^* \varphi$ means that φ is true everywhere in a model whenever ψ is true everywhere in the model. We can capture this by introducing another modal operator $⊡$ with the intended meaning 'everywhere in the model:'

$$(\mathfrak{M}, x) \models ⊡\psi \quad \text{iff} \quad (\mathfrak{M}, y) \models \psi \text{ for all points } y \text{ in } \mathfrak{M}.$$

Its dual $⟐$ means 'somewhere in the model:'

$$(\mathfrak{M}, x) \models ⟐\psi \quad \text{iff} \quad (\mathfrak{M}, y) \models \psi \text{ for some point } y \text{ in } \mathfrak{M}.$$

The relation $\mathfrak{M} \models ⊡\psi \to \varphi$ would then read 'if ψ is true everywhere in $\mathfrak{M}$ then φ is also true everywhere in $\mathfrak{M}$.' The modal operators $⊡$ and $⟐$ are known as the *universal modalities*. They were introduced and investigated by Goranko and Passy (1992). Denote by $\mathcal{ML}_n^u$ the language $\mathcal{ML}_n$ enriched with $⊡$ (and its dual $⟐$).

Suppose now that we have an n-modal logic L and want to introduce in it the universal modalities with their intended interpretation. As $\boxdot$ and $\diamonddot$ should behave like the **S5** box and diamond, the most natural way to do this is to take the $n+1$-modal logic

$$L_u = L \oplus \{\text{axioms of } \mathbf{S5} \text{ for } \diamonddot \text{ and } \boxdot\} \oplus \{\boxdot\, p_0 \to \Box_i p_0 \mid 1 \leq i \leq n\}$$

in the language $\mathcal{ML}_n^u$. It is not hard to check that in *any* rooted $n+1$-frame for L_u the accessibility relation corresponding to $\boxdot$ should be universal. The connection between L and L_u is established by the following two results of (Goranko and Passy 1992):

Lemma 1.24. *For every n-modal logic L and all $\mathcal{ML}_n$-formulas φ and ψ,*

$$\psi \vdash_L^* \varphi \quad \textit{iff} \quad \boxdot\, \psi \to \varphi \in L_u.$$

Theorem 1.25. *For every n-modal logic L,*

(i) *L_u is Kripke complete iff L is globally Kripke complete;*

(ii) *L_u has the fmp iff $\vdash_L^*$ is determined by finite frames.*

In general, L_u does not inherit 'good' properties from L. For example, Spaan (1993) constructs a unimodal logic L such that L has the polynomial fmp and is decidable in coNP, while L_u is undecidable. Using the filtration technique (see e.g. Chagrov and Zakharyaschev 1997), one can prove the following:

Theorem 1.26. *Let L be any of the logics* **K**, **S4**, **S5**, **KD45**, **K4.3**, **S4.3**, **GL**, **K4**, **T** *and* **D**. *Then L_u has the exponential fmp.*

Some complexity results for modal logics with the universal modalities follow from (Hemaspaandra 1996) and (Areces *et al.* 2000):

Theorem 1.27. (i) *The logics $\mathbf{K}_u$, $\mathbf{T}_u$ and $\mathbf{D}_u$ are* EXPTIME-*complete.*

(ii) *The logics $\mathbf{K4}_u$ and $\mathbf{S4}_u$ are* PSPACE-*complete.*

However, a detailed complexity analysis of the decision problem for these kinds of logics seems to be missing.

Another way of proving decidability of a multimodal logic L is to reduce the decision problem for L to some known decidable problem, say, to a decidable set Γ of formulas, written in some language $\mathcal{L}$. The task then is to find a recursive function f mapping $\mathcal{ML}_n$-formulas to $\mathcal{L}$-formulas and such that, for every $\mathcal{ML}_n$-formula φ,

$$\varphi \in L \quad \text{iff} \quad f(\varphi) \in \Gamma.$$

An example of a very expressive formalism is *monadic second-order logic*, where various classes of structures are known to have decidable theories. As we will use reductions to such theories in Part III, here we briefly present all the required definitions and results.

The *monadic second-order language* $\mathcal{MSOL}$ is based on the alphabet of first-order logic $\mathcal{QL}^=$ extended with a countably infinite list $P_0, P_1, \ldots$ of *unary* (or *monadic*) *predicate variables*. In what follows we consider the first-order signature consisting of one binary predicate symbol $<$ only. The formula formation rules of $\mathcal{MSOL}$ are those of $\mathcal{QL}^=$ plus the following two:

- if x is an individual variable and P is a monadic predicate variable then $P(x)$ is an (atomic) $\mathcal{MSOL}$-formula;
- if φ is an $\mathcal{MSOL}$-formula and P is a monadic predicate variable then $\forall P\varphi$ and $\exists P\varphi$ are $\mathcal{MSOL}$-formulas.

An $\mathcal{MSOL}$-*sentence* is an $\mathcal{MSOL}$-formula without occurrences of free individual variables or free monadic predicate variables.

$\mathcal{MSOL}$ is interpreted in usual first-order structures $I = \langle D^I, <^I \rangle$. However, this time an *assignment in* I is a function $\mathfrak{a}$ mapping each individual variable x to an element $\mathfrak{a}(x) \in D^I$ and each monadic predicate variable P to a subset $\mathfrak{a}(P) \subseteq D^I$. The *truth-relation* $I \models^{\mathfrak{a}} \varphi$, for an $\mathcal{MSOL}$-formula φ, is defined by induction on the construction of φ in the same way as for $\mathcal{QL}^=$-formulas in Section 1.3; the only missing clauses are:

- $I \models^{\mathfrak{a}} P(x)$ iff $\mathfrak{a}(x) \in \mathfrak{a}(P)$, P a predicate variable;
- $I \models^{\mathfrak{a}} \forall P\psi$ iff $I \models^{\mathfrak{b}} \psi$ for every assignment $\mathfrak{b}$ in I such that $\mathfrak{a}$ and $\mathfrak{b}$ agree on all individual variables and on all predicate variables different from P;
- $I \models^{\mathfrak{a}} \exists P\psi$ iff $I \models^{\mathfrak{b}} \psi$ for some assignment $\mathfrak{b}$ in I such that $\mathfrak{a}$ and $\mathfrak{b}$ agree on all individual variables and on all predicate variables save P.

An $\mathcal{MSOL}$-formula φ is said to be *true in* I, if $I \models^{\mathfrak{a}} \varphi$ for all assignments $\mathfrak{a}$ in I. Given a class $\mathcal{C}$ of first-order structures, the *monadic second-order theory of* $\mathcal{C}$ is the set of $\mathcal{MSOL}$-sentences that are true in each $I \in \mathcal{C}$.

The following theorem is a consequence of results of Büchi (1962) and Rabin (1969); for details consult (Gabbay *et al.* 1994):

Theorem 1.28. *Let $\mathcal{C}$ be one of the following classes of first-order structures: $\{\langle \mathbb{N}, < \rangle\}$, $\{\langle \mathbb{Z}, < \rangle\}$, $\{\langle \mathbb{Q}, < \rangle\}$, the class of all finite strict linear orders. Then the monadic second-order theory of $\mathcal{C}$ is decidable.*[15]

[15]However, the monadic second-order theories of $\{\langle \mathbb{R}, < \rangle\}$ and of the class of all strict linear orders are undecidable according to results of Shelah (1975) and Gurevich and Shelah (1982) (see also Gabbay *et al.* 1994).

It is to be noted, however, that the price of proving decidability via a reduction to these decidable monadic second-order theories is that the resulting decision algorithm is non-elementary (see Robertson 1974, Meyer 1974, Rabin 1977).

We conclude this section with a brief discussion of how to prove that a logic is not recursively axiomatizable. As was observed by Craig (1953), every recursively enumerable logic L is recursively axiomatizable as well. Thus, to show that L is not recursively axiomatizable, it suffices to reduce a problem, whose complement is not recursively enumerable, to the satisfiability problem for L. It is well-known from computability theory that Σ^1_1-hard problems are such. Roughly, the membership problem '$x \in X$?', for a set X of natural numbers, is in Σ^1_1 if there is a formula

$$\exists P_0 \ldots \exists P_k \; \varphi(y)$$

of monadic second-order arithmetic such that, for all natural numbers n, we have

$$n \in X \quad \text{iff} \quad \langle \mathbb{N}, 0, +, \cdot \rangle \models \exists P_0 \ldots \exists P_k \; \varphi[n].$$

(Here $P_0, \ldots, P_k$ are the only monadic predicate variables in φ, y its only free individual variable, and no quantification over predicate variables occurs in φ.) Then we say that a problem A is Σ^1_1-*hard* if every problem in Σ^1_1 can be recursively reduced to A. A problem is Σ^1_1-*complete* if it is Σ^1_1-hard and belongs to Σ^1_1.

Chapter 2

Applied modal logic

So far we were considering modal logics with rather 'abstract' necessity and possibility operators. Let us now concentrate on logical formalisms specially designed for reasoning about certain concrete application domains, such as time, space, knowledge, etc., and show how they can be related to modal logics.

2.1 Temporal logic

Perhaps the most natural and intuitive use of modal logic is reasoning about time. There are many different models of time. In the framework of the possible world semantics we can imagine, for instance, that the *flow of time* is represented as a frame $\mathfrak{F} = \langle T, < \rangle$ in which T is a set of *moments of time* and $<$ a binary *precedence relation* between them. If time is regarded to be linear then we may assume that $<$ is a strict linear order on T, i.e., $<$ is transitive, irreflexive and connected:

$$\forall x, y \in T \ (x < y \vee y < x \vee x = y)$$

(see Section 1.4). The necessity and possibility operators interpreted in such frames can be understood then as 'always in the future' and 'some time in the future,' respectively.

According to Theorem 1.12, the logic determined by the class of *all* strict linear orders is **K4.3**. If we regard time to be *infinite* and *discrete* (in the sense that between any two points there are only finitely many other points) or, on the contrary, *dense* (that is, there is a third point between any two distinct points) then we may need temporal logics determined by the flows of time $\langle \mathbb{N}, < \rangle$, $\langle \mathbb{Z}, < \rangle$, $\langle \mathbb{Q}, < \rangle$ or $\langle \mathbb{R}, < \rangle$, where $\mathbb{N}$, $\mathbb{Z}$, $\mathbb{Q}$, and $\mathbb{R}$ are the sets of natural, integer, rational and real numbers, respectively.

Theorem 2.1. *The following equalities hold true:*

$$\mathsf{Log}\{\langle \mathbb{N}, <\rangle\} = \mathsf{Log}\{\langle \mathbb{Z}, <\rangle\} = \mathbf{K4.3} \oplus \Diamond\top \oplus \Box(\Box p \to p) \to (\Diamond\Box p \to \Box p),$$

$$\mathsf{Log}\{\langle \mathbb{Q}, <\rangle\} = \mathsf{Log}\{\langle \mathbb{R}, <\rangle\} = \mathbf{K4.3} \oplus \Diamond\top \oplus \Box\Box p \to \Box p.$$

For proofs see, e.g., (Segerberg 1970, Goldblatt 1982). As both of these logics have the fmp (for instance, $\mathsf{Log}\{\langle \mathbb{N}, <\rangle\}$ is determined by the class of finite 'balloons,' i.e., finite strict linear orders ending with nondegenerate finite clusters), they are decidable. Moreover, it is easy to show (see, e.g., Ono and Nakamura 1980, Sistla and Clarke 1985) that we have the following:

Theorem 2.2. $\mathsf{Log}\{\langle \mathbb{N}, <\rangle\}$ *and* $\mathsf{Log}\{\langle \mathbb{Q}, <\rangle\}$ *have the polynomial fmp and are* coNP*-complete.*

One can understand future as 'from now on' (i.e., including the present moment) and consider logics determined by classes of (reflexive) linear orders. Again by Theorem 1.12, the logic determined by the class of all linear orders is **S4.3**. For other reflexive flows of time we have:

$$\mathsf{Log}\{\langle \mathbb{N}, \leq\rangle\} = \mathsf{Log}\{\langle \mathbb{Z}, \leq\rangle\} = \mathbf{S4.3} \oplus \Box(\Box(p \to \Box p) \to p) \to (\Diamond\Box p \to \Box p),$$

$$\mathsf{Log}\{\langle \mathbb{Q}, \leq\rangle\} = \mathsf{Log}\{\langle \mathbb{R}, \leq\rangle\} = \mathbf{S4.3}.$$

Similarly to the case of strict linear orders, both of these logics have the polynomial fmp and are coNP-complete.

The unimodal language $\mathcal{ML}$ is able to speak only about the future, but we can easily extend it to deal with the past by adding one more pair of necessity and possibility operators interpreted by the converse $<^{-1}$ of $<$ (that is, by $>$). Thus we come to the bimodal temporal language $\mathcal{ML}_2$ with two boxes $\Box_F$ ('always in the future') and $\Box_P$ ('always in the past'), together with their duals $\Diamond_F$ and $\Diamond_P$, respectively. Although frames for this language are of the form $\langle T, <, >\rangle$, we can safely write $\mathfrak{F} = \langle T, <\rangle$ keeping in mind that $\Box_P$ is interpreted by the converse of $<$. The truth-relation for $\Box_F$ and $\Box_P$ in a model based on a flow of time $\mathfrak{F} = \langle T, <\rangle$ is defined as follows:

$$t \models \Box_F\psi \quad \text{iff} \quad (\forall t' > t)\ t' \models \psi;$$

$$t \models \Box_P\psi \quad \text{iff} \quad (\forall t' < t)\ t' \models \psi.$$

Even this simple language is enough to say that some property φ will take place infinitely often in the future or that φ is always caused by another property ψ:

$$\Box_F\Diamond_F\varphi,$$

$$\Box_F(\varphi \to \Diamond_P\psi) \wedge \Box_P(\varphi \to \Diamond_P\psi) \wedge (\varphi \to \Diamond_P\psi).$$

Statements of this sort are used in the field of verification and specification of reactive systems (such as operating systems); see, e.g., (Manna and Pnueli 1992, 1995).

We can also define operators $\boxdot$ and $\diamondplus$ which behave on linear frames like the universal modalities:

$$\boxdot\varphi = \varphi \wedge \Box_F\varphi \wedge \Box_P\varphi,$$
$$\diamondplus\varphi = \varphi \vee \Diamond_F\varphi \vee \Diamond_P\varphi.$$

Moreover, we can say that some formula φ holds precisely at one point of a flow of time:

$$\diamondplus!\varphi = \diamondplus(\varphi \wedge \neg\Diamond_F\varphi \wedge \neg\Diamond_P\varphi).$$

Denote by **Lin** the bimodal logic determined by the class of all strict linear orders, i.e.,

$$\mathbf{Lin} = \{\varphi \in \mathcal{ML}_2 \mid \mathfrak{F} \models \varphi,\ \mathfrak{F} \text{ a strict linear order}\}.$$

This logic can be axiomatized as

$$\begin{aligned}\mathbf{Lin} = \mathbf{K4}_2 &\oplus p \to \Box_F\Diamond_P p\\ &\oplus p \to \Box_P\Diamond_F p\\ &\oplus \Diamond_F\Diamond_P p \vee \Diamond_P\Diamond_F p \to p \vee \Diamond_F p \vee \Diamond_P p\end{aligned}$$

(see, e.g., Gabbay *et al.* 1994). Roughly speaking, if R_F and R_P are (transitive) accessibility relations interpreting $\Box_F$ and $\Box_P$, then the first two axioms describe the conditions $R_F \subseteq R_P^{-1}$ and $R_P \subseteq R_F^{-1}$, respectively, and the third one says that these relations are connected.

Given a class $\mathcal{C}$ of flows of time, we denote by $\mathsf{Log}_{FP}\mathcal{C}$ the bimodal logic (with the operators $\Box_F$ and $\Box_P$) determined by $\mathcal{C}$. (Recall that $\mathsf{Log}\mathcal{C}$ denotes the unimodal logic determined by $\mathcal{C}$.) To simplify notation, we will write $\mathsf{Log}_{FP}(\mathbb{T})$ instead of $\mathsf{Log}_{FP}\{\langle\mathbb{T}, <\rangle\}$, for $\mathbb{T} \in \{\mathbb{N}, \mathbb{Z}, \mathbb{Q}, \mathbb{R}\}$.

Theorem 2.3. *The following equalities hold true:*

$$\begin{aligned}\mathsf{Log}_{FP}(\mathbb{N}) &= \mathbf{Lin} \oplus \Diamond_F\top \oplus \Box_P(\Box_P p \to p) \to \Box_P p\\ &\qquad \oplus \Box_F(\Box_F p \to p) \to (\Diamond_F\Box_F p \to \Box_F p),\\ \mathsf{Log}_{FP}(\mathbb{Z}) &= \mathbf{Lin} \oplus \Diamond_F\top \oplus \Diamond_P\top\\ &\qquad \oplus \Box_F(\Box_F p \to p) \to (\Diamond_F\Box_F p \to \Box_F p)\\ &\qquad \oplus \Box_P(\Box_P p \to p) \to (\Diamond_P\Box_P p \to \Box_P p),\\ \mathsf{Log}_{FP}(\mathbb{Q}) &= \mathbf{Lin} \oplus \Diamond_F\top \oplus \Diamond_P\top \oplus \Box_F\Box_F p \to \Box_F p,\\ \mathsf{Log}_{FP}(\mathbb{R}) &= \mathsf{Log}_{FP}(\mathbb{Q}) \oplus \boxdot(\Box_P p \to \Diamond_F\Box_P p) \to (\Box_P p \to \Box_F p).\end{aligned}$$

We refer the reader to (Goldblatt 1982) for a proof of this theorem; see also (Bull 1968, Segerberg 1970, Wolter 1996).

Theorem 2.4. *The decision problem for the logics* **Lin**, $\mathsf{Log}_{FP}(\mathbb{N})$, $\mathsf{Log}_{FP}(\mathbb{Z})$, $\mathsf{Log}_{FP}(\mathbb{Q})$ *and* $\mathsf{Log}_{FP}(\mathbb{R})$ *is* coNP*-complete.*

The results for **Lin**, $\mathsf{Log}_{FP}(\mathbb{Q})$ and $\mathsf{Log}_{FP}(\mathbb{R})$ follow from the fact that these logics have the polynomial fmp. For $\mathsf{Log}_{FP}(\mathbb{N})$ and $\mathsf{Log}_{FP}(\mathbb{Z})$ (which do not have the fmp), the complexity results can be obtained by proving that these logics are determined by some special models of polynomial size (Wolter 1996).

'Always in the future' and 'always in the past' are just one type of possible temporal operators. When reasoning about the behavior of programs, we quite often need to say that if at some moment of time the program is at state φ, then at the next moment it passes to a state ψ. This behavior can be captured by the *next-time operator* denoted by $\bigcirc$ and semantically defined in linear orders $\mathfrak{F} = \langle T, < \rangle$ in the following way:

$$t \models \bigcirc\varphi \quad \text{iff} \quad \text{there is an immediate } <\text{-successor } t+1 \text{ of } t \text{ and } t+1 \models \varphi.$$

The statement above about programs can be represented then as

$$\boxed{\forall}\, (\varphi \to \bigcirc\psi).$$

More expressive are the *binary* temporal operators 'since' and 'until' with their natural meaning:

- φ since ψ: 'φ has been the case since ψ';
- φ until ψ: 'φ will be the case until ψ'.

We will denote these operators by $\mathcal{S}$ and $\mathcal{U}$, respectively. Let $\mathcal{ML}_{\mathcal{SU}}$ be the temporal language which results from the language $\mathcal{L}$ of classical propositional logic by extending it with the binary connectives $\mathcal{S}$, $\mathcal{U}$ and the corresponding formula formation clause: if φ and ψ are $\mathcal{ML}_{\mathcal{SU}}$-formulas then so are $\varphi\mathcal{S}\psi$ and $\varphi\mathcal{U}\psi$.

The semantics of the new operators is defined as follows. Let $\mathfrak{F} = \langle T, < \rangle$ be a strict linear order and $t_1, t_2 \in T$. Denote by (t_1, t_2) the *open interval* $\{t \in T \mid t_1 < t < t_2\}$. Then

$$t_1 \models \varphi\mathcal{S}\psi \text{ iff there is } t_2 < t_1 \text{ such that } t_2 \models \psi \text{ and } t \models \varphi \text{ for all } t \in (t_2, t_1),$$
$$t_1 \models \varphi\mathcal{U}\psi \text{ iff there is } t_2 > t_1 \text{ such that } t_2 \models \psi \text{ and } t \models \varphi \text{ for all } t \in (t_1, t_2).$$

Note that the operators $\Diamond_F$ and $\Diamond_P$ can be defined in $\mathcal{ML}_{\mathcal{SU}}$ as abbreviations

$$\Diamond_F \varphi = \top \mathcal{U} \varphi, \quad \Diamond_P \varphi = \top \mathcal{S} \varphi, \quad \bigcirc \varphi = \bot \mathcal{U} \varphi.$$

In this language we can say, for instance, that a system will respond (p will be true) only when it gets a request (q is true):

$$\Diamond_F p \rightarrow (\neg p) \mathcal{U} q.$$

The expressive power of the temporal language $\mathcal{ML}_{\mathcal{SU}}$ over the flows of time $\langle \mathbb{N}, < \rangle$, $\langle \mathbb{Z}, < \rangle$ and $\langle \mathbb{R}, < \rangle$ can be characterized in terms of the first-order sublanguage $\mathcal{QL}_t$ of $\mathcal{QL}$ having one binary predicate symbol $<$ and countably infinitely many unary predicate symbols $P_0, P_1, \ldots$. (Here we denote individual variables by t, t', etc.) The language $\mathcal{QL}_t$ is interpreted in flows of time $\langle T, < \rangle$: the interpretation of $<$ is given by the flow, and P_i are interpreted by arbitrary subsets of T. Consider the following *standard translation* $\cdot^\star$ of $\mathcal{ML}_{\mathcal{SU}}$ into $\mathcal{QL}_t$:

$$\begin{aligned}
p_i^\star &= P_i(t), \quad p_i \text{ a propositional variable,} \\
(\varphi \wedge \psi)^\star &= \varphi^\star \wedge \psi^\star, \\
(\neg \varphi)^\star &= \neg \varphi^\star, \\
(\varphi \mathcal{U} \psi)^\star &= \exists t' > t \, (\psi^\star\{t'/t\} \wedge \forall t''(t < t'' < t' \rightarrow \varphi^\star\{t''/t\})), \\
(\varphi \mathcal{S} \psi)^\star &= \exists t' < t \, (\psi^\star\{t'/t\} \wedge \forall t''(t' < t'' < t \rightarrow \varphi^\star\{t''/t\})).
\end{aligned}$$

Observe that, for any $\mathcal{ML}_{\mathcal{SU}}$-formula φ, $\varphi^\star$ is a $\mathcal{QL}_t$-formula with precisely one free variable t Let $\mathfrak{M} = \langle \mathfrak{F}, \mathfrak{V} \rangle$ be a model for $\mathcal{ML}_{\mathcal{SU}}$ based on a flow of time $\mathfrak{F} = \langle T, < \rangle$. Define a first-order structure

$$I(\mathfrak{M}) = \left\langle T, <, P_0^{I(\mathfrak{M})}, P_1^{I(\mathfrak{M})}, \ldots \right\rangle$$

by taking $P_i^{I(\mathfrak{M})} = \mathfrak{V}(p_i)$, for all $i < \omega$. Then we clearly have, for every $t \in T$ and every $\mathcal{ML}_{\mathcal{SU}}$-formula φ:

$$(\mathfrak{M}, t) \models \varphi \quad \text{iff} \quad I(\mathfrak{M}) \models \varphi^\star[t].$$

In other words, $\mathcal{ML}_{\mathcal{SU}}$ can be regarded as a fragment of $\mathcal{QL}_t$. Say that $\mathcal{ML}_{\mathcal{SU}}$ is *expressively complete* for a class $\mathcal{C}$ of flows of time if, for any $\mathcal{QL}_t$-formula $\phi(t)$ with one free variable, there exists an $\mathcal{ML}_{\mathcal{SU}}$-formula φ such that, for all models $\mathfrak{M}$ based on some $\mathfrak{F} \in \mathcal{C}$,

$$I(\mathfrak{M}) \models \forall t \, (\phi \leftrightarrow \varphi^\star).$$

This means that, modulo the flows of time in $\mathcal{C}$, the temporal language $\mathcal{ML}_{\mathcal{SU}}$ has the same expressive power as the fragment of the first-order language $\mathcal{QL}_t$ consisting of formulas with one free variable.

The following result is known as *Kamp's theorem*; for proofs see (Kamp 1968) and (Gabbay *et al.* 1994).

Theorem 2.5. *$\mathcal{ML}_{\mathcal{SU}}$ is expressively complete for the flows of time $\{\langle \mathbb{N}, < \rangle\}$, $\{\langle \mathbb{Z}, < \rangle\}$, $\{\langle \mathbb{R}, < \rangle\}$, and the class of all finite strict linear orders.*

Note that $\mathcal{ML}_{\mathcal{SU}}$ is expressively complete neither for $\{\langle \mathbb{Q}, < \rangle\}$ nor for the class of all strict linear orders. Expressive completeness for these classes can be obtained by adding the so-called *Stavi connectives* to $\mathcal{ML}_{\mathcal{SU}}$; we refer the reader to (Gabbay *et al.* 1994) for more information.

Given a class $\mathcal{C}$ of strict linear orders, we denote by $\mathsf{Log}_{\mathcal{SU}}(\mathcal{C})$ the temporal logic in the language $\mathcal{ML}_{\mathcal{SU}}$ determined by $\mathcal{C}$. To simplify notation, if $\mathcal{C}$ consists of a single flow of time $\mathfrak{F}$ then we will write $\mathsf{Log}_{\mathcal{SU}}(\mathfrak{F})$ instead of $\mathsf{Log}_{\mathcal{SU}}(\{\mathfrak{F}\})$.

We denote by $\mathbf{Lin}_{\mathcal{SU}}$ the temporal logic determined by the class of all strict linear orders. $\mathsf{Log}_{\mathcal{SU}}(\mathbb{N})$, $\mathsf{Log}_{\mathcal{SU}}(\mathbb{Z})$, $\mathsf{Log}_{\mathcal{SU}}(\mathbb{Q})$ and $\mathsf{Log}_{\mathcal{SU}}(\mathbb{R})$ are the logics of $\langle \mathbb{N}, < \rangle$, $\langle \mathbb{Z}, < \rangle$, etc. All these logics are known to be finitely axiomatizable and PSPACE-complete; see (Goldblatt 1982, Sistla and Clarke 1985, Gabbay and Hodkinson 1990, Reynolds 1992, 1999, 2003). We present here an axiomatization of a somewhat simpler logic which has found many applications as a program verification and specification formalism (see Manna and Pnueli 1992, 1995). The logic is known as **PTL**, *propositional temporal logic*.[1] It is formulated in the $\mathcal{S}$-free reduct $\mathcal{ML}_{\mathcal{U}}$ of the language $\mathcal{ML}_{\mathcal{SU}}$ and has $\langle \mathbb{N}, < \rangle$ as its intended flow of time, i.e.,

$$\mathbf{PTL} = \mathsf{Log}_{\mathcal{SU}}(\mathbb{N}) \cap \mathcal{ML}_{\mathcal{U}}.$$

Theorem 2.6. **PTL** *can be axiomatized by the following Hilbert-style system:*

Axioms:

$$\Box_F(p \to q) \to (\Box_F p \to \Box_F q),$$

$$\bigcirc(p \to q) \to (\bigcirc p \to \bigcirc q),$$

$$\bigcirc\neg p \leftrightarrow \neg\bigcirc p, \tag{2.1}$$

$$\Box_F p \leftrightarrow \bigcirc p \land \bigcirc\Box_F p, \tag{2.2}$$

$$\Box_F(p \to \bigcirc p) \to \bigcirc(p \to \Box_F p), \tag{2.3}$$

$$p\mathcal{U}q \to \Diamond_F q, \tag{2.4}$$

$$p\mathcal{U}q \leftrightarrow \bigcirc q \lor \bigcirc(p \land p\mathcal{U}q). \tag{2.5}$$

Inference rules: *MP, Subst, and RN for* $\Box_F$.

Theorem 2.7. *The decision problem for* **PTL** *is* PSPACE-*complete.*

For proofs see e.g. (Segerberg 1989, Sistla and Clarke 1985, Gabbay *et al.* 1994) or Corollary 11.36.

[1] This logic is also called **LTL** (*linear temporal logic*); see, e.g., (Clarke *et al.* 2000).

It may be of interest to note that as far as decidability and computational complexity are concerned, it is enough to deal with **PTL** instead of $\mathsf{Log}_{\mathcal{SU}}(\mathbb{N})$, for we have the following (see Gabbay *et al.* 1980):

Proposition 2.8. *$\mathsf{Log}_{\mathcal{SU}}(\mathbb{N})$ is polynomially reducible to* **PTL**.

Proof. It is sufficient to show that, given a $\mathcal{ML}_{\mathcal{SU}}$-formula φ, one can effectively construct an $\mathcal{S}$-free formula $\bar{\varphi}$ (the length of which is linear in the length of φ) such that $\bar{\varphi}$ is satisfied in a model based on $\langle \mathbb{N}, < \rangle$ iff φ is satisfied in a model based on $\langle \mathbb{N}, < \rangle$. Suppose φ is given. Clearly, without loss of generality we may assume that $\varphi = \vartheta \vee \Diamond_F \vartheta$, so that φ is satisfiable iff it is satisfiable at time point 0. Given a subformula of φ of the form $\psi_1 \mathcal{S} \psi_2$ with $\mathcal{S}$-free ψ_1 and ψ_2, we introduce a fresh propositional variable $p_{\psi_1 \mathcal{S} \psi_2}$. Let φ' be the result of replacing every occurrence of $\psi_1 \mathcal{S} \psi_2$ in φ with $p_{\psi_1 \mathcal{S} \psi_2}$, and let

$$\varphi'' = \varphi' \wedge \neg p_{\psi_1 \mathcal{S} \psi_2} \wedge \Box_F^+ \Big(\bigcirc p_{\psi_1 \mathcal{S} \psi_2} \leftrightarrow \big(\psi_2 \vee (\psi_1 \wedge p_{\psi_1 \mathcal{S} \psi_2}) \big) \Big),$$

where $\Box_F^+ \chi$ denotes $\chi \wedge \Box_F \chi$. We claim that φ'' is satisfiable at 0 iff φ is satisfiable at 0. Indeed, suppose first that $(\mathfrak{M}, 0) \models \varphi''$, for some model $\mathfrak{M}$. By induction on $n \in \mathbb{N}$ one can show that $(\mathfrak{M}, n) \models p_{\psi_1 \mathcal{S} \psi_2} \leftrightarrow \psi_1 \mathcal{S} \psi_2$, for every $n \in \mathbb{N}$. It follows that $(\mathfrak{M}, 0) \models \varphi$.

Conversely, if $(\mathfrak{M}, 0) \models \varphi$ for some $\mathfrak{M} = \langle \langle \mathbb{N}, < \rangle, \mathfrak{V} \rangle$, then we can extend $\mathfrak{V}$ to $\mathfrak{V}'$ by taking

$$\mathfrak{V}'(p_{\psi_1 \mathcal{S} \psi_2}) = \{ n \mid (\mathfrak{M}, n) \models \psi_1 \mathcal{S} \psi_2 \}.$$

It should be clear that φ'' is true at 0 in the resulting model $\langle \langle \mathbb{N}, < \rangle, \mathfrak{V}' \rangle$.

By iterating this process sufficiently many times we end up with an $\mathcal{S}$-free formula $\bar{\varphi}$. As $\bar{\varphi}$ is $\mathcal{S}$-free, it is satisfiable at 0 iff it is satisfiable at all. Thus, $\bar{\varphi}$ is as required. ❑

As a consequence we obtain:

Theorem 2.9. *The decision problem for $\mathsf{Log}_{\mathcal{SU}}(\mathbb{N})$ is* PSPACE*-complete.*

We will also be considering the bimodal fragment $\mathbf{PTL}_{\Box\bigcirc}$ of **PTL** having only $\Box_F$ and $\bigcirc$ as its temporal operators, i.e.,

$$\mathbf{PTL}_{\Box\bigcirc} = \mathsf{Log}\{\langle \mathbb{N}, <, +1 \rangle\}.$$

It is not hard to see that if we omit axioms (2.4) and (2.5), then the resulting Hilbert-type system axiomatizes $\mathbf{PTL}_{\Box\bigcirc}$. Actually, again it turns out that in a sense $\mathbf{PTL}_{\Box\bigcirc}$ has the same expressive power as full **PTL**. For we have the following:

Proposition 2.10. **PTL** *is polynomially reducible to* $\mathbf{PTL}_{\Box\bigcirc}$.

Proof. Given an $\mathcal{ML}_{\mathcal{U}}$-formula φ, denote by φ^U the result of replacing every subformula of the form $\psi\mathcal{U}\chi$ in φ with a fresh propositional variable $p_{\psi\mathcal{U}\chi}$. Let $\mathcal{R}_U(\varphi)$ be the union of the sets

$$\{p_{\psi\mathcal{U}\chi} \to \Diamond_F \chi^U \mid \psi\mathcal{U}\chi \in sub\,\varphi\} \quad \text{and}$$

$$\{p_{\psi\mathcal{U}\chi} \leftrightarrow (\bigcirc\chi^U \vee (\bigcirc\psi^U \wedge \bigcirc p_{\psi\mathcal{U}\chi})) \mid \psi\mathcal{U}\chi \in sub\,\varphi\}.$$

We will show that, for every $\mathcal{ML}_{\mathcal{U}}$-formula φ,

$$\varphi \in \mathbf{PTL} \quad \text{iff} \quad \Box_F^+ \bigwedge \mathcal{R}_U(\varphi) \to \varphi^U \in \mathbf{PTL}_{\Box\bigcirc}.$$

Suppose first that

$$(\mathfrak{M}, 0) \models \Box_F^+ \bigwedge \mathcal{R}_U(\varphi) \wedge \neg\varphi^U$$

for some model $\mathfrak{M}$ based on $\langle \mathbb{N}, <, +1\rangle$. (This $\mathfrak{M}$ can also be considered as a model based on $\langle \mathbb{N}, <\rangle$.) We claim that for every subformula α of φ and every $n \in \mathbb{N}$,

$$(\mathfrak{M}, n) \models \alpha \quad \text{iff} \quad (\mathfrak{M}, n) \models \alpha^U. \tag{2.6}$$

The proof is by induction on the construction of α, where the only non-trivial case is $\alpha = \psi\mathcal{U}\chi$.

($\Rightarrow$) If $(\mathfrak{M}, n) \models \psi\mathcal{U}\chi$ then there is an $m > n$ such that $(\mathfrak{M}, m) \models \chi$ and $(\mathfrak{M}, k) \models \psi$ for all $k \in (n, m)$. It follows by the induction hypothesis that $(\mathfrak{M}, m) \models \chi^U$, whence $(\mathfrak{M}, m-1) \models \bigcirc\chi^U$, and so $(\mathfrak{M}, m-1) \models p_{\psi\mathcal{U}\chi}$, since we have $(\mathfrak{M}, i) \models \bigwedge \mathcal{R}_U(\varphi)$ for all $i \in \mathbb{N}$. If $n < m-1$ then we have $(\mathfrak{M}, m-2) \models \bigcirc p_{\psi\mathcal{U}\chi}$ and $(\mathfrak{M}, m-1) \models \bigcirc\psi^U$, from which we obtain $(\mathfrak{M}, m-2) \models p_{\psi\mathcal{U}\chi}$. By repeating this argument sufficiently many times we shall end up with $(\mathfrak{M}, n) \models p_{\psi\mathcal{U}\chi}$, as required.

($\Leftarrow$) If $(\mathfrak{M}, n) \models p_{\psi\mathcal{U}\chi}$ then $(\mathfrak{M}, n) \models \bigcirc\chi^U \vee (\bigcirc\psi^U \wedge \bigcirc p_{\psi\mathcal{U}\chi})$. We either have $(\mathfrak{M}, n+1) \models \chi^U$, whence by the induction hypothesis $(\mathfrak{M}, n+1) \models \chi$, and so $(\mathfrak{M}, n) \models \psi\mathcal{U}\chi$, or $(\mathfrak{M}, n+1) \models \psi^U \wedge p_{\psi\mathcal{U}\chi}$. Since $(\mathfrak{M}, n) \models p_{\psi\mathcal{U}\chi} \to \Diamond_F\chi^U$, there is an $m > n$ such that $(\mathfrak{M}, m) \models \chi^U$, and so by the induction hypothesis, $(\mathfrak{M}, m) \models \chi$. Now, by using the above argument at most m times, we can show that ψ holds 'everywhere in between,' i.e., $(\mathfrak{M}, n) \models \psi\mathcal{U}\chi$, which completes the proof of (2.6).

Now it follows from (2.6) that $(\mathfrak{M}, 0) \not\models \varphi$, as required.

Conversely, assume that φ is refuted in some model $\mathfrak{M} = \langle\langle \mathbb{N}, <\rangle, \mathfrak{V}\rangle$. Extend $\mathfrak{V}$ to a valuation $\mathfrak{V}^+$ by taking, for all $\psi\mathcal{U}\chi \in sub\,\varphi$,

$$\mathfrak{V}^+(p_{\psi\mathcal{U}\chi}) = \{n \in \mathbb{N} \mid (\mathfrak{M}, n) \models \psi\mathcal{U}\chi\},$$

and let $\mathfrak{M}^+ = \langle\langle \mathbb{N}, <, +1\rangle, \mathfrak{V}^+\rangle$. We leave it to the reader to show that for all $n \in \mathbb{N}$ and all $\psi \in sub\,\varphi$,

$$(\mathfrak{M}^+, n) \models \bigwedge \mathcal{R}_U(\varphi), \qquad \text{and}$$
$$(\mathfrak{M}, n) \models \psi \quad \text{iff} \quad (\mathfrak{M}^+, n) \models \psi^U.$$

Therefore, $\Box_F^+ \bigwedge \mathcal{R}_U(\varphi) \to \varphi^U$ is refuted in $\mathfrak{M}^+$. ❑

It follows from Theorem 2.7 that $\mathbf{PTL}_{\Box\bigcirc}$ is also PSPACE-complete. A discussion of some related topics can be found in (Sistla and Zuck 1987).

Remark 2.11. It is worth mentioning that there exist rooted frames for $\mathbf{PTL}_{\Box\bigcirc}$ different from $\langle \mathbb{N}, <, +1\rangle$. However, *all* of them satisfy two important properties. First, by (2.1), the accessibility relation $R_\bigcirc$ interpreting $\bigcirc$ (as a box-like operator) is a function (i.e., $\forall x \exists! y\; xR_\bigcirc y$) and, by (2.3) and (2.2), the relation corresponding to $\Box_F$ is the transitive closure of $R_\bigcirc$ (for a proof see, e.g., Blackburn *et al.* 2001). Second, every rooted frame for $\mathsf{Log}\{\langle \mathbb{N}, <\rangle\}$ (and for **PTL**) different from $\langle \mathbb{N}, <\rangle$ is a *balloon*—a finite strict linear order followed by a (possibly uncountably infinite) nondegenerate cluster (see, e.g., Goldblatt 1987). So every rooted frame for $\mathbf{PTL}_{\Box\bigcirc}$ different from $\langle \mathbb{N}, <, +1\rangle$ is of the form $\langle W, R, f\rangle$, where $\langle W, R\rangle$ is a balloon and f is a function on W that is the R-successor on the 'finite linear order part' and arbitrary otherwise. In particular, every countable rooted frame for $\mathbf{PTL}_{\Box\bigcirc}$ is in fact a p-morphic image of $\langle \mathbb{N}, <, +1\rangle$.

2.2 Interval temporal logic

In Section 2.1 we considered temporal logics interpreted in Kripke models the points of which are linearly ordered and represent moments of time. In this section we discuss another approach to temporal reasoning which takes as primitive *temporal intervals* rather than points. Interval-based temporal logics originate from the same areas as modal logic in general: philosophy, linguistics, computer science and artificial intelligence. They arise from the observation that time-dependent assertions can be of different kinds. Some of them describe instant situations and can be evaluated at single moments (points) of time, for example:

My temperature is 37.3 C°.

But there are also assertions that can be evaluated only at some period (interval) of time, say:

Mary often visits her mother.

And of course there are cases when we can regard temporal assertions as both point-dependent and interval-dependent, perhaps with some difference in meaning, for instance:

William Shakespeare is an actor.

The interplay between these types of propositions has been studied thoroughly in linguistical semantics, see e.g. (Dowty 1979, Kamp 1979, Bennett 1977, Nishimura 1980). Similar considerations can be found in works on computational logic (Lamport 1985, Kowalski and Sergot 1985). We address the reader to (van Benthem 1995) for further discussion and references.

Now let us consider some specific systems of interval logic. Allen (1983) observed that relative positions of any two intervals i and j can be described by precisely one of the following thirteen basic interval relations: $\mathsf{before}(i,j)$, $\mathsf{meets}(i,j)$, $\mathsf{overlaps}(i,j)$, $\mathsf{during}(i,j)$, $\mathsf{starts}(i,j)$, $\mathsf{finishes}(i,j)$, their inverses (i.e., $\mathsf{before}(j,i)$, $\mathsf{meets}(j,i)$, etc.), and $\mathsf{equal}(i,j)$. Let us denote by $\mathcal{A}\ell\ell$-13 the language whose alphabet contains these thirteen binary predicate symbols, a sufficient supply of *interval variables* i, j, etc., and the Booleans. Formulas of $\mathcal{A}\ell\ell$-13 are just Boolean combinations of the above listed atomic ones.

In order to provide a semantics for $\mathcal{A}\ell\ell$-13 formulas, suppose that the flow of time is represented as a strict linear order $\mathfrak{F} = \langle W, < \rangle$. (Often the intended flow of time in interval logic is not just an arbitrary strict linear order, but a dense order like $\langle \mathbb{Q}, < \rangle$ or $\langle \mathbb{R}, < \rangle$.) An *assignment in* $\mathfrak{F}$ is a function $\mathfrak{a}$ mapping the interval variables into *temporal intervals in* $\mathfrak{F}$. There may be different views on what the temporal intervals in $\mathfrak{F}$ should be. First we take perhaps the most 'liberal' version by defining them as arbitrary non-empty convex sets in $\mathfrak{F}$. In other words, a temporal interval $\mathfrak{a}(i)$ in $\mathfrak{F}$ is a non-empty subset of W such that

$$\forall x, y \in \mathfrak{a}(i)\ \forall z \in W (x < z < y \rightarrow z \in \mathfrak{a}(i)).$$

For example, if $u < v$ then the open interval (u, v), the sets

$$(u, v] = \{w \in W \mid u < w \leq v\},$$
$$[u, v) = \{w \in W \mid u \leq w < v\},$$

and if $u \leq v$ (i.e., $u < v$ or $u = v$) then the set

$$[u, v] = \{w \in W \mid u \leq w \leq v\}$$

are temporal intervals. Now, following (Allen 1984), the *truth-relation* $\mathfrak{F} \models^{\mathfrak{a}} \varphi$ for atomic $\mathcal{A}\ell\ell$-13 formulas can be defined as follows (see Fig. 2.1):

$\mathfrak{F} \models^{\mathfrak{a}} \mathsf{equals}(i,j)$ iff $\mathfrak{a}(i) = \mathfrak{a}(j)$,

$\mathfrak{F} \models^{\mathfrak{a}} \mathsf{before}(i,j)$ iff $\forall x, y\ \big(x \in \mathfrak{a}(i) \wedge y \in \mathfrak{a}(j) \rightarrow x < y \wedge \exists z (x < z < y \wedge z \notin \mathfrak{a}(i) \wedge z \notin \mathfrak{a}(j))\big)$,

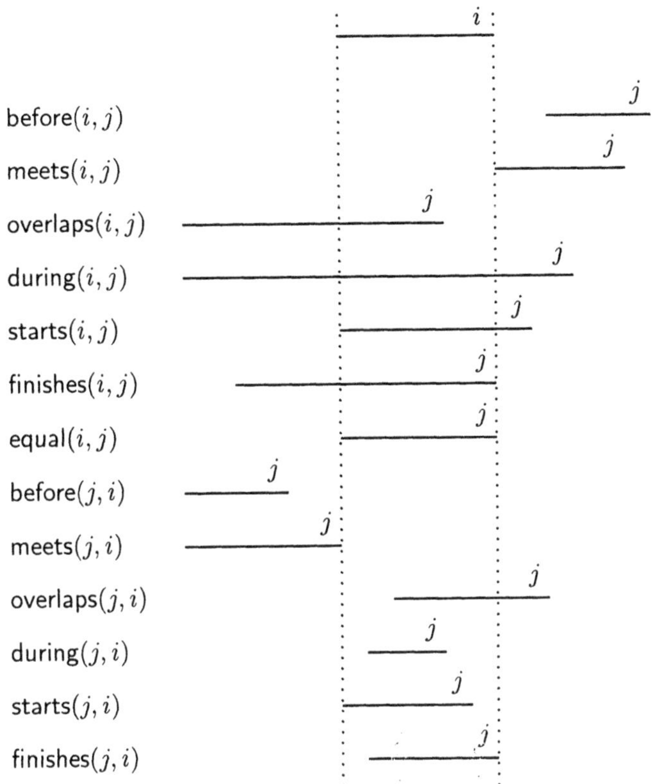

Figure 2.1: The atomic formulas of $\mathcal{A}\ell\ell$-13.

$$\begin{array}{lll}
\mathfrak{F} \models^{\mathfrak{a}} \mathsf{meets}(i,j) & \text{iff} & \forall x, y\, \big(x \in \mathfrak{a}(i) \wedge y \in \mathfrak{a}(j) \to x < y \wedge \forall z (x < z < y \\
& & \to\; z \in \mathfrak{a}(i) \vee z \in \mathfrak{a}(j))\big), \\
\mathfrak{F} \models^{\mathfrak{a}} \mathsf{overlaps}(i,j) & \text{iff} & \mathfrak{a}(i) \cap \mathfrak{a}(j) \neq \emptyset \wedge \exists x, y\, \big(x < y \\
& & \wedge\, x \in \mathfrak{a}(j) \wedge x \notin \mathfrak{a}(i) \wedge y \in \mathfrak{a}(i) \wedge y \notin \mathfrak{a}(j)\big), \\
\mathfrak{F} \models^{\mathfrak{a}} \mathsf{starts}(i,j) & \text{iff} & \mathfrak{a}(i) \subseteq \mathfrak{a}(j) \wedge \mathfrak{a}(i) \neq \mathfrak{a}(j) \wedge \forall x, y\, \big(x < y \\
& & \wedge\, x \in \mathfrak{a}(j) \wedge y \in \mathfrak{a}(i) \to x \in \mathfrak{a}(i)\big), \\
\mathfrak{F} \models^{\mathfrak{a}} \mathsf{during}(i,j) & \text{iff} & \exists x, y, z\, \big(x < y < z \wedge x \in \mathfrak{a}(j) \wedge x \notin \mathfrak{a}(i) \\
& & \wedge\, y \in \mathfrak{a}(i) \wedge z \in \mathfrak{a}(j) \wedge z \notin \mathfrak{a}(i)\big), \\
\mathfrak{F} \models^{\mathfrak{a}} \mathsf{finishes}(i,j) & \text{iff} & \mathfrak{a}(i) \subseteq \mathfrak{a}(j) \wedge \mathfrak{a}(i) \neq \mathfrak{a}(j) \wedge \forall x, y\, \big(x < y \\
& & \wedge\, y \in \mathfrak{a}(j) \wedge x \in \mathfrak{a}(i) \to y \in \mathfrak{a}(i)\big).
\end{array}$$

The truth-conditions for the Booleans are the same as in **Cl**. We say that φ *is satisfied in* $\mathfrak{F}$ if $\mathfrak{F} \models^{\mathfrak{a}} \varphi$ holds for some assignment $\mathfrak{a}$ in $\mathfrak{F}$; φ *is satisfiable in a class* $\mathcal{C}$ of flows of time if φ is satisfied in $\mathfrak{F}$ for some $\mathfrak{F} \in \mathcal{C}$.

Using formulas of this language we can express various constraints on time intervals, which can be used for representing qualitative temporal information, for instance, in planning (see, e.g., Allen *et al.* 1991, Allen and Koomen 1983). Usually $\mathcal{A}\ell\ell$-13 serves as a basis for more complex languages which, besides temporal constraints, use other predicates such as $\mathsf{HOLDS}(P, i)$ (property P holds during interval i), $\mathsf{OCCURS}(E, i)$ (event E happens over interval i). Some examples will be provided in Section 3.2.

The reader may now be wondering why such a formalism as $\mathcal{A}\ell\ell$-13 is discussed in a book on modal logic. First, because it can be regarded as a fragment of a suitable point-based temporal logic. Indeed, let us define a translation $\cdot^*$ from $\mathcal{A}\ell\ell$-13 into the temporal language $\mathcal{ML}_2$ of the previous section (cf. Blackburn 1992). For atomic formulas we take:

$$
\begin{aligned}
(\mathsf{equals}(i,j))^* &= \boxdot(p_i \leftrightarrow p_j),\\
(\mathsf{before}(i,j))^* &= \boxdot(p_i \to \neg p_j \wedge \Diamond_F p_j) \wedge \diamondplus(\neg p_i \wedge \neg p_j \wedge \Diamond_P p_i \wedge \Diamond_F p_j),\\
(\mathsf{meets}(i,j))^* &= \boxdot(p_i \to \neg p_j \wedge \Diamond_F p_j) \wedge \neg\diamondplus(\neg p_i \wedge \neg p_j \wedge \Diamond_P p_i \wedge \Diamond_F p_j),\\
(\mathsf{overlaps}(i,j))^* &= \diamondplus(p_i \wedge p_j) \wedge \diamondplus(p_i \wedge \neg p_j \wedge \Diamond_P p_j) \wedge \diamondplus(p_j \wedge \neg p_i \wedge \Diamond_F p_i),\\
(\mathsf{starts}(i,j))^* &= \boxdot(p_i \to p_j) \wedge \diamondplus(p_j \wedge \neg p_i) \wedge \boxdot(p_i \wedge p_j \to \Box_P(p_j \to p_i)),\\
(\mathsf{during}(i,j))^* &= \boxdot(p_i \to p_j) \wedge \diamondplus(p_j \wedge \neg p_i \wedge \Diamond_P p_i) \wedge \diamondplus(p_j \wedge \neg p_i \wedge \Diamond_F p_i),\\
(\mathsf{finishes}(i,j))^* &= \boxdot(p_i \to p_j) \wedge \diamondplus(p_j \wedge \neg p_i) \wedge \boxdot(p_i \wedge p_j \to \Box_F(p_j \to p_i)).
\end{aligned}
$$

Now, given an $\mathcal{A}\ell\ell$-13 formula φ, we replace in it all atomic subformulas ψ with ψ^* and add to the result the conjunct

$$
\diamondplus p_i \wedge \boxdot(\Diamond_P p_i \wedge \Diamond_F p_i \to p_i), \tag{2.7}
$$

for every interval variable i occurring in φ (which ensures that the propositional variable p_i associated with the interval variable i is interpreted by a nonempty convex set). The resultant formula is denoted by φ^*. It is not hard to see that, for every $\mathcal{A}\ell\ell$-13 formula φ and every flow $\mathfrak{F}$ of time, φ is satisfiable in $\mathfrak{F}$ iff φ^* is satisfiable in $\mathfrak{F}$.

By appropriately changing the formulas of this translation one can capture different understandings of the nature of temporal intervals. For example, if we want temporal intervals to contain at least two points then the first conjunct of (2.7) should be replaced by the formula $\diamondplus(p_i \wedge \Diamond_F p_i)$. The reader can try to define modal formulas describing Allen's relations between only closed intervals $[u, v]$ or between only open ones (u, v).

As $\mathcal{A\ell\ell}$-13 clearly contains **Cl** (a propositional variable p_i can be translated to $\mathsf{equals}(i,j)$, for some interval variable $j \neq i$), the satisfiability problem for $\mathcal{A\ell\ell}$-13 formulas in any class of flows of time is NP-hard. Moreover, as follows from (Vilain and Kautz 1986), we have:

Theorem 2.12. *The satisfiability problem for $\mathcal{A\ell\ell}$-13 formulas in any class of strict linear flows of time is* NP*-complete.*

For the interested reader, here we give a sketch of a simple proof of how satisfiability of an $\mathcal{A\ell\ell}$-13 formula φ in an arbitrary infinite strict linear flow of time $\mathfrak{F} = \langle W, < \rangle$ can be reduced to satisfiability in the finite linear order

$$\mathfrak{G} = \langle \{0, \dots, 4n-1\}, < \rangle,$$

where n is the number of interval variables in φ. (Theorem 2.12 will follow immediately.)

Suppose φ is satisfied in $\mathfrak{F}$. Without loss of generality we may assume $\mathfrak{F}$ to be *Dedekind-complete*. (This means that every convex set in $\mathfrak{F}$ can be represented as one of the four types of intervals: (u,v), $(u,v]$, $[u,v)$ and $[u,v]$, where $u,v \in W \cup \{-\infty, +\infty\}$, with the standard interpretation of the infinity symbols: $(u,+\infty) = \{v \in W \mid v > u\}$, $(-\infty,u) = \{v \in W \mid v < u\}$, etc. Examples of Dedekind-complete orders are $\langle \mathbb{R}, < \rangle$, $\langle \mathbb{N}, < \rangle$, $\langle \mathbb{Z}, < \rangle$; however, $\langle \mathbb{Q}, < \rangle$ is not Dedekind-complete.) If our $\mathfrak{F}$ is not Dedekind-complete, then we take its completion $\mathfrak{F}'$ (the smallest Dedekind-complete order containing $\mathfrak{F}$). It is readily seen that if φ is satisfied in $\mathfrak{F}$ then it is satisfied in $\mathfrak{F}'$. So let $\mathfrak{F}$ be Dedekind-complete and let $x_0 < \cdots < x_m$, $m < 2n$, be all the endpoints of the intervals interpreting the variables in φ. If a variable i is interpreted by (x_k, x_ℓ) ($[x_k, x_\ell)$, $(x_k, x_\ell]$, $[x_k, x_\ell]$) in $\mathfrak{F}$ then we interpret it as $[2k+1, 2\ell-1]$ (respectively, $[2k, 2\ell-1)$, $[2k+1, 2\ell]$, $[2k, 2\ell]$) in $\mathfrak{G}$. It is not hard to check that $\mathfrak{G}$ satisfies φ under this assignment.

Now suppose that φ is satisfied in $\mathfrak{G}$. As $\mathfrak{G}$ is finite, all of its intervals can be regarded as closed (e.g. $(k,\ell) = [k+1, \ell-1]$). Select points $x_0 < \cdots < x_{4n}$ in $\mathfrak{F}$. Now, if i is interpreted as $[k,\ell]$ in $\mathfrak{G}$ then we interpret it as $[x_k, x_{\ell+1})$ in $\mathfrak{F}$. It is readily checked that φ holds in $\mathfrak{F}$ under this assignment.

The second reason for considering interval temporal logic in this book is that one can construct rather expressive *modal* logics of intervals (see, e.g., Humberstone 1979, van Benthem 1983, Allen and Hayes 1985, Halpern and Shoham 1991). Here we present a variant of the *Halpern–Shoham logic* **HS** following (Marx and Venema 1997). The language of **HS** is $\mathcal{ML}_4$ with four diamonds $\Diamond_s$, $\Diamond_f$, $\Diamond_s^{-1}$, $\Diamond_f^{-1}$, and the corresponding boxes $\Box_s$, $\Box_f$, $\Box_s^{-1}$, $\Box_f^{-1}$. Frames for **HS**, or simply **HS**-*frames*, contain *closed intervals* of the form $[u,v]$, $u \leq v$, of some strict linear order $\mathfrak{F} = \langle W, < \rangle$ as their worlds and interpret $\Diamond_s$, $\Diamond_f$, $\Diamond_s^{-1}$, $\Diamond_f^{-1}$ by the relations iSj, iFj saying that interval i

starts j, interval i *finishes* j, and their converses, respectively. More precisely, let

$$In(\mathfrak{F}) = \{[u,v] \mid u,v \in W,\ u \leq v\}.$$

Then an **HS**-*frame* (corresponding to $\mathfrak{F}$) is the triple $\mathfrak{I}(\mathfrak{F}) = \langle In(\mathfrak{F}), S, F\rangle$, where

$$\begin{array}{lll} [u_1,v_1]S[u_2,v_2] & \text{iff} & u_1 = u_2 \text{ and } v_1 < v_2, \\ [u_1,v_1]F[u_2,v_2] & \text{iff} & u_1 > u_2 \text{ and } v_1 = v_2. \end{array}$$

Such a frame is called an **HS**-frame *over* $\mathfrak{F}$. Thus, for a model $\mathfrak{M} = \langle \mathfrak{I}(\mathfrak{F}), \mathfrak{V}\rangle$ based on $\mathfrak{I}(\mathfrak{F})$, we have

$$\begin{array}{lll} (\mathfrak{M},[u,v]) \models \Diamond_s\varphi & \text{iff} & \exists v' > v\ (\mathfrak{M},[u,v']) \models \varphi, \\ (\mathfrak{M},[u,v]) \models \Diamond_s^{-1}\varphi & \text{iff} & \exists u'\ \big(u \leq u' < v \wedge (\mathfrak{M},[u,u']) \models \varphi\big), \end{array}$$

i.e., $\Diamond_s\varphi$ is true in $[u,v]$ iff φ is true in an interval which has $[u,v]$ as a starting subinterval, and $\Diamond_s^{-1}\varphi$ is true in $[u,v]$ iff φ is true in a starting subinterval of $[u,v]$. The meaning of the other two diamonds is defined analogously.

This language is quite expressive. For example, $[u,v] \models \Box_f^{-1}\Box_s^{-1}\varphi$ iff 'φ is true at all subintervals of $[u,v]$.' Further, the modal operator $\Diamond_s$ represents the basic relation starts of $\mathcal{A\ell\ell}$-13 in the following sense:

$$[u,v] \models \Diamond_s\varphi \ \text{ iff } \ [u,v'] \models \varphi \text{ for some } v' \text{ such that } \mathsf{starts}([u,v],[u,v']) \text{ holds.}$$

In fact, we can define modal operators representing all the thirteen basic relations of $\mathcal{A\ell\ell}$-13 in the same sense. For instance, here is a formula representing meets:

$$\Diamond_m\varphi = (\Box_f^{-1}\bot \wedge \Diamond_s\varphi) \vee \Diamond_f^{-1}(\Box_f^{-1}\bot \wedge \Diamond_s\varphi).$$

Indeed, we have

$$[u,v] \models \Diamond_m\varphi \ \text{ iff } \ [v,w] \models \varphi \text{ for some } w \text{ such that } \mathsf{meets}([u,v],[v,w]) \text{ holds.}$$

One can also characterize many standard properties of linear orders using **HS**-formulas. Say, the formula $\neg(\Diamond_s^{-1}\top \wedge \Box_s^{-1}\Box_s^{-1}\bot)$ is valid in $\mathfrak{I}(\mathfrak{F})$ iff $\mathfrak{F}$ is dense. For more examples consult (Halpern and Shoham 1986, Marx and Venema 1997).

Various classes of strict linear orders give rise to different **HS**-logics. For such a class $\mathcal{C}$, let

$$\mathbf{HS}_{\mathcal{C}} = \{\varphi \in \mathcal{ML}_4 \mid \mathfrak{I}(\mathfrak{F}) \models \varphi, \text{ for all } \mathfrak{F} \in \mathcal{C}\}.$$

Halpern and Shoham (1986) show that the decision problem for $\mathbf{HS}_{\mathcal{C}}$ is very complex for almost all interesting classes $\mathcal{C}$ of linear orders:

Theorem 2.13. *Let $\mathcal{C}$ be any class of linear orders such that at least one member of $\mathcal{C}$ contains an infinite ascending chain of distinct points. Then* $\mathbf{HS}_{\mathcal{C}}$ *is undecidable.*

Note that (Halpern and Shoham 1986) contains many other results concerning the high complexity of **HS**-logics. Explicit axiomatizations of some **HS**-logics can be found in (Marx and Venema 1997). However, they are not finite in the sense of Section 1.4, because they use the irreflexivity rule of Gabbay (1981a). We will return to **HS**-logics in Section 3.9, where it will be considered from a two-dimensional perspective, and in Section 7.1, where Theorem 2.13 will be obtained as a consequence of a more general result.

We have defined only those temporal languages that will be used later on in this book. For other kinds of temporal logics designed for various applications in philosophy, computer science, artificial intelligence, computational linguistics and other fields, for instance, branching time temporal logics, or computation tree logics, we refer the reader to (Clarke and Emerson 1981, 1982, Emerson 1990, Emerson and Halpern 1985, Thomason 1984, Zanardo 1990, 1996, Gabbay *et al.* 1994, 2000) and references therein.

2.3 Epistemic logic

Epistemic logics, or logics of knowledge, have been studied in philosophy with the aim of analyzing formal properties of reasoning about knowledge and belief since the 1950s (see, e.g., Hintikka 1962, Lenzen 1978). Over the last 20 years, however, epistemic logic has found applications in various other disciplines. Here are some of them:

- in game theory, it is used for an epistemic analysis of games with incomplete information (Aumann 1976, Bacharach 1994, Kaneko and Nagashima 1997);

- in artificial intelligence, epistemic logic is applied in order to find out what an agent has to know (in particular, about what it knows) to show intelligent behavior (Laux and Wansing 1995, Meyer and van der Hoek 1995, Halpern and Moses 1992, Fagin *et al.* 1995);

- in computer science, it is employed to analyze the behavior of multi-agent systems; see (Fagin *et al.* 1995) and references therein.

This list is by no means complete; other applications can be found in (Fagin *et al.* 1995, Meyer and van der Hoek 1995).

In all these cases the use of the multimodal language $\mathcal{ML}_n$ for capturing properties of knowledge and belief seems quite natural. Suppose, for instance, that we have a group of n agents called $1, \ldots, n$. For each of them we introduce a modal operator $\Box_i$ which is read as 'agent i knows' or 'agent i believes.' The axioms

(1) $\Box_i(p_0 \to p_1) \to (\Box_i p_0 \to \Box_i p_1)$,

(2) $\Box_i p_0 \to p_0$,

(3) $\Box_i p_0 \to \Box_i \Box_i p_0$,

(4) $\neg\Box_i p_0 \to \Box_i \neg\Box_i p_0$

mean then that

(1) agent i knows all the logical consequences of its knowledge (this phenomenon is known in the literature as *logical omniscience*),

(2) everything that i knows is true,

(3) agent i knows what it knows (*positive introspection*), and

(4) agent i knows what it does not know (*negative introspection*).

Recall now that (1) is an axiom of every normal modal logic, (2) an axiom of **T**, (3) an axiom of **K4**, and (2)–(4) are axioms of **S5**. All the epistemic logics to be considered in this book contain axiom (1) for every agent and are closed under the necessitation rules $\varphi/\Box_i\varphi$, which mean that agents know what is valid; in particular, they know all the tautologies of classical logic. Of course, this assumption gives a somewhat idealized model of knowledge for human agents (and perhaps for robots as well), but for many purposes of modeling the behavior of multi-agent systems in artificial intelligence this simplification seems to be justified or at least the best possible approximation. For a philosophical discussion of principles which can be acceptable under this or that interpretation of knowledge and belief the reader is referred to (Lenzen 1978).

The basic *epistemic logics* in the language $\mathcal{ML}_n$ are the following multimodal variants of the systems **K**, **T**, **K4**, **S4**, **KD45** and **S5** defined in Section 1.4:

- $\mathbf{K}_n$: no property different from logical omniscience of all agents is assumed,
- $\mathbf{T}_n$: besides logical omniscience, it is assumed that what is known is true,
- $\mathbf{K4}_n$: besides logical omniscience, positive introspection is assumed,

- $\mathbf{S4}_n$: besides the properties of $\mathbf{T}_n$, we have positive introspection,
- $\mathbf{KD45}_n$: besides logical omniscience and positive introspection, negative introspection, and consistency of what is known is assumed,
- $\mathbf{S5}_n$: besides the properties of $\mathbf{S4}_n$, we have negative introspection.

It was shown in (Halpern and Moses 1992) that all these logics are decidable and their decision problems are PSPACE-complete; see Theorem 1.17.

Having postulated (1) and the necessitation rules for epistemic logics, we get again into the class of normal multimodal logics which can be interpreted in Kripke models. The question is how these models fit into the epistemic context. According to the possible world semantics, the meaning of the formula $\Box_i\varphi$ is analyzed as follows: $\Box_i\varphi$ is true in a world w if and only if φ is true in every world (or situation) which agent i regards as possible. And a world v is regarded as possible by i in w if v is accessible from w via the relation interpreting $\Box_i$. It follows that i does not know φ iff there exists a world, which is considered possible by i, where φ is false. The following example[2] illustrates how elegant this analysis is.

Example 2.14. (The wise men puzzle.) Imagine that there are three wise men and a king who has two white and three red hats, and that all wise men know that he only has these hats. The king puts a hat on the head of each of the three wise men. Each of them sees the colors of the hats of the other two men, but not the color of his own hat. Now the king asks whether any of them knows the color of his hat. No one says he does. The king asks again—and again none of them knows. But having been asked the third time, all of them say that they know the color. How did the wise men solve the puzzle?

We analyze this puzzle in the framework of the possible world semantics. Assume that the three wise men are called A, B, and C. The following seven situations (alias worlds) are possible:

- all wise men have red hats; this situation is represented by the triple $\langle r,r,r\rangle$;
- A has a white hat, B and C have red hats, or $\langle w,r,r\rangle$, in symbols;
- B has a white hat, A and C have red hats, i.e., $\langle r,w,r\rangle$;
- C has a white hat, A and B have red hats, i.e., $\langle r,r,w\rangle$;
- A and B have white hats, C has a red hat, i.e., $\langle w,w,r\rangle$;

[2] The reader can find numerous examples of this sort in the literature, cf. (Fagin *et al.* 1995).

- B and C have white hats, A has a red hat, i.e., $\langle r, w, w\rangle$;
- A and C have white hats, B has a red hat, i.e., $\langle w, r, w\rangle$.

Denote this collection of triples by W. Since they describe all possible situations, no world outside W is possible.

Now, A sees the hats of the two other wise men. So he knows their colors. Therefore, only one the following four sets of worlds can be regarded as possible by A:

- $V_{?rr} = \{\langle c_1, c_2, c_3\rangle \in W \mid c_2 = r,\ c_3 = r\}$,
- $V_{?wr} = \{\langle c_1, c_2, c_3\rangle \in W \mid c_2 = w,\ c_3 = r\}$,
- $V_{?ww} = \{\langle c_1, c_2, c_3\rangle \in W \mid c_2 = w,\ c_3 = w\}$,
- $V_{?rw} = \{\langle c_1, c_2, c_3\rangle \in W \mid c_2 = r,\ c_3 = w\}$.

As A does not know the color of his hat, the set of worlds he considers possible must contain at least one world in which he has a white hat and at least one world in which his hat is red. This excludes $V_{?ww}$. Similarly, the sets of worlds considered possible by B and C are $V_{r?r}$, $V_{r?w}$, or $V_{w?r}$, and $V_{rr?}$, $V_{rw?}$, or $V_{wr?}$, respectively.

After the wise men have stated that they do not know the colors of their hats, it is *common knowledge* that none of them knows the color of his hat. Thus, it is common knowledge that at least two of them have red hats. So, now the set of worlds A considers possible belongs to the list $V_{?rr}$, $\{\langle r, w, r\rangle\}$, and $\{\langle r, r, w\rangle\}$. Similarly, the sets of worlds considered possible by B and C are among $V_{r?r}$, $\{\langle r, r, w\rangle\}$, $\{\langle w, r, r\rangle\}$ and $V_{rr?}$, $\{\langle r, w, r\rangle\}$, $\{\langle w, r, r\rangle\}$, respectively.

In the second round each of the three wise men again says that he does not know the color of his hat. This means that the set of worlds A considers possible contains a world in which he has a red hat and a world in which he has a white hat. The same holds for B and C. It follows that the sets of worlds A, B and C consider possible are $V_{?rr}$, $V_{r?r}$ and $V_{rr?}$, respectively. This is common knowledge after all of them have stated that they don't know the colors of their own hats. But then the only remaining possible world is $\langle r, r, r\rangle$.

Observe that a number of assumptions have been made to derive this conclusion. For example, we assumed that the three wise men are logically omniscient (and that each of them knows that the other wise men are logically omniscient). Moreover, we used the facts that (i) at the beginning all wise men know that there are three red hats and two white hats, that (ii) every wise man knows that every wise men knows that there are three red hats and two white hats, and that (iii) every wise man knows that every wise men knows that every wise men knows that there are three red hats and two white hats.

This phenomenon—the need to take a potentially infinite iteration of epistemic operators—turns out to be fundamental for various representations of multi-agent systems in game theory, artificial intelligence and computer science. In a finitary language like $\mathcal{ML}_n$ we are not able to express directly the infinite conjunction saying that φ is *common knowledge* among a group M of agents, that is

$$\bigwedge_{k<\omega} \mathsf{E}_M^k \varphi,$$

where

$$\mathsf{E}_M \varphi = \bigwedge_{i \in M} \Box_i \varphi, \quad \mathsf{E}_M^0 \varphi = \varphi, \quad \mathsf{E}_M^{k+1} \varphi = \mathsf{E}_M \mathsf{E}_M^k \varphi.$$

The standard solution to this problem is to take the *common knowledge operators* C_M, 'it is common knowledge among the agents in M,' as primitive and interpret them by the transitive and reflexive closure of the relations $\bigcup_{i \in M} R_i$, i.e., by $\left(\bigcup_{i \in M} R_i\right)^*$, where the R_i interpret the operators $\Box_i$, for $i \in M$. In other words, we define

$$\begin{array}{lll} w \models \mathsf{C}_M \varphi & \text{iff} & \forall v \in W \left(w \left(\bigcup_{i \in M} R_i\right)^* v \text{ implies } v \models \varphi\right) \\ & \text{iff} & \forall k < \omega \ w \models \mathsf{E}_M^k \varphi. \end{array}$$

Remark 2.15. An alternative way would be to interpret C_M by the transitive (but not reflexive) closure of $\bigcup_{i \in M} R_i$ as is done, e.g., in (Fagin *et al.* 1995). From the technical point of view these two ways are equivalent. Indeed, let C_M^+ denote the operator interpreted by the transitive closure of $\bigcup_{i \in M} R_i$. Then $\mathsf{C}_M \varphi$ can be defined as $\varphi \wedge \mathsf{C}_M^+ \varphi$ and $\mathsf{C}_M^+ \varphi$ as $\mathsf{E}_M \mathsf{C}_M \varphi$.

Let $\mathcal{ML}_n^C$ denote the language that results from $\mathcal{ML}_n$ by extending it with the common knowledge operator C_M for every nonempty subset M of $\{1, \dots, n\}$ (and the corresponding formula formation rules).

Given a normal modal logic L in the language $\mathcal{ML}_n$, denote by L^C the $(n + 2^n - 1)$-modal logic formulated in $\mathcal{ML}_n^C$ and determined by the class of all frames of the form

$$\left\langle W, R_1, \dots, R_n, \{(\bigcup_{i \in M} R_i)^* \mid M \subseteq \{1, \dots, n\},\ M \neq \emptyset\} \right\rangle, \tag{2.8}$$

where $\langle W, R_1, \dots, R_n \rangle$ is a frame for L and the common knowledge operators C_M are interpreted by $(\bigcup_{i \in M} R_i)^*$.

Remark 2.16. It is to be noted that this kind of semantic definition leaves a possibility for L^C to have nonstandard (or not intended) frames that are different from those above (for example, the operation of the transitive reflexive closure is not first-order definable). Fortunately, this is not the case. As follows from the axiomatization given in Theorem 2.17 below, *all* frames for L^C are standard frames of the form (2.8) (and the operation of the transitive reflexive closure is modally definable). That is why, when dealing with frames for L^C, we need to know only the relations R_i. So, to simplify notation, we will usually represent frames for L^C as $\langle W, R_1, \ldots, R_n \rangle$.

Note that in $\mathbf{S4}_n^C$ and $\mathbf{S5}_n^C$ the operators $\mathsf{C}_{\{i\}}$ have the same interpretation as the operators $\Box_i$, while in $\mathbf{K}_n^C$, $\mathbf{T}_n^C$, $\mathbf{K4}_n^C$, and $\mathbf{KD45}_n^C$ their behavior is different. In particular, $\mathbf{S4}_1^C$ and $\mathbf{S5}_1^C$ are just notational variants of **S4** and **S5**, respectively.

The following theorem summarizes the most important facts about epistemic logics with common knowledge operators:

Theorem 2.17. *Suppose that either* $n \geq 1$ *and* $L \in \{\mathbf{K}_n, \mathbf{T}_n\}$, *or* $n > 1$ *and* $L \in \{\mathbf{K4}_n, \mathbf{S4}_n, \mathbf{KD45}_n, \mathbf{S5}_n\}$. *Then*

- L^C *can be axiomatized by adding the following axioms and inference rules to those of* L, *for all nonempty sets* $M \subseteq \{1, \ldots, n\}$:

$$\mathsf{C}_M p_0 \leftrightarrow (p_0 \wedge \mathsf{E}_M \mathsf{C}_M p_0), \tag{2.9}$$

$$\textit{given } p_0 \rightarrow p_1 \wedge \mathsf{E}_M p_0, \textit{ derive } p_0 \rightarrow \mathsf{C}_M p_1; \tag{2.10}$$

- *the decision problem for* L^C *is* EXPTIME-*complete;*
- L^C *has the finite model property.*

Remark 2.18. An alternative axiomatization for any epistemic logic L^C above can be obtained by omitting rule (2.10) and adding the following axioms and inference rules to those of L and (2.9), for all nonempty sets $M \subseteq \{1, \ldots, n\}$:

- $\mathsf{C}_M(p_0 \rightarrow p_1) \rightarrow (\mathsf{C}_M p_0 \rightarrow \mathsf{C}_M p_1)$,
- $\mathsf{C}_M(p_0 \rightarrow \mathsf{E}_M p_0) \rightarrow (p_0 \rightarrow \mathsf{C}_M p_0)$,
- given p_0, derive $\mathsf{C}_M p_0$.

(Observe the similarities with the axiomatization of **PTL** in Theorem 2.6.) We leave it to the reader to show that the two axiomatizations are interderivable.

Axioms for common knowledge appear in (Lehmann 1984, Milgrom 1981, McCarthy *et al.* 1979), although in these papers only the operator C expressing common knowledge of all agents is used. A completeness proof, based on the ideas of Kozen and Parikh (1981), can be found in (Halpern and Moses 1992). The decidability and complexity results are based on the fact that—as will be shown in Section 2.8—the above logics are embeddable into propositional dynamic logics **PDL** and **CPDL** (Halpern and Moses 1992).

2.4 Dynamic logic

Propositional dynamic logic **PDL** was designed for reasoning—at a rather abstract level—about the behavior of programs. The field of computer science which is concerned with formal languages that are able to express various properties of programs, in particular their correctness, is known as program verification and specification. One of the most influential approaches to verification of ordinary sequential programs (e.g., programs for sorting lists of integers) proposed by Floyd (1967) and Hoare (1967) uses correctness assertions of the form

$$\{\varphi\}\,\alpha\,\{\psi\}$$

which state that any execution of program (command or action type) α starting from a state where φ holds reaches a state where ψ holds. The formulas φ and ψ are called the *pre-* and *post-conditions* of this assertion. The idea of using such assertions is based on the fact that the program's underlying semantics can be described in terms of a transformation from an initial state to a final state.[3] The transition graph representing this transformation can be regarded as a Kripke frame whose accessibility relations are labeled with commands. So if we associate with every program α the modal operator $[\alpha]$ with the intended meaning '$w \models [\alpha]\psi$ iff every possible execution of α at state w arrives at a state in which ψ holds,' then the correctness assertion above can be represented as an ordinary modal formula:

$$\varphi \to [\alpha]\psi.$$

(Note that $[\alpha]\psi$ is the *weakest pre-condition* for which any execution of α reaches a ψ-state.)

As computer programs are usually composed from commands, our 'abstract' programs can also be complex entities composed from primitive ones. Our operations on programs are *sequencing* (or *composition*) ';', *nondeterministic choice* '$\cup$', *iteration* '$*$', and *test* '?' (see Fig. 2.2). For example, for programs α, β and a statement φ, we can represent the compound program 'if φ then α else β' as $(\varphi?\,;\alpha) \cup (\neg\varphi?\,;\beta)$. The programs 'while φ do α' and 'repeat α until φ' can be represented as $(\varphi?\,;\alpha)^*\,;\neg\varphi?$ and $\alpha\,;(\neg\varphi?,\alpha)^*\,;\varphi?$, respectively.

Before turning to the precise definitions of the syntax and semantics of **PDL**, it may be worth noting that apart from its ability to describe abstract properties of programs, **PDL** and its extensions turned out to be useful for at least two other reasons as well.

[3] Observe that assertions of this form are not appropriate for the verification and specification of continuously operating reactive programs which are usually nonterminating. Since there is no final state, post-conditions are of no use to describe the behavior of such programs. In this case temporal logic provides an appropriate formalism (see, e.g., Clarke and Emerson 1981, Emerson and Halpern 1985, Manna and Pnueli 1992, 1995).

$\alpha ; \beta$	do α followed by β,
$\alpha \cup \beta$	do either α or β, nondeterministically,
α^*	repeat α a finite number of times,
$\varphi?$	proceed if φ is true, else fail.

Figure 2.2: The intended reading of operations on programs.

First, various important modal logics can be embedded into propositional dynamic logics, and so inherit some of their properties, say, decidability or upper bounds for their computational complexity. We will discuss the embedding of expressive epistemic logics with the common knowledge operator (Halpern and Moses 1992) in Section 2.8. Fischer and Immerman (1987) embedded temporalized epistemic logics into **CPDL**—an extension of **PDL** with the 'converse operator.' A variant of their embedding can be found in Section 6.3. Description logics have also been analyzed by means of embeddings in propositional dynamic logics (Schild 1991, De Giacomo and Lenzerini 1994, De Giacomo and Lenzerini 1996); see Section 2.5. And second, in artificial intelligence and philosophy, propositional dynamic logics are often taken as a basis for constructing deontic logics and logics intended for reasoning about actions; see, e.g., (Segerberg 1980, Prendinger and Schurz 1996, De Giacomo and Lenzerini 1995, Meyer 1988, Fischer and Immerman 1987).

Besides the alphabet of classical propositional logic (where $\wedge$ and $\neg$ are regarded as the only primitive connectives), the alphabet of the language $\mathcal{PDL}$ contains

- a countably infinite set $\alpha_0, \alpha_1, \ldots$ of *atomic actions* (or *atomic programs*),
- the symbols ;, $\cup$, * and ?.

The sets of *$\mathcal{PDL}$-formulas* and *action terms* are defined by simultaneous induction as follows:

- every propositional variable is a formula,
- every atomic action is an action term,
- if φ and ψ are formulas and α is an action term, then $\varphi \wedge \psi$, $\neg\varphi$ and $[\alpha]\varphi$ are formulas,
- if α and β are action terms and φ is a formula, then $\alpha \cup \beta$, $\alpha ; \beta$, α^* and $\varphi?$ are action terms.

As before, we define $\langle\alpha\rangle\varphi$ as an abbreviation for $\neg[\alpha]\neg\varphi$. Observe that the internal structure of the modal operators is the only difference between the language $\mathcal{PDL}$ and the standard multimodal language $\mathcal{ML}_\omega$ with infinitely many boxes.

The language $\mathcal{PDL}$ is interpreted in *$\mathcal{PDL}$-structures* which are frames of the form

$$\mathfrak{F} = \langle W, T_{\alpha_0}, T_{\alpha_1}, \ldots \rangle,$$

where W is a (nonempty) set of *states* and the T_{α_i} are binary relations—this time called *transition relations*—on W, one for each atomic action α_i. Unlike the possible world semantics, now $wT_{\alpha_i}v$ reads as 'there is an execution of α_i which starts at state w and ends at state v.'

As usual, a *valuation* $\mathfrak{V}$ *in* $\mathfrak{F}$ is a map from the set of propositional variables into the set of all subsets of W. Given a *model* $\mathfrak{M} = \langle \mathfrak{F}, \mathfrak{V}\rangle$, we define the *truth-relation* $(\mathfrak{M}, w) \models \varphi$ (or $w \models \varphi$, if understood) and the *compound transition relations* $T_\alpha^{\mathfrak{M}}$ (or T_α) by parallel induction, for any state w, formula φ and action term α:

- $w \models p$ iff $w \in \mathfrak{V}(p)$,
- $w \models \varphi \wedge \psi$ iff $w \models \varphi$ and $w \models \psi$,
- $w \models \neg\varphi$ iff not $w \models \varphi$,
- $w \models [\alpha]\varphi$ iff $v \models \varphi$ for every $v \in W$ such that $wT_\alpha v$,
- $T_{\alpha\cup\beta} = T_\alpha \cup T_\beta$ (i.e., $x(T_\alpha \cup T_\beta)y$ iff $xT_\alpha y$ or $xT_\beta y$),
- $T_{\alpha\,;\,\beta}$ is the *composition* (or *relative product*) $T_\alpha \circ T_\beta$ of T_α and T_β (i.e., $x(T_\alpha \circ T_\beta)y$ iff $\exists z \in W\ xT_\alpha zT_\beta y$),
- $T_{\alpha^*} = (T_\alpha)^*$ (i.e., T_{α^*} is the reflexive and transitive closure of T_α),
- $T_{\varphi?} = \{\langle x, x\rangle \mid x \models \varphi\}$.

Observe that T_α depends on the valuation $\mathfrak{V}$ only if α contains test, otherwise it is completely determined by $\mathfrak{F}$.

We say that φ is *true in* $\mathfrak{M}$ if $(\mathfrak{M}, w) \models \varphi$ for all $w \in W$ and define the logic **PDL** as the set of all $\mathcal{PDL}$-formulas that are true in all models based on $\mathcal{PDL}$-structures. Note that the fragment of **PDL** with only atomic action terms is just a syntactic variant of the multimodal logic $\mathbf{K}_\omega$ with infinitely many **K**-boxes. One can define (along the lines of Section 1.5) an algebraic semantics for **PDL**. These modal algebras are studied in the literature under the name of *dynamic algebras*; see (Kozen 1981, Pratt 1991).

Syntactically, **PDL** can be characterized as follows (see Berman 1979, Gabbay 1977a, Nishimura 1979, Parikh 1978, Pratt 1978, Segerberg 1977).

Theorem 2.19. **PDL** *is the smallest set of $\mathcal{PDL}$-formulas containing classical propositional logic* **Cl**, *the axioms*

$$[\alpha](p \to q) \to ([\alpha]p \to [\alpha]q), \tag{2.11}$$
$$[\alpha\,;\beta]p \leftrightarrow [\alpha][\beta]p, \tag{2.12}$$
$$[\alpha \cup \beta]p \leftrightarrow [\alpha]p \wedge [\beta]p, \tag{2.13}$$
$$[\alpha^*]p \leftrightarrow p \wedge [\alpha][\alpha^*]p, \tag{2.14}$$
$$[\alpha^*](p \to [\alpha]p) \to (p \to [\alpha^*]p), \tag{2.15}$$
$$[q?]p \leftrightarrow (q \to p), \tag{2.16}$$

for all action terms α, β, and closed under modus ponens, substitution, and the necessitation rules

'given φ, derive $[\alpha]\varphi$,'

for all action terms α.

Theorem 2.20. **PDL** *has the fmp and is decidable, with the decision problem being* EXPTIME-*complete.*

The fmp is shown by filtration using the Fischer–Ladner closure in (Fischer and Ladner 1977). The decidability of **PDL** and the exponential lower bound is proved in (Fischer and Ladner 1979). The exponential upper bound was established in (Pratt 1979).

A more expressive language, called $\mathcal{CPDL}$ (*converse* $\mathcal{PDL}$), is obtained by extending $\mathcal{PDL}$ with a constructor for representing backward executions of programs. Namely, we add to the alphabet of $\mathcal{PDL}$ the *converse operator* $^-$ on action terms, so that α^- is an action term of $\mathcal{CPDL}$ whenever α is an action term, and associate with α^- the transition relation

- $T_{\alpha^-} = T_\alpha^{-1}$ (i.e., $xT_\alpha^{-1}y$ iff $yT_\alpha x$).

(Note that $[\alpha^-]\varphi$ is the *strongest post-condition* satisfied after any execution of α starting from a state at which φ holds.)

The logic **CPDL** is defined to be the set of all valid $\mathcal{CPDL}$-formulas. It is not hard to see that the following identities always hold:

$$T_{\alpha^{--}} = T_\alpha, \qquad T_{(\alpha\cup\beta)^-} = T_{\alpha^-\cup\beta^-},$$
$$T_{(\alpha\,;\beta)^-} = T_{\beta^-\,;\,\alpha^-}, \qquad T_{(\alpha^*)^-} = T_{(\alpha^-)^*},$$
$$T_{(\varphi?)^-} = T_{\varphi?}.$$

Thus we have:

Proposition 2.21. *Every $\mathcal{CPDL}$-formula is equivalent in* **CPDL** *to a formula in which the converse operator is applied only to atomic actions.*

The above axiomatization of **PDL** can be extended to an axiomatization of **CPDL** by adding the axioms (see Parikh 1978):

$$p \to [\alpha]\langle\alpha^-\rangle p \quad \text{and} \quad p \to [\alpha^-]\langle\alpha\rangle p. \tag{2.17}$$

The filtration for **PDL** goes through for **CPDL** as well, and as was shown in (Pratt 1979, Vardi 1985, Vardi and Wolper 1986), the complexity of the extended logic does not increase:

Theorem 2.22. **CPDL** *has the fmp, and the decision problem for* **CPDL** *is* EXPTIME-*complete.*

Remark 2.23. Observe that the *test-free fragment* $\mathbf{CPDL}^{-?}$ of **CPDL** (i.e., those formulas in **CPDL** that do not contain action terms of the form φ?) is in fact a Kripke complete multimodal logic. Indeed, the language of this fragment has a modal operator $[\alpha]$ for every test-free action term α. So, strictly speaking, a frame interpreting this multimodal language is not a $\mathcal{PDL}$-structure as introduced above, but any structure of the form

$$\mathfrak{F} = \langle W, T_\alpha, \ldots \rangle, \tag{2.18}$$

where W is a (nonempty) set and the T_α are binary relations on W, one for each test-free action term α (not only for atomic actions). Then $\mathbf{CPDL}^{-?}$ is the multimodal logic determined by the class $\mathcal{C}$ of all frames of this kind such that the relations T_α for nonatomic test-free action terms α are obtained as above. In principle, there can be frames for $\mathbf{CPDL}^{-?}$ that are not in $\mathcal{C}$. It can be shown, however, that by omitting (2.16) from the axiomatization of **CPDL**, we obtain an axiomatization for $\mathbf{CPDL}^{-?}$. So, actually, in *every* frame for $\mathbf{CPDL}^{-?}$ of the form (2.18), the relation T_{α^*} is the reflexive and transitive closure of T_α, $T_{\alpha\cup\beta} = T_\alpha \cup T_\beta$, $T_{\alpha;\beta} = T_\alpha \circ T_\beta$, and $T_{\alpha^-} = T_\alpha^{-1}$, for all test-free action terms α, β (see Remarks 2.11, 2.16 and observe the similarities between the axiomatizations for **PDL**, $\mathbf{PTL}_{\Box\bigcirc}$ and epistemic logics L^C given in Remark 2.18).

Different variants of **PDL** as well as first-order dynamic logic can be found in (Goldblatt 1987, Harel *et al.* 2000); see also Section 3.6. The reader may find useful surveys of other dynamic formalisms in (van Benthem 1996, van Eijck and Visser 1994, Goldblatt 1982, Harel 1984, Kozen and Tiuryn 1990, Ponse *et al.* 1996); see also Section 3.10.

2.5 Description logic

Description logic is not a modal logic. It was created at the beginning of the 1980s as a formalism for knowledge representation and reasoning in artificial intelligence. And only ten years later it was observed that description logic and modal logic are more than close: to a large extent they are simply notational variants of each other.

Briefly, the history of description logic is as follows. After the straightforward attack on knowledge representation with the help of the heavy artillery of first-order logic failed in the 1960s, a number of ideas were proposed the essence of which was to treat knowledge in a more structural, visual, object-oriented way (see Quillian 1967, 1968, Minsky 1975 and the collection Brachman and Levesque 1985) *without using logic*.

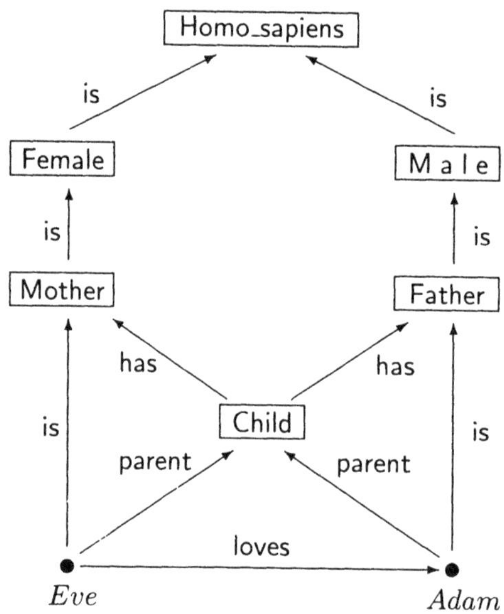

Figure 2.3: Semantic network.

Figure 2.3 shows a simple example of representing some information about human relationships in the form of a *semantic network* of Quillian (1967, 1968) and Raphael (1968). The application domain in this example—human beings—is divided into (not necessarily disjoint) classes (Homo_sapiens, Female, Male, Father, Mother, Child), concrete individuals (*Eve*, *Adam*), and the relations between them (is, has, parent, loves) are depicted in the form of labeled arrows.

The main deficit of such representations was the lack of semantics, and as a consequence, ambiguities. (For how can we be sure that a reasoning program our company has bought provides us with a complete set of correct answers, if it was not even precisely formulated in the manual what a correct answer is?) In the depicted network it is not clear, for instance, whether *all* members of the class Child are children of *Eve* or only *some* of them.

Description logic appeared as a sort of compromise between the above mentioned features of semantic networks and Minsky frames, on the one hand, and logic- (and so semantic-) based formalisms, on the other. It originated from the KL-ONE system of Brachman and Schmolze (1985), which combined in itself many ideas of its predecessors.

Like modal logic, description logic consists of a wide spectrum of languages. Since our road in this book comes from modal logic, as the basis of our description language we choose the language $\mathcal{ALC}$ of Schmidt-Schauß and Smolka (1991) which, as we shall see, is closely related to multimodal **K**.

The alphabet of $\mathcal{ALC}$ consists of

- *concept names* $C_0, C_1, \dots$;
- *role names* $R_0, R_1, \dots$ (or $R, S, \dots$);
- *object names* $a_0, a_1, \dots$ (or $a, b, \dots$);
- the *Boolean concept constructors* $\sqcap$, $\neg$;
- the *existential quantifier* $\exists$;
- the *Boolean formula constructors* $\wedge$, $\neg$;
- the symbols . (dot), : (colon) and =.

Concept names are supposed to denote classes of objects in a certain domain Δ (say, Mother, Male, etc. in the example above), role names are intended for denoting binary relations between elements of Δ (has, loves), and object names stand for some concrete elements in Δ (*Eve*, *Adam*).

Now we define by induction (complex) *concepts* and *formulas* of $\mathcal{ALC}$. Every concept name is an (*atomic*) *concept*. If C and D are concepts, a and b object names, and R is a role name, then

- $C \sqcap D$, $\neg C$ and $\exists R.C$ are *concepts*,
- $a : C$, aRb, $C = D$ are *atomic formulas*, and
- Boolean combination of atomic formulas are *formulas*.

The intended meaning of $C \sqcap D$ is simply the intersection of C and D; $\neg C$ means the complement (in the domain under consideration) of C; $\exists R.C$ denotes the class of all objects from which at least one object in C is accessible via R. In the usual way we can also define concepts $\forall R.C$, $C \sqcup D$, $C \to D$, $C \leftrightarrow D$, $\top$, $\bot$: e.g., $\forall R.C$ is $\neg\exists R.\neg C$, $C \sqcup D$ is $\neg(\neg C \sqcap \neg D)$, $\top$ is $C \sqcup \neg C$. The formulas $a : C$ and aRb mean that object a belongs to concept C, and that a and b are related by role R, respectively; $C = D$ says that concepts C

and D contain the same elements (a precise definition of semantics of $\mathcal{ALC}$ is given below).

Traditionally, in 'standard' description logic the Booleans are not among the formula constructors; all formulas are atomic. Instead of equality $C = D$ often inclusion (subsumption) $C \sqsubseteq D$ is preferred. (Note that $C \sqsubseteq D$ can be expressed as $(C \sqcap \neg D) = \bot$. Conversely, $C = D$ is defined via $\sqsubseteq$ as $(C \sqsubseteq D) \wedge (D \sqsubseteq C)$.)

An $\mathcal{ALC}$ *knowledge base* is just a finite set of $\mathcal{ALC}$ formulas. As usual in knowledge representation, we distinguish between knowledge bases containing only terminological knowledge and those containing only assertional knowledge. More precisely, we call a knowledge base Σ a *TBox* (*terminological box*) if it contains formulas of the form $C = D$ only; Σ is an *ABox* (*assertion box*) if it contains only formulas of the form $a : C$ or aRb. Note that without loss of generality we may assume all concept equations to be of the form $C = \top$, since $C = D$ is equivalent to $(C \leftrightarrow D) = \top$.

Example 2.24. The following $\mathcal{ALC}$ knowledge base represents the semantic network in Fig. 2.3:

$$\left.\begin{array}{l} \mathsf{Female} \sqcup \mathsf{Male} \sqsubseteq \mathsf{Homo\_sapiens} \\ \mathsf{Mother} \sqsubseteq \mathsf{Female} \\ \mathsf{Father} \sqsubseteq \mathsf{Male} \\ \mathsf{Child} \sqsubseteq \exists\mathsf{has.Mother} \sqcap \exists\mathsf{has.Father} \end{array}\right\} \text{TBox}$$

$$\left.\begin{array}{l} \mathit{Eve} : \mathsf{Mother} \\ \mathit{Adam} : \mathsf{Father} \\ \mathit{Eve}\ \mathsf{loves}\ \mathit{Adam} \\ \mathit{Eve} : \exists\mathsf{parent.Child} \\ \mathit{Adam} : \exists\mathsf{parent.Child} \end{array}\right\} \text{ABox}$$

Observe that the relation is in Fig. 2.3 is represented in the form of $C \sqsubseteq D$ if it connects concepts (like Mother and Female) and $a : C$ if it holds between an object name and a concept (like *Eve* and Mother).

Formally, the semantics of $\mathcal{ALC}$ is defined in the following way. A *model* for $\mathcal{ALC}$ is a structure of the form

$$I = \langle \Delta, R_0^I, \dots, C_0^I, \dots, a_0^I, \dots \rangle, \tag{2.19}$$

where Δ is a nonempty set, the *domain* of I (elements of which are often called *objects*), and for all $i = 0, 1, \dots,$ R_i^I are binary relations on Δ (interpreting the role names), C_i^I subsets of Δ (interpreting the concept names), and a_i^I are elements of Δ (interpreting the object names).

The *value* C^I *of a concept* C *in* I, and the *truth-relation* $I \models \varphi$, φ a formula, are defined inductively as follows:

- $(C \sqcap D)^I = C^I \cap D^I$;
- $(\neg C)^I = \Delta - C^I$;
- $(\exists R_i.C)^I = \{x \in \Delta \mid \exists y \in C^I \; xR_i^I y\}$;
- $I \models a : C$ iff $a^I \in C^I$;
- $I \models aR_ib$ iff $a^I R_i^I b^I$;
- $I \models C = D$ iff $C^I = D^I$;
- $I \models \varphi \wedge \psi$ iff $I \models \varphi$ and $I \models \psi$;
- $I \models \neg\varphi$ iff not $I \models \varphi$.

A formula φ is said to be *true in* I if $I \models \varphi$; we then also say that I is a *model for* φ. A formula φ is called *satisfiable* if there exists a model for φ. A concept C is *satisfiable* if there is a model I in which $C^I \neq \emptyset$.

Suppose we are given (or have constructed) a set Σ of $\mathcal{ALC}$-formulas describing an application domain. This is our knowledge base. How can it be used? There are several typical reasoning tasks we should be able to solve. We formulate them in terms of the consequence relation $\Sigma \models \varphi$ defined as follows.

Say that an $\mathcal{ALC}$-formula φ is a *logical consequence* of the knowledge base Σ and write $\Sigma \models \varphi$ if φ is true in every $\mathcal{ALC}$-model where all formulas in Σ are true.

For instance, let Σ be the knowledge base of Example 2.24. Then we clearly have

$$\Sigma \models \mathsf{Mother} \sqsubseteq \mathsf{Homo\_sapiens}, \qquad \Sigma \models \mathit{Eve} : \mathsf{Female}.$$

The main *reasoning tasks* for a knowledge base Σ are:

- *Concept satisfiability*: $\Sigma \not\models C = \bot$. (Is there is a model I for Σ such that $C^I \neq \emptyset$?)
- *Subsumption*: $\Sigma \models C \sqsubseteq D$. (Does $C^I \subseteq D^I$ hold in every model I for Σ?)
- *Consistency*: $\Sigma \not\models \bot$. (Is there a model for Σ?)
- *Instance checking*: $\Sigma \models a : C$, a an object name. (Does a^I belong to C^I in every model I for Σ?)

Since knowledge bases are supposed to be finite, all the listed reasoning tasks are reducible to the

- *Satisfiability problem*: given an $\mathcal{ALC}$-formula φ, determine whether it is satisfiable.

Indeed, we have $\Sigma \models \psi$ iff the formula $\bigwedge_{\chi\in\Sigma} \chi \wedge \neg\psi$ is not satisfiable. Note also that concept satisfiability, subsumption, consistency and instance checking are reducible to each other: for example

$$\Sigma \models C \sqsubseteq D \quad \text{iff} \quad \Sigma \models C \sqcap \neg D = \bot, \tag{2.20}$$

and

$$\Sigma \not\models C = \bot \quad \text{iff} \quad \Sigma \not\models C \sqsubseteq \bot$$

(see Table 2.1, where $A \to B$ means that problem A is reducible to problem B).

The reader must have already observed that the concept fragment of $\mathcal{ALC}$ is just a notational variant of multimodal **K**.[4] Indeed, assuming that $\mathcal{ALC}$ contains n role names $R_0, \dots, R_{n-1}$, we can define a translation $\cdot^\dagger$ from the set of $\mathcal{ML}_n$-formulas *onto* the set of $\mathcal{ALC}$-concepts by taking:

$$\begin{aligned} p_i^\dagger &= C_i, \\ (\varphi \wedge \psi)^\dagger &= \varphi^\dagger \sqcap \psi^\dagger, \\ (\neg\varphi)^\dagger &= \neg\varphi^\dagger, \\ (\Diamond_i\varphi)^\dagger &= \exists R_i.\varphi^\dagger. \end{aligned}$$

Every $\mathcal{ML}_n$-model $\mathfrak{M} = \langle \mathfrak{F}, \mathfrak{V} \rangle$ with $\mathfrak{F} = \langle W, S_0, \dots, S_{n-1} \rangle$ can be transformed to an $\mathcal{ALC}$-model

$$I_{\mathfrak{M}} = \left\langle W, R_0^{I_{\mathfrak{M}}}, \dots, R_{n-1}^{I_{\mathfrak{M}}}, C_0^{I_{\mathfrak{M}}}, \dots, a_0^{I_{\mathfrak{M}}}, \dots \right\rangle,$$

where $R_i^{I_{\mathfrak{M}}} = S_i$, $C_i^{I_{\mathfrak{M}}} = \mathfrak{V}(p_i)$ and $a_i^{I_{\mathfrak{M}}} \in W$ arbitrary. Then it should be clear that for every $\mathcal{ML}_n$-formula φ, every $\mathcal{ML}_n$-model $\mathfrak{M}$, and every world w in $\mathfrak{M}$,

$$(\mathfrak{M}, w) \models \varphi \quad \text{iff} \quad w \in (\varphi^\dagger)^{I_{\mathfrak{M}}}.$$

Conversely, every $\mathcal{ALC}$-model of the form (2.19) gives rise to an $\mathcal{ML}_n$-model $\mathfrak{M}_I = \langle \mathfrak{F}_I, \mathfrak{V}_I \rangle$, where $\mathfrak{F}_I = \langle \Delta, R_0^I, \dots, R_{n-1}^I \rangle$ and $\mathfrak{V}_I(p_i) = C_i^I$, for all

[4] Note, however, that a modal logic is intended to represent schemes of correct reasoning that involves modal operators; formally, it is a set of formulas containing some basic axioms and closed under certain inference rules (in particular, substitution). Description logic was designed to represent knowledge about some application domains rather than universal logical truth. In mathematical logic such 'knowledge bases' are known as theories.

$i = 0, 1, \ldots$. Take the inverse $\ddagger$ of translation $\dagger$ from $\mathcal{ALC}$-concepts onto $\mathcal{ML}_n$-formulas. Then clearly $\mathfrak{M}$ is isomorphic to $\mathfrak{M}_{I_{\mathfrak{M}}}$, and for every $\mathcal{ALC}$-concept C, every $\mathcal{ALC}$-model I and every object w in I,

$$w \in C^I \quad \text{iff} \quad (\mathfrak{M}_I, w) \models C^{\ddagger}.$$

As a consequence we have that the problem of $\mathcal{ALC}$ concept satisfiability with empty knowledge base is equivalent to the satisfiability problem for $\mathbf{K}_n$. Thus by Theorem 1.17 we obtain:

Proposition 2.25. *The problem of $\mathcal{ALC}$ concept satisfiability and the subsumption problem, both with empty knowledge bases, are* PSPACE*-complete.*

On the other hand, $\mathcal{ALC}$-formulas can refer explicitly to the names of objects (worlds) in the models and express some facts about these objects. Thus, the formula part of $\mathcal{ALC}$ is more expressive than multimodal $\mathbf{K}$. Moreover, even pure terminological reasoning is more complex than reasoning in $\mathbf{K}_n$ because the global consequence relation $\vdash^*_{\mathbf{K}_n}$ is equivalent to the concept satisfiability problem relative to TBoxes, i.e., to the problem '$\Sigma \not\models C = \bot$?,' where Σ is a TBox (see Table 2.1). First, for all $\mathcal{ML}_n$-formulas φ and ψ,

$$\begin{aligned} \varphi \vdash^*_{\mathbf{K}_n} \psi \quad & \text{iff} \quad \varphi^{\dagger} = \top \models \psi^{\dagger} = \top \\ & \text{iff} \quad \neg\psi^{\dagger} \text{ is not satisfiable in a model for } \varphi^{\dagger} = \top. \end{aligned}$$

Conversely, given a TBox Σ and a concept C, we have

$$\Sigma \not\models C = \bot \quad \text{iff} \quad \{C^{\ddagger} \leftrightarrow D^{\ddagger} \mid C = D \in \Sigma\} \not\vdash^*_{\mathbf{K}_n} \neg C^{\ddagger}.$$

As a consequence of Theorem 1.23 we obtain then the following result, which was first proved by Schild (1991) who embedded (an extension of) $\mathcal{ALC}$ (without assertion formulas) into **PDL**.

Theorem 2.26. *The $\mathcal{ALC}$ concept satisfiability problem relative to TBoxes is* EXPTIME*-complete.*

It is worth mentioning that standard tableau procedures for TBox-reasoning in $\mathcal{ALC}$—as implemented, for example, by Horrocks (1998)—do not run in exponential time, but are double-exponential. Only recently Donini and Masacci (2000) have presented a satisfiability checking tableau algorithm for TBoxes running in exponential time.

As follows from Theorem 2.26, the satisfiability problem for $\mathcal{ALC}$-formulas is EXPTIME-hard. Moreover, we have a matching upper bound:

Theorem 2.27. *The satisfiability problem for $\mathcal{ALC}$-formulas is* EXPTIME*-complete.*

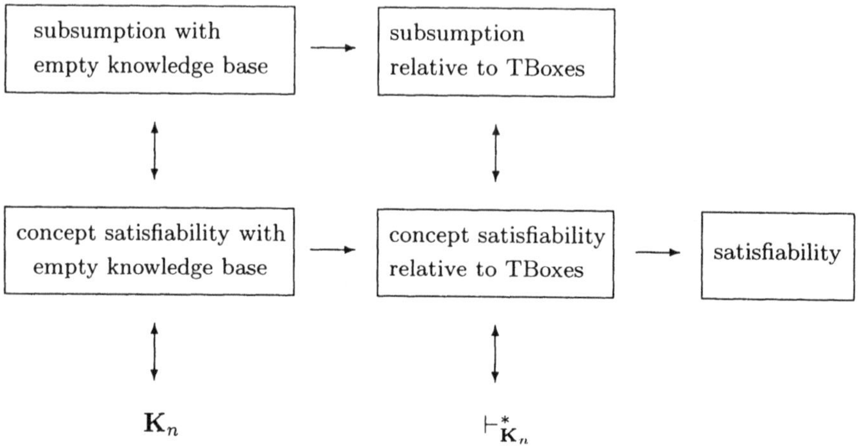

Table 2.1: Reasoning tasks in $\mathcal{ALC}$.

As this result does not seem to appear explicitly in the existing literature, we show here a satisfiability checking algorithm for $\mathcal{ALC}$-formulas running in exponential time. An alternative proof would be a generalization of the proof of the exponential upper bound for $\vdash^*_{\mathbf{K}_n}$.

Suppose φ is an $\mathcal{ALC}$-formula. Let $ob\,\varphi$ be the set of all object names in φ and let $con\,\varphi$ and $sub\,\varphi$ denote the closure under negation of, respectively, the set of all concepts in φ and the set of all subformulas in φ. By identifying E and $\neg\neg E$, for every concept or formula E, we have

$$|ob\,\varphi| \leq \ell(\varphi), \quad |con\,\varphi| \leq 2\ell(\varphi), \quad \text{and} \quad |sub\,\varphi| \leq 2\ell(\varphi),$$

where $\ell(\varphi)$ is the length of φ, i.e., the number of symbol occurrences in φ.

We call a *concept type for* φ any subset $\mathbf{c}$ of $con\,\varphi$ such that

- $C \sqcap D \in \mathbf{c}$ iff $C, D \in \mathbf{c}$, for every $C \sqcap D \in con\,\varphi$;
- $\neg C \in \mathbf{c}$ iff $C \notin \mathbf{c}$, for every $C \in con\,\varphi$.

A *formula type* $\boldsymbol{f}$ *for* φ is a subset of $sub\,\varphi$ such that

- $\psi \wedge \chi \in \boldsymbol{f}$ iff $\psi, \chi \in \boldsymbol{f}$, for every $\psi \wedge \chi \in sub\,\varphi$;
- $\neg\psi \in \boldsymbol{f}$ iff $\psi \notin \boldsymbol{f}$, for every $\psi \in sub\,\varphi$.

Clearly, there are at most $2^{|con\,\varphi|}$ concept types and at most $2^{|sub\,\varphi|}$ formula types for φ. We are going to use these types to construct a model for φ, if any.

Let us call a *model candidate for* φ a triple $\langle T, o, \boldsymbol{f}\rangle$ such that T is a set of concept types for φ, o is a function from $ob\,\varphi$ to T, $\boldsymbol{f}$ a formula type for φ, and $\langle T, o, \boldsymbol{f}\rangle$ satisfies the conditions:

(a) $\varphi \in \boldsymbol{f}$;

(b) $(a : C) \in \boldsymbol{f}$ implies $C \in o(a)$;

(c) $aRb \in \boldsymbol{f}$ implies $\{\neg C \mid \neg\exists R.C \in o(a)\} \subseteq o(b)$.

A model candidate $\langle T, o, \boldsymbol{f}\rangle$ for φ is said to be a *quasimodel for* φ if the following conditions hold:

(d) for every concept type $\boldsymbol{c} \in T$ and every concept $\exists R.C \in \boldsymbol{c}$, there is a $\boldsymbol{c}' \in T$ such that $\{\neg D \mid \neg\exists R.D \in \boldsymbol{c}\} \cup \{C\} \subseteq \boldsymbol{c}'$;

(e) for every concept type $\boldsymbol{c} \in T$ and every concept C, if $\neg C \in \boldsymbol{c}$ then $(C = \top) \notin \boldsymbol{f}$;

(f) for every concept C, if $\neg(C = \top) \in \boldsymbol{f}$ then there is a $\boldsymbol{c} \in T$ such that $C \notin \boldsymbol{c}$;

(g) T is not empty.

We now show that our formula φ is satisfiable iff there is a quasimodel for φ. Suppose first that we have found a quasimodel $\langle T, o, \boldsymbol{f}\rangle$ for φ. Define a model $I = \langle \Delta, R_0^I, \dots, C_0^I, \dots, a_0^I, \dots\rangle$ by taking

- $\Delta = T \cup ob\,\varphi$;
- $a^I = a$, for $a \in ob\,\varphi$;
- $C_i^I = \{\boldsymbol{c} \in T \mid C_i \in \boldsymbol{c}\} \cup \{a \in ob\,\varphi \mid C_i \in o(a)\}$;
- $\boldsymbol{c}R^I\boldsymbol{c}'$ iff $\{\neg C \mid \neg\exists R.C \in \boldsymbol{c}\} \subseteq \boldsymbol{c}'$, for $\boldsymbol{c}, \boldsymbol{c}' \in T$;
- aR^Ib iff $aRb \in \boldsymbol{f}$, for $a, b \in ob\,\varphi$;
- $aR^I\boldsymbol{c}$ iff $\{\neg C \mid \neg\exists R.C \in o(a)\} \subseteq \boldsymbol{c}$, for $a \in ob\,\varphi$ and $\boldsymbol{c} \in T$.

It is readily proved by induction that

$$C^I = \{\boldsymbol{c} \in T \mid C \in \boldsymbol{c}\} \cup \{a \in ob\,\varphi \mid C \in o(a)\},$$

for every $C \in con\,\varphi$, and that $I \models \boldsymbol{f}$. Therefore, $I \models \varphi$.

Conversely, suppose that $I \models \varphi$ for some model

$$I = \left\langle \Delta, R_0^I, \dots, C_0^I, \dots, a_0^I, \dots \right\rangle.$$

Define a triple $\langle T, o, \boldsymbol{f}\rangle$ as follows:

- $T = \{\boldsymbol{c}(x) \mid x \in \Delta\}$, where $\boldsymbol{c}(x) = \{C \in con\,\varphi \mid x \in C^I\}$;
- $o(a) = \boldsymbol{c}(a^I)$;
- $\boldsymbol{f} = \{\chi \in sub\,\varphi \mid I \models \chi\}$.

It is easily seen that $\langle T, o, \boldsymbol{f}\rangle$ is a quasimodel for φ.

Our exponential time satisfiability-checking algorithm runs as follows. Given a formula φ, we first enumerate all model candidates $\langle T, o, \boldsymbol{f}\rangle$ for φ in which T contains *all* concept types for φ; denote these candidates by $\mathfrak{C}_1, \dots, \mathfrak{C}_N$. It should be clear that

$$N \leq 2^{|con\,\varphi||ob\,\varphi|} \cdot 2^{|sub\,\varphi|} \leq 2^{2\ell(\varphi)^2 + 2\ell(\varphi)},$$

and so this enumeration can be performed in exponential time. Set $i = 1$ and consider $\mathfrak{C}_i = \langle T, o, \boldsymbol{f}\rangle$.

Step 1. Enumerate all pairs $\langle \boldsymbol{c}, D\rangle$, where $\boldsymbol{c} \in T$ and $D \in \boldsymbol{c}$. Call such a pair a *defect (in T)* if either (i) D is of the form $\exists R.C$ and there is no $\boldsymbol{c}' \in T$ such that $\{\neg D \mid \neg\exists R.D \in \boldsymbol{c}\} \cup \{C\} \subseteq \boldsymbol{c}'$; or (ii) D is of the form $\neg C$ and $(C = \top) \in \boldsymbol{f}$. If we find such a defect $\langle \boldsymbol{c}, D\rangle$ in T and $\boldsymbol{c}$ is not in the range of o, then we set $T := T - \{\boldsymbol{c}\}$ and then proceed further with Step 1. If $\boldsymbol{c}$ belongs to the range of o, then we stop considering $\mathfrak{C}_i$ and go to Step 3. When all defects are exhausted, we go to Step 2.

Step 2. Check whether the resulting triple $\langle T', o, \boldsymbol{f}\rangle$ satisfies (f) and (g). If it does, then we stop with a verdict: $\langle T', o, \boldsymbol{f}\rangle$ is a quasimodel for φ. Otherwise we go to Step 3.

Step 3. Set $i := i + 1$. If $i \leq N$ then we go to Step 1. Otherwise we stop with a verdict: there is no quasimodel for φ.

Clearly, if the algorithm says that $\langle T', o, \boldsymbol{f}\rangle$ is a quasimodel for φ, then this is indeed the case. On the other hand, if $\langle T', o, \boldsymbol{f}\rangle$ is a quasimodel for φ then there is $\mathfrak{C}_i = \langle T, o, \boldsymbol{f}\rangle$ and no concept type from T' can ever occur in a defect. So the algorithm will stop at Step 2 producing a quasimodel for φ.

Actually, from the application point of view we may be interested not in arbitrary models satisfying a given formula, but only in *finite* ones. For logics like $\mathcal{ALC}$ there is no difference between these two variants of the satisfiability problem:

Proposition 2.28. *$\mathcal{ALC}$ has the fmp: every satisfiable $\mathcal{ALC}$-formula can be satisfied in a finite model.*

(This fact follows immediately from the proof above: given a satisfiable formula φ, our algorithm constructs a model for φ of size $\leq 2^{|con\,\varphi|} + |ob\,\varphi|$.)

However, there are much more expressive description languages that do not enjoy the fmp; some of them will be discussed later on in this section.

There are several ways of reducing the complexity of the reasoning tasks. Of course, all of them mean reducing the expressive power of the language as well. One can restrict the use of some concept constructors. For instance, by allowing applications of $\neg$ only to atomic concepts we cannot form the union

($\sqcup$) of concepts. The subsumption problem for such a restricted language (with empty knowledge base) becomes NP-complete (see Donini *et al.* 1995 and references therein).

Another way is to impose restrictions on formulas that may be used in our knowledge bases. Suppose a TBox Σ consists of statements of the form

$$A = C,$$

where A is a concept name. The equation $A = C$ can be regarded as a *definition* of A. Say that Σ is a *simple TBox* if, for every concept name A, there is at most one definition of A in Σ. Thus, to define a concept A in a simple TBox means to single out necessary and sufficient conditions for an object to be in A. In simple TBoxes, one can distinguish between *defined* concepts—those which appear in the left-hand side of an equation—and *atomic* ones, i.e., those that are not defined. Of course, in order to obtain explicit definitions of defined concepts one has to require that no defined concept name occurs in its own definition. To make this more precise, let us define a binary relation $\prec$ on the set of concept names occurring in Σ by taking $A \prec B$ if A is defined and the definition of A in Σ contains an occurrence of B. Now call Σ *acyclic* if the relation $\prec$ contains no cycles (i.e., no sequences of the form $A_0 \prec A_1 \prec \dots \prec A_0$), otherwise it is *cyclic*.

Acyclic simple TBoxes are an important type of knowledge bases for applications. The reason for this is that for a simple acyclic TBox Σ, the subsumption problem

$$\Sigma \models C \sqsubseteq D$$

reduces to the subsumption problem with empty knowledge base

$$\models C' \sqsubseteq D',$$

where C' and D' are obtained from C and D by replacing recursively every defined concept by its definition, so that the resultant C' and D' contain only atomic concept names. Unfortunately, as was shown by Nebel (1990), this 'unfolding' technique may result in an exponential blowup of the concept size, and so we can't use Proposition 2.25 to obtain a PSPACE algorithm. Nevertheless, such an algorithm exists: Lutz (1999) presents a PSPACE tableau procedure for checking satisfiability of $\mathcal{ALC}$-concepts with respect to acyclic simple TBoxes. (We remind the reader that by Theorem 2.26, both concept satisfiability and subsumption for arbitrary TBoxes are EXPTIME-complete.)

Suppose now that our knowledge base Σ is a cyclic simple TBox. Then a statement of the form

$$A = T(A)$$

is contained in Σ, where $T(A)$ denotes a concept with an occurrence of A. According to the interpretation given above, $A = T(A)$ is understood as a constraint stating that an object belongs to A^I iff it belongs to $T(A)^I$. There are two other interpretations of such equations in the literature, known as the *least* and the *greatest fixed point interpretations*: according to them, A^I is understood as the least (respectively, greatest) solution of the equation $A = T(A)$ if it exists; see (Baader 1990, Nebel 1991). However, this topic is beyond the scope of this book.

The natural desire to improve the computational behavior of description logics comes across the need to increase their expressive power. For instance, in the knowledge base of Example 2.24 we might want to refine the information by stating that

- *Eve* and *Adam* have only two children;
- every child has only one mother; and
- all children have *Eve* and *Adam* as their ancestors.

This desire may lead to richer description languages, say, to the one introduced by De Giacomo and Lenzerini (1996) and De Giacomo (1995) under the name $\mathcal{CQ}$.

The language of $\mathcal{CQ}$ is an extension of that of $\mathcal{ALC}$ with a number of role and concept constructors. First, by a *basic role* we mean any role name R_i. Now, if R, S are roles, B is a basic role, C, D are concepts and n a natural number, then

- $R \sqcup S$, $R \circ S$, R^* are *roles*, and
- $C \sqcap D$, $\neg C$, $\exists R.C$, $\exists_{\geq n} B.C$ are *concepts*.

The intended meaning of the introduced constructors will be clear from the following definition (which extends the corresponding definition for $\mathcal{ALC}$). Let I be a model of the form (2.19). Then

- $(R \sqcup S)^I = R^I \cup S^I$;
- $(R \circ S)^I = R^I \circ S^I$ (the composition of R^I and S^I);
- $(R^*)^I = (R^I)^*$ (the transitive and reflexive closure of R^I);
- $x \in (\exists_{\geq n} R.C)^I$ iff $\left|\{y \in C^I \mid xR^I y\}\right| \geq n$.

Concepts of the form $\exists_{\geq n} R.C$ are called in description logic *qualified number restrictions*; in modal logic they appeared under the name of *graded modalities* in (Fine 1972b, van der Hoek 1992). Observe that

$$x \in (\neg \exists_{\geq n} R.C)^I \quad \text{iff} \quad \left|\{y \in C^I \mid xR^I y\}\right| < n.$$

So we denote $\neg\exists_{\geq n}R.C$ by $\exists_{\leq n-1}R.C$, and $\exists_{\geq n}R.C \wedge \exists_{\leq n}R.C$ by $\exists_{=n}R.C$.

It is easy to extend the translation $\cdot^{\dagger}$ from multimodal **K** onto $\mathcal{ALC}$-concepts above to a translation of **PDL** into $\mathcal{CQ}$-concepts (see Section 14.1). On the other hand, De Giacomo (1995) showed that the satisfiability problem for $\mathcal{CQ}$-concepts relative to $\mathcal{CQ}$ TBoxes is polynomially reducible to the satisfiability problem for **CPDL**. This reduction is easily generalized to a reduction of $\mathcal{CQ}$ formula satisfiability to **PDL**. By Theorem 2.22, we obtain:

Proposition 2.29. *The satisfiability problem for $\mathcal{CQ}$-formulas is* EXPTIME-*complete.*

In $\mathcal{CQ}$, we can extend the knowledge base of Example 2.24 with the following formulas:

$$Eve : \exists_{=2}\mathsf{parent.Child}$$

(Eve has two children),

$$\mathsf{Child} \sqsubseteq \exists_{=1}\mathsf{has.Mother}$$

(every child has one mother),

$$Eve : \mathsf{First\_Parent}, \quad Adam : \mathsf{First\_Parent}$$

(Eve and Adam are first parents),

$$\mathsf{First\_Parent} \sqsubseteq \exists(\mathsf{parent} \circ \mathsf{parent}^*).\exists\mathsf{drives.Car}$$

(the first parents have a descendent who drives a car). Note, however, that we cannot express in $\mathcal{CQ}$ that Eve and Adam are the only first parents. To be able to do this we need a constructor allowing us to form concepts $\{a\}$ out of object names a. The concept $\{a\}$ is interpreted in a model I in a straightforward way:

$$\{a\}^I = \{a^I\}.$$

Such concepts are closely related to *nominals* in modal and hybrid logics; see, e.g., (Blackburn 1993). Using this construct we can define

$$\mathsf{First\_Parent} = \{Eve\} \sqcup \{Adam\}.$$

The extension $\mathcal{CQO}$ of $\mathcal{CQ}$ with the constructor of nominals above was introduced by De Giacomo (1995). Observe that having concepts of the form $\{a\}$, there is no need to define $a : C$ and aRb as atomic formulas: they are equivalent to $\{a\} \to C = \top$ and $\{a\} \to \exists R.\{b\} = \top$, respectively. It is shown in (De Giacomo 1995) that the satisfiability problem for $\mathcal{CQO}$ is EXPTIME-complete.

The language $\mathcal{CQ}$ and its extensions are not available yet in implemented systems. A less expressive but important extension of $\mathcal{ALC}$, which is part of almost all working systems, adds to the syntax of $\mathcal{ALC}$

(i) a set of *transitive role names* $T_0, T_1, \ldots$ interpreted by *transitive* binary relations, and

(ii) the possibility to use *role inclusion axioms* of the form

$$S_1 \sqsubseteq S_2$$

in TBoxes, where S_1 and S_2 are transitive or standard role names. Such a role inclusion axiom is satisfied in a model I iff $S_1^I \subseteq S_2^I$.

Now, instead of defining the role descendent as the transitive closure of the role parent in the example above, one can approximate the properties of descendent by introducing it from the very beginning as a transitive role name and adding the role inclusion axiom

$$\mathsf{parent} \sqsubseteq \mathsf{descendent}$$

to the TBox. Of course, some information is lost, since now the interpretation of descentent does not coincide with but only contains the transitive closure of parent, but in implemented systems the computational behavior of transitive role names is much better than that of transitive closures. $\mathcal{ALC}$ extended with transitive roles was introduced under the name $\mathcal{ALC}_{R^+}$ in (Sattler 1996), but now is usually called $\mathcal{S}$; see e.g. (Horrocks *et al.* 2000b). As $\mathcal{S}$ is already reserved for the temporal operator 'Since,' in what follows we will use the original name $\mathcal{ALC}_{R^+}$. $\mathcal{ALC}_{R^+}$ with role inclusion axioms is called $\mathcal{ALCH}_{R^+}$.

Since $\mathcal{ALCH}_{R^+}$ can obviously be embedded into $\mathcal{CQ}$, we immediately obtain:

Proposition 2.30. *The satisfiability problem for $\mathcal{ALC}_{R^+}$- and $\mathcal{ALCH}_{R^+}$-formulas is* EXPTIME*-complete.*

For more information about description logic we refer the reader to the *Description Logic Handbook* (Baader *et al.* 2003) and the surveys (Donini *et al.* 1996, Calvanese *et al.* 2001).

2.6 Spatial logic

'Spatial logic' is a collective name for various logical languages and systems describing topological and geometric sets and relations. Some of them have been motivated by applications in computer science and artificial intelligence, such as image processing, visual databases, geographical information systems, robotics, etc. Others come from pure mathematics and mathematical physics (in particular, topology, projective geometry, relativity theory). Of the enormous number of spatial formalisms developed in these diverse fields, we concentrate in this book only on those that were devised within the knowledge representation and reasoning branch of artificial intelligence. Most of these logics are of *qualitative* rather than *quantitative* character because quite often precise numerical information is either not available or not appropriate for (common sense) reasoning about spatial structures in knowledge representation systems (Cohn 1997).[5]

Even within the field of knowledge representation and reasoning there exist different approaches to logical description of spatial structures; see, e.g., the collection (Stock 1997) and references therein, and monographs (Casati and Varzi 1999, Galton 2000). In this book, we will consider only some of them which—explicitly or implicitly—are based on the formalism of modal logic.

Let us begin by discussing a 'naïve' approach to representing space in the framework of possible world semantics.

Compass relations on the plane

In human everyday practice, most spatial structures are attached to coordinate systems; such are, for example, maps (geographical, celestial, anatomical, etc.) or images (fixed or moving). This observation suggests the following straightforward use of Kripke frames to represent coordinates. Consider the real plane $\mathbb{R} \times \mathbb{R}$ as an infinite map. The *compass relations* between points $\langle x, y \rangle$ and $\langle x', y' \rangle$ are defined by taking:

$$
\begin{array}{lll}
\langle x,y\rangle\, R_E\, \langle x',y'\rangle & \text{iff} & x < x',\ y = y'\ (\langle x',y'\rangle \text{ is to the East of } \langle x,y\rangle);\\
\langle x,y\rangle\, R_N\, \langle x',y'\rangle & \text{iff} & x = x',\ y < y'\ (\langle x',y'\rangle \text{ is to the North of } \langle x,y\rangle);\\
\langle x,y\rangle\, R_S\, \langle x',y'\rangle & \text{iff} & x = x',\ y > y'\ (\langle x',y'\rangle \text{ is to the South of } \langle x,y\rangle);\\
\langle x,y\rangle\, R_W\, \langle x',y'\rangle & \text{iff} & x > x',\ y = y'\ (\langle x',y'\rangle \text{ is to the West of } \langle x,y\rangle).
\end{array}
$$

The plane with these relations can be regarded as the 4-frame

$$\langle \mathbb{R} \times \mathbb{R}, R_E, R_W, R_N, R_S \rangle,$$

and one can consider the corresponding modal logic with four necessity operators: $\Box_E$, $\Box_W$, $\Box_N$ and $\Box_S$. Instead of $\mathbb{R}$ we can take any other linearly ordered set, for instance $\mathbb{Z}$, thus obtaining a grid-like map. Of course, the change of the basic set may affect the resulting logic. But some formulas are valid in any plane of this sort, say,

$$\Box_E \Box_N p \leftrightarrow \Box_N \Box_E p$$

[5]Traditionally, spatial structures are investigated by many mathematical disciplines from different viewpoints. The closest ones to the modal logic ideology are those studying qualitative properties and behavior of space structures. A typical example is the mathematical theory of dynamical systems (with its more recent parts, such as catastrophe theory and chaos theory). The basic concept here is a 'phase space' consisting of 'states,' the coordinates of which are parameters of a certain system. This allows one to represent various structures (mechanical, biological, economical) as spatial.

and the other commutativity axioms. Instead of the whole plane one can restrict attention only to a certain subset, say, the North-Western half-plane

$$\{\langle x, y\rangle \mid x < y\}.$$

The formula

$$\Box_N \Box_E p \to \Box_E \Box_N p$$

is valid in this half-plane, while its converse

$$\Box_E \Box_N p \to \Box_N \Box_E p$$

is not. Note that the logic of this half-plane can also be regarded as a variant of interval temporal logic—it will be considered in Section 3.9.

There are at least two big flaws in this simple approach to spatial representation and reasoning. First, the resulting 'spatial logics' often turn out to be undecidable and even not recursively axiomatizable; see Chapter 7. And second, the language of the compass logic speaks only about points, but not about *spatial regions*, that is the space occupied by physical bodies, say, countries, which are much more important for applications.

Region connection calculus

$\mathcal{RCC}$—*Region Connection Calculus*—is a first-order theory devised by Randell, Cui and Cohn (1992) for qualitative spatial representation and reasoning. The signature of $\mathcal{RCC}$ contains only one binary predicate symbol C. Atomic formulas of the form $\mathsf{C}(X, Y)$ are read as 'region X *is connected* with region Y.' (We denote individual variables of $\mathcal{RCC}$ by X, Y, Z, etc.) Using C one can define other relations between spatial regions. Here are some of them:

$\mathsf{DC}(X, Y)$	—	'X and Y are disconnected,'
$\mathsf{P}(X, Y)$	—	'X is a part of Y,'
$\mathsf{EQ}(X, Y)$	—	'X is identical with Y,'
$\mathsf{O}(X, Y)$	—	'X overlaps Y,'
$\mathsf{PO}(X, Y)$	—	'X partially overlaps Y,'
$\mathsf{EC}(X, Y)$	—	'X is externally connected to Y,'
$\mathsf{PP}(X, Y)$	—	'X is a proper part of Y,'
$\mathsf{TPP}(X, Y)$	—	'X is a tangential proper part of Y,'
$\mathsf{TPPi}(X, Y)$	—	'Y is a tangential proper part of X,'
$\mathsf{NTPP}(X, Y)$	—	'X is a nontangential proper part of Y,'
$\mathsf{NTPPi}(X, Y)$	—	'Y is a nontangential proper part of X.'

$\mathsf{DC}(X,Y)$	$=$	$\neg\mathsf{C}(X,Y)$
$\mathsf{P}(X,Y)$	$=$	$\forall Z\ (\mathsf{C}(Z,X) \to \mathsf{C}(Z,Y))$
$\mathsf{EQ}(X,Y)$	$=$	$\mathsf{P}(X,Y) \land \mathsf{P}(Y,X)$
$\mathsf{O}(X,Y)$	$=$	$\exists Z\ (\mathsf{P}(Z,X) \land \mathsf{P}(Z,Y))$
$\mathsf{PO}(X,Y)$	$=$	$\mathsf{O}(X,Y) \land \neg\mathsf{P}(X,Y) \land \neg\mathsf{P}(Y,X)$
$\mathsf{EC}(X,Y)$	$=$	$\mathsf{C}(X,Y) \land \neg\mathsf{O}(X,Y)$
$\mathsf{PP}(X,Y)$	$=$	$\mathsf{P}(X,Y) \land \neg\mathsf{P}(Y,X)$
$\mathsf{TPP}(X,Y)$	$=$	$\mathsf{PP}(X,Y) \land \exists Z\ (\mathsf{EC}(Z,X) \land \mathsf{EC}(Z,Y))$
$\mathsf{NTPP}(X,Y)$	$=$	$\mathsf{PP}(X,Y) \land \neg\exists Z\ (\mathsf{EC}(Z,X) \land \mathsf{EC}(Z,Y))$

Figure 2.4: Some relations between spatial regions, defined in terms of C.

Their definitions via C are given in Fig. 2.4. The axioms of $\mathcal{RCC}$ can be found in (Randell *et al.* 1992). We will not use them in this book.

From the computational point of view $\mathcal{RCC}$ turns out to be too expressive: as was observed by Gotts (1996b) (and actually follows from Grzegorczyk 1951), the full first-order theory of $\mathcal{RCC}$ is undecidable. Fortunately, there are various decidable (and even tractable) fragments of $\mathcal{RCC}$. One of the most important is known as $\mathcal{RCC}$-8. It was constructed (independently and almost simultaneously) by two parallel research streams of spatial knowledge representation and reasoning: in the framework of geographical information systems (Egenhofer and Franzosa 1991, Egenhofer and Mark 1995, Bennett *et al.* 1997, Haarslev *et al.* 1999) and as an effective fragment of $\mathcal{RCC}$ (Randell *et al.* 1992).

$\mathcal{RCC}$-8

If we are interested only in relationships between spatial regions without taking into account their topological shape, then the eight predicates in Fig. 2.5 are enough: they turn out to be jointly exhaustive and pairwise disjoint, which means that any two (non-empty) regions stand precisely in one of these eight relations. Moreover, according to the experiments reported in (Renz and Nebel 1998), the eight predicates turn out to be conceptually cognitive adequate in the sense that people indeed distinguish between these relations.

Formally, the language of $\mathcal{RCC}$-8 consists of a countably infinite set of individual variables $X_0, X_1, \ldots$ (or $X, Y, Z, \ldots$), called *region variables*, eight binary predicate symbols DC, EQ, PO, EC, TPP, TPPi, NTPP, NTPPi and the Booleans out of which we construct in the usual way *spatial formulas*.

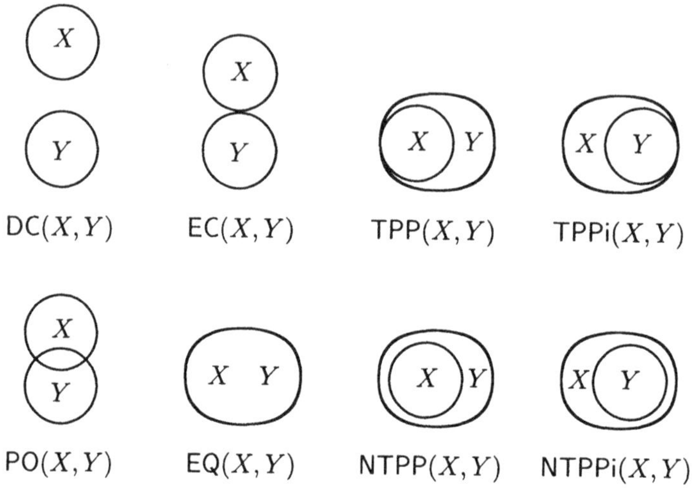

Figure 2.5: The $\mathcal{RCC}$-8 predicates.

For example, using the language of $\mathcal{RCC}$-8 we can compose spatial knowledge bases like

$$\begin{aligned}
&\mathsf{EC}(\textit{Catalunya, France}),\\
&\mathsf{TPP}(\textit{Catalunya, Spain}) \vee \mathsf{NTPP}(\textit{Catalunya, Spain}),\\
&\mathsf{DC}(\textit{Spain, France}) \vee \mathsf{EC}(\textit{Spain, France}),\\
&\mathsf{NTPP}(\textit{Paris, France}).
\end{aligned}$$

Then the formulas

$$\mathsf{EC}(\textit{Spain, France}), \quad \mathsf{TPP}(\textit{Catalunya, Spain}), \quad \mathsf{DC}(\textit{Spain, Paris})$$

should be consequences of this knowledge base.

Note that the other relations in Fig. 2.4 can be expressed as Boolean combinations of the $\mathcal{RCC}$-8 predicates as follows:

$$\begin{aligned}
\mathsf{P}(X,Y) &= \mathsf{TPP}(X,Y) \vee \mathsf{EQ}(X,Y) \vee \mathsf{NTPP}(X,Y),\\
\mathsf{P}(Y,X) &= \mathsf{TPPi}(X,Y) \vee \mathsf{EQ}(X,Y) \vee \mathsf{NTPPi}(X,Y),\\
\mathsf{O}(X,Y) &= \mathsf{PO}(X,Y) \vee \mathsf{P}(X,Y) \vee \mathsf{P}(Y,X).
\end{aligned}$$

Spatial formulas can be interpreted in *topological spaces*. We remind the reader that a *topological space* is a pair $\mathfrak{T} = \langle U, \mathbb{I}\rangle$ in which U is a nonempty set, the *universe* of the space, and $\mathbb{I}$ is the *interior operator* on U satisfying the following *Kuratowski axioms*: for all $X, Y \subseteq U$,

$$\mathbb{I}(X \cap Y) = \mathbb{I}X \cap \mathbb{I}Y, \quad \mathbb{I}X \subseteq \mathbb{I}\mathbb{I}X, \quad \mathbb{I}X \subseteq X, \quad \mathbb{I}U = U.$$

Figure 2.6: Regular closure.

The operator dual to $\mathbb{I}$ is called the *closure operator* and denoted by $\mathbb{C}$. Thus $\mathbb{C}X = U - \mathbb{I}(U - X)$, for all $X \subseteq U$. A set $X \subseteq U$ is called *open* if $\mathbb{I}X = X$, and *closed* if $\mathbb{C}X = X$. We will also consider some special topological spaces, such as the *connected spaces* (which are not unions of two disjoint nonempty open sets), and the *Euclidean spaces* $\langle \mathbb{R}^n, \mathbb{I} \rangle$ for $n \geq 1$ (where a point $x \in \mathbb{R}^n$ belongs to $\mathbb{I}X$ if, for some $\varepsilon > 0$, all points in the ε-neighborhood[6] of x belong to X).

Region variables range over *regular closed sets* of the topological space $\mathfrak{T}$, i.e., an *assignment in* $\mathfrak{T}$ is a map $\mathfrak{a}$ associating with every variable X a set $\mathfrak{a}(X) \subseteq U$ such that $\mathfrak{a}(X) = \mathbb{C}\mathbb{I}\mathfrak{a}(X)$. (For instance, $\emptyset$ and U are regular closed sets. Examples of sets that are not regular closed in, say, the two-dimensional Euclidean space are balloons—circles with attached threads (1D lines)—or sets with isolated points, etc., which can hardly be regarded as regions; see Fig. 2.6 where the region $\mathbb{C}\mathbb{I}X$ consists of two disconnected parts, with one of them containing a 'hole.') Often it is also assumed that regions are nonempty, i.e., $\mathfrak{a}(X) \neq \emptyset$. However, this constraint can be expressed in $\mathcal{RCC}$-8 explicitly: for instance, $\neg\mathsf{DC}(X, X)$, according to the interpretation below, guarantees that region X is not empty.

The *truth-relation* $\mathfrak{T} \models^{\mathfrak{a}} \varphi$ for atomic formulas of $\mathcal{RCC}$-8 is defined in the following way:[7]

$$\begin{array}{lll}
\mathfrak{T} \models^{\mathfrak{a}} \mathsf{DC}(X_1, X_2) & \text{iff} & \neg\exists x\; x \in \mathfrak{a}(X_1) \cap \mathfrak{a}(X_2), \\
\mathfrak{T} \models^{\mathfrak{a}} \mathsf{EQ}(X_1, X_2) & \text{iff} & \forall x\; (x \in \mathfrak{a}(X_1) \leftrightarrow x \in \mathfrak{a}(X_2)), \\
\mathfrak{T} \models^{\mathfrak{a}} \mathsf{PO}(X_1, X_2) & \text{iff} & \exists x\; x \in \mathbb{I}\mathfrak{a}(X_1) \cap \mathbb{I}\mathfrak{a}(X_2) \\
 & & \quad \wedge\ \exists x\; x \in \mathfrak{a}(X_1) \cap (U - \mathfrak{a}(X_2)) \\
 & & \quad \wedge\ \exists x\; x \in (U - \mathfrak{a}(X_1)) \cap \mathfrak{a}(X_2), \\
\mathfrak{T} \models^{\mathfrak{a}} \mathsf{EC}(X_1, X_2) & \text{iff} & \exists x\; x \in \mathfrak{a}(X_1) \cap \mathfrak{a}(X_2) \\
 & & \quad \wedge\ \neg\exists x\; x \in \mathfrak{a}(X_1) \cap \mathbb{I}\mathfrak{a}(X_2) \\
 & & \quad \wedge\ \neg\exists x\; x \in \mathbb{I}\mathfrak{a}(X_1) \cap \mathfrak{a}(X_2),
\end{array}$$

[6]The *ε-neighborhood* of $x = \langle x_1, \ldots, x_n \rangle$ in $\mathbb{R}^n$ consists of all points $y = \langle y_1, \ldots, y_n \rangle$ such that $\sum_{i=1}^{n} |x_i - y_i|^2 < \varepsilon^2$.

[7]Note that since we allow regions to be empty, the $\mathcal{RCC}$-8 predicates are no longer pairwise disjoint. For instance, both $\mathsf{DC}(\emptyset, X)$ and $\mathsf{NTPP}(\emptyset, X)$ hold in every topological space whenever $X \neq \emptyset$, as well as $\mathsf{DC}(\emptyset, \emptyset)$ and $\mathsf{EQ}(\emptyset, \emptyset)$.

$$\begin{aligned}
\mathfrak{I} \models^{\mathfrak{a}} \mathsf{TPP}(X_1, X_2) \quad &\text{iff} \quad \forall x\, x \in (U - \mathfrak{a}(X_1)) \cup \mathfrak{a}(X_2) \\
&\qquad \wedge \exists x\, x \in \mathfrak{a}(X_1) \cap \mathfrak{a}(X_2) \cap (U - \mathbb{I}\mathfrak{a}(X_2)) \\
&\qquad \wedge \exists x\, x \in (U - \mathfrak{a}(X_1)) \cap \mathfrak{a}(X_2), \\
\mathfrak{I} \models^{\mathfrak{a}} \mathsf{NTPP}(X_1, X_2) \quad &\text{iff} \quad \forall x\, x \in (U - \mathfrak{a}(X_1)) \cup \mathbb{I}\mathfrak{a}(X_2) \\
&\qquad \wedge \exists x\, x \in (U - \mathfrak{a}(X_1)) \cap \mathfrak{a}(X_2).
\end{aligned}$$

Note that although the full $\mathcal{RCC}$ formalism was originally presented as a naïve theory without any specific models, Gotts (1996a) and Bennett (1998) showed that it can also be interpreted in classical point-set topology. The truth-definition for $\mathsf{C}(X, Y)$ is formalized then as follows:

$$\mathfrak{I} \models^{\mathfrak{a}} \mathsf{C}(X, Y) \quad \text{iff} \quad \exists x \in \mathfrak{a}(X) \cap \mathfrak{a}(Y).$$

As was proved by Gotts (1996a), the syntactical definitions of Fig. 2.4 are correct in Euclidean spaces, and these spaces are models of the $\mathcal{RCC}$ axioms as well.

It is not hard to see that the above truth-definition and the Kuratowski axioms together yield the following equivalences (which one might consider as more natural truth-definitions):

$$\begin{aligned}
\mathfrak{I} \models^{\mathfrak{a}} \mathsf{PO}(X_1, X_2) \quad &\text{iff} \quad \exists x\, x \in \mathbb{I}\mathfrak{a}(X_1) \cap \mathbb{I}\mathfrak{a}(X_2) \\
&\qquad \wedge \exists x\, x \in \mathbb{I}\mathfrak{a}(X_1) \cap (U - \mathfrak{a}(X_2)) \\
&\qquad \wedge \exists x\, x \in (U - \mathfrak{a}(X_1)) \cap \mathbb{I}\mathfrak{a}(X_2), \\
\mathfrak{I} \models^{\mathfrak{a}} \mathsf{EC}(X_1, X_2) \quad &\text{iff} \quad \exists x\, x \in \mathfrak{a}(X_1) \cap \mathfrak{a}(X_2) \\
&\qquad \wedge \neg\exists x\, x \in \mathbb{I}\mathfrak{a}(X_1) \cap \mathbb{I}\mathfrak{a}(X_2), \\
\mathfrak{I} \models^{\mathfrak{a}} \mathsf{TPP}(X_1, X_2) \quad &\text{iff} \quad \forall x\, x \in (U - \mathfrak{a}(X_1)) \cup \mathfrak{a}(X_2), \\
&\qquad \wedge \exists x\, x \in \mathfrak{a}(X_1) \cap (U - \mathbb{I}\mathfrak{a}(X_2)) \\
&\qquad \wedge \exists x\, x \in (U - \mathfrak{a}(X_1)) \cap \mathfrak{a}(X_2).
\end{aligned}$$

Indeed, it is readily checked that for any sets A, B,

$$A \cap B = \emptyset \quad \text{implies} \quad \mathbb{C}A \cap \mathbb{I}B = \emptyset. \tag{2.21}$$

Now, in order to prove that the two definitions of PO are equivalent, it is enough to show that for all regular closed sets V and W,

$$\mathbb{I}V - W \neq \emptyset \quad \text{iff} \quad V - W \neq \emptyset.$$

One direction is obvious. For the other, suppose that $V - W \neq \emptyset$. Since $V - W = \mathbb{C}\mathbb{I}V \cap \mathbb{I}(U - W)$, by (2.21) we have $\mathbb{I}V - W \neq \emptyset$.

In the case of EC, we have to show that for all regular closed sets V and W,

$$\mathbb{I}V \cap \mathbb{I}W = \emptyset \quad \text{iff} \quad V \cap \mathbb{I}W = \emptyset \text{ and } \mathbb{I}V \cap W = \emptyset.$$

The implication ($\Leftarrow$) is obvious. For the converse, suppose that $\mathbb{I}V \cap \mathbb{I}W = \emptyset$. Then by (2.21) we have

$$\emptyset = \mathbb{C}\mathbb{I}V \cap \mathbb{I}\mathbb{I}\mathbb{I}W = V \cap \mathbb{I}W,$$
$$\emptyset = \mathbb{I}\mathbb{I}\mathbb{I}V \cap \mathbb{C}\mathbb{I}W = \mathbb{I}V \cap W.$$

Finally, in the case of TPP we have:

$$\mathfrak{a}(X_1) \cap \mathfrak{a}(X_2) \cap (U - \mathbb{I}\mathfrak{a}(X_2)) = \mathfrak{a}(X_1) \cap (U - \mathbb{I}\mathfrak{a}(X_2)),$$

because $\mathfrak{a}(X_1) \subsetneq \mathfrak{a}(X_2)$.

The main reasoning task for $\mathcal{RCC}$-8 is the following *entailment problem*: given a finite set Σ of spatial formulas and a formula φ, decide whether φ is a *logical consequence of* Σ (or Σ *entails* φ), i.e., for every topological space $\mathfrak{T}$ and every assignment $\mathfrak{a}$ in it, we have $\mathfrak{T} \models^{\mathfrak{a}} \varphi$ whenever $\mathfrak{T} \models^{\mathfrak{a}} \psi$ for all $\psi \in \Sigma$. If φ is a logical consequence of Σ, then we write $\Sigma \models \varphi$. It should be clear that the entailment problem is reducible to the *satisfiability problem*: given a spatial formula φ, decide whether φ is *satisfiable* (or *realizable*) in a topological space, i.e., whether there exists a topological space $\mathfrak{T}$ and an assignment $\mathfrak{a}$ in it such that $\mathfrak{T} \models^{\mathfrak{a}} \varphi$. Indeed, we have $\Sigma \models \varphi$ iff the formula $\bigwedge \Sigma \wedge \neg\varphi$ is not satisfiable in any topological space. Sometimes satisfiability in more restricted classes of topological spaces is considered, say, only in connected spaces or in the Euclidean spaces $\langle \mathbb{R}^n, \mathbb{I} \rangle$, for $n \geq 1$.

That the satisfiability (and so entailment) problem for $\mathcal{RCC}$-8 formulas in topological spaces is decidable was observed by Bennett (1994, 1996). Renz and Nebel (1999) showed the NP-completeness of the satisfiability problem and described maximal tractable fragments of $\mathcal{RCC}$-8, i.e., those that belong to P.

Bennett (1994, 1996) embedded $\mathcal{RCC}$-8 into the bimodal logic $\mathbf{S4}_u$, i.e., Lewis's **S4** with the universal modality, using the fact that $\mathbf{S4}_u$ is complete with respect to topological spaces. But before considering this connection in more detail, let us extend $\mathcal{RCC}$-8 with Boolean operations on regions.

$\mathcal{BRCC}$-8

One apparent 'deficit' of $\mathcal{RCC}$-8 is that it operates only with *atomic* regions. We cannot form unions ($\sqcup$) or intersections ($\sqcap$) of regions to say, for instance, that

$$\mathsf{EQ}(EU, Spain \sqcup Italy \sqcup \dots)$$

('the EU consists of Spain, Italy, etc.'),

$$\mathsf{P}(Alps, Italy \sqcup France \sqcup \dots)$$

('the Alps are located in Italy, France, etc.'),

$$\mathsf{EC}(Austria, Alps \sqcap Italy)$$

('Austria is externally connected to the alpine part of Italy'), and deduce from these that there is a country Z such that $\mathsf{TPP}(Z, EU)$ (i.e., 'Z is a tangential proper part of the EU'), or that if $\mathsf{EC}(X, EU)$, for some country X, then $\mathsf{EC}(X, Y)$ for some country Y in the EU. Note, by the way, that $\mathsf{TPP}(Z, EU)$ is a correct conclusion only if we interpret our formulas in Euclidean (or, more generally, connected topological spaces (and if there are non-EU countries): in a discrete topological space (where all sets are open) the EU may be an open set with empty boundary. This simple observation and the result of (Renz 1998), according to which every satisfiable $\mathcal{RCC}$-8 formula is satisfiable in all Euclidean spaces $\langle \mathbb{R}^n, \mathbb{I} \rangle$, $n \geq 1$, show that the Boolean operations on region terms indeed increase the expressive power of $\mathcal{RCC}$-8.

Denote by $\mathcal{BRCC}$-8 the extension of $\mathcal{RCC}$-8 which allows the use of *Boolean region terms*, i.e., combinations of region variables using the Boolean operators $\sqcup$, $\sqcap$ and $\neg$, as arguments of the $\mathcal{RCC}$-8 predicates. The *value* $\mathfrak{a}(t)$ of a Boolean region term t in a topological space $\mathfrak{T} = \langle U, \mathbb{I} \rangle$ under assignment $\mathfrak{a}$ is defined inductively as follows:

$$\begin{aligned} \mathfrak{a}(t \sqcup t') &= \mathbb{CI}(\mathfrak{a}(t) \cup \mathfrak{a}(t')) = \mathfrak{a}(t) \cup \mathfrak{a}(t'), \\ \mathfrak{a}(t \sqcap t') &= \mathbb{CI}(\mathfrak{a}(t) \cap \mathfrak{a}(t')), \\ \mathfrak{a}(\neg t) &= \mathbb{CI}(U - \mathfrak{a}(t)). \end{aligned}$$

As the Boolean operators do not in general preserve the property of being regular closed, we need the prefix $\mathbb{CI}$ in the right-hand parts of these definitions. Thus, every region term is interpreted as a regular closed set of $\mathfrak{T}$. Note that $\mathfrak{a}(X \sqcap \neg X) = \emptyset$ and $\mathfrak{a}(X \sqcup \neg X) = U$ for any $\mathfrak{a}$ and $\mathfrak{T}$. We denote the region terms $X \sqcap \neg X$ and $X \sqcup \neg X$ by $\bot$ and $\top$, respectively. The constraint $\neg\mathsf{EQ}(X, \bot)$ asserts that X is a nonempty region.

$\mathbf{S4}_u$ as a spatial formalism

In the late 1930s and early 1940s several logicians (Stone 1937, Tarski 1938, Tsao Chen 1938, McKinsey 1941) noticed that **S4** can be interpreted in topological spaces. Actually, there is a striking similarity between the axioms of **S4** and Kuratowski's axioms for the interior operator. (Axiom (K) and rule (RN) of **S4** can be replaced with $\Box(p_1 \wedge p_2) \leftrightarrow (\Box p_1 \wedge \Box p_2)$ and $\Box\top$, corresponding to the first and the last topological axioms above.) Using this observation, it is readily seen that every topological space $\mathfrak{T} = \langle U, \mathbb{I} \rangle$ gives rise to the modal algebra

$$\mathfrak{T}^+ = \langle 2^U, \cap, -, \mathbb{I}, \emptyset, U \rangle$$

which is an algebra for **S4** (see Section 1.5). Moreover, one can show that **S4** is complete with respect to algebras of this sort. This follows, in particular, from the fact that given a Kripke frame $\mathfrak{F} = \langle W, R\rangle$ for **S4**, we can construct the topological space $\mathfrak{T}_{\mathfrak{F}} = \langle W, \mathbb{I}_{\mathfrak{F}}\rangle$, where for any $X \subseteq W$,

$$\mathbb{I}_{\mathfrak{F}}X = \{x \in X \mid \forall y \in W\ (xRy \to y \in X)\}.$$

Moreover, $\mathfrak{F}$ and $\mathfrak{T}_{\mathfrak{F}}^{+}$ validate the same modal formulas, i.e., $\mathsf{Log}\mathfrak{F} = \mathsf{Log}\mathfrak{T}_{\mathfrak{F}}^{+}$. Therefore,

$$\mathbf{S4} = \{\varphi \in \mathcal{ML} \mid \mathfrak{T} \models \varphi \text{ for every (finite) topological space } \mathfrak{T}\},$$

where the relation $\mathfrak{T} \models \varphi$ is defined as follows. Given a topological space $\mathfrak{T} = \langle U, \mathbb{I}\rangle$, a *valuation* $\mathfrak{V}$ *in* $\mathfrak{T}$ maps each propositional variable to a subset of U. The pair $\langle \mathfrak{T}, \mathfrak{V}\rangle$ is then called a *topological model* (*based on* $\mathfrak{T}$). The valuation $\mathfrak{V}$ can be extended to all $\mathcal{ML}$-formulas by interpreting $\Box$ as $\mathbb{I}$, $\Diamond$ as $\mathbb{C}$, $\wedge$ as $\cap$, and $\neg$ as $-$. Now we say that φ is *satisfiable in* $\mathfrak{T}$ if $\mathfrak{V}(\varphi) \neq \emptyset$, for some topological model $\langle \mathfrak{T}, \mathfrak{V}\rangle$; φ is *valid in* $\mathfrak{T}$ ($\mathfrak{T} \models \varphi$ in symbols) if $\mathfrak{V}(\varphi) = U$ for all topological models based on $\mathfrak{T}$. Thus, **S4** can be regarded as the *logic of topological spaces*.

We can increase the expressive power of $\mathcal{ML}$ by enriching it with the universal box $\boxdot$ and diamond $\diamondplus$ (see Section 1.6), the topological meaning of which is 'for all points in the space' and 'for some point in the space,' respectively. More precisely, for every formula φ in the language of $\mathcal{ML}^u$ and for every topological model $\langle \mathfrak{T}, \mathfrak{V}\rangle$ based on $\mathfrak{T} = \langle U, \mathbb{I}\rangle$, we have:

$$\mathfrak{V}(\boxdot\varphi) = \begin{cases} U, & \text{if } \mathfrak{V}(\varphi) = U,\\ \emptyset, & \text{otherwise;}\end{cases} \qquad \mathfrak{V}(\diamondplus\varphi) = \begin{cases} U, & \text{if } \mathfrak{V}(\varphi) \neq \emptyset,\\ \emptyset, & \text{otherwise.}\end{cases}$$

In view of the connection between **S4**-frames and topological spaces mentioned above and Theorem 1.26, we have:

$$\mathbf{S4}_u = \{\varphi \in \mathcal{ML}^u \mid \mathfrak{T} \models \varphi \text{ for every (finite) topological space } \mathfrak{T}\}.$$

$\mathbf{S4}_u$ is expressive enough to encode the topological meaning of the $\mathcal{RCC}$-8 predicates and that of Boolean region terms.[8] Indeed, let us denote the box and the diamond of **S4** by, respectively, **I** and **C** (to emphasize their topological interpretation as the interior and closure operators). For a Boolean region term t, define inductively a modal formula $t^{\bowtie}$ by taking:

$$\begin{aligned}
X_i^{\bowtie} &= \mathbf{CI}p_i, \ (X_i \text{ is a region variable, } p_i \text{ a propositional variable}),\\
(t_1 \sqcap t_2)^{\bowtie} &= \mathbf{CI}(t_1^{\bowtie} \wedge t_2^{\bowtie}),\\
(t_1 \sqcup t_2)^{\bowtie} &= \mathbf{CI}(t_1^{\bowtie} \vee t_2^{\bowtie}),\\
(\neg t)^{\bowtie} &= \mathbf{CI}\neg t^{\bowtie}.
\end{aligned}$$

[8]Recently, the expressive power of the language of $\mathbf{S4}_u$ has been characterized in terms of bisimulations by Aiello and van Benthem (2000). The associated topo-games have been used in (Aiello 2001) to measure the difference between spatial regions.

Then, with every atomic $\mathcal{BRCC}$-8 formula $P(s,t)$ we associate a modal formula $(P(s,t))^{\bowtie}$ defined by:

$$
\begin{aligned}
(\mathsf{DC}(s,t))^{\bowtie} &= \neg\Diamond\!\!\!\Box(s^{\bowtie} \wedge t^{\bowtie}),\\
(\mathsf{EQ}(s,t))^{\bowtie} &= \boxdot(s^{\bowtie} \leftrightarrow t^{\bowtie}),\\
(\mathsf{PO}(s,t))^{\bowtie} &= \Diamond\!\!\!\Box(\mathbf{I}s^{\bowtie} \wedge \mathbf{I}t^{\bowtie}) \wedge \Diamond\!\!\!\Box(s^{\bowtie} \wedge \neg t^{\bowtie}) \wedge \Diamond\!\!\!\Box(\neg s^{\bowtie} \wedge t^{\bowtie}),\\
(\mathsf{EC}(s,t))^{\bowtie} &= \Diamond\!\!\!\Box(s^{\bowtie} \wedge t^{\bowtie}) \wedge \neg\Diamond\!\!\!\Box(\mathbf{I}s^{\bowtie} \wedge \mathbf{I}t^{\bowtie}),\\
(\mathsf{TPP}(s,t))^{\bowtie} &= \boxdot(\neg s^{\bowtie} \vee t^{\bowtie}) \wedge \Diamond\!\!\!\Box(s^{\bowtie} \wedge \neg\mathbf{I}t^{\bowtie}) \wedge \Diamond\!\!\!\Box(\neg s^{\bowtie} \wedge t^{\bowtie}),\\
(\mathsf{NTPP}(s,t))^{\bowtie} &= \boxdot(\neg s^{\bowtie} \vee \mathbf{I}t^{\bowtie}) \wedge \Diamond\!\!\!\Box(\neg s^{\bowtie} \wedge t^{\bowtie}).
\end{aligned}
$$

Finally, given a $\mathcal{BRCC}$-8 formula φ, denote by $\varphi^{\bowtie}$ the result of replacing all occurrences of atomic formulas $P(s,t)$ in φ by $(P(s,t))^{\bowtie}$. Note that in view of $\mathbf{CICI}p \leftrightarrow \mathbf{CI}p \in \mathbf{S4}$, the translations of all region variables and terms in $\varphi^{\bowtie}$ are interpreted in topological spaces as regular closed sets.

Since the definition of the translation $\cdot^{\bowtie}$ mimics the truth-definition of the $\mathcal{RCC}$-8 predicates, and since $\mathbf{S4}_u$ has the fmp (see Theorem 1.26), we immediately obtain:

Theorem 2.31. *For every $\mathcal{BRCC}$-8 formula φ, the following conditions are equivalent:*

(i) *φ is satisfiable in a topological space,*
(ii) *$\varphi^{\bowtie}$ is satisfiable in a topological space,*
(iii) *$\varphi^{\bowtie}$ is satisfiable in a finite Kripke frame for $\mathbf{S4}_u$,*
(iv) *$\varphi^{\bowtie}$ is satisfiable in a finite topological space,*
(v) *φ is satisfiable in a finite topological space.*

As a corollary we have:

Theorem 2.32. *The satisfiability problem for $\mathcal{BRCC}$-8 formulas is decidable.*

Actually, $\mathbf{S4}_u$ makes it possible to express much more complex relations among regions than those available in $\mathcal{BRCC}$-8. For example, we can define a ternary relation

$$\mathsf{EC3}(X,Y,Z) = \boxdot\neg(\mathbf{I}X \wedge \mathbf{I}Y) \wedge \boxdot\neg(\mathbf{I}Y \wedge \mathbf{I}Z) \wedge \boxdot\neg(\mathbf{I}Z \wedge \mathbf{I}X) \wedge \Diamond\!\!\!\Box(X \wedge Y \wedge Z)$$

and write $\mathsf{EC3}(\mathit{Russia}, \mathit{Poland}, \mathit{Lithuania})$ to say that Russia, Poland and Lithuania have a common border, but no common interior point. Unlike $\mathcal{RCC}$-8 and $\mathcal{BRCC}$-8, where regions are usually assumed to be regular closed, $\mathbf{S4}_u$ gives more flexibility. In the extreme, we can express such 'pathological' properties of sets as 'X is dense in Y, but has no interior:'

$$\boxdot\neg\mathbf{I}X \wedge \boxdot(\mathbf{C}X \leftrightarrow Y).$$

Embedding $\mathcal{BRCC}$-8 into S5

The modal translation $\varphi^{\bowtie}$ of a $\mathcal{BRCC}$-8 formula φ has a rather special form. Renz (1998) used this form to show that satisfiable $\mathcal{RCC}$-8 formulas can be satisfied in very simple topological spaces, namely in those determined by $\mathbf{S4}_u$-frames that we call quasisaws. (Renz used this result to show that all satisfiable $\mathcal{RCC}$-8 formulas can be satisfied in $\langle \mathbb{R}^n, \mathbb{I} \rangle$ for any $n \geq 1$. Note, however, that $\mathbf{S4}_u$ is not complete with respect to $\{\langle \mathbb{R}^n, \mathbb{I} \rangle \mid n \geq 1\}$; for a counterexample see Proposition 16.20.)

A *quasisaw* is a 2-frame $\mathfrak{F} = \langle W, R, R_U \rangle$ such that R_U is the universal relation on W and $\langle W, R \rangle$ is a partial order of *depth* ≤ 1 and *width* ≤ 2 (that is, no R-chain has more than two distinct points, and no point has more than two distinct proper successors). An example of a quasisaw is shown in Fig. 2.7. A *fork* is a frame $\mathfrak{f} = \langle W_{\mathfrak{f}}, R_{\mathfrak{f}} \rangle$ such that $W_{\mathfrak{f}} = \{b_{\mathfrak{f}}, l_{\mathfrak{f}}, r_{\mathfrak{f}}\}$ and $R_{\mathfrak{f}}$

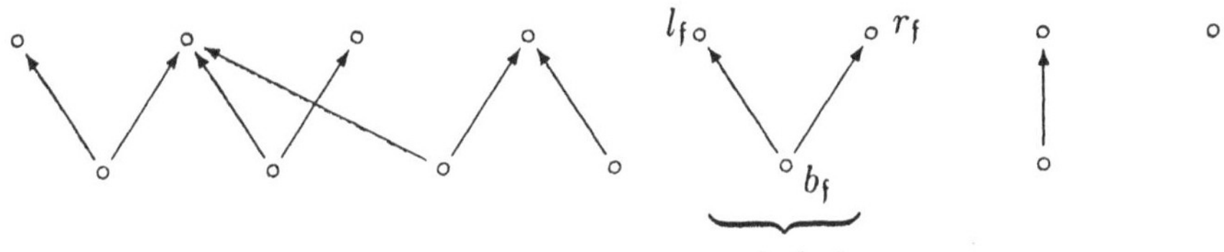

Figure 2.7: Quasisaw.

is the reflexive closure of $\{\langle b_{\mathfrak{f}}, l_{\mathfrak{f}} \rangle, \langle b_{\mathfrak{f}}, r_{\mathfrak{f}} \rangle\}$. Thus, $b_{\mathfrak{f}}$ is the root of $\mathfrak{f}$ with two immediate successors $l_{\mathfrak{f}}$ and $r_{\mathfrak{f}}$. It should be clear that if an $\mathbf{S4}_u$-formula is satisfied in a quasisaw then it is satisfied in a disjoint union of forks (equipped with the universal relation) as well.

The following generalization of Renz's result was proved in (Wolter and Zakharyaschev 2000a); see Theorem 16.4 for a further generalization:

Theorem 2.33. *A $\mathcal{BRCC}$-8 formula φ is satisfiable in a topological space iff $\varphi^{\bowtie}$ is satisfiable in a quasisaw containing $\leq \ell(\varphi^{\bowtie})$ forks.*

Thus, the satisfiability problem for $\mathcal{BRCC}$-8 formulas φ in topological spaces reduces to the satisfiability problem for their translations $\varphi^{\bowtie}$ in quasisaws which are disjoint unions of forks. We can make one step further by observing that the latter problem can be reduced to the satisfiability of propositional unimodal formulas in **S5**-models. The idea behind this reduction is to represent every subformula ψ of $\varphi^{\bowtie}$ by means of three **S5**-formulas ψ^b, ψ^l, ψ^r which encode the 'behavior' of ψ at the three points of a fork.

Given such a formula $\varphi^{\bowtie}$, we define inductively three translations $\cdot^b$, $\cdot^l$

and $\cdot^r$ by taking

$$\begin{aligned}
(p)^i &= p^i,\ p \text{ and } p^i \text{ propositional variables, } i \in \{b,l,r\},\\
(\psi \wedge \chi)^i &= \psi^i \wedge \chi^i,\ \text{for } i \in \{b,l,r\},\\
(\psi \vee \chi)^i &= \psi^i \vee \chi^i,\ \text{for } i \in \{b,l,r\},\\
(\neg\psi)^i &= \neg\psi^i,\ \text{for } i \in \{b,l,r\},\\
(\mathbf{I}\psi)^b &= \psi^b \wedge \psi^l \wedge \psi^r,\\
(\mathbf{I}\psi)^i &= \psi^i,\ \text{for } i \in \{l,r\},\\
(\mathbf{C}\psi)^b &= \psi^b \vee \psi^l \vee \psi^r,\\
(\diamondplus\psi)^i &= \Diamond(\psi^b \vee \psi^l \vee \psi^r),\ \text{for } i \in \{b,r,l\},\\
(\boxtimes\psi)^i &= \Box(\psi^b \wedge \psi^l \wedge \psi^r),\ \text{for } i \in \{b,r,l\}.
\end{aligned}$$

Finally, we define the **S5**-*translation* $\varphi^{\odot}$ of a $\mathcal{BRCC}$-8 formula φ as $(\varphi^{\bowtie})^b$. It should be clear that the length of $\varphi^{\odot}$ is polynomial in the length of φ.

Theorem 2.34. *For every $\mathcal{BRCC}$-8 formula φ, $\varphi^{\bowtie}$ is satisfiable in a quasisaw iff $\varphi^{\odot}$ is* **S5**-*satisfiable.*

Proof. ($\Rightarrow$) Suppose that $\varphi^{\bowtie}$ is satisfiable in an $\mathbf{S4}_u$-model $\mathfrak{M} = \langle \mathfrak{G}, \mathfrak{V}\rangle$ based on a frame $\mathfrak{G} = \langle V, R, R_U\rangle$, where R_U is the universal relation on V and $\langle V, R\rangle$ is a disjoint union of forks. Without loss of generality we can clearly assume that $\varphi^{\bowtie}$ is satisfied at the bottom point $b_{\mathfrak{f}}$ of some fork $\mathfrak{f}$.

Construct an **S5**-model $\mathfrak{N} = \langle \mathfrak{F}, \mathfrak{U}\rangle$ by taking $\mathfrak{F} = \langle U, S\rangle$, where U consists of all forks in $\mathfrak{G}$, $S = U \times U$, and for every propositional variable p in $\varphi^{\bowtie}$,

$$\begin{aligned}
\mathfrak{U}(p^b) &= \{\mathfrak{f} \in U \mid (\mathfrak{M}, b_{\mathfrak{f}}) \models p\},\\
\mathfrak{U}(p^l) &= \{\mathfrak{f} \in U \mid (\mathfrak{M}, l_{\mathfrak{f}}) \models p\},\\
\mathfrak{U}(p^r) &= \{\mathfrak{f} \in U \mid (\mathfrak{M}, r_{\mathfrak{f}}) \models p\}.
\end{aligned}$$

Now, by induction on the construction of a subformula ψ of $\varphi^{\bowtie}$ we show that, for every fork $\mathfrak{f}$ in $\mathfrak{G}$ and every $i \in \{b,l,r\}$,

$$(\mathfrak{N}, \mathfrak{f}) \models \psi^i \quad \text{iff} \quad (\mathfrak{M}, i_{\mathfrak{f}}) \models \psi. \tag{2.22}$$

The basis of induction follows from the definition of $\mathfrak{N}$, and the case of the Boolean connectives is trivial.

Suppose $\psi = \mathbf{I}\chi$. If $i \in \{l,r\}$ then (2.22) holds by the induction hypothesis, since $(\mathfrak{M}, i_{\mathfrak{f}}) \models \mathbf{I}\chi \leftrightarrow \chi$ and $(\mathbf{I}\chi)^i = \chi^i$. And if $i = b$ then, on the one hand,

$$(\mathfrak{M}, b_{\mathfrak{f}}) \models \mathbf{I}\chi \quad \text{iff} \quad (\mathfrak{M}, b_{\mathfrak{f}}) \models \chi,\ (\mathfrak{M}, l_{\mathfrak{f}}) \models \chi,\ (\mathfrak{M}, r_{\mathfrak{f}}) \models \chi,$$

and on the other, by the definition of the translation,

$$(\mathfrak{N},\mathfrak{f}) \models (\mathbf{I}\chi)^b \quad \text{iff} \quad (\mathfrak{N},\mathfrak{f}) \models \chi^b,\ (\mathfrak{N},\mathfrak{f}) \models \chi^l,\ (\mathfrak{N},\mathfrak{f}) \models \chi^r,$$

which yields (2.22) by the induction hypothesis.

Suppose now that $\psi = \Diamond\chi$. Then

$$\begin{aligned}(\mathfrak{N},\mathfrak{f}) \models \psi^i \quad & \text{iff} \quad \exists \mathfrak{f}' \in U\ \exists j \in \{b,l,r\}\ (\mathfrak{N},\mathfrak{f}') \models \chi^j \\ & \text{iff} \quad \exists \mathfrak{f}' \in U\ \exists j \in \{b,l,r\}\ (\mathfrak{M}, j_{\mathfrak{f}'}) \models \chi \\ & \text{iff} \quad (\mathfrak{M}, i_{\mathfrak{f}}) \models \Diamond\chi.\end{aligned}$$

The remaining cases are considered analogously.

It follows that $\varphi^{\odot}$ is satisfied in $\mathfrak{N}$.

($\Leftarrow$) Assume that $\varphi^{\odot}$ is satisfied in an **S5**-model $\mathfrak{N} = \langle \mathfrak{F}, \mathfrak{U}\rangle$, $\mathfrak{F} = \langle U, S\rangle$. With every point $x \in U$ we associate a fork $\mathfrak{f}_x = \langle W_x, R_x\rangle$ so that the sets W_x, for $x \in U$, are pairwise disjoint. Construct an $\mathbf{S4}_u$-model $\mathfrak{M} = \langle \mathfrak{G}, \mathfrak{V}\rangle$ by taking $\mathfrak{G} = \langle V, R, R_U\rangle$,

- $V = \bigcup_{x\in U} W_x$, $R_U = V \times V$,
- uRv iff $u = v$ or $\exists x \in U\ (u = b_{\mathfrak{f}_x} \wedge (v = l_{\mathfrak{f}_x} \vee v = r_{\mathfrak{f}_x}))$,
- $\mathfrak{V}(p) = \{i_{\mathfrak{f}_x} \in V \mid (\mathfrak{N}, x) \models p^i,\ i = b,l,r\}$.

Then $\mathfrak{G}$ is clearly a quasisaw. By a straightforward induction one can show that for every $x \in U$, every subformula ψ of $\varphi^{\bowtie}$, and every $i = b,l,r$, we have

$$(\mathfrak{N}, x) \models \psi^i \quad \text{iff} \quad (\mathfrak{M}, i_{\mathfrak{f}_x}) \models \psi.$$

For example,

$$\begin{aligned}(\mathfrak{N},x) \models (\mathbf{I}\chi)^b \quad & \text{iff} \quad (\mathfrak{N},x) \models \chi^i,\ \text{for } i \in \{b,l,r\} \\ & \text{iff} \quad (\mathfrak{M}, i_{\mathfrak{f}_x}) \models \chi,\ \text{for } i \in \{b,l,r\} \\ & \text{iff} \quad (\mathfrak{M}, b_{\mathfrak{f}_x}) \models \mathbf{I}\chi.\end{aligned}$$

It follows that $\varphi^{\bowtie}$ is satisfied in $\mathfrak{M}$. ❑

As **S5** is NP-complete and $\mathcal{RCC}$-8 can encode propositional classical logic (using the predicate EQ), we immediately obtain that the computational behavior of $\mathcal{BRCC}$-8 in arbitrary topological spaces is precisely the same as that of $\mathcal{RCC}$-8:

Theorem 2.35. *The satisfiability problem for $\mathcal{BRCC}$-8 formulas in topological spaces is* NP*-complete.*

However, if only Euclidean (or even connected) topological spaces are regarded as possible interpretations, the satisfiability problem for $\mathcal{BRCC}$-8 formulas becomes PSPACE-complete (for details consult Wolter and Zakharyaschev 2000a).

2.7 Intuitionistic logic

Intuitionistic logic is yet another type of logic which can be embedded in **S4**; actually, as we have already said, to provide such an embedding was the main reason for constructing **S4** by Gödel (1933) and Orlov (1928).

Intuitionistic logic, and more generally *intuitionism* as the trend in the foundations of mathematics initiated by Brouwer (1907, 1908), aimed to single out and describe the principles of 'constructive' mathematical reasoning, constructive in the sense that it provides (at least) an algorithm constructing an object the existence of which is proved. Classical logic **Cl**, as well as all other logics having **Cl** as their fragment, are not constructive: using the law of the excluded middle (A10) we can establish the existence of objects by *reductio ad absurdum* without even giving a hint of how to find them (mathematical textbooks abound with proofs of this sort[9]).

Intuitionistic propositional logic **Int** was first constructed syntactically by Kolmogorov (1925), Glivenko (1929) and Heyting (1930). It has the same language $\mathcal{L}$ as **Cl**, and an $\mathcal{L}$-formula φ belongs to **Int** iff φ can be derived from the axioms (A1)–(A9) using MP and Subst. In other words, **Int** is obtained from **Cl** by discarding axiom (A10). (It should be noted, however, that unlike **Cl**, the connectives $\wedge$, $\vee$, $\rightarrow$ and $\bot$ are independent: they cannot be expressed via each other.)

The intended meaning of the intuitionistic connectives was explained first in terms of the *proof interpretation* due to Brouwer, Kolmogorov and Heyting:

- a proof of a proposition $\varphi \wedge \psi$ consists of a proof of φ and a proof of ψ;
- a proof of $\varphi \vee \psi$ is given by presenting either a proof of φ or a proof of ψ;
- a proof of $\varphi \rightarrow \psi$ is a construction which, given a proof of φ, returns a proof of ψ;
- $\bot$ has no proof and a proof of $\neg\varphi$ is a construction which, given a proof of φ, would return a proof of $\bot$.

According to this interpretation, **Int** contains only those formulas that have proofs. The existence of open mathematical problems (e.g., 'P = NP?') shows that the formula $p \vee \neg p$ has no proof, and so cannot be accepted as an intuitionistically valid principle.

[9]Here is a well-known example: to prove that there exists an irrational number x such that $x^{\sqrt{2}}$ is rational, we observe first that, by (A10), $\sqrt{2}^{\sqrt{2}}$ is either rational or irrational; if it is rational then we take $x = \sqrt{2}$, otherwise $x = \sqrt{2}^{\sqrt{2}}$.

Various more formal semantics have been constructed for **Int** (see, e.g., Kleene 1945, Gödel 1958, Kreisel 1962, Medvedev 1962, Skvortsov 1979, Artemov 2001). Here we briefly consider three of them: the topological, the algebraic and the relational (or possible world) semantics.

Stone (1937) and Tarski (1938) discovered that **Int** can be interpreted in topological spaces $\mathfrak{T} = \langle U, \mathbb{I} \rangle$ by associating with each variable p an open set $\mathfrak{V}(p) \subseteq U$, the *value* of p in $\mathfrak{T}$ under the *valuation* $\mathfrak{V}$. The values of arbitrary $\mathcal{L}$-formulas in $\mathfrak{T}$ are defined inductively as follows:

$$\begin{aligned} \mathfrak{V}(\bot) &= \emptyset, \\ \mathfrak{V}(\varphi \wedge \psi) &= \mathfrak{V}(\varphi) \cap \mathfrak{V}(\psi), \\ \mathfrak{V}(\varphi \vee \psi) &= \mathfrak{V}(\varphi) \cup \mathfrak{V}(\psi), \\ \mathfrak{V}(\varphi \to \psi) &= \mathbb{I}\big((U - \mathfrak{V}(\varphi)) \cup \mathfrak{V}(\psi)\big). \end{aligned}$$

If $\mathfrak{V}(\varphi) = U$ for every valuation $\mathfrak{V}$ in $\mathfrak{T}$, then we say that φ is *valid* in $\mathfrak{T}$ and write $\mathfrak{T} \models \varphi$. It turns out that $\varphi \in$ **Int** iff φ is valid in all topological spaces iff φ is valid in $\mathbb{R}^n$, for any $n \geq 1$; see, e.g., (Rasiowa and Sikorski 1963).

A more general algebraic semantics was constructed by McKinsey and Tarski (1944, 1946). A *Heyting* (or *pseudo-Boolean*) *algebra* is a structure of the form

$$\mathfrak{A} = \langle A, \wedge^{\mathfrak{A}}, \vee^{\mathfrak{A}}, \to^{\mathfrak{A}}, 0^{\mathfrak{A}}, 1^{\mathfrak{A}} \rangle$$

such that $\wedge^{\mathfrak{A}}$, $\vee^{\mathfrak{A}}$ and $\to^{\mathfrak{A}}$ are binary operations on A, $0^{\mathfrak{A}}, 1^{\mathfrak{A}} \in A$, $\wedge^{\mathfrak{A}}$ and $\vee^{\mathfrak{A}}$ are commutative, associative and have the absorption property (like in Boolean algebras, see Section 1.5) and for all $a, b, c \in A$,

- $c \wedge^{\mathfrak{A}} a \leq^{\mathfrak{A}} b$ iff $c \leq^{\mathfrak{A}} a \to^{\mathfrak{A}} b$ ($a \to^{\mathfrak{A}} b$ is the greatest element in the set $\{c \in A \mid c \wedge^{\mathfrak{A}} a \leq^{\mathfrak{A}} b\}$);
- $0^{\mathfrak{A}} \leq^{\mathfrak{A}} a \leq^{\mathfrak{A}} 1^{\mathfrak{A}}$ ($0^{\mathfrak{A}}$ and $1^{\mathfrak{A}}$ are the least and greatest elements in $\mathfrak{A}$, respectively),

where the binary relation $\leq^{\mathfrak{A}}$ on A is defined by taking

$$a \leq^{\mathfrak{A}} b \quad \text{iff} \quad a \wedge^{\mathfrak{A}} b = a.$$

Note that one can also define Heyting algebras as the algebras of open elements of modal algebras for **S4** (see Sections 1.5 and 2.6).

Int is sound and complete with respect to the class of all Heyting algebras; moreover, every extension of **Int** (closed under MP and Subst)—these extensions are known as *intermediate* or *superintuitionistic logics*—is characterized by the class of Heyting algebras validating its formulas; see, e.g., (Chagrov and Zakharyaschev 1997).

The possible world semantics for **Int** defined by Beth (1956) and Kripke (1965b) (see also Grzegorczyk 1964) reflects the epistemic character of intuitionistic logic, namely that it takes into account the development of knowledge.

Let us imagine that our knowledge is developing discretely, nondeterministically passing from one state to another. Being at some state of knowledge (or information) x, we can say which facts are known at x and which are not established yet. Besides, we know what states of information are possible in the future (i.e., do not contradict the knowledge at x). This does not mean, however, that we shall reach all these possible states (for instance, we can imagine now not only a course of events under which the equality P = NP will be proved, but also situations when it will remain unproved or will be refuted). It is also reasonable to assume that when passing to a new state, all the facts known at x are preserved, and some new facts can possibly be established. The propositions established at x are regarded as true at x; they will remain true at all further possible states. But a proposition which is not true at x cannot be said to be false, because it may become true at one of the subsequent states.

Possible states of information are represented as Kripke frames $\mathfrak{F} = \langle W, R\rangle$ in which R is a partial order on W, i.e., R is reflexive, transitive and antisymmetric ($\forall x, y\ (xRy \wedge yRx \rightarrow x = y)$). A *valuation* $\mathfrak{V}$ in $\mathfrak{F}$ indicates which atomic propositions hold true in each state $x \in W$. Thus $\mathfrak{V}$ is a map from the set of propositional variables into the set $Up\mathfrak{F}$ of upward closed subsets of W ($X \in Up\mathfrak{F}$ iff $\forall x \in X \forall y \in W\ (xRy \rightarrow y \in X)$). The pair $\mathfrak{M} = \langle \mathfrak{F}, \mathfrak{V}\rangle$ is called an *intuitionistic (Kripke) model* of the language $\mathcal{L}$. The *truth-relation* $(\mathfrak{M}, x) \models \varphi$ (or simply $x \models \varphi$) is defined inductively as follows:

$$
\begin{array}{lll}
(\mathfrak{M}, x) \models p & \text{iff} & x \in \mathfrak{V}(p); \\
(\mathfrak{M}, x) \not\models \bot; & & \\
(\mathfrak{M}, x) \models \psi \wedge \chi & \text{iff} & (\mathfrak{M}, x) \models \psi \text{ and } (\mathfrak{M}, x) \models \chi; \\
(\mathfrak{M}, x) \models \psi \vee \chi & \text{iff} & (\mathfrak{M}, x) \models \psi \text{ or } (\mathfrak{M}, x) \models \chi; \\
(\mathfrak{M}, x) \models \psi \rightarrow \chi & \text{iff} & \text{for all } y \in W \text{ such that } xRy, \\
& & (\mathfrak{M}, y) \models \psi \text{ implies } (\mathfrak{M}, y) \models \chi.
\end{array}
$$

It follows from this definition that

$$(\mathfrak{M}, x) \models \neg\psi \quad \text{iff} \quad \text{for all } y \in W \text{ such that } xRy,\ (\mathfrak{M}, y) \not\models \psi.$$

(Observe that an intuitionistic model based on the single-point frame is nothing else but a standard model for **Cl**.) For example, Fig. 2.8 shows an intuitionistic model refuting axiom (A10).

Int is sound and complete with respect to the class of all intuitionistic frames. Moreover, it has the exponential fmp, and the decidability problem

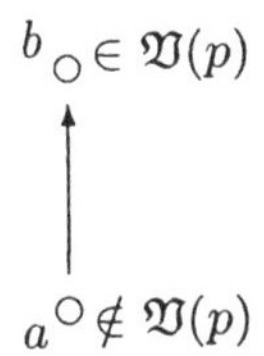

$$b \models p$$
$$b \models p \vee (p \to \bot)$$
$$b \not\models \bot$$
$$b \not\models p \to \bot$$

$$a \not\models p$$
$$a \not\models p \to \bot$$
$$a \not\models p \vee (p \to \bot)$$

Figure 2.8: An intuitionistic model refuting $p \vee (p \to \bot)$.

for **Int** is PSPACE-complete. (It should be noted that the problem of whether a given intuitionistic formula is satisfiable is NP-complete: it is enough to check satisfiability in single-point—i.e., classical—models.)

The constructive character of **Int** is reflected by the fact that it has the so-called *disjunction property*: for all $\mathcal{L}$-formulas φ and ψ,

$$\varphi \vee \psi \in \mathbf{Int} \quad \text{iff} \quad \varphi \in \mathbf{Int} \ \text{ or } \ \psi \in \mathbf{Int}.$$

Note, however, that this property is not characteristic for **Int**: there are proper extensions of **Int** having the disjunction property.

We conclude this section with the definition of the *Gödel translation* T which embeds **Int** into **S4**:

- $\mathsf{T}(p) = \Box p$, p a propositional variable;
- $\mathsf{T}(\bot) = \Box\bot$;
- $\mathsf{T}(\varphi \wedge \psi) = \mathsf{T}(\varphi) \wedge \mathsf{T}(\psi)$;
- $\mathsf{T}(\varphi \vee \psi) = \mathsf{T}(\varphi) \vee \mathsf{T}(\psi)$;
- $\mathsf{T}(\varphi \to \psi) = \Box(\mathsf{T}(\varphi) \to \mathsf{T}(\psi))$.

If we understand the **S4**-box as 'it is provable' then the intuitionistic connectives are transformed by T into the corresponding classical ones, but they are understood now in the context of 'provability.' One can show that for every $\mathcal{L}$-formula φ,

$$\varphi \in \mathbf{Int} \quad \text{iff} \quad \mathsf{T}(\varphi) \in \mathbf{S4}.$$

For more information about intuitionistic logic we refer the reader to (van Dalen 1986) or (Chagrov and Zakharyaschev 1997).

2.8 'Model level' reductions between logics

We conclude Chapter 2 by establishing a number of useful polynomial reductions between modal, epistemic, dynamic and temporal logics, summarized in Table 2.2. On the one hand, that such reductions exist follows immediately from the complexity results presented in this chapter. For example, $\mathbf{K}_1^C$ and $\mathbf{K4}_2^C$ (introduced in Section 2.3) are polynomially reducible to each other simply because they are both EXPTIME-complete. However, such reductions via Turing machines usually do not give any information on how models of the two logics are connected. In contrast, our reductions below work on the 'model level,' and this will enable us to generalize the results to many-dimensional logics in Sections 6.3 and 6.5.

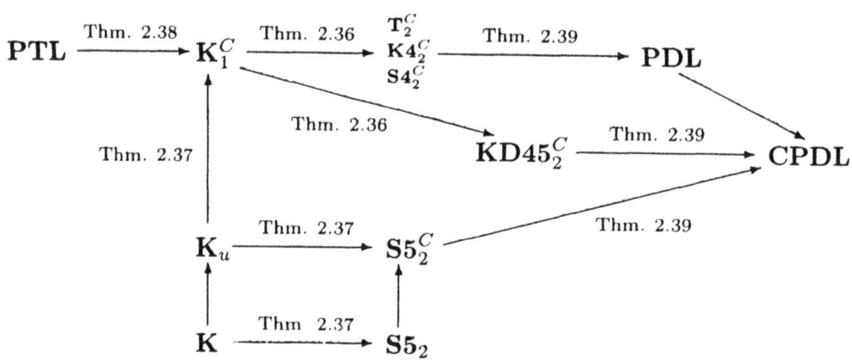

Table 2.2: 'Model level' reductions between modal, epistemic, dynamic, and temporal logics.

Theorem 2.36. $\mathbf{K}_1^C$ *is polynomially reducible to* $\mathbf{T}_2^C$, $\mathbf{K4}_2^C$, $\mathbf{S4}_2^C$ *and* $\mathbf{KD45}_2^C$.

Proof. First, we show that $\mathbf{K}_1^C$ is polynomially reducible to $\mathbf{D}_1^C$. Fix a fresh propositional variable p and define a translation ^{r} from $\mathcal{ML}_1^C$-formulas (with modal operators $\Box$ and C) into $\mathcal{ML}_1^C$ by taking

$$\begin{aligned} q^r &= p \wedge q, \qquad (q \text{ a propositional variable}) \\ (\psi_1 \wedge \psi_2)^r &= \psi_1^r \wedge \psi_2^r, \\ (\neg\psi)^r &= p \wedge \neg\psi^r, \\ (\Box\psi)^r &= p \wedge \Box(p \to \psi^r), \\ (\mathsf{C}\,\psi)^r &= p \wedge \mathsf{C}\,(p \to \psi^r). \end{aligned}$$

Our aim is to show that for all $\mathcal{ML}_1^C$-formulas φ,

$$\varphi \in \mathbf{K}_1^C \qquad \text{iff} \qquad p \wedge \mathsf{C}(\neg p \to \mathsf{C}\,\neg p) \to \varphi^r \in \mathbf{D}_1^C.$$

Suppose first that $\varphi \notin \mathbf{K}_1^C$. Then there is a model $\mathfrak{M} = \langle \mathfrak{F}, \mathfrak{V} \rangle$ based on a frame $\mathfrak{F} = \langle W, R \rangle$ with root r and such that $(\mathfrak{M}, r) \not\models \varphi$. Let X be the set of all points in W having no R-successors. For each $x \in X$, take a fresh point x^+, and define a new model $\mathfrak{M}' = \langle \mathfrak{F}', \mathfrak{V}' \rangle$ based on a frame $\mathfrak{F}' = \langle W', R' \rangle$ by setting

- $W' = W \cup \{x^+ \mid x \in X\}$,
- $R' = R \cup \{\langle x, x^+ \rangle \mid x \in X\} \cup \{\langle x^+, x^+ \rangle \mid x \in X\}$,
- $\mathfrak{V}'(p) = W$,
- $\mathfrak{V}'(q) = \mathfrak{V}(q)$, for any other propositional variable q.

Then clearly R' is serial and $\mathfrak{M}' \models \neg p \to \mathsf{C}\,\neg p$. An easy induction shows that for all $\mathcal{ML}_1^C$-formulas ψ and all $x \in W$,

$$(\mathfrak{M}, x) \models \psi \qquad \text{iff} \qquad (\mathfrak{M}', x) \models \psi^r.$$

It follows that $(\mathfrak{M}', r) \not\models \varphi^r$.

Conversely, suppose that $p \wedge \mathsf{C}(\neg p \to \mathsf{C}\,\neg p) \wedge \neg\varphi^r$ is satisfied at the root of a model $\mathfrak{M} = \langle \mathfrak{F}, \mathfrak{V} \rangle$ based on $\mathfrak{F} = \langle W, R \rangle$. Define a model $\mathfrak{M}' = \langle \mathfrak{F}', \mathfrak{V}' \rangle$ based on $\mathfrak{F}' = \langle W', R' \rangle$ by taking

- $W' = \mathfrak{V}(p)$,
- $R' = R \cap (W' \times W')$,
- $\mathfrak{V}'(q) = \mathfrak{V}(p) \cap \mathfrak{V}(q)$, for any variable q.

We leave it to the reader to show that $(\mathfrak{M}', r) \not\models \varphi$.

Thus, it suffices to construct the reductions we need from $\mathbf{D}_1^C$ instead of $\mathbf{K}_1^C$. Take a fresh variable p and define a translation $\sharp$ from $\mathcal{ML}_1^C$-formulas into $\mathcal{ML}_2^C$-formulas (with $\Box_1$, $\Box_2$ and $\mathsf{C}_{\{1,2\}}$) as follows:

$$\begin{aligned}
q^\sharp &= p \wedge q, \qquad (q \text{ a propositional variable})\\
(\psi_1 \wedge \psi_2)^\sharp &= \psi_1^\sharp \wedge \psi_2^\sharp,\\
(\neg\psi)^\sharp &= p \wedge \neg\psi^\sharp,\\
(\Box\psi)^\sharp &= p \wedge \Box_1(\neg p \to \Box_2(p \to \psi^\sharp)),\\
(\mathsf{C}\,\psi)^\sharp &= p \wedge \mathsf{C}_{\{1,2\}}(p \to \psi^\sharp).
\end{aligned}$$

Given an $\mathcal{ML}_1^C$-formula φ, we set:

$$\chi_{\mathbf{S4}}^{\varphi} = p \wedge \mathsf{C}_{\{1,2\}}(p \rightarrow \Diamond_1 \neg p) \wedge \mathsf{C}_{\{1,2\}}(\neg p \rightarrow \Diamond_2 p) \wedge \tag{2.23}$$

$$\mathsf{C}_{\{1,2\}}(p \rightarrow \Box_2 p) \wedge \mathsf{C}_{\{1,2\}}(\neg p \rightarrow \Box_1 \neg p) \wedge \tag{2.24}$$

$$\mathsf{C}_{\{1,2\}}\Big(\bigwedge_{\psi \in sub\, \varphi^{\sharp}} (p \wedge \psi \rightarrow \Box_1(p \rightarrow \psi)) \wedge (p \wedge \psi \rightarrow \Box_2 \psi) \wedge \tag{2.25}$$

$$(\neg p \wedge \psi \rightarrow \Box_1 \psi) \wedge (\neg p \wedge \psi \rightarrow \Box_2(\neg p \rightarrow \psi))\Big). \tag{2.26}$$

Our aim is to show that

(i) if $\chi_{\mathbf{S4}}^{\varphi} \rightarrow \varphi^{\sharp} \in \mathbf{S4}_2^C$ then $\varphi \in \mathbf{D}_1^C$;

(ii) if $\varphi \in \mathbf{D}_1^C$ then $\chi_{\mathbf{S4}}^{\varphi} \rightarrow \varphi^{\sharp} \in \mathbf{K}_2^C$.

It will follow then that $\varphi \in \mathbf{D}_1^C$ iff $\chi_{\mathbf{S4}}^{\varphi} \rightarrow \varphi^{\sharp} \in L^C$, for all Kripke complete modal logics L between $\mathbf{K}_2$ and $\mathbf{S4}_2$, in particular, for the logics mentioned in the theorem—save $\mathbf{KD45}_2^C$.

To prove (i), suppose that $\varphi \notin \mathbf{D}_1^C$. Using Proposition 1.7, it is not hard to see that $(\mathfrak{M}, r) \not\models \varphi$, for some model $\mathfrak{M} = \langle \mathfrak{F}, \mathfrak{V} \rangle$ based on an intransitive tree $\mathfrak{F} = \langle W, R \rangle$ without endpoints and with root r. Define a new 2-frame $\mathfrak{F}' = \langle W', R_1, R_2 \rangle$ by taking

- $W' = W \cup (W \times \{1\})$,
- xR_1y iff either $x \in W$ and $y = \langle x, 1 \rangle$ or $x = y$,
- xR_2y iff either there is $z \in W$ such that $x = \langle z, 1 \rangle$ and zRy or $x = y$

(see Fig. 2.9). Obviously, both R_1 and R_2 are partial orders, which means that $\mathfrak{F}' = \langle W', R_1, R_2 \rangle$ is a frame for $\mathbf{S4}_2$. Further, it is straightforward to see that, for all $x, y \in W$,

$$xR^*y \qquad \text{iff} \qquad x(R_1 \cup R_2)^*y.$$

Now define a model $\mathfrak{M}' = \langle \mathfrak{F}', \mathfrak{V}' \rangle$ by taking

- $\mathfrak{V}'(p) = W$,
- $\mathfrak{V}'(q) = \mathfrak{V}(q)$, for any other propositional variable q.

An easy induction shows that, for all $x \in W$ and all subformulas ψ of φ,

$$(\mathfrak{M}, x) \models \psi \qquad \text{iff} \qquad (\mathfrak{M}', x) \models \psi^{\sharp}.$$

Thus, $(\mathfrak{M}', r) \not\models \varphi^{\sharp}$. It is not hard to check also that $(\mathfrak{M}', r) \models \chi_{\mathbf{S4}}^{\varphi}$, and so we have

$$(\mathfrak{M}', r) \not\models \chi_{\mathbf{S4}}^{\varphi} \rightarrow \varphi^{\sharp},$$

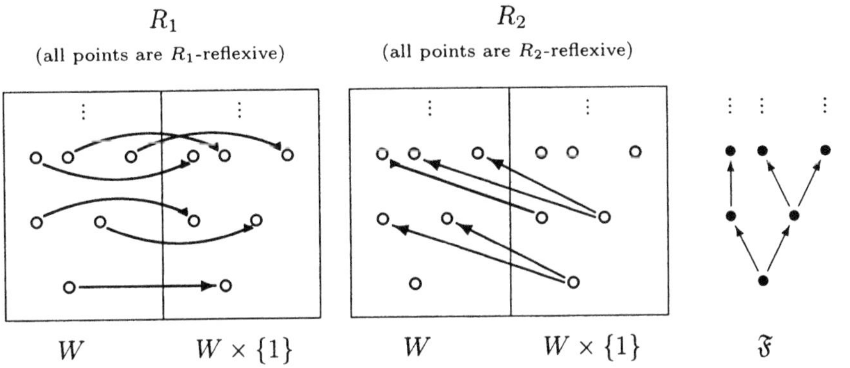

Figure 2.9: $\langle W', R_1, R_2\rangle$ is a frame for $\mathbf{S4}_2$.

from which $\chi^{\varphi}_{\mathbf{S4}} \to \varphi^{\sharp} \notin \mathbf{S4}_2^C$.

Let us now show (ii). Suppose that $\chi^{\varphi}_{\mathbf{S4}} \wedge \neg\varphi^{\sharp}$ is satisfied at the root of a model $\mathfrak{M} = \langle \mathfrak{F}, \mathfrak{V}\rangle$ based on some frame $\mathfrak{F} = \langle W, R_1, R_2\rangle$. For $i = 1, 2$, let

$$R_i^p = R_i \cap (\mathfrak{V}(p) \times \mathfrak{V}(p)) \quad \text{and} \quad R_i^{\neg p} = R_i \cap \big((W - \mathfrak{V}(p)) \times (W - \mathfrak{V}(p))\big).$$

Define a new frame $\mathfrak{F}' = \langle W', R\rangle$ by taking $W' = \mathfrak{V}(p)$ and, for all $x, y \in W'$, xRy iff there are $x', y' \in W$ and $z', z'' \in W - W'$, such that

$$x(R_1^p \cup R_2^p)^* x', \quad x' R_1 z', \quad z'(R_1^{\neg p} \cup R_2^{\neg p})^* z'', \quad z'' R_2 y', \quad y'(R_1^p \cup R_2^p)^* y$$

(see Fig. 2.10). By (2.23), R is serial. Clearly, for all $x, y \in W'$,

$$\text{if } xR^*y \text{ then } x(R_1 \cup R_2)^* y. \tag{2.27}$$

On the other hand, it is not hard to show using (2.24) that, for all $x, y \in W'$,

$$\text{if } x(R_1 \cup R_2)^* y \text{ then either } x(R_1^p \cup R_2^p)^* y \text{ or } xR^*y. \tag{2.28}$$

Now define a model $\mathfrak{M}' = \langle \mathfrak{F}', \mathfrak{V}'\rangle$ by taking $\mathfrak{V}'(q) = \mathfrak{V}(q) \cap W'$. We claim that, for every $x \in W'$ and every subformula ψ of φ,

$$(\mathfrak{M}', x) \models \psi \quad \text{iff} \quad (\mathfrak{M}, x) \models \psi^{\sharp}.$$

We show only the induction steps for $\psi = \Box\chi$ and $\psi = \mathsf{C}\chi$. First, suppose that $(\mathfrak{M}, x) \models \Box_1(\neg p \to \Box_2(p \to \chi^{\sharp}))$ and let xRy. We need to show that $(\mathfrak{M}', y) \models \chi$. By the definition of R, there are x', z', z'', y' as above. Since $\Box_1(\neg p \to \Box_2(p \to \chi^{\sharp}))$ is a subformula of $\varphi^{\sharp}$, by (2.25) we have

$$(\mathfrak{M}, x') \models \Box_1(\neg p \to \Box_2(p \to \chi^{\sharp})),$$

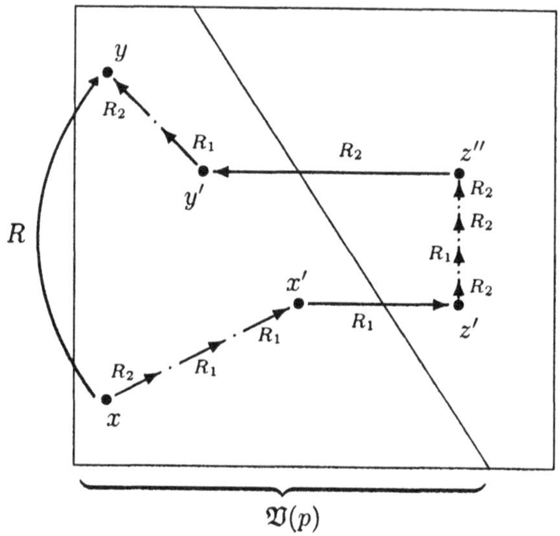

Figure 2.10: The accessibility relation R.

and so $(\mathfrak{M}, z') \models \Box_2(p \to \chi^\sharp)$. By (2.26), we have $(\mathfrak{M}, z'') \models \Box_2(p \to \chi^\sharp)$, and so $(\mathfrak{M}, y') \models \chi^\sharp$. Finally, again by (2.25) we obtain $(\mathfrak{M}, y) \models \chi^\sharp$. Thus, by the induction hypothesis, we have $(\mathfrak{M}', y) \models \chi$, as required. The other direction for $\Box$-formulas is straightforward.

Now suppose $(\mathfrak{M}', x) \models \mathsf{C}\chi$ and let $y \in W'$ be such that $x(R_1 \cup R_2)^* y$. We need to show that $(\mathfrak{M}, y) \models \chi^\sharp$. By the induction hypothesis, we have $(\mathfrak{M}, z) \models \chi^\sharp$, for all z with xR^*z. By (2.28), we have either $x(R_1^p \cup R_2^p)^* y$ or xR^*y, so in the latter case we have $(\mathfrak{M}, y) \models \chi^\sharp$. If $x(R_1^p \cup R_2^p)^* y$ holds then we obtain this by (2.25). The other direction for C-formulas follows from (2.27).

Finally, as the root of $\mathfrak{F}$ belongs to $\mathfrak{F}'$, it follows that $\mathfrak{M}'$ refutes φ.

In the case of $\mathbf{KD45}_2^C$ we need another reduction. Define a translation $\natural$ from $\mathcal{ML}_1^C$-formulas into $\mathcal{ML}_2^C$-formulas as follows. First, we associate with each $\mathcal{ML}_1^C$-formula of the form $\mathsf{C}\,\psi$ a new propositional variable $p_{\mathsf{C}\psi}$. Take a fresh variable p. Then define inductively:

$$
\begin{aligned}
q^\natural &= p \wedge q, \qquad (q \text{ a propositional variable})\\
(\psi_1 \wedge \psi_2)^\natural &= \psi_1^\natural \wedge \psi_2^\natural,\\
(\neg\psi)^\natural &= p \wedge \neg\psi^\natural,\\
(\Box\psi)^\natural &= p \wedge \Box_1(\neg p \to \Box_2(p \to \psi^\natural)),\\
(\mathsf{C}\,\psi)^\natural &= p \wedge p_{\mathsf{C}\psi}.
\end{aligned}
$$

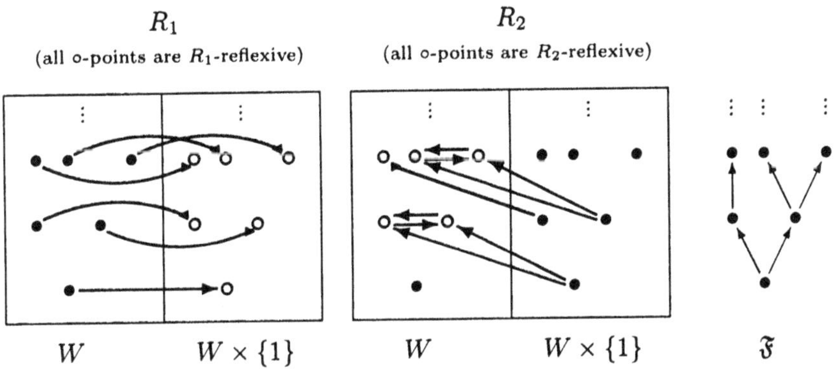

Figure 2.11: $\langle W', R_1, R_2\rangle$ is a frame for $\mathbf{KD45}_2$.

Finally,[10] given an $\mathcal{ML}_1^C$-formula φ, we set

$$\chi^{\varphi}_{sim} = \bigwedge_{\mathsf{C}\psi \in sub\,\varphi} \Big(p_{\mathsf{C}\psi} \leftrightarrow \psi^{\natural} \wedge \Box_1(\neg p \to \Box_2(p \to \mathsf{C}_{\{1,2\}}(p \to \psi^{\natural})))\Big),$$

$$\chi^{\varphi}_{\mathbf{KD45}} = p \wedge \mathsf{C}_{\{1,2\}}\big((p \to (\chi^{\varphi}_{sim} \wedge \Box_1\neg p)) \wedge (\neg p \to \Box_2 p)\big).$$

Our aim is to show that

$$\varphi \in \mathbf{D}_1^C \qquad \text{iff} \qquad \chi^{\varphi}_{\mathbf{KD45}} \to \varphi^{\natural} \in \mathbf{KD45}_2^C. \tag{2.29}$$

To prove the ($\Leftarrow$) direction, suppose $\varphi \notin \mathbf{D}_1^C$. Then $(\mathfrak{M}, r) \not\models \varphi$, for some model $\mathfrak{M} = \langle \mathfrak{F}, \mathfrak{V}\rangle$ based on an intransitive tree $\mathfrak{F} = \langle W, R\rangle$ without endpoints and with root r. Define a 2-frame $\mathfrak{F}' = \langle W', R_1, R_2\rangle$ by taking

- $W' = W \cup (W \times \{1\})$,
- xR_1y iff either $x \in W$ and $y = \langle x, 1\rangle$, or $x, y \in W \times \{1\}$ and $x = y$,
- R_2 to be the closure of R_2' under the rule '$xSy \wedge xSz \Rightarrow ySz$' (i.e., the Euclidean closure), where $xR_2'y$ iff there exists $z \in W$ such that $x = \langle z, 1\rangle$ and zRy

(see Fig. 2.11). It is not hard to check that both R_1 and R_2 are transitive, serial and Euclidean. So $\mathfrak{F}' = \langle W', R_1, R_2\rangle$ is a frame for $\mathbf{KD45}_2$. Observe that, for all $x, y \in W$,

$$xRy \text{ iff there is a } z \in W' - W \text{ such that } xR_1z \text{ and } zR_2y. \tag{2.30}$$

[10] It would be more natural to define $(\mathsf{C}\psi)^{\natural} = \psi^{\natural} \wedge \Box_1(\neg p \to \Box_2(p \to \mathsf{C}_{\{1,2\}}(p \to \psi^{\natural})))$. However, this would not be a polynomial translation, since $\psi^{\natural}$ would occur twice in the right-hand side.

Now define a model $\mathfrak{M}' = \langle \mathfrak{F}', \mathfrak{V}' \rangle$ by taking

- $\mathfrak{V}'(p) = W$,
- $\mathfrak{V}'(p_{\mathsf{C}\psi}) = \{x \in W \mid (\mathfrak{M}, x) \models \mathsf{C}\,\psi\}$,
- $\mathfrak{V}'(q) = \mathfrak{V}(q)$, for any other propositional variable q.

An easy induction shows that, for all $x \in W$ and all subformulas ψ of φ,

$$(\mathfrak{M}, x) \models \psi \quad \text{iff} \quad (\mathfrak{M}', x) \models \psi^{\natural}. \tag{2.31}$$

Thus, $(\mathfrak{M}', r) \not\models \varphi^{\natural}$. Further, we claim that $(\mathfrak{M}', r) \models \chi^{\varphi}_{\mathbf{KD45}}$. Indeed, we clearly have $(\mathfrak{M}', r) \models \mathsf{C}_{\{1,2\}}\big((p \to \Box_1 \neg p) \wedge (\neg p \to \Box_2 p)\big)$. Now let $x \in W$. We need to show that $(\mathfrak{M}', x) \models \chi^{\varphi}_{sim}$. Take a subformula of φ of the form $\mathsf{C}\,\psi$. Suppose first that

$$(\mathfrak{M}', x) \models \psi^{\natural} \wedge \Box_1\big(\neg p \to \Box_2(p \to \mathsf{C}_{\{1,2\}}(p \to \psi^{\natural}))\big),$$

and let xR^*y. Then, by (2.30), we have $(\mathfrak{M}', y) \models \psi^{\natural}$, and so, by (2.31), $(\mathfrak{M}, y) \models \psi$. It follows that $(\mathfrak{M}, x) \models \mathsf{C}\,\psi$, from which $(\mathfrak{M}', x) \models p_{\mathsf{C}\psi}$.

Conversely, suppose that $(\mathfrak{M}', x) \models p_{\mathsf{C}\psi}$, and so $(\mathfrak{M}, x) \models \mathsf{C}\,\psi$. By (2.31), we have $(\mathfrak{M}', x) \models \psi^{\natural}$. Now let $y \in W' - W$ and $z, u \in W$ be such that xR_1y, yR_2z, and $z(R_1 \cup R_2)^*u$. Using the transitivity of R_2 it is not hard to show that in this case u can be reached from x via an alternating chain of nonreflexive R_1- and R_2-arrows. Then, by (2.30), xR^*u, and so $(\mathfrak{M}, u) \models \psi$. Using (2.31) once again, we finally obtain $(\mathfrak{M}', u) \models \psi^{\natural}$, as required.

So we have $(\mathfrak{M}', r) \not\models \chi^{\varphi}_{\mathbf{KD45}} \to \varphi^{\natural}$, whence $\chi^{\varphi}_{\mathbf{KD45}} \to \varphi^{\natural} \notin \mathbf{KD45}^{C}_{2}$.

Let us now show the $(\Rightarrow)$ direction of (2.29). Suppose that $\chi^{\varphi}_{\mathbf{KD45}} \wedge \neg\varphi^{\natural}$ is satisfied at the root r of a model $\mathfrak{M} = \langle \mathfrak{F}, \mathfrak{V} \rangle$ based on a $\mathbf{KD45}_2$-frame $\mathfrak{F} = \langle W, R_1, R_2 \rangle$. Define a model $\mathfrak{M}' = \langle \mathfrak{F}', \mathfrak{V}' \rangle$ based on $\mathfrak{F}' = \langle W', R \rangle$ by taking

- $W' = \mathfrak{V}(p)$,
- xRy iff there exists $z \in W - \mathfrak{V}(p)$ such that xR_1z and zR_2y,
- $\mathfrak{V}'(q) = \mathfrak{V}(q) \cap \mathfrak{V}(p)$.

Then clearly R is serial. We claim that, for every $x \in W'$ and every subformula ψ of φ,

$$(\mathfrak{M}', x) \models \psi \quad \text{iff} \quad (\mathfrak{M}, x) \models \psi^{\natural}.$$

We show only the induction step for $\psi = \mathsf{C}\chi$. Suppose first that $(\mathfrak{M}', x) \models \mathsf{C}\chi$. According to the definition of $\chi^{\varphi}_{\mathbf{KD45}}$, we need to show that

$$(\mathfrak{M}, x) \models \chi^{\natural} \wedge \Box_1\big(\neg p \to \Box_2(p \to \mathsf{C}_{\{1,2\}}(p \to \chi^{\natural}))\big).$$

We have $(\mathfrak{M}, x) \models \chi^{\natural}$ by the induction hypothesis. Now let $x' \in W' - W$, $x'', y \in W$ be such that xR_1x', $x'R_2x''$ and $x''(R_1 \cup R_2)^*y$. Using that $(\mathfrak{M}, r) \models \chi^{\varphi}_{\mathbf{KD45}}$ and the transitivity of R_1 and R_2, it is not hard to see that then xR^*y holds. So by the induction hypothesis, we have $(\mathfrak{M}, y) \models \chi^{\natural}$ as required. The other direction for C-formulas is straightforward.

As the root of $\mathfrak{F}$ belongs to $\mathfrak{F}'$, it follows that $\mathfrak{M}'$ refutes φ. ❑

Theorem 2.37. (1) $\mathbf{K}_u$ *is polynomially reducible to* $\mathbf{K}_1^C$. *Further,*

(2) $\mathbf{K}$ *is polynomially reducible to* L, *and*

(3) $\mathbf{K}_u$ *is polynomially reducible to* L^C,

for any bimodal logic L *between* $\mathbf{K}_2$ *and* $\mathbf{S5}_2$.

Proof. First, observe that the decision problem for $\mathbf{K}_u$ can be polynomially reduced to the decision problem for $\mathcal{ML}_1^u$-formulas in which no $\boxtimes$ occurs in the scope of a modal operator ($\Box$ or $\boxtimes$). Indeed, given an $\mathcal{ML}_1^u$-formula φ, denote by φ^u the result of replacing every subformula of the form $\chi = \boxtimes\,\psi$ with a fresh propositional variable p_χ. Let

$$\mathcal{R}_u(\varphi) = \{(\boxtimes\,\psi^u \leftrightarrow \boxtimes\,p_\chi) \wedge (\Diamond\!\!\!\!\diamond p_\chi \leftrightarrow \boxtimes\,p_\chi) \mid \chi = \boxtimes\,\psi \in sub\,\varphi\}.$$

Then it is not hard to see that

$$\varphi \in \mathbf{K}_u \qquad \text{iff} \qquad \bigwedge \mathcal{R}_u(\varphi) \to \varphi^u \in \mathbf{K}_u,$$

and the formula in the right-hand side is as required.

In order to show (1), we take a fresh propositional variable p and define a (polynomial) translation ^{c} from $\mathcal{ML}_1^u$ into $\mathcal{ML}_1^C$ as follows:

$$\begin{aligned} q^c &= q, \quad (q \text{ a propositional variable}) \\ (\neg\psi)^c &= \neg\psi^c, \\ (\chi \wedge \psi)^c &= \chi^c \wedge \psi^c, \\ (\Box\psi)^c &= \Box(p \to \psi^c), \\ (\boxtimes\,\psi)^c &= \mathsf{C}\,\psi^c. \end{aligned}$$

Let us show that for every $\mathcal{ML}_1^u$-formula φ without occurrences of $\boxtimes$ in the scope of a modal operator,

$$\varphi \in \mathbf{K}_u \quad \text{iff} \quad \varphi^c \in \mathbf{K}_1^C.$$

First, suppose $\varphi \notin \mathbf{K}_u$. By a straightforward generalization of Proposition 1.7, we may assume that $(\mathfrak{M}, r) \not\models \varphi$ for some model $\mathfrak{M}$ based on a frame

$\mathfrak{F} = \langle W, R, R_u \rangle$, where $\langle W, R \rangle$ is a disjoint union of intransitive trees with r being the root of one of them, and R_u is the universal relation on W.

Now extend the relation R to a relation R' by connecting the roots of the trees with each other. Define a new model $\mathfrak{M}'$ based on $\langle W, R', R'^* \rangle$ by taking p to be true everywhere but at the roots of the above trees. One can first prove by induction that, for all $x \in W$ and all subformulas ψ of φ without the operator $\boxtimes$,

$$(\mathfrak{M}, x) \models \psi \qquad \text{iff} \qquad (\mathfrak{M}', x) \models \psi^c.$$

Now, since no $\boxtimes$ occurs within the scope of a modal operator, we derive for all subformulas ψ of φ:

$$(\mathfrak{M}, r) \models \psi \qquad \text{iff} \qquad (\mathfrak{M}', r) \models \psi^c.$$

It follows that $(\mathfrak{M}', r) \not\models \varphi^c$.

Conversely, suppose that $\varphi^c \notin \mathbf{K}_1^C$. Then $(\mathfrak{M}, r) \not\models \varphi$ for a model $\mathfrak{M}$ based on a rooted frame with root r. Remove from this frame all arrows leading to points where $\neg p$ holds in $\mathfrak{M}$, and define the accessibility relation interpreting $\boxtimes$ as the universal one. Since φ has no occurrences of $\boxtimes$ in the scope of a modal operator, it is not hard to see that φ is refuted at r in the resulting model.

Claims (2) and (3) are proved simultaneously. Define a translation $\clubsuit$ from $\mathcal{ML}_1^u$-formulas into $\mathcal{ML}_2^C$-formulas (with $\Box_1$, $\Box_2$, and $\mathsf{C}_{\{1,2\}}$) by taking

$$\begin{aligned} q^\clubsuit &= p \wedge q, \qquad (q \text{ a propositional variable}) \\ (\psi_1 \wedge \psi_2)^\clubsuit &= \psi_1^\clubsuit \wedge \psi_2^\clubsuit, \\ (\neg\psi)^\clubsuit &= p \wedge \neg\psi^\clubsuit, \\ (\Box\psi)^\clubsuit &= p \wedge \Box_1(\neg p \wedge \neg e \rightarrow \Box_2(p \rightarrow \psi^\clubsuit)), \\ (\boxtimes\psi)^\clubsuit &= p \wedge \mathsf{C}_{\{1,2\}}(p \rightarrow \psi^\clubsuit), \end{aligned}$$

where p and e are fresh variables. Note that if ψ is an $\mathcal{ML}_1$-formula then $\psi^\clubsuit$ is an $\mathcal{ML}_2$-formula.

We now show that, for every $\mathcal{ML}_1^u$-formula φ without occurrences of $\boxtimes$ in the scope of a modal operator,

(i) if $p \rightarrow \varphi^\clubsuit \in \mathbf{S5}_2^C$ then $\varphi \in \mathbf{K}_u$,

(ii) if $\varphi \in \mathbf{K}_u$ then $p \rightarrow \varphi^\clubsuit \in \mathbf{K}_2^C$.

We will then have, for any bimodal logic L between $\mathbf{K}_2$ and $\mathbf{S5}_2$,

$$\begin{aligned} \varphi \in \mathbf{K} \qquad &\text{iff} \qquad p \rightarrow \varphi^\clubsuit \in L, \\ \varphi \in \mathbf{K}_u \qquad &\text{iff} \qquad p \rightarrow \varphi^\clubsuit \in L^C. \end{aligned}$$

To prove (i), suppose that $\varphi \notin \mathbf{K}_u$. Then $(\mathfrak{M}, r) \not\models \varphi$ for some model $\mathfrak{M} = \langle \mathfrak{F}, \mathfrak{V} \rangle$ based on a frame $\mathfrak{F} = \langle W, R, R_u \rangle$, with R_u being the universal relation on W. Define a model $\mathfrak{M}' = \langle \mathfrak{F}', \mathfrak{V}' \rangle$ based on $\mathfrak{F}' = \langle W', R_1, R_2 \rangle$ by taking

- $W' = W \cup (W \times W)$,
- R_1 to be the reflexive, transitive and symmetric closure of R_1', where $xR_1'y$ iff $x \in W$ and $y = \langle x, z \rangle$, for some $z \in W$,
- R_2 to be the reflexive, transitive and symmetric closure of R_2', where $xR_2'y$ iff $y \in W$ and $x = \langle z, y \rangle$, for some $z \in W$,
- $\mathfrak{V}'(p) = W$,
- $\mathfrak{V}'(e) = \{\langle x, y \rangle \in W \times W \mid \langle x, y \rangle \notin R\}$,
- $\mathfrak{V}'(q) = \mathfrak{V}(q)$, for any other propositional variable q.

Thus, $\mathfrak{F}' = \langle W', R_1, R_2 \rangle$ is a frame for $\mathbf{S5}_2$. Further, it is not hard to see that $(R_1 \cup R_2)^*$ is the universal relation on W'. An easy induction shows that, for all $x \in W$ and all subformulas ψ of φ, we have

$$(\mathfrak{M}, x) \models \psi \qquad \text{iff} \qquad (\mathfrak{M}', x) \models \psi^{\clubsuit}.$$

It follows that $(\mathfrak{M}', r) \not\models p \to \varphi^{\clubsuit}$, and so $p \to \varphi^{\clubsuit} \notin \mathbf{S5}_2^C$.

Now let us prove (ii). Suppose that $p \wedge \neg\varphi^{\clubsuit}$ is satisfied at the root r of a model $\mathfrak{M} = \langle \mathfrak{F}, \mathfrak{V} \rangle$ based on $\mathfrak{F} = \langle W, R_1, R_2 \rangle$. Define a model $\mathfrak{M}' = \langle \mathfrak{F}', \mathfrak{V}' \rangle$ based on $\mathfrak{F}' = \langle W', R, R_u \rangle$ by taking

- $W' = \mathfrak{V}(p)$,
- xRy iff there exists $z \in W - (\mathfrak{V}(p) \cup \mathfrak{V}(e))$ such that xR_1z and zR_2y,
- R_u to be the universal relation on W',
- $\mathfrak{V}'(q) = \mathfrak{V}(q) \cap \mathfrak{V}(p)$.

One can first prove by induction that, for all $x \in W'$ and all subformulas ψ of φ without the operator $\boxtimes$,

$$(\mathfrak{M}', x) \models \psi \qquad \text{iff} \qquad (\mathfrak{M}, x) \models \psi^{\clubsuit}.$$

Now, since no $\boxtimes$ occurs within the scope of a modal operator, we derive for all subformulas ψ of φ:

$$(\mathfrak{M}', r) \models \psi \qquad \text{iff} \qquad (\mathfrak{M}, r) \models \psi^{\clubsuit}.$$

It follows that $(\mathfrak{M}', r) \not\models \varphi$. ❑

Theorem 2.38. **PTL** *is polynomially reducible to* $\mathbf{K}_1^C$.

Proof. By Proposition 2.10, **PTL** is polynomially reducible to $\mathbf{PTL}_{\Box\bigcirc}$. Hence, it suffices to show that $\mathbf{PTL}_{\Box\bigcirc}$ is polynomially reducible to $\mathbf{K}_1^C$.

For a formula φ of the bimodal language $\mathcal{ML}_2$ (with $\Box_F$ and $\bigcirc$), we put

$$sub^{\bigcirc}\varphi = sub\,\varphi \cup \{\bigcirc\chi \mid \chi \in sub\,\varphi\}$$

and denote by $\varphi^\bullet$ the result of replacing all occurrences of $\bigcirc$ and $\Diamond_F$ in φ with $\Diamond_1$ and $\Diamond_1\neg\mathsf{C}\neg$, respectively. Let

$$\mathcal{R}(\varphi) = \{\Diamond_1\chi^\bullet \to \Box_1\chi^\bullet \mid \chi \in sub^{\bigcirc}\varphi\}.$$

We show now that for every $\mathcal{ML}_2$-formula φ,

$$\varphi \in \mathbf{PTL}_{\Box\bigcirc} \quad \text{iff} \quad \mathsf{C}(\Diamond_1\top \wedge \bigwedge\mathcal{R}(\varphi)) \to \varphi^\bullet \in \mathbf{K}_1^C.$$

The implication ($\Leftarrow$) should be clear. Conversely, suppose that

$$(\mathfrak{M}, w_0) \models \neg\varphi^\bullet \wedge \mathsf{C}(\Diamond_1\top \wedge \bigwedge\mathcal{R}(\varphi)) \tag{2.32}$$

for some model $\mathfrak{M} = \langle\mathfrak{F}, \mathfrak{V}\rangle$ based on a rooted intransitive tree $\mathfrak{F} = \langle W, R_1\rangle$ with root w_0. First we construct a countable sequence $w_0, w_1, \dots$ of distinct points in W such that $w_i R_1 w_{i+1}$, for all $i \in \mathbb{N}$. This sequence will then be used as the flow of time in which we refute φ.

Suppose that a sequence $\sigma = \langle w_0, \dots, w_n\rangle$ has already been constructed. Call a pair $\left\langle m, \Diamond_F\psi\right\rangle$ a *σ-defect* if $m < n$, $\Diamond_F\psi \in sub\,\varphi$, and

- $(\mathfrak{M}, w_m) \models \Diamond_1\neg\mathsf{C}\neg\psi^\bullet$, but
- for all i with $m + 1 \leq i \leq n$, we have $(\mathfrak{M}, w_i) \not\models \psi^\bullet$.

If there are no σ-defects then we take some $w_{n+1} \in W$ such that $w_n R_1 w_{n+1}$ (w_{n+1} exists because $(\mathfrak{M}, w_n) \models \Diamond_1\top$) and continue with the new sequence $\sigma' = \langle w_0, \dots, w_n, w_{n+1}\rangle$.

Otherwise, we list all the σ-defects (there are finitely many of them). Take the first σ-defect $\left\langle m, \Diamond_F\psi\right\rangle$ in the list. One can prove by induction that for all $i = m, \dots, n$,

$$(\mathfrak{M}, w_i) \models \Diamond_1\neg\mathsf{C}\neg\psi^\bullet. \tag{2.33}$$

Indeed, for $i = m$ this holds by the definition of a σ-defect. Now suppose that (2.33) holds for some i with $m \leq i < n$. Then either

(a) $(\mathfrak{M}, w_i) \models \Diamond_1\psi^\bullet$, or

(b) $(\mathfrak{M}, w_i) \models \Diamond_1 \Diamond_1 \neg \mathsf{C} \neg \psi^\bullet$.

In the former case, in view of (2.32), we have $(\mathfrak{M}, w_i) \models \Box_1 \psi^\bullet$, contrary to $\left\langle m, \Diamond_F \psi \right\rangle$ being a σ-defect. So case (b) must hold. Since $\bigcirc \Diamond_F \psi \in sub^{\bigcirc} \varphi$,

$$(\mathfrak{M}, w_i) \models \Diamond_1 \Diamond_1 \neg \mathsf{C} \neg \psi^\bullet \to \Box_1 \Diamond_1 \neg \mathsf{C} \neg \psi^\bullet,$$

and so

$$(\mathfrak{M}, w_i) \models \Box_1 \Diamond_1 \neg \mathsf{C} \neg \psi^\bullet,$$

from which $(\mathfrak{M}, w_{i+1}) \models \Diamond_1 \neg \mathsf{C} \neg \psi^\bullet$.

We have shown that $(\mathfrak{M}, w_n) \models \Diamond_1 \neg \mathsf{C} \neg \psi^\bullet$. So we can find distinct points $w_{n+1}, \dots, w_{k_1}$ in W such that

- $w_i R_1 w_{i+1}$ for all i with $n \leq i < k_1$, and
- $(\mathfrak{M}, w_{k_1}) \models \psi^\bullet$.

Now consider the sequence $\sigma_1 = \langle w_0, \dots, w_n, w_{n+1}, \dots, w_{k_1} \rangle$, and take the second σ-defect from the previous list σ (if any). If it is also a σ_1-defect then, by repeating the above argument, we can extend σ_1 to some

$$\sigma_2 = \langle w_0, \dots, w_n, \dots, w_{k_1}, \dots, w_{k_2} \rangle,$$

and so on. After fixing all the σ-defects this way, we obtain a new sequence σ'. Then we list all the σ'-defects, 'fix' them, and so forth.

In the limit we obtain a sequence $\langle w_i \mid i \in \mathbb{N} \rangle$. Define a valuation $\mathfrak{V}'$ in the frame $\langle \mathbb{N}, <, +1 \rangle$ by taking

$$\mathfrak{V}'(p) = \{ n \in \mathbb{N} \mid w_n \in \mathfrak{V}(p) \},$$

for every propositional variable p, and let $\mathfrak{M}' = \langle \langle \mathbb{N}, <, +1 \rangle, \mathfrak{V}' \rangle$. It can be shown by induction that for all $\psi \in sub\, \varphi$ and all $n \in \mathbb{N}$,

$$(\mathfrak{M}, w_n) \models \psi^\bullet \quad \text{iff} \quad (\mathfrak{M}', n) \models \psi.$$

Hence, by (2.32), we have $(\mathfrak{M}', 0) \not\models \varphi$, as required. ❑

We conclude this section by showing that all epistemic logics L_n^C, introduced in Section 2.3, as well as the temporal logic **PTL**, can be embedded into dynamic logics **PDL** and **CPDL**. First, with every operator $\Box_i$ of $\mathcal{ML}_n$ $(1 \leq i \leq n)$ we associate an action term $\mathsf{t}_j(\Box_i)$, $j \leq 5$. To this end, we fix action variables $\alpha_1, \dots, \alpha_n$, $\beta_1, \dots, \beta_n$ and put, for $i \le n$,

$$\begin{aligned}
\mathsf{m}_1(\Box_i) &= \alpha_i, \\
\mathsf{m}_2(\Box_i) &= \alpha_i \,; \alpha_i^*, \\
\mathsf{m}_3(\Box_i) &= \alpha_i^*, \\
\mathsf{m}_4(\Box_i) &= (\alpha_i \cup \alpha_i^-)^*, \\
\mathsf{m}_5(\Box_i) &= \alpha_i \cup (\alpha_i^- \,; \alpha_i) \cup (\beta_i \cup \beta_i^-)^*.
\end{aligned}$$

Define translations $\mathsf{t}_1, \dots, \mathsf{t}_6$ from the language $\mathcal{ML}_n^C$ into the language $\mathcal{CPDL}$ by taking for every non-empty set $M = \{i_1, \dots, i_k\} \subseteq \{1, \dots, n\}$ and every $j = 1, \dots, 6$:

$$\begin{aligned}
\mathsf{t}_j(p_i) &= p_i, \\
\mathsf{t}_j(\varphi \wedge \psi) &= \mathsf{t}_j(\varphi) \wedge \mathsf{t}_j(\psi), \\
\mathsf{t}_j(\neg\varphi) &= \neg \mathsf{t}_j(\varphi), \\
\mathsf{t}_j(\Box_i \varphi) &= \begin{cases} [\mathsf{m}_j(\Box_i)]\, \mathsf{t}_j(\varphi), & \text{if } j \neq 6; \\ \mathsf{t}_6(\varphi) \wedge [\mathsf{m}_1(\Box_i)]\, \mathsf{t}_6(\varphi), & \text{otherwise,} \end{cases} \\
\mathsf{t}_j(\mathsf{C}_M \varphi) &= [(\mathsf{m}_j(\Box_{i_1}) \cup \dots \cup \mathsf{m}_j(\Box_{i_k}))^*]\, \mathsf{t}_j(\varphi).
\end{aligned}$$

Theorem 2.39. *If $L \in \{\mathbf{K}_n, \mathbf{T}_n, \mathbf{K4}_n, \mathbf{S4}_n, \mathbf{KD45}_n, \mathbf{S5}_n\}$ then the epistemic logic L^C is polynomially reducible to* **CPDL**. *More precisely, for every $\mathcal{ML}_n^C$-formula φ, we have*

(i) $\varphi \in \mathbf{K}_n^C$ *iff* $\mathsf{t}_1(\varphi) \in \mathbf{PDL}$,

(ii) $\varphi \in \mathbf{T}_n^C$ *iff* $\mathsf{t}_6(\varphi) \in \mathbf{PDL}$,

(iii) $\varphi \in \mathbf{K4}_n^C$ *iff* $\mathsf{t}_2(\varphi) \in \mathbf{PDL}$,

(iv) $\varphi \in \mathbf{S4}_n^C$ *iff* $\mathsf{t}_3(\varphi) \in \mathbf{PDL}$,

(v) $\varphi \in \mathbf{S5}_n^C$ *iff* $\mathsf{t}_4(\varphi) \in \mathbf{CPDL}$,

(vi) $\varphi \in \mathbf{KD45}_n^C$ *iff* $[\gamma^*]\,\chi \to \mathsf{t}_5(\varphi) \in \mathbf{CPDL}$, *where*

$$\begin{aligned}
\gamma &= \alpha_1 \cup \alpha_1^- \cup \ldots \cup \alpha_n \cup \alpha_n^- \cup \beta_1 \cup \beta_1^- \cup \ldots \cup \beta_n \cup \beta_n^-, \\
\chi &= \bigwedge_{i \le n} [\alpha_i \,; \alpha_i] \bot \wedge \bigwedge_{i \le n} \langle \alpha_i \cup \alpha_i^- \rangle \top \to [\beta_i \cup \beta_i^-] \bot.
\end{aligned}$$

Proof. The proofs are rather straightforward. Here we only sketch the proof of (vi). Suppose a $\mathbf{KD45}_n$-frame $\mathfrak{F} = \langle W, R_1, \dots, R_n \rangle$ refutes φ in a world w under some valuation. Without loss of generality we may assume that w is the root of $\mathfrak{F}$. Define a $\mathcal{PDL}$-structure $\mathfrak{G} = \langle W, T_{\alpha_1}, \dots, T_{\beta_1}, \dots \rangle$ by taking, for all $u, v \in W$, $1 \le i \le n$,

- $uT_{\alpha_i}v$ iff uR_iv and not vR_iu;
- $uT_{\beta_i}v$ iff uR_iv, vR_iu and there is no $x \in W$ such that xR_iu and not uR_ix.

It is easy to show that $\mathfrak{G}$ refutes $[\gamma^*]\,\chi \rightarrow \mathrm{t}_5(\varphi)$.

Conversely, suppose that a $\mathcal{PDL}$-structure $\mathfrak{G} = \langle W, T_{\alpha_1}, \ldots, T_{\beta_1}, \ldots\rangle$ refutes $[\gamma^*]\,\chi \rightarrow \mathrm{t}_5(\varphi)$ at its root w. Define an n-frame $\mathfrak{F} = \langle W, R_1, \ldots, R_n\rangle$ by taking, for all $u, v \in W$,

- uR_iv iff either $uT_{\alpha_i}v$ or $uT_{\alpha_i^-;\alpha_i}v$, or $uT_{(\beta_i \cup \beta_i^-)^*}v$.

One can readily show that $\mathfrak{F}$ is a frame for $\mathbf{KD45}_n^C$ refuting φ. ❑

Observe that the translations above embed the epistemic logics in question into the *test-free fragments* of **PDL** and **CPDL**.

Since the composition of two polynomial reductions is a polynomial reduction, Theorems 2.38 and 2.39 yield:

Theorem 2.40. *The temporal logic* **PTL** *is polynomially reducible to* **PDL**.

Chapter 3

Many-dimensional modal logics

So far we have been considering modal formalisms intended for reasoning about time, knowledge, beliefs, actions, space independently of each other. We have completely abstracted from the fact that in reality all these entities exist in close interaction: knowledge, beliefs and spatial regions can change over time and under actions, agents in a multi-agent system may have their own knowledge bases, and so forth. In this chapter we discuss possible ways of constructing *many-dimensional* (or *combined*) *modal logics* which are able to capture such interactions. Computational properties of these logics will be investigated in Parts II–IV.

3.1 Fusions

The formation of *fusions*, or *independent joins*, is the simplest and perhaps most frequently used way of combining logics. Let L_1 and L_2 be two multimodal[1] logics formulated in languages $\mathcal{L}_1$ and $\mathcal{L}_2$, both containing the language $\mathcal{L}$ of classical propositional logic, but having disjoint sets of modal operators. Denote by $\mathcal{L}_1 \otimes \mathcal{L}_2$ the union of $\mathcal{L}_1$ and $\mathcal{L}_2$. Then the *fusion* $L_1 \otimes L_2$ *of* L_1 *and* L_2 is the smallest multimodal logic L in the language $\mathcal{L}_1 \otimes \mathcal{L}_2$ containing $L_1 \cup L_2$. In particular, if L_1 is axiomatized by a set of axioms Ax_1 and L_2 is axiomatized by Ax_2, then $L_1 \otimes L_2$ is axiomatized by the union $Ax_1 \cup Ax_2$. This means that *no* axiom containing modal operators from both languages $\mathcal{L}_1$ and $\mathcal{L}_2$ is required to axiomatize the fusion of L_1 and

[1] Here by a multimodal logic we mean a logic formulated in any of the languages $\mathcal{ML}_n$, $\mathcal{ML}_n^C$, $\mathcal{ML}_\mathcal{U}$, or $\mathcal{ML}_{\mathcal{SU}}$.

L_2. The modal operators in $\mathcal{L}_1$ and $\mathcal{L}_2$ remain 'independent,' they 'do not interact;' however, both L_1 and L_2 contain classical propositional logic **Cl**. Note that the formation of fusions is clearly an associative binary operation on logics. Thus, one can define the fusion $L_1 \otimes L_2 \otimes \cdots \otimes L_n$ of n logics in a straightforward way, for any natural number $n \geq 2$. For example, we have

$$\mathbf{K}_n = \underbrace{\mathbf{K} \otimes \cdots \otimes \mathbf{K}}_{n}, \qquad \mathbf{S5}_n = \underbrace{\mathbf{S5} \otimes \cdots \otimes \mathbf{S5}}_{n}, \qquad \text{etc.}$$

(see Section 1.4).

Fusions of modal logics have been studied for a relatively long time. The first explicit result about fusions was obtained by Thomason (1980), who proved that fusions of consistent modal logics turn out to be conservative extensions of their components. Further results showing that many important properties of logics are preserved under fusions were obtained by Kracht and Wolter (1991), Fine and Schurz (1996), Goranko and Passy (1992), Spaan (1993), Gabbay (1996) and Wolter (1998).

So far we have considered fusions only from the syntactical point of view. However, fusions have a very natural semantical interpretation as well, at least for logics which are Kripke complete. Consider two classes $\mathcal{C}_1$ and $\mathcal{C}_2$ of m- and n-frames, respectively, that are closed under disjoint unions and isomorphic copies. The *fusion* $\mathcal{C}_1 \otimes \mathcal{C}_2$ *of* $\mathcal{C}_1$ *and* $\mathcal{C}_2$ is the class of all $n+m$-frames of the form

$$\langle W, R_1, \ldots, R_m, S_1, \ldots, S_n \rangle$$

such that $\langle W, R_1, \ldots, R_m \rangle \in \mathcal{C}_1$ and $\langle W, S_1, \ldots, S_n \rangle \in \mathcal{C}_2$.

Thus, $\mathcal{C}_1 \otimes \mathcal{C}_2$ consists of arbitrary combinations of frames from $\mathcal{C}_1$ and $\mathcal{C}_2$ sharing the same set of worlds. It should be clear that if $\mathcal{C}_1$ and $\mathcal{C}_2$ determine logics L_1 and L_2, respectively, then all frames in $\mathcal{C}_1 \otimes \mathcal{C}_2$ validate the fusion $L_1 \otimes L_2$. However, it is rather nontrivial to prove that actually the converse also holds, i.e., $\mathcal{C}_1 \otimes \mathcal{C}_2$ characterizes $L_1 \otimes L_2$.

Another important preservation theorem shows that the fusion of two decidable logics is decidable as well. Thus, modulo decidability, fusions can be reduced to their components. This result heavily relies upon the fact that we combine *propositional modal logics* rather than, say, first-order theories, where such a result does not hold. For example, the first-order theory of one equivalence relation $\sim$ has the finite model property and is decidable. However, the first-order theory of two equivalence relations $\sim_1$ and $\sim_2$ is undecidable (Janiczak 1953, Ershov *et al.* 1965).

These results as well as other preservation theorems concerning fusions are proved in Chapter 4.

We conclude this introductory section by illustrating the role of fusions with some simple examples.

Example 3.1. First we explain in more detail why it is natural to consider the basic *epistemic logics*, introduced in Section 2.3, as fusions. Take an agent A and an epistemic logic L_A with the modal operator $\Box_A$ ('agent A knows') intended for reasoning about the knowledge of A. Assume now that L_B is another epistemic logic formalizing the knowledge of another agent B by means of the operator $\Box_B$ ('agent B knows'). If agents A and B are supposed to interact, we may need a formalism which is able to represent not only the knowledge about A's and B's 'objects,' but also their knowledge about each other's knowledge.

Naturally, we then take the bimodal epistemic language with both operators $\Box_A$ and $\Box_B$. But what are the principles (axioms) of the logic intended for reasoning in the combined language? Of course, it should contain $L_A \cup L_B$, since the principles governing a single knowledge operator should remain the same in the combined logic. Thus, the logic will contain the fusion $L_A \otimes L_B$. If no information about the relation between A and B is available, then we have no grounds to add any axioms containing both boxes $\Box_A$ and $\Box_B$. So in this case the fusion $L_A \otimes L_B$ is the epistemic logic which can serve for reasoning about the knowledge of two agents A and B.

It is not hard to imagine various situations when interaction axioms are required, for instance, when A knows everything that B knows. Then we should extend the fusion with the axiom

$$\Box_B p \to \Box_A p.$$

Another example: A knows about B's knowledge (when, say, A has constructed B). Then we need the extra axiom

$$\Box_B p \to \Box_A \Box_B p.$$

But in any case the formation of fusions is the first basic step towards constructing multi-agent logics of knowledge.

Example 3.2. Epistemic logics are used in order to formalize reasoning about knowledge of agents having incomplete information. However, such logics are able to describe only *static* pictures. They do not have enough expressive power to reason, for instance, about changes of knowledge when new information becomes available or certain facts are forgotten. To construct a language which can capture various dynamic features of knowledge, a new *temporal* 'dimension' should be added to the epistemic one. Suppose, for example, that an epistemic logic is extended by means of the temporal operator $\mathcal{U}$ (until). The resulting *temporal epistemic* language will then contain the modal operators $\Box_i$ ('agent i knows') and $\mathcal{U}$, so that we can express conditions like

$$(\neg \Box_i p)\, \mathcal{U} (\Box_j p)$$

saying that agent j will know property p, and not later than agent i.

Again, it seems natural to start constructing an axiomatization of the desirable combination of the epistemic and temporal logics by taking their fusion. When doing that, we assume no interaction between time and knowledge, so agents may forget, learn, etc. Actually, such a fusion is the basic temporal epistemic logic introduced in (Fagin *et al.* 1995). In Section 3.4 we discuss this temporal epistemic logic as well as some other logics having interactions between time and knowledge.

Example 3.3. The nice behavior of fusions of modal logics is particularly useful in *description logic*. For instance, having a decidable description logic with one transitive role, another decidable description logic with one functional role, and one more decidable description logic with one ordinary role, and taking a suitable fusion of them, we can construct a decidable description logic with arbitrarily many transitive, functional, and ordinary roles. More advanced applications of fusions in description logic are explored in (Baader *et al.* 2002).

From the semantical point of view, the formation of fusions does not change the 'dimension' of logics: worlds in their frames are still regarded as points without any 'many-dimensional feature' (cf., however, Section 9.1). Let us see now what happens when we combine logics whose modal operators are supposed to interact. Perhaps the most intuitively transparent is the combination of temporal and spatial logics.

3.2 Spatio-temporal logics

Suppose that we need a logical formalism which is able to represent knowledge and reason about spatial regions changing over time. We can then choose a spatial logic and a temporal logic that reflect our views on space and time (and satisfy the required effectiveness and expressiveness parameters), say, $\mathcal{BRCC}$-8 and $\mathsf{Log}_{\mathcal{SU}}(\mathcal{C})$, for some class $\mathcal{C}$ of flows of time, and try to combine them into a single spatio-temporal system.[2]

This choice (together with common sense considerations) almost uniquely determines the semantical paradigm of the hybrid under construction. As we saw in Sections 2.6 and 2.1, static spatial regions are interpreted in a topological space $\mathfrak{T} = \langle U, \mathbb{I} \rangle$, and the flow of time is represented by a frame $\mathfrak{F} = \langle W, < \rangle$, where $<$ is a strict linear order on W. It is reasonable to assume that space with its topology always remains the same. However, the spatial regions occupied by the objects under consideration may move with time

[2]Our choice is motivated mainly by the fact that both components are 'modal' logics considered in this book.

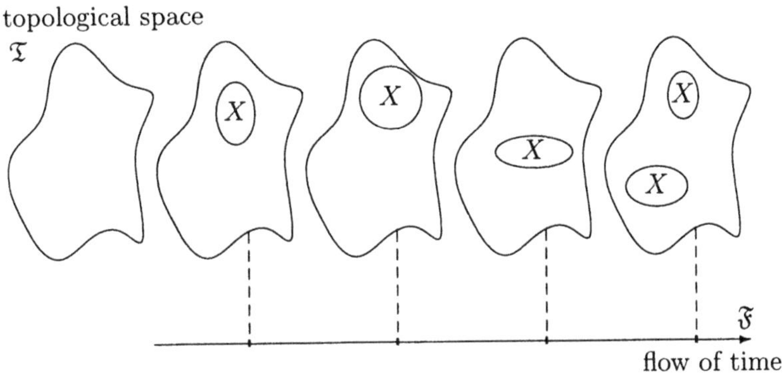

Figure 3.1: Spatial regions moving in time.

passing by (see Fig. 3.1). This naïve picture can be formalized by means of the following concept of topological temporal model.

A *topological temporal model* (or *tt-model*, for short) based on a flow of time $\mathfrak{F} = \langle W, < \rangle$ and a topological space $\mathfrak{T} = \langle U, \mathbb{I} \rangle$ is a triple $\mathfrak{M} = \langle \mathfrak{F}, \mathfrak{T}, \mathfrak{a} \rangle$, where $\mathfrak{a}$, an *assignment in* $\mathfrak{M}$, associates with every region variable X and every moment of time $w \in W$ a regular closed set $\mathfrak{a}(X, w) \subseteq U$ (that is, a set $\mathfrak{a}(X) \subseteq U$ such that $\mathfrak{a}(X) = \mathbb{C}\mathbb{I}\mathfrak{a}(X))$, the *state* of X at w. Thus, tt-models can be regarded as *two-dimensional* structures. Having fixed a moment of time, we can move in the 'spatial dimension' representing the states of regions at this moment. Having fixed a spatial region, we can move along the 'temporal dimension' tracing the evolution of this region in time.

Let us turn now to the syntactical parameters of spatio-temporal hybrids. Actually, there are different ways of introducing a temporal dimension into the syntax of $\mathcal{BRCC}$-8, which give rise to a hierarchy of possible spatio-temporal languages

$$\mathcal{ST}_0 \subseteq \mathcal{ST}_1 \subseteq \mathcal{ST}_2.$$

The spatio-temporal language $\mathcal{ST}_0$. The most obvious one allows applications of the temporal operators $\mathcal{S}$ and $\mathcal{U}$ only to spatial formulas of $\mathcal{BRCC}$-8. More precisely, the *spatio-temporal language* $\mathcal{ST}_0$ is defined as follows. Every formula of $\mathcal{BRCC}$-8 is also an $\mathcal{ST}_0$-formula, and if φ and ψ are $\mathcal{ST}_0$-formulas then so are $\varphi\mathcal{S}\psi$, $\varphi\mathcal{U}\psi$, $\varphi \wedge \psi$, and $\neg\varphi$. As usual, we use abbreviations $\bigcirc\varphi = \bot\mathcal{U}\varphi$, $\Diamond_F\varphi = \top\mathcal{U}\varphi$, $\Box_F\varphi = \neg\Diamond_F\neg\varphi$; a new one is $\varphi\mathcal{W}\psi = \Box_F\varphi \vee (\varphi\mathcal{U}\psi)$, where $\mathcal{W}$ stands for 'waiting for' (it is also known as 'unless;' see Manna and Pnueli 1992).

For a tt-model $\mathfrak{M} = \langle \mathfrak{F}, \mathfrak{T}, \mathfrak{a} \rangle$, an $\mathcal{ST}_0$-formula φ, and $w \in W$, define the *truth-relation* $(\mathfrak{M}, w) \models \varphi$ ('φ holds in $\mathfrak{M}$ at moment w') by induction on the construction of φ. Denote by $\mathfrak{a}_w$ the assignment in $\mathfrak{T}$ defined by $\mathfrak{a}_w(X) = \mathfrak{a}(X, w)$, for every region variable X and every $w \in W$. (Recall that the truth-relation $\mathfrak{T} \models^{\mathfrak{a}_w} \varphi$ was introduced in Section 2.6.) Now,

- if φ contains no temporal operators, then $(\mathfrak{M}, w) \models \varphi$ iff $\mathfrak{T} \models^{\mathfrak{a}_w} \varphi$;
- $(\mathfrak{M}, w) \models \varphi \mathcal{U} \psi$ iff there is $v > w$ such that $(\mathfrak{M}, v) \models \psi$ and $(\mathfrak{M}, u) \models \varphi$ for every u in the interval $w < u < v$;
- $(\mathfrak{M}, w) \models \varphi \mathcal{S} \psi$ iff there is $v < w$ such that $(\mathfrak{M}, v) \models \psi$ and $(\mathfrak{M}, u) \models \varphi$ for every u in the interval $v < u < w$.

Although the interaction between time and space in $\mathcal{ST}_0$ is rather weak, the language $\mathcal{ST}_0$ is expressive enough to capture some aspects of *continuity of changes* (see, e.g., Cohn 1997):

$$\begin{aligned}
&\mathsf{DC}(X,Y) \to \mathsf{DC}(X,Y)\, \mathcal{W}\, \mathsf{EC}(X,Y),\\
&\mathsf{EC}(X,Y) \to \mathsf{EC}(X,Y)\, \mathcal{W}\, (\mathsf{DC}(X,Y) \vee \mathsf{PO}(X,Y)),\\
&\mathsf{PO}(X,Y) \to \mathsf{PO}(X,Y)\, \mathcal{W}\, (\mathsf{EC}(X,Y) \vee\\
&\qquad\qquad \mathsf{TPP}(X,Y) \vee \mathsf{EQ}(X,Y) \vee \mathsf{TPPi}(X,Y)),\\
&\text{etc.}
\end{aligned}$$

The first of these formulas, for instance, says that if two regions are disconnected at some moment, then either they will remain disconnected forever or they are disconnected until they become externally connected. If the flow of time is discrete then these conditions can be rewritten as:

$$\begin{aligned}
&\mathsf{DC}(X,Y) \to \bigcirc(\mathsf{DC}(X,Y) \vee \mathsf{EC}(X,Y)),\\
&\mathsf{EC}(X,Y) \to \bigcirc(\mathsf{EC}(X,Y) \vee \mathsf{DC}(X,Y) \vee \mathsf{PO}(X,Y)),\\
&\mathsf{PO}(X,Y) \to \bigcirc(\mathsf{PO}(X,Y) \vee \mathsf{EC}(X,Y) \vee\\
&\qquad\qquad \mathsf{TPP}(X,Y) \vee \mathsf{EQ}(X,Y) \vee \mathsf{TPPi}(X,Y)),\\
&\text{etc.}
\end{aligned}$$

The spatio-temporal language $\mathcal{ST}_1$. Of course, the expressive power of $\mathcal{ST}_0$ is rather limited. In particular, we can compare regions only at one moment of time, but we are not able to connect a region as it is 'today' with its state 'tomorrow' to say, for example, that it is expanding or remains the same. In other words, we can express the dynamics of relations between regions, say,

$$\neg\Box_F \mathsf{P}(\mathit{Kosovo}, \mathit{Yugoslavia})$$

('it is not true that Kosovo will always be part of Yugoslavia'), but not the dynamics of regions themselves, for instance, that

$$\Box_F^+ \mathsf{P}(EU, \bigcirc EU),$$

where $\bigcirc EU$ at moment n intends to denote the space occupied by the EU at the next moment (so for the flow of time $\langle \mathbb{N}, < \rangle$ the last formula means: 'the EU will never contract'). This new constructor may also be important to refine the continuity assumption by requiring that

$$\Box_F^+ (\mathsf{EQ}(X, \bigcirc X) \vee \mathsf{O}(X, \bigcirc X)),$$

i.e., 'regions X and $\bigcirc X$ either coincide or overlap.' (Recall from Section 2.6 that the predicates P and O are expressible in $\mathcal{BRCC}$-8.)

To capture this dynamics, we extend $\mathcal{ST}_0$ by allowing applications of the next-time operator $\bigcirc$ not only to formulas but also to Boolean region terms. Thus, arguments of the predicate symbols in $\mathcal{BRCC}$-8 can be now arbitrary region $\bigcirc$-*terms* which are constructed from region variables using the Booleans and $\bigcirc$. For instance, $\bigcirc\bigcirc X$ represents region X as it will be 'the day after tomorrow.' Denote the resulting language by $\mathcal{ST}_1$. If $\mathfrak{M} = \langle \mathfrak{F}, \mathfrak{T}, \mathfrak{a} \rangle$ is a tt-model and t a $\bigcirc$-term, then put

$$\mathfrak{a}(\bigcirc t, w) = \begin{cases} \mathfrak{a}(t, w'), & \text{if } w' \text{ is an immediate successor of } w \text{ in } \mathfrak{F}, \\ \emptyset, & \text{if } w \text{ has no immediate successor in } \mathfrak{F}. \end{cases}$$

Note that for every $\bigcirc$-term and every time point w, $\mathfrak{a}(t, w)$ is a regular closed set in $\mathfrak{T}$. Using $\mathcal{ST}_1$ we can express over $\langle \mathbb{N}, < \rangle$ that region X will always be the same (i.e., X is *rigid*):

$$\Box_F^+ \mathsf{EQ}(X, \bigcirc X),$$

or that it has at most two distinct states, one on 'even days,' another on 'odd ones:'

$$\Box_F^+ \mathsf{EQ}(X, \bigcirc\bigcirc X).$$

Note, by the way, that the $\mathcal{ST}_1$-formula

$$\Box_F^+ \mathsf{NTPP}(X, \bigcirc X)$$

is satisfiable only in models based on infinite topological spaces—in contrast to $\mathcal{BRCC}$-8 formulas, for which finite topological spaces are enough (see Theorem 2.31).

It may appear that $\mathcal{ST}_1$ is able to compare regions only within fixed time intervals. However, using an auxiliary rigid variable X we can write, for instance,

$$\Box_F^+ \mathsf{EQ}(X, \bigcirc X) \wedge \Diamond_F \mathsf{EQ}(X, EU) \wedge \mathsf{P}(\mathit{Russia}, X).$$

This formula is satisfiable iff 'some day in the future the *present* territory of Russia will be part of the EU.' Note that the formula

$$\Diamond_F \mathsf{P}(\mathit{Russia}, \mathit{EU})$$

means that there will be a day when Russia—its territory on that day (say, without Chechnya but with Byelorussia)—becomes part of the EU.

The spatio-temporal language $\mathcal{ST}_2$. Imagine now that we want to express in our spatio-temporal language that all countries in Europe will pass through the euro-zone, but only Germany (in its present territory) will use the euro forever. Unfortunately, we do not know which countries will be formed in Europe in the future, so we cannot simply write down all formulas of the form

$$\Diamond_F \mathsf{P}(X, \mathit{Euro\text{-}zone}).$$

What we actually need is the possibility of constructing regions $\Diamond_F X$ and $\Box_F X$ which contain all the points that will belong to region X in the future and only common points of all future states of X, respectively. Then we can write:

$$\mathsf{EQ}(\mathit{Europe}, \Diamond_F^+ \mathit{Euro\text{-}zone})$$

and

$$\mathsf{EQ}(\mathit{Germany}, \Box_F^+ \mathit{Euro\text{-}zone}).$$

The formula

$$\mathsf{P}(\mathit{Russia}, \Diamond_F \mathit{EU})$$

says that all points of the present territory of Russia will belong to the EU in the future (but perhaps at different moments of time).

So let us extend $\mathcal{ST}_0$ by allowing the use of *temporal region terms*, constructed from region variables, the Booleans, and the temporal operators $\mathcal{U}$ and $\mathcal{S}$ with all their derivatives, as arguments of the $\mathcal{RCC}$-8 predicates. In other words, every region variable is a temporal region term, and if t_1 and t_2 are temporal region terms then so are $t_1 \sqcap t_2$, $t_1 \sqcup t_2$, $\neg t_1$, $\Diamond_F t_1$, $\Box_F t_1$, $\Diamond_P t_1$, $\Box_P t_1$, $\bigcirc t_1$, $t_1 \mathcal{U} t_2$ and $t_1 \mathcal{S} t_2$. The resulting language will be denoted by $\mathcal{ST}_2$.

The intended semantics of temporal region terms is as follows. Suppose $\mathfrak{M} = \langle \mathfrak{F}, \mathfrak{T}, \mathfrak{a} \rangle$ is a tt-model. Define inductively the *value* $\mathfrak{a}(t, w)$ *of a temporal region term* t *under* $\mathfrak{a}$ *at* w *in* $\mathfrak{M}$ by taking:

$$\mathfrak{a}(\Diamond_F t, w) = \mathbb{CI} \bigcup_{v>w} \mathfrak{a}(t, v),$$

$$\mathfrak{a}(\Box_F t, w) = \mathbb{CI} \bigcap_{v>w} \mathfrak{a}(t, v),$$

$$\mathfrak{a}(t_1 \mathcal{U} t_2, w) = \mathbb{CI}\{x \mid \exists v > w (x \in \mathfrak{a}(t_2, v) \land \forall u (w < u < v \to x \in \mathfrak{a}(t_1, u)))\},$$

$$\mathfrak{a}(t_1 \mathcal{S} t_2, w) = \mathbb{CI}\{x \mid \exists v < w (x \in \mathfrak{a}(t_2, v) \land \forall u (w > u > v \to x \in \mathfrak{a}(t_1, u)))\},$$

and the corresponding clauses for $\Diamond_P$, $\Box_P$ and the Booleans. For example, the formula

$$\mathsf{DC}(Russia\ \mathcal{S}\ Russian\_Empire,\ Russia\ \mathcal{S}\ Germany)$$

can be used to say that the part of Russia that has been remaining Russian since 1917 is not connected to the part of Germany (Königsberg) that became Russian after the Second World War.

Note that the operators $\Diamond_F$ and $\Box_F$ on temporal region terms are dual in the sense that for every assignment $\mathfrak{a}$, every region term t, and every moment w we have

$$\mathfrak{a}(\Diamond_F t, w) = \mathfrak{a}(\neg\Box_F\neg t, w).$$

Indeed, suppose that $\mathfrak{a}(t, v) = \mathbb{CI}X_v$ for $v > w$. Using the duality of $\mathbb{C}$, $\mathbb{I}$ and $\cup$, $\cap$, it is easy to see that the equality above is equivalent to the following one

$$\mathbb{CI}\bigcup_{v>w}\mathbb{CI}X_v = \mathbb{CIC}\bigcup_{v>w}\mathbb{ICI}X_v$$

which holds in any topological space.

The inclusion $\supseteq$ follows from

$$\mathbb{CIC}\bigcup_{v>w}\mathbb{ICI}X_v = \mathbb{CICI}\bigcup_{v>w}\mathbb{ICI}X_v = \mathbb{CI}\bigcup_{v>w}\mathbb{ICI}X_v.$$

To show $\subseteq$, it suffices to observe that for every $v > w$, we have $\mathbb{CI}X_v \subseteq \mathbb{CICI}X_v$, from which $\mathbb{CI}X_v \subseteq \mathbb{C}\bigcup_{v>w}\mathbb{ICI}X_v$, whence $\mathbb{I}\bigcup_{v>w}\mathbb{CI}X_v \subseteq \mathbb{C}\bigcup_{v>w}\mathbb{ICI}X_v$, and so $\mathbb{CI}\bigcup_{v>w}\mathbb{CI}X_v \subseteq \mathbb{C}\bigcup_{v>w}\mathbb{ICI}X_v = \mathbb{CI}\bigcup_{v>w}\mathbb{ICI}X_v$ because $\mathbb{C}X$ is the smallest closed set containing X and every union of open sets is open.

Further, $\Diamond_F$ and $\bigcirc$ can be defined via $\mathcal{U}$ as usual:

$$\Diamond_F t = \top\mathcal{U}t, \quad \bigcirc t = \bot\mathcal{U}t.$$

(So $\mathcal{ST}_2$ is in fact an extension of $\mathcal{ST}_1$.)

It is also worth noting that in the definition above we have to use the prefix $\mathbb{CI}$ in the right-hand parts because infinite unions and intersections of regular closed sets are not necessarily regular closed, while all temporal region terms are supposed to be interpreted by 'regions' of topological spaces. For example, an infinite union of closed intervals in $\mathbb{R}$ can be open and an infinite intersection of closed intervals can be just a single point, the regular closure of which is empty:

$$\bigcup_{n=1}^{\infty}[1/n, 1-1/n] = (0,1), \qquad \bigcap_{n=1}^{\infty}[-1/n, 1/n] = \{0\}.$$

Actually, as we shall see below, infinite operations bring various semantical complications. To avoid this problem, we can try to restrict assignments in models in such a way that infinite intersections and unions can be reduced to finite ones. There are different ways of doing this.

One idea would be to accept the *Finite Change Assumption*:

FCA *No region can change its spatial configuration infinitely often.*

This means that under **FCA** we consider only those tt-models $\mathfrak{M} = \langle \mathfrak{F}, \mathfrak{T}, \mathfrak{a} \rangle$ that satisfy the following condition: for every temporal region term t there are pairwise disjoint convex sets $I_1, \dots, I_n$ of points in $\mathfrak{F} = \langle W, < \rangle$ such that

$$W = I_1 \cup \cdots \cup I_n$$

and the state of t remains constant on each I_j (i.e., $\mathfrak{a}(t, u) = \mathfrak{a}(t, v)$ for all $u, v \in I_j$). Note that for the flow $\mathfrak{F} = \langle \mathbb{N}, < \rangle$ **FCA** can be captured by the $\mathcal{ST}_2$-formulas $\Diamond_F \Box_F \mathsf{EQ}(t, \bigcirc t)$.

Of course, **FCA** excludes some mathematically interesting cases. Yet, it is absolutely adequate for many applications,[3] for example, when we are planning a job which eventually must be completed (consider a robot painting a wall). Optimists would accept **FCA** to describe the geography of Europe in the examples above. In temporal databases the time line is often assumed to be finite, though arbitrarily long, which corresponds to **FCA**.

Another, more general, way of reducing infinite unions and intersections to finite ones is to adopt the *Finite State Assumption*:

FSA *Every region can have only finitely many possible states (although it may change its states infinitely often).*

Say that a tt-model $\mathfrak{M} = \langle \mathfrak{F}, \mathfrak{T}, \mathfrak{a} \rangle$ satisfies **FSA**, or is an **FSA**-*model*, if for every temporal region term t there are finitely many regular closed sets $A_1, \dots, A_m \subseteq U$ such that $\{\mathfrak{a}(t, w) \mid w \in W\} = \{A_1, \dots, A_m\}$.

Example 3.4. We illustrate possible applications of the language introduced above by showing a toy spatio-temporal knowledge base. Consider the following scenario of how the foot and mouth epidemic spreads across a country. Assume that the country consists of disjoint regions: farms, towns, forests, rivers, etc. The map of the country can clearly be represented as a database of $\mathcal{RCC}$-8 formulas. Besides, we require that all these regions are rigid, i.e., $\Box_F^+ \mathsf{EQ}(X, \bigcirc X)$ (as quantification over regions is not allowed, we have to write such formulas for all regions X on the map). Now, suppose that at moment 0 foot and mouth has been detected only at one farm X_0:

$$\mathsf{EQ}(F\&M, X_0) \wedge \mathsf{P}(X_0, \mathit{Farm}).$$

[3]'What has been is what will be and what has been done is what will be done; there is nothing new under the sun.' (Ecclesiastes)

The region $F\&M$, representing the current contaminated part of the country, is not rigid. Nor is the region *Stock* representing the farms with live-stock. Let $X_0, \dots, X_n$ be all the farms in the country. We then should clearly have, for all $i \leq n$:

$$\Box_F^+(\mathsf{O}(X_i, \mathit{Stock}) \to \mathsf{P}(X_i, \mathit{Stock})).$$

$$\Box_F^+\mathsf{P}(\mathit{Stock}, X_0 \sqcup \dots \sqcup X_n).$$

$$\Box_F^+((\mathsf{O}(X_i, F\&M) \to \mathsf{P}(X_i, F\&M)).$$

$$\Box_F^+\mathsf{P}(F\&M, \mathit{Stock}).$$

Suppose also that if one farm suffers from foot and mouth, then at the next moment the disease will spread to all neighboring farms with stock, but not further, i.e., for all $i, j \leq n$,

$$\Box_F^+(\mathsf{P}(X_i, F\&M) \wedge \mathsf{EC}(X_i, X_j) \wedge \mathsf{P}(X_j, \mathit{Stock}) \to \bigcirc\mathsf{P}(X_j, F\&M)).$$

$$\Box_F^+(\neg\mathsf{EC}(X_i, F\&M) \to \bigcirc\neg\mathsf{P}(X_i, F\&M)).$$

As the government takes proper measures against the disease, in a few moments (say, two for definiteness), a farm with foot and mouth will have no live-stock. On the other hand, the government is going to help the farmers to continue their business, so eventually new stock will be purchased (but nobody knows when):

$$\Box_F^+\Big(\mathsf{P}(X_i, F\&M) \to \bigcirc\bigcirc(\neg\mathsf{O}(X_i, F\&M) \wedge \neg\mathsf{O}(X_i, \mathit{Stock}))\Big).$$

$$\Box_F^+(\mathsf{P}(X_i, \mathit{Stock}) \to \Diamond_F\mathsf{P}(X_i, \mathit{Stock})).$$

Denote the resulting knowledge base by Σ. We can use it to answer queries like 'how much time the government needs to get rid of the disease' or 'when it is safe to buy new animals,' for instance, by checking whether formulas of the form

$$\bigcirc \dots \bigcirc\mathsf{EQ}(F\&M, \bot), \qquad \bigcirc \dots \bigcirc(\neg\Diamond_F\mathsf{P}(X_i, F\&M))$$

are logical consequences of Σ.

It is worth noting that in this example we have a typical mixture of 'a sort of' model checking and deduction: while the map of the country is simulated by taking all $\mathcal{RCC}$-8 relations which hold true between farms, towns, forests, etc., knowledge about regions like $F\&M$ and *Stock* is incomplete, since it depends on the future development. So to decide whether $\Sigma \models \varphi$ holds or not *proper deduction* (or *theorem proving*) is required; cf. (Halpern and Vardi 1991).

Modal formalisms for spatio-temporal reasoning

As we saw in Section 2.6, $\mathcal{BRCC}$-8 can be embedded into the bimodal logic $\mathbf{S4}_u$. Similarly, the constructed temporalizations of $\mathcal{BRCC}$-8 can be translated into the language

$$\mathcal{PST} = \mathcal{ML}_{\mathcal{SU}} \otimes \mathcal{ML}^u,$$

or *propositional spatio-temporal language*, which contains the temporal operators $\mathcal{S}$ and $\mathcal{U}$, and the modal operators of $\mathbf{S4}_u$ (which we denote, to emphasize their topological interpretation, by $\mathbf{I}$, $\mathbf{C}$, and $\boxtimes$, $\diamondsuit$) as well. The intended models of $\mathcal{PST}$, called *topological $\mathcal{PST}$-models*, are triples of the form $\mathfrak{M} = \langle \mathfrak{F}, \mathfrak{T}, \mathfrak{U} \rangle$, in which $\mathfrak{F} = \langle W, < \rangle$ is a flow of time, $\mathfrak{T} = \langle U, \mathbb{I} \rangle$ a topological space, and $\mathfrak{U}$, a *valuation*, is a map associating with every propositional variable p and every $w \in W$ a set $\mathfrak{U}(p, w) \subseteq U$. $\mathfrak{U}$ is then extended to arbitrary $\mathcal{PST}$-formulas in the following way:

- $\mathfrak{U}(\psi \wedge \chi, w) = \mathfrak{U}(\psi, w) \cap \mathfrak{U}(\chi, w)$;
- $\mathfrak{U}(\neg\psi, w) = U - \mathfrak{U}(\psi, w)$;
- $\mathfrak{U}(\boxtimes\psi, w) = U$ if $\mathfrak{U}(\psi, w) = U$, and $\mathfrak{U}(\boxtimes\psi, w) = \emptyset$ otherwise;
- $\mathfrak{U}(\mathbf{I}\psi, w) = \mathbb{I}\mathfrak{U}(\psi, w)$;
- $x \in \mathfrak{U}(\psi\mathcal{U}\chi, w)$ iff there is $v > w$ such that $x \in \mathfrak{U}(\chi, v)$ and $x \in \mathfrak{U}(\psi, u)$ for all u in the interval $w < u < v$;
- $x \in \mathfrak{U}(\psi\mathcal{S}\chi, w)$ iff there is $v < w$ such that $x \in \mathfrak{U}(\chi, v)$ and $x \in \mathfrak{U}(\psi, u)$ for all u in the interval $v < u < w$.

In particular,

$$\mathfrak{U}(\Diamond_F\psi, w) = \bigcup_{v>w} \mathfrak{U}(\psi, v), \qquad \mathfrak{U}(\Box_F\psi, w) = \bigcap_{v>w} \mathfrak{U}(\psi, v),$$

$$\mathfrak{U}(\bigcirc\psi, w) = \begin{cases} \mathfrak{U}(\psi, w'), & \text{if } w' \text{ is an immediate successor of } w \text{ in } \mathfrak{F}, \\ \emptyset, & \text{if } w \text{ has no immediate successor in } \mathfrak{F}. \end{cases}$$

A $\mathcal{PST}$-formula φ is *satisfied in* $\mathfrak{M}$ if $\mathfrak{U}(\varphi, w) \neq \emptyset$, for some $w \in W$. We say that a topological $\mathcal{PST}$-model $\mathfrak{M} = \langle \mathfrak{F}, \mathfrak{T}, \mathfrak{U} \rangle$ *satisfies* **FSA** if for every variable p there are finitely many sets $U_1, \dots, U_n \subseteq U$ such that

$$\{\mathfrak{U}(p, w) \mid w \in W\} = \{U_1, \dots, U_n\}.$$

The following toy example illustrates the expressive power of $\mathcal{PST}$:

$$\Box_F \boxtimes(\neg\mathbf{I}\textit{cockroach} \wedge (\mathbf{C}\textit{cockroach} \leftrightarrow \textit{habitat})),$$

$$\Box_F \boxtimes(\textit{habitat} \to \bigcirc\textit{habitat}),$$

$$\boxtimes\mathbf{C}\Diamond_F \textit{cockroach}.$$

These formulas say that (a) cockroaches form a dense set in their habitat (but for humans they are invisible), (b) the cockroach habitat will never contract, and (c) sooner or later, cockroaches will appear in the neighborhood of every place on Earth.

Let us encode now $\mathcal{ST}_i$-formulas in the language $\mathcal{PST}$ by extending the translation $\cdot^{\bowtie}$ of Section 2.6. For a temporal region term t, define a $\mathcal{PST}$-formula $t^{\bowtie}$ by taking:

$$X_i^{\bowtie} = \mathbf{CI}p_i, \ (X_i \text{ is a region variable}), \qquad (t_1 \sqcap t_2)^{\bowtie} = \mathbf{CI}(t_1^{\bowtie} \wedge t_2^{\bowtie}),$$
$$(t_1 \sqcup t_2)^{\bowtie} = \mathbf{CI}(t_1^{\bowtie} \vee t_2^{\bowtie}), \qquad (t_1 \mathcal{U} t_2)^{\bowtie} = \mathbf{CI}(t_1^{\bowtie} \mathcal{U} t_2^{\bowtie}),$$
$$(\neg t)^{\bowtie} = \mathbf{CI}\neg t^{\bowtie}, \qquad (t_1 \mathcal{S} t_2)^{\bowtie} = \mathbf{CI}(t_1^{\bowtie} \mathcal{S} t_2^{\bowtie}).$$

Note that we also have

$$(\bigcirc t)^{\bowtie} = \mathbf{CI}\bigcirc t^{\bowtie}, \qquad (\Diamond_F t)^{\bowtie} = \mathbf{CI}\Diamond_F t^{\bowtie}, \qquad (\Box_F t)^{\bowtie} = \mathbf{CI}\Box_F t^{\bowtie}.$$

For atomic $\mathcal{ST}_2$-formulas, let

$$\begin{aligned}
(\mathsf{DC}(t_1,t_2))^{\bowtie} &= \neg \diamondplus (t_1^{\bowtie} \wedge t_2^{\bowtie}),\\
(\mathsf{EQ}(t_1,t_2))^{\bowtie} &= \boxplus (t_1^{\bowtie} \leftrightarrow t_2^{\bowtie}),\\
(\mathsf{PO}(t_1,t_2))^{\bowtie} &= \diamondplus(\mathbf{I}t_1^{\bowtie} \wedge \mathbf{I}t_2^{\bowtie}) \wedge \diamondplus(t_1^{\bowtie} \wedge \neg t_2^{\bowtie}) \wedge \diamondplus(\neg t_1^{\bowtie} \wedge t_2^{\bowtie}),\\
(\mathsf{EC}(t_1,t_2))^{\bowtie} &= \diamondplus(t_1^{\bowtie} \wedge t_2^{\bowtie}) \wedge \neg\diamondplus(\mathbf{I}t_1^{\bowtie} \wedge \mathbf{I}t_2^{\bowtie}),\\
(\mathsf{TPP}(t_1,t_2))^{\bowtie} &= \boxplus(\neg t_1^{\bowtie} \vee t_2^{\bowtie}) \wedge \diamondplus(t_1^{\bowtie} \wedge \neg\mathbf{I}t_2^{\bowtie}) \wedge \diamondplus(\neg t_1^{\bowtie} \wedge t_2^{\bowtie}),\\
(\mathsf{NTPP}(t_1,t_2))^{\bowtie} &= \boxplus(\neg t_1^{\bowtie} \vee \mathbf{I}t_2^{\bowtie}) \wedge \diamondplus(\neg t_1^{\bowtie} \wedge t_2^{\bowtie}).
\end{aligned}$$

Suppose now that φ is an arbitrary $\mathcal{ST}_2$-formula. Then $\varphi^{\bowtie}$ denotes the result of replacing all occurrences of atomic formulas $R(t_1,t_2)$ in φ with $(R(t_1,t_2))^{\bowtie}$.

It should be clear from the definition that we have:

Theorem 3.5. *An $\mathcal{ST}_2$-formula φ is satisfiable in a tt-model (with* **FSA***) based on a flow of time $\mathfrak{F}$ iff $\varphi^{\bowtie}$ is satisfiable in a topological $\mathcal{PST}$-model (with* **FSA***) based on $\mathfrak{F}$.*

The two-dimensional character of spatio-temporal logics becomes even more apparent if we interpret spatial formulas of $\mathcal{BRCC}$-8 in rooted Kripke frames $\mathfrak{G} = \langle V, R_{\mathbf{I}}, R_{\forall}\rangle$ for $\mathbf{S4}_u$ (i.e., where $\langle V, R_{\mathbf{I}}\rangle$ is a quasi-order and $R_{\forall}$ is the universal relation on V). Let $\mathfrak{F} = \langle W, <\rangle$ be a flow of time. Then $\mathcal{PST}$-formulas are interpreted in 3-frames of the form

$$\mathfrak{F} \times \mathfrak{G} = \left\langle W \times V, \overline{<}, \overline{R}_{\mathbf{I}}, \overline{R}_{\forall}\right\rangle,$$

where $W \times V$ is the Cartesian product of W and V, i.e., the set of all pairs $\langle w, x\rangle$, for $w \in W$ and $x \in V$, and the relations $\overline{<}$, $\overline{R}_{\mathbf{I}}$ and $\overline{R}_{\forall}$ are defined coordinate-wise: for all $\langle w_1, x_1\rangle$ and $\langle w_2, x_2\rangle$ in $W \times V$,

$$\begin{array}{lll} \langle w_1, x_1\rangle \,\overline{<}\, \langle w_2, x_2\rangle & \text{iff} & w_1 < w_2 \text{ and } x_1 = x_2, \\ \langle w_1, x_1\rangle \,\overline{R}_{\mathbf{I}}\, \langle w_2, x_2\rangle & \text{iff} & w_1 = w_2 \text{ and } x_1 R_{\mathbf{I}} x_2, \\ \langle w_1, x_1\rangle \,\overline{R}_{\forall}\, \langle w_2, x_2\rangle & \text{iff} & w_1 = w_2. \end{array}$$

The temporal operators $\mathcal{S}$ and $\mathcal{U}$ are interpreted by means of the relation $\overline{<}$, while the interior and closure operators of $\mathbf{S4}_u$ (into which we embed $\mathcal{BRCC}$-8) are interpreted by $\overline{R}_{\mathbf{I}}$, and the universal modalities of $\mathbf{S4}_u$ by $\overline{R}_{\forall}$. The frame $\mathfrak{F} \times \mathfrak{G}$ is known as the *product of Kripke frames* $\mathfrak{F} = \langle W, <\rangle$ *and* $\mathfrak{G} = \langle V, R_{\mathbf{I}}, R_{\forall}\rangle$. (Products of frames and the corresponding many-dimensional modal logics are among the main topics of this book; we will introduce them in Section 3.3.)

As we saw in Section 2.6, every Kripke frame $\mathfrak{G}$ for $\mathbf{S4}_u$ gives rise to a topological space $\mathfrak{T}_{\mathfrak{G}}$. Similarly, every Kripke model $\langle \mathfrak{F} \times \mathfrak{G}, \mathfrak{V}\rangle$, where $\mathfrak{F} = \langle W, <\rangle$ is a flow of time and $\mathfrak{G} = \langle V, R_{\mathbf{I}}, R_{\forall}\rangle$ is a rooted $\mathbf{S4}_u$-frame, can be transformed into a topological $\mathcal{PST}$-model $\langle \mathfrak{F}, \mathfrak{T}_{\mathfrak{G}}, \mathfrak{U}\rangle$ in which, for every propositional variable p, every $w \in W$ and every $v \in V$,

$$v \in \mathfrak{U}(p, w) \quad \text{iff} \quad \langle w, v\rangle \in \mathfrak{V}(p).$$

Now it is straightforward to prove the following:

Proposition 3.6. *For every $\mathcal{PST}$-formula φ, if φ is satisfied in the Kripke model $\langle \mathfrak{F} \times \mathfrak{G}, \mathfrak{V}\rangle$, then φ is satisfied in the topological $\mathcal{PST}$-model $\langle \mathfrak{F}, \mathfrak{T}_{\mathfrak{G}}, \mathfrak{U}\rangle$.*

It is worth noting, however, that the sets of $\mathcal{PST}$-formulas satisfiable in the above Kripke models and in topological $\mathcal{PST}$-models turn out to be different. Consider, for example, the formula

$$\Diamond_F \mathbf{C} p \leftrightarrow \mathbf{C} \Diamond_F p.$$

It is clearly valid in every Kripke model based on the product of a flow of time and a rooted $\mathbf{S4}_u$-frame. On the other hand, we can refute this formula in a topological $\mathcal{PST}$-model: it suffices to take $\mathfrak{T} = \langle \mathbb{R}, \mathbb{I}\rangle$ with the standard interior operator on the real line and the flow of time $\mathfrak{F} = \langle \mathbb{N}, <\rangle$, then select a sequence X_n of closed sets such that $\bigcup_{n \in \mathbb{N}} X_n$ is not closed, and put $\mathfrak{U}(p, n) = X_n$. As we shall see in Section 16.2, the two types of models are equivalent with respect to the modal translations of $\mathcal{ST}_2$-formulas under the finite state assumption **FSA**.

$\mathcal{BRCC}$-8 + $\mathcal{A\ell\ell}$-13

We conclude this section by showing how one can design a temporal extension of $\mathcal{BRCC}$-8 based on the interval approach to temporal representation and reasoning. Such a combination may appear to be rather natural because the region-based approach to spatial reasoning closely mirrors the interval-based approach to temporal reasoning—they both take extended entities rather than points as primitives.

Following (Allen 1984) we write $\mathsf{HOLDS}(\varphi, i)$ to say that a formula φ holds during a time interval i. For example, $\mathsf{HOLDS}(\mathsf{PO}(X, Y), i)$ means that during interval i regions X and Y partially overlap. Let us call an $\mathcal{ARCC}$-8 *formula* any Boolean combination of atomic $\mathcal{A}\ell\ell$-13 formulas and formulas of the form $\mathsf{HOLDS}(\varphi, i)$, where φ is a $\mathcal{BRCC}$-8 formula.

$\mathcal{ARCC}$-8 formulas are interpreted in *interval topological models* which are triples of the form $\mathfrak{M} = \langle \mathfrak{F}, \mathfrak{T}, \mathfrak{a} \rangle$, where $\mathfrak{F} = \langle W, < \rangle$ is a strict linear order, $\mathfrak{T} = \langle U, \mathbb{I} \rangle$ a topological space, and assignment $\mathfrak{a}$ associates with every interval variable i a non-empty convex set $\mathfrak{a}(i)$ in $\mathfrak{F}$, and with every region variable X and every moment of time u it associates a regular closed set $\mathfrak{a}(X, u)$ in $\mathfrak{T}$. Now, the truth-relation for the $\mathcal{A}\ell\ell$-13 atomic formulas is defined as in Section 2.2, and $\mathsf{HOLDS}(\varphi, i)$ is true in $\mathfrak{M}$ iff for every point $u \in \mathfrak{a}(i)$ we have $\mathfrak{T} \models^{\mathfrak{a}_u} \varphi$ (as defined in Section 2.6). Here is a simple example of a formula of this unsophisticated language which holds in every interval topological model:

$$
\begin{aligned}
&\mathsf{meets}(i, j) \wedge \mathsf{during}(i, k) \wedge \mathsf{during}(j, k)\\
&\wedge \mathsf{HOLDS}(\mathsf{TPP}(\mathit{Hong\_Kong}, \mathit{UK}) \wedge \mathsf{EC}(\mathit{Hong\_Kong}, \mathit{China}), i)\\
&\wedge \mathsf{HOLDS}(\mathsf{DC}(\mathit{Hong\_Kong}, \mathit{UK}), j)\\
&\wedge \mathsf{HOLDS}(\mathsf{EC}(\mathit{UK}, \mathit{China}) \vee \mathsf{DC}(\mathit{UK}, \mathit{China}), k)\\
&\qquad \rightarrow\ \mathsf{HOLDS}(\mathsf{EC}(\mathit{UK}, \mathit{China}), i).
\end{aligned}
$$

By combining the translations $\cdot^*$ of Section 2.2 and $\cdot^{\bowtie}$ of Section 2.6, it is not hard to embed $\mathcal{ARCC}$-8 into the language $\mathcal{PST}$ interpreted in topological $\mathcal{PST}$-models based on arbitrary flows of time. Moreover, a combination of satisfiability-checking algorithms for $\mathcal{A}\ell\ell$-13 and $\mathcal{BRCC}$-8 yields a satisfiability-checking algorithm for $\mathcal{ARCC}$-8, also showing that the satisfiability problem for $\mathcal{ARCC}$-8 is in NP. We leave details to the reader as an exercise.

3.3 Products

In the previous section we saw how spatio-temporal logics can be interpreted in products of certain frames. The formation of Cartesian products of various structures—vector and topological spaces, algebras, etc.—is a standard mathematical way of capturing the multidimensional character of our world. In modal logic, products of Kripke frames are natural constructions allowing us to reflect interactions between modal operators representing time, space, knowledge, actions, etc. Products of modal logics (i.e., the sets of multimodal formulas valid in products of Kripke frames for those logics) have been studied in both pure modal logic (see, e.g., Segerberg 1973, Shehtman 1978, Gabbay and Shehtman 1998, Marx 1999) and applications in computer science and artificial intelligence (see, e.g., Reif and Sistla 1985, Fagin *et al.* 1995, Baader and Ohlbach 1995, Reynolds 1997, Finger and Reynolds 1999) since the 1970s.

Two-dimensional products

We define the *product* of an n-frame $\mathfrak{F}_1 = \langle W_1, R_1^1, \dots, R_1^n \rangle$ and an m-frame $\mathfrak{F}_2 = \langle W_2, R_2^1, \dots, R_2^m \rangle$ as the $n+m$-frame of the form

$$\mathfrak{F}_1 \times \mathfrak{F}_2 = \langle W_1 \times W_2, R_h^1, \dots, R_h^n, R_v^1, \dots, R_v^m \rangle$$

in which, for all $u_1, u_2 \in W_1$ and $v_1, v_2 \in W_2$,

$$\begin{array}{lcl} \langle u_1, v_1 \rangle R_h^i \langle u_2, v_2 \rangle & \text{iff} & u_1 R_1^i u_2 \text{ and } v_1 = v_2 \quad (1 \le i \le n), \\ \langle u_1, v_1 \rangle R_v^j \langle u_2, v_2 \rangle & \text{iff} & u_1 = u_2 \text{ and } v_1 R_2^j v_2 \quad (1 \le j \le m). \end{array}$$

Such a frame will be called a *product frame*. The subscripts h and v appeal to the geometrical intuition of considering the R_h^i as 'horizontal' accessibility relations in $\mathfrak{F}_1 \times \mathfrak{F}_2$ and the R_v^j as 'vertical' ones; see Fig. 3.2 for an illustration.

Given a class $\mathcal{C}_1$ of n-frames and a class $\mathcal{C}_2$ of m-frames, we define their *product* $\mathcal{C}_1 \times \mathcal{C}_2$ by taking

$$\mathcal{C}_1 \times \mathcal{C}_2 = \{\mathfrak{F}_1 \times \mathfrak{F}_2 \mid \mathfrak{F}_1 \in \mathcal{C}_1, \mathfrak{F}_2 \in \mathcal{C}_2\}.$$

Let L_1 and L_2 be two Kripke complete multimodal logics formulated in languages $\mathcal{L}_1$ and $\mathcal{L}_2$. As in Section 3.1, denote by $\mathcal{L}_1 \otimes \mathcal{L}_2$ the smallest multimodal language containing the language $\mathcal{L}$ of classical propositional logic together with the disjoint union of the modal operators of $\mathcal{L}_1$ and $\mathcal{L}_2$. (For example, if $\mathcal{L}_1 = \mathcal{ML}_n$ and $\mathcal{L}_2 = \mathcal{ML}_m$ then $\mathcal{L}_1 \otimes \mathcal{L}_2 = \mathcal{ML}_{n+m}$.) We define the *product of* L_1 *and* L_2 as the multimodal logic

$$L_1 \times L_2 = \mathsf{Log}(\mathsf{Fr}L_1 \times \mathsf{Fr}L_2)$$

in the language $\mathcal{L}_1 \otimes \mathcal{L}_2$. In other words, $L_1 \times L_2$ is the set of $\mathcal{L}_1 \otimes \mathcal{L}_2$-formulas that are valid in all product frames $\mathfrak{F}_1 \times \mathfrak{F}_2$, where $\mathfrak{F}_1$ is a frame for L_1 and $\mathfrak{F}_2$ a frame for L_2. For example, $\mathbf{K}_n \times \mathbf{K}_m$ is the $n+m$-modal logic determined by all product frames $\mathfrak{F}_1 \times \mathfrak{F}_2$, where $\mathfrak{F}_1$ is an n-frame and $\mathfrak{F}_2$ an m-frame; $\mathbf{S4} \times \mathbf{S5}$ is the bimodal logic determined by all product frames $\mathfrak{F}_1 \times \mathfrak{F}_2$ such that $\mathfrak{F}_1 \models \mathbf{S4}$ and $\mathfrak{F}_2 \models \mathbf{S5}$.

It is worth emphasizing that in the definition of $L_1 \times L_2$ we take the classes of *all* Kripke frames for L_1 and L_2. The reason is that the equalities $\mathsf{Log}\mathcal{C}_1 = \mathsf{Log}\mathcal{C}_1'$ and $\mathsf{Log}\mathcal{C}_2 = \mathsf{Log}\mathcal{C}_2'$ do not necessarily imply that

$$\mathsf{Log}(\mathcal{C}_1 \times \mathcal{C}_2) = \mathsf{Log}(\mathcal{C}_1' \times \mathcal{C}_2')$$

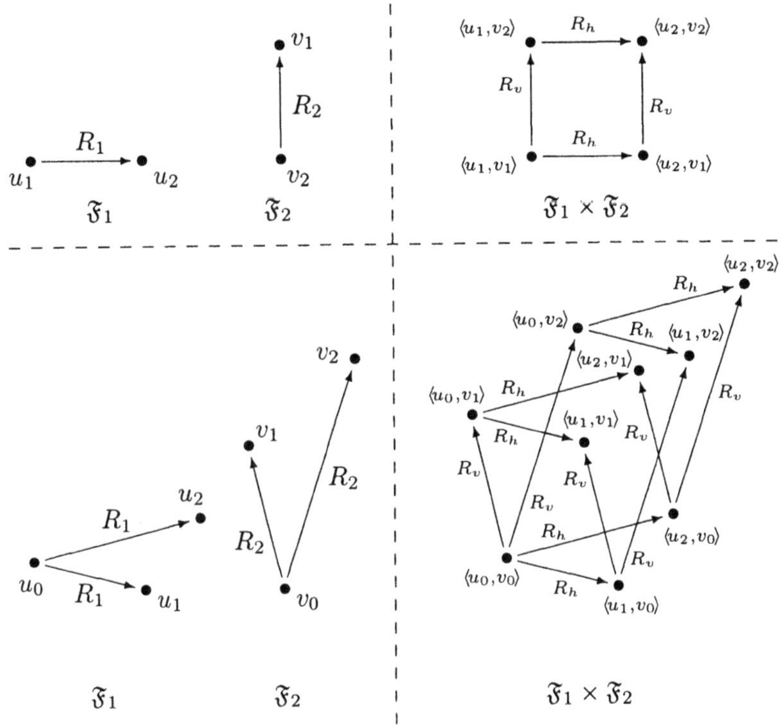

Figure 3.2: Product frames.

(an example will be given in Theorem 7.11 of Section 7.2). Actually, instead of the classes of all frames for L_1 and L_2 in this definition we can take the classes $\mathsf{Fr}^r L_1$ and $\mathsf{Fr}^r L_2$ of *rooted* frames for L_1 and L_2. Indeed, the inclusion

$$L_1 \times L_2 \subseteq \mathsf{Log}(\mathsf{Fr}^r L_1 \times \mathsf{Fr}^r L_2)$$

is clear. To show the converse, suppose $\varphi \notin L_1 \times L_2$, i.e., φ is refuted at a point $\langle u, v\rangle$ in some $\mathfrak{F}_1 \times \mathfrak{F}_2 \in \mathsf{Fr}L_1 \times \mathsf{Fr}L_2$ under some valuation. Let $\mathfrak{G}_1$ and $\mathfrak{G}_2$ be the subframes of $\mathfrak{F}_1$ and $\mathfrak{F}_2$ generated by u and v, respectively. Then by Theorem 1.13 $\mathfrak{G}_i \models L_i$, for $i = 1, 2$. On the other hand, it is readily checked that $\mathfrak{G}_1 \times \mathfrak{G}_2$ is isomorphic to the subframe of $\mathfrak{F}_1 \times \mathfrak{F}_2$ generated by $\langle u, v\rangle$. It follows that $\varphi \notin \mathsf{Log}(\mathsf{Fr}^r L_1 \times \mathsf{Fr}^r L_2)$. Thus we obtain the following:

Proposition 3.7. *For all Kripke complete modal logics L_1 and L_2,*

$$L_1 \times L_2 = \mathsf{Log}(\mathsf{Fr}^r L_1 \times \mathsf{Fr}^r L_2).$$

Products of logics always contain their fusions. Indeed, given a product frame

$$\mathfrak{F}_1 \times \mathfrak{F}_2 = \langle W_1 \times W_2, R_h^1, \ldots, R_h^n, R_v^1, \ldots, R_v^m \rangle$$

and points $x \in W_1$, $y \in W_2$, we define

$$\begin{aligned} W_1^y &= \{\langle w, y\rangle \mid w \in W_1\}, \\ W_2^x &= \{\langle x, v\rangle \mid v \in W_2\}, \\ S_h^{i,y} &= R_h^i \cap (W_1^y \times W_1^y) \quad (1 \le i \le n), \\ S_v^{j,x} &= R_v^j \cap (W_2^x \times W_2^x) \quad (1 \le j \le m), \end{aligned}$$

and the 'coordinate-wise' frames

$$\mathfrak{F}_1^y = \left\langle W_1^y, S_h^{1,y}, \ldots, S_h^{n,y} \right\rangle, \qquad \mathfrak{F}_2^x = \left\langle W_2^x, S_v^{1,x}, \ldots, S_v^{m,x} \right\rangle.$$

Then for all $x \in W_1$, $y \in W_2$, the frames $\mathfrak{F}_1^y$ and $\mathfrak{F}_2^x$ are isomorphic to $\mathfrak{F}_1$ and $\mathfrak{F}_2$, respectively, and

$$\langle W_1 \times W_2, R_h^1, \ldots, R_h^n \rangle = \sum_{y \in W_2} \mathfrak{F}_1^y, \quad \langle W_1 \times W_2, R_v^1, \ldots, R_v^m \rangle = \sum_{x \in W_1} \mathfrak{F}_2^x.$$

Now suppose that $\mathfrak{F}_i \models L_i$ $(i = 1, 2)$. Then, by Theorem 1.13, $\mathfrak{F}_1 \times \mathfrak{F}_2$ is a frame for the fusion $L_1 \otimes L_2$ of L_1 and L_2. Thus we have proved the following:

Proposition 3.8. *For all Kripke complete modal logics L_1 and L_2,*

$$L_1 \otimes L_2 \subseteq L_1 \times L_2.$$

As we shall see in Section 5.1, this inclusion is proper: product logics always include certain *interactions* between the modal operators of their components. Note, however, that the modal operators within each component are not affected by these interactions. More precisely, we have:

Proposition 3.9. *For any two consistent Kripke complete modal logics L_1 and L_2, their product $L_1 \times L_2$ is a conservative extension of both L_1 and L_2.*

Proof. Let φ be a formula in the language of L_1 such that $\varphi \notin L_1$. Then $\mathfrak{F}_1 \not\models \varphi$ for some $\mathfrak{F}_1 \models L_1$. Take any frame $\mathfrak{F}_2$ for L_2. It should be clear that $\mathfrak{F}_1 \times \mathfrak{F}_2 \not\models \varphi$, and so $\varphi \notin L_1 \times L_2$. ❑

The following simple result showing that the product construction commutes with the three basic operations on frames (see Section 1.4) will be often used in Part II. We leave the proof to the reader as an exercise.

Proposition 3.10. *For all frames $\mathfrak{F}$, $\mathfrak{G}$, $\mathfrak{H}$, $\mathfrak{H}_i$, $i \in I$, the following hold:*

(i) *If $\mathfrak{F}$ is a p-morphic image of $\mathfrak{H}$, then $\mathfrak{F} \times \mathfrak{G}$ is a p-morphic image of $\mathfrak{H} \times \mathfrak{G}$.*

(ii) *If $\mathfrak{F}$ is a generated subframe of $\mathfrak{H}$, then $\mathfrak{F} \times \mathfrak{G}$ is a generated subframe of $\mathfrak{H} \times \mathfrak{G}$.*

(iii) *If $\mathfrak{F}$ is a disjoint union of $\mathfrak{H}_i$, $i \in I$, then $\mathfrak{F} \times \mathfrak{G}$ is isomorphic to the disjoint union of $\mathfrak{H}_i \times \mathfrak{G}$, $i \in I$.*

Similarly to products of logics, one can also define *products of consequence relations.* Given Kripke complete modal logics L_1 and L_2 formulated in languages $\mathcal{L}_1$ and $\mathcal{L}_2$, respectively, define the consequence relation $\vdash^*_{L_1} \times \vdash^*_{L_2}$ between formulas in the language $\mathcal{L}_1 \otimes \mathcal{L}_2$ by taking

$$\varphi(\vdash^*_{L_1} \times \vdash^*_{L_2})\psi \quad \text{iff} \quad \text{for all models } \mathfrak{M} \text{ based on a frame in } \mathsf{Fr}L_1 \times \mathsf{Fr}L_2, \quad \mathfrak{M} \models \psi \text{ whenever } \mathfrak{M} \models \varphi.$$

A natural question arises then as to how $\vdash^*_{L_1} \times \vdash^*_{L_2}$ relates to the global consequence relation $\vdash^*_{L_1 \times L_2}$. Clearly, if $L_1 \times L_2$ is globally Kripke complete then $\vdash^*_{L_1} \times \vdash^*_{L_2}$ always contains $\vdash^*_{L_1 \times L_2}$. In fact, as we shall see in Theorem 5.12, in many cases they coincide.

The reader should have no difficulties with defining products of logics in the languages $\mathcal{ML}_\mathcal{U}$ and $\mathcal{ML}_{\mathcal{SU}}$ (say, **PTL** $\times$ **PTL**, **PTL** $\times$ **S5**, $\mathsf{Log}_{\mathcal{SU}}(\mathbb{N}) \times \mathbf{S4}_2$) by extending the definitions above in a straightforward way.

Higher-dimensional products

In principle, there are two ways of defining products of three or more modal logics. First, we can generalize in a straightforward way the definitions of the previous subsection. (To simplify notation, we consider here products of unimodal logics.) The *product* $\mathfrak{F}_1 \times \cdots \times \mathfrak{F}_n$ of frames $\mathfrak{F}_i = \langle W_i, R_i \rangle$, $i = 1, \dots, n$, is the n-frame

$$\langle W_1 \times \cdots \times W_n, \bar{R}_1, \dots, \bar{R}_n \rangle$$

where, for each $i = 1, \dots, n$, $\bar{R}_i$ is a binary relation on $W_1 \times \cdots \times W_n$ such that

$$\langle u_1, \dots, u_n \rangle \, \bar{R}_i \, \langle v_1, \dots, v_n \rangle \quad \text{iff} \quad u_i R_i v_i \quad \text{and} \quad u_k = v_k, \quad \text{for } k \neq i.$$

Then, given Kripke complete (unimodal) logics L_i ($i = 1, \dots, n$), we define the *product logic* $L_1 \times \cdots \times L_n$ as the set of all n-modal formulas that are valid in all product frames $\mathfrak{F}_1 \times \cdots \times \mathfrak{F}_n$ such that $\mathfrak{F}_i \models L_i$ for every $i = 1, \dots, n$. For example, $\mathbf{K}^n$ is the logic determined by all n-dimensional product frames; $\mathbf{S5}^n$ is the logic determined by all product frames $\mathfrak{F}_1 \times \cdots \times \mathfrak{F}_n$, where $\mathfrak{F}_i \models \mathbf{S5}$ for each $i = 1, \dots, n$.

The second way would be to define $L_1 \times \cdots \times L_n$ as

$$(((L_1 \times L_2) \times L_3) \times \cdots \times L_{n-1}) \times L_n.$$

The easily established fact that the frame $\mathfrak{F}_1 \times \cdots \times \mathfrak{F}_n$ is isomorphic to

$$(((\mathfrak{F}_1 \times \mathfrak{F}_2) \times \mathfrak{F}_3) \times \cdots \times \mathfrak{F}_{n-1}) \times \mathfrak{F}_n$$

might seem to suggest that the two definitions are equivalent. However, the situation is not that simple. For example, it is an open question (asked by V. Shehtman) whether the equalities

$$\mathbf{K}^4 = \mathbf{K}^3 \times \mathbf{K} \quad \text{and} \quad \mathbf{S5}^4 = \mathbf{S5}^3 \times \mathbf{S5}$$

hold. The problem here is that $\mathbf{K}^4$ is characterized by the class of products of four 1-frames, while $\mathbf{K}^3 \times \mathbf{K}$ by the class of products of *arbitrary* 3-frames for $\mathbf{K}^3$ and 1-frames for $\mathbf{K}$. Now, the thing is that these arbitrary $\mathbf{K}^3$-frames are not necessarily isomorphic to product frames (in fact, we do not even know what they look like; see Theorem 8.29).

For this reason, we take as the only 'official' definition of $L_1 \times \cdots \times L_n$ the equality

$$L_1 \times \cdots \times L_n = \mathsf{Log}(\mathsf{Fr}L_1 \times \cdots \times \mathsf{Fr}L_n).$$

Note, however, that in Section 5.1 we provide a characterization of arbitrary (countable) frames for $\mathbf{K} \times \mathbf{K}$ and $\mathbf{S5} \times \mathbf{S5}$ (among many other 2D logics), and prove—with the help of this characterization—that for many three-dimensional products the two definitions coincide. For instance,

$$\begin{aligned} \mathbf{K}^3 &= (\mathbf{K} \times \mathbf{K}) \times \mathbf{K}, \\ \mathbf{S5}^3 &= (\mathbf{S5} \times \mathbf{S5}) \times \mathbf{S5} \end{aligned}$$

(see Corollary 5.11).

Similarly to Proposition 3.7, we have:

Proposition 3.11. *For all Kripke complete modal logics $L_1, \ldots, L_n$,*

$$L_1 \times \cdots \times L_n = \mathsf{Log}(\mathsf{Fr}^r L_1 \times \cdots \times \mathsf{Fr}^r L_n).$$

In particular, $\mathbf{S5}^n$ is determined by products of frames $\langle W_i, R_i \rangle$ where $R_i = W_i \times W_i$ is the universal relation on W_i, for every $i = 1, \ldots, n$. Product frames of this kind will be called *universal product* $\mathbf{S5}^n$*-frames*. We denote such a frame by $\langle W_1, \ldots, W_n \rangle$ and sometimes call it the *universal product frame on* $W_1 \times \cdots \times W_n$. It is to be noted that each universal product frame $\langle W_1, \ldots, W_n \rangle$ is a p-morphic image of a *cubic universal product frame*, i.e., a frame of the form $\langle W, \ldots, W \rangle$. Indeed, it is easy to see that if a set W is such that there are surjections $f_i : W \to W_i$, for $i = 1, \dots, n$, then the map f defined by

$$f(w_1, \dots, w_n) = \langle f_1(w_1), \dots, f_n(w_n) \rangle$$

is a p-morphism from the frame $\langle W, \dots, W \rangle$ onto $\langle W_1, \dots, W_n \rangle$. Such a set and surjections can be found, for example, by taking the disjoint union of the W_i as W and defining f_i so that it is the identity map on W_i and arbitrary otherwise. Thus we obtain:

Proposition 3.12. $\mathbf{S5}^n$ *is determined by the cubic universal product frames.*

Observe that the n-dimensional analogs of Propositions 3.8 and 3.9 hold:

Proposition 3.13. *For all Kripke complete modal logics* $L_1, \dots, L_n$,

$$L_1 \otimes L_2 \otimes \dots \otimes L_n \subseteq L_1 \times L_2 \times \dots \times L_n.$$

Proposition 3.14. *The product* $L_1 \times \dots \times L_n$ *of consistent Kripke complete logics* $L_1, \dots, L_n$ *is a conservative extension of each of them.*

Moreover, we also have:

Proposition 3.15. *Let* $L_1, \dots, L_n, L_{n+1}$ *be consistent Kripke complete unimodal logics. Then the logic* $L_1 \times \dots \times L_n \times L_{n+1}$ *is a conservative extension of* $L_1 \times \dots \times L_n$, *i.e., for every* $\mathcal{ML}_n$*-formula* φ,

$$\varphi \in L_1 \times \dots \times L_n \quad \textit{iff} \quad \varphi \in L_1 \times \dots \times L_n \times L_{n+1}.$$

Proof. We prove this only for $L_i = L$, $i = 1, \dots, n, n+1$; the general case is considered in a similar way. First, it is readily checked that for any $n+1$-dimensional product frame

$$\mathfrak{F} = \left\langle W_1 \times \dots \times W_n \times W_{n+1}, \bar{R}_1, \dots, \bar{R}_n, \bar{R}_{n+1} \right\rangle,$$

the projection map

$$f(w_1, \dots, w_n, w_{n+1}) = \langle w_1, \dots, w_n \rangle$$

is a p-morphism from the 'n-reduct'

$$\mathfrak{F}_{(n)} = \left\langle W_1 \times \dots \times W_n \times W_{n+1}, \bar{R}_1, \dots, \bar{R}_n \right\rangle$$

of $\mathfrak{F}$ onto the n-dimensional product frame

$$\mathfrak{F}^- = \left\langle W_1 \times \dots \times W_n, \bar{R}_1, \dots, \bar{R}_n \right\rangle.$$

Now suppose that $\varphi \in L^{n+1}$ and $\mathfrak{G}$ is an n-dimensional product frame for L^n. As L is consistent and Kripke complete, there exists a frame $\mathfrak{H}$ for L. Then the product $\mathfrak{F} = \mathfrak{G} \times \mathfrak{H}$ is a frame for L^{n+1}, and so $\mathfrak{F} \models \varphi$. Since $\mathfrak{F}^- = \mathfrak{G}$, by the p-morphism theorem we finally obtain $\mathfrak{G} \models \varphi$.

Conversely, suppose that $\varphi \in L^n$, and let $\mathfrak{F}$ be an $n+1$-dimensional product frame for L^{n+1}. Then clearly $\mathfrak{F}^-$ is a frame for L^n, and so $\mathfrak{F} \models \varphi$. ❑

Product logics were defined as sets of modal formulas that are valid in classes of product frames. It is important to stress that in general there are frames for product logics which are *not* product frames. Thus, in the case of product logics, it is meaningful to speak not only about the finite model property, but also about *product finite model property*: a product logic L has the *product fmp* if L is characterized by the class of its finite product frames. Note that by Proposition 3.13, for every product frame $\mathfrak{F} = \mathfrak{F}_1 \times \cdots \times \mathfrak{F}_n$ and product logic $L = L_1 \times \cdots \times L_n$,

$$\mathfrak{F} \models L \quad \text{iff} \quad \mathfrak{F}_i \models L_i, \text{ for all } 1 \leq i \leq n.$$

Obviously, the product fmp implies the fmp. However, the converse does not hold: we shall see a number of counterexamples in Section 8.4.

We can enumerate the formulas that are not in a product logic L (and thereby obtain a decision algorithm for L whenever L is recursively enumerable) if

- L has the product fmp, and
- finite product frames for L are recursively enumerable (up to isomorphism).

The latter property clearly holds if L is a product of finitely axiomatizable Kripke complete logics such as **K**, **K4**, **S5**, etc. However, not so many product logics enjoy the product fmp.

We can say much more about *countable* product frames:

Theorem 3.16. *Let L_i be a Kripke complete unimodal logic such that $\mathsf{Fr}L_i$ is first-order definable in the language having equality and a binary predicate symbol R_i, for each $i = 1, \dots, n$. Then $L_1 \times \cdots \times L_n$ is determined by the class of its countable product frames.*

Proof. For each i, let Γ_i denote the first-order theory defining $\mathsf{Fr}L_i$. Extend our first-order language having equality and $R_1, \dots, R_n$ with n unary function symbols $f_1, \dots, f_n$. For each $\phi \in \Gamma_i$, denote by ϕ' the formula obtained by substituting $f_i(x)$ for all occurrences of each variable x in ϕ ($i = 1, \dots, n$). Let

$$\Sigma = \{\phi' \mid \phi \in \Gamma_i,\ i = 1, \dots, n\} \cup \{\pi\},$$

where π is the following sentence:

$$\begin{aligned} &\forall x \forall y\ \big(f_1(x) = f_1(y) \land \cdots \land f_n(x) = f_n(y) \ \rightarrow\ x = y\big) \land \\ &\qquad \forall x_1 \dots \forall x_n \exists y\ \big(f_1(y) = x_1 \land \cdots \land f_n(y) = x_n\big) \land \\ &\qquad \bigwedge_{i=1}^{n} \forall x \forall y \Big(x R_i y \ \leftrightarrow\ \big(f_i(x) R_i f_i(y) \land \bigwedge_{\substack{j=1 \\ j \neq i}}^{n} f_j(x) = f_j(y)\big)\Big). \end{aligned}$$

Now suppose that $\varphi \notin L_1 \times \cdots \times L_n$, for some $\mathcal{ML}_n$-formula φ. Then φ is not true in a model $\mathfrak{M} = \langle \mathfrak{F}, \mathfrak{V} \rangle$ based on the product $\mathfrak{F}_1 \times \cdots \times \mathfrak{F}_n$ of frames $\mathfrak{F}_i = \langle W_i, S_i \rangle$ such that $\mathfrak{F}_i \models \Gamma_i$ for $i = 1, \ldots, n$. Take the first-order language having equality and $R_1, \ldots, R_n, f_1, \ldots, f_n$ as above and also countably many unary predicate symbols $P_0, P_1, \ldots$. Define a first-order structure I of this language by taking

$$I = \langle W_1 \times \cdots \times W_n, \bar{S}_1, \ldots, \bar{S}_n, pr_1, \ldots, pr_n, \mathfrak{V}(p_0), \mathfrak{V}(p_1), \ldots \rangle ,$$

where $pr_i : W_1 \times \cdots \times W_n \to W_i$ are the projection functions. It is readily checked that $I \models \Sigma$. Since without the projections I is nothing but the modal model $\mathfrak{M}$ considered as a first-order structure (see Section 1.3), we also have $I \not\models \forall x \varphi^\star(x)$ (where $\varphi^\star$ is the standard translation of φ). In other words, $\Sigma' = \Sigma \cup \{\exists x \neg \varphi^\star(x)\}$ is true I. By the downward Löwenheim–Skolem–Tarski theorem, there is a countable first-order structure

$$J = \langle W, R_1^J, \ldots, R_n^J, f_1^J, \ldots, f_n^J, P_0^J, P_1^J, \ldots \rangle$$

such that $J \models \Sigma'$. For each $i = 1, \ldots, n$, define

$$\begin{aligned} U_i &= \{f_i^J(w) \mid w \in W\}, \\ Q_i &= R_i^J \cap (U_i \times U_i), \end{aligned}$$

and for each $j < \omega$,

$$P_j^{I'} = \{\langle f_1^J(w), \ldots, f_n^J(w) \rangle \mid w \in P_j^J\}.$$

Since $J \models \pi$, the map $h(w) = \langle f_1^J(w), \ldots, f_n^J(w) \rangle$ is an isomorphism between J and the first-order structure

$$I' = \left\langle U_1 \times \cdots \times U_n, \bar{Q}_1, \ldots, \bar{Q}_n, pr_1, \ldots, pr_n, P_0^{I'}, P_1^{I'}, \ldots \right\rangle .$$

Thus, $I' \models \Sigma$ and $I' \not\models \forall x \varphi^\star(x)$. Let $\mathfrak{G}_i = \langle U_i, Q_i \rangle$, $i = 1, \ldots, n$. Define a valuation $\mathfrak{W}$ in the (countable) product frame

$$\mathfrak{G} = \mathfrak{G}_1 \times \cdots \times \mathfrak{G}_n$$

by taking $\mathfrak{W}(p_j) = P_j^{I'}$ for $j < \omega$. Then I' without the projections is just the modal model $\mathfrak{N} = \langle \mathfrak{G}, \mathfrak{W} \rangle$ considered as a first-order structure, and so φ is not true in $\mathfrak{N}$.

Note that in fact we have also proved that

$$\varphi \in L_1 \times \cdots \times L_n \quad \text{iff} \quad \Sigma \models \forall x \varphi^\star(x), \tag{3.1}$$

for any $\mathcal{ML}_n$-formula φ. ❑

In many cases the product construction preserves recursive enumerability of the components:

Theorem 3.17. *Let L_i be a Kripke complete unimodal logic such that $\mathsf{Fr}L_i$ is definable by a recursive set of first-order sentences in the language having equality and a binary predicate symbol R_i, for each $i = 1, \dots, n$. Then the product logic $L_1 \times \dots \times L_n$ is recursively enumerable.*

Proof. We use the notation of the proof of Theorem 3.16. Since now the sets Γ_i are recursive, Σ is recursive as well. And since the consequence relation of first-order logic **QCl** is recursively enumerable, it follows from (3.1) that $L_1 \times \dots \times L_n$ is recursively enumerable. ❑

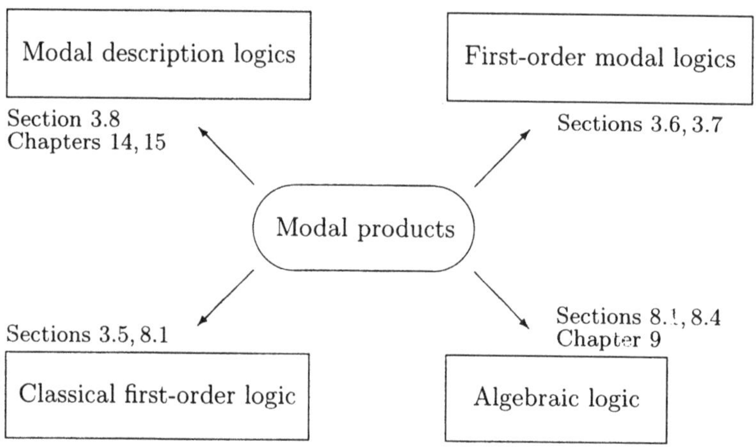

Figure 3.3: Products and other many-dimensional formalisms.

Besides their obvious connection to fusions, products of modal logics are related to other many-dimensional formalisms considered in this book. We saw in Section 3.2 how they show up in spatio-temporal representation and reasoning. In the next section we shall see a family of temporal epistemic logics ranging from fusions to products. Fig. 3.3 indicates some other connections which will be discussed later on in the book. Product logics themselves will be investigated in detail in Chapters 5–8.

3.4 Temporal epistemic logics

A large family of combined modal logics has been constructed with the aim of formalizing the behavior of multi-agent systems; see, e.g., (Ladner and

Reif 1986, Lehmann 1984, Parikh and Ramanujam 1985, Sato 1977). In this section we briefly discuss the approach proposed by Fagin *et al.* (1995), which gave rise to various combinations of propositional temporal and epistemic logics ranging from fusions to products of these logics.

Consider a certain system $\mathfrak{S}$ about which we know only that the state of $\mathfrak{S}$ at each moment of time belongs to some set S of states. Suppose further that the flow of time is $\mathfrak{F} = \langle T, < \rangle$. Then every possible *evolution* of $\mathfrak{S}$ over $\mathfrak{F}$ can be represented by means of a function f associating with each moment $t \in T$ the state $f(t) \in S$ of $\mathfrak{S}$ at t. Such an f will be called a *run over* $\mathfrak{F}$. Thus, the collection of all possible runs over $\mathfrak{F}$ is the set of all functions from T to S.

Example 3.18. We illustrate these concepts using the 'multi-agent system' of three wise men from the 'wise men puzzle' analyzed in Section 2.3. The 'agents' of the system are the three wise men, denoted by A, B and C. Each of them wears either a red or a white hat. Thus, for each $D \in \{A, B, C\}$ we can define the set of (relevant) local states S_D of D as

$$S_D = \{r, w\}.$$

The meaning of 'D is in state r' or 'D is in state w' is 'D's hat is red' or 'D's hat is white,' respectively. The set S of states of the whole multi-agent system is then the Cartesian product

$$S = S_A \times S_B \times S_C.$$

A run f in this example is a function which associates with every moment of time t the distribution of the red and white hats among the wise men at t. As the king does not change the location of the hats, we may assume that each run in the wise men puzzle is a constant function associating with every $t \in T$ the same triple $\langle c_1, c_2, c_3 \rangle$ of colors. We will come back to this example later on in this section.

Assume now that the states $s \in S$ come equipped with the set of classical propositional (i.e., nontemporal) formulas that are true in s. In other words, assume that there is a valuation $\mathfrak{V}$ which associates with every propositional variable p the set of states $\mathfrak{V}(p) \subseteq S$ in which p is true. Now, for each run f over $\mathfrak{F}$, we can define a valuation $\mathfrak{U}_f$ over $\mathfrak{F}$ by taking

$$\mathfrak{U}_f(p) = \{t \in T \mid f(t) \in \mathfrak{V}(p)\}$$

for each propositional variable p. Then, for every $\mathcal{ML}_{\mathcal{SU}}$-formula φ, every moment $t \in T$, and every run f over $\mathfrak{F}$, we can define the truth-relation $\langle t, f \rangle \models \varphi$:

$$\langle t, f \rangle \models \varphi \quad \text{iff} \quad \langle \mathfrak{F}, \mathfrak{U}_f, t \rangle \models \varphi.$$

This formalism is nothing else but a special representation of the temporal logics discussed in Section 2.1.

Example 3.18 (cont.) From now on, we assume that the flow of time T consists of the natural numbers $\mathbb{N}$, i.e., $\mathfrak{F} = \langle \mathbb{N}, < \rangle$. At moment 0 the wise men do not answer questions: they observe the hats of each other. The first round of answers to the king's question happens at moment 1, the second round at moment 2, and so on.

Assume that the respective colors of wise men A, B and C are h_1, h_2 and h_3. Take a propositional variable p which intends to mean this. In other words, we have a valuation $\mathfrak{V}$ in $S_A \times S_B \times S_C$ with

$$\mathfrak{V}(p) = \{\langle h_1, h_2, h_3 \rangle\}.$$

Now, if f_p denotes the run which is constantly $\langle h_1, h_2, h_3 \rangle$ then $\langle n, f_p \rangle \models p$, for every $n \in \mathbb{N}$.

As we see, the pure temporal perspective does not enable us to model the interesting part of the three wise men puzzle. Recall that the main ingredients in the analysis of this puzzle were statements of the form 'agent A knows that agent B knows ....' So the question is how to represent within the temporal framework the fact that an agent A_i knows φ at a moment $t \in T$ under the assumption that the evolution is represented by a run f. To put it another way, if we mix the temporal and epistemic languages then how shall we define the truth-relation $\langle t, f \rangle \models \Box_i \varphi$?

In epistemic logic we defined $\Box_i \varphi$ to be true in a world u iff φ is true in every world which is considered possible by agent A_i. In the current framework this means: $\Box_i \varphi$ is true in $\langle t, f \rangle$ iff φ is true in $\langle t', f' \rangle$ for every moment t' and every run f' that are regarded possible by agent A_i. Thus, in order to define a truth condition for $\langle t, f \rangle \models \Box_i \varphi$, we require accessibility relations R_i between pairs $\langle t, f \rangle$ and $\langle t', f' \rangle$.

The following definition of the semantics for temporal epistemic logics with n agents should appear natural now. Suppose S is a non-empty set (of states) and $\mathfrak{F} = \langle T, < \rangle$ is a strict linear order. Suppose also that $\mathcal{R}$ is a non-empty set of functions from T to S (the available runs over $\mathfrak{F}$), and let $R_1, \dots, R_n$ be binary relations on $T \times \mathcal{R}$. Then the tuple

$$\mathfrak{S} = \langle T \times \mathcal{R}, <, R_1, \dots, R_n \rangle$$

is called a *temporal epistemic structure*. A *valuation* $\mathfrak{V}$ *in* $\mathfrak{S}$ is a function from the set of propositional variables into the set 2^S of all subsets of S. The pair $\mathfrak{M} = \langle \mathfrak{S}, \mathfrak{V} \rangle$ is called a *model based on* $\mathfrak{S}$.

We will consider two modal languages interpreted in temporal epistemic structures: the language $\mathcal{ML}_{SU} \otimes \mathcal{ML}_n$ consisting of modal formulas with the temporal operators $\mathcal{S}$ and $\mathcal{U}$, and the epistemic boxes $\Box_1, \ldots, \Box_n$, and its extension $\mathcal{ML}_{\mathcal{SU}} \otimes \mathcal{ML}_n^C$ with the common knowledge operators.

Suppose $\mathfrak{M} = \langle \mathfrak{G}, \mathfrak{V} \rangle$ is a model based on a temporal epistemic structure $\mathfrak{G} = \langle T \times \mathcal{R}, <, R_1, \ldots, R_n \rangle$. Define the truth-relation $\models$ between elements of $T \times \mathcal{R}$ and $\mathcal{ML}_{\mathcal{SU}} \otimes \mathcal{ML}_n^C$-formulas as follows:

- $(\mathfrak{M}, \langle t, f \rangle) \models p$ iff $f(t) \in \mathfrak{V}(p)$,
- $(\mathfrak{M}, \langle t, f \rangle) \models \varphi \wedge \psi$ iff $(\mathfrak{M}, \langle t, f \rangle) \models \varphi$ and $\langle t, f \rangle \models \psi$,
- $(\mathfrak{M}, \langle t, f \rangle) \models \neg\varphi$ iff not $(\mathfrak{M}, \langle t, f \rangle) \models \varphi$,
- $(\mathfrak{M}, \langle t, f \rangle) \models \varphi \mathcal{S} \psi$ iff there exists $t' < t$ such that $(\mathfrak{M}, \langle t', f \rangle) \models \psi$ and $(\mathfrak{M}, \langle s, f \rangle) \models \varphi$ for every s in the interval $t' < s < t$,
- $(\mathfrak{M}, \langle t, f \rangle) \models \varphi \mathcal{U} \psi$ iff there exists $t' > t$ such that $(\mathfrak{M}, \langle t', f \rangle) \models \psi$ and $(\mathfrak{M}, \langle s, f \rangle) \models \varphi$ for every s in the interval $t < s < t'$,
- $(\mathfrak{M}, \langle t, f \rangle) \models \Box_i \varphi$ iff $(\mathfrak{M}, \langle t', f' \rangle) \models \varphi$ whenever $\langle t, f \rangle \, R_i \, \langle t', f' \rangle$,
- $(\mathfrak{M}, \langle t, f \rangle) \models \mathsf{C}_M \varphi$ iff $(\mathfrak{M}, \langle t', f' \rangle) \models \varphi$ when $\langle t, f \rangle \, (\bigcup_{i \in M} R_i)^* \, \langle t', f' \rangle$.

As usual, we say that φ *is true in* $\mathfrak{M}$ (in symbols: $\mathfrak{M} \models \varphi$) if $(\mathfrak{M}, \langle t, f \rangle) \models \varphi$ holds, for every $\langle t, f \rangle \in T \times \mathcal{R}$.

For any epistemic logic L from the list $\mathbf{K}_n$, $\mathbf{T}_n$, $\mathbf{K4}_n$, $\mathbf{S4}_n$, $\mathbf{KD45}_n$, $\mathbf{S5}_n$ and any class $\mathcal{C}$ of strict linear orders, we let $\mathcal{TE}_{L,\mathcal{C}}$ denote the class of all temporal epistemic structures of the form

$$\langle T \times \mathcal{R}, <, R_1, \ldots, R_n \rangle$$

such that $\langle T, < \rangle \in \mathcal{C}$ and $\langle T \times \mathcal{R}, R_1, \ldots, R_n \rangle \models L$. If $\mathcal{C}$ consists of a single flow of time $\mathfrak{F}$, then we write $\mathcal{TE}_{L,\mathfrak{F}}$ instead of $\mathcal{TE}_{L,\mathcal{C}}$. The *temporal epistemic logic* determined by a class $\mathcal{K}$ of temporal epistemic structures,

$$\mathsf{ELog}_{\mathcal{SU}}(\mathcal{K})$$

in symbols, is the set of all $\mathcal{ML}_{\mathcal{SU}} \otimes \mathcal{ML}_n$-formulas that are true in every model based on a structure in $\mathcal{K}$. The *common knowledge logic* $\mathsf{ELog}_{\mathcal{SU}}^C(\mathcal{K})$ is defined analogously.

The following result is a consequence of Theorems 4.1 and 4.12 stating the transfer of some properties under the formation of fusions.

Theorem 3.19. *Let L be one of the epistemic logics $\mathbf{K}_n$, $\mathbf{T}_n$, $\mathbf{K4}_n$, $\mathbf{S4}_n$, $\mathbf{KD45}_n$, $\mathbf{S5}_n$, and let $\mathfrak{F} = \langle T, < \rangle$ be a strict linear order. Then the following holds:*

- *The temporal epistemic logic* $\mathsf{ELog}_{\mathcal{SU}}(\mathcal{TE}_{L,\mathfrak{F}})$ *coincides with the fusion of the temporal logic* $\mathsf{Log}_{\mathcal{SU}}(\mathfrak{F})$ *and* L. *That is to say,* $\mathsf{ELog}_{\mathcal{SU}}(\mathcal{TE}_{L,\mathfrak{F}})$ *can be axiomatized by putting together the sets of axioms and inference rules for* $\mathsf{Log}_{\mathcal{SU}}(\mathfrak{F})$ *and* L.

- $\mathsf{ELog}_{\mathcal{SU}}(\mathcal{TE}_{L,\mathfrak{F}})$ *is decidable whenever* $\mathfrak{F}$ *is one of* $\langle\mathbb{N},<\rangle$, $\langle\mathbb{Z},<\rangle$, $\langle\mathbb{Q},<\rangle$ *or* $\langle\mathbb{R},<\rangle$.

The same results hold for the common knowledge extensions $\mathsf{ELog}^{C}_{\mathcal{SU}}(\mathcal{TE}_{L,\mathfrak{F}})$ *of these logics.*

Theorem 3.19 does not necessarily hold for classes $\mathcal{C}$ containing more than one flow of time. For example, while the formula

$$\Box_F\bot \to \Box_1(\Box_F\bot \vee \Diamond_F\Box_F\bot)$$

does not belong to the fusion $\mathbf{Lin}_{\mathcal{SU}} \otimes \mathbf{S5}$, it is easy to see that it belongs to $\mathsf{ELog}_{\mathcal{SU}}(\mathcal{TE}_{\mathbf{S5},\mathcal{C}})$, where $\mathcal{C}$ is the class of all strict linear orders.

By imposing various constraints on temporal epistemic structures, we can reflect some interesting features of agents; see (Fagin *et al.* 1995). Here are some examples.

Synchronous systems

A temporal epistemic structure $\mathfrak{S}$ models agents who *know the time* if, for all $t,t' \in T$, $f,f' \in \mathcal{R}$, and $i \le n$,

$$\langle t,f\rangle\, R_i\, \langle t',f'\rangle \text{ implies } t = t'.$$

In other words, if A_i believes that at moment t relative to an evolution f the pair $\langle t',f'\rangle$ represents a possible state of affairs, then $t = t'$. So at each moment t the agents are assumed to know that the clock is at t. Systems represented by structures of this type are known as *synchronous*.

In Section 13.1 we will show that many temporal epistemic logics determined by classes of synchronous systems are decidable by embedding them into decidable fragments of first-order temporal logics.

Agents who know the time and neither forget nor learn

A temporal epistemic structure models agents who *do not learn* if, for all agents A_i, $f,f' \in \mathcal{R}$ and $t,t' \in T$, we have

$$\langle t,f\rangle\, R_i\, \langle t',f'\rangle \text{ implies } \forall s \ge t\ \exists s' \ge t'\ \ \langle s,f\rangle\, R_i\, \langle s',f'\rangle\,.$$

Intuitively, an agent A_i does not learn if, whenever it regards w as a possible state of affairs at moment t, then it regards w as a possible state of affairs at every moment $s \geq t$ as well. Under the condition that agents know the time, this means that if agent A_i regards an evolution f' as possible at t then it regards f' as possible at every $s > t$.

A temporal epistemic structure models agents who *do not forget* if, for all A_i, $t, t' \in T$ and $f, f' \in \mathcal{R}$, we have

$$\langle t, f\rangle \, R_i \, \langle t', f'\rangle \text{ implies } \forall s \leq t \; \exists s' \leq t' \;\; \langle s, f\rangle \, R_i \, \langle s', f'\rangle \, .$$

The intuition behind this definition is dual to that behind the models for agents who do not learn. Systems of this type are known also as *systems with perfect recall*.

Observe that if a temporal epistemic structure models agents who know time, do not forget and do not learn, then, for all agents A_i, $t, t' \in T$ and $f, f' \in \mathcal{R}$, we have

$$\langle t, f\rangle \, R_i \, \langle t', f'\rangle \text{ implies } t = t' \text{ and } \forall s \;\; \langle s, f\rangle \, R_i \, \langle s, f'\rangle \, .$$

Thus, $\mathfrak{G}$ is isomorphic to the product of frames $\mathfrak{F} = \langle T, <\rangle$ and $\langle \mathcal{R}, S_1, \dots, S_n\rangle$, where

$$f S_i f' \quad \text{iff} \quad \exists t, t' \in T \;\; \langle t, f\rangle \, R_i \, \langle t', f'\rangle \quad \text{iff} \quad \forall t \in T \;\; \langle t, f\rangle \, R_i \, \langle t, f'\rangle \, .$$

Example 3.18 (cont.) Let us complete the analysis of the 'wise men puzzle' by collecting first the information we already have. We have a temporal epistemic structure

$$\mathfrak{G} = \langle \mathbb{N} \times \mathcal{R}, <, R_A, R_B, R_C\rangle \, ,$$

where $\mathcal{R}$ is the set of all constant functions from $\mathbb{N}$ to $\{r, w\} \times \{r, w\} \times \{r, w\}$. (We will identify such a run f with its only value.) But what are the accessibility relations R_A, R_B and R_C? There is also some model $\mathfrak{M} = \langle \mathfrak{G}, \mathfrak{V}\rangle$ with

$$\mathfrak{V}(p) = \{\langle h_1, h_2, h_3\rangle\},$$

for some triple $\langle h_1, h_2, h_3\rangle$ of colors. For $D \in \{A, B, C\}$, we denote by $\Box_D$ the knowledge operator for agent D. Then the following should hold in $\mathfrak{M}$, for all $f \in \mathcal{R}$:

$$\langle 0, f\rangle \not\models \Box_A p \vee \Box_B p \vee \Box_C p, \tag{3.2}$$

$$\langle 0, f\rangle \not\models \bigcirc(\Box_A p \vee \Box_B p \vee \Box_C p). \tag{3.3}$$

We show how to define—using some of our implicit assumptions—the relations R_A, R_B and R_C in order to find out what $\langle h_1, h_2, h_3\rangle$ should be. We assume that the wise men are logically omniscient, capable of positive and negative introspection and that they know only true things. In other words, $\Box_A$, $\Box_B$ and $\Box_C$ are **S5**-boxes, and so R_A, R_B and R_C must be equivalence relations. Further, we assume that the wise men know the time and do not forget. Therefore, for every $D \in \{A, B, C\}$,

$$R_D = R_D^0 \cup R_D^1 \cup R_D^2 \cup \dots$$

where R_D^n ($n < \omega$) is a binary relation on the set $\{\langle n, f\rangle \mid f \in \mathcal{R}\}$ and

$$R_D^0 \supseteq R_D^1 \supseteq R_D^2 \supseteq \dots$$

holds.

Consider first the R_D^0. Since all the three wise men see the other two, we have

$$\begin{aligned}
\langle 0, \langle c_1, c_2, c_3\rangle\rangle\, R_A^0\, \langle 0, \langle c_1', c_2', c_3'\rangle\rangle \quad &\text{iff} \quad c_2 = c_2' \text{ and } c_3 = c_3', \\
\langle 0, \langle c_1, c_2, c_3\rangle\rangle\, R_B^0\, \langle 0, \langle c_1', c_2', c_3'\rangle\rangle \quad &\text{iff} \quad c_1 = c_1' \text{ and } c_3 = c_3', \\
\langle 0, \langle c_1, c_2, c_3\rangle\rangle\, R_C^0\, \langle 0, \langle c_1', c_2', c_3'\rangle\rangle \quad &\text{iff} \quad c_1 = c_1' \text{ and } c_2 = c_2'.
\end{aligned}$$

Now state $\langle r, w, w\rangle$ has only one R_A^0-successor (itself), $\langle w, r, w\rangle$ has only one R_B^0-successor, and $\langle w, w, r\rangle$ has only one R_C^0-successor. Thus by (3.2), $\mathfrak{V}(p)$ cannot have any of these states as its only element. Since all three wise men have this knowledge, R_D^1 is defined as follows, for all $D \in \{A, B, C\}$:

$$f R_D^1 f' \;\text{ iff }\; f = f' \text{ or } (f R_D^0 f' \text{ and } f \notin \{\langle r, w, w\rangle, \langle w, r, w\rangle, \langle w, w, r\rangle\})$$

(see Fig. 3.4). Therefore, by (3.3), the only possibility which remains for

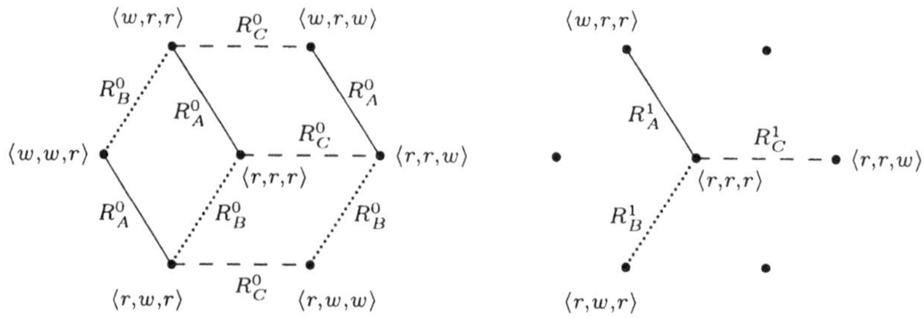

Figure 3.4: The relations R_D^0 and R_D^1.

$\mathfrak{V}(p)$ is $\{\langle r, r, r\rangle\}$, since every other state has only one R_D^1-successor for some $D \in \{A, B, C\}$. This also shows that

$$f R_D^n f' \;\text{ iff }\; f = f'$$

must hold for all $n \geq 2$. Note that while the wise men do not forget, they do learn (at least at the beginning) because

$$R_D^0 \supsetneq R_D^1 \supsetneq R_D^2$$

for all $D \in \{A, B, C\}$.

3.5 Classical first-order logic as a propositional multimodal logic

As we saw in Section 1.3, the standard translation, mapping the modal operators to the corresponding first-order quantifiers, embeds propositional modal logic **S5** into classical first-order logic. Moreover, the inverse map is an embedding of the one-variable fragment of first-order logic into **S5**. A natural question arising in this situation is whether we can generalize the inverse translation by considering quantification over each variable as a new modal operator and thereby representing full first-order logic as a propositional modal logic. The idea of such a 'modal approach' to first-order logic was suggested by Quine (1971) and Kuhn (1980), and fully realized by Venema (1991). On the other hand, 'approximating' first-order logic with logical systems of propositional character was an important motive in the *algebraic* treatment of classical first-order logic; see the work of Tarski and his school (Halmos 1962, Henkin *et al.* 1971, 1985, Craig 1974, Blok and Pigozzi 1989, Németi 1991, Andréka *et al.* 2000).

In this section we exploit this idea to establish connections between classical first-order logic and products of propositional **S5**.

Let us fix a natural number $n > 0$ and consider the sublanguage $r\mathcal{QL}^n$ of the n-variable fragment of $\mathcal{QL}$ which contains no individual constants and whose atomic formulas are of the form $P(x_0, \dots, x_{n-1})$, where P is an n-ary predicate symbol and $x_0, \dots, x_{n-1}$ are the first n individual variables (r in $r\mathcal{QL}^n$ stands for 'restricted').

Note that by allowing atomic formulas of the form $P(x_0, \dots, x_{n-1})$ only, we restrict the expressive power of the n-variable fragment of $\mathcal{QL}$. As was observed by Tarski, if we extend the language with equality then variable substitutions like $P(x_0, x_0, x_2, \dots, x_{n-1})$ become expressible in $r\mathcal{QL}^n$:

$$P(x_0, x_0, x_2, \dots, x_{n-1}) \leftrightarrow \exists x_1 (x_0 = x_1 \wedge P(x_0, x_1, x_2, \dots, x_{n-1})).$$

However, even with the help of equality, variable interchanges like

$$P(x_1, x_0, x_2, \dots, x_{n-1})$$

are expressible only by using an extra $n + 1$st variable; see (Henkin *et al.* 1985).

Define a translation $\cdot^\bullet$ of $r\mathcal{QL}^n$-formulas into the multimodal language $\mathcal{ML}_n$ by taking

$$\begin{aligned}
P_i(x_0,\dots,x_{n-1})^\bullet &= p_i\\
(\varphi\wedge\psi)^\bullet &= \varphi^\bullet\wedge\psi^\bullet,\\
(\neg\varphi)^\bullet &= \neg\varphi^\bullet,\\
(\forall x_i\psi)^\bullet &= \Box_{i+1}\psi^\bullet \quad (i<n),\\
(\exists x_i\psi)^\bullet &= \Diamond_{i+1}\psi^\bullet \quad (i<n).
\end{aligned}$$

Every $r\mathcal{QL}^n$-structure $I=\langle D^I,P_0^I,\dots\rangle$ can be considered then as a modal model $\mathfrak{M}(I)=\langle\langle W,R_0,\dots\rangle,\mathfrak{V}\rangle$, where

- W is the set of all variable assignments in I, i.e., the set of all functions from the variables $x_0,\dots,x_{n-1}$ into D^I;
- $\mathfrak{a}R_i\mathfrak{b}$ iff $\mathfrak{a}(x_j)=\mathfrak{b}(x_j)$ for all variables x_j different from x_i, $i<n$;
- $\mathfrak{V}(p_i)=P_i^I$.

It is not hard to see that for all $r\mathcal{QL}^n$-formulas φ, $r\mathcal{QL}^n$-structures I, and all assignments $\mathfrak{a}$ in I, we have

$$I\models^{\mathfrak{a}}\varphi \quad\text{iff}\quad (\mathfrak{M}(I),\mathfrak{a})\models\varphi^\bullet. \tag{3.4}$$

The set W of all assignments in I can be regarded as the nth Cartesian power of the domain D^I. The underlying frame of $\mathfrak{M}(I)$ then turns into a product frame for $\mathbf{S5}^n$: the nth power of the Kripke frame $\langle D^I,S\rangle$, where S is the universal relation on D^I.

Conversely, we can turn every modal model $\mathfrak{M}=\langle\mathfrak{F},\mathfrak{V}\rangle$ based on a cubic universal product $\mathbf{S5}^n$-frame $\mathfrak{F}=\langle W,W,\dots,W\rangle$ into the first-order structure

$$I(\mathfrak{M})=\left\langle W,\dots,P_i^{I(\mathfrak{M})},\dots\right\rangle,$$

where $P_i^{I(\mathfrak{M})}=\mathfrak{V}(p_i)$ for each i. Then for all $r\mathcal{QL}^n$-formulas φ and all worlds $\langle w_1,\dots,w_n\rangle$ in $\mathfrak{F}$ we clearly have:

$$(\mathfrak{M},\langle w_1,\dots,w_n\rangle)\models\varphi^\bullet \quad\text{iff}\quad I(\mathfrak{M})\models\varphi[w_1,\dots,w_n]. \tag{3.5}$$

According to Proposition 3.12, $\mathbf{S5}^n$ is determined by the class of cubic universal product frames. Thus, by (3.4) and (3.5), for every $r\mathcal{QL}^n$-formula φ, we obtain

$$\varphi\in\mathbf{QCl} \quad\text{iff}\quad \varphi^\bullet\in\mathbf{S5}^n.$$

This equivalence shows that, since the translation $^\bullet$ is clearly onto the set of $\mathcal{ML}_n$-formulas, the logic $\mathbf{S5}^n$ can be regarded as the *n-variable 'substitution free' fragment* of classical first-order logic. To put it in another way, the following inverse translation, mapping $\mathcal{ML}_n$-formulas to $\mathcal{QL}$-formulas, extends Wajsberg's map † (see the end of Section 1.3) and embeds $\mathbf{S5}^n$ into **QCl**:

$$\begin{aligned}
p_i^\dagger &= P_i(x_0, \dots, x_{n-1}),\\
(\varphi \wedge \psi)^\dagger &= \varphi^\dagger \wedge \psi^\dagger,\\
(\neg\varphi)^\dagger &= \neg\varphi^\dagger,\\
(\Box_i \psi)^\dagger &= \forall x_{i-1}\psi^\dagger \quad (i = 1, \dots, n),\\
(\Diamond_i \psi)^\dagger &= \exists x_{i-1}\psi^\dagger \quad (i = 1, \dots, n).
\end{aligned}$$

We shall return to interconnections between classical first-order logic and modal product logics in Sections 8.1 and 9.1. The reader can find more information on algebraization and modalization of other versions of first-order logic in (Andréka *et al.* 2000, Blok and Pigozzi 1989, Marx and Venema 1997).

3.6 First-order modal logics

After the previous section it should not come as a surprise that we introduce first-order modal logics here, in the chapter on many-dimensional systems, rather than in Chapter 1 dealing with basic modal logics; the more so that first-order modal logics can be regarded as combinations of propositional modal logics with classical first-order logic. Many interesting features of the resulting systems arise because of subtle interactions between the quantifiers and the modal operators independently of the underlying modal logic. It is in fact the 'combined system' aspect that makes first-order modal logic so exciting. To illustrate this claim, let us consider two formulas

$$\Box\exists x\, \varphi(x) \quad \text{and} \quad \exists x\, \Box\varphi(x).$$

Under the epistemic reading of $\Box$, the former formula means that the agent knows that there exists an x to which φ applies, while the latter means that there exists an x for which the agent knows that φ applies to x. For example, suppose $\varphi(x)$ stands for

'x is the telephone number of Mary.'

Then the former formula is true if the agent knows that Mary has a telephone, while the latter one is true if the agent knows the telephone number of Mary. (In the former case $\Box$ is called a modality *de dicto* and in the latter a modality *de re*.)

Our *first-order* (or *quantified*) *modal language* $\mathcal{QML}_l$ is based on the alphabet of $\mathcal{QL}$ (Section 1.3) extended with the *necessity operators* $\Box_1, \dots, \Box_l$, for $l \geq 1$. The formulas of $\mathcal{QML}_l$ are defined using the formula-formation rules of $\mathcal{QL}$ together with the rule for the $\Box_i$: if φ is a $\mathcal{QML}_l$-formula then so is $\Box_i\varphi$, for every i, $1 \leq i \leq l$. As in propositional modal logic, we regard $\Diamond_i\varphi$ as an abbreviation for $\neg\Box_i\neg\varphi$ and write $\mathcal{QML}$ for $\mathcal{QML}_1$.

Another reason to consider first-order modal logics in this chapter is that their models are in a sense two-dimensional. Actually, there is a spectrum of different semantics for first-order modal logics. In this book we will be considering perhaps the simplest one of them. It was first introduced by Kripke (1963b) and is characterized by 'constant (or common) domains' and 'rigid designators.' More precisely, we interpret $\mathcal{QML}_l$ in *first-order Kripke models* which are structures of the form $\mathfrak{M} = \langle \mathfrak{F}, D, I \rangle$, where

- $\mathfrak{F} = \langle W, R_1, \ldots, R_l \rangle$ is an l-frame (the R_i being binary relations on a nonempty set of worlds W),
- D is a nonempty set, the *domain* of $\mathfrak{M}$, and
- I is a function associating with every world $w \in W$ a first-order $\mathcal{QL}$-structure
$$I(w) = \left\langle D, P_0^{I(w)}, \ldots, c_0^{I(w)}, \ldots \right\rangle$$
such that $P_i^{I(w)}$, for each i, is a relation on D of the same arity as P_i, and $c_i^{I(w)}$ is an element in D such that $c_i^{I(u)} = c_i^{I(v)}$ for all $u, v \in W$.

(As before, we say that $\mathfrak{M}$ is *based on* $\mathfrak{F}$ or that $\mathfrak{F}$ is the underlying frame of $\mathfrak{M}$.) To simplify notation we will omit the superscript I and write P_i^w, c_i^w, etc., if this does not cause ambiguity.

An *assignment in* D is a function $\mathfrak{a}$ from the set of individual variables to D. The *value* $\tau^{\mathfrak{M},\mathfrak{a}}$ of a term τ in $\mathfrak{M}$ under the assignment $\mathfrak{a}$ is $\mathfrak{a}(x)$ if τ is a variable x, and (the unique) $c^{I(w)}$ if τ is a constant c.

According to the given definition, our models have *rigid designators* in the sense that they interpret each term (a constant or a variable) by the same element of D in all worlds of W. Under the temporal interpretation (see Section 3.7) of the modal operators this means that the names of objects do not vary in time so that we can refer to an object by its name even if it does not exist yet (or does not exist any more). Under the epistemic interpretation, rigid designators mean, in particular, that we assume all agents to *know* which object a constant denotes. It is to be noted that, from the technical point of view, not too much will change if we consider models with nonrigid *constants* (but not variables)—to allow names like the Queen to denote different objects at different moments of time.

The *truth-relation* $(\mathfrak{M}, w) \models^{\mathfrak{a}} \varphi$ (or simply $w \models^{\mathfrak{a}} \varphi$, if $\mathfrak{M}$ is understood) in the model $\mathfrak{M}$ under the assignment $\mathfrak{a}$ is defined by induction on the construction of φ in the following way:

- $w \models^{\mathfrak{a}} P_i(\tau_1, \ldots, \tau_n)$ iff $\left\langle \tau_1^{\mathfrak{M},\mathfrak{a}}, \ldots, \tau_n^{\mathfrak{M},\mathfrak{a}} \right\rangle \in P_i^{I(w)}$ (this fact will also be written as $I(w) \models P_i[\tau_1^{\mathfrak{M},\mathfrak{a}}, \ldots, \tau_n^{\mathfrak{M},\mathfrak{a}}]$);
- $w \models^{\mathfrak{a}} \psi \wedge \chi$ iff $w \models^{\mathfrak{a}} \psi$ and $w \models^{\mathfrak{a}} \chi$;
- $w \models^{\mathfrak{a}} \neg\psi$ iff not $w \models^{\mathfrak{a}} \psi$;
- $w \models^{\mathfrak{a}} \forall x\psi$ iff $w \models^{\mathfrak{b}} \psi$ for every assignment $\mathfrak{b}$ in D that may differ from $\mathfrak{a}$ only on x;
- $w \models^{\mathfrak{a}} \Box_i\varphi$ iff $v \models^{\mathfrak{a}} \varphi$ for all $v \in W$ such that wR_iv.

We say that a formula φ is *true in* $\mathfrak{M}$ if $(\mathfrak{M}, w) \models^{\mathfrak{a}} \varphi$ holds for all assignments $\mathfrak{a}$ in D and all worlds w in W. The set of $\mathcal{QML}_l$-formulas that are true in all models is denoted by $\mathbf{QK}_l$ (*quantified* $\mathbf{K}_l$). In general, given an l-modal logic L, we denote by $\mathbf{Q}L$ the set of $\mathcal{QML}_l$-formulas that are true in all models based on frames for L. For instance, $\mathbf{QT}_l$, $\mathbf{QK4}_l$ and $\mathbf{QS4}_l$ are the sets of $\mathcal{QML}_l$-formulas that are true in all models based on reflexive, transitive and quasi-ordered frames, respectively.

The models introduced above are known as *models with constant domains*. In other words, we make the *constant domain assumption*. Under this assumption all constants and variables do denote some objects, and the quantifiers range over the same domain everywhere in the model. It follows immediately that the resulting logic is a conservative extension of classical predicate logic **QCl**. (Note that under the epistemic interpretation of the modal operators the constant domain assumption says that the domain is common knowledge.)

However, the defined semantics is just one of a dozen possible alternatives. Imagine, for example, that we deal with a temporal interpretation of the modal operators. Then our everyday life experience suggests the following:

1. The domains of $I(w)$ can all be *different* for different w, because their elements can 'die' and 'be born.'
2. When we name an element x then its name is a rigid designator whenever x exists.
3. Predicates at moment w can apply to elements not existing at w. We are all familiar with young expectant parents talking about their babies (yet to be born), buying things for them, or similarly talking about their dead parents, etc.

This leads us to *models with varying* (or *changing*) *domains* which contain one more function $\mathfrak{d}$ associating with every world $w \in W$ a nonempty set $\mathfrak{d}(w) \subseteq D$—the existing elements in w—such that $D = \bigcup_{w \in W} \mathfrak{d}(w)$. The only difference in the definition of the truth-relation above is in the truth-condition for $\forall x\psi$, which now looks as follows:

- $w \models^{\mathfrak{a}} \forall x\psi$ iff $w \models^{\mathfrak{b}} \psi$ for every assignment $\mathfrak{b}$ that may differ from $\mathfrak{a}$ only on x, provided that $\mathfrak{b}(x) \in \mathfrak{d}(w)$.

Thus, $\mathfrak{d}(w)$ is regarded as the 'true' domain of $I(w)$. The set of $\mathcal{QML}_l$-formulas that are true in all models with varying domains under all assignments will be denoted by $\mathbf{Q}_v\mathbf{K}_l$, quantified $\mathbf{K}_l$ with varying domains. It is worth noting that $\mathbf{Q}_v\mathbf{K}_l$ has a number of 'unorthodox' properties. For example, neither

$$\forall x P(x) \to P(c) \quad \text{nor} \quad \forall x P(x) \to P(y)$$

belongs to $\mathbf{Q}_v\mathbf{K}_l$ simply because there may be a world w such that $P^w = \mathfrak{d}(w)$, but $c^w \notin \mathfrak{d}(w)$ and $\mathfrak{a}(y) \notin \mathfrak{d}(w)$. This means, in particular, that $\mathbf{Q}_v\mathbf{K}_l$ does not obey the principles of classical first-order logic. Various authors have regarded this as an argument against the semantics defined above (see, e.g., Garson 1984). The interested reader can find various alternative approaches to the semantics of first-order modal logics in (Garson 1984, Hughes and Cresswell 1996, Fitting and Mendelson 1998).

One way to 'repair' $\mathbf{Q}_v\mathbf{K}_l$ is to require that

- $c_i^{I(w)} \in \mathfrak{d}(w)$ for every $w \in W$ and every constant c_i

and to modify the notion of truth in a model by saying that a formula φ is *true* (*satisfied*) in a model $\mathfrak{M}$ with varying domains if $(\mathfrak{M}, w) \models^{\mathfrak{a}} \varphi$ holds for every (some) $w \in W$ and every (respectively, some) assignment $\mathfrak{a}$ in D such that $\mathfrak{a}(x) \in \mathfrak{d}(w)$ for all individual variables x. The set of $\mathcal{QML}_l$-formulas that are true in all 'repaired' models with varying domains will be denoted by $\mathbf{Q}^v\mathbf{K}_l$. By definition,

$$\varphi \in \mathbf{Q}^v\mathbf{K}_l \quad \text{iff} \quad \forall x_1 \dots \forall x_n \varphi \in \mathbf{Q}^v\mathbf{K}_l,$$

for any $\mathcal{QML}_l$-formula φ and list $x_1, \dots, x_n$ of all variables which occur free in φ. It is easy to see that $\mathbf{Q}^v\mathbf{K}_l$ is a conservative extension of $\mathbf{QCl}$ and that $\mathbf{Q}^v\mathbf{K}_l$ and $\mathbf{Q}_v\mathbf{K}_l$ contain precisely the same constant-free sentences; however, $\mathbf{Q}_v\mathbf{K}_l \subsetneq \mathbf{Q}^v\mathbf{K}_l$.

Two other important classes of models consist of *models with expanding domains* and *with decreasing domains*, i.e., models with varying domains in which $\mathfrak{d}(u) \subseteq \mathfrak{d}(v)$ or $\mathfrak{d}(u) \supseteq \mathfrak{d}(v)$ whenever uR_iv, respectively. Under the 'old' understanding of truth, these models also give rise to some unorthodox properties. For instance, the formula $\forall y \Box(\forall x P(x) \to P(y))$ is true in all models with expanding domains, while $\Box(\forall x P(x) \to P(y))$ is not (contrary to the classical principle $\varphi \in \mathbf{QCl}$ iff $\forall x \varphi \in \mathbf{QCl}$). As mentioned above, this does not happen under the new definition of truth, which will be considered as the only 'official' definition from now on.

Actually, later on in this section we will show that both varying and expanding domains can be reduced to constant ones, at least as far as the decidability of (fragments of) the logics in question is concerned. For that reason we will mostly be considering models with constant domains.

On the syntactic level, the difference between the domain assumptions can be captured by the *Barcan formulas*

$$\forall x \Box_i \varphi \rightarrow \Box_i \forall x \varphi$$

and the *converse Barcan formulas*

$$\Box_i \forall x \varphi \rightarrow \forall x \Box_i \varphi.$$

It is not hard to see that the Barcan formulas are true in all models with decreasing domains (but refuted in a model with nondecreasing domains), while the converse Barcan formulas are true in all models with expanding domain (and refuted in a model with nonexpanding domains). So, both types of Barcan formulas are true in models with constant domains.

The (converse) Barcan formulas can be used to axiomatize $\mathbf{QK}_l$: it can be represented by the calculus containing all the axiom schemata and inference rules of classical predicate calculus, the Barcan and converse Barcan formulas, the modal schemata

$$\Box_i(\varphi \rightarrow \psi) \rightarrow (\Box_i \varphi \rightarrow \Box_i \psi),$$

for $1 \leq i \leq l$, and the necessitation rules $\varphi / \Box_i \varphi$.

By adding to **QK** the standard modal axiom schemata of **T**, **K4**, **S4** we obtain modal predicate logics **QT**, **QK4**, **QS4** (see, e.g., Hughes and Cresswell 1996).

Let us now see how satisfiability in models with varying and expanding domains can be reduced to satisfiability in models with constant domains. Let φ be a $\mathcal{QML}_l$-formula, and let $E(x)$ be a unary predicate symbol which does not occur in φ. By induction on the construction of φ we define its *relativization* $\varphi \downarrow E$:

$$\begin{aligned}
P_i(\tau_1, \ldots, \tau_n) \downarrow E &= P_i(\tau_1, \ldots, \tau_n), \\
(\psi \wedge \chi) \downarrow E &= (\psi \downarrow E) \wedge (\chi \downarrow E), \\
(\neg \psi) \downarrow E &= \neg(\psi \downarrow E), \\
(\forall x \psi) \downarrow E &= \forall x (E(x) \rightarrow (\psi \downarrow E)), \\
(\Box_i \psi) \downarrow E &= \Box_i (\psi \downarrow E) \qquad (i = 1, \ldots, l).
\end{aligned}$$

As before, we denote by $md(\varphi)$ the *modal depth* of φ, i.e., the maximal number of nested modal operators in φ.

Proposition 3.20. *Let φ be a $\mathcal{QML}_l$-sentence, $c_1, \dots, c_n$ all the constants occurring in φ and $E(\bar{c}) = E(c_1) \wedge \dots \wedge E(c_n)$. Then for any Kripke frame $\mathfrak{F} = \langle W, R_1, \dots, R_l \rangle$, we have*

(i) *φ is satisfied in a model based on $\mathfrak{F}$ and having varying domains iff*

$$\varphi \downarrow E \wedge \boldsymbol{M}_{(l)}^{\leq md(\varphi)}(\exists x E(x) \wedge E(\bar{c}))$$

is satisfied in a model based on $\mathfrak{F}$ and having constant domains;

(ii) *φ is satisfied in a model based on $\mathfrak{F}$ and having expanding domains iff*

$$\varphi' = \varphi \downarrow E \wedge \exists x E(x) \wedge E(\bar{c}) \wedge \boldsymbol{M}_{(l)}^{\leq md(\varphi)} \forall x (E(x) \rightarrow \bigwedge_{i=1}^{l} \Box_i E(x))$$

is satisfied in a model based on $\mathfrak{F}$ and having constant domains.

Proof. We prove only (ii), leaving the simpler case (i) to the reader. Assuming that φ is satisfied in a model $\mathfrak{M} = \langle \mathfrak{F}, D, \mathfrak{d}, I \rangle$ with expanding domains and that

$$I(w) = \left\langle D, P_0^{I(w)}, \dots, c_0^{I(w)}, \dots \right\rangle$$

for $w \in W$, we construct a model $\mathfrak{N} = \langle \mathfrak{F}, D, J \rangle$ with constant domains by taking

$$J(w) = \left\langle D, E^{J(w)}, P_0^{I(w)}, \dots, c_0^{I(w)}, \dots \right\rangle,$$

where $E^{J(w)} = \mathfrak{d}(w)$. It is readily checked by induction that $(\mathfrak{M}, w) \models^{\mathfrak{a}} \psi$ iff $(\mathfrak{N}, w) \models^{\mathfrak{a}} \psi \downarrow E$, for every $w \in W$, every subformula ψ of φ, and every assignment $\mathfrak{a}$ in $\mathfrak{d}(w)$. It follows that φ' is satisfied in $\mathfrak{N}$.

Conversely, suppose φ' is satisfied at root v of a model $\mathfrak{N} = \langle \mathfrak{F}, D, J \rangle$ with constant domains and

$$J(w) = \left\langle D, E^{J(w)}, P_0^{J(w)}, \dots, c_0^{J(w)}, \dots \right\rangle,$$

for $w \in W$. Consider the model $\mathfrak{M} = \langle \mathfrak{F}, D, \mathfrak{d}, I \rangle$ such that

$$I(w) = \left\langle D, P_0^{J(w)}, \dots, c_0^{I(w)}, \dots \right\rangle$$

for all $w \in W$, $\mathfrak{d}(w) = E^{J(w)}$ whenever w is accessible in $\leq md(\varphi)$ steps from v (via the relation $\bigcup_{1 \leq i \leq l} R_i$) and $\mathfrak{d}(w) = D$ otherwise, $c_i^{I(w)} = c_i^{J(v)}$ if c_i occurs in φ and $c_i^{I(w)}$ is an arbitrary element of $\mathfrak{d}(v)$ otherwise. By the fourth conjunct of φ', $\mathfrak{M}$ has expanding domains. Now, using that the truth-value of φ at v depends only on the worlds accessible in $\leq md(\varphi)$ steps from v, one can easily show by induction that φ is satisfied at v in $\mathfrak{M}$. ❑

The connection between products $\mathbf{S5}^n$ and classical first-order logic $\mathbf{QCl}$ we established in Section 3.5 suggests that modal product logics of the form

$$L \times \overbrace{\mathbf{S5} \times \cdots \times \mathbf{S5}}^{n}$$

can be reduced to the n-variable fragments of first-order modal logics $\mathbf{Q}L$ (with constant domains). Indeed, fix some natural number $n > 0$ and take the sublanguage $r\mathcal{QML}_l^n$ of the n-variable fragment of $\mathcal{QML}_l$ which contains no constant symbols and whose only atomic formulas are of the form $P(x_0, \dots, x_{n-1})$, where P is an n-ary predicate symbol and $x_0, \dots, x_{n-1}$ are the first n individual variables. The translation † from $\mathcal{ML}_n$ onto $r\mathcal{QL}^n$ defined in Section 3.5 can be extended to a translation from $\mathcal{ML}_{l+n}$ onto $r\mathcal{QML}_l^n$-formulas by taking

$$\begin{aligned}
p_i^\dagger &= P_i(x_0, \dots, x_{n-1}),\\
(\varphi \wedge \psi)^\dagger &= \varphi^\dagger \wedge \psi^\dagger,\\
(\neg\varphi)^\dagger &= \neg\varphi^\dagger,\\
(\Box_i \psi)^\dagger &= \Box_i \psi^\dagger, && \text{for } i = 1, \dots, l,\\
(\Box_j \psi)^\dagger &= \forall x_{j-l-1} \psi^\dagger, && \text{for } j = l+1, \dots, l+n.
\end{aligned}$$

An argument similar to the proof of Proposition 3.12 shows that the product logic $L \times \mathbf{S5} \times \cdots \times \mathbf{S5}$ is determined by product frames of the form

$$\mathfrak{G} = \mathfrak{F} \times \langle D, D \times D \rangle \times \cdots \times \langle D, D \times D \rangle,$$

where $\mathfrak{F} = \langle W, R_1, \dots, R_l \rangle$ is a frame for L and D is a nonempty set. Now, (propositional) Kripke models $\langle \mathfrak{G}, \mathfrak{V} \rangle$ based on such a product frame $\mathfrak{G}$ and first-order Kripke models of the form $\langle \mathfrak{F}, D, I \rangle$ are in one-to-one correspondence with each other:

$$\langle w, a_1, \dots, a_n \rangle \in \mathfrak{V}(p_i) \quad \text{iff} \quad \langle a_1, \dots, a_n \rangle \in P_i^{I(w)},$$

for all propositional variables p_i, $w \in W$ and $a_1, \dots, a_n \in D$. It should be clear that in fact, for all $\mathcal{ML}_{l+n}$-formulas φ, we have

$$(\langle \mathfrak{G}, \mathfrak{V} \rangle, \langle w, a_1, \dots, a_n \rangle) \models \varphi \quad \text{iff} \quad (\langle \mathfrak{F}, D, I \rangle, w) \models^{\mathfrak{a}} \varphi^\dagger,$$

where $\mathfrak{a}$ is the assignment in D such that $\mathfrak{a}(x_i) = a_{i+1}$ $(i < n)$. As a consequence we obtain the following:

Theorem 3.21. *Let L be a Kripke complete l-modal logic. Then for every $\mathcal{ML}_{l+n}$-formula φ,*

$$\varphi \in L \times \overbrace{\mathbf{S5} \times \cdots \times \mathbf{S5}}^{n} \quad \textit{iff} \quad \varphi^\dagger \in \mathbf{Q}L.$$

This observation will be used in Section 8.4 (Theorem 8.35) to show the undecidability of the two-variable fragment of any logic between **QK** and **QS5**.

First-order epistemic logics

Let us now have a closer look at the interaction between modal operators and quantifiers in epistemic logics with common knowledge operators. Denote by $\mathcal{QML}_n^C$ the language of modal predicate logic with epistemic operators $\Box_i$, $1 \le i \le n$, and common knowledge operators C_M for all nonempty subsets M of $\{1, \dots, n\}$ (see Section 2.3 for the propositional case). For a propositional logic $L \in \{\mathbf{K}_n, \mathbf{T}_n, \mathbf{K4}_n, \mathbf{S4}_n, \mathbf{KD45}_n, \mathbf{S5}_n\}$, let $\mathbf{Q}L^C$ be the first-order epistemic logic which consists of those $\mathcal{QML}_n^C$-formulas that are true in all first-order Kripke models based on frames for L and having constant domains.

Note that unlike standard first-order modal logics like $\mathbf{QK}_n$ and $\mathbf{QS4}_n$, which can be axiomatized in a natural way by putting together the axioms of their propositional fragments and those of **QCl**, first-order modal logics with common knowledge operators behave quite differently. The following result of (Wolter 2000a) will be partly proved in Section 12.1:

Theorem 3.22. *Let* $L \in \{\mathbf{K}_n, \mathbf{T}_n, \mathbf{K4}_n, \mathbf{S4}_n, \mathbf{KD45}_n, \mathbf{S5}_n\}$, *where* $n > 1$. *Then* $\mathbf{Q}L^C$ *is not recursively enumerable.*

First-order epistemic logic has found interesting applications in game theory; see, e.g., (Kaneko and Nagashima 1997) and references therein. A static noncooperative strategic normal form 2-person game G consists of two agents (or players), say, 1 and 2. The players have finite sets $S_1 = \{s_1^1, \dots, s_{l(1)}^1\}$ and $S_2 = \{s_1^2, \dots, s_{l(2)}^2\}$ of actions, respectively. Payoff functions u_i, $i = 1, 2$, from $S = S_1 \times S_2$ into the set of rational numbers determine the payoff of the players: $u_i(s_1, s_2)$ is the payoff of player i when player 1 performs action $s_1 \in S_1$ and player 2 performs action $s_2 \in S_2$.

Here is a variant of a game known as *Prisoner's Dilemma* (Gibbons 1992). Two partners in a crime (the players in this game) have been captured, placed in separate cells and offered an opportunity to confess. Their actions can be '*confess*' and '*not confess*.' The payoff functions are defined as follows. If neither suspect confesses, they go free and split the proceeds of the crime (which we represent by, say, 5 units of utility). If one player confesses and the other does not, the one who confesses testifies against the other, and so goes free and gets the entire 10 units of utility. The other prisoner goes to prison and gets nothing. If both prisoners confess, then both are given a reduced term, but both are convicted (which we represent by 1 unit of utility). The following table summarizes the definition:

	not confess	confess
not confess	(5,5)	(0,10)
confess	(10,0)	(1,1)

For an intelligent individual it is always best to confess: if his partner does not confess, he receives 10 units (instead of 5) and if the partner confesses as well, he receives 1 (instead of 0). (Note however, that for the common good it would be better if neither of them confessed. This conflict between the pursuit of individual goals and common good is the driving force behind many game theoretic problems.)

The argument above is formalized by the notion of Nash equilibrium (Nash 1991): the strategy (s_1, s_2) is a *Nash equilibrium* for G if

$$u_1(s_1, s_2) \geq u_1(s, s_2) \quad \text{and} \quad u_2(s_1, s_2) \geq u_2(s_1, s')$$

for all actions s and s' of players 1 and 2, respectively. The Prisoner's Dilemma has precisely one Nash equilibrium: (confess, confess).

Not all games have a Nash equilibrium in this sense. However, extended to mixed strategies (which are probability distributions over the sets of actions S_i—modeling, for example, that you flip a coin to choose an action), Nash equilibria (which are now defined via the *expected payoff*) exist for any game. In fact, for every game G with strategies $S_1 = \{s^1_1, \ldots, s^1_{l(1)}\}$ and $S_2 = \{s^2_1, \ldots, s^2_{l(2)}\}$, one can construct a formula denoted by $\mathsf{Nash}_G(\overline{x}, \overline{y})$, where $(\overline{x}, \overline{y}) = (x_1, \ldots, x_{l(1)}, y_1, \ldots, y_{l(2)})$, such that

$$\langle \mathbb{R}, \ldots \rangle \models \mathsf{Nash}_G[a_1, \ldots, a_{l(1)}, b_1, \ldots, b_{l(2)}]$$

holds iff the probability distributions $P_1(s^1_j) = a_j$ and $P_2(s^2_j) = b_j$ define a Nash equilibrium for G. Here $\mathsf{Nash}_G(\overline{x}, \overline{y})$ is a first-order formula in the language of real closed fields, and $\langle \mathbb{R}, \ldots \rangle$ is the standard model for this language based on the real numbers. (Our first-order language $\mathcal{QL}$ does not contain function symbols of arity ≥ 1, so operations like '+' should be represented by appropriate predicate symbols.) The reader can consult (Kaneko and Nagashima 1997, Wolter 2000a) for details of the construction.

Now, in the epistemic analysis of games one has to be aware of the difference between $\mathsf{C}_{\{1,2\}} \exists \overline{x} \exists \overline{y} \mathsf{Nash}_G(\overline{x}, \overline{y})$, which states that it is common knowledge among the two players 1 and 2 that game G has a Nash equilibrium (knowledge *de dicto*), and $\exists \overline{x} \exists \overline{y} \mathsf{C}_{\{1,2\}} \mathsf{Nash}_G(\overline{x}, \overline{y})$, which says that at least one Nash equilibrium for G is common knowledge among 1 and 2 (knowledge *de re*). Knowledge *de re* is useful for playing the game, while knowledge *de dicto* is not. Actually, it turns out that the relation between these two assertions depends on the formal representation. For example, assume that mathematics is common knowledge and that both players know the game.

One possibility to formalize this assumption is to accept

$$\mathsf{C}_{\{1,2\}}\forall\overline{x}\left(\mathsf{C}_{\{1,2\}}P(\overline{x}) \leftrightarrow P(\overline{x})\right),$$

for every 'mathematical predicate' P (say, the ternary predicate for '+') and to assume that the constant symbols representing the payoffs $u_i(s_1, s_2)$ are interpreted globally. The last condition need not be added explicitly, since it is 'built into' the semantics of constants. What are the consequences for 'common knowledge about Nash equilibria'? Since all relevant predicates are global, there is no difference between *de re* and *de dicto* knowledge! The two formulas are equivalent.

The outcome is completely different if another natural interpretation of the phrase 'mathematics is common knowledge' is chosen. This time we formalize this by the assumption that the theory of real closed fields is common knowledge without requiring that the mathematical predicates are global. So, we just accept

$$\{\mathsf{C}_{\{1,2\}}\psi \mid \psi \in \Phi\},$$

where Φ is an axiomatization of the theory of real closed fields. Under this formalization, it is common knowledge that every game has a Nash equilibrium (since, according to (Nash 1991), every game has a Nash equilibrium in $\mathbb{R}$ and the theory of real closed fields is complete (Tarski 1948)), but it does not follow that a Nash equilibrium is common knowledge; see (Wolter 2000a) for details. The following result illustrates formally the fact that common knowledge about theories implies common knowledge about objects only if these objects are denoted by global constant symbols:

Proposition 3.23. *Let $L \in \{\mathbf{K}_2, \mathbf{T}_2, \mathbf{K4}_2, \mathbf{S4}_2, \mathbf{KD45}_2, \mathbf{S5}_2\}$. Suppose that $\varphi(x)$ and ψ are $\mathcal{QL}$-formulas (without epistemic operators), x is the only free variable in φ, and ψ is a sentence. Then $\mathsf{C}_{\{1,2\}}\psi \to \exists x\mathsf{C}_{\{1,2\}}\varphi(x) \in \mathbf{Q}L^C$ iff there exists a constant c such that $\psi \to \varphi(c)$ is in classical logic $\mathbf{QCl}$.*

Proof. The implication ($\Leftarrow$) is clear because constants are interpreted globally. Conversely, suppose there is no constant c such that $\psi \to \varphi(c) \in \mathbf{QCl}$. Then for each constant c we have a $\mathcal{QL}$-structure

$$I(c) = \left\langle D, P_0^{I(c)}, \dots, c_0^{I(c)}, \dots \right\rangle$$

such that $I(c) \not\models \psi \to \varphi(c)$, where D is the set of all constants and $c_i^{I(c)} = c_i$ for all $c \in D$ (such a term model exists, since $\mathcal{QL}$ does not contain equality). Define $\mathfrak{M} = \langle \mathfrak{F}, D, I\rangle$ by taking $\mathfrak{F} = \langle D, R_1, R_2\rangle$, where $R_1 = R_2 = D \times D$. Then $c \models \psi$ and $c \not\models \varphi(c)$ holds for all $c \in D$. So $c \models \mathsf{C}_{\{1,2\}}\psi \wedge \neg\exists x\mathsf{C}_{\{1,2\}}\varphi(x)$ for all $c \in D$. ❑

A detailed discussion of alternative formalizations and the connection with issues in the philosophy of mathematics lies outside the scope of this book. We just wanted to show that the interaction between quantifiers and epistemic operators is not as simple as it may appear when only 'telephone numbers' are considered.

First-order dynamic logics

First-order dynamic logic has a flavor that is quite different from first-order modal logic and even propositional dynamic logic: its modal operators are not constructed from abstract atomic programs, but from concrete programs of the form $x := \tau$ which assign the value of a term τ to a variable x. The worlds (or states) of models of standard dynamic logics consist of structures interpreting first-order logic together with assignments of values to variables. The accessibility relation interpreting the program $x := \tau$ consists of all pairs $(\mathfrak{a}_1, \mathfrak{a}_2)$ of assignments such that $\mathfrak{a}_2$ is obtained from $\mathfrak{a}_1$ by taking $\mathfrak{a}_2(x) = \mathfrak{a}_1(\tau)$. Obviously, this language allows for natural representations of many concrete programs; we refer the reader to (Harel *et al.* 2000) for details. First-order dynamic logic in this sense is outside the scope of this book.

The languages $\mathcal{QDL}$ ('quantified $\mathcal{PDL}$') and $\mathcal{CQDL}$ ('quantified $\mathcal{CPDL}$') we consider here extend $\mathcal{PDL}$ in the same manner as first-order modal logics extend propositional modal logics. We will use logics based on these languages as expressive formalisms into which other logics (like first-order epistemic or temporal logics) can be embedded. For simplicity, we will not even allow for the test operator '?'. Thus, the modal operators of $\mathcal{QDL}$ are composed from abstract atomic programs $\alpha_0, \alpha_1, \dots$ by means of ;, $\cup$, and *. In $\mathcal{CQDL}$ we allow for the converse operator as well. Now the syntax and semantics of $\mathcal{QDL}$ and $\mathcal{CQDL}$ are defined in the obvious manner. By **QDL** and **CQDL** we denote the respective sets of valid formulas. As in the propositional case (Theorem 2.39), all first-order epistemic logics can be embedded into **CQDL**:

Theorem 3.24. *Let* $L \in \{\mathbf{K}_n, \mathbf{T}_n, \mathbf{K4}_n, \mathbf{S4}_n, \mathbf{KD45}_n, \mathbf{S5}_n\}$. *Then* $\mathbf{Q}L^C$ *is polynomially reducible to* **CQDL**.

First-order intuitionistic logic

As its propositional fragment **Int**, first-order intuitionistic logic **QInt** was originally constructed by Heyting (1930) in the form of an axiomatic system reflecting the constructive proof interpretation of the propositional connectives $\to$, $\wedge$, $\vee$, $\bot$ (see Section 2.7) and the quantifiers:

- a proof of $\exists x \varphi(x)$ is a construction presenting an object a together with a proof of $\varphi(a)$;

- a proof of $\forall x \varphi(x)$ is a construction which, given an object a as an input, returns a proof of $\varphi(a)$.

Similar to the propositional case, such a system can be obtained from the classical first-order calculus of Section 1.3 by deleting the law of the excluded middle (A10).

Intuitionistic first-order Kripke models can be defined as a special case of first-order modal Kripke models: they are of the form

$$\mathfrak{M} = \langle \mathfrak{F}, D, \mathfrak{d}, I \rangle ,$$

where

- $\mathfrak{F} = \langle W, R \rangle$ is an intuitionistic frame, i.e., R is a partial order on W,
- I is a function associating with every $w \in W$ a first-order $\mathcal{QL}$-structure

$$I(w) = \left\langle D, P_0^{I(w)}, \dots, c_0^{I(w)}, \dots \right\rangle$$

  such that $c_i^{I(u)} = c_i^{I(v)}$ for all $u, v \in W$,

- $\mathfrak{M}$ has expanding domains, i.e., $\mathfrak{d}(u) \subseteq \mathfrak{d}(v)$ whenever uRv,
- $c_i^{I(u)} \in \mathfrak{d}(u)$ for every $u \in W$,
- the truth of predicates is preserved in all accessible worlds, i.e., for every n-ary predicate symbol P, if uRv then $P^{I(u)} \subseteq P^{I(v)}$.

An *assignment in* D is a function $\mathfrak{a}$ from the set of individual variables to D. The *value* $\tau^{\mathfrak{M},\mathfrak{a}}$ of a term τ in $\mathfrak{M}$ under the assignment $\mathfrak{a}$ is $\mathfrak{a}(x)$ if τ is a variable x, and (the unique) $c^{I(w)}$ if τ is a constant c. The *truth-relation* $(\mathfrak{M}, w) \models^{\mathfrak{a}} \varphi$ (or simply $w \models^{\mathfrak{a}} \varphi$) is defined as follows:

- $w \models^{\mathfrak{a}} P_i(\tau_1, \dots, \tau_n)$ iff $\left\langle \tau_1^{\mathfrak{M},\mathfrak{a}}, \dots, \tau_n^{\mathfrak{M},\mathfrak{a}} \right\rangle \in P_i^{I(w)}$;
- $w \models^{\mathfrak{a}} \psi \wedge \chi$ iff $w \models^{\mathfrak{a}} \psi$ and $w \models^{\mathfrak{a}} \chi$;
- $w \models^{\mathfrak{a}} \psi \vee \chi$ iff $w \models^{\mathfrak{a}} \psi$ or $w \models^{\mathfrak{a}} \chi$;
- $w \models^{\mathfrak{a}} \psi \to \chi$ iff for all v such that wRv, $v \models^{\mathfrak{a}} \psi$ implies $v \models^{\mathfrak{a}} \chi$;
- $w \not\models^{\mathfrak{a}} \bot$;
- $w \models^{\mathfrak{a}} \forall x \psi$ iff $v \models^{\mathfrak{b}} \psi$ for every $v \in W$ with wRv and every assignment $\mathfrak{b}$ in D such that $\mathfrak{b}(x) \in \mathfrak{d}(v)$ and $\mathfrak{a}(y) = \mathfrak{b}(y)$ for all variables $y \neq x$;
- $w \models^{\mathfrak{a}} \exists x \psi$ iff $w \models^{\mathfrak{b}} \psi$ for some assignment $\mathfrak{b}$ in D such that $\mathfrak{b}(x) \in \mathfrak{d}(w)$ and $\mathfrak{a}(y) = \mathfrak{b}(y)$ for all variables $y \neq x$.

(Note that this definition generalizes the truth-conditions for classical first-order formulas and intuitionistic propositional formulas: the clauses for the propositional connectives are the same as in Section 2.7, and if the underlying frame $\mathfrak{F}$ consists of a single point, $\mathfrak{M}$ is simply a $\mathcal{QL}$-structure.) We say that a formula φ is *true in* $\mathfrak{M}$ if $(\mathfrak{M}, w) \models^{\mathfrak{a}} \varphi$ holds for every world $w \in W$ and every assignment $\mathfrak{a}$ in D such that $\mathfrak{a}(x) \in \mathfrak{d}(w)$ for all individual variables x.

As in the propositional case, an intuitionistic first-order Kripke model can be understood as a dynamic database with the set of states W: $\mathfrak{d}(w)$ is the set of all objects available at the state w, and $w \models^{\mathfrak{a}} \varphi$ means that the truth of φ is established at w, given the values of parameters of φ according to $\mathfrak{a}$. Thus, the above truth-definition says that $\exists x \varphi(x)$ is true at w iff at the state w we have an object a such that the truth of $\varphi(x)$ is established at w for $x = a$. On the other hand, $\forall x \varphi(x)$ is true at w iff for every state v accessible from w we can guarantee the truth of $\varphi(x)$ for every replacement of x by any object available at v, i.e., iff the universal truth of φ is predictable at the state w. For example, from the intuitionistic viewpoint, we have

$$\text{10 May 2000} \not\models \exists x \,(\text{‘}x \text{ is a planet’} \wedge \text{‘}x \neq \text{Earth’} \wedge \text{‘there is life on } x\text{’}),$$

but

$$\text{10 May 2000} \not\models \forall x \,(\text{‘}x \text{ is a planet} \wedge \text{‘}x \neq \text{Earth’} \rightarrow \text{‘there is no life on } x\text{’}).$$

As was shown by Kripke (1965) (see also Schütte 1968, Gabbay 1981b), the following completeness theorem holds:

Theorem 3.25. *A $\mathcal{QL}$-formula is in* **QInt** *iff it is true in all intuitionistic Kripke models.*

Using this result it is not hard to see that **QInt** has both the disjunction and the existence properties demonstrating its ‘constructive’ character, viz.,

$$\begin{array}{lll} \varphi \vee \psi \in \mathbf{QInt} & \text{iff} & \varphi \in \mathbf{QInt} \text{ or } \psi \in \mathbf{QInt}; \\ \exists x \varphi(x) \in \mathbf{QInt} & \text{iff} & \varphi(\tau) \in \mathbf{QInt} \text{ for some term } \tau. \end{array}$$

It is worth noting also that the formulas

$$\neg \forall x \neg P(x) \rightarrow \exists x P(x), \quad \neg \exists x \neg P(x) \rightarrow \forall x P(x)$$

do not belong to **QInt**. Indeed, they are refuted in a model with two worlds uRv each of which has one object, say a, such that

$$u \not\models P(x) \text{ and } v \models P(x).$$

Thus, the quantifiers $\forall$ and $\exists$ are not dual in **QInt** as they are in **QCl**.

Similarly to the propositional case, first-order intuitionistic logic can be interpreted in first-order (classical) modal logic using the Gödel translation T which prefixes $\Box$ to every subformula of a $\mathcal{QL}$-formula. This translation turns out to be an embedding of **QInt** into the modal logic $\mathbf{Q}^e\mathbf{S4}$ determined by quasi-ordered models with expanding domains. Namely, for every $\mathcal{QL}$-formula φ,

$$\varphi \in \mathbf{QInt} \quad \text{iff} \quad \mathsf{T}(\varphi) \in \mathbf{Q}^e\mathbf{S4}.$$

A proof can be found in (Rasiowa and Sikorski 1963).

Denote by **QIntCD** the extension of **QInt** with the axiom schema

$$\boldsymbol{cd} = \forall x\ (\varphi(x) \vee \psi) \rightarrow \forall x\ \varphi(x) \vee \psi.$$

The following result was obtained by Görnemann (1971) (see also Gabbay 1981b): for every $\mathcal{QL}$-formula φ,

$\varphi \in \mathbf{QIntCD}$ iff φ is true in all intuitionistic Kripke models with constant domains.

As a consequence we have:

$$\varphi \in \mathbf{QIntCD} \quad \text{iff} \quad \mathsf{T}(\varphi) \in \mathbf{QS4}.$$

Thus, semantically the formula $\boldsymbol{cd}$ can be viewed as an intuitionistic analog of the Barcan formula $\forall x \Box P(x) \rightarrow \Box \forall x P(x)$. Note, however, that the Barcan formula cannot be derived from $\mathsf{T}(\boldsymbol{cd})$ and that the extension of $\mathbf{Q}^e\mathbf{S4}$ with $\mathsf{T}(\boldsymbol{cd})$ turns out to be a proper sublogic of **QS4**, which means, in particular, that this extension is incomplete with respect to the above Kripke semantics; see (Shehtman and Skvortsov 1990).

We conclude this section by noting that satisfiability in arbitrary intuitionistic Kripke models reduces to satisfiability in models with constant domains. This can be shown in the same way as in the modal case. Namely, let E be a unary predicate symbol which has no occurrences in φ. Define the relativization $\varphi \downarrow E$ of φ by taking

$$\begin{aligned}
P_i(\tau_1, \ldots, \tau_n) \downarrow E &= P_i(\tau_1, \ldots, \tau_n), \\
(\psi \odot \chi) \downarrow E &= (\psi \downarrow E) \odot (\chi \downarrow E), \text{ where } \odot \in \{\vee, \wedge, \rightarrow\}, \\
(\bot) \downarrow E &= \bot, \\
(\forall x \psi) \downarrow E &= \forall x\ (E(x) \rightarrow (\psi \downarrow E)), \\
(\exists x \psi) \downarrow E &= \exists x\ (E(x) \wedge (\psi \downarrow E)).
\end{aligned}$$

The reader can readily prove by induction that, for every $\mathcal{QL}$-sentence φ, φ is satisfied in an intuitionistic model based on a frame $\mathfrak{F}$ and having expanding domains iff the sentence $\varphi \downarrow E \wedge \exists x E(x) \wedge E(\bar{c})$ is satisfied in a model based on $\mathfrak{F}$ and having constant domains, where $E(\bar{c}) = E(c_1) \wedge \cdots \wedge E(c_k)$ and $c_1, \ldots, c_k$ are all the constants occurring in φ.

For more information on first-order intuitionistic logic and its extensions we refer the reader to (Dummett 1977, van Dalen 1986, Gabbay *et al.* 2000).

3.7 First-order temporal logics

Let us turn now to first-order extensions of temporal logics. These kinds of first-order modal logics are probably the most interesting from the viewpoint of possible applications in computer science and artificial intelligence: they have been used in program specification and verification (Pnueli 1986, Manna and Pnueli 1992, 1995), temporal databases and e-commerce (Chomicki 1994, Abiteboul *et al.* 1996, Chomicki and Toman 1998, Spielmann 2000, Chomicki *et al.* 2001), knowledge representation and reasoning (Fagin *et al.* 1995, Schild 1993, Wolter and Zakharyaschev 2000b, Artale and Franconi 2003) and some other fields.

For instance, in temporal databases first-order temporal logic can be used as both a query language and a language capable of formalizing temporal integrity constraints. Here are two simple examples (more information and further references can be found in Chomicki and Toman 1998). The query

> *'find all people who have been unemployed since their graduation and married before graduation'*

can be represented as the formula

$$\mathsf{unemployed}(x)\ \mathcal{S}\ \big(\mathsf{grad}(x) \wedge \Diamond_P \mathsf{married}(x)\big).$$

The temporal integrity constraint

> *'a student cannot graduate without attending a course in logic'*

can be rephrased as

$$\neg\exists x\ \Big(\mathsf{grad}(x) \wedge \Box_P \neg\exists y\ \big(\mathsf{attend}(x, y) \wedge \mathsf{logic\_course}(y)\big)\Big).$$

(The reader should appreciate the elegance and readability of these formulas. Later on in this section we shall see first-order translations which are much more artificial.)

Formally, the language of first-order temporal logic and its models are defined as follows. Denote by $\mathcal{QTL}$ the *first-order temporal language* constructed in the standard way from the alphabet of $\mathcal{QML}_l$ in which the boxes of $\mathcal{ML}_l$ are replaced with the (binary) *temporal operators* $\mathcal{S}$ (since) and $\mathcal{U}$ (until) of the language $\mathcal{ML}_{\mathcal{SU}}$. As before, we use the standard abbreviations:

$$\Diamond_F \varphi = \top \mathcal{U} \varphi, \qquad \Box_F \varphi = \neg \Diamond_F \neg \varphi,$$

$$\Diamond_P \varphi = \top \mathcal{S} \varphi, \qquad \Box_P \varphi = \neg \Diamond_P \neg \varphi,$$

$$\bigcirc \varphi = \bot \mathcal{U} \varphi.$$

$\mathcal{QTL}$ is interpreted in *first-order temporal models* of the form $\mathfrak{M} = \langle \mathfrak{F}, D, I \rangle$, where $\mathfrak{F} = \langle W, < \rangle$ is a strict linear order representing the flow of time, D is a nonempty set, the *domain* of $\mathfrak{M}$, and I is a function associating with every moment of time $w \in W$ a first-order $\mathcal{QL}$-structure

$$I(w) = \left\langle D, P_0^{I(w)}, \dots, c_0^{I(w)}, \dots \right\rangle,$$

the *state* of $\mathfrak{M}$ at moment w, such that $c_i^{I(u)} = c_i^{I(v)}$ for all $u, v \in W$. Given an assignment $\mathfrak{a}$ in D, we define the *truth-relation* $(\mathfrak{M}, w) \models^{\mathfrak{a}} \varphi$ (or simply $w \models^{\mathfrak{a}} \varphi$) as in Section 3.6, adding the standard temporal clauses:

- $w \models^{\mathfrak{a}} \varphi \mathcal{S} \psi$ iff there is $v < w$ such that $v \models^{\mathfrak{a}} \psi$ and $u \models^{\mathfrak{a}} \varphi$ for every $u \in (v, w)$;
- $w \models^{\mathfrak{a}} \varphi \mathcal{U} \psi$ iff there is $v > w$ such that $v \models^{\mathfrak{a}} \psi$ and $u \models^{\mathfrak{a}} \varphi$ for every $u \in (w, v)$.

As before, we say that a formula φ is *true in* $\mathfrak{M}$ if $(\mathfrak{M}, w) \models^{\mathfrak{a}} \varphi$ holds for all assignments $\mathfrak{a}$ in D and all time points w in W. (We do not consider here models with expanding, varying, or decreasing domains: the discussion on domain assumptions of Section 3.6 can be translated to the temporal context in a straightforward way. In particular, the satisfiability problem for $\mathcal{QTL}$-formulas in first-order temporal models with varying and expanding domains is reducible to the same problem in models with constant domains.)

For a class $\mathcal{C}$ of strict linear orders, we denote by $\mathsf{QLog}_{\mathcal{SU}}(\mathcal{C})$ the set of $\mathcal{QTL}$-formulas that are true in all models based on frames in $\mathcal{C}$:

$$\mathsf{QLog}_{\mathcal{SU}}(\mathcal{C}) = \{\varphi \in \mathcal{QTL} \mid (\mathfrak{M}, w) \models^{\mathfrak{a}} \varphi \text{ for all } \mathfrak{M} = \langle \mathfrak{F}, D, I \rangle \text{ with } \mathfrak{F} \in \mathcal{C}, \text{ all } w \text{ in } \mathfrak{F}, \text{ and all assignments } \mathfrak{a} \text{ in } D\}.$$

$\mathsf{QLog}_{\mathcal{SU}}(\mathcal{C})$ is said to be the (*first-order* or *quantified*) *temporal logic of the class* $\mathcal{C}$. We will also be considering the $\mathcal{S}$-free sublanguage $\mathcal{QTL}_{\mathcal{U}}$ of $\mathcal{QTL}$, and the logic

$$\mathsf{QLog}_{\mathcal{U}}(\mathcal{C}) = \mathsf{QLog}_{\mathcal{SU}}(\mathcal{C}) \cap \mathcal{QTL}_{\mathcal{U}}.$$

Instead of $\mathsf{QLog}_{\mathcal{SU}}(\{\langle \mathbb{N} < \rangle\})$ and $\mathsf{QLog}_{\mathcal{U}}(\{\langle \mathbb{N}, < \rangle\})$ we write $\mathsf{QLog}_{\mathcal{SU}}(\mathbb{N})$ and $\mathsf{QLog}_{\mathcal{U}}(\mathbb{N})$, respectively; similar notation is used for $\langle \mathbb{Z}, < \rangle$, $\langle \mathbb{Q}, < \rangle$, and $\langle \mathbb{R}, < \rangle$.

Unfortunately, the temporal logics of many natural and useful flows of time turn out to be not recursively enumerable, and so not presentable as axiomatic systems with finitely many axiom schemata. Such are, for instance, $\mathsf{QLog}_{\mathcal{SU}}(\mathbb{N})$, $\mathsf{QLog}_{\mathcal{SU}}(\mathbb{Z})$ or $\mathsf{QLog}_{\mathcal{SU}}(\mathbb{R})$ (some proofs can be found in Chapter 11; for a general result consult (Gabbay *et al.* 1994)). Here we show the axiomatizations of Reynolds (1996) for the temporal logics $\mathsf{QLog}_{\mathcal{SU}}(\mathcal{LO})$, where $\mathcal{LO}$ is the class of all strict linear orders, and $\mathsf{QLog}_{\mathcal{SU}}(\mathbb{Q})$.

Theorem 3.26. (i) $\mathsf{QLog}_{\mathcal{SU}}(\mathcal{LO})$ *is axiomatized by the following axiom schemata and inference rules:*

Axiom schemata: *those of* **QCl** *plus*

$$\psi\mathcal{U}(\exists x\varphi) \to \exists x(\psi\mathcal{U}\varphi),$$
$$\Box_F(\varphi \to \psi) \to (\chi\mathcal{U}\varphi \to \chi\mathcal{U}\psi),$$
$$\Box_F(\varphi \to \psi) \to (\varphi\mathcal{U}\chi \to \psi\mathcal{U}\chi),$$
$$\varphi \wedge (\chi\mathcal{U}\psi) \to \chi\mathcal{U}(\psi \wedge (\chi\mathcal{S}\varphi)),$$
$$\psi\mathcal{U}\varphi \to (\psi \wedge (\psi\mathcal{U}\varphi))\mathcal{U}\varphi,$$
$$\psi\mathcal{U}(\psi \wedge (\psi\mathcal{U}\varphi)) \to \psi\mathcal{U}\varphi,$$
$$(\psi\mathcal{U}\varphi) \wedge (\beta\mathcal{U}\alpha) \to (\psi \wedge \beta)\mathcal{U}(\varphi \wedge \alpha) \vee (\psi \wedge \beta)\mathcal{U}(\varphi \wedge \beta) \vee (\psi \wedge \beta)\mathcal{U}(\psi \wedge \alpha),$$

and their past counterparts.

Inference rules: *those of* **QCl** *plus RN for both* $\Box_F$ *and* $\Box_P$.

(ii) $\mathsf{QLog}_{\mathcal{SU}}(\mathbb{Q})$ *can be axiomatized by extending* $\mathsf{QLog}_{\mathcal{SU}}(\mathcal{LO})$ *with two extra axioms:* $\Diamond_F\top \wedge \Diamond_P\top$ *and* $\neg(\bot\mathcal{U}\top) \wedge \neg(\bot\mathcal{S}\top)$ *(saying that the flow of time has no end points and is dense).*

According to Theorem 2.5, the propositional temporal language $\mathcal{ML}_{\mathcal{SU}}$ is expressively complete for the flows of time $\langle\mathbb{N},<\rangle$, $\langle\mathbb{Z},<\rangle$, $\langle\mathbb{R},<\rangle$. In this connection it would be interesting to find out whether this characterization of the expressive power of $\mathcal{ML}_{\mathcal{SU}}$ can be lifted to $\mathcal{QTL}$. Actually, this question was raised in the context of temporal databases; see (Abiteboul *et al.* 1996, Chomicki 1994, Chomicki and Niwinski 1995). $\mathcal{QTL}$ provides only 'implicit' access to time: quantification over points in time in the sense of first-order logic is not permitted, and the only means of expressing temporal properties is by the operators $\mathcal{S}$ and $\mathcal{U}$. An obvious alternative is to reason about time explicitly, using the full power of first-order logic. This leads us to a *two-sorted first-order language*, called $\mathcal{TS}$ in what follows, one sort of which refers to points in time and the other to the first-order domain. In $\mathcal{TS}$, every predicate P has precisely one 'temporal argument' so that $P(t, x_1, \ldots, x_n)$ means that P applies to $x_1, \ldots, x_n$ at moment t. (This reflects the *timestamp* view of temporal databases, see Chomicki 1994.) The query and the constraint above can be reformulated in $\mathcal{TS}$ as the formulas

$$\exists t_1 \Big(t_1 < t_0 \wedge \mathsf{grad}(t_1, x) \wedge \exists t_2 \ (t_2 < t_1 \wedge \mathsf{married}(t_2, x) \wedge \\ \forall t_3(t_1 < t_3 < t_0 \rightarrow \mathsf{unemployed}(t_3, x)))\Big)$$

and

$$\forall t \ \Big(\neg\exists x\big(\mathsf{grad}(t, x) \wedge \forall t' \ (t' < t \rightarrow \neg\exists y \ (\mathsf{attend}(t', x, y) \wedge \\ \mathsf{logic\_course}(t', y))))\big)\Big),$$

respectively.

Let us define the syntax and semantics of $\mathcal{TS}$ more precisely. $\mathcal{TS}$ is based on the following alphabet:

- individual variables $x_0, x_1, \dots$ (or $x, y, z, \dots$) and constants $c_0, c_1, \dots$ of *domain sort* (we also call them *domain terms*),
- individual variables $t_0, t_1, \dots$ (or $t, t', t'', \dots$) of *temporal sort*,
- the binary predicate symbol $<$ of sort 'temporal $\times$ temporal,'
- predicate symbols $P_0, P_1, \dots$ of sort 'temporal $\times$ domain^n,' $n < \omega$,
- the Boolean logical connectives $\neg$ and $\wedge$,
- the universal quantifier $\forall$.

Formulas of $\mathcal{TS}$ are defined inductively:

- $t_i < t_j$ is an (atomic) formula, for temporal variables t_i, t_j,
- $P(t, \tau_1, \dots, \tau_n)$ is an (atomic) formula, for a predicate symbol P of sort 'temporal $\times$ domain^n,' a temporal variable t, and domain terms $\tau_1, \dots, \tau_n$,
- if φ and ψ are formulas, t a temporal variable, and x a domain variable, then $\neg\varphi$, $\varphi \wedge \psi$, $\forall t\varphi$ and $\forall x\varphi$ are formulas.

$\mathcal{TS}$ is interpreted in the same kind of *first-order temporal models* as $\mathcal{QTL}$, i.e., structures of the form $\mathfrak{M} = \langle \mathfrak{F}, D, I \rangle$, where $\mathfrak{F} = \langle W, < \rangle$ is a flow of time, D is a nonempty set, the *domain* of $\mathfrak{M}$, and I a function associating with every moment of time $w \in W$ a first-order $\mathcal{QL}$-structure

$$I(w) = \Big\langle D, P_0^{I(w)}, \dots, c_0^{I(w)}, \dots \Big\rangle,$$

in which $P_i^{I(w)}$ is an n-ary relation on D whenever P_i is a predicate symbol of arity $n+1$, and $c_i^{I(w)} \in D$ with $c_i^{I(u)} = c_i^{I(v)}$ for all $u, v \in W$.

An *assignment in* $\mathfrak{M}$ is a function $\mathfrak{a} = \mathfrak{a}_1 \cup \mathfrak{a}_2$ such that $\mathfrak{a}_1$ associates with every temporal variable t a moment of time $\mathfrak{a}_1(t) \in W$ and $\mathfrak{a}_2$ associates with every domain variable x an element $\mathfrak{a}_2(x)$ of D. The *value* $\tau^{\mathfrak{M},\mathfrak{a}}$ of a domain term τ in $\mathfrak{M}$ under the assignment $\mathfrak{a}$ is $\mathfrak{a}_2(x)$ if τ is a domain variable x, and (the time independent) $c^{I(w)}$ if τ is a constant c.

The *truth-relation* $\mathfrak{M} \models^{\mathfrak{a}} \varphi$ is defined inductively as follows:

- $\mathfrak{M} \models^{\mathfrak{a}} t_i < t_j$ iff $\mathfrak{a}_1(t_i) < \mathfrak{a}_1(t_j)$ in $\mathfrak{F}$,
- $\mathfrak{M} \models^{\mathfrak{a}} P(t, \tau_1, \dots, \tau_n)$ iff $\left\langle \tau_1^{\mathfrak{M},\mathfrak{a}}, \dots, \tau_n^{\mathfrak{M},\mathfrak{a}} \right\rangle \in P^{I(\mathfrak{a}_1(t))}$,
- $\mathfrak{M} \models^{\mathfrak{a}} \forall t \varphi$ iff $\mathfrak{M} \models^{\mathfrak{b}} \varphi$ for every assignment $\mathfrak{b}$ that may differ from $\mathfrak{a}$ only on t,
- $\mathfrak{M} \models^{\mathfrak{a}} \forall x \varphi$ iff $\mathfrak{M} \models^{\mathfrak{b}} \varphi$ for every assignment $\mathfrak{b}$ that may differ from $\mathfrak{a}$ only on x,

and the standard clauses for the Booleans.

It is fairly easy to see that everything expressible in $\mathcal{QTL}$ can be expressed in $\mathcal{TS}$ as well. Indeed, suppose that each n-ary predicate symbol Q_i of $\mathcal{QTL}$ is associated with the $(n+1)$-ary predicate symbol P_i of $\mathcal{TS}$. Define a translation $\ddagger$ from $\mathcal{QTL}$ into $\mathcal{TS}$ by taking, for some fixed temporal variable t,

$$
\begin{aligned}
Q_i(\tau_1, \dots, \tau_n)^\ddagger &= P_i(t, \tau_1, \dots, \tau_n), \\
(\varphi \wedge \psi)^\ddagger &= \varphi^\ddagger \wedge \psi^\ddagger, \\
(\neg \varphi)^\ddagger &= \neg(\varphi^\ddagger), \\
(\forall x \varphi)^\ddagger &= \forall x (\varphi^\ddagger), \\
(\varphi \mathcal{S} \psi)^\ddagger &= \exists t' (t' < t \wedge \psi^\ddagger\{t'/t\} \wedge \forall t''(t' < t'' < t \rightarrow \varphi^\ddagger\{t''/t\})), \\
(\varphi \mathcal{U} \psi)^\ddagger &= \exists t' (t < t' \wedge \psi^\ddagger\{t'/t\} \wedge \forall t''(t < t'' < t' \rightarrow \varphi^\ddagger\{t''/t\})),
\end{aligned}
$$

where t' and t'' are fresh temporal variables. Let $\mathfrak{M} = \langle \mathfrak{F}, D, I \rangle$ be a first-order temporal structure such that, for all w in $\mathfrak{F}$,

$$
I(w) = \left\langle D, P_0^{I(w)}, \dots, c_0^{I(w)}, \dots \right\rangle .
$$

Then the relations $P_i^{I(w)}$ can be regarded as interpretations of the n-ary predicate symbols Q_i of $\mathcal{QTL}$, as well as interpretations of the 'domain parts' of the corresponding $(n+1)$-ary predicate symbols P_i of $\mathcal{TS}$. Thus, we have the following:

Lemma 3.27. *For every $\mathcal{QTL}$-formula φ, every moment of time w and every assignment $\mathfrak{a}$ in D,*

$$(\mathfrak{M}, w) \models^{\mathfrak{a}} \varphi \qquad \textit{iff} \qquad \mathfrak{M} \models^{\mathfrak{b}} \varphi^{\ddagger},$$

where $\mathfrak{b} = \mathfrak{b}_1 \cup \mathfrak{a}$ and $\mathfrak{b}_1(t) = w$.

We are now in a position to formulate a natural first-order version of the expressive completeness definition for propositional temporal logic. Let $\mathcal{C}$ be a class of flows of time, $\mathcal{L}'$ a sublanguage of $\mathcal{QTL}$ and $\mathcal{L}''$ a sublanguage of $\mathcal{TS}$. We say that *$\mathcal{L}'$ is expressively complete for $\mathcal{L}''$ over $\mathcal{C}$* if for every $\mathcal{L}''$-formula $\varphi(t)$ with at most one free temporal variable, there exists an $\mathcal{L}'$-formula $\widehat{\varphi}$ such that for all models $\mathfrak{M}$ based on flows of time in $\mathcal{C}$,

$$\mathfrak{M} \models \forall t \left(\varphi \leftrightarrow (\widehat{\varphi})^{\ddagger}\right).$$

In this case we also say that *$\widehat{\varphi}$ expresses φ over $\mathcal{C}$.*

Having the expressive completeness result for $\mathcal{ML}_{\mathcal{SU}}$, it may seem plausible to conjecture that there are interesting classes $\mathcal{C}$ of flows of time over which $\mathcal{QTL}$ is expressively complete for $\mathcal{TS}$ itself. Unfortunately, this conjecture turns out to be wrong: the sentence

$$\exists t_1 \exists t_2 \left(t_1 < t_2 \wedge \forall x \; (P(t_1, x) \leftrightarrow P(t_2, x))\right)$$

is not expressible in $\mathcal{QTL}$ over any interesting class of flows of time; see (Kamp 1971), where this is proved for $\{\langle \mathbb{Q}, < \rangle\}$, and (Abiteboul *et al.* 1996), where this is proved for $\{\langle \mathbb{N}, < \rangle\}$ and the class of all finite linear orders.

We now define a natural fragment of $\mathcal{TS}$ for which $\mathcal{QTL}$ is expressively complete over every class of flows of time $\mathcal{C}$ for which $\mathcal{ML}_{\mathcal{SU}}$ is expressively complete.

Denote by $\mathcal{TS}_{1t}$ the set of all $\mathcal{TS}$-formulas φ which do not contain subformulas of the form $\forall x \psi$ such that ψ has more than one free temporal variable.

Note that for every $\mathcal{QTL}$-formula φ, we have $\varphi^{\ddagger} \in \mathcal{TS}_{1t}$. The following result was obtained in (Hodkinson *et al.* 2000):

Theorem 3.28. *Let $\mathcal{C}$ be any class of flows of time for which $\mathcal{ML}_{\mathcal{SU}}$ is expressively complete (for example, the class*

$$\{\langle \mathbb{N}, < \rangle, \langle \mathbb{Z}, < \rangle, \langle \mathbb{R}, < \rangle\} \cup \{\mathfrak{F} \mid \mathfrak{F} \text{ a finite strict linear order}\}).$$

Then $\mathcal{QTL}$ is expressively complete for $\mathcal{TS}_{1t}$ over $\mathcal{C}$.

Proof. Suppose that $\mathcal{ML}_{\mathcal{SU}}$ is expressively complete for $\mathcal{C}$. So for any formula $\varphi(t, P_1, \ldots, P_k)$ of the first-order language $\mathcal{QL}_t$ with one free variable t

and unary predicates $P_1, \dots, P_k$, we may fix a propositional temporal formula $\overline{\varphi}(p_1, \dots, p_k)$ such that for every first-order structure

$$\mathfrak{M} = \langle W, <, P_0^{\mathfrak{M}}, P_1^{\mathfrak{M}}, \dots \rangle$$

based on a flow of time $\mathfrak{F} = \langle W, < \rangle \in \mathcal{C}$, and every $\mathcal{ML}_{\mathcal{SU}}$-model $\mathfrak{N} = \langle \mathfrak{F}, \mathfrak{V} \rangle$ with $\mathfrak{V}(p_i) = P_i^{\mathfrak{M}}$, we have

$$(\mathfrak{N}, w) \models \overline{\varphi} \quad \text{iff} \quad \mathfrak{M} \models \varphi[w/t], \text{ for all } w \in W.$$

Suppose now that $\chi = \chi(t, Q_1, \dots, Q_k)$ is a $\mathcal{TS}_{1t}$-formula. We prove that for every subformula ψ of χ with at most one free temporal variable, there is a $\mathcal{QTL}$-formula $\widehat{\psi}$ that expresses ψ. The proof is by induction on the construction of ψ.

Case 1: ψ is atomic. If $\psi = t < t$, then put $\widehat{\psi} = \bot$. If $\psi = Q_i(t, x_1, \dots, x_n)$, then put $\widehat{\psi} = P_i(x_1, \dots, x_n)$.

Case 2: $\psi = \forall x \psi_1$. By the induction hypothesis, there exists $\widehat{\psi_1}$ that expresses ψ_1. But then $\widehat{\psi} = \forall x \widehat{\psi_1}$ expresses ψ.

Case 3: otherwise. Let $\psi_1, \dots, \psi_l$ be a list of all subformulas of ψ of the form either $Q_i(t', y_1, \dots, y_n)$ or $\forall z \psi'$ that have an occurrence in ψ that is not within the scope of a domain quantifier $\forall y$. (This means that ψ is constructed from $\psi_1, \dots, \psi_l$ using the Booleans and quantification over temporal variables.) Since $\psi \in \mathcal{TS}_{1t}$, every ψ_i of the form $\forall z \psi_i'$ has at most one free temporal variable. Thus, by the induction hypothesis, there exists a $\mathcal{QTL}$-formula $\widehat{\psi_i}$ that expresses ψ_i, for each $i \leq l$.

Now replace in ψ every occurrence of a $\psi_i(t')$ that is not within the scope of a $\forall y$ by a unary predicate $Q_{\psi_i}(t')$. Denote the resulting $\mathcal{QL}_t$-formula by $\psi'(t, Q_{\psi_1}, \dots, Q_{\psi_l})$ (note that it contains no free variables different from t). Take the propositional temporal formula $\overline{\psi'}(q_{\psi_1}, \dots, q_{\psi_l})$ expressing ψ', and in it, replace every propositional variable q_{ψ_i} by $\widehat{\psi_i}$. The resulting formula $\widehat{\psi}$ clearly expresses ψ.

This completes the induction. So there is a $\mathcal{QTL}$-formula $\widehat{\chi}$ expressing χ, which proves the claim of the theorem. ❑

We conclude this section by establishing a connection between products of propositional temporal logics with **S5** and first-order temporal logics. Similarly to the first-order modal case, the translation † from $\mathcal{ML}_n$ into the n-variable fragment of $\mathcal{QL}$ defined in Section 3.5 can be extended to a translation from $\mathcal{ML}_{\mathcal{SU}} \otimes \mathcal{ML}_n$ into the n-variable fragment of $\mathcal{QTL}$ by taking

$$\begin{aligned} p_i^\dagger &= P_i(x_0, \dots, x_{n-1}), \\ (\varphi \wedge \psi)^\dagger &= \varphi^\dagger \wedge \psi^\dagger, \\ (\neg\varphi)^\dagger &= \neg\varphi^\dagger, \end{aligned}$$

$$
\begin{aligned}
(\varphi \mathcal{S} \psi)^\dagger &= \varphi^\dagger \mathcal{S} \psi^\dagger, \\
(\varphi \mathcal{U} \psi)^\dagger &= \varphi^\dagger \mathcal{U} \psi^\dagger, \\
(\Box_j \psi)^\dagger &= \forall x_{j-1} \psi^\dagger, \qquad \text{for } j = 1, \dots, n.
\end{aligned}
$$

Then the following theorem can be proved in a way similar to Theorem 3.21:

Theorem 3.29. *Let $\mathcal{C}$ be a class of strict linear orders. Then for every $\mathcal{ML}_{\mathcal{SU}} \otimes \mathcal{ML}_n$-formula φ,*

$$\varphi \in \mathsf{Log}_{\mathcal{SU}}(\mathcal{C} \times \overbrace{\mathsf{FrS5} \times \cdots \times \mathsf{FrS5}}^{n}) \quad \textit{iff} \quad \varphi^\dagger \in \mathsf{QLog}_{\mathcal{SU}}(\mathcal{C}).$$

In particular, by Theorem 6.29 to be proved in Section 6.4, we shall have:

Theorem 3.30. *For every $\mathcal{ML}_{\mathcal{U}} \otimes \mathcal{ML}$-formula φ,*

$$\varphi \in \mathbf{PTL} \times \mathbf{S5} \quad \textit{iff} \quad \varphi^\dagger \in \mathsf{QLog}_{\mathcal{U}}(\mathbb{N}).$$

This observation will be used in Section 6.5 for establishing an upper bound for the complexity of $\mathbf{PTL} \times \mathbf{S5}$, and in Section 11.4 for establishing lower bounds for the complexity of fragments of $\mathsf{QLog}_{\mathcal{U}}(\mathbb{N})$.

3.8 Description logics with modal operators

Description logics have been designed and used as a formalism for knowledge representation and reasoning only in *static* application domains. They are not able to express such dynamic aspects of knowledge as time- or action-dependence, beliefs of different agents, obligations, etc., which are regarded to be important ingredients in modeling intelligent agents.

Imagine, for instance, a car salesman who, trying to understand the development of the car market, implements a knowledge base about his customers. Besides standard ABox and TBox of the form

$$
\begin{aligned}
&\mathit{John} : \exists \mathsf{has.Car} \\
&\mathit{John} \ \mathsf{likes} \ \mathit{Golf} \\
&\mathit{Golf} : \mathsf{VW} \\
&\mathsf{VW} \sqsubseteq \mathsf{Car} \\
&\mathsf{Male\_customer} = \mathsf{Male} \sqcap \mathsf{Customer} \\
&\mathsf{Modern\_car} = \mathsf{Car} \sqcap \exists \mathsf{has.Computer}
\end{aligned}
$$

it may also contain 'modalized' formulas, e.g.,

$$
\begin{aligned}
&\mathsf{Customer} = \mathsf{Homo\_sapiens} \sqcap \langle \text{sometime in the past} \rangle \, \exists \mathsf{buys.Car} \\
&\mathsf{Potential\_customer} = \langle \text{eventually} \rangle \, \mathsf{Customer}
\end{aligned}
$$

$$\mathsf{Faithful\_customer} = \mathsf{Customer} \sqcap \exists[\mathsf{always}]\mathsf{buys.Car}$$

$$\langle\text{John believes}\rangle\,\langle\text{next year}\rangle\,(\mathsf{Male\_customer} \sqsubseteq \exists\mathsf{buys.Modern\_car})$$

The meaning of the first two modalized formulas should be clear. The third one means that a faithful customer always buys a car of the same type, say, Golf. And the fourth formula says that according to John's beliefs, next year every male customer will buy a modern car.

To provide such a language with a reasonable semantics, we obviously come to many-dimensional structures. First, we need an object dimension—a usual model of the underlying description language. To capture beliefs of agents, every such model may have a number of alternatives. And to reflect the development of the knowledge base in time we need a time axis. The whole model thus has at least three dimensions.

There are several many-dimensional approaches in the literature to the design of 'dynamic' description logics (see, e.g., Schmiedel 1990, Schild 1993, Laux 1994, Graber *et al.* 1995, Baader and Ohlbach 1995, Artale and Franconi 1998, Baader and Laux 1995, Wolter and Zakharyaschev 1998, 2000a, 2000c). Perhaps the most general perspective was proposed by Baader and Ohlbach (1995). Roughly speaking, each dimension (object, time, belief, etc.) is represented by a set D_i (of objects, moments of time, possible worlds, etc.), concepts are interpreted as subsets of the Cartesian product $\prod_{i=1}^{n} D_i$ and roles of dimension i as binary relations between n-tuples that may differ only in the ith coordinate. And one can quantify over roles not only to obtain concepts, but also roles themselves and concept equations (like in the example above). However, the constructed language turned out to be too expressive: the satisfiability problem in such models is undecidable.

Trying to simplify this semantics, Baader and Laux (1995) noticed that different dimensions may have different status. For instance, time should probably be the same for all objects inhabiting the object dimension of our knowledge base. This observation led to a somewhat more transparent semantics: models now consist of worlds (or states) which represent—in terms of some standard description logic—the 'current states of affairs;' these worlds may change with time passing by or under certain actions, or they may have a number of alternative worlds reflecting the beliefs of agents, and the connection between concepts and roles from different worlds is described by means of the corresponding temporal, dynamic, epistemic, or some other 'modal' operators.

There are several 'degrees of freedom' within this semantical paradigm:

1. The worlds in models may have arbitrary, expanding or constant domains. Of course, the choice depends on the application we deal with. However, as in first-order modal logic, technically the most important is the constant domain assumption: we shall see that if the satisfiab-

ility problem is decidable in models with constant domains then it is decidable in models with expanding or varying domains as well.

2. The concept, role and object names of the underlying description language may be *local* or *global*. Global names have the same values in all worlds, while local ones may have different values. However, technically local object names present no difficulty as compared with global ones, and it will be shown that global concepts are expressible via local concepts and modal operators. On the other hand, we shall see in Chapter 14 that it does make a difference whether we use local or global role names.

3. As we saw in the example above, in general we may need modal operators applicable to all syntactic terms of the language: concepts, roles and formulas. However, sometimes only some of them require modal 'quantification.'

4. And finally, depending on the application domain we may choose between various kinds of modal operators (e.g., temporal, epistemic, action, etc.), the corresponding accessibility relations (say, linear for time, universal for knowledge, arbitrary for actions), and between the underlying pure description logics.

We begin by introducing a modal description language $\mathcal{ML}_{\mathcal{ALC}}$ whose alphabet consists of the alphabet of $\mathcal{ALC}$ (where we distinguish between *global role names* $R_0, R_1, \ldots$ and *local role names* $S_0, S_1, \ldots$) and the necessity operators $\Box_1, \ldots, \Box_n$ together with their duals $\Diamond_1, \ldots, \Diamond_n$ of the modal language $\mathcal{ML}_n$.

Starting from this alphabet, we construct compound *concepts* and *roles* in the following way:

- all role names are *roles*, and all concept names are *concepts*;
- if R is a role then so are $\Box_i R$ and $\Diamond_i R$, for every $i = 1, \ldots, n$ (we will call such roles *modalized*);
- if C, D are concepts and R is a role then $C \sqcap D$, $\neg C$, $\exists R.C$ and $\Box_i C$ are concepts, $i = 1, \ldots, n$.

Atomic $\mathcal{ML}_{\mathcal{ALC}}$-formulas are expressions of the form $C = D$, $a : C$, aRb, where a and b are object names. If φ and ψ are *$\mathcal{ML}_{\mathcal{ALC}}$-formulas* then so are $\varphi \wedge \psi$, $\neg\varphi$ and $\Box_i\varphi$.

The intended semantics of $\mathcal{ML}_{\mathcal{ALC}}$ is defined as follows. Suppose that $\mathfrak{F} = \langle W, \lhd_1, \ldots, \lhd_n \rangle$ is an n-frame. An *$\mathcal{ML}_{\mathcal{ALC}}$-model based on $\mathfrak{F}$* is a pair

$\mathfrak{M} = \langle \mathfrak{F}, I \rangle$ in which I is a function associating with each $w \in W$ an $\mathcal{ALC}$-model

$$I(w) = \left\langle \Delta, R_0^{I(w)}, \ldots, S_0^{I(w)}, \ldots, C_0^{I(w)}, \ldots, a_0^{I(w)}, \ldots \right\rangle,$$

where

- Δ is a nonempty set, called the *domain* of $\mathfrak{M}$,
- $R_i^{I(w)}$ are binary relations on Δ such that $R_i^{I(u)} = R_i^{I(v)}$ for all $u, v \in W$ (they interpret the global role names),
- $S_i^{I(w)}$ are arbitrary binary relations on Δ (interpreting the local role names),
- $C_i^{I(w)}$ are subsets of Δ (interpreting the concept names), and
- $a_i^{I(w)}$ are elements of Δ such that $a_i^{I(u)} = a_i^{I(v)}$ for any $u, v \in W$ (they interpret the (global) object names).

The definition of a model given above presupposes that we accept the *constant domain assumption*, that the object names are *rigid designators* (i.e., they are global), and that all concepts are local. Later on in this section we shall see that we do not lose too much by imposing these restrictions.

To define models with *varying domains*, one should replace everywhere in the above definition the common domain Δ with an individual nonempty domain Δ^w, for each world w. In models with *expanding domains*, we have $\Delta^u \subseteq \Delta^v$ whenever $u \lhd_i v$, for some $i = 1, \ldots, n$. Note that the values $R_i^{I(w)}$ of global role names must be given so that, for all $u, v \in W$ and all $x, y \in \bigcup_{w \in W} \Delta^w$, we have $xR_i^{I(u)}y$ iff $xR_i^{I(v)}y$ whenever $x, y \in \Delta^u \cap \Delta^v$. Besides, all the $a_i^{I(w)}$ must belong to Δ^w for every $w \in W$.

Now, for a model $\mathfrak{M} = \langle \mathfrak{F}, I \rangle$ and a world w in $\mathfrak{F}$, we define the *values* $C^{I(w)}$ *of a concept* C and $R^{I(w)}$ *of a role* R *in* w, and the *truth-relation* $(\mathfrak{M}, w) \models \varphi$ (or simply $w \models \varphi$) for a formula φ by taking:

- $x(\Box_i R)^{I(w)}y$ iff $\forall v \rhd_i w\; xR^{I(v)}y$;
- $x(\Diamond_i R)^{I(w)}y$ iff $\exists v \rhd_i w\; xR^{I(v)}y$;
- $(C \sqcap D)^{I(w)} = C^{I(w)} \cap D^{I(w)}$;
- $(\neg C)^{I(w)} = \Delta - C^{I(w)}$;
- $x \in (\exists R.C)^{I(w)}$ iff $\exists y \in C^{I(w)}\; xR^{I(w)}y$;
- $x \in (\Box_i C)^{I(w)}$ iff $\forall v \rhd_i w\; x \in C^{I(v)}$;

- $w \models C = D$ iff $C^{I(w)} = D^{I(w)}$;
- $w \models aRb$ iff $a^{I(w)} R^{I(w)} b^{I(w)}$;
- $w \models a : C$ iff $a^{I(w)} \in C^{I(w)}$;
- $w \models \varphi \wedge \psi$ iff $w \models \varphi$ and $w \models \psi$;
- $w \models \neg\varphi$ iff not $w \models \varphi$;
- $w \models \Box_i \varphi$ iff $\forall v \rhd_i w\ v \models \varphi$.

(We recommend the reader to analyze the semantical meaning of the concept Faithful_customer defined above.) A concept C is *satisfied in* $\mathfrak{M}$ if there is a world w in $\mathfrak{F}$ such that $C^{I(w)} \neq \emptyset$. A formula φ is *satisfied in* $\mathfrak{M}$ if there is a world w in $\mathfrak{F}$ such that $w \models \varphi$.

Note that in the same way as we have combined $\mathcal{ALC}$ and $\mathcal{ML}_n$, one can construct hybrids of other description and modal logics, say, $\mathcal{CQ}$ and temporal logics, or $\mathcal{ALC}$ and **CPDL**. Such combinations will be considered in Chapter 14. Here we only show examples of the use of the 'temporal' and 'action' description languages.

Example 3.31. The following is a definition of a concept 'mortal:'

$$\begin{aligned}\mathsf{Mortal} = {} & \mathsf{Living\_being} \sqcap (\exists \mathsf{lives\_in.Place}) \sqcap \\ & (\mathsf{Living\_being}\ \mathcal{U}\ \Box_F \neg \mathsf{Living\_being}) \sqcap \\ & (\mathsf{Living\_being}\ \mathcal{S}\ \Box_P \neg \mathsf{Living\_being}).\end{aligned}$$

In other words, a mortal is a living being who lives in a certain place, remains alive until it dies and was born some time in the past.

Another example: suppose we have two atomic actions submit and accept. Then we can specify concepts published_paper and submitted_paper using the formulas

$$\begin{aligned}&\mathsf{published\_paper} \sqsubseteq \langle \mathsf{accept}^- \rangle\, \mathsf{submitted\_paper},\\ &\mathsf{submitted\_paper} \sqsubseteq \langle \mathsf{submit}^- \rangle\, \mathsf{manuscript}.\end{aligned}$$

The former inclusion, for instance, says that if an object is currently a published paper then there is a state in which it is a submitted paper and from which the current state is reachable via the action 'accept.' Suppose now that author_of is a global role name. Then we can derive from these two formulas that

$$\exists \mathsf{author\_of}.\ \mathsf{published\_paper} \sqsubseteq \langle (\mathsf{submit};\mathsf{accept})^- \rangle\, \exists \mathsf{author\_of}.\ \mathsf{manuscript},$$

i.e., if somebody is currently the author of a published paper, then there is a state in which she is the author of a manuscript and from which the current state is accessible via the action 'submit' followed by the action 'accept.'

Now we show how to reduce satisfiability of $\mathcal{ML}_{\mathcal{ALC}}$-concepts and formulas in models with expanding and varying domains to satisfiability in models with constant domains.

Given a concept C or a formula φ, let ex be a 'fresh' concept name (not occurring in C and φ) the intended meaning of which is to contain in each world precisely those objects that are assumed to exist (under the varying or expanding domain assumption) in this world. By relativizing all concepts and formulas to ex, one can simulate varying and expanding domains using constant ones. For simplicity, we will do this for the unimodal language $\mathcal{ML}_1$ with the operators $\Box$ and $\Diamond$. The results are readily generalized to the multimodal case.

By induction on the construction of C define its *relativization* $C\!\downarrow\!\mathsf{ex}$:

$$
\begin{aligned}
C_i\!\downarrow\!\mathsf{ex} &= C_i, \quad C_i \text{ a concept name},\\
(D \sqcap E)\!\downarrow\!\mathsf{ex} &= (D\!\downarrow\!\mathsf{ex}) \sqcap (E\!\downarrow\!\mathsf{ex}),\\
(\neg D)\!\downarrow\!\mathsf{ex} &= \neg(D\!\downarrow\!\mathsf{ex}),\\
(\exists R.D)\!\downarrow\!\mathsf{ex} &= \exists R.(\mathsf{ex} \sqcap D\!\downarrow\!\mathsf{ex}),\\
(\Box D)\!\downarrow\!\mathsf{ex} &= \Box(D\!\downarrow\!\mathsf{ex}).
\end{aligned}
$$

The relativization $\varphi\!\downarrow\!\mathsf{ex}$ of φ is then defined inductively as follows:

$$
\begin{aligned}
(C = D)\!\downarrow\!\mathsf{ex} &= \big((C\!\downarrow\!\mathsf{ex}) = (D\!\downarrow\!\mathsf{ex})\big),\\
(a : C)\!\downarrow\!\mathsf{ex} &= a : (C\!\downarrow\!\mathsf{ex}),\\
(aRb)\!\downarrow\!\mathsf{ex} &= aRb,\\
(\chi \wedge \psi)\!\downarrow\!\mathsf{ex} &= (\chi\!\downarrow\!\mathsf{ex}) \wedge (\psi\!\downarrow\!\mathsf{ex}),\\
(\neg\psi)\!\downarrow\!\mathsf{ex} &= \neg(\psi\!\downarrow\!\mathsf{ex}),\\
(\Box\psi)\!\downarrow\!\mathsf{ex} &= \Box(\psi\!\downarrow\!\mathsf{ex}).
\end{aligned}
$$

To formulate the statement, we require the notion of modal depth $md(\cdot)$ of roles, concepts and formulas, and the notion of role depth $rd(\cdot)$ of concepts. The *modal depth* of roles, concepts and formulas is defined inductively in the following way:

$$
\begin{aligned}
md(R_i) &= md(S_i) = md(C_i) = 0, & md(\Box R) &= md(\Diamond R) = md(R) + 1,\\
md(C \sqcap D) &= \max\{md(C), md(D)\}, & md(\neg C) &= md(C),\\
md(\exists R.C) &= \max\{md(R), md(C)\}, & md(\Box C) &= md(C) + 1,\\
md(C = D) &= \max\{md(C), md(D)\}, & md(a : C) &= md(C),\\
md(aRb) &= md(R), & md(\varphi \wedge \psi) &= \max\{md(\varphi), md(\psi)\},\\
md(\neg\varphi) &= md(\varphi), & md(\Box\varphi) &= md(\varphi) + 1.
\end{aligned}
$$

The *role depth* $rd(C)$ of a concept C is defined analogously:

$$\begin{aligned} rd(C_i) &= 0, \\ rd(C \sqcap D) &= \max\{rd(C), rd(D)\}, \\ rd(\neg C) &= rd(C), \\ rd(\exists R.C) &= rd(C) + 1, \\ rd(\Box C) &= rd(C). \end{aligned}$$

Given concepts C, D and a formula φ, define inductively the concepts $\Box^{\leq n}C$, $\boldsymbol{R}_D^n C$ and $\boldsymbol{A}_D^n C$, and the formula $\Box^{\leq n}\varphi$ by taking

$$\Box^0 C = \Box^{\leq 0} C = C \qquad \boldsymbol{R}_D^0 C = C$$
$$\boldsymbol{A}_D^0 C = C \qquad \Box^0 \varphi = \Box^{\leq 0}\varphi = \varphi,$$

and for $k \geq 0$,

$$\begin{aligned} \Box^{k+1} C &= \Box\Box^k C, \\ \Box^{\leq k+1} C &= \Box^{\leq k} C \sqcap \Box^{k+1} C, \\ \boldsymbol{R}_D^{k+1} C &= \boldsymbol{R}_D^k C \sqcap \textstyle\bigsqcap\{\forall R.\boldsymbol{R}_D^k C \mid R \text{ occurs in } D\}, \\ \boldsymbol{A}_D^{k+1} C &= \boldsymbol{A}_D^k C \sqcap \boldsymbol{R}_D^{rd(D)}(C \to \Box^{\leq md(D)} \boldsymbol{A}_D^k C), \end{aligned}$$

and

$$\begin{aligned} \Box^{k+1}\varphi &= \Box\Box^k \varphi, \\ \Box^{\leq k+1}\varphi &= \Box^{\leq k}\varphi \wedge \Box^{k+1}\varphi. \end{aligned}$$

Proposition 3.32. *For all Kripke frames $\mathfrak{F}$ the following hold:*

(i) *An $\mathcal{ML}_{\mathcal{ALC}}$-concept C is satisfied in a model based on $\mathfrak{F}$ and having varying domains iff the concept $C{\downarrow}\mathsf{ex} \sqcap \mathsf{ex}$ is satisfied in a model based on $\mathfrak{F}$ and having constant domains.*

(ii) *An $\mathcal{ML}_{\mathcal{ALC}}$-concept C is satisfied in a model based on $\mathfrak{F}$ and having expanding domains iff the concept*

$$C' = C{\downarrow}\mathsf{ex} \sqcap \mathsf{ex} \sqcap \Box^{\leq md(C)} \boldsymbol{A}_C^{rd(C)}\mathsf{ex}$$

is satisfied in a model based on $\mathfrak{F}$ and having constant domains.

(iii) *An $\mathcal{ML}_{\mathcal{ALC}}$-formula φ containing object names $b_1, \dots, b_m$ is satisfied in a model based on $\mathfrak{F}$ and having varying domains iff the formula*

$$\varphi{\downarrow}\mathsf{ex} \wedge \Box^{\leq md(\varphi)}\Big(\neg(\mathsf{ex} = \bot) \wedge \bigwedge_{i=1}^{m}(b_i : \mathsf{ex})\Big)$$

is satisfied in a model based on $\mathfrak{F}$ and having constant domains.

(iv) *An $\mathcal{ML}_{\mathcal{ALC}}$-formula φ containing object names $b_1, \ldots, b_m$ is satisfied in a model based on $\mathfrak{F}$ and having expanding domains iff the formula*

$$\varphi{\downarrow}\mathsf{ex} \wedge \neg(\mathsf{ex} = \bot) \wedge \bigwedge_{i=1}^{m} (b_i : \mathsf{ex}) \wedge \Box^{\leq md(\varphi)}(\mathsf{ex} \sqsubseteq \Box\mathsf{ex})$$

is satisfied in a model based on $\mathfrak{F}$ and having constant domains.

Proof. We show here only (ii), leaving (i), (iii) and (iv) to the reader as an exercise. Suppose that $\mathfrak{F} = \langle W, \lhd \rangle$ is a frame and $C^{I(v)} \neq \emptyset$ for a world v in a model $\mathfrak{M} = \langle \mathfrak{F}, I \rangle$ with expanding domains such that

$$I(w) = \left\langle \Delta^w, R_0^{I(w)}, \ldots, S_0^{I(w)}, \ldots, C_0^{I(w)}, \ldots, a_0^{I(w)}, \ldots \right\rangle,$$

for every $w \in W$ (so that $\Delta^u \subseteq \Delta^{u'}$ whenever $u \lhd u'$). We construct a model $\mathfrak{N} = \langle \mathfrak{F}, J \rangle$ with constant domains by taking

$$J(w) = \left\langle \Delta, R_0^J, \ldots, S_0^{I(w)}, \ldots, C_0^{I(w)}, \ldots, \mathsf{ex}^{J(w)}, a_0^{I(w)}, \ldots \right\rangle,$$

where $\Delta = \bigcup_{w \in W} \Delta^w$, $R_i^J = \bigcup_{w \in W} R_i^{I(w)}$ (which can be done by the definition of global roles) and $\mathsf{ex}^{J(w)} = \Delta^w$, for all $w \in W$. Then it is not hard to see that

$$\left(\mathsf{ex} \sqcap \Box^{\leq md(C)} \boldsymbol{A}_C^{rd(C)}\mathsf{ex}\right)^{J(v)} = \Delta^v = \mathsf{ex}^{J(v)}.$$

And it is readily checked by induction that for every subconcept D of C and every $w \in W$,

$$D^{I(w)} = (D{\downarrow}\mathsf{ex} \sqcap \mathsf{ex})^{J(w)}.$$

It follows that $(C')^{J(v)} = C^{I(v)}$, i.e., C' is satisfied in $\mathfrak{N}$.

Conversely, suppose that $(C')^{J(v)} \neq \emptyset$ at root v of a model $\mathfrak{N} = \langle \mathfrak{F}, J \rangle$ with constant domains such that for all $w \in W$,

$$J(w) = \left\langle \Delta, R_0^{J(w)}, \ldots, S_0^{J(w)}, \ldots, C_0^{J(w)}, \ldots, \mathsf{ex}^{J(w)}, a_0^{J(w)}, \ldots \right\rangle.$$

Choose a point $x \in (C')^{J(v)}$. For any $n < \omega$ and any world $w \in W$, we say that a point $y \in \Delta$ is *role-accessible from x in n steps in w* if there exist points $x_0, \ldots, x_n \in \Delta$, worlds $w_0, \ldots, w_{n-1} \in W$, natural numbers $m_0, \ldots, m_n$ and roles $Q_0, \ldots, Q_{n-1}$ occurring in C such that the following hold:

$$x_0 = x, \qquad x_n = y, \tag{3.6}$$

$$x_i Q_i^{J(w_i)} x_{i+1} \qquad \text{for } i < n, \tag{3.7}$$

$$x_{i+1} \in \mathsf{ex}^{J(w_i)} \qquad \text{for } i < n, \tag{3.8}$$

$$w_0 \text{ is } \lhd\text{-accessible from } v \text{ in } m_0 \text{ steps,} \tag{3.9}$$

$$w_{i+1} \text{ is } \lhd\text{-accessible from } w_i \text{ in } m_{i+1} \text{ steps, for } i < n, \tag{3.10}$$

$$\sum_{i=0}^{n} m_i \le md(C), \quad \text{and} \tag{3.11}$$

$$w \text{ is } \lhd^*\text{-accessible from } w_{n-1}. \tag{3.12}$$

Now, for any world $w \in W$, if w is $\lhd$-accessible from v in $\le md(C)$ steps then let

$$\Delta^w = \{y \in \Delta \mid y \text{ is role-accessible from } x \text{ in } n \text{ steps in } w, \text{ for some } n \le rd(C)\}.$$

For all other $w \in W$, let $\Delta^w = \Delta$. By (3.12), we have that whenever y is role-accessible from x in n steps in w then it is role-accessible from x in n steps in w' as well, for all w' with $w \lhd w'$. Thus $\Delta^w \subseteq \Delta^{w'}$ holds whenever $w \lhd w'$. Moreover, for every w which is $\lhd$-accessible from v in $\le md(C)$ steps, we have

$$\Delta^w \subseteq \mathsf{ex}^{J(w)}. \tag{3.13}$$

Indeed, suppose that $y \in \Delta^w$ and that $x_0, \dots, x_n$ and $w_0, \dots, w_{n-1}$ satisfy (3.6)–(3.12) for some $n \le rd(C)$. By assumption,

$$x = x_0 \in \left(\Box^{\le md(C)} \boldsymbol{A}_C^{rd(C)} \mathsf{ex}\right)^{J(v)}.$$

Thus, by (3.7)–(3.9),

$$x_1 \in \left(\Box^{\le md(C)} \boldsymbol{A}_C^{rd(C)-1} \mathsf{ex}\right)^{J(w_0)}.$$

It can be shown by induction on n (using (3.7)–(3.11)) that in fact for all $i < n$, we have

$$x_{i+1} \in \left(\Box^{\le md(C)} \boldsymbol{A}_C^{rd(C)-i-1} \mathsf{ex}\right)^{J(w_i)}.$$

So, by (3.12),

$$y \in \left(\boldsymbol{A}_C^{rd(C)-n} \mathsf{ex}\right)^{J(w)}.$$

Since, by the definition, $k < k'$ implies $(\boldsymbol{A}_C^{k'} \mathsf{ex})^{J(w)} \subseteq (\boldsymbol{A}_C^{k} \mathsf{ex})^{J(w)}$, we finally obtain that $y \in (\boldsymbol{A}_C^{0} \mathsf{ex})^{J(w)} = \mathsf{ex}^{J(w)}$.

Now consider the model $\mathfrak{M} = \langle \mathfrak{F}, I \rangle$

$$I(w) = \left\langle \Delta^w, R_0^{I(w)}, \dots, S_0^{I(w)}, \dots, C_0^{I(w)}, \dots, a_0^{I(w)}, \dots \right\rangle,$$

where $a_i^{I(w)}$ are arbitrary elements of Δ^v, and $R_i^{I(w)}$, $S_i^{I(w)}$ and $C_i^{I(w)}$ are the restrictions of $R_i^{J(w)}$, $S_i^{J(w)}$ and $C_i^{J(w)}$ to Δ^w, respectively, for every $w \in W$. It can be shown by induction that for every subconcept D of C, every world

w which is $\lhd$-accessible from v in $\leq md(C) - md(D)$ steps, and every $y \in \Delta^w$ such that y is role-accessible from x in $rd(C) - rd(D)$ steps in w,

$$y \in D^{I(w)} \quad \text{iff} \quad y \in (D \downarrow \mathsf{ex})^{J(w)} \cap \Delta^w.$$

We consider only the case of $D = \exists R.E$. Suppose that $y \in D^{I(w)}$ and y is role-accessible from x in $rd(C) - rd(D)$ steps in w. Then there is a $z \in \Delta^w$ such that $yR^{I(w)}z$ and $z \in E^{I(w)}$. So $yR^{J(w)}z$ and, by (3.13), $z \in \mathsf{ex}^{J(w)}$ and z is role-accessible from x in

$$rd(C) - rd(D) + 1 = rd(C) - rd(E)$$

steps in w. Then, by the induction hypothesis, we have $z \in (E \downarrow \mathsf{ex})^{J(w)}$, and so $z \in (\mathsf{ex} \sqcap E \downarrow \mathsf{ex})^{J(w)}$. It follows that $y \in (D \downarrow \mathsf{ex})^{J(w)} \cap \Delta^w$. Conversely, suppose that $y \in (D \downarrow \mathsf{ex})^{J(w)} \cap \Delta^w$ and y is role-accessible from x in $rd(C) - rd(D)$ steps in w. Then there is a z such that $yR^{J(w)}z$, $z \in (E \downarrow \mathsf{ex})^{J(w)}$ and $z \in \mathsf{ex}^{J(w)}$. This means that z is role-accessible from x in $rd(C) - rd(E)$ steps in w, and hence $z \in \Delta^w$. Therefore, $yR^{I(w)}z$. By the induction hypothesis we have $z \in E^{I(w)}$, and so $y \in D^{I(w)}$.

Since $x \in \Delta^v$ and x is role-accessible from x in 0 steps in v, this is enough to show that C is satisfied in $\mathfrak{M}$. ❑

We have justified our acceptance of the constant domain assumption. And the following proposition shows that the addition of global concepts does not increase the expressive power of our languages. (We again consider the language with only one pair of modal operators.)

Proposition 3.33. (i) *A concept C is satisfied in a model based on a frame $\mathfrak{F}$ and interpreting concept names $C_1, \ldots, C_k$ globally iff*

$$C \sqcap \Box^{\leq md(C)} \boldsymbol{A}_C^{rd(C)} \prod_{1 \leq i \leq k} ((C_i \to \Box C_i) \sqcap (\neg C_i \to \Box \neg C_i))$$

is satisfied in a model based on $\mathfrak{F}$ with the local interpretation of concepts.

(ii) *An $\mathcal{ML}_{\mathcal{ALC}}$-formula φ is satisfied in a model based on $\mathfrak{F}$ and interpreting concept names $C_1, \ldots, C_k$ globally iff*

$$\varphi \wedge \Box^{\leq md(\varphi)} \prod_{1 \leq i \leq k} ((C_i \to \Box C_i) \sqcap (\neg C_i \to \Box \neg C_i)) = \top$$

is satisfied in a model based on $\mathfrak{F}$ with the local interpretation of concepts.

Proof. Exercise. ❑

Let us see now how the reasoning tasks formulated in Section 2.5 modify for modal description logics. Suppose $\mathcal{C}$ is a class of n-frames (modeling time, beliefs, actions, or something else). The *satisfiability problem* for $\mathcal{ML}_{\mathcal{ALC}}$-formulas in $\mathcal{C}$ can be formulated as follows:

- Given an $\mathcal{ML}_{\mathcal{ALC}}$-formula φ, decide whether there is a model based on a frame in $\mathcal{C}$ and satisfying φ.

Usually we consider the classes of all frames for some multimodal logic L in the language $\mathcal{ML}_n$. By the *formula satisfiability problem for* $L_{\mathcal{ALC}}$ we mean then the satisfiability problem for $\mathcal{ML}_{\mathcal{ALC}}$-formulas in $\mathsf{Fr}L$.

Given a multimodal logic L and a knowledge base Σ, we now have two variants of reasoning tasks for Σ and L, which reflect the two ways of understanding the consequence relation in L.

- *Concept satisfiability*: Are there a model $\langle \mathfrak{F}, I\rangle$ and a world v in $\mathfrak{F}$ such that $C^{I(v)} \neq \emptyset$, $\mathfrak{F}$ is a frame for L and $v \models \Sigma$?

- *Global concept satisfiability*: Are there a model $\langle \mathfrak{F}, I\rangle$ and a world v in $\mathfrak{F}$ such that $C^{I(v)} \neq \emptyset$, $\mathfrak{F}$ is a frame for L and $w \models \Sigma$ for all w in $\mathfrak{F}$?

- *Subsumption*: Does $C^{I(v)} \subseteq D^{I(v)}$ hold for every model $\langle \mathfrak{F}, I\rangle$ and every world v in $\mathfrak{F}$ such that $\mathfrak{F}$ is a frame for L and $v \models \Sigma$?

- *Global subsumption*: Does $C^{I(v)} \subseteq D^{I(v)}$ hold for every model $\langle \mathfrak{F}, I\rangle$ and every world v in $\mathfrak{F}$ such that $\mathfrak{F}$ is a frame for L and $w \models \Sigma$ for all w in $\mathfrak{F}$?

- *Instance checking*: Does $a^{I(v)}$ belong to $C^{I(v)}$ for every model $\langle \mathfrak{F}, I\rangle$ and every world v in $\mathfrak{F}$ such that $\mathfrak{F}$ is a frame for L and $v \models \Sigma$?

- *Global instance checking*: Does $a^{I(v)}$ belong to $C^{I(v)}$ for every model $\langle \mathfrak{F}, I\rangle$ and every world v in $\mathfrak{F}$ such that $\mathfrak{F}$ is a frame for L and $w \models \Sigma$ for all w in $\mathfrak{F}$?

In the 'global cases,' the knowledge base is assumed to be applied to all worlds in the model, while in the 'local' ones only to a single world. As we shall see below, the decidability of a local problem does not imply in general the decidability of the corresponding global one. However, as in Section 2.5, we again have obvious reductions between the various reasoning tasks; see Table 3.1.

We explain those which connect modal description logics and products of modal logics.

Modal description logics and products

Consider first the fragment of the language $\mathcal{ML}_{\mathcal{ALC}}$ containing m *global* role names $R_0, \dots, R_{m-1}$ and neither local role names nor modalized roles at all. In this case the translation † from the modal language $\mathcal{ML}_n$ onto $\mathcal{ALC}$-concepts defined in Section 2.5 can be extended to a translation from $\mathcal{ML}_{n+m}$

onto $\mathcal{ML}_{\mathcal{ALC}}$-concepts by taking

$$\begin{aligned}
p_i^\dagger &= C_i, \\
(\varphi \wedge \psi)^\dagger &= \varphi^\dagger \sqcap \psi^\dagger, \\
(\neg\varphi)^\dagger &= \neg\varphi^\dagger, \\
(\Diamond_i\varphi)^\dagger &= \Diamond_i\varphi^\dagger, && \text{for } 1 \le i \le n, \\
(\Diamond_i\varphi)^\dagger &= \exists R_{i-n-1}.\varphi^\dagger, && \text{for } n+1 \le i \le n+m.
\end{aligned}$$

Since all the role names are global, every modal model $\mathfrak{M} = \langle \mathfrak{F} \times \mathfrak{G}, \mathfrak{V} \rangle$ based on the product of an n-frame $\mathfrak{F} = \langle W, \lhd_1, \dots, \lhd_n \rangle$ and an m-frame $\mathfrak{G} = \langle \Delta, T_0, \dots, T_{m-1} \rangle$ can be transformed into an $\mathcal{ML}_{\mathcal{ALC}}$-model $\langle \mathfrak{F}, I_{\mathfrak{M}} \rangle$ such that, for each $w \in W$,

$$I_{\mathfrak{M}}(w) = \left\langle \Delta, R_0^{I(w)}, \dots, R_{m-1}^{I(w)}, C_0^{I(w)}, \dots, a_0^{I(w)}, \dots \right\rangle,$$

where $R_i^{I(w)} = T_i$, $C_i^{I(w)} = \{u \in \Delta \mid \langle w, u \rangle \in \mathfrak{V}(p_i)\}$, and $a_i^{I(w)} \in \Delta$ arbitrary. Then it should be clear that for every $\mathcal{ML}_{n+m}$-formula φ and every world $\langle w, u \rangle$ in $\mathfrak{M}$, we have

$$(\mathfrak{M}, \langle w, u \rangle) \models \varphi \quad \text{iff} \quad u \in (\varphi^\dagger)^{I(w)}.$$

Conversely, every $\mathcal{ML}_{\mathcal{ALC}}$-model $\langle \mathfrak{F}, I \rangle$ with $\mathfrak{F} = \langle W, \lhd_1, \cdots \lhd_n \rangle$ and

$$I(w) = \left\langle \Delta, R_0^{I(w)}, \dots, R_{m-1}^{I(w)}, C_0^{I(w)}, \dots, a_0^{I(w)}, \dots \right\rangle,$$

gives rise to a modal model $\mathfrak{M} = \langle \mathfrak{F} \times \mathfrak{G}_I, \mathfrak{V}_I \rangle$, where

$$\mathfrak{G}_I = \left\langle \Delta, R_0^{I(w)}, \dots, R_{m-1}^{I(w)} \right\rangle \quad \text{and} \quad \mathfrak{V}_I(p_i) = \bigcup_{w \in W} C_i^{I(w)}.$$

By taking the inverse $\ddagger$ of the translation $\dagger$, we obtain

$$u \in C^{I(w)} \quad \text{iff} \quad (\mathfrak{M}_I, \langle w, u \rangle) \models C^\ddagger,$$

for every $\mathcal{ML}_{\mathcal{ALC}}$-concept C, every object $u \in \Delta$ and every world w in $\mathfrak{F}$.

It follows that if the knowledge base is empty, then the $L_{\mathcal{ALC}}$-satisfiability problem for concepts containing neither modalized roles nor local role names is equivalent to the satisfiability problem for the product logic $L \times \mathbf{K}_m$. (Note that the same problem for models with expanding domains corresponds to the satisfiability problem for 'expanding relativized products,' see Section 9.1.)

We know (see page 71) that the global consequence relation $\vdash^*_{\mathbf{K}_m}$ corresponds to the satisfiability problem for $\mathcal{ALC}$-concepts relative to TBoxes. We now lift this correspondence to modalized $\mathcal{ALC}$. Following the standard

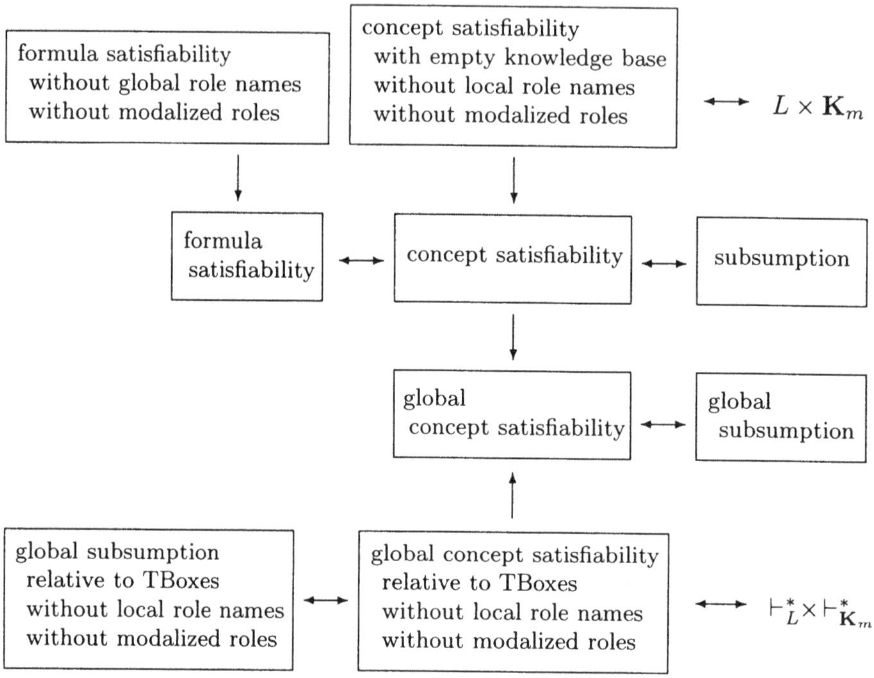

Table 3.1: Reasoning tasks in $L_{\mathcal{ALC}}$.

terminology of description logic, by a *TBox* we mean a knowledge base Σ consisting only of equations $C = D$ of $\mathcal{ML}_{\mathcal{ALC}}$-concepts. For $\mathcal{ML}_{\mathcal{ALC}}$ without local role names and modalized roles, the consequence relation $\vdash^*_L \times \vdash^*_{\mathbf{K}_m}$ (cf. Section 3.3) and the global concept satisfiability problem relative to TBoxes are reducible to each other. Indeed, it is enough to observe that

$$\varphi(\vdash^*_L \times \vdash^*_{\mathbf{K}_m})\psi \text{ iff } \neg\psi^\dagger \text{ is not globally } L_{\mathcal{ALC}}\text{-satisfiable relative to } \{\varphi^\dagger = \top\}.$$

As will be shown in Theorem 5.36, $\vdash^*_{\mathbf{K}} \times \vdash^*_{\mathbf{K}}$ is undecidable. Thus, we have:

Proposition 3.34. *The global $\mathbf{K}_{\mathcal{ALC}}$-satisfiability problem for concepts containing neither local role names nor modalized roles is undecidable (even relative to TBoxes).*

On the other hand, we shall see (Theorem 14.8) that the formula satisfiability problem (and thus the concept satisfiability problem) for $\mathbf{K}_{\mathcal{ALC}}$ is decidable. Unlike $\mathbf{K}_{\mathcal{ALC}}$, for many other modal description logics the global concept satisfiability problem can easily be reduced to the formula satisfiability problem. In particular, this is the case for **CPDL**, **S4**, **K4**, $\mathbf{S5}^C_n$ and

PTL. Here, for example, is a reduction for **S4**. Given any finite knowledge base Σ, we have:

$$C \text{ is globally satisfiable relative to } \Sigma \text{ iff } (\Box \bigwedge_{\varphi \in \Sigma} \varphi) \wedge \neg(C = \bot) \text{ is satisfiable.}$$

The connections between modal description logics and products of modal logics established so far are based on what one may call 'semantic equivalence:' we just have to replace concepts by propositional variables and roles by accessibility relations. In contrast, the following polynomial reduction is invariant only under satisfiability. Yet, it will be very useful for the transfer of decidability and complexity results.

Theorem 3.35. *Let L be a Kripke complete modal logic. Then the satisfiability problem for $L \times$ **S5** is polynomially reducible to the formula satisfiability problem for $L_{\mathcal{ALC}}$ (without any roles at all).*

Proof. Assume for simplicity that L is a unimodal logic with modal operator $\Box_1$ and that **S5** has modal operator $\Box_2$. For every $\varphi \in \mathcal{ML}_2$ we define an $\mathcal{ML}_{\mathcal{ALC}}$-formula which is $L_{\mathcal{ALC}}$-satisfiable iff φ is $L \times$ **S5**-satisfiable. Let C_p and C_χ be fresh concept names for every propositional variable p and every $\mathcal{ML}_2$-formula χ of the form $\chi = \Box_2\chi'$, respectively. Define inductively a translation $\psi^\sharp$ of an $\mathcal{ML}_2$-formula ψ by taking

$$\begin{aligned} p^\sharp &= C_p, \\ (\psi_1 \wedge \psi_2)^\sharp &= \psi_1^\sharp \sqcap \psi_2^\sharp, \\ (\neg\psi)^\sharp &= \neg\psi^\sharp, \\ (\Box_1\psi)^\sharp &= \Box_1\psi^\sharp, \\ (\Box_2\psi)^\sharp &= C_{\Box_2\psi}. \end{aligned}$$

Now, denote by χ the formula

$$\bigwedge_{\Box_2\psi \in sub\varphi} ((C_{\Box_2\psi} = \top) \vee (C_{\Box_2\psi} = \bot)) \wedge \bigwedge_{\Box_2\psi \in sub\varphi} ((C_{\Box_2\psi} = \top) \leftrightarrow (\psi^\sharp = \top)).$$

The first conjunct of χ says that $C_{\Box_2\psi}$ applies to all objects iff it applies to some object. The second one says that this is the case iff $\psi^\sharp$ applies to all objects. Now one can readily show by induction that $\neg(\varphi^\sharp = \bot) \wedge \Box_1^{\leq md(\varphi)}\chi$ is satisfiable in $L_{\mathcal{ALC}}$ iff φ is $L \times$ **S5**-satisfiable. ❑

The theorem we have just proved states that the $L_{\mathcal{ALC}}$-satisfiability problem for formulas without global role names and modalized roles is at least as complex as the satisfiability problem for $L \times$ **S5**. We will use this result in Chapter 14 to obtain lower bounds of the computational complexity for modal description logics from lower bounds of the computational complexity for products with **S5** obtained in Section 6.5.

The following result is proved similarly; it will be used in Chapter 14 to obtain undecidability results for description logics with modal operators.

Theorem 3.36. *Let L be a Kripke complete modal logic. Then the satisfiability problem for $L \times \mathbf{K}_u$ is polynomially reducible to the formula satisfiability problem for $L_{\mathcal{ALC}}$ with global role names (but without local role names and modalized roles).*

Modal description logics and first-order modal logics

Now we show that modalized $\mathcal{ALC}$ can be regarded as a fragment of first-order modal logic. This embedding will be used in Chapter 14 to obtain decidability and complexity results for modal description logics from corresponding results for certain fragments of first-order modal logics established in Chapters 11 and 12. To simplify notation, we assume again that there is only one pair $\Box$ and $\Diamond$ of modal operators.

Fix two different individual variables, say, x, y. The translation R^T of a role R is a formula with two free variables x, y defined by taking

$$R_i^T = R_i(x,y), \qquad S_i^T = S_i(x,y),$$
$$(\Box R)^T = \Box R^T, \qquad (\Diamond R)^T = \Diamond R^T.$$

The translation C^T of a concept C is a formula with one free variable x:

$$\begin{aligned} C_i^T &= C_i(x), & & \\ (C \sqcap D)^T &= C^T \wedge D^T, & (\neg C)^T &= \neg C^T, \\ (\exists R.C)^T &= \exists y\,(R^T \wedge C^T\{y/x\}), & (\Box C)^T &= \Box C^T. \end{aligned}$$

The translation φ^T of an $\mathcal{ML}_{\mathcal{ALC}}$-formula φ is a $\mathcal{QML}$-sentence defined as follows (we assume that the object names of $\mathcal{ALC}$ coincide with the constant symbols of first-order logic):

$$\begin{aligned} (C = D)^T &= \forall x(C^T \leftrightarrow D^T), \\ (a : C)^T &= C^T\{a/x\}, \\ (aRb)^T &= R^T\{a/x, b/y\}, \\ (\varphi \wedge \psi)^T &= \varphi^T \wedge \psi^T, \\ (\neg\varphi)^T &= \neg\varphi^T, \\ (\Box\varphi)^T &= \Box\varphi^T. \end{aligned}$$

It is readily checked that the following theorem holds:

Theorem 3.37. (i) *Let C be an $\mathcal{ML}_{\mathcal{ALC}}$-concept with global role names $R_1, \ldots, R_k$. Then C is satisfied in an $\mathcal{ML}_{\mathcal{ALC}}$-model based on a frame $\mathfrak{F}$ and having domain Δ iff the first-order modal sentence*

$$\exists x\, C^T \wedge$$
$$\Box^{\leq md(C)} \bigwedge_{1\leq i\leq k} \forall x, y\Big(\big(R_i(x,y) \to \Box R_i(x,y)\big) \wedge \big(\neg R_i(x,y) \to \Box\neg R_i(x,y)\big)\Big)$$

is satisfied in a first-order Kripke model based on $\mathfrak{F}$ and having domain Δ.

(ii) *Let φ be an $\mathcal{ML}_{\mathcal{ALC}}$-formula with global role names $R_1, \ldots, R_k$. Then φ is satisfied in an $\mathcal{ML}_{\mathcal{ALC}}$-model based on a frame $\mathfrak{F}$ and having domain Δ iff the first-order modal sentence*

$$\varphi^T \wedge$$
$$\Box^{\leq md(\varphi)} \bigwedge_{1\leq i\leq k} \forall x, y\Big(\big(R_i(x,y) \to \Box R_i(x,y)\big) \wedge \big(\neg R_i(x,y) \to \Box\neg R_i(x,y)\big)\Big)$$

is satisfied in a first-order Kripke model based on $\mathfrak{F}$ and having domain Δ.

3.9 HS as a two-dimensional logic

In Section 2.2 we interpreted formulas of the temporal logic **HS** of intervals in frames of the form

$$\mathfrak{I}(\mathfrak{F}) = \langle In(\mathfrak{F}), S, F\rangle,$$

where $\mathfrak{F} = \langle W, <\rangle$ is a strict linear order, $In(\mathfrak{F})$ the set of all closed intervals $[u, v]$ in $\mathfrak{F}$ (for $u, v \in W$, $u \leq v$) and S, F are the 'starts' and 'finishes' relations between intervals.

As every interval $[u, v]$ is determined by its two end-points $u, v \in W$, we can represent it as a pair $\langle u, v\rangle$, i.e., as an element of the Cartesian product $W \times W$. Let us define the 'North-Western' subset of $W \times W$ as

$$nw(W \times W) = \{\langle u, v\rangle \in W \times W \mid u \leq v\},$$

and then the *compass relations* R_E, R_W, R_N, R_S on $nw(W \times W)$ in precisely the same way as we did in Section 2.6 for $\mathbb{R} \times \mathbb{R}$, but using the relation $<$ of $\mathfrak{F}$. For example,

$$\langle u, v\rangle\, R_E\, \langle u', v'\rangle \quad \text{iff} \quad u < u' \text{ and } v = v'.$$

The compass relations can represent the relations S, F of $\mathfrak{I}(\mathfrak{F})$ and their converses, viz.,

$$[u, v]S[u', v'] \quad \text{iff} \quad \langle u, v\rangle\, R_N\, \langle u', v'\rangle,$$
$$[u, v]F[u', v'] \quad \text{iff} \quad \langle u, v\rangle\, R_W\, \langle u', v'\rangle,$$

$$[u,v]S^{-1}[u',v'] \quad \text{iff} \quad \langle u,v\rangle\, R_S\, \langle u',v'\rangle\,,$$
$$[u,v]F^{-1}[u',v'] \quad \text{iff} \quad \langle u,v\rangle\, R_E\, \langle u',v'\rangle\,.$$

In other words, we can view the frame $\mathfrak{I}(\mathfrak{F})$ as a 'map' on which the intervals of $In(\mathfrak{F})$ correspond to the points on and above the diagonal coming from 'South-West' to 'North-East.' (The points on the diagonal correspond to intervals $[u,u]$ without duration.) Note that by using the compass relations, we can represent all the $\mathcal{A}\ell\ell$-13 relations between intervals; see Fig. 3.5.

Observe that the frame

$$nw_{\mathfrak{F}} = \langle nw(W \times W), R_E, R_W, R_N, R_S\rangle$$

we obtain this way is in fact a subframe of the product frame

$$\langle W, <, >\rangle \times \langle W, <, >\rangle\,.$$

So, for any class $\mathcal{C}$ of strict linear orders, the logic $\mathbf{HS}_{\mathcal{C}}$ can be defined as

$$\mathbf{HS}_{\mathcal{C}} = \mathsf{Log}\{nw_{\mathfrak{F}} \mid \mathfrak{F} \in \mathcal{C}\}.$$

As we shall see in Section 9.1, the two-dimensional **HS**-frames introduced above are examples of *expanding* (and *decreasing*) *relativized product frames*.

3.10 Modal transition logics

The structures of action terms in **PDL**-like logics have been represented in algebraic form as *Kleene algebras*, *action algebras* (Kozen 1981, 1990, Pratt 1990), as well as *process algebras* intended for describing the behavior of parallel processes (Bergstra and Klop 1984, Hoare 1985, Milner 1980). Some of these algebraic formalisms allow for the Boolean operations on action terms, which has led to the idea of considering them in the modal perspective as *modal transition logics*: multimodal logics reasoning about *transitions* (or procedures, programs, actions, preferences, etc.). Formulas of these logics are similar to action terms of **PDL**, they are evaluated in frames having transitions as their worlds, and **PDL**-operations like composition become modal operators. A successful research program in this direction was initiated by van Benthem (1991) and Venema (1991) under the name of *arrow logic*, with arrows being abstractions of transitions. Here we give a brief survey of this approach; for more information the reader is referred to (Marx *et al.* 1996, Marx and Venema 1997).

The language $\mathcal{AL}$ of arrow logic has three modal operators. However, unlike $\mathcal{ML}_3$, not all of them are unary. Namely, we have

- a binary operator ; (for taking the composition of arrows),

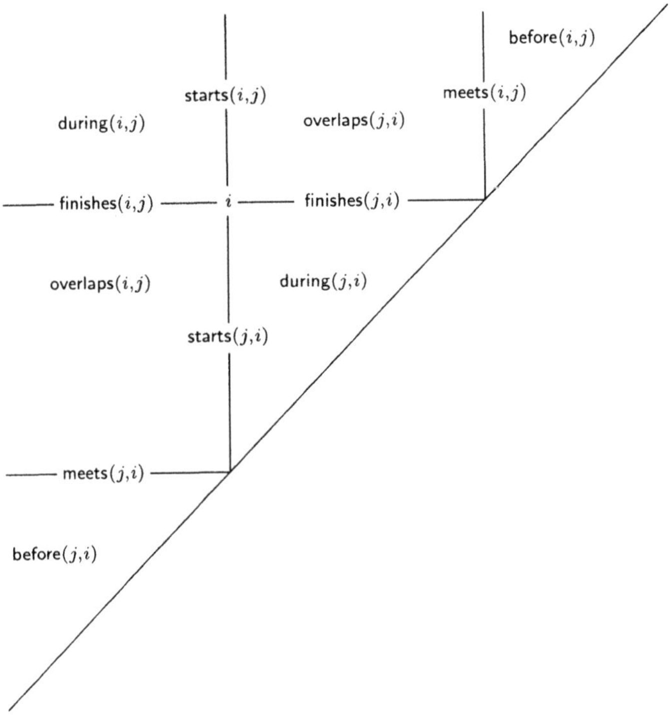

Figure 3.5: Representation of the $\mathcal{All}$-13 relations.

- a unary operator $^-$ (for taking the converse of an arrow), and
- a constant Id (for identity arrows that are similar to ? in $\mathcal{PDL}$).[4]

The corresponding formula-formation rules are as follows:

- Id is an $\mathcal{AL}$-formula,
- if φ and ψ are $\mathcal{AL}$-formulas, then so are $\varphi ; \psi$ and φ^-.

$\mathcal{AL}$-formulas are interpreted in *arrow frames* which are structures of the form

$$\mathfrak{F} = \langle W, C, R, E \rangle ,$$

where W is a nonempty set (of *arrows*), C is a ternary, R is a binary and E is a unary relations on W. An *arrow model* is a pair $\mathfrak{M} = \langle \mathfrak{F}, \mathfrak{V} \rangle$, where $\mathfrak{F}$ is an arrow frame and $\mathfrak{V}$ a *valuation in* $\mathfrak{F}$, i.e., a function mapping the

[4]Sometimes ; is denoted by $\circ$, $\bullet$ or $|$, $^-$ by $\breve{}$ or $\otimes$, and Id by $\iota\delta$ or $1'$.

propositional variables to subsets of W. The *truth-relation* $(\mathfrak{M}, x) \models \varphi$, x an arrow in W and φ an $\mathcal{AL}$-formula, is defined as follows:

$$\begin{array}{lll}
(\mathfrak{M}, x) \models p & \text{iff} & x \in \mathfrak{V}(p) \ (p \text{ a propositional variable}),\\
(\mathfrak{M}, x) \models \varphi \wedge \psi & \text{iff} & (\mathfrak{M}, x) \models \varphi \text{ and } (\mathfrak{M}, x) \models \psi,\\
(\mathfrak{M}, x) \models \neg\varphi & \text{iff} & \text{not } (\mathfrak{M}, x) \models \varphi,\\
(\mathfrak{M}, x) \models \varphi ; \psi & \text{iff} & \exists y, z \in W \ \big(C(x, y, z) \ \& \ (\mathfrak{M}, y) \models \varphi \ \& \ (\mathfrak{M}, z) \models \psi\big),\\
(\mathfrak{M}, x) \models \varphi^{-} & \text{iff} & \exists y \in W \ \big(R(x, y) \ \& \ (\mathfrak{M}, y) \models \varphi\big),\\
(\mathfrak{M}, x) \models Id & \text{iff} & E(x).
\end{array}$$

We say that an $\mathcal{AL}$-formula φ is *valid* in an arrow frame $\mathfrak{F}$ if $(\mathfrak{M}, x) \models \varphi$ holds for every arrow model $\mathfrak{M}$ based on $\mathfrak{F}$ and every arrow x in $\mathfrak{F}$. Note that according to the given definition the operators ; and $^{-}$ are diamond-like modalities. They satisfy the following analogs of (the dual version of) axiom (K):

$$\begin{array}{ll}
(\mathrm{K})_{;} & (((p_0 \vee p_1) ; p_2) \leftrightarrow ((p_0 ; p_2) \vee (p_1 ; p_2)))\\
& \wedge \ ((p_0 ;(p_1 \vee p_2)) \leftrightarrow ((p_0 ; p_1) \vee (p_0 ; p_2)))\\
(\mathrm{K})_{-} & (p_0 \vee p_1)^{-} \leftrightarrow (p_0^{-} \vee p_1^{-})
\end{array}$$

A set of $\mathcal{AL}$-formulas is called an *arrow logic* if it contains the axioms (A1)–(A10) of classical propositional logic **Cl**, the formulas $(\mathrm{K})_{;}$ and $(\mathrm{K})_{-}$ and is closed under the rules of MP, Subst and the following analogs of RN:

$$\begin{array}{ll}
(\mathrm{RN})_{;} & \text{given } \varphi, \text{ derive } \neg(\neg\varphi ; \neg\psi) \text{ and } \neg(\neg\psi ; \neg\varphi),\\
(\mathrm{RN})_{-} & \text{given } \varphi, \text{ derive } \neg(\neg\varphi)^{-}.
\end{array}$$

The smallest arrow logic is denoted by $\mathbf{AL}_{min}$.

This 'abstract' approach to arrows permits various more concrete 'representations.' Perhaps the most natural way of depicting an arrow w is to consider it as a pair

$$\langle \text{start-point of } w, \text{ end-point of } w \rangle ,$$

i.e., as an edge in a directed graph. More generally, arrows can be regarded as edges of a directed multi-graph (where more than one edge between two nodes is allowed); see (Vakarelov 1996a, 1996b, 1997). This view can be put into a two-dimensional perspective: a set W of arrows is the Cartesian square $U \times U$ of some nonempty set U of points, and C, R and E are the following relations on $U \times U$:

$$\begin{aligned}
C_U &= \{\langle \langle a, b\rangle , \langle a, c\rangle , \langle c, b\rangle \rangle \mid a, b, c \in U\},\\
R_U &= \{\langle \langle a, b\rangle , \langle b, a\rangle \rangle \mid a, b \in U\},\\
E_U &= \{\langle a, a\rangle \mid a \in U\}.
\end{aligned}$$

We call an arrow frame of the form $\langle U \times U, C_U, R_U, E_U \rangle$ the *square arrow frame* over U. Subframes of such a frame are called *pair arrow frames*. Models which are based on square (pair) arrow frames are called *square (pair) arrow models*. The set of arrows satisfying an $\mathcal{AL}$-formula in a pair arrow model is simply a binary relation. This observation connects arrow logics with the algebraic theory of relations which originated in the 19th century from the work of De Morgan (1860), Peirce (1870) and Schröder (1895). The modern development of this theory—a branch of algebraic logic—started with Tarski and his colleagues. Along the lines of Section 1.5, one can easily generalize all the concepts involved in the algebraic semantics of modal logics to arrow logics. Then the class RRA of *representable relation algebras*, defined by Tarski (1941), turns out to be the class of modal algebras for the arrow logic $\mathbf{AL}_{sq}$, which is the set of all $\mathcal{AL}$-formulas that are valid in every square arrow frame. In an attempt to extend the finite equational axiomatization of Boolean algebras to RRA, Tarski (1941) gave a list of natural properties of binary relations formulated here as $\mathcal{AL}$-formulas:

(AL$_1$) $(p_0^-)^- \leftrightarrow p_0$,

(AL$_2$) $(p_0 \,; Id) \leftrightarrow p_0$,

(AL$_3$) $(p_0 \,; p_1)^- \leftrightarrow (p_1^- \,; p_0^-)$,

(AL$_4$) $(p_0^- \,; \neg(p_0 \,; p_1)) \rightarrow \neg p_1$,

(AL$_5$) $((p_0 \,; p_1) \,; p_2) \leftrightarrow (p_0 \,;(p_1 \,; p_2))$.

It is readily seen that all these formulas are valid in every square arrow frame and so belong to $\mathbf{AL}_{sq}$. Moreover, even over 'abstract' arrow frames they express certain properties in the sense that a formula φ is valid in an arrow frame $\mathfrak{F} = \langle W, C, R, E \rangle$ iff $\mathfrak{F}$ has the corresponding property. Roughly, (AL$_1$) says that

(1) R is an idempotent total function on W,

that is, if $r(x)$ denotes the unique y such that $R(x, y)$, then $r(r(x)) = x$. (AL$_2$) says that

(2) $\forall x \exists y \big(E(y) \wedge C(x, x, y)\big) \wedge \forall x, y, z \big(C(x, y, z) \wedge E(z) \rightarrow x = y\big)$;

(AL$_3$) and (AL$_4$) mean that

(3) $$\forall x, y, z \Big(\big(C(r(x), y, z) \rightarrow C(x, r(z), r(y))\big) \wedge \big(C(x, y, z) \rightarrow C(z, r(y), x)\big)\Big);$$

and (AL$_5$) corresponds to

(4) $$\forall x, y, z, u, v\Big(C(x, y, z) \wedge C(z, u, v) \rightarrow \exists w\big(C(x, w, v) \wedge C(w, y, u)\big)\Big).$$

Arrow logics having axiom (AL_5) are called *associative*. For proofs of these correspondences see, e.g., (Marx and Venema 1997). Given an arrow logic L and a set Γ of $\mathcal{AL}$-formulas, we write $L \oplus \Gamma$ to denote the smallest arrow logic containing $L \cup \Gamma$. Let

$$\mathbf{AL}_{RA} = \mathbf{AL}_{min} \oplus \{(AL_1), (AL_2), (AL_3), (AL_4), (AL_5)\}.$$

(The modal algebras of $\mathbf{AL}_{RA}$ are what Tarski (1941) called *relation algebras*.) As was mentioned above, we have $\mathbf{AL}_{RA} \subsetneq \mathbf{AL}_{sq}$. However, the converse inclusion does not hold (i.e., the axioms (AL_1)–(AL_5) do not give a complete axiomatization of $\mathbf{AL}_{sq}$), as the following consequence of the algebraic results of Lyndon (1950) and Monk (1964) show:

Theorem 3.38. *The arrow logics $\mathbf{AL}_{sq}$ and $\mathbf{AL}_{RA}$ are different. In fact, $\mathbf{AL}_{sq}$ is not finitely axiomatizable.*

Various strengthenings of this theorem can be found in (Andréka 1997). Marx and Venema (1997) give an axiomatization of $\mathbf{AL}_{sq}$ using an irreflexivity rule. Actually, $\mathbf{AL}_{sq}$ is 'quite far' from the finitely axiomatizable logic $\mathbf{AL}_{RA}$. In Section 8.4 we will discuss a remarkable result of Hirsch and Hodkinson (2001) (Theorem 8.32) which has the following consequence:

Corollary 3.39. *It is undecidable whether a finite arrow frame for $\mathbf{AL}_{RA}$ is a frame for $\mathbf{AL}_{sq}$.*

Tarski showed (see Tarski 1941, Tarski and Givant 1987) that strong set and number theories can be developed within the framework of relation algebras. As a consequence we obtain:

Theorem 3.40. *Every arrow logic L in the interval $\mathbf{AL}_{RA} \subseteq L \subseteq \mathbf{AL}_{sq}$ is undecidable. In particular, $\mathbf{AL}_{sq}$ is recursively enumerable but undecidable.*

Many more associative arrow logics are proven to be undecidable in (Andréka *et al.* 1994, Kurucz *et al.* 1995), see also (Andréka *et al.* 1997). These proofs use reductions to undecidable word problems for semigroups.

The undecidability of relation algebras led to the study of various weakenings of the associativity axiom (AL_5); see, e.g., (Hirsch and Hodkinson 2002, Maddux 1978) and references therein. For instance, let

$$\begin{aligned}\mathbf{AL}_{NA} &= \mathbf{AL}_{min} \oplus \{(AL_1), (AL_2), (AL_3), (AL_4)\},\\ \mathbf{AL}_{WA} &= \mathbf{AL}_{min} \oplus \{(AL_1), (AL_2), (AL_3), (AL_4), (AL_6)\},\end{aligned}$$

where (AL_6) is the following axiom of *weak associativity*:

(AL_6) $(((p_0 \wedge Id) ; p_1) ; p_2) \leftrightarrow ((p_0 \wedge Id) ;(p_1 ; p_2))$.

(The corresponding classes of modal algebras are the *nonassociative relation algebras* (NA) and the *weakly associative relation algebras* (WA), respectively.) It follows from the algebraic results of Németi (1987) that these logics are decidable:

Theorem 3.41. *The arrow logics* $\mathbf{AL}_{NA}$ *and* $\mathbf{AL}_{WA}$ *are decidable.*

Marx (2002) provides a detailed complexity analysis of various arrow logics. Here we mention his result concerning $\mathbf{AL}_{WA}$ only:

Theorem 3.42. *The decision problem for* $\mathbf{AL}_{WA}$ *is* EXPTIME-*complete.*

There are intuitive properties of arrows (e.g., that composition of arrows is a partial function) that the language $\mathcal{AL}$ is not able to express. This gave rise to extensions of $\mathcal{AL}$ with various new (i.e., not expressible in $\mathcal{AL}$) connectives. In particular, all extensions of $\mathbf{AL}_{WA}$ with the *universal modality*, the *difference operator*, *counting* or *graded modalities*, *polyadic compositions* or the *iteration* * turned out to be decidable; see, e.g., (van Benthem 1994, Kurucz 2000b, Marx 1995, Marx and Venema 1997, Mikulás 1995, Stebletsova 2000a, 2000b). On the other hand, the extensions of $\mathbf{AL}_{WA}$ with *coordinate-wise difference* or *counting operators* are undecidable (Marx 2002); those with *projection elements* or the *fork operation* are not even recursively enumerable (Kurucz 1997).

There is also an interesting semantical characterization of $\mathbf{AL}_{WA}$. Recall that a pair arrow frame is a subframe of a square arrow frame, that is, a structure of the form

$$\mathfrak{F}_U^{(W)} = \left\langle W, C_U^{(W)}, R_U^{(W)}, E_U^{(W)} \right\rangle,$$

where $\emptyset \neq W \subseteq U \times U$ for some set U and $C_U^{(W)}$, $R_U^{(W)}$ and $E_U^{(W)}$ are the respective restrictions of C_U, R_U and E_U to W, i.e.,

$$\begin{aligned} C_U^{(W)} &= \{\langle\langle a,b\rangle, \langle a,c\rangle, \langle c,b\rangle\rangle \mid \langle a,b\rangle, \langle a,c\rangle, \langle c,b\rangle \in W\}, \\ R_U^{(W)} &= \{\langle\langle a,b\rangle, \langle b,a\rangle\rangle \mid \langle a,b\rangle, \langle b,a\rangle \in W\}, \\ E_U^{(W)} &= \{\langle a,a\rangle \mid \langle a,a\rangle \in W\} = E_U \cap W. \end{aligned}$$

The following theorem (in its algebraic form) was proved by Maddux (1982):

Theorem 3.43. *The arrow logic* $\mathbf{AL}_{WA}$ *is characterized by the class of all pair arrow frames* $\mathfrak{F}_U^{(W)}$, *with* W *being a reflexive and symmetric relation on* U.

For a proof and further completeness theorems concerning weak arrow logics consult (Marx and Venema 1997). Note that according to Theorem 3.43,

$\mathbf{AL}_{WA}$ can be considered as an example of the *relativization* technique to be discussed in Section 9.1.

A remarkable feature of the language $\mathcal{AL}$ is that its expressive power over square arrow frames is the same as that of the first-order language with equality $\mathcal{QL}_2^{3,=}$ having only binary predicate symbols $R_0, R_1, \dots$ and three individual variables x, y, z. Indeed, define a translation $\cdot^\dagger$ of $\mathcal{AL}$-formulas into $\mathcal{QL}_2^{3,=}$-formulas (having free variables among x and y) by taking

$$\begin{aligned} p_i^\dagger &= R_i(x,y) \quad (p_i \text{ a propositional variable}), \\ (\varphi \wedge \psi)^\dagger &= \varphi^\dagger \wedge \psi^\dagger, \\ (\neg\varphi)^\dagger &= \neg\varphi^\dagger, \\ (\varphi\,;\psi)^\dagger &= \exists z\,(\varphi^\dagger\{z/y\} \wedge \psi^\dagger\{z/x\}), \\ (\varphi^-)^\dagger &= \varphi^\dagger\{x/y, y/x\}, \\ Id^\dagger &= (x = y). \end{aligned}$$

On the semantical level, one may regard any arrow model $\mathfrak{M} = \langle \mathfrak{F}, \mathfrak{V} \rangle$ based on a square arrow frame $\mathfrak{F}$ over some set U as a first-order structure

$$J(\mathfrak{M}) = \left\langle U, R_0^{J(\mathfrak{M})}, R_1^{J(\mathfrak{M})}, \dots \right\rangle$$

of the language $\mathcal{QL}_2^{3,=}$, where $R_i^{J(\mathfrak{M})} = \mathfrak{V}(p_i)$ for all i. Similarly, one can regard a first-order $\mathcal{QL}_2^{3,=}$-structure J with domain U as a square arrow model $\mathfrak{M}(J)$ over U. It is not hard to see the following equivalence:

Theorem 3.44. *For every $\mathcal{AL}$-formula φ, every square arrow model $\mathfrak{M}$ and every arrow $\langle a, b \rangle$ in $\mathfrak{M}$,*

$$(\mathfrak{M}, \langle a, b \rangle) \models \varphi \quad \textit{iff} \quad J(\mathfrak{M}) \models \varphi^\dagger[a, b].$$

The translation in the other direction is more complicated. Nevertheless, the following theorem was proved by Tarski in an algebraic setting (see Tarski and Givant 1987):

Theorem 3.45. *There is a recursive function $^\bullet$ from $\mathcal{QL}_2^{3,=}$ to $\mathcal{AL}$ such that, for any $\mathcal{QL}_2^{3,=}$-formula $\phi(x, y)$, any first-order $\mathcal{QL}_2^{3,=}$-structure J with domain U and any points $a, b \in U$,*

$$J \models \phi[a, b] \quad \textit{iff} \quad (\mathfrak{M}(J), \langle a, b \rangle) \models \phi^\bullet.$$

The reader can find an 'arrow logic' proof of this result in (Marx and Venema 1997). In Section 8.1 we will give an embedding of $\mathcal{QL}_2^{3,=}$ into the logic $\mathbf{S5} \times \mathbf{S5} \times \mathbf{S5}$ which provides a connection between $\mathbf{AL}_{RA}$ and three-dimensional product logics.

To conclude this section, we briefly discuss some connections between arrow logics and other modal transition formalisms. In action logics (Pratt 1990), the binary *residuals* \ and / of the composition connective ; of $\mathcal{AL}$ are also considered, with the following semantical meaning (in arrow frames):

$$(\mathfrak{M}, x) \models \varphi \backslash \psi \quad \text{iff} \quad \forall y, z\ (\text{if } C(z, y, x) \text{ and } (\mathfrak{M}, y) \models \varphi \text{ then } (\mathfrak{M}, z) \models \psi),$$
$$(\mathfrak{M}, x) \models \varphi / \psi \quad \text{iff} \quad \forall y, z\ (\text{if } C(z, x, y) \text{ and } (\mathfrak{M}, y) \models \psi \text{ then } (\mathfrak{M}, z) \models \varphi).$$

It is readily checked that these connectives are expressible in $\mathbf{AL}_{NA}$, viz., the following equivalences are valid in all arrow frames for $\mathbf{AL}_{NA}$:

$$(\varphi \backslash \psi) \leftrightarrow \neg(\varphi^- ; \neg\psi),$$
$$(\varphi / \psi) \leftrightarrow \neg(\neg\varphi ; \psi^-).$$

The residuals \ and / are also connectives of implicational calculi of *categorial grammars*[5] (where they are called *pre-* and *post-implications* and often denoted by $\rightarrow$ and $\leftarrow$, respectively). It is not hard to see that the above semantical definition gives us the following implications:

- if $\varphi \backslash \psi$ follows from χ, then ψ follows from $\varphi ; \chi$,
- if φ / ψ follows from χ, then one φ follows from $\chi ; \psi$.

These are the rules of one of the best-known implicational inference systems—the *Lambek calculus* (Lambek 1958). It follows that associative arrow frames provide a sound semantics for this calculus (van Benthem 1991). Moreover, it is shown in (Andréka and Mikulás 1994) that the Lambek calculus is complete for the class of pair arrow frames $\mathfrak{F}_U^{(W)}$, with W being a transitive relation on U. Implicational calculi and their embeddings into modal logics are studied in detail in (Kurtonina 1995, Roorda 1991). Other semantics for implicational calculi are reviewed in (van Benthem 1991).

3.11 Intuitionistic modal logics

Intuitionistic modal logics, i.e., combinations of modal logics and (extensions of) intuitionistic logic, originate from several distinguishable sources and have various areas of application. They include:

- intuitionistic analysis of modalities in philosophy (see, e.g., Prior 1957, Ewald 1986, Williamson 1992);
- modal analysis of formal constructive provability in the foundations of mathematics (Kuznetsov 1985, Kuznetsov and Muravitskij 1986);

[5]The reader can find a detailed treatment of categorial grammars in (Moortgat 1996); a brief survey is in (van Benthem 1996).

- possible applications in computer science (Plotkin and Stirling 1986, Stirling 1987, Wijesekera 1990);
- modalities are added to intuitionistic logic in the framework of studying 'new intuitionistic connectives' (Bessonov 1977, Gabbay 1977, Yashin 1994) and
- to simulate the monadic fragment of intuitionistic first-order logic (Bull 1966, Ono 1977, Ono and Suzuki 1988, Bezhanishvili 1997).

There are different ways of defining intuitionistic analogs of classical modal logics. One approach is to use the fact that classical **S5** and **K** can be regarded as fragments of classical first-order logic **QCl** and introduce their intuitionistic counterparts as 'solutions' x and y to the equations

$$\frac{\mathbf{QCl}}{\mathbf{S5}} = \frac{\mathbf{QInt}}{x}, \qquad \frac{\mathbf{QCl}}{\mathbf{K}} = \frac{\mathbf{QInt}}{y}.$$

More precisely, let us recall that in Section 1.2 we defined two maps $\varphi \to \varphi^{\star}$ and $\varphi \to \varphi^{\dagger}$ from $\mathcal{ML}$ into the language of classical first-order logic in such a way that for all $\varphi \in \mathcal{ML}$ we have:

- $\varphi \in \mathbf{S5}$ iff $\varphi^{\dagger} \in \mathbf{QCl}$,
- $\varphi \in \mathbf{K}$ iff $\varphi^{\star} \in \mathbf{QCl}$.

So the 'solutions' x and y to the equations above can be written as

- $\mathbf{MIPC} = \{\varphi \in \mathcal{ML} \mid \varphi^{\dagger} \in \mathbf{QInt}\}$,
- $\mathbf{FS} = \{\varphi \in \mathcal{ML} \mid \varphi^{\star} \in \mathbf{QInt}\}$,

respectively.

These two logics can be regarded as 'intuitionistic' because their non-modal fragments coincide with propositional intuitionistic logic **Int**, and their modal operators $\Box$ and $\Diamond$ reflect the behavior of the intuitionistic quantifiers $\forall$ and $\exists$ in the same way as the classical $\Box$ and $\Diamond$ of **S5** and **K** simulate the classical quantifiers. Moreover, **MIPC** and **FS** turn out to be of interest from one more point of view. The former logic corresponds to the one-variable fragment of **QInt** and the latter to a natural fragment of the two-variable sublanguage of **QInt**. As both of them are decidable (this will be shown in Section 10.2), we obtain expressive and natural fragments of the undecidable first-order logic **QInt**.[6]

[6]The decidability of **FS** is of particular interest, since no more expressive decidable fragments of **QInt** are known. In contrast to classical predicate logic, the two-variable fragment of **QInt** is undecidable, at least under the constant domain assumption (Gabbay and Shehtman 1993). While the decidability of **K** can be 'explained' by the fact that it is embeddable into the two-variable fragment of classical predicate logic, the decidability of **FS** cannot be justified by the observation that it is embedded into the two-variable fragment of **QInt**.

MIPC and **FS** were introduced—axiomatically, not as fragments of first-order intuitionistic logic—by Prior (1957) and Fischer Servi (1980, 1984). But before defining their syntactical and semantical characterizations, it is worth considering another, more general and abstract, approach to constructing intuitionistic modal logics.

Let M be a nonempty subset of $\{\Box, \Diamond\}$. Denote by $\mathcal{L}_{\mathsf{M}}$ the standard propositional intuitionistic language extended with the connectives in M. In particular, $\mathcal{L}_{\{\Box,\Diamond\}} = \mathcal{ML}$. We stick to the new notation, however, in order to emphasize that $\mathcal{L}_{\{\Box,\Diamond\}}$ on the intuitionistic basis is usually more expressive than $\mathcal{L}_{\{\Box\}}$. Instead of $\mathcal{L}_{\{\Box\}}$ and $\mathcal{L}_{\{\Box,\Diamond\}}$, we will write $\mathcal{L}_{\Box}$ and $\mathcal{L}_{\Box\Diamond}$, respectively. By an *intuitionistic modal logic* in the language $\mathcal{L}_{\mathsf{M}}$ (*im-logic*, for short) we understand any subset of $\mathcal{L}_{\mathsf{M}}$ containing intuitionistic logic **Int** and closed under MP, Subst and the *regularity rule*

$$\frac{\varphi \to \psi}{\odot\varphi \to \odot\psi}$$

for every $\odot \in \mathsf{M}$. Given such a logic L and a set Γ of $\mathcal{L}_{\mathsf{M}}$-formulas we denote by $L \oplus \Gamma$ the smallest im-logic containing $L \cup \Gamma$.

Within this approach, there are three obvious ways of defining intuitionistic analogs of classical modal logics. First, one can take the family of logics extending the basic system $\mathbf{IntK}_{\Box}$ in the language $\mathcal{L}_{\Box}$, which is axiomatized by adding to **Int** the axioms of **K**, say,

- $\Box(p \wedge q) \leftrightarrow \Box p \wedge \Box q$ and
- $\Box\top$.

An example of a logic in this family is Kuznetsov's (1985) *intuitionistic provability logic* $\mathbf{I}^{\Delta}$ (Kuznetsov used Δ instead of $\Box$), the intuitionistic analog of the Gödel–Löb classical provability logic **GL**:

$$\mathbf{I}^{\Delta} = \mathbf{IntK}_{\Box} \oplus p \to \Box p \oplus (\Box p \to p) \to p \oplus ((p \to q) \to p) \to (\Box q \to p).$$

A possibility operator $\Diamond$ in logics of this sort can be defined in the classical way by taking $\Diamond\varphi = \neg\Box\neg\varphi$. Note, however, that in general this $\Diamond$ does not distribute over disjunction and that the connection via negation between $\Box$ and $\Diamond$ is too strong from the intuitionistic standpoint. The situation here is similar to that in intuitionistic predicate logic where $\exists$ and $\forall$ are not dual. Consequently, neither **MIPC** nor **FS** are axiomatic extensions of $\mathbf{IntK}_{\Box}$.

Another family of im-logics can be defined in the language $\mathcal{L}_{\Diamond}$ by taking as the basis system the smallest logic in $\mathcal{L}_{\Diamond}$ to contain the axioms

- $\Diamond(p \vee q) \leftrightarrow \Diamond p \vee \Diamond q$ and

- $\neg\Diamond\bot$.

This logic will be denoted by $\mathbf{IntK}_\Diamond$. However, again neither **FS** nor **MIPC** are axiomatic extensions of $\mathbf{IntK}_\Diamond$ because $\Box$ cannot be defined by means of $\Diamond$ in those logics.

A family of logics containing **FS** and **MIPC** can be obtained if we consider im-logics with independent $\Box$ and $\Diamond$. These are extensions of the system $\mathbf{IntK}_{\Box\Diamond}$ which is the smallest im-logic in the language $\mathcal{L}_{\Box\Diamond}$ containing both $\mathbf{IntK}_\Box$ and $\mathbf{IntK}_\Diamond$. Then we have:

$$\begin{aligned}
\mathbf{FS} &= \mathbf{IntK}_{\Box\Diamond} \oplus \Diamond(p \to q) \to (\Box p \to \Diamond q) \oplus \\
&\qquad (\Diamond p \to \Box q) \to \Box(p \to q), \\
\mathbf{MIPC} &= \mathbf{FS} \oplus \Box p \to p \oplus \Box p \to \Box\Box p \oplus \Diamond p \to \Box\Diamond p \oplus \\
&\qquad p \to \Diamond p \oplus \Diamond\Diamond p \to \Diamond p \oplus \Diamond\Box p \to \Box p.
\end{aligned}$$

The axiomatization of **MIPC** was first given by Bull (1966) and, according to Simpson (1994), the axiomatization of **FS** was determined by C. Stirling; see also (Grefe 1998). We will prove these equations in Chapter 10.

Now let us consider the relational semantics for some of the im-logics introduced above. All the semantical concepts to be defined turn out to be natural combinations of the corresponding notions developed for classical modal and intuitionistic logics.

We begin with the semantics for logics in the language $\mathcal{L}_\Box$. Consider frames $\mathfrak{F} = \langle W, R, R_\Box \rangle$ in which R is a partial order on W interpreting the intuitionistic connectives and $R_\Box$ is an arbitrary binary relation interpreting $\Box$ in the standard manner. Say that a map $\mathfrak{V}$ is a *valuation* in $\mathfrak{F}$ if it associates with every propositional variable p an *R-closed* subset $\mathfrak{V}(p) \in Up\mathfrak{F}$ of W. The *truth-relation* $(\mathfrak{M}, x) \models \varphi$ (or simply $x \models \varphi$) is defined inductively as follows:

$$\begin{array}{lll}
(\mathfrak{M}, x) \models p & \text{iff} & x \in \mathfrak{V}(p); \\
(\mathfrak{M}, x) \models \psi \wedge \chi & \text{iff} & (\mathfrak{M}, x) \models \psi \text{ and } (\mathfrak{M}, x) \models \chi; \\
(\mathfrak{M}, x) \models \psi \vee \chi & \text{iff} & (\mathfrak{M}, x) \models \psi \text{ or } (\mathfrak{M}, x) \models \chi; \\
(\mathfrak{M}, x) \models \psi \to \chi & \text{iff} & \text{for all } y \in W \text{ such that } xRy, \\
 & & (\mathfrak{M}, y) \models \psi \text{ implies } (\mathfrak{M}, y) \models \chi; \\
(\mathfrak{M}, x) \not\models \bot; & & \\
(\mathfrak{M}, x) \models \Box\varphi & \text{iff} & \forall y \in W\ (xR_\Box y \to y \models \varphi).
\end{array}$$

In accordance with the principles of intuitionistic semantics, we want the set of states $\{x \in W \mid x \models \Box\varphi\}$ to be *R-closed* whenever the set $\{x \in W \mid x \models \varphi\}$ is R-closed. This will be the case if

$$\{x \in W \mid \forall y\ (xR_\Box y \to y \in X)\}$$

is R-closed whenever $X \subseteq W$ is R-closed. Equivalently this condition can be represented as

$$R \circ R_{\Box} \subseteq R_{\Box}.$$

We shall see in Section 10.1 that $\mathbf{IntK}_{\Box}$ is sound and complete with respect to frames satisfying this condition.[7]

Another possibility of constructing a semantics for $\mathbf{IntK}_{\Box}$ is to manipulate the inductive definition of the truth-relation for $\Box$ in such a way that the set $\{x \mid x \models \Box\varphi\}$ becomes R-closed for arbitrary frames of the form $\langle W, R, R_{\Box}\rangle$, where R is a partial order. This can be achieved by using the truth-relation $\models'$ which is obtained from $\models$ by replacing the clause for $\Box$ with the following one:

$$(\mathfrak{M}, x) \models' \Box\varphi \quad \text{iff} \quad \forall y \forall z\, (xRyR_{\Box}z \rightarrow z \models \varphi).$$

With the help of the completeness result formulated above it is not difficult to see that $\mathbf{IntK}_{\Box}$ is sound and complete with respect to arbitrary frames $\langle W, R, R_{\Box}\rangle$ under the truth-relation $\models'$.

We arrive at a similar adjustment of the truth-relation for $\Box$ if we try to construct a semantics for **FS** and **MIPC** starting from their definitions as fragments of **QInt** and considering proper reducts of first-order intuitionistic frames.

Consider a set of information states W with a partial order $\lhd$ on it and a function $\mathfrak{d}$ which 'describes' the states $w \in W$ by associating with them structures $\mathfrak{d}(w) = \langle \Delta^w, S^w\rangle$ such that Δ^w is nonempty and S^w is a binary relation on Δ^w satisfying the following monotonicity conditions:

- $\Delta^w \subseteq \Delta^v$ whenever $w \lhd v$,
- $S^w \subseteq S^v$ whenever $w \lhd v$.

The triple $\mathfrak{F} = \langle W, \lhd, \mathfrak{d}\rangle$ will be called a *standard* **FS**-*frame*. A *valuation* $\mathfrak{V}$ in it is a map which associates with every propositional variable p and every state w a set $\mathfrak{d}(w,p) \subseteq \Delta^w$ satisfying the condition

- $\mathfrak{d}(w,p) \subseteq \mathfrak{d}(v,p)$ whenever $w \lhd v$.

The pair $\langle \mathfrak{F}, \mathfrak{V}\rangle$ is called a *standard* **FS**-*model.*.

The *truth-relation* $\models$ between pairs (w, x), for $w \in W$, $x \in \Delta^w$, and $\mathcal{L}_{\Box\Diamond}$-formulas is defined inductively as follows:

[7] Actually, we will show that $\mathbf{IntK}_{\Box}$ is determined by the class of frames satisfying the stronger condition $R \circ R_{\Box} \circ R = R_{\Box}$; see Proposition 10.4.

$(w,x) \models p$	iff	$x \in \mathfrak{V}(w,p)$;
$(w,x) \models \psi \wedge \chi$	iff	$(w,x) \models \psi$ and $(w,x) \models \chi$;
$(w,x) \models \psi \vee \chi$	iff	$(w,x) \models \psi$ or $(w,x) \models \chi$;
$(w,x) \models \psi \to \chi$	iff	for all $v \in W$ such that $w \lhd v$, $(w,x) \models \psi$ implies $(w,x) \models \chi$;
$(w,x) \not\models \bot$;		
$(w,x) \models \Box\varphi$	iff	$\forall v\ (w \lhd v \to \forall y\ (xS^v y \to (v,y) \models \varphi))$;
$(w,x) \models \Diamond\varphi$	iff	$\exists y\ (xS^w y \to (w,y) \models \varphi)$;

A direct inspection of the inductive definitions shows that standard **FS**-models indeed simulate first-order intuitionistic models. It follows that **FS** coincides with the set of $\mathcal{L}_{\Box\Diamond}$-formulas that are true in every standard **FS**-frame under every valuation. It should also be clear that **MIPC** is the set of formulas that are valid in all standard **FS**-frames in which $S^w = \Delta^w \times \Delta^w$, for all $w \in W$.

We close our introduction to intuitionistic modal logic with the following observation.

Proposition 3.46. *Neither* **FS** *nor* **MIPC** *have the finite model property with respect to standard* **FS***-models.*

Proof. We show that the formula

$$\varphi = \Box\neg\neg p \to \neg\neg\Box p$$

is valid in all finite standard **FS**-frames, but is refuted in an infinite standard **FS**-frame validating **MIPC** (this example was provided by Ono and Suzuki 1988). First, suppose that $\mathfrak{F} = \langle W, \lhd, \mathfrak{d} \rangle$ is a finite standard **FS**-frame, $\mathfrak{V}$ a valuation in it, and $(w,x) \models \Box\neg\neg p$. Consider a maximal $v \in W$ with $w \lhd v$ and any $y \in \mathfrak{d}(v)$ with $xS^v y$. Then $(v,y) \models \neg\neg p$, and so $(v,y) \models p$ because v is maximal. Hence $(v,x) \models \Box p$ (again because v is maximal). It follows that $(v,x) \models \Box p$ for all maximal $v \in W$ with $w \lhd v$. Therfore, since W is finite, $(w,x) \models \neg\neg\Box p$. Thus, $\mathfrak{F}$ validates φ.

Now we construct a standard **FS**-frame $\langle W, \lhd, \mathfrak{d} \rangle$ by taking

- $W = \mathbb{N}$,
- $n \lhd m$ iff $n \leq m$, for all $n, m \in W$,
- $\mathfrak{d}(n) = \mathbb{N}$,
- $S^m = \mathbb{N} \times \mathbb{N}$, for $m \in W$.

Let $\mathfrak{V}(n,p) = \{0, \dots, n\}$. Then it is easily checked that $(0,0) \not\models \varphi$. ❑

We have discussed two different types of im-logics: those in which only one primitive modal operator is available (viz., extensions of $\mathbf{IntK}_\Box$ and $\mathbf{IntK}_\Diamond$) and those comprising two primitive operators $\Box$ and $\Diamond$ connected by the principles of **FS**. Intuitively, the former logics are much simpler; they can hardly be described as 'many-dimensional.' On the contrary, the frames for **FS** and its extensions have a clear two-dimensional flavor. This intuition will be made more precise in Chapter 10, where we show that $\mathbf{IntK}_\Box$ can be embedded into *fusions* of classical modal logics, while **FS** lies embedded into *products* of them.

Part II

Fusions and products

Chapter 4

Fusions of modal logics

We begin our study of the combined systems introduced in the previous chapter by considering the simplest and most fundamental kind of combination—the formation of fusions. Unlike other, more sophisticated, combinations to be investigated in the subsequent chapters, the fusion operation has a very nice feature: it preserves properties of the fused logics. In particular, the following properties are transferred from the component logics to their fusion:

- Kripke completeness (Theorem 4.1),
- the fmp (Theorem 4.2),
- decidability (Theorem 4.12),
- decidability of the global consequence relation (Theorem 4.10),
- (uniform) interpolation property (Theorem 4.18).

We prove these remarkable theorems in the first five sections of this chapter. Section 4.6 provides a brief survey of known complexity results concerning fusions.

4.1 Preserving Kripke completeness and the finite model property

In this section we outline a proof of the following two transfer theorems concerning fusions of multimodal[1] logics due to Kracht and Wolter (1991) and Fine and Schurz (1996):

[1]As in Section 3.1, by a multimodal logic we mean a logic formulated in any of the languages $\mathcal{ML}_n$, $\mathcal{ML}_n^C$, $\mathcal{ML}_\mathcal{U}$, or $\mathcal{ML}_{\mathcal{SU}}$.

Theorem 4.1. *If multimodal logics L_1 and L_2 are characterized by classes of frames $\mathcal{C}_1$ and $\mathcal{C}_2$, respectively, and if $\mathcal{C}_1$ and $\mathcal{C}_2$ are closed under the formation of disjoint unions and isomorphic copies, then the fusion $L_1 \otimes L_2$ of L_1 and L_2 is characterized by $\mathcal{C}_1 \otimes \mathcal{C}_2$. In particular,*

$$L_1 \otimes L_2 = \mathsf{Log}(\mathsf{Fr}L_1 \otimes \mathsf{Fr}L_2).$$

Theorem 4.2. *If both L_1 and L_2 are multimodal logics having the finite model property, then their fusion $L_1 \otimes L_2$ has the finite model property as well.*

Actually, Theorem 4.2 follows from the proof of Theorem 4.1, since the closure under *finite* disjoint unions is enough when we work with finite frames. So we concentrate on the proof of Theorem 4.1. To simplify notation, we assume that L_1 and L_2 are unimodal logics with the boxes $\Box_1$ and $\Box_2$, respectively. The fusion $L = L_1 \otimes L_2$ is then a bimodal logic in the language $\mathcal{ML}_2$.

With each $\mathcal{ML}_2$-formula φ of the form $\Box_i\psi$ ($i = 1, 2$) we associate a new variable q_φ which will be called the *surrogate of* φ. For an $\mathcal{ML}_2$-formula φ containing no surrogate variables, denote by $\varphi^{\mathbf{1}}$ the formula that results from φ by replacing all its subformulas of the form $\Box_2\psi$, which are not within the scope of other $\Box_2$, with their surrogate variables $q_{\Box_2\psi}$. So $\varphi^{\mathbf{1}}$ is a unimodal formula containing only $\Box_1$. Let

$$\Theta^{\mathbf{1}}(\varphi) = \{p \mid p \text{ is a variable in } \varphi\} \cup \{\chi \in sub\,\Box_2\psi \mid \Box_2\psi \in sub\,\varphi\}.$$

The formula $\varphi^{\mathbf{2}}$ and the set $\Theta^{\mathbf{2}}(\varphi)$ are defined symmetrically.

Suppose now that φ is satisfiable in a model based on a frame for L. To prove that L is characterized by $\mathcal{C}_1 \otimes \mathcal{C}_2$, we have to construct a frame in $\mathcal{C}_1 \otimes \mathcal{C}_2$ satisfying φ. As we know only how to build frames for the unimodal fragments of L, the frame is constructed step-by-step alternating between $\Box_1$ and $\Box_2$.

Note first that since L_1 is characterized by $\mathcal{C}_1$, there is a model $\mathfrak{M}$ based on a frame in $\mathcal{C}_1$ and satisfying $\varphi^{\mathbf{1}}$ at a point r. Our aim now is to ensure that the formulas of the form $\Box_2\psi$ have the same truth-values as their surrogates $q_{\Box_2\psi}$. To do this, with each point x in $\mathfrak{M}$ we can associate the formula

$$\varphi_x = \bigwedge\{\psi \in \Theta^{\mathbf{1}}(\varphi) \mid (\mathfrak{M}, x) \models \psi^{\mathbf{1}}\} \wedge \bigwedge\{\neg\psi \mid \psi \in \Theta^{\mathbf{1}}(\varphi),\ (\mathfrak{M}, x) \not\models \psi^{\mathbf{1}}\},$$

construct a model $\mathfrak{M}_x$ based on a frame in $\mathcal{C}_2$ and satisfying $\varphi_x^{\mathbf{2}}$ in a world y, and then hook $\mathfrak{M}_x$ to $\mathfrak{M}$ by identifying x and y. After that we can switch to $\Box_1$ and in the same manner ensure that formulas $\Box_1\psi$ have the same truth-values as $q_{\Box_1\psi}$ at all points in every $\mathfrak{M}_x$. And so forth; see Fig. 4.1 for an example. In this construction we use the fact that $\mathcal{C}_1$ and $\mathcal{C}_2$ are closed under isomorphic copies and disjoint unions: the $\mathfrak{M}_x$ should be mutually disjoint and the final model is the union of the models constructed at each step. Note also that this construction is a special case of *fibring semantics* that is called iterated *dovetailing* (Gabbay 1996, 1999).

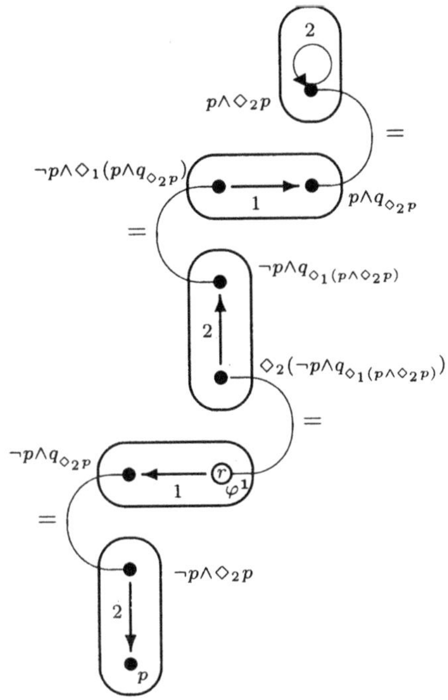

Figure 4.1: Satisfying $\varphi = p \wedge \Diamond_1(\neg p \wedge \Diamond_2 p) \wedge \Diamond_2\big(\neg p \wedge \Diamond_1(p \wedge \Diamond_2 p)\big)$ at r.

However, to realize this quite obvious scheme, we must be sure that $\varphi_x^{\mathbf{2}}$ is really satisfiable in a frame for L_2, which may impose some restrictions on the models we choose. First, in the construction above it is enough to deal with points x accessible from r in at most $md(\varphi)$ steps; no other point has any influence on the truth of φ at r. Let X be the set of all such points. Now, a sufficient and necessary condition for φ_x to be satisfiable in a frame for L (and so for $\varphi_x^{\mathbf{2}}$ to be satisfiable in a frame for L_2) can be formulated using the following general description of formulas of type φ_x.

Suppose Γ is a finite set of formulas closed under subformulas. Define the *consistency-set* $C(\Gamma)$ *of* Γ by taking

$$C(\Gamma) = \{\psi_\Delta \mid \Delta \subseteq \Gamma\},$$

where for $\Delta \subseteq \Gamma$,

$$\psi_\Delta = \bigwedge\{\chi \mid \chi \in \Delta\} \wedge \bigwedge\{\neg\chi \mid \chi \in \Gamma - \Delta\}.$$

In particular, for all $x \in X$, we have $\varphi_x \in C(\Theta^1(\varphi))$. Given a formula φ, define

$$\Sigma_1(\varphi) = \{\psi \in C(\Theta^1(\varphi)) \mid \neg\psi \notin L\},$$
$$\Sigma_2(\varphi) = \{\psi \in C(\Theta^2(\varphi)) \mid \neg\psi \notin L\}.$$

The formulas in $\Sigma_i(\varphi)$ can be regarded as 'state descriptions' of the points in the possible models with respect to the formulas in $\Theta^i(\varphi)$. In particular, for all $x \in X$, φ_x is satisfiable in a frame for L iff $\varphi_x \in \Sigma_1(\varphi)$. In other words, we should start with a model $\mathfrak{M}$ satisfying $\varphi^1 \wedge \Box_1^{\leq md(\varphi)}(\bigvee \Sigma_1(\varphi))^1$ at a point r. Of course, the subsequent models $\mathfrak{M}_x$ must satisfy $\varphi_x^2 \wedge \Box_2^{\leq md(\varphi)}(\bigvee \Sigma_2(\varphi_x))^2$ at all points $x \in X$, etc.

We hope this sketch is enough to illustrate the basic idea of the proof. We will not go further into the technical details of the inductive construction: they can be either found in (Kracht and Wolter 1991) or restored from the algebraic proof of Theorem 4.10 below.

Note that although frames for fusions of n unimodal logics have n different accessibility relations, these frames—unlike product frames—can hardly be regarded as 'genuinely many-dimensional' in the geometric sense. However, in Section 9.1 we show that in many cases fusions can be characterized by certain classes of *subframes* of product frames.

4.2 Algebraic preliminaries

This section is intended to provide the reader with the algebraic prerequisites that are necessary for the proofs of the other transfer theorems.

The algebraic characterization of fusions is quite natural. Suppose L_1 is an n-modal logic (with the boxes $\Box_1, \dots, \Box_n$) and L_2 an m-modal logic (with the boxes $\Box_{n+1}, \dots, \Box_{n+m}$). Given classes $\mathcal{C}_1$ and $\mathcal{C}_2$ of n-modal and m-modal algebras, respectively, denote by $\mathcal{C}_1 \otimes \mathcal{C}_2$ the class of $n+m$-modal algebras of the form

$$\Big\{ \big\langle A, \wedge, \neg, 0, 1, \Box_1, \dots, \Box_{n+m} \big\rangle \mid \big\langle A, \wedge, \neg, 0, 1, \Box_1, \dots, \Box_n \big\rangle \in \mathcal{C}_1 \text{ and } \big\langle A, \wedge, \neg, 0, 1, \Box_{n+1}, \dots, \Box_{n+m} \big\rangle \in \mathcal{C}_2 \Big\}.$$

It should be clear that

$$\mathsf{Alg}(L_1 \otimes L_2) = \mathsf{Alg}L_1 \otimes \mathsf{Alg}L_2.$$

Let $\mathfrak{A} = \langle A, \wedge^{\mathfrak{A}}, \neg^{\mathfrak{A}}, 0^{\mathfrak{A}}, 1^{\mathfrak{A}} \rangle$ be a Boolean algebra. For all $a, b \in A$, put

$$a \leq^{\mathfrak{A}} b \quad \text{iff} \quad a \wedge^{\mathfrak{A}} b = a.$$

It is easy to see that $\leq^{\mathfrak{A}}$ is a partial order on A with least element $0^{\mathfrak{A}}$ and greatest element $1^{\mathfrak{A}}$. As usual, we write $a <^{\mathfrak{A}} b$ if $a \leq^{\mathfrak{A}} b$ and $a \neq b$. An element a of $\mathfrak{A}$ is called an *atom* if $a \neq 0^{\mathfrak{A}}$ and

$$\{x \in A \mid x \leq^{\mathfrak{A}} a\} = \{0^{\mathfrak{A}}, a\}.$$

In other words, an atom of $\mathfrak{A}$ is an immediate successor of the zero element of $\mathfrak{A}$. The algebra $\mathfrak{A}$ is called *atomic* if for every nonzero $x \in A$ there exists an atom a such that $a \leq^{\mathfrak{A}} x$. The algebra $\mathfrak{A}$ is said to be *atomless* if $\mathfrak{A}$ contains no atoms. A proof of the following theorem can be found in (Koppelberg 1988).

Theorem 4.3. *Any two countably infinite atomless Boolean algebras are isomorphic.*[2]

We say that a modal algebra $\mathfrak{A}$ is a *c.i.a.-algebra* if the Boolean reduct of $\mathfrak{A}$ is a countably infinite atomless Boolean algebra. For a multimodal logic L, denote by $\mathsf{Atg}L$ the class of all c.i.a.-algebras in $\mathsf{Alg}L$.

The following result generalizes Theorem 1.14 by stating that all modal logics L as well as the global consequence relations $\vdash_L^*$ are characterized by their c.i.a.-algebras.

Theorem 4.4. *Let L be a n-modal logic. Then*

(i) *for any two formulas φ and ψ, we have*

$$\varphi \vdash_L^* \psi \quad \textit{iff} \quad \mathfrak{V}(\varphi) = 1^{\mathfrak{A}} \textit{ implies } \mathfrak{V}(\psi) = 1^{\mathfrak{A}},$$

for every $\mathfrak{A} \in \mathsf{Atg}L$ and every model $\mathfrak{M} = \langle \mathfrak{A}, \mathfrak{V} \rangle$;

(ii) $L = \mathsf{LogAtg}L$.

Proof. To simplify notation, we assume L to be a unimodal logic formulated in the language $\mathcal{ML}$. Suppose that $\varphi \not\vdash_L^* \psi$. Define an equivalence relation $\sim$ on $\mathcal{ML}$ by taking

$$\chi_1 \sim \chi_2 \quad \text{iff} \quad \varphi \vdash_L^* \chi_1 \leftrightarrow \chi_2.$$

Denote by $[\chi]$ the $\sim$-equivalence class generated by χ. We define an algebra

$$\mathfrak{A} = \left\langle A, \wedge^{\mathfrak{A}}, \neg^{\mathfrak{A}}, 0^{\mathfrak{A}}, 1^{\mathfrak{A}}, \Box^{\mathfrak{A}} \right\rangle$$

[2] For readers not familiar with atomless Boolean algebras: this theorem can be proved in the same way (namely, using a back-and-forth construction) as Cantor proved that any two countably infinite dense linear orders without endpoints are isomorphic.

by taking $A = \{[\chi] \mid \chi \in \mathcal{ML}\}$ and

$$\begin{aligned} [\chi_1] \wedge^{\mathfrak{A}} [\chi_2] &= [\chi_1 \wedge \chi_2], \\ \neg^{\mathfrak{A}}[\chi] &= [\neg\chi], \\ 0^{\mathfrak{A}} &= [\bot], \\ 1^{\mathfrak{A}} &= [\top], \\ \Box^{\mathfrak{A}}[\chi] &= [\Box\chi]. \end{aligned}$$

The reader can readily check that $\mathfrak{A}$ is a well-defined modal algebra for L.

Define a valuation $\mathfrak{V}$ in $\mathfrak{A}$ by taking $\mathfrak{V}(p) = [p]$ for every propositional variable p. It can be proved by induction that $\mathfrak{V}(\chi) = [\chi]$, for every $\mathcal{ML}$-formula χ. Since $\varphi \sim \top$ and $\psi \not\sim \top$, we then have $\mathfrak{V}(\varphi) = 1^{\mathfrak{A}}$ and $\mathfrak{V}(\psi) \neq 1^{\mathfrak{A}}$.

It remains to show that $\mathfrak{A}$ is countably infinite and atomless. Clearly, $\mathfrak{A}$ is countably infinite whenever it is atomless. So it suffices to prove that $\mathfrak{A}$ is atomless. Suppose χ is an arbitrary formula such that $[\chi] \neq 0^{\mathfrak{A}}$. Take any propositional variable p which does not occur in χ and φ. Then

$$0^{\mathfrak{A}} <^{\mathfrak{A}} [\chi] \wedge^{\mathfrak{A}} [p] <^{\mathfrak{A}} [\chi],$$

and so $[\chi]$ is not an atom of $\mathfrak{A}$. It follows that $\mathfrak{A}$ is atomless.

This proves (i), and (ii) is its obvious consequence. ❑

The second part of the following theorem was first proved in (Thomason 1980):

Theorem 4.5. *For all consistent multimodal logics L_1 and L_2,*

(i) $\vdash^*_{L_1 \otimes L_2}$ *is a conservative extension of both* $\vdash^*_{L_1}$ *and* $\vdash^*_{L_2}$;

(ii) $L_1 \otimes L_2$ *is a conservative extension of both L_1 and L_2.*

Proof. As before, we assume that L_1 and L_2 are unimodal logics with the boxes $\Box_1$ and $\Box_2$, respectively. And again it suffices to prove only (i): (ii) follows immediately. Suppose $\varphi \not\vdash^*_{L_1} \psi$. We show that $\varphi \not\vdash^*_{L_1 \otimes L_2} \psi$. By Theorem 4.4, there exist

$$\mathfrak{A} = \left\langle A, \wedge^{\mathfrak{A}}, \neg^{\mathfrak{A}}, 0^{\mathfrak{A}}, 1^{\mathfrak{A}}, \Box_1^{\mathfrak{A}} \right\rangle \in \mathsf{Atg}L_1$$

and a valuation $\mathfrak{V}$ in $\mathfrak{A}$ such that $\mathfrak{V}(\varphi) = 1^{\mathfrak{A}}$ and $\mathfrak{V}(\psi) \neq 1^{\mathfrak{A}}$. Take any

$$\mathfrak{B} = \left\langle B, \wedge^{\mathfrak{B}}, \neg^{\mathfrak{B}}, 0^{\mathfrak{B}}, 1^{\mathfrak{B}}, \Box_2^{\mathfrak{B}} \right\rangle \in \mathsf{Atg}L_2.$$

(Such an algebra exists, since L_2 is consistent.) By Theorem 4.3, the Boolean reducts of $\mathfrak{A}$ and $\mathfrak{B}$ are isomorphic. Hence we may assume that $A = B$ and that the Boolean operations of $\mathfrak{A}$ and $\mathfrak{B}$ coincide, i.e., $\wedge^{\mathfrak{A}} = \wedge^{\mathfrak{B}}$, $\neg^{\mathfrak{A}} = \neg^{\mathfrak{B}}$, $0^{\mathfrak{A}} = 0^{\mathfrak{B}}$ and $1^{\mathfrak{A}} = 1^{\mathfrak{B}}$. But then

$$\mathfrak{D} = \left\langle A, \wedge^{\mathfrak{A}}, \neg^{\mathfrak{A}}, 0^{\mathfrak{A}}, 1^{\mathfrak{A}}, \Box_1^{\mathfrak{A}}, \Box_2^{\mathfrak{B}} \right\rangle$$

is in $\mathsf{Atg}(L_1 \otimes L_2)$ and $\langle \mathfrak{D}, \mathfrak{V} \rangle$ is an algebraic model in which φ is true, but ψ is not. ❑

We now remind the reader of some basic facts about Boolean algebras. (Detailed proofs can be found in (Koppelberg 1988).)

Recall that the domain of a direct product $\mathfrak{A} = \prod_{i \in I} \mathfrak{A}_i$ of algebras consists of all functions f from I into $\bigcup_{i \in I} A_i$ with $f(i) \in A_i$, for all $i \in I$. In other words, it consists of all sequences $\langle a_i \mid i \in I \rangle$ with $a_i \in A_i$ for all $i \in I$. The operations are defined component-wise. For example,

$$\langle a_i \mid i \in I \rangle \wedge^{\mathfrak{A}} \langle b_i \mid i \in I \rangle = \left\langle a_i \wedge^{\mathfrak{A}_i} b_i \mid i \in I \right\rangle .$$

Let us recall next that given a Boolean algebra $\mathfrak{A} = \left\langle A, \wedge^{\mathfrak{A}}, \neg^{\mathfrak{A}}, 0^{\mathfrak{A}}, 1^{\mathfrak{A}} \right\rangle$ and a nonzero element $a \in A$, we can construct the Boolean algebra

$$\mathfrak{A}_a = \left\langle \{a \wedge^{\mathfrak{A}} b \mid b \in A\}, \wedge^{\mathfrak{A}}, \neg^a, 0^{\mathfrak{A}}, a \right\rangle ,$$

where for all $b \in A$,

$$\neg^a b = a \wedge^{\mathfrak{A}} \neg^{\mathfrak{A}} b.$$

This algebra is called the *relativization of* $\mathfrak{A}$ *by* a.

A finite set $\{a_i \mid i \in I\}$ is said to be a *partition of* $\mathfrak{A}$ if

- $a_i \neq 0^{\mathfrak{A}}$, for all $i \in I$,
- $a_i \wedge^{\mathfrak{A}} a_j = 0^{\mathfrak{A}}$, for all distinct $i, j \in I$, and
- $\bigvee^{\mathfrak{A}}_{i \in I} a_i = 1^{\mathfrak{A}}$.

Lemma 4.6. *Suppose that* $\{a_i \mid i \in I\}$ *is a partition of* $\mathfrak{A}$. *Then the map*

$$\sigma : \mathfrak{A} \to \prod_{i \in I} \mathfrak{A}_{a_i},$$

defined by taking $\sigma(a) = \left\langle a \wedge^{\mathfrak{A}} a_i \mid i \in I \right\rangle$, *for* $a \in A$, *is an isomorphism from* $\mathfrak{A}$ *onto* $\prod_{i \in I} \mathfrak{A}_{a_i}$.

Proof. First we observe that σ is a bijection. Indeed, suppose $\sigma(a) = \sigma(b)$. Then

$$a \wedge^{\mathfrak{A}} a_i = b \wedge^{\mathfrak{A}} a_i,$$

for all $i \in I$, from which

$$\bigvee_{i\in I}^{\mathfrak{A}} (a \wedge^{\mathfrak{A}} a_i) = \bigvee_{i\in I}^{\mathfrak{A}} (b \wedge^{\mathfrak{A}} a_i),$$

and so

$$a \wedge^{\mathfrak{A}} \bigvee_{i\in I}^{\mathfrak{A}} a_i = b \wedge^{\mathfrak{A}} \bigvee_{i\in I}^{\mathfrak{A}} a_i.$$

As $\bigvee_{i\in I}^{\mathfrak{A}} a_i = 1^{\mathfrak{A}}$, we then obtain $a = b$.

To prove that σ is a surjection, suppose that $\langle b_i \mid i \in I\rangle$ is an element of $\prod_{i\in I} \mathfrak{A}_{a_i}$. Then

$$\sigma(\bigvee_{i\in I}^{\mathfrak{A}} b_i) = \left\langle a_i \wedge^{\mathfrak{A}} \bigvee_{i\in I}^{\mathfrak{A}} b_i \mid i \in I \right\rangle = \langle b_i \mid i \in I\rangle\,,$$

since $b_i \leq^{\mathfrak{A}} a_i$ and $a_i \wedge^{\mathfrak{A}} a_j = 0^{\mathfrak{A}}$, for all $i, j \in I$, $i \neq j$.

It remains to show that σ respects the Booleans. For $\mathfrak{B} = \prod_{i\in I} \mathfrak{A}_{a_i}$ and all a in the universe of $\mathfrak{A}$, we have:

$$\begin{aligned}
\sigma(\neg^{\mathfrak{A}} a) &= \langle \neg^{\mathfrak{A}} a \wedge^{\mathfrak{A}} a_i \mid i \in I\rangle \\
&= \langle \neg^{a_i} (a \wedge^{\mathfrak{A}} a_i) \mid i \in I\rangle \\
&= \neg^{\mathfrak{B}} \langle a \wedge^{\mathfrak{A}} a_i \mid i \in I\rangle \\
&= \neg^{\mathfrak{B}} \sigma(a).
\end{aligned}$$

The other operations are considered analogously. ❑

Since an element $b \in A$ such that $b \leq^{\mathfrak{A}} a$ is an atom in $\mathfrak{A}$ iff it is an atom in $\mathfrak{A}_a$, we clearly have:

Lemma 4.7. *If $\mathfrak{A}$ is atomless, then $\mathfrak{A}_a$ is atomless for each nonzero $a \in A$.*

Given maps $\sigma_i : A_i \to B_i$, for $i \in I$, we denote by

$$\sigma = \prod_{i\in I} \sigma_i$$

the map from $\prod_{i\in I} A_i$ into $\prod_{i\in I} B_i$ defined by

$$\sigma \langle a_i \mid i \in I\rangle = \langle \sigma_i(a_i) \mid i \in I\rangle\,.$$

If the σ_i are isomorphisms from $\mathfrak{A}_i$ onto $\mathfrak{B}_i$, then clearly $\prod_{i\in I} \sigma$ is an isomorphism from $\prod_{i\in I} \mathfrak{A}_i$ onto $\prod_{i\in I} \mathfrak{B}_i$.

Proposition 4.8. *Suppose that $\mathfrak{A}$ and $\mathfrak{B}$ are countably infinite atomless Boolean algebras, with $\{a_i \mid i \in I\}$ and $\{b_i \mid i \in I\}$ being partitions of $\mathfrak{A}$ and $\mathfrak{B}$, respectively. Then there exists an isomorphism σ from $\mathfrak{A}$ onto $\mathfrak{B}$ such that $\sigma(a_i) = b_i$ for all $i \in I$.*

Proof. By Lemma 4.7, $\mathfrak{A}_{a_i}$ and $\mathfrak{B}_{b_i}$ $(i \in I)$ are countably infinite atomless algebras. Hence, by Theorem 4.3, there are isomorphisms σ_i from $\mathfrak{A}_{a_i}$ onto $\mathfrak{B}_{b_i}$. So

$$\sigma = \prod_{i \in I} \sigma_i$$

is an isomorphism from $\prod_{i \in I} \mathfrak{A}_{a_i}$ onto $\prod_{i \in I} \mathfrak{B}_{b_i}$. Lemma 4.6 supplies an isomorphism ρ_0 from $\mathfrak{A}$ onto $\prod_{i \in I} \mathfrak{A}_{a_i}$ such that $\rho_0(a_i) = \langle 0^{\mathfrak{A}}, \dots, a_i, \dots, 0^{\mathfrak{A}} \rangle$ for all $i \in I$. It also supplies an isomorphism ρ_1 from $\mathfrak{B}$ onto $\prod_{i \in I} \mathfrak{B}_{b_i}$ such that $\rho_1(b_i) = \langle 0^{\mathfrak{B}}, \dots, b_i, \dots, 0^{\mathfrak{B}} \rangle$, for $i \in I$. The composition $\rho_0 \circ \sigma \circ \rho_1^{-1}$ is then the required isomorphism from $\mathfrak{A}$ onto $\mathfrak{B}$ (here ρ_1^{-1} is the inverse of ρ_1). ❑

Given a sequence of valuations $\mathfrak{V}_i$ in algebras $\mathfrak{A}_i$, $i \in I$, we denote by

$$\mathfrak{V} = \prod_{i \in I} \mathfrak{V}_i$$

the valuation in $\prod_{i \in I} \mathfrak{A}_i$ defined by

$$\mathfrak{V}(p) = \langle \mathfrak{V}_i(p) \mid i \in I \rangle$$

for every propositional variable p.

Lemma 4.9. *Suppose $\mathfrak{V}_i$ is a valuation in a modal algebra $\mathfrak{A}_i$, for $i \in I$. Then for all formulas ψ, we have*

$$\left(\prod_{i \in I} \mathfrak{V}_i \right) (\psi) = \langle \mathfrak{V}_i(\psi) \mid i \in I \rangle .$$

Proof. The easy inductive proof is left to the reader. ❑

4.3 Preserving decidability of global consequence

In this section we prove the following result of (Wolter 1998):

Theorem 4.10. *Suppose that L_1 and L_2 are consistent multimodal logics and $L = L_1 \otimes L_2$. Then $\vdash_L^*$ is decidable iff both $\vdash_{L_1}^*$ and $\vdash_{L_2}^*$ are decidable.*

The implication ($\Rightarrow$) follows immediately from Theorem 4.5. The converse implication is a consequence of Lemma 4.11 below and the fact that the size of the set $C(\Theta^1(\varphi) \cup \Theta^1(\psi))$ can be bounded by a recursive function in the lengths of φ and ψ.

Again, let us assume for simplicity that L_1 and L_2 are unimodal logics with the boxes $\Box_1$ and $\Box_2$, respectively. Then $L = L_1 \otimes L_2$ is a bimodal logic in the language $\mathcal{ML}_2$.

Lemma 4.11. *For any two $\mathcal{ML}_2$-formulas φ and ψ, the following conditions* (i)–(iii) *are equivalent:*

(i) $\varphi \not\vdash^*_L \psi$;

(ii) *there exists* $D \subseteq C(\Theta^1(\varphi) \cup \Theta^1(\psi))$ *such that*

- $\varphi^1 \wedge (\bigvee D)^1 \not\vdash^*_{L_1} \psi^1$,
- $\varphi^1 \wedge (\bigvee D)^1 \not\vdash^*_{L_1} \neg\chi^1$ *and* $(\bigvee D)^2 \not\vdash^*_{L_2} \neg\chi^2$, *for every* $\chi \in D$;

(iii) *there exists* $D \subseteq C(\Theta^2(\varphi) \cup \Theta^2(\psi))$ *such that*

- $\varphi^2 \wedge (\bigvee D)^2 \not\vdash^*_{L_2} \psi^2$,
- $\varphi^2 \wedge (\bigvee D)^2 \not\vdash^*_{L_2} \neg\chi^2$ *and* $(\bigvee D)^1 \not\vdash^*_{L_1} \neg\chi^1$, *for every* $\chi \in D$.

Proof. (ii)$\Rightarrow$(i). Take any $D \subseteq C(\Theta^1(\varphi) \cup \Theta^1(\psi))$ satisfying (ii). By Theorem 4.4, for each $\chi \in D$ there exist $\mathfrak{A}_\chi \in \mathsf{Atg}L_1$ and a valuation $\mathfrak{V}_\chi$ in $\mathfrak{A}_\chi$ such that

$$\mathfrak{V}_\chi(\varphi^1 \wedge (\bigvee D)^1) = 1^{\mathfrak{A}_\chi} \quad \text{and} \quad \mathfrak{V}_\chi(\chi^1) \neq 0^{\mathfrak{A}_\chi}.$$

Further, we can find $\mathfrak{A}_\psi \in \mathsf{Atg}L_1$ and $\mathfrak{V}_\psi$ such that

$$\mathfrak{V}_\psi(\varphi^1 \wedge (\bigvee D)^1) = 1^{\mathfrak{A}_\psi} \quad \text{and} \quad \mathfrak{V}_\psi(\psi^1) \neq 1^{\mathfrak{A}_\psi}.$$

Put

$$\mathfrak{A} = \prod_{\chi \in D \cup \{\psi\}} \mathfrak{A}_\chi$$

and define a valuation $\mathfrak{V}^1$ in $\mathfrak{A}$ by taking

$$\mathfrak{V}^1 = \prod_{\chi \in D \cup \{\psi\}} \mathfrak{V}_\chi.$$

We then have:

$$\mathfrak{A} \in \mathsf{Atg}L_1, \tag{4.1}$$
$$\mathfrak{V}^1(\varphi^1) = 1^{\mathfrak{A}}, \tag{4.2}$$
$$\mathfrak{V}^1(\psi^1) \neq 1^{\mathfrak{A}}, \tag{4.3}$$
$$\{\mathfrak{V}^1(\chi^1) \mid \chi \in D\} \text{ is a partition of } \mathfrak{A}. \tag{4.4}$$

(4.1) follows from the fact that any direct product of finitely many c.i.a.-algebras is clearly a c.i.a.-algebra, and the class of algebras for a logic is closed under the formation of direct products (see Theorem 1.15). For (4.2)–(4.4), observe first that, by Lemma 4.9,

$$\mathfrak{V}^1(\alpha) = \langle \mathfrak{V}_\chi(\alpha) \mid \chi \in D \cup \{\psi\}\rangle,$$

for every $\mathcal{ML}_1$-formula α. Now (4.2) follows from $\mathfrak{V}_\chi(\varphi^1) = 1^{\mathfrak{A}_\chi}$ for all $\chi \in D \cup \{\psi\}$. (4.3) follows from $\mathfrak{V}_\psi(\psi^1) \neq 1^{\mathfrak{A}_\psi}$. To show (4.4), note that for any two distinct $\chi_1, \chi_2 \in C(\Theta^1(\varphi) \cup \Theta^1(\psi))$, there exists a formula α such that either α is a conjunct of χ_1 and $\neg\alpha$ a conjunct of χ_2 or vice versa. Hence $\mathfrak{V}^1(\chi_1^1) \wedge^{\mathfrak{A}} \mathfrak{V}^1(\chi_2^1) = 0^{\mathfrak{A}}$ for any two distinct $\chi_1, \chi_2 \in D$. We have $\bigvee^{\mathfrak{A}}_{\chi \in D} \mathfrak{V}^1(\chi^1) = 1^{\mathfrak{A}}$ because $\mathfrak{V}_\chi((\bigvee D)^1) = 1^{\mathfrak{A}_\chi}$ for every $\chi \in D \cup \{\psi\}$. Finally, $\mathfrak{V}^1(\chi^1) \neq 0^{\mathfrak{A}}$ because $\mathfrak{V}_\chi(\chi^1) \neq 0^{\mathfrak{A}_\chi}$, for all $\chi \in D$.

On the other hand, using the fact that $(\bigvee D)^2 \not\vdash^*_{L_2} \neg\chi^2$ for all $\chi \in D$, we obtain in a similar way an algebra $\mathfrak{B} \in \mathsf{Atg}L_2$ with a valuation $\mathfrak{V}^2$ such that the set

$$\{\mathfrak{V}^2(\chi^2) \mid \chi \in D\}$$

is a partition of $\mathfrak{B}$. By Proposition 4.8, there exists an isomorphism σ from the Boolean reduct of $\mathfrak{B}$ onto the Boolean reduct of $\mathfrak{A}$ such that

$$\sigma(\mathfrak{V}^2(\chi^2)) = \mathfrak{V}^1(\chi^1)$$

for all $\chi \in D$. By identifying the Boolean reducts of $\mathfrak{B}$ and $\mathfrak{A}$, we can therefore assume that we have an algebra

$$\mathfrak{D} = \left\langle A, \wedge^{\mathfrak{D}}, \neg^{\mathfrak{D}}, 0^{\mathfrak{D}}, 1^{\mathfrak{D}}, \Box_1^{\mathfrak{A}}, \Box_2^{\mathfrak{B}} \right\rangle \in \mathsf{Atg}L$$

with two valuations $\mathfrak{V}^1$ and $\mathfrak{V}^2$ satisfying

$$\mathfrak{V}^1(\chi^1) = \mathfrak{V}^2(\chi^2)$$

for all $\chi \in D$ and such that $\mathfrak{V}^1$ still has properties (4.1)–(4.4).

Now, taking into account the definition of $C(\Theta^1(\varphi) \cup \Theta^1(\psi))$, one can easily show that the following equations hold for every $\alpha \in \Theta^1(\varphi) \cup \Theta^1(\psi)$:

$$\begin{aligned}
\mathfrak{V}^1(\alpha^1) &= \bigvee\nolimits^{\mathfrak{D}} \{\mathfrak{V}^1(\chi^1) \mid \chi \in D,\ \alpha \text{ is a conjunct of } \chi\} \\
&= \bigvee\nolimits^{\mathfrak{D}} \{\mathfrak{V}^2(\chi^2) \mid \chi \in D,\ \alpha \text{ is a conjunct of } \chi\} \\
&= \mathfrak{V}^2(\alpha^2).
\end{aligned}$$

Thus, we can define a new valuation $\mathfrak{V}$ in $\mathfrak{D}$ by taking, for all (nonsurrogate) variables p in φ and ψ,

$$\mathfrak{V}(p) = \mathfrak{V}^1(p) = \mathfrak{V}^2(p).$$

By induction on the construction of α we show that

$$\mathfrak{V}(\alpha) = \mathfrak{V}^1(\alpha^1) = \mathfrak{V}^2(\alpha^2)$$

for every $\alpha \in \Theta^1(\varphi) \cup \Theta^1(\psi)$. The only nontrivial steps are $\alpha = \Box_1\beta$ and $\alpha = \Box_2\beta$:

$$\mathfrak{V}(\Box_1\beta) = \Box_1^{\mathfrak{D}}(\mathfrak{V}(\beta)) = \Box_1^{\mathfrak{A}}(\mathfrak{V}^1(\beta^1)) = \mathfrak{V}^1((\Box_1\beta)^1). \tag{4.5}$$

The proof of $\mathfrak{V}(\Box_2\beta) = \mathfrak{V}^2((\Box_2\beta)^2)$ is similar. Since φ and ψ are built up from formulas in $\Theta^1(\varphi) \cup \Theta^1(\psi)$ using only the Booleans and $\Box_1$, an argument similar to (4.5) shows that $\mathfrak{V}(\varphi) = \mathfrak{V}^1(\varphi^1) = 1^{\mathfrak{D}}$ and $\mathfrak{V}(\psi) = \mathfrak{V}^1(\psi^1) \neq 1^{\mathfrak{D}}$. Hence $\varphi \not\vdash_L^* \psi$, as required.

(i)$\Rightarrow$(ii). Suppose that $\varphi \not\vdash_L^* \psi$. So we have an algebra

$$\mathfrak{A} = \left\langle A, \wedge^{\mathfrak{A}}, \neg^{\mathfrak{A}}, 0^{\mathfrak{A}}, 1^{\mathfrak{A}}, \Box_1^{\mathfrak{A}}, \Box_2^{\mathfrak{A}} \right\rangle$$

and a valuation $\mathfrak{V}$ in $\mathfrak{A}$ such that $\mathfrak{A} \in \mathsf{Alg}L$, $\mathfrak{V}(\varphi) = 1^{\mathfrak{A}}$ and $\mathfrak{V}(\psi) \neq 1^{\mathfrak{A}}$. But then

$$D = \{\chi \in C(\Theta^1(\varphi) \cup \Theta^1(\psi)) \mid \mathfrak{V}(\chi) \neq 0^{\mathfrak{A}}\}$$

satisfies (ii). Indeed, let

$$\mathfrak{A}_1 = \left\langle A, \wedge^{\mathfrak{A}}, \neg^{\mathfrak{A}}, 0^{\mathfrak{A}}, 1^{\mathfrak{A}}, \Box_1^{\mathfrak{A}} \right\rangle$$

and

$$\mathfrak{V}^1(p) = \begin{cases} \mathfrak{V}(p), & \text{if } p \text{ is a nonsurrogate variable,} \\ \mathfrak{V}(\Box_2\chi), & \text{if } p = q_{\Box_2\chi} \text{ for some } \Box_2\chi \in \Theta^1(\varphi) \cup \Theta^1(\psi). \end{cases}$$

The algebra $\mathfrak{A}_2$ and the valuation $\mathfrak{V}^2$ in $\mathfrak{A}_2$ are defined analogously. Then we clearly have $\mathfrak{V}^1(\varphi^1) = \mathfrak{V}^1((\bigvee D)^1) = \mathfrak{V}^2((\bigvee D)^2) = 1^{\mathfrak{A}}$, $\mathfrak{V}^1(\psi^1) \neq 1^{\mathfrak{A}}$, and $\mathfrak{V}^1(\chi^1) \neq 0^{\mathfrak{A}}$, $\mathfrak{V}^2(\chi^2) \neq 0^{\mathfrak{A}}$ for any $\chi \in D$.

The equivalence of (i) and (iii) is proved in the same way. ❑

4.4 Preserving decidability

As we saw in Section 1.5, from the algebraic point of view every modal logic L can be regarded as the equational theory of modal algebras generated by the equations $\{\varphi = 1 \mid \varphi \in L\}$. Thus, the problem of whether decidability is preserved under the formation of fusions of modal logics is an instance of the more general question: under which conditions does the decidability of two equational theories T_1 and T_2 imply the decidability of the union $T_1 \cup T_2$? So it would seem to be natural to begin investigating decidability of fusions by trying to take advantage of the available results concerning unions of equational theories. But unfortunately, none of them is applicable to modal logics. For instance, the first rather general sufficient condition found by Pigozzi (1974) says that decidability is preserved whenever the languages of T_1 and T_2 are disjoint. However, we cannot use this result to prove decidability of fusions of modal logics because the Boolean operators are always shared by the equational theories of modal algebras. There are a number of preservation results for joins of equational theories with shared symbols; see, e.g., (Baader and Tinelli 1997, Baader and Tinelli 2002, Domenjoud *et al.* 1994). But again the special conditions they impose on the equational theories make these results nonapplicable to fusions.

Here we prove the following theorem due to Wolter (1998):

Theorem 4.12. *Suppose that L_1 and L_2 are multimodal logics. Then $L_1 \otimes L_2$ is decidable whenever both L_1 and L_2 are decidable.*

Proof. We again assume that L_1 and L_2 are unimodal logics with the boxes $\Box_1$ and $\Box_2$, respectively. Let $L = L_1 \otimes L_2$.

It is natural to begin the proof by trying to use the criterion of Lemma 4.11. Observe that for any consistency-set $C(\Gamma)$, modal algebra $\mathfrak{A}$ and valuation $\mathfrak{V}$ in $\mathfrak{A}$, we have

$$\mathfrak{V}(\bigvee C(\Gamma)) = 1^{\mathfrak{A}}.$$

Therefore, taking into account the definition of $\Sigma_1(\varphi)$, we obtain for all $\mathcal{ML}_2$-formulas φ that

$$\bigvee \Sigma_1(\varphi) \in L, \tag{4.6}$$

and so

$$\bigvee \Sigma_1(\varphi) \not\vdash_L^* \neg\chi \quad \text{for all } \chi \in \Sigma_1(\varphi).$$

Thus, for all $\chi \in \Sigma_1(\varphi)$, we have

$$(\bigvee \Sigma_1(\varphi))^{\mathbf{1}} \not\vdash_{L_1}^* \neg\chi^{\mathbf{1}}, \qquad (\bigvee \Sigma_1(\varphi))^{\mathbf{2}} \not\vdash_{L_2}^* \neg\chi^{\mathbf{2}}. \tag{4.7}$$

Now, if $\varphi \notin L$, then by (4.6)

$$(\bigvee \Sigma_1(\varphi))^{\mathbf{1}} \not\vdash_{L_1}^* \varphi^{\mathbf{1}}.$$

If $\varphi \in L$, then by taking $D = \Sigma_1(\varphi)$ and using (4.7) and Lemma 4.11 we obtain

$$(\bigvee \Sigma_1(\varphi))^{\mathbf{1}} \vdash_{L_1}^* \varphi^{\mathbf{1}}.$$

Thus, we arrive at the following corollary (in which the equivalence (i) $\Leftrightarrow$ (iii) is proved in the same manner as (i) $\Leftrightarrow$ (ii)):

Corollary 4.13. *Suppose that L_1 and L_2 are consistent and φ is an $\mathcal{ML}_2$-formula. Then the following conditions are equivalent:*

(i) $\varphi \in L$,

(ii) $(\bigvee \Sigma_1(\varphi))^{\mathbf{1}} \vdash^*_{L_1} \varphi^{\mathbf{1}}$,

(iii) $(\bigvee \Sigma_2(\varphi))^{\mathbf{2}} \vdash^*_{L_2} \varphi^{\mathbf{2}}$.

Define $a^1(\varphi)$ to be the length of the longest chain $\Box_2, \Box_1, \Box_2, \dots$ of boxes starting with $\Box_2$ and such that a subformula of the form

$$\Box_2(\dots \Box_1(\dots \Box_2(\dots)))$$

occurs in φ. The function $a^2(\varphi)$ is defined analogously by swapping $\Box_1$ and $\Box_2$. The sum $a(\varphi) = a^1(\varphi) + a^2(\varphi)$ will be called the *alternation depth of* φ. It is readily seen that the following lemma holds:

Lemma 4.14. *For every $\mathcal{ML}_2$-formula φ with at least one box, either*

$$a(\bigvee C(\Theta^{\mathbf{1}}(\varphi))) < a(\varphi)$$

or

$$a(\bigvee C(\Theta^{\mathbf{2}}(\varphi))) < a(\varphi).$$

Now Lemma 4.14 together with Corollary 4.13 provide us with a decision algorithm for L under the condition that $\vdash^*_{L_1}$ and $\vdash^*_{L_2}$ are decidable. We proceed by induction on $a(\varphi)$. Suppose that we already know how to decide whether $\alpha \in L$, for every α with $a(\alpha) < a(\varphi)$. By Lemma 4.14, we may assume, say, that $a(\chi) < a(\varphi)$ for all $\chi \in C(\Theta^{\mathbf{1}}(\varphi))$. Hence $\Sigma_1(\varphi)$ can be constructed effectively and, according to Corollary 4.13, it remains to check whether $(\bigvee \Sigma_1(\varphi))^{\mathbf{1}} \vdash^*_{L_1} \varphi^{\mathbf{1}}$ holds, which can be done effectively because $\vdash^*_{L_1}$ is decidable. The case when $a(\chi) < a(\varphi)$ for all $\chi \in C(\Theta^{\mathbf{2}}(\varphi))$ is similar.

Unfortunately, $\vdash^*_{L_i}$ is not necessarily decidable when L_i is decidable; for a counterexample see Section 5.4. So this argument cannot be used to prove Theorem 4.12. However, it indicates a path we shall follow to conduct our proof. In fact, if we find recursive functions which for every φ give two natural numbers n_1, n_2 such that

$$(\bigvee \Sigma_1(\varphi))^{\mathbf{1}} \vdash^*_{L_1} \varphi^{\mathbf{1}} \quad \text{iff} \quad \Box_{\mathbf{1}}^{\leq n_1}(\bigvee \Sigma_1(\varphi))^{\mathbf{1}} \to \varphi^{\mathbf{1}} \in L_1,$$

and

$$(\bigvee \Sigma_2(\varphi))^{\mathbf{2}} \vdash^*_{L_2} \varphi^{\mathbf{2}} \quad \text{iff} \quad \Box_{\mathbf{2}}^{\leq n_2}(\bigvee \Sigma_2(\varphi))^{\mathbf{2}} \to \varphi^{\mathbf{2}} \in L_2,$$

then we shall have a decision procedure for L provided that both L_1 and L_2 are decidable. It turns out that the 1-*depth* $d^1(\varphi)$ and the 2-*depth* $d^2(\varphi)$ of φ

defined below can be used as the required n_1 and n_2:

$$\begin{aligned} d^1(p) &= 0 \\ d^1(\varphi \wedge \psi) &= \max\{d^1(\varphi), d^1(\psi)\} \\ d^1(\neg\varphi) &= d^1(\varphi) \\ d^1(\Box_1\varphi) &= d^1(\varphi) + 1 \\ d^1(\Box_2\varphi) &= d^1(\varphi); \end{aligned}$$

$d^2(\varphi)$ is defined analogously.

Proposition 4.15. *For every $\mathcal{ML}_2$-formula φ, the following conditions are equivalent:*

(i) $\varphi \in L$,

(ii) $\Box_1^{\leq d^1(\varphi)}(\bigvee \Sigma_1(\varphi))^{\mathbf{1}} \to \varphi^{\mathbf{1}} \in L_1$,

(iii) $\Box_2^{\leq d^2(\varphi)}(\bigvee \Sigma_2(\varphi))^{\mathbf{2}} \to \varphi^{\mathbf{2}} \in L_2$.

Proof. The implications (ii) $\Rightarrow$ (i) and (iii) $\Rightarrow$ (i) are clear, since L_1 and L_2 are modal logics. To prove (i) $\Rightarrow$ (ii), suppose that

$$\Box_1^{\leq d^1(\varphi)}(\bigvee \Sigma_1(\varphi))^{\mathbf{1}} \to \varphi^{\mathbf{1}} \notin L_1.$$

Then there exist $\mathfrak{A}^- \in \mathsf{Atg}L_1$ and a valuation $\mathfrak{W}$ in $\mathfrak{A}^-$ such that

$$\mathfrak{W}(\neg\varphi^{\mathbf{1}} \wedge \Box_1^{\leq d^1(\varphi)}(\bigvee \Sigma_1(\varphi))^{\mathbf{1}}) \neq 0^{\mathfrak{A}^-}.$$

We will construct an algebra $\mathfrak{D} \in \mathsf{Atg}L$ and a valuation $\mathfrak{V}$ in it such that

$$\mathfrak{V}(\neg\varphi) \neq 0^{\mathfrak{D}}.$$

We may assume that

$$\mathfrak{W}((\bigvee \Sigma_1(\varphi))^{\mathbf{1}}) \neq 1^{\mathfrak{A}^-}, \tag{4.8}$$

for otherwise we would have $(\bigvee \Sigma_1(\varphi))^{\mathbf{1}} \not\models^*_{L_1} \varphi^{\mathbf{1}}$, and so $\varphi \notin L$ by Corollary 4.13.

Lemma 4.16. *For each $i = 1, 2$, there exist an algebra $\mathfrak{A}_i \in \mathsf{Atg}L_i$ and a valuation $\mathfrak{W}^i$ in $\mathfrak{A}_i$ such that*

$$\{\mathfrak{W}^i(\chi^{\mathbf{i}}) \mid \chi \in \Sigma_1(\varphi)\}$$

is a partition of $\mathfrak{A}_i$.

Proof. By the definition of $\Sigma_1(\varphi)$, for each $\chi \in \Sigma_1(\varphi)$ there exist an algebra $\mathfrak{A}_\chi \in \mathsf{Atg}L$ and a valuation $\mathfrak{W}_\chi$ in $\mathfrak{A}_\chi$ such that $\mathfrak{W}_\chi(\chi) \neq 0^{\mathfrak{A}_\chi}$. By (4.6), $\mathfrak{W}_\chi(\bigvee \Sigma_1(\varphi)) = 1^{\mathfrak{A}_\chi}$ for all $\chi \in \Sigma_1(\varphi)$. Given any two distinct formulas $\psi_1, \psi_2 \in \Sigma_1(\varphi)$, we can find a formula α such that either α is a conjunct of ψ_1 and $\neg\alpha$ a conjunct of ψ_2 or vice versa. Hence $\mathfrak{W}_\chi(\psi_1) \wedge^{\mathfrak{A}_\chi} \mathfrak{W}_\chi(\psi_2) = 0^{\mathfrak{A}_\chi}$.

Now, for each $\chi \in \Sigma_1(\varphi)$, let $\mathfrak{A}^1_\chi$ be the $\Box_2$-free reduct of $\mathfrak{A}_\chi$. Define a valuation $\mathfrak{W}^1_\chi$ in $\mathfrak{A}^1_\chi$ by taking

$$\mathfrak{W}^1_\chi(p) = \begin{cases} \mathfrak{W}_\chi(p), & \text{if } p \text{ is a nonsurrogate variable,} \\ \mathfrak{W}_\chi(\Box_2\psi), & \text{if } p = q_{\Box_2\psi} \text{ for some } \Box_2\psi \in \Theta^1(\varphi). \end{cases}$$

Put

$$\mathfrak{A}_1 = \prod_{\chi \in \Sigma_1(\varphi)} \mathfrak{A}^1_\chi \quad \text{and} \quad \mathfrak{W}^1 = \prod_{\chi \in \Sigma_1(\varphi)} \mathfrak{W}^1_\chi.$$

Then $\mathfrak{A}_1 \in \mathsf{Atg}L_1$. It should be clear that for all $\alpha \in \Theta^1(\varphi)$ and all $\chi \in \Sigma_1(\varphi)$ we have

$$\mathfrak{W}_\chi(\alpha) = \mathfrak{W}^1_\chi(\alpha^1).$$

Thus, $\{\mathfrak{W}^1(\chi^1) \mid \chi \in \Sigma_1(\varphi)\}$ is a partition of $\mathfrak{A}_1$. $\mathfrak{A}_2$ and $\mathfrak{W}^2$ can be defined analogously. ❑

Lemma 4.17. *Let $m = d^1(\varphi)$ and suppose that $\Box_1^{\leq m}(\bigvee \Sigma_1(\varphi))^1 \to \varphi^1 \notin L_1$. Then there exist an algebra $\mathfrak{A} = \langle A, \wedge, \neg, 0, 1, \Box_1 \rangle$ in $\mathsf{Atg}L_1$, a valuation $\mathfrak{V}^1$ in $\mathfrak{A}$, and elements $a_0, \dots, a_m \in A$ such that*

(a1) *$0 < a_m < a_{m-1} < \dots < a_0 < 1$ (so $\{\neg a_0, a_0 \wedge \neg a_1, \dots, a_m\}$ is a partition of $\mathfrak{A}$);*

(a2) *$a_{n+1} \leq a_n \wedge \Box_1 a_n$, for all $n < m$;*

(a3) *$a_m \wedge \mathfrak{V}^1(\neg\varphi^1) \neq 0$;*

(a4) *$\{a_m \wedge \mathfrak{V}^1(\chi^1) \mid \chi \in \Sigma_1(\varphi)\}$ is a partition of $\mathfrak{A}_{a_m}$;*

(a5) *for every $n < m$, $\{(a_n \wedge \neg a_{n+1}) \wedge \mathfrak{V}^1(\chi^1) \mid \chi \in \Sigma_1(\varphi)\}$ is a partition of $\mathfrak{A}_{(a_n \wedge \neg a_{n+1})}$.*

Proof. By assumption, there exist $\mathfrak{A}^- \in \mathsf{Atg}L_1$ and a valuation $\mathfrak{W}$ in $\mathfrak{A}^-$ such that

$$\mathfrak{W}(\neg\varphi^1 \wedge \Box_1^{\leq m}(\bigvee \Sigma_1(\varphi))^1) \neq 0^{\mathfrak{A}^-}. \tag{4.9}$$

For each $n \leq m$, we take an algebra $\mathfrak{A}_n \in \mathsf{Atg}L_1$ with a valuation $\mathfrak{W}_n$ such that

$$\{\mathfrak{W}_n(\chi^1) \mid \chi \in \Sigma_1(\varphi)\} \text{ is a partition of } \mathfrak{A}_n. \tag{4.10}$$

(Such algebras exist by Lemma 4.16.) Put

$$\mathfrak{A} = \mathfrak{A}^- \times \prod_{n \le m} \mathfrak{A}_n, \qquad \mathfrak{V}^1 = \mathfrak{W} \times \prod_{n \le m} \mathfrak{W}_n,$$

and, for every $n \le m$,

$$a_n = \left\langle \mathfrak{W}(\Box_1^{\le n}(\bigvee \Sigma_1(\varphi))^1), 0^{\mathfrak{A}_0}, \ldots, 0^{\mathfrak{A}_{n-1}}, 1^{\mathfrak{A}_n}, \ldots, 1^{\mathfrak{A}_m} \right\rangle.$$

Recall that we denote the constants and operations of $\mathfrak{A}$ by 0, 1, $\wedge$, $\neg$, $\Box_1$. We show that the sequence $a_0, \ldots, a_m$ and the valuation $\mathfrak{V}^1$ are as required. By (4.8), $a_0 < 1$, and so (a1) clearly follows from the definition. Condition (a2) follows from the fact that $a_n \wedge \Box_1 a_n$ is equal to

$$\left\langle \mathfrak{W}(\Box_1^{\le n+1}(\bigvee \Sigma_1(\varphi))^1), \Box_1^{\mathfrak{A}_0} 0^{\mathfrak{A}_0}, \ldots, \Box_1^{\mathfrak{A}_{n-1}} 0^{\mathfrak{A}_{n-1}}, 1^{\mathfrak{A}_n}, \ldots, 1^{\mathfrak{A}_m} \right\rangle.$$

Condition (a3) follows from (4.9). For (a4), observe that for all $\chi \in \Sigma_1(\varphi)$

$$\mathfrak{V}^1(\chi^1) = \langle \mathfrak{W}(\chi^1), \mathfrak{W}_0(\chi^1), \ldots, \mathfrak{W}_m(\chi^1) \rangle.$$

Thus, by the definition of $\Sigma_1(\varphi)$, $a_m \wedge \mathfrak{V}^1(\chi_1^1)$ and $a_m \wedge \mathfrak{V}^1(\chi_2^1)$ are disjoint for distinct χ_1 and χ_2. By (4.10),

$$\bigvee_{\chi \in \Sigma_1(\varphi)} \mathfrak{V}^1(\chi^1) = \left\langle \mathfrak{W}((\bigvee \Sigma_1(\varphi))^1), 1^{\mathfrak{A}_0}, \ldots, 1^{\mathfrak{A}_m} \right\rangle = a_0,$$

and so, by (a1), we have

$$\bigvee_{\chi \in \Sigma_1(\varphi)} a_m \wedge \mathfrak{V}^1(\chi^1) = a_m = 1^{\mathfrak{A}_{a_m}}.$$

Finally, by the definition of a_m, we have $a_m \wedge \mathfrak{V}^1(\chi^1) > 0$, for all $\chi \in \Sigma_1(\varphi)$. Condition (a5) is proved similarly to (a4). ❑

Now we can complete the proof of Proposition 4.15 as follows. For each $n = 1, \ldots, m+1$, take a $\mathfrak{B}_n \in \mathsf{Atg}L_2$ and a valuation $\mathfrak{V}_n$ such that

$$\{\mathfrak{V}_n(\chi^2) \mid \chi \in \Sigma_1(\varphi)\}$$

is a partition of $\mathfrak{B}_n$. Take also an arbitrary $\mathfrak{B}_0 \in \mathsf{Atg}L_2$ and any valuation $\mathfrak{V}_0$ in $\mathfrak{B}_0$. Define a valuation $\mathfrak{V}^2$ in the direct product

$$\mathfrak{B} = \prod_{n \le m+1} \mathfrak{B}_n$$

by taking

$$\mathfrak{V}^2 = \prod_{n \le m+1} \mathfrak{V}_n.$$

Next, choose $\mathfrak{A} \in \mathsf{Atg}L_1$ with a valuation $\mathfrak{V}^1$ and a sequence $a_0, \dots, a_m$ satisfying (a1)–(a5) of Lemma 4.17. Let $b_0 = \neg a_0$, $b_{n+1} = a_n \wedge \neg a_{n+1}$, for $n < m$, and $b_{m+1} = a_m$. Then by (a1) and Lemma 4.6, there is a Boolean isomorphism

$$\varrho : \mathfrak{A} \to \prod_{n \le m+1} \mathfrak{A}_{b_n}.$$

By (a4)–(a5) and Proposition 4.8, for all $n = 1, \dots, m+1$ there are Boolean isomorphisms

$$\sigma_n : \mathfrak{A}_{b_n} \to \mathfrak{B}_n$$

such that

$$\sigma_n(b_n \wedge \mathfrak{V}^1(\chi^1)) = \mathfrak{V}_n(\chi^2), \tag{4.11}$$

for all $\chi \in \Sigma_1(\varphi)$. Take an arbitrary Boolean isomorphism σ_0 from $\mathfrak{A}_{b_0}$ onto $\mathfrak{B}_0$, and put

$$\sigma = \prod_{n \le m+1} \sigma_n.$$

Then $\varrho \circ \sigma$ is a Boolean isomorphism from $\mathfrak{A}$ onto $\mathfrak{B}$. Using this isomorphism we can identify the Boolean reducts of $\mathfrak{A}$ and $\mathfrak{B}$ in the usual way and obtain an algebra

$$\mathfrak{D} = \left\langle A, \wedge, \neg, 0, 1, \Box_1, \Box_2^{\mathfrak{B}} \right\rangle$$

such that $\mathfrak{D} \in \mathsf{Atg}L$.

Observe that using the isomorphisms ϱ and σ we obtain

$$\begin{aligned} a_n &\stackrel{\varrho}{=} \left\langle 0, \dots, 0, 1^{\mathfrak{A}_{b_{n+1}}}, \dots, 1^{\mathfrak{A}_{b_{m+1}}} \right\rangle \\ &\stackrel{\sigma}{=} \langle 0^{\mathfrak{B}_0}, \dots, 0^{\mathfrak{B}_n}, 1^{\mathfrak{B}_{n+1}}, \dots, 1^{\mathfrak{B}_{m+1}} \rangle, \end{aligned} \tag{4.12}$$

for all $n \le m$. Therefore, by (4.11),

$$\begin{aligned} a_n \wedge \mathfrak{V}^1(\chi^1) &\stackrel{\varrho\circ\sigma}{=} \langle 0^{\mathfrak{B}_0}, \dots, 0^{\mathfrak{B}_n}, \mathfrak{V}_{n+1}(\chi^2), \dots, \mathfrak{V}_{m+1}(\chi^2) \rangle \\ &= a_n \wedge \mathfrak{V}^2(\chi^2), \end{aligned}$$

for all $n \le m$ and all $\chi \in \Sigma_1(\varphi)$. Now using the properties of $\Sigma_1(\varphi)$, one can easily show that, for all $n \le m$ and all $\alpha \in \Theta^1(\varphi)$,

$$\begin{aligned} a_n \wedge \mathfrak{V}^1(\alpha^1) &= a_n \wedge \bigvee \{\mathfrak{V}^1(\chi^1) \mid \chi \in \Sigma_1(\varphi),\ \alpha \text{ is a conjunct of } \chi\} \\ &= a_n \wedge \bigvee \{\mathfrak{V}^2(\chi^2) \mid \chi \in \Sigma_1(\varphi),\ \alpha \text{ is a conjunct of } \chi\} \\ &= a_n \wedge \mathfrak{V}^2(\alpha^2). \end{aligned} \tag{4.13}$$

Define a new valuation $\mathfrak{V}$ in $\mathfrak{D}$ by taking, for all variables p in φ,

$$\mathfrak{V}(p) = \mathfrak{V}^1(p) = \mathfrak{V}^2(p).$$

We claim that for every $n \leq m$ and every $\alpha \in \Theta^{\mathbf{1}}(\varphi) \cup sub\,\neg\varphi$ such that $d^1(\alpha) \leq n$, we have:

$$a_k \wedge \mathfrak{V}(\alpha) = a_k \wedge \mathfrak{V}^1(\alpha^{\mathbf{1}}). \tag{4.14}$$

The proof is by induction on n. The basis of induction, i.e., the case $n = 0$, is proved by induction on the subformulas of α with $d^1(\alpha) = 0$. For propositional variables this follows from the definition of $\mathfrak{V}$. The case of the Booleans is trivial. So suppose $\alpha = \Box_2\beta$. Then $\alpha \in \Theta^{\mathbf{1}}(\varphi)$. Notice that $\Box_1$ does not occur in β, since $d^1(\alpha) = 0$. Consequently, $\alpha = \alpha^{\mathbf{2}}$ and the equality $a_0 \wedge \mathfrak{V}(\alpha) = a_0 \wedge \mathfrak{V}^2(\alpha^{\mathbf{2}})$ follows immediately. Hence, $a_0 \wedge \mathfrak{V}(\alpha) = a_0 \wedge \mathfrak{V}^1(\alpha^{\mathbf{1}})$ by (4.13).

The induction step is also proved by induction on the subformulas of α with $d^1(\alpha) \leq n+1$. The only interesting cases are: $\alpha = \Box_1\beta$ and $\alpha = \Box_2\beta$. Let us first assume that $\alpha = \Box_1\alpha_1$. By the induction hypothesis,

$$a_n \wedge \mathfrak{V}(\beta) = a_n \wedge \mathfrak{V}^1(\beta^{\mathbf{1}}).$$

Hence,

$$\begin{aligned} a_n \wedge \Box_1 a_n \wedge \mathfrak{V}(\alpha) &= a_n \wedge \Box_1 a_n \wedge \Box_1\mathfrak{V}(\beta) = a_n \wedge \Box_1(a_n \wedge \mathfrak{V}(\beta)) \\ &= a_n \wedge \Box_1(a_n \wedge \mathfrak{V}^1(\beta^{\mathbf{1}})) = a_n \wedge \Box_1 a_n \wedge \Box_1\mathfrak{V}^1(\beta^{\mathbf{1}}) \\ &= a_n \wedge \Box_1 a_n \wedge \mathfrak{V}^1(\alpha^{\mathbf{1}}). \end{aligned}$$

Now $a_{n+1} \wedge \mathfrak{V}(\alpha) = a_{n+1} \wedge \mathfrak{V}^1(\alpha^{\mathbf{1}})$ follows from $a_{n+1} \leq a_n \wedge \Box_1 a_n$, i.e., condition (a2).

Let $\alpha = \Box_2\beta$. Then $\alpha, \beta \in \Theta^{\mathbf{1}}(\varphi)$. We know, by the induction hypothesis and (4.13), that $a_{n+1} \wedge \mathfrak{V}(\beta) = a_{n+1} \wedge \mathfrak{V}^2(\beta^{\mathbf{2}})$. Hence

$$\begin{aligned} \Box_2^{\mathfrak{B}} a_{n+1} \wedge \mathfrak{V}(\alpha) &= \Box_2^{\mathfrak{B}} a_{n+1} \wedge \Box_2^{\mathfrak{B}}\mathfrak{V}(\beta) = \Box_2^{\mathfrak{B}}(a_{n+1} \wedge \mathfrak{V}(\beta)) \\ &= \Box_2^{\mathfrak{B}}(a_{n+1} \wedge \mathfrak{V}^2(\beta^{\mathbf{2}})) = \Box_2^{\mathfrak{B}} a_{n+1} \wedge \Box_2^{\mathfrak{B}}\mathfrak{V}^2(\beta^{\mathbf{2}}) \\ &= \Box_2^{\mathfrak{B}} a_{n+1} \wedge \mathfrak{V}^2(\alpha^{\mathbf{2}}). \end{aligned}$$

On the other hand, by (4.12) we have $a_{n+1} \leq \Box_2^{\mathfrak{B}} a_{n+1}$. Thus, we can conclude that

$$a_{n+1} \wedge \mathfrak{V}(\alpha) = a_{n+1} \wedge \mathfrak{V}^2(\alpha^{\mathbf{2}}),$$

which, by (4.13), yields

$$a_{n+1} \wedge \mathfrak{V}(\alpha) = a_{n+1} \wedge \mathfrak{V}^1(\alpha^{\mathbf{1}}).$$

To complete the proof of Proposition 4.15, it remains to observe that, by (4.14) and (a3), we have $a_m \wedge \mathfrak{V}(\neg\varphi) \neq 0$, and so $\mathfrak{V}(\varphi) \neq 1$.

The implication (i) $\Rightarrow$ (iii) can be proved in the same way. ❑

As was shown above, Proposition 4.15 provides us with a decision procedure for L whenever both L_1 and L_2 are decidable. This completes the proof of Theorem 4.12. ❑

4.5 Preserving interpolation

Denote by $var\varphi$ the set of all propositional variables in φ. We remind the reader that a logic L has the *interpolation property* if whenever $\varphi \to \psi \in L$ then there is a formula χ with $var\chi \subseteq var\varphi \cap var\psi$ such that $\varphi \to \chi \in L$ and $\chi \to \psi \in L$. The formula χ is then called an *interpolant for* $\varphi \to \psi$ *in* L. The interpolation property was introduced and investigated by Craig (1957) who discovered that classical logic enjoys interpolation. For information about interpolation in modal logic we refer the reader to (Maksimova 1979) and (Chagrov and Zakharyaschev 1997).

Say that a modal logic L has *uniform interpolation* if for every formula φ and every set $Q = \{q_1, \dots, q_k\}$ of variables, there exists a *uniform interpolant* φ_Q *for* φ with the following properties:

- $\varphi \to \varphi_Q \in L$,
- $var\varphi_Q \subseteq var\varphi - Q$,
- for every ψ, if $\varphi \to \psi \in L$ and $var\psi \cap Q = \emptyset$ then $\varphi_Q \to \psi \in L$.

Pitts (1992) proved that intuitionistic propositional logic **Int** has uniform interpolation. Ghilardi (1995) and Visser (1996) showed that **K**, the provability logic **GL** and the Grzegorczyk logic **Grz** also have this property. However, **S4** does not have uniform interpolation, although it enjoys Craig interpolation; see also (Ghilardi and Zawadowski 1995). It is easy to see that a modal logic L has uniform interpolation whenever it has interpolation and $\mathsf{Alg}L$ is *locally finite*, i.e., each finitely generated algebra in $\mathsf{Alg}L$ is finite. (Take as the uniform interpolant for φ the conjunction of all interpolants for $\varphi \to \psi$ in L.) It follows, for instance, that **S5** has uniform interpolation, since it has interpolation and $\mathsf{Alg}\mathbf{S5}$ is locally finite.

In this section we prove the following result of Wolter (1998). (Note that the transfer of interpolation can be proved similarly; see Kracht and Wolter 1991).

Theorem 4.18. *Suppose* L_1 *and* L_2 *are multimodal logics. Then* $L = L_1 \otimes L_2$ *has the uniform interpolation property iff both* L_1 *and* L_2 *have this property.*

Proof. Let

$$\nabla(\varphi,\psi) = \{\varphi_1 \to \neg\psi_1 \mid \varphi_1 \in \Sigma_1(\varphi),\ \psi_1 \in \Sigma_1(\psi),\ \varphi_1 \to \neg\psi_1 \in L\}.$$

Then $\bigvee \Sigma_1(\varphi \to \psi)$ is equivalent (modulo Boolean transformations) to

$$\bigvee \Sigma_1(\varphi) \wedge \bigvee \Sigma_1(\psi) \wedge \bigwedge \nabla(\varphi,\psi).$$

Now, the proof of Proposition 4.15 can be easily extended to show that, for any two formulas φ and ψ, we have

$$\varphi \to \psi \in L \quad \text{iff} \quad \gamma \in L_1,$$

where γ is

$$\Box_1^{\leq d^1(\varphi)}(\bigvee \Sigma_1(\varphi))^1 \wedge \Box_1^{\leq d^1(\psi)}(\bigvee \Sigma_1(\psi))^1 \wedge \Box_1^{\leq d^1(\varphi)}(\bigwedge \nabla(\varphi,\psi))^1 \to (\varphi \to \psi)^1.$$

Suppose that L_1 and L_2 have uniform interpolation. Fix $Q = \{q_1,\dots,q_k\}$. We prove by induction on $a(\varphi)$ that there exists a uniform interpolant φ_Q for φ. This is clear if φ contains no modal operators. Assume that φ contains them and that uniform interpolants exist for all χ with $a(\chi) < a(\varphi)$. We may assume that $a(\chi) < a(\varphi)$ for all $\chi \in C(\Theta^1(\varphi))$, and take uniform interpolants χ_Q for these χ. Let

$$\beta = \varphi \wedge \Box_1^{\leq d^1(\varphi)} \bigvee \Sigma_1(\varphi) \wedge \Box_1^{\leq d^1(\varphi)} \bigwedge \{\chi \to \chi_Q \mid \chi \in \Sigma_1(\varphi)\},$$

and

$$R = Q \cup \{q_{\Box_2\alpha} \mid \Box_2\alpha \in \Theta^1(\varphi),\ Q \cap var\alpha \neq \emptyset\}.$$

As L_1 has uniform interpolation, we can take a uniform interpolant β^1_R for β^1 in L_1. There exists a (uniquely determined) formula $\bar\varphi$ such that

$$\beta^1_R = (\bar\varphi)^1.$$

By the definition of R, we have $Q \cap var\bar\varphi = \emptyset$. We show that $\bar\varphi$ is equivalent to a uniform interpolant φ_Q for φ. Indeed, we have $\beta^1 \to \beta^1_R \in L_1$. Thus $\beta \to \bar\varphi \in L$, and so $\varphi \to \bar\varphi \in L$, since $\beta \leftrightarrow \varphi \in L$. Assume now that $\varphi \to \psi \in L$ and $var\psi \cap Q = \emptyset$. We show that $\bar\varphi \to \psi \in L$. It follows from $\gamma \in L_1$ that $\beta^1 \to \delta^1 \in L_1$, where δ is

$$\Box_1^{\leq d^1(\psi)} \bigvee \Sigma_1(\psi) \wedge \Box_1^{\leq d^1(\varphi)} \bigwedge \{(\varphi_1)_Q \to \neg\psi_1 \mid \varphi_1 \to \neg\psi_1 \in \nabla(\varphi,\psi)\} \to \psi.$$

So $\beta^1_R \to \delta^1 \in L_1$, since $R \cap var\delta^1 = \emptyset$. But then $\bar\varphi \to \delta \in L$, from which $\bar\varphi \to \psi \in L$, since $\psi \leftrightarrow \delta \in L$. ❑

It follows, for example, that $\mathbf{K}_n$ and $\mathbf{S5} \otimes \mathbf{S5}$ have uniform interpolation. (For $\mathbf{K}_n$ this was first proved by D'Agostino and Hollenberg 1998.)

4.6 On the computational complexity of fusions

Unlike the properties considered above, the upper bounds for the computational complexity do not always transfer under the formation of fusions (obviously, the lower bounds are inherited as long as we take fusions of consistent logics). We already know that the validity problems for **S5** and **KD45** are coNP-complete, while for $\mathbf{S5}_2 = \mathbf{S5} \otimes \mathbf{S5}$ and $\mathbf{KD45}_2 = \mathbf{KD45} \otimes \mathbf{KD45}$ these problems become PSPACE-complete; see Theorems 1.16 and 1.17. On the other hand, we know that in many cases PSPACE-completeness transfers under the formation of fusions: examples are the fusions of **K**, **T**, **S4**, etc. (see Theorem 1.17). In fact, the proof of Theorem 1.17 from (Halpern and Moses 1992) can be easily modified so as to obtain the following result on the computational complexity of fusions of basic modal logics:

Theorem 4.19. *Let* $n > 1$ *and* $L_i \in \{\mathbf{K}, \mathbf{T}, \mathbf{K4}, \mathbf{S4}, \mathbf{KD45}, \mathbf{S5}\}$, *for all* $1 \leq i \leq n$. *Then* $L_1 \otimes \cdots \otimes L_n$ *is* PSPACE-*complete.*

These observations lead to the following question:

(1) Given a complexity class $\mathcal{C}$, is it the case that the validity problem for the fusion $L_1 \otimes L_2$ is in $\mathcal{C}$ whenever the validity problem for both L_1 and L_2 is in $\mathcal{C}$?

According to the example above, the answer to this question is negative for the class coNP. But then we are facing the following problem:

(2) Give a criterion describing when the validity problem for the fusion of two coNP-complete modal logics is also in coNP.

Unfortunately, except the observation that coNP does not transfer, nothing nontrivial is known about question (1). For example, it is an open problem whether PSPACE- or EXPTIME-completeness transfer under the formation of fusions.

The second problem, however, has been solved by Spaan (1993). To formulate her classification theorem, we require the following notion. Say that a frame $\langle W', R' \rangle$ is a *skeleton subframe* of a frame $\langle W, R \rangle$ if $W' \subseteq W$ and $R' \subseteq R$. Recall also that we used to denote by $\circ$ reflexive points and by $\bullet$ irreflexive ones.

Theorem 4.20. *Suppose that* $\mathcal{C}_1$ *and* $\mathcal{C}_2$ *are classes of unimodal frames that are closed under the formation of isomorphic copies and disjoint unions. Then the validity problem for* $L = \mathsf{Log}(\mathcal{C}_1 \otimes \mathcal{C}_2) = \mathsf{Log}(\mathcal{C}_1) \otimes \mathsf{Log}(\mathcal{C}_2)$ *is* PSPACE-*hard whenever one of the following six cases holds (here* $\{n, \overline{n}\} = \{1, 2\}$*):*

(i) $\bullet\!\leftarrow\!\bullet\!\rightarrow\!\bullet$ *is a skeleton subframe of a frame in* $\mathcal{C}_n$ *and* $\bullet\!\rightarrow\!\bullet$ *is a skeleton subframe of a frame in* $\mathcal{C}_{\overline{n}}$;

(ii) $\circ \to \bullet \to \bullet$ *is a skeleton subframe of a frame in* $\mathcal{C}_n$ *and* $\bullet \to \bullet$ *is a skeleton subframe of a frame in* $\mathcal{C}_{\overline{n}}$*;*

(iii) $\bullet \to \circ \to \bullet$ *is a skeleton subframe of a frame in* $\mathcal{C}_n$ *and* $\bullet \to \bullet$ *is a skeleton subframe of a frame in* $\mathcal{C}_{\overline{n}}$*;*

(iv) $\bullet \to \bullet \to \bullet$ *is a skeleton subframe of a frame in* $\mathcal{C}_n$ *and* $\circ \to \bullet$ *is a skeleton subframe of a frame in* $\mathcal{C}_{\overline{n}}$*;*

(v) $\bullet \to \bullet \to \bullet$ *and* $\circ \to \bullet$ *are skeleton subframes of a frame in* $\mathcal{C}_n$ *and* $\bullet \leftrightarrow \bullet$ *is a skeleton subframe of a frame in* $\mathcal{C}_{\overline{n}}$*;*

(vi) $\bullet \to \bullet \to \bullet$ *and* $\circ \to \bullet$ *are skeleton subframes of a frame in* $\mathcal{C}_n$ *and* $\bullet \to \circ$ *is a skeleton subframe of a frame in* $\mathcal{C}_{\overline{n}}$*.*

Otherwise, either $\mathcal{C}_n$*, for some* $n \in \{1, 2\}$*, consists of disjoint unions of singleton frames—in which case* L *is polynomially reducible to* $\mathsf{Log}(\mathcal{C}_{\overline{n}})$*—or* L *is* coNP*-complete.*

A close inspection of this result shows that almost all interesting fusions are PSPACE-hard. Besides those already mentioned in Theorem 1.17, this lower bound holds, for example, for **K4.3** $\otimes$ **K4.3** and **S4.3** $\otimes$ **S4.3**. The only interesting exceptions are the fusions **Alt** $\otimes$ **Alt** and **DAlt** $\otimes$ **DAlt**, which by Theorem 4.20 are coNP-complete.

Chapter 5

Products of modal logics: introduction

Unlike fusions, where modal operators of the fused logics do not interact, products of modal logics do involve a rather strong interaction, which makes them much more complex than fusions. In particular, no general transfer theorem comparable with those we saw in the previous chapter can be proved for products. Their computational behavior subtly 'feels' various frame properties of logics (transitivity, linearity, etc.): it strongly depends on the dimensionality of the product. All this as well as the connections with other many-dimensional formalisms makes the theory of products of modal logics challenging and exciting.

We begin our study of decidability, complexity and axiomatizability problems for products of standard modal logics with an introductory chapter, where, to keep the exposition as transparent as possible, we consider only two-dimensional products of unimodal logics. (However, all the definitions and many of the results can be easily generalized to two-dimensional products of multimodal logics; see Gabbay and Shehtman 1998). Thus, we will be dealing with product logics formulated in the bimodal language $\mathcal{ML}_2$. To reflect the geometrical intuition behind product frames (see Section 3.3), we denote the boxes and diamonds of $\mathcal{ML}_2$ by $\boxminus$, $\diamondminus$, and $\boxbar$, $\diamondbar$; the former pair is interpreted in 2-frames $\mathfrak{F} = \langle W, R_h, R_v \rangle$ by the 'horizontal' accessibility relation R_h and the latter one by the 'vertical' R_v.

The aim of the chapter is to consider the interaction axioms between $\boxminus$ and $\boxbar$, which will result in rather simple axiomatizations of certain kinds of products, and then, using the example of relatively simple $\mathbf{S5} \times \mathbf{S5}$, to gently introduce the methods of obtaining decidability and complexity results we will be applying later on to more complex products.

5.1 Axiomatizing products

Recall that the *product logic* $L_1 \times L_2$ of Kripke complete modal logics L_1 and L_2 is defined as

$$L_1 \times L_2 = \mathsf{Log}\{\mathfrak{F}_1 \times \mathfrak{F}_2 \mid \mathfrak{F}_1 \in \mathsf{Fr}L_1,\ \mathfrak{F}_2 \in \mathsf{Fr}L_2\},$$

where the product $\mathfrak{F}_1 \times \mathfrak{F}_2$ of frames $\mathfrak{F}_1 = \langle W_1, R_1 \rangle$ and $\mathfrak{F}_2 = \langle W_2, R_2 \rangle$ is the frame $\langle W_1 \times W_2, R_h, R_v \rangle$ in which, for all $u, u' \in W_1$, $v, v' \in W_2$,

$$\begin{array}{lll} \langle u, v \rangle\, R_h\, \langle u', v' \rangle & \text{iff} & uR_1u' \text{ and } v = v', \\ \langle u, v \rangle\, R_v\, \langle u', v' \rangle & \text{iff} & vR_2v' \text{ and } u = u'. \end{array}$$

Product logics are defined in a semantical way: they are logics determined by classes of product frames. Thus, a good start to understand their behavior is to find properties that hold in every product frame. The most obvious are the three diagrams in Fig. 5.1 the meaning of which can be described by the following first-order sentences:

- *left commutativity*: $\forall x \forall y \forall z\, \big(xR_vy \wedge yR_hz \rightarrow \exists u\, (xR_hu \wedge uR_vz)\big)$,
- *right commutativity*: $\forall x \forall y \forall z\, \big(xR_hy \wedge yR_vz \rightarrow \exists u\, (xR_vu \wedge uR_hz)\big)$,
- *Church–Rosser property*: $\forall x \forall y \forall z\, \big(xR_vy \wedge xR_hz \rightarrow \exists u\, (yR_hu \wedge zR_vu)\big)$.

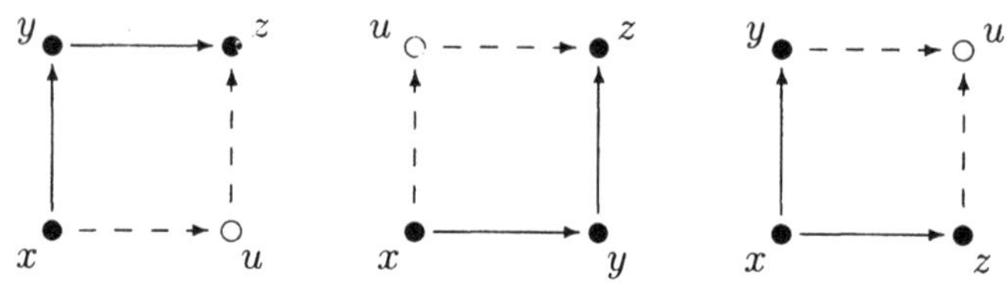

Figure 5.1: Left and right commutativity and Church–Rosser properties.

These properties can also be expressed by modal formulas. One can easily check that a 2-frame is left commutative iff it validates the formula

$$\boldsymbol{com^l} = \Diamond_v \Diamond_h p \rightarrow \Diamond_h \Diamond_v p,$$

it is right commutative iff it validates

$$\boldsymbol{com^r} = \Diamond_h \Diamond_v p \rightarrow \Diamond_v \Diamond_h p,$$

and it is Church–Rosser iff it validates

$$\boldsymbol{chr} = \Diamond_h \Box_v p \rightarrow \Box_v \Diamond_h p.$$

The left and right commutativity axioms can be combined into a single *commutativity* axiom

$$\boldsymbol{com} = \boldsymbol{com}^l \wedge \boldsymbol{com}^r.$$

Are these axioms enough to characterize the product frames? The answer is negative: there are commutative and Church–Rosser 2-frames that are not (isomorphic to) products of any frames. A simple example of such a frame is shown in Fig. 5.2. Anyway, it is tempting to conjecture that every product

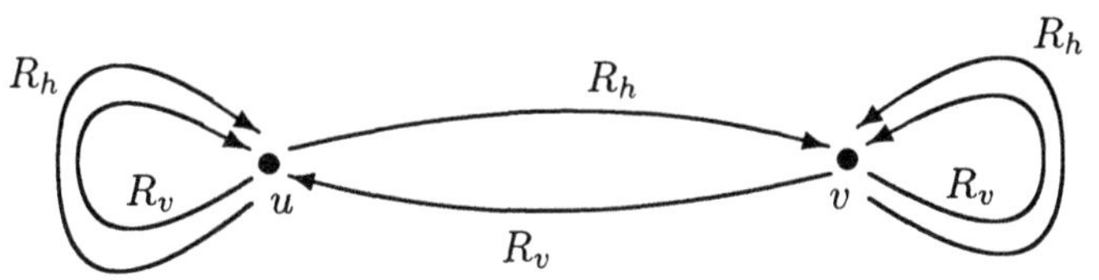

Figure 5.2: Commutative and Church–Rosser but not product frame.

logic $L_1 \times L_2$ can be represented as

$$[L_1, L_2] = (L_1 \otimes L_2) \oplus \boldsymbol{com} \oplus \boldsymbol{chr}.$$

Logics L_1 and L_2 for which this is the case, i.e., $L_1 \times L_2 = [L_1, L_2]$, will be called *product-matching*. Of course, by Proposition 3.8, we always have the inclusion

$$[L_1, L_2] \subseteq L_1 \times L_2. \tag{5.1}$$

The question is whether the converse holds. It turns out that many pairs of standard modal logics are indeed product-matching; however, there are many counterexamples as well. The results we are about to present were obtained by Gabbay and Shehtman (1998).

Axiomatizing K × K

First, since both $\boldsymbol{com}$ and $\boldsymbol{chr}$ are *Sahlqvist formulas*, the logic $[\mathbf{K}, \mathbf{K}]$ is *canonical*, and so we have:

Proposition 5.1. *$[\mathbf{K}, \mathbf{K}]$ is Kripke complete. In particular,* $\mathsf{Fr}\,[\mathbf{K}, \mathbf{K}]$ *is the class of all 2-frames having the commutativity and Church–Rosser properties.*

For definitions of and classical results on Sahlqvist formulas and canonicity; see, e.g., (Sahlqvist 1975, Chagrov and Zakharyaschev 1997, Blackburn *et al.* 2001).

Now we show that **K** and **K** are product-matching. The heart of the proof is the following:

Lemma 5.2. *Every countable rooted* 2*-frame validating* $\boldsymbol{com}$ *and* $\boldsymbol{chr}$ *is a p-morphic image of a product frame (in fact, of the product of two countable intransitive trees).*

Proof. Let $\mathfrak{G} = \langle W, R_h, R_v \rangle$ be a countable rooted frame validating $\boldsymbol{com}$ and $\boldsymbol{chr}$. We will build, step-by-step, frames $\mathfrak{F}_1 = \langle U, R_1 \rangle$ and $\mathfrak{F}_2 = \langle V, R_2 \rangle$ and a p-morphism f from $\mathfrak{F}_1 \times \mathfrak{F}_2$ onto $\mathfrak{G}$. One way of formalizing this straightforward step-by-step argument is by defining a *game* $G(\mathfrak{G})$ between two players $\forall$ (male) and $\exists$ (female) over $\mathfrak{G}$. (Our game and its properties are similar to those of (Hirsch and Hodkinson 1997), where games are played over relation algebras. For a detailed treatment of games over many-dimensional structures see (Hirsch and Hodkinson 2002).)

We define a $\mathfrak{G}$-*network* to be a tuple

$$N = \left\langle U^N, V^N, R_1^N, R_2^N, f^N \right\rangle,$$

where $\mathfrak{F}_1^N = \left\langle U^N, R_1^N \right\rangle$ and $\mathfrak{F}_2^N = \left\langle V^N, R_2^N \right\rangle$ are finite irreflexive and intransitive trees, and f^N is a homomorphism from $\mathfrak{F}_1^N \times \mathfrak{F}_2^N$ to $\mathfrak{G}$, that is, for all $u, u' \in U^N$ and $v, v' \in V^N$,

$$\begin{array}{lll} \text{if } uR_1^N u' & \text{then} & f^N(u,v) R_h f^N(u',v), \\ \text{if } vR_2^N v' & \text{then} & f^N(u,v) R_v f^N(u,v'). \end{array}$$

The players $\forall$ and $\exists$ build a countable sequence of finite $\mathfrak{G}$-networks

$$N_0 \subseteq N_1 \subseteq \ldots \subseteq N_i \subseteq \ldots$$

where $N_{i-1} \subseteq N_i$ means that $U^{N_{i-1}} \subseteq U^{N_i}$, $V^{N_{i-1}} \subseteq V^{N_i}$, $R_\ell^{N_{i-1}} \subseteq R_\ell^{N_i}$, for $\ell = 1, 2$, and $f^{N_{i-1}} \subseteq f^{N_i}$ if we consider functions as sets of pairs.

In round 0, $\forall$ picks the root r of $\mathfrak{G}$. $\exists$ responds with some $\mathfrak{G}$-network N_0 such that both U^{N_0} and V^{N_0} are singleton sets, the accessibility relations $R_1^{N_0}$ and $R_2^{N_0}$ are empty, and f^{N_0} maps the only pair in $U^{N_0} \times V^{N_0}$ to r.

Suppose now that in round i, $0 < i < \omega$, the players built a sequence $N_0 \subseteq \cdots \subseteq N_{i-1}$ of $\mathfrak{G}$-networks. Player $\forall$'s aim is to challenge $\exists$ with possible *defects* of N_{i-1} which indicate that the homomorphism $f^{N_{i-1}}$ is not a p-morphism onto $\mathfrak{G}$ yet. $\forall$ picks such a defect which consists of

- a pair $\langle u, v \rangle \in U^{N_{i-1}} \times V^{N_{i-1}}$,
- a 'direction' $d \in \{h, v\}$,
- a world w in $\mathfrak{G}$ such that $f^{N_{i-1}}(u,v) R_d w$.

Player $\exists$ can respond in two ways. Assume that $\forall$ has picked direction h. If there is some $u' \in U^{N_{i-1}}$ such that $uR_1^{N_{i-1}}u'$ and $f^{N_{i-1}}(u', v) = w$, then she responds with $N_i = N_{i-1}$. Otherwise, she responds (if she can) with some $\mathfrak{G}$-network N_i extending N_{i-1} in such a way that

- $U^{N_i} = U^{N_{i-1}} \cup \{u^+\}$, u^+ being a fresh point, $R_1^{N_i} = R_1^{N_{i-1}} \cup \{\langle u, u^+ \rangle\}$,
- $\mathfrak{F}_2^{N_i} = \mathfrak{F}_2^{N_{i-1}}$ and
- $f^{N_i}(u^+, v) = w$.

If $\forall$ picked direction v, her move is analogous, possibly extending $\mathfrak{F}_2^{N_{i-1}}$. If $\exists$ can respond in each round $i < \omega$ then *she wins the play*. Say that $\exists$ *has a winning strategy in the game* $G(\mathfrak{G})$ if she can win all plays. (We assume that in each round of a play, $\exists$ possesses the information about all the previous moves of $\forall$ and remembers her answers.)

Claim 5.3. *If $\exists$ has a winning strategy in $G(\mathfrak{G})$ then there are countable intransitive trees $\mathfrak{F}_1$, $\mathfrak{F}_2$ such that $\mathfrak{G}$ is a p-morphic image of $\mathfrak{F}_1 \times \mathfrak{F}_2$.*

Proof. Consider a play of the game $G(\mathfrak{G})$ when $\forall$ eventually picks all possible defects (he can do this because $\mathfrak{G}$ is countable and $\mathfrak{G}$-networks are always finite). If $\exists$ uses her winning strategy in this play, then she succeeds in constructing a countable ascending chain of $\mathfrak{G}$-networks whose union gives the required p-morphism. ❑

Thus, it remains to define a winning strategy for $\exists$ in $G(\mathfrak{G})$. In round 0, her response is determined by the rules of the game. In round i $(0 < i < \omega)$, some sequence $N_0 \subseteq \cdots \subseteq N_{i-1}$ of $\mathfrak{G}$-networks is already constructed. Assume that $\forall$ picks the defect which consists of a pair $\langle u, v \rangle \in U^{N_{i-1}} \times V^{N_{i-1}}$, a direction d and a world w in $\mathfrak{G}$ such that $f^{N_{i-1}}(u, v) R_d w$.

Let for definiteness $d = h$ (the case of $d = v$ is similar). By the rules of the game, if there is $u' \in U^{N_{i-1}}$ such that $uR_1^{N_{i-1}}u'$ and $f^{N_{i-1}}(u', v) = w$, then $\exists$ must respond with $N_i = N_{i-1}$. Otherwise she has to add a fresh point u^+ to $U^{N_{i-1}}$ and to respond with a $\mathfrak{G}$-network N_i satisfying the above conditions. The value of $f^{N_i}(u^+, v)$ is defined to be w by the rules. What remains to be done is to define f^{N_i} on all pairs of the form $\langle u^+, v' \rangle$, where $v' \in V^{N_i} = V^{N_{i-1}}$ and $v' \neq v$. These pairs will be called *new pairs*.

Claim 5.4. *There is an enumeration $\{v_0, v_1, \dots, v_M\}$ of $V^{N_{i-1}}$ such that $v_0 = v$ and, for all k, $0 < k \leq M$, there is a unique index $pred(k) < k$ for which either $v_{pred(k)} R_2^{N_{i-1}} v_k$ or $v_k R_2^{N_{i-1}} v_{pred(k)}$.*

Proof. Take the unique $R_2^{N_{i-1}}$-path starting from the root of the tree $\mathfrak{F}_2^{N_{i-1}}$ and ending with v, and enumerate it backwards; then proceed with all the other points in $V^{N_{i-1}}$ by enumerating them in the order of their 'creation' in the current play of the game. ❑

In order to define f^{N_i} on the new pairs, suppose that $0 < k \leq M$ and that we have already defined $f^{N_i}(u^+, v_\ell)$ for all $\ell < k$ in such a way that

$$f^{N_i}(u, v_\ell) R_h f^{N_i}(u^+, v_\ell), \tag{5.2}$$

$$f^{N_i}(u^+, v_\ell) R_v f^{N_i}(u^+, v_{pred(\ell)}), \text{ if } \ell > 0 \text{ and } v_\ell R_2^{N_{i-1}} v_{pred(\ell)}, \tag{5.3}$$

$$f^{N_i}(u^+, v_{pred(\ell)}) R_v f^{N_i}(u^+, v_\ell), \text{ if } \ell > 0 \text{ and } v_{pred(\ell)} R_2^{N_{i-1}} v_\ell. \tag{5.4}$$

Let us now define $f^{N_i}(u^+, v_k)$. By Claim 5.4, we have $pred(k) < k$, and either $v_{pred(k)} R_2^{N_{i-1}} v_k$ or $v_k R_2^{N_{i-1}} v_{pred(k)}$. Consider two possible cases.

Case 1: $v_{pred(k)} R_2^{N_{i-1}} v_k$. As $f^{N_{i-1}}$ is a homomorphism from $\mathfrak{F}_1^{N_{i-1}} \times \mathfrak{F}_2^{N_{i-1}}$,

$$f^{N_{i-1}}(u, v_{pred(k)}) R_v f^{N_{i-1}}(u, v_k),$$

and by (5.2) we have also

$$f^{N_{i-1}}(u, v_{pred(k)}) R_h f^{N_{i-1}}(u^+, v_{pred(k)})$$

(see Fig. 5.3). Since $\mathfrak{G}$ validates ***chr***, there is $s \in W$ such that $f^{N_{i-1}}(u, v_k) R_h s$

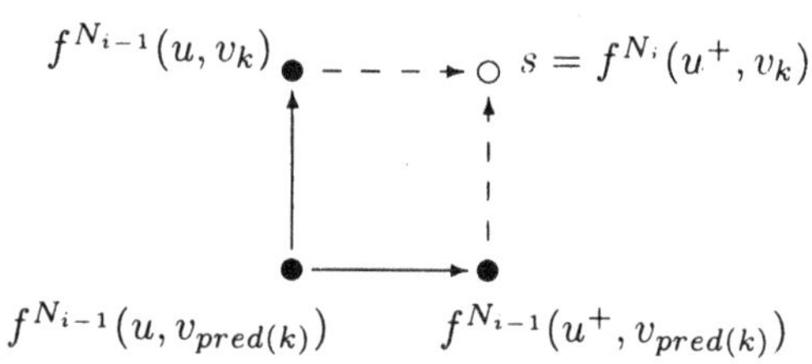

Figure 5.3: Using ***chr***.

and $f^{N_{i-1}}(u^+, v_{pred(k)}) R_v s$. Take any such s and define $f^{N_i}(u^+, v_k) = s$. Then $f^{N_i}(u^+, v_k)$ satisfies (5.2) and (5.4), that is,

$$f^{N_i}(u, v_k) R_h f^{N_i}(u^+, v_k),$$
$$f^{N_i}(u^+, v_{pred(k)}) R_v f^{N_i}(u^+, v_k).$$

Case 2: $v_k R_2^{N_{i-1}} v_{pred(k)}$. We then have

$$f^{N_{i-1}}(u, v_k) R_v f^{N_{i-1}}(u, v_{pred(k)}),$$
$$f^{N_{i-1}}(u, v_{pred(k)}) R_h f^{N_{i-1}}(u^+, v_{pred(k)}),$$

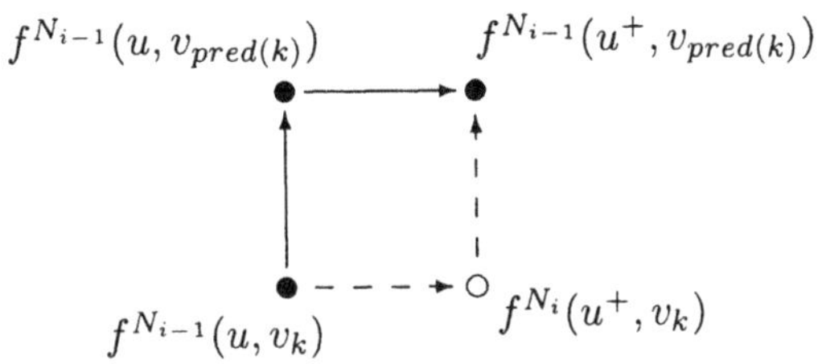

Figure 5.4: Using $\boldsymbol{com^l}$.

and so we can use $\mathfrak{G} \models \boldsymbol{com^l}$ to define $f^{N_i}(u^+, v_k)$; see Fig. 5.4. (If player $\forall$ chooses $d = v$, then we use $\mathfrak{G} \models \boldsymbol{com^r}$.) Then $f^{N_i}(u^+, v_k)$ satisfies (5.2) and (5.3), that is,

$$f^{N_i}(u, v_k) R_h f^{N_i}(u^+, v_k),$$
$$f^{N_i}(u^+, v_k) R_v f^{N_i}(u^+, v_{pred(k)}).$$

Now we prove that the defined function f^{N_i} is a homomorphism from $\mathfrak{F}_1^{N_i} \times \mathfrak{F}_2^{N_i}$ to $\mathfrak{G}$. Suppose first that $x, y \in U^{N_i}$, $x R_1^{N_i} y$ and $z \in V^{N_i}$. We show that

$$f^{N_i}(x, z) R_h f^{N_i}(y, z). \tag{5.5}$$

Indeed, if $y \neq u^+$ then $x \in U^{N_{i-1}}$, and so (5.5) holds because f^{N_i} coincides with $f^{N_{i-1}}$ on the 'old' pairs. And if $y = u^+$ then we have $x = u$, and then (5.5) follows from (5.2).

Now suppose that $x, y \in V^{N_i}$, $x R_2^{N_i} y$ and $z \in U^{N_i}$. We show that

$$f^{N_i}(z, x) R_v f^{N_i}(z, y). \tag{5.6}$$

Again, if $z \neq u^+$ then (5.5) is clear. Let $z = u^+$. There are $k, \ell \leq M$ such that $k \neq \ell$, $x = v_\ell$ and $y = v_k$. Suppose first that $k > \ell$. Then, by Claim 5.4, we have $\ell = pred(k)$, and so (5.6) follows from (5.4). Similarly, if $k < \ell$ then $k = pred(\ell)$ and (5.6) holds by (5.3). ❑

We are now in a position to prove the following:

Theorem 5.5. $\mathbf{K} \times \mathbf{K} = [\mathbf{K}, \mathbf{K}]$.

Proof. By Proposition 5.1, $[\mathbf{K}, \mathbf{K}]$ is determined by the class of all commutative and Church–Rosser frames. This class is first-order definable in the language having equality and two binary predicate symbols. Let $\varphi \notin [\mathbf{K}, \mathbf{K}]$. Then, by Theorem 1.6, we have a countable rooted 2-frame $\mathfrak{F}$ for $[\mathbf{K}, \mathbf{K}]$ refuting φ. Now, using Lemma 5.2, we can find a product frame $\mathfrak{G}$ having $\mathfrak{F}$

as its p-morphic image. By Theorem 1.13 (i), it follows that $\mathfrak{G} \not\models \varphi$, and so $\varphi \notin \mathbf{K} \times \mathbf{K}$. Therefore, we obtain $\mathbf{K} \times \mathbf{K} \subseteq [\mathbf{K}, \mathbf{K}]$. The converse inclusion has already been shown above. ❑

Product-matching logics

Actually, Theorem 5.5 can be generalized to many other pairs of standard modal logics. Consider the first-order language with equality and a binary predicate R. A formula ψ in this language is called *positive* if it is built up from atoms using only $\wedge$ and $\vee$. A sentence of the form

$$\forall x \forall y \forall \bar{z} \left(\psi(x, y, \bar{z}) \to R(x, y)\right)$$

is said to be a *universal Horn sentence* if $\psi(x, y, \bar{z})$ is a positive formula. We call an $\mathcal{ML}$-formula φ a *Horn formula*, if there is a universal Horn sentence φ_H such that, for all frames $\mathfrak{F}$,

$$\mathfrak{F} \models \varphi \quad \text{iff} \quad \mathfrak{F} \models \varphi_H.$$

An $\mathcal{ML}$-formula is called *variable free* if it contains no propositional variables, i.e., all its atomic subformulas are constants $\bot$ or $\top$.

Lemma 5.6. (i) *For every variable-free $\mathcal{ML}$-formula φ, the class $\mathsf{Fr}\{\varphi\}$ is first-order definable. In fact, for all frames $\mathfrak{F}$,*

$$\mathfrak{F} \models \varphi \quad \textit{iff} \quad \mathfrak{F} \models \varphi^\star,$$

where $\varphi^\star$ is the standard translation of φ (see Section 1.3).

(ii) *If φ is a variable-free formula and $\mathfrak{F}$ is a p-morphic image of $\mathfrak{G}$ then*

$$\mathfrak{F} \models \varphi \quad \textit{iff} \quad \mathfrak{G} \models \varphi.$$

Proof. An easy induction is left to the reader as an exercise. ❑

We call a unimodal logic *Horn axiomatizable* if it is axiomatizable by only Horn and variable-free formulas. Examples of Kripke complete Horn axiomatizable logics are $\mathbf{K}$, $\mathbf{D} = \mathbf{K} \oplus \Diamond\top$, $\mathbf{K4}$, $\mathbf{S4}$, $\mathbf{KD45}$, $\mathbf{T}$, $\mathbf{S5}$.

Clearly, if L is a Kripke complete and Horn axiomatizable logic then $\mathsf{Fr}L$ is defined by the set

$$\begin{aligned}\Gamma_L = &\{\varphi_H \mid \varphi \text{ is a Horn axiom of } L\} \cup \\ &\qquad \{\varphi^\star \mid \varphi \text{ is a variable-free axiom of } L\}\end{aligned} \tag{5.7}$$

of first-order formulas.

Proposition 5.7. *Let L_1 and L_2 be Kripke complete and Horn axiomatizable unimodal logics. Then $[L_1, L_2]$ is Kripke complete. In particular, $\mathsf{Fr}\,[L_1, L_2]$ is determined by the class of frames in $\mathsf{Fr}(L_1 \otimes L_2)$ having the commutativity and Church–Rosser properties.*

Proof. Since both $\mathsf{Fr}L_1$ and $\mathsf{Fr}L_2$ are first-order definable, $\mathsf{Fr}(L_1 \otimes L_2)$ is first-order definable as well. By Theorem 4.1, $L_1 \otimes L_2$ is complete with respect to $\mathsf{Fr}(L_1 \otimes L_2)$, and so, by Fine's (1975b) theorem, $L_1 \otimes L_2$ is canonical (see also Chagrov and Zakharyaschev 1997, Blackburn *et al.* 2001). Using the fact that $[\mathbf{K}, \mathbf{K}]$ is canonical and that the sum of two canonical logics is canonical (see, e.g., Chagrov and Zakharyaschev 1997), we can conclude that $[L_1, L_2]$ is canonical as well, and therefore Kripke complete. ❑

Lemma 5.8. *Let L_1 and L_2 be Kripke complete and Horn axiomatizable unimodal logics. Then every countable rooted 2-frame for $[L_1, L_2]$ is a p-morphic image of a product frame for $L_1 \times L_2$.*

Proof. For each $i = 1, 2$, define the set Γ_{L_i} of first-order formulas as in (5.7). Then, by Proposition 5.7, $\mathsf{Fr}[L_1, L_2]$ is defined by Γ_{L_1} (for R_h), Γ_{L_2} (for R_v), plus the commutativity and Church–Rosser properties. Suppose that $\mathfrak{G} = \langle W, R_h, R_v \rangle$ is a countable rooted 2-frame for $[L_1, L_2]$. Then $\mathfrak{G} \models \boldsymbol{com} \wedge \boldsymbol{chr}$, $\langle W, R_h \rangle \models \Gamma_{L_1}$ and $\langle W, R_v \rangle \models \Gamma_{L_2}$. Therefore, by Lemma 5.2, there are frames $\mathfrak{F}_1 = \langle U, R_1 \rangle$ and $\mathfrak{F}_2 = \langle V, R_2 \rangle$ and a p-morphism f from $\mathfrak{F}_1 \times \mathfrak{F}_2$ onto $\mathfrak{G}$. However, $\mathfrak{F}_1$ and $\mathfrak{F}_2$ can refute some axioms of L_1 and L_2. By Lemma 5.6 (ii), these can only be some of the Horn axioms.

To 'repair' the $\mathfrak{F}_i$, we will form the 'L_i-closure' of $\mathfrak{F}_i$ by extending, step-by-step, their accessibility relations R_i in the following way. First, let $i = 1$. Define an infinite ascending chain

$$R_1^0 \subseteq R_1^1 \subseteq \dots \subseteq R_1^n \subseteq \dots$$

of binary relations on U by taking $R_1^0 = R_1$, $\mathfrak{F}_1^0 = \mathfrak{F}_1$ and, for $n < \omega$,

$$R_1^{n+1} = R_1^n \cup \{\langle a, b \rangle \in U \times U \mid \mathfrak{F}_1^n \models \exists \bar{z} \psi(a, b, \bar{z}), \text{ for some } \psi \text{ such that } \forall x \forall y \forall \bar{z} (\psi(x, y, \bar{z}) \to R(x, y)) \in \Gamma_{L_1}\},$$

and $\mathfrak{F}_1^{n+1} = \langle U, R_1^{n+1} \rangle$. R_2^n and $\mathfrak{F}_2^n$ are defined similarly, using universal Horn sentences in Γ_{L_2}. We claim that, for each $n < \omega$, f is a p-morphism from $\mathfrak{F}_1^n \times \mathfrak{F}_2^n$ onto $\mathfrak{G}$. Indeed, the 'backward' condition always holds after extending the accessibility relation of the pre-image. Let us assume inductively that f is a homomorphism from $\mathfrak{F}_1^n \times \mathfrak{F}_2^n$ to $\mathfrak{G}$ and let $aR_1^{n+1}b$, for $a, b \in U$, and $c \in V$. If $aR_1^n b$ also holds then $f(a, c) R_h f(b, c)$ by the induction hypothesis. Otherwise, there are a positive formula $\psi(x, y, z_1, \dots, z_m)$ and $d_1, \dots, d_m \in U$ such that

$$\mathfrak{F}_1^n \models \psi(a, b, d_1, \dots, d_m) \quad \text{and} \quad \forall x \forall y \forall \bar{z} (\psi(x, y, \bar{z}) \to R(x, y)) \in \Gamma_{L_1}.$$

Since f is a homomorphism and ψ is positive, we have

$$\mathfrak{G} \models \psi(f(a,c), f(b,c), f(d_1,c), \dots, f(d_m,c))$$

when R is interpreted as R_h. It follows that

$$\langle W, R_h \rangle \models \exists \bar{z} \psi(f(a,c), f(b,c), \bar{z}),$$

and so, by $\langle W, R_h \rangle \models \Gamma_{L_1}$, we obtain $f(a,c) R_h f(b,c)$.

Finally, let

$$R_1^\infty = \bigcup_{n<\omega} R_1^n, \qquad \mathfrak{F}_1^\infty = \langle U, R_1^\infty \rangle,$$
$$R_2^\infty = \bigcup_{n<\omega} R_2^n, \qquad \mathfrak{F}_2^\infty = \langle V, R_2^\infty \rangle.$$

It is easy to see that $\mathfrak{F}_i^\infty \models \Gamma_{L_i}$, $i = 1, 2$, and f is a p-morphism from $\mathfrak{F}_1^\infty \times \mathfrak{F}_2^\infty$ onto $\mathfrak{G}$. ❑

Now we obtain:

Theorem 5.9. *Let L_1 and L_2 be Kripke complete and Horn axiomatizable unimodal logics. Then $L_1 \times L_2 = [L_1, L_2]$.*

Proof. By Proposition 5.7, $[L_1, L_2]$ is determined by the class of commutative and Church–Rosser frames from $\mathsf{Fr}(L_1 \otimes L_2)$. This class is first-order definable in the language with equality and two binary predicate symbols. Let $\varphi \notin [L_1, L_2]$. Then, by Theorem 1.6, we have a countable rooted 2-frame $\mathfrak{F}$ for $[L_1, L_2]$ refuting φ. Now, using Lemma 5.8, we can find a product frame $\mathfrak{G}$ for $L_1 \times L_2$ having $\mathfrak{F}$ as its p-morphic image. By Theorem 1.13 (i), it follows that $\mathfrak{G} \not\models \varphi$, and so $\varphi \notin L_1 \times L_2$. Therefore, $L_1 \times L_2 \subseteq [L_1, L_2]$. The converse inclusion has already been shown as (5.1). ❑

Corollary 5.10. *Let L_1 and L_2 be any logics from the following list:* **K**, **D**, **K4**, **S4**, **KD45**, **T**, **S5**. *Then $L_1 \times L_2 = [L_1, L_2]$.*

More axiomatizability results about products of temporal, dynamic, and epistemic logics with **S5** will be obtained in Sections 6.5, 11.7 and 12.2.

Corollary 5.11. *Let L_1, L_2 and L_3 be Kripke complete and Horn axiomatizable unimodal logics. Then*

$$L_1 \times L_2 \times L_3 = (L_1 \times L_2) \times L_3 = L_1 \times (L_2 \times L_3).$$

Proof. Follows from Theorem 3.16 and Lemma 5.8. ❑

Note that according to the definition above, the sentence

$$\forall x \forall y \forall z (R(x,y) \wedge R(x,z) \to y = z)$$

is not regarded as a universal Horn sentence, so Lemma 5.8 (and hence Theorem 5.9 and Corollary 5.11) does not seem to apply to products of **Alt** (and **DAlt**) logics. However, in Section 8.5 we shall show how to modify the proof of Lemma 5.8 in order to obtain these results for products of **Alt** (and **DAlt**) logics of *any* finite dimension.

In Section 3.3 we introduced the product $\vdash^*_{L_1} \times \vdash^*_{L_2}$ of global consequence relations $\vdash^*_{L_1}$ and $\vdash^*_{L_2}$: for any formulas φ and ψ, we have $\psi(\vdash^*_{L_1} \times \vdash^*_{L_2})\varphi$ iff $\mathfrak{M} \models \varphi$ whenever $\mathfrak{M} \models \psi$, for every model $\mathfrak{M}$ based on a frame in $\mathsf{Fr}L_1 \times \mathsf{Fr}L_2$. In general, not too much is known about how $\vdash^*_{L_1} \times \vdash^*_{L_2}$ relates to the global consequence relation $\vdash^*_{L_1 \times L_2}$ (see the questions below). But in case both L_1 and L_2 are Kripke complete and Horn axiomatizable, we have the following:

Theorem 5.12. *Suppose L_1 and L_2 are Kripke complete and Horn axiomatizable unimodal logics. Then $L_1 \times L_2$ is globally Kripke complete, and $\vdash^*_{L_1 \times L_2}$ coincides with $\vdash^*_{L_1} \times \vdash^*_{L_2}$.*

Proof. By Proposition 5.7 and Theorem 5.9, $L_1 \times L_2$ is determined by a first-order definable class of frames. So, by Theorem 1.19, it is globally Kripke complete, and thus we have $\vdash^*_{L_1 \times L_2} \subseteq \vdash^*_{L_1} \times \vdash^*_{L_2}$.

Suppose now that $\psi \not\vdash^*_{L_1 \times L_2} \varphi$. Then, by Theorem 1.20, we can find a countable 2-frame $\mathfrak{F}$ for $L_1 \times L_2$ (and so for $[L_1, L_2]$) such that $\mathfrak{M} \models \psi$ and $\mathfrak{M} \not\models \varphi$, for some model $\mathfrak{M}$ based on $\mathfrak{F}$. By Lemma 5.8, $\mathfrak{F}$ is a p-morphic image of a product frame $\mathfrak{G}$ from $\mathsf{Fr}L_1 \times \mathsf{Fr}L_2$. But then $\mathfrak{M}' \models \psi$ and $\mathfrak{M}' \not\models \varphi$ for some model $\mathfrak{M}'$ based on $\mathfrak{G}$, and so $\psi(\vdash^*_{L_1} \times \vdash^*_{L_2})\varphi$ does not hold. ❑

Note that Theorem 5.12 can also be regarded as a syntactical characterization of the semantically defined consequence relation $\vdash^*_{L_1} \times \vdash^*_{L_2}$: φ is a consequence of ψ if φ can be derived from ψ using the theorems of $[L_1, L_2]$, modus ponens, and the rules of necessitation.

Question 5.13. Is it true that $L_1 \times L_2$ is globally Kripke complete whenever

(i) both L_1 and L_2 are globally Kripke complete;

(ii) both L_1 and L_2 are determined by first-order definable classes of frames?

Question 5.14. Give an example of a globally Kripke complete product logic $L_1 \times L_2$ such that $\vdash^*_{L_1 \times L_2}$ does not coincide with $\vdash^*_{L_1} \times \vdash^*_{L_2}$.

Logics that are not product-matching

Of course, there are numerous cases when Theorem 5.9 does not apply. In Chapter 7 we shall see many examples of finitely axiomatizable modal logics whose products are not even recursively enumerable. Such are, for instance, $\mathsf{Log}\{\langle \mathbb{N}, < \rangle\} \times \mathsf{Log}\{\langle \mathbb{N}, < \rangle\}$ (cf. (Segerberg 1970, Spaan 1993) and Corollary 7.14), $\mathbf{GL.3} \times \mathbf{GL.3}$ and $\mathbf{Grz.3} \times \mathbf{Grz.3}$ (see Corollary 7.16). Here we prove that Theorem 5.9 cannot be generalized even to logics whose classes of frames are definable by *universal* first-order formulas. As the following theorem shows, for many transitive logics L, the pair of $\mathbf{K4.3}$ and L is not product-matching:

Theorem 5.15. *Let L be any Kripke complete logic containing* $\mathbf{K4}$ *and having the two-element reflexive chain as its frame. Then* $\mathbf{K4.3} \times L \neq [\mathbf{K4.3}, L]$.

Proof. Let $\mathfrak{F}$ be the frame in Fig. 5.5. It is readily seen that $\mathfrak{F} \models [\mathbf{K4.3}, L]$.

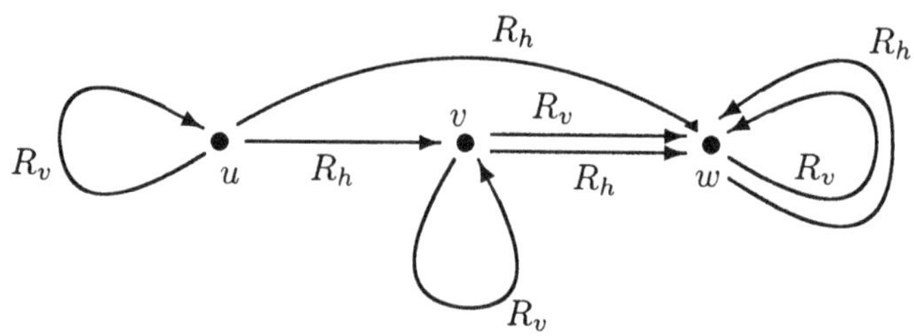

Figure 5.5: The frame $\mathfrak{F}$.

Now, with each world p in $\mathfrak{F}$ we associate a propositional variable, denoted also by p. The following formula $\varphi_{\mathfrak{F}}$ is an analog of the *frame* (or *Jankov–Fine*) *formula for* $\mathfrak{F}$ (see Chagrov and Zakharyaschev 1997):

$$u \wedge \Box^{+}\Big[\bigvee_{p=u,v,w} (p \wedge \neg \bigvee_{p' \neq p} p') \wedge \bigwedge_{\substack{p,p'=u,v,w \\ pR_h p'}} (p \rightarrow \diamondminus p') \wedge \bigwedge_{\substack{p,p'=u,v,w \\ \neg(pR_h p')}} (p \rightarrow \neg \diamondminus p') \wedge$$

$$\bigwedge_{\substack{p,p'=u,v,w \\ pR_v p'}} (p \rightarrow \diamondbar p') \wedge \bigwedge_{\substack{p,p'=u,v,w \\ \neg(pR_v p')}} (p \rightarrow \neg \diamondbar p')\Big].$$

Here $\Box^{+}\psi$ abbreviates $\psi \wedge \boxminus\psi \wedge \boxbar\psi \wedge \boxminus\boxbar\psi$.

The formula $\varphi_{\mathfrak{F}}$ is clearly satisfiable in $\mathfrak{F}$: it is enough to take the model $\mathfrak{M} = \langle \mathfrak{F}, \mathfrak{V} \rangle$ with $\mathfrak{V}(p) = \{p\}$ for every p in $\mathfrak{F}$, and then $(\mathfrak{M}, u) \models \varphi_{\mathfrak{F}}$. So $\neg\varphi_{\mathfrak{F}} \notin [\mathbf{K4.3}, L]$. (Note that we do not assume here that $[\mathbf{K4.3}, L]$ is Kripke complete, which in general—say, for $L = \mathbf{Grz}$—we do not know.)

On the other hand, it is easy to see that, for every two-dimensional product $\mathfrak{G}$ of transitive frames,

$$\varphi_{\mathfrak{F}} \text{ is satisfiable in } \mathfrak{G} \quad \text{iff} \quad \mathfrak{F} \text{ is a p-morphic image of a generated subframe of } \mathfrak{G} \tag{5.8}$$

(see (Chagrov and Zakharyaschev 1997) or Claim 8.36 below). Since a generated subframe of a product frame is a product frame itself, it suffices to show that $\mathfrak{F}$ is not a p-morphic image of any product frame $\mathfrak{G}_1 \times \mathfrak{G}_2$, where $\mathfrak{G}_1 \models$ **K4.3**. Then $\neg\varphi_{\mathfrak{F}} \in$ **K4.3** $\times L$ follows by (5.8).

Suppose otherwise, i.e., there exists a p-morphism f from a product frame $\mathfrak{G} = \langle W, S_h, S_v \rangle$ with weakly connected S_h onto $\mathfrak{F}$. Then there are points $x_u, x_v, x_w \in W$ such that $f(x_u) = u$, $f(x_v) = v$, $f(x_w) = w$ and $x_u S_h x_v S_v x_w$. Since $\mathfrak{G}$ is a product frame, there is a $y_u \in W$ such that $x_u S_v y_u S_h x_w$:

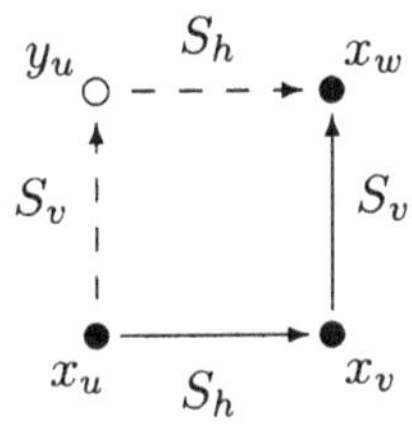

Then $f(y_u) = u$ must hold. Next, there is a $y_v \in W$ such that $f(y_v) = v$ and $y_u S_h y_v$. Since S_h is weakly connected and $f(x_w) = w$, we have $y_v S_h x_w$. And since $\mathfrak{G}$ is a product frame, there has to be a point $z \in W$ such that $x_u S_h z S_h x_v$:

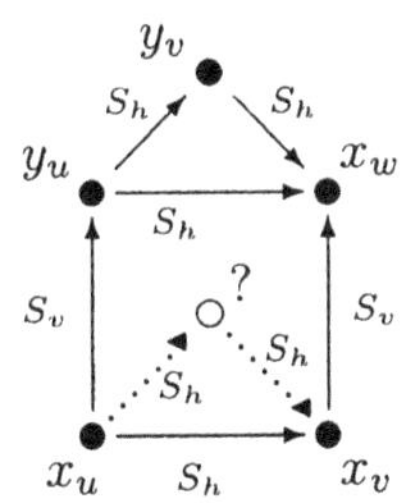

But then $u R_h f(z) R_h v$ should hold, which is a contradiction. ❑

Corollary 5.16. *Let L be any logic from the list* **K4**, **S4**, **Grz**, **K4.3**, **S4.3**, **Grz.3**. *Then* **K4.3** $\times L \neq$ [**K4.3**, L].

Next, we prove a theorem of Gabbay and Shehtman (1998) from which it follows that, for many transitive logics L, the pair of **Grz.3** and L is not product-matching either.

Theorem 5.17. *Let L_1 be any Kripke complete logic containing* **Grz** *and having the two-element reflexive chain as its frame. Let L_2 be any Kripke complete logic containing* **S4** *and having either* (i) *the two-element reflexive chain or* (ii) *the two-element cluster as its frame. Then $L_1 \times L_2 \neq [L_1, L_2]$.*

Proof. Consider the formula

$$\psi = \Box_1\Box_2(\Box_1(p \to \Box_2 p) \to p) \to p.$$

First, we show that $\psi \in L_1 \times L_2$. Suppose otherwise. Then there is a model $\mathfrak{M}$ based on the product $\mathfrak{F}_1 \times \mathfrak{F}_2$ of a frame $\mathfrak{F}_1 = \langle U, R_1 \rangle$ for L_1 and a frame $\mathfrak{F}_2 = \langle V, R_2 \rangle$ for L_2 and such that

$$(\mathfrak{M}, \langle u_0, v_0 \rangle) \models \neg p \land \Box_1\Box_2(\neg p \to \Diamond_1(p \land \Diamond_2 \neg p)) \tag{5.9}$$

for some point $\langle u_0, v_0 \rangle \in U \times V$. To derive a contradiction, we are going to construct an infinite ascending chain of at least two distinct points in $\mathfrak{F}_1$, thereby showing that it cannot be a frame for **Grz**, and so for L_1 either.

Let $0 < n < \omega$ and assume inductively that we have already defined points $\langle u_k, v_k \rangle \in U \times V$, for all $k < n$, such that:

$$(\mathfrak{M}, \langle u_k, v_k \rangle) \models \neg p, \tag{5.10}$$

$$u_0 R_1 u_k \text{ and } v_0 R_2 v_k, \tag{5.11}$$

$$u_k \neq u_{k-1}, \text{ if } k > 0. \tag{5.12}$$

By (5.10), (5.11) and (5.9),

$$(\mathfrak{M}, \langle u_{n-1}, v_{n-1} \rangle) \models \Diamond_1(p \land \Diamond_2 \neg p),$$

and so there are $u_n \in U$, $v_n \in V$ such that

$$u_{n-1} R_1 u_n \text{ and } v_{n-1} R_2 v_n, \tag{5.13}$$

$$(\mathfrak{M}, \langle u_{n-1}, v_n \rangle) \models p \text{ and } (\mathfrak{M}, \langle u_n, v_n \rangle) \models \neg p. \tag{5.14}$$

Then $\langle u_n, v_n \rangle$ clearly satisfies (5.10) and (5.12), and (5.11) follows from (5.13) and the transitivity of R_1 and R_2. Now, consider the points $u_n \in U$, $n < \omega$. Two cases are possible: either there are m, n such that $m > n + 1$ and $u_m = u_n$, or all the u_n are distinct. In the former case $\mathfrak{F}_1$ contains a proper cluster and in the latter an infinite ascending chain of distinct points, which is a contradiction.

It remains to show that $\psi \notin [L_1, L_2]$. Let $\mathfrak{G}$ be the frame in Fig. 5.2, if case (i) in the formulation of our theorem holds, and that in Fig. 5.6, if (ii) holds. In either case we define a valuation on $\mathfrak{G}$ in such a way that $u \models \neg p$ and $v \models p$. Then it is readily checked that $u \not\models \psi$ and $\mathfrak{G} \models [L_1, L_2]$. (Note that we do not assume that $[L_1, L_2]$ is Kripke complete. In fact, we do not know this.) ❑

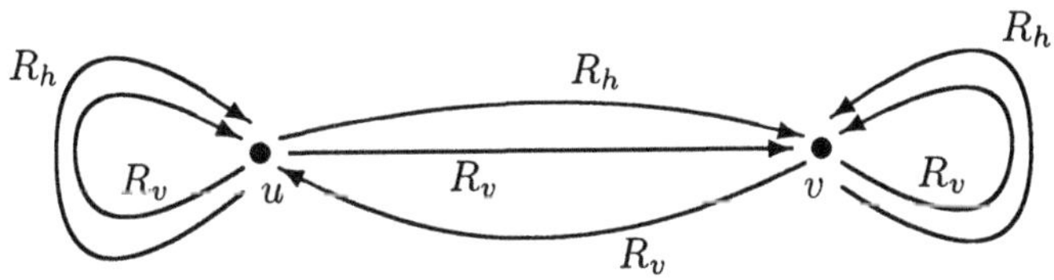

Figure 5.6: The frame $\mathfrak{G}$ in case (ii).

It is worth noting that in fact each of the Theorems 5.15 and 5.17 gives a continuum of non-product-matching pairs of logics (see, for instance, Theorem 11.19 of (Chagrov and Zakharyaschev 1997)).

There are still many pairs of logics that are beyond the scope of Theorems 5.15 and 5.17. For instance, the following question is open:

Question 5.18. Are any of the logics $\mathbf{K4.3} \times \mathbf{S5}$, $\mathbf{K4.3} \times \mathbf{K}$, $\mathbf{S4.3} \times \mathbf{S4.3}$, $\mathbf{GL} \times \mathbf{S4}$, $\mathsf{Log}\{\langle \mathbb{N}, < \rangle\} \times \mathbf{K}$, $\mathsf{Log}\{\langle \mathbb{Q}, < \rangle\} \times \mathbf{S5}$ product-matching? Are any of them finitely axiomatizable?

And for pairs that are known to be not product-matching, no finite axiomatization is known either:

Question 5.19. Let L be any logic from the list $\mathbf{K4}$, $\mathbf{S4}$, $\mathbf{Grz}$, $\mathbf{K4.3}$, $\mathbf{S4.3}$, $\mathbf{Grz.3}$. Is $\mathbf{K4.3} \times L$ finitely axiomatizable?

Many of these logics are recursively enumerable by Theorem 3.17. An axiomatization of $\mathbf{K4.3} \times \mathbf{K4.3}$ using Gabbay-style irreflexivity rules can be found in (Reynolds and Zakharyaschev 2001).

$[\mathbf{K4.3}, \mathbf{K4.3}]$ looks rather 'harmless' (maybe not?), whereas, as we show in Chapter 7, $\mathbf{K4.3} \times \mathbf{K4.3}$ is undecidable. The following interesting problems are also open:

Question 5.20. Do there exist Kripke complete logics L_1 and L_2 such that only one of $[L_1, L_2]$ and $L_1 \times L_2$ is decidable?

Question 5.21. Give an example of Kripke complete logics L_1 and L_2 such that $[L_1, L_2]$ is Kripke incomplete?

5.2 Proving decidability with quasimodels

This section introduces the main technique we will use to prove decidability of product logics and other two-dimensional logical formalisms—the *method of quasimodels*.[1]

There are three basic approaches to establishing decidability of one-dimensional modal logics; see, e.g., (Gabbay *et al.* 1994, Chagrov and Zakharyaschev 1997, Zakharyaschev *et al.* 2001). Given such a logic L, we can try to prove that it has the fmp (and that the class of finite frames for L is recursively enumerable, which is the case if L is finitely axiomatizable). This is the most popular approach. If L does not enjoy the fmp, then we can try to show that it is characterized by (in general) infinite models having a certain 'regular structure,' say, constructed from repeating finite pieces. The third approach is to try to reduce the decision problem for L to another problem that is already known to be decidable (say, to the decision problem for a suitable monadic second-order theory or the emptiness problem for a certain tree automaton (Vardi and Wolper 1986)).

In principle, the same approaches can be applied to many-dimensional modal logics. At first sight, proving decidability by establishing the (product) fmp may appear as very promising. By definition, the product logic $L_1 \times L_2$ is determined by product frames $\mathfrak{F}_1 \times \mathfrak{F}_2$ such that $\mathfrak{F}_i \models L_i$, $i = 1, 2$. If both L_1 and L_2 are finitely axiomatizable, then clearly the finite product frames for $L_1 \times L_2$ are recursively enumerable. So to prove that $L_1 \times L_2$ is decidable, it is enough 'just' to show that it has the *product fmp*. Unfortunately, many product logics do *not* have this property; see Theorem 6.21.

There still remains a possibility that our $L_1 \times L_2$ is complete with respect to the class of its *all* (not necessarily product) finite frames—i.e., it enjoys the '*abstract*' *fmp*. However, on the one hand, very few product logics are known to have the fmp (some examples are given in Sections 5.3 and 8.3). And on the other hand, now we are facing the problem of enumerating the finite 'abstract' frames for $L_1 \times L_2$. This problem can be much harder than enumerating its finite product frames. In fact, the standard approach to enumerating the finite frames by first providing a finite axiomatization works only in a limited number of cases, because, as we saw in the previous section, not too many product logics are known to be finitely axiomatizable. The complexity of the structure of finite abstract frames for product logics is also illustrated by results of Section 8.4, where we shall show examples of finitely axiomatizable (and decidable) logics L_i, $i = 1, 2, 3$, such that the property of being a finite frame for $L_1 \times L_2 \times L_3$ is undecidable.

Thus, if we want to develop a reasonably general machinery for proving decidability of product and other many-dimensional logics, we may be bound to deal with *infinite* models. The question then is how to represent these

[1] The method of quasimodels was first developed in the series of papers on description logics with various modal and temporal operators in (Wolter and Zakharyaschev 1998, 1999b, 2000c, 2001b) and then extended to products in (Wolter 2000b) and to fragments of first-order modal and temporal logics in (Hodkinson *et al.* 2000, Wolter and Zakharyaschev 2001a, 2002).

infinite models as 'regular structures of repeating finite pieces,' if this is at all possible. Consider, for example, the product $\mathfrak{F} \times \mathfrak{G}$ of a frame $\mathfrak{F} = \langle W, R \rangle$ for a logic L_1 and a frame $\mathfrak{G} = \langle \Delta, S \rangle$ for a logic L_2, and let $\mathfrak{V}$ be a valuation in $\mathfrak{F} \times \mathfrak{G}$. We can then represent the model $\mathfrak{M} = \langle \mathfrak{F} \times \mathfrak{G} \rangle$ in the following way. Let $\boldsymbol{q}'$ be a function associating with each $w \in W$ the L_2-model $\boldsymbol{q}'(w) = \langle \mathfrak{G}, \mathfrak{V}_w \rangle$, where

$$\mathfrak{V}_w(p) = \{x \in \Delta \mid \langle w, x \rangle \in \mathfrak{V}(p)\},$$

for all variables p. As the valuation $\mathfrak{V}$ is clearly restored from $\boldsymbol{q}'$ by taking

$$\mathfrak{V}(p) = \{\langle w, x \rangle \in W \times \Delta \mid x \in \mathfrak{V}_w(p)\},$$

we can think of $\mathfrak{M}$ as the pair $\langle \mathfrak{F}, \boldsymbol{q}' \rangle$. In other words, we may assume that product models for $L_1 \times L_2$ consist of 'slices' $\boldsymbol{q}'(w)$, $w \in W$, of L_2-models (which take care of the 'vertical' modal operators $\boxdot$ and $\diamondsuit$). In fact, this kind of representation was originally built in the definitions of some other two-dimensional formalisms, say, description or first-order temporal logics.

The next step is to 'finitize' the slices $\boldsymbol{q}'(w)$. There are different ways of doing this depending on the logics we deal with. However, the standard starting observation is as follows. If we are interested, say, in satisfiability of a formula φ, then there are only finitely many formulas that may influence the truth-value of φ, for example, the set $sub\,\varphi$ of all its subformulas or a certain closure of $sub\,\varphi$, say, under $\neg$. Suppose that we have fixed a set Γ of such 'relevant' formulas, with its size being bounded by some computable function of $\ell(\varphi)$. For every point $x \in \Delta$, we can then define the *type* of $\langle w, x \rangle$ as

$$\boldsymbol{t}_w(x) = \{\psi \in \Gamma \mid (\mathfrak{M}, \langle w, x \rangle) \models \psi\}$$

and think of $\boldsymbol{q}'(w)$ as populated by these types. More precisely, instead of $\boldsymbol{q}'(w) = \langle \mathfrak{G}, \mathfrak{V}_w \rangle$ we consider now the pair $\boldsymbol{q}''(w) = \langle \mathfrak{G}, \boldsymbol{t}_w \rangle$, where $\boldsymbol{t}_w$ 'labels' every element of $\mathfrak{G}$ with a type. The fact that the number of pairwise distinct types cannot exceed $2^{|\Gamma|}$ opens a way to various finitizations of $\boldsymbol{q}''(w)$, for instance, filtration, by identifying different points of the same type; see e.g., (Blackburn *et al.* 2001, Chagrov and Zakharyaschev 1997). We call the obtained finite type structures $\boldsymbol{q}(w) = \langle \mathfrak{G}_w, \boldsymbol{t}'_w \rangle$ *quasistates*—they still take care of $\boxdot$ and $\diamondsuit$—and the pair $\langle \mathfrak{F}, \boldsymbol{q} \rangle$ a *basic structure*. (The reader will find various kinds of quasistates and basic structures later on in the book, the simplest ones being those for **S5** $\times$ **S5** defined below in this section.)

As the slices $\boldsymbol{q}''(u)$ and $\boldsymbol{q}''(v)$ are in general different for different u and v, although sharing the same frame $\mathfrak{G}$, their finitizations $\boldsymbol{q}(u) = \langle \mathfrak{G}_u, \boldsymbol{t}'_u \rangle$ and $\boldsymbol{q}(v) = \langle \mathfrak{G}_v, \boldsymbol{t}'_v \rangle$ may have nonisomorphic frames $\mathfrak{G}_u$ and $\mathfrak{G}_v$, i.e., we are losing the product structure of the original model. To put it another way, we no longer know what happens with the point $\langle u, x \rangle$—or the type representing it—when we move to a successor v of u in $\mathfrak{F}$. To restore this lost information, we require functions r which trace the evolution of each point (or type) in $\mathfrak{G}_w$ along the horizontal axis (and thereby take care of the horizontal operators $\boxminus$ and $\diamondminus$); see Fig. 5.7. Such functions r are called *runs*: they map every $w \in W$ to a point $r(w)$ in the underlying frame $\mathfrak{G}_w$ of $\boldsymbol{q}(w)$. In order to reconstruct the product structure, in some cases certain conditions should be imposed on the runs. A basic structure together with an appropriate (structured) set $\mathfrak{R}$ of runs is called a *quasimodel for* φ.

Given concrete L_1 and L_2, our first aim will be to find a proper notion of quasimodel for which the following 'quasimodel lemma' holds: *a formula* φ *is satisfiable in a model based on a frame for* $L_1 \times L_2$ *iff there is a quasimodel for* φ.

Although quasistates in quasimodels are always finite, quasimodels themselves are usually infinite (since the frame $\mathfrak{F}$ can be infinite). How can we use them to prove decidability? Depending on the logics in question, there may be several different ways:

(1) In certain cases it is easy to find a finite quasimodel for φ and then to construct a finite product model out of it, thereby showing that the logic has the product fmp. This will be done in the decidability proofs for $\mathbf{S5} \times \mathbf{S5}$ (later on in this section) and $\mathbf{K} \times \mathbf{K}$ in Section 6.1.

(2) It is shown that there is a quasimodel for φ iff there exists a *finite* set $\mathcal{S}$ of *finite* 'partial' quasimodels (called *blocks*) satisfying some effectively checkable conditions and that the cardinality of $\mathcal{S}$ as well as the size of each block in it do not exceed a number effectively computable from φ. The 'effectively checkable conditions' are supposed to guarantee that blocks can be used as 'small mosaic pieces' to construct the quasimodel we need.[2] (This technique is used in Sections 6.2, 6.4, 6.5, 6.6, 14.2, and—in a somewhat 'degenerate' form—in Sections 11.4 and 11.5.)

(3) In some cases, the statement that a quasimodel exists can be translated into monadic second-order logic or reduced to other known decidable problems. (This approach is taken in Sections 11.3, 11.8, and 13.2.)

(4) In Section 11.6 we also decompose quasimodels into 'partial' quasimodels, but neither the 'pieces' nor their collection is finite. Nevertheless, the existence of an appropriate set of partial quasimodels can be checked effectively using a reduction to a decidable problem in monadic second-order logic.

(5) Tableau type decision procedures building quasimodels are developed in Chapter 15.

[2]The mosaic method of (Németi 1995, Venema and Marx 1999) is of similar flavor: it also builds big structures out of small pieces.

(Quasimodels will also be used for axiomatizing many-dimensional logics in Sections 11.7 and 12.2.)

We illustrate the method of quasimodels by proving the following well-known theorem:

Theorem 5.22. **S5** $\times$ **S5** *is decidable.*

As we saw in Section 3.5, products of **S5** can be embedded into (finite variable fragments of) first-order logic. Thus, this theorem follows from Scott's (1962) result on the decidability of the two-variable fragment of first-order logic. An algebraic proof (in the setting of diagonal-free cylindric algebras of dimension 2) was found by Henkin in (Henkin *et al.* 1985). Segerberg (1973) uses filtration to prove the fmp of **S5** $\times$ **S5**. A mosaic type proof (also in the algebraic setting) can be found in (Marx and Mikulás 1999).

Proof. Let us fix an $\mathcal{ML}_2$-formula φ and see how to construct quasimodels for **S5** $\times$ **S5**. First, we define a *type* for φ as a subset $\boldsymbol{t}$ of $sub\,\varphi$ which is *Boolean-saturated* in the sense that

- $\psi \wedge \chi \in \boldsymbol{t}$ iff $\psi \in \boldsymbol{t}$ and $\chi \in \boldsymbol{t}$, for every $\psi \wedge \chi \in sub\,\varphi$,
- $\neg\psi \in \boldsymbol{t}$ iff $\psi \notin \boldsymbol{t}$, for every $\neg\psi \in sub\,\varphi$.

A *quasistate* for φ is a set $\boldsymbol{T}$ of distinct types for φ which is $\Diamond$-*saturated*, i.e.,

(qm1) $\quad \forall \boldsymbol{t} \in \boldsymbol{T}\, \forall \Diamond\psi \in sub\,\varphi \; \big(\Diamond\psi \in \boldsymbol{t} \;\leftrightarrow\; \exists \boldsymbol{t}' \in \boldsymbol{T}\; \psi \in \boldsymbol{t}'\big).$

It follows that if $\Diamond\psi \in \boldsymbol{t}$, for some type $\boldsymbol{t}$ in $\boldsymbol{T}$, then $\Diamond\psi \in \boldsymbol{t}'$ for all other types $\boldsymbol{t}'$ in $\boldsymbol{T}$. Note that we can consider a quasistate as a 'cluster of types:' a universal frame each world in which is labeled by a type. Clearly, the cardinality of a quasistate for φ (i.e., the number of distinct types in it) does not exceed $2^{|sub\varphi|}$.

A *basic structure* for φ is a pair $\langle W, \boldsymbol{q}\rangle$ such that W is a nonempty set and $\boldsymbol{q}$ a function from W into the set of quasistates for φ. (In other words, we can think of a basic structure as a multiset of type-clusters.)

A *run through* $\langle W, \boldsymbol{q}\rangle$ is a function r from W to the set of types for φ such that

$$\forall w \in W \; r(w) \in \boldsymbol{q}(w).$$

That is, a run is a 'choice-function' which, for every $w \in W$, chooses a type from the type-cluster $\boldsymbol{q}(w)$.

A run r is called *coherent* if

$$\forall w \in W\, \forall \Diamond\psi \in sub\,\varphi \; \big((\exists v \in W\; \psi \in r(v)) \;\rightarrow\; \Diamond\psi \in r(w)\big),$$

and *saturated* if

$$\forall w \in W\, \forall \Diamond\psi \in sub\,\varphi \; \big(\Diamond\psi \in r(w) \;\rightarrow\; \exists v \in W\; \psi \in r(v)\big).$$

We say that a triple $\mathfrak{Q} = \langle W, \boldsymbol{q}, \mathfrak{R} \rangle$ is an **S5** $\times$ **S5**-*quasimodel for* φ if $\langle W, \boldsymbol{q} \rangle$ is a basic structure for φ and $\mathfrak{R}$ is a set of coherent and saturated runs through $\langle W, \boldsymbol{q} \rangle$ such that

(qm2) φ belongs to a type occurring in a quasistate $\boldsymbol{q}(w)$ for some $w \in W$;

(qm3) for every $w \in W$ and every $\boldsymbol{t} \in \boldsymbol{q}(w)$, there is an $r \in \mathfrak{R}$ such that $r(w) = \boldsymbol{t}$.

This definition is illustrated by Fig. 5.7 in which types are represented by • and quasistates by the framed sets of types on the same vertical line.

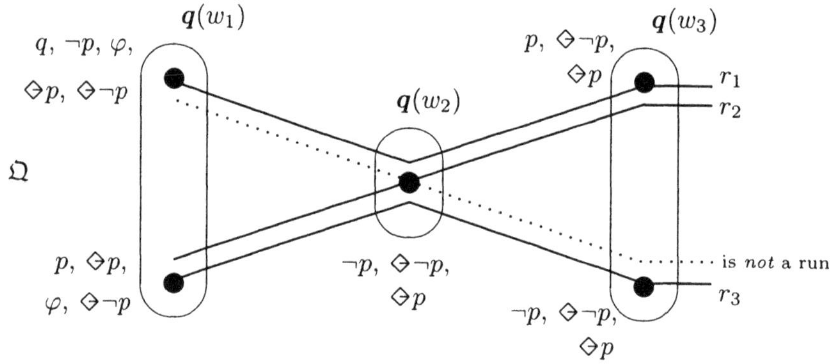

Figure 5.7: An **S5** $\times$ **S5**-quasimodel for $\varphi = \diamond(\diamond p \wedge \diamond \neg p \wedge q)$.

Remark 5.23. It is to be noted that the quasimodel $\mathfrak{Q} = \langle W, \boldsymbol{q}, \mathfrak{R} \rangle$ can be regarded as an abstract model for **S5** $\times$ **S5** satisfying φ. The set of points in this model is

$$U = \{r(w) \mid w \in W, \ r \in \mathfrak{R}\},$$

the accessibility relations R_v and R_h are defined by taking

$$\begin{aligned} r(u) R_v r'(v) \quad &\text{iff} \quad u = v, \\ r(u) R_h r'(v) \quad &\text{iff} \quad r = r'. \end{aligned}$$

(It is easily checked that they satisfy the commutativity and Church–Rosser conditions.) And the valuation of the model is determined by the types:

$$\mathfrak{V}(p) = \{r(w) \in U \mid p \in r(w)\}.$$

An important property of this abstract model is that it can be reconstructed into a product model; see Lemma 5.24.

Recall from Section 3.3 that the product of universal frames $\langle W_1, W_1 \times W_1 \rangle$ and $\langle W_2, W_2 \times W_2 \rangle$ is denoted by $\langle W_1, W_2 \rangle$. The following statement is the 'quasimodel lemma' mentioned above.

Lemma 5.24. *An $\mathcal{ML}_2$-formula φ is satisfiable in a model based on a universal product* **S5** $\times$ **S5***-frame $\langle W_1, W_2 \rangle$ iff there is an* **S5** $\times$ **S5***-quasimodel $\langle W_1, \boldsymbol{q}, \mathfrak{R} \rangle$ for φ.*

Proof. Suppose that we have a model $\mathfrak{M}$ based on $\langle W_1, W_2 \rangle$ and satisfying φ. With every pair $\langle x, y \rangle \in W_1 \times W_2$ we associate the type

$$\boldsymbol{t}(x, y) = \{\psi \in sub\,\varphi \mid (\mathfrak{M}, \langle x, y \rangle) \models \psi\},$$

and with every $x \in W_1$ we associate the quasistate ('vertical type-cluster')

$$\boldsymbol{q}(x) = \{\boldsymbol{t}(x, y) \mid y \in W_2\}.$$

For every $y \in W_2$, define a function r_y by taking, for $x \in W_1$,

$$r_y(x) = \boldsymbol{t}(x, y).$$

Put $\mathfrak{R} = \{r_y \mid y \in W_2\}$. Then clearly $\langle W_1, \boldsymbol{q}, \mathfrak{R} \rangle$ is a quasimodel for φ.

Conversely, suppose $\langle W_1, \boldsymbol{q}, \mathfrak{R} \rangle$ is a quasimodel for φ. Take the universal relations on W_1 and on the set $\mathfrak{R}$ of runs, and let $\mathfrak{F} = \langle W_1 \times \mathfrak{R}, R_h, R_v \rangle$ be the product of these two universal frames: for all $x, x' \in W_1$ and $r, r' \in \mathfrak{R}$,

$$\begin{aligned} \langle x, r \rangle\, R_h\, \langle x', r' \rangle \quad &\text{iff} \quad r = r', \\ \langle x, r \rangle\, R_v\, \langle x', r' \rangle \quad &\text{iff} \quad x = x'. \end{aligned}$$

Observe that

$$\text{if } \langle W_1, \boldsymbol{q}, \mathfrak{R} \rangle \text{ is finite then the product frame } \mathfrak{F} \text{ is finite as well.} \tag{5.15}$$

Let $\mathfrak{V}$ be a valuation in $\mathfrak{F}$ defined by

$$\mathfrak{V}(p) = \{\langle x, r \rangle \mid p \in r(x)\}$$

for every propositional variable p. Put $\mathfrak{M} = \langle \mathfrak{F}, \mathfrak{V} \rangle$. By induction on the construction of $\psi \in sub\,\varphi$ one can readily show that, for every $\langle x, r \rangle$ in $\mathfrak{M}$, we have

$$(\mathfrak{M}, \langle x, r \rangle) \models \psi \quad \text{iff} \quad \psi \in r(x).$$

The basis of induction and the case of Booleans are trivial (here we use the fact that types are Boolean saturated). Let $\psi = \diamondsuit \chi$. We then have

$$\begin{aligned} (\mathfrak{M}, \langle x, r \rangle) \models \diamondsuit \chi \quad &\Longleftrightarrow \quad \exists x' \in W_1\ (\mathfrak{M}, \langle x', r \rangle) \models \chi \\ &\Longleftrightarrow \quad \exists x' \in W_1\ \chi \in r(x') \qquad \text{[by the induction hypothesis]} \\ &\Longleftrightarrow \quad \diamondsuit \chi \in r(x) \qquad \text{[since } r \text{ is coherent and saturated].} \end{aligned}$$

Now let $\psi = \Diamond\chi$. Then

$$\begin{aligned} (\mathfrak{M}, \langle x, r\rangle) \models \Diamond\chi \quad &\Longrightarrow \quad \exists r' \in \mathfrak{R}\ (\mathfrak{M}, \langle x, r'\rangle) \models \chi \\ &\Longrightarrow \quad \exists r' \in \mathfrak{R}\ \chi \in r'(x) \qquad \text{[by the induction hypothesis]} \\ &\Longrightarrow \quad \Diamond\chi \in r(x) \qquad \text{[by } \mathbf{(qm1)}\text{]}. \end{aligned}$$

Conversely,

$$\begin{aligned} \Diamond\chi \in r(x) \quad &\Longrightarrow \quad \exists \boldsymbol{t} \in \boldsymbol{q}(x)\ \chi \in \boldsymbol{t} \qquad \text{[by } \mathbf{(qm1)}\text{]} \\ &\Longrightarrow \quad \exists r' \in \mathfrak{R}\ \chi \in r'(x) \qquad \text{[by } \mathbf{(qm3)}\text{]} \\ &\Longrightarrow \quad \exists r' \in \mathfrak{R}\ (\mathfrak{M}, \langle x, r'\rangle) \models \chi \qquad \text{[by the induction hypothesis]} \\ &\Longrightarrow \quad (\mathfrak{M}, \langle x, r\rangle) \models \Diamond\chi. \end{aligned}$$

It now follows from **(qm2)** that $\mathfrak{M}$ satisfies φ. ❑

We show now that for every $\mathbf{S5} \times \mathbf{S5}$-satisfiable formula φ there is a quasimodel the size of which is effectively bounded in the length of φ. Let $\mathfrak{Q} = \langle W, \boldsymbol{q}, \mathfrak{R}\rangle$ be a quasimodel for φ. Without loss of generality we may assume that each point $w \in W$ has a *twin in* $\mathfrak{Q}$, i.e., a point $w' \in W$ such that $\boldsymbol{q}(w) = \boldsymbol{q}(w')$ and, for all runs $r \in \mathfrak{R}$, we have $r(w) = r(w')$. (Such a 'duplication' of points clearly does not change $\mathfrak{Q}$ being a quasimodel for φ.) We construct a smaller quasimodel $\mathfrak{Q}' = \langle W', \boldsymbol{q}', \mathfrak{R}'\rangle$ out of $\mathfrak{Q}$ in the following way. To begin with, we put in W' a point $w_\varphi \in W$ such that $\boldsymbol{q}(w_\varphi)$ contains a type with φ. Then, for every $\boldsymbol{t} \in \boldsymbol{q}(w_\varphi)$, we fix a run $r_{\boldsymbol{t}}$ such that $r_{\boldsymbol{t}}(w_\varphi) = \boldsymbol{t}$ and, for each $\Diamond\psi \in \boldsymbol{t}$, select a $v \in W$ such that $\psi \in r_{\boldsymbol{t}}(v)$ and put v into W' together with its twin v'. Thus, the resulting W' contains at most

$$2^{|sub\,\varphi|} \cdot 2\,|sub\,\varphi| \tag{5.16}$$

elements. Let $\boldsymbol{q}'$ be the restriction of $\boldsymbol{q}$ to W'. It should be clear from the construction that, for all types $\boldsymbol{t} \in \boldsymbol{q}(w_\varphi)$, the restriction $r'_{\boldsymbol{t}}$ of $r_{\boldsymbol{t}}$ to W' is a coherent and saturated run through $\langle W', \boldsymbol{q}'\rangle$. Let $\mathfrak{S} = \{r'_{\boldsymbol{t}} \mid \boldsymbol{t} \in \boldsymbol{q}(w_\varphi)\}$. Although for every $\boldsymbol{t} \in \boldsymbol{q}(w_\varphi)$ we have a run coming through $\boldsymbol{t}$, $\mathfrak{S}$ is not necessarily big enough to satisfy **(qm3)**, i.e., to contain runs coming through *all* types in $\langle W', \boldsymbol{q}'\rangle$.

To fix this problem, we extend $\mathfrak{S}$ to a larger set $\mathfrak{R}'$ in the following way. We know that for every $v \in W'$ different from w_φ and every type $\boldsymbol{t} \in \boldsymbol{q}'(v)$, there is a run $r_{v,\boldsymbol{t}} \in \mathfrak{R}$ such that $r_{v,\boldsymbol{t}}(v) = \boldsymbol{t}$. By the construction of $\mathfrak{S}$, we have a run $s \in \mathfrak{S}$ with $s(w_\varphi) = r_{v,\boldsymbol{t}}(w_\varphi)$. Define the run $r_{v,\boldsymbol{t}} + s$ through $\langle W', \boldsymbol{q}'\rangle$ by taking, for every $w \in W'$,

$$(r_{v,\boldsymbol{t}} + s)(w) = \begin{cases} \boldsymbol{t}, & \text{if } w = v, \\ s(w), & \text{otherwise.} \end{cases}$$

It is easy to see that $r_{v,t} + s$ is coherent and saturated. Indeed, suppose that $\diamondminus\psi \in sub\,\varphi$ and that there is $w' \in W'$ such that $\psi \in (r_{v,t} + s)(w')$. As $t = r_{v,t}(v)$, $r_{v,t}(w_\varphi) = s(w_\varphi)$ and both $r_{v,t}$ and s are coherent, we have $\diamondminus\psi \in (r_{v,t} + s)(w)$ for all $w \in W'$. Now, suppose that $\diamondminus\psi \in (r_{v,t} + s)(w)$ for some $w \in W'$. Then $\diamondminus\psi \in s(w_\varphi)$. Since $s \in \mathfrak{S}$ and v has a twin in W', it does not really matter that we have changed the saturated run s at v: there is still some $w' \in W'$ such that $w' \neq v$ and $\psi \in s(w') = (r_{v,t} + s)(w')$ (see Fig. 5.8). Thus, we can take $\mathfrak{R}'$ to be the set of all coherent and saturated runs though $\langle W', \boldsymbol{q}' \rangle$.

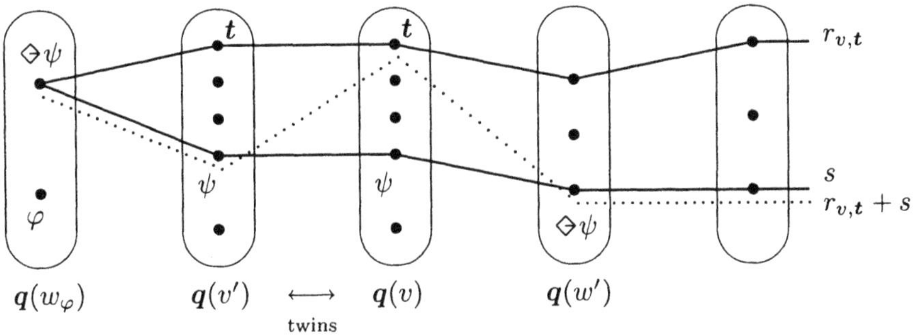

Figure 5.8: Constructing run $r_{v,t} + s$ using twins.

The number of runs in $\mathfrak{R}'$ is at most

$$2^{2|sub\,\varphi|} \cdot 2\,|sub\,\varphi|\,, \tag{5.17}$$

and so we get a quasimodel for φ of effectively bounded size. Since, by Proposition 3.7, **S5** × **S5** is determined by universal product frames and in view of Lemma 5.24, we can conclude that **S5** × **S5** is decidable. ❑

Observe that by (5.15), (5.16) and (5.17) we obtain the following:

Theorem 5.25. **S5** × **S5** *has the product fmp. In particular, each* **S5** × **S5***-satisfiable formula* φ *is satisfiable in a universal product* **S5** × **S5***-frame containing at most*

$$2^{3|sub\,\varphi|} \cdot 4\,|sub\,\varphi|^2$$

points.

The product fmp of **S5** × **S5** follows from Mortimer's (1975) result on the fmp of the two-variable fragment of first-order logic; the exponential fmp of this fragment is shown in (Grädel *et al.* 1997). A short algebraic proof is given in (Andréka and Németi 1994).

Theorem 5.25 provides us with a nondeterministic exponential time algorithm for satisfiability checking in universal product **S5** × **S5**-frames. Indeed, given a formula φ, we first guess an **S5** × **S5**-model of exponential size in the length of φ and then check whether φ is satisfied in it. So we have:

Theorem 5.26. *The satisfiability problem for* **S5** × **S5** *is in* NEXPTIME, *and so the decision problem for* **S5** × **S5** *is in* coNEXPTIME.

Compared with the polynomial (in fact, linear) upper bound for the size of satisfying **S5**-frames (see Theorem 1.16), the upper bound obtained in Theorem 5.25 may appear too high. In Section 5.5 we will show that actually it cannot be significantly reduced.

5.3 The finite model property

We begin by showing that a variant of the 'good old' *filtration method*, known in modal logic since the 1940s (for history and references consult, e.g., Chagrov and Zakharyaschev 1997), can also be used to establish decidability (and the fmp) of some products with **S5**. The following theorem is due to Gabbay and Shehtman (1998):

Theorem 5.27. *For every logic L in the list* $\mathbf{K}_n$, $\mathbf{T}_n$, $\mathbf{D}_n$, $\mathbf{K4}_n$, $\mathbf{S4}_n$, $\mathbf{KD45}_n$, $\mathbf{S5}_n$, *and every $n \geq 1$, the product $L \times$* **S5** *has the 2-exponential (abstract) fmp.*

Proof. We illustrate the method by giving a proof for **K4** × **S5**. The generalization to $\mathbf{K4}_n \times \mathbf{S5}$ for $n > 1$, as well as the other cases, is similar and left to the reader.

By Theorem 5.9, we know that $\mathbf{K4} \times \mathbf{S5} = [\mathbf{K4}, \mathbf{S5}]$. Suppose $\varphi \notin [\mathbf{K4}, \mathbf{S5}]$ for some $\mathcal{ML}_2$-formula φ. Then, by Proposition 5.7, there exists a model $\mathfrak{M} = \langle \mathfrak{F}, \mathfrak{V} \rangle$ refuting φ and based on a 2-frame $\mathfrak{F} = \langle W, R_h, R_v \rangle$ such that R_h is transitive, R_v is an equivalence relation, and R_h and R_v commute (the Church–Rosser property follows from commutativity in this case).

For each world x in W, let

$$\Sigma(x) = \{\psi \in sub\,\varphi \mid (\mathfrak{M}, x) \models \psi\}.$$

Define an equivalence relation $\sim$ on W by taking, for all $x, y \in W$,

$$x \sim y \quad \text{iff} \quad \Sigma(x) = \Sigma(y) \ \text{ and } \ \{\Sigma(z) \mid xR_v z\} = \{\Sigma(z) \mid yR_v z\}.$$

Now we construct a new model $\mathfrak{M}^\sim = \langle \mathfrak{F}^\sim, \mathfrak{V}^\sim \rangle$ based on $\mathfrak{F}^\sim = \langle W^\sim, R_h^\sim, R_v^\sim \rangle$ as follows:

- $W^\sim = \{[x] \mid x \in W\}$, where $[x]$ denotes the $\sim$-equivalence class of x;

- for all $x, y \in W$,

$$[x]R_v^\sim[y] \quad \text{iff} \quad \exists x'\exists y'\ (x' \sim x,\ y' \sim y \text{ and } x'R_v y');$$

- $R_h^\sim$ is the transitive closure of the relation $R_h^\bullet$ defined by taking, for all $x, y \in W$,

$$[x]R_h^\bullet[y] \quad \text{iff} \quad \exists x'\exists y'\ (x' \sim x,\ y' \sim y \text{ and } x'R_h y');$$

- $\mathfrak{V}^\sim(p) = \{[x] \mid x \in \mathfrak{V}(p)\}$, for all $p \in sub\,\varphi$, and $\mathfrak{V}^\sim(q) = \emptyset$, for all other propositional variables q.

Observe first that each world $[x]$ in $\mathfrak{M}^\sim$ is uniquely determined by the pair $\langle \Sigma(x), \{\Sigma(z) \mid xR_v z\}\rangle$ of sets. So we have

$$|W^\sim| \leq 2^{|sub\,\varphi|} \cdot 2^{2^{|sub\,\varphi|}}.$$

We will show now that

(1) $\mathfrak{M}^\sim$ refutes φ, and

(2) $\langle W^\sim, R_h^\sim, R_v^\sim\rangle$ is a frame for $[\mathbf{K4}, \mathbf{S5}]$.

Claim (1) follows from the fact that $\mathfrak{M}^\sim$ is a *filtration* of $\mathfrak{M}$ in the sense that, for all $x, y \in W$, the following two conditions hold:

(f1) if $xR_h y$ then $[x]R_h^\sim[y]$, and if $xR_v y$ then $[x]R_v^\sim[y]$,

(f2) if $[x]R_h^\sim[y]$ then, for all ψ,

$$\text{if } \boxminus\psi \in sub\,\varphi \text{ and } (\mathfrak{M}, x) \models \boxminus\psi \text{ then } (\mathfrak{M}, y) \models \psi, \text{ and}$$

if $[x]R_v^\sim[y]$ then, for all ψ,

$$\text{if } \Box\psi \in sub\,\varphi \text{ and } (\mathfrak{M}, x) \models \Box\psi \text{ then } (\mathfrak{M}, y) \models \psi.$$

The proofs are straightforward and left to the reader. ($R_v^\sim$ and $R_h^\sim$ are known as the *least filtration* and the *Lemmon* (or *least transitive*) *filtration*, respectively; see e.g., (Chagrov and Zakharyaschev 1997, Goldblatt 1987).) By induction on the construction of ψ, the reader can readily check that for every $\psi \in sub\,\varphi$ and every $x \in W$,

$$(\mathfrak{M}, x) \models \psi \quad \text{iff} \quad (\mathfrak{M}^\sim, [x]) \models \psi,$$

which yields (1).

To prove (2), observe first that $R_h^\sim$ is transitive by definition. Further, it follows easily from the definition of $R_v^\sim$ and the corresponding properties of R_v that $R_v^\sim$ is reflexive and symmetric. In order to show transitivity of $R_v^\sim$, we first prove that the equivalence relations $\sim$ and R_v commute, that is, for all x, y in W,

$$\exists z\; x \sim zR_v y \qquad \text{iff} \qquad \exists u\; xR_v u \sim y. \tag{5.18}$$

Clearly, it is enough to show only one direction, since the other follows by taking the converse and using the fact that both $\sim$ and R_v are symmetric. So suppose $x \sim zR_v y$. By the definition of $\sim$, there exists a u such that $xR_v u$ and $\Sigma(u) = \Sigma(y)$. We claim that $u \sim y$ holds. Now we make essential use of the fact that R_v is an equivalence relation (actually, a transitive and symmetric, or a transitive and Euclidean R_v would suffice here):

$$\{w \mid uR_v w\} = \{w \mid xR_v w\} \qquad \text{and} \qquad \{w \mid zR_v w\} = \{w \mid yR_v w\}.$$

Since $x \sim z$, we have that $\{\Sigma(w) \mid uR_v w\} = \{\Sigma(w) \mid yR_v w\}$, as required.

Now we can easily obtain the transitivity of $R_v^\sim$. Suppose $[x]R_v^\sim[y]R_v^\sim[z]$. So there are $x', y', y'', z' \in W$ such that

$$x \sim \underline{x'R_v y' \sim y \sim y''}R_v z' \sim z.$$

Then $y' \sim y''$ and, by (5.18), there is a u such that $x' \sim uR_v y''$, which implies $[x]R_v^\sim[z]$ by the transitivity of R_v.

Finally, we prove that $R_h^\sim$ and $R_v^\sim$ commute. Since $R_h^\sim$ is the transitive closure of $R_h^\bullet$, it clearly suffices to show that $R_h^\bullet$ and $R_v^\sim$ commute. Suppose first that $[x]R_h^\bullet[y]R_v^\sim[z]$. Then there are $x', y', y'', z' \in W$ such that

$$x \sim x'R_h \underline{y' \sim y \sim y''R_v z'} \sim z.$$

By (5.18), there is a u such that $y'R_v u \sim z'$, and so we obtain

$$x \sim x'R_v wR_h u \sim z' \sim z$$

for some w, since R_h and R_v commute. Thus we have $[x]R_v^\sim[w]R_h^\bullet[z]$, as required. The other direction is similar and left to the reader. ❑

As a consequence we immediately obtain:

Theorem 5.28. *Suppose $L \in \{\mathbf{K}_n, \mathbf{T}_n, \mathbf{D}_n, \mathbf{K4}_n, \mathbf{S4}_n, \mathbf{KD45}_n, \mathbf{S5}_n\}$. Then the decision problem for $L \times \mathbf{S5}$ is in* coN2EXPTIME.

The interested reader can find various generalizations of Theorem 5.27 in (Gabbay and Shehtman 1998).[3] Further generalizations will be discussed in

[3] In (Gabbay and Shehtman 1998) the related theorems are stated for products with $\mathbf{S5}_m$ (not only with **S5**). However, for $m > 1$, there is a gap in the proof of Proposition 12.5, item (2.1). In fact, the theorem itself does not seem to hold for $m > 1$, cf. Theorems 6.71 and 6.72, and the proof of Lemma 6.47 below.

Section 6.5. Note, however, that all these are about products where one of the components is **S5**. No filtration argument is known to the authors that works for other types of products. In particular, the following problem is open:

Question 5.29. Find 'natural' unimodal logics L_1 and L_2 such that both L_1 and L_2 are finitely axiomatizable, have the fmp (and hence are decidable), their product $L_1 \times L_2$ also has the fmp, but is undecidable. (This would mean that there is no algorithm capable of deciding whether a finite frame is a frame for $L_1 \times L_2$. In particular, $L_1 \times L_2$ would not be finitely axiomatizable.)

Note that the bimodal $L_1 = \mathbf{K} \times \mathbf{K}$ and $L_2 = \mathbf{K}$ satisfy these properties (see Corollary 5.11, Theorems 5.5, 8.24 and 8.28). Using the results of (Kracht and Wolter 1999) on Thomason's (1974b, 1975) reduction of bimodal logics to unimodal ones, it is not difficult to construct a pair of appropriate unimodal logics. However, the resulting logics are rather 'artificial.' The reader will find some related open questions in Chapters 6 and 7.

On the other hand, there exist logics L such that L has the fmp but $L \times \mathbf{S5}$ does not, for instance, $L = \mathbf{S5} \times \mathbf{S5}$ (see Corollary 5.11, Theorems 5.9, 5.25 and 8.12). Another example was given by Reynolds (1997) who proved that $\mathbf{Lin} \times \mathbf{S5}$ has no fmp. Here we use Reynold's idea to show the following more general result:

Theorem 5.30. *Suppose that $\mathcal{C}$ is a class of linear orders containing either $\langle \mathbb{N}, < \rangle$ or $\langle \mathbb{Z}, < \rangle$. Suppose also that L is a Kripke complete unimodal logic*[4] *having an infinite frame $\langle W, R \rangle$ with a point $x \in W$ such that xRy, for all $y \in W$, $y \neq x$. Then $\mathsf{Log}_{FP}\,\mathcal{C} \times L$ does not have the (abstract) fmp.*

Proof. Consider the formulas ψ_n, $n < \omega$, defined inductively by taking

$$\psi_0 = q \wedge \Box_P \neg q,$$

$$\psi_{n+1} = \Diamond_P \psi_n \wedge \Box_P \Box_P \neg \psi_n.$$

It is not hard to see that, for every $n < \omega$, we have

$$\psi_n \to \Box_F \neg \psi_n \in \mathsf{Log}_{FP}\,\mathcal{C} \times L. \tag{5.19}$$

Let

$$\chi = \psi_0 \wedge \Diamond_F \psi_1 \wedge \Box_F \diamondsuit(\psi_0 \wedge \Diamond_F \psi_1).$$

Let $\mathfrak{F}$ be either $\langle \mathbb{N}, < \rangle$ or $\langle \mathbb{Z}, < \rangle$, and let $\mathfrak{G} = \langle W, R \rangle$ be an infinite frame for L such that there is an $x_0 \in W$ with $x_0 R y$, for all $y \in W$, $y \neq x_0$. Take an arbitrary enumeration $\{x_1, x_2, \dots\}$ of a countably infinite subset of $W - \{x_0\}$,

[4] Almost all logics considered in this book satisfy the condition on L, e.g., **K**, **K4**, **S5**, **S4.3**, $\mathsf{Log}\{\langle \mathbb{N}, < \rangle\}$, **GL**, **Grz.3**, etc. (a counterexample is **Alt**).

and define a valuation $\mathfrak{V}$ in $\mathfrak{F} \times \mathfrak{G}$ by taking $\mathfrak{V}(q) = \{\langle n, x_n \rangle \mid n \in \mathbb{N}\}$. Put $\mathfrak{M} = \langle \mathfrak{F} \times \mathfrak{G}, \mathfrak{V} \rangle$. It is not hard to compute that $(\mathfrak{M}, \langle 0, x_0 \rangle) \models \chi$.

On the other hand, assume that $\mathfrak{F} = \langle W, R_h, R_v \rangle$ is a frame for $\mathsf{Log}_{FP}\, \mathcal{C} \times L$ and that $\mathfrak{M}$ is a model based on $\mathfrak{F}$ such that $(\mathfrak{M}, w) \models \chi$ for some $w \in W$. We show that then $\mathfrak{F}$ must be infinite.

Let $w_0 = w$. Define inductively, for every $n > 0$, $n \in \mathbb{N}$, a world w_n in $\mathfrak{F}$ such that the following hold:

(i) $w_0 R_h w_n$, and

(ii) $(\mathfrak{M}, w_n) \models \psi_n \wedge \bigwedge_{k<n} \neg\psi_k$.

Since, by (ii), all w_n should be different, this will prove the infinity of $\mathfrak{F}$. To begin with, as $(\mathfrak{M}, w_0) \models \Diamond_F \psi_1$, there is a w_1 such that $w_0 R_h w_1$ and $(\mathfrak{M}, w_1) \models \psi_1$. By (5.19), we also have $(\mathfrak{M}, w_1) \models \neg\psi_0$. Now assume that, for all $k \leq n$, points w_k satisfying (i) and (ii) have already been defined. Then, by the induction hypothesis, we have

$$(\mathfrak{M}, w_n) \models \diamondsuit(\psi_0 \wedge \Diamond_F \psi_1) \wedge \psi_n. \tag{5.20}$$

Claim 5.31. $\diamondsuit(\psi_0 \wedge \Diamond_F \psi_1) \wedge \psi_n \to \Diamond_F \psi_{n+1} \in \mathsf{Log}_{FP}\, \mathcal{C} \times L$.

Proof. Suppose $(\mathfrak{N}, \langle u, v \rangle) \models \diamondsuit(\psi_0 \wedge \Diamond_F \psi_1) \wedge \psi_n$ for some model $\mathfrak{N}$ based on the product of a rooted frame $\langle U, < \rangle$ for $\mathsf{Log}_{FP}\, \mathcal{C}$ and a frame $\langle V, S \rangle$ for L. Then there are $v' \in V$, $u' \in U$ such that vSv', $u < u'$, $(\mathfrak{N}, \langle u, v' \rangle) \models \psi_0$ and

$$(\mathfrak{N}, \langle u', v' \rangle) \models \Diamond_P \psi_0 \wedge \Box_P \Box_P \neg\psi_0.$$

It follows that

$$\text{there is no } x \in U \text{ such that } u < x < u'. \tag{5.21}$$

We claim that $(\mathfrak{N}, \langle u', v \rangle) \models \psi_{n+1}$. Clearly, $(\mathfrak{N}, \langle u', v \rangle) \models \Diamond_P \psi_n$. And, by (5.19) and (5.21), we also have $(\mathfrak{N}, \langle u', v \rangle) \models \Box_P \Box_P \neg\psi_n$ (here we use the fact that $<$ is transitive and weakly connected). ❑

Thus, by (5.20), we have $(\mathfrak{M}, w_n) \models \Diamond_F \psi_{n+1}$, and so there is some w_{n+1} such that $w_n R_h w_{n+1}$ and $(\mathfrak{M}, w_{n+1}) \models \psi_{n+1}$. Finally, by the induction hypothesis, the transitivity of R_h and (5.19), we obtain

$$(\mathfrak{M}, w_{n+1}) \models \bigwedge_{k<n+1} \neg\psi_k,$$

as required. ❑

We have already seen in Section 5.2 that by the quasimodel technique one can show not only the fmp, but the stronger product fmp for **S5** × **S5**, and also obtain a better, coNEXPTIME, upper bound for its complexity. In Section 6.5 similar results will be proved for **K** × **S5** and **K** × **KD45** as well.

On the other hand, the next theorem shows that products like **K4** × **S5** and **S4** × **S5** do not enjoy the product fmp.

Given a frame $\langle W, R\rangle$, we call a sequence $\langle x_n \mid n < \omega\rangle$ of distinct points from W an *infinite ascending chain* if $x_0 R x_1 R x_2 R \dots$. If in addition we have $\langle x_i, x_j\rangle \notin R$, whenever $j < i$, then we call $\langle x_n \mid n < \omega\rangle$ an *ascending ω-type chain*. For instance, Fr**S5** contains frames having infinite ascending chains, but none of them have ascending ω-type chains.

Theorem 5.32. *Let $\mathcal{C}$ be a class of transitive frames at least one of which contains an ascending ω-type chain. Suppose also that L is a Kripke complete unimodal logic having an infinite frame $\langle W, R\rangle$ with a point $x \in W$ such that xRy, for all $y \in W$, $y \neq x$. Then* $\mathsf{Log}\,\mathcal{C} \times L$ *does not have the product fmp.*

Proof. Consider the formula

$$\varphi = \boxminus^+ \diamondvert p \wedge \boxminus^+ \boxbar (p \to \diamondminus \boxminus^+ \neg p).$$

Let $\mathfrak{F}$ be a frame in $\mathcal{C}$ containing an ascending ω-type chain $\langle x_n \mid n < \omega\rangle$. Let $\mathfrak{G} = \langle W, R\rangle$ be an infinite frame for L such that there is an $x \in W$ with xRy, for all $y \in W$, $y \neq x$. Take an arbitrary enumeration $\{y_0, y_1, \dots\}$ of a countably infinite subset of $W - \{x\}$, and define a valuation $\mathfrak{V}$ in $\mathfrak{F} \times \mathfrak{G}$ by taking $\mathfrak{V}(p) = \{\langle x_n, y_n\rangle \mid n < \omega\}$. Put $\mathfrak{M} = \langle \mathfrak{F} \times \mathfrak{G}, \mathfrak{V}\rangle$. It is not hard to compute that $(\mathfrak{M}, \langle x_0, y_0\rangle) \models \varphi$. On the other hand, it is readily checked that φ is not satisfiable in any finite product frame where the first component is transitive; see Fig. 5.9. ❑

There are products without the product fmp for which Theorem 5.32 does not apply. The following two theorems cover more cases.

Theorem 5.33. *Let $\mathcal{C}$ be a class of transitive frames at least one of which contains an infinite ascending chain. Then neither of the logics* $\mathsf{Log}\,\mathcal{C} \times$ **GL.3** *and* $\mathsf{Log}\,\mathcal{C} \times$ **Grz.3** *has the product fmp.*

Proof. Consider the formula

$$\psi = \boxminus^+ \diamondvert p \wedge \boxminus^+ \boxbar (p \to \diamondminus \boxbar^+ \neg p).$$

Let $\mathfrak{F}$ be a frame in $\mathcal{C}$ containing an infinite ascending chain $\langle x_n \mid n < \omega\rangle$ and let $\mathfrak{G}$ be either $\langle\{0, 1, \dots, \omega\}, >\rangle$ or $\langle\{0, 1, \dots, \omega\}, \geq\rangle$. Define a valuation $\mathfrak{V}$ in $\mathfrak{F} \times \mathfrak{G}$ by taking $\mathfrak{V}(p) = \{\langle x_n, n\rangle \mid n \in \mathbb{N}\}$. Put $\mathfrak{M} = \langle \mathfrak{F} \times \mathfrak{G}, \mathfrak{V}\rangle$. It is not hard to compute that $(\mathfrak{M}, \langle x_0, \omega\rangle) \models \psi$. On the other hand, it is readily

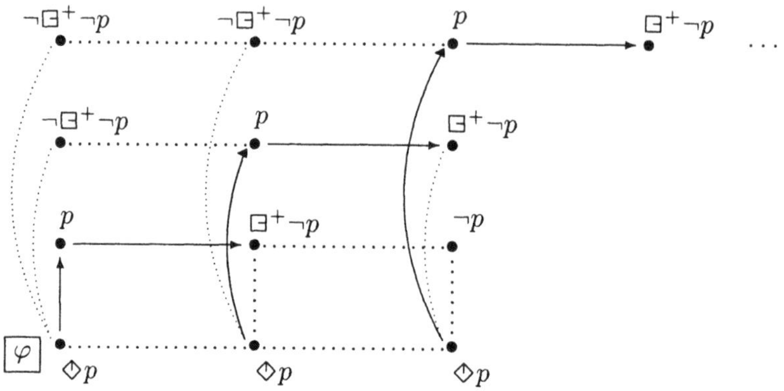

Figure 5.9: φ is not satisfiable in any finite product frame where the first component is transitive.

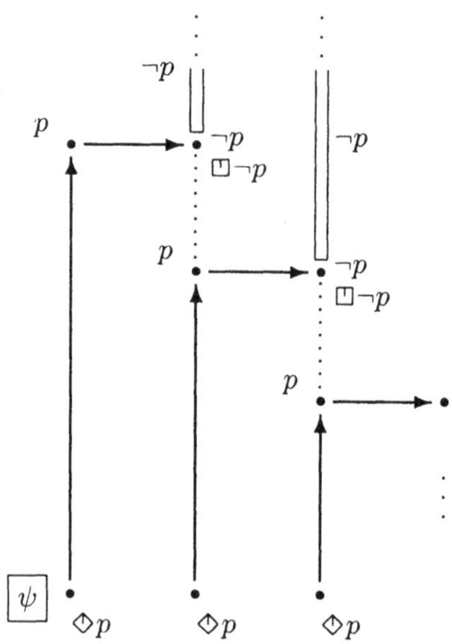

Figure 5.10: ψ is not satisfiable in any finite product frame where the first component is transitive and the second component is weakly connected.

checked that ψ is not satisfiable in any finite product frame, where the first component is transitive and the second component is weakly connected; see Fig. 5.10. ❑

Note that none of the logics **GL.3** $\times$ **GL.3**, **Grz.3** $\times$ **Grz.3**, **GL.3** $\times$ **Grz.3** has the product fmp either, see Theorem 7.10.

Theorem 5.34. *Let L be any Kripke complete unimodal logic having an infinite frame $\langle W, R \rangle$ with a point $x \in W$ such that xRy, for all $y \in W$, $y \neq x$. Then $\mathbf{K}_u \times L$ does not have the product fmp.*

Proof. Take the formula

$$\boxed{\forall}\Diamond_v p \wedge \boxed{\forall}\Box_v(p \to \Diamond_h \neg p) \wedge \boxed{\forall}\Box_v(\neg p \to \Box_h \neg p),$$

and repeat the previous proof. ❑

Note again that almost all unimodal logics we consider in the book satisfy the condition on L formulated in Theorems 5.30, 5.32 and 5.34.

We conclude this section with the following observation which will be used in Section 14.4.

Proposition 5.35. *Suppose that L_1 and L_2 are Kripke complete unimodal logics and L_1 has the fmp. Then $L_1 \times L_2$ has the product fmp iff $L_1 \times L_2$ is determined by product frames of the form $\mathfrak{F}_1 \times \mathfrak{F}_2$, where $\mathfrak{F}_1$ is a frame for L_1 and $\mathfrak{F}_2$ is a finite frame for L_2.*

Proof. Obviously, if $L_1 \times L_2$ is determined by finite product frames, then it is determined by product frames in which the second component is finite. Conversely, assume that L_1 has the fmp and that $L_1 \times L_2$ is determined by product frames the second component of which is finite. Suppose that $\varphi \in \mathcal{ML}_2$ and $\mathfrak{G} = \langle W, R \rangle$ is a finite frame. We are going to encode 'the behavior of $\mathfrak{G}$ as a second component of a product frame' by means of an $\mathcal{ML}$-formula (having modal operator $\Diamond_h$). To this end, for every $w \in W$ and every $\psi \in sub\,\varphi \cup \{\neg\varphi\}$, introduce a new propositional variable $p_{w,\psi}$. Denote by $\mathsf{code}_{\mathfrak{G}}$ the conjunction of the following $\mathcal{ML}$-formulas, for all $w \in W$:

$$\bigwedge_{\psi_1 \wedge \psi_2 \in sub\,\varphi} p_{w,\psi_1 \wedge \psi_2} \leftrightarrow (p_{w,\psi_1} \wedge p_{w,\psi_2}),$$

$$\bigwedge_{\neg\psi \in sub\,\varphi \cup \{\neg\varphi\}} p_{w,\neg\psi} \leftrightarrow \neg p_{w,\psi},$$

$$\bigwedge_{\Diamond_h \psi \in sub\,\varphi} p_{w,\Diamond_h \psi} \leftrightarrow \Diamond_h p_{w,\psi},$$

$$\bigwedge_{\Diamond_v \psi \in sub\,\varphi} (p_{w,\Diamond_v \psi} \leftrightarrow \bigvee_{wRv} p_{v,\psi}).$$

We claim that, for all frames $\mathfrak{F}$, the following conditions are equivalent:

(i) $\neg\varphi$ is satisfiable in $\mathfrak{F} \times \mathfrak{G}$;

(ii) there exists a $w \in W$ such that $p_{w,\neg\varphi} \land \boxminus^{\leq md(\varphi)}\mathsf{code}_{\mathfrak{G}}$ is satisfiable in $\mathfrak{F}$.

Indeed, suppose $(\mathfrak{M}, \langle v_0, w_0\rangle) \models \varphi$ for some model $\mathfrak{M}$ based on $\mathfrak{F} \times \mathfrak{G}$. Define a valuation $\mathfrak{V}$ in $\mathfrak{F} = \langle V, S\rangle$ by setting, for all $w \in W$ and $\psi \in sub\,\varphi \cup \{\neg\varphi\}$,

$$\mathfrak{V}(p_{w,\psi}) = \{v \in V \mid (\mathfrak{M}, \langle v, w\rangle) \models \psi\},$$

and let $\mathfrak{M}' = \langle \mathfrak{F}, \mathfrak{V}\rangle$. It is readily seen that

$$(\mathfrak{M}', v_0) \models p_{w_0,\neg\varphi} \land \boxminus^{\leq md(\varphi)}\mathsf{code}_{\mathfrak{G}}.$$

Conversely, suppose $p_{w,\neg\varphi} \land \boxminus^{\leq md(\varphi)}\mathsf{code}_{\mathfrak{G}}$ is satisfied in a model $\mathfrak{M} = \langle \mathfrak{F}, \mathfrak{V}\rangle$, for some $w \in W$. Define a valuation $\mathfrak{V}'$ in $\mathfrak{F} \times \mathfrak{G}$ by setting, for all propositional variables $p \in sub\,\varphi$,

$$\mathfrak{V}'(p) = \{\langle v, w\rangle \mid v \in \mathfrak{V}(p_{w,p})\}.$$

It is readily checked that $\neg\varphi$ is satisfied in the model $\langle \mathfrak{F} \times \mathfrak{G}, \mathfrak{V}'\rangle$.

Now suppose $\varphi \notin L_1 \times L_2$. By assumption, we find a frame $\mathfrak{F} \in \mathsf{Fr}L_1$ and a finite frame $\mathfrak{G} \in \mathsf{Fr}L_2$ such that $\mathfrak{F} \times \mathfrak{G}$ refutes φ. Then $p_{w,\neg\varphi} \land \boxminus^{\leq md(\varphi)}\mathsf{code}_{\mathfrak{G}}$ is satisfied in $\mathfrak{F}$, for some point w in $\mathfrak{G}$. Since L_1 has the fmp, there is a finite frame $\mathfrak{F}' \in \mathsf{Fr}L_1$ which satisfies $p_{w,\neg\varphi} \land \boxminus^{\leq md(\varphi)}\mathsf{code}_{\mathfrak{G}}$. Therefore, φ is refuted in the finite product frame $\mathfrak{F}' \times \mathfrak{G}$. ❑

5.4 Proving undecidability

The aim of this section is to demonstrate on simple examples the three basic techniques of establishing undecidability we shall use later on in this book.

Undecidability by tiling

The following $\mathbb{N} \times \mathbb{N}$ *tiling problem* is known to be undecidable (see Berger 1966, Robinson 1971, Börger *et al.* 1997). Given a finite set T of *tile types*, which are 4-tuples of colors

$$t = \langle \mathit{left}(t), \mathit{right}(t), \mathit{up}(t), \mathit{down}(t)\rangle\,,$$

decide whether T *tiles* the grid $\mathbb{N} \times \mathbb{N}$, i.e., whether there exists a function (called a *tiling*) τ from $\mathbb{N} \times \mathbb{N}$ to T such that, for all $i, j \in \mathbb{N}$,

- $up(\boldsymbol{\tau}(i,j)) = down(\boldsymbol{\tau}(i,j+1))$ and
- $right(\boldsymbol{\tau}(i,j)) = left(\boldsymbol{\tau}(i+1,j))$.

If we think of a tile as a physical 1×1-square with colors along its four edges, then a tiling $\boldsymbol{\tau}$ of $\mathbb{N} \times \mathbb{N}$ is just a way of placing tiles, each of a type from T, together to cover the $\mathbb{N} \times \mathbb{N}$ grid, with no rotation of the tiles allowed and the colors on adjacent edges of adjacent tiles matching. The reader may find a useful survey of various tiling problems in (van Emde Boas 1997).

We are going to use this tiling problem to prove the following result of Marx (1999):

Theorem 5.36. *The consequence relation* $\vdash^*_{\mathbf{K}} \times \vdash^*_{\mathbf{K}}$ *(and so, by Theorem 5.12, the global consequence relation* $\vdash^*_{\mathbf{K}\times\mathbf{K}}$*) is undecidable.*

Proof. Given a finite set T of tile types, we associate with every $t \in T$ a propositional variable p_t. Using these variables, we then construct a formula φ_T as the conjunction of the following formulas:

$$\bigvee_{t\in T} p_t \wedge \bigwedge_{\substack{t,t'\in T\\ t\neq t'}} \neg(p_t \wedge p_{t'}), \tag{5.22}$$

$$\bigwedge_{t\in T} \Big(p_t \to \bigvee_{up(t)=down(t')} \boxminus p_{t'}\Big), \tag{5.23}$$

$$\bigwedge_{t\in T} \Big(p_t \to \bigvee_{right(t)=left(t')} \boxminus p_{t'}\Big), \tag{5.24}$$

$$\diamondminus\top \wedge \diamondminus\top. \tag{5.25}$$

We show that φ_T is true in a model $\mathfrak{M}$ based on a frame of the form $\mathfrak{F}_1 \times \mathfrak{F}_2$ iff T tiles $\mathbb{N} \times \mathbb{N}$.

($\Rightarrow$) Suppose that $\mathfrak{M}$ is based on a product frame $\mathfrak{F} = \langle W, R_h, R_v\rangle$ and $\mathfrak{M} \models \varphi_T$. By (5.22), for every $x \in W$ there is precisely one variable p_t, $t \in T$, such that $(\mathfrak{M}, x) \models p_t$. And in view of (5.25) every point has both R_h- and R_v-successors.

Now, fix some $x_{00} \in W$ and take infinite sequences

$$x_{00} R_h x_{10} R_h \dots R_h x_{n0} R_h \dots$$
$$x_{00} R_v x_{01} R_v \dots R_v x_{0n} R_v \dots$$

By the Church–Rosser property, we then have points $x_{ij} \in W$, for all $i,j \in \mathbb{N}$, such that

$$x_{ij} R_v x_{i(j+1)} R_h x_{(i+1)(j+1)} \quad \text{and} \quad x_{ij} R_h x_{(i+1)j} R_v x_{(i+1)(j+1)}.$$

Define a map τ from $\mathbb{N} \times \mathbb{N}$ to T by taking

$$\tau(i,j) = t \qquad \text{iff} \qquad (\mathfrak{M}, x_{ij}) \models p_t.$$

Formulas (5.23) and (5.24) ensure color matching, so that τ is a tiling of $\mathbb{N} \times \mathbb{N}$.

($\Leftarrow$) Suppose τ is a tiling of $\mathbb{N} \times \mathbb{N}$ with T. The reader can readily check that φ_T is true in the model $\mathfrak{M} = \langle \langle \mathbb{N} \times \mathbb{N}, R_h, R_v \rangle, \mathfrak{V} \rangle$, where

$$\begin{array}{lcl} \langle i,j \rangle \, R_h \, \langle k,l \rangle & \text{iff} & k = i+1, \ l = j, \\ \langle i,j \rangle \, R_v \, \langle k,l \rangle & \text{iff} & k = i, \ l = j+1, \end{array}$$

and

$$\mathfrak{V}(p_t) = \{ \langle i,j \rangle \mid \tau(i,j) = t \}.$$

Thus, we have proved:

$$T \text{ tiles } \mathbb{N} \times \mathbb{N} \qquad \text{iff} \qquad \text{not } \varphi_T \, (\vdash^*_{\mathbf{K}} \times \vdash^*_{\mathbf{K}}) \, \bot.$$

Since φ_T is effectively constructed from T, it follows that $\vdash^*_{\mathbf{K}} \times \vdash^*_{\mathbf{K}}$ is undecidable. ❑

As we shall see in Section 6.1, the validity problem (and thus the local consequence relation) for $\mathbf{K} \times \mathbf{K}$ is decidable. Note also that, by Lemma 1.24, $\mathbf{K} \times \mathbf{K}$ enriched with the universal modalities (that is, $(\mathbf{K} \times \mathbf{K})_u$) is undecidable.

In Chapter 14 we will need the following consequence of Theorem 5.36:

Theorem 5.37. $\mathbf{K}_u \times \mathbf{K}_u$ *is undecidable.*

Proof. Denote the two universal boxes of $\mathbf{K}_u \times \mathbf{K}_u$ by $\boxed{\forall}_1$ and $\boxed{\forall}_2$. It is not hard to show that, for any two formulas φ and ψ in the language of $\mathbf{K} \times \mathbf{K}$,

$$\varphi \, (\vdash^*_{\mathbf{K}} \times \vdash^*_{\mathbf{K}}) \, \psi \qquad \text{iff} \qquad \boxed{\forall}_1 \boxed{\forall}_2 \varphi \to \psi \in \mathbf{K}_u \times \mathbf{K}_u.$$

Details are left to the reader. ❑

Undecidability by Turing machines

We assume that the reader knows (at least at an intuitive level) what a Turing machine is, so we just fix the notation and terminology to be used later on.

A *single-tape right-infinite deterministic Turing machine* $\boldsymbol{A}$ is given by

- a finite set S of *states* containing, in particular, the *initial state* s_0 and the *halt state* s_1 (such that $s_0 \neq s_1$),

- a *tape alphabet* A ($b \in A$ stands for blank), and
- a *transition function* ϱ (or a set of instructions).

Configurations of $\boldsymbol{A}$ will be represented by infinite sequences (words) of the form

$$\langle \pounds, a_1, \ldots, a_i, \ldots, a_n, b, \ldots \rangle,$$

where $\pounds \notin A$ is a symbol marking the left end of the tape, all $a_1, \ldots, a_n$ save one, say a_i, are in A, while a_i belongs to $S \times A$ and represents the active cell and the current state (all cells of the tape located to the right of a_n are blank, i.e., contain b). If the machine starts on the empty tape (all cells of which are blank), then the *start configuration* is represented by the word

$$\langle \pounds, \langle s_0, b \rangle, b, \ldots \rangle.$$

The transition function

$$\varrho : (S - \{s_1\}) \times (A \cup \{\pounds\}) \to S \times (A \cup \{\mathsf{L}, \mathsf{R}\})$$

transforms each pair of the form $\langle s, a \rangle$ into one of the following pairs:

- $\langle s', a' \rangle$ (write a' and come to state s'),
- $\langle s', \mathsf{L} \rangle$ (move one cell left and come to state s'),
- $\langle s', \mathsf{R} \rangle$ (move one cell right and come to state s'),

where L and R are fresh symbols. If $a = \pounds$ (i.e., the leftmost cell of the tape is active) then we assume that $\varrho(s, a) = \langle s', \mathsf{R} \rangle$ (that is, having reached the leftmost cell, the machine always moves to the right). According to this definition, the machine always makes another step whenever the current state is different from s_1.

Another important observation, which will help us to simulate the behavior of Turing machines via modal formulas, is that only the active cell and its neighbors can be changed by the transition to the next configuration, while all other cells remain the same. To make this property of Turing machines explicit, we represent the transition function ϱ as a function δ defined on triples of the form $\langle a_i, \langle s, a_j \rangle, a_k \rangle$, for $a_i \in A \cup \{\pounds\}$, $a_j, a_k \in A$, $s \in S - \{s_1\}$, by taking

$$\delta(a_i, \langle s, a_j \rangle, a_k) = \begin{cases} \langle a_i, \langle s', a_j' \rangle, a_k \rangle, & \text{if } \varrho(s, a_j) = \langle s', a_j' \rangle, \\ \langle \langle s', a_i \rangle, a_j, a_k \rangle, & \text{if } \varrho(s, a_j) = \langle s', \mathsf{L} \rangle \text{ and } a_i \neq \pounds, \\ \langle \pounds, \langle s', a_j \rangle, a_k \rangle, & \text{if } \varrho(s, a_j) = \langle s', \mathsf{L} \rangle \text{ and } a_i = \pounds, \\ \langle a_i, a_j, \langle s', a_k \rangle \rangle, & \text{if } \varrho(s, a_j) = \langle s', \mathsf{R} \rangle. \end{cases}$$

We call a (finite or infinite) sequence

$$c_0, c_1, \ldots, c_k, \ldots$$

of configurations a *computation* of $\boldsymbol{A}$, if the state of c_0 is s_0 and, for all k, c_{k+1} (if it exists) is obtained from c_k by replacing the triple

$$\langle \text{left neighbor of the active cell, active cell, right neighbor of the active cell} \rangle$$

of c_k by its δ-image. We say that $\boldsymbol{A}$ *halts* (starting with the empty tape), if there is a finite computation $c_0, \ldots, c_k$ such that c_0 is the start configuration and the state of c_k is s_1.

It is well known (see, e.g., Barwise 1977, Enderton 1972, Shoenfield 1967) that the *halting problem* for Turing machines is undecidable: no algorithm can decide, given a Turing machine $\boldsymbol{A}$, whether $\boldsymbol{A}$ comes to a stop having started from the empty tape. A Turing machine $\boldsymbol{A}$ is called *recurrent* if, having started from the empty tape, it works forever and reenters the start state s_0 infinitely many times. It is known (see Harel *et al.* 1983) that the problem 'given $\boldsymbol{A}$, decide whether it is recurrent' is Σ_1^1-complete. This means, in particular, that if we recursively enumerate all Turing machines $\boldsymbol{A}_0, \boldsymbol{A}_1, \ldots$ then the set

$$\{n \mid \boldsymbol{A}_n \text{ is not recurrent}\} \text{ is not recursively enumerable.} \tag{5.26}$$

Recall from Section 2.1 the fragment $\mathbf{PTL}_{\Box\bigcirc}$ of propositional temporal logic **PTL** having only $\Box_F$ and $\bigcirc$ as its temporal operators. We will use (5.26) to prove the following:

Theorem 5.38. *The product logic* $\mathbf{PTL}_{\Box\bigcirc} \times \mathbf{PTL}_{\Box\bigcirc}$ *is not recursively enumerable.*

Proof. Given a Turing machine $\boldsymbol{A}$, we construct a formula $\varphi_{\boldsymbol{A}}$ (in the language with $\boxminus$, $\Box\!\!\!|$, $\ominus$ and $\bigcirc\!\!\!|$) such that

$$\varphi_{\boldsymbol{A}} \text{ is } \mathbf{PTL}_{\Box\bigcirc} \times \mathbf{PTL}_{\Box\bigcirc}\text{-satisfiable} \quad \text{iff} \quad \boldsymbol{A} \text{ is recurrent.} \tag{5.27}$$

Let $A' = A \cup \{\pounds\} \cup (S \times A)$. With each $x \in A'$ we associate a propositional variable p_x. We also use three extra variables q_s, q_l and q_r the meaning of which will be clear from the formulas below. Define $\varphi_{\boldsymbol{A}}$ to be the conjunction of the following formulas, for all instructions $\delta(\alpha, \beta, \gamma) = \langle \alpha', \beta', \gamma' \rangle$ of $\boldsymbol{A}$:

$$\Box\!\!\!|^{+} \boxminus^{+} \bigwedge_{\substack{x, x' \in A' \\ x \neq x'}} \neg(p_x \wedge p_{x'}), \tag{5.28}$$

$$p_\pounds \wedge \ominus(p_{\langle s_0, b \rangle} \wedge \boxminus p_b), \tag{5.29}$$

$$\boxbar^+\boxminus^+\Big((q_s \leftrightarrow \bigvee_{\langle s,a\rangle\in S\times A} p_{\langle s,a\rangle}) \wedge (q_l \leftrightarrow \ominus q_s) \wedge (q_s \leftrightarrow \ominus q_r)\Big), \tag{5.30}$$

$$\boxbar^+\Big(\diamondminus^+(q_l \wedge p_\alpha) \wedge \diamondminus(q_s \wedge p_\beta) \wedge \diamondminus(q_r \wedge p_\gamma) \rightarrow \tag{5.31}$$
$$\boxminus^+((q_l \rightarrow \obar p_{\alpha'}) \wedge (q_s \rightarrow \obar p_{\beta'}) \wedge (q_r \rightarrow \obar p_{\gamma'}))\Big),$$

$$\boxbar^+\boxminus^+ \bigwedge_{a\in A\cup\{\pounds\}} (\neg q_l \wedge \neg q_s \wedge \neg q_r \wedge p_a \rightarrow \obar p_a), \tag{5.32}$$

$$\neg\diamondbar\diamondminus \bigvee_{a\in A} p_{\langle s_1,a\rangle}, \tag{5.33}$$

$$\boxbar\diamondbar\diamondminus \bigvee_{a\in A} p_{\langle s_0,a\rangle}, \tag{5.34}$$

where $\boxbar^+\chi = \chi \wedge \boxbar\chi$, $\diamondbar^+\chi = \chi \vee \diamondbar\chi$, and similarly for $\boxminus^+$ and $\diamondminus^+$.

Now suppose that $\varphi_{\boldsymbol{A}}$ is satisfiable in a model $\mathfrak{M}$ based on a frame for $\mathbf{PTL}_{\Box\bigcirc} \times \mathbf{PTL}_{\Box\bigcirc}$. By Theorem 6.29, we may assume that this frame is $\langle \mathbb{N}, <, +1\rangle \times \langle \mathbb{N}, <, +1\rangle$. So we have

$$(\mathfrak{M}, \langle 0, 0\rangle) \models \varphi_{\boldsymbol{A}}.$$

We may think of horizontal sequences

$$r_j = \langle \langle 0, j\rangle, \langle 1, j\rangle, \langle 2, j\rangle, \dots\rangle$$

of pairs as representations of configurations c_j of $\boldsymbol{A}$ (in the sense that the ith cell of c_j contains a iff $(\mathfrak{M}, \langle i, j\rangle) \models p_a$). Then

- (5.28) says that for all $i, j < \omega$, there is at most one $x \in A'$ such that $\langle i, j\rangle$ is marked by p_x;
- (5.29) says that $\boldsymbol{A}$ starts with an empty tape;
- (5.30) marks with q_s, q_l and q_r the active cell and its left and right neighbors, respectively;
- (5.31) and (5.32) ensure that the sequence $r_0, r_1, \dots, r_j, \dots$ represents a computation of $\boldsymbol{A}$;
- (5.33) says that $\boldsymbol{A}$ never halts and
- (5.34) that it reenters the start state s_0 infinitely often.

Conversely, suppose that $\boldsymbol{A}$ is a recurrent Turing machine and $c_0, \dots, c_k, \dots$ is its computation starting with the empty tape. Define a valuation $\mathfrak{V}$ in the frame $\langle \mathbb{N}, <, +1\rangle \times \langle \mathbb{N}, <, +1\rangle$ by taking, for all $x \in A'$,

$$\mathfrak{V}(p_x) = \{\langle i, j\rangle \in \mathbb{N} \times \mathbb{N} \mid \text{the } i\text{th cell of } c_j \text{ contains } x\}$$

and

$$\begin{aligned}\mathfrak{V}(q_s) &= \{\langle i,j\rangle \in \mathbb{N}\times\mathbb{N} \mid \text{the active cell of } c_j \text{ is the } i\text{th one}\}\\ \mathfrak{V}(q_l) &= \{\langle i-1,j\rangle \mid \langle i,j\rangle \in \mathfrak{V}(q_s)\}\\ \mathfrak{V}(q_r) &= \{\langle i+1,j\rangle \mid \langle i,j\rangle \in \mathfrak{V}(q_s)\}.\end{aligned}$$

It is now readily checked that $\varphi_{\boldsymbol{A}}$ is satisfied at point $\langle 0,0\rangle$ in this model, which yields (5.27).

Thus we obtain that

$$\neg\varphi_{\boldsymbol{A}} \in \mathbf{PTL}_{\Box\bigcirc} \times \mathbf{PTL}_{\Box\bigcirc} \quad \text{iff} \quad \boldsymbol{A} \text{ is not recurrent.}$$

So $\mathbf{PTL}_{\Box\bigcirc} \times \mathbf{PTL}_{\Box\bigcirc}$ cannot be recursively enumerable. ❑

As a consequence we also have:

Corollary 5.39. $\mathbf{PTL} \times \mathbf{PTL}$ *is not recursively enumerable.*

Undecidability by Post's correspondence problem

The third undecidable 'master problem' we use in this book is known as *Post's correspondence problem* or *PCP*, for short (Post 1946). It is formulated as follows. Given a finite alphabet $A = \{a_1, \ldots, a_m\}$ and a finite set P of pairs $\langle v_1, w_1\rangle, \ldots, \langle v_k, w_k\rangle$ of nonempty finite sequences (*words*) v_i, w_i over A, decide whether there exist an $N \geq 1$ and a sequence $i_1, \ldots, i_N$ of indices such that

$$v_{i_1} * \cdots * v_{i_N} = w_{i_1} * \cdots * w_{i_N} \tag{5.35}$$

(here $*$ denotes the *concatenation* of sequences). A proof showing the undecidability of PCP (via a reduction of the halting problem for Turing machines) can be found, e.g., in (Hopcroft *et al.* 2001). Here we use this fact to prove the following:

Theorem 5.40. *The product logic* $\mathbf{PTL}_{\Box\bigcirc} \times \mathbf{K4}$ *is undecidable.*

Proof. Given a finite alphabet A and a set $P = \{\langle v_1, w_1\rangle, \ldots, \langle v_k, w_k\rangle\}$ of pairs of words over A, we construct a formula $\varphi_{A,P}$ (in the language with $\boxminus$, $\ominus$ and $\boxdot$) which is $\mathbf{PTL}_{\Box\bigcirc} \times \mathbf{K4}$-satisfiable iff there exist an $N \geq 1$ and a sequence $i_1, \ldots, i_N$ of indices such that (5.35) holds. The formula $\varphi_{A,P}$ is built from the propositional variables:

- pair_i, for every pair $\langle v_i, w_i\rangle$, $1 \leq i \leq k$,
- left_a and right_a, for every $a \in A$,
- left and right.

For each $1 \leq i \leq k$, let l_i and r_i be the lengths of words v_i and w_i, respectively, and let

$$\begin{aligned} v_i &= \left\langle b_0^i, \dots, b_{l_i}^i \right\rangle, \\ w_i &= \left\langle c_0^i, \dots, c_{r_i}^i \right\rangle. \end{aligned}$$

The formula $\varphi_{A,P}$ is defined as the conjunction

$$\varphi_{A,P} = \varphi_1 \wedge \varphi_2 \wedge \varphi_{left} \wedge \varphi_{right}$$

in which

$$\begin{aligned} \varphi_1 &= \boxminus^+\Big(\bigvee_{1 \leq i \leq k} \mathsf{pair}_i \wedge \bigwedge_{\substack{i \neq j \\ 1 \leq i,j \leq k}} \neg(\mathsf{pair}_i \wedge \mathsf{pair}_j)\Big), \\ \varphi_2 &= \Diamond \Box^+ \bigwedge_{a \in A} (\mathsf{left}_a \leftrightarrow \mathsf{right}_a), \end{aligned}$$

φ_{left} is a conjunction of (5.36)–(5.42), for all i with $1 \leq i \leq k$ and for all $j < l_i$,

$$\boxminus^+ \Box^+ \Big(\bigwedge_{\substack{a \neq b \\ a,b \in A}} \neg(\mathsf{left}_a \wedge \mathsf{left}_b) \wedge \big(\mathsf{left} \leftrightarrow \bigvee_{a \in A} \mathsf{left}_a\big)\Big), \tag{5.36}$$

$$\boxminus^+ \Box^+ \bigwedge_{a \in A} (\mathsf{left}_a \to \boxminus \mathsf{left}_a), \tag{5.37}$$

$$\neg\mathsf{left} \wedge \boxminus^+ \Box^+ (\neg\mathsf{left} \to \Box\neg\mathsf{left}), \tag{5.38}$$

$$\boxminus^+ (\mathsf{pair}_i \to \Box^+(\neg\mathsf{left} \to \ominus \Box^{l_i} \neg\mathsf{left})), \tag{5.39}$$

$$\boxminus^+ (\mathsf{pair}_i \to \Box^+ \ominus(\Diamond^j \mathsf{left} \wedge \neg\Diamond^{j+1} \mathsf{left} \to \mathsf{left}_{b^i_{l_i - j}})), \tag{5.40}$$

$$\mathsf{pair}_i \to \ominus\Big(\mathsf{left}_{b_1^i} \wedge \Diamond\big(\mathsf{left}_{b_2^i} \wedge \Diamond(\mathsf{left}_{b_3^i} \wedge \dots \wedge \Diamond \mathsf{left}_{b^i_{l_i}})\dots\big)\Big), \tag{5.41}$$

$$\boxminus\Big(\mathsf{pair}_i \to \Box^+(\mathsf{left} \wedge \Box\neg\mathsf{left} \to \ominus\Diamond(\mathsf{left}_{b_1^i} \wedge \Diamond(\mathsf{left}_{b_2^i} \wedge \dots \wedge \Diamond \mathsf{left}_{b^i_{l_i}})\dots))\Big). \tag{5.42}$$

The conjunct φ_{right} is defined by replacing in φ_{left} all occurrences of left with right, left_a with right_a (for $a \in A$), l_i with r_i and the sequence of $\mathsf{left}_{b_j^i}$ (for $1 \leq j \leq l_i$) with $\mathsf{right}_{c_j^i}$ ($1 \leq j \leq r_i$). (Note that pair_i occurs in both φ_{left} and φ_{right}.)

We prove now that $\varphi_{A,P}$ is as required. Suppose first that $\varphi_{A,P}$ is satisfiable in a model $\mathfrak{M}$ based on a frame for $\mathbf{PTL}_{\Box\bigcirc} \times \mathbf{K4}$. By Theorem 6.29, we may assume that this frame is the product of $\langle \mathbb{N}, <, +1 \rangle$ and some frame $\langle V, S \rangle$ for $\mathbf{K4}$. Then we have

$$(\mathfrak{M}, \langle 0, y_0 \rangle) \models \varphi_{A,P},$$

for some $y_0 \in V$. Since $(\mathfrak{M}, \langle 0, y_0\rangle) \models \varphi_2$, we can find an N, $1 \leq N < \omega$, such that

$$(\mathfrak{M}, \langle N, y_0\rangle) \models \Box^+ \bigwedge_{a\in A} (\mathsf{left}_a \leftrightarrow \mathsf{right}_a). \tag{5.43}$$

Let $i_1, \dots, i_N$ be the sequence of indices such that, for $1 \leq j \leq N$, we have $(\mathfrak{M}, \langle j-1, y_0\rangle) \models \mathsf{pair}_{i_j}$ (φ_1 ensures that there is a unique sequence of this sort). We claim that (5.35) holds.

For every j with $1 \leq j \leq N$, let

$$\mathfrak{V}_j(\mathsf{left}) = \{y \in V \mid (\mathfrak{M}, \langle j, y\rangle) \models \mathsf{left}\}.$$

Given a sequence $z_1, \dots, z_l$ of points from $\mathfrak{V}_j(\mathsf{left})$, define

$$\mathit{leftword}_j(z_1, \dots, z_l) = \langle c_1, \dots, c_l\rangle,$$

where $c_i = a$ for the (uniquely determined by (5.36)) $a \in A$ such that $(\mathfrak{M}, \langle j, z_i\rangle) \models \mathsf{left}_a$. Call a sequence $\langle y_0, \dots, y_{l-1}\rangle$ of (not necessarily distinct) points from V an *S-path in* $\mathfrak{V}_j(\mathsf{left})$ if $y_0, \dots, y_{l-1} \in \mathfrak{V}_j(\mathsf{left})$ and $y_0 S y_1 S \dots S y_{l-1}$. The number l is called the *length* of the S-path. We will show that, for every $1 \leq j \leq N$, the following holds:

(i) there exists an S-path $\langle y_0, \dots, y_{n_j-1}\rangle$ in $\mathfrak{V}_j(\mathsf{left})$ of length

$$n_j = l_{i_1} + \dots + l_{i_j}$$

such that

$$\mathit{leftword}_j(y_0, \dots, y_{n_j-1}) = v_{i_1} * \dots * v_{i_j};$$

(ii) every S-path in $\mathfrak{V}_j(\mathsf{left})$ is of length $\leq n_j$;

(iii) for every S-path $\langle y_0, \dots, y_{n_j-1}\rangle$ in $\mathfrak{V}_j(\mathsf{left})$, we have

$$\mathit{leftword}_j(y_0, \dots, y_{n_j-1}) = v_{i_1} * \dots * v_{i_j}.$$

Indeed, for $j = 1$, we have (i) by $(\mathfrak{M}, \langle 0, y_0\rangle) \models \mathsf{pair}_{i_1}$ and (5.41), (ii) by (5.38) and (5.39), and (iii) by (5.40). Now assume inductively that (i)–(iii) hold for some $1 \leq j < N$. Let $\langle y_0, \dots, y_{n_j-1}\rangle$ be a maximal S-path in $\mathfrak{V}_j(\mathsf{left})$. First, by (5.37), we have $y_0, \dots, y_{n_j-1} \in \mathfrak{V}_{j+1}(\mathsf{left})$. Second, since $(\mathfrak{M}, \langle j, y_{n_j-1}\rangle) \models \mathsf{left} \land \Box\neg\mathsf{left}$ and $(\mathfrak{M}, \langle j, y_0\rangle) \models \mathsf{pair}_{i_{j+1}}$, (5.42) now implies that there exist $y_{n_j}, \dots, y_{n_j+l_{i_{j+1}}-1}$ such that $\left\langle y_0, \dots, y_{n_j+l_{i_{j+1}}-1}\right\rangle$ is an S-path in $\mathfrak{V}_{j+1}(\mathsf{left})$, as required in (i). For (ii) and (iii), observe first that for every S-path $\langle y_0, \dots, y_{l-1}\rangle$ in $\mathfrak{V}_{j+1}(\mathsf{left})$, $\left\langle y_0, \dots, y_{l-l_{i_{j+1}}-1}\right\rangle$ is an S-path in $\mathfrak{V}_j(\mathsf{left})$, by (5.39). So $l \leq n_{j+1}$ must hold. If $l = n_{j+1}$ then

$\mathit{leftword}_j(y_0, \dots, y_{l-l_{i_{j+1}}-1}) = v_{i_1} * \dots * v_{i_j}$ by the induction hypothesis, and so $\mathit{leftword}_{j+1}(y_0, \dots, y_{l-l_{i_{j+1}}-1}) = v_{i_1} * \dots * v_{i_j}$ by (5.37). On the other hand, $\mathit{leftword}_{j+1}(y_{l-l_{i_{j+1}}}, \dots, y_{l-1}) = v_{i_{j+1}}$ by (5.40), and therefore we have $\mathit{leftword}_{j+1}(y_0, \dots, y_{l-1}) = v_{i_1} * \dots * v_{i_{j+1}}$, as required.

We can repeat the argument above for the 'right side' as well. Take, for $1 \le j \le N$,

$$\mathfrak{V}_j(\mathsf{right}) = \{y \in V \mid (\mathfrak{M}, \langle j, y\rangle) \models \mathsf{right}\},$$

and, for every sequence $z_1, \dots, z_l$ of points from $\mathfrak{V}_j(\mathsf{right})$, define

$$\mathit{rightword}_n(z_1, \dots, z_l) = \langle c_1, \dots, c_l\rangle,$$

where $c_i = a$ for the uniquely determined $a \in A$ with $(\mathfrak{M}, \langle j, z_i\rangle) \models \mathsf{right}_a$. We then have, for every $1 \le j \le N$:

(i)′ there exists an S-path $\langle y_0, \dots, y_{m_j-1}\rangle$ in $\mathfrak{V}_j(\mathsf{right})$ of length

$$m_j = r_{i_1} + \dots + r_{i_j}$$

such that

$$\mathit{rightword}_j(y_0, \dots, y_{m_j-1}) = w_{i_1} * \dots * w_{i_j};$$

(ii)′ every S-path in $\mathfrak{V}_j(\mathsf{right})$ is of length $\le m_j$;

(iii)′ for every S-path $\langle y_0, \dots, y_{m_j-1}\rangle$ in $\mathfrak{V}_j(\mathsf{right})$, we have

$$\mathit{rightword}_j(y_0, \dots, y_{m_j-1}) = w_{i_1} * \dots * w_{i_j}.$$

Now, by (5.36) and (5.43), we have $\mathfrak{V}_N(\mathsf{left}) = \mathfrak{V}_N(\mathsf{right})$. By (i), there exists an S-path $\langle y_0, \dots, y_{l-1}\rangle$ in $\mathfrak{V}_N(\mathsf{left})$ such that $l = n_N$ and

$$\mathit{leftword}_N(y_0, \dots, y_{l-1}) = v_{i_1} * \dots * v_{i_N}.$$

By (ii)′, we have $n_N \le m_N$. Similarly, using (i)′ and (ii), we obtain $m_N \le n_N$, from which $n_N = m_N$. Hence, by (iii)′,

$$\mathit{rightword}_N(y_0, \dots, y_{l-1}) = w_{i_1} * \dots * w_{i_N}.$$

Since, by (5.43),

$$\mathit{leftword}_N(y_0, \dots, y_{l-1}) = \mathit{rightword}_N(y_0, \dots, y_{l-1}),$$

we finally obtain $v_{i_1} * \dots * v_{i_N} = w_{i_1} * \dots * w_{i_N}$, as required.

Conversely, suppose that there is an $N \ge 1$ and a sequence $i_1, \dots, i_N$ of indices such that (5.35) holds. Our aim is to show that $\varphi_{A,P}$ is satisfiable in the product frame $\langle \mathbb{N}, <, +1\rangle \times \langle \mathbb{N}, <\rangle$. For each $j > N$, choose an arbitrary pair $\langle v_{i_j}, w_{i_j} \rangle$ from P. For every j with $1 \leq j < \omega$, let l_{i_j} and r_{i_j} be the lengths of words v_{i_j} and w_{i_j}, respectively, and let

$$\begin{aligned} v_{i_1} * \ldots * v_{i_j} &= \langle b_0, \ldots, b_{n_j - 1} \rangle, \\ w_{i_1} * \ldots * w_{i_j} &= \langle c_0, \ldots, c_{m_j - 1} \rangle, \end{aligned}$$

where $n_j = l_{i_1} + \cdots + l_{i_j}$ and $m_j = r_{i_1} + \cdots + r_{i_j}$. Note that, by our assumption, $n_N = m_N$ and $b_j = c_j$, for every $j < n_N$.

Define a valuation $\mathfrak{V}$ in $\langle \mathbb{N}, <, +1 \rangle \times \langle \mathbb{N}, < \rangle$ by taking

- $\mathfrak{V}(\mathsf{pair}_i) = \{\langle j-1, 0 \rangle \mid i = i_j, \text{ for some } j \geq 1\}$, for $1 \leq i \leq k$,
- $\mathfrak{V}(\mathsf{left}_a) = \{\langle j, l \rangle \mid j \geq 1,\ l < n_j,\ b_l = a\}$, for $a \in A$,
- $\mathfrak{V}(\mathsf{right}_a) = \{\langle j, l \rangle \mid j \geq 1,\ l < m_j,\ c_l = a\}$, for $a \in A$,
- $\mathfrak{V}(\mathsf{left}) = \bigcup_{a \in A} \mathfrak{V}(\mathsf{left}_a)$, $\qquad \mathfrak{V}(\mathsf{right}) = \bigcup_{a \in A} \mathfrak{V}(\mathsf{right}_a)$.

One can easily check that under this valuation we have $\langle 0, 0 \rangle \models \varphi_{A,P}$. ❑

5.5 Proving complexity with tilings

In this section we demonstrate how bounded tiling problems can be used to establish NEXPTIME and EXPSPACE lower bounds of computational complexity.

Proving NEXPTIME lower bounds

First we show that the NEXPTIME upper bound of the satisfiability problem for **S5** × **S5**, obtained in Section 5.2, is optimal. To prove this, we will reduce a NEXPTIME-complete problem to the satisfiability problem for **S5** × **S5**.

In the spectrum of NEXPTIME-complete problems, the most suitable for dealing with many-dimensional logics seems to be the following $k \times k$*-bounded tiling problem*: given $k < \omega$, a finite set T of tile types (see Section 5.4) and a $t_0 \in T$, decide whether T can tile the $k \times k$ grid in such a way that t_0 is placed onto $\langle 0, 0 \rangle$. In other words, the problem is to decide whether there exists a function $\boldsymbol{\tau}$ from the set $\{\langle i, j \rangle \mid i, j < k\}$ to T such that

- $up(\boldsymbol{\tau}(i, j)) = down(\boldsymbol{\tau}(i, j+1))$, for all $i < k$, $j < k - 1$,
- $right(\boldsymbol{\tau}(i, j)) = left(\boldsymbol{\tau}(i+1, j))$, for all $i < k - 1$, $j < k$,
- $\boldsymbol{\tau}(0, 0) = t_0$.

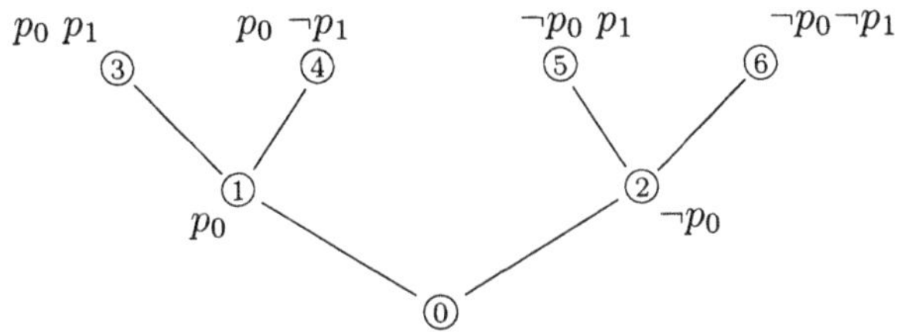

Figure 5.11: The binary tree $\mathfrak{H}_1$ of depth 2.

If k is given in its binary representation then this problem is known to be NEXPTIME-complete; see e.g., (Levin 1973, Lewis 1978, Lewis and Papadimitriou 1981, van Emde Boas 1997).

Suppose we are given a finite set T of tile types, a $t_0 \in T$ and a natural number k in its binary form. Without loss of generality we may assume that $k = 2^n$ for some $n < \omega$. Our aim is to construct a formula $\varphi_{n,T}$ such that

(i) the length of $\varphi_{n,T}$ is a polynomial function of $|T|$ and n;

(ii) T tiles $2^n \times 2^n$ grid, with t_0 being placed onto $\langle 0, 0\rangle$, iff $\varphi_{n,T}$ is **S5** $\times$ **S5**-satisfiable.

At first sight it should not be too hard to reduce $k \times k$-tiling to **S5** $\times$ **S5**-satisfiability: the $k \times k$ grid looks like a perfect universal product **S5** $\times$ **S5**-frame. The problem, however, is that we are not able to refer in the language $\mathcal{ML}_2$ to 'my right neighbor,' 'my neighbor above,' etc., in order to ensure color matching simply because both R_h and R_v are equivalence relations.

The idea of encoding $k \times k$-tiling in an **S5** $\times$ **S5**-model proposed by Marx (1999) is as follows. It is known (see, e.g., Halpern and Moses 1992 or Chagrov and Zakharyaschev 1997) that there is a modal formula of length $O(n^2)$ which is satisfied in a **K**-model $\mathfrak{M}$ iff $\mathfrak{M}$ contains as a submodel a binary tree $\mathfrak{H}_n = \langle\langle W, S\rangle, \mathfrak{W}\rangle$ of depth $2n$ with the valuation $\mathfrak{W}$ depicted in Fig. 5.11 for the case $n = 1$. Here is such a formula:

$$\chi_n = \bigwedge_{\ell<2n} \Box^\ell \Big((\Diamond p_\ell \wedge \Diamond \neg p_\ell) \wedge \bigwedge_{i<\ell} \big((p_i \to \Box p_i) \wedge (\neg p_i \to \Box \neg p_i) \big) \Big).$$

The 2^{2n} leaves of the tree $\mathfrak{H}_n$ are labeled by $2n$-tuples containing either p_i or $\neg p_i$, for each $i < 2n$. By replacing in such a $2n$-tuple every p_i with 1 and every $\neg p_j$ with 0, we obtain a pair $\langle \ell, m\rangle$, where ℓ and m are decimal numbers whose binary representations are the first and the last n bits in the resulting word of 1s and 0s, respectively. The pair $\langle \ell, m\rangle$ determined by a leaf x of $\mathfrak{H}_n$ will be denoted by *grid*(x). For example, if

$$(\mathfrak{H}_n, x) \models \neg p_0 \wedge \cdots \wedge \neg p_{n-1} \wedge p_n \wedge \cdots \wedge p_{2n-1}$$

then $grid(x) = \langle 0, 2^n - 1\rangle$. This way we get all the pairs $\langle \ell, m\rangle$, for $\ell, m < 2^n$. Moreover, it is easily seen that for each pair of leaves x and y, we have $grid(x) = \langle \ell, m\rangle$ and $grid(y) = \langle \ell, m+1\rangle$ iff the following conditions (5.44)–(5.47) hold:

$$(\mathfrak{H}_n, x) \models p_i \quad \text{iff} \quad (\mathfrak{H}_n, y) \models p_i, \quad \text{for all } i < n, \tag{5.44}$$

and there exists an i, $n \leq i < 2n$, such that

$$(\mathfrak{H}_n, x) \models p_j \quad \text{iff} \quad (\mathfrak{H}_n, y) \models p_j, \quad \text{for all } n \leq j < i, \tag{5.45}$$

$$(\mathfrak{H}_n, x) \models \neg p_i \quad \text{and} \quad (\mathfrak{H}_n, y) \models p_i, \tag{5.46}$$

$$(\mathfrak{H}_n, x) \models p_j \quad \text{and} \quad (\mathfrak{H}_n, y) \models \neg p_j, \quad \text{for all } i < j < 2n. \tag{5.47}$$

Similar conditions hold for 'horizontal neighbors' of the $2^n \times 2^n$ grid, which makes it possible to use the leaves of the tree $\mathfrak{H}_n$ to encode the grid.

The problem, however, is that in **S5** × **S5**-models we do not have the modal operators of **K** which are required to 'grow' such a tree. But we can simulate it using the two **S5**-boxes in the following way. We represent the nodes of $\mathfrak{H}_n$ by points which are marked with a special variable d (for 'diagonal'), and use another variable s as a pointer to the S-successors of a given node. The **K**-diamond and **K**-box will be simulated as

$$\Diamond\psi = \Diamond_v(s \wedge \Diamond_h(d \wedge \psi)), \qquad \Box\psi = \neg\Diamond\neg\psi. \tag{5.48}$$

Then the points representing the leaves of $\mathfrak{H}_n$ will validate the formula $d \wedge \Box\bot$ (see Fig. 5.12).

We are now in a position to define the formula $\varphi_{n,T}$. For each tile type $t \in T$, take (with a slight abuse of notation) three propositional variables t, t^h, t^v, and for each $i < 2n$, take two variables p_i^h and p_i^v. Then the formula $\varphi_{n,T}$ is the conjunction of the formulas (5.49)–(5.58), in which $\Diamond$ and $\Box$ are defined by (5.48):

$$d \wedge \Box^{2n+1}\bot \wedge \chi_n, \tag{5.49}$$

$$\Box_v\Box_h(d \wedge \Box\bot \leftrightarrow \bigvee_{t\in T} t), \tag{5.50}$$

$$\Box_v\Box_h \bigwedge_{t\in T}(t \rightarrow \bigwedge_{\substack{t,t'\in T\\ t'\neq t}} \neg t'), \tag{5.51}$$

$$\bigwedge_{i<2n} \left(\Box_v(\Diamond_h p_i \rightarrow \Box_h p_i^h) \wedge \Box_v(\Diamond_h\neg p_i \rightarrow \Box_h\neg p_i^h)\right), \tag{5.52}$$

$$\bigwedge_{t\in T} \left(\Box_v(\Diamond_h t \rightarrow \Box_h t^h) \wedge \Box_v(\Diamond_h\neg t \rightarrow \Box_h\neg t^h)\right), \tag{5.53}$$

$$\bigwedge_{i<2n} \left(\Box_h(\Diamond_v p_i \rightarrow \Box_v p_i^v) \wedge \Box_h(\Diamond_v\neg p_i \rightarrow \Box_v\neg p_i^v)\right), \tag{5.54}$$

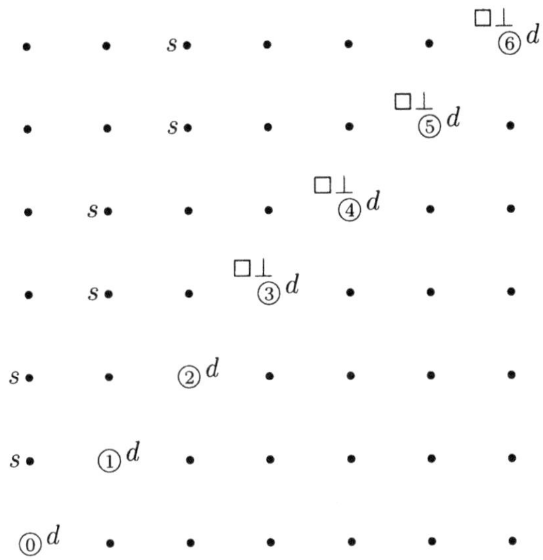

Figure 5.12: Coding $\mathfrak{H}_1$ in an $\mathbf{S5} \times \mathbf{S5}$-model.

$$\bigwedge_{t \in T} \Big(\boxminus(\diamondbar t \to \boxbar t^v) \wedge \boxminus(\diamondbar \neg t \to \boxbar \neg t^v) \Big), \tag{5.55}$$

$$\boxbar\boxminus \bigwedge_{t_1 \in T} \Big(\alpha \wedge \beta_1 \wedge t_1^v \wedge \bigvee_{t \in T} t^h \to \bigvee_{\substack{t \in T \\ up(t_1) = down(t)}} t^h \Big), \tag{5.56}$$

$$\boxbar\boxminus \bigwedge_{t_1 \in T} \Big(\beta \wedge \alpha_1 \wedge t_1^v \wedge \bigvee_{t \in T} t^h \to \bigvee_{\substack{t \in T \\ right(t_1) = left(t)}} t^h \Big), \tag{5.57}$$

$$\boxbar\boxminus \Big(d \wedge \Box\bot \wedge \bigwedge_{i < 2n} \neg p_i \to t_0 \Big), \tag{5.58}$$

where

$$\alpha = \bigwedge_{i<n} (p_i^h \leftrightarrow p_i^v), \qquad \beta = \bigwedge_{n \le i < 2n} (p_i^h \leftrightarrow p_i^v),$$

$$\alpha_1 = \bigvee_{i<n} \Big(\bigwedge_{j<i} (p_j^h \leftrightarrow p_j^v) \wedge p_i^h \wedge \neg p_i^v \wedge \bigwedge_{i<j<n} (\neg p_j^h \wedge p_j^v) \Big),$$

$$\beta_1 = \bigvee_{n \le i < 2n} \Big(\bigwedge_{n \le j < i} (p_j^h \leftrightarrow p_j^v) \wedge p_i^h \wedge \neg p_i^v \wedge \bigwedge_{i<j<2n} (\neg p_j^h \wedge p_j^v) \Big).$$

Clearly, the length of $\varphi_{n,T}$ is polynomial in n and $|T|$. Let us show that $\varphi_{n,T}$ really does the job.

Suppose first that T tiles $2^n \times 2^n$, and let $\mathfrak{H}_n = \langle\langle W, S\rangle, \mathfrak{W}\rangle$ be the binary tree model as above. Define a valuation $\mathfrak{V}$ on the universal product $\mathbf{S5} \times \mathbf{S5}$-frame $\langle W, W\rangle$ by taking:

$$\begin{aligned}
\mathfrak{V}(d) &= \{\langle x, x\rangle \mid x \in W\},\\
\mathfrak{V}(s) &= \{\langle x, y\rangle \mid xSy\},\\
\mathfrak{V}(p_i) &= \{\langle x, x\rangle \mid (\mathfrak{H}_n, x) \models p_i\},\\
\mathfrak{V}(p_i^v) &= \{\langle x, y\rangle \mid \langle x, x\rangle \in \mathfrak{V}(p_i)\},\\
\mathfrak{V}(p_i^h) &= \{\langle x, y\rangle \mid \langle y, y\rangle \in \mathfrak{V}(p_i)\},\\
\mathfrak{V}(t) &= \{\langle x, x\rangle \mid x \text{ is a leaf in } \mathfrak{H}_n \text{ and } t \text{ tiles } grid(x)\},\\
\mathfrak{V}(t^v) &= \{\langle x, y\rangle \mid \langle x, x\rangle \in \mathfrak{V}(t)\},\\
\mathfrak{V}(t^h) &= \{\langle x, y\rangle \mid \langle y, y\rangle \in \mathfrak{V}(t)\}.
\end{aligned}$$

Let $\mathfrak{N} = \langle\langle W, W\rangle, \mathfrak{V}\rangle$. It is not hard to check that $\varphi_{n,T}$ is true in $\mathfrak{N}$ at $\langle u, u\rangle$, where u is the root of $\mathfrak{H}_n$. Indeed, the meaning of (5.49) was explained above. (5.50) says that only the diagonal points representing the leaves of $\mathfrak{H}_n$ (i.e., those where $\Box\bot$ holds) validate at least one 'tile variable,' and (5.51) ensures that there is only one such variable for every leaf. Formulas (5.52)–(5.58) say in effect that the colors on adjacent edges of adjacent tiles match and that t_0 is placed onto $\langle 0, 0\rangle$. That (5.52)–(5.55) and (5.58) are true at $\langle u, u\rangle$ follows directly from the definition of the valuation $\mathfrak{V}$. Of the remaining two formulas, we check only (5.56). Assume that for some $\langle x, y\rangle \in W \times W$ and $t_1 \in T$

$$(\mathfrak{N}, \langle x, y\rangle) \models \alpha \wedge \beta_1 \wedge t_1^v \wedge \bigvee_{t \in T} t^h.$$

By the definition of $\mathfrak{V}(t^h)$ and $\mathfrak{V}(t^v)$, both x and y must be leaves in $\mathfrak{H}_n$. Since $(\mathfrak{N}, \langle x, y\rangle) \models t_1^v$, we have $\langle x, x\rangle \in \mathfrak{V}(t_1)$, which means that t_1 tiles $grid(x)$. Let t_2 tile $grid(y)$, that is, $\langle y, y\rangle \in \mathfrak{V}(t_2)$ and so $(\mathfrak{N}, \langle x, y\rangle) \models t_2^h$. As $(\mathfrak{N}, \langle x, y\rangle) \models \alpha \wedge \beta_1$, we have by (5.44)–(5.47) that $grid(x) = \langle \ell, m\rangle$ and $grid(y) = \langle \ell, m+1\rangle$ for some ℓ, m. It follows that $up(t_1) = down(t_2)$.

Conversely, suppose that $\varphi_{n,T}$ is $\mathbf{S5} \times \mathbf{S5}$-satisfiable. By Proposition 3.11, we may assume that the formula $\varphi_{n,T}$ is satisfied at a point $\langle x_0, y_0\rangle$ in a model $\mathfrak{M} = \langle\langle U_1, U_2\rangle, \mathfrak{V}\rangle$ based on a universal product $\mathbf{S5} \times \mathbf{S5}$-frame $\langle U_1, U_2\rangle$. Our aim is to show that T tiles $2^n \times 2^n$ as required.

For a set $U \subseteq U_1 \times U_2$, let $\mathfrak{V}_U$ denote the restriction of $\mathfrak{V}$ to U, that is, $\mathfrak{V}_U(p) = \mathfrak{V}(p) \cap U$, for all variables p. As $(\mathfrak{M}, \langle x_0, y_0\rangle) \models \chi_n$, there is some $U \subseteq U_1 \times U_2$ and a binary relation R on U such that the model $\mathfrak{M}_U = \langle\langle U, R\rangle, \mathfrak{V}_U\rangle$ is isomorphic to the binary tree model $\mathfrak{H}_n$ (see Fig. 5.11), with $\langle x_0, y_0 \rangle$ being the root of $\mathfrak{M}_U$. Moreover, since $(\mathfrak{M}, \langle x_0, y_0 \rangle) \models \Box^{2n+1}\bot$ and by (5.48), $d \wedge \Box\bot$ is true at all leaves in $\mathfrak{M}_U$. So, in view of (5.50) and (5.51), precisely one tile variable is true at each such leaf. Therefore, the following map τ from $\{\langle \ell, m \rangle \mid \ell, m < n\}$ to T is well defined:

$$\tau(\ell, m) = t \quad \text{iff} \quad \mathit{grid}(x) = \langle \ell, m \rangle \text{ and } (\mathfrak{M}, x) \models t, \text{ for some leaf } x \text{ in } \mathfrak{M}_U.$$

To show that τ is in fact a tiling, we have to check that the colors on adjacent edges of adjacent tiles match.

Suppose, for instance, that $\tau(\ell, m) = t_1$, $\tau(\ell, m+1) = t_2$, and for some leaves x, y of $\mathfrak{M}_U$, $\mathit{grid}(x) = \langle \ell, m \rangle$ and $\mathit{grid}(y) = \langle \ell, m+1 \rangle$. Let $x = \langle u_1, u_2 \rangle$ and $y = \langle u_1', u_2' \rangle$. Take the point $\langle u_1, u_2' \rangle$. Then by (5.53) and (5.55) we have

$$(\mathfrak{M}, \langle u_1, u_2' \rangle) \models t_1^v \wedge t_2^h$$

and by (5.52), (5.54) and (5.44)–(5.47), $(\mathfrak{M}, \langle u_1, u_2' \rangle) \models \alpha \wedge \beta_1$. In view of (5.51) and (5.56), we then obtain that $\mathit{up}(t_1) = \mathit{down}(t_2)$, as required.

We have proved that the satisfiability problem for **S5** $\times$ **S5** is NEXPTIME-hard. Together with Theorem 5.26 this yields:

Theorem 5.41. *The satisfiability problem for* **S5** $\times$ **S5** *is* NEXPTIME-*complete, and so the decision problem for* **S5** $\times$ **S5** *is* coNEXPTIME-*complete.*

Actually, almost the same proof can establish NEXPTIME-hardness of the satisfiability problem for many other bimodal logics. The following theorem was also proved by Marx (1999):

Theorem 5.42. *Let L be a Kripke complete bimodal logic between* **K** $\times$ **K** *and* **S5** $\times$ **S5**. *Then the satisfiability problem for L is* NEXPTIME-*hard, and so the decision problem for L is* coNEXPTIME-*hard.*

Proof. One has to replace in $\varphi_{n,T}$ the modal operators $\boxminus$, $\diamondminus$, $\boxbar$, $\diamondbar$ with $\boxminus^{\leq 2n}$, $\diamondminus^{\leq 2n}$, $\boxbar^{\leq 2n}$ and $\diamondbar^{\leq 2n}$, respectively (but leave those in the definition of $\Box$ and $\Diamond$ untouched). Using the fact that every frame for L must validate the commutativity and Church–Rosser axioms, it is not hard to see that this formula does the job. ❑

Proving EXPSPACE lower bounds

Now we will use the 2^n-*corridor tiling problem* which is EXPSPACE-complete (see van Emde Boas 1997 and references therein): given a finite set T of tile types, two tile types $t_0, t_1 \in T$ and $n \in \mathbb{N}$ in binary, decide whether there is an $m \in \mathbb{N}$ such that T tiles the $m \times 2^n$-corridor in such a way that t_0 is placed onto $\langle 0, 0 \rangle$, t_1 is placed onto $\langle m-1, 0 \rangle$, and the top and bottom sides of the corridor are of some fixed color, say, *white*. We are about to prove the following:

Theorem 5.43. *The satisfiability problem for* **PTL**$\times$**S5** *is* EXPSPACE-*hard.*

Proof. Suppose that a finite set T of tile types, $t_0, t_1 \in T$ and a natural number n are given. Our aim is to construct a formula $\varphi_{n,T}$ (in the language with $\mathcal{U}_h$, $\boxminus$, $\diamondminus$, $\ominus$ and $\Box$, $\Diamond$) such that (i) its length is a polynomial function of $|T|$ and n, and (ii) $\varphi_{n,T}$ is **PTL** $\times$ **S5**-satisfiable iff there is an $m \in \mathbb{N}$ such that T tiles the $m \times 2^n$-corridor as described above. Moreover, we will see that $\varphi_{n,T}$ is **PTL** $\times$ **S5**-satisfiable iff it is satisfied in a model based on the product of $\langle \mathbb{N}, < \rangle$ and a *finite* **S5**-frame.[5]

Suppose our formula $\varphi_{n,T}$ is satisfied in a model $\mathfrak{M}$ based on a frame for **PTL**$\times$**S5**. By Theorem 6.29, we may assume that this frame is the product of $\langle \mathbb{N}, < \rangle$ and a universal **S5**-frame $\langle W, R \rangle$. Our first step in the construction of $\varphi_{n,T}$ (which will contain, among many others, propositional variables t for all $t \in T$) is to write down formulas forcing a finite sequence $y_0, y_1, \ldots, y_{m \cdot 2^n - 1}$ of distinct points from W for some $m \in \mathbb{N}$ such that for each $i < m \cdot 2^n$, $\langle i, y_i \rangle \models t$ for a unique tile type t. If $i = k \cdot 2^n + j$ for some $k < m$, $j < 2^n$ then we will use the point $\langle i, y_i \rangle$ to encode the pair $\langle k, j \rangle$ of the $m \times 2^n$-grid. Thus the up neighbor $\langle k, j+1 \rangle$ of $\langle k, j \rangle$ will be coded by the point $\langle i+1, y_{i+1} \rangle$, and its right neighbor $\langle k+1, j \rangle$ by $\langle i + 2^n, y_{i+2^n} \rangle$.

Let $q_0, \ldots, q_{n-1}$ be pairwise distinct propositional variables, and $q_i^1 = q_i$, $q_i^0 = \neg q_i$, for $i < n$. Set

$$\sigma_j = q_0^{d_0} \wedge \cdots \wedge q_{n-1}^{d_{n-1}},$$

where $d_{n-1} \ldots d_0$ is the binary representation of $j < 2^n$. The formula

$$\boxminus^+ \bigwedge_{i<n} (\Box q_i \vee \Box \neg q_i) \tag{5.59}$$

says that the truth-values of the q_i (and so those of the σ_j) do not change along the vertical axis. We force subsequent columns to satisfy the infinitely repeating sequence

$$\sigma_0, \sigma_1, \ldots, \sigma_{2^n-1}, \sigma_0, \sigma_1, \ldots$$

by the following 'counting' formulas (the length of which is polynomial in n):

$$\sigma_0 \wedge \boxminus^+ \bigwedge_{k<n} \Big(\big(\bigwedge_{i<k} q_i \wedge \neg q_k\big) \to \big(\bigwedge_{j=k+1}^{n-1} (q_j \leftrightarrow \ominus q_j)\big) \wedge \ominus\big(\bigwedge_{i<k} \neg q_i \wedge q_k\big)\Big), \tag{5.60}$$

$$\boxminus^+ \Big(\bigwedge_{i<n} q_i \to \ominus\big(\bigwedge_{i<n} \neg q_i\big)\Big). \tag{5.61}$$

[5]The finiteness of the **S5**-component will be used in Theorems 11.33 and 11.52 below.

Now let $p_0, \ldots, p_{n-1}$ be a fresh n-tuple of distinct variables such that their truth-values do not change along the horizontal axis. This requirement can be ensured by the formula

$$\boxminus^+ \boxdot \bigwedge_{i<n} (p_i \leftrightarrow \ominus p_i). \tag{5.62}$$

Let $\pi_j = p_0^{d_0} \wedge \cdots \wedge p_{n-1}^{d_{n-1}}$, where $d_{n-1} \ldots d_0$ is the binary representation of $j < 2^n$, and let

$$\mathsf{equ} = \bigwedge_{i<n} (p_i \leftrightarrow q_i).$$

We also require

$$\mathsf{mark} = \bigvee_{t \in T} t,$$

$$\mathsf{tile} = \mathsf{equ} \wedge \mathsf{mark} \wedge \boxminus\neg\mathsf{mark}.$$

Now we can generate the required sequence of points using the following formulas:

$$(\diamondsuit\mathsf{mark})\, \mathcal{U}_h\, (\sigma_0 \wedge \boxminus^+ \boxdot \neg\mathsf{mark}), \tag{5.63}$$

$$\mathsf{tile} \wedge \boxminus\big(\diamondsuit\mathsf{mark} \to \diamondsuit\mathsf{tile}\big). \tag{5.64}$$

Indeed, suppose that the conjunction of (5.59)–(5.64) holds at $\langle 0, y_0 \rangle$, for some $y_0 \in W$. Then

$$\langle 1, y_0 \rangle \models \diamondsuit\mathsf{mark} \to \diamondsuit\mathsf{tile}.$$

Since, by (5.60) and (5.63), we have (if $n > 0$)

$$\langle 1, y_0 \rangle \models \diamondsuit\mathsf{mark},$$

there is a point $y_1 \in W$ such that $\langle 1, y_1 \rangle \models \mathsf{tile}$. In particular, we have:

(a) $\langle 1, y_1 \rangle \models \mathsf{equ}$, and so $\langle k, y_1 \rangle \models \pi_1$ for all $k \in \mathbb{N}$;

(b) no point of the form $\langle k, y_1 \rangle$ with $k > 1$ makes mark true.

Note that $y_1 \neq y_0$, since $\langle 0, y_0 \rangle \models \boxminus\neg\mathsf{mark}$, by (5.64). Now we consider $\langle 1, y_0 \rangle$ and by the same argument find a point y_2 (which is different from y_1 by (b)), and so forth; see Fig. 5.13. By (5.63), this construction cannot go on forever, that is, there is some $k \in \mathbb{N}$ such that

$$\langle k, y_0 \rangle \models \sigma_0 \wedge \boxminus^+ \boxdot \neg\mathsf{mark},$$

and so, by (5.60) and (5.61), $k = m \cdot 2^n$ must hold, for some $m \in \mathbb{N}$. Thus we have 'generated' distinct points

$$y_0, y_1, \dots, y_{m \cdot 2^n - 1}$$

from W.

Our next aim is to write down formulas that could serve as pointers to the up and right neighbors of a given pair in the corridor (at this moment we do not bother about its top border). Let

$$\mathsf{up} = \ominus \mathsf{tile},$$
$$\mathsf{right} = \mathsf{equ} \wedge (\neg \mathsf{equ})\,\mathcal{U}_h\, \mathsf{tile}.$$

It is easy to see that:

- for all $i < m \cdot 2^n - 1$, $\langle i, y_{i+1}\rangle \models \mathsf{up}$ and $\langle i, y_j\rangle \not\models \mathsf{up}$ for all $j \neq i+1$,
- for all $i < (m-1) \cdot 2^n$, $\langle i, y_{i+2^n}\rangle \models \mathsf{right}$ and $\langle i, y_j\rangle \not\models \mathsf{right}$ for all $j \neq i + 2^n$.

Finally, the formulas below ensure that $\langle 0, 0\rangle$ is covered by t_0, $\langle m-1, 0\rangle$ is covered by t_1, every point of the $m \times 2^n$-corridor is covered by at most one tile, the top and bottom sides of the corridor are *white* and the colors on adjacent edges of adjacent tiles match:

$$t_0 \wedge \boxminus^+ \boxdot \bigwedge_{\substack{t,t' \in T, \\ t \neq t'}} \neg(t \wedge t'), \tag{5.65}$$

$$\boxminus^+ \boxdot\big(\sigma_0 \wedge \mathsf{mark} \wedge \boxminus(\sigma_0 \to \boxdot\neg\mathsf{mark}) \to t_1\big), \tag{5.66}$$

$$\boxminus^+ \boxdot\Big(\sigma_0 \wedge \mathsf{mark} \to \bigvee_{\substack{t \in T, \\ down(t) = white}} t\Big), \tag{5.67}$$

$$\boxminus^+ \boxdot\Big(\sigma_{2^n - 1} \wedge \mathsf{mark} \to \bigvee_{\substack{t \in T, \\ up(t) = white}} t\Big), \tag{5.68}$$

$$\boxminus^+ \boxdot\Big(\neg\sigma_{2^n - 1} \to \bigwedge_{\substack{t,t' \in T, \\ up(t) \neq down(t')}} \big(t \to \boxdot(\mathsf{up} \to \boxminus\neg t')\big)\Big), \tag{5.69}$$

$$\boxminus^+ \boxdot\Big(\bigwedge_{\substack{t,t' \in T, \\ right(t) \neq left(t')}} \big(t \to \boxdot(\mathsf{right} \to \boxminus\neg t')\big)\Big). \tag{5.70}$$

Let $\varphi_{n,T}$ be the conjunction of (5.59)–(5.70). Suppose that

$$(\mathfrak{M}, \langle 0, y_0\rangle) \models \varphi_{n,T}.$$

Then we define a map $\boldsymbol{\tau} : m \times 2^n \to T$ by taking

$$\boldsymbol{\tau}(k, j) = t \quad \text{iff} \quad (\mathfrak{M}, \langle k \cdot 2^n + j, y_{k \cdot 2^n + j} \rangle) \models t.$$

We leave it to the reader to check that $\boldsymbol{\tau}$ is indeed a tiling of $m \times 2^n$ as required.

For the other direction, Fig. 5.13 shows that $\varphi_{n,T}$ is satisfiable in a product of $\langle \mathbb{N}, < \rangle$ and a universal **S5**-frame having $m \cdot 2^n$ points. ❑

As **PTL** $\times$ **S5** is polynomially reducible to $\mathbf{PTL}_{\Box\bigcirc} \times \mathbf{S5}$ (see Claim 6.25), we also obtain the EXPSPACE-hardness of the satisfiability problem for $\mathbf{PTL}_{\Box\bigcirc} \times \mathbf{S5}$. We give a generalization of Theorem 5.43 in Section 6.5 (see Theorem 6.63).

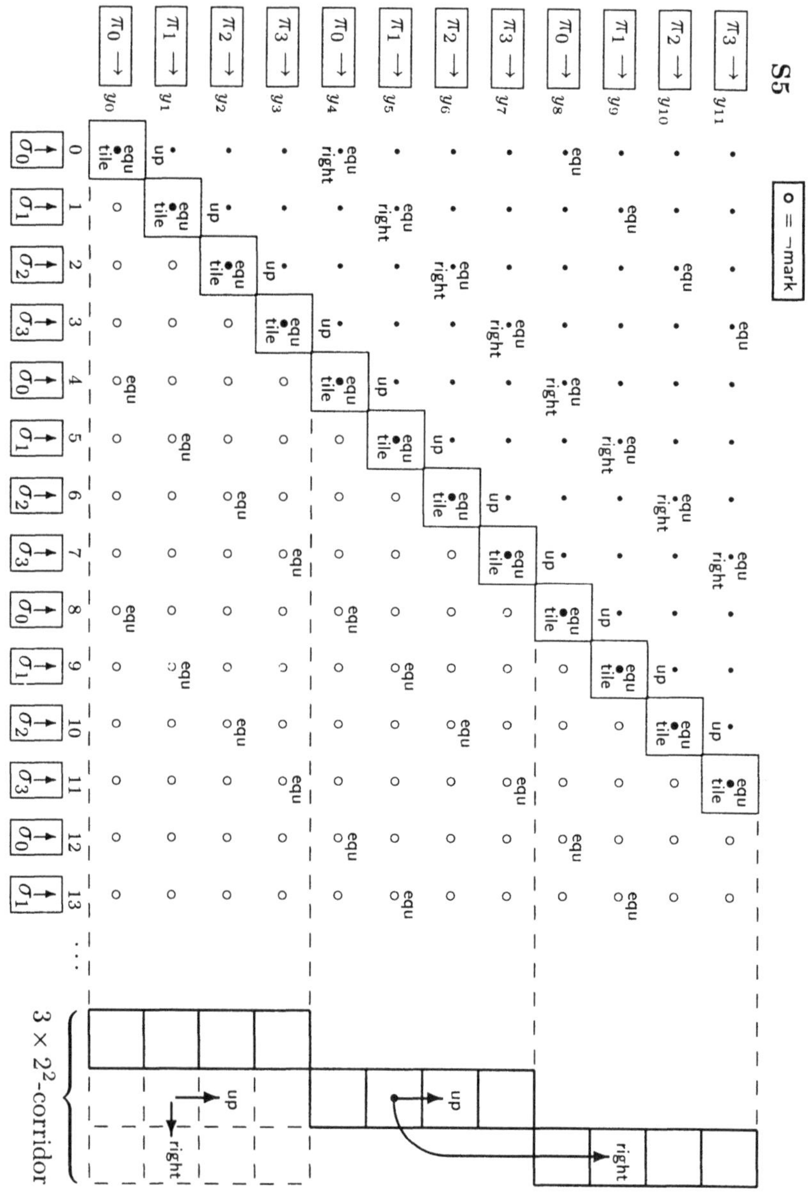

Figure 5.13: Satisfying $\varphi_{2,T}$ in the product of $\langle \mathbb{N}, <, +1 \rangle$ and an **S5**-frame having $3 \cdot 2^2$ elements.

Chapter 6

Decidable products

The landscape of decidable product logics known so far can be roughly described as follows: these are products with $\mathbf{K}_n$- and $\mathbf{S5}_n$-type logics.

We begin this chapter by proving the decidability of products of various expressive multimodal logics with $\mathbf{K}_m$. First, in Section 6.1, we show on the example of $\mathbf{K}_n \times \mathbf{K}_m$ how the method of quasimodels, introduced in Section 5.2, can be used to prove the decidability of products with $\mathbf{K}_m$. Then, in Section 6.2, we generalize the method to establish the decidability of $\mathbf{CPDL} \times \mathbf{K}_m$. In Section 6.3, we draw as consequences the decidability of products of epistemic logics (with common knowledge operators) with $\mathbf{K}_m$. In Section 6.4, we consider products of temporal logics with $\mathbf{K}_m$. In particular, we prove the decidability of $\mathbf{PTL} \times \mathbf{K}_m$ by means of a reduction to $\mathbf{K}_1^C \times \mathbf{K}_m$. We also show how to modify the quasimodel proofs to obtain the decidability of product logics like $\mathbf{K4.3} \times \mathbf{K}_m$, $\mathbf{Lin} \times \mathbf{K}_m$, and $\mathsf{Log}_{FP}(\mathbb{Q}) \times \mathbf{K}_m$.

None of the decision procedures for products with $\mathbf{K}$ we present in this chapter runs in ELEM. Although it is still a challenging open problem whether the product logic $\mathbf{K} \times \mathbf{K}$ is elementary, in Section 6.4 we show that the decision problem for $\mathbf{PTL} \times \mathbf{K}$ (and so for $\mathbf{CPDL} \times \mathbf{K}$ and most of the products of epistemic logics with $\mathbf{K}$) does not belong to ELEM.

None of these results depends on whether we consider products with unimodal $\mathbf{K}$ or multi-modal $\mathbf{K}_m$, $m > 1$. The situation changes drastically if we deal with $\mathbf{S5}$ instead of $\mathbf{K}$. Products with $\mathbf{S5}$ turn out to be computationally simpler than products with $\mathbf{K}$, while products with $\mathbf{S5}_m$, for $m > 1$, behave similarly to products with $\mathbf{K}$. In particular, we show that the filtration technique used in Section 5.3 can be extended to prove that $\mathbf{CPDL} \times \mathbf{S5}$ is decidable in N2EXPTIME. One can also 'mix' the quasimodel techniques used in the proofs of the decidability of $\mathbf{CPDL} \times \mathbf{K}_m$ and $\mathbf{S5} \times \mathbf{S5}$ (Theorem 5.22) to obtain another proof of the decidability of $\mathbf{CPDL} \times \mathbf{S5}$. Product logics like $\mathbf{K4.3} \times \mathbf{S5}$, $\mathbf{Lin} \times \mathbf{S5}$, and $\mathsf{Log}_{FP}(\mathbb{Q}) \times \mathbf{S5}$ are decidable in 2EXPTIME. And finally, **PTL** $\times$ **S5** is EXPSPACE-hard, which matches the upper bound to be established in Section 11.4. On the other hand, in Section 6.6, we consider products with $\mathbf{S5}_m$, $m > 1$, and show that both the 'positive' decidability results and the 'negative' nonelementarity results proved for products with **K** can be generalized to these logics as well.

Properties of a representative family of product logics as well as open questions are summarized in Tables 6.2–6.4 at the end of this chapter.

6.1 Warming up: $\mathbf{K}_n \times \mathbf{K}_m$

Let us begin by using the method of quasimodels to prove the following result of Gabbay and Shehtman (1998):[1]

Theorem 6.1. $\mathbf{K}_n \times \mathbf{K}_m$ *is decidable.*

Proof. To simplify notation, we confine ourselves only to the case of $\mathbf{K} \times \mathbf{K}$; the reader should have no problems with generalizing the proof to the multimodal case. Thus, as in the previous chapter, here we also work with the language $\mathcal{ML}_2$ the modal operators of which are denoted by $\boxminus$, $\boxed{\vert}$ and $\diamondminus$, $\Diamond\!\!\vert$.

Let us fix an $\mathcal{ML}_2$-formula φ and try to define a suitable notion of $\mathbf{K}\times\mathbf{K}$-quasimodel for φ following the pattern of Section 5.2.

Again, by a *type* for φ we mean any Boolean-saturated subset of the set $sub\,\varphi$ of all subformulas in φ. However, clusters of types cannot be used as quasistates for $\mathbf{K} \times \mathbf{K}$. More promising structures are suggested by Proposition 1.8, viz., finite intransitive trees of depth not exceeding the modal depth $md(\varphi)$ of φ.

A *quasistate candidate* for φ is a pair $\langle\langle T, <\rangle, \boldsymbol{t}\rangle$, where $\langle T, <\rangle$ is a finite intransitive tree of depth $\leq md(\varphi)$ and $\boldsymbol{t}$ a labeling function associating with each $x \in T$ a type $\boldsymbol{t}(x)$ for φ. (So we can think of a quasistate candidate as a tree of types.) Two quasistate candidates $\langle\langle T, <\rangle, \boldsymbol{t}\rangle$ and $\langle\langle T', <'\rangle, \boldsymbol{t}'\rangle$ are called *isomorphic* if there is an isomorphism f between the trees $\langle T, <\rangle$ and $\langle T', <'\rangle$ such that $\boldsymbol{t}(x) = \boldsymbol{t}'(f(x))$, for all $x \in T$.

A quasistate candidate $\langle\langle T, <\rangle, \boldsymbol{t}\rangle$ is called a *quasistate* for φ if the following conditions hold:

(qm1) ($\Diamond\!\!\vert$*-saturation*) For all $x \in T$ and $\Diamond\!\!\vert\,\psi \in sub\,\varphi$,

$$\Diamond\!\!\vert\,\psi \in \boldsymbol{t}(x) \quad \text{iff} \quad \exists y \in T\ \big(x < y\ \wedge\ \psi \in \boldsymbol{t}(y)\big).$$

[1] In fact, Gabbay and Shehtman (1998) give two proofs of the theorem which are different from ours: one shows that $\mathbf{K}_n \times \mathbf{K}_m$ has the fmp (cf. Theorem 8.24 below), the other uses the method of normal forms due to Fine (1975a). Marx and Mikulás (2001) obtain the same result using a kind of filtration.

(qm1′) (*smallness*) For all $x, x_1, x_2 \in T$ such that $x < x_1$, $x < x_2$ and $x_1 \neq x_2$, the structures $\langle\langle T^{x_1}, <^{x_1}\rangle, \boldsymbol{t}^{x_1}\rangle$ and $\langle\langle T^{x_2}, <^{x_2}\rangle, \boldsymbol{t}^{x_2}\rangle$ are not isomorphic,

where $\langle T^{x_i}, <^{x_i}\rangle$ is the subtree of $\langle T, <\rangle$ generated by x_i, and $\boldsymbol{t}^{x_i}$ is the restriction of $\boldsymbol{t}$ to T^{x_i}, $i = 1, 2$.

As the number of different types for φ does not exceed $2^{|sub\,\varphi|}$, the number of pairwise nonisomorphic quasistates for φ of depth 0 is at most $2^{|sub\,\varphi|}$ as well. Now define inductively

$$n_0(\varphi) = 2^{|sub\,\varphi|}, \qquad n_{k+1}(\varphi) = 2^{|sub\,\varphi|} \cdot 2^{n_k(\varphi)}.$$

Clearly, $n_k(\varphi)$ is an upper bound for the number of nonisomorphic quasistates for φ of depth k, and so

$$b(\varphi) = \sum_{k=0}^{md(\varphi)} n_k(\varphi) \tag{6.1}$$

is an upper bound for the number of different quasistates for φ. The number of points in any quasistate for φ is bounded by

$$n_0(\varphi) + \sum_{k=1}^{md(\varphi)} \prod_{j=1}^{k} n_{md(\varphi)-j}(\varphi) \leq b(\varphi)^{md(\varphi)} = p(\varphi).$$

In what follows we assume that nonisomorphic quasistates are disjoint and that isomorphic quasistates actually coincide.

A *basic structure* of depth m for φ is a pair $\langle \mathfrak{F}, \boldsymbol{q}\rangle$ such that $\mathfrak{F} = \langle W, R\rangle$ is a frame and $\boldsymbol{q}$ a function associating with each $w \in W$ a quasistate

$$\boldsymbol{q}(w) = \langle\langle T_w, <_w\rangle, \boldsymbol{t}_w\rangle$$

for φ such that the depth of each $\langle T_w, <_w\rangle$ is m.

Let $\langle \mathfrak{F}, \boldsymbol{q}\rangle$ be a basic structure for φ of depth m and let $k \leq m$. A *k-run through* $\langle \mathfrak{F}, \boldsymbol{q}\rangle$ is a function r giving for each $w \in W$ a point $r(w) \in T_w$ of co-depth[2] k. (That is, a run 'goes along' the frame $\mathfrak{F}$ and chooses a (location of a) type of the same co-depth from each type-tree $\langle T_w, <_w\rangle$.) Given a set $\mathfrak{R}$ of runs, we denote by $\mathfrak{R}_k$ the set of all k-runs from $\mathfrak{R}$. Clearly, if $\mathfrak{R}_0$ is not empty, then it is a singleton set, with its only member r_0 being the run through the roots of the quasistates.

A run r is called *coherent* if

$$\forall w \in W\, \forall \Diamond\psi \in sub\,\varphi \left(\exists v \in W\, (wRv \wedge \psi \in \boldsymbol{t}_v(r(v))) \rightarrow \Diamond\psi \in \boldsymbol{t}_w(r(w))\right)$$

[2]The notions of depth and co-depth were defined in Section 1.4.

and *w-saturated* for $w \in W$ if

$$\forall \diamondminus\psi \in sub\,\varphi \left(\diamondminus\psi \in \boldsymbol{t}_w(r(w)) \rightarrow \exists v \in W\left(wRv \wedge \psi \in \boldsymbol{t}_v(r(v))\right)\right).$$

A run is *saturated* if it is w-saturated for all $w \in W$.

Finally, we say that a quadruple $\mathfrak{Q} = \langle \mathfrak{F}, \boldsymbol{q}, \mathfrak{R}, \lhd \rangle$ is a $\mathbf{K} \times \mathbf{K}$-*quasimodel for* φ *(based on* $\mathfrak{F}$*)* if $\langle \mathfrak{F}, \boldsymbol{q}\rangle$ is a basic structure for φ of depth $m \leq md(\varphi)$ such that

(qm2) $\exists w_0 \in W\ \varphi \in \boldsymbol{t}_{w_0}(x_0)$, where x_0 is the root of $\langle T_{w_0}, <_{w_0}\rangle$,

$\mathfrak{R}$ is a set of coherent and saturated runs through $\langle \mathfrak{F}, \boldsymbol{q}\rangle$, and $\lhd$ is a binary relation on $\mathfrak{R}$ satisfying the following conditions:

(qm3) for all $r, r' \in \mathfrak{R}$, if $r \lhd r'$ then $r(w) <_w r'(w)$ for all $w \in W$;

(qm4) $\mathfrak{R}_0 \neq \emptyset$, and for all $k < m$, $r \in \mathfrak{R}_k$, $w \in W$ and $x \in T_w$, if $r(w) <_w x$ then there is $r' \in \mathfrak{R}_{k+1}$ such that $r'(w) = x$ and $r \lhd r'$.

The notion of quasimodel has been defined, and now we have to prove the 'quasimodel lemma:'

Lemma 6.2. *An* $\mathcal{ML}_2$*-formula* φ *is satisfiable in a product frame* $\mathfrak{F} \times \mathfrak{G}$ *iff there is a* $\mathbf{K} \times \mathbf{K}$*-quasimodel for* φ *based on* $\mathfrak{F}$.

Proof. ($\Leftarrow$) Suppose $\langle \mathfrak{F}, \boldsymbol{q}, \mathfrak{R}, \lhd\rangle$ is a quasimodel for φ and $\mathfrak{F} = \langle W, R\rangle$. Take the product frame $\mathfrak{F} \times \langle \mathfrak{R}, \lhd\rangle$ and define a valuation $\mathfrak{V}$ in it as follows:

$$\mathfrak{V}(p) = \{\langle w, r\rangle \mid p \in \boldsymbol{t}_w(r(w))\}$$

for every propositional variable p. Let $\mathfrak{M} = \langle \mathfrak{F} \times \langle \mathfrak{R}, \lhd\rangle, \mathfrak{V}\rangle$. By induction on the construction of $\psi \in sub\,\varphi$ one can show that for every $\langle w, r\rangle$ in $\mathfrak{M}$ we have

$$(\mathfrak{M}, \langle w, r\rangle) \models \psi \quad \text{iff} \quad \psi \in \boldsymbol{t}_w(r(w)).$$

For variables this is just the definition of $\mathfrak{V}$, and the case of Booleans follows from the fact that types are Boolean-saturated. Let $\psi = \diamondminus\chi$. We then have:

$$\begin{aligned}
(\mathfrak{M}, \langle w, r\rangle) \models \diamondminus\chi &\iff \exists w' \in W\left(wRw' \wedge (\mathfrak{M}, \langle w', r\rangle) \models \chi\right)\\
&\iff \exists w' \in W\left(wRw' \wedge \chi \in \boldsymbol{t}_{w'}(r(w'))\right) \text{ [by the induction hypothesis]}\\
&\iff \diamondminus\chi \in \boldsymbol{t}_w(r(w)) \qquad \text{[since } r \text{ is coherent and saturated]}.
\end{aligned}$$

Suppose $\psi = \diamond\!\!\mid\chi$. Then

$$\begin{aligned}
(\mathfrak{M}, \langle w, r\rangle) \models \diamond\!\!\mid\chi &\Longrightarrow \exists r' \in \mathfrak{R}\left(r \lhd r' \wedge (\mathfrak{M}, \langle w, r'\rangle) \models \chi\right)\\
&\Longrightarrow \exists r' \in \mathfrak{R}\left(r \lhd r' \wedge \chi \in \boldsymbol{t}_{w'}(r'(w))\right) \text{ [by the induction hypothesis]}\\
&\Longrightarrow \exists r' \in \mathfrak{R}\left(r(w) <_w r'(w) \wedge \chi \in \boldsymbol{t}_w(r'(w))\right) \qquad \text{[by (qm3)]}\\
&\Longrightarrow \diamond\!\!\mid\chi \in \boldsymbol{t}_w(r(w)) \qquad \text{[by (qm1)]}.
\end{aligned}$$

Conversely,

$$\diamondsuit\chi \in \boldsymbol{t}_w(r(w)) \implies \exists x \in T_w \left(r(w) <_w x \wedge \chi \in \boldsymbol{t}_w(x)\right) \qquad \text{[by } \mathbf{(qm1)}\text{]}$$

Therefore, $r \in \mathfrak{R}_k$ for some $k < m$ (where m is the depth of $\langle \mathfrak{F}, \boldsymbol{q} \rangle$), and so

$$\begin{aligned} & \exists r' \in \mathfrak{R} \left(r \lhd r' \wedge \chi \in \boldsymbol{t}_w(r'(w))\right) && \text{[by } \mathbf{(qm4)}\text{]} \\ \implies & \exists r' \in \mathfrak{R} \left(r \lhd r' \wedge (\mathfrak{M}, \langle w, r' \rangle) \models \chi\right) && \text{[by the induction hypothesis]} \\ \implies & (\mathfrak{M}, \langle w, r \rangle) \models \diamondsuit\chi. \end{aligned}$$

In view of **(qm2)** and $\mathfrak{R}_0 \neq \emptyset$ (which we have by **(qm4)**), it follows that φ is satisfied in $\mathfrak{M}$.

($\Rightarrow$) Suppose that φ is satisfied in a model $\mathfrak{M}$ based on the product $\mathfrak{F} \times \mathfrak{G}$ of frames $\mathfrak{F} = \langle W, R \rangle$ and $\mathfrak{G} = \langle \Delta, < \rangle$. By Propositions 1.7 and 3.10, we may assume that $\mathfrak{G}$ is an intransitive tree of depth $m \leq md(\varphi)$ and that

$$(\mathfrak{M}, \langle w_0, x_0 \rangle) \models \varphi$$

for some $w_0 \in W$, with x_0 being the root of $\mathfrak{G}$. With every pair $\langle w, x \rangle \in W \times \Delta$ we associate the type

$$\boldsymbol{t}(w, x) = \{\psi \in sub\,\varphi \mid (\mathfrak{M}, \langle w, x \rangle) \models \psi\}.$$

Now we have to construct a quasistate $\langle \langle T_w, <_w \rangle, \boldsymbol{t}_w \rangle$ for each $w \in W$. The obvious choice of $T_w = \Delta$, $<_w = <$ and $\boldsymbol{t}_w(x) = \boldsymbol{t}(w, x)$ does not work, because Δ can be infinite. So let us make it finite in such a way that the resulting structure still satisfies **(qm1)** and also complies with the smallness condition **(qm1′)**. Fix a $w \in W$ and define a binary relation $\sim_w$ on Δ as follows. If $x, y \in \Delta$ are of depth 0 (i.e., they are leaves of $\mathfrak{G}$) then

$$x \sim_w y \quad \text{iff} \quad \boldsymbol{t}(w, x) = \boldsymbol{t}(w, y).$$

For $x, y \in \Delta$ of depth k ($0 < k \leq md(\varphi)$), let

$$\begin{aligned} x \sim_w y \qquad \text{iff} \qquad & \boldsymbol{t}(w, x) = \boldsymbol{t}(w, y) \\ & \wedge \forall z \in \Delta \left(x < z \rightarrow \exists z' \in \Delta \, (y < z' \wedge z \sim_w z')\right) \\ & \wedge \forall z \in \Delta \left(y < z \rightarrow \exists z' \in \Delta \, (x < z' \wedge z \sim_w z')\right). \end{aligned}$$

Clearly $\sim_w$ is an equivalence relation on Δ. Denote by $[x]_w$ the $\sim_w$-equivalence class of x and put

$$\begin{aligned} & \Delta_w = \{[x]_w \mid x \in \Delta\}, \\ & [x]_w R_w [y]_w \quad \text{iff} \quad \exists y' \in [y]_w \; x < y', \\ & \boldsymbol{l}_w([x]_w) = \boldsymbol{t}(w, x). \end{aligned}$$

Then, by the definition of $\sim_w$, R_w is well-defined and the structure

$$\langle\langle\Delta_w, R_w\rangle, \boldsymbol{l}_w\rangle$$

clearly satisfies **(qm1′)**. Observe that the map $f_w : x \mapsto [x]_w$ is a p-morphism from $\langle\Delta, <\rangle$ onto $\langle\Delta_w, R_w\rangle$, and so it also satisfies **(qm1)**. However, $\langle\Delta_w, R_w\rangle$ is not necessarily a tree. The tree $\langle T_w, <_w\rangle$ we need can be obtained by unraveling $\langle\Delta_w, R_w\rangle$:

$$\begin{aligned}
T_w &= \{\langle[x_0]_w, \dots, [x_k]_w\rangle \mid k \le m,\ [x_0]_w R_w [x_1]_w R_w \dots R_w [x_k]_w\},\\
u <_w v \quad &\text{iff} \quad u = \langle[x_0]_w, \dots, [x_k]_w\rangle,\ v = \langle[x_0]_w, \dots, [x_k]_w, [x_{k+1}]_w\rangle\\
&\qquad\text{and}\quad [x_k]_w R_w [x_{k+1}]_w.
\end{aligned}$$

Let

$$\boldsymbol{t}_w(\langle[x_0]_w, \dots, [x_k]_w\rangle) = \boldsymbol{l}_w([x_k]_w) = \boldsymbol{t}(w, x_k).$$

It is not hard to see that, for any $w \in W$, $\langle\langle T_w, <_w\rangle, \boldsymbol{t}_w\rangle$ is a quasistate for φ. Moreover, $\varphi \in \boldsymbol{t}_{w_0}(\langle[x_0]_{w_0}\rangle)$. So, by taking

$$\boldsymbol{q}(w) = \langle\langle T_w, <_w\rangle, \boldsymbol{t}_w\rangle$$

for each $w \in W$ we obtain a basic structure $\langle\mathfrak{F}, \boldsymbol{q}\rangle$ for φ satisfying **(qm2)**.

It remains to define appropriate runs through $\langle\mathfrak{F}, \boldsymbol{q}\rangle$. To this end, for each $k \le m$ and each sequence $\langle x_0, \dots, x_k\rangle$ of points in Δ such that $x_0 < \dots < x_k$, take the map

$$r : w \mapsto \langle[x_0]_w, \dots, [x_k]_w\rangle.$$

It is easy to check that r is a coherent and saturated k-run. Let $\mathfrak{R}$ be the set of all such runs. For $r, r' \in \mathfrak{R}$, let $r \lhd r'$ iff $r(w) <_w r'(w)$ for all $w \in W$. Then **(qm3)** holds by definition. It remains to prove **(qm4)**. Let $r \in \mathfrak{R}_k$, $v \in W$ and $z \in T_v$ be such that $r(v) <_v z$. We have to show that there is $r' \in \mathfrak{R}_{k+1}$ such that $r \lhd r'$ and $r'(v) = z$. Since $r(v) <_v z$, we have $r(v) = \langle[x_0]_v, \dots, [x_k]_v\rangle$ and $z = \langle[x_0]_v, \dots, [x_k]_v, [x_{k+1}]_v\rangle$, for some $x_1, \dots, x_k, x_{k+1}$ with $x_0 < x_1 < \dots < x_k$ and $[x_k]_v R_v [x_{k+1}]_v$. By the definition of R_v, there is $y \in [x_{k+1}]_v$ such that $x_k < y$. But then the map

$$r' : w \mapsto \langle[x_0]_w, \dots, [x_k]_w, [y]_w\rangle$$

is in $\mathfrak{R}$. Thus, $\langle\mathfrak{F}, \boldsymbol{q}, \mathfrak{R}, \lhd\rangle$ is a quasimodel for φ. ❑

Our next task is to provide an algorithm for deciding whether there exists a $\mathbf{K} \times \mathbf{K}$-quasimodel for φ. In fact, we will show that instead of finding such a quasimodel, it is enough to find a finite set of finite 'building blocks' out of which a quasimodel for φ can be constructed, with the size of the set and the size of blocks in it being effectively computable.

A *block for* φ *with root* w is a quadruple $\mathfrak{B} = \langle\mathfrak{F}, \boldsymbol{q}, \mathfrak{R}, \lhd\rangle$ such that

- $\mathfrak{F} = \langle \Delta, < \rangle$ is a tree of depth ≤ 1 with root w,
- $\langle \mathfrak{F}, \boldsymbol{q} \rangle$ is a basic structure for φ of depth m, for some $m \leq md(\varphi)$,
- $\mathfrak{R}$ is a set of coherent and w-saturated runs through $\langle \mathfrak{F}, \boldsymbol{q} \rangle$,
- $\lhd$ is a binary relation on $\mathfrak{R}$ satisfying **(qm3)** and **(qm4)**.

Such a block is 'almost' a quasimodel for φ: what is missing is that the runs are not necessarily leaf-saturated and that **(qm2)** may not hold, i.e., φ may not belong to the root of the type-tree at w.

A set $\mathcal{S}$ of blocks for φ is called *satisfying* if

- all blocks in $\mathcal{S}$ are of the same depth m, for some $m \leq md(\varphi)$,
- $\mathcal{S}$ contains a block satisfying **(qm2)** and
- for every block $\mathfrak{B} = \langle \mathfrak{F}, \boldsymbol{q}, \mathfrak{R}, \lhd \rangle$ in $\mathcal{S}$ with $\mathfrak{F} = \langle \Delta, < \rangle$ and every $v \in \Delta$ there exists a block $\mathfrak{B}' = \langle \mathfrak{F}', \boldsymbol{q}', \mathfrak{R}', \lhd' \rangle$ in $\mathcal{S}$ such that $\boldsymbol{q}(v) = \boldsymbol{q}'(w')$ for the root w' of $\mathfrak{B}'$.

Lemma 6.3. *There is a* $\mathbf{K} \times \mathbf{K}$*-quasimodel for* φ *iff there is a satisfying set of blocks for* φ *such that the number of quasistates in each block does not exceed*

$$M(\varphi) = 1 + (md(\varphi) + 1) \cdot p(\varphi) \cdot |sub\, \varphi|.$$

Proof. ($\Leftarrow$) First we show how a quasimodel for φ can be constructed from a satisfying set $\mathcal{S}$ of blocks for φ. To begin with, we call a quadruple $\langle \mathfrak{F}, \boldsymbol{q}, \mathfrak{R}, \lhd \rangle$ a *weak quasimodel for* φ if the following conditions hold:

(wq1) $\mathfrak{F} = \langle W, R \rangle$ is a finite frame and $\langle \mathfrak{F}, \boldsymbol{q} \rangle$ is a basic structure for φ satisfying **(qm2)**;

(wq2) $\mathfrak{R}$ is a set of runs through $\langle \mathfrak{F}, \boldsymbol{q} \rangle$ and $\lhd$ is a binary relation on $\mathfrak{R}$ satisfying **(qm3)** and **(qm4)**;

(wq3) for all $w, v \in W$ such that $w \neq v$ and wRv, there exists a block $\mathfrak{B}^{wv} = \langle \mathfrak{F}^{wv}, \boldsymbol{q}^{wv}, \mathfrak{R}^{wv}, \lhd^{wv} \rangle$ in $\mathcal{S}$ with $\mathfrak{F}^{wv} = \langle \Delta, < \rangle$ such that

- $\Delta \subseteq W$ and $w, v \in \Delta$,
- for all $u \in \Delta$, $\boldsymbol{q}(u) = \boldsymbol{q}^{wv}(u)$,
- for all $u, u' \in \Delta$, if uRu' then $u < u'$,
- for all $r \in \mathfrak{R}$, the restriction r^{wv} of r to Δ is a run in $\mathfrak{R}^{wv}$.

We construct by induction a sequence $\langle \mathfrak{Q}_n \mid n < \omega \rangle$ of weak quasimodels that 'converges' to a quasimodel for φ. Let $\mathfrak{Q}_0 = \langle \mathfrak{F}_0, \boldsymbol{q}_0, \mathfrak{R}_0, \lhd_0 \rangle$ be a block in $\mathcal{S}$ with root w_0 for which **(qm2)** holds. Clearly, it is a weak quasimodel for φ

as well. Suppose now that we have already constructed $\mathfrak{Q}_n = \langle \mathfrak{F}_n, \boldsymbol{q}_n, \mathfrak{R}_n, \lhd_n \rangle$ with $\mathfrak{F}_n = \langle W_n, R_n \rangle$. For each $w \in W_n - W_{n-1}$ (here and in what follows $W_{-1} = \{w_0\}$) select a block $\mathfrak{B}^w = \langle \mathfrak{F}^w, \boldsymbol{q}^w, \mathfrak{R}^w, \lhd^w \rangle$ from $\mathcal{S}$ with root w and $\mathfrak{F}^w = \langle \Delta^w, <^w \rangle$ such that $\boldsymbol{q}_n(w) = \boldsymbol{q}^w(w)$. (The existence of such a block follows from **(wq3)**.) We may assume that all the selected blocks are pairwise disjoint and $\Delta^w \cap W_n = \{w\}$. Define $\langle \mathfrak{F}_{n+1}, \boldsymbol{q}_{n+1} \rangle$ by taking

$$
\begin{aligned}
W_{n+1} &= W_n \cup \bigcup \{\Delta^w \mid w \in W_n - W_{n-1}\}, \\
R_{n+1} &= R_n \cup \bigcup \{<^w \mid w \in W_n - W_{n-1}\}, \\
\mathfrak{F}_{n+1} &= \langle W_{n+1}, R_{n+1} \rangle, \\
\boldsymbol{q}_{n+1}(v) &= \begin{cases} \boldsymbol{q}^w(v), & \text{if } v \in \Delta^w, \ w \in W_n - W_{n-1}, \\ \boldsymbol{q}_n(v), & \text{if } v \in W_n. \end{cases}
\end{aligned}
$$

In other words, we 'glue together' the basic structures $\langle \mathfrak{F}_n, \boldsymbol{q}_n \rangle$ and $\langle \mathfrak{F}^w, \boldsymbol{q}^w \rangle$ at point w.

Next we define $\mathfrak{R}_{n+1}$ and $\lhd_{n+1}$. Suppose that we have $r \in \mathfrak{R}_n$ and a sequence $\bar{s} = \langle s^w \in \mathfrak{R}^w \mid w \in W_n - W_{n-1} \rangle$ such that $r(w) = s^w(w)$, for all $w \in W_n - W_{n-1}$. Define the *extension* $r \cup \bar{s}$ of r by taking, for all $v \in W_{n+1}$,

$$
r \cup \bar{s}(v) = \begin{cases} s^w(v), & \text{if } v \in \Delta^w, \ w \in W_n - W_{n-1}, \\ r(v), & \text{if } v \in W_n. \end{cases}
$$

Let $\mathfrak{R}_{n+1}$ be the set of all such extensions and let

$$
(r_1 \cup \bar{s}_1) \lhd_{n+1} (r_2 \cup \bar{s}_2) \quad \text{iff} \quad r_1 \lhd_n r_2 \text{ and } s_1^w \lhd^w s_2^w, \text{ for all } w \in W_n - W_{n-1}.
$$

It can be readily checked that $\mathfrak{R}_{n+1}$ and $\lhd_{n+1}$ satisfy **(qm3)** and **(qm4)**, and so $\mathfrak{Q}_{n+1} = \langle \mathfrak{F}_{n+1}, \boldsymbol{q}_{n+1}, \mathfrak{R}_{n+1}, \lhd_{n+1} \rangle$ is a weak quasimodel.

The 'limit quasimodel' is defined as follows: let $\mathfrak{F} = \langle W, R \rangle$, where

$$
W = \bigcup_{n<\omega} W_n, \qquad R = \bigcup_{n<\omega} R_n,
$$

and let

$$
\boldsymbol{q} = \bigcup_{n<\omega} \boldsymbol{q}_n.
$$

For each sequence of runs $\langle r_n \in \mathfrak{R}_n \mid n < \omega \rangle$ such that r_{n+1} is an extension of r_n take $r = \bigcup_{n<\omega} r_n$. Let $\mathfrak{R}$ be the set of all such runs. For $r, r' \in \mathfrak{R}$, define

$$
r \lhd r' \qquad \text{iff} \qquad r_n \lhd_n r'_n \text{ for all } n < \omega
$$

(where $r' = \bigcup_{n<\omega} r'_n$).

It is not hard to see, using **(wq1)**–**(wq3)**, that $\langle \mathfrak{F}, \boldsymbol{q}, \mathfrak{R}, \lhd \rangle$ is a quasimodel for φ. Here we show only that all runs in $\mathfrak{R}$ are coherent and saturated, i.e., for all $r \in \mathfrak{R}$, $w \in W$ and $\Diamond\psi \in sub\,\varphi$,

$$\Diamond\psi \in \boldsymbol{t}_w(r(w)) \quad \text{iff} \quad \exists v \in W \left(wRv \;\wedge\; \psi \in \boldsymbol{t}_v(r(v))\right).$$

Suppose that $\Diamond\psi \in \boldsymbol{t}_w(r(w))$, and let n be such that $w \in W_n - W_{n-1}$. Then $\Diamond\psi \in \boldsymbol{t}_w(r_n(w))$ and, by the definition of $\mathfrak{Q}_{n+1}$, there exists $v \in W_{n+1}$ for which $wR_{n+1}v$ and $\psi \in \boldsymbol{t}_v(r_{n+1}(v))$. Conversely, suppose wR_nv and $\psi \in \boldsymbol{t}_v(r_n(v))$. Then it follows from **(wq3)** that $\Diamond\psi \in \boldsymbol{t}_w(r_n(w))$.

($\Rightarrow$) Now we have to show how to extract a set of 'small' blocks from a given quasimodel $\mathfrak{Q} = \langle \mathfrak{F}, \boldsymbol{q}, \mathfrak{R}, \lhd \rangle$ for φ of depth $m \leq md(\varphi)$ with $\mathfrak{F} = \langle W, R \rangle$.

Note first that we may assume each world w in $\mathfrak{F}$ to have arbitrarily many indistinguishable copies in $\mathfrak{Q}$ in the following sense. Say that two distinct worlds $w, w' \in W$ are *twins* (*in* $\mathfrak{Q}$) if

- $\boldsymbol{q}(w) = \boldsymbol{q}(w')$;
- for all $v \in W$, vRw iff vRw', and wRv iff $w'Rv$;
- and for all runs $r \in \mathfrak{R}$, $r(w) = r(w')$.

To construct a satisfying set $\mathcal{S}$ of blocks, we will associate with each $w \in W$ a block $\mathfrak{B}^w = \langle \mathfrak{F}^w, \boldsymbol{q}^w, \mathfrak{R}^w, \lhd^w \rangle$ with root w such that $\boldsymbol{q}^w(w) = \boldsymbol{q}(w)$, and put

$$\mathcal{S} = \{\mathfrak{B}^w \mid w \in W\}.$$

The resulting $\mathcal{S}$ will clearly be a satisfying set of blocks for φ.

So, let $w \in W$. First we define inductively sets of runs $\mathfrak{S}_k \subseteq \mathfrak{R}_k$, $k \leq m$:

- $\mathfrak{S}_0 = \{r_0\}$.
- Given $\mathfrak{S}_k$, we construct $\mathfrak{S}_{k+1}$ as follows. For every run $r \in \mathfrak{S}_k$ and every $x \in T_w$ with $r(w) <_w x$, select an $r' \in \mathfrak{R}_{k+1}$ such that $r \lhd r'$ and $r'(w) = x$, and put it into $\mathfrak{S}_{k+1}$. (Such a run r' exists by **(qm4)**.)

Finally, let $\mathfrak{S} = \bigcup_{k \leq m} \mathfrak{S}_k$. Clearly, $|\mathfrak{S}| \leq p(\varphi)$.

For every $r \in \mathfrak{S}$ and every $\Diamond\psi \in \boldsymbol{t}_w(r(w))$ we then let

$$Sat(r, \Diamond\psi) = \{v \in W \mid wRv,\ \psi \in \boldsymbol{t}_v(r(v))\}.$$

As r is saturated, $Sat(r, \Diamond\psi) \neq \emptyset$. We select a finite subset $\Delta^w(r, \Diamond\psi)$ of $Sat(r, \Diamond\psi)$ in the following way. If $Sat(r, \Diamond\psi) = \{w\}$ then $\Delta^w(r, \Diamond\psi) = \{w\}$ as well. Otherwise, let $\Delta^w(r, \Diamond\psi)$ consist of a $v \neq w$ from $Sat(r, \Diamond\psi)$ together with $m+1$ twins of v. We may assume that the obtained sets $\Delta^w(r, \Diamond\psi)$ are pairwise disjoint.

Now we define

- $\Delta^w = \{w\} \cup \bigcup\{\Delta^w(r, \diamondminus\psi) \mid r \in \mathfrak{S},\ \diamondminus\psi \in \boldsymbol{t}_w(r(w))\}$,
- for all $v, v' \in \Delta^w$, $vR^w v'$ iff $v = w$ and vRv',
- $\mathfrak{F}^w = \langle \Delta^w, R^w \rangle$ and
- for all $v \in \Delta^w$, $\boldsymbol{q}^w(v) = \boldsymbol{q}(v)$.

Then $\mathfrak{F}^w$ is a tree of depth ≤ 1 and $\langle \mathfrak{F}^w, \boldsymbol{q}^w \rangle$ is a basic structure for φ. The cardinality of Δ^w is clearly bounded by $1 + (md(\varphi) + 1) \cdot p(\varphi) \cdot |sub\,\varphi|$.

It remains to define a set $\mathfrak{R}^w$ of coherent and w-saturated runs through $\langle \mathfrak{F}^w, \boldsymbol{q}^w \rangle$ and a binary relation $\lhd^w$ on $\mathfrak{R}^w$ such that **(qm3)** and **(qm4)** hold. A natural candidate for $\mathfrak{R}^w$ seems to be the set $\mathfrak{S}^-$ of all restrictions of runs in $\mathfrak{S}$ to Δ^w. These runs are clearly coherent and w-saturated. However, there may not be enough runs in $\mathfrak{S}^-$: although there is a run in $\mathfrak{S}^-$ coming through every $x \in T_w$, there may exist $x \in T_v$, for some $v \in \Delta^w$, $v \neq w$, which is not in the range of any run r in $\mathfrak{S}^-$. Another choice might be the set $\mathfrak{R}^-$ of the restrictions of all runs in $\mathfrak{R}$ to Δ^w. They clearly go through all points in $\langle \mathfrak{F}^w, \boldsymbol{q}^w \rangle$, they are coherent, but, alas, not necessarily w-saturated. So, to construct the required set $\mathfrak{R}^w$, we have to compose new runs out of old ones.

Let $v \in \Delta^w$, $v \neq w$, and suppose that r and r' are functions whose domain contains Δ^w and $r(w) = r'(w)$. Define a function $r +_v r'$ with domain Δ^w by taking, for all $z \in \Delta^w$,

$$(r +_v r')(z) = \begin{cases} r(z), & \text{if } z = v, \\ r'(z), & \text{if } z \neq v. \end{cases}$$

(Note that a similar operation was used in the proof of Theorem 5.22.) Using this 'addition' function, we now define sets $\mathfrak{R}^w_k$ of runs, for every $k \leq m$. Let $\mathfrak{R}^w_0$ consist of the restriction of r_0 to Δ^w. For $k > 0$, we put all the restrictions of runs from $\mathfrak{S}_k$ into $\mathfrak{R}^w_k$ (i.e., $\mathfrak{S}^-_k \subseteq \mathfrak{R}^w_k$) and also add there the functions

$$r_1 +_{v_1} (r_2 +_{v_2} (\dots (r_l +_{v_l} r) \dots)),$$

where $1 \leq l \leq k$, $r \in \mathfrak{S}_k$, $r_1, \dots, r_l \in \mathfrak{R}_k$ such that $r(w) = r_i(w)$, for $1 \leq i \leq l$, and $v_1, \dots, v_l$ are pairwise distinct points in Δ^w different from w.

Obviously every run $s \in \mathfrak{R}^w$ is coherent. We show that it is w-saturated. This is clear if s belongs to $\mathfrak{S}^-$. Otherwise, s is of the form

$$r_1 +_{v_1} (r_2 +_{v_2} (\dots (r_k +_{v_k} r) \dots))$$

for some $k \leq m$. So, we modified the w-saturated run r at $\leq m$ places. Take some formula $\diamondminus\psi \in \boldsymbol{t}_w(s(w))$. Since we selected for Δ^w $m+1$ twins for each point in $Sat(r, \diamondminus\psi)$, there is still at least one v left to 'saturate s with respect to $\diamondminus\psi$,' that is such that $\psi \in \boldsymbol{t}_v(s(v))$.

Finally, let

$$s = r_1 +_{v_1} (r_2 +_{v_2} (\dots (r_l +_{v_l} r) \dots)), \qquad (6.2)$$
$$s' = r'_1 +_{v'_1} (r'_2 +_{v'_2} (\dots (r'_n +_{v'_n} r') \dots))$$

be two runs in $\mathfrak{R}^w$. (If either s or s' belongs to $\mathfrak{S}^-$ then we consider l or n as 0, respectively.) We let $s \lhd s'$ iff the following hold:

- $s \in \mathfrak{R}^w_k$ and $s' \in \mathfrak{R}^w_{k+1}$, for some $k < m$,
- $r \lhd r'$,
- $l \leq n$ and $v_i = v'_i$, for all $1 \leq i \leq l$,
- for all $z \in \Delta^w$, $r_i(z) <_z r'_i(z)$ whenever $1 \leq i \leq l$, and $r(z) <_z r'_i(z)$ whenever $l+1 \leq i \leq n$.

Then **(qm3)** holds by definition. We show that **(qm4)** also holds. Suppose that s is of the form (6.2), $z \in \Delta^w$, $x \in T_z$ and $s(z) <_z x$. We need a run s' in $\mathfrak{R}^w$ such that $s \lhd^w s'$ and $s'(z) = x$.

Case 1: $z = v_j$ for some $1 \leq j \leq l$. Then $s(z) = r_j(z) = v_j$ for some $r_i \in \mathfrak{R}$. As the original quasimodel $\mathfrak{Q}$ satisfies **(qm4)**, we have a run $r'_j \in \mathfrak{R}$ such that $r_j \lhd r'_j$ and $r'_j(z) = x$. Similarly, for all $i \neq j$, $1 \leq i \leq l$, take a run r'_i from $\mathfrak{R}$ such that $r_i \lhd r'_i$ and $r'_i(w) = r'_j(w)$. Finally, take a run r' from $\mathfrak{S}$ such that $r \lhd r'$ and $r'(w) = r'_j(w)$. (Such a run exists by the definition of $\mathfrak{S}$.) Then

$$s' = r'_1 +_{v_1} (r'_2 +_{v_2} (\dots (r'_l +_{v_l} r') \dots))$$

is a run in $\mathfrak{R}^w$ as required.

Case 2: $z \neq v_j$ for any $1 \leq j \leq l$. Then $s(z) = r(z)$. Select a run r'_{l+1} from $\mathfrak{R}$ such that $r \lhd r'_{l+1}$ and $r'_{l+1}(z) = x$. For each i, $1 \leq i \leq l$, take a run r'_i from $\mathfrak{R}$ such that $r_i \lhd r'_i$ and $r'_i(w) = r'_{l+1}(w)$. Finally, take a run r' from $\mathfrak{S}$ such that $r \lhd r'$ and $r'(w) = r'_{l+1}(w)$. Then

$$s' = r'_1 +_{v_1} (r'_2 +_{v_2} (\dots (r'_{l+1} +_z r') \dots))$$

is a run in $\mathfrak{R}^w$ as required.

Thus, $\langle \mathfrak{F}^w, \boldsymbol{q}^w, \mathfrak{R}^w, \lhd^w \rangle$ is indeed a block with root w. ❑

To complete the proof of Theorem 6.1, it remains to observe that we can effectively construct all possible quasistates for φ, compose out of them, also effectively, the set of all blocks for φ with $\leq M(\varphi)$ quasistates, and then decide whether this set contains a satisfying subset. ❑

As an almost immediate consequence of the proof above, we obtain the following result of (Gabbay and Shehtman 2000); a different proof can be found in (Marx and Mikulás 2002):

Theorem 6.4. $\mathbf{K}_n \times \mathbf{K}_m$ *has the product fmp.*

Proof. We again confine ourselves to the case of $\mathbf{K} \times \mathbf{K}$. Suppose φ is $\mathbf{K} \times \mathbf{K}$-satisfiable. Then, by Propositions 1.7 and 3.10, it is satisfiable in a product $\mathfrak{H} \times \mathfrak{G}$ of two intransitive trees of depths $\leq md(\varphi)$. By Lemma 6.2, there exists a quasimodel $\mathfrak{Q}$ for φ based on $\mathfrak{H}$. Let $\mathfrak{Q}'$ result from $\mathfrak{Q}$ by adding the twins required for constructing blocks. Although, in general, the underlying frame $\mathfrak{H}'$ of $\mathfrak{Q}'$ is not a tree, it is still intransitive. So all the blocks of the satisfying set that are constructed out of $\mathfrak{Q}'$ in the proof of Lemma 6.3 are based on intransitive trees.

The quasimodel for φ built from this satisfying set is based on an intransitive tree $\mathfrak{F} = \langle W, R \rangle$, it satisfies φ at its root w_0, and every point in $\mathfrak{F}$ has at most $M(\varphi)$ R-successors. Now, if we stop the construction of this quasimodel after $md(\varphi)$ steps, then we get a structure $\mathfrak{Q}^- = \langle \mathfrak{F}^-, \boldsymbol{q}^-, \mathfrak{R}^-, \lhd^- \rangle$ based on a finite frame $\mathfrak{F}^-$. This structure satisfies all the required properties of quasimodels for φ save only one: it is not necessarily leaf-saturated. However, we can use the proof of Lemma 6.2 to convert $\mathfrak{Q}^-$ into a model based on the finite product frame $\mathfrak{F}^- \times \langle \mathfrak{R}^-, \lhd^- \rangle$, and this model will clearly satisfy φ at its root. ❑

Observe that the upper bound $b(\varphi)$ for the number of different quasistates (see (6.1) in the proof of Theorem 6.1) is a nonelementary function of the modal depth of φ. So the above decision procedure, as well as all other known decision algorithms for $\mathbf{K} \times \mathbf{K}$, is nonelementary. (It is shown in (Marx and Mikulás 2001) that $\mathbf{K} \times \mathbf{K}$-satisfiability of $\mathcal{ML}_2$-formulas of modal depth ≤ 2 is already NEXPTIME-hard.) However, the following challenging question is still waiting for a solution:

Question 6.5. What is the computational complexity of $\mathbf{K} \times \mathbf{K}$?

As a (relatively) easy exercise the reader can prove the following:

Theorem 6.6. (i) *The decision problem for* $\mathbf{K} \times \mathbf{Alt}$ *is in* EXPTIME.

(ii) $\mathbf{K}_n \times \mathbf{Alt}_m$ *has the product fmp.*.

Hint: since in this case a quasistate is just a $\diamondsuit$-saturated chain of types of length $\leq md(\varphi) + 1$, the upper bound $b(\varphi)$ for the number of different quasistates for a formula φ can be defined as

$$b(\varphi) = 2^{|sub\,\varphi| \cdot (md(\varphi)+1)}.$$

An EXPTIME satisfiability-checking algorithm for $\mathbf{K} \times \mathbf{Alt}$ can be constructed similarly to that in the proof of Theorem 2.27.

We do not know, however, whether this algorithm is optimal:

Question 6.7. Is $\mathbf{K} \times \mathbf{Alt}$ EXPTIME-complete?

6.2 CPDL $\times$ $\mathbf{K}_m$

Now we show how to generalize the constructions of the previous section in order to prove the decidability of the product of **CPDL** (propositional dynamic logic with the converse operator) and $\mathbf{K}_m$. To simplify notation, we again consider products only with unimodal **K**; all the definitions and proofs are easily generalized to multimodal $\mathbf{K}_m$.

To begin with, we briefly explain how the definitions of the syntax and semantics for the products of modal logics introduced above can be extended to products with **CPDL**. Formulas and action terms of $\mathcal{CPDL} \otimes \mathcal{ML}$ are defined by parallel induction as in Section 2.4. (We only note that $\psi?$ is an action term of $\mathcal{CPDL} \otimes \mathcal{ML}$ whenever ψ is a $\mathcal{CPDL} \otimes \mathcal{ML}$-formula.) The modal operators of $\mathcal{CPDL} \otimes \mathcal{ML}$ are $[\alpha]$ and $\langle\alpha\rangle$, for every action term α, as well as $\boxminus$ and $\diamondminus$.

Formulas of $\mathcal{CPDL} \otimes \mathcal{ML}$ are interpreted in $\mathcal{CPDL} \otimes \mathcal{ML}$-*structures*, that is, frames of the form

$$\mathfrak{F} = \langle U, T_{\alpha_1}, T_{\alpha_2}, \ldots, R\rangle,$$

where $\langle U, T_{\alpha_1}, T_{\alpha_2}, \ldots\rangle$ is a $\mathcal{PDL}$-structure and $\langle U, R\rangle$ is a Kripke frame. As usual, a valuation $\mathfrak{V}$ in $\mathfrak{F}$ is a map from the set of propositional variables into subsets of U, and the pair $\mathfrak{M} = \langle \mathfrak{F}, \mathfrak{V}\rangle$ is a model based on $\mathfrak{F}$. Given such a model $\mathfrak{M}$, we define the truth-relation $(\mathfrak{M}, u) \models \varphi$ and the compound transition relations $T_\alpha^{\mathfrak{M}}$ by parallel induction as we did for $\mathcal{CPDL}$ in Section 2.4. In particular, the following clause defines $T_{\varphi?}^{\mathfrak{M}}$ for a $\mathcal{CPDL} \otimes \mathcal{ML}$-formula φ:

- $T_{\varphi?}^{\mathfrak{M}} = \{\langle u, u\rangle \mid (\mathfrak{M}, u) \models \varphi\}$.

Observe that if α does not contain test then $T_\alpha^{\mathfrak{M}}$ is determined only by the $\mathcal{PDL}$-structure $\langle U, T_{\alpha_1}, T_{\alpha_2}, \ldots\rangle$.

The *product*

$$\mathfrak{F} \times \mathfrak{G} = \langle W \times \Delta, \bar{T}_{\alpha_1}, \bar{T}_{\alpha_2}, \ldots, R_v\rangle$$

of a $\mathcal{PDL}$-structure $\mathfrak{F} = \langle W, T_{\alpha_1}, T_{\alpha_2}, \ldots\rangle$ and a frame $\mathfrak{G} = \langle \Delta, R\rangle$ is a special kind of $\mathcal{CPDL} \otimes \mathcal{ML}$-structure defined by taking for all $u_1, u_2 \in W$, all $x_1, x_2 \in \Delta$, and all atomic actions α_i,

$$\begin{aligned} \langle u_1, x_1\rangle \, \bar{T}_{\alpha_i} \, \langle u_2, x_2\rangle \quad &\text{iff} \quad u_1 T_{\alpha_i} u_2 \ \text{ and } \ x_1 = x_2, \\ \langle u_1, x_1\rangle \, R_v \, \langle u_2, x_2\rangle \quad &\text{iff} \quad x_1 R x_2 \ \text{ and } \ u_1 = u_2. \end{aligned} \tag{6.3}$$

Note that if α contains test then (6.3) does not necessarily hold for $\bar{T}_\alpha$. However, for any model $\mathfrak{M}$ based on $\mathfrak{F} \times \mathfrak{G}$ and any transition relation $\bar{T}_\alpha^{\mathfrak{M}}$, we have the following:

$$\text{if } \ \langle u_1, x_1\rangle \, \bar{T}_\alpha^{\mathfrak{M}} \, \langle u_2, x_2\rangle \ \text{ then } \ x_1 = x_2. \tag{6.4}$$

The logic **CPDL** $\times$ **K** is defined as the set of all $\mathcal{CPDL}\otimes\mathcal{ML}$-formulas that are valid in all product $\mathcal{CPDL}\otimes\mathcal{ML}$-structures. Similarly to Proposition 2.21, one can show that every $\mathcal{CPDL}\otimes\mathcal{ML}$-formula is equivalent in **CPDL** $\times$ **K** to a formula in which the converse operator is applied only to action variables. So in what follows we consider only formulas of this form.

As to finding an *axiomatization* for **CPDL** $\times$ **K**, first observe that all the axioms of **CPDL** (see Section 2.4) hold in every model based on a product $\mathcal{CPDL}\otimes\mathcal{ML}$-structure. Further, the formulas

$$\Diamond\langle\alpha_i\rangle p \leftrightarrow \langle\alpha_i\rangle\Diamond p \quad \text{and} \quad \langle\alpha_i\rangle\Box p \to \Box\langle\alpha_i\rangle p \tag{6.5}$$

with atomic actions α_i hold in such models as well. (Note that commutativity and Church–Rosser properties for all action terms α not containing test follow. On the other hand, it is easy to find models based on product frames where (6.5) does not hold for some action term having test.) So, a natural candidate for an axiomatization of **CPDL** $\times$ **K** could be obtained by putting together the **CPDL**-axioms and (6.5). It is not known, however, whether the resulting logic is complete with respect to 'standard' models, that is, models based on $\mathcal{CPDL}\otimes\mathcal{ML}$-structures, with each pair T_{α_i} and R having the commutativity and Church–Rosser properties. So the following question is open:

Question 6.8. Give an axiomatization for **CPDL** $\times$ **K**.

An axiomatization for **CPDL** $\times$ **S5** is given in Section 6.5.

Remark 6.9. We can define the *test-free fragment of* **CPDL** $\times$ **K** as the set of those formulas in **CPDL** $\times$ **K** that do not contain action terms of the form φ?. The language of the test-free fragment of **CPDL** $\times$ **K** has the $\Box$ of $\mathcal{ML}$ and a modal operator $[\alpha]$ for every test-free action term α. So strictly speaking a frame interpreting this multimodal language is not a $\mathcal{CPDL}\otimes\mathcal{ML}$-structure as introduced above, but any structure of the form

$$\mathfrak{F} = \langle U, T_\alpha, \ldots, R\rangle,$$

where U is a (nonempty) set and the T_α are binary relations on U, one for each test-free action term α (not only for atomic actions).

It is easy to see that the test-free fragment of **CPDL** $\times$ **K** in fact coincides with the usual product $\mathbf{CPDL}^{-?} \times \mathbf{K}$, where $\mathbf{CPDL}^{-?}$ is the test-free fragment of **CPDL** (which is a Kripke complete multimodal logic, see Remark 2.23). Since all the axioms of $\mathbf{CPDL}^{-?}$ hold in $\mathbf{CPDL}^{-?} \times \mathbf{K}$, we obtain that in *all* (not just in the product) frames (of the above form) for $\mathbf{CPDL}^{-?} \times \mathbf{K}$, the relation T_{α^*} is the reflexive and transitive closure of T_α, $T_{\alpha\cup\beta} = T_\alpha \cup T_\beta$, $T_{\alpha;\beta} = T_\alpha \circ T_\beta$, and $T_{\alpha^-} = T_\alpha^{-1}$, for all test-free action terms α, β (cf. Remark 2.23).

In the remaining part of this section we prove the following result of Wolter (2000b):

Theorem 6.10. **CPDL × $\mathbf{K}_m$** *is decidable.*

Proof. To begin with, let us fix a $\mathcal{CPDL} \otimes \mathcal{ML}$-formula φ and define a notion of a **CPDL × K**-quasimodel for φ.

As is well known, when treating logics related to **PDL**, it is not enough to consider only subformulas of φ: a somewhat larger set of formulas, known as the *Fischer–Ladner closure* of φ, is required. This set, denoted here by $flc(\varphi)$, is the smallest set of formulas containing φ and satisfying the following conditions:

- if $\psi \wedge \chi \in flc(\varphi)$ then $\psi \in flc(\varphi)$ and $\chi \in flc(\varphi)$,
- if $\neg\psi \in flc(\varphi)$ then $\psi \in flc(\varphi)$,
- if $\Diamond\psi \in flc(\varphi)$ then $\psi \in flc(\varphi)$,
- if $\langle\alpha\rangle\,\psi \in flc(\varphi)$ then $\psi \in flc(\varphi)$,
- if $\langle\alpha;\beta\rangle\,\psi \in flc(\varphi)$ then $\langle\alpha\rangle\langle\beta\rangle\,\psi \in flc(\varphi)$,
- if $\langle\alpha \cup \beta\rangle\,\psi \in flc(\varphi)$ then $\langle\alpha\rangle\,\psi \in flc(\varphi)$ and $\langle\beta\rangle\,\psi \in flc(\varphi)$,
- if $\langle\alpha^*\rangle\,\psi \in flc(\varphi)$ then $\psi \in flc(\varphi)$ and $\langle\alpha\rangle\,\langle\alpha^*\rangle\,\psi \in flc(\varphi)$,
- if $\langle\alpha_i^-\rangle\,\psi \in flc(\varphi)$ then $\langle\alpha_i\rangle\,\psi \in flc(\varphi)$,
- if $\langle\psi?\rangle\,\chi \in flc(\varphi)$ then $\psi \in flc(\varphi)$ and $\chi \in flc(\varphi)$.

Note that $|flc(\varphi)|$ is linear in the length (i.e., the number of symbols) of φ (for a proof see, e.g., Harel *et al.* 2000).

Now, a *type* for φ is a Boolean saturated subset $\boldsymbol{t}$ of $flc(\varphi)$ satisfying the following conditions:

(t1) $\langle\alpha;\beta\rangle\,\psi \in \boldsymbol{t}$ iff $\langle\alpha\rangle\langle\beta\rangle\,\psi \in \boldsymbol{t}$, for all $\langle\alpha;\beta\rangle\,\psi \in flc(\varphi)$,

(t2) $\langle\alpha^*\rangle\,\psi \in \boldsymbol{t}$ iff either $\psi \in \boldsymbol{t}$ or $\langle\alpha\rangle\langle\alpha^*\rangle\,\psi \in \boldsymbol{t}$, for all $\langle\alpha^*\rangle\,\psi \in flc(\varphi)$,

(t3) $\langle\alpha \cup \beta\rangle\,\psi \in \boldsymbol{t}$ iff either $\langle\alpha\rangle\,\psi \in \boldsymbol{t}$ or $\langle\beta\rangle\,\psi \in \boldsymbol{t}$, for all $\langle\alpha \cup \beta\rangle\,\psi \in flc(\varphi)$,

(t4) $\langle\psi?\rangle\,\chi \in \boldsymbol{t}$ iff $\psi \in \boldsymbol{t}$ and $\chi \in \boldsymbol{t}$, for all $\langle\psi?\rangle\,\chi \in flc(\varphi)$.

The number of pairwise distinct types for φ does not exceed $2^{|flc(\varphi)|}$.

A *quasistate* for φ is defined word by word as in the previous section, but using the above definitions of types, $flc(\varphi)$ instead of $sub\,\varphi$, and the following modified definition of $md(\varphi)$.

The *modal depth* $md(\varphi)$ of a $\mathcal{CPDL}\otimes\mathcal{ML}$-formula φ and the *modal depth* $md(\alpha)$ of a $\mathcal{CPDL}\otimes\mathcal{ML}$-action term α are defined inductively as follows:

$$\begin{aligned} md(p) &= 0, & md(\varphi\wedge\psi) &= \max\{md(\varphi), md(\psi)\},\\ md(\neg\varphi) &= md(\varphi), & md(\Diamond\varphi) &= md(\varphi)+1,\\ md(\langle\alpha\rangle\varphi) &= md(\alpha)+md(\varphi), & md(\alpha_i) &= md(\alpha_i^-) = 0,\\ md(\alpha;\beta) &= \max\{md(\alpha), md(\beta)\}, & md(\alpha\cup\beta) &= \max\{md(\alpha), md(\beta)\},\\ md(\alpha^*) &= md(\alpha), & md(\psi?) &= md(\psi). \end{aligned}$$

The upper bound $b(\varphi)$ for the number of different (i.e., nonisomorphic) quasistates for φ and the upper bound $p(\varphi)$ for the number of points in a quasistate for φ are computed as in the previous section using $flc(\varphi)$ in place of $sub\,\varphi$.

A *basic structure* of depth m for φ is a pair $\langle\mathfrak{F}, \boldsymbol{q}\rangle$ such that

$$\mathfrak{F} = \langle W, T_{\alpha_1}, \dots, T_{\alpha_n}\rangle$$

is an n-frame, where $\alpha_1, \dots, \alpha_n$ is an enumeration of all action variables in φ, and $\boldsymbol{q}$ is a function associating with each $w\in W$ a quasistate

$$\boldsymbol{q}(w) = \langle\langle T_w, <_w\rangle, \boldsymbol{t}_w\rangle$$

for φ of depth m.

As before, for any $k\le m$, a *k-run through* $\langle\mathfrak{F}, \boldsymbol{q}\rangle$ is a function r associating with each $w\in W$ a point $r(w)\in T_w$ of co-depth k. Now we need to define analogs of the coherency and saturation conditions for the new runs. To this end, for each run r through $\langle\mathfrak{F}, \boldsymbol{q}\rangle$, we define first a binary relation T^r_α on W as follows:

- $wT^r_{\alpha_i}v$ iff $wT_{\alpha_i}v$,
- $wT^r_{\alpha_i^-}v$ iff $vT^r_{\alpha_i}w$,
- $wT^r_{\alpha;\beta}v$ iff $w(T^r_\alpha\circ T^r_\beta)v$,
- $wT^r_{\alpha\cup\beta}v$ iff $w(T^r_\alpha\cup T^r_\beta)v$,
- $wT^r_{\alpha^*}v$ iff $w(T^r_\alpha)^*v$,
- $wT^r_{\psi?}v$ iff $w=v$ and $\psi\in\boldsymbol{t}_w(r(w))$.

The relation T^r_α depends on r only when α contains test.

Now, a run r is called *coherent* if

$$\forall w\in W\,\forall\langle\alpha\rangle\psi\in flc(\varphi)$$
$$\Big(\exists v\in W\,\big(wT^r_\alpha v\wedge\psi\in\boldsymbol{t}_v(r(v))\big)\rightarrow\langle\alpha\rangle\psi\in\boldsymbol{t}_w(r(w))\Big).$$

It is called *w-saturated*, for $w \in W$, if

$$\forall \langle\alpha\rangle\,\psi \in \mathit{flc}(\varphi)\ \Big(\langle\alpha\rangle\,\psi \in \boldsymbol{t}_w(r(w)) \ \rightarrow\ \exists v \in W\ \big(wT^r_\alpha v\ \wedge\ \psi \in \boldsymbol{t}_v(r(v))\big)\Big).$$

A run is *saturated* if it is w-saturated for all $w \in W$.

Finally, **CPDL** × **K**-*quasimodels for* φ are defined precisely as **K** × **K**-quasimodels in the previous section, using the new definitions of basic structures and runs.

The following lemma, like Lemma 6.2, establishes a connection between models based on product frames and quasimodels:

Lemma 6.11. *A* $\mathcal{CPDL} \otimes \mathcal{ML}$*-formula* φ *is satisfied in a model based on a product* $\mathcal{CPDL} \otimes \mathcal{ML}$*-structure iff there is a* **CPDL** × **K**-*quasimodel for* φ.

Proof. In principle, the proof follows the lines of the proof of Lemma 6.2, but of course it is a bit more tiresome.

($\Leftarrow$) Consider a quasimodel $\langle\mathfrak{F}, \boldsymbol{q}, \mathfrak{R}, \lhd\rangle$ for φ with

$$\mathfrak{F} = \langle W, T_{\alpha_1}, \ldots, T_{\alpha_n}\rangle,$$

and let $\mathfrak{F}' = \big\langle W, T_{\alpha_1}, \ldots, T_{\alpha_n}, T_{\alpha_{n+1}}, \ldots\big\rangle$ be any $\mathcal{PDL}$-structure 'extending' $\mathfrak{F}$. Define a valuation $\mathfrak{V}$ in the product $\mathcal{CPDL} \otimes \mathcal{ML}$-structure $\mathfrak{F}' \times \langle\mathfrak{R}, \lhd\rangle$ by taking

$$\mathfrak{V}(p) = \{\langle w, r\rangle \mid p \in \boldsymbol{t}_w(r(w))\}$$

for every propositional variable p. Put $\mathfrak{M} = \langle\mathfrak{F}' \times \langle\mathfrak{R}, \lhd\rangle, \mathfrak{V}\rangle$. The following two equivalences can be proved by parallel induction for all $w, v \in W$:

- for every $\psi \in \mathit{flc}(\varphi)$ and every $r \in \mathfrak{R}$,

$$(\mathfrak{M}, \langle w, r\rangle) \models \psi \quad \text{iff} \quad \psi \in \boldsymbol{t}_w(r(w));$$

- for every action term α occurring in $\mathit{flc}(\varphi)$ and every $r \in \mathfrak{R}$,

$$wT^r_\alpha v \quad \text{iff} \quad \langle w, r\rangle\,\bar{T}^{\mathfrak{M}}_\alpha\,\langle v, r\rangle\,.$$

We show only the induction steps for $\psi = \langle\alpha\rangle\,\chi$ and $\alpha = \chi?$:

$$\begin{aligned}
&(\mathfrak{M}, \langle w, r\rangle) \models \langle\alpha\rangle\,\chi\\
&\quad\Longleftrightarrow\quad \exists v \in W, s \in \mathfrak{R}\ \big(\langle w, r\rangle\,\bar{T}^{\mathfrak{M}}_\alpha\,\langle v, s\rangle\ \wedge\ (\mathfrak{M}, \langle v, s\rangle) \models \chi\big)\\
&\quad\Longleftrightarrow\quad \exists v \in W\ \big(wT^r_\alpha v\ \wedge\ \chi \in \boldsymbol{t}_v(r(v))\big) \quad \text{[by (6.4) and IH]}\\
&\quad\Longleftrightarrow\quad \langle\alpha\rangle\,\chi \in \boldsymbol{t}_w(r(w)) \qquad \text{[since } r \text{ is coherent and saturated];}
\end{aligned}$$

$$\begin{aligned}
wT^r_{\chi?}v \quad&\Longleftrightarrow\quad w = v \ \text{ and }\ \chi \in \boldsymbol{t}_w(r(w))\\
&\Longleftrightarrow\quad w = v \ \text{ and }\ (\mathfrak{M}, \langle w, r\rangle) \models \chi \qquad \text{[by the induction hypothesis]}\\
&\Longleftrightarrow\quad \langle w, r\rangle\,\bar{T}^{\mathfrak{M}}_{\chi?}\,\langle v, r\rangle \qquad \text{[by the definition of } \bar{T}^{\mathfrak{M}}_{\chi?}\text{].}
\end{aligned}$$

It follows then from **(qm2)** that φ is satisfied in $\mathfrak{M}$.

($\Rightarrow$) Suppose that φ is satisfied in a model $\mathfrak{M}$ based on the product $\mathfrak{H} \times \mathfrak{G}$ of a $\mathcal{PDL}$-structure $\mathfrak{H} = \langle W, T_{\alpha_1}, \dots \rangle$ and a frame $\mathfrak{G} = \langle \Delta, < \rangle$. By Proposition 1.7 and a straightforward generalization of Proposition 3.10, we may also assume that $\mathfrak{G}$ is an intransitive tree of depth $m \leq md(\varphi)$ and that

$$(\mathfrak{M}, \langle w_0, x_0 \rangle) \models \varphi,$$

for some $w_0 \in W$ and root x_0 of $\mathfrak{G}$. As before, with each pair $\langle w, x \rangle$ in $W \times \Delta$ we associate the type

$$\boldsymbol{t}(w, x) = \{\psi \in \mathit{flc}(\varphi) \mid (\mathfrak{M}, \langle w, x \rangle) \models \psi\}.$$

Let $\alpha_1, \dots, \alpha_n$ be an enumeration of the action variables occurring in φ. A quasimodel for φ based on the frame

$$\mathfrak{F} = \langle W, T_{\alpha_1}, \dots, T_{\alpha_n} \rangle$$

can be then constructed in precisely the same way as in the proof of Lemma 6.2. ❑

We now show how to extend the proof of Theorem 6.1 to obtain a decidability proof for **CPDL** $\times$ **K**.

Note first that trees of depth ≤ 1 are no longer enough for constructing blocks. Now a *block for* φ *with root* w_0 is a quadruple $\mathfrak{B} = \langle \mathfrak{F}, \boldsymbol{q}, \mathfrak{R}, \lhd \rangle$ satisfying the following properties:

(b1) $\mathfrak{F} = \langle \Delta, T_{\alpha_1}, \dots, T_{\alpha_n} \rangle$ is a finite n-frame with a simple 'tree-like' structure:

- for all $w, v \in \Delta$ with $w \neq v$ and $\beta_1, \beta_2 \in \{\alpha_j, \alpha_j^- \mid 1 \leq j \leq n\}$, if $wT_{\beta_1}v$ and $wT_{\beta_2}v$ then $\beta_1 = \beta_2$;
- for every $w \in \Delta$ such that $w \neq w_0$,

$$\left|\{v \mid \exists \beta \in \{\alpha_i, \alpha_i^- \mid i \leq n\}\ wT_\beta v\}\right| \leq 2;$$

- for every $v \in W$, there exists a unique sequence $\langle v_0, v_1, \dots, v_m \rangle$ of distinct points in Δ such that $w_0 = v_0$, $v = v_m$ and, for any $i < m$, there exists $\beta \in \{\alpha_j, \alpha_j^- \mid 1 \leq j \leq n\}$ with $v_i T_\beta v_{i+1}$;

(b2) $\langle \mathfrak{F}, \boldsymbol{q} \rangle$ is a basic structure for φ;

(b3) $\mathfrak{R}$ is a set of coherent and w_0-saturated runs through $\langle \mathfrak{F}, \boldsymbol{q} \rangle$;

(b4) $\lhd$ is a binary relation on $\mathfrak{R}$ satisfying **(qm3)** and **(qm4)**.

The definition of a *satisfying set of blocks* remains precisely the same as in the previous section.

Our aim is to show that φ is satisfiable iff there is a satisfying set $\mathcal{S}$ for φ such that the size of each block in $\mathcal{S}$ is at most $N(\varphi)$, for some natural number $N(\varphi)$ effectively computable from the length of φ. In contrast to the upper bound $M(\varphi)$ found for $\mathbf{K} \times \mathbf{K}$, now the size of the blocks depends also on the number of nestings of action terms in φ. In order to compute the necessary upper bound, we first have to make every action term in φ 'iteration-free.' Namely, for every natural number n and every action term α, we define an action term $\alpha(n)$ as follows:

- $\alpha(n) = \alpha$, if $\alpha = \alpha_i$, $\alpha = \alpha_i^-$, or $\alpha = \psi?$,
- $(\beta \cup \gamma)(n) = \beta(n) \cup \gamma(n)$,
- $(\beta \,;\gamma)(n) = \beta(n) \,; \gamma(n)$,
- $\beta^*(n) = \beta^{\leq n}(n)$, where

$$\beta^{\leq n} = \top? \cup \beta \cup (\beta \,;\beta) \cup \cdots \cup \beta^n \quad \text{and} \quad \beta^n = \overbrace{\beta\,;\ldots;\beta}^{n}.$$

In other words, $\alpha(n)$ results from α by replacing every occurrence of an action term of the form β^* (which is not in the scope of a test $\psi?$) with $\beta^{\leq n}$. Therefore, $\alpha(n)$ contains no occurrence of *.

The *length* $|\alpha|$ of an action term α without iteration is defined as follows:

- $|\psi?| = 0$,
- $|\alpha_i| = |\alpha_i^-| = 1$,
- $|\beta \cup \gamma| = \max\{|\beta|, |\gamma|\}$,
- $|\beta \,;\gamma| = |\beta| + |\gamma|$.

Finally, we put

$$l(\varphi) = \max\left\{ \left|\alpha(b(\varphi)\cdot p(\varphi))\right| \mid \langle\alpha\rangle\,\psi \in \mathit{flc}(\varphi)\right\}.$$

(Recall that $b(\varphi)$ and $p(\varphi)$ are the upper bounds for the number of different quasistates for φ and the number of different points in a quasistate for φ, respectively.) We are now in a position to formulate and prove a satisfiability criterion.

Lemma 6.12. *There is a* **CPDL** $\times$ **K**-*quasimodel for* φ *iff there is a satisfying set of blocks for* φ *in which the size of each block is at most*

$$N(\varphi) = 1 + (md(\varphi) + 1)\cdot l(\varphi)\cdot p(\varphi)\cdot |\mathit{flc}(\varphi)|.$$

Proof. ($\Leftarrow$) Suppose that $\mathcal{S}$ is a satisfying set of blocks for φ. The construction of the limit quasimodel $\langle \mathfrak{F}, \boldsymbol{q}, \mathfrak{R}, \lhd \rangle$ is analogous to that in the proof of Lemma 6.3. The only point where the proof gets more complicated is the argument showing that all the runs in $\mathfrak{R}$ are coherent. Let $\mathfrak{F} = \langle W, T_{\alpha_1}, \dots, T_{\alpha_n} \rangle$. We claim that for all $r \in \mathfrak{R}$, $w \in W$ and $\langle \alpha \rangle \psi \in \mathit{flc}(\varphi)$,

$$\exists w' \in W \left(w T^r_\alpha w' \ \wedge \ \psi \in \boldsymbol{t}_{w'}(r(w')) \right) \ \rightarrow \ \langle \alpha \rangle \psi \in \boldsymbol{t}_w(r(w)).$$

The proof is by induction on the construction of α.

Case 1: $\alpha = \alpha_i$, where α_i an action variable. Suppose that there is a $w' \in W$ such that $w T_{\alpha_i} w'$ and $\psi \in \boldsymbol{t}_{w'}(r(w'))$. Then by **(wq3)**, we have $\langle \alpha_i \rangle \psi \in \boldsymbol{t}_w(r(w))$.

Case 2: $\alpha = \alpha_i^-$, α_i an action variable. This case is analogous to Case 1.

Case 3: $\alpha = \chi?$. Then we have

$$\begin{aligned}
&\exists w' \in W \left(w T^r_{\chi?} w' \ \wedge \ \psi \in \boldsymbol{t}_{w'}(r(w')) \right) \\
&\quad \Longrightarrow \quad \chi \in \boldsymbol{t}_w(r(w)) \ \wedge \ \psi \in \boldsymbol{t}_w(r(w)) \qquad \text{[by the definition of } T^r_{\chi?}] \\
&\quad \Longrightarrow \quad \langle \chi? \rangle \psi \in \boldsymbol{t}_w(r(w)) \qquad \text{[by } \mathbf{(t4)}].
\end{aligned}$$

Case 4: $\alpha = \beta\,;\gamma$. Then

$$\begin{aligned}
&\exists w' \in W \left(w T^r_{\beta\,;\gamma} w' \ \wedge \ \psi \in \boldsymbol{t}_{w'}(r(w')) \right) \\
&\quad \Longrightarrow \quad \exists w' \in W \left(w (T^r_\beta \circ T^r_\gamma) w' \ \wedge \ \psi \in \boldsymbol{t}_{w'}(r(w')) \right) \quad \text{[by def. of } T^r_{\beta\,;\gamma}] \\
&\quad \Longrightarrow \quad \langle \beta \rangle \langle \gamma \rangle \psi \in \boldsymbol{t}_w(r(w)) \qquad \text{[by IH]} \\
&\quad \Longrightarrow \quad \langle \beta\,;\gamma \rangle \psi \in \boldsymbol{t}_w(r(w)) \qquad \text{[by } \mathbf{(t1)}].
\end{aligned}$$

Case 5: $\alpha = \beta \cup \gamma$. Then

$$\begin{aligned}
&\exists w' \in W \Big(w T^r_{\beta \cup \gamma} w' \ \wedge \ \psi \in \boldsymbol{t}_{w'}(r(w')) \Big) \\
&\quad \Longrightarrow \quad \exists w' \in W \left((w T^r_\beta w' \vee w T^r_\gamma w') \wedge \psi \in \boldsymbol{t}_{w'}(r(w')) \right) \quad \text{[by def. of } T^r_{\beta \cup \gamma}] \\
&\quad \Longrightarrow \quad \langle \beta \rangle \psi \in \boldsymbol{t}_w(r(w)) \ \vee \ \langle \gamma \rangle \psi \in \boldsymbol{t}_w(r(w)) \quad \text{[by IH]} \\
&\quad \Longrightarrow \quad \langle \beta \cup \gamma \rangle \psi \in \boldsymbol{t}_w(r(w)) \qquad \text{[by } \mathbf{(t3)}].
\end{aligned}$$

Case 6: $\alpha = \beta^*$. Suppose that $w T^r_{\beta^*} w'$ and $\psi \in \boldsymbol{t}_{w'}(r(w'))$, for some $w' \in W$. Then, by the definition of $T^r_{\beta^*}$, there are $w_0, \dots, w_k \in W$ such that $w_0 = w$, $w_k = w'$ and $w_i T^r_\beta w_{i+1}$ for all $i < k$. By **(t2)**, we have $\langle \beta^* \rangle \psi \in \boldsymbol{t}_{w_k}(r(w_k))$. If $k = 0$ then we are done. Otherwise, by the induction hypothesis,

$$\langle \beta \rangle \langle \beta^* \rangle \psi \in \boldsymbol{t}_{w_{k-1}}(r(w_{k-1})),$$

and so again by **(t2)**, we have $\langle \beta^* \rangle \psi \in \boldsymbol{t}_{w_{k-1}}(r(w_{k-1}))$. By repeating this argument we obtain $\langle \beta^* \rangle \psi \in \boldsymbol{t}_{w_0}(r(w_0))$.

This proves the implication ($\Leftarrow$) of Lemma 6.12.

($\Rightarrow$) Suppose $\mathfrak{Q} = \langle \mathfrak{F}, \boldsymbol{q}, \mathfrak{R}, \lhd \rangle$ is a quasimodel for φ of depth $m \leq md(\varphi)$ and $\mathfrak{F} = \langle W, T_{\alpha_1}, \dots, T_{\alpha_n} \rangle$. We may again assume that each world w in $\mathfrak{F}$ has arbitrarily many indistinguishable copies in $\mathfrak{Q}$ in the following sense. Two distinct worlds $w, w' \in W$ are called *twins (in $\mathfrak{Q}$)* if

- $\boldsymbol{q}(w) = \boldsymbol{q}(w')$;
- for all $v \in W$ and i, $1 \leq i \leq n$, $vT_{\alpha_i}w$ iff $vT_{\alpha_i}w'$, and $wT_{\alpha_i}v$ iff $w'T_{\alpha_i}v$;
- for all runs $r \in \mathfrak{R}$, $r(w) = r(w')$.

We will construct a satisfying set $\mathcal{S}$ of blocks by associating with each $w_0 \in W$ a block $\mathfrak{B}^{w_0} = \langle \mathfrak{F}^{w_0}, \boldsymbol{q}^{w_0}, \mathfrak{R}^{w_0}, \lhd^{w_0} \rangle$ with root w such that $\boldsymbol{q}^w(w) = \boldsymbol{q}(w)$.

Fix a $w_0 \in W$. Similarly to the previous section, define first a set $\mathfrak{S}$ of 'auxiliary' runs as follows:

- $\mathfrak{S}_0 = \{r_0\}$;
- to construct $\mathfrak{S}_{k+1}$, for every $r \in \mathfrak{S}_k$ and every $x \in T_w$ with $r(w) <_w x$ we take an $r' \in \mathfrak{R}_{k+1}$ such that $r \lhd r'$ and $r'(w) = x$, and put it in $\mathfrak{S}_{k+1}$ (such a run r' exists by **(qm4)**).

Finally, let $\mathfrak{S} = \bigcup_{k \leq m} \mathfrak{S}_k$. Then we clearly have $|\mathfrak{S}| \leq p(\varphi)$.

Next, we define $\mathfrak{F}^{w_0} = \langle \Delta^{w_0}, S_{\alpha_1}, \dots, S_{\alpha_n} \rangle$. Recall the definition of Δ^w from the proof of Lemma 6.3: in order to make the runs in $\mathfrak{S}$ root-saturated, we put $md(\varphi) + 1$ points to Δ^w, for each r in $\mathfrak{S}$ and $\Diamond\psi \in \boldsymbol{t}_w(r(w))$. Now it is not enough to choose points; we have to choose 'α-paths' whenever we have $\langle \alpha \rangle \psi \in \boldsymbol{t}_w(r(w_0))$.

To this end, for every run r in $\mathfrak{R}$ and every α occurring in $flc(\varphi)$, we define by induction the set $path_r(\alpha)$:

$$\begin{aligned}
path_r(\alpha_i) &= \{\langle w \rangle \mid wT_{\alpha_i}w\} \cup \{\langle w, \alpha_i, v \rangle \mid w \neq v,\ wT_{\alpha_i}v\} \\
path_r(\alpha_i^-) &= \{\langle w \rangle \mid wT_{\alpha_i}w\} \cup \{\langle w, \alpha_i^-, v \rangle \mid w \neq v,\ wT_{\alpha_i^-}v\} \\
path_r(\alpha \cup \beta) &= path_r(\alpha) \cup path_r(\beta) \\
path_r(\alpha\,;\beta) &= \{\langle w, \dots, v, \dots, u \rangle \mid \\
&\qquad \langle w, \dots, v \rangle \in path_r(\alpha),\ \langle v, \dots, u \rangle \in path_r(\beta)\} \\
path_r(\alpha^*) &= \{\langle w \rangle \mid w \in W\} \cup \bigcup \{path_r(\alpha^n) \mid n > 0\} \\
path_r(\psi?) &= \{\langle w \rangle \mid \psi \in \boldsymbol{t}_w(r(w))\}.
\end{aligned}$$

A path of the form $\langle w \rangle$ is called *degenerate*. For a path

$$\bar{w} = \langle w_0, \beta_0, \dots, \beta_{k-1}, w_k \rangle$$

we put $start(\bar{w}) = w_0$, $end(\bar{w}) = w_k$ and call k the *length* of $\bar{w}$. Given two paths $\bar{v}_1 = \langle w_0, \beta_0, \dots, \beta_{k-1}, w_k\rangle$ and $\bar{v}_2 = \langle w_k, \beta_k, \dots, \beta_{\ell-1}, w_\ell\rangle$, we put

$$\bar{v}_1 * \bar{v}_2 = \langle w_0, \beta_0, \dots, \beta_{k-1}, w_k, \beta_k, \dots, \beta_{\ell-1}, w_\ell\rangle .$$

Two paths $\bar{w} = \langle w_0, \beta_0, \dots, \beta_{k-1}, w_k\rangle$ and $\bar{v} = \langle v_0, \gamma_0, \dots, \gamma_{\ell-1}, v_\ell\rangle$ are called *twins* if $k = \ell$, $w_0 = v_0$, $\beta_i = \gamma_i$ $(i < k)$, and w_j is a twin of v_j $(1 \le j \le k)$. Since each point in our quasimodel can have arbitrarily many twins, we may assume that in fact each path has arbitrarily many twins as well.

A straightforward induction shows the following: for all $u, v \in W$, all runs $r \in \mathfrak{R}$ and all action terms occurring in $flc(\varphi)$,

$$uT^r_\alpha v \quad \text{iff} \quad \exists \bar{w} \in path_r(\alpha)\ \big(start(\bar{w}) = u\ \&\ end(\bar{w}) = v\big). \tag{6.6}$$

Observe that, for any action term α without iteration, the length of paths $\bar{w} \in path_r(\alpha)$ is bounded by the length $|\alpha|$ of α. However, if α contains iteration, then these lengths are not necessarily bounded. To solve this problem, we define the 'truncated' version $tr^r_\alpha(\bar{w})$ of each path $\bar{w} \in path_r(\alpha)$ by induction on the complexity of α. If α does not contain an occurrence of * then we put $tr^r_\alpha(\bar{w}) = \bar{w}$.

Suppose now that α contains iteration. If $\alpha = \psi?$ then let $tr^r_\alpha(\bar{w}) = \bar{w}$. If $\alpha = \beta \cup \gamma$, then

$$tr^r_\alpha(\bar{w}) = \begin{cases} tr^r_\beta(\bar{w}), & \text{if } \bar{w} \in path_r(\beta), \\ tr^r_\gamma(\bar{w}), & \text{if } \bar{w} \in path_r(\gamma). \end{cases}$$

If $\alpha = \beta ; \gamma$, then $\bar{w} = \bar{w}_1 * \bar{w}_2$, where $\bar{w}_1 \in path_r(\beta)$ and $\bar{w}_2 \in path_r(\gamma)$. Then we put

$$tr^r_\alpha(\bar{w}) = tr^r_\beta(\bar{w}_1) * tr^r_\gamma(\bar{w}_2).$$

Let $\alpha = \beta^*$. Then there are a natural number k and $\bar{w}_1, \dots \bar{w}_k \in path_r(\beta)$ such that

$$\bar{w} = \bar{w}_1 * \dots * \bar{w}_k.$$

If $k \le b(\varphi) \cdot p(\varphi)$, then we put

$$tr^r_\alpha(\bar{w}) = tr^r_\beta(\bar{w}_1) * \dots * tr^r_\beta(\bar{w}_k).$$

Otherwise there must be i, j, $1 \le i < j \le k$, such that

- $end(\bar{w}_i) \ne start(\bar{w}_j)$,
- $\boldsymbol{q}(end(\bar{w}_i)) = \boldsymbol{q}(start(\bar{w}_j))$ and $r(end(\bar{w}_i)) = r(start(\bar{w}_j))$.

In this case we choose the largest such i and j, and put

$$tr(\bar{w}) = tr^r_\beta(\bar{w}_1) * \dots * tr^r_\beta(\bar{w}_i) * tr^r_\beta(\bar{w}_j) * \dots * tr^r_\beta(\bar{w}_k).$$

If $k-(j-i-1) \leq b(\varphi)\cdot p(\varphi)$, then we put $tr^r_\alpha(\bar{w}) = tr(\bar{w})$. Otherwise we continue truncating $tr(\bar{w})$ in the same manner.

It should be clear that in this way we always obtain a path $tr^r_\alpha(\bar{w})$ the length of which is bounded by $l(\varphi)$. If $\bar{w}$ and $\bar{v}$ are twins then $tr^r_\alpha(\bar{w})$ and $tr^r_\alpha(\bar{v})$ are twins as well, with

$$start(tr^r_\alpha(\bar{w})) = start(\bar{w}) \quad \text{and} \quad end(tr^r_\alpha(\bar{w})) = end(\bar{w}).$$

We are now in a position to define Δ^{w_0} for the block $\mathfrak{B}^{w_0}$. For every $r \in \mathfrak{S}$ and every $\langle\alpha\rangle\,\psi \in \boldsymbol{t}_{w_0}(r(w_0))$ let

$$Sat(r, \langle\alpha\rangle\,\psi) = \\ \{tr^r_\alpha(\bar{w}) \mid \bar{w} \in path_r(\alpha),\ start(\bar{w}) = w_0,\ \psi \in \boldsymbol{t}_{end(\bar{w})}(r(end(\bar{w})))\}.$$

By (6.6), $Sat(r, \langle\alpha\rangle\,\psi) \neq \emptyset$, since r is saturated. We select a finite subset $Sel(r, \langle\alpha\rangle\,\psi)$ of $Sat(r, \langle\alpha\rangle\,\psi)$ as follows. If $Sat(r, \langle\alpha\rangle\,\psi) = \{\langle w_0\rangle\}$ then let $Sel(r, \langle\alpha\rangle\,\psi) = \{\langle w_0\rangle\}$ as well. Otherwise, let $Sel(r, \langle\alpha\rangle\,\psi)$ consist of a nondegenerate path $tr^r_\alpha(\bar{w})$ from $Sat(r, \langle\alpha\rangle\,\psi)$ together with its $m+1$ twins. Define $\Delta^{w_0}(r, \langle\alpha\rangle\,\psi)$ as the set of points different from w_0 which occur in a path in $Sel(r, \langle\alpha\rangle\,\psi)$. Clearly, the cardinality of $\Delta^{w_0}(r, \langle\alpha\rangle\,\psi)$ is bounded by

$$(md(\varphi)+1)\cdot|\alpha(b(\varphi)\cdot p(\varphi))|.$$

We may assume that the sets $\Delta^{w_0}(r, \langle\alpha\rangle\,\psi)$ defined this way are pairwise disjoint. Now put

- $\Delta^{w_0} = \{w_0\} \cup \bigcup\{\Delta^{w_0}(r, \langle\alpha\rangle\,\psi) \mid r \in \mathfrak{S},\ \langle\alpha\rangle\,\psi \in \boldsymbol{t}_{w_0}(r(w_0))\}$,
- for all $v, v' \in \Delta^{w_0}$ and $1 \leq i \leq n$, $vS_{\alpha_i}v'$ iff there are $\langle\alpha\rangle\,\psi \in \boldsymbol{t}_{w_0}(r(w_0))$ and $tr^r_\alpha(\bar{w}) \in Sel(r, \langle\alpha\rangle\,\psi)$ such that

$$tr^r_\alpha(\bar{w}) = \langle w_0, \dots, v, \alpha_i, v', \dots, w_k\rangle,$$

- $\mathfrak{F}^{w_0} = \langle\Delta^{w_0}, S_{\alpha_1}, \dots, S_{\alpha_n}\rangle$, and
- for all $v \in \Delta^{w_0}$, $\boldsymbol{q}^{w_0}(v) = \boldsymbol{q}(v)$.

It is not hard to check that $\langle\mathfrak{F}^{w_0}, \boldsymbol{q}^{w_0}\rangle$ is a basic structure for φ. The cardinality of Δ^{w_0} does not exceed $l(\varphi)$.

It remains to define a set $\mathfrak{R}^{w_0}$ of coherent and w_0-saturated runs through $\langle\mathfrak{F}^{w_0}, \boldsymbol{q}^{w_0}\rangle$ and a binary relation $\lhd^{w_0}$ on $\mathfrak{R}^{w_0}$ satisfying **(qm3)** and **(qm4)**. This is done in the same way as in the proof of Lemma 6.3 using the following modified definition of the 'run addition' function. Suppose that $\bar{w} \in path_r(\alpha)$, for some α occurring in $flc(\varphi)$, and that r and r' are functions whose domains contain Δ^{w_0} such that $r(w_0) = r'(w_0)$. Define a function $r +_{\bar{w}} r'$ with domain Δ^{w_0} by taking, for $v \in \Delta^{w_0}$,

$$r +_{\bar{w}} r'(v) = \begin{cases} r(v), & \text{if } v \text{ occurs in } tr^r_\alpha(\bar{w}), \\ r'(v), & \text{otherwise.} \end{cases}$$

Then the definitions of $\mathfrak{R}^{w_0}$ and $\lhd^{w_0}$, as well as the proofs that all runs in $\mathfrak{R}^{w_0}$ are coherent and w_0-saturated, and that $\mathfrak{R}^{w_0}$ and $\lhd^{w_0}$ satisfy **(qm3)** and **(qm4)** follow the lines of the proof of Lemma 6.3.

Thus, $\langle \mathfrak{F}^{w_0}, \boldsymbol{q}^{w_0}, \mathfrak{R}^{w_0}, \lhd^{w_0} \rangle$ is a block (of appropriate size) with root w_0, which proves Lemma 6.12. ❑

The decidability of **CPDL** $\times$ **K** follows immediately. ❑

Straightforward modifications of this proof show that **CPDL** $\times$ $\mathbf{T}_m$ and **CPDL** $\times$ $\mathbf{D}_m$ are also decidable.

Note that, unlike in the case of **K** $\times$ **K**, we cannot use the above proof for constructing a finite product model satisfying a given formula:

Theorem 6.13. **CPDL** $\times$ **K** *does not have the product fmp.*

Proof. By Theorem 5.32, **PTL** $\times$ **K** lacks the product fmp. This logic is reducible to **CPDL** $\times$ **K** by Theorems 6.18 and 6.24 below. Since these reductions turn finite product models to finite product models, it follows that **CPDL** $\times$ **K** lacks the product fmp as well.

(Alternatively, one can use the formula

$$\varphi = [\alpha_1^*]\Diamond p \wedge [\alpha_1^*]\Box(p \to \langle \alpha_1^* \rangle [\alpha_1^*]\neg p)$$

like in the proof of Theorem 5.32.) ❑

However, the following problem remains open:

Question 6.14. Does **CPDL** $\times$ **K** have the (abstract) fmp?

The decision procedure we have obtained is clearly nonelementary, and the following theorem says that no elementary algorithm can be found:

Theorem 6.15. *The satisfiability problems for* **PDL** $\times$ **K** *and* **CPDL** $\times$ **K** *do not belong to* ELEM.

Proof. By Theorems 6.18 and 6.24 below, **PTL** $\times$ **K** is polynomially reducible to **PDL** $\times$ **K**. On the other hand, by Theorem 6.37 below, **PTL** $\times$ **K** is not elementary. ❑

6.3 Products of epistemic logics with $\mathbf{K}_m$

In this section we consider products of epistemic logics—with and without common knowledge operators—and multimodal **K**.

Since for every Kripke complete multimodal logic L, its 'common knowledge extension' L^C is a Kripke complete multimodal logic as well, we do not need new definitions to introduce the product logic $L^C \times \mathbf{K}_m$:

$$L^C \times \mathbf{K}_m = \mathsf{Log}(\mathsf{Fr}L^C \times \mathsf{Fr}\mathbf{K}_m).$$

Before turning to the decision and complexity problems, we notice first that Theorem 3.16 holds for this kind of product as well:

Theorem 6.16. *Let L and L' be Kripke complete multimodal logics such that $\mathsf{Fr}L$ and $\mathsf{Fr}L'$ are first-order definable. Then $L^C \times L'$ is determined by the class of its countable product frames.*

Proof. Suppose $\varphi \notin L^C \times L'$. Then, by Proposition 3.7, φ is refuted in a model $\mathfrak{M}$ based on a product of a rooted frame for L^C and a rooted frame for L'. Starting from $\mathfrak{M}$, we define a first-order structure I as in the proof of Theorem 3.16. When applying the downward Löwenheim–Skolem–Tarski theorem, we take a countable *elementary substructure* J of I.

Now let $R_1^I, \dots, R_n^I$ be the relations in I interpreting the $\Box_i$ of L and let R_M^I be the relation interpreting the common knowledge operator C_M, for nonempty $M \subseteq \{1, \dots, n\}$ (we use a similar notation for J as well). By Remark 2.16, R_M^I is the reflexive and transitive closure of $\bigcup_{i \in M} R_i^I$. Although the operation of taking the reflexive and transitive closure is not first-order definable, we can still deduce that R_M^J is the reflexive and transitive closure of $\bigcup_{i \in M} R_i^J$. Indeed, suppose $u R_M^J v$. Then $u R_M^I v$, and so there is a first-order formula $\eta(x, y)$ of the form

$$\exists z_0 \dots \exists z_k \, (x R_{i_0} x_0 \wedge x_0 R_{i_1} x_1 \wedge \dots \wedge x_k R_{i_k} y)$$

such that $i_j \in M$ and $I \models \eta(x, y)[u, v]$. It follows that $J \models \eta(x, y)[u, v]$ as well, which means that there is a chain of $R_{i_j}^J$-arrows from u to v. Turning J into a modal model $\mathfrak{N}$ as in the proof of Theorem 3.16, we end up with a model refuting φ and based on a product of countable rooted frames for L^C and L', as required. ❑

As concerns finding an *axiomatization* for a logic of the form $L^C \times \mathbf{K}_m$, a natural candidate could be obtained by putting together the axioms of L^C (see Theorem 2.17) and the commutativity and Church–Rosser axioms between the modal operators of L and $\mathbf{K}_m$. It is not known, however, whether the resulting logic is Kripke complete (cf. the discussion before Question 6.8). So the following question is open:

Question 6.17. Suppose that either $n \geq 1$ and $L \in \{\mathbf{K}_n, \mathbf{T}_n\}$, or $n > 1$ and $L \in \{\mathbf{K4}_n, \mathbf{S4}_n, \mathbf{KD45}_n, \mathbf{S5}_n\}$. Is $L^C \times \mathbf{K}_m$ finitely axiomatizable?

The decidability of products of $\mathbf{K}_m$ ($\mathbf{T}_m$ and $\mathbf{D}_m$) with the standard epistemic logics can be easily obtained from the decidability results of the previous section if we can 'lift' the embeddings of epistemic logics into **CPDL**, given in Theorem 2.39, to products (all the reductions between product logics used in this chapter are shown in Table 6.1).

Theorem 6.18. *Suppose that* $L \in \{\mathbf{K}_n, \mathbf{T}_n, \mathbf{K4}_n, \mathbf{S4}_n, \mathbf{KD45}_n, \mathbf{S5}_n\}$ *and that* L' *is a Kripke complete* m*-modal logic. Then* $L^C \times L'$ *is polynomially reducible to* $\mathbf{CPDL} \times L'$.

Proof. We extend the translations t_j, $1 \leq j \leq 6$, of Theorem 2.39 (from $\mathcal{ML}_n^C$ into $\mathcal{CPDL}$) to translations

$$\mathsf{t}'_j : \mathcal{ML}_n^C \otimes \mathcal{ML}_m \to \mathcal{CPDL} \otimes \mathcal{ML}_m$$

by taking

- $\mathsf{t}'_j(\Box_i \varphi) = \Box_i \mathsf{t}'_j(\varphi)$, for all boxes $\Box_i$ of $\mathcal{ML}_m$.

It is pretty easy to extend the proof of Theorem 2.39 to show that, for every $\mathcal{ML}_n^C \otimes \mathcal{ML}_m$-formula φ,

$$\begin{array}{lll}
\varphi \in \mathbf{K}_n^C \times L' & \text{iff} & \mathsf{t}'_1(\varphi) \in \mathbf{PDL} \times L', \\
\varphi \in \mathbf{T}_n^C \times L' & \text{iff} & \mathsf{t}'_6(\varphi) \in \mathbf{PDL} \times L', \\
\varphi \in \mathbf{K4}_n^C \times L' & \text{iff} & \mathsf{t}'_2(\varphi) \in \mathbf{PDL} \times L', \\
\varphi \in \mathbf{S4}_n^C \times L' & \text{iff} & \mathsf{t}'_3(\varphi) \in \mathbf{PDL} \times L', \\
\varphi \in \mathbf{S5}_n^C \times L' & \text{iff} & \mathsf{t}'_4(\varphi) \in \mathbf{CPDL} \times L', \\
\varphi \in \mathbf{KD45}_n^C \times L' & \text{iff} & [\gamma^*]\chi \to \mathsf{t}'_5(\varphi) \in \mathbf{CPDL} \times L'.
\end{array}$$

(The translations t'_j are similar to those defined and used in (Fischer and Immerman 1987).) ❑

Remark 6.19. Note that in general it is not the case that the existence of a polynomial reduction of L_1 to L'_1 implies that $L_1 \times L$ is polynomially reducible to $L'_1 \times L$. Consider, for example, $\mathsf{Log}\{\langle \mathbb{N}, < \rangle\}$ and **S5**. Both logics are coNP-complete, so $\mathsf{Log}\{\langle \mathbb{N}, < \rangle\}$ is polynomially reducible to **S5**. On the other hand, according to Corollary 7.13, $\mathsf{Log}\{\langle \mathbb{N}, < \rangle\} \times \mathsf{Log}\{\langle \mathbb{N}, < \rangle\}$ is not even recursively enumerable, while $\mathbf{S5} \times \mathsf{Log}\{\langle \mathbb{N}, < \rangle\}$ is in EXPSPACE by Theorem 6.60.

As a consequence of Theorems 6.10 and 6.18 we obtain:

Theorem 6.20. *The logics* $\mathbf{K}_n^C \times \mathbf{K}_m$, $\mathbf{T}_n^C \times \mathbf{K}_m$, $\mathbf{K4}_n^C \times \mathbf{K}_m$, $\mathbf{S4}_n^C \times \mathbf{K}_m$, $\mathbf{KD45}_n^C \times \mathbf{K}_m$, *and* $\mathbf{S5}_n^C \times \mathbf{K}_m$ *are decidable.*

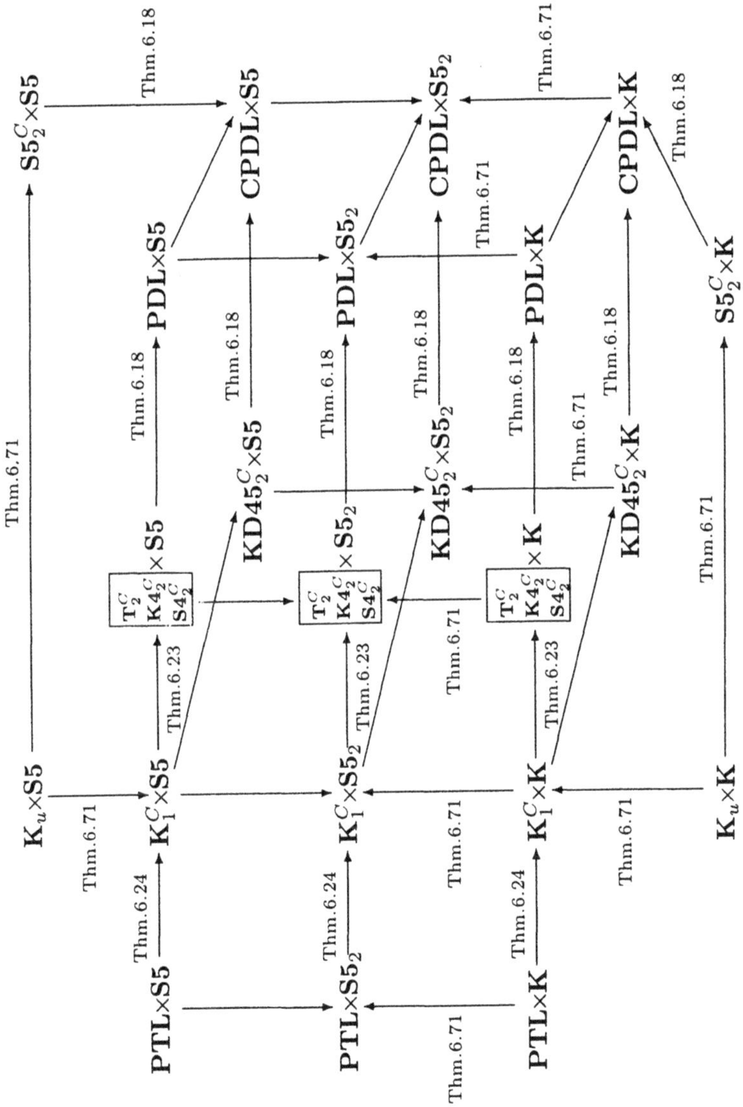

Table 6.1: Reductions between decidable product logics.

Note that $\mathbf{T} \times \mathbf{K}_m$ and $\mathbf{KD45}_1^C \times \mathbf{K}_m$ have the product fmp (cf. Theorems 6.4 and 6.56, respectively). On the other hand, we have:

Theorem 6.21. *No logic in the following list has the product fmp:*

$$\mathbf{K}_1^C \times \mathbf{K},\ \mathbf{T}_1^C \times \mathbf{K},\ \mathbf{K4}_2^C \times \mathbf{K},\ \mathbf{S4}_2^C \times \mathbf{K},\ \mathbf{KD45}_2^C \times \mathbf{K},\ \mathbf{S5}_2^C \times \mathbf{K}.$$

Proof. By Theorem 5.34, $\mathbf{K}_u \times \mathbf{K}$ does not have the product fmp. According to Theorem 6.71 below, $\mathbf{K}_u \times \mathbf{K}$ is reducible to all of the listed logics. Since these reductions work on the 'model' level (turning finite product models to finite product models), none of the listed logics can have the product fmp. (Alternatively, for $\mathbf{K}_1^C \times \mathbf{K}$ and $\mathbf{T}_1^C \times \mathbf{K}$ one can use the formula

$$\mathsf{C}\Diamond p \wedge \mathsf{C}\,\Box(p \to \Diamond\mathsf{C}\neg p),$$

for $\mathbf{KD45}_2^C \times \mathbf{K}$

$$\begin{aligned}&\Diamond(p \wedge q) \wedge \mathsf{C}_{\{1,2\}}(\Diamond\neg q \to \Diamond(p \wedge q)) \wedge \\ &\quad \mathsf{C}_{\{1,2\}}\,\Box(p \wedge q \to \Diamond_1(\neg p \wedge q \wedge \Diamond_2\mathsf{C}_{\{1,2\}}\neg q)),\end{aligned}$$

and we leave it to the reader to find a suitable formula for showing that $\mathbf{S5}_2^C \times \mathbf{K}$ lacks the product fmp.) ❑

Yet, some of these logics may still have the abstract fmp. In particular, it would be interesting to find a solution to the following problem:

Question 6.22. Do the products $\mathbf{K4} \times \mathbf{K}$ and $\mathbf{S4} \times \mathbf{K}$ have the fmp?

As to the complexity of products of epistemic logics with $\mathbf{K}_m$, we first 'lift' the reduction of Theorem 2.36 to the product level:

Theorem 6.23. *For every Kripke complete multimodal logic L, $\mathbf{K}_1^C \times L$ is polynomially reducible to any of $\mathbf{T}_2^C \times L$, $\mathbf{K4}_2^C \times L$, $\mathbf{S4}_2^C \times L$ and $\mathbf{KD45}_2^C \times L$.*

Proof. We prove the theorem only for unimodal L; the proof can be easily generalized to the multimodal case. First we show that $\mathbf{K}_1^C \times L$ is polynomially reducible to $\mathbf{D}_1^C \times L$. Denote the modal operator of the language $\mathcal{ML}$ of L by $\Box_3$. We extend the translation ^{r} defined in the proof of Theorem 2.36 (from $\mathcal{ML}_1^C$ into $\mathcal{ML}_1^C$) to a translation

$$^{r'} : \mathcal{ML}_1^C \otimes \mathcal{ML} \to \mathcal{ML}_1^C \otimes \mathcal{ML}$$

by taking $(\Box_3\varphi)^{r'} = \Box_3\varphi^{r'}$. It is easy to extend the proof of Theorem 2.36 to show that, for all $\mathcal{ML}_1^C \otimes \mathcal{ML}$-formulas φ,

$$\begin{aligned}&\varphi \in \mathbf{K}_1^C \times L \quad \text{iff} \\ &\quad p \wedge \Box_3^{\leq md(\varphi)}\mathsf{C}((p \to \Box_3 p) \wedge (\neg p \to (\Box_3\neg p \wedge \mathsf{C}\,\neg p))) \to \varphi^{r'} \in \mathbf{D}_1^C \times L.\end{aligned}$$

Next, we extend the translation $^\sharp$ in the proof of Theorem 2.36 (from $\mathcal{ML}_1^C$ into $\mathcal{ML}_2^C$) to a translation

$$^{\sharp'} : \mathcal{ML}_1^C \otimes \mathcal{ML} \to \mathcal{ML}_2^C \otimes \mathcal{ML}$$

by taking $(\Box_3\varphi)^{\sharp'} = \Box_3\varphi^{\sharp'}$. Now, given an $\mathcal{ML}_1^C \otimes \mathcal{ML}$-formula φ, define the formula $\chi_{\mathbf{S4}}^{\varphi'}$ as the result of replacing each occurrence of $\mathsf{C}_{\{1,2\}}$ in the formula $\chi_{\mathbf{S4}}^{\varphi}$ in the proof of Theorem 2.36 with $\Box_3^{\leq md(\varphi)}\mathsf{C}_{\{1,2\}}$, and adding the conjunct

$$\Box_3^{\leq md(\varphi)}\mathsf{C}_{\{1,2\}}\big((p \to \Box_3 p) \land (\neg p \to \Box_3\neg p)\big).$$

It is straightforward to extend the proof of Theorem 2.36 to show that

(i) if $\chi_{\mathbf{S4}}^{\varphi'} \to \varphi^{\sharp'} \in \mathbf{S4}_2^C \times L$ then $\varphi \in \mathbf{D}_1^C \times L$;

(ii) if $\varphi \in \mathbf{D}_1^C \times L$ then $\chi_{\mathbf{S4}}^{\varphi'} \to \varphi^{\sharp'} \in \mathbf{K}_2^C \times L$.

To obtain a reduction to $\mathbf{KD45}_2^C \times L$, we extend the translation $^\natural$ in the proof of Theorem 2.36 to a translation

$$^{\natural'} : \mathcal{ML}_1^C \otimes \mathcal{ML} \to \mathcal{ML}_1^C \otimes \mathcal{ML}$$

by again taking $(\Box_3\varphi)^{\natural'} = \Box_3\varphi^{\natural'}$. Given an $\mathcal{ML}_1^C \otimes \mathcal{ML}$-formula φ, we define the formula $\chi_{\mathbf{KD45}}^{\varphi'}$ as

$$p \land \Box_3^{\leq md(\varphi)}\mathsf{C}_{\{1,2\}}\big((p \to (\Box_3 p \land \chi_{sim}^{\varphi} \land \Box_1\neg p)) \land (\neg p \to (\Box_3\neg p \land \Box_2 p))\big),$$

where χ_{sim}^{φ} is as in the proof of Theorem 2.36. One can extend the proof of that theorem to show that

$$\varphi \in \mathbf{D}_1^C \times L \quad \text{iff} \quad \chi_{\mathbf{KD45}}^{\varphi'} \to \varphi^{\natural'} \in \mathbf{KD45}_2^C \times L$$

as required. ❑

The reduction of Theorem 2.38 can also be generalized to product logics:

Theorem 6.24. *Let L be any Kripke complete m-modal logic such that* $\mathsf{Fr}L$ *is first-order definable in the language having equality and m binary predicate symbols. Then* $\mathbf{PTL} \times L$ *is polynomially reducible to* $\mathbf{K}_1^C \times L$.

Proof. To simplify notation, we confine ourselves only to the case of a unimodal L. We denote the box of the language $\mathcal{ML}$ of L by $\Box$ and, as before, the modal operators of $\mathbf{K}_1^C$ by $\Box_1$ and C.

First we 'get rid of' the $\mathcal{U}$ operator:

Claim 6.25. *$\mathbf{PTL} \times L$ is polynomially reducible to $\mathbf{PTL}_{\Box\bigcirc} \times L$.*

Proof. Given an $\mathcal{ML}_\mathcal{U} \otimes \mathcal{ML}$-formula φ, denote by φ^U the result of replacing every subformula of φ of the form $\chi = \chi_1 \mathcal{U} \chi_2$ with a variable p_χ. Let $\mathcal{R}_U(\varphi)$ be defined as in the proof of Proposition 2.10. Then it is straightforward to show (cf. the proof of Proposition 2.10) that for every $\mathcal{ML}_\mathcal{U} \otimes \mathcal{ML}$-formula φ,

$$\varphi \in \mathbf{PTL} \times L \qquad \text{iff} \qquad \Box^{\leq md(\varphi)} \Box_F^+ \bigwedge \mathcal{R}_U(\varphi) \to \varphi^U \in \mathbf{PTL}_{\Box\bigcirc} \times L,$$

as required. ❑

Now, for every $\mathcal{U}$-free $\mathcal{ML}_\mathcal{U} \otimes \mathcal{ML}$-formula φ, define the set $\mathcal{R}(\varphi)$ and the formula $\varphi^\bullet$ as in the proof of Theorem 2.38. We claim that

$$\varphi \in \mathbf{PTL}_{\Box\bigcirc} \times L \qquad \text{iff} \qquad \Box^{\leq md(\varphi)} \mathsf{C}(\Diamond_1 \top \wedge \bigwedge \mathcal{R}(\varphi)) \to \varphi^\bullet \in \mathbf{K}_1^C \times L.$$

The implication ($\Leftarrow$) follows from Theorem 6.29 below. Conversely, suppose that we have a model $\mathfrak{M} = \langle \mathfrak{F} \times \mathfrak{G}, \mathfrak{V} \rangle$ based on a product frame $\mathfrak{F} \times \mathfrak{G}$ for $\mathbf{K}_1^C \times L$ and such that

$$(\mathfrak{M}, \langle w_0, x_0 \rangle) \models \neg\varphi^\bullet \wedge \Box^{\leq md(\varphi)} \mathsf{C}(\Diamond_1 \top \wedge \bigwedge \mathcal{R}(\varphi)).$$

By Theorem 6.16, we may assume that $\mathfrak{F} = \langle W, R_1, R_1^* \rangle$ and $\mathfrak{G} = \langle \Delta, R \rangle$ are countable rooted frames with roots w_0 and x_0, respectively, and $\langle W, R_1 \rangle$ is an intransitive tree.

We construct a countable sequence $w_0, w_1, \dots$ of distinct points in W such that $w_i R_1 w_{i+1}$, for all $i \in \mathbb{N}$. The construction is similar to the one in the proof of Theorem 2.38; the only difference is in the kind of defects we have to 'fix.'

Suppose that a sequence $\sigma = \langle w_0, \dots, w_n \rangle$ has already been constructed. Define

$$\Delta' = \{x_0\} \cup \{y \in \Delta \mid x_0 R \dots R x_k = y,\ x_0, \dots, x_k \in \Delta,\ k \leq md(\varphi)\}.$$

(Thus, $\Delta' = \{y \in \Delta \mid dp(y) \leq k\}$.) We call a triple $\left\langle x, m, \Diamond_F \psi \right\rangle$ a *σ-defect* if $x \in \Delta'$, $m < n$, $\Diamond_F \psi \in sub\, \varphi$, and

- $(\mathfrak{M}, \langle w_m, x \rangle) \models \Diamond_1 \neg \mathsf{C} \neg \psi^\bullet$, but
- for all i with $m + 1 \leq i \leq n$, we have $(\mathfrak{M}, \langle w_i, x \rangle) \not\models \psi^\bullet$.

Since for each finite sequence σ there can be only countably many σ-defects, after fixing all defects in the limit we obtain a sequence $\langle w_i \mid i \in \mathbb{N} \rangle$ as required.

Define a valuation $\mathfrak{V}'$ in the frame $\langle \mathbb{N}, < \rangle \times \langle \Delta, R \rangle$ by taking

$$\mathfrak{V}'(p) = \{\langle n, x \rangle \in \mathbb{N} \times \Delta \mid \langle w_n, x \rangle \in \mathfrak{V}(p)\},$$

for every propositional variable p, and let $\mathfrak{M}' = \langle \langle \mathbb{N}, < \rangle \times \Delta, \mathfrak{V}' \rangle$. It can be shown by induction that for all $\psi \in sub\,\varphi$, $n \in \mathbb{N}$, and $x \in \Delta'$,

$$(\mathfrak{M}, \langle w_n, x \rangle) \models \psi^\bullet \qquad \text{iff} \qquad (\mathfrak{M}', \langle n, x \rangle) \models \psi.$$

Hence, we have $(\mathfrak{M}', \langle 0, x_0 \rangle) \not\models \varphi$, as required. ❑

In Theorem 6.37 we will show that the satisfiability problem for $\mathbf{PTL} \times \mathbf{K}$ is not elementary. So Theorems 6.23, 6.24 (cf. Table 6.1) and 6.37 yield:

Theorem 6.26. *The satisfiability problem for $L \times \mathbf{K}$ does not belong to* ELEM, *whenever $L \in \{\mathbf{K}_1^C, \mathbf{T}_2^C, \mathbf{K4}_2^C, \mathbf{S4}_2^C, \mathbf{KD45}_2^C\}$.*

Question 6.27. Is $\mathbf{S5}_2^C \times \mathbf{K}$ elementary?

We will discuss the complexity of $\mathbf{S5} \times \mathbf{K}$ in Section 6.5. The following question is also open:

Question 6.28. What is the complexity of $\mathbf{T} \times \mathbf{K}$, $\mathbf{K4} \times \mathbf{K}$ and $\mathbf{S4} \times \mathbf{K}$?

Note that, by Theorem 5.42, the satisfiability problem for these logics is NEXPTIME-hard.

6.4 Products of temporal logics with $\mathbf{K}_m$

In temporal logic, we are often interested not in the class $\mathsf{Fr}L$ of all frames for a logic L, but only in some class of the *intended flows of time*. For example, $\mathsf{Fr}\,\mathsf{Log}\{\langle \mathbb{N}, < \rangle\}$ contains all finite strict linear orders followed by clusters with one or more reflexive points, which are certainly not the intended models of time. However, according to the definition of products we have

$$\mathsf{Log}\{\langle \mathbb{N}, < \rangle\} \times L = \mathsf{Log}\left(\mathsf{Fr}\,\mathsf{Log}\{\langle \mathbb{N}, < \rangle\} \times \mathsf{Fr}L\right),$$

for any Kripke complete modal logic L. The question important for applications of products to temporal reasoning is whether this product logic is determined by products with the intended flow of time $\langle \mathbb{N}, < \rangle$ only. The following theorem shows that this is often indeed the case. We formulate it not only for products with $\mathsf{Log}\{\langle \mathbb{N}, < \rangle\}$, but also with $\mathsf{Log}_{FP}(\mathbb{N})$, $\mathbf{PTL}$ and $\mathbf{PTL}_{\Box\bigcirc}$.

Theorem 6.29. (i) *Let L be any of the logics $\mathsf{Log}\{\langle \mathbb{N}, < \rangle\}$, $\mathsf{Log}_{FP}(\mathbb{N})$, $\mathbf{PTL}$, and let L' be any Kripke complete m-modal logic such that $\mathsf{Fr}L'$ is first-order* *definable in the language with equality and m binary predicate symbols. Then $L \times L'$ is determined by the class of frames*

$$\{\langle \mathbb{N}, < \rangle \times \mathfrak{F} \mid \mathfrak{F} \textit{ is a countable frame for } L'\},$$

and $L \times L'^{C}$ is determined by the class of frames

$$\{\langle \mathbb{N}, < \rangle \times \mathfrak{F} \mid \mathfrak{F} \textit{ is a countable frame for } L'^{C}\}.$$

If L' is also $\mathsf{Log}\{\langle \mathbb{N}, < \rangle\}$, $\mathsf{Log}_{FP}(\mathbb{N})$, or **PTL** *then $L \times L'$ is determined by the sole frame $\langle \mathbb{N}, < \rangle \times \langle \mathbb{N}, < \rangle$.*

(ii) *Let L' be as in* (i). *Then $\mathbf{PTL}_{\Box\bigcirc} \times L'$ is determined by the class of frames*

$$\{\langle \mathbb{N}, <, +1 \rangle \times \mathfrak{F} \mid \mathfrak{F} \textit{ is a countable frame for } L'\},$$

and $\mathbf{PTL}_{\Box\bigcirc} \times L'^{C}$ is determined by the class of frames

$$\{\langle \mathbb{N}, <, +1 \rangle \times \mathfrak{F} \mid \mathfrak{F} \textit{ is a countable frame for } L'^{C}\}.$$

$\mathbf{PTL}_{\Box\bigcirc} \times \mathbf{PTL}_{\Box\bigcirc}$ *is determined by the sole frame $\langle \mathbb{N}, <, +1 \rangle \times \langle \mathbb{N}, <, +1 \rangle$.*

Proof. We prove the theorem only for $L = \mathsf{Log}\{\langle \mathbb{N}, < \rangle\}$. The remaining cases are considered analogously. According to Remark 2.11, the class of rooted frames for $\mathsf{Log}\{\langle \mathbb{N}, < \rangle\}$ consists of $\langle \mathbb{N}, < \rangle$ and all finite strict linear orders followed by a (possibly uncountably infinite) cluster of reflexive points. Observe that this class of frames is closed under taking elementary substructures.

Suppose that some formula φ is refuted in the product of a rooted frame for $\mathsf{Log}\{\langle \mathbb{N}, < \rangle\}$ and a rooted frame for L' (or for L'^{C}). Now we can follow the proof of Theorem 3.16, but take a countable elementary substructure when applying the downward Löwenheim–Skolem–Tarski theorem. This shows that φ is refuted in the product of a countable rooted frame for $\mathsf{Log}\{\langle \mathbb{N}, < \rangle\}$ and a countable rooted frame for L'. (In the case of L'^{C} we also need the argument from the proof of Theorem 6.16.) It remains to notice that any countable rooted frame for $\mathsf{Log}\{\langle \mathbb{N}, < \rangle\}$ is a p-morphic image of $\langle \mathbb{N}, < \rangle$. Hence, by Proposition 3.10 (i), $\mathsf{Log}\{\langle \mathbb{N}, < \rangle\} \times L'$ (or $\mathsf{Log}\{\langle \mathbb{N}, < \rangle\} \times L'^{C}$) is determined by the required class of frames. ❑

As to flows of time different from $\langle \mathbb{N}, < \rangle$, we consider here products with **K4.3** and $\mathsf{Log}\{\langle \mathbb{Q}, < \rangle\}$ as well as their bimodal temporal variants **Lin** and $\mathsf{Log}_{FP}(\mathbb{Q})$. To begin with, we describe classes of product logics which are determined by their intended flows of time.

Theorem 6.30. *If L' is a Kripke complete multimodal logic then both product logics $\mathbf{K4.3} \times L'$ and $\mathbf{Lin} \times L'$ are determined by the class of all frames $\mathfrak{F} \times \mathfrak{F}'$, where $\mathfrak{F}$ is a strict linear order and $\mathfrak{F}' \in \mathsf{Fr}L'$.*

Proof. Note that for any transitive connected frame $\langle W_0, <_0\rangle$ we can find a strict linear order $\langle W_1, <_1\rangle$ such that the bimodal frame $\langle W_0, <_0, <_0^{-1}\rangle$ is a p-morphic image of the bimodal frame $\langle W_1, <_1, <_1^{-1}\rangle$. (If $\langle W_0, <_0\rangle$ is countable, then $\langle W_1, <_1\rangle$ can be obtained from $\langle W_0, <_0\rangle$ by replacing each of its nondegenerate clusters $C_w = \{v \in W_0 \mid w <_0 v \wedge v <_0 w\}$ with a copy of the integers $\langle \mathbb{Z}, <\rangle$. For uncountable clusters take a sufficiently large well-ordering and replace the cluster by a copy of the converse of the well-ordering followed by a copy of the well-ordering.) Now it is readily checked that the class of rooted frames for **K4.3** is contained in the class of transitive connected frames and that all rooted frames for **Lin** are transitive and connected. The theorem follows immediately from Proposition 3.10 (i). ❑

Theorem 6.31. (i) *Let L' be a Kripke complete m-modal logic such that $\mathsf{Fr}L'$ is first-order definable in the language with equality and m binary predicate symbols. Then $\mathsf{Log}\{\langle \mathbb{Q}, <\rangle\} \times L'$ and $\mathsf{Log}_{FP}(\mathbb{Q}) \times L'$ are determined by the class of frames*

$$\{\langle \mathbb{Q}, <\rangle \times \mathfrak{F} \mid \mathfrak{F} \textit{ is a countable frame for } L'\},$$

and $\mathsf{Log}\{\langle \mathbb{Q}, <\rangle\} \times L'^C$ and $\mathsf{Log}_{FP}(\mathbb{Q}) \times L'^C$ are determined by the class of frames

$$\{\langle \mathbb{Q}, <\rangle \times \mathfrak{F} \mid \mathfrak{F} \textit{ is a countable frame for } L'^C\}.$$

Proof. The class of rooted frames for $\mathsf{Log}\{\langle \mathbb{Q}, <\rangle\}$ consists of dense transitive and connected frames without right-endpoints but with left-endpoints. The class of rooted frames for $\mathsf{Log}_{FP}(\mathbb{Q})$ consists of dense transitive and connected frames without endpoints. Both of these classes are known to be first-order definable, so they are closed under taking elementary substructures. Hence, following the proof of Theorem 6.29 we can show that all the logics $\mathsf{Log}\{\langle \mathbb{Q}, <\rangle\} \times L'$, $\mathsf{Log}_{FP}(\mathbb{Q}) \times L'$, $\mathsf{Log}\{\langle \mathbb{Q}, <\rangle\} \times L'^C$, and $\mathsf{Log}_{FP}(\mathbb{Q}) \times L'^C$ are determined by countable rooted product frames. It is not difficult to show that any rooted countable frame in $\mathsf{FrLog}\{\langle \mathbb{Q}, <\rangle\}$ is a p-morphic image of a generated subframe of $\langle \mathbb{Q}, <\rangle$. This holds for the corresponding bimodal frames in $\mathsf{FrLog}_{FP}(\mathbb{Q})$ as well. Hence, by Proposition 3.10 (i) and (ii), we obtain the classes we need. ❑

It is to be noted, however, that not all products of temporal logics are determined by the intended product frames.

Example 6.32. Consider, for instance, the formula $\Box_1 \Diamond_2 (p \wedge \Box_1 \neg p)$. It is clearly satisfied in the product of $\langle \mathbb{R}, <\rangle$ and a cluster with continuum-many points, but not in $\langle \mathbb{R}, <\rangle \times \langle \mathbb{N}, <\rangle$. It follows that

$$\mathsf{Log}\{\langle \mathbb{R}, <\rangle\} \times \mathsf{Log}\{\langle \mathbb{N}, <\rangle\} \subsetneq \mathsf{Log}\{\langle \mathbb{R}, <\rangle \times \langle \mathbb{N}, <\rangle\}.$$

In Section 7.3 we shall see that the logic $\mathsf{Log}\{\langle\mathbb{R},<\rangle\times\langle\mathbb{Q},<\rangle\}$ is not recursively axiomatizable, while

$$\mathsf{Log}\{\langle\mathbb{R},<\rangle\}\times\mathsf{Log}\{\langle\mathbb{Q},<\rangle\}=\mathsf{Log}\{\langle\mathbb{Q},<\rangle\}\times\mathsf{Log}\{\langle\mathbb{Q},<\rangle\}$$

is recursively enumerable by Theorem 3.17, because the class of frames for $\mathsf{Log}\{\langle\mathbb{Q},<\rangle\}$ is definable by a finite set of first-order formulas in the language with one binary predicate and equality (Segerberg 1970, Goldblatt 1987).

Now, returning to products of temporal logics with $\mathbf{K}_m$, first observe that, by Theorems 6.20 and 6.24, we have (cf. Table 6.1):

Theorem 6.33. **PTL** $\times\,\mathbf{K}_m$ *(and so* $\mathsf{Log}\{\langle\mathbb{N},<\rangle\}\times\mathbf{K}_m$*) is decidable.*

In Section 13.2 we give another proof for this result by a reduction to the monadic second-order theory of $\langle\mathbb{N},<\rangle$ (see Theorem 13.6).

Let us turn to the complexity of **PTL** $\times$ **K**. In what follows, we denote the modal operators of **PTL** by $\mathcal{U}_h$, $\boxminus$, $\diamondminus$, $\ominus$, and the modal operators of **K** by $\boxbar$ and $\diamondbar$. Given a formula φ in this language, we denote by $vmd(\varphi)$ the maximal number of nested 'vertical' modal operators (i.e., $\boxbar$ and $\diamondbar$) in φ. For example, $vmd(p)=0$ and $vmd(\diamondbar(p\wedge\boxbar p))=2$. Further, for each natural number d, we define the functions $\exp_d:\mathbb{N}\rightarrow\mathbb{N}$ by taking inductively for all $m\in\mathbb{N}$:

$$\exp_0(m)=m,\quad \exp_{d+1}(m)=\exp_d(m)\cdot 2^{\exp_d(m)}.$$

We prove the following result:[3]

Theorem 6.34. *Let $d>0$. Then any problem '$x\in X$?' which is solvable by a deterministic algorithm in space bounded by $\exp_d(|x|)$ on input x is polynomially reducible to the* **PTL** $\times$ **K**-*satisfiability problem for formulas φ with $vmd(\varphi)\leq d$.*

Proof. The proof is conducted in two steps. First, we show that 'yardsticks' of the type used in (Stockmeyer 1974) can be encoded by **PTL** $\times$ **K**-formulas:

Lemma 6.35. *For all natural numbers $d>0$ and $d'>1$, there exist a formula $\delta_{d,d'}$ with a propositional variable p_d such that $vmd(\delta_{d,d'})=d-1$, the length of $\delta_{d,d'}$ is linear in $d+d'$, and the following hold:*

(a) *for every model $\mathfrak{M}$ based on the product of $\langle\mathbb{N},<\rangle$ and some frame $\langle W,R\rangle$ and all $n\in\mathbb{N}$, $x\in W$, if $(\mathfrak{M},\langle n,x\rangle)\models p_d\wedge\delta_{d,d'}$, then for each $m\geq n$,*

$$(\mathfrak{M},\langle m,x\rangle)\models p_d \quad \textit{iff} \quad m=n+j\cdot\exp_d(d') \textit{ for some } j\in\mathbb{N}. \tag{6.7}$$

[3]Our proof is very close to the one given in (Halpern and Vardi 1989) and showing that the satisfiability problem for **PTL** $\times$ $\mathbf{S5}_2$ is nonelementary.

(b) $\delta_{d,d'}$ *is* **PTL** $\times$ **K**-*satisfiable; moreover, for every* $k < \exp_d(d')$ *there exist a model* $\mathfrak{M}_k$ *based on a product frame* $\langle \mathbb{N}, < \rangle \times \langle W_k, R_k \rangle$ *and a point* $x_k \in W_k$ *such that*

- $(\mathfrak{M}_k, \langle n, x_k \rangle) \models \delta_{d,d'}$ *for all* $n \in \mathbb{N}$;
- $(\mathfrak{M}_k, \langle k, x_k \rangle) \models p_d$.

Proof. The construction of $\delta_{d,d'}$ is by induction on d. To begin with, let $\delta_{1,d'}$ be the conjunction of the following formulas:

$$\Diamond^+ p_0 \wedge \boxminus^+ \big((p_0 \leftrightarrow \ominus^{d'} p_0) \wedge (p_0 \rightarrow \bigwedge_{1 \le i < d'} \ominus^i \neg p_0) \big), \tag{6.8}$$

$$\boxminus^+ \big(q_1 \mathcal{U}_h p_0 \rightarrow (q_1 \leftrightarrow \ominus^{d'} \neg q_1) \big), \tag{6.9}$$

$$\boxminus^+ \big(\neg (q_1 \mathcal{U}_h p_0) \rightarrow (q_1 \leftrightarrow \ominus^{d'} q_1) \big), \tag{6.10}$$

$$\boxminus^+ (p_1 \leftrightarrow (p_0 \wedge \neg q_1 \wedge \neg q_1 \mathcal{U}_h p_0)). \tag{6.11}$$

Suppose that $\mathfrak{M}$ is a model based on the product of $\langle \mathbb{N}, < \rangle$ and some frame $\langle W, R \rangle$, and $n \in \mathbb{N}$, $x \in W$ are such that

$$(\mathfrak{M}, \langle n, x \rangle) \models p_1 \wedge \delta_{1,d'}.$$

By (6.11), $(\mathfrak{M}, \langle n, x \rangle) \models p_0$, and so by (6.8), the time points $m \ge n$ such that $(\mathfrak{M}, \langle m, x \rangle) \models p_0$ are precisely d' steps apart from each other.

Let $a < 2^{d'}$ and $a_0 a_1 \dots a_{d'-1}$ be the d'-bit binary representation of a (say, 1 is represented as $00\dots01$, and $2^{d'-1}$ as $100\dots00$). We say that an interval $[n + j \cdot d', n + (j+1) \cdot d' - 1]$, for some $j \in \mathbb{N}$, *simulates* a if for every $i \le d' - 1$,

$$(\mathfrak{M}, \langle n + j \cdot d' + i, x \rangle) \models q_1 \quad \text{iff} \quad a_i = 1.$$

Recall that for two d'-bit binary numbers $a = a_0 \dots a_{d'-1}$ and $b = b_0 \dots b_{d'-1}$ we have

$$b = a + 1 \pmod{2^{d'}} \quad \text{iff} \quad \forall i < d' \; \big(a_i = b_i \leftrightarrow (\exists j > i \;\; a_j = 0) \big).$$

It is not hard to see that if an interval $[n + j \cdot d', n + (j+1) \cdot d' - 1]$ simulates a number a, then formulas (6.9) and (6.10) force the next interval

$$[n + (j+1) \cdot d', n + (j+2) \cdot d' - 1]$$

to simulate $a + 1 \pmod{2^{d'}}$.

Finally, by (6.11) we have that the interval $[n, n + d' - 1]$ simulates the number 0, and for all $m \ge n$,

$$(\mathfrak{M}, \langle m, x \rangle) \models p_1 \quad \text{iff} \quad m = n + j \cdot d' \cdot 2^{d'} \text{ for some } j \in \mathbb{N},$$

as required in (a).

To show (b), fix a number $k < d' \cdot 2^{d'}$. Then there are unique numbers $k' < d'$, $k'' < 2^{d'}$ such that $k = k' + k'' \cdot d'$. Let $\mathfrak{G}$ be a frame with a single irreflexive point x. Define a model $\mathfrak{M}_k = \langle\langle \mathbb{N}, <\rangle \times \mathfrak{G}, \mathfrak{W}_k\rangle$ by taking

- $\mathfrak{W}_k(p_0) = \{k' + j \cdot d' \mid j \in \mathbb{N}\} \times \{x\}$;
- $\mathfrak{W}_k(p_1) = \{k + j \cdot d' \cdot 2^{d'} \mid j \in \mathbb{N}\} \times \{x\}$;
- for all $n \in \mathbb{N}$,

$$\langle n, x\rangle \in \mathfrak{W}_k(q_1) \quad \text{iff} \quad \exists i < d', j \in \mathbb{N} \; \big(n = k' + i + j \cdot d' \text{ and the } i\text{th bit of the } d'\text{-bit binary representation of the number } (2^{d'} - k'') + j \ (\bmod\ (2^{d'})) \text{ equals } 1\big).$$

The reader can readily check that $(\mathfrak{M}_k, \langle n, x\rangle) \models \delta_{1,d'}$, for all $n \in \mathbb{N}$, and $(\mathfrak{M}_k, \langle k, x\rangle) \models p_1$.

Assume now that we have constructed $\delta_{d,d'}$ such that (a) and (b) hold. Our aim is to construct $\delta_{d+1,d'}$. First, let $\psi_{d,d'}$ be the conjunction of the formulas

$$\boxminus^+(\Box\delta_{d,d'} \wedge \Diamond p_d),$$
$$\boxminus^+\Big(\big(p_d \leftrightarrow \Diamond(r_d \wedge p_d)\big) \wedge \big(p_d \leftrightarrow \Box(r_d \rightarrow p_d)\big)\Big),$$
$$\boxminus^+\Box\big((r_d \leftrightarrow \boxminus r_d) \wedge (r_d \leftrightarrow \diamondminus r_d)\big).$$

Suppose that $\mathfrak{M}$ is a model based on the product of $\langle \mathbb{N}, <\rangle$ and some frame $\langle W, R\rangle$. It is straightforward to show that the following claim holds:

Claim 6.36. *If $(\mathfrak{M}, \langle n, x\rangle) \models \psi_{d,d'}$ for some $n \in \mathbb{N}$, $x \in W$ then*

(i) *$(\mathfrak{M}, \langle m, y\rangle) \models \delta_{d,d'}$, for all $m \geq n$ and $y \in W$ such that xRy;*

(ii) *for each $m \geq n$ there is $y_m \in W$ such that xRy_m and $(\mathfrak{M}, \langle m, y_m\rangle) \models p_d$;*

(iii) *there is a $y \in W$ such that xRy and for every $m \geq n$,*

$$(\mathfrak{M}, \langle m, x\rangle) \models p_d \quad \textit{iff} \quad (\mathfrak{M}, \langle m, y\rangle) \models p_d.$$

Define $\delta_{d+1,d'}$ to be the conjunction of $\psi_{d,d'}$ and the following formulas:

$$\boxminus^+\big((q_{d+1} \leftrightarrow \Diamond q_{d+1}) \wedge (q_{d+1} \leftrightarrow \Box q_{d+1})\big), \tag{6.12}$$

$$\boxminus^+\Big(q_{d+1}\mathcal{U}_h p_d \rightarrow \Big(q_{d+1} \leftrightarrow \Box\big(p_d \rightarrow (\neg p_d)\mathcal{U}_h(p_d \wedge \neg q_{d+1})\big)\Big)\Big), \tag{6.13}$$

$$\boxminus^+\Big(\neg(q_{d+1}\mathcal{U}_h p_d) \to \Big(q_{d+1} \leftrightarrow \Box\big(p_d \to (\neg p_d)\mathcal{U}_h(p_d \wedge q_{d+1})\big)\Big)\Big), \tag{6.14}$$

$$\boxminus^+\Big(p_{d+1} \leftrightarrow \big(p_d \wedge \neg q_{d+1} \wedge (\neg q_{d+1})\mathcal{U}_h p_d\big)\Big). \tag{6.15}$$

Now suppose that $\mathfrak{M}$ is as above and

$$(\mathfrak{M}, \langle n, x\rangle) \models p_{d+1} \wedge \delta_{d+1,d'}$$

for some $n \in \mathbb{N}$, $x \in W$. By (6.15), we have $(\mathfrak{M}, \langle n, x\rangle) \models p_d$. Although we do not know whether $\delta_{d,d'}$ holds at $\langle n, x\rangle$, still we claim that the time points $m \geq n$ such that $(\mathfrak{M}, \langle m, x\rangle) \models p_d$ are precisely $\exp_d(d')$ steps apart from each other. Indeed, choose a $y \in W$ as in Claim 6.36 (iii). Then, by Claim 6.36 (i), we have $(\mathfrak{M}, \langle n, y\rangle) \models p_d \wedge \delta_{d,d'}$, and so, by the induction hypothesis, the time points $m \geq n$ such that $(\mathfrak{M}, \langle m, y\rangle) \models p_d$ are $\exp_d(d')$ steps apart from each other. The choice of y ensures that we also have, for all $m \geq n$,

$$(\mathfrak{M}, \langle m, x\rangle) \models p_d \quad \text{iff} \quad m = n + j \cdot \exp_d(d') \text{ for some } j \in \mathbb{N}$$

(see Fig. 6.1).

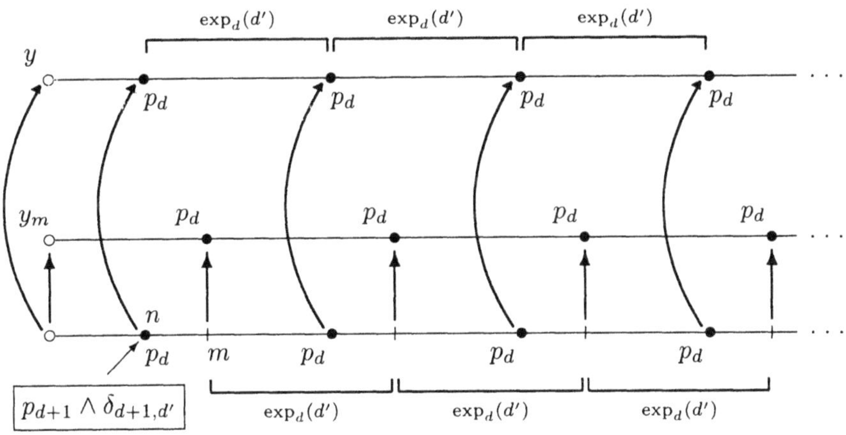

Figure 6.1: Yardsticks of length $\exp_d(d')$.

Similarly to the case $d = 1$, we want to use the intervals

$$[m, m + \exp_d(d') - 1]$$

such that $m \geq n$ and $(\mathfrak{M}, \langle m, x\rangle) \models p_d$ to simulate $< 2^{\exp_d(d')}$ numbers in such a way that consecutive intervals simulate consecutive (mod $(2^{\exp_d(d')})$) numbers. The variable q_{d+1} is used to encode the bits of the $\exp_d(d')$-bit binary representation of these numbers: q_{d+1} encodes 1, while $\neg q_{d+1}$ encodes 0.

Since we aimed to have a formula $\delta_{d+1,d'}$ of length linear in $d+1+d'$, we cannot simply use formulas similar to (6.9) and (6.10) to simulate the modulo $2^{\exp_d(d')}$ successor function. However, we already have 'yardsticks' of length $\exp_d(d')$, so we use them as follows. Suppose that the jth bit of some number $a < 2^{\exp_d(d')}$ is 'stored' at a point $\langle m, x\rangle$. Then by Claim 6.36 (i) and (ii), there is a $y_m \in W$ such that xRy_m and $(\mathfrak{M}, \langle m, y_m\rangle) \models p_d \wedge \delta_{d,d'}$. Formula (6.12) ensures that, for all $m \geq n$, q_{d+1} is 'uniform' among $\langle m, x\rangle$ and all $\langle m, y\rangle$ with xRy, so $\langle m, y_m\rangle$ also stores the jth bit of our number a. Now by the induction hypothesis, the next $m' > m$ with $(\mathfrak{M}, \langle m', y_m\rangle) \models p_d$ is $m' = m + \exp_d(d')$. So formulas (6.13) and (6.14) force $\langle m + \exp_d(d'), y_m\rangle$ to store the jth bit of $a + 1 \pmod{(2^{\exp_d(d')})}$. Then, again by (6.12), $\langle m + \exp_d(d'), x\rangle$ stores the same bit as well (see Fig. 6.1).

Finally, (6.15) guarantees that p_{d+1} holds at $\langle m, x\rangle$ iff the number simulated by the interval $[m, m + \exp_d(d') - 1]$ equals 0. So these numbers m are $\exp_d(d') \cdot 2^{\exp_d(d')} = \exp_{d+1}(d')$ steps apart from each other, as required in (a).

For (b), take a number $k < \exp_{d+1}(d')$. Then there are unique numbers $k' < \exp_d(d')$, $k'' < 2^{\exp_d(d')}$ such that $k = k' + k'' \cdot \exp_d(d')$. By the induction hypothesis, for each $l < \exp_d(d')$, there exist a model $\mathfrak{M}_l = \langle \mathfrak{F}_l, \mathfrak{V}_l\rangle$ based on a product frame $\mathfrak{F}_l = \langle \mathbb{N}, <\rangle \times \langle W_l, R_l\rangle$ and a point $x_l \in W_l$ such that

- $(\mathfrak{M}_l, \langle n, x_l\rangle) \models \delta_{d,d'}$ for all $n \in \mathbb{N}$;
- $(\mathfrak{M}_l, \langle l, x_l\rangle) \models p_d$.

We may assume that the sets W_l are pairwise disjoint. Take a fresh point x and put

$$W = \{x\} \cup \bigcup_{l < \exp_d(d')} W_l,$$

$$R = \{\langle x, x_l\rangle \mid l < \exp_d(d')\} \cup \bigcup_{l < \exp_d(d')} R_l,$$

$$\mathfrak{F} = \langle \mathbb{N}, <\rangle \times \langle W, R\rangle .$$

For each number k as above, we define a new model $\mathfrak{N}_k = \langle \mathfrak{F}, \mathfrak{W}_k\rangle$ by taking

- $\mathfrak{W}_k(p_i) = \bigcup_{l < \exp_d(d')} \mathfrak{V}_l(p_i)$, for all $i < d$;
- $\mathfrak{W}_k(p_d) = \bigcup_{l < \exp_d(d')} \mathfrak{V}_l(p_d) \cup \{\langle n, x\rangle \mid n = k' + j \cdot \exp_d(d'),\ j \in \mathbb{N}\}$;
- $\mathfrak{W}_k(p_{d+1}) = \{\langle n, x\rangle \mid n = k + j \cdot \exp_{d+1}(d'),\ j \in \mathbb{N}\}$;

- $\mathfrak{W}_k(r_i) = \bigcup_{l<\exp_d(d')} \mathfrak{V}_l(r_i)$, for all $1 \leq i < d$;
- $\mathfrak{W}_k(r_d) = \mathbb{N} \times \{x_{k'}\}$;
- $\mathfrak{W}_k(q_i) = \bigcup_{l<\exp_d(d')} \mathfrak{V}_l(q_i)$, for all $1 \leq i \leq d$;
- for all $n \in \mathbb{N}$, $y \in W$,

$$\langle n, y\rangle \in \mathfrak{W}_k(q_{d+1}) \quad \text{iff} \quad \exists i < \exp_d(d'),\ j \in \mathbb{N}\ \big(n = k' + i + j \cdot \exp_d(d') \text{ and the } i\text{th bit of the } \exp_d(d')\text{-bit binary representation of } (2^{\exp_d(d')} - k'') + j \ (\bmod\ (2^{\exp_d(d')})) \text{ equals to } 1\big).$$

It is not difficult to see now that $(\mathfrak{N}_k, \langle n, x\rangle) \models \delta_{d+1,d'}$, for all $n \in \mathbb{N}$, and $(\mathfrak{N}_k, \langle k, x\rangle) \models p_{d+1}$, which completes the proof of Lemma 6.35. ❑

We now come back to the proof of Theorem 6.34. First, let us define briefly what it means to say that a single-tape right-infinite deterministic Turing machine solves a problem '$x \in X$?' in bounded space. Such Turing machines were defined in Section 5.4. Here we use a slight variation of that definition: instead of one halt state s_1, a Turing machine $\boldsymbol{A}$ has two halt states, s_{yes} (the *accepting state*), and s_{no} (the *rejecting state*). Otherwise we use the same notation as in Section 5.4.

Let Y be any finite set having more than one element, and let Y^* denote the set of all finite sequences (words) over Y. Let $\boldsymbol{A}$ be a Turing machine with tape alphabet $A = Y \cup \{b\}$. Given an $x = \langle x_1, x_2, \dots, x_n\rangle$ in Y^*, the *computation of* $\boldsymbol{A}$ *on input* x is the (unique) computation of $\boldsymbol{A}$ starting with the configuration

$$\langle \pounds, \langle s_0, x_1\rangle, x_2, \dots, x_n, b, b, \dots\rangle .$$

Clearly, for each configuration $c = \langle \pounds, a_0, a_1, \dots\rangle$ in the computation of $\boldsymbol{A}$ on x there is a number N_c such that $a_m = b$ for every $m > N_c$. If $\boldsymbol{A}$ halts on x, then define the *space used by* $\boldsymbol{A}$ *on* x as the maximum of these numbers N_c.

Now, let X be a subset of Y^*, for some set Y as above. Take a function $f : \mathbb{N} \to \mathbb{N}$. We say that a Turing machine $\boldsymbol{A}$ with tape alphabet $A = Y \cup \{b\}$ *solves the problem '$x \in X$?' in space bounded by* f, if for all x in Y^*,

- $\boldsymbol{A}$ halts on input x,
- the space used by $\boldsymbol{A}$ on x is $f(|x|)$, and
- $x \in X$ iff $\boldsymbol{A}$ halts on x at state s_{yes}.

Fix a $d > 0$. Given any set X such that $X \subseteq Y^*$ for some Y and the problem '$x \in X$?' is solved by a Turing machine $\boldsymbol{A}$ in space bounded by $\exp_d$, we will construct a family of formulas $\varphi_{\boldsymbol{A},x}$ $(x \in Y^*)$ such that for every $x \in Y^*$,

- $vmd(\varphi_{\boldsymbol{A},x}) = d$,
- the length of $\varphi_{\boldsymbol{A},x}$ is linear in $|x|$,

and

$$\varphi_{\boldsymbol{A},x} \text{ is } \mathbf{PTL} \times \mathbf{K}\text{-satisfiable iff } \boldsymbol{A} \text{ halts on input } x \text{ at state } s_{yes}. \tag{6.16}$$

To this end, we introduce a propositional variable t_a for each a in the alphabet $A' = A \cup \{\pounds\} \cup (S \times A)$. We also use three extra variables q_s, q_l and q_r the meaning of which will be clear from the formulas below. Fix an $x = \langle x_1, \dots, x_{d'} \rangle \in Y^*$. We define $\varphi_{\boldsymbol{A},x}$ as the conjunction of $p_d \wedge \psi_{d,d'}$ (see the proof of Lemma 6.35) and the following formulas, for all instructions $\delta(\alpha, \beta, \gamma) = \langle \alpha', \beta', \gamma' \rangle$ of $\boldsymbol{A}$:

$$\boxminus^+ \bigvee_{a \in A'} (t_a \wedge \bigwedge_{a' \in A',\, a' \neq a} \neg t_{a'}), \tag{6.17}$$

$$\boxminus^+ \bigwedge_{a \in A'} ((t_a \leftrightarrow \Diamond t_a) \wedge (t_a \leftrightarrow \Box t_a)), \tag{6.18}$$

$$\boxminus^+ (t_\pounds \leftrightarrow p_d), \tag{6.19}$$

$$t_\pounds \wedge \ominus(t_{\langle s_0, x_1 \rangle} \wedge \ominus(t_{x_2} \wedge \ominus(\cdots \wedge \ominus(t_{x_{d'}} \wedge (t_b \mathcal{U}_h t_\pounds)) \dots))), \tag{6.20}$$

$$\boxminus^+ \Big(t_\alpha \wedge \ominus t_\beta \wedge \ominus\ominus t_\gamma \rightarrow$$

$$\Box\big(p_d \rightarrow (\neg p_d)\mathcal{U}_h(p_d \wedge t_{\alpha'} \wedge \ominus t_{\beta'} \wedge \ominus\ominus t_{\gamma'})\big)\Big), \tag{6.21}$$

$$\boxminus^+ ((q_s \leftrightarrow \bigvee_{\langle s,a \rangle \in S \times A} t_{\langle s,a \rangle}) \wedge (q_l \leftrightarrow \ominus q_s) \wedge (q_s \leftrightarrow \ominus q_r)), \tag{6.22}$$

$$\boxminus^+ \bigwedge_{a \in A \cup \{\pounds\}} \Big(\neg q_l \wedge \neg q_s \wedge \neg q_r \wedge t_a \rightarrow \Box\big(p_d \rightarrow (\neg p_d)\mathcal{U}_h(p_d \wedge t_a)\big)\Big), \tag{6.23}$$

$$\diamondminus^+ \bigvee_{a \in A} t_{\langle s_{yes}, a \rangle} \wedge \neg \diamondminus^+ \bigvee_{a \in A} t_{\langle s_{no}, a \rangle}. \tag{6.24}$$

Suppose first that $\varphi_{\boldsymbol{A},x}$ is $\mathbf{PTL} \times \mathbf{K}$-satisfiable. By Theorem 6.29, we may assume that

$$(\mathfrak{M}, \langle n, x \rangle) \models \varphi_{\boldsymbol{A},x}, \tag{6.25}$$

for some model $\mathfrak{M}$ based on the product of $\langle \mathbb{N}, < \rangle$ and some (countable) frame $\langle W, R \rangle$. We show that $\boldsymbol{A}$ halts on input x at state s_{yes}.

To begin with, we know that the space used by $\boldsymbol{A}$ on input x is $\exp_d(d')$. So we can represent each configuration of the computation of $\boldsymbol{A}$ on x as a finite word

$$c = \langle \pounds, a_0, \ldots, a_l, b, \ldots, b\rangle$$

of length $\exp_d(d')$. We will use the 'yardsticks' provided by Lemma 6.35 to encode these configurations by means of the intervals $[m, m + \exp_d(d') - 1]$ for which $(\mathfrak{M}, \langle m, x\rangle) \models p_d$ for some $m \geq n$. Given such an m, we will say that the interval $[m, m + \exp_d(d') - 1]$ *encodes* the configuration c if, for all $k < \exp_d(d')$ and $a \in A'$,

$$(\mathfrak{M}, \langle m + k, x\rangle) \models t_a \quad \text{iff} \quad a \text{ occupies the } k\text{th cell of } c. \tag{6.26}$$

As we shall see, $\varphi_{\boldsymbol{A},x}$ also ensures that consecutive configurations in the computation of $\boldsymbol{A}$ on x are encoded by consecutive intervals.

By (6.25), we have $(\mathfrak{M}, \langle n, x\rangle) \models p_d \wedge \psi_{d,d'}$. Choose a $y \in W$ as in Claim 6.36 (iii). Then by Claim 6.36, $(\mathfrak{M}, \langle n, y\rangle) \models p_d \wedge \delta_{d,d'}$. So by Lemma 6.35, the numbers $m \geq n$ for which $(\mathfrak{M}, \langle m, y\rangle) \models p_d$ holds are located $\exp_d(d')$ steps apart from each other. By the choice of y, we obtain that for the same numbers we actually have $(\mathfrak{M}, \langle m, x\rangle) \models p_d$ (see Fig. 6.1).

Now the formula (6.17) says that for every $m \geq n$, $\langle n, x\rangle$ validates precisely one t_a. The formula (6.18) ensures that, for every $m \geq n$, the t_a are uniform among $\langle m, x\rangle$ and all $\langle m, y\rangle$ with xRy. By (6.19), we have that the delimiters of the configurations coincide with the delimiters of the intervals. Formula (6.20) ensures that the start configuration

$$\langle \pounds, \langle s_0, x_1\rangle, x_2, \ldots, x_{d'}, b, \ldots, b\rangle$$

is encoded by the interval $[n, n + \exp_d(d') - 1]$. It is not hard to see that formula (6.21) forces the correct transitions for the active cell and its left and right neighbors (for each $m \geq n$, make use of the y_m provided by Claim 6.36 (ii)). Formula (6.22) marks with q_s, q_l and q_r the active cell and its left and right neighbors, respectively, and (6.23) ensures that at each step of $\boldsymbol{A}$ only the active cell and its left and right neighbors are changed. Finally, (6.24) says that $\boldsymbol{A}$ halts on x at state s_{yes}.

Conversely, suppose that $\boldsymbol{A}$ halts on x at state s_{yes}. We need to show that $\varphi_{\boldsymbol{A},x}$ is $\mathbf{PTL} \times \mathbf{K}$-satisfiable. Since $\psi_{d,d'}$ is actually a conjunct of $\delta_{d+1,d'}$, by Lemma 6.35 we know that there exists a model $\mathfrak{M}$ based on a product frame $\langle \mathbb{N}, <\rangle \times \langle W, R\rangle$ such that

$$(\mathfrak{M}, \langle 0, x\rangle) \models p_{d+1} \wedge \psi_{d,d'},$$

for some $x \in W$. By (6.15), we also have $(\mathfrak{M}, \langle 0, x\rangle) \models p_d$. It is not hard to see that by encoding the configurations of the computation of $\boldsymbol{A}$ on x as in (6.26), the remaining conjuncts of $\varphi_{\boldsymbol{A},x}$ are also satisfied at $\langle 0, x\rangle$. ❑

As a consequence of Theorem 6.34 we obtain that *any* problem in ELEM can be polynomially reduced to the satisfiability problem for **PTL** $\times$ **K**. So we have:

Theorem 6.37. *The satisfiability problem for* **PTL** $\times$ **K** *does not belong to* ELEM.

The following question still remains open:

Question 6.38. What is the complexity of the products $\mathsf{Log}\{\langle \mathbb{N}, < \rangle\} \times \mathbf{K}$ and $\mathsf{Log}_{FP}(\mathbb{N}) \times \mathbf{K}$?

Note that by Theorem 5.32, $\mathsf{Log}\{\langle \mathbb{N}, < \rangle\} \times \mathbf{K}$ does not have the product fmp.

Question 6.39. Does $\mathsf{Log}\{\langle \mathbb{N}, < \rangle\} \times \mathbf{K}$ have the (abstract) fmp?

Now let us consider products of other temporal logics with $\mathbf{K}_m$. Using the ideas of (Wolter and Zakharyaschev 2000c), we show the following result:

Theorem 6.40. *The modal product logics* $\mathbf{K4.3} \times \mathbf{K}_m$ *and* $\mathsf{Log}\{\langle \mathbb{Q}, < \rangle\} \times \mathbf{K}_m$ *and the corresponding temporal variants* $\mathbf{Lin} \times \mathbf{K}_m$ *and* $\mathsf{Log}_{FP}(\mathbb{Q}) \times \mathbf{K}_m$ *are decidable.*

Proof. We use the fact (established by Propositions 6.30 and 6.31) that $\mathbf{K4.3} \times \mathbf{K}_m$ and $\mathsf{Log}\{\langle \mathbb{Q}, < \rangle\} \times \mathbf{K}_m$ are determined by their intended frames (i.e., products with strict linear orders and with $\langle \mathbb{Q}, < \rangle$, respectively).

We confine ourselves to considering **K** instead of arbitrary $\mathbf{K}_m$. Let us show first how to modify the quasimodel proof for $\mathbf{K} \times \mathbf{K}$ to obtain the decidability of $\mathbf{K4.3} \times \mathbf{K}$. The definition of a quasimodel and the proof of the corresponding 'quasimodel lemma' are almost the same as in the $\mathbf{K} \times \mathbf{K}$-proof. The only difference is that the underlying frames $\mathfrak{F} = \langle W, R \rangle$ are not arbitrary: now they must be strict linear orders.

Blocks, however, are significantly different from blocks in the $\mathbf{K} \times \mathbf{K}$-proof, not only in their shape, but also in that they are not necessarily root-saturated. A *block for* φ is a quadruple $\mathfrak{B}^{uv} = \langle \mathfrak{F}^{uv}, \boldsymbol{q}^{uv}, \mathfrak{R}^{uv}, \lhd^{uv} \rangle$ such that

- $\mathfrak{F}^{uv} = \langle \{u, v\}, < \rangle$ is a 2-element strict linear order with $u < v$,
- $\langle \mathfrak{F}^{uv}, \boldsymbol{q}^{uv} \rangle$ is a basic structure for φ of depth m, for some $m \leq md(\varphi)$,
- $\mathfrak{R}^{uv}$ is a set of runs through $\langle \mathfrak{F}^{uv}, \boldsymbol{q}^{uv} \rangle$ such that, for all $r \in \mathfrak{R}^{uv}$ and $\diamondsuit\psi \in sub\,\varphi$,

$$\text{if } \psi \in \boldsymbol{t}_v(r(v)) \text{ or } \diamondsuit\psi \in \boldsymbol{t}_v(r(v)) \text{ then } \diamondsuit\psi \in \boldsymbol{t}_u(r(u)),$$

- $\lhd^{uv}$ is a binary relation on $\mathfrak{R}^{uv}$ satisfying **(qm3)** and **(qm4)**.

(We remind the reader that quasistates occurring in such a block are denoted by $\boldsymbol{q}^{uv}(u) = \langle\langle T_u, <_u\rangle, \boldsymbol{t}_u\rangle$ and $\boldsymbol{q}(v)^{uv} = \langle\langle T_v, <_v\rangle, \boldsymbol{t}_v\rangle$.)

A set $\mathcal{S}$ of blocks for φ is called *satisfying* if the following properties hold:

(ssb1) all blocks in $\mathcal{S}$ are of the same depth m, for some $m \leq md(\varphi)$;

(ssb2) $\mathcal{S}$ contains a block satisfying **(qm2)**;

(ssb3) for every $\mathfrak{B}^{uv}$ in $\mathcal{S}$, if $\diamond\psi \in \boldsymbol{t}_v(r(v))$ for some run $r \in \mathfrak{R}^{uv}$ then there exist a block $\mathfrak{B}^{vw}$ in $\mathcal{S}$ and a sequence $\langle x_s \in T_w \mid s \in \mathfrak{R}^{uv}\rangle$ of points in T_w such that

(i) $\boldsymbol{q}^{uv}(v) = \boldsymbol{q}^{vw}(v)$,

(ii) for every $s \in \mathfrak{R}^{uv}$, the function p defined by $p(v) = s(v)$, $p(w) = x_s$ is a run in $\mathfrak{R}^{vw}$,

(iii) for all $s, s' \in \mathfrak{R}^{uv}$, if $s \lhd^{uv} s'$ then $x_s <_w x_{s'}$,

(iv) $\psi \in \boldsymbol{t}_w(x_r)$;

(ssb4) for every block $\mathfrak{B}^{uv}$ in $\mathcal{S}$, if $\diamond\psi \in \boldsymbol{t}_u(r(u))$, $\psi \notin \boldsymbol{t}_v(r(v))$ and $\diamond\psi \notin \boldsymbol{t}_v(r(v))$ for some run $r \in \mathfrak{R}^{uv}$ then there are blocks $\mathfrak{B}^{uw}$ and $\mathfrak{B}^{wv}$ in $\mathcal{S}$ and a sequence $\langle x_s \in T_w \mid s \in \mathfrak{R}^{uv}\rangle$ of points in T_w such that

(i) $\boldsymbol{q}^{uv}(u) = \boldsymbol{q}^{uw}(u)$, $\boldsymbol{q}^{uw}(w) = \boldsymbol{q}^{wv}(w)$, $\boldsymbol{q}^{wv}(v) = \boldsymbol{q}^{uv}(v)$,

(ii) for every $s \in \mathfrak{R}^{uv}$, the function p' defined by $p'(u) = s(u)$, $p'(w) = x_s$ is a run in $\mathfrak{R}^{uw}$, and the function p'' defined by $p''(w) = x_s$, $p''(v) = s(v)$ is a run in $\mathfrak{R}^{wv}$,

(iii) for all $s, s' \in \mathfrak{R}^{uv}$, if $s \lhd^{uv} s'$ then $x_s <_w x_{s'}$,

(iv) $\psi \in \boldsymbol{t}_w(x_r)$.

Clearly, one can effectively check whether there exists a satisfying set of blocks for φ. As satisfiability in a single element strict linear order is trivially decidable, to establish the decidability of $\mathbf{K4.3} \times \mathbf{K}$, it is enough to prove the following 'block lemma:'

Lemma 6.41. *There is a* $\mathbf{K4.3} \times \mathbf{K}$*-quasimodel for* φ *based on a strict linear order with* ≥ 2 *elements iff there is a satisfying set of blocks for* φ.

Proof. The construction of a satisfying set from a quasimodel is easy. Suppose that $\mathfrak{Q} = \langle \mathfrak{F}, \boldsymbol{q}, \mathfrak{R}, \lhd\rangle$ is a quasimodel for φ, with $\mathfrak{F} = \langle W, R\rangle$ being a strict linear order with ≥ 2 elements. For all $u, v \in W$ such that uRv, define the *restriction* $\mathfrak{Q}^{uv}$ of $\mathfrak{Q}$ to the 2-element strict linear order on $\{u, v\}$ in the natural way. It is straightforward to check that these $\mathfrak{Q}^{uv}$ are blocks and that the collection $\mathcal{S}$ of them is a satisfying set.

Now we show how a quasimodel for φ can be constructed from a satisfying set $\mathcal{S}$ of blocks for φ. Similarly to the $\mathbf{K} \times \mathbf{K}$-proof, we call a quadruple $\mathfrak{Q} = \langle \mathfrak{F}, \boldsymbol{q}, \mathfrak{R}, \lhd \rangle$ a *weak quasimodel for* φ if the following conditions hold:

(wq1′) $\mathfrak{F} = \langle W, R \rangle$ is a finite strict linear order, $W = \{w_0, w_1, \dots, w_m\}$ for some $m > 0$, $w_0 R w_1 R \dots R w_m$, and $\langle \mathfrak{F}, \boldsymbol{q} \rangle$ is a basic structure for φ satisfying **(qm2)**;

(wq2′) $\mathfrak{R}$ is a set of runs through $\langle \mathfrak{F}, \boldsymbol{q} \rangle$ such that for all $i < j \leq m$, $r \in \mathfrak{R}$ and $\Diamond\psi \in sub\,\varphi$,

$$\text{if } \psi \in \boldsymbol{t}_{w_j}(r(w_j)) \text{ or } \Diamond\psi \in \boldsymbol{t}_{w_j}(r(w_j)) \text{ then } \Diamond\psi \in \boldsymbol{t}_{w_i}(r(w_i)),$$

(wq2″) $\lhd$ is a binary relation on $\mathfrak{R}$ satisfying **(qm4)** and such that, for all $r, s \in \mathfrak{R}$,

$$r \lhd s \qquad \text{iff} \qquad r(w_i) <_{w_i} s(w_i) \;\text{ for all } i \leq m$$

(this property is a bit stronger than **(qm3)**),

(wq3′) for every $i < m$, the restriction of $\mathfrak{Q}$ to the two-element strict linear order on $\{w_i, w_{i+1}\}$ is a block in $\mathcal{S}$.

A weak quasimodel is almost a quasimodel. What is missing is that runs are not necessarily saturated. To fix this, take a triple $\left\langle i, r, \Diamond\psi \right\rangle$ such that $i \leq m$, $r \in \mathfrak{R}$ and $\Diamond\psi \in sub\,\varphi$. Such a triple is called a *defect in* $\mathfrak{Q}$ if $\Diamond\psi \in \boldsymbol{t}_{w_i}(r(w_i))$ and for all j such that $i < j \leq m$, $\psi \notin \boldsymbol{t}_{w_j}(r(w_j))$ and $\Diamond\psi \notin \boldsymbol{t}_{w_j}(r(w_j))$. If $i = m$ then such a defect is called an *end-defect*, otherwise it is a *middle-defect*.

We construct a sequence $\langle \mathfrak{Q}_n \mid n < \omega \rangle$ of weak quasimodels which 'converges' to a real quasimodel for φ. Take a block $\mathfrak{Q}_0 = \langle \mathfrak{F}_0, \boldsymbol{q}_0, \mathfrak{R}_0, \lhd_0 \rangle$ in $\mathcal{S}$ satisfying **(qm2)**. Clearly, it is a weak quasimodel for φ as well. Suppose now that we have already constructed $\mathfrak{Q}_n = \langle \mathfrak{F}_n, \boldsymbol{q}_n, \mathfrak{R}_n, \lhd_n \rangle$ such that $\mathfrak{F}_n = \langle W_n, R_n \rangle$, $W_n = \{w_0, w_1, \dots, w_m\}$ and $w_0 R_n w_1 R_n \dots R_n w_m$. If the set D_n of all defects in $\mathfrak{Q}_n$ is empty then we are done: $\mathfrak{Q}_n$ is obviously a quasimodel for φ. Otherwise, we take some $d = \left\langle i, r, \Diamond\psi \right\rangle$ from D_n.

Case 1: d is a middle-defect, that is, $i < m$. By **(wq3′)**, the restriction $\mathfrak{Q}^{w_i w_{i+1}}$ of $\mathfrak{Q}_n$ to the two-element strict linear order on $\{w_i, w_{i+1}\}$ is a block in $\mathcal{S}$. Choose two blocks $\mathfrak{B}^{w_i w}$ and $\mathfrak{B}^{w w_{i+1}}$ according to **(ssb4)** (with $u = w_i$ and $v = w_{i+1}$). We may assume that $w \notin W_n$. Define a basic structure

$\langle \mathfrak{F}_n^d, \boldsymbol{q}_n^d \rangle$ by taking

$$W_n^d = W_n \cup \{w\},$$
$$R_n^d = R_n \cup \{\langle w_j, w \rangle \mid j \le i,\, w_j \in W_n\} \cup \{\langle w, w_j \rangle \mid i < j \le m,\, w_j \in W_n\},$$
$$\mathfrak{F}_n^d = \langle W_n^d, R_n^d \rangle,$$
$$\boldsymbol{q}_n^d(v) = \begin{cases} \boldsymbol{q}^{w_i w}(v) = \boldsymbol{q}^{w w_{i+1}}(v), & \text{if } v = w, \\ \boldsymbol{q}_n(v), & \text{if } v \in W_n. \end{cases}$$

For all runs $s, p \in \mathfrak{R}_n$, $s' \in \mathfrak{R}^{w_i w}$, $s'' \in \mathfrak{R}^{w w_{i+1}}$, such that $s(w_i) = s'(w_i)$, $s'(w) = s''(w)$, $s''(w_{i+1}) = p(w_{i+1})$, define the function $s \cup s' \cup s'' \cup p$ on W_n^d by taking, for all $v \in W_n^d$,

$$(s \cup s' \cup s'' \cup p)(v) = \begin{cases} s(v), & \text{if } v = w_j,\ j \le i, \\ s'(v) = s''(v), & \text{if } v = w, \\ p(v), & \text{if } v = w_j,\ i < j \le m. \end{cases}$$

Let $\mathfrak{R}_n^d$ be the set of all such functions. Elements in $\mathfrak{R}_n^d$ of the form $s \cup s' \cup s'' \cup s$, for some $s \in \mathfrak{R}_n$, are called *extensions of* s. We call an extension $s \cup s' \cup s'' \cup s$ *good*, if $s'(w) = s''(w) = x_s$; cf. **(ssb4)**. Observe that every $s \in \mathfrak{R}_n$ has a unique good extension in $\mathfrak{R}_n^d$.

For all $s, s' \in \mathfrak{R}_n^d$, define

$$s \lhd_n^d s' \quad \text{iff} \quad s(v) \lhd_v s'(v) \text{ for all } v \in W_n^d.$$

In other words, we 'glue together' the blocks $\mathfrak{B}^{w_i w}$ and $\mathfrak{B}^{w w_{i+1}}$ at w, and then 'insert' the resulting piece into $\mathfrak{Q}_n$ between w_i and w_{i+1}. It can be readily checked that $\mathfrak{Q}_n^d = \langle \mathfrak{F}_n^d, \boldsymbol{q}_n^d, \mathfrak{R}_n^d, \lhd_n^d \rangle$ is a weak quasimodel. Moreover, the defect d in $\mathfrak{Q}_n^d$ is 'cured' in the sense that (by **(ssb4)**) the good extension r^+ of r is such that $\psi \in \boldsymbol{t}_w(r^+(w))$.

Case 2: d is an end-defect. This case is analogous to Case 1, but we have to use **(ssb3)** instead of **(ssb4)** for 'gluing together' $\mathfrak{Q}_n$ and a block $\mathfrak{B}^{w_m w}$ at w_m.

Next we turn the remaining defects in $\mathfrak{Q}_n$ to a subset D_n^d of the set of defects in $\mathfrak{Q}_n^d$ as follows. Suppose $\langle j, s, \Diamond\chi \rangle$ is a defect in D_n different from d. Let s^+ be the good extension of s and let $k = j$ if $j \le i$ and $k = j + 1$ otherwise. If $\langle k, s^+, \Diamond\chi \rangle$ is a defect in $\mathfrak{Q}_n^d$ then we put it into D_n^d. Clearly, $|D_n^d| \le |D_n| - 1$. If $D_n^d \ne \emptyset$ then we take a defect $d' \in D_n^d$, construct $\mathfrak{Q}_n^{dd'}$, and so on. When all the finitely many defects in D_n are cured, we obtain a weak quasimodel $\mathfrak{Q}_{n+1}$. Note that every run $r_n \in \mathfrak{R}_n$ has a unique extension $r_{n+1} \in \mathfrak{R}_{n+1}$ obtained by taking at every step the good extension of the previous run. We call this r_{n+1} the *good extension of* r_n *in* $\mathfrak{Q}_{n+1}$.

The limit quasimodel is defined as follows. Let $\mathfrak{F} = \langle W, R\rangle$, where

$$W = \bigcup_{n<\omega} W_n, \qquad R = \bigcup_{n<\omega} R_n$$

and

$$\boldsymbol{q} = \bigcup_{n<\omega} \boldsymbol{q}_n.$$

Then clearly $\mathfrak{F}$ is a strict linear order and $\langle \mathfrak{F}, \boldsymbol{q}\rangle$ is a basic structure for φ.

For every $i < \omega$ and every sequence of runs $\langle r_n \in \mathfrak{R}_n \mid n \geq i\rangle$ such that r_{n+1} is the good extension of r_n in $\mathfrak{Q}_{n+1}$ for all $n \geq i$, take $r = \bigcup\{r_n \mid n \geq i\}$. Let $\mathfrak{R}$ be the set of such runs. For $r = \bigcup\{r_n \mid n \geq i\}$ and $r' = \bigcup\{r'_n \mid n \geq j\}$ in $\mathfrak{R}$, define

$$r \lhd r' \qquad \text{iff} \qquad r_n \lhd_n r'_n, \quad \text{for all } n \geq \max(i,j).$$

We show that $\mathfrak{R}$ and $\lhd$ satisfy **(qm3)** and **(qm4)**. Indeed, suppose that r and r' are of the above form and $r \lhd r'$. Take a $w \in W$. There is an $n \geq \max(i,j)$ such that $w \in W_n$. Then $r(w) = r_n(w)$, $r'(w) = r'_n(w)$ and $r_n \lhd_n r'_n$, which implies $r(w) <_w r'(w)$ by **(wq2″)**. For **(qm4)**, suppose that $r = \bigcup\{r_n \mid n \geq i\}$ and $r(w) <_w x$ for some $x \in T_w$. Then there is an $n \geq i$ such that $w \in W_n$, and so $r(w) = r_n(w)$. Since $\mathfrak{Q}_n$ satisfies **(qm4)**, there is an $s_n \in \mathfrak{R}_n$ such that $s_n(w) = x$ and $r_n \lhd_n s_n$. Let $s = \bigcup\{s_m \mid m \geq n\}$, where s_{m+1} is the good extension of s_m in $\mathfrak{Q}_{m+1}$ for all $m \geq n$. Then $s(w) = s_n(w) = x$, and it is not hard to see that, by **(ssb3)**, **(ssb4)** and **(wq2″)**, $r_m \lhd_m s_m$ hold for all $m \geq n$, from which $r \lhd s$.

Finally, we show that all the runs in $\mathfrak{R}$ are coherent and saturated. Indeed, suppose that $r = \bigcup\{r_n \mid n \geq i\}$ and $\Diamond\psi \in \boldsymbol{t}_w(r(w))$ for some $w \in W$. Then there is an $n \geq i$ such that $w \in W_n$, and so $r(w) = r_n(w)$. If $\left\langle w, r_n, \Diamond\psi\right\rangle$ is not a defect in $\mathfrak{Q}_n$ then there is a $v \in W_n$ such that wRv, $r_n(v) = r(v)$ and $\psi \in \boldsymbol{t}_v(r_n(v))$. And if $\left\langle w, r_n, \Diamond\psi\right\rangle$ is a defect in $\mathfrak{Q}_n$ then it is cured in its good extension r_{n+1} in $\mathfrak{Q}_{n+1}$: there is $v \in W_{n+1}$ such that wRv, $r_{n+1}(v) = r(v)$ and $\psi \in \boldsymbol{t}_v(r_{n+1}(v))$. Conversely, assume that $\psi \in \boldsymbol{t}_w(r(w))$ and let vRw. Then there is an $n \geq i$ such that $v, w \in W_n$. Thus $r(w) = r_n(w)$, $r(v) = r_n(v)$ and vR_nw, and so $\Diamond\psi \in \boldsymbol{t}_v(r(v))$ follows by **(wq2′)**.

Therefore, $\mathfrak{Q} = \langle \mathfrak{F}, \boldsymbol{q}, \mathfrak{R}, \lhd\rangle$ is a quasimodel for φ, as required. ❑

The case of $\mathsf{Log}\{\langle \mathbb{Q}, <\rangle\} \times \mathbf{K}$ is similar. The only difference is that one has to extend the definition of a satisfying set of blocks with the following three properties:

(ssb5) for every $\mathfrak{B}^{uv}$ in $\mathcal{S}$, there exist a block $\mathfrak{B}^{vw}$ in $\mathcal{S}$ and a sequence $\langle x_s \in T_w \mid s \in \mathfrak{R}^{uv}\rangle$ of points in T_w such that **(ssb3)**(i)–(iii) hold;

(ssb6) for every $\mathfrak{B}^{uv}$ in $\mathcal{S}$, there exist a block $\mathfrak{B}^{wu}$ in $\mathcal{S}$ and a sequence $\langle x_s \in T_w \mid s \in \mathfrak{R}^{uv}\rangle$ of points in T_w such that

(i) $\boldsymbol{q}^{uv}(u) = \boldsymbol{q}^{wu}(u)$,

(ii) for every $s \in \mathfrak{R}^{uv}$, the function p defined by $p(w) = x_s$, $p(u) = s(u)$ is a run in $\mathfrak{R}^{wu}$,

(iii) for all $s, s' \in \mathfrak{R}^{uv}$, if $s \lhd^{uv} s'$ then $x_s <_w x_{s'}$;

(ssb7) for every $\mathfrak{B}^{uv}$ in $\mathcal{S}$, there are blocks $\mathfrak{B}^{uw}$ and $\mathfrak{B}^{wv}$ in $\mathcal{S}$ and a sequence $\langle x_s \in T_w \mid s \in \mathfrak{R}^{uv}\rangle$ of points in T_w such that **(ssb4)**(i)–(iii) hold.

Then in the construction of the sequence of weak quasimodels, after having cured all defects of $\mathfrak{Q}_n$ and constructed a weak quasimodel $\mathfrak{Q}'_{n+1}$ based on a finite strict linear order $\mathfrak{F}'_{n+1} = \langle W'_{n+1}, R'_{n+1}\rangle$, where

$$W'_{n+1} = \{w_0, w_1, \dots, w_m\} \text{ and } w_0 R'_{n+1} w_1 R'_{n+1} \dots R'_{n+1} w_m,$$

we define (with the help of **(ssb5)**–**(ssb7)**) a weak quasimodel $\mathfrak{Q}_{n+1}$ based on a finite strict linear order $\mathfrak{F}_{n+1} = \langle W_{n+1}, R_{n+1}\rangle$, where

$$W_{n+1} = W'_{n+1} \cup \{u_0, u_1, \dots, u_m, u_{m+1}\}, \text{ and}$$
$$u_0 R_{n+1} w_0 R_{n+1} u_1 R_{n+1} w_1 R_{n+1} \dots R_{n+1} u_m R_{n+1} w_m R_{n+1} u_{m+1}.$$

As a result, we construct a quasimodel for φ which is based on a linear order isomorphic to $\langle \mathbb{Q}, <\rangle$.

The decidability of **Lin** $\times$ $\mathbf{K}_m$ and $\mathsf{Log}_{FP}(\mathbb{Q}) \times \mathbf{K}_m$ can be established by mixing the 'tricks' introduced so far. ❑

Observe that the decision procedures given in the above proof are nonelementary. In Section 13.2 we give another proof with the help of reductions to monadic second-order theories of certain linear orders (see Theorem 13.6).

Question 6.42. What is the complexity of **K4.3** $\times$ **K**, $\mathsf{Log}\{\langle \mathbb{Q}, <\rangle\} \times \mathbf{K}$, **Lin** $\times$ **K**, and $\mathsf{Log}_{FP}(\mathbb{Q}) \times \mathbf{K}$? Are these logics in ELEM?

Note that, by Theorem 5.32, none of the logics listed in Question 6.42 have the product fmp.

Question 6.43. Do **K4.3** $\times$ **K** or $\mathsf{Log}\{\langle \mathbb{Q}, <\rangle\} \times \mathbf{K}$ have the (abstract) fmp?

6.5 Products with S5

Products with **S5** are usually not so complex as products with **K** or $\mathbf{S5}_m$, for $m \geq 2$. In this section we justify this claim by providing elementary upper bounds for the computational complexity of logics like **CPDL** $\times$ **S5** and **Lin** $\times$ **S5**. On the other hand, we also show that the elementary decision procedures are still of considerable complexity by proving that almost all products with **S5** are at least coNEXPTIME- or EXPSPACE-hard.

First, observe that the definition of **CPDL** $\times$ **K** given in Section 6.2 can be extended to define the product of **CPDL** and any Kripke complete logic L, in particular, to define **CPDL** $\times$ **S5** and **CPDL** $\times$ **KD45**. Now the quasimodel techniques for establishing the decidability of **CPDL** $\times$ **K** (Theorem 6.10) and **S5** $\times$ **S5** (Theorem 5.22) can be 'mixed' to prove the following:

Theorem 6.44. *The product logics* **CPDL** $\times$ **S5** *and* **CPDL** $\times$ **KD45** *are decidable.*

Proof. We give a sketch for **CPDL** $\times$ **S5**, emphasizing the important steps. The **CPDL** component suggests that *types* should be again Boolean saturated subsets of $\mathit{flc}(\varphi)$; however, the **S5** component makes it possible to define *quasistates* as just $\diamondsuit$-saturated subsets of types. So the number of points in a quasistate is now bounded by

$$p(\varphi) = 2^{|\mathit{flc}(\varphi)|}$$

and the number of different quasistates by

$$b(\varphi) = 2^{2^{|\mathit{flc}(\varphi)|}}.$$

Besides, no ordering of the runs is needed. Thus, a **CPDL**$\times$**S5**-*quasimodel* for φ is a triple $\langle \mathfrak{F}, \boldsymbol{q}, \mathfrak{R} \rangle$ with $\mathfrak{F} = \langle W, T_{\alpha_1}, \dots, T_{\alpha_n} \rangle$, which satisfies conditions **(qm2)** of the **CPDL** $\times$ **K**-proof, and **(qm1)** and **(qm3)** of the **S5** $\times$ **S5**-proof, where the coherency and saturation conditions on runs are taken from the **CPDL** $\times$ **K**-proof. Then the proof of the 'quasimodel lemma' is straightforward:

Lemma 6.45. *A* $\mathcal{CPDL} \otimes \mathcal{ML}$*-formula* φ *is satisfiable in a product frame for* **CPDL** $\times$ **S5** *iff there is a* **CPDL** $\times$ **S5**-*quasimodel for* φ.

A *block for* φ *with root* w_0 is a triple $\langle \mathfrak{F}, \boldsymbol{q}, \mathfrak{R} \rangle$ satisfying **(b1)**–**(b3)** of the **CPDL** $\times$ **K**-proof and **(qm3)** of the **S5** $\times$ **S5**-proof. A set $\mathcal{S}$ of blocks for φ is *satisfying* if

- $\mathcal{S}$ contains a block satisfying **(qm2)** and
- for every block $\mathfrak{B} = \langle \mathfrak{F}, \boldsymbol{q}, \mathfrak{R} \rangle$ in $\mathcal{S}$ and every world v in $\mathfrak{F}$, there exists a block $\mathfrak{B}' = \langle \mathfrak{F}', \boldsymbol{q}', \mathfrak{R}' \rangle$ with root w' in $\mathcal{S}$ such that $\boldsymbol{q}(v) = \boldsymbol{q}'(w')$.

Now we have the 'block lemma' as well:

Lemma 6.46. *There is a* **CPDL**×**S5**-*quasimodel for* φ *iff there is a satisfying set* $\mathcal{S}$ *of blocks for* φ *such that*

- *the number of quasistates in each block in* $\mathcal{S}$ *is at most*

$$N(\varphi) = 1 + 2 \cdot l(\varphi) \cdot p(\varphi) \cdot |\mathit{flc}(\varphi)|,$$

- *the number of runs in each block in* $\mathcal{S}$ *is at most*

$$R(\varphi) = p(\varphi) \cdot \big(1 + N(\varphi) \cdot |\mathit{flc}(\varphi)| \cdot b(\varphi)\big),$$

where $l(\varphi)$ *is defined as in the* **CPDL** × **K**-*proof (that is,* $l(\varphi)$ *is a polynomial function of* $b(\varphi) \cdot p(\varphi)$*).*

Observe that the factor $md(\varphi) + 1$ in $N(\varphi)$ is now replaced by 2 due to the fact that in the construction of small blocks out of a quasimodel it is enough to take one twin copy of each path (cf. the **S5** × **S5**-proof). Now the decidability of **CPDL** × **S5** follows from Lemmas 6.45 and 6.46. ❑

Observe that this proof provides us with a 3EXPTIME algorithm deciding **CPDL** × **S5** in the following way. Call a block $\mathfrak{B} = \langle \mathfrak{F}, \boldsymbol{q}, \mathfrak{R} \rangle$ for φ *small* if $|\mathfrak{F}| \leq N(\varphi)$ and $|\mathfrak{R}| \leq R(\varphi)$. Take the set of all small blocks for φ (a straightforward computation shows that the cardinality of this set is at most 3-exponential in the length of φ). Eliminate iteratively those blocks $\mathfrak{B} = \langle \mathfrak{F}, \boldsymbol{q}, \mathfrak{R} \rangle$ for which there is a world v in $\mathfrak{F}$ such that for all the 'noneliminated' blocks $\mathfrak{B}' = \langle \mathfrak{F}', \boldsymbol{q}', \mathfrak{R}' \rangle$, we have $\boldsymbol{q}(v) \neq \boldsymbol{q}'(w')$ for the root w' of $\mathfrak{B}'$. This elimination procedure stops after at most 3-exponentially many steps in the length of φ. Obviously, the set $\mathcal{R}$ of remaining blocks forms a satisfying set for φ if it contains a block satisfying **(qm2)**. Conversely, if there exists a satisfying set $\mathcal{S}$ of blocks for φ, then $\mathcal{R} \supseteq \mathcal{S}$ and $\mathcal{R}$ is a satisfying set for φ itself. Hence, φ is satisfiable iff $\mathcal{R}$ contains a block satisfying **(qm2)**.

However, it is shown by Schmidt and Tishkovsky (2003) that by 'mixing' the filtration techniques for **CPDL** with the one discussed in Section 5.3, we can obtain a slightly better, nondeterministic 2-exponential, upper bound of the satisfiability problem for **CPDL** × **S5**. Here we give a sketch of this argument.

To begin with, define **CPDL** · **S5** as the set of all $\mathcal{CPDL} \otimes \mathcal{ML}$-formulas that are true in every model based on a (not necessarily product) $\mathcal{CPDL} \otimes \mathcal{ML}$-structure $\langle U, T_{\alpha_1}, T_{\alpha_2}, \dots, R \rangle$ such that R is an equivalence relation and T_{α_i} commutes with R, for each atomic action α_i.

The following analog of Theorem 5.27 was shown by Schmidt and Tishkovsky (2003):

Lemma 6.47. **CPDL** · **S5** *has the 2-exponential fmp.*

Proof. Suppose a $\mathcal{CPDL} \otimes \mathcal{ML}$-formula φ is refuted in a model $\mathfrak{M} = \langle \mathfrak{F}, \mathfrak{V} \rangle$ based on a $\mathcal{CPDL} \otimes \mathcal{ML}$-structure $\mathfrak{F} = \langle W, T_{\alpha_1}, T_{\alpha_2}, \dots, R \rangle$ such that R is an equivalence relation and each T_{α_i} commutes with R. Take the Fischer–Ladner closure $flc(\varphi)$ of φ (see Section 6.2). Similarly to the proof of Theorem 5.27, we define, for each world x in W,

$$\Sigma(x) = \{\psi \in flc(\varphi) \mid (\mathfrak{M}, x) \models \psi\},$$

and let $\sim$ be the following equivalence relation on W:

$$x \sim y \qquad \text{iff} \qquad \Sigma(x) = \Sigma(y) \;\text{ and }\; \{\Sigma(z) \mid xRz\} = \{\Sigma(z) \mid yRz\}.$$

Now define a $\mathcal{CPDL} \otimes \mathcal{ML}$-structure $\mathfrak{F}^{\sim} = \langle W^{\sim}, T^{\sim}_{\alpha_1}, T^{\sim}_{\alpha_2}, \dots, R^{\sim} \rangle$ as the *smallest $\sim$-filtration* of $\mathfrak{F}$, that is, by taking,

- $W^{\sim} = \{[x] \mid x \in W\}$, where $[x]$ denotes the $\sim$-equivalence class of x;
- for all $x, y \in W$ and all atomic actions α_i,

$$[x] T^{\sim}_{\alpha_i} [y] \qquad \text{iff} \qquad \exists x' \exists y' \; (x' \sim x, \; y' \sim y \text{ and } x' T_{\alpha_i} y');$$

- for all $x, y \in W$,

$$[x] R^{\sim} [y] \qquad \text{iff} \qquad \exists x' \exists y' \; (x' \sim x, \; y' \sim y \text{ and } x' R y').$$

It is not hard to show that $R^{\sim}$ is an equivalence relation and each $T^{\sim}_{\alpha_i}$ commutes with $R^{\sim}$ (cf. the proof of Theorem 5.27). Define a valuation $\mathfrak{V}^{\sim}$ in $\mathfrak{F}^{\sim}$ by taking

- $\mathfrak{V}^{\sim}(p) = \{[x] \mid x \in \mathfrak{V}(p)\}$, for all $p \in flc(\varphi)$, and $\mathfrak{V}^{\sim}(q) = \emptyset$, for all other propositional variables q,

and let $\mathfrak{M}^{\sim} = \langle \mathfrak{F}^{\sim}, \mathfrak{V}^{\sim} \rangle$ (the compound transition relations in $\mathfrak{M}^{\sim}$ are defined as usual). A straightforward induction (see, e.g., Harel *et al.* 2000) shows that for all worlds x in W and all $\mathcal{CPDL} \otimes \mathcal{ML}$-formulas φ,

$$(\mathfrak{M}, x) \models \varphi \qquad \text{iff} \qquad (\mathfrak{M}^{\sim}, [x]) \models \varphi,$$

and so $\mathfrak{M}^{\sim}$ refutes φ as well. Since $|flc(\varphi)|$ is linear in the length $\ell(\varphi)$ of φ, the size of $W^{\sim}$ is 2-exponential in $\ell(\varphi)$ (cf. the proof of Theorem 5.27). ❑

Now we can show that **CPDL** $\cdot$ **S5** in fact coincides with **CPDL** $\times$ **S5**:

Lemma 6.48. **CPDL** $\cdot$ **S5** $=$ **CPDL** $\times$ **S5**.

Proof. The inclusion **CPDL** $\cdot$ **S5** $\subseteq$ **CPDL** $\times$ **S5** is obvious.

To prove the converse, first we observe that, by Lemma 6.47, it is enough to consider models for **CPDL** $\cdot$ **S5** based on finite $\mathcal{CPDL} \otimes \mathcal{ML}$-structures $\mathfrak{F} = \langle U, T_{\alpha_1}, T_{\alpha_2}, \dots, R\rangle$ such that R is an equivalence relation and each T_{α_i} commutes with R. Given such a model $\mathfrak{M} = \langle \mathfrak{F}, \mathfrak{V}\rangle$, we can construct step-by-step, like in the proofs of Lemmas 5.2 and 5.8, a p-morphism f from the product $\mathfrak{F}'$ of a $\mathcal{PDL}$-structure and a frame for **S5** onto $\mathfrak{F}$. Now define a valuation $\mathfrak{V}'$ in $\mathfrak{F}'$ by taking

$$\mathfrak{V}'(p) = \{x \mid f(x) \in \mathfrak{V}(p)\}.$$

Let $\mathfrak{M}' = \langle \mathfrak{F}', \mathfrak{V}'\rangle$. Define the compound transition relations in $\mathfrak{M}'$ the usual way. It is not hard to show that f is still a p-morphism with respect to the compound relations, and for all x in $\mathfrak{F}'$ and all $\mathcal{CPDL} \otimes \mathcal{ML}$-formulas φ,

$$(\mathfrak{M}', x) \models \varphi \quad \text{iff} \quad (\mathfrak{M}, f(x)) \models \varphi$$

(for formulas not containing test it is straightforward; for those with test the proof is by induction on the nesting of tests). ❑

Now, by Lemmas 6.47 and 6.48, we obtain:

Theorem 6.49. **CPDL** $\times$ **S5** *has the 2-exponential fmp, and so it is decidable in* coN2EXPTIME.

Using the reductions of Theorems 6.18 (see Table 6.1), we obtain:

Theorem 6.50. *Suppose* $L \in \{\mathbf{PDL}, \mathbf{K}_n^C, \mathbf{T}_n^C, \mathbf{K4}_n^C, \mathbf{S4}_n^C, \mathbf{KD45}_n^C, \mathbf{S5}_n^C\}$. *Then* $L \times$ **S5** *is decidable in* coN2EXPTIME.

Since by Theorem 5.34, $\mathbf{K}_u \times \mathbf{S5}$ lacks the product fmp, and the reductions in Theorems 6.18 and 6.71 (cf. Table 6.1) all turn finite product models to finite product models, we also have the following:

Theorem 6.51. *Suppose* $L \in \{\mathbf{CPDL}, \mathbf{PDL}, \mathbf{K}_n^C, \mathbf{T}_n^C, \mathbf{K4}_n^C, \mathbf{S4}_n^C, \mathbf{KD45}_n^C, \mathbf{S5}_n^C\}$. *Then* $L \times$ **S5** *does not have the product fmp.*

However, these reductions do not necessarily preserve the (abstract) fmp. One can repeat the above filtration argument for products of epistemic logics with **S5** and obtain the following:

Theorem 6.52. *Suppose* $L \in \{\mathbf{K}_n^C, \mathbf{T}_n^C, \mathbf{K4}_n^C, \mathbf{S4}_n^C, \mathbf{KD45}_n^C, \mathbf{S5}_n^C\}$. *Then* $L \times$ **S5** *has the 2-exponential fmp.*

The above filtration also helps us to axiomatize products of dynamic and epistemic logics with **S5**. Define the logic $[\mathbf{CPDL}, \mathbf{S5}]_w$ as the logic axiomatized by putting together the **CPDL**-axioms, the **S5**-axioms and the axioms

$$\Diamond\langle\alpha_i\rangle p \leftrightarrow \langle\alpha_i\rangle\Diamond p, \tag{6.27}$$

for all *atomic* actions α_i. More precisely, let $[\mathbf{CPDL}, \mathbf{S5}]_w$ be the smallest set of $\mathcal{CPDL} \otimes \mathcal{ML}$-formulas containing classical propositional logic **Cl**, the axioms (2.11)–(2.17) (for all action terms in $\mathcal{CPDL} \otimes \mathcal{ML}$), the **S5**-axioms for $\boxminus$, the axioms (6.27) (for all atomic actions α_i), and closed under modus ponens, substitution, and the necessitation rules (for all $[\alpha]$ and $\boxminus$). The following analog of Proposition 5.7 was shown in (Schmidt and Tishkovsky 2003):

Proposition 6.53. $[\mathbf{CPDL}, \mathbf{S5}]_w = \mathbf{CPDL} \cdot \mathbf{S5}$.

Proof. The inclusion $[\mathbf{CPDL}, \mathbf{S5}]_w \subseteq \mathbf{CPDL} \cdot \mathbf{S5}$ is clear. To show the converse, we observe that in the *canonical model* for $[\mathbf{CPDL}, \mathbf{S5}]_w$ the **S5**-accessibility relation commutes with each of the atomic transition relations. Further, it can be shown (see, e.g., Harel *et al.* 2000) that by filtrating the canonical model in the same way as in the proof of Lemma 6.47, we obtain a 'standard' model, that is, a model based on a $\mathcal{CPDL} \otimes \mathcal{ML}$-structure $\langle U, T_{\alpha_1}, T_{\alpha_2}, \dots, R\rangle$ where R is an equivalence relation and each T_{α_i} commutes with R. ❑

So, by Lemma 6.48 and Proposition 6.53, we obtain that $\mathbf{CPDL} \times \mathbf{S5}$ is 'kind of' product-matching:

Theorem 6.54. $\mathbf{CPDL} \times \mathbf{S5} = [\mathbf{CPDL}, \mathbf{S5}]_w$.

Note that, for action terms α containing test, the commutativity axioms (6.27) do not belong to the logic $\mathbf{CPDL} \times \mathbf{S5}$. In fact, as is shown in (Schmidt and Tishkovsky 2003), if we add these axioms to $[\mathbf{CPDL}, \mathbf{S5}]_w$, then the resulting logic is (linearly) reducible to **CPDL**.

As concerns axiomatization of products of epistemic logics L with **S5**, we can define the logic $[L, \mathbf{S5}]$ by putting together the axioms of the epistemic logic L (see Theorem 2.17), those of **S5**, and commutativity. By repeating the above filtration argument, we then obtain:

Theorem 6.55. *Suppose* $L \in \{\mathbf{K}_n^C, \mathbf{T}_n^C, \mathbf{K4}_n^C, \mathbf{S4}_n^C, \mathbf{KD45}_n^C, \mathbf{S5}_n^C\}$. *Then* $L \times \mathbf{S5} = [L, \mathbf{S5}]$.

Note that Theorems 12.8 and 12.14 below can also be used to obtain the above axiomatization results.

In some cases there are even better complexity bounds. As was shown in Theorem 5.41, $\mathbf{S5} \times \mathbf{S5}$ is coNEXPTIME-complete. Similarly, by 'simplifying' the proof of Theorem 6.44 for the case of **K** in place of **CPDL** we can obtain:

Theorem 6.56. *Every* $\mathbf{K} \times \mathbf{S5}$*-satisfiable formula* φ *can be satisfied in a product* $\mathbf{K} \times \mathbf{S5}$*-frame of size exponential in the length of* φ.

Together with Theorem 5.42, this yields the following result of (Marx 1999):

Theorem 6.57. *The satisfiability problem for* $\mathbf{K}\times\mathbf{S5}$ (*and* $\mathbf{K}\times\mathbf{KD45}$) *is* NEXPTIME-*complete, and so the decision problem is* coNEXPTIME-*complete.*

This theorem follows also from the tableau algorithm for the modal description logic $\mathbf{K}_{\mathcal{ALC}}$ (see Theorem 15.15) and the reduction of Theorem 3.35.

Moreover, the following holds:

Theorem 6.58. *The global consequence relation* $\vdash^*_{\mathbf{K}\times\mathbf{S5}}$ *is decidable.*

Proof. By Theorem 14.8 below and the reduction of Theorem 3.36, we obtain that $\mathbf{K}_u \times \mathbf{S5}$ is decidable.

Denote the universal box of $\mathbf{K}_u$ by $\boxtimes_1$ and the box of **S5** by $\Box_2$. It is not hard to show that, for any two formulas φ and ψ in the language $\mathcal{ML}_2$,

$$\varphi\,(\vdash^*_{\mathbf{K}} \times \vdash^*_{\mathbf{S5}})\,\psi \qquad \text{iff} \qquad \Box_2\boxtimes_1\varphi \to \psi \in \mathbf{K}_u \times \mathbf{S5}.$$

Since by Theorem 5.12, $\vdash^*_{\mathbf{K}\times\mathbf{S5}}$ is the same as $\vdash^*_{\mathbf{K}} \times \vdash^*_{\mathbf{S5}}$, the decidability of $\vdash^*_{\mathbf{K}\times\mathbf{S5}}$ follows. ❑

$\mathbf{Alt}\times\mathbf{S5}$ is even simpler than $\mathbf{K}\times\mathbf{S5}$. The proof of the following theorem is left to the reader as an exercise:

Theorem 6.59. *Every* $\mathbf{Alt}\times\mathbf{S5}$-*satisfiable formula* φ *can be satisfied in a product* $\mathbf{Alt}\times\mathbf{S5}$-*frame of size polynomial in the length of* φ. *So the decision problem for* $\mathbf{Alt}\times\mathbf{S5}$ *is* coNP-*complete.*

Let us now consider products of temporal logics with **S5**. First, Theorem 11.31 below provides an EXPSPACE decision algorithm for the one-variable fragment of temporal first-order logic $\mathsf{QLog}_{\mathcal{U}}(\mathbb{N})$ and so, by Theorem 3.29, for $\mathbf{PTL}\times\mathbf{S5}$ and $\mathsf{Log}\{\langle\mathbb{N},<\rangle\}\times\mathbf{S5}$ as well. Thus we obtain:

Theorem 6.60. *The decision problems for* $\mathsf{Log}\{\langle\mathbb{N},<\rangle\}\times\mathbf{S5}$ *and* $\mathbf{PTL}\times\mathbf{S5}$ *are in* EXPSPACE.

Note that in Section 11.7 (see Theorem 11.78) we show that $\mathbf{PTL}\times\mathbf{S5}$ is kind of product-matching.

Next, by 'mixing' the quasimodel proofs of Theorems 6.40 and 5.22, we can show the following:

Theorem 6.61. *Suppose* L *is one of* $\mathbf{K4.3}$, $\mathsf{Log}\{\langle\mathbb{Q},<\rangle\}$, $\mathbf{Lin}$, $\mathsf{Log}_{FP}(\mathbb{Q})$. *Then* $L\times\mathbf{S5}$ *and* $L\times\mathbf{KD45}$ *are decidable in* 2EXPTIME.

Proof. We only give a sketch of how to modify the quasimodels used in the proof of Theorem 6.40. Given a formula φ, the **S5** component makes it possible to define *quasistates* for φ as just $\Diamond$-saturated subsets of types for φ. So the number of points in a quasistate is now bounded by

$$p(\varphi) = 2^{|sub\,\varphi|}$$

and the number of different quasistates by

$$b(\varphi) = 2^{2^{|sub\,\varphi|}}.$$

A *basic structure* for φ is a pair $\langle \mathfrak{F}, \boldsymbol{q} \rangle$, where $\mathfrak{F} = \langle W, < \rangle$ is a strict linear order, and $\boldsymbol{q}$ is a function associating with each point w in W a quasistate $\boldsymbol{q}(w)$. A *run* through such a basic structure is a function r associating with each point w in W a type $r(w)$ from $\boldsymbol{q}(w)$.

A *block for* φ in this case is a triple $\mathfrak{B}^{uv} = \langle \mathfrak{F}^{uv}, \boldsymbol{q}^{uv}, \mathfrak{R}^{uv} \rangle$ such that

- $\mathfrak{F}^{uv} = \langle \{u, v\}, < \rangle$ is a 2-element strict linear order with $u < v$,
- $\langle \mathfrak{F}^{uv}, \boldsymbol{q}^{uv} \rangle$ is a basic structure for φ,
- $\mathfrak{R}^{uv}$ is a set of runs through $\langle \mathfrak{F}^{uv}, \boldsymbol{q}^{uv} \rangle$ such that
  - for all $r \in \mathfrak{R}^{uv}$ and $\Diamond\psi \in sub\,\varphi$,

    $$\text{if } \psi \in r(v) \text{ or } \Diamond\psi \in r(v) \text{ then } \Diamond\psi \in r(u),$$

  - for each $w \in \{u, v\}$ and each $\boldsymbol{t} \in \boldsymbol{q}^{uv}(w)$, there is an $r \in \mathfrak{R}^{uv}$ such that $r(w) = \boldsymbol{t}$.

A set $\mathcal{S}$ of blocks for φ is called *satisfying* if $\mathcal{S}$ contains a block satisfying **(qm2)**, and for every $\mathfrak{B}^{uv}$ in $\mathcal{S}$ the following properties hold:

(ssb3′) if $\Diamond\psi \in r(v)$, for some run $r \in \mathfrak{R}^{uv}$, then there exist a block $\mathfrak{B}^{vw}$ in $\mathcal{S}$ and a sequence $\langle \boldsymbol{t}_s \mid s \in \mathfrak{R}^{uv} \rangle$ of types in $\boldsymbol{q}^{vw}(w)$ such that

- $\boldsymbol{q}^{uv}(v) = \boldsymbol{q}^{vw}(v)$,
- for every $s \in \mathfrak{R}^{uv}$, the function p defined by $p(v) = s(v)$, $p(w) = \boldsymbol{t}_s$ is a run in $\mathfrak{R}^{vw}$,
- $\psi \in \boldsymbol{t}_r$;

(ssb4′) if $\Diamond\psi \in r(u)$, $\psi \notin r(v)$ and $\Diamond\psi \notin r(v)$ for some run $r \in \mathfrak{R}^{uv}$ then there are blocks $\mathfrak{B}^{uw}$ and $\mathfrak{B}^{wv}$ in $\mathcal{S}$ and a sequence $\langle \boldsymbol{t}_s \mid s \in \mathfrak{R}^{uv} \rangle$ of types in $\boldsymbol{q}^{uw}(w)$ such that

- $\boldsymbol{q}^{uv}(u) = \boldsymbol{q}^{uw}(u)$, $\boldsymbol{q}^{uw}(w) - \boldsymbol{q}^{wv}(w)$, $\boldsymbol{q}^{wv}(v) = \boldsymbol{q}^{uv}(v)$,
- for every $s \in \mathfrak{R}^{uv}$, the function p' defined by $p'(u) = s(u)$, $p'(w) = \boldsymbol{t}_s$ is a run in $\mathfrak{R}^{uw}$, and the function p'' defined by $p''(w) = \boldsymbol{t}_s$, $p''(v) = s(v)$ is a run in $\mathfrak{R}^{wv}$, and
- $\psi \in \boldsymbol{t}_r$.

Then, as usual, one can prove the 'quasimodel' and 'block' lemmas (cf. the proof of Theorem 6.40), and the decidability of **K4.3** × **S5** follows. For the remaining logics the proof has to be modified similarly to that of Theorem 6.40.

A 2EXPTIME decision algorithm is obtained as follows. Take the set of all blocks for φ (a straightforward computation shows that the cardinality of this set is at most 2-exponential in the length of φ). Eliminate iteratively those blocks for which there are no 'noneliminated' blocks satisfying **(ssb3′)** and **(ssb4′)**. This elimination procedure stops after at most 2-exponentially many steps in the length of φ. Now it is not hard to show that φ is satisfiable iff the set $\mathcal{S}$ of remaining blocks contains a block satisfying **(qm2)**. ❑

In Section 13.2 we give another proof of the decidability of **PTL** × **KD45**, **PTL** × **S5** and the logics in Theorem 6.61 with the help of reductions to monadic second-order theories of certain linear orders (see Theorem 13.6).

Note that a 2EXPTIME decision algorithm for **Lin** × **S5** was first given by Reynolds (1997), who also showed that **Lin** × **S5** has no fmp (see Theorem 5.30). By Theorem 5.32, none of the logics **K4.3** × **S5**, Log{⟨ℕ, <⟩} × **S5**, Log{⟨ℚ, <⟩} × **S5** have the product fmp.

Question 6.62. Do any of the logics Log{⟨ℕ, <⟩} × **S5**, Log{⟨ℚ, <⟩} × **S5** or **K4.3** × **S5** have the fmp?

So far our main concern was to obtain upper bounds for the computational complexity of products with **S5**. Now we prove an EXPSPACE-hardness result of (Hodkinson *et al.* 2003) generalizing Theorem 5.43:

Theorem 6.63. *Let $\mathcal{C}$ be any class of strict linear orders at least one of which contains an infinite ascending chain. Then the satisfiability problem for* Log($\mathcal{C}$ × FrS5) *(and for* Log($\mathcal{C}$ × Fr**KD45**)) *is* EXPSPACE-*hard. The same lower bound holds if we consider satisfiability in products of frames from $\mathcal{C}$ and finite* **S5**-*frames.*

Proof. Trying to modify the proof of Theorem 5.43, we are facing two main problems. First, the strict linear orders in $\mathcal{C}$ are not necessarily discrete, unlike

$\langle \mathbb{N}, < \rangle$. So, having generated an infinite sequence of 'horizontal' points, we cannot ensure that the 'horizontal diamond' $\Diamond$ actually refers to one of them. Second, even in the case of $\langle \mathbb{N}, < \rangle$ we do not have the next-time operator $\ominus$.

In order to solve the first problem, here we will use a reduction of the following *infinite version* of the 2^n-corridor tiling problem, which is also EXPSPACE-complete (the proof is left to the reader as an easy exercise): given a finite set T of tile types, a tile type $t_0 \in T$ and $n \in \mathbb{N}$ in binary, decide whether T tiles the $\mathbb{N} \times 2^n$-corridor in such a way that t_0 is placed onto $\langle 0, 0 \rangle$ and the top and bottom sides of the corridor are of some fixed color, say, *white*.

Suppose that a finite set T of tile types, $t_0 \in T$ and a natural number n are given. Our first aim is to construct an $\mathcal{ML}_2$-formula $\varphi_{n,T}$ such that (i) its length is a polynomial function of $|T|$ and n, and (ii) $\varphi_{n,T}$ is satisfiable in a frame from $\mathcal{C} \times \mathsf{FrS5}$ iff T tiles the $\mathbb{N} \times 2^n$-corridor such that its top and bottom sides are *white* and t_0 is placed onto $\langle 0, 0 \rangle$. Later on we will modify $\varphi_{n,T}$ in such a way that the resulting formula $\psi_{n,T}$ is satisfiable in a frame from $\mathcal{C} \times \mathsf{FrS5}$ iff it is satisfied in a model based on the product of a frame from $\mathcal{C}$ and a *finite* **S5**-frame.

Take a strict linear order $\langle U, < \rangle \in \mathcal{C}$, a universal frame $\langle W, R \rangle$ for **S5**, and suppose that a model $\mathfrak{M}$ is based on $\mathfrak{F} = \langle U, < \rangle \times \langle W, R \rangle$. Our first step in the construction of $\varphi_{n,T}$ (which will again contain, among many others, propositional variables t for all $t \in T$) is to write down formulas forcing not only an infinite sequence $y_0, y_1, \dots$ of distinct points from W, but at the same time an infinite sequence $x_0 < x_1 < x_2 < \dots$ of points from U such that, for each $i \in \mathbb{N}$, $\langle x_i, y_i \rangle \models t$ for a unique tile type t. As before, if $i = k \cdot 2^n + j$ for some $j < 2^n$ then we will use the point $\langle x_i, y_i \rangle$ to encode the pair $\langle k, j \rangle$ of the $\mathbb{N} \times 2^n$-grid. Thus the up neighbor $\langle k, j+1 \rangle$ of $\langle k, j \rangle$ will be coded by the point $\langle x_{i+1}, y_{i+1} \rangle$, and its right neighbor $\langle k+1, j \rangle$ by $\langle x_{i+2^n}, y_{i+2^n} \rangle$.

Let $q_0, \dots, q_{n-1}$ be pairwise distinct propositional variables, and $q_i^1 = q_i$, $q_i^0 = \neg q_i$, for $i < n$. Set

$$\sigma_j = q_0^{d_0} \wedge \dots \wedge q_{n-1}^{d_{n-1}},$$

where $d_{n-1} \dots d_0$ is the binary representation of $j < 2^n$. The formula

$$\boxminus^+ \bigwedge_{i<n} (\Box q_i \vee \Box \neg q_i) \tag{6.28}$$

again says that the truth-values of the q_i (and so that of the σ_j) do not change along the vertical axis. We call the set $\{\langle u, w \rangle \mid w \in W\}$, for $u \in U$, a *slice* of $\mathfrak{F}$, the *u-slice*, to be more precise, and say that *the u-slice is of type* j, for $u \in U$, $j < 2^n$, if

$$\langle u, w \rangle \models \sigma_j, \quad \text{for all } w \in W.$$

Let $p_0, \dots, p_{n-1}$ be a fresh n-tuple of distinct variables such that their truth-values do not change along the horizontal axis. This requirement can be ensured by the formula

$$\boxbar \bigwedge_{i<n} (\boxminus^+ p_i \vee \boxminus^+ \neg p_i). \tag{6.29}$$

Let $\pi_j = p_0^{d_0} \wedge \dots \wedge p_{n-1}^{d_{n-1}}$, where $d_{n-1} \dots d_0$ is the binary representation of $j < 2^n$, and let

$$\mathsf{equ} = \bigwedge_{i<n} (p_i \leftrightarrow q_i).$$

It should be clear that, for all $u \in U$, $w \in W$, if $\langle u, w \rangle \models \mathsf{equ}$ and the u-slice is of type j (i.e., it makes σ_j true) then $\langle u', w \rangle \models \pi_j$ for all $u' \in U$.

We can now define 'counting' formulas of length polynomial in n. Suppose that succ is a propositional variable and that formulas (6.28), (6.29),

$$\boxminus^+ \boxbar \bigwedge_{k<n} \Big(\big(\bigwedge_{i<k} q_i \wedge \neg q_k\big) \to \big(\mathsf{succ} \leftrightarrow \bigwedge_{i<k} \neg p_i \wedge p_k \wedge \bigwedge_{j=k+1}^{n-1} (p_j \leftrightarrow q_j)\big) \Big), \tag{6.30}$$

$$\boxminus^+ \boxbar \Big(\bigwedge_{i<n} q_i \to \big(\mathsf{succ} \leftrightarrow \bigwedge_{i<n} \neg p_i\big) \Big) \tag{6.31}$$

hold at some point $\langle x_0, y_0 \rangle$ of $\mathfrak{M}$. If $u \geq x_0$, $\langle u, w \rangle \models \mathsf{succ}$ and the u-slice is of type j for some $j < 2^n$, then $\langle u', w \rangle \models \pi_{j+1(\mathrm{mod}\ 2^n)}$ for all $u' \in U$, $u' \geq x_0$.

Now we can generate the required infinite sequences of points using the formula

$$\sigma_0 \wedge \mathsf{equ} \wedge \mathsf{tile} \wedge \neg \diamondminus \mathsf{tile} \wedge$$
$$\boxminus^+ \Big[\diamondbar \mathsf{tile} \to \diamondbar \big(\mathsf{succ} \wedge \diamondminus(\mathsf{equ} \wedge \mathsf{tile}) \wedge \boxminus(\diamondminus \mathsf{tile} \to \neg \diamondbar \mathsf{tile})\big)\Big], \tag{6.32}$$

where $\mathsf{tile} = \bigvee_{t \in T} t$. Indeed, suppose that the conjunction of (6.28)–(6.32) holds at some point $\langle x_0, y_0 \rangle$ of $\mathfrak{M}$. Then $\langle u, y_0 \rangle \models \pi_0$ for all $u \in U$, and the x_0-slice is of type 0. Since

$$\langle x_0, y_0 \rangle \models \diamondbar \big(\mathsf{succ} \wedge \diamondminus(\mathsf{equ} \wedge \mathsf{tile}) \wedge \boxminus(\diamondminus \mathsf{tile} \to \neg \diamondbar \mathsf{tile})\big),$$

there are points $y_1 \in W$ and $x_1 > x_0$ in U such that

- $\langle x_0, y_1 \rangle \models \mathsf{succ}$ (that is, $\langle u, y_1 \rangle \models \pi_1$ for all $u \in U$, and so $y_1 \neq y_0$),
- $\langle x_1, y_1 \rangle \models \mathsf{equ}$ (that is, the x_1-slice is of type 1),
- $\langle x_1, y_1 \rangle \models \mathsf{tile}$,

- no point of the form $\langle u, y_1 \rangle$ with $u > x_1$ makes tile true, and
- no point belonging to a u-slice such that $x_0 < u < x_1$ makes tile true.

Now we consider $\langle x_1, y_0 \rangle$ and by the same argument find points $y_2 \notin \{y_0, y_1\}$ and $x_2 > x_1$ such that the x_2-slice is of type 2, etc., and so forth till we get to a point x_{2^n-1} whose slice is of type $2^n - 1$, and then to a x_{2^n}-slice of type 0 again; see Fig. 6.2.

Our next aim is to write down formulas that could serve as pointers to the up and right neighbors of a given pair in the corridor (at this moment we do not bother about its top border). Let

$$\mathsf{up} = \mathsf{succ} \wedge \diamondminus\mathsf{tile} \wedge \boxminus(\diamondminus\mathsf{tile} \to \neg\diamondbar\mathsf{tile})$$

$$\mathsf{right} = \mathsf{equ} \wedge \diamondminus\mathsf{tile} \wedge \boxminus\big(\diamondminus\mathsf{tile} \wedge \diamondbar\mathsf{tile} \to \neg\mathsf{equ}\big).$$

It is easy to see that for all $i \in \mathbb{N}$,

- $\langle x_i, y_{i+1} \rangle \models \mathsf{up}$ and $\langle x_i, y_j \rangle \not\models \mathsf{up}$ for all $j \neq i+1$,
- $\langle x_i, y_{i+2^n} \rangle \models \mathsf{right}$ and $\langle x_i, y_j \rangle \not\models \mathsf{right}$ for all $j \neq i + 2^n$.

Finally, the formulas below ensure that $\langle 0, 0 \rangle$ is covered by t_0, every point of the $\mathbb{N} \times 2^n$-corridor is covered by at most one tile, the top and bottom sides of the corridor are *white* and the colors on adjacent edges of adjacent tiles match:

$$t_0 \wedge \boxminus^+\boxbar \bigwedge_{\substack{t,t' \in T,\\ t \neq t'}} \neg(t \wedge t'), \tag{6.33}$$

$$\boxminus^+\boxbar\Big(\sigma_0 \wedge \mathsf{tile} \to \bigvee_{\substack{t \in T,\\ down(t) = white}} t\Big), \tag{6.34}$$

$$\boxminus^+\boxbar\Big(\sigma_{2^n-1} \wedge \mathsf{tile} \to \bigvee_{\substack{t \in T,\\ up(t) = white}} t\Big), \tag{6.35}$$

$$\boxminus^+\boxbar\Bigg(\neg\sigma_{2^n-1} \to \bigwedge_{\substack{t,t' \in T,\\ up(t) \neq down(t')}} (t \to \boxbar(\mathsf{up} \to \boxminus\neg t'))\Bigg), \tag{6.36}$$

$$\boxminus^+\boxbar\Bigg(\bigwedge_{\substack{t,t' \in T,\\ right(t) \neq left(t')}} (t \to \boxbar(\mathsf{right} \to \boxminus\neg t'))\Bigg). \tag{6.37}$$

Let $\varphi_{n,T}$ be the conjunction of (6.28)–(6.37). Suppose that

$$(\mathfrak{M}, \langle x_0, y_0 \rangle) \models \varphi_{n,T}.$$

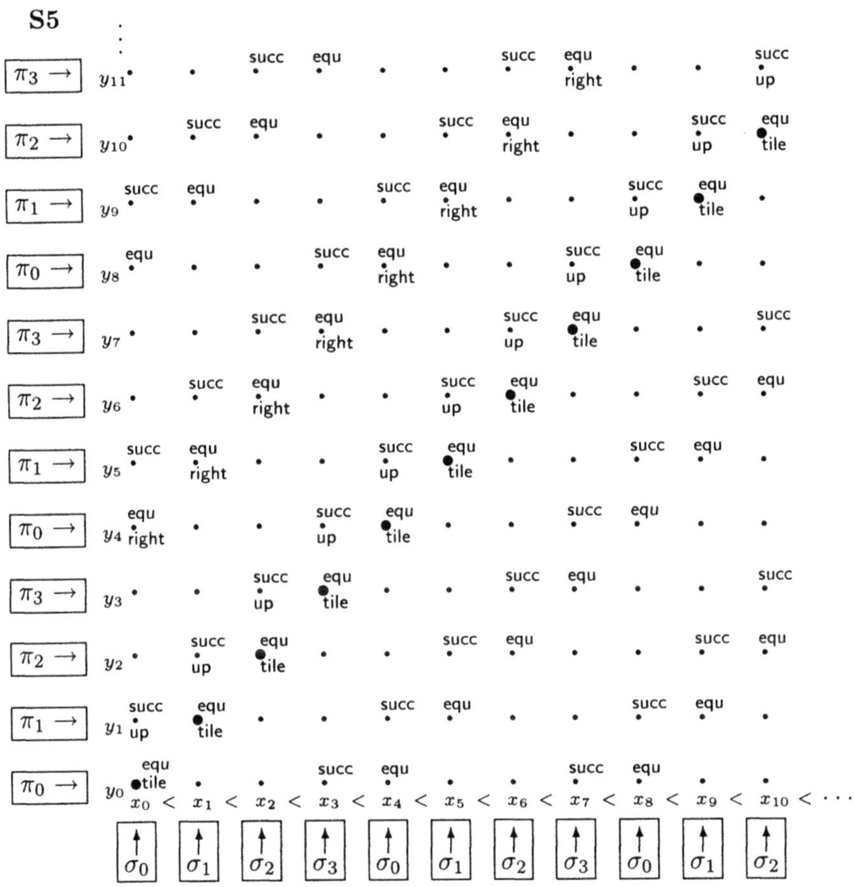

Figure 6.2: Satisfying $\varphi_{2,T}$ in a frame from $\mathcal{C} \times \mathsf{Fr}\mathbf{S5}$.

Then we define a map $\boldsymbol{\tau} : \mathbb{N} \times 2^n \to T$ by taking

$$\boldsymbol{\tau}(k, j) = t \quad \text{iff} \quad \langle x_{k \cdot 2^n + j}, y_{k \cdot 2^n + j} \rangle \models t.$$

We leave it to the reader to check that $\boldsymbol{\tau}$ is indeed a tiling of $\mathbb{N} \times 2^n$ as required.

For the other direction, take a strict linear order $\mathfrak{F}$ from $\mathcal{C}$ having an infinite ascending chain of distinct points x_i. Figure 6.2 shows then that $\varphi_{n,T}$ is satisfiable in a product of $\mathfrak{F}$ and an arbitrary infinite universal **S5**-frame.

Let us now turn to satisfiability in products of strict linear orders from $\mathcal{C}$ and *finite* **S5**-frames. By the pigeon-hole principle, any tiling of $\mathbb{N} \times 2^n$ by T has two identical columns X, Y, so it can be converted into an eventually periodic tiling by iterating the 'interval' $[X, Y)$ between the columns. In

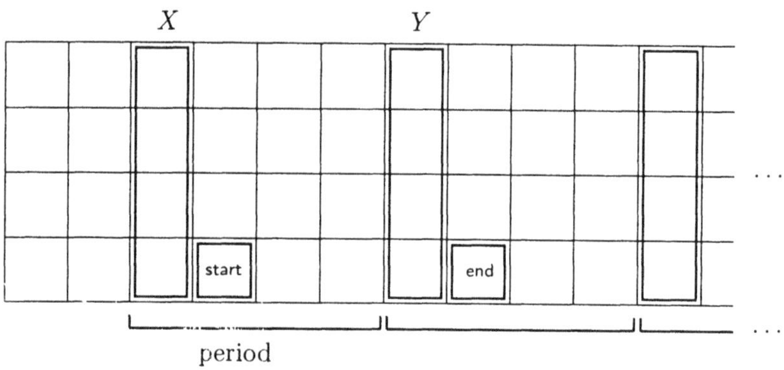

Figure 6.3: Marking identical columns with start and end.

order to force such a tiling, we modify $\varphi_{n,T}$ as follows. First, we introduce new propositional variables start and end (which are intended to mark the bottom tile of the columns following X and Y, respectively; see Fig. 6.3), and add the following conjuncts to $\varphi_{n,T}$:

$$\Diamond_h(\mathsf{start} \land \sigma_0 \land \Diamond_v\mathsf{tile} \land \Diamond_h(\mathsf{end} \land \sigma_0 \land \Diamond_v\mathsf{tile})), \tag{6.38}$$

$$\Box_h^+\big(\mathsf{start} \to (\Box_v\mathsf{start} \land \Box_h\neg\mathsf{start})\big) \land \Box_h^+\big(\mathsf{end} \to (\Box_v\mathsf{end} \land \Box_h\neg\mathsf{end})\big). \tag{6.39}$$

Then we replace the 'grid-generating' formula (6.32) with

$$\sigma_0 \land \mathsf{equ} \land \mathsf{tile} \land \neg\Diamond_h\mathsf{tile} \land \Box_h^+\Big[\Diamond_v\mathsf{tile} \land \Diamond_h\mathsf{end} \to \Diamond_v\big(\mathsf{succ} \land \Diamond_h(\mathsf{equ} \land \mathsf{tile}) \land \Box_h(\Diamond_h\mathsf{tile} \to \neg\Diamond_v\mathsf{tile})\big)\Big]. \tag{6.40}$$

Finally, to guarantee that the tiling is periodic, we add the conjunct

$$\boxminus^{+}\Box \bigwedge_{t\in T} \Big[t \wedge \diamondminus\mathsf{start} \wedge \neg\diamondminus(\sigma_0 \wedge \Diamond\mathsf{tile} \wedge \diamondminus\mathsf{start}) \rightarrow$$

$$\Box\Big(\mathsf{equ} \rightarrow \boxminus\big[\mathsf{equ} \wedge \mathsf{tile} \wedge \diamondminus\mathsf{end} \wedge \neg\diamondminus(\sigma_0 \wedge \Diamond\mathsf{tile} \wedge \diamondminus\mathsf{end}) \rightarrow t\big]\Big)\Big]. \qquad (6.41)$$

Denote the resulting formula by $\psi_{n,T}$.

Suppose first that $\psi_{n,T}$ is satisfiable in a frame from $\mathcal{C} \times \mathsf{Fr}\mathbf{S5}$. By (6.40), we have points x_0 and y_0 as before. By (6.38) and (6.39), there are unique points $x_{start} > x_0$ and $x_{end} > x_{start}$ such that, for every $y \in W$,

$$\langle x_{start}, y\rangle \models \mathsf{start} \quad \text{and} \quad \langle x_{end}, y\rangle \models \mathsf{end}.$$

So, as before, we can generate $x_1, y_1, \ldots, x_i, y_i$—as long as $\diamondminus\mathsf{end}$ holds at $\langle x_{i-1}, y_0\rangle$. Two cases are possible.

Case 1: there is an $i < \omega$ such that $x_{end} \leq x_i$. Then (6.38) and (6.40) guarantee that $x_{end} = x_i$ and $i = k \cdot 2^n$ must hold for some k, $0 < k < \omega$. The same applies to x_{start} as well: we have $x_{start} = l \cdot 2^n$ for some l, $0 < l < k$. Define a periodic tiling $\tau : \mathbb{N} \times 2^n \rightarrow T$ (repeating the pattern between columns $(l-1)$ and $(k-1)$) by taking

$$\tau(i,j) = t \quad \text{iff} \quad \langle x_{f(i)\cdot 2^n + j}, y_{f(i)\cdot 2^n + j}\rangle \models t,$$

where

$$f(i) = \begin{cases} i, & \text{if } i < l, \\ m + l, & \text{if } i \geq l \text{ and } i - l =_{\mathrm{mod}(k-l)} m. \end{cases}$$

Case 2: $x_{end} > x_i$ for all $i < \omega$. Then we can recover the tiling τ as in the 'infinite case' above. Note that now formula (6.41) has no effect: it is satisfied simply because

$$\langle x_i, y_i\rangle \not\models \mathsf{equ} \wedge \mathsf{tile} \wedge \diamondminus\mathsf{end} \wedge \neg\diamondminus(\sigma_0 \wedge \Diamond\mathsf{tile} \wedge \diamondminus\mathsf{end}),$$

for all $i < \omega$.

Conversely, as we said above, if T tiles the $\mathbb{N} \times 2^n$-corridor as required, then we can always assume that the tiling is eventually periodic. We leave it to the reader to check that $\psi_{n,T}$ is satisfiable in the product of a (finite) strict linear order and a finite universal **S5**-frame. ❑

As a consequence, we obtain the following theorem (which confirms a conjecture of Reynolds 1997):

Theorem 6.64. *Suppose that L is one of the logics* $\mathsf{Log}\{\langle\mathbb{N}, <\rangle\}$, **Lin**, **K4.3**, $\mathsf{Log}\{\langle\mathbb{Q}, <\rangle\}$, $\mathsf{Log}_{FP}(\mathbb{Q})$. *Then the satisfiability problem for $L \times$ **S5** (and for $L \times$ **KD45**) is* EXPSPACE-*hard.*

Proof. By Theorems 6.29, 6.30 and 6.31, $L \times \mathbf{S5}$ is always determined by a class of product frames each of which is the product of a strict linear order and a frame for **S5**. ❑

So, by Theorem 6.60, we have:

Theorem 6.65. *Both* $\mathsf{Log}\{\langle \mathbb{N}, < \rangle\} \times \mathbf{S5}$ *and* $\mathbf{PTL} \times \mathbf{S5}$ *are* EXPSPACE-*complete.*

Using the reductions of Theorems 6.18, 6.23, and 6.24, we also obtain the following (cf. Table 6.1):

Theorem 6.66. *Suppose* $L \in \{\mathbf{PDL}, \mathbf{CPDL}, \mathbf{K}_1^C, \mathbf{T}_2^C, \mathbf{K4}_2^C, \mathbf{S4}_2^C, \mathbf{KD45}_2^C\}$. *Then the satisfiability problem for* $L \times \mathbf{S5}$ *(and for* $L \times \mathbf{KD45}$*) is* EXPSPACE-*hard.*

However, the exact complexity of these logics is not known. In particular, the following question is open:

Question 6.67. What is the complexity of $\mathbf{K4} \times \mathbf{S5}$ and $\mathbf{S4} \times \mathbf{S5}$?

M. Marx conjectures that these logics are also EXPSPACE-complete.

6.6 Products with multimodal S5

First, we show how to generalize the quasimodel technique used for establishing the decidability of $\mathbf{CPDL} \times \mathbf{K}$ (Theorem 6.10) in order to prove the following result:

Theorem 6.68. $\mathbf{CPDL} \times \mathbf{S5}_m$ *and* $\mathbf{CPDL} \times \mathbf{KD45}_m$ *are decidable.*

Proof. The proof is similar to the decidability proofs for $\mathbf{CPDL} \times \mathbf{K}$ and $\mathbf{K} \times \mathbf{K}$. We only define the new notions of quasistates and quasimodels required in the proof of the decidability of $\mathbf{K} \times \mathbf{S5}_2$. The extension of these notions to the case of $\mathbf{CPDL} \times \mathbf{S5}_m$ as well as the remaining steps of the proof are straightforward and left to the reader.

Given an $\mathcal{ML}_3$-formula φ (in the language with boxes $\boxminus$, $\Box_1$ and $\Box_2$), define $a^1(\varphi)$ to be the length of the longest chain $\Box_2, \Box_1, \Box_2, \dots$ of boxes starting with $\Box_2$ and such that a subformula of the form $\Box_2(\dots\Box_1(\dots\Box_2(\dots)))$ occurs in φ. The function $a^2(\varphi)$ is defined analogously by swapping $\Box_1$ and $\Box_2$ (cf. Section 4.4).

A *quasistate candidate* for φ is a tuple $\langle\langle T, <_1, <_2\rangle, \boldsymbol{t}\rangle$, where

- $\langle T, <_1 \cup <_2\rangle$ is a finite intransitive tree of depth $\leq \max(a^1(\varphi), a^2(\varphi))$;
- $<_1$ and $<_2$ are disjoint;

- there are no $x, y, z \in T$ and $i \in \{1,2\}$ such that $x <_i y <_i z$;
- $\boldsymbol{t}$ is a function associating with each $x \in T$ a type $\boldsymbol{t}(x)$ for φ.

Two quasistate candidates $\langle\langle T, <_1, <_2\rangle, \boldsymbol{t}\rangle$ and $\langle\langle T', <_1', <_2'\rangle, \boldsymbol{t}'\rangle$ are called *isomorphic* if there is an isomorphism f between the graphs $\langle T, <_1, <_2\rangle$ and $\langle T', <_1', <_2'\rangle$ such that $\boldsymbol{t}(x) = \boldsymbol{t}'(f(x))$ for all $x \in T$.

Intuitively, every quasistate candidate $\langle\langle T, <_1, <_2\rangle, \boldsymbol{t}\rangle$ corresponds to an $\mathbf{S5}_2$-model $\mathfrak{M} = (\langle T, R_1, R_2\rangle, \mathfrak{V})$, where R_i is the transitive, reflexive and symmetric closure of $<_i$, $i = 1, 2$, and $\mathfrak{V}(p) = \{x \in T \mid p \in \boldsymbol{t}(x)\}$. It follows from the proof of Theorem 4.1 that $\mathbf{S5}_2$ is determined by frames obtained in this way from quasistate candidates.

Given a quasistate candidate $\boldsymbol{q} = \langle\langle T, <_1, <_2\rangle, \boldsymbol{t}\rangle$, we define, for every $i = 1, 2$, and every $x \in T$,

$$d_i^{\boldsymbol{q}}(x) = \begin{cases} 1, & \text{if there is a } y \in T \text{ with } x <_i y, \\ 0, & \text{otherwise} \end{cases}$$

and

$$cd_i^{\boldsymbol{q}}(x) = \begin{cases} 1, & \text{if there is a } y \in T \text{ with } y <_i x, \\ 0, & \text{otherwise.} \end{cases}$$

Note that for no point $x \in T$ do we have both $d_i^{\boldsymbol{q}}(x) = 1$ and $cd_i^{\boldsymbol{q}}(x) = 1$. A quasistate candidate $\boldsymbol{q} = \langle\langle T, <_1, <_2\rangle, \boldsymbol{t}\rangle$ is called a *quasistate* for φ if the following conditions hold:

(qm1) ($\Diamond$-*saturation*) For all $x \in T$, $i = 1, 2$ and $\Diamond_i\psi \in sub\,\varphi$,

$$\Diamond_i\psi \in \boldsymbol{t}(x) \quad \text{iff} \quad \exists y \in T\ \big(xR_iy \wedge \psi \in \boldsymbol{t}(y)\big),$$

and if in addition $d_i^{\boldsymbol{q}}(x) = 1$, then

$$\Diamond_i\psi \in \boldsymbol{t}(x) \quad \text{iff} \quad \exists y \in T\ \big(x <_i y \wedge \psi \in \boldsymbol{t}(y)\big).$$

(qm1′) (*smallness*) For all $i = 1, 2$ and all $x, x_1, x_2 \in T$ such that $x <_i x_1$, $x <_i x_2$ and $x_1 \neq x_2$, the quasistate candidates $\langle\langle T^{x_1}, <_1^{x_1}, <_2^{x_2}\rangle, \boldsymbol{t}^{x_1}\rangle$ and $\langle\langle T^{x_2}, <_1^{x_2}, <_2^{x_2}\rangle, \boldsymbol{t}^{x_2}\rangle$ are not isomorphic.

Observe that the number of nonisomorphic quasistates for φ is not bounded by an elementary function in the length of φ.

A $\mathbf{K} \times \mathbf{S5}_2$-*basic structure* of depth m for φ is a pair $\langle\mathfrak{F}, \boldsymbol{q}\rangle$ such that $\mathfrak{F} = \langle W, R\rangle$ is a frame and $\boldsymbol{q}$ is a function associating with each $w \in W$ a quasistate $\boldsymbol{q}(w) = \langle\langle T_w, <_1^w, <_2^w\rangle, \boldsymbol{t}_w\rangle$ for φ such that the depth of each tree $\langle T_w, <_1^w \cup <_2^w\rangle$ is m.

Let $\langle \mathfrak{F}, \boldsymbol{q} \rangle$ be a basic structure for φ of depth m and let $k \leq m$. A *k-run through* $\langle \mathfrak{F}, \boldsymbol{q} \rangle$ is a function r giving for each $w \in W$ a point $r(w) \in T_w$ such that

- for every $w \in W$, the co-depth of $r(w)$ in $\langle T_w, <_1^w \cup <_2^w \rangle$ is k;
- for all $w_1, w_2 \in W$ and $i = 1, 2$, $d_i^{\boldsymbol{q}(w_1)}(r(w_1)) = d_i^{\boldsymbol{q}(w_2)}(r(w_2))$ and $cd_i^{\boldsymbol{q}(w_1)}(r(w_1)) = cd_i^{\boldsymbol{q}(w_2)}(r(w_2))$.

Coherent and saturated runs are defined as in the proof of Theorem 6.1. Finally, we say that $\mathfrak{Q} = \langle \mathfrak{F}, \boldsymbol{q}, \mathfrak{R}, \lhd_1, \lhd_2 \rangle$ is a $\mathbf{K} \times \mathbf{S5}_2$*-quasimodel for φ (based on $\mathfrak{F}$)* if $\langle \mathfrak{F}, \boldsymbol{q} \rangle$ is a basic structure for φ of depth $m \leq \max(a^1(\varphi), a^2(\varphi))$ such that

(qm2) $\exists w_0 \in W \; \varphi \in \boldsymbol{t}_{w_0}(x_0)$, where x_0 is the root of $\langle T_{w_0}, <_1^{w_0} \cup <_2^{w_0} \rangle$,

$\mathfrak{R}$ is a set of coherent and saturated runs through $\langle \mathfrak{F}, \boldsymbol{q} \rangle$, and $\lhd_i$, for $i = 1, 2$, are binary relations on $\mathfrak{R}$ satisfying the following conditions:

(qm3) for all $r, r' \in \mathfrak{R}$, if $r \lhd_i r'$ then $r(w) <_i^w r'(w)$ for every $w \in W$;

(qm4) for all $w \in W$, $i = 1, 2$, $x \in T_w$ and $r \in \mathfrak{R}$, if $r(w) <_i^w x$ ($x <_i^w r(w)$) then there is $r' \in \mathfrak{R}$ such that $r'(w) = x$ and $r \lhd_i r'$ (respectively, $r' \lhd_i r$).

Being equipped with these definitions, one can then proceed as in the proof of Theorem 6.1. ❑

Using the reductions of Theorems 6.18 and 6.24 (see Table 6.1), we then obtain:

Theorem 6.69. *Suppose L is one of* $\mathbf{K}_n^C$, $\mathbf{T}_n^C$, $\mathbf{K4}_n^C$, $\mathbf{S4}_n^C$, $\mathbf{KD45}_n^C$, $\mathbf{S5}_n^C$, **PDL**, **PTL**, $\mathsf{Log}\{\langle \mathbb{N}, < \rangle\}$. *Then $L \times \mathbf{S5}_m$ (and $L \times \mathbf{KD45}_m$) are decidable.*

A similar generalization of the proof of Theorem 6.40 yields:

Theorem 6.70. *Suppose L is one of* **K4.3**, $\mathsf{Log}\{\langle \mathbb{Q}, < \rangle\}$, **Lin**, $\mathsf{Log}_{FP}(\mathbb{Q})$. *Then $L \times \mathbf{S5}_m$ and $L \times \mathbf{KD45}_m$ are decidable.*

In Section 13.2 we give another proof of the decidability of $\mathbf{PTL} \times \mathbf{KD45}_m$, $\mathbf{PTL} \times \mathbf{S5}_m$, and that of the logics in Theorem 6.70 with the help of reductions to monadic second-order theories of certain linear orders (see Theorem 13.6). Similarly to products with $\mathbf{K}$, none of these decision procedures for products with $\mathbf{S5}_m$, $m > 1$, runs in elementary time. We now show that in most cases we cannot do better. Actually, this is easily done by 'lifting' the reduction of Theorem 2.37 to products:

Theorem 6.71. *Suppose L is either a Kripke complete multimodal logic, or $L \in \{$***PTL, PDL, CPDL***$\}$. Then*

(1) *$L \times \mathbf{K}_u$ is polynomially reducible to $L \times \mathbf{K}_1^C$,*

and, for every bimodal logic L' between $\mathbf{K}_2$ and $\mathbf{S5}_2$,

(2) *$L \times \mathbf{K}$ is polynomially reducible to $L \times L'$,*

(3) *$L \times \mathbf{K}_u$ is polynomially reducible to $L \times L'^C$.*

Proof. We give a sketch of a proof for the case when L is a Kripke complete unimodal logic; the other cases are similar. Denote by $\Box_3$ the modal operator of the language $\mathcal{ML}$ of L. To begin with, we claim that the decision problem for $L \times \mathbf{K}_u$ can be polynomially reduced to the decision problem for $\mathcal{ML} \otimes \mathcal{ML}_1^u$-formulas in which no $\boxdot$ occurs in the scope of another modal operator ($\Box$, $\boxdot$ or $\Box_3$). Indeed, given an $\mathcal{ML} \otimes \mathcal{ML}_1^u$-formula φ, denote by φ^u the result of replacing every subformula of the form $\chi = \boxdot\psi$ in φ with a fresh propositional variable p_χ. Define the set $\mathcal{R}_u(\varphi)$ as in the proof of Theorem 2.37. Then it is not hard to see that

$$\varphi \in L \times \mathbf{K}_u \qquad \text{iff} \qquad \Box_3^{\leq md(\varphi)} \bigwedge \mathcal{R}_u(\varphi) \to \varphi^u \in L \times \mathbf{K}_u,$$

and the formula in the right-hand side is as required.

Next, we extend the translations ^{c} and $^\clubsuit$ of Theorem 2.37 to translations

$$^{c'} : \mathcal{ML} \otimes \mathcal{ML}_1^u \to \mathcal{ML} \otimes \mathcal{ML}_1^C$$
$$^{\clubsuit'} : \mathcal{ML} \otimes \mathcal{ML}_1^u \to \mathcal{ML} \otimes \mathcal{ML}_2^C$$

by taking $(\Box_3\varphi)^{c'} = \Box_3\varphi^{c'}$ and $(\Box_3\varphi)^{\clubsuit'} = \Box_3\varphi^{\clubsuit'}$. Note that if φ is an $\mathcal{ML}\otimes\mathcal{ML}_1$-formula then $\varphi^{\clubsuit'}$ is an $\mathcal{ML}\otimes\mathcal{ML}_2$-formula. It is straightforward to extend the proof of Theorem 2.37 to show that for every $\mathcal{ML} \otimes \mathcal{ML}_1^u$-formula φ without occurrences of $\boxdot$ in the scope of another modal operator,

- $\varphi \in L \times \mathbf{K}_u$ iff $\Box_3^{\leq md(\varphi)}\big((p \to \Box_3 p) \wedge (\neg p \to \Box_3\neg p)\big) \to \varphi^{c'} \in L \times \mathbf{K}_1^C$;
- if $p \wedge \Box_3^{\leq md(\varphi)}\big((p \to \Box_3 p) \wedge (\neg p \to \Box_3\neg p)\big) \to \varphi^{\clubsuit'} \in L \times \mathbf{S5}_2^C$ then $\varphi \in L \times \mathbf{K}_u$;
- if $\varphi \in L \times \mathbf{K}_u$ then

$$p \wedge \Box_3^{\leq md(\varphi)}\big((p \to \Box_3 p) \wedge (\neg p \to \Box_3\neg p)\big) \to \varphi^{\clubsuit'} \in L \times \mathbf{K}_2^C,$$

as required. ❑

As a consequence of Theorems 6.15, 6.37 and 6.26 we then obtain the following (see Table 6.1):

Theorem 6.72. *Let* $L \in \{\mathbf{PTL}, \mathbf{K}_1^C, \mathbf{T}_2^C, \mathbf{K4}_2^C, \mathbf{S4}_2^C, \mathbf{KD45}_2^C, \mathbf{PDL}, \mathbf{CPDL}\}$. *Then the satisfiability problem for* $L \times \mathbf{S5}_2$ *(and for* $L \times \mathbf{KD45}_2$*) does not belong to* ELEM.

For $\mathbf{PTL} \times \mathbf{S5}_2$ this was first shown in (Halpern and Vardi 1989).

Question 6.73. What is the complexity of $\mathbf{K4.3} \times \mathbf{S5}_2$, $\mathsf{Log}\{\langle\mathbb{Q}, <\rangle\} \times \mathbf{S5}_2$, $\mathbf{Lin} \times \mathbf{S5}_2$, and $\mathsf{Log}_{FP}(\mathbb{Q}) \times \mathbf{S5}_2$? Are these logics in ELEM?

	finitely axiomatizable	has fmp	has product fmp	decidable	complexity	global cons. is dec.
$\mathbf{K} \times \mathbf{K}$	yes (Thm. 5.5)	yes	yes (Thm. 6.4)	yes (Thm. 6.1)	not in ELEM?	no (Thm. 5.36)
$\mathbf{K4} \times \mathbf{K}$	yes (Thm. 5.9)	?	no (Thm. 5.32)	yes (Thm. 6.20)	coNEXPTIME-hard (Thm. 5.42) ?	?
$\mathbf{S5} \times \mathbf{K}$	yes (Thm. 5.9)	yes	yes (Thm. 6.56)	yes (Thm. 6.20)	coNEXPTIME-compl. (Thm. 6.57)	yes (Thm. 6.58)
$\mathbf{K4.3} \times \mathbf{K}$	?	?	no (Thm. 5.32)	yes (Thm. 6.40)	not in ELEM?	?
$\mathsf{Log}\{\langle \mathbb{N}, < \rangle\} \times \mathbf{K}$	?	?	no (Thm. 5.32)	yes (Thm. 6.33)	not in ELEM?	?
$\mathsf{Log}\{\langle \mathbb{Q}, < \rangle\} \times \mathbf{K}$	?	?	no (Thm. 5.32)	yes (Thm. 6.40)	not in ELEM?	?
$\mathbf{Alt} \times \mathbf{K}$	yes (Thm. 8.55)	yes	yes (Thm. 6.6)	yes (Thm. 6.6)	in EXPTIME (Thm. 6.6) ?	no (Thm. 8.54)

Table 6.2: Products of unimodal logics with **K**.

	finitely axiomatizable	has fmp	has product fmp	decidable	complexity
K4 × **S5**	yes (Thm. 5.9)	yes (Thm. 5.27)	no (Thm. 5.32)	yes (Thm. 5.28)	coNEXPTIME-hard in coN2EXPTIME (Thms. 5.42, 5.28) [?]
S5 × **S5**	yes (Thm. 5.9)	yes	yes (Thm. 5.25)	yes (Thm. 5.22)	coNEXPTIME-compl. (Thm. 5.41)
K4.3 × **S5**	[?] r.e. (Thm. 3.17)	[?]	no (Thm. 5.32)	yes (Thm. 6.61)	EXPSPACE-hard in 2EXPTIME (Thms. 6.64, 6.61) [?]
$\mathsf{Log}\{\langle \mathbb{N}, < \rangle\}$ × **S5**	[?]	[?]	no (Thm. 5.32)	yes (Thm. 6.50)	EXPSPACE-complete (Thm. 6.65)
$\mathsf{Log}\{\langle \mathbb{Q}, < \rangle\}$ × **S5**	[?] r.e. (Thm. 3.17)	[?]	no (Thm. 5.32)	yes (Thm. 6.61)	EXPSPACE-hard in 2EXPTIME (Thms. 6.64, 6.61) [?]
Alt × **S5**	yes (Thm. 8.55)	yes	yes (Thm. 6.59)	yes (Thm. 6.59)	coNP-complete (Thm. 6.59)

Table 6.3: Products of unimodal logics with **S5**.

$\times$	**K**	$\mathbf{S5}_n$, $n \geq 2$	**S5**
PTL	not in ELEM (Thm. 6.37)	not in ELEM (Thm. 6.72)	EXPSPACE-complete (Thm. 6.65)
Lin	[not in ELEM?]	[not in ELEM?]	EXPSPACE-hard in 2EXPTIME (Thms. 6.66, 6.61) [?]
PDL	not in ELEM (Thm. 6.15)	not in ELEM (Thm. 6.72)	EXPSPACE-hard in coN2EXPTIME (Thms. 6.66, 6.50) [?]
$\mathbf{K}_1^C$	not in ELEM (Thm. 6.26)	not in ELEM (Thm. 6.72)	EXPSPACE-hard in coN2EXPTIME (Thms. 6.66, 6.50) [?]
$\mathbf{K4}_2^C$	not in ELEM (Thm. 6.26)	not in ELEM (Thm. 6.72)	EXPSPACE-hard in coN2EXPTIME (Thms. 6.66, 6.50) [?]
$\mathbf{KD45}_2^C$	not in ELEM (Thm. 6.26)	not in ELEM (Thm. 6.72)	EXPSPACE-hard in coN2EXPTIME (Thms. 6.66, 6.50) [?]
$\mathbf{S5}_2^C$	[?]	[?]	[?] in coN2EXPTIME (Thm. 6.50)

Table 6.4: Complexity of decidable products of multimodal logics with **K** and $\mathbf{S5}_n$.

Chapter 7

Undecidable products

The method of proving the decidability of two-dimensional products developed in Chapter 6 was essentially based on the fact that every rooted frame can be unraveled into an intransitive tree (see Proposition 1.7). Since these trees are not frames for transitive modal logics, i.e., those containing **K4**, such logics need a different approach.

Modal logics determined by transitive linear frames—in other words, extensions of **K4.3**—seem to be a good starting point for analyzing products of transitive logics. All of them are known to be Kripke complete (Fine 1974b). All finitely axiomatizable extensions of **K4.3** are decidable (Zakharyaschev and Alekseev 1995). All 'linear' modal logics determined by reflexive linear frames, i.e., extensions of **S4.3**, are finitely axiomatizable and have the finite model property (Bull 1966, Fine 1971). The satisfiability problem in many natural classes of linear frames (say, arbitrary ones) is NP-complete (Ono and Nakamura 1980). So, what about products like $\mathbf{K4.3} \times \mathbf{K4.3}$ or $\mathbf{S4.3} \times \mathbf{S4.3}$? In this chapter we show that these logics—among many other products of 'linear' modal logics—are undecidable (some of them are not even recursively enumerable). Note that the transitivity of frames is essential for obtaining these kinds of undecidability results. For instance, the logics **Alt** and **DAlt** can be considered as the logics of some *intransitive* linear frames containing infinite ascending chains; yet the product logics $\mathbf{Alt} \times \mathbf{Alt}$ and $\mathbf{DAlt} \times \mathbf{DAlt}$, and in fact all $\mathbf{Alt}^n$ and $\mathbf{DAlt}^n$, for $n > 0$, do have the product finite model property and are decidable, as will be shown in Section 8.5.

Not too much is known about the computational properties of product logics whose both components are transitive and at least one of them is not necessarily linear (or of a fixed finite width). We discuss these logics in Section 7.5 and show that $\mathsf{Log}\{\langle \mathbb{N}, < \rangle\} \times \mathbf{K4}$ and $\mathsf{Log}\{\langle \mathbb{N}, < \rangle\} \times \mathbf{S4}$ are undecidable. We also prove that many products with $\mathbf{K}_u$ are undecidable, and those with $\mathbf{K}_1^C$ are not even recursively enumerable.

Similarly to the previous chapters, the product of two frames $\mathfrak{F}_1 = \langle W_1, <_1 \rangle$ and $\mathfrak{F}_2 = \langle W_2, <_2 \rangle$ is denoted by

$$\mathfrak{F}_1 \times \mathfrak{F}_2 = \langle W_1 \times W_2, <_h, <_v \rangle .$$

The modal operators of product logics are $\boxminus$, $\boxbar$, $\diamondminus$, and $\diamondbar$.

7.1 Products of linear orders with infinite ascending chains

Let us recall from Sections 1.2 and 5.3 that a frame $\langle W, R \rangle$ is called *weakly connected* if

$$\forall x, y, z \in W \ (xRy \wedge xRz \rightarrow yRz \vee y = z \vee zRy).$$

We call a sequence $\langle x_n \mid n < \omega \rangle$ of distinct points from W an *ascending ω-type chain* if $x_0 R x_1 R x_2 R \dots$ and $\langle x_i, x_j \rangle \notin R$ whenever $j < i$ (i.e., different points in the sequence belong to different clusters).

Our main aim in this section is to prove the following general theorem due to Reynolds and Zakharyaschev (2001):

Theorem 7.1. *Let $\mathcal{C}_1$ and $\mathcal{C}_2$ be classes of transitive weakly connected frames such that at least one frame in each of these classes contains an ascending ω-type chain. Then* $\mathsf{Log}(\mathcal{C}_1 \times \mathcal{C}_2)$ *(and so* $\mathsf{Log}\mathcal{C}_1 \times \mathsf{Log}\mathcal{C}_2$*) is undecidable.*

As a consequence we shall have, in particular, the following results:

Theorem 7.2. (i) *Let L_1 and L_2 be any logics from the list:*

$$\mathbf{K4.3},\ \mathbf{S4.3},\ \mathsf{Log}\{\langle \mathbb{O}, < \rangle\},\ \mathsf{Log}\{\langle \mathbb{O}, \leq \rangle\},\ \textit{for } \mathbb{O} \in \{\mathbb{N}, \mathbb{Z}, \mathbb{Q}\}.$$ [1]

Then $L_1 \times L_2$ is undecidable.

(ii) *The logics* $\mathsf{Log}\{\langle \mathbb{O}, < \rangle \times \langle \mathbb{O}', < \rangle\}$ *and* $\mathsf{Log}\{\langle \mathbb{O}, \leq \rangle \times \langle \mathbb{O}', \leq \rangle\}$*, where* $\mathbb{O}, \mathbb{O}' \in \{\mathbb{N}, \mathbb{Z}, \mathbb{Q}, \mathbb{R}\}$*, are undecidable.*

We will see in Section 7.3 that many of the logics in Theorem 7.2 are in fact not recursively enumerable.

Let us turn now to the proof of Theorem 7.1.

Proof. The idea is to reduce the undecidable tiling problem for $\mathbb{N} \times \mathbb{N}$ (see Section 5.4) to the satisfiability problem in $\mathcal{C}_1 \times \mathcal{C}_2$. Without loss of generality we will assume all frames in the classes $\mathcal{C}_1$ and $\mathcal{C}_2$ to be rooted. (Note that

[1] Note that $\mathsf{Log}\{\langle \mathbb{R}, < \rangle\} = \mathsf{Log}\{\langle \mathbb{Q}, < \rangle\}$ and $\mathsf{Log}\{\langle \mathbb{R}, \leq \rangle\} = \mathsf{Log}\{\langle \mathbb{Q}, \leq \rangle\} = \mathbf{S4.3}$. In fact, all the listed logics are finitely axiomatizable, see Section 2.1.

a rooted, transitive and weakly connected frame can be viewed as a *chain of clusters*.)

As we saw in Section 5.4, such a reduction would be pretty simple if our frames were intransitive (or the language contained the 'next-time' operators); for then we would be able to refer to the tiles on the right and above directly. As this is not the case, we will use the idea of Marx and Reynolds (1999) to enumerate the pairs of natural numbers and refer to the right and above neighbors of a pair indirectly via special pointers. Another problem is that our frames are in general not irreflexive (so the diamond operators cannot say 'here but not later') and that apart from the existence of an ascending ω-type chain we know nothing of the order type of these frames (for instance, they can be of type $\omega + 1$).

First we attack the latter problem by using the following trick (cf. Spaan 1993), which makes it possible to deal with frames that may contain nondegenerate clusters. Suppose that we have a product $\mathfrak{F} = \langle W_1 \times W_2, <_h, <_v \rangle$ of two rooted, transitive and weakly connected frames $\mathfrak{F}_1 = \langle W_1, <_1 \rangle$ and $\mathfrak{F}_2 = \langle W_2, <_2 \rangle$. The relations $<_h$ and $<_v$ are not necessarily irreflexive, and our task is to 'simulate irreflexivity' in both directions of $\mathfrak{F}$.

To this end, we partition a part of $W_1 \times W_2$ into 'black' and 'white' squares using the following formulas with propositional variables h_0, h_1 (horizontal) and v_0, v_1 (vertical):

$$h_0 \wedge v_0,$$

$$\Box^*(h_0 \to \neg h_1),$$

$$\Box^*(v_0 \to \neg v_1),$$

$$\boxminus\big((h_0 \to \diamondminus h_1) \wedge (h_1 \to \diamondminus h_0)\big),$$

$$\boxbar\big((v_0 \to \diamondbar v_1) \wedge (v_1 \to \diamondbar v_0)\big),$$

$$\Box^*\big((h_i \to \boxbar h_i) \wedge (\neg h_i \to \boxbar \neg h_i)\big),$$

$$\Box^*\big((v_i \to \boxminus v_i) \wedge (\neg v_i \to \boxminus \neg v_i)\big),$$

$$\boxminus^+(\diamondminus h_0 \to h_0 \vee h_1) \wedge \boxbar^+(\diamondbar v_0 \to v_0 \vee v_1),$$

where $i \in \{0, 1\}$ and

$$\Box^* \psi = \psi \wedge \boxminus \psi \wedge \boxbar \psi \wedge \boxbar \boxminus \psi.$$

The conjunction of these formulas will be denoted by $\mathsf{Chessboard}$. Let

$$\mathsf{w} = (h_1 \wedge v_0) \vee (h_0 \wedge v_1), \qquad \mathsf{b} = (h_0 \wedge v_0) \vee (h_1 \wedge v_1).$$

Say that a point x in $\mathfrak{F}$ (under some valuation) is *white* (or *black*) if $x \models \mathsf{w}$ (respectively, $x \models \mathsf{b}$). A point that is neither black nor white will be called a *cloud point.* The cloud points validate the formula

$$\mathsf{cloud} = (\neg h_1 \wedge \neg h_0) \vee (\neg v_0 \wedge \neg v_1).$$

A maximal set S of points in $\mathfrak{F}$ will be called a *square* if the following conditions are satisfied:

- all points in S are of the same color (i.e., either black or white) and
- S is connected in the sense that, for any two distinct points $x, y \in S$, either there is a path of $<_v$ and $<_h$-arrows between these points and every such path entirely belongs to S, or S contains the two points u, v such that $x <_v u$, $y <_h u$, $v <_h x$, $v <_v y$ (or symmetrically, $x <_h u$, $y <_v u$, $v <_v x$, $v <_h y$).

It is not hard to check that if $\mathsf{Chessboard}$ is true at the root r of $\mathfrak{F}$ (under some valuation) then the noncloud part of $\mathfrak{F}$ can be viewed as a chess-board that is either infinite or finite 'circular' in each direction: this part of $\mathfrak{F}$ is divided into columns and rows of black and white squares in such a way that r belongs to a black square and every square has a horizontal and a vertical (not necessarily immediate) successor of different color (in particular, our chess-board may look like the Euclidean plane $\mathbb{R}^2$ all points in which are squares).

For squares S_1 and S_2, we write

$$\begin{array}{lll} S_1 <_h S_2 & \text{iff} & \forall x \in S_1\, \exists y \in S_2\ (x \neq y \text{ and } x <_h y), \\ S_1 <_v S_2 & \text{iff} & \forall x \in S_1\, \exists y \in S_2\ (x \neq y \text{ and } x <_v y). \end{array}$$

Now we can define new possibility operators ⬘ and ⬗ by taking, for any formula ψ,

$$⬘\psi = \big(h_0 \to \Diamond_h(h_1 \wedge \Diamond_h^+\psi)\big) \wedge \big(h_1 \to \Diamond_h(h_0 \wedge \Diamond_h^+\psi)\big),$$
$$⬗\psi = \big(v_0 \to \Diamond_v(v_1 \wedge \Diamond_v^+\psi)\big) \wedge \big(v_1 \to \Diamond_v(v_0 \wedge \Diamond_v^+\psi)\big).$$

Let ⬒ and ◨ be the duals of ⬘ and ⬗, respectively.

For each noncloud point x in $\mathfrak{F}$, let *square*(x) denote the square containing x. It should be clear that for every x on the chess-board we have:

$$\begin{array}{lll} x \models ⬘\psi & \text{iff} & \exists y\ (\mathit{square}(x) <_h \mathit{square}(y) \text{ and } y \models \psi), \\ x \models ⬗\psi & \text{iff} & \exists y\ (\mathit{square}(x) <_v \mathit{square}(y) \text{ and } y \models \psi). \end{array}$$

Note that $<_h$ and $<_v$ are not necessarily irreflexive on squares either: if, say, $\{x \mid \exists y\ \langle x, y\rangle \in S\}$ is a $<_1$-cluster with at least two elements, then $S <_h S$ holds.

However, we can force certain squares to be 'irreflexive' with respect to $<_h$ and $<_v$. Given a propositional variable p, a square S on the chess-board is called a *p-square* if the following three conditions hold:

- $\forall x \in S\ x \models p$;
- $\forall x \in S \forall y \notin S(x <_h y \vee x <_v y \to y \not\models p)$;
- S is *irreflexive*, i.e., $S \not<_h S$ and $S \not<_v S$.

Let p-square be the conjunction of the following formulas (in which q' and q'' are auxiliary variables different from p):

$$\Box^*(p \to \neg\mathsf{cloud}),$$
$$\Box^*(p \to \neg\text{⬗}p \wedge \neg\text{⬘}p),$$
$$\Box^*(\text{⬙}p \wedge \neg\text{⬘}p \to p),$$
$$\Box^*(\text{⬖}p \wedge \neg\text{⬗}p \to p),$$
$$\Box^*(p \to \text{⬘}q' \wedge \text{⬗}q'' \wedge \neg\text{⬘}\text{⬘}q' \wedge \neg\text{⬗}\text{⬗}q''),$$
$$\Box^*(q' \to \neg\text{⬘}q'),$$
$$\Box^*(q'' \to \neg\text{⬗}q''),$$
$$\Box^*(p \to \boxminus(\neg q' \wedge \text{⬘}q' \to p)),$$
$$\Box^*(p \to \boxbar(\neg q'' \wedge \text{⬗}q'' \to p)).$$

The reader can readily check that if p-square holds at the root of $\mathfrak{F}$ and $x \models p$ then *square*(x) is a p-square: the first conjunct guarantees that x is on the chess-board, the second ensures that the second and third p-square conditions hold; the third and fourth conjuncts of p-square ensure that where p holds in a square it also holds to the left and downwards within that square; the last five formulas say that the auxiliary variables hold in the immediate right and upwards neighbors of a p-square and use this to ensure that where p holds in a square it also holds to the right and upwards in that square. (It is to be noted that there may be infinitely many different p-squares on the chess-board.)

Given a square S, from now on we will write $S \models p$ whenever $x \models p$ hold for all $x \in S$.

Let *pair* : $\mathbb{N} \to \mathbb{N} \times \mathbb{N}$ be the enumeration of the points $\langle m, n\rangle$ in $\mathbb{N} \times \mathbb{N}$ defined recursively by taking:

- *pair*$(0) = \langle 0, 0\rangle$,
- if *pair*$(n) = \langle 0, j\rangle$ then *pair*$(n+1) = \langle j+1, 0\rangle$,
- otherwise, if *pair*$(n) = \langle i+1, j\rangle$ then *pair*$(n+1) = \langle i, j+1\rangle$;

see Fig. 7.1. Let *right*(n) denote the number of the pair to the right of *pair*(n)

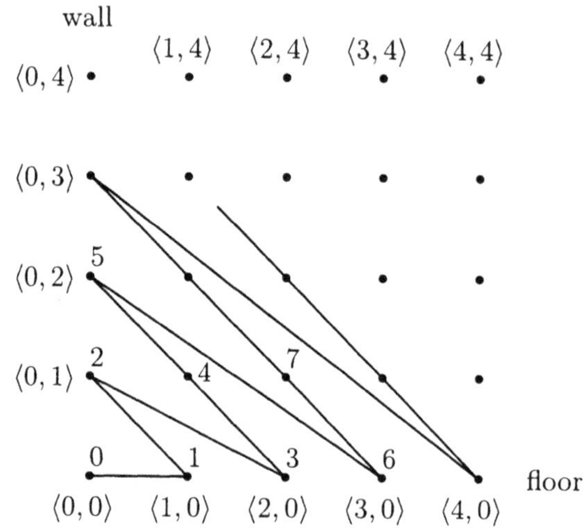

Figure 7.1: The enumeration *pair*.

and *above*(n) the number of the pair above *pair*(n). For instance, *right*$(3) = 6$, *above*$(3) = 7$. An important property of the enumeration is that

$$right(n+1) = \begin{cases} above(n), & \text{if } pair(n) \text{ is not on the wall;} \\ above(n)+1, & \text{if } pair(n) \text{ is on the wall.} \end{cases} \tag{7.1}$$

Given a set T of tile types, we can now write down a formula φ_T which is satisfiable in $\mathfrak{F}$ iff T tiles $\mathbb{N} \times \mathbb{N}$. The formula φ_T will contain the propositional variables

- t, for every tile type $t \in T$,
- tile $(= \bigvee\{t \mid t \in T\})$,
- next (a pointer to the next tile according to the enumeration),
- right (a pointer to the right-neighbor of a tile),
- above (a pointer to the above-neighbor of a tile),
- wall (marking the wall, i.e., the pairs $\langle 0, n\rangle$),
- floor (marking the floor, i.e., the pairs $\langle n, 0\rangle$).

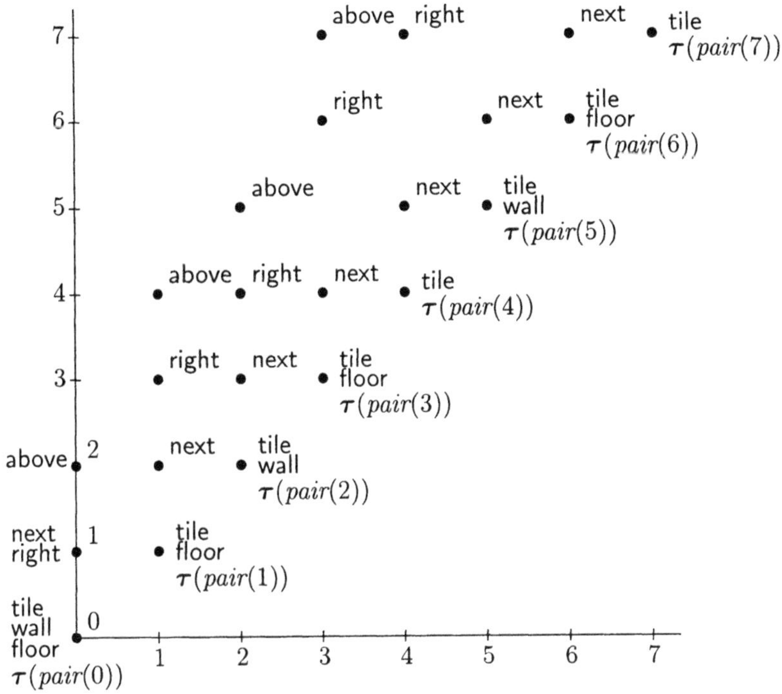

Figure 7.2: The formula φ_T in the product of ascending ω-type chains.

Let Tiling be the conjunction of the following formulas:

$$\square^*(\mathsf{tile} \leftrightarrow \bigvee_{t\in T} t),$$

$$\square^* \bigwedge_{\substack{t,t'\in T\\ t\neq t'}} \neg(t \wedge t'),$$

$$\square^* \bigwedge_{\substack{t,t'\in T\\ up(t)\neq down(t')}} \big(t \to ◨(\mathsf{above} \to \neg ⬘ t')\big),$$

$$\square^* \bigwedge_{\substack{t,t'\in T\\ right(t)\neq left(t')}} \big(t \to ◨(\mathsf{right} \to \neg ⬘ t')\big).$$

The intended meaning of these formulas should be clear.

Define φ_T to be the conjunction of Chessboard, Tiling, t-square, for all

$t \in T$, tile, next-square, right-square, above-square, wall-square, floor-square,[2] and the following formulas as well:

$$\Box^*(\mathsf{tile} \to ⬗\mathsf{next}), \tag{7.2}$$

$$\Box^*(\mathsf{next} \to ⬘\mathsf{tile}), \tag{7.3}$$

$$\Box^*(\mathsf{tile} \to ⬗\mathsf{right}), \tag{7.4}$$

$$\Box^*(\mathsf{right} \to ⬘\mathsf{tile}), \tag{7.5}$$

$$\Box^*\neg(⬗\mathsf{next} \wedge ⬘\mathsf{tile}), \tag{7.6}$$

$$\Box^*\big(\mathsf{next} \to ⬒(⬘\mathsf{tile} \to \neg⬗(\mathsf{right} \vee \mathsf{above}))\big), \tag{7.7}$$

$$\Box^*(\mathsf{right} \to ⬗\mathsf{above}), \tag{7.8}$$

$$\Box^*(\mathsf{above} \to ⬘\mathsf{tile}), \tag{7.9}$$

$$\mathsf{floor} \wedge \mathsf{wall} \wedge \neg⬘⬗(\mathsf{floor} \wedge \mathsf{wall}), \tag{7.10}$$

$$\Box^*\big(\mathsf{wall} \to ◨(\mathsf{next} \to ⬒(\mathsf{tile} \to \mathsf{floor}))\big), \tag{7.11}$$

$$\Box^*\big(\mathsf{wall} \to ◨(\mathsf{above} \to ⬒(\mathsf{tile} \to \mathsf{wall}))\big), \tag{7.12}$$

$$\Box^*\big(\mathsf{tile} \wedge \neg\mathsf{wall} \to ◨(\mathsf{above} \to ⬒(\mathsf{tile} \to \neg\mathsf{wall}))\big), \tag{7.13}$$

$$\neg⬗(\mathsf{next} \wedge \neg\mathsf{right}), \tag{7.14}$$

$$\Box^*(⬗\mathsf{above} \wedge ⬘\mathsf{tile} \to \mathsf{right} \vee ⬗\mathsf{right}), \tag{7.15}$$

$$◨(⬘(\mathsf{tile} \wedge \neg\mathsf{wall}) \to (\mathsf{right} \vee ⬘\mathsf{right}), \tag{7.16}$$

$$\Box^*\neg(⬗\mathsf{right} \wedge ⬘\mathsf{right}), \tag{7.17}$$

$$\neg⬗(⬘\mathsf{wall} \wedge ⬘\mathsf{right}). \tag{7.18}$$

Figure 7.2 gives the reader some general intuition about these formulas. They will be explained in detail in the proof of the next lemma.

Lemma 7.3. *T tiles $\mathbb{N} \times \mathbb{N}$ iff φ_T is satisfiable in a frame $\mathfrak{F} \in \mathcal{C}_1 \times \mathcal{C}_2$.*

Proof. ($\Rightarrow$) Suppose $\boldsymbol{\tau} : \mathbb{N} \times \mathbb{N} \to T$ is a tiling and $\mathfrak{F} \in \mathcal{C}_1 \times \mathcal{C}_2$ is the product of two rooted frames $\mathfrak{F}_1$ and $\mathfrak{F}_2$ with ascending ω-type chains and roots x_0, y_0, respectively. Take ascending ω-type chains

$$x_0 \lneqq_1 x_1 \lneqq_1 x_2 \lneqq_1 \dots \quad \text{and} \quad y_0 \lneqq_2 y_1 \lneqq_2 y_2 \lneqq_2 \dots$$

in $\mathfrak{F}_1$ and $\mathfrak{F}_2$. Define a valuation $\mathfrak{V}$ of h_0, h_1, v_0 and v_1 in $\mathfrak{F} = \mathfrak{F}_1 \times \mathfrak{F}_2$ by

[2] We always take distinct pairs of auxiliary variables q' and q'' in the square-formulas.

taking:

$$\begin{aligned}
\mathfrak{V}(h_0) &= \{\langle x, y\rangle \mid x_n \le_1 x <_1 x_{n+1},\ n < \omega,\ n \text{ is even}\};\\
\mathfrak{V}(h_1) &= \{\langle x, y\rangle \mid x_n \le_1 x <_1 x_{n+1},\ n < \omega,\ n \text{ is odd}\};\\
\mathfrak{V}(v_0) &= \{\langle x, y\rangle \mid y_n \le_2 y <_2 y_{n+1},\ n < \omega,\ n \text{ is even}\};\\
\mathfrak{V}(v_1) &= \{\langle x, y\rangle \mid y_n \le_2 y <_2 y_{n+1},\ n < \omega,\ n \text{ is odd}\}.
\end{aligned}$$

The formula Chessboard is satisfied at the root of $\mathfrak{F}$ under this valuation. For any $m, n < \omega$, let

$$(m, n) = \{\langle x, y\rangle \mid x_m \le_1 x <_1 x_{m+1},\ y_n \le_2 y <_2 y_{n+1}\}.$$

Clearly, the (m, n) are all the squares on the chess-board. Now we extend $\mathfrak{V}$ to the other variables in φ_T by taking

$$\begin{aligned}
\mathfrak{V}(t) &= \bigcup\{(n, m) \mid m = n \text{ and } \tau(\mathit{pair}(n)) = t\};\\
\mathfrak{V}(\mathsf{tile}) &= \bigcup\{(n, m) \mid n = m\};\\
\mathfrak{V}(\mathsf{next}) &= \bigcup\{(n, m) \mid m = n + 1\};\\
\mathfrak{V}(\mathsf{right}) &= \bigcup\{(n, m) \mid m = \mathit{right}(n)\};\\
\mathfrak{V}(\mathsf{above}) &= \bigcup\{(n, m) \mid m = \mathit{above}(n)\};\\
\mathfrak{V}(\mathsf{wall}) &= \bigcup\{(n, m) \mid m = n \text{ and } \mathit{pair}(n) \text{ is on the wall}\};\\
\mathfrak{V}(\mathsf{floor}) &= \bigcup\{(n, m) \mid m = n \text{ and } \mathit{pair}(n) \text{ is on the floor}\}.
\end{aligned}$$

(This situation is depicted in Fig. 7.2.) It is not hard to check that under this valuation we have $(0, 0) \models \varphi_T$.

($\Leftarrow$) Suppose φ_T is satisfied at the root x_0 of some $\mathfrak{F} \in \mathcal{C}_1 \times \mathcal{C}_2$ under some valuation. Then x_0 belongs to a tile-square; let us denote this irreflexive square by $(0, 0)$. By (7.2) and (7.3), we have an infinite sequence of squares

$$(0, 0) <_v (0, 1) <_h (1, 1) <_v (1, 2) <_h \cdots <_h (i, i) <_v (i, i + 1) <_h \dots,$$

in which every (n, n), $n \in \mathbb{N}$, is a tile-square and every $(n, n + 1)$ is a next-square, so they are all irreflexive by the formulas next-square and t-square ($t \in T$). In fact, it is not hard to see that, by (7.6), all the squares in the sequence are distinct. For all $m, n \in \mathbb{N}$, denote by (m, n) the square located in the same column with (m, m) and the same row with (n, n). (Note that there can be squares on the chess-board other than those of the form (m, n).) Given a square S, we write $\mathit{row}(S) < \mathit{row}(n, n)$ (or $\mathit{row}(S) > \mathit{row}(n, n)$) if for all $x \in S$ there are points y in $\mathfrak{F}$ such that $x \le_h y$ and $\mathit{square}(y) <_v (n, n)$ (respectively, $x \le_h y$ and $(n, n) <_v \mathit{square}(y)$).

Consider an arbitrary square (n, n). By (7.4) and right-square, there is a unique right-square r_n such that $(n, n) <_v r_n$. By (7.5) and tile-square, there is a unique tile-square s such that $r_n <_h s$. In this case we say that r_n *points to* s. Using (7.2) and (7.6), one can easily show that

$$\text{if } row(r_n) < row(k,k) \text{ then } r_n = (n,i) \text{ for some } i < k. \tag{7.19}$$

In other words, (7.19) means that r_n points to tile-square (i,i).

By (7.4), (7.8) and above-square, there is a unique above-square a_n such that $(n,n) <_v a_n$. By (7.9) and tile-square, there is a unique tile-square s such that $a_n <_h s$. We will again say that a_n *points to* s. In the same way as above we can show that

$$\text{if } row(a_n) < row(k,k) \text{ then } a_n = (n,i) \text{ for some } i < k. \tag{7.20}$$

Now, we will prove by induction on $n \geq 0$ that

(i) if $pair(right(n))$ is on the floor, then $(right(n), right(n)) \models \mathsf{floor}$;

(ii) $(right(n), right(n)) \models \neg\mathsf{wall}$;

(iii) $r_n = (n, right(n))$, i.e., r_n points to $(right(n), right(n))$;

(iv) $a_n = (n, above(n))$, i.e., a_n points to $(above(n), above(n))$;

(v) if $pair(above(n))$ is on the wall then $(above(n), above(n)) \models \mathsf{wall}$.

For the base case $n = 0$, observe that by (7.10), $(0,0) \models \mathsf{floor} \wedge \mathsf{wall}$. By (7.11) we then have $(1,1) \models \mathsf{floor}$ and by (7.10) again, $(1,1) \models \neg\mathsf{wall}$. By (7.14) $r_0 = (0,1)$, and so r_0 points to $(1,1)$.

Notice that by (7.8), (7.20), (7.15) and right-square, $a_0 = (0,2)$ points to $(2,2)$: if a_0 was further above $(0,2)$ then (7.15) would imply that $r_0 = (0,1)$ is not the unique right-square above $(0,0)$, contrary to right-square. By (7.12), we have $(2,2) \models \mathsf{wall}$.

Now consider the induction step for $n > 0$ (below IH stands for 'induction hypothesis').

(i) If $pair(right(n))$ is on the floor, then $pair(n)$ is on the floor as well, hence $pair(n-1)$ is on the wall. By (7.1), we have $right(n) - 1 = above(n-1)$ and so

$$pair(right(n) - 1) = pair(above(n-1))$$

is on the wall too, whence by IH (v), $(right(n) - 1, right(n) - 1) \models \mathsf{wall}$. It follows directly by (7.11) that we must have $(right(n), right(n)) \models \mathsf{floor}$.

(ii) If $pair(right(n))$ is on the floor then $(right(n), right(n)) \models \neg\mathsf{wall}$ by (7.10) and (i). Otherwise $right(n) = above(n-1)$ and $n - 1 = right(k)$ for some k with $0 \leq k < n$. By IH (ii), we have $(right(k), right(k)) \models \neg\mathsf{wall}$, hence $(n-1, n-1) \models \neg\mathsf{wall}$. Since IH (iv) tells us that a_{n-1} points to $(above(n-1), above(n-1))$, by (7.13) we obtain

$$(above(n-1), above(n-1)) \models \neg\mathsf{wall}$$

which is $(right(n), right(n)) \models \neg\mathsf{wall}$ as required.

(iii) It follows from (ii) and (7.16) that there is a right-square r of the form $(m, right(n))$. We will show that $r = (n, right(n))$.

We cannot have $m \geq right(n)$ by (7.5) and tile-square. By (7.7), there is some i such that $0 \leq i < right(n)$ with $(i, i) <_v r$. It follows that $r = r_i$. If $i < n$ then by IH $r = (i, right(i))$, which is impossible. Suppose that $i > n$. Let us then examine r_n. By right-square, r_n cannot be of the form $(k, right(n))$ for some k. The case when $row(r_n) > row(right(n), right(n))$ is impossible by (7.17). If $row(r_n) < row(right(n), right(n))$ then, by (7.19) $r_n = (n, j)$ for some $n < j < right(n)$. If $j = right(k)$ then $k < n$, and by IH (iii) $r_k = (k, j)$, contrary to right-square. And if $j \neq right(k)$ for any $k < n$ then $pair(j)$ is on the wall, i.e., $j = above(k)$ for some $k < n$, from which by IH (v) $(j, j) \models \mathsf{wall}$, contrary to (7.18). Thus $i = n$, that is, $r_n = (n, right(n))$.

(iv) By (7.8), (iii), (7.15), (7.20) and right-square we have $a = (n, above(n))$.

(v) If $pair(above(n))$ is on the wall then $pair(n)$ is on the wall as well. Since $n > 0$, $n = above(k)$ for some $k < n$, and so by IH (v) $(n, n) \models \mathsf{wall}$. By (iv) $a_n = (n, above(n))$ and by (7.12), $(above(n), above(n)) \models \mathsf{wall}$.

A tiling τ of $\mathbb{N} \times \mathbb{N}$ is defined now as follows. Given a pair $\langle m, n \rangle$ of natural numbers, let $\tau(m, n)$ be the unique $t \in T$ such that the tile-square $(pair^{-1}(m, n), pair^{-1}(m, n))$ is a t-square. Using the formula Tiling it is readily seen that τ is indeed a tiling. ❑

To complete the proof of Theorem 7.1, it remains only to observe that, for any set T of tile types, we clearly have: $\neg\varphi_T \in \mathsf{Log}(\mathcal{C}_1 \times \mathcal{C}_2)$ iff φ_T is not satisfiable in $\mathcal{C}_1 \times \mathcal{C}_2$ iff, by Lemma 7.3, T does not tile $\mathbb{N} \times \mathbb{N}$. It follows that $\mathsf{Log}(\mathcal{C}_1 \times \mathcal{C}_2)$ is undecidable. ❑

We conclude this section by proving an undecidability result concerning **HS** interval temporal logics (see Section 2.2). Given a class $\mathcal{C}$ of strict linear orders, the *forward fragment* $\mathbf{HS}_{\mathcal{C}}^{fw}$ of logic $\mathbf{HS}_{\mathcal{C}}$ consists of those formulas of $\mathbf{HS}_{\mathcal{C}}$ that contain only modal operators $\Box_s$, $\Box_f^{-1}$ and their duals (and do not contain $\Box_s^{-1}$ and $\Box_f$). The following result is a generalization of Theorem 2.13:

Theorem 7.4. *Let $\mathcal{C}$ be a class of strict linear orders at least one of which contains an infinite ascending chain of distinct points. Then $\mathbf{HS}_{\mathcal{C}}^{fw}$ is undecidable.*

Proof. Having recalled the two-dimensional representation of **HS**-logics from Section 3.9, it is straightforward to see that the forward fragment $\mathbf{HS}_{\mathcal{C}}^{fw}$ is determined by the class

$$\{\mathfrak{G} \mid \mathfrak{G} \text{ is the 'North-Western subframe' of } \mathfrak{F} \times \mathfrak{F}, \text{ for some } \mathfrak{F} \in \mathcal{C}\}.$$

(We remind the reader that the North-Western subframe of $\mathfrak{F} \times \mathfrak{F}$, where $\mathfrak{F} = \langle W, < \rangle$, consists of all points $\langle u, v \rangle$ such that $u \leq v$.)

Now, the undecidability of $\mathbf{HS}_{\mathcal{C}}^{fw}$ follows from two facts. First, a close inspection of the formula φ_T, constructed in the proof of Theorem 7.1, shows that if φ_T is satisfiable in the North-Western subframe of $\mathfrak{F} \times \mathfrak{F}$ under some valuation, then this valuation can be extended to the whole product frame $\mathfrak{F} \times \mathfrak{F}$ without changing the truth values of φ_T in the North-Western subframe. And second, if φ_T is satisfied in $\mathfrak{F} \times \mathfrak{F}$, for $\mathfrak{F} \in \mathcal{C}$, then we can always modify the valuation in such a way that squares are just singletons, the tile-squares are points above the 'diagonal' of $\mathfrak{F} \times \mathfrak{F}$, and φ_T is satisfied in the North-Western subframe of $\mathfrak{F} \times \mathfrak{F}$. Details are left to the reader. ❑

7.2 Products of linear orders with infinite descending chains

In this section we prove the following result of Reynolds and Zakharyaschev (2001):

Theorem 7.5. *Let $\mathcal{C}_1$ and $\mathcal{C}_2$ be classes of transitive and weakly connected frames such that each of them contains a rooted Noetherian linear order having an infinite descending chain of distinct points. Then $\mathsf{Log}(\mathcal{C}_1 \times \mathcal{C}_2)$ (and so $\mathsf{Log}\mathcal{C}_1 \times \mathsf{Log}\mathcal{C}_2$) is undecidable.*

Proof. Given a finite set T of tile types, we are again going to construct a formula χ_T which is satisfiable in a frame from $\mathcal{C}_1 \times \mathcal{C}_2$ iff T tiles $\mathbb{N} \times \mathbb{N}$.

To this end, we again have to solve the problem of having not necessarily irreflexive frames. As in Section 7.1, we partition the frames into 'black' and 'white' squares. However, now this should be done in a somewhat different way. We will use the formulas:

$$\Box^*((h \vee \text{⟠} h \to \text{◫} h) \wedge (\neg h \vee \text{⟠} \neg h \to \text{◫} \neg h)), \tag{7.21}$$

$$\Box^*((v \vee \text{⟐} v \to \boxminus v) \wedge (\neg v \vee \text{⟐} \neg v \to \boxminus \neg v)), \tag{7.22}$$

$$\neg h \wedge \neg v \wedge \text{⟠}\text{⟐}\Box^*(h \wedge v), \tag{7.23}$$

$$\Box^*(\text{◨}\bot \wedge \text{⬒}\bot \to p), \tag{7.24}$$

$$\text{◫}\text{⬘}(\neg p \wedge \text{⬒} p), \tag{7.25}$$

$$\boxminus\text{⬗}(p \wedge \text{◨}\neg p), \tag{7.26}$$

$$\Box^*(p \to \text{⬒}(p \wedge \text{⬗} p)), \tag{7.27}$$

$$\Box^*(\neg p \to \text{◨}(\neg p \wedge \text{⬘}\neg p)), \tag{7.28}$$

where

$$\blacklozenge_h\psi = (h \to \Diamond_h(\neg h \land \Diamond_h^+\psi)) \land (\neg h \to \Diamond_h(h \land \Diamond_h^+\psi)),$$
$$\blacklozenge_v\psi = (v \to \Diamond_v(\neg v \land \Diamond_v^+\psi)) \land (\neg v \to \Diamond_v(v \land \Diamond_v^+\psi)).$$

Denote the conjunction of (7.21)–(7.28) by Diagonal.

Lemma 7.6. *Let $\mathfrak{F}_1 = \langle W_1, <_1\rangle$ and $\mathfrak{F}_2 = \langle W_2, <_2\rangle$ be rooted Noetherian linear orders such that each of them contains an infinite descending chain of distinct points. Then* Diagonal *is satisfied in $\mathfrak{F}_1 \times \mathfrak{F}_2$.*

Proof. Take infinite descending chains

$$x_0 \gneq_1 x_1 \gneq_1 x_2 \gneq_1 \dots \quad \text{and} \quad y_0 \gneq_2 y_1 \gneq_2 y_2 \gneq_2 \dots$$

in $\mathfrak{F}_1$ and $\mathfrak{F}_2$, respectively. Define a valuation $\mathfrak{V}$ in $\mathfrak{F}_1 \times \mathfrak{F}_2$ by taking:

$$\begin{aligned}
\mathfrak{V}(h) &= \{\langle x, y\rangle \mid x_0 <_1 x\} \cup \{\langle x, y\rangle \mid x_{n+1} \lneq_1 x \leq_1 x_n,\ n < \omega,\ n \text{ is even}\};\\
\mathfrak{V}(v) &= \{\langle x, y\rangle \mid y_0 <_2 y\} \cup \{\langle x, y\rangle \mid y_{n+1} \lneq_2 y \leq_2 y_n,\ n < \omega,\ n \text{ is even}\};\\
\mathfrak{V}(p) &= \{\langle x, y\rangle \mid x_1 \lneq_1 x\} \cup \{\langle x, y\rangle \mid x_{n+1} \lneq_1 x,\ y \leq_2 y_n,\ n > 0\}
\end{aligned}$$

(see Fig. 7.3). Since $\mathfrak{F}_1$ is rooted and Noetherian, there is a $<_1$-greatest point z_1 in $\mathfrak{F}_1$ such that $z_1 <_1 x_n$ for all $n < \omega$. Similarly, there is a $<_2$-greatest point z_2 in $\mathfrak{F}_2$ such that $z_2 <_2 y_n$ for all $n < \omega$. The reader can easily check that under the valuation $\mathfrak{V}$, we have $\langle z_1, z_2\rangle \models$ Diagonal. ❑

Next, let us recall from Section 7.1 the notion of square and the extensions of the relations $<_h$ and $<_v$ to squares.

Lemma 7.7. *If $x \models$ Diagonal in a frame $\mathfrak{F} \in \mathcal{C}_1 \times \mathcal{C}_2$ under some valuation, then there are squares (m, n) in $\mathfrak{F}$ ($n, m < \omega$) such that for all $k, n, m < \omega$,*

(a) *(m, n) is irreflexive,*

(b) *$(m, n) \lneq_h (k, n)$ and $(n, m) \lneq_v (n, k)$, if $m > k$,*

(c) *$(m, n) \models p$ iff $m \leq n$.*

Proof. Suppose $\langle z_1, z_2\rangle \models$ Diagonal for some $\langle z_1, z_2\rangle$ in $\mathfrak{F} = \mathfrak{F}_1 \times \mathfrak{F}_2$. By induction on n we define squares (k, m), for $k, m \leq n$, and show that for all $k, \ell, m \leq n$,

(i) (k, m) is irreflexive,

(ii) $(k, m) \lneq_h (\ell, m)$ and $(m, k) \lneq_v (m, \ell)$, if $k > \ell$,

(iii) $(k, m) \models p$ iff $k \leq m$.

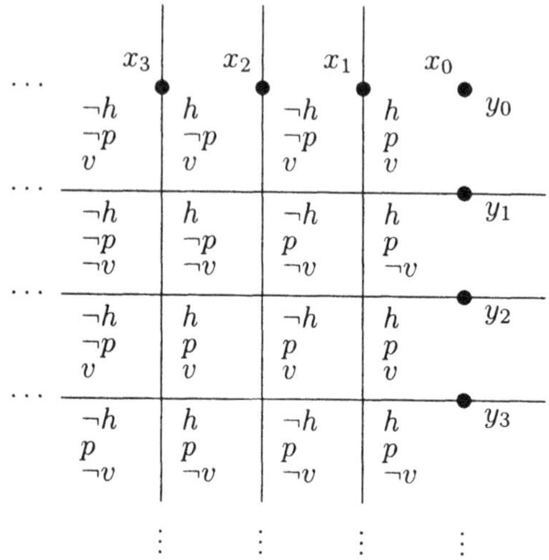

Figure 7.3: Satisfying Diagonal.

First, let $n = 0$. By (7.23), there are some $x_0 \gneqq_1 z_1$ and $y_0 \gneqq_2 z_2$ such that $\langle x_0, y_0 \rangle \models \Box^*(h \wedge v)$. Define $(0,0)$ as *square*$(\langle x_0, y_0 \rangle)$. (For every point z in $\mathfrak{F}$, *square*(z) exists by (7.21) and (7.22).) Then $(0,0)$ is irreflexive, and $(0,0) \models p$ holds by (7.24).

Assume now that $n > 0$ and squares (k, m) satisfying (i)–(iii) have been defined for all $k, m < n$. For each $k < n$, choose some x_k, y_k such that $\langle x_k, y_k \rangle$ belongs to (k, k). By IH and (7.25), there is an x_n such that $z_1 \lneqq_1 x_n \lneqq_1 x_{n-1}$ and $\langle x_n, y_{n-1} \rangle \models \neg p \wedge \text{⬒}p$. Let $(n, k) =$ *square*$(\langle x_n, y_k \rangle)$, for all $k < n$. Then by IH and (7.25), (n, k) is irreflexive, and by IH and (7.27), we have $(n, k) \models \neg p$, for all $k < n$. Next, by IH and (7.26), there is a y_n such that $z_2 \lneqq_2 y_n \lneqq_2 y_{n-1}$ and $\langle x_n, y_n \rangle \models p \wedge \text{◨}\neg p$. Let $(k, n) =$ *square*$(\langle x_k, y_n \rangle)$, for all $k \le n$. Then by IH and (7.26), (k, n) is irreflexive, and by IH and (7.28), we have $(k, n) \models p$ for all $k \le n$, as required. ❑

Given a finite set T of tile types, we now construct a formula χ_T which is satisfiable in a $\mathfrak{F} \in \mathcal{C}_1 \times \mathcal{C}_2$ iff T tiles $\mathbb{N} \times \mathbb{N}$. As in the previous section, we will use the enumeration *pair* : $\mathbb{N} \to \mathbb{N} \times \mathbb{N}$ and represent the tiles on $\mathbb{N} \times \mathbb{N}$ by squares in $\mathfrak{F}$ of the form (n, n), using the variables tile and t, for $t \in T$. Let *left*(n) denote the number of the pair to the left of *pair*(n) and *below*(n) the number of the pair below *pair*(n). As before, wall and floor will indicate that a certain pair is on the wall or on the floor. But instead of the pointers next, right and above we will now use the variables prev, left and below which

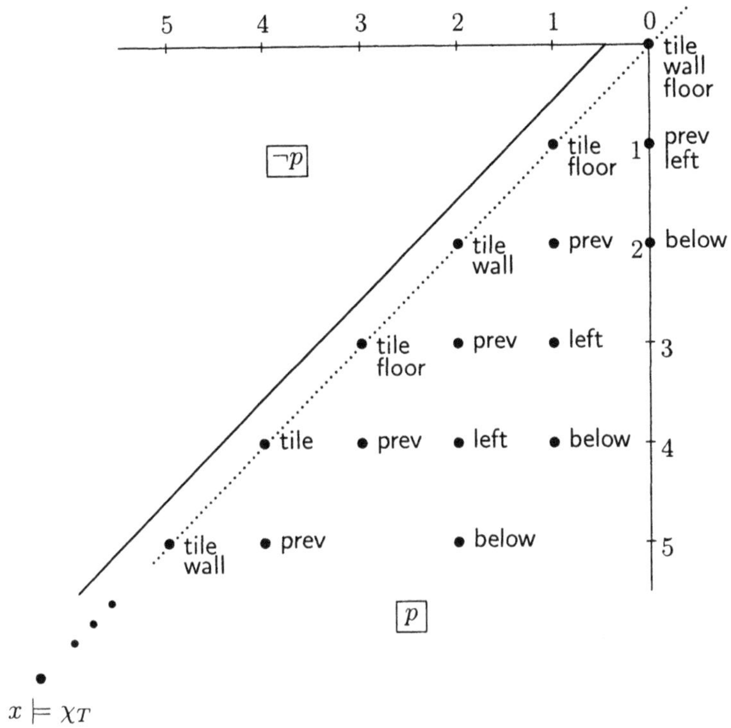

Figure 7.4: The formula χ_T in the product of infinite descending chains.

are supposed to point to the previous pair in the enumeration, to that on the left and below, respectively (see Fig. 7.4).

Let Tiling be the conjunction of the following formulas:

$$\Box^*(\mathsf{tile} \leftrightarrow \bigvee_{t \in T} t),$$

$$\Box^* \bigwedge_{\substack{t,t' \in T \\ t \neq t'}} \neg(t \wedge t'),$$

$$\Box^* \bigwedge_{\substack{t,t' \in T \\ down(t) \neq up(t')}} \big(t \to \blacksquare(\mathsf{below} \to \neg\Diamond t')\big),$$

$$\Box^* \bigwedge_{\substack{t,t' \in T \\ left(t) \neq right(t')}} \big(t \to \blacksquare(\mathsf{left} \to \neg\Diamond t')\big).$$

Given a propositional variable r, we define a formula r-square in almost the same way as in Section 7.1: it is the conjunction of the formulas

$$\Box^*(r \to \neg ⬗ r \wedge \neg ⬘ r),$$

$$\Box^*(\diamondminus r \wedge \neg ⬘ r \to r),$$

$$\Box^*(\diamondbar r \wedge \neg ⬗ r \to r),$$

$$\Box^*(r \wedge ⬘ \top \to ⬘ q' \wedge \neg ⬘ ⬘ q'),$$

$$\Box^*(r \wedge ⬗ \top \to ⬗ q'' \wedge \neg ⬗ ⬗ q''), r$$

$$\Box^*((q' \to \neg ⬘ q') \wedge (q'' \to \neg ⬗ q'')),$$

$$\Box^*(r \to \boxminus(\neg q' \wedge ⬘ q' \to r)),$$

$$\Box^*(r \to \boxbar(\neg q'' \wedge ⬗ q'' \to r)),$$

$$\Box^*((r \wedge ⬒ \bot \to \boxminus r) \wedge (r \wedge ◨ \bot \to \boxbar r)).$$

As before, it is readily checked that if r-square holds at the root of $\mathfrak{F}$ and $x \models r$ then *square*(x) is a r-square.

Now, we define χ_T to be the conjunction of Tiling, Diagonal, t-square, for all $t \in T$, prev-square, left-square, below-square, wall-square, floor-square and the formulas:

$$\Box^*(\mathsf{tile} \leftrightarrow p \wedge \neg ⬗ p), \tag{7.29}$$

$$\Box^*(\mathsf{prev} \leftrightarrow ⬗ \mathsf{tile} \wedge \neg ⬗ ⬗ \mathsf{tile}), \tag{7.30}$$

$$\Box^*(\mathsf{tile} \wedge \neg \mathsf{wall} \to ⬘ \mathsf{left}), \tag{7.31}$$

$$\Box^*(\mathsf{left} \to ⬗ \mathsf{tile}), \tag{7.32}$$

$$\Box^*(\mathsf{tile} \wedge \neg \mathsf{floor} \to ⬘ \mathsf{below}), \tag{7.33}$$

$$\Box^*(\mathsf{below} \to ⬗ \mathsf{tile}), \tag{7.34}$$

$$\Box^*(\mathsf{tile} \wedge \neg \mathsf{floor} \wedge \neg \mathsf{wall} \to ⬘(\mathsf{left} \wedge ⬘ \mathsf{below} \wedge \neg ⬘ ⬘ \mathsf{below})), \tag{7.35}$$

$$\Box^*(⬒ \bot \wedge ◨ \bot \leftrightarrow \mathsf{floor} \wedge \mathsf{wall}), \tag{7.36}$$

$$\Box^*((⬒ \bot \wedge ◨ \bot) \vee (\mathsf{tile} \wedge ⬘(\mathsf{prev} \wedge ⬗ \mathsf{wall})) \leftrightarrow \mathsf{floor}), \tag{7.37}$$

$$\Box^*(⬗ \top \to (\mathsf{tile} \wedge ⬘(\mathsf{below} \wedge ⬗ \mathsf{wall}) \leftrightarrow \mathsf{wall})), \tag{7.38}$$

$$\Box^*(\mathsf{tile} \wedge \neg \mathsf{floor} \to ⬒(\mathsf{below} \to ⬗ \mathsf{left} \wedge \neg ⬗ ⬗ \mathsf{left})), \tag{7.39}$$

$$\Box^*(\mathsf{tile} \wedge \mathsf{floor} \to \neg ⬘(⬗ \mathsf{floor} \wedge ⬘ \mathsf{left})), \tag{7.40}$$

$$\Box^*(\mathsf{floor} \to ⬒(\mathsf{left} \to ⬗ \mathsf{floor})). \tag{7.41}$$

Lemma 7.8. *T tiles $\mathbb{N} \times \mathbb{N}$ iff χ_T is satisfiable in a frame $\mathfrak{F} \in \mathcal{C}_1 \times \mathcal{C}_2$.*

Proof. ($\Rightarrow$) Take rooted Noetherian frames $\mathfrak{F}_1 \in \mathcal{C}_1$ and $\mathfrak{F}_2 \in \mathcal{C}_2$ having infinite descending chains of distinct points. Define a valuation $\mathfrak{V}$ of h, v and

p as in the proof of Lemma 7.6. Observe that under this valuation we have the following squares (m,n):

$$\{\langle x,y\rangle \mid x_1 \lneqq_1 x,\ y_1 \lneqq_2 y\} = (0,0),$$
$$\{\langle x,y\rangle \mid x_{n+2} \lneqq_1 x \leq_1 x_{n+1},\ y_1 \lneqq_2 y\} = (n+1,0),\ \text{ for } n<\omega,$$
$$\{\langle x,y\rangle \mid x_1 \lneqq_1 x,\ y_{m+2} \lneqq_2 y \leq_2 y_{m+1},\} = (0,m+1),\ \text{ for } m<\omega,$$
$$\{\langle x,y\rangle \mid x_{n+1} \lneqq_1 x \leq_1 x_n,\ y_{m+1} \lneqq_2 y \leq_2 y_m\} = (n,m),\ \text{ for } 0<n,m<\omega.$$

Suppose $\boldsymbol{\tau} : \mathbb{N}\times\mathbb{N} \to T$ is a tiling. Extend the valuation $\mathfrak{V}$ to the remaining propositional variables by taking

$$\begin{aligned}
\mathfrak{V}(\mathsf{tile}) &= \bigcup\{(n,n) \mid n<\omega\},\\
\mathfrak{V}(t) &= \bigcup\{(n,n) \mid n<\omega,\ \boldsymbol{\tau}(\mathit{pair}(n)) = t\},\ t\in T,\\
\mathfrak{V}(\mathsf{prev}) &= \bigcup\{(n,n+1) \mid n<\omega\},\\
\mathfrak{V}(\mathsf{left}) &= \bigcup\{(m,n) \mid m,n<\omega,\ m=\mathit{left}(n)\},\\
\mathfrak{V}(\mathsf{below}) &= \bigcup\{(m,n) \mid m,n<\omega,\ m=\mathit{below}(n)\},\\
\mathfrak{V}(\mathsf{wall}) &= \bigcup\{(n,n) \mid n<\omega,\ \mathit{pair}(n) \text{ is on the wall}\},\\
\mathfrak{V}(\mathsf{floor}) &= \bigcup\{(n,n) \mid n<\omega,\ \mathit{pair}(n) \text{ is on the floor}\}
\end{aligned}$$

(see Fig. 7.4). Let $\langle z_1, z_2\rangle$ be the 'limit' point in $\mathfrak{F}_1 \times \mathfrak{F}_2$ as in the proof of Lemma 7.6. It is a matter of routine to check that under the defined valuation χ_T holds at $\langle z_1, z_2\rangle$.

($\Leftarrow$) Suppose χ_T is true at some point x in $\mathfrak{F}$ under some valuation. Then $x \models \mathsf{Diagonal}$, and so we have squares (m,n) $(n,m<\omega)$ satisfying conditions (a)–(c) of Lemma 7.7.

By (7.29), tile holds in (n,n), for all $n<\omega$, and is false in (n,m) whenever $n\neq m$. And by (7.30), prev holds in (m,n) iff $n=m+1$. (Note that tile and prev may be true at some other points that do not belong to the depicted grid, but they are of no concern to us.)

Now by induction on n we show that

(i) $\mathit{pair}(n)$ is on the floor iff $(n,n)\models \mathsf{floor}$;

(ii) if $\mathit{pair}(n)$ is not on the floor then $(\mathit{below}(n),n)\models \mathsf{below}$;

(iii) $\mathit{pair}(n)$ is on the wall iff $(n,n)\models \mathsf{wall}$;

(iv) if $\mathit{pair}(n)$ is not on the wall then $(\mathit{left}(n),n)\models \mathsf{left}$.

By (7.36), $(0,0)\models \mathsf{floor}\wedge\mathsf{wall}$. By (7.37), $(1,1)\models\mathsf{floor}$ and, in view of (7.36), $(1,1)\models\neg\mathsf{wall}$. By (7.31), $(0,1)\models\mathsf{left}$. Now suppose $n>1$. Observe that by (7.32) and (7.34), if $(m,n)\models\mathsf{left}$ or $(m,n)\models\mathsf{below}$ then $m<n$.

Suppose $\mathit{pair}(n)$ is on the floor. Then $\mathit{pair}(n-1)$ is on the wall, and so by IH $(n-1,n-1)\models\mathsf{wall}$. It follows that $(n,n)\models \mathsf{tile}\wedge \blacklozenge(\mathsf{prev}\wedge \lozenge\mathsf{wall})$. By

(7.37) we then have $(n, n) \models \mathsf{floor}$. The converse implication also follows by IH from (7.37). This proves (i).

Assume now that $pair(n)$ is not on the floor, and so, by (i), $(n, n) \models \neg\mathsf{floor}$. By (7.33), $(k, n) \models \mathsf{below}$ for some $k < n$. By (7.39), we have $(k, n-1) \models \mathsf{left}$. Since by IH we have $(left(n-1), n-1) \models \mathsf{left}$, $k = left(n-1) = below(n)$ follows by left-square. This yields (ii).

Suppose $pair(n)$ is on the wall. Then by (i), $(n, n) \models \neg\mathsf{floor}$ and, by (ii), $(below(n), n) \models \mathsf{below}$. Since $below(n) < n$ and $pair(below(n))$ is also on the wall, by IH we have $(n, n) \models \mathsf{tile} \wedge \Diamond(\mathsf{below} \wedge \Diamond\mathsf{wall})$. And since $(n, n) \models \Diamond\top$, we obtain by (7.38) that $(n, n) \models \mathsf{wall}$. The converse implication follows from IH and (7.38). Thus we have (iii).

Finally, to prove (iv), suppose that $pair(n)$ is not on the wall. Then by (iii), $(n, n) \models \neg\mathsf{wall}$. By (7.31), $(k, n) \models \mathsf{left}$ for some $k < n$. If $pair(n)$ is not on the floor, then by (i), $(n, n) \models \neg\mathsf{floor}$. So by (7.35) and left-square, we have $(k-1, n) \models \mathsf{below}$. But then by (ii) and below-square, $k - 1 = below(n)$, from which $k = below(n) + 1 = left(n)$ as required. If $pair(n)$ is on the floor, then again by (i), $(n, n) \models \mathsf{floor}$. Since $left(n) < n$ and $pair(left(n))$ is also on the floor, by IH $left(n)$ is the largest $m < n$ such that $(m, m) \models \mathsf{floor}$. But then, by (7.40) and (7.41), we have $k = left(n)$.

Now define a map $\boldsymbol{\tau} : \mathbb{N} \times \mathbb{N} \to T$ by taking

$$\boldsymbol{\tau}(i, j) = t \quad \text{iff} \quad t \in T,\ pair(n) = \langle i, j \rangle \text{ and } (n, n) \models t.$$

It follows from left-square, below-square, (ii), (iv) and $x \models \mathsf{Tiling}$ that $\boldsymbol{\tau}$ is well-defined and a tiling of $\mathbb{N} \times \mathbb{N}$. ❑

Theorem 7.5 follows immediately from Lemma 7.8. ❑

Observe that as a consequence of Lemmas 7.6 and 7.7 above we also obtain:

Theorem 7.9. *Let $\mathcal{C}_1$ and $\mathcal{C}_2$ be classes of transitive and weakly connected frames such that each of them contains a rooted Noetherian linear order having an infinite descending chain of distinct points. Then* $\mathsf{Log}(\mathcal{C}_1 \times \mathcal{C}_2)$ *(and so* $\mathsf{Log}\mathcal{C}_1 \times \mathsf{Log}\mathcal{C}_2$*) does not have the product finite model property.*

Since all frames for the logics **GL.3** and **Grz.3** are transitive and weakly connected (see Section 1.2), and since the addition of a root to $\langle \mathbb{N}, > \rangle$ and to $\langle \mathbb{N}, \geq \rangle$ results in frames for **GL.3** and **Grz.3**, respectively, our theorems have the following corollaries:

Theorem 7.10. *The logics* **GL.3** $\times$ **GL.3**, **Grz.3** $\times$ **Grz.3**, **GL.3** $\times$ **Grz.3** *are undecidable and do not have the product finite model property.*

By Theorem 1.12, **GL.3** and **Grz.3** are also characterized by classes of finite frames. Similarly, it is not hard to show that both $\mathsf{Log}\{\langle \mathbb{N}, > \rangle \times \langle \mathbb{N}, > \rangle\}$

and $\mathsf{Log}\{\langle \mathbb{N}, \geq \rangle \times \langle \mathbb{N}, \geq \rangle\}$ are characterized by (recursive) classes of finite product frames, that is, they have the product fmp. Thus we obtain:

Theorem 7.11.
$$\mathbf{GL.3} \times \mathbf{GL.3} \neq \mathsf{Log}\{\langle \mathbb{N}, > \rangle \times \langle \mathbb{N}, > \rangle\},$$
$$\mathbf{Grz.3} \times \mathbf{Grz.3} \neq \mathsf{Log}\{\langle \mathbb{N}, \geq \rangle \times \langle \mathbb{N}, \geq \rangle\}.$$

We will see in Sections 7.3 and 7.4 that in fact all the logics mentioned in Theorem 7.11 are not recursively enumerable.

7.3 Products of Dedekind complete linear orders

Harel (1986) proved that the following problem is Σ_1^1-complete:

- Given a finite set T of tile types and a $t_0 \in T$, can T tile $\mathbb{N} \times \mathbb{N}$ in such a way that t_0 appears infinitely often in the first column?

We will use this result to show that logics of certain classes of products of linear frames are not recursively enumerable, and so not recursively axiomatizable.

Say that a transitive and weakly connected frame $\mathfrak{F} = \langle W, < \rangle$ is *Dedekind complete* if every bounded (with respect to $<$) subset $V \subseteq W$ has a *least upper bound* in $\mathfrak{F}$, i.e., the set

$$\{w \in W \mid \forall v \in V \; v < w\}$$

has a least element. It is not hard to see that both $\langle \mathbb{N}, \leq \rangle$ and $\langle \mathbb{R}, \leq \rangle$ are Dedekind complete, but $\langle \mathbb{Q}, \leq \rangle$ is not. Also, all Noetherian linear orders, like $\langle \mathbb{N}, > \rangle$ and $\langle \mathbb{N}, \geq \rangle$, are Dedekind complete.

Theorem 7.12. *Let $\mathcal{C}$ be a class of products of transitive and weakly connected frames satisfying the following conditions:*

min *there is a frame $\langle W_1, <_1 \rangle \times \langle W_2, <_2 \rangle \in \mathcal{C}$ with each $\langle W_i, <_i \rangle$, for $i = 1, 2$, containing an ascending ω-type chain;*

max *if $\langle W_1, <_1 \rangle \times \langle W_2, <_2 \rangle \in \mathcal{C}$, then $\langle W_1, <_1 \rangle$ is Dedekind complete.*

Then the satisfiability problem for $\mathcal{ML}_2$-formulas in $\mathcal{C}$ is Σ_1^1-hard, and so $\mathsf{Log}\,\mathcal{C}$ is not recursively enumerable.

Proof. Given a set T of tile types and a $t_0 \in T$, let ψ_T be the conjunction of φ_T defined in Section 7.1 and the following three formulas:

$$\Box^*(\mathsf{tile} \to \text{⬗}\text{⬘}(t_0 \wedge \mathsf{wall})), \tag{7.42}$$

$$\text{⬒}((\neg\text{⬗}\mathsf{tile} \wedge \neg\text{⬗}\text{⬘}\mathsf{tile}) \vee \text{⬗}\mathsf{tile} \vee \text{⬗}(\text{⬘}\mathsf{tile} \wedge \neg\mathsf{next} \wedge \neg\text{⬘}\mathsf{next})), \tag{7.43}$$

$$\text{◨}(\text{⬘}\mathsf{tile} \to \mathsf{next} \vee \text{⬘}\mathsf{next}). \tag{7.44}$$

We will show that ψ_T is satisfiable in $\mathcal{C}$ iff T tiles $\mathbb{N} \times \mathbb{N}$ with t_0 appearing infinitely often on the wall.

If we have a tiling with t_0 appearing infinitely often on the wall then it is clear that ψ_T has a model based on a frame in $\mathcal{C}$: just define a valuation as in the proof of Lemma 7.3 ($\Rightarrow$) in a product frame in $\mathcal{C}$ whose component frames have ascending ω-type chains in both dimensions (see Fig. 7.2). The three new conjuncts are obviously satisfied.

Conversely, if ψ_T has a model based on a frame $(W_1, <_1) \times (W_2, <_2)$ from $\mathcal{C}$ then this is also a model of φ_T. So in the same way as in the proof of Lemma 7.3 ($\Leftarrow$) we can construct a tiling of $\mathbb{N} \times \mathbb{N}$. (Recall that we use a sequence of tile-squares $(0,0), (1,1), \ldots$, that is, squares in which tile holds.)

We will show that the only points at which tile $\wedge$ wall holds are those in squares (i,i) for which *pair*(i) is on the wall. (7.42) then tells us that t_0 appears infinitely often as a tile on the wall in the tiling, which is precisely what we need. In fact it is sufficient to show that there are no tile-squares apart from (n,n), $n \in \mathbb{N}$. This is because in Section 7.1 we had shown that (n,n) is a wall-square iff *pair*(n) is on the wall (i.e., not a right-neighbor of any other *pair*(k)).

To this end, for each $n \in \mathbb{N}$, choose $x_n \in W_1$ and $y_n \in W_2$ such that $\langle x_n, y_n \rangle$ belongs to the square (n,n). If the ascending ω-type chain

$$x_0 <_1 x_1 <_1 \ldots$$

is unbounded in $\langle W_1, <_1 \rangle$ then we are done as tile-square, (7.2) and (7.6) can easily be used to show that there are no other tile-squares.

So consider the situation in which $x_0 <_1 x_1 <_1 \ldots$ is bounded in $\langle W_1, <_1 \rangle$. As $\langle W_1, <_1 \rangle$ is Dedekind complete, there is some least upper bound z to the sequence. We will show that 'from z on' there are no more tile-squares, i.e.,

$$\forall u \in W_2 \, \forall z' \in W_1 \, (z' \geq_1 z \to \langle z', u \rangle \not\models \mathsf{tile}).$$

By (7.43), there are three possibilities for the pair $\langle z, y_0 \rangle$:

case 1: We may have $\neg \Diamond_h \mathsf{tile} \wedge \neg \Diamond_h \Diamond_v \mathsf{tile}$ true at $\langle z, y_0 \rangle$. But then indeed tile is false from z on. Let us show that the other cases cannot occur.

case 2: $\langle z, y_0 \rangle$ makes $\Diamond_h \mathsf{tile}$ true. Then tile is true at $\langle z, u \rangle$ for some $u \geq_2 y_0$. By (7.44), next is true at $\langle z', u \rangle$ for some $z' \geq_1 x_0$. By (7.3) and tile-square, we have $z' <_1 z$. As z is the least upper bound of the sequence of x_ns, we know there is some x_n such that $z' <_1 x_n$. There are three cases depending on the ordering of u and y_n. We cannot have $u <_2 y_n$ by (7.2) and (7.6). We cannot have $u >_2 y_n$ by (7.6). And finally, we cannot have $u = y_n$ by tile-square.

case 3: $⬗(⬘\mathsf{tile} \wedge \neg\mathsf{next} \wedge \neg⬘\mathsf{next})$ is true at $\langle z, y_0\rangle$. This case is similar to case 2 and cannot happen.

So all the tile-squares lie before the least upper bound z. But then we can use (7.2) and (7.6) to show that there are no other tile-squares. ❑

Corollary 7.13. *None of the logics*

$$\mathsf{Log}\{\langle \mathbb{O}, <\rangle \times \langle \mathbb{O}', <\rangle\}, \qquad \mathsf{Log}\{\langle \mathbb{O}, \leq\rangle \times \langle \mathbb{O}', \leq\rangle\},$$

for $\mathbb{O} \in \{\mathbb{N}, \mathbb{Z}, \mathbb{R}\}$ *and* $\mathbb{O}' \in \{\mathbb{N}, \mathbb{Z}, \mathbb{R}, \mathbb{Q}\}$, *is recursively enumerable.*

It is not hard to see that all frames for $\mathsf{Log}\{\langle \mathbb{N}, <\rangle\}$ and $\mathsf{Log}\{\langle \mathbb{N}, \leq\rangle\}$ are Dedekind complete (see the axiomatizations in Section 2.1). So we also have the following:

Corollary 7.14. *Let* $L_1 \in \{\mathsf{Log}\{\langle \mathbb{N}, <\rangle\}, \mathsf{Log}\{\langle \mathbb{N}, \leq\rangle\}\}$ *and let* $L_2 = \mathsf{Log}\, \mathcal{C}$, *where* $\mathcal{C}$ *is a class of transitive and weakly connected frames at least one of which contains an ascending* ω*-type chain of distinct points (e.g.,* L_2 *is any logic mentioned in Theorem 7.2). Then* $L_1 \times L_2$ *is not recursively enumerable.*

By using an equally devious set of extra formulas we can also produce a similar theorem for classes of frames with infinite descending chains.

Theorem 7.15. *Let* $\mathcal{C}$ *be a class of products of transitive and weakly connected frames satisfying the following conditions:*

min *there is a frame* $\langle W_1, <_1\rangle \times \langle W_2, <_2\rangle \in \mathcal{C}$ *with each* $\langle W_i, <_i\rangle$, *for* $i = 1, 2$, *being rooted, Noetherian and containing an infinite descending chain;*

max *if* $\langle W_1, <_1\rangle \times \langle W_2, <_2\rangle \in \mathcal{C}$, *then both* $\langle W_1, >_1\rangle$ *and* $\langle W_2, >_2\rangle$ *are Dedekind complete.*

Then the satisfiability problem for $\mathcal{ML}_2$*-formulas in* $\mathcal{C}$ *is* Σ^1_1*-hard, and so* $\mathsf{Log}\, \mathcal{C}$ *is not recursively enumerable.*

Proof. As before, suppose that we have a finite set T of tile types and a $t_0 \in T$. We extend χ_T of Section 7.2 to ξ_T by choosing a new propositional variable r and adding the following conjuncts to χ_T:

$$\neg ⬗⬘\chi_T, \tag{7.45}$$

$$\Box^*(\neg p \to \neg r), \tag{7.46}$$

$$\Box^*(t_0 \wedge \mathsf{wall} \to r \wedge ⊟ r), \tag{7.47}$$

$$◫(⟠ r \to ⟠(t_0 \wedge \mathsf{wall})), \tag{7.48}$$

$$⬒⬗ r. \tag{7.49}$$

These are designed to ensure that:

- ξ_T can be true only at the 'Dedekind limit' of the diagonal squares (n, n), $n \in \mathbb{N}$, (7.45) (for χ_T holds at this limit),
- r is false above the diagonal squares (7.46),
- r is true
  - in every t_0-square which is a wall-square, and at the points which are to the right of such a square (7.47),
  - and only there (7.48),
- if one moves even a bit to the right from the square where ξ_T holds, one sees r above (7.49).

Thus, if ξ_T is satisfied in some square then r (and so t_0 on the wall) must be infinitely close to this square.

Condition **min** guarantees that if T tiles $\mathbb{N} \times \mathbb{N}$ so that t_0 appears infinitely often on the wall, then ξ_T is satisfiable in $\mathcal{C}$. ❑

It is not hard to see that all Noetherian frames are Dedekind complete (see the axiomatizations for **GL.3** and **Grz.3** in Section 1.2), so we have the following:

Corollary 7.16. **GL.3** × **GL.3**, **Grz.3** × **Grz.3**, *and* **GL.3** × **Grz.3** *are not recursively enumerable.*

7.4 Products of finite linear orders

This section proves the following theorem due to Reynolds and Zakharyaschev (2001):

Theorem 7.17. *If* $\mathcal{C}_1$ *and* $\mathcal{C}_2$ *are classes of finite (strict) linear orders both containing arbitrarily long (but finite) chains, then the logic* $\mathsf{Log}(\mathcal{C}_1 \times \mathcal{C}_2)$ *is undecidable.*

Proof. For simplicity we will assume here that both $\mathcal{C}_1$ and $\mathcal{C}_2$ contain only *strict* linear orders. If this is not the case, one can use variables h and v and the formulas (7.21)–(7.22) to partition frames into black and white squares, and then use the modal operators $\blacklozenge$ and $\lozenge$ as in the previous sections.

We are going to reduce the undecidable halting problem for Turing machines (see Section 5.4 for definitions and notation) to the satisfiability problem in $\mathcal{C}_1 \times \mathcal{C}_2$. Given a Turing machine $\boldsymbol{A}$, we construct a formula $\varphi_{\boldsymbol{A}}$ which is satisfiable in a frame from $\mathcal{C}_1 \times \mathcal{C}_2$ iff $\boldsymbol{A}$ comes to a stop having started from the configuration $\langle \pounds, \langle s_0, b \rangle, b, b, \ldots \rangle$.

The proof consists of two basic steps. First, using the enumeration depicted in Fig. 7.1, we generate a sequence of 'diagonal points' to represent (now a finite part of) the $\mathbb{N} \times \mathbb{N}$ grid. The formulas doing this job are similar to formulas (7.2)–(7.18) but much simpler, since now we are dealing with finite (and so discrete) linear orders:

$$\mathsf{tm} \wedge \Box^*(\mathsf{tm} \wedge \Diamond_v \top \to \Diamond_v \mathsf{next} \wedge \neg \Diamond_v \Diamond_v \mathsf{next}), \tag{7.50}$$

$$\Box^*(\mathsf{next} \wedge \Diamond_h \top \to \Diamond_h \mathsf{tm} \wedge \neg \Diamond_h \Diamond_h \mathsf{tm}), \tag{7.51}$$

$$\mathsf{floor} \wedge \mathsf{wall} \wedge \neg \Diamond_v \Diamond_h (\mathsf{floor} \wedge \mathsf{wall}), \tag{7.52}$$

$$\Box_v(\mathsf{next} \to \mathsf{right}), \tag{7.53}$$

$$\Box^*(\mathsf{right} \wedge \Diamond_v \top \to \Diamond_v \mathsf{above} \wedge \neg \Diamond_v \Diamond_v \mathsf{above} \wedge \neg \Diamond_h \mathsf{above}), \tag{7.54}$$

$$\Box^*(\neg \mathsf{wall} \wedge \mathsf{tm} \to \Box_v(\mathsf{above} \wedge \Diamond_h \top \to \Diamond_h \mathsf{right} \wedge \neg \Diamond_h \Diamond_h \mathsf{right})), \tag{7.55}$$

$$\Box^*(\mathsf{wall} \wedge \mathsf{tm} \to \Box_v(\mathsf{above} \wedge \Diamond_h \top \to \Diamond_h q \wedge \neg \Diamond_h \Diamond_h q \wedge \neg \Diamond_h \mathsf{right})), \tag{7.56}$$

$$\Box^*(q \wedge \Diamond_v \top \to \Diamond_v \mathsf{right} \wedge \neg \Diamond_v \Diamond_v \mathsf{right}), \tag{7.57}$$

$$\Box^*\big(\mathsf{wall} \to \Box_v(\mathsf{next} \to \Box_h(\mathsf{tm} \to \mathsf{floor}))\big), \tag{7.58}$$

$$\Box^*\big(\mathsf{wall} \to \Box_v(\mathsf{above} \to \Box_h(\mathsf{tm} \to \mathsf{wall}))\big), \tag{7.59}$$

$$\Box^*(\mathsf{above} \to \neg \Diamond_h \mathsf{floor}), \tag{7.60}$$

$$\Box^*(\mathsf{right} \to \neg \Diamond_h \mathsf{wall}), \tag{7.61}$$

$$\Box^* \neg (p \wedge (\Diamond_v p \vee \Diamond_h p)), \tag{7.62}$$

where p ranges over all the variables occurring in (7.50)–(7.61).

Next, we represent a run of the Turing machine $\boldsymbol{A}$ as a sequence of consecutive rows on the grid, each of which represents a configuration of $\boldsymbol{A}$. (Note that although our grid is finite, its rows are suitable for representing infinite configurations because any configuration in a computation of $\boldsymbol{A}$ starting from $\langle \pounds, \langle s_0, b\rangle, b, b, \dots \rangle$ can contain only finitely many symbols different from b.) This can be done by the conjunction of the following formulas, for all instructions $\delta(\alpha, \beta, \gamma) = \langle \alpha', \beta', \gamma' \rangle$ of $\boldsymbol{A}$:

$$\Box^*(\mathsf{tm} \leftrightarrow \bigvee_{x \in A'} p_x), \tag{7.63}$$

$$\Box^* \bigwedge_{\substack{x, x' \in A' \\ x \neq x'}} \neg (p_x \wedge p_{x'}), \tag{7.64}$$

$$p_\pounds \wedge \Diamond_v\big(\mathsf{right} \wedge \Diamond_h(p_{\langle s_0, b\rangle} \wedge \Diamond_v(\mathsf{right} \wedge \Diamond_h p_b))\big), \tag{7.65}$$

$$\Box^*(q_l \leftrightarrow \mathsf{tm} \wedge \Diamond_v(\mathsf{right} \wedge \Diamond_h q_s)), \tag{7.66}$$

$$\Box^*(q_s \leftrightarrow \bigvee_{\langle s,a\rangle \in S\times A} p_{\langle s,a\rangle}), \tag{7.67}$$

$$\Box^*\big(q_s \leftrightarrow \Diamond(\mathsf{right} \wedge \Diamond(\mathsf{tm} \wedge q_r))\big), \tag{7.68}$$

$$\Box^*\big[p_\alpha \wedge q_l \wedge \Diamond(\mathsf{right} \wedge \Diamond(p_\beta \wedge \Diamond(\mathsf{right} \wedge \Diamond p_\gamma))) \rightarrow \Diamond\big(\mathsf{above} \wedge \Diamond(p_{\alpha'} \wedge \Diamond(\mathsf{right} \wedge \Diamond(p_{\beta'} \wedge \Diamond(\mathsf{right} \wedge \Diamond p_{\gamma'})))))\big], \tag{7.69}$$

$$\Box^* \bigwedge_{a\in A\cup\{\pounds\}} \big(\neg q_l \wedge \neg q_s \wedge \neg q_r \wedge p_a \rightarrow \Box(\mathsf{above} \rightarrow \Box(\mathsf{tm} \rightarrow p_a))\big). \tag{7.70}$$

Define $\varphi_{\boldsymbol{A}}$ to be the conjunction of (7.50)–(7.70). Suppose that frames $\mathfrak{F}_1 \in \mathcal{C}_1$ and $\mathfrak{F}_2 \in \mathcal{C}_2$ are given, and $\varphi_{\boldsymbol{A}}$ is true at the root of the frame $\mathfrak{F}_1 \times \mathfrak{F}_2$ under some valuation. Then there are points $x_0, \dots, x_\ell$ in $\mathfrak{F}_1$ and $y_0, \dots, y_\ell$ in $\mathfrak{F}_2$, for $\ell < \omega$, such that $\langle x_i, y_i\rangle \models \mathsf{tm}$ for all $i \leq \ell$. It is not hard to show that, for $i, j \leq \ell$, we have

- $\langle x_i, y_j\rangle \models \mathsf{next}$ iff $j = i + 1$,
- $\langle x_i, y_j\rangle \models \mathsf{right}$ iff $right(i) = j$,
- $\langle x_i, y_j\rangle \models \mathsf{above}$ iff $above(i) = j$, and
- the variables wall and floor mark the wall and the floor of the grid, respectively.

Every point marked by tm is also marked by a propositional variable p_x, for some $x \in A'$. Configurations are represented by tuples

$$\langle\langle x_{i_0}, y_{i_0}\rangle, \dots, \langle x_{i_k}, y_{i_k}\rangle\rangle$$

such that $\langle x_{i_0}, y_{i_0}\rangle \models \mathsf{wall}$ and $\langle x_{i_j}, y_{i_{j+1}}\rangle \models \mathsf{right}$, for all $j < k$. We also have $\langle x_{i_j}, y_{i_j}\rangle \models q_s$ iff $\langle x_{i_j}, y_{i_j}\rangle \models p_{\langle s,a\rangle}$ for some $\langle s, a\rangle \in S \times A$, $\langle x_{i_{j-1}}, y_{i_{j-1}}\rangle \models q_l$ and $\langle x_{i_{j+1}}, y_{i_{j+1}}\rangle \models q_r$. The formula (7.65) says that $\boldsymbol{A}$ starts working on the empty tape. Finally, the formulas (7.69) and (7.70) describe the effect of the transition function δ: we move to the next row above and change the active cell and its left and right neighbors, leaving other cells intact.

With this explanation, it is not hard to show that $\varphi_{\boldsymbol{A}}$ is as required. We leave this to the reader. ❑

Corollary 7.18. $\mathsf{Log}\{\langle\mathbb{N}, >\rangle \times \langle\mathbb{N}, >\rangle\}$ *and* $\mathsf{Log}\{\langle\mathbb{N}, \geq\rangle \times \langle\mathbb{N}, \geq\rangle\}$ *are not recursively enumerable.*

Proof. As we already mentioned in Section 7.2, it is not hard to see that these logics are determined by recursive classes of finite frames. So if they were recursively enumerable, then they would be decidable, contrary to Theorem 7.17. ❑

7.5 More undecidable products

First we use a modification of the 'chess-board' technique of the previous sections to show undecidability of certain product logics, where one of the components is **K** enriched with either the universal modality or the common knowledge operator. The results below will be applied in Part IV to show undecidability and nonaxiomatizability of certain temporal epistemic and modal description logics.

Theorem 7.19. *Let $\mathcal{C}$ be a class of transitive and weakly connected frames such that at least one frame in $\mathcal{C}$ contains an ascending ω-type chain of distinct points. Then $\mathsf{Log}(\mathcal{C} \times \mathsf{Fr}\,\mathbf{K}_u)$ and $\mathsf{Log}\,\mathcal{C} \times \mathbf{K}_u$ are undecidable, and the logics $\mathsf{Log}(\mathcal{C} \times \mathsf{Fr}\,\mathbf{K}_1^{\mathcal{C}})$ and $\mathsf{Log}\,\mathcal{C} \times \mathbf{K}_1^{\mathcal{C}}$ are not even recursively enumerable.*

Proof. First we show the undecidability of $\mathsf{Log}(\mathcal{C} \times \mathsf{Fr}\,\mathbf{K}_u)$ and then explain how to modify the proof for $\mathsf{Log}(\mathcal{C} \times \mathsf{Fr}\,\mathbf{K}_1^{\mathcal{C}})$. The statements for product logics $\mathsf{Log}\,\mathcal{C} \times \mathbf{K}_u$ and $\mathsf{Log}\,\mathcal{C} \times \mathbf{K}_1^{\mathcal{C}}$ will clearly follow. As before, we use $\diamondminus$ and $\boxminus$ to denote the 'horizontal' modal operators. In the 'vertical' dimension, the **K**-modalities are denoted by $\diamondbar$ and $\boxbar$, and the universal box by $\boxed{\forall}$.

The undecidable problem we are going to reduce to the satisfiability problem for $\mathsf{Log}(\mathcal{C} \times \mathsf{Fr}\,\mathbf{K}_u)$ is the halting problem for Turing machines (see Section 5.4). Given a Turing machine $\boldsymbol{A}$, our aim is to construct a formula $\psi_{\boldsymbol{A}}$ which is satisfiable in a frame from $\mathcal{C} \times \mathsf{Fr}\,\mathbf{K}_u$ iff $\boldsymbol{A}$ does not come to a stop having started from the configuration $\langle \pounds, \langle s_0, b\rangle, b, b, \dots\rangle$.

Suppose that $\mathfrak{F} = \langle W, <\rangle$ is a frame in $\mathcal{C}$ and $\mathfrak{G} = \langle U, R, R_u\rangle$ is a frame for $\mathbf{K}_u$. Without loss of generality we may assume $\mathfrak{F}$ and $\mathfrak{G}$ to be rooted (i.e., $R_u = U \times U$). First, we generate consecutive disjoint 'slices' in $\mathfrak{F} \times \mathfrak{G}$ by generating consecutive disjoint 'intervals' in $\mathfrak{F}$. (We call a nonempty subset I of W an interval if, for all $u, v, w \in W$, whenever $u, v \in I$ and $u < w < v$ then $w \in I$.) This can be done using propositional variables h_0 and h_1 and the following formulas (which are similar to the conjuncts of $\mathsf{Chessboard}$ in Section 7.1, cf. also (Spaan 1993)):

$$\boxminus^+(h_0 \to \neg h_1),$$

$$h_0 \wedge \boxminus^+((h_0 \to \diamondminus h_1) \wedge (h_1 \to \diamondminus h_0)),$$

$$\boxminus^+(\diamondminus h_0 \to h_0 \vee h_1),$$

$$\boxminus^+\big((h_0 \to \boxed{\forall} h_0) \wedge (\neg h_0 \to \boxed{\forall}\neg h_0)\big),$$

$$\boxminus^+\big((h_1 \to \boxed{\forall} h_1) \wedge (\neg h_1 \to \boxed{\forall}\neg h_1)\big).$$

The conjunction of these formulas will be denoted by Scale. Say that a point x in $\mathfrak{F}$ (under some valuation in $\mathfrak{F} \times \mathfrak{G}$) is *white* (or *black*) if $\langle x, y\rangle \models h_0$ (respectively, $\langle x, y\rangle \models h_1$) for all y in $\mathfrak{G}$. A point that is neither black nor white will be called a *cloud point*. If Scale is true at a point $\langle x_0, y_0\rangle$ in $\mathfrak{F} \times \mathfrak{G}$ under some valuation, with x_0 being the root of $\mathfrak{F}$, then the noncloud part of $\mathfrak{F}$ can be viewed as a scale that is either infinite or finite 'circular:' this part of $\mathfrak{F}$ is divided into intervals in such a way that x_0 belongs to a white one and every interval has (not necessarily immediate) successors of different color.

For intervals I_1 and I_2, we write

$$I_1 < I_2 \quad \text{iff} \quad \forall x \in I_1\, \exists y \in I_2\ (x \neq y \text{ and } x < y).$$

As before, we can define a new possibility operator $\blacklozenge$ by taking, for any formula ψ,

$$\blacklozenge\psi = \big(h_0 \to \diamondminus(h_1 \wedge \diamondminus^+\psi)\big) \wedge \big(h_1 \to \diamondminus(h_0 \wedge \diamondminus^+\psi)\big);$$

$\blacksquare$ is the dual of $\blacklozenge$. For any noncloud point x in $\mathfrak{F}$, let *interval*(x) denote the interval containing x. It should be clear that, for every y in $\mathfrak{G}$, we have:

$$\langle x, y\rangle \models \blacklozenge\psi \quad \text{iff} \quad \exists w\ \big(\mathit{interval}(x) < \mathit{interval}(w) \text{ and } \langle w, y\rangle \models \psi\big).$$

Note that $<$ is not necessarily irreflexive on intervals either: if I is a $<$-cluster having at least two elements, then $I < I$ holds.

However, as before, we can force certain intervals to be irreflexive. Given propositional variables p and q, define a formula $\mathsf{next}(p, q)$ as the conjunction of the following formulas:

$$\begin{gathered}
\boxminus^+(p \vee q \to (h_0 \vee h_1)),\\
\boxminus^+(p \to \neg\blacklozenge p),\\
\boxminus^+(\diamondminus p \wedge \neg\blacklozenge p \to p),\\
\boxminus^+(p \to \blacklozenge q \wedge \neg\blacklozenge\blacklozenge q),\\
\boxminus^+(q \to \neg\blacklozenge q),\\
\boxminus^+(p \to \boxminus(\neg q \wedge \blacklozenge q \to p))
\end{gathered}$$

(cf. the p-square formulas in Section 7.1). It is easy to see that if

$$\langle x_0, y\rangle \models \diamondminus^+ p \wedge \mathsf{next}(p, q)$$

then there are points x and $w > x$ such that $\langle x', y\rangle \models p$ for all $x' \in \mathit{interval}(x)$, $\langle w, y\rangle \models q$, $\mathit{interval}(x) < \mathit{interval}(w)$, both $\mathit{interval}(x)$ and $\mathit{interval}(w)$ are irreflexive, and for all u with $x < u < w$ we have either $u \in \mathit{interval}(x)$ or $u \in \mathit{interval}(w)$. It follows in particular that if

$$\langle x_0, y\rangle \models \diamondminus^+ q_0 \wedge \mathsf{next}(q_0, q_1) \wedge \mathsf{next}(q_1, q_2) \wedge \cdots \wedge \mathsf{next}(q_{n-1}, q_n)$$

then there exist n consecutive irreflexive intervals $I_0, \dots, I_{n-1}$ such that we have $\langle w, y\rangle \models q_j$ for all $w \in I_j$, $j < n$, and $\langle u, y\rangle \models q_n$ for some point u such that $u \notin I_{n-1}$ and $u > w$ for all $w \in I_{n-1}$.

Now we can encode runs of the Turing machine $\boldsymbol{A}$ as consecutive rows of $\mathfrak{F} \times \mathfrak{G}$, each of which represents a configuration. Consider the following formulas, for all instructions $\delta(\alpha, \beta, \gamma) = \langle \alpha', \beta', \gamma' \rangle$ of $\boldsymbol{A}$ (here d and d' are dummy variables):

$$\boxed{\forall}\boxminus^+ \bigwedge_{\substack{x,x' \in A' \\ x \neq x'}} \neg(p_x \wedge p_{x'}), \tag{7.71}$$

$$p_{\pounds} \wedge \mathsf{next}(p_{\pounds}, p_{\langle s_0, b\rangle}) \wedge \mathsf{next}(p_{\langle s_0, b\rangle}, d) \wedge \boxminus^+(p_{\pounds} \vee p_{\langle s_0, b\rangle} \vee p_b), \tag{7.72}$$

$$\boxed{\forall}\boxminus^+(q_s \leftrightarrow \bigvee_{\langle s,a\rangle \in S \times A} p_{\langle s,a\rangle}), \tag{7.73}$$

$$\boxed{\forall}(\diamondminus^+ q_l \wedge \mathsf{next}(q_l, q_s) \wedge \mathsf{next}(q_s, q_r) \wedge \mathsf{next}(q_r, d')), \tag{7.74}$$

$$\boxed{\forall}\Big(\diamondminus^+(q_l \wedge p_\alpha) \wedge \diamondminus(q_s \wedge p_\beta) \wedge \diamondminus(q_r \wedge p_\gamma) \rightarrow \diamondbar\top \wedge \\ \boxminus^+((q_l \rightarrow \boxbar p_{\alpha'}) \wedge (q_s \rightarrow \boxbar p_{\beta'}) \wedge (q_r \rightarrow \boxbar p_{\gamma'}))\Big), \tag{7.75}$$

$$\boxed{\forall}\boxminus^+ \bigwedge_{a \in A \cup \{\pounds\}} (\neg q_l \wedge \neg q_s \wedge \neg q_r \wedge p_a \rightarrow \boxbar p_a), \tag{7.76}$$

$$\boxed{\forall}\neg\diamondminus \bigvee_{a \in A} p_{\langle s_1, a\rangle}. \tag{7.77}$$

Let $\psi_{\boldsymbol{A}}$ be the conjunction of Scale and (7.71)–(7.77). We leave it to the reader to check that $\psi_{\boldsymbol{A}}$ is satisfied in a model based on a frame for $\mathsf{Log}(\mathcal{C} \times \mathsf{Fr}\,\mathbf{K}_u)$ iff $\boldsymbol{A}$ has an infinite computation which starts from the empty tape.

That $\mathsf{Log}(\mathcal{C} \times \mathsf{Fr}\,\mathbf{K}_1^C)$ is not recursively enumerable can be proved as follows. Replace all occurrences of $\boxed{\forall}$ in $\psi_{\boldsymbol{A}}$ by $\mathsf{C}_{\{1\}}$, and add the conjunct

$$\mathsf{C}_{\{1\}}\neg\mathsf{C}_{\{1\}}\neg\diamondminus \bigvee_{a \in A} p_{\langle s_0, a\rangle}.$$

It is not hard to see that the resulting formula $\chi_{\boldsymbol{A}}$ is satisfied in a model based on a frame for $\mathsf{Log}(\mathcal{C} \times \mathsf{Fr}\,\mathbf{K}_1^C)$ iff $\boldsymbol{A}$ is recurrent (see Section 5.4). ❑

As a consequence of the reductions of Chapter 6 (see Table 7.1) we obtain:

Theorem 7.20. *Suppose* $L \in \{$**PDL**, **CPDL**, $\mathbf{K}_1^C$, $\mathbf{T}_2^C$, $\mathbf{K4}_2^C$, $\mathbf{S4}_2^C$, $\mathbf{KD45}_2^C$, $\mathbf{S5}_2^C\}$. *Then* $L \times \mathbf{K}_u$ *and* $L \times \mathbf{K}_1^C$ *are undecidable.*

Proof. By Theorem 7.19, we know that $\mathbf{PTL} \times \mathbf{K}_u$ is undecidable. The undecidability of the remaining product logics of the form $L \times \mathbf{K}_u$—save

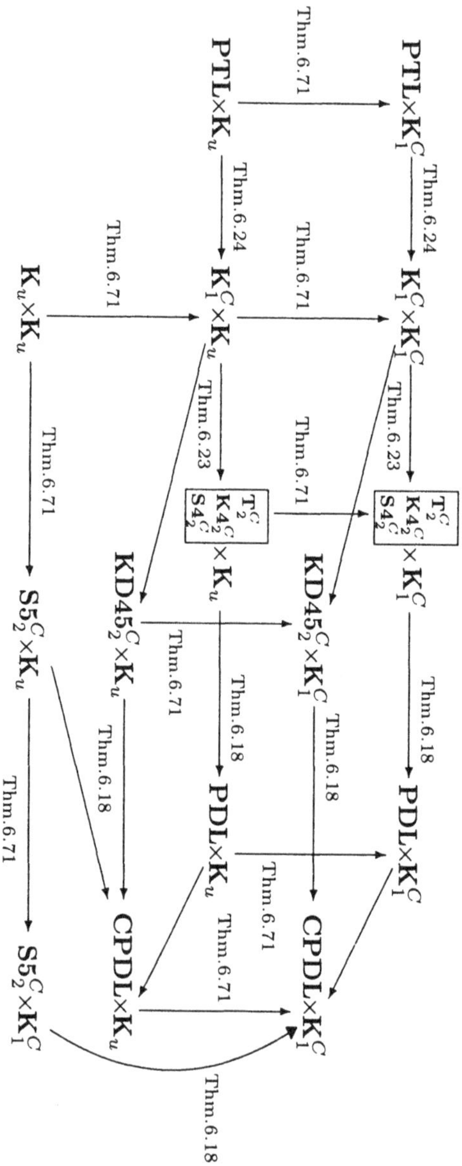

Table 7.1: Reductions between undecidable product logics.

$\mathbf{S5}_2^C \times \mathbf{K}_u$—now follows from the reductions of Table 7.1. Finally, $\mathbf{S5}_2^C \times \mathbf{K}_u$ is undecidable, because $\mathbf{K}_u \times \mathbf{K}_u$ is undecidable (Theorem 5.37) and polynomially reducible to $\mathbf{S5}_2^C \times \mathbf{K}_u$ (Table 7.1).

For any L listed in the theorem, $L \times \mathbf{K}_u$ is polynomially reducible to $L \times \mathbf{K}_1^C$, see Theorem 6.71 (1). Hence, $L \times \mathbf{K}_1^C$ is undecidable. ❑

It seems that neither the undecidability proofs of this chapter nor the method of quasimodels developed in Chapter 6 can be applied to 'nontrivial' products of unimodal logics whose both components are transitive and at least one of them is not necessarily linear (or of a fixed finite width); see Table 7.2. In fact, the only known result concerning this kind of logic is Theorem 7.24 below.

In particular, the following challenging problems are open:

Question 7.21. Are any of the products $\mathbf{K4.3} \times \mathbf{K4}$, $\mathbf{K4} \times \mathbf{K4}$, $\mathbf{S4} \times \mathbf{S4}$ decidable?

It is worth noting that $\mathbf{K4.3} \times \mathbf{K4.3}$ is known to be undecidable (see Theorem 7.2). Since **K4.3** is in coNP-complete and **K4** is PSPACE-complete, **K4.3** is polynomially reducible to **K4**. However, without knowing how such a reduction works, it is not clear how to 'lift' it to the product level, and deduce, for instance, the undecidability of $\mathbf{K4.3} \times \mathbf{K4}$ from the undecidability of $\mathbf{K4.3} \times \mathbf{K4.3}$ (cf. Remark 6.19).

As concerns the fmp of these logics, by Theorem 5.32 we know that $L_1 \times L_2$ does not have the product fmp, whenever L_1 and L_2 are any logics from the list

$$\mathbf{K4},\ \mathbf{S4},\ \mathbf{K4.3},\ \mathbf{S4.3},\ \mathsf{Log}\{\langle \mathbb{O}, <\rangle\},\ \mathsf{Log}\{\langle \mathbb{O}, \leq\rangle\},\ \text{for } \mathbb{O} \in \{\mathbb{N}, \mathbb{Z}, \mathbb{Q}\}. \quad (7.78)$$

However, these logics may still enjoy the (abstract) finite model property:

Question 7.22. Do any of the logics in (7.78) have the fmp?

A positive answer to this question for $\mathbf{K4} \times \mathbf{K4}$ and $\mathbf{S4} \times \mathbf{S4}$ would also solve affirmatively the corresponding parts of Question 7.21, since both these logics are finitely axiomatizable by Corollary 5.10. Note that even solutions to the following problems are not known:

Question 7.23. Are any of the logics $(\mathbf{K4} \otimes \mathbf{K4}) \oplus (\Box_1\Box_2 p \leftrightarrow \Box_2\Box_1 p)$ and $(\mathbf{S4} \otimes \mathbf{S4}) \oplus (\Box_1\Box_2 p \leftrightarrow \Box_2\Box_1 p)$ decidable? Do they have the fmp?

Note that, by Theorem 5.40, we do know that $\mathbf{PTL} \times \mathbf{K4}$ is undecidable. Here we prove the following generalization of this result:

Theorem 7.24. $\mathsf{Log}\{\langle \mathbb{N}, <\rangle\} \times \mathbf{K4}$ *and* $\mathsf{Log}\{\langle \mathbb{N}, <\rangle\} \times \mathbf{S4}$ *are undecidable.*

Proof. First we show that $\mathsf{Log}\{\langle\mathbb{N},<\rangle\} \times \mathbf{K4}$ is undecidable by modifying the proof of Theorem 5.40. Throughout, we will use the notation of that proof.

Given a finite alphabet A and a set $P = \{\langle v_1, w_1\rangle, \dots, \langle v_k, w_k\rangle\}$ of pairs of words over A, we construct a formula $\psi_{A,P}$ (in the language with $\boxminus$ and $\boxbar$) which is $\mathsf{Log}\{\langle\mathbb{N},<\rangle\} \times \mathbf{K4}$-satisfiable iff there exist an $N \geq 1$ and a sequence $i_1, \dots, i_N$ of indices such that

$$v_{i_1} * \dots * v_{i_N} = w_{i_1} * \dots * w_{i_N}. \tag{7.79}$$

Let the formula ψ_{left} be φ_{left} as defined in the proof of Theorem 5.40, but (5.39) replaced by

$$\boxminus^+\big(\mathsf{pair}_i \to \boxbar^+(\neg\mathsf{left} \to \diamondminus\boxbar^{l_i}\neg\mathsf{left})\big), \tag{7.80}$$

(5.40) replaced by

$$\boxminus^+\Big(\mathsf{pair}_i \to \boxbar^+\big(\neg\mathsf{left} \land \boxminus\mathsf{left} \to \boxminus(\diamondbar^j\mathsf{left} \land \neg\diamondbar^{j+1}\mathsf{left} \to \mathsf{left}_{b^i_{l_i-j}})\big)\Big), \tag{7.81}$$

(5.41) replaced by

$$\mathsf{pair}_i \to \boxminus\Big(\mathsf{left}_{b^i_1} \land \diamondbar\big(\mathsf{left}_{b^i_2} \land \diamondbar(\mathsf{left}_{b^i_3} \land \dots \land \diamondbar\mathsf{left}_{b^i_{l_i}})\dots\big)\Big), \tag{7.82}$$

and (5.42) replaced by

$$\boxminus\Big(\mathsf{pair}_i \to \boxbar^+\big(\mathsf{left} \land \boxbar\neg\mathsf{left} \to$$
$$\boxminus\diamondbar(\mathsf{left}_{b^i_1} \land \diamondbar(\mathsf{left}_{b^i_2} \land \dots \land \diamondbar\mathsf{left}_{b^i_{l_i}})\dots)\big)\Big). \tag{7.83}$$

Define ψ_{right} from φ_{right} in a similar way, and let

$$\psi_{A,P} = \varphi_1 \land \varphi_2 \land \psi_{left} \land \psi_{right},$$

where φ_1 and φ_2 are as in the proof of Theorem 5.40.

We show that $\psi_{A,P}$ is as required. First, if there exist an $N \geq 1$ and a sequence $i_1, \dots, i_N$ of indices such that (7.79) holds, then $\psi_{A,P}$ is satisfied in the same model based on $\langle\mathbb{N},<\rangle \times \langle\mathbb{N},<\rangle$ as in the proof of Theorem 5.40. Conversely, suppose that $\psi_{A,P}$ is $\mathsf{Log}\{\langle\mathbb{N},<\rangle\} \times \mathbf{K4}$-satisfiable. By Theorem 6.29, we may assume that

$$(\mathfrak{M}, \langle 0, y_0\rangle) \models \psi_{A,P}$$

for a model $\mathfrak{M}$ based on the product of $\langle\mathbb{N},<\rangle$ and a frame $\langle V, S\rangle$ for $\mathbf{K4}$. By φ_2, we can find an N, $1 \leq N < \omega$, such that

$$(\mathfrak{M}, \langle N, y_0\rangle) \models \boxbar^+ \bigwedge_{a \in A}(\mathsf{left}_a \leftrightarrow \mathsf{right}_a).$$

Let $i_1, \dots, i_N$ be the sequence of indices such that, for $1 \le j \le N$, we have $(\mathfrak{M}, \langle j-1, y_0 \rangle) \models \mathsf{pair}_{i_j}$ (by φ_1 we have such a sequence and it is unique). Now one can almost repeat the proof of Theorem 5.40 with the points $0, 1, 2, \dots$ in place of $x_0, x_1, x_2, \dots$. The only difference is that the (inductive) proofs of statements (i)–(iii) and (i)′–(iii)′ are a bit more complex. We show how to prove (i)–(iii).

For $j = 1$, we have (i) by $(\mathfrak{M}, \langle 0, y_0 \rangle) \models \mathsf{pair}_{i_1}$ and (7.82), (ii) by (5.38), (7.80) and (5.37), and (iii) by (7.81) and again (5.37). Now assume inductively that (i)–(iii) hold for some $1 \le j < N$. Let $\langle y_0, \dots, y_{n_j - 1} \rangle$ be a maximal S-path in $\mathfrak{V}_j(\mathsf{left})$. First, we have $y_0, \dots, y_{n_j-1} \in \mathfrak{V}_{j+1}(\mathsf{left})$ by (5.37). Second, $(\mathfrak{M}, \langle j, y_{n_j - 1} \rangle) \models \mathsf{left} \wedge \Box\neg\mathsf{left}$ and $(\mathfrak{M}, \langle j, y_0 \rangle) \models \mathsf{pair}_{i_{j+1}}$, so (7.83) now implies that there exist $y_{n_j}, \dots, y_{n_j + l_{i_{j+1}} - 1}$ such that $\left\langle y_0, \dots, y_{n_j + l_{i_{j+1}} - 1} \right\rangle$ is an S-path in $\mathfrak{V}_{j+1}(\mathsf{left})$, as required in (i). For (ii) and (iii), observe first that for any S-path $\langle y_0, \dots, y_{l-1} \rangle$ in $\mathfrak{V}_{j+1}(\mathsf{left})$, $\left\langle y_0, \dots, y_{l - l_{i_{j+1}} - 1} \right\rangle$ is an S-path in $\mathfrak{V}_j(\mathsf{left})$, by (7.80) and (5.37). So $l \le n_{j+1}$ must hold. If $l = n_{j+1}$ then $\mathit{leftword}_j(y_0, \dots, y_{l - l_{i_{j+1}} - 1}) = v_{i_1} * \dots * v_{i_j}$ by the induction hypothesis, so $\mathit{leftword}_{j+1}(y_0, \dots, y_{l - l_{i_{j+1}} - 1}) = v_{i_1} * \dots * v_{i_j}$ by (5.37). On the other hand, by the induction hypothesis and (5.37), we have

$$(\mathfrak{M}, \langle j, y_{l - l_{i_{j+1}}} \rangle) \models \neg\mathsf{left} \wedge \boxminus\mathsf{left}.$$

Now (7.81) implies that $\mathit{leftword}_{j+1}(y_{l - l_{i_{j+1}}}, \dots, y_{l-1}) = v_{i_{j+1}}$, so

$$\mathit{leftword}_{j+1}(y_0, \dots, y_{l-1}) = v_{i_1} * \dots * v_{i_{j+1}},$$

as required.

To prove the undecidability of $\mathsf{Log}\{\langle \mathbb{N}, < \rangle\} \times \mathbf{S4}$, we need to modify the formula $\psi_{A,P}$ constructed above. We apply a trick similar (but simpler) to the one in Sections 7.1–7.4: we use an extra variable s to imitate the **K4**-modalities on **S4**-frames.

Let

$$\varphi_0 = \Box\big((s \to \boxminus s) \wedge (\neg s \to \boxminus\neg s)\big).$$

Introduce a 'strict' possibility operator $\blacklozenge$ by taking, for any formula χ,

$$\blacklozenge\chi = \big(s \to \Diamond(\neg s \wedge \Diamond\chi)\big) \wedge \big(\neg s \to \Diamond(s \wedge \Diamond\chi)\big),$$

and let $\blacksquare$ be the dual of $\blacklozenge$. Now, the formula $\sigma_{A,P}$ is obtained from $\psi_{A,P}$ above by replacing each occurrence of $\Box$ or $\Diamond$ with $\blacksquare$ or $\blacklozenge$, respectively, and taking the conjunct of the resulting formula and φ_0.

We show that $\sigma_{A,P}$ is as required. Suppose first that there is a sequence of indices $i_1, \dots, i_N$, $N \ge 1$, such that (7.79) holds. Then $\sigma_{A,P}$ is satisfiable in the product frame $\langle \mathbb{N}, < \rangle \times \langle \mathbb{N}, \leq \rangle$. Indeed, define a valuation $\mathfrak{V}$ in this frame as in the proof of Theorem 5.40 and extend it to s by taking

$$\mathfrak{V}(s) = \{\langle n, 2m \rangle \mid n, m \in \mathbb{N}\}.$$

One can readily check that under this valuation we have $\langle 0, 0 \rangle \models \sigma_{A,P}$.

Conversely, suppose that $(\mathfrak{M}, \langle 0, y_0 \rangle) \models \sigma_{A,P}$ for some model $\mathfrak{M}$ based on a product frame $\langle \mathbb{N}, < \rangle \times \langle V, R \rangle$, where R is a reflexive and transitive relation on V. Define a new relation S on V by taking, for all $x, y \in V$, xSy iff one of the following conditions hold:

- $(\mathfrak{M}, \langle 0, x \rangle) \models s$ and there is a z in V such that $(\mathfrak{M}\, \langle 0, z \rangle) \models \neg s$ and $xRzRy$, or
- $(\mathfrak{M}, \langle 0, x \rangle) \models \neg s$ and there is a z in V such that $(\mathfrak{M}\, \langle 0, z \rangle) \models s$ and $xRzRy$.

Clearly, S is a transitive relation and (since φ_0 is a conjunct of $\sigma_{A,P}$), the operator $\Diamond$ is nothing but the modal operator interpreted by the 'vertical' relation of the product frame $\langle \mathbb{N}, < \rangle \times \langle V, S \rangle$. Now one can repeat the above proof given for $\psi_{A,P}$. ❑

Now by the reductions in Theorems 6.18, 6.23 and 6.24 we obtain:

Theorem 7.25. *Suppose* $L \in \{$**PTL**, **PDL**, **CPDL**, $\mathbf{K}_1^C$, $\mathbf{T}_2^C$, $\mathbf{K4}_2^C$, $\mathbf{S4}_2^C$, $\mathbf{KD45}_2^C\}$. *Then* $L \times$ **K4** *and* $L \times$ **S4** *are undecidable.*

	finitely axiomatizable	decidable	has fmp	has product fmp
$\mathbf{K4} \times \mathbf{K4}$	yes (Thm. 5.9)	?	?	no (Thm. 5.32)
$\mathbf{S4} \times \mathbf{S4}$	yes (Thm. 5.9)	?	?	no (Thm. 5.32)
$\mathsf{Log}\{\langle \mathbb{N}, < \rangle\} \times \mathbf{S4}$	?	no (Thm. 7.24)	?	no (Thm. 5.32)
$\mathsf{Log}\{\langle \mathbb{N}, < \rangle\} \times \mathsf{Log}\{\langle \mathbb{N}, < \rangle\}$	no (not r.e.) (Cor. 7.13)	no	?	no (Thm. 5.32)
$\mathbf{K4.3} \times \mathbf{K4.3}$	? r.e. (Thm. 3.17)	no (Thm. 7.2)	?	no (Thm. 5.32)
$\mathbf{K4.3} \times \mathbf{K4}$	? r.e. (Thm. 3.17)	?	?	no (Thm. 5.32)

Table 7.2: Products of transitive unimodal logics.

Chapter 8

Higher-dimensional products

As we saw in the previous three chapters, the computational complexity of two-dimensional product logics may grow dramatically as compared with the complexity of their components. This suggests that we can hardly expect 'decent' computational behavior from higher-dimensional products. The main aim of this chapter is to show that actually *no* n-modal logic between $\mathbf{K}^n$ and $\mathbf{S5}^n$ is decidable if $n \geq 3$ and that *no* logic in this interval is finitely axiomatizable. Examples of finitely axiomatizable and decidable higher-dimensional product logics are given in Section 8.5. For the reader's convenience Table 8.1 summarizes the properties of higher-dimensional products of some standard logics.

Let us begin by recalling the basic definitions of Section 3.3. The (*n-dimensional*) *product* of Kripke frames $\mathfrak{F}_1 = \langle W_1, R_1 \rangle, \dots, \mathfrak{F}_n = \langle W_n, R_n \rangle$ is the n-frame

$$\mathfrak{F}_1 \times \cdots \times \mathfrak{F}_n = \langle W_1 \times \cdots \times W_n, \bar{R}_1, \dots, \bar{R}_n \rangle,$$

where, for each $i = 1, \dots, n$, $\bar{R}_i$ is a binary relation on $W_1 \times \cdots \times W_n$ such that

$$\langle u_1, \dots, u_n \rangle \bar{R}_i \langle v_1, \dots, v_n \rangle \quad \text{iff} \quad u_i R_i v_i \ \text{ and } \ u_k = v_k \ \text{ for } k \neq i.$$

The (*n-dimensional*) *product* of Kripke complete modal logics L_i, $i = 1, \dots, n$, is the n-modal logic

$$L_1 \times \cdots \times L_n = \mathsf{Log}\{\mathfrak{F}_1 \times \cdots \times \mathfrak{F}_n \mid \mathfrak{F}_i \in \mathsf{Fr} L_i,\ i = 1, \dots, n\}.$$

	finitely axiomatizable	has fmp	has product fmp	decidable
$\mathbf{S5}^n$ $(n \geq 3)$	no (Thm. 8.2)	no (Thm. 8.12)	no	no (Thm. 8.6)
$\mathbf{K4}^n$ $(n \geq 3)$	no (Thm. 8.30)	?	no (Thm. 8.31)	no (Thm. 8.28)
$\mathbf{K}^n$ $(n \geq 3)$	no (Thm. 8.30)	yes (Thm. 8.24)	no (Thm. 8.31)	no (Thm. 8.28)
$\mathbf{Alt}^n$	yes (Thm. 8.46)	yes	yes (Thm. 8.52)	yes coNP-complete (Thm. 8.53)

Table 8.1: Some higher-dimensional product logics.

Similarly to the two-dimensional case, any two coordinates of a product frame satisfy the property of left and right commutativity, as well as the Church–Rosser property. To put it another way, all product frames of any dimension $n \geq 2$ validate the formulas:

$$\begin{aligned} \boldsymbol{com}^l_{ij} &= \Diamond_j\Diamond_i p \to \Diamond_i\Diamond_j p, \\ \boldsymbol{com}^r_{ij} &= \Diamond_i\Diamond_j p \to \Diamond_j\Diamond_i p, \\ \boldsymbol{chr}_{ij} &= \Diamond_i\Box_j p \to \Box_j\Diamond_i p \qquad (i,j = 1,\dots,n,\ i \neq j). \end{aligned}$$

As before, the logic that results by extending the fusion of the L_i with these axioms will be denoted by $[L_1,\dots,L_n]$ and called the *commutator* of $L_1,\dots,L_n$. In other words,

$$[L_1,\dots,L_n] = (L_1 \otimes \dots \otimes L_n) \oplus \bigoplus_{\substack{1\leq i,j\leq n \\ i\neq j}} (\boldsymbol{com}_{ij} \oplus \boldsymbol{chr}_{ij}),$$

where $\boldsymbol{com}_{ij} = \boldsymbol{com}^l_{ij} \wedge \boldsymbol{com}^r_{ij}$. By Proposition 3.13, we always have

$$[L_1,\dots,L_n] \subseteq L_1 \times \dots \times L_n.$$

8.1 S5 × S5 × ⋯ × S5

We begin our investigation of higher-dimensional products by considering the products of **S5**. These logics—their algebraic counterparts, to be more precise—have been thoroughly studied in *algebraic logic* (Henkin *et al.* 1971, 1985, Andréka *et al.* 2000). As we know from Section 1.5, every n-frame $\mathfrak{F} = \langle W, R_1, \ldots, R_n \rangle$ gives rise to the n-modal algebra $\mathfrak{F}^+$ of all subsets in W, where for every $X \subseteq W$ and every $i = 1, \ldots, n$,

$$\Diamond_i^{\mathfrak{F}^+} X = \{w \in W \mid \exists u \in X \; wR_i u\}.$$

Thus, elements of the algebraic dual $\mathfrak{F}^+$ of a universal product $\mathbf{S5}^n$-frame $\mathfrak{F}$ are all subsets of some Cartesian product $W_1 \times \cdots \times W_n$ and, for each such subset X and each $i = 1, \ldots, n$,

$$\Diamond_i^{\mathfrak{F}^+} X = \{\langle w_1, \ldots, w_n \rangle \mid \exists u \in W_i \; \langle w_1, \ldots, w_{i-1}, u, w_{i+1}, \ldots, w_n \rangle \in X\}.$$

In algebraic logic, these kinds of algebras are called *full diagonal-free cylindric set algebras of dimension n*.[1] By Proposition 3.11, these algebras generate the variety (equational class) $\mathsf{Alg}\mathbf{S5}^n$ of n-modal algebras for $\mathbf{S5}^n$ which is known in the algebraic logic literature as the variety RDf_n of *representable diagonal-free cylindric algebras of dimension n*. The class $\mathsf{Alg}[\mathbf{S5}, \mathbf{S5}, \ldots, \mathbf{S5}]$ is known in algebraic logic as the class Df_n of *diagonal-free cylindric algebras of dimension n*; see (Henkin *et al.* 1985).

Axiomatization

As we know, $[\mathbf{S5}, \mathbf{S5}, \ldots, \mathbf{S5}] \subseteq \mathbf{S5}^n$. However, unlike the 2D case, now this inclusion is proper. To show this, we note first that all n-dimensional product frames satisfy the following 'cubifying' properties whenever $n \geq 3$ and $i, j, k \in \{1, \ldots, n\}$ are distinct:

$$\Phi_{cub}^{ijk} = \forall x, y, z, v \; \big(xR_i v \wedge xR_j y \wedge xR_k z \;\rightarrow\; \exists a, b, c, d \; (vR_j c \wedge vR_k b \wedge yR_i c \wedge yR_k a \wedge zR_i b \wedge zR_j a \wedge aR_i d \wedge bR_j d \wedge cR_k d)\big).$$

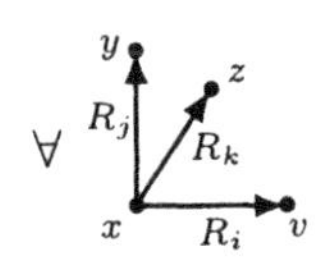

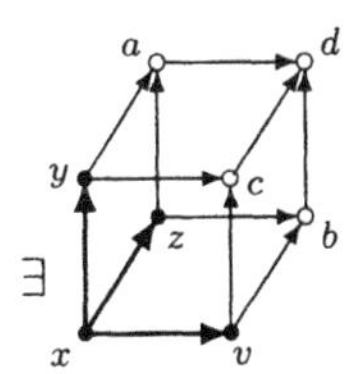

[1]Note that we enumerate 'dimensions' starting from 1, while the standard algebraic logic convention is to start from 0.

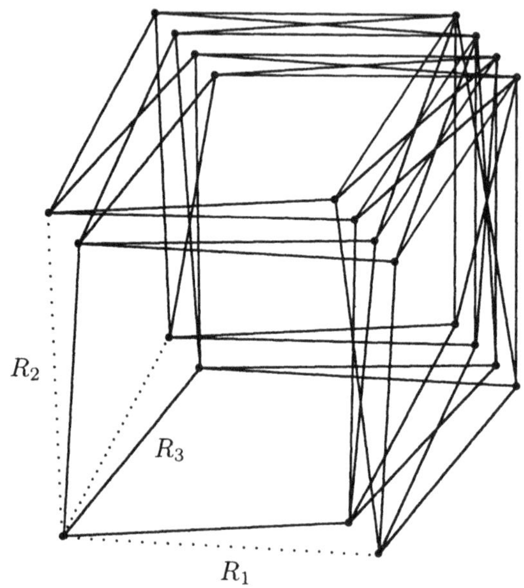

Figure 8.1: An $[\mathbf{S5}, \mathbf{S5}, \mathbf{S5}]$-frame refuting $\boldsymbol{cub}^{123}$.

It is not hard to check that a $[\mathbf{S5}, \mathbf{S5}, \dots, \mathbf{S5}]$-frame $\mathfrak{F}$ satisfies this property iff the following modal formula $\boldsymbol{cub}^{ijk}$ is valid in $\mathfrak{F}$ (cf. Henkin *et al.* 1985 [3.2.67]):

$$\boldsymbol{cub}^{ijk} = \Diamond_i p \wedge \Diamond_j q \wedge \Diamond_k r \rightarrow \\ \Diamond_i \Diamond_j \Diamond_k \big(\Diamond_k(\Diamond_j p \wedge \Diamond_i q) \wedge \Diamond_j(\Diamond_k p \wedge \Diamond_i r) \wedge \Diamond_i(\Diamond_k q \wedge \Diamond_j r)\big).$$

Thus $\boldsymbol{cub}^{ijk}$ belongs to $\mathbf{S5}^n$. On the other hand, Fig. 8.1 shows a 23-element $[\mathbf{S5}, \mathbf{S5}, \mathbf{S5}]$-frame refuting $\boldsymbol{cub}^{123}$ (see again Henkin *et al.* 1985 [3.2.67]). So $[\mathbf{S5}, \mathbf{S5}, \mathbf{S5}]$ and $\mathbf{S5}^3$ must be different. But the situation is even worse:

Theorem 8.1. (Johnson 1969) *The equational theory of* RDf_n *is not finitely axiomatizable, whenever* $n \geq 3$.

Translating this result into the language of modal logic, we obtain:

Theorem 8.2. *For no* $n \geq 3$ *is the logic* $\mathbf{S5}^n$ *finitely axiomatizable.*

(This result also follows from Theorem 8.30 below.) The only consolation can be the following consequence of Theorem 3.17:

Corollary 8.3. $\mathbf{S5}^n$ *is recursively enumerable.*

The interested reader can find an infinite recursive axiomatization of RDf_n (and thereby of $\mathbf{S5}^n$) in (Hirsch and Hodkinson 1997).

Question 8.4. Is it possible to axiomatize $\mathbf{S5}^n$ using only finitely many propositional variables? (For related questions in algebraic logic consult (Andréka 1997).)

Undecidability

That $\mathbf{S5}^n$ is undecidable for any $n \geq 3$ was first proved (in the algebraic setting) by Maddux (1980), who used a reduction of the word problem of semigroups.

Theorem 8.5. (Maddux 1980) *Let V be a variety such that $\mathsf{RDf}_n \subseteq V \subseteq \mathsf{Df}_n$. Then the equational theory of V is undecidable whenever $n \geq 3$.*

Reformulating this result in the language of modal logic, we obtain:

Theorem 8.6. *Any n-modal logic L in the interval*

$$[\mathbf{S5}, \mathbf{S5}, \ldots, \mathbf{S5}] \subseteq L \subseteq \mathbf{S5}^n$$

is undecidable whenever $n \geq 3$.

Here we show a different proof which uses a reduction of a tiling problem. To make the proof more transparent and to illustrate the important connection between $\mathbf{S5}^n$ and a fragment of classical first-order logic, we consider only the case of $\mathbf{S5}^n$. (However, the proof can be generalized to cover all logics mentioned in Theorem 8.6.)

Note first that by Proposition 3.15, it is enough to show that $\mathbf{S5}^3$ is undecidable. We will reduce an undecidable fragment of first-order logic to $\mathbf{S5}^3$. This fragment—$\mathcal{QL}_2^{3,=}$—was introduced in Section 3.10. It consists of all first-order formulas with equality that contain only binary predicate symbols and at most three distinct individual variables x, y, z. The undecidability of this fragment follows from Theorems 3.40 and 3.44. Since Theorem 3.40 is not proved in this book, below we give a direct proof by reducing the undecidable $\mathbb{N} \times \mathbb{N}$ tiling problem (see Section 5.4) to the satisfiability problem for $\mathcal{QL}_2^{3,=}$-formulas.

Proposition 8.7. *The satisfiability problem for $\mathcal{QL}_2^{3,=}$-formulas is undecidable.*

Proof. Suppose we are given a finite set T of tile types. With every $t \in T$ we associate a binary predicate P_t, and let $<_1$, $<_2$ be two extra binary predicates.

Using these we construct a $\mathcal{QL}_2^{3,=}$-formula ϕ_T as the conjunction of (8.1)–(8.6):

$$\forall x \exists y\, x <_1 y \;\wedge\; \forall x \exists y\, x <_2 y, \tag{8.1}$$

$$\forall x \forall y \forall z\, \big(x <_1 y \wedge x <_2 z \to \exists x (z <_1 x \wedge y <_2 x)\big), \tag{8.2}$$

$$\forall x \bigvee_{t \in T} P_t(x, x), \tag{8.3}$$

$$\forall x \bigwedge_{t \neq t'} \neg (P_t(x,x) \wedge P_{t'}(x,x)), \tag{8.4}$$

$$\forall x \forall y \Big(x <_1 y \to \bigwedge_{right(t) \neq left(t')} \neg (P_t(x,x) \wedge P_{t'}(y,y))\Big), \tag{8.5}$$

$$\forall x \forall y \Big(x <_2 y \to \bigwedge_{up(t) \neq down(t')} \neg (P_t(x,x) \wedge P_{t'}(y,y))\Big). \tag{8.6}$$

It should be clear that ϕ_T has a model whenever T tiles $\mathbb{N} \times \mathbb{N}$ ($<_1$ and $<_2$ are interpreted as the horizontal and vertical successor functions in the $\mathbb{N} \times \mathbb{N}$ grid, and $P_t(x,x)$ holds iff t tiles x, for any $x \in \mathbb{N} \times \mathbb{N}$ and $t \in T$).

Conversely, suppose that ϕ_T has a model. Take any point y_{00} in it. By (8.1), we have two infinite ascending chains of (not necessarily distinct) points:

$$y_{00} <_1 y_{10} <_1 y_{20} <_1 \dots,$$
$$y_{00} <_2 y_{01} <_2 y_{02} <_2 \dots.$$

By (8.2), there is y_{11} such that $y_{01} <_1 y_{11}$, $y_{10} <_2 y_{11}$. For the same reason, if we have already constructed points $y_{i(j+1)}$ and $y_{(i+1)j}$ for which $y_{ij} <_1 y_{(i+1)j}$ and $y_{ij} <_2 y_{i(j+1)}$, then we can find a point $y_{(i+1)(j+1)}$ such that

$$y_{i(j+1)} <_1 y_{(i+1)(j+1)} \quad \text{and} \quad y_{(i+1)j} <_2 y_{(i+1)(j+1)}.$$

Define a map τ from $\mathbb{N} \times \mathbb{N}$ to T by taking $\tau(i,j) = t$ iff $P_t(y_{ij}, y_{ij})$ holds in the model under consideration. It follows then from (8.3)–(8.6) that τ is well-defined and gives a tiling of $\mathbb{N} \times \mathbb{N}$. ❑

Now we construct a recursive translation ‡ of $\mathcal{QL}_2^{3,=}$-formulas into the language $\mathcal{ML}_3$ of $\mathbf{S5}^3$. Let $\Box\varphi$ denote $\Box_1\Box_2\Box_3\varphi$, and let d_{ij} and p_ℓ be propositional variables, for $1 \le i < j \le 3$ and $\ell < \omega$. Note first that without loss of generality we may assume that the only atoms in $\mathcal{QL}_2^{3,=}$-formulas are equalities $x_i = x_j$, for $1 \le i < j \le 3$, and predicates of the form $P_\ell(x,y)$, $\ell < \omega$ (cf. Section 3.5).

Given such a formula ϕ, we replace in it every occurrence of

- $\exists x_i$ with $\Diamond_i$, $\forall x_i$ with $\Box_i$, for $i = 1, 2, 3$,

- $x_i = x_j$ with d_{ij}, for $1 \le i < j \le 3$,
- $P_\ell(x, y)$ with $\Diamond_3 p_\ell$, for $\ell < \omega$.

The resulting 3-modal formula is denoted by $\phi^\dagger$. Let δ_ϕ be the conjunction (8.7)–(8.13)

$$\Box \Diamond_i d_{ij}, \tag{8.7}$$

$$\Box \Diamond_j d_{ij}, \tag{8.8}$$

$$\Box(\Diamond_k d_{ij} \to d_{ij}), \tag{8.9}$$

$$\Box(d_{12} \wedge d_{23} \to d_{13}), \tag{8.10}$$

$$\Box(d_{12} \wedge d_{13} \to d_{23}), \tag{8.11}$$

$$\Box(d_{13} \wedge d_{23} \to d_{12}), \tag{8.12}$$

$$\Box(d_{ij} \wedge \Diamond_i(d_{ij} \wedge \Diamond_3 p) \to \Diamond_3 p), \tag{8.13}$$

in which $1 \le i < j \le 3$, $i, j \ne k$, $1 \le k \le 3$, and p ranges over the propositional variables occurring in $\phi^\dagger$. Finally, we put

$$\phi^\ddagger = \phi^\dagger \wedge \delta_\phi.$$

Proposition 8.8. *A $\mathcal{QL}_2^{3,=}$-formula ϕ is satisfiable iff its translation $\phi^\ddagger$ is satisfiable in an $\mathbf{S5}^3$-frame.*

Proof. ($\Rightarrow$) Suppose ϕ is satisfied in a first-order structure I with domain D. Consider the model $\mathfrak{M}_I = \langle \langle D, D, D \rangle, \mathfrak{V} \rangle$ based on the universal product frame $\langle D, D, D \rangle$ (cf. Section 3.3) and such that

$$\begin{aligned} \mathfrak{V}(p_\ell) &= \{\langle a_1, a_2, a_3 \rangle \in D^3 \mid I \models P_\ell(x, y)[a_1, a_2]\}, \\ \mathfrak{V}(d_{ij}) &= \{\langle a_1, a_2, a_3 \rangle \in D^3 \mid a_i = a_j\} \qquad (1 \le i < j \le 3). \end{aligned}$$

It should be clear that $\mathfrak{M}_I \models \delta_\phi$, and by a straightforward induction on the construction of ϕ one can easily show that, for all $a_1, a_2, a_3 \in D$,

$$I \models \phi(x, y, z)[a_1, a_2, a_3] \quad \text{iff} \quad (\mathfrak{M}_I, \langle a_1, a_2, a_3 \rangle) \models \phi^\dagger.$$

($\Leftarrow$) Suppose now that $\phi^\ddagger$ is satisfied in a model $\mathfrak{M} = \langle \mathfrak{F}, \mathfrak{V} \rangle$ based on an $\mathbf{S5}^3$-frame $\mathfrak{F}$. By Proposition 3.11, we may assume that $\mathfrak{F}$ is a universal product frame $\langle W_1, W_2, W_3 \rangle$. Further, we may assume that, for all $\langle w_1, w_2, w_3 \rangle \in W_1 \times W_2 \times W_3$ and all variables p_ℓ in $\phi^\ddagger$,

$$(\mathfrak{M}, \langle w_1, w_2, w_3 \rangle) \models p_\ell \quad \text{iff} \quad \forall v \in W_3\ (\mathfrak{M}, \langle w_1, w_2, v \rangle) \models p_\ell, \tag{8.14}$$

$$(\mathfrak{M}, \langle w_1, w_2, w_3 \rangle) \models d_{12} \quad \text{iff} \quad \forall v \in W_3\ (\mathfrak{M}, \langle w_1, w_2, v \rangle) \models d_{12}, \tag{8.15}$$

$$(\mathfrak{M}, \langle w_1, w_2, w_3 \rangle) \models d_{13} \quad \text{iff} \quad \forall v \in W_2\ (\mathfrak{M}, \langle w_1, v, w_3 \rangle) \models d_{13}, \tag{8.16}$$

$$(\mathfrak{M}, \langle w_1, w_2, w_3 \rangle) \models d_{23} \quad \text{iff} \quad \forall v \in W_1\ (\mathfrak{M}, \langle v, w_2, w_3 \rangle) \models d_{23}, \tag{8.17}$$

and $\mathfrak{V}(q) = \emptyset$ for all other propositional variables q. Indeed, (8.14) follows from the fact that p_ℓ occurs in $\phi^\ddagger$ only in the context of $\Diamond_3 p_\ell$, and (8.15)–(8.17) from (8.9). Such a model $\mathfrak{M}$ will be called *binary generated* to reflect that the truth-values of variables in $\mathfrak{M}$ depend only on at most two coordinates of the worlds.

Our aim is to construct from $\mathfrak{M}$ an $\mathbf{S5}^3$-model $\mathfrak{N} = \langle\langle V, V, V\rangle, \mathfrak{U}\rangle$, based on a cubic universal product frame, such that $\phi^\ddagger$ is satisfied in $\mathfrak{N}$ and, for all i, j with $1 \leq i < j \leq 3$,

$$\mathfrak{U}(d_{ij}) = \{\langle v_1, v_2, v_3\rangle \in V^3 \mid v_i = v_j\}.$$

Then by taking

$$I_\mathfrak{N} \models P_\ell(x, y)[v_1, v_2] \quad \text{iff} \quad \exists v_3 \in V \; (\mathfrak{N}, \langle v_1, v_2, v_3\rangle) \models p_\ell,$$

we will obtain a first-order structure $I_\mathfrak{N}$ (with domain V) satisfying ϕ.

That such a model $\mathfrak{N}$ exists is a consequence of Lemmas 8.9 and 8.10 to be proved below. These lemmas are due to Johnson (1969) (who used ideas of Halmos (1957)) and were proved for representable diagonal-free cylindric (and polyadic) algebras; see also (Henkin *et al.* 1985).

Lemma 8.9. (Johnson, Halmos) *Let* $\mathfrak{M} = \langle\langle W_1, W_2, W_3\rangle, \mathfrak{V}\rangle$ *be an* $\mathbf{S5}^3$-*model and let* U *be the disjoint union of the sets* W_i, $i = 1, 2, 3$. *Then there is an* $\mathbf{S5}^3$-*model* $\mathfrak{M}^+ = \langle\langle U, U, U\rangle, \mathfrak{V}^+\rangle$ *having* $\mathfrak{M}$ *as its p-morphic image. Moreover, if formulas* (8.7)–(8.12) *are true in* $\mathfrak{M}$, *then* $\mathfrak{V}^+$ *can be chosen so that*

$$\mathfrak{V}^+(d_{ij}) \supseteq \{\langle u_1, u_2, u_3\rangle \in U^3 \mid u_i = u_j\} \tag{8.18}$$

hold for $1 \leq i < j \leq 3$.

Proof. As we saw in the proof of Proposition 3.12, if a set U is such that there are surjections $f_i : U \to W_i$, for $i = 1, 2, 3$, then the map f defined by

$$f(u_1, u_2, u_3) = \langle f_1(u_1), f_2(u_2), f_3(u_3)\rangle$$

is a p-morphism from the model $\mathfrak{M}^+ = \langle\langle U, U, U\rangle, \mathfrak{V}^+\rangle$ onto $\mathfrak{M}$, where

$$\mathfrak{V}^+(p) = \{\langle u_1, u_2, u_3\rangle \in U^3 \mid f(u_1, u_2, u_3) \in \mathfrak{V}(p)\}.$$

Assume now that δ_ϕ is true in $\mathfrak{M}$. Take U to be the disjoint union of W_1, W_2 and W_3. We will define surjections $f_i : U \to W_i$ in such a way that (8.18) holds. But first let us define an auxiliary function g on U.

We claim that for every $u \in W_1$ there is a world $g(u) = \langle u, v, w\rangle$ in $W_1 \times W_2 \times W_3$ such that $(\mathfrak{M}, g(u)) \models d_{12} \wedge d_{13} \wedge d_{23}$. Indeed, take some $\langle u, v', w'\rangle$. By (8.8), there is a $v \in W_2$ with $(\mathfrak{M}, \langle u, v, w'\rangle) \models d_{12}$, and there is a $w \in W_3$ with $(\mathfrak{M}, \langle u, v, w\rangle) \models d_{23}$. By (8.9), $(\mathfrak{M}, \langle u, v, w\rangle) \models d_{12}$, and so, by (8.10), $(\mathfrak{M}, \langle u, v, w\rangle) \models d_{13}$. In the same way one can show that for every $v \in W_2$ (every $w \in W_3$) there is $g(v) = \langle u, v, w\rangle$ (respectively, $g(w) = \langle u, v, w\rangle$) in $W_1 \times W_2 \times W_3$ such that $(\mathfrak{M}, g(v)) \models d_{12} \wedge d_{13} \wedge d_{23}$ (and $(\mathfrak{M}, g(w)) \models d_{12} \wedge d_{13} \wedge d_{23}$).

Construct maps f_i from U onto W_i $(i = 1, 2, 3)$ by taking $f_i(u)$ to be the i-th coordinate of $g(u)$, for every $u \in U$. (Since f_i is identical on W_i, f_i is surjective.) Let $\mathfrak{M}^+$ be the model as defined above. We show that it satisfies (8.18). Suppose, for instance, that $\langle u, u, v\rangle \in U^3$ and $u \in W_3$. Then

$$f(u, u, u) = g(u) \in \mathfrak{V}(d_{12}),$$

and so $\langle u, u, u\rangle \in \mathfrak{V}^+(d_{12})$. In view of (8.9), it follows that $\langle u, u, v\rangle \in \mathfrak{V}^+(d_{12})$. Other cases are treated analogously. ❑

As was mentioned above, we may assume that the constructed model $\mathfrak{M}^+$ is binary generated. By the p-morphism theorem, $\mathfrak{M}^+$ satisfies $\phi^{\ddagger}$. Since $\mathfrak{M}^+$ is based on a universal product $\mathbf{S5}^3$-frame and δ_ϕ is a conjunction of formulas prefixed by $\Box_1\Box_2\Box_3$, this actually means that $\mathfrak{M}^+ \models \delta_\phi$.

Lemma 8.10. (Johnson, Halmos) *Let $\mathfrak{M}^+ = \langle\langle U, U, U\rangle, \mathfrak{V}^+\rangle$ be a binary generated $\mathbf{S5}^3$-model satisfying condition (8.18) and such that $\mathfrak{M}^+ \models \delta_\phi$. Then there is a p-morphism from $\mathfrak{M}^+$ onto an $\mathbf{S5}^3$-model $\mathfrak{N} = \langle\langle V, V, V\rangle, \mathfrak{U}\rangle$ such that $|V| \le |U|$ and*

$$\mathfrak{U}(d_{ij}) = \{\langle v_1, v_2, v_3\rangle \in V^3 \mid v_i = v_j\},$$

for $1 \le i < j \le 3$.

Proof. For every pair i, j such that $1 \le i < j \le 3$, we define a relation $R_{ij} \subseteq U \times U$ by taking

$$R_{ij} = \{\langle u, v\rangle \in U \times U \mid \exists \langle w_1, w_2, w_3\rangle \in \mathfrak{V}^+(d_{ij})(w_i = u \wedge w_j = v)\}.$$

In fact, these three relations coincide. Let us check, for instance, that we have $R_{12} \subseteq R_{13}$. Suppose that $(\mathfrak{M}^+, \langle u, v, w\rangle) \models d_{12}$. By (8.9), we have $(\mathfrak{M}^+, \langle u, v, v\rangle) \models d_{12}$ and by (8.18), $(\mathfrak{M}^+, \langle u, v, v\rangle) \models d_{23}$. It follows then from (8.10) that $(\mathfrak{M}^+, \langle u, v, v\rangle) \models d_{13}$.

So we denote R_{ij} by R and prove that it is an equivalence relation on $U \times U$. By (8.18), it is reflexive. Let us show that it is symmetric. Suppose $(\mathfrak{M}^+, \langle u, v, w\rangle) \models d_{12}$. Then, as we saw above, $(\mathfrak{M}^+, \langle u, v, v\rangle) \models d_{13}$. By (8.9), $(\mathfrak{M}^+, \langle u, u, v\rangle) \models d_{13}$ and by (8.18), $(\mathfrak{M}^+, \langle u, u, v\rangle) \models d_{12}$, from which by (8.11), $(\mathfrak{M}^+, \langle u, u, v\rangle) \models d_{23}$. In view of (8.9), we have $(\mathfrak{M}^+, \langle v, u, v\rangle) \models d_{23}$. Then, by (8.18), $(\mathfrak{M}^+, \langle v, u, v\rangle) \models d_{13}$ and by (8.12), $(\mathfrak{M}^+, \langle v, u, v\rangle) \models d_{12}$.

Thus, $\langle v, u\rangle \in R_{12}$ holds. To prove transitivity, suppose $uR_{12}v$ and $vR_{23}w$. This means that we have $(\mathfrak{M}^+, \langle u, v, x\rangle) \models d_{12}$ and $(\mathfrak{M}^+, \langle y, v, w\rangle) \models d_{23}$ for some x and y. It follows from (8.9) that $(\mathfrak{M}^+, \langle u, v, w\rangle) \models d_{12} \wedge d_{23}$, and by (8.10) we obtain $(\mathfrak{M}^+, \langle u, v, w\rangle) \models d_{13}$, i.e., $uR_{13}w$.

Denote by $[u]$ the R-equivalence class containing u. Let $V = \{[u] \mid u \in U\}$. Define a valuation $\mathfrak{U}$ on $V \times V \times V$ by taking

$$\mathfrak{U}(p) = \{\langle [u_0], [u_1], [u_2]\rangle \mid \langle u_1, u_2, u_3\rangle \in \mathfrak{V}^+(p)\}.$$

This definition does not depend on the choice of $\langle u_1, u_2, u_3\rangle$. Indeed, suppose that u_iRv_i, for each $i = 1, 2, 3$. We show that in this case

$$\langle u_1, u_2, u_3\rangle \in \mathfrak{V}^+(p) \quad \text{iff} \quad \langle v_1, v_2, v_3\rangle \in \mathfrak{V}^+(p),$$

for every variable p in $\phi^{\ddagger}$.

Suppose first that p does not depend on coordinate 3, that is, p is either some p_ℓ or d_{12}. Let $\langle u_1, u_2, u_3\rangle \in \mathfrak{V}^+(p)$. We show that $\langle u_1, v_2, u_3\rangle \in \mathfrak{V}^+(p)$. By (8.8), we have $(\mathfrak{M}^+, \langle u_1, u_2, w\rangle) \models d_{23}$ for some w. It follows that $u_2R_{23}w$ and $(\mathfrak{M}^+, \langle u_1, u_2, w\rangle) \models p$ (since $\mathfrak{V}^+(p)$ does not depend on coordinate 3). So $(\mathfrak{M}^+, \langle u_1, v_2, w\rangle) \models \Diamond_2(\Diamond_3 p \wedge d_{23})$. By the transitivity of R, we have v_1Rw, and so there is u such that $(\mathfrak{M}^+, \langle u, v_2, w\rangle) \models d_{23}$. In view of (8.9), we then have $(\mathfrak{M}^+, \langle u_1, v_2, w\rangle) \models d_{23}$. Therefore, by (8.13), $(\mathfrak{M}^+, \langle u_1, v_2, w\rangle) \models \Diamond_3 p$, from which $(\mathfrak{M}^+, \langle u_1, v_2, u_3\rangle) \models p$. Starting from this, in a similar way we can show that $(\mathfrak{M}^+, \langle v_1, v_2, u_3\rangle) \models p$. But then $(\mathfrak{M}^+, \langle v_1, v_2, v_3\rangle) \models p$ (see Fig. 8.2).

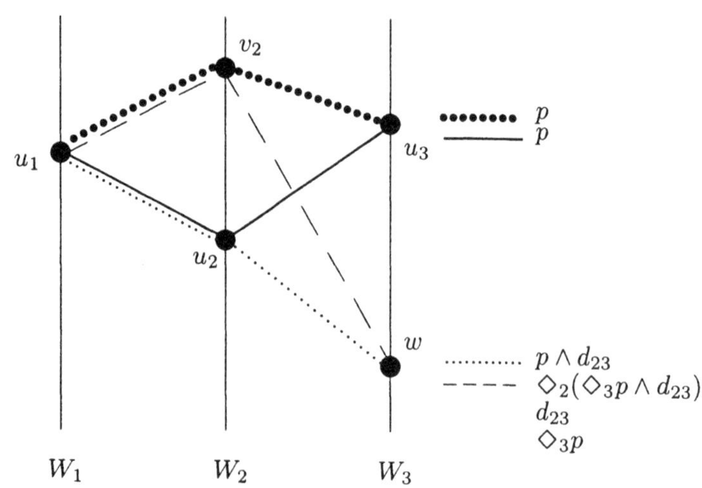

Figure 8.2: $\mathfrak{U}(p)$ is well-defined.

Now suppose that $p = d_{13}$, i.e., $\langle u_1, u_2, u_3 \rangle \in \mathfrak{V}^+(d_{13})$ and $u_i R v_i$, for each $i = 1, 2, 3$. Then $u_1 R u_3$ and, since R is an equivalence relation, $v_1 R v_3$, from which $\langle v_1, w, v_3 \rangle \in \mathfrak{V}^+(d_{13})$, for some w. Using (8.9), we then obtain $\langle v_1, v_2, v_3 \rangle \in \mathfrak{V}^+(d_{13})$. The case of $p = d_{23}$ is treated similarly.

It follows from the definition of $\mathfrak{U}$ that

$$\mathfrak{U}(d_{ij}) = \{\langle v_1, v_2, v_3 \rangle \in V^3 \mid v_i = v_j\}.$$

It remains to observe that the map f from U^3 onto V^3 defined by

$$f(u_1, u_2, u_3) = \langle [u_1], [u_2], [u_3] \rangle$$

is a p-morphism from $\mathfrak{M}^+$ onto $\mathfrak{N} = \langle V \times V \times V, \mathfrak{U} \rangle$, which completes the proof of Lemma 8.10. ❑

Returning to the proof of Proposition 8.8, we see that if $\phi^\ddagger$ is satisfied in an $\mathbf{S5}^3$-model $\mathfrak{M}$ then $\phi^\ddagger$ is satisfied in $\mathfrak{M}^+$ (by Lemma 8.9), and so it is also satisfied in $\mathfrak{N}$ (by Lemma 8.10), as required. ❑

As a consequence of Propositions 8.7 and 8.8 we obtain that $\mathbf{S5}^3$ is undecidable.

Corollary 8.11. $\mathbf{S5}^n$ *does not have the product fmp whenever* $n \geq 3$.

Proof. By Corollary 8.3, $\mathbf{S5}^n$ is recursively enumerable. It is easy to see that finite product $\mathbf{S5}^n$-frames are recursively enumerable as well. Thus, by Theorem 8.6, $\mathbf{S5}^n$ cannot have the product fmp. ❑

As $[\mathbf{S5}, \dots, \mathbf{S5}]$ is finitely axiomatizable and undecidable, it does not even have the (abstract) fmp. But actually a much more general result holds.

Lack of the finite model property

Theorem 8.12. *No n-modal logic L in the interval*

$$[\mathbf{S5}, \mathbf{S5}, \dots, \mathbf{S5}] \subseteq L \subseteq \mathbf{S5}^n$$

has the finite model property, whenever $n \geq 3$.

Proof. The proof we present here is based on an idea of Németi (1984) (for another proof using a property of finite semigroups see (Kurucz 2002)).

Let Φ be the conjunction of formulas (8.7)–(8.13) above and the following formulas:

(a) $\Box(p \wedge \Diamond_1(p \wedge d_{13}) \to d_{13})$ ('p^{-1} is a function'),

(b) $\Box(p \leftrightarrow \Diamond_3 p)$ ('p is binary'),

(c) $\Box\Diamond_2 p$ ('$Dom\, p = \top$'),

(d) $\neg\Diamond_2(\Diamond_1 p \wedge d_{12})$ ('$Rng\, p \neq \top$').

Lemma 8.13. *Φ is* $\mathbf{S5}^3$*-satisfiable.*

Proof. Consider a model $\mathfrak{M} = \langle \mathfrak{F}, \mathfrak{V} \rangle$, where $\mathfrak{F}$ is the universal product frame on $\mathbb{N} \times \mathbb{N} \times \mathbb{N}$ and

$$\begin{aligned} \mathfrak{V}(p) &= \{\langle x, x+1, z \rangle \mid x, z \in \mathbb{N}\}, \\ \mathfrak{V}(d_{ij}) &= \{\langle x_1, x_2, x_3 \rangle \mid x_1, x_2, x_3 \in \mathbb{N},\ x_i = x_j\} \quad (1 \leq i < j \leq 3). \end{aligned}$$

Then it is readily seen that $(\mathfrak{M}, \langle 0,0,0 \rangle) \models \Phi$. ❑

Lemma 8.14. *Φ is not satisfiable in any finite frame for* $[\mathbf{S5}, \mathbf{S5}, \mathbf{S5}]$.

Proof. Suppose that $\mathfrak{F} = \langle W, R_1, R_2, R_3 \rangle$ is a frame for $[\mathbf{S5}, \mathbf{S5}, \mathbf{S5}]$ (i.e., the R_i are commuting equivalence relations on W) and that $\mathfrak{M}$ is a model on $\mathfrak{F}$ such that $(\mathfrak{M}, x) \models \Phi$ for some x. We show that then $\mathfrak{F}$ must be infinite.

For each $n < \omega$, we define a formula φ_n and worlds x_n, y_n in $\mathfrak{F}$ as follows:

$$\begin{aligned} \varphi_0 &= \neg \Diamond_2 (\Diamond_1 p \wedge d_{12}), \\ \varphi_{n+1} &= \Diamond_2 (\Diamond_1 (\varphi_n \wedge p) \wedge d_{12}). \end{aligned}$$

Let $x_0 = x$. Assume that x_k has already been defined. By (c) and (8.7), there are y_k, x_{k+1} such that $x_k R_2 y_k R_1 x_{k+1}$, $y_k \models p$ and $x_{k+1} \models d_{12}$.

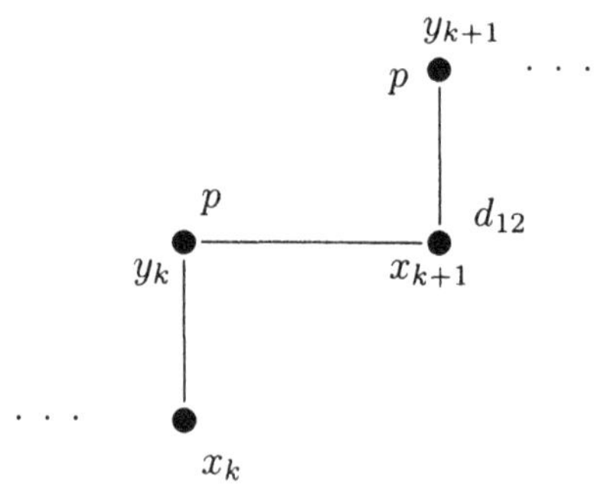

Claim 8.15. $\forall n < \omega\ y_n \models \varphi_n$.

Proof. An easy induction on n is left to the reader. ❑

Claim 8.16. $\forall n < \omega\ \big(w \models \varphi_n \ \&\ w R_3 w' \rightarrow w' \models \varphi_n\big)$.

Proof. The proof is also by induction on n. Let $w' \not\models \varphi_0$. Then, by commutativity, there are u, v such that

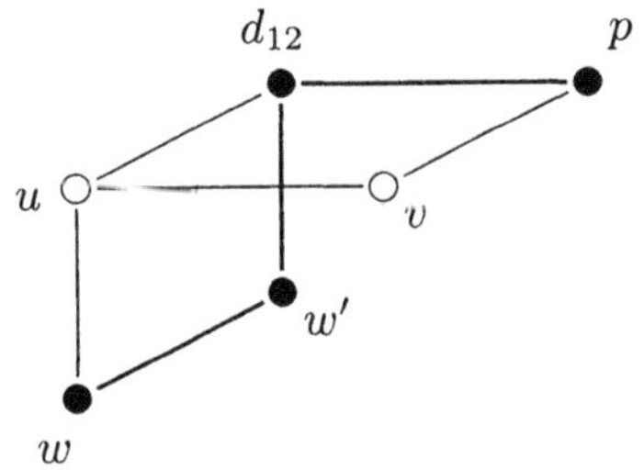

Then, by (8.9), $u \models d_{12}$, by (b), $v \models p$, and so $w \not\models \varphi_0$.

Now let $w \models \varphi_{n+1}$. Then, by commutativity, there are u, v such that:

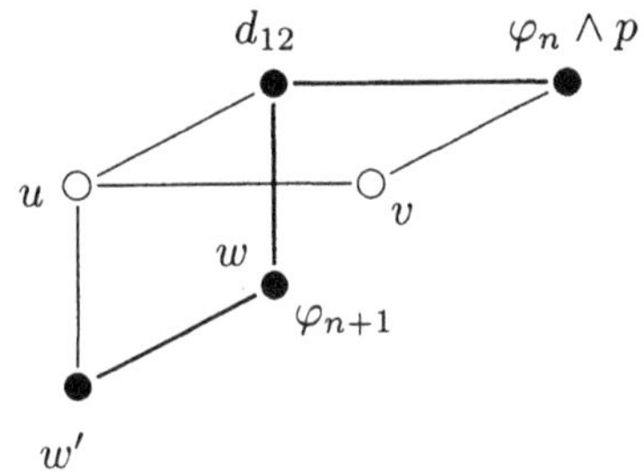

Then $u \models d_{12}$ by (8.9), $v \models \varphi_n \wedge p$ by (b) and the induction hypothesis, from which $w' \models \varphi_{n+1}$. ❑

Claim 8.17. $\forall n < \omega\ \mathfrak{M} \models \Box(d_{13} \wedge \Diamond_1(d_{13} \wedge \varphi_n) \rightarrow \varphi_n)$.

Proof. *Case* $n = 0$. Suppose that $w \models d_{13} \wedge \Diamond_1(d_{13} \wedge \neg\Diamond_2(\Diamond_1 p \wedge d_{12}))$ and $w \models \Diamond_2(\Diamond_1 p \wedge d_{12})$. Then there are u, v and (by commutativity) w' such that:

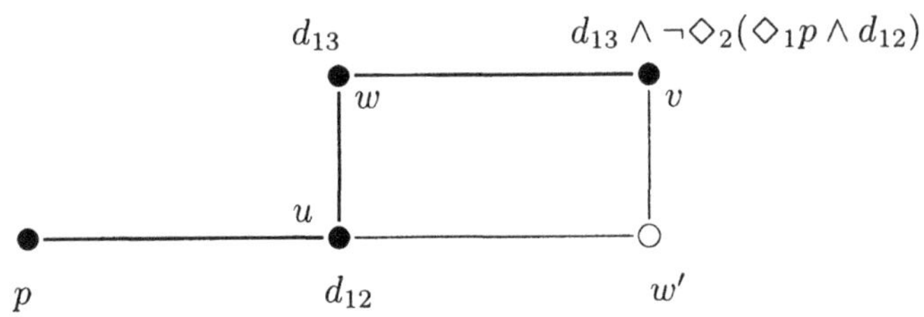

Then $u \models d_{13}$ and $w' \models d_{13}$, by (8.9). So $u \models d_{23}$, by (8.11), and $w' \models d_{23}$, by (8.9). It follows from (8.12) that $w' \models d_{12}$, contrary to $v \models \neg\Diamond_2(\Diamond_1 p \wedge d_{12})$.

Case $n + 1$. Let $w \models d_{13} \wedge \Diamond_1[d_{13} \wedge \Diamond_2(\Diamond_1(\varphi_n \wedge p) \wedge d_{12})]$. Then there are u and (by commutativity) w' such that:

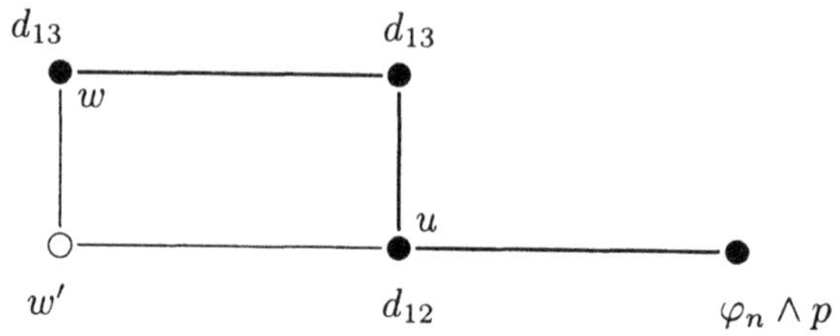

Then $u \models d_{13}$ and $w' \models d_{13}$, by (8.9). So by (8.11), $u \models d_{23}$. Then $w' \models d_{23}$, by (8.9), and $w' \models d_{12}$, by (8.12). Thus $w' \models \Diamond_1(\varphi_0 \wedge p) \wedge d_{12}$, from which $w \models \Diamond_2(\Diamond_1(\varphi_n \wedge p) \wedge d_{12})$. ❑

Claim 8.18. $\forall k, n < \omega \, \forall w \, \big(k < n \to w \not\models \varphi_k \wedge \varphi_n\big)$.

Proof. The proof is by induction on k. Let $n > 0$ and $k = 0$. If $w \models \varphi_n$ then w R_2-sees a d_{12}-world which R_1-sees a p-world. On the other hand, if $w \models \varphi_0$ then w does not R_2-see a d_{12}-world which R_1-sees a p-world.

Suppose now that $w \models \varphi_{k+1} \wedge \varphi_{n+1}$. Then there is a world w' such that $w' \models d_{12} \wedge \Diamond_1(\varphi_n \wedge p)$. By (8.8), there is u for which $w' R_3 u$ and $u \models d_{13}$. Hence, by commutativity, there are v, x, w'', y such that:

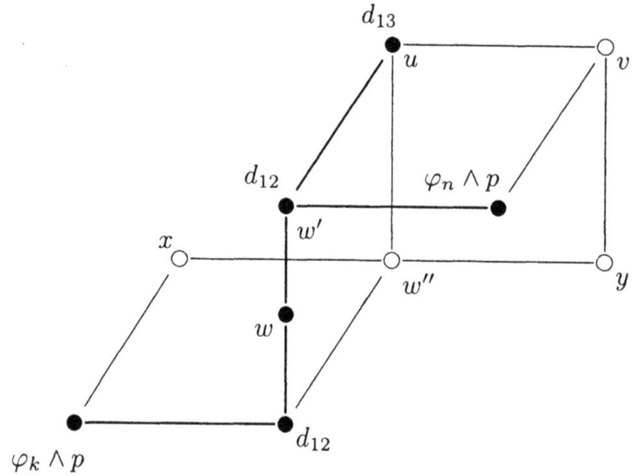

Then $u \models d_{12}$, by (8.9), $u \models d_{23}$, by (8.11), and $v \models d_{23}$ again by (8.9). Further, $v \models \varphi_n$, by Claim 8.16, and $y \models \varphi_n$, by the definition of φ_n. On the other hand, $w'' \models d_{12} \wedge d_{13}$, by (8.9), and so $w'' \models d_{23}$, by (8.11). Therefore, $y \models d_{23}$, by (8.9). Finally, $y \models p$, by (8.13). Since $x \models \varphi_k \wedge p$, by Claim 8.16 and (b), we obtain that x and y are such that

$$xR_1y, \quad x \models \varphi_k \wedge p \quad \text{and} \quad y \models \varphi_n \wedge p.$$

By (8.8), there is some s such that yR_3s and $s \models d_{13}$, and by commutativity, there is w''' for which:

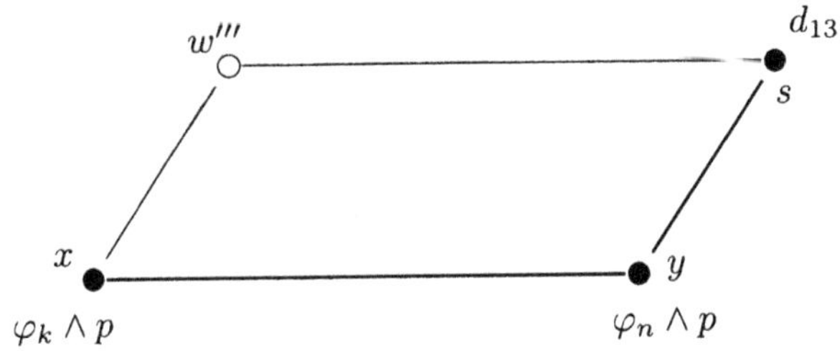

By (b) and Claim 8.16, $w''' \models \varphi_k \wedge p$ and $s \models \varphi_n \wedge p$. It follows from (a) that $w''' \models d_{13}$. Then, by Claim 8.17, $w''' \models \varphi_n$. Thus $w''' \models \varphi_k \wedge \varphi_n$, contrary to the induction hypothesis. ❑

Lemma 8.14 follows immediately from Claims 8.15 and 8.18. ❑

Finally, Theorem 8.12 follows from Lemmas 8.13 and 8.14. ❑

8.2 Products between $\mathbf{K4}^n$ and $\mathbf{S5}^n$

In this section we show that $\mathbf{S5}^n$ can be easily reduced to any product logic between $\mathbf{K4}^n$ and $\mathbf{S5}^n$, and thus all these logics are undecidable.

Theorem 8.19. *Let $n \geq 3$ and, for each $i = 1, \ldots, n$, let L_i be a Kripke complete unimodal logic from the interval $\mathbf{K4} \subseteq L_i \subseteq \mathbf{S5}$. Then the product logic $L_1 \times \cdots \times L_n$ is undecidable and does not have the product fmp.*

Proof. By Proposition 3.15, it is enough to prove the theorem for $n = 3$. So we define a reduction of $\mathbf{S5}^3$ to any logic $L_1 \times L_2 \times L_3$ such that $\mathbf{K4} \subseteq L_i \subseteq \mathbf{S5}$, $i = 1, 2, 3$.

Given an $\mathcal{ML}_3$-formula φ, we construct two $\mathcal{ML}_3$-formulas φ^r and φ^+ as follows. First, φ^r is obtained from φ by inductively replacing each subformula of φ of the form $\Diamond_i\psi$ (or $\Box_i\psi$) with $\Diamond_i^+\psi = \psi \vee \Diamond_i\psi$ (or $\Box_i^+\psi = \psi \wedge \Box_i\psi$, respectively,). Second, with every subset Σ of the set $sub\,\varphi$ of all subformulas in φ we associate the formula

$$\alpha_\varphi^\Sigma = \bigwedge_{\psi \in \Sigma} \psi \wedge \bigwedge_{\psi \in sub\,\varphi - \Sigma} \neg\psi.$$

Further, for every collection S of subsets of $sub\,\varphi$ and every $i = 1, 2, 3$, we construct the formula

$$\alpha_\varphi^{S,i} = \bigwedge_{\Sigma \in S} \Diamond_i^+(\alpha_\varphi^\Sigma)^r \wedge \bigwedge_{\Sigma \in 2^{sub\,\varphi} - S} \neg\Diamond_i^+(\alpha_\varphi^\Sigma)^r.$$

Finally, we define

$$\alpha_\varphi = \bigwedge_{|\{i,j,k\}|=3} \left(\Box_i^+ \Box_j^+ \bigvee_{S \subseteq 2^{sub\,\varphi}} \Box_k^+ \alpha_\varphi^{S,k} \right)$$

and take

$$\varphi^+ = \varphi^r \wedge \alpha_\varphi.$$

Lemma 8.20. *For every $\mathcal{ML}_3$-formula φ, the following conditions are equivalent:*

(i) *φ is satisfiable in* $\mathbf{S5}^3$,

(ii) *φ^+ is satisfiable in $L_1 \times L_2 \times L_3$.*

Proof. The implication (i) $\Rightarrow$ (ii) is easy. If φ is satisfied in a product frame for $\mathbf{S5}^3$ then α_φ is also satisfied in this frame. As φ and φ^r are equivalent in reflexive models, φ^r is satisfied in the same product frame too. And since $L_i \subseteq \mathbf{S5}$ for all $i = 1, 2, 3$, this product frame is a frame for $L_1 \times L_2 \times L_3$.

(ii) $\Rightarrow$ (i). By Proposition 3.11, we may assume that φ^+ is satisfied at the root $\overline{r} = \langle r_1, r_2, r_3 \rangle$ of a model $\mathfrak{M} = \langle \mathfrak{F}, \mathfrak{V} \rangle$ based on a rooted product $L_1 \times L_2 \times L_3$-frame $\mathfrak{F} = \langle W_1, R_1 \rangle \times \langle W_2, R_2 \rangle \times \langle W_3, R_3 \rangle$. Let

$$\mathfrak{F}^+ = \langle W_1, W_2, W_3 \rangle$$

be the universal product $\mathbf{S5}^3$-frame based on $W = W_1 \times W_2 \times W_3$ and let $\mathfrak{M}^+ = \langle \mathfrak{F}^+, \mathfrak{V} \rangle$. We show by induction on the construction of $\psi \in sub\,\varphi$ that for every $\overline{x} = \langle x_1, x_2, x_3 \rangle \in W$,

$$(\mathfrak{M}, \overline{x}) \models \psi^r \quad \text{iff} \quad (\mathfrak{M}^+, \overline{x}) \models \psi.$$

The basis of the induction and the case of the Booleans are trivial. Suppose that, for some $i \in \{1, 2, 3\}$, $(\mathfrak{M}, \overline{x}) \models (\Diamond_i \chi)^r$, that is, $(\mathfrak{M}, \overline{x}) \models \chi^r \vee \Diamond_i \chi^r$. Then there is $\overline{y} = \langle y_1, y_2, y_3 \rangle$ in W such that either $\overline{y} = \overline{x}$ or $x_i R_i y_i$ and $x_j = y_j$ for $j \neq i$, and $(\mathfrak{M}, \overline{y}) \models \chi^r$. By the induction hypothesis, $(\mathfrak{M}^+, \overline{y}) \models \chi$ and so $(\mathfrak{M}^+, \overline{x}) \models \Diamond_i \chi$.

Conversely, let $i = 1$ and let $(\mathfrak{M}^+, \overline{x}) \models \Diamond_1 \chi$, i.e., $(\mathfrak{M}^+, \overline{y}) \models \chi$, for some $\overline{y} = \langle y_1, y_2, y_3 \rangle \in W$ such that $y_2 = x_2$ and $y_3 = x_3$. By the induction hypothesis, we have $(\mathfrak{M}, \overline{y}) \models \chi^r$, but the relation $x_1 R_1 y_1$ may not hold.

Let $\Sigma = \{\psi \in sub\,\varphi \mid (\mathfrak{M}, \overline{y}) \models \psi^r\}$. Then clearly $\chi \in \Sigma$ and

$$(\mathfrak{M}, \overline{y}) \models (\alpha_\varphi^\Sigma)^r. \tag{8.19}$$

Now, by assumption, $(\mathfrak{M}, \overline{r}) \models \alpha_\varphi$, and so there is a collection S of subsets of $sub\,\varphi$ such that $(\mathfrak{M}, \langle r_1, x_2, x_3 \rangle) \models \Box_1^+ \alpha_\varphi^{S,1}$. As $\mathbf{K4} \subseteq L_1$, $\langle W_1, R_1 \rangle$ is a transitive frame with root r_1. So either $r_1 R_1 x_1$ or $r_1 = x_1$. In any

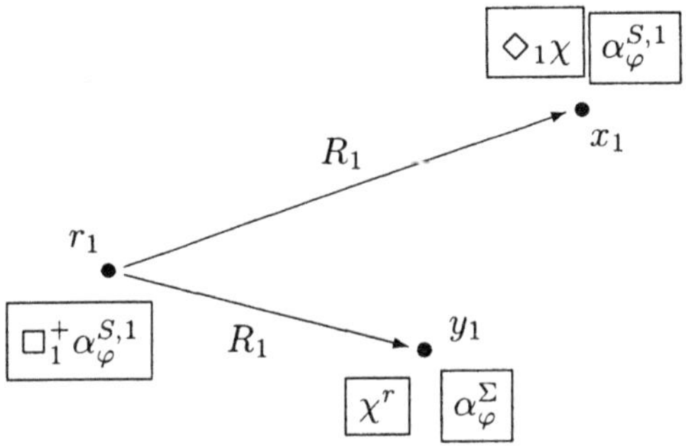

Figure 8.3: The induction step for $\Diamond_1\chi$.

case, $(\mathfrak{M},\overline{x}) \models \alpha_\varphi^{S,1}$ holds. Note that $\Sigma \in S$, for otherwise we would have $(\mathfrak{M},\langle r_1,x_2,x_3\rangle) \models \neg\Diamond_1^+(\alpha_\varphi^\Sigma)^r$, contrary to (8.19) and $\langle W_1,R_1\rangle$ being transitive with root r_1. It follows that $(\mathfrak{M},\overline{x}) \models \Diamond_1^+(\alpha_\varphi^\Sigma)^r$. And since $\chi \in \Sigma$, we obtain $(\mathfrak{M},\overline{x}) \models \chi^r \vee \Diamond_1\chi^r$. The cases of $i = 2,3$ are considered analogously (see Fig. 8.3). ❑

Theorem 8.19 is an immediate consequence of Lemma 8.20 and Theorem 8.6. ❑

Note that the proof above transforms finite product $L_1 \times L_2 \times L_3$-frames into finite product $\mathbf{S5}^3$-frames. So, as a consequence of Corollary 8.11 we obtain the following:

Corollary 8.21. *Let L_i be any Kripke complete unimodal logic in the interval $\mathbf{K4} \subseteq L_i \subseteq \mathbf{S5}$ $(i = 1,2,3)$. Then the product logic $L_1 \times L_2 \times L_3$ does not have the product fmp.*

The proof of Lemma 8.20 can be extended to so-called *weakly transitive* logics. These are logics of the form

$$\mathbf{K4}(k) = \mathbf{K} \oplus \Box p \wedge \cdots \wedge \Box^k p \to \Box^{k+1} p,$$

where $k > 0$. It is easily checked that $\mathbf{K4}(k)$ is determined by the class of *k-transitive* frames $\langle W,R\rangle$ which satisfy the condition

$$\forall x,y \in W(xR^{k+1}y \to xRy \vee \cdots \vee xR^k y).$$

Theorem 8.22. *Let $k > 0$ and let L_i be any Kripke complete logic from the interval $\mathbf{K4}(k) \subseteq L_i \subseteq \mathbf{S5}$ $(i = 1,2,3)$. Then the logic $L_1 \times L_2 \times L_3$ is undecidable.*

Proof. It suffices to use the following modification of the translation defined in the proof of Lemma 8.20. For each $\mathcal{ML}_3$-formula φ, let $\varphi^{(k)}$ be the result of replacing in φ each subformula of the form $\Box_i\psi$ with $\Box_i^{\leq k}\psi$. The formulas α_φ^Σ, $\alpha_\varphi^{S,i}$, α_φ and φ^+ are defined in the same way as above, but using $\Box_i^{\leq k}\psi$ instead of $\Box_i^+\psi$ and $\psi^{(k)}$ instead of ψ^r. The remaining steps of the proof are left to the reader. ❑

Question 8.23. Do product logics like $\mathbf{K4}^3$, $\mathbf{S4}^3$, $\mathbf{K4} \times \mathbf{S4} \times \mathbf{S5}$ have the fmp?

8.3 Products with the fmp

In this section we prove the following theorem of Gabbay and Shehtman (1998):

Theorem 8.24. *Any product of* **Alt** *and* **K** *has the fmp. In particular,* $\mathbf{K}^n$, $\mathbf{Alt}^n$ *and* $\mathbf{K}^n \times \mathbf{Alt}^m$ *have the fmp, for any* $n, m \geq 1$.

Proof. We obtain Theorem 8.24 as a consequence of Lemmas 8.25 and 8.26 to be proved below.

As in Section 1.4, we say that an arbitrary n-frame $\mathfrak{F} = \langle W, R_1, \dots, R_n\rangle$ is of *depth* k if k is the length of the longest path in $\mathfrak{F}$. An n-modal logic L is said to have the *finite depth property* if L is determined by a class of frames of finite depth. Note that the depth of a frame $\mathfrak{F}$ does not exceed $k < \omega$ iff $\mathfrak{F} \models M_{(n)}^{k+1}\bot$.

Lemma 8.25. *Any product* $L_1 \times \cdots \times L_n$ *of* **Alt** *and* **K** *has the finite depth property.*

Proof. Suppose $\varphi \notin L_1 \times \cdots \times L_n$ for some n-modal formula φ. Then there are rooted frames $\mathfrak{F}_i$, $i = 1, \dots, n$, such that $\mathfrak{F}_i \models L_i$ and φ is refuted at the root of $\mathfrak{F}_1 \times \cdots \times \mathfrak{F}_n$. By Proposition 1.7, for each $i = 1, \dots, n$, there is an intransitive tree $\mathfrak{T}_i$ and a p-morphism h_i from $\mathfrak{T}_i$ onto $\mathfrak{F}_i$. Note that if $L_i = \mathbf{Alt}$ then the unraveling $\mathfrak{T}_i$ of $\mathfrak{F}_i$ is just a chain of irreflexive points. So we always have $\mathfrak{T}_i \models L_i$. It is straightforward to check that the function h defined by

$$h(x_1, \dots, x_n) = \langle h_1(x_1), \dots, h_n(x_n)\rangle$$

is a p-morphism from $\mathfrak{T}_1 \times \cdots \times \mathfrak{T}_n$ onto $\mathfrak{F}_1 \times \cdots \times \mathfrak{F}_n$ (cf. Proposition 3.10(i)). Now we prune all the trees $\mathfrak{T}_i$ down to the modal depth $md(\varphi)$ of φ. Clearly, the resulting product frame $\mathfrak{T}_1^- \times \cdots \times \mathfrak{T}_n^-$ is of depth $n \cdot md(\varphi)$; it validates $L_1 \times \cdots \times L_n$ and refutes φ at its root. ❑

Lemma 8.26. *If an* n*-modal logic* L *has the finite depth property, then it has the fmp as well.*

Proof. Suppose that $\mathfrak{M} \not\models \varphi$, for some $\mathcal{ML}_n$-formula φ and some model $\mathfrak{M} = \langle \mathfrak{F}, \mathfrak{V} \rangle$ based on an n-frame $\mathfrak{F} = \langle W, R_1, \dots, R_n \rangle$ such that $\mathfrak{F} \models L$ and the depth of $\mathfrak{F}$ is $k < \omega$. Suppose also that the propositional variables of φ are among $p_1, \dots, p_m$, and so without loss of generality we may assume that

$$\mathfrak{V}(p_j) = \mathfrak{V}(p_m) \quad \text{for all } j \geq m. \tag{8.20}$$

Let Σ be the set of all $\mathcal{ML}_n$-formulas built up from $p_1, \dots, p_m$. Consider any *filtration* $\mathfrak{M}^\Sigma = \langle \mathfrak{F}^\Sigma, \mathfrak{V}^\Sigma \rangle$ of $\mathfrak{M}$ through Σ, which is defined as follows. The worlds of $\mathfrak{F}^\Sigma$ are the equivalence classes $[x]$ of the equivalence relation $\sim_\Sigma$ defined by taking, for all $x, y \in W$,

$$x \sim_\Sigma y \quad \text{iff} \quad \forall \psi \in \Sigma \; \big((\mathfrak{M}, x) \models \psi \iff (\mathfrak{M}, y) \models \psi \big).$$

The accessibility relation R_i^Σ, for each $i = 1, \dots, n$, is any relation between points in $\mathfrak{F}^\Sigma$ satisfying two conditions:

- if xR_iy then $[x]R_i^\Sigma[y]$, and
- if $[x]R_i^\Sigma[y]$ then, for all ψ,

$$\text{if } \Box_i\psi \in \Sigma \text{ and } (\mathfrak{M}, x) \models \Box_i\psi \text{ then } (\mathfrak{M}, y) \models \psi.$$

The valuation $\mathfrak{V}^\Sigma$ is defined by taking

$$\mathfrak{V}^\Sigma(p) = \big\{ [x] \mid (\mathfrak{M}, x) \models p \big\},$$

for every propositional variable p; cf. (8.20). By induction on the construction of ψ, it is readily checked that

$$\forall \psi \in \Sigma \, \forall x \in W \; \big((\mathfrak{M}, x) \models \psi \iff (\mathfrak{M}^\Sigma, [x]) \models \psi \big). \tag{8.21}$$

It follows that $\mathfrak{M}^\Sigma \not\models \varphi$, $\mathfrak{M}^\Sigma \models L$ and $\mathfrak{M}^\Sigma \models \boldsymbol{M}_{(n)}^{k+1}\bot$. Therefore, $\mathfrak{M}^\Sigma$ is of depth k.

We will show now that not only its depth, but $\mathfrak{M}^\Sigma$ itself is finite. To this end, observe first that each world in $\mathfrak{M}^\Sigma$ is uniquely determined by the set of propositional variables that are true in it and the set of worlds accessible from it:

Claim 8.27. *Let $[x]$ and $[y]$ be such that*

(i) $(\mathfrak{M}^\Sigma, [x]) \models p_i$ *iff* $(\mathfrak{M}^\Sigma, [y]) \models p_i$, *for every* $1 \leq i \leq m$;

(ii) $[x]R_j^\Sigma[z]$ *iff* $[y]R_j^\Sigma[z]$, *for every world* $[z]$ *and* $1 \leq j \leq n$.

Then $[x] = [y]$.

Proof. We prove by induction that for all $\psi \in \Sigma$,

$$(\mathfrak{M}, x) \models \psi \quad \text{iff} \quad (\mathfrak{M}, y) \models \psi.$$

The basis of induction follows from (i), and the case of Booleans is trivial. For $\Box_i \chi$ we have:

$$(\mathfrak{M}, x) \models \Box_i \chi \overset{\text{by (8.21)}}{\Longleftrightarrow} (\mathfrak{M}^\Sigma, [x]) \models \Box_i \chi \overset{\text{by (ii)}}{\Longleftrightarrow}$$
$$(\mathfrak{M}^\Sigma, [y]) \models \Box_i \chi \overset{\text{by (8.21)}}{\Longleftrightarrow} (\mathfrak{M}, y) \models \Box_i \chi.$$

❑

Now, for each $j \leq k$, let N_j denote the number of worlds $[x]$ in $\mathfrak{F}^\Sigma$ such that the length of the longest path in $\mathfrak{F}^\Sigma$ starting from $[x]$ is j. By Claim 8.27, we have the following upper bound for N_j:

$$N_0 \leq 2^m, \qquad N_{j+1} \leq 2^{n(N_0 + \cdots + N_j)} \cdot 2^m.$$

Therefore, N_j is finite for every $j \leq k$, and so $\mathfrak{M}^\Sigma$ is finite as well, which completes the proof of Lemma 8.26. ❑

Theorem 8.24 follows immediately. ❑

It is to be noted that even for recursively enumerable product logics the fmp does not always imply decidability. Although $\mathbf{Alt}^n$ is decidable for any n (Theorem 8.53), $\mathbf{K}^n$ turns out to be undecidable if $n \geq 3$ (Theorem 8.28). The reason for this is that we do not necessarily have an algorithm deciding whether a finite frame is a frame for the logic in question (cf. Theorem 8.29).

8.4 Between $\mathbf{K}^n$ and $\mathbf{S5}^n$

The fact that $\mathbf{K}^n$ has the fmp (Theorem 8.24), while $\mathbf{S5}^n$ does not (for $n \geq 3$, Theorem 8.12), might give some hope that the computational properties of $\mathbf{K}^n$ are 'better' than those of $\mathbf{S5}^n$. In this section we show that this is not the case: in higher dimensions all logics between $\mathbf{K}^n$ and $\mathbf{S5}^n$ are quite complex.

To begin with, we note that

$$[\mathbf{K}, \mathbf{K}, \ldots, \mathbf{K}] \subsetneq \mathbf{K}^n.$$

Indeed, one can generalize the first-order 'cubifying' property Φ^{ijk}_{cub} of Section 8.1 as follows. For each $\ell > 0$ and distinct $i, j \in \{1, \ldots, n\}$, let

$$\begin{aligned}
&\Phi^{ijk}_{cub}(\ell) = \\
&\quad \forall x_1, \ldots, x_\ell, x, y, z, \ \big(xR_i x_1 \wedge x_1 R_i x_2 \wedge \cdots \wedge x_{\ell-1} R_i x_\ell \wedge xR_j y \wedge xR_k z \to \\
&\quad \exists u, y_1, \ldots y_\ell, z_1, \ldots, z_\ell, u_1, \ldots, u_\ell \ (yR_k u \wedge zR_j u \wedge yR_i y_1 \wedge \\
&\quad y_1 R_i y_2 \wedge \cdots \wedge y_{k-1} R_i y_\ell \wedge zR_i z_1 \wedge z_1 R_i z_2 \wedge \cdots \wedge z_{k-1} R_i z_\ell \wedge uR_i u_1 \wedge \\
&\quad u_1 R_i u_2 \wedge \cdots \wedge u_{k-1} R_i u_\ell \wedge x_\ell R_j y_\ell \wedge x_\ell R_k z_\ell \wedge y_\ell R_k u_\ell \wedge z_\ell R_j u_\ell)\big);
\end{aligned}$$

see Fig. 8.4.

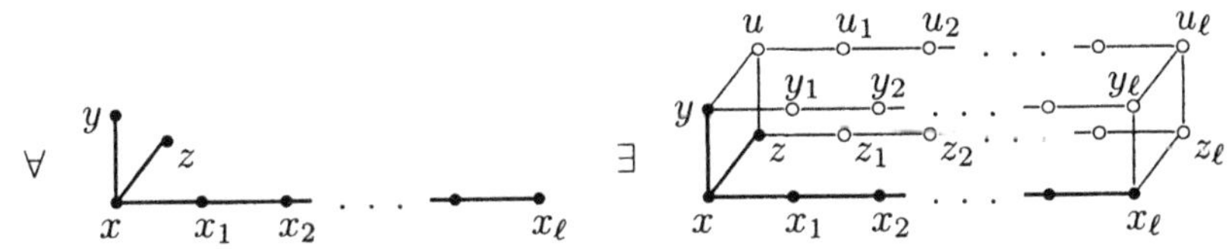

Figure 8.4: The property $\Phi_{cub}^{ijk}(\ell)$.

Then clearly $\Phi_{cub}^{ijk} = \Phi_{cub}^{ijk}(1)$ and n-dimensional product frames satisfy $\Phi_{cub}^{ijk}(\ell)$ for all $\ell > 0$. On the other hand, it is routine to check that a $[\mathbf{K}, \mathbf{K}, \mathbf{K}]$-frame $\mathfrak{F}$ satisfies $\Phi_{cub}^{123}(\ell)$ iff the following $\mathcal{ML}_3$-formula $\boldsymbol{cub}_\ell$ is valid in $\mathfrak{F}$:

$$\begin{aligned} \boldsymbol{cub}_\ell = \Big[& \Diamond_1^\ell(\Box_2 p_{12} \wedge \Box_3 p_{13}) \ \wedge \ \Diamond_2(\Box_1^\ell p_{21} \wedge \Box_3 p_{23}) \ \wedge \ \Diamond_3(\Box_1^\ell p_{31} \wedge \Box_2 p_{32}) \\ & \wedge \ \Box_1^\ell \Box_2(p_{12} \wedge p_{21} \to \Box_3 q_3) \ \wedge \ \Box_1^\ell \Box_3(p_{13} \wedge p_{31} \to \Box_2 q_2) \ \wedge \\ & \wedge \ \Box_2 \Box_3(p_{23} \wedge p_{32} \to \Box_1^\ell q_1) \Big] \ \longrightarrow \ \Diamond_1^\ell \Diamond_2 \Diamond_3(q_1 \wedge q_2 \wedge q_3) \, . \end{aligned}$$

It is proved in (Kurucz 2000b) that, for every $\ell > 0$,

$$\boldsymbol{cub}_\ell \notin [\mathbf{K}, \mathbf{K}, \mathbf{K}] \oplus \bigoplus_{0<k<\ell} \boldsymbol{cub}_k$$

and that these formulas can be used to show that $\mathbf{K}^n$ is not finitely axiomatizable whenever $n \geq 3$. Here we prove this in different way, using the following general results of (Hirsch *et al.* 2002).

From now on let $n \geq 3$ and let L be any n-modal logic such that

$$\mathbf{K}^n \subseteq L \subseteq \mathbf{S5}^n.$$

Theorem 8.28. *L is undecidable.*

Theorem 8.29. *It is undecidable whether a finite frame is a frame for L.*

Theorem 8.30. *L is not finitely axiomatizable.*

Theorem 8.31. *L does not have the product finite model property in the following strong sense: there is an $\mathcal{ML}_3$-formula which does not belong to L, but which is valid in all finite k-dimensional product frames for all $k \geq 3$.*

In the proofs we use the following result of Hirsch and Hodkinson (2001) about relation algebras (all the necessary definitions will be given below):

Theorem 8.32. *It is undecidable whether a finite simple relation algebra is representable.*[2]

Given a natural number $n \geq 3$ and a finite simple relation algebra $\mathfrak{A}$, we will define below a finite n-frame $\mathfrak{F}_{\mathfrak{A},n}$ and an $\mathcal{ML}_3$-formula $\varphi_{\mathfrak{A}}$ such that the following lemmas hold:

Lemma 8.33. *The following conditions are equivalent:*

(i) $\mathfrak{F}_{\mathfrak{A},n}$ *is a frame for* L*;*

(ii) *the formula* $\neg\varphi_{\mathfrak{A}}$ *does not belong to* L*;*

(iii) $\mathfrak{F}_{\mathfrak{A},3}$ *is a p-morphic image of some universal product* $\mathbf{S5}^3$*-frame.*

Lemma 8.34. $\mathfrak{A}$ *is representable iff* $\mathfrak{F}_{\mathfrak{A},3}$ *is a p-morphic image of some universal product* $\mathbf{S5}^3$*-frame. Moreover,* $\mathfrak{A}$ *is representable with a finite base iff* $\mathfrak{F}_{\mathfrak{A},3}$ *is a p-morphic image of some finite universal product* $\mathbf{S5}^3$*-frame.*

Now, Theorems 8.28 and 8.29 follow immediately from Theorem 8.32 and Lemmas 8.33, 8.34. Theorem 8.30 follows from Theorem 8.29, since if L were finitely axiomatizable then there would be an effective test for finite frames being frames for L. Note that if L is recursively enumerable and finite product frames for L are also recursively enumerable (such as, e.g., for $\mathbf{K}^n$, $\mathbf{K4}^n$, $\mathbf{S5}^n$) then the fact that L does not have the product fmp follows already from Theorem 8.28.

Note also that as a consequence of Theorems 3.21 and 8.28 we obtain the following result of Gabbay and Shehtman (1993):

Theorem 8.35. *For any unimodal logic* L *between* $\mathbf{K}$ *and* $\mathbf{S5}$*, the two-variable fragment of the first-order modal logic* $\mathbf{Q}L$ *is undecidable.*

We are about to prove Lemmas 8.33, 8.34 and Theorem 8.31.

Frame formulas in product frames

In this subsection we establish a connection between arbitrary product frames and product frames for $\mathbf{S5}^3$. This connection (Claim 8.37 below) is the heart of the proof of Lemma 8.33.

Let $\mathfrak{F} = \langle F, R_1, R_2, R_3 \rangle$ be a finite 3-frame with the following property:

$$\forall p, p' \in F \; \exists s_1, s_2 \in F \; \big(pR_1s_1 \;\&\; s_1R_2s_2 \;\&\; s_2R_3p'\big). \tag{8.22}$$

For example, all universal product $\mathbf{S5}^3$-frames have this property. With each point $p \in F$ we associate a propositional variable, denoted also by p. The

[2] Although Hirsch and Hodkinson (2001) formulated this theorem for arbitrary finite relation algebras, the algebras constructed in their proof are in fact simple.

following formula $\varphi_{\mathfrak{F}}$ can be regarded as (a variant of) the *frame* (or *Jankov–Fine*) *formula for* $\mathfrak{F}$ (see Chagrov and Zakharyaschev 1997):

$$\begin{aligned}
\varphi_{\mathfrak{F}} \quad = \quad & \Box^+ \bigvee_{p\in F} (p \wedge \neg \bigvee_{p'\in F-\{p\}} p') && (8.23)\\
& \wedge \Box^+ \bigwedge_{\substack{i=1,2,3,\\ p,p'\in F,\, pR_ip'}} (p \to \Diamond_i p') && (8.24)\\
& \wedge \Box^+ \bigwedge_{\substack{i=1,2,3,\\ p,p'\in F,\, \neg(pR_ip')}} (p \to \neg\Diamond_i p'). && (8.25)
\end{aligned}$$

Here and below, $\Box_i^+\psi$ abbreviates $\psi \wedge \Box_i\psi$, and $\Box^+\psi$ stands for $\Box_1^+\Box_2^+\Box_3^+\psi$.

It is easy to see that $\varphi_{\mathfrak{F}}$ is satisfiable in $\mathfrak{F}$: it is enough to take the model $\mathfrak{M} = \langle \mathfrak{F}, \mathfrak{V}\rangle$ with $\mathfrak{V}(p) = \{p\}$, and then $(\mathfrak{M}, q) \models \varphi_{\mathfrak{F}}$, for every $q \in F$. Moreover, we have the following claim:

Claim 8.36. *Let* $\mathfrak{F} = \langle F, R_1, R_2, R_3\rangle$ *and* $\mathfrak{H}$ *be 3-frames satisfying* (8.22), *with* $\mathfrak{F}$ *being finite. If* $\mathfrak{H}$ *satisfies* $\varphi_{\mathfrak{F}}$ *then* $\mathfrak{F}$ *is a p-morphic image of* $\mathfrak{H}$.

Proof. Suppose that $\varphi_{\mathfrak{F}}$ is satisfied in some model $\mathfrak{M}$ based on a 3-frame $\mathfrak{H} = \langle U, S_1, S_2, S_3\rangle$ satisfying (8.22). Define a function $h : U \to F$ by taking, for all $u \in U$,

$$h(u) = p \quad \text{iff} \quad (\mathfrak{M}, u) \models p.$$

Then, by (8.23), h is well-defined. By (8.23), (8.24) and since $\mathfrak{F}$ satisfies (8.22), h is 'onto.' Finally, (8.24) and (8.25) guarantee that h is a p-morphism from $\mathfrak{H}$ onto $\mathfrak{F}$. ❑

Now, given an n-dimensional product frame ($n \geq 3$)

$$\mathfrak{H} = \langle U_1, S_1\rangle \times \langle U_2, S_2\rangle \times \langle U_3, S_3\rangle \times \cdots \times \langle U_n, S_n\rangle$$

and a world $\bar{u} = \langle u_1, u_2, \dots, u_n\rangle$ in it, define the sets

$$U_i(\bar{u}) = \{v \in U_i \mid v = u_i \text{ or } u_iS_iv\}, \qquad \text{for } i = 1, 2, 3.$$

Define a universal product $\mathbf{S5}^3$-frame $\mathfrak{H}(\bar{u})$ by taking

$$\mathfrak{H}(\bar{u}) = \langle U_1(\bar{u}), U_2(\bar{u}), U_3(\bar{u})\rangle .$$

Clearly, $\langle u_1, u_2, u_3\rangle$ is in $\mathfrak{H}(\bar{u})$, and $\mathfrak{H}(\bar{u})$ (like any universal product $\mathbf{S5}^3$-frame) satisfies (8.22). Observe that if $\mathfrak{H}$ is finite then $\mathfrak{H}(\bar{u})$ is finite as well.

Claim 8.37. *Let* $\mathfrak{F} = \langle F, R_1, R_2, R_3\rangle$ *be a finite 3-frame such that the* R_i *are equivalence relations and* (8.22) *holds in* $\mathfrak{F}$. *If* $\varphi_{\mathfrak{F}}$ *is satisfied at some point* $\bar{u}$ *in an* n*-dimensional product frame* $\mathfrak{H}$ *for some* $n \geq 3$, *then* $\varphi_{\mathfrak{F}}$ *is satisfiable in the universal product* $\mathbf{S5}^3$*-frame* $\mathfrak{H}(\bar{u})$.

Proof. Suppose $\mathfrak{M}$ is a model based on an n-dimensional product frame $\mathfrak{H}$ such that

$$(\mathfrak{M}, \bar{u}) \models \varphi_{\mathfrak{F}} \tag{8.26}$$

holds. Define a model $\mathfrak{M}'$ based on $\mathfrak{H}(\bar{u})$ by taking, for all $\langle v_1, v_2, v_3 \rangle$ in $U_1(\bar{u}) \times U_2(\bar{u}) \times U_3(\bar{u})$ and $p \in F$,

$$(\mathfrak{M}', \langle v_1, v_2, v_3 \rangle) \models p \quad \text{iff} \quad (\mathfrak{M}, \langle v_1, v_2, v_3, u_4, \dots, u_n \rangle) \models p. \tag{8.27}$$

We claim that $(\mathfrak{M}', \langle u_1, u_2, u_3 \rangle) \models \varphi_{\mathfrak{F}}$. Indeed, (8.23) clearly holds by (8.27) and (8.26). To prove (8.25), we show that if $i = 1, 2, 3$, $\langle v_1, v_2, v_3 \rangle$, $\langle w_1, w_2, w_3 \rangle$ are in $U_1(\bar{u}) \times U_2(\bar{u}) \times U_3(\bar{u})$, $v_j = w_j$ for $j \neq i$, $(\mathfrak{M}', \langle v_1, v_2, v_3 \rangle) \models p$ and $(\mathfrak{M}', \langle w_1, w_2, w_3 \rangle) \models p'$, then pR_ip'. Without loss of generality we may assume that $i = 1$. By the definition of $U_1(\bar{u})$, either $v_1 = u_1$ or $u_1 S_1 v_1$, and similarly, either $w_1 = u_1$ or $u_1 S_1 w_1$. By (8.23), there is a unique $p'' \in F$ such that $(\mathfrak{M}', \langle u_1, v_2, v_3 \rangle) \models p''$. So, by (8.27), we have

$$(\mathfrak{M}, \langle v_1, v_2, v_3, u_4, \dots, u_n \rangle) \models p \quad \text{and} \quad (\mathfrak{M}, \langle u_1, v_2, v_3, u_4, \dots, u_n \rangle) \models p''.$$

We claim that $p'' R_1 p$ and $p'' R_1 p'$. Indeed, if $v_1 = u_1$ then $p = p''$, and so $p'' R_1 p$ holds by the reflexivity of R_1. If $u_1 S_1 v_1$ then

$$(\mathfrak{M}, \langle u_1, v_2, v_3, u_4, \dots, u_n \rangle) \models p'' \land \Diamond_1 p$$

which, by (8.26), implies $p'' R_1 p$. Similarly, one can show that $p'' R_1 p'$. Therefore, we must have pR_1p', because R_1 is symmetric and transitive.

For (8.24), we show that if $\langle v_1, v_2, v_3 \rangle \in U_1(\bar{u}) \times U_2(\bar{u}) \times U_3(\bar{u})$, pR_1p' and $(\mathfrak{M}', \langle v_1, v_2, v_3 \rangle) \models p$ then there is a $w \in U_1(\bar{u})$ such that $(\mathfrak{M}', \langle w, v_2, v_3 \rangle) \models p'$. Similar statements hold for 2 and $U_2(\bar{u})$, and 3 and $U_3(\bar{u})$, respectively. As we saw in the previous paragraph, $p'' R_1 p$ for the unique $p'' \in F$ such that $(\mathfrak{M}', \langle u_1, v_2, v_3 \rangle) \models p''$. As R_1 is transitive, $p'' R_1 p'$. By (8.27) and (8.26), we have

$$(\mathfrak{M}, \langle u_1, v_2, v_3, u_4, \dots, u_n \rangle) \models \Diamond_1 p'.$$

Hence, there is some $w \in U_1$ such that $u_1 S_1 w$ and

$$(\mathfrak{M}, \langle w, v_2, v_3, u_4, \dots, u_n \rangle) \models p'.$$

Since such a w is in $U_1(\bar{u})$ and in view of (8.27), we finally obtain that $(\mathfrak{M}', \langle w, v_2, v_3 \rangle) \models p'$, as required. ❑

Relation algebras and product frames

Recall from Section 3.10 that a *relation algebra* is a modal algebra for arrow logic $\mathbf{AL}_{RA}$. In other words, a relation algebra is a structure of the form

$$\mathfrak{A} = \langle A, \land, \neg, 0, 1, ;, ^{-}, Id \rangle$$

satisfying the following properties, for all $x, y, z \in A$:

- $\langle A, \wedge, \neg, 0, 1\rangle$ is a Boolean algebra,
- $x\,;(y\,;z) = (x\,;y)\,;z$,
- $x^{--} = x$ and $x\,;Id = Id; x = x$,
- ; and $^-$ distribute over $\vee$ (so they are monotone with respect to the Boolean $\leq$),
- the *cycle law* holds, i.e.,

$$x \wedge (y\,;z) = 0 \quad \text{iff} \quad y \wedge (x\,;z^-) = 0 \quad \text{iff} \quad z \wedge (y^-\,;x) = 0.$$

It is not hard to see that these two definitions of relation algebras are equivalent; consult (Maddux 1991) and (Hirsch and Hodkinson 2002) for a discussion and a detailed introduction to relation algebras.[3]

A relation algebra is *atomic* if its Boolean reduct is an atomic Boolean algebra (see Section 4.2). Thus, all finite relation algebras are atomic. A relation algebra is *simple* if it has no nontrivial homomorphic images. It is well-known (see e.g., Maddux 1991, Theorem 17) that a relation algebra $\mathfrak{A}$ is simple iff $1\,;a\,;1 = 1$ holds for all $a \neq 0$ in $\mathfrak{A}$.

A natural example is the (simple) relation algebra of all subsets of $U \times U$, for some nonempty set U. Here ; is the composition (relative product) of binary relations, $^-$ is the converse, and Id the identity relation on U.

A simple relation algebra is called *representable with base* U if it is embeddable into the relation algebra of all subsets of $U \times U$. As we already mentioned, it follows from the main result of (Hirsch and Hodkinson 2001) that there is no algorithm deciding whether a finite simple relation algebra is representable.

Now, take some finite simple relation algebra $\mathfrak{A}$. Call a triple $\langle t_1, t_2, t_3\rangle$ of atoms of $\mathfrak{A}$ *consistent* if $t_3^- \leq t_1\,;t_2$.

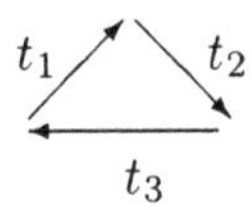

Note that, by the cycle law, if a triple $\langle t_1, t_2, t_3\rangle$ is consistent then $\langle t_2, t_3, t_1\rangle$, $\langle t_3, t_1, t_2\rangle$, $\langle t_1^-, t_3^-, t_2^-\rangle$, $\langle t_3^-, t_2^-, t_1^-\rangle$ and $\langle t_2^-, t_1^-, t_3^-\rangle$ are also consistent.

We are now in a position to define the n-frame $\mathfrak{F}_{\mathfrak{A},n}$ and the $\mathcal{ML}_3$-formula $\varphi_{\mathfrak{A}}$. With each consistent triple $\langle t_1, t_2, t_3\rangle$ of atoms of $\mathfrak{A}$ we associate a point $t = t_1t_2t_3$. The set of all such points will be denoted by $\mathcal{T}_{\mathfrak{A}}$. For $t, t' \in \mathcal{T}_{\mathfrak{A}}$

[3]Note that in the algebraic logic literature, $^-$ is usually denoted by $\breve{}$ and Id by $1'$. We use $^-$ and Id to be consistent with our arrow logic notation.

and $i = 1, 2, 3$, define tR_it' iff $t_i = t'_i$. For $4 \leq i \leq n$, let R_i be the identity on $\mathcal{T}_{\mathfrak{A}}$. Finally, set

$$\mathfrak{F}_{\mathfrak{A},n} = \langle \mathcal{T}_{\mathfrak{A}}, R_1, R_2, R_3, \ldots, R_n \rangle .$$

Clearly, $\mathfrak{F}_{\mathfrak{A},n}$ is finite and the R_i are equivalence relations.

Claim 8.38. $\mathfrak{F}_{\mathfrak{A},3}$ *satisfies* (8.22).

Proof. Take some $t, s \in \mathcal{T}_{\mathfrak{A}}$. Since ; and $^-$ are monotone and $\mathfrak{A}$ is simple, there are atoms x, y of $\mathfrak{A}$ such that $t_1^- \leq x^- ; s_3^- ; y$. It follows that there is an atom z for which $t_1^- \leq z ; y$ and $z \leq x^- ; s_3^-$. So the following chain of consistent triples

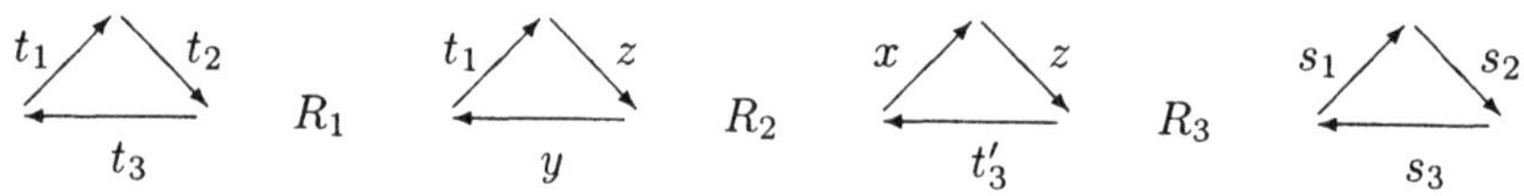

is as required. ❑

Thus, we can define $\varphi_{\mathfrak{A}}$ as the frame formula for $\mathfrak{F}_{\mathfrak{A},3}$.

Proof of Lemma 8.33. (i) $\Rightarrow$ (ii). Suppose $\mathfrak{F}_{\mathfrak{A},n} \models L$. Since $\varphi_{\mathfrak{A}}$ is an $\mathcal{ML}_3$-formula satisfiable in $\mathfrak{F}_{\mathfrak{A},3}$, it is satisfiable in $\mathfrak{F}_{\mathfrak{A},n}$, for any $n \geq 3$. Therefore, $\neg\varphi_{\mathfrak{A}}$ is not valid in $\mathfrak{F}_{\mathfrak{A},n}$, and so does not belong to L.

(iii) $\Rightarrow$ (i). Suppose $\mathfrak{F}_{\mathfrak{A},3}$ is a p-morphic image of some universal product $\mathbf{S5}^3$-frame $\mathfrak{G}_1 \times \mathfrak{G}_2 \times \mathfrak{G}_3$. Then clearly $\mathfrak{F}_{\mathfrak{A},n}$ is a p-morphic image of the universal product $\mathbf{S5}^n$-frame $\mathfrak{G}_1 \times \mathfrak{G}_2 \times \mathfrak{G}_3 \times \cdots \times \mathfrak{G}_n$, where $\mathfrak{G}_i$ is the one-point reflexive frame, for each $4 \leq i \leq n$. Since $L \subseteq \mathbf{S5}^n$, $\mathfrak{F}_{\mathfrak{A},n}$ is a frame for L.

Finally, if (ii) holds, that is, if $\neg\varphi_{\mathfrak{A}} \notin L$ then, in view of $\mathbf{K}^n \subseteq L$, $\varphi_{\mathfrak{A}}$ is satisfiable in an n-dimensional product frame, and so (iii) is an immediate consequence of Claims 8.36 and 8.37. ❑

Proof of Lemma 8.34. The proof is obtained from a chain of known results in algebraic logic using the duality between Kripke frames and Boolean algebras with operators. Here we present the arguments in the modal logic setting, as 'modal mirror images' of the algebraic proofs of Halmos (1957), Johnson (1969) and Monk (1961).

Claim 8.39. (Monk) *If the (finite and simple) relation algebra* $\mathfrak{A}$ *is representable with base* U *then the 3-frame* $\mathfrak{F}_{\mathfrak{A},3}$ *is a p-morphic image of the universal product* $\mathbf{S5}^3$*-frame* $\langle U, U, U \rangle$.

Proof. Suppose that there is a function *rep* embedding $\mathfrak{A}$ into the relation algebra of all subsets of $U \times U$. Define a function h from $U \times U \times U$ to the set $\mathcal{T}$ of consistent triples of atoms of $\mathfrak{A}$ by taking

$$h(u_1, u_2, u_3) = t_1 t_2 t_3 \quad \text{iff } \langle u_1, u_2\rangle \in rep(t_3),\ \langle u_3, u_1\rangle \in rep(t_2),\ \langle u_2, u_3\rangle \in rep(t_1).$$

It is easy to check that h is a well-defined p-morphism onto $\mathfrak{F}_{\mathfrak{A},3}$. ❑

Take a finite simple relation algebra $\mathfrak{A}$ and define, for $1 \le i < j \le 3$, a subset E_{ij} of $\mathcal{T}_{\mathfrak{A}}$ as follows. Let $k \in \{1, 2, 3\}$ be different from both i and j. Then

$$E_{ij} = \{t \in \mathcal{T}_{\mathfrak{A}} \mid t_k \le Id\}.$$

(Recall that Id denotes the identity element of $\mathfrak{A}$.) It is not hard to see that the following properties hold whenever $1 \le i < j \le 3$, $1 \le k \le 3$ and $k \ne i, j$:

$$\forall t \in \mathcal{T}_{\mathfrak{A}}\ \exists t', t'' \in E_{ij}\ (tR_i t' \ \&\ tR_j t''), \tag{8.28}$$

$$\forall t, t' \in \mathcal{T}_{\mathfrak{A}}\ (t \in E_{ij} \ \&\ tR_k t' \to t' \in E_{ij}), \tag{8.29}$$

$$E_{12} \cap E_{13} \subseteq E_{23},\quad E_{12} \cap E_{23} \subseteq E_{13},\quad E_{13} \cap E_{23} \subseteq E_{12}, \tag{8.30}$$

$$\forall t, t' \in E_{ij}\ (tR_i t' \vee tR_j t' \to t = t'). \tag{8.31}$$

The following claim is a consequence of Lemma 8.9.

Claim 8.40. *Assume that h is a p-morphism from a universal product $\mathbf{S5}^3$-frame $\langle U_1, U_2, U_3\rangle$ onto $\mathfrak{F}_{\mathfrak{A},3}$. Let U be the disjoint union of the sets U_i, $i = 1, 2, 3$. Then there is a p-morphism f from the universal product $\mathbf{S5}^3$-frame $\langle U, U, U\rangle$ onto $\mathfrak{F}_{\mathfrak{A},3}$ such that*

$$\text{for all } u_1, u_2, u_3 \in U,\ 1 \le i < j \le 3, \quad \text{if } u_i = u_j \text{ then } f(u_1, u_2, u_3) \in E_{ij}. \tag{8.32}$$

Proof. Suppose our propositional variables are d_{12}, d_{13}, d_{23} and a, for each atom a of $\mathfrak{A}$. Define a model $\mathfrak{M} = \langle\langle U_1, U_2, U_3\rangle, \mathfrak{V}\rangle$ by taking

$$\mathfrak{V}(a) = \{\langle u_1, u_2, u_3\rangle \mid h(u_1, u_2, u_3)_3 = a\},$$
$$\mathfrak{V}(d_{ij}) = \{\langle u_1, u_2, u_3\rangle \mid h(u_1, u_2, u_3) \in E_{ij}\},$$

and $\mathfrak{V}(p) = \emptyset$ for all other variables p. Using (8.28)–(8.30), it is readily checked that all formulas (8.7)–(8.12) are true in $\mathfrak{M}$. Therefore, by Lemma 8.9, there is an $\mathbf{S5}^3$-model $\mathfrak{M}^+ = \langle\langle U, U, U\rangle, \mathfrak{V}^+\rangle$ and a p-morphism k from $\mathfrak{M}^+$ onto $\mathfrak{M}$ such that

$$\mathfrak{V}^+(d_{ij}) \supseteq \{\langle u_1, u_2, u_3\rangle \mid u_i = u_j\}.$$

Define a function f from $U \times U \times U$ to $\mathcal{T}_{\mathfrak{A}}$ by taking, for all $u_1, u_2, u_3 \in U$,

$$f(u_1, u_2, u_3) = h(k(u_1, u_2, u_3)).$$

It should be clear that f is a p-morphism from $\langle U, U, U\rangle$ onto $\mathfrak{F}_{\mathfrak{A},3}$ satisfying (8.32). ❑

Our next claim is a consequence of Lemma 8.10:

Claim 8.41. *Suppose f is a p-morphism from a universal product $\mathbf{S5}^3$-frame $\langle U, U, U\rangle$ onto $\mathfrak{F}_{\mathfrak{A},3}$ satisfying* (8.32). *Then there is a set V with $|V| \leq |U|$ and a p-morphism g from $\langle V, V, V\rangle$ onto $\mathfrak{F}_{\mathfrak{A},3}$ such that*

$$\begin{aligned} &\textit{for all } v_1, v_2, v_3 \in V,\ 1 \leq i < j \leq 3, \\ &\qquad v_i = v_j \ \textit{iff}\ g(v_1, v_2, v_3) \in E_{ij}. \end{aligned} \tag{8.33}$$

Proof. Define a model $\mathfrak{M}^+ = \langle\langle U, U, U\rangle, \mathfrak{V}^+\rangle$ by taking

$$\begin{aligned} \mathfrak{V}^+(a) &= \{\langle u_1, u_2, u_3\rangle \mid f(u_1, u_2, u_3)_3 = a\}, \\ \mathfrak{V}^+(d_{ij}) &= \{\langle u_1, u_2, u_3\rangle \mid f(u_1, u_2, u_3) \in E_{ij}\}, \end{aligned}$$

and $\mathfrak{V}^+(p) = \emptyset$ for all other variables p. By (8.28)–(8.30), all formulas (8.7)–(8.12) are true in $\mathfrak{M}^+$, and by (8.31), (8.13) holds in $\mathfrak{M}^+$ as well. Moreover, by the definition of $\mathfrak{V}^+$ and (8.29), $\mathfrak{M}^+$ is binary generated. Therefore, by Lemma 8.10, there is an $\mathbf{S5}^3$-model $\mathfrak{N} = \langle\langle V, V, V\rangle, \mathfrak{U}\rangle$ and a p-morphism ℓ from $\mathfrak{M}^+$ onto $\mathfrak{N}$ such that $|V| \leq |U|$ and

$$\mathfrak{U}(d_{ij}) = \{\langle v_1, v_2, v_3\rangle \mid v_i = v_j\}.$$

Define a function g from $V \times V \times V$ to $T_{\mathfrak{A}}$ by taking, for all $v_1, v_2, v_3 \in V$,

$$g(v_1, v_2, v_3) = f(\ell^{-1}(v_1, v_2, v_3)).$$

(Here ℓ^{-1} denotes the inverse of ℓ.) Since ℓ is a p-morphism and by the definition of $\mathfrak{V}^+(a)$, the function g is well-defined. It is readily checked that g is a p-morphism from $\langle V, V, V\rangle$ onto $\mathfrak{F}_{\mathfrak{A},3}$ satisfying (8.33). ❑

Claim 8.42. (Monk) *Suppose g is a p-morphism from a universal product $\mathbf{S5}^3$-frame $\langle V, V, V\rangle$ onto $\mathfrak{F}_{\mathfrak{A},3}$ satisfying* (8.33). *Then the relation algebra $\mathfrak{A}$ is representable with base V, that is, $\mathfrak{A}$ is embeddable into the set relation algebra of all subsets of $V \times V$.*

Proof. Recall that the points of $\mathfrak{F}_{\mathfrak{A},3}$ are the consistent triples of atoms of $\mathfrak{A}$. Define the representation *rep* of $\mathfrak{A}$ with base V as follows. For each atom c of $\mathfrak{A}$, take

$$rep(c) = \{\langle u, v\rangle \in V \times V \mid \exists w \in V\ g(u, v, w)_3 = c\}.$$

Then, by the definition of $\mathfrak{F}_{\mathfrak{A},3}$, $rep(c_2)$ and $rep(c_3)$ are disjoint whenever $c_2 \neq c_3$. Extend *rep* to an arbitrary element x of $\mathfrak{A}$ by taking

$$rep(x) = \bigcup\{rep(c) \mid c \text{ is an atom of } \mathfrak{A} \text{ and } c \leq x\}.$$

It is straightforward to check that *rep* is a Boolean embedding. We show that it is a relation algebra homomorphism. First, $rep(Id) = \{\langle u, u\rangle \mid u \in V\}$ holds because of (8.33). Since ; and $^-$ distribute over the Boolean join $\vee$, it is enough to show that *rep* preserves ; and $^-$ for atoms. To this end, we need the following claim: for all $u, v, w \in V$ and atoms a, b, c of $\mathfrak{A}$,

$$g(u, v, w) = abc \text{ iff } \langle u, v\rangle \in rep(c),\ \langle v, w\rangle \in rep(a),\ \langle w, u\rangle \in rep(b). \quad (8.34)$$

We use the following property of $\mathfrak{F}_{\mathfrak{A},3}$: for all $t \in \mathcal{T}_{\mathfrak{A}}$, $1 \le i < j \le 3$ and $1 \le k \le 3$ with $k \ne i, j$,

$$t \in E_{ij} \implies t_k \le Id \implies t_i = t_j^- . \quad (8.35)$$

Suppose that $g(u, v, w) = abc$. Then $\langle u, v\rangle \in rep(c)$ by definition. In order to prove $\langle v, w\rangle \in rep(a)$, we show—with the help of (8.33) and (8.35)—that $g(v, w, u) = bca$:

$g(u, v, w) = abc$	$g(u, v, w) = abc$	$g(u, v, w) = abc$
R_2	R_3	R_1
$g(u, w, w) = {*}bb^-$	$g(u, v, u) = c^- {*} c$	$g(v, v, w) = aa^-{*}$
R_3	R_1	R_2
$g(u, w, u) = b{*}b^-$	$g(v, v, u) = c^-c{*}$	$g(v, w, w) = {*}a^-a$
R_1	R_2	R_3
$g(v, w, u) = b{*}{*}$	$g(v, w, u) = {*}c{*}$	$g(v, w, u) = {*}{*}a.$

In the same way one can show $g(w, u, v) = cab$. So $\langle w, u\rangle \in rep(b)$.

For the other direction, we know by (8.35) that $g(w, u, w) = b^- {*} b$ and $g(v, w, w) = {*}a^-a$. So again an argument similar to that above proves $g(u, v, w) = abc$.

Using (8.35) and (8.34), it is not hard to check that $rep(c)^- = rep(c^-)$ and $rep(c_2 ; c_3) = rep(c_2) ; rep(c_3)$ hold for all atoms c, c_2, c_3. ❑

Lemma 8.34 is a direct consequence of Claims 8.39–8.42. ❑

Lack of product fmp

First we show how Theorem 8.31 follows from what we have so far and then give a concrete, relatively simple, formula which forces an infinite product frame.

Proof of Theorem 8.31. Take some finite, simple, representable relation algebra $\mathfrak{A}$ which is representable only with an infinite base (e.g., the *linear* or *point relation algebra*; see (Maddux 1991)), and consider the 3-frame $\mathfrak{F}_{\mathfrak{A},3}$ and the $\mathcal{ML}_3$-formula $\varphi_{\mathfrak{A}}$. Then, by Lemmas 8.33 and 8.34, $\neg\varphi_{\mathfrak{A}}$ is not in L. We show that $\neg\varphi_{\mathfrak{A}}$ is valid in all finite k-dimensional product frames, for any $k \geq 3$. Suppose that there is a finite product frame satisfying $\varphi_{\mathfrak{A}}$. Then, by Claims 8.36 and 8.37, $\mathfrak{F}_{\mathfrak{A},3}$ is a p-morphic image of some finite universal product $\mathbf{S5}^3$-frame, contrary to Lemma 8.34, since $\mathfrak{A}$ is representable only with an infinite base. ❑

Now we construct a 6-element 3-frame $\mathfrak{F}$ and show that the frame formula for $\mathfrak{F}$ can be satisfied only in an infinite product frame. This $\mathfrak{F}$ is a simplification of the 3-frame $\mathfrak{F}_{\mathfrak{A},3}$ obtained from the linear (point) relation algebra used in the proof of Theorem 8.31.

Let F consist of all permutations of the set $\{1,2,3\}$. For $i = 1,2,3$, define R_i as 'forgetting about i in the triples,' that is, for $p, q \in F$, let pR_iq iff

$$p(j) < p(k) \iff q(j) < q(k), \quad \text{whenever } \{i,j,k\} = \{1,2,3\},$$

and let $\mathfrak{F} = \langle F, R_1, R_2, R_3 \rangle$. To simplify notation, given some $p \in F$, we write p_i for $p^{-1}(i)$ and identify p with the triple $p_1p_2p_3$; see Fig. 8.5. We also write $p = *i*j*$ whenever $p(i) < p(j)$ holds.

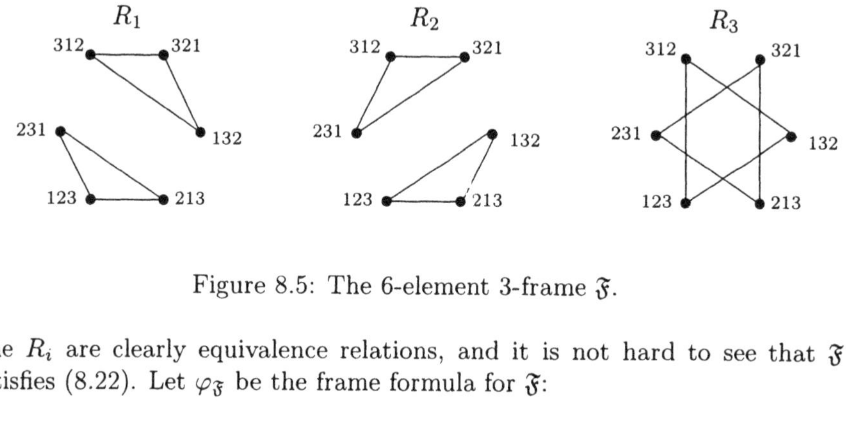

Figure 8.5: The 6-element 3-frame $\mathfrak{F}$.

The R_i are clearly equivalence relations, and it is not hard to see that $\mathfrak{F}$ satisfies (8.22). Let $\varphi_{\mathfrak{F}}$ be the frame formula for $\mathfrak{F}$:

$$\varphi_{\mathfrak{F}} = \Box^+ \bigvee_{p \in F} (p \wedge \neg \bigvee_{p' \neq p} p') \wedge \Box^+ \bigwedge_{\substack{i=1,2,3,\\ p,p' \in F,\, pR_ip'}} (p \to \Diamond_i p') \wedge \Box^+ \bigwedge_{\substack{i=1,2,3,\\ p,p' \in F,\, \neg(pR_ip')}} (p \to \neg\Diamond_i p').$$

Claim 8.43. *There is a product frame satisfying $\varphi_{\mathfrak{F}}$.*

Proof. Let Q_1, Q_2 and Q_3 be three pairwise disjoint dense subsets of the rationals. Take the universal product $\mathbf{S5}^3$-frame $\langle Q_1, Q_2, Q_3 \rangle$ and define a valuation $\mathfrak{V}$ in it as follows:

$$\mathfrak{V}(p) = \{\langle x_1, x_2, x_3 \rangle \in Q_1 \times Q_2 \times Q_3 \mid x_{p_1} < x_{p_2} < x_{p_3}\}.$$

Let $\mathfrak{M} = \langle\langle Q_1, Q_2, Q_3\rangle, \mathfrak{V}\rangle$. It is not hard to check that

$$(\mathfrak{M}, \langle x_1, x_2, x_3\rangle) \models \varphi_{\mathfrak{F}},$$

for any $\langle x_1, x_2, x_3\rangle$. ❑

Claim 8.44. *Any product frame satisfying $\varphi_{\mathfrak{F}}$ is infinite.*

Proof. Let $\mathfrak{M}$ be a model on the product frame $\langle U, S_U\rangle \times \langle V, S_V\rangle \times \langle W, S_W\rangle$. For simplicity, points $\langle x, y, z\rangle$ in $\mathfrak{M}$ will be denoted by xyz. Suppose that $x_0 \in U$, $y_0 \in V$, $z_0 \in W$ are such that

$$(\mathfrak{M}, x_0y_0z_0) \models \varphi_{\mathfrak{F}} \text{ and, say, } (\mathfrak{M}, x_0y_0z_0) \models 312. \tag{8.36}$$

We will show then that both U and V must be infinite sets. Let $0 < n < \omega$ and assume inductively that we have already defined points $x_i \in U$ and $y_i \in V$ for each $i < n$ such that

$$\begin{aligned}
&x_0S_Ux_i \text{ and } y_0S_Vy_i, \text{ for } 0 < i < n,\\
&x_i \neq x_j \text{ and } y_i \neq y_j, \text{ for } i,j < n,\ i \neq j,\\
&(\mathfrak{M}, x_iy_jz_0) \models 312, \text{ for } i \leq j < n, && (8.37)\\
&(\mathfrak{M}, x_iy_jz_0) \models 321, \text{ for } j < i < n. && (8.38)
\end{aligned}$$

Define x_n and y_n. We have $312R_1321$ and, by (8.37), $(\mathfrak{M}, x_0y_{n-1}z_0) \models 312$. By (8.36), there is some $x_n \in U$ such that

$$x_0S_Ux_n \text{ and } (\mathfrak{M}, x_ny_{n-1}z_0) \models 321. \tag{8.39}$$

By (8.36) and (8.37), $x_n \neq x_i$, for $i < n$. We show that

$$(\mathfrak{M}, x_ny_iz_0) \models 321 \text{ for all } i < n-1 \tag{8.40}$$

(see Fig. 8.6). To this end we first prove the following claim: there are no points $u_0, u_1 \in U$ and $v_0, v_1 \in V$ such that

- $(\mathfrak{M}, u_0v_0z_0) \models 321$ and $(\mathfrak{M}, u_1v_1z_0) \models 321$,
- $(\mathfrak{M}, u_0v_1z_0) \models 312$ and $(\mathfrak{M}, u_1v_0z_0) \models 312$,

and, for each $i < 2$,

- either $u_i = x_0$ or $x_0S_Uu_i$, and
- either $v_i = y_0$ or $y_0S_Vv_i$.

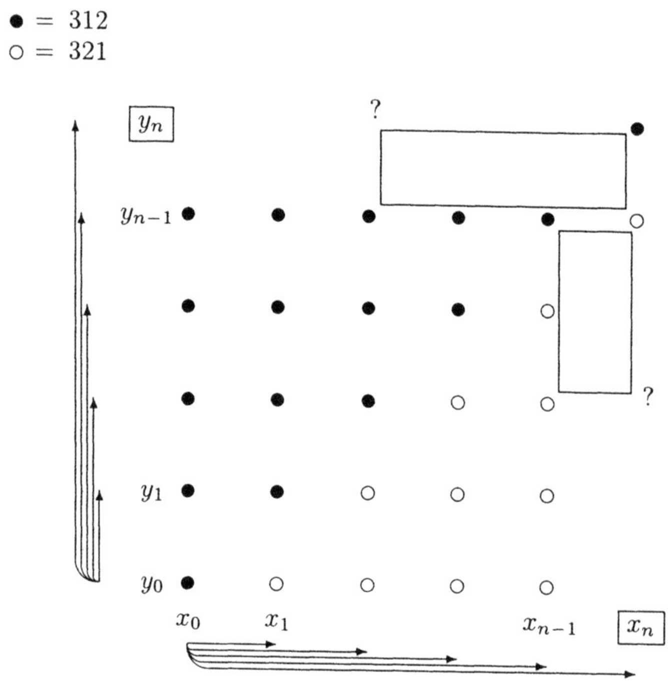

Figure 8.6: The points x_n and y_n.

Suppose that such points u_0, u_1, v_0, v_1 do exist; see Fig. 8.7. Since we have $(\mathfrak{M}, u_0v_1z_0) \models 312$ and $312R_3132$, there is a $z \in W$ such that z_0S_Wz and

$$(\mathfrak{M}, u_0v_1z) \models 132. \tag{8.41}$$

Then $(\mathfrak{M}, u_0y_0z) \models a$, for some $a \in F$ with $a = *1*3*$, from which

$$(\mathfrak{M}, u_0v_0z) \models b, \quad \text{for some } b \in F \text{ with } b = *1*3*. \tag{8.42}$$

On the other hand, $(\mathfrak{M}, u_0v_0z_0) \models 321$ by assumption. Thus $b = *2*1*$, which, by (8.42), means that

$$b = 213. \tag{8.43}$$

By (8.41), $(\mathfrak{M}, x_0v_1z) \models c$ for some $c \in F$ with $c = *3*2*$. So $(\mathfrak{M}, u_1v_1z) \models d$, for some $d \in F$ with $d = *3*2*$. On the other hand, since by assumption $(\mathfrak{M}, u_1v_1z_0) \models 321$, we have $d = *2*1*$, and so $d = 321$. Therefore, $(\mathfrak{M}, u_1y_0z) \models e$ for some $e \in F$ with $e = *3*1*$. Hence $(\mathfrak{M}, u_1v_0z) \models f$, for some $f \in F$ with $f = *3*1*$. By (8.43), we have $(\mathfrak{M}, x_0v_0z) \models g$ for some $g \in F$ with $g = *2*3*$, whence $f = *2*3*$, and so $f = 231$. It follows that

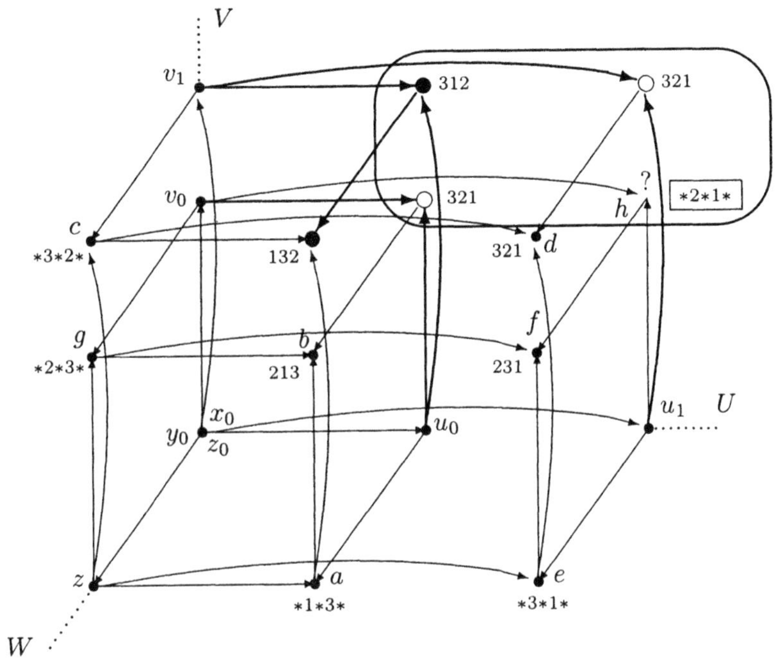

Figure 8.7: The points u_0, u_1, v_0, v_1.

$(\mathfrak{M}, u_1 v_0 z_0) \models h$ for some $h \in F$ with $h = *2*1*$, contrary to the assumption $(\mathfrak{M}, u_1 v_0 z_0) \models 312$.

Now one can prove (8.40) as follows. Take some $i < n-1$. By (8.38), we then have $(\mathfrak{M}, x_{n-1} y_i z_0) \models 321$. Therefore, $(\mathfrak{M}, x_0 y_i z_0) \models k$ for some $k \in F$ with $k = *3*2*$. Thus $(\mathfrak{M}, x_n y_i z_0) \models \ell$ for some $\ell \in F$ with $\ell = *3*2*$. On the other hand, by (8.39), $(\mathfrak{M}, x_n y_0 z_0) \models m$ for some $m \in F$ with $m = *3*1*$, and so $\ell = *3*1*$. Hence, either $\ell = 312$ or $\ell = 321$. Finally, we use the claim above with $u_0 = x_{n-1}$, $u_1 = x_n$, $v_0 = y_i$ and $v_1 = y_{n-1}$ to obtain $\ell = 321$.

Now we can define y_n. We know that $321 R_2 312$ and we have just shown that $(\mathfrak{M}, x_n y_0 z_0) \models 321$. By (8.36), there is some $y_n \in V$ such that

$$y_0 S_V y_n \quad \text{and} \quad (\mathfrak{M}, x_n y_n z_0) \models 312. \tag{8.44}$$

By (8.36), (8.39) and (8.40), $y_n \neq y_i$, for $i < n$. It remains to show that, for all $i < n$, $(\mathfrak{M}, x_i y_n z_0) \models 312$ holds as well. To this end, take some $i < n$. By (8.44), we have $(\mathfrak{M}, x_0 y_n z_0) \models p$ for some $p \in F$ with $p = *3*2*$. Thus, $(\mathfrak{M}, x_i y_n z_0) \models q$ for some $q \in F$ with $q = *3*2*$. On the other hand, by (8.37) and (8.38), $q = *3*1*$, and so either $q = 312$ or $q = 321$. Now apply the above claim with $u_0 = x_i$, $u_1 = x_n$, $v_0 = y_n$ and $v_1 = y_{n-1}$ to obtain $q = 312$. Thus, we have shown that both U and V are infinite, which completes the proof of Claim 8.44. ❑

We conclude this section with three open problems:

Question 8.45. Let $n \geq 3$.

(1) Give an (infinite) axiomatization of $\mathbf{K}^n$.

(2) Is it possible to axiomatize $\mathbf{K}^n$ (or any product logic between $\mathbf{K}^n$ and $\mathbf{S5}^n$) using only finitely many propositional variables?

(3) Is $\mathbf{S5}^n$ finitely axiomatizable over $\mathbf{K}^n$?

8.5 Finitely axiomatizable and decidable products

Although the previous section shows that many higher-dimensional products are neither decidable nor finitely axiomatizable, there are some (not completely trivial) examples of product logics with a better computational behavior. In this section, we prove the result of Gabbay and Shehtman (1998) according to which all (finite) products of the logics **Alt** and **DAlt** are finitely axiomatizable (in fact, product-matching) and decidable.

Theorem 8.46. *Any product of* **Alt** *and* **DAlt** *is finitely axiomatizable. In particular,* $\mathbf{Alt}^n$, $\mathbf{DAlt}^n$ *and* $\mathbf{Alt}^n \times \mathbf{DAlt}^m$ *are finitely axiomatizable, for all* $n, m \geq 1$, *namely,*

$$\begin{aligned} \mathbf{Alt}^n &= [\mathbf{Alt}, \mathbf{Alt}, \dots, \mathbf{Alt}], \\ \mathbf{DAlt}^n &= [\mathbf{DAlt}, \mathbf{DAlt}, \dots, \mathbf{DAlt}], \\ \mathbf{Alt}^n \times \mathbf{DAlt}^m &= [\underbrace{\mathbf{Alt}, \dots, \mathbf{Alt}}_{n}, \underbrace{\mathbf{DAlt}, \dots, \mathbf{DAlt}}_{m}]. \end{aligned}$$

Proof. We remind the reader that every rooted Kripke frame for **Alt** is the p-morphic image of an intransitive chain, i.e., an intransitive tree with a single branch. Rooted frames for **DAlt** are p-morphic images of infinite intransitive chains. The proof of the theorem is based on the fact that for products of **Alt** and **DAlt** the following higher-dimensional analog of Lemmas 5.2 and 5.8 holds:

Lemma 8.47. *Every countable rooted n-frame for* $[\mathbf{Alt}, \mathbf{Alt}, \dots, \mathbf{Alt}]$ *is a p-morphic image of the product of n countable intransitive chains.*

Proof. We consider only the case of $n = 3$; for $n \geq 3$ the proof is similar. Like in the proof of Lemma 5.2, we formalize the step-by-step argument with the help of two-player games. Suppose that

$$\mathfrak{G} = \langle G, S_1, S_2, S_3 \rangle$$

is a countable rooted frame for $[\mathbf{Alt}, \mathbf{Alt}, \mathbf{Alt}]$. The game $G(\mathfrak{G})$ over $\mathfrak{G}$ is a modification of the game in the proof of Lemma 5.2. Namely, a $\mathfrak{G}$-*network* is a tuple

$$N = \langle U^N, V^N, W^N, R_1^N, R_2^N, R_3^N, f^N \rangle,$$

where $\mathfrak{F}_1^N = \langle U^N, R_1^N \rangle$, $\mathfrak{F}_2^N = \langle V^N, R_2^N \rangle$ and $\mathfrak{F}_3^N = \langle W^N, R_3^N \rangle$ are finite intransitive chains and f^N is a homomorphism from $\mathfrak{F}_1^N \times \mathfrak{F}_2^N \times \mathfrak{F}_3^N$ to $\mathfrak{G}$.

As before, the players $\forall$ and $\exists$ build a countable sequence of finite $\mathfrak{G}$-networks

$$N_0 \subseteq N_1 \subseteq \ldots \subseteq N_i \subseteq \ldots .$$

In round 0, $\forall$ picks the root r of $\mathfrak{G}$. $\exists$ responds with some $\mathfrak{G}$-network N_0 such that U^{N_0}, V^{N_0} and W^{N_0} are all one-element sets, the accessibility relations $R_i^{N_0}$ are empty, for all $i = 1, 2, 3$, and f^{N_0} takes the only triplet to r.

In round i $(0 < i < \omega)$, some sequence $N_0 \subseteq \cdots \subseteq N_{i-1}$ of $\mathfrak{G}$-networks is already built. $\forall$ picks

- a triplet $\langle u, v, w \rangle \in U^{N_{i-1}} \times V^{N_{i-1}} \times W^{N_{i-1}}$,
- a 'direction' d such that $1 \leq d \leq 3$,
- a world g in $\mathfrak{G}$ such that $f^{N_{i-1}}(u, v, w) S_d g$.

Player $\exists$ can respond in two ways. Assume that $\forall$ picked direction $d = 1$. If there is some $u' \in U^{N_{i-1}}$ with $u R_1^{N_{i-1}} u'$ then $f^{N_{i-1}}(u', v, w) = g$ must hold, since $f^{N_{i-1}}$ is a homomorphism and $\langle G, S_1 \rangle \models \mathbf{Alt}$. In this case $\exists$ responds with $N_i = N_{i-1}$. Otherwise, she responds (if she can) with some $\mathfrak{G}$-network N_i extending N_{i-1} in such a way that

- $U^{N_i} = U^{N_{i-1}} \cup \{u^+\}$ (where u^+ is a fresh point),
- $R_1^{N_i} = R_1^{N_{i-1}} \cup \{\langle u, u^+ \rangle\}$,
- $\mathfrak{F}_k^{N_i} = \mathfrak{F}_k^{N_{i-1}}$, for $k = 2, 3$, and
- $f^{N_i}(u^+, v, w) = g$.

If $\forall$ picked direction 2 or 3, $\exists$'s move is similar, possibly extending $\mathfrak{F}_2^{N_{i-1}}$ or $\mathfrak{F}_3^{N_{i-1}}$. Note that in any case the frames $\mathfrak{F}_k^{N_i}$ are finite intransitive chains again, for all $k = 1, 2, 3$.

$\exists$ *has a winning strategy in the game* $G(\mathfrak{G})$ if she can respond in each round $i < \omega$, whatever moves $\forall$ chooses to make. Similarly to Claim 5.3, one can prove the following:

Claim 8.48. *If $\exists$ has a winning strategy in $G(\mathfrak{G})$ then there are countable intransitive chains $\mathfrak{F}_1$, $\mathfrak{F}_2$, $\mathfrak{F}_3$ such that $\mathfrak{G}$ is a p-morphic image of $\mathfrak{F}_1 \times \mathfrak{F}_2 \times \mathfrak{F}_3$.*

Using the fact that $\mathfrak{G}$ is a frame for $[\mathbf{Alt}, \mathbf{Alt}, \mathbf{Alt}]$, it remains to define a winning strategy for $\exists$ in the game $G(\mathfrak{G})$. In round 0, her response is determined by the rules of the game. In round i $(0 < i < \omega)$, some sequence $N_0 \subseteq \cdots \subseteq N_{i-1}$ of $\mathfrak{G}$-networks is already constructed. Assume that $\forall$ picks the triplet $\langle u, v, w\rangle \in U^{N_{i-1}} \times V^{N_{i-1}} \times W^{N_{i-1}}$, direction d and world g in $\mathfrak{G}$ such that $f^{N_{i-1}}(u, v, w) S_d g$.

Suppose $d = 1$ (the cases of $d = 2, 3$ are similar). By the rules of the game, if there is $u' \in U^{N_{i-1}}$ such that $u R_1^{N_{i-1}} u'$ then $\exists$ responds with $N_i = N_{i-1}$. Otherwise, she has to add a fresh point u^+ to $U^{N_{i-1}}$ and respond with some $\mathfrak{G}$-network N_i satisfying the above conditions. $f^{N_i}(u^+, v, w)$ is defined to be g by the rules. The remaining task is to define f^{N_i} on all the triplets of the form $\langle u^+, v', w'\rangle$, where $v' \in V^{N_i} = V^{N_{i-1}}$, $w' \in W^{N_i} = W^{N_{i-1}}$, and $\langle v', w'\rangle \neq \langle v, w\rangle$.

Claim 8.49. *There are enumerations $\{v_0, v_1, \dots, v_{M_1}\}$ and $\{w_0, w_1, \dots, w_{M_2}\}$ of $V^{N_{i-1}}$ and $W^{N_{i-1}}$, respectively, satisfying the following properties:*

- *$v_0 = v$ and $w_0 = w$;*
- *for all k, $0 < k \leq M_1$, there is a unique index $pred(k) < k$ such that either $v_{pred(k)} R_2^{N_{i-1}} v_k$ or $v_k R_2^{N_{i-1}} v_{pred(k)}$;*
- *for all ℓ $0 < \ell \leq M_2$, there is a unique index $pred(\ell) < \ell$ such that either $w_{pred(\ell)} R_3^{N_{i-1}} w_\ell$ or $w_\ell R_3^{N_{i-1}} w_{pred(\ell)}$.*

Figure 8.8: Enumerating $\mathfrak{F}_2^{N_{i-1}}$.

Proof. Take, for example, the unique $R_2^{N_{i-1}}$-path, starting from the root of the chain $\mathfrak{F}_2^{N_{i-1}}$ and ending with v, and enumerate it backwards; then continue with enumerating the chain starting with the $R_2^{N_{i-1}}$-successor of v; see Fig. 8.8. Do the same for $\mathfrak{F}_3^{N_{i-1}}$ and w. ❑

In order to define f^{N_i} on the new triplets, first define $\prec$ to be the lexicographic ordering on the pairs of numbers induced by the above enumerations:

$$\langle j, \ell\rangle \prec \langle k, m\rangle \quad \text{iff} \quad \text{either } j < k \text{ or } j = k \text{ and } \ell < m.$$

Now let $\langle 0,0\rangle \prec \langle k,m\rangle \prec \langle M_1+1, M_2+1\rangle$ and assume inductively that we have already defined $f^{N_i}(u^+, v_j, w_\ell)$, for all $\langle j,\ell\rangle \prec \langle k,m\rangle$ such that

$$\begin{array}{ll} f^{N_i}(u, v_j, w_\ell) S_1 f^{N_i}(u^+, v_j, w_\ell), & \\ f^{N_i}(u^+, v_j, w_\ell) S_2 f^{N_i}(u^+, v_{pred(j)}, w_\ell), & \text{if } j>0 \text{ and } v_j R_2^{N_{i-1}} v_{pred(j)}, \\ f^{N_i}(u^+, v_{pred(j)}, w_\ell) S_2 f^{N_i}(u^+, v_j, w_\ell), & \text{if } j>0 \text{ and } v_{pred(j)} R_2^{N_{i-1}} v_j, \\ f^{N_i}(u^+, v_j, w_\ell) S_3 f^{N_i}(u^+, v_j, w_{pred(\ell)})), & \text{if } \ell>0 \text{ and } w_\ell R_3^{N_{i-1}} w_{pred(\ell)}, \\ f^{N_i}(u^+, v_j, w_{pred(\ell)}) S_3 f^{N_i}(u^+, v_j, w_\ell), & \text{if } \ell>0 \text{ and } w_{pred(\ell)} R_3^{N_{i-1}} w_\ell. \end{array}$$

We will now define $f^{N_i}(u^+, v_k, w_m)$. By Claim 8.49, we have to consider the following cases:

(1) $m=0$, i.e., $w_m = w$ and either
(1a) $v_{pred(k)} R_2^{N_{i-1}} v_k$ or
(1b) $v_k R_2^{N_{i-1}} v_{pred(k)}$.
(2) $k=0$, i.e., $v_k = v$ and either
(2a) $w_{pred(m)} R_3^{N_{i-1}} w_m$ or
(2b) $w_m R_3^{N_{i-1}} w_{pred(m)}$.
(3) $m, k > 0$ and either
(3a) $v_k R_2^{N_{i-1}} v_{pred(k)}$ and $w_m R_3^{N_{i-1}} w_{pred(m)}$, or
(3b) $v_{pred(k)} R_2^{N_{i-1}} v_k$ and $w_m R_3^{N_{i-1}} w_{pred(m)}$, or
(3c) $v_k R_2^{N_{i-1}} v_{pred(k)}$ and $w_{pred(m)} R_3^{N_{i-1}} w_m$, or
(3d) $v_{pred(k)} R_2^{N_{i-1}} v_k$ and $w_{pred(m)} R_3^{N_{i-1}} w_m$.

Cases (1a)–(2b) are similar to cases 1 and 2 in the proof of Lemma 5.2. In order to define $f^{N_i}(u^+, v_k, w_m)$, one has to use properties $\boldsymbol{chr}_{12}$, $\boldsymbol{com}_{12}$, $\boldsymbol{chr}_{13}$ and $\boldsymbol{com}_{13}$ of $\mathfrak{G}$, respectively. Consider case (3a). Figure 8.9 shows the relevant points of $\mathfrak{G}$ and the relations among them which hold by the induction hypothesis. Since $\langle G, S_1\rangle \models \mathbf{Alt}$ and $\boldsymbol{com}_{12} \wedge \boldsymbol{com}_{13}$ holds in $\mathfrak{G}$, there is a (unique) $s \in G$ such that

$$f^{N_i}(u, v_k, w_m) S_1 s, \;\; s S_2 f^{N_i}(u^+, v_{pred(k)}, w_m) \text{ and } s S_3 f^{N_i}(u^+, v_k, w_{pred(m)}).$$

Put $f^{N_i}(u^+, v_k, w_m) = s$. The cases (3b)–(3d) are similar; see Fig. 8.9. It is straightforward to see that F^{N_i} satisfies the induction hypothesis. Finally, as in the proof of Lemma 5.2, the induction hypothesis can be used to show that f^{N_i} is a homomorphism from $\mathfrak{F}_1^{N_i} \times \mathfrak{F}_2^{N_i} \times \mathfrak{F}_3^{N_i}$ to $\mathfrak{G}$. ❑

To complete the proof of Theorem 8.46, let

$$L = [\underbrace{\mathbf{Alt}, \dots, \mathbf{Alt}}_{n}, \underbrace{\mathbf{DAlt}, \dots, \mathbf{DAlt}}_{m}].$$

Case (3a):

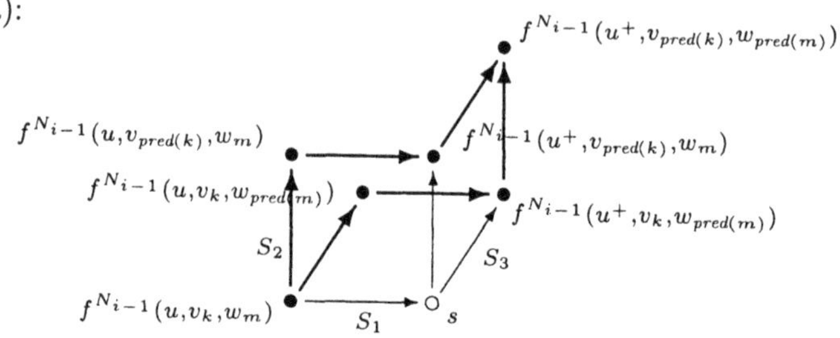

Case (3b):

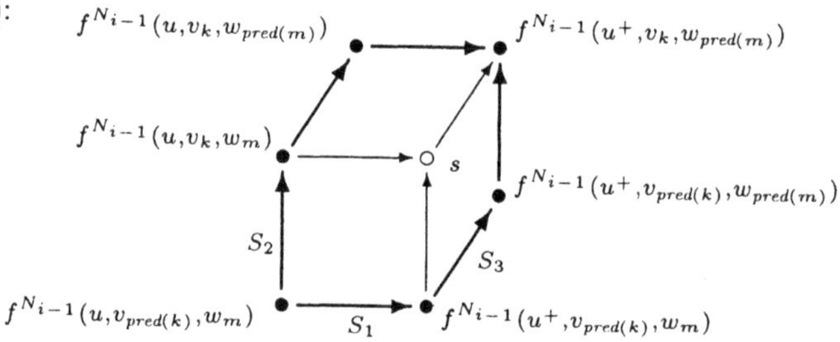

Case (3c):

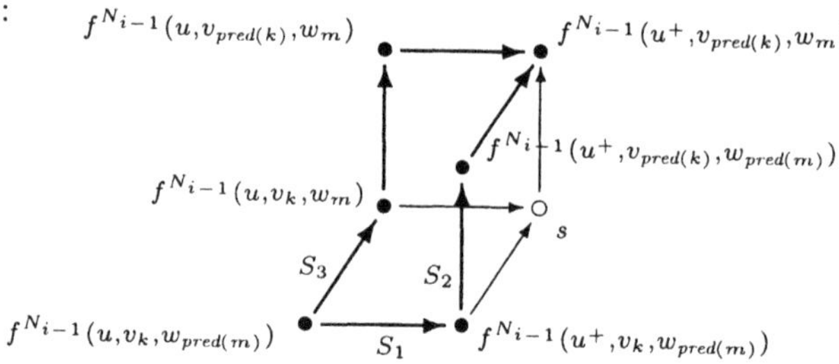

Case (3d):

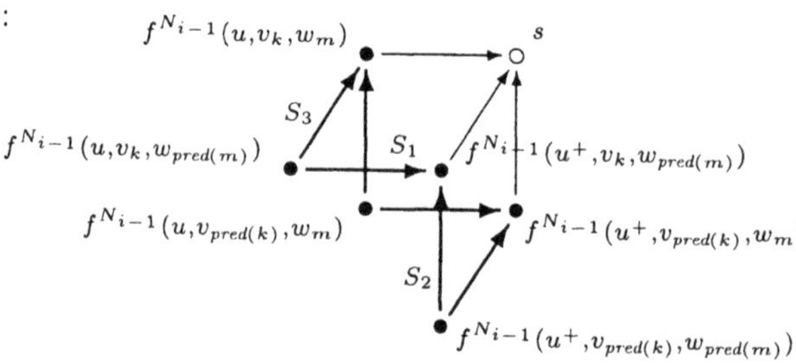

Figure 8.9: Case (3).

As we know, $\mathsf{Fr}L$ is the class of all commutative and Church–Rosser frames with n functional and m functional and serial accessibility relations. Thus, it is first-order definable in the language having $n+m$ binary predicate symbols. Therefore, by Theorem 1.6, $\varphi \notin L$ means that there is a countable $n+m$-frame $\mathfrak{F}$ for L and a model $\mathfrak{M}$ based on $\mathfrak{F}$ such that $(\mathfrak{M}, u) \not\models \varphi$ for some point u. Then we also have $\mathfrak{G} \not\models \varphi$, for the countable subframe $\mathfrak{G}$ of $\mathfrak{F}$ generated by u. Now $\varphi \notin \mathbf{Alt}^n \times \mathbf{DAlt}^m$ follows from Lemma 8.47, since infinite intransitive chains are frames for **DAlt**. ❑

As consequences of Theorem 8.46 and Lemma 8.47 we obtain:

Corollary 8.50. *Let* $L_1, L_2, L_3 \in \{\mathbf{Alt}, \mathbf{DAlt}\}$. *Then*

$$L_1 \times L_2 \times L_3 = (L_1 \times L_2) \times L_3 = L_1 \times (L_2 \times L_3).$$

Corollary 8.51. *Let* $L_1, L_2 \in \{\mathbf{Alt}, \mathbf{DAlt}\}$. *Then* $L_1 \times L_2$ *is globally Kripke complete, and* $\vdash^*_{L_1 \times L_2}$ *coincides with* $\vdash^*_{L_1} \times \vdash^*_{L_2}$.

Given a formula $\varphi \notin \mathbf{Alt}^n \times \mathbf{DAlt}^m$, one can cut the component chains of a product frame refuting φ at the modal depth $md(\varphi)$ of φ. In the **DAlt**-components, the last point of the corresponding chain should be made reflexive. The resulting product frame is still a frame for $\mathbf{Alt}^n \times \mathbf{DAlt}^m$, it refutes φ and its size is polynomial in the length of φ. Thus, we obtain the following theorem (which is also a consequence of Theorem 8.24):

Theorem 8.52. $\mathbf{Alt}^n$, $\mathbf{DAlt}^n$ *and* $\mathbf{Alt}^n \times \mathbf{DAlt}^m$ *have the polynomial product fmp, for all* $n, m \geq 1$.

Putting together the results obtained earlier in this section, we arrive at the following:

Theorem 8.53. *The decision problem for* $\mathbf{Alt}^n$, $\mathbf{DAlt}^n$ *and* $\mathbf{Alt}^n \times \mathbf{DAlt}^m$ *is* coNP*-complete, for all* $n, m \geq 1$.

On the other hand, the proof of Theorem 5.36 also yields the following:

Theorem 8.54. *The global consequence relations for logics like* $\mathbf{Alt} \times \mathbf{K}$, $\mathbf{D} \times \mathbf{D}$, $\mathbf{Alt} \times \mathbf{Alt}$ *and* $\mathbf{DAlt} \times \mathbf{DAlt}$ *are undecidable.*

Note that by Lemma 1.24 we also obtain the undecidability of $(\mathbf{D} \times \mathbf{D})_u$, $(\mathbf{Alt} \times \mathbf{Alt})_u$ and $(\mathbf{DAlt} \times \mathbf{DAlt})_u$. A proof similar to that of Theorem 5.37 gives the undecidability of $\mathbf{D}_u \times \mathbf{D}_u$, $\mathbf{Alt}_u \times \mathbf{Alt}_u$ and $\mathbf{DAlt}_u \times \mathbf{DAlt}_u$.

Finally, we observe that by 'mixing' the proofs of Lemmas 8.47 and 5.8 one can show that every countable rooted 2-frame for $[\mathbf{Alt}, L]$ is a p-morphic image of a product frame for $\mathbf{Alt} \times L$, whenever L is a Kripke complete and Horn axiomatizable logic. From this we obtain:

Theorem 8.55. *Let L be a Kripke complete and Horn axiomatizable unimodal logic. Then* $\mathbf{Alt} \times L = [\mathbf{Alt}, L]$.

Note that this theorem together with Theorem 8.24 gives another proof of the decidability of $\mathbf{Alt} \times \mathbf{K}$ (cf. Theorem 6.6).

Chapter 9

Variations on products

So far in Part II we have been considering two ways of combining modal logics: fusions and products.

Fusions (which can actually be defined for a wide range of knowledge representation formalisms called in (Baader *et al.* 2002) *abstract description systems*) are used to speak about different but not interacting aspects of application domains. For example, we may take the fusion of n copies of **S5**, each of which representing knowledge of a single agent, and of n copies of **KD45** representing their beliefs. The resulting combination

$$\mathbf{S5}_n \otimes \mathbf{KD45}_n$$

is capable of reasoning about knowledge and beliefs of the n agents 'living independently and knowing nothing of each other.' Moreover, it provides no connection between what is known and what is believed by agent i whatsoever, say, the formula

$$\Box_i^{\mathbf{S5}} \varphi \to \Box_i^{\mathbf{KD45}} \varphi$$

('if agent i knows φ then i believes that φ holds') does not belong to the fusion.

It is the absence of any interaction between the modal operators of the fused logics that ensures good algorithmic behavior of the fusions, as was shown in Chapter 4.[1]

Products of logics do provide such interactions, which makes them a good tool for constructing formalisms suitable for, say, spatio-temporal representation and reasoning; see Section 3.2 and Chapter 16. However, as we saw earlier

[1]Note, however, that the classes of models of the fused logics must be closed under disjoint unions, which is not the case when we form fusions of, say, description logics with nominals or negations of roles. To overcome this difficulty, another method of combining logics, called *e-connections*, was introduced in (Kutz *et al.* 2002).

in this part, the formation of products can dramatically increase the computational complexity of logics (remember, the compass logic of Section 2.6—i.e., the product of two NP-complete logics $\mathsf{Log}\{\langle \mathbb{N}, < \rangle\}$—is not even recursively enumerable; see Corollary 7.13).

Some examples considered above—say, modal description or modal first-order logics with expanding, decreasing, and arbitrary domains, or spatio-temporal logics with the finite state assumption—suggest two possible ways of reducing the expressive power of product logics in the hope of obtaining more 'user-friendly' and still useful many-dimensional formalisms.

First, in Section 9.1 we consider sublogics of product logics determined by classes of certain (not necessarily generated) *subframes* of their product frames. This kind of restriction on the 'domains' of modal operators is similar to 'relativizations' of the quantifiers in first-order logic and algebraic logic, where it indeed results in improving the bad algorithmic behavior of logics, cf. (Németi 1995, Marx and Venema 1997).

And second, in Section 9.2 we impose various restrictions on possible valuations in product frames.

Unfortunately, neither of these ways has been studied systematically yet. The modest aim of this chapter is only to give a few (sometimes nontrivial) observations and, perhaps, some warnings.

9.1 Relativized products

Product logics are determined by classes of product frames. The attractive feature of product frames is their geometrically intuitive many-dimensional structure: worlds are tuples and the accessibility relations act coordinatewise. However, this nice structure results in strong interaction, like commutativity, between the different modal operators. A natural way of loosening this strong connection but keeping the transparent many-dimensional structure is to consider *subframes* of product frames. Worlds are still tuples, the relations still act coordinatewise, but not all tuples of the Cartesian product are available, so the commutativity and Church–Rosser properties do not necessarily hold.

This idea gives rise to the following 'product-like' combinations of logics. First, we choose a class of 'desirable' subframes of product frames. This can be any class: the class of all such subframes, the so-called 'locally cubic' frames, frames that 'expand' along one of the coordinates (see below for precise definitions), a class of frames satisfying some (modal or first-order) formulas, etc. Having chosen such a class $\mathcal{K}$, we then take the logic determined by those subframes of the appropriate product frames that belong to $\mathcal{K}$. Thus, each choice of $\mathcal{K}$ defines a new product-like operator on logics. As we shall see, in many cases the resulting logics are indeed located between the fusions and the products of the components.

Formally, let n be a positive natural number and $\mathcal{K}$ a class of subframes of n-ary product frames. Given Kripke complete unimodal logics $L_1, \ldots, L_n$, the *$\mathcal{K}$-relativized product* $(L_1 \times \cdots \times L_n)^{\mathcal{K}}$ of $L_1, \ldots, L_n$ is defined by taking

$$(L_1 \times \cdots \times L_n)^{\mathcal{K}} = \mathsf{Log}\{\mathfrak{G} \in \mathcal{K} \mid \mathfrak{G} \subseteq \mathfrak{F} \text{ for some } \mathfrak{F} \in \mathsf{Fr}L_1 \times \cdots \times \mathsf{Fr}L_n\}.$$

The usual product logics are then special cases of $\mathcal{K}$-relativized product logics:

$$L_1 \times \cdots \times L_n = (L_1 \times \cdots \times L_n)^{\mathsf{Fr}L_1 \times \cdots \times \mathsf{Fr}L_n}.$$

Relativized products were first suggested as a modification of the product construction by Mikulás and Marx (2000). The results of this section were obtained in (Kurucz and Zakharyaschev 2003).

Arbitrary relativizations

We begin by considering the product operator determined by the class SF_n of *all* subframes of n-ary product frames. SF_n-relativized products of logics will be called *arbitrarily relativized products*. Clearly, for all classes $\mathcal{K}$ such that $\mathsf{Fr}L_1 \times \cdots \times \mathsf{Fr}L_n \subseteq \mathcal{K} \subseteq \mathsf{SF}_n$, we have

$$(L_1 \times \cdots \times L_n)^{\mathcal{K}} \subseteq L_1 \times \cdots \times L_n.$$

Note that if $n \geq 2$ and $\mathcal{K}$ contains a frame which does not satisfy either commutativity or the Church–Rosser property (e.g., $\mathcal{K} = \mathsf{SF}_n$) then this inclusion is proper.

On the other hand, unlike product logics, arbitrarily relativized products of logics do not necessarily contain the fusion of the components. For example, the formula $\Diamond_2\top$ clearly belongs to the fusion $\mathbf{K} \otimes \mathbf{D}$, but is refuted in any *finite* subframe of, say, $\langle \omega, < \rangle \times \langle \omega, < \rangle$, and so $\Diamond_2\top \notin (\mathbf{K} \times \mathbf{D})^{\mathsf{SF}_2}$. However, as we shall see below, for a large class of natural logics, arbitrarily relativized products do contain the fusions.

A Kripke complete modal logic L is called a *subframe logic* if for all $\mathfrak{F} \in \mathsf{Fr}L$ and $\mathfrak{G} \subseteq \mathfrak{F}$, we have $\mathfrak{G} \in \mathsf{Fr}L$ as well (for a general theory of subframe logics consult (Fine 1985, Chagrov and Zakharyaschev 1997, Wolter 1997, Zakharyaschev *et al.* 2001) and references therein). Typical examples of subframe logics are those determined by classes of Kripke frames that are definable by universal first-order formulas. The reader can easily check that all logics in Fig. 1.1 except **D**, **DAlt**, and **KD45** are subframe logics. (Note that **GL**, **GL.3** and **Grz** are subframe logics but not first-order definable.)

Proposition 9.1. *If $L_1, \ldots, L_n$ are subframe logics then*

$$L_1 \otimes \cdots \otimes L_n \subseteq (L_1 \times \cdots \times L_n)^{\mathsf{SF}_n}. \tag{9.1}$$

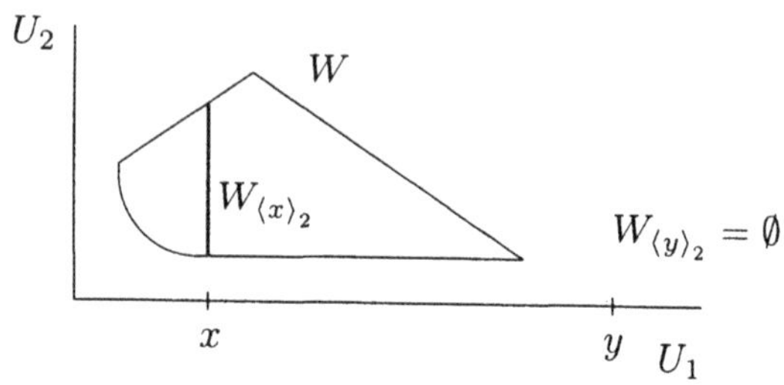

Figure 9.1: 'Coordinatewise' subframes.

Proof. The proof is similar to that of Proposition 3.8. Suppose that an n-frame $\mathfrak{G} = \langle W, S_1, \dots, S_n\rangle$ is a subframe of some product frame

$$\langle U_1, R_1\rangle \times \dots \times \langle U_n, R_n\rangle \in \mathsf{Fr}L_1 \times \dots \times \mathsf{Fr}L_n.$$

Fix some i, $1 \le i \le n$. For every $n-1$-tuple $\overline{u}_i = \langle u_1, \dots, u_{i-1}, u_{i+1}, \dots, u_n\rangle$ with $u_j \in U_j$, for $j \ne i$, we take the set

$$W_{\overline{u}_i} = \{\langle u_1, \dots, u_n\rangle \in W \mid u_i \in U_i,\ \langle u_1, \dots, u_{i-1}, u_{i+1}, \dots, u_n\rangle = \overline{u}_i\},$$

and let $S_{\overline{u}_i}$ be the restriction of S_i to $W_{\overline{u}_i}$, i.e., $S_{\overline{u}_i} = S_i \cap (W_{\overline{u}_i} \times W_{\overline{u}_i})$ (see Fig. 9.1). Then clearly we have the following:

- if $W_{\overline{u}_i}$ is not empty then $\langle W_{\overline{u}_i}, S_{\overline{u}_i}\rangle$ is isomorphic to a subframe of $\langle U_i, R_i\rangle$;
- $\langle W, S_i\rangle$ is the disjoint union of the frames $\langle W_{\overline{u}_i}, S_{\overline{u}_i}\rangle$, for all possible $n-1$-tuples $\overline{u}_i$ with nonempty $W_{\overline{u}_i}$.

Therefore, since L_i is a subframe logic, $\langle W, S_i\rangle \models L_i$. ❑

As we shall see below, the converse of inclusion (9.1) does not always hold. However, as the following theorem shows, for many standard subframe logics, their arbitrarily relativized product coincides with their fusion. Thus, 'arbitrary relativization' can be regarded as a 'many-dimensional' semantical characterization of fusions of these logics.

Theorem 9.2. *Let $L_i \in \{\mathbf{K}, \mathbf{T}, \mathbf{K4}, \mathbf{S4}, \mathbf{S5}, \mathbf{S4.3}\}$, for $i = 1, \dots, n$. Then*

$$(L_1 \times \dots \times L_n)^{\mathsf{SF}_n} = L_1 \otimes \dots \otimes L_n.$$

Proof. According to Proposition 1.11 and Theorems 1.16, 4.1, 4.2, all fusions $L_1 \otimes \dots \otimes L_n$ mentioned in the formulation of the theorem are characterized by countable (in fact, finite) rooted n-frames $\mathfrak{G} = \langle W, S_1, \dots, S_n\rangle$, where $\langle W, S_i\rangle$ is a frame for L_i, $i = 1, \dots, n$. We now prove the following analog of Lemma 5.8:

Lemma 9.3. *Suppose that $L_i \in \{\mathbf{K}, \mathbf{T}, \mathbf{K4}, \mathbf{S4}, \mathbf{S5}, \mathbf{S4.3}\}$, $i = 1, \dots, n$, and let $\mathfrak{G} = \langle W, S_1, \dots, S_n \rangle$ be a countable rooted n-frame such that $\langle W, S_i \rangle \models L_i$ for all $i = 1, \dots, n$. Then $\mathfrak{G}$ is a p-morphic image of a subframe of some product frame for $L_1 \times \cdots \times L_n$.*

Proof. First we show that every countable rooted n-frame

$$\mathfrak{G} = \langle W, S_1, \dots, S_n \rangle$$

is a p-morphic image of a subframe of some product frame. Similarly to the proof of Lemma 5.2, we will construct, step-by-step, frames $\mathfrak{F}_i = \langle U_i, R_i \rangle$ $(i = 1, \dots, n)$, a subframe $\mathfrak{H} \subseteq \mathfrak{F}_1 \times \cdots \times \mathfrak{F}_n$, and a p-morphism f from $\mathfrak{H}$ onto $\mathfrak{G}$. As before, we formalize this step-by-step argument by defining a game $G(\mathfrak{G})$ between two players $\forall$ and $\exists$ over $\mathfrak{G}$.

Define a $\mathfrak{G}$-*network* to be a tuple

$$N = \langle U_1^N, \dots, U_n^N, V^N, R_1^N, \dots, R_n^N, f^N \rangle$$

such that $\mathfrak{F}_i^N = \langle U_i^N, R_i^N \rangle$ are finite intransitive trees for all $i = 1, \dots, n$, $V^N \subseteq U_1^N \times \cdots \times U_n^N$, and f^N is a homomorphism from the subframe $\mathfrak{H}^N$ of $\mathfrak{F}_1^N \times \cdots \times \mathfrak{F}_n^N$, having V^N as its set of worlds, to $\mathfrak{G}$. In other words, for all $u_1 \in U_1, \dots, u_n \in U_n$, $i = 1, \dots, n$, and $u_i' \in U_i$,

$$\text{if } \langle u_1, \dots, u_n \rangle \in V^N,\ \langle u_1, \dots, u_{i-1}, u_i', u_{i+1}, \dots, u_n \rangle \in V^N \text{ and } u_i R_i^N u_i'$$
$$\text{then } f^N(u_1, \dots, u_i, \dots, u_n) S_i f^N(u_1, \dots, u_i', \dots, u_n)$$

The players $\forall$ and $\exists$ build a countable 'expanding' sequence of finite $\mathfrak{G}$-networks as follows.

In round 0, $\forall$ picks the root r of $\mathfrak{G}$. $\exists$ responds with a $\mathfrak{G}$-network N_0 such that all the $U_i^{N_0}$ are singleton sets, $V^{N_0} = U_1^{N_0} \times \cdots \times U_n^{N_0}$, the relations $R_i^{N_0}$ are all empty, and f^{N_0} maps the only n-tuple in V^{N_0} to r.

Suppose now that in round j, $0 < j < \omega$, the players have already built a finite $\mathfrak{G}$-network N_{j-1}. Now player $\forall$ challenges player $\exists$ with a possible *defect* of N_{j-1} which indicates that the homomorphism $f^{N_{j-1}}$ is not a p-morphism onto $\mathfrak{G}$ yet. $\forall$ picks such a defect which consists of

- an n-tuple $\langle u_1, \dots, u_n \rangle \in V^{N_{j-1}}$,
- a coordinate $i \in \{1, \dots, n\}$, and
- a world w in $\mathfrak{G}$ such that $f^{N_{j-1}}(u_1, \dots, u_n) R_i w$.

Player $\exists$ can respond in two ways. If there is some u_i' such that

$$\langle u_1, \dots, u_i', \dots, u_n \rangle \in V^{N_{j-1}},\ u_i R_i^{N_{j-1}} u_i' \text{ and } f^{N_{j-1}}(u_1, \dots, u_i', \dots, u_n) = w,$$

then she responds with $N_j = N_{j-1}$. Otherwise, she responds with the following $\mathfrak{G}$-network N_j extending N_{j-1}:

- $U_i^{N_j} = U_i^{N_{j-1}} \cup \{u^+\}$, u^+ being a fresh point, $R_i^{N_j} = R_i^{N_{j-1}} \cup \{\langle u_i, u^+\rangle\}$,
- $V^{N_j} = V^{N_{j-1}} \cup \{\langle u_1, \dots, u_{i-1}, u^+, u_{i+1}, \dots, u_n\rangle\}$,
- $\mathfrak{F}_k^{N_j} = \mathfrak{F}_k^{N_{j-1}}$ for all $k \neq i$, and
- $f^{N_j}(u_1, \dots, u^+, \dots, u_n) = w$.

Observe that $\exists$ can *always* respond this way. In other words, she always has a winning strategy in the ω-long game $G(\mathfrak{G})$. It is straightforward to see that the union (in the natural sense) of the constructed $\mathfrak{G}$-networks gives the required p-morphism f from a subframe $\mathfrak{H} = \langle V, \dots\rangle$ of a product frame $\mathfrak{F}_1 \times \dots \times \mathfrak{F}_n$ onto $\mathfrak{G}$. This proves the lemma for $L_i = \mathbf{K}$, $i = 1, \dots, n$.

However, in the other cases nothing guarantees that the 'coordinate' frames $\mathfrak{F}_i = \langle U_i, R_i\rangle$ are actually frames for L_i. In what follows we fix some i with $1 \leq i \leq n$ and try to transform $\mathfrak{F}_i$ into a frame for L_i and keep all other frames $\mathfrak{F}_j$ for $j \neq i$ and the set V intact. Without loss of generality we may assume that $i = 1$.

To begin with, we show that the frames $\mathfrak{F}_i$ and the subframe $\mathfrak{H} = \langle V, \dots\rangle$ have some useful properties. First, it should be clear from the construction that

$$\text{for each } i = 1, \dots, n, \text{ the frame } \mathfrak{F}_i \text{ is an intransitive tree.} \tag{9.2}$$

To formulate another property, we require an auxiliary definition.

Given an odd natural number k, a sequence $\langle v^0, \dots, v^k\rangle$ of distinct n-tuples $v^\ell = \langle v_1^\ell, \dots, v_n^\ell\rangle$, $\ell \leq k$, from V is called a *path in V between v^0 and v^k* if the following two conditions hold:

- for each even number $\ell < k$, $v_j^\ell = v_j^{\ell+1}$ whenever $j \neq 1$, and
- for each odd number $\ell < k$, $v_1^\ell = v_1^{\ell+1}$

(see Fig. 9.2). We call k the *length* of such a path. If in addition $v_1^0 = v_1^k$ also holds then we call $\langle v^0, \dots, v^k\rangle$ a *circle in* V (since all the n-tuples are distinct in a path, this can happen only if $k \geq 3$; see Fig. 9.3).

Observe that if $\langle v^0, \dots, v^k\rangle$ is a circle then, for every $\ell \leq k$, and every i, $1 \leq i \leq n$, there exists an $\ell' \leq k$, $\ell' \neq \ell$, such that $v_i^\ell = v_i^{\ell'}$.

The second important property is that

$$\text{there are no circles in } V. \tag{9.3}$$

For suppose otherwise. Take a circle $\langle v^0, \dots, v^k\rangle$ in V and enumerate all of its n-tuples according to their 'creation time' in the game. Let v^ℓ be the last one in this list. By the rules of the game, one of the coordinates of v^ℓ should be fresh, contrary to the observation above.

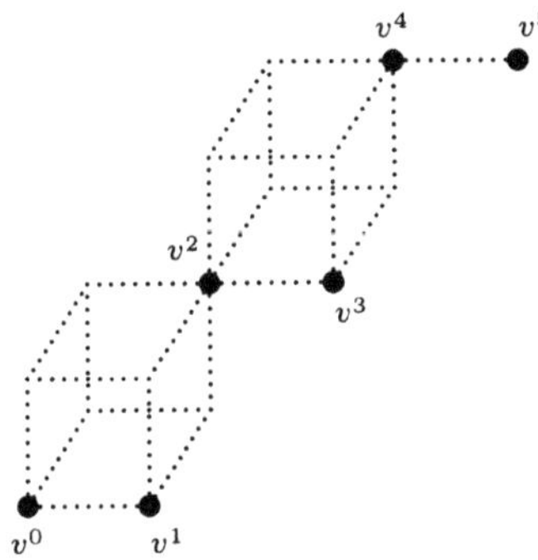

Figure 9.2: A three-dimensional path of length 5.

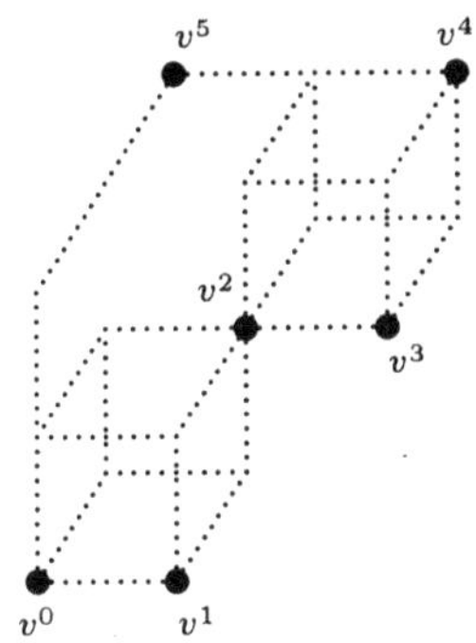

Figure 9.3: A three-dimensional circle.

Note that as a special case of (9.3) we conclude that there are no *squares* in V, i.e., four distinct n-tuples of the form $\langle x, w_2, \dots, w_n\rangle$, $\langle x, w_2', \dots, w_n'\rangle$, $\langle y, w_2, \dots, w_n\rangle$, and $\langle y, w_2', \dots, w_n'\rangle$.

Now in order to transform $\mathfrak{F}_1 = \langle U_1, R_1\rangle$ into a frame for L_1, we will extend, step-by-step (like in the proof of Lemma 5.8), the accessibility relation R_1 (but always leave the sets U_1, V and the frames $\mathfrak{F}_j$ for $j \neq 1$ unchanged).

First let $L_1 = \mathbf{K4}$. Define an infinite ascending chain

$$R_1^0 \subseteq R_1^1 \subseteq \cdots \subseteq R_1^m \subseteq \dots$$

of binary relations on U_1 by taking $R_1^0 = R_1$ and, for $m < \omega$,

$$R_1^{m+1} = R_1^m \cup \{\langle x_1, y_1\rangle \in U_1 \times U_1 \mid x_1 R_1^m z_1 \text{ and } z_1 R_1^m y_1 \text{ for some } z_1 \in U_1\}.$$

For every $m < \omega$, let $\mathfrak{F}_1^m = \langle U_1, R_1^m\rangle$ and let $\mathfrak{H}^m$ be the subframe of

$$\mathfrak{F}_1^m \times \mathfrak{F}_2 \times \cdots \times \mathfrak{F}_n$$

with V as its set of worlds. Finally, let

$$R_1^\infty = \bigcup_{m<\omega} R_1^m, \qquad \mathfrak{F}_1^\infty = \langle U_1, R_1^\infty \rangle,$$

and let $\mathfrak{H}^\infty$ be the corresponding subframe of $\mathfrak{F}_1^\infty \times \mathfrak{F}_2 \times \cdots \times \mathfrak{F}_n$.

Clearly, $\mathfrak{F}_1^\infty$ is a frame for **K4**. We are about to show that f is still a p-morphism from $\mathfrak{H}^\infty$ onto $\mathfrak{G}$. Since the 'backward' p-morphism condition always holds after extending the accessibility relation of the pre-image, it is enough to show that f is a homomorphism from $\mathfrak{H}^\infty$ onto $\mathfrak{G}$. We will prove by parallel induction on m that the following two statements hold, for all $m < \omega$ and for $x_1, y_1 \in U_1$:

(1) If $x_1 R_1^m y_1$ then there are $x_2, \ldots, x_n, y_2, \ldots, y_n$ such that there is a path in V between $x = \langle x_1, x_2, \ldots, x_n \rangle$ and $y = \langle y_1, y_2, \ldots, y_n \rangle$.

(2) If $x_1 R_1^m y_1$ and both $w^x = \langle x_1, w_2, \ldots, w_n \rangle$ and $w^y = \langle y_1, w_2, \ldots, w_n \rangle$ are in V for some $w_j \in U_j$, $j = 2, \ldots, n$, then $f(w^x) S_1 f(w^y)$. In other words, f is a homomorphism from $\mathfrak{H}^m$ onto $\mathfrak{G}$.

Assume first that $m = 0$. Then by the definition of $\mathfrak{H}$, (2) holds and there exist $w_2, \ldots, w_n$ such that both $w^x = \langle x_1, w_2, \ldots, w_n \rangle$ and $w^y = \langle y_1, w_2, \ldots, w_n \rangle$ are in V. By (9.2), $x_1 \neq y_1$, and so the sequence $\langle w^x, w^y \rangle$ is a path in V as required.

Let us assume inductively that (1) and (2) hold for some $m < \omega$, and let $x_1, y_1 \in U_1$ be such that $x_1 R_1^{m+1} y_1$, but $x_1 R_1^m y_1$ does not hold. Then there is a $z_1 \in U_1$ such that $x_1 R_1^m z_1$ and $z_1 R_1^m y_1$. It is not hard to see that, by (9.2), x_1, y_1 and z_1 should be all distinct. By item (1) of the induction hypothesis, there are x_j, z_j, z_j', y_j, for $j = 2, \ldots, n$, such that

- there is a path in V between $x = \langle x_1, x_2, \ldots, x_n \rangle$ and $z = \langle z_1, z_2, \ldots, z_n \rangle$;
- there is a path in V between $z' = \langle z_1, z_2', \ldots, z_n' \rangle$ and $y = \langle y_1, y_2, \ldots, y_n \rangle$.

If $z \neq z'$ then the concatenation of these two paths gives a path between x and y. If $z = z'$ then leave out z from the concatenated sequence, and the rest gives a path as required in (1).

For (2), suppose that $w^x = \langle x_1, w_2, \ldots, w_n \rangle$ and $w^y = \langle y_1, w_2, \ldots, w_n \rangle$ are in V for some $w_j \in U_j$, $j = 2, \ldots, n$. Let $w^z = \langle z_1, w_2, \ldots, w_n \rangle$. Consider the n-tuples x, y, z, z' given above. We claim that

$$x = w^x, \ y = w^y \text{ and } z = z' = w^z. \tag{9.4}$$

Suppose otherwise. Then several cases are possible. We are going to show that any of them means that there is a circle in V, contrary to (9.3). Let $\rho = \langle x, v^1, \ldots, v^k, y \rangle$ denote the path in V between x and y (which exists because of (1)).

- Suppose first that $x \neq w^x$ and $y \neq w^y$. Then the concatenation of $\langle w^y, w^x \rangle$ and ρ is a circle in V.

- Suppose $x = w^x$ and $y \neq w^y$. Then the length of ρ is ≥ 3, so $\langle w^y, v^1, \dots, v^k, y \rangle$ is a circle in V. The case when $x \neq w^x$ and $y = w^y$ is similar.

- Finally, suppose $x = w^x$ and $y = w^y$. If $z \neq w^z$ and $z' \neq w^z$ then the length of ρ should be ≥ 5, and $\langle v^k, v^1, \dots, v^{k-1} \rangle$ is a circle in V. The cases when one of z and z' coincides with w^z but the other does not are similar.

As a consequence of (9.4), we obtain that w^z is in V. So by item (2) of the induction hypothesis,

$$f(w^x) S_1 f(w^z) \quad \text{and} \quad f(w^z) S_1 f(w^y).$$

Since S_1 is transitive, we have $f(w^x) S_1 f(w^y)$, which completes the proof of Lemma 9.3 for $L_1 = \mathbf{K4}$.

If $L_1 = \mathbf{S4}$ or $L_1 = \mathbf{T}$, we simply make all worlds of $\mathfrak{F}_1$ reflexive and f is still a p-morphism. In the case of $L_1 = \mathbf{S5}$, we have to 'close' $\mathfrak{F}_1$ under both transitivity and symmetry. It is not hard to see that this causes no problem, since there are no squares in V.

For $L_1 = \mathbf{S4.3}$ we need a slight modification of the above proof for $\mathbf{K4}$. We have to turn $\mathfrak{F}_1$ to a reflexive, transitive and weakly connected frame. To this end, we modify the definition of the accessibility relation R_1^{m+1} ($m < \omega$). First, we make all the points in U_1 reflexive. Then for all distinct $x_1, y_1 \in U_1$, we define $\langle x_1, y_1 \rangle$ to be in R_1^{m+1} iff one of the following three conditions hold:

- $x_1 R_1^m y_1$;

- there is a $z_1 \in U_1$ such that $x_1 R_1^m z_1$ and $z_1 R_1^m y_1$;

- there is a $z_1 \in U_1$ such that $z_1 R_1^m x_1$, $z_1 R_1^m y_1$, and
  - either there are no $w_2, \dots, w_n$ such that both $w^x = \langle x_1, w_2, \dots, w_n \rangle$ and $w^y = \langle y_1, w_2, \dots, w_n \rangle$ are in V,
  - or there exist $w_2, \dots, w_n$ such that both w^x and w^y are in V, and $f(w^x) S_1 f(w^y)$ holds. (Note that although $w^x \neq w^y$, it can happen that $f(w^x) = f(w^y)$.)

Since there are no squares in V, R_1^{m+1} is well-defined. The very same inductive proof as above shows that the frame $\mathfrak{F}_1^\infty$ obtained this way is reflexive, transitive and weakly connected, and f is still a p-morphism from $\mathfrak{H}^\infty$ onto $\mathfrak{G}$. ❑

Now we can complete the proof of Theorem 9.2. Let $\varphi \notin L_1 \otimes \cdots \otimes L_n$. Take a countable rooted n-frame $\mathfrak{G} = \langle W, S_1, \ldots, S_n \rangle$ refuting φ and such that, for every $i = 1, \ldots, n$, $\langle W, S_i \rangle$ is a frame for L_i. Now, using Lemma 9.3, we can find a subframe $\mathfrak{H}$ of a product frame for $L_1 \times \cdots \times L_n$ having $\mathfrak{G}$ as its p-morphic image. It follows that $\mathfrak{H} \not\models \varphi$, and so $\varphi \notin (L_1 \times \cdots \times L_n)^{\mathsf{SF}_n}$. Therefore, $(L_1 \times \cdots \times L_n)^{\mathsf{SF}_n} \subseteq L_1 \otimes \cdots \otimes L_n$. Proposition 9.1 gives the converse inclusion. ❑

It is not clear how far Theorem 9.2 can be generalized. On the one hand, we conjecture that it holds for $L_i \in \{\mathbf{K4.3}, \mathbf{Grz}, \mathbf{GL}, \mathbf{GL.3}\}$ as well. For **K4.3** even Lemma 9.3 may hold, although a somewhat different, 'more careful' proof would be needed. However, it is not true that every countable (even finite) frame for, say, $\mathbf{Grz} \otimes \mathbf{Grz}$ is a p-morphic image of a subframe of a product of two **Grz**-frames. Consider, for instance, the 2-frame $\langle \{x, y, z, w\}, R_1, R_2 \rangle$ with $xR_1yR_2zR_1wR_2x$. It is not hard to see that if this frame is a p-morphic image of a subframe of $\mathfrak{F}_1 \times \mathfrak{F}_2$ then both $\mathfrak{F}_1$ and $\mathfrak{F}_2$ must contain infinite strictly ascending chains of points, and so cannot be frames for **Grz**.

On the other hand, Theorem 9.2 does not hold for all subframe logics, not even for those of them that (unlike **Grz**) are characterized by universally first-order definable classes of frames. Take, for instance, the logic

$$\mathbf{K5} = \mathbf{K} \oplus \Diamond\Box p \to \Box p.$$

It is well-known (see, e.g., Chagrov and Zakharyaschev 1997) that **K5** is Kripke complete and characterized by the class of Euclidean frames, i.e., frames $\langle W, R \rangle$ satisfying the universal (Horn) sentence

$$\forall x \forall y \forall u \big(R(u,x) \wedge R(u,y) \to R(x,y)\big).$$

In particular, frames for **K5** have the property

$$\forall x \forall u \big(R(u,x) \to R(x,x)\big).$$

Now consider the formula

$$\varphi = \Diamond_1\big(p \wedge \Diamond_2(q \wedge \neg p)\big) \wedge \Box_1\Box_2(q \to \neg\Diamond_1 q).$$

It is clearly satisfiable in the following frame for $\mathbf{K5} \otimes \mathbf{K}$:

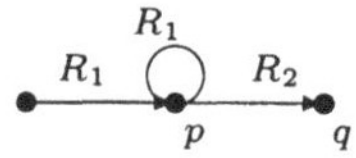

On the other hand, it is not hard to see that φ is not satisfiable in any subframe of a product frame for $\mathbf{K5} \times \mathbf{K}$. Therefore,

$$\mathbf{K5} \otimes \mathbf{K} \subsetneq (\mathbf{K5} \times \mathbf{K})^{\mathsf{SF}_2} \subsetneq \mathbf{K5} \times \mathbf{K}.$$

In fact, a similar statement holds for any logic $\mathbf{K} \oplus \Diamond^i \Box p \to \Box^i p$ $(i \geq 1)$ in place of **K5**. Further, the same argument shows that

$$\mathbf{K45} \otimes \mathbf{K4} \subsetneq (\mathbf{K45} \times \mathbf{K4})^{\mathsf{SF}_2} \subsetneq \mathbf{K45} \times \mathbf{K4},$$

where $\mathbf{K45} = \mathbf{K4} \oplus \Diamond\Box p \to \Box p$.

Other kinds of logics for which Theorem 9.2 does not hold are those having frames with a finite bound on their branching, e.g. **Alt**. Recall that $\langle W, R\rangle$ is a frame for **Alt** iff every point in W has at most one R-successor. Now consider the formula

$$\psi = p \wedge \Diamond_1(\neg p \wedge \Diamond_2 q) \wedge \Diamond_2(\neg p \wedge \Diamond_1 r) \wedge \Box_1\Box_2(q \to \neg r).$$

ψ is clearly satisfiable in the $\mathbf{Alt} \otimes \mathbf{Alt}$-frame

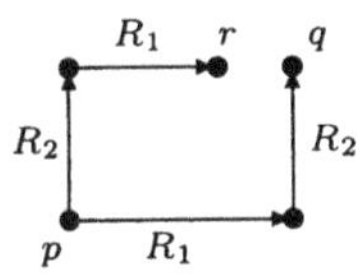

On the other hand, it should be clear that ψ is not satisfiable in any subframe of a frame for $\mathbf{Alt} \times \mathbf{Alt}$. Thus,

$$\mathbf{Alt} \otimes \mathbf{Alt} \subsetneq (\mathbf{Alt} \times \mathbf{Alt})^{\mathsf{SF}_2} \subsetneq \mathbf{Alt} \times \mathbf{Alt}.$$

However, in general the behavior of arbitrarily relativized products remains unexplored. It would be of interest, for instance, to find solutions to the following problems.

Question 9.4. Are arbitrarily relativized products of finitely axiomatizable logics also finitely axiomatizable (in those cases when they differ from the fusions)?

Question 9.5. Are arbitrarily relativized products of decidable logics also decidable?

Question 9.6. Find a general characterization of those arbitrarily relativized products of logics that coincide with their fusions.

Cubic and locally cubic relativizations

To motivate another kind of relativization, let us briefly discuss a possible way of creating new, more expressive logics from products. Given the product of n unimodal logics, one may want to add new operations to $\Box_1, \ldots, \Box_n$ that 'connect' the different dimensions. Perhaps the simplest and most natural operations of this sort are the *diagonal constants* d_{ij} which have already showed up in various disguises in this book. Given two natural numbers i and j with $1 \leq i, j \leq n$, the truth-relation for the constant d_{ij} in models over subframes of n-ary product frames is defined as follows:

$$(\mathfrak{M}, \langle u_1, \dots, u_n \rangle) \models \mathsf{d}_{ij} \quad \text{iff} \quad u_i = u_j.$$

The set of n-tuples satisfying d_{ij} is usually called the (i, j)-*diagonal element*.

Actually, the main reason for introducing such constants is to give a 'modal treatment' of equality of classical first-order logic: one can extend the translation $^\bullet$ of Section 3.5 by taking

$$(x_i = x_j)^\bullet = \mathsf{d}_{ij},$$

for all variables x_i, x_j. Let $(\mathbf{S5}^n)^=$ denote the logic (in the language $\mathcal{ML}_n$ with the diagonal constants) determined by the class of cubic universal product $\mathbf{S5}^n$-frames extended with the diagonal elements (interpreting the d_{ij}). Modal algebras for this logic are called *representable cylindric algebras* and are extensively studied in the algebraic logic literature; see, e.g., (Henkin *et al.* 1981, 1985, Hirsch and Hodkinson 2002) and the references therein. By the algebraic results of (Monk 1969) and (Maddux 1980), $(\mathbf{S5}^n)^=$ is neither finitely axiomatizable nor decidable. Note also that $(\mathbf{S5}^n)^=$ is not a conservative extension of $\mathbf{S5}^n$ (Henkin *et al.* 1985).

Another natural way of connecting dimensions is via so-called 'jump' modalities. Given a function $\pi : \{1, \dots, n\} \to \{1, \dots, n\}$ (such a map can be called a *jump*), define the truth-relation for the unary modal operator s_π in models over subframes of product frames as follows:

$$(\mathfrak{M}, \langle u_1, \dots, u_n \rangle) \models \mathsf{s}_\pi \varphi \quad \text{iff} \quad (\mathfrak{M}, \langle u_{\pi(1)}, \dots, u_{\pi(n)} \rangle) \models \varphi.$$

These modal operators are often called (*generalized*) *substitutions*, since by taking

$$P(x_{\pi(1)-1}, \dots, x_{\pi(n)-1})^\bullet = \mathsf{s}_\pi P(x_0, \dots, x_{n-1}) \qquad (P \text{ an atomic formula})$$

one can extend the translation $^\bullet$ of Section 3.5 from formulas with a fixed order of the variables to arbitrary first-order formulas. Note that in cubic universal product $\mathbf{S5}^n$-frames certain substitutions are expressible with the help of the boxes and the diagonal constants (Henkin *et al.* 1985). Various versions of modal algebras corresponding to products of $\mathbf{S5}$ logics with substitutions and with or without diagonal constants (e.g., *polyadic* and *substitution algebras*) are studied in (Halmos 1957, 1962, Pinter 1973, 1975); see also (Daigneault and Monk 1963, Németi 1991, Sági 2002, Sain and Thompson 1991). Again, the algebraic results show that most of these logics are nonfinitely axiomatizable and undecidable.

Arbitrary relativizations of these extensions of **S5**-products do result in new, decidable many-dimensional logics; see (Németi 1995, Venema and Marx 1999). Moreover, both the diagonal constants and the substitutions can 'detect' some properties of the set of worlds, so it makes sense to consider, for example, those frames whose sets of worlds are closed under jumps. A nonempty set W of n-tuples is called a *local n-cube* if for all maps

$$\pi : \{1, \ldots, n\} \to \{1, \ldots, n\}$$

and all $\langle u_1, \ldots, u_n \rangle \in W$, we have $\langle u_{\pi(1)}, \ldots, u_{\pi(n)} \rangle \in W$. It is easy to see that W is a local n-cube iff for every $\langle u_1, \ldots, u_n \rangle \in W$, the Cartesian power $\{u_1, \ldots, u_n\}^n$ is a subset of W, that is, W is the union of 'n-dimensional cubes.' In particular, local 2-cubes are just the reflexive and symmetric binary relations.

A set W such that $W = U^n$, for some nonempty set U, will be called an *n-cube*. Clearly, n-cubes are special cases of local n-cubes. Let

$$\begin{aligned} \mathsf{LC}_n &= \{\langle W, S_1, \ldots, S_n \rangle \in \mathsf{SF}_n \mid W \text{ is a local } n\text{-cube}\}, \\ \mathsf{C}_n &= \{\langle W, S_1, \ldots, S_n \rangle \in \mathsf{SF}_n \mid W \text{ is an } n\text{-cube}\}. \end{aligned}$$

Note that cubic universal product frames belong to C_n. In general, we will refer to frames whose sets of worlds are n-cubes as *cubic*.

Locally cubic relativizations of the above extensions of **S5**-products again give new logics that are also different from the arbitrarily relativized versions. Moreover, all these 'extended relativized **S5**-products' turn out to be decidable and often finitely axiomatizable. A comprehensive treatment of relativized versions of $(\mathbf{S5}^n)^=$ and products of **S5** logics extended with substitutions can be found in (Marx and Venema 1997) under the respective names of *cylindric modal logics* and *modal logics of relations*.

Note that one can also establish connections between different dimensions by introducing polyadic modal operators on product frames. This is the road taken by arrow logics (see Section 3.10), where a binary modal operator is considered. Relativized versions of arrow logics are among the main topics of (Marx and Venema 1997); see also references therein and in Section 3.10 above.

Question 9.7. What can be said about extensions with diagonals and/or substitutions of arbitrarily and locally cubic relativized products of modal logics other than **S5**?

As mentioned in (Mikulás and Marx 2000), decidability of these extensions of relativized $\mathbf{K}^n$ can be proved by a reduction to the $n+1$-variable packed fragment of first-order logic. According to (Mikulás 2000), the mosaic method (which has been so successful for extensions of relativized $\mathbf{S5}^n$) can also be used to show decidability of extensions of

$$(L_1 \times \cdots \times L_n)^{\mathsf{LC}_n} \quad \text{and} \quad (L_1 \times \cdots \times L_n)^{\mathsf{SF}_n},$$

whenever $L_i \in \{\mathbf{K}, \mathbf{T}, \mathbf{K4}, \mathbf{S4}, \mathbf{S5}\}$.

The following two propositions show that if we do not enrich the language $\mathcal{ML}_n$, then locally cubic and cubic relativizations do not yield anything new.

Proposition 9.8. *For all Kripke complete unimodal logics $L_1, \ldots, L_n$ and all classes $\mathcal{K}$ such that $\mathsf{LC}_n \subseteq \mathcal{K} \subseteq \mathsf{SF}_n$,*

$$(L_1 \times \cdots \times L_n)^{\mathsf{LC}_n} = (L_1 \times \cdots \times L_n)^{\mathcal{K}} = (L_1 \times \cdots \times L_n)^{\mathsf{SF}_n}.$$

Proof. The inclusions

$$(L_1 \times \cdots \times L_n)^{\mathsf{LC}_n} \supseteq (L_1 \times \cdots \times L_n)^{\mathcal{K}} \supseteq (L_1 \times \cdots \times L_n)^{\mathsf{SF}_n}$$

are obvious. To prove the converse ones, we show that any rooted n-frame $\mathfrak{H}$ in $(\mathsf{Fr}L_1 \times \cdots \times \mathsf{Fr}L_n)^{\mathsf{SF}_n}$ is isomorphic to a generated subframe of some $\mathfrak{G}$ in $(\mathsf{Fr}L_1 \times \cdots \times \mathsf{Fr}L_n)^{\mathsf{LC}_n}$. Indeed, suppose that $\mathfrak{H} \subseteq \mathfrak{F}$ for some $\mathfrak{F}$ in $\mathsf{Fr}L_1 \times \cdots \times \mathsf{Fr}L_n$ of the form

$$\langle U_1, R_1 \rangle \times \cdots \times \langle U_n, R_n \rangle .$$

Take an isomorphic copy of $\mathfrak{F}$ such that the U_i are pairwise disjoint. By Makinson's theorem (see Section 1.2), for each Kripke complete unimodal logic L, either the one-element reflexive frame ($\circ$) or the one-element irreflexive frame ($\bullet$) is a frame for L. For all $i, j \in \{1, \ldots, n\}$, we define binary relations R_i^j on U_j by taking

$$R_i^j = \begin{cases} R_i, & \text{if } i = j, \\ \emptyset, & \text{if } \bullet \models L_i, \\ \{\langle u, u \rangle \mid u \in U_j\}, & \text{if } \circ \models L_i. \end{cases}$$

Now let $U = \bigcup_{1 \le i \le n} U_i$. For every $i \in \{1, \ldots, n\}$, set $R_i^+ = \bigcup_{1 \le j \le n} R_i^j$, and take

$$\mathfrak{F}_i = \langle U, R_i^+ \rangle .$$

Since each $\mathfrak{F}_i$ is a disjoint union of L_i-frames, the product frame

$$\mathfrak{F}^+ = \mathfrak{F}_1 \times \cdots \times \mathfrak{F}_n$$

is then a frame for $L_1 \times \cdots \times L_n$. Let W denote the set of worlds of $\mathfrak{H}$. Define W^+ as the smallest local n-cube containing W, that is,

$$W^+ = \bigcup \{ \{u_1, \ldots, u_n\}^n \mid \langle u_1, \ldots, u_n \rangle \in W \},$$

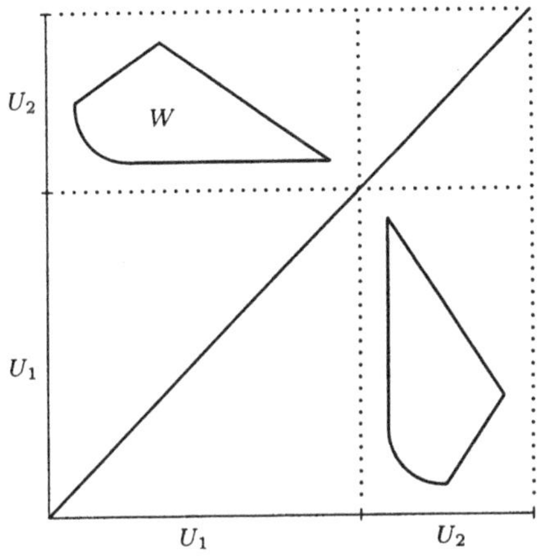

Figure 9.4: The smallest local 2-cube containing W.

and let $\mathfrak{G}$ be the subframe of $\mathfrak{F}^+$ with W^+ as its set of worlds (see Fig. 9.4 for the case $n = 2$). Then clearly

$$\mathfrak{G} \in (\mathsf{Fr}L_1 \times \cdots \times \mathsf{Fr}L_n)^{\mathsf{LC}_n}$$

and $\mathfrak{H} \subseteq \mathfrak{G}$. It is not hard to see that $\mathfrak{H}$ is in fact a generated subframe of $\mathfrak{G}$, because $W^+ \cap (U_1 \times \cdots \times U_n) = W$. ❑

Proposition 9.9. *For all subframe logics $L_1, \dots, L_n$,*

$$(L_1 \times \cdots \times L_n)^{\mathsf{C}_n} = L_1 \times \cdots \times L_n.$$

Proof. The inclusion $L_1 \times \cdots \times L_n \subseteq (L_1 \times \cdots \times L_n)^{\mathsf{C}_n}$ is easy, since the L_i are subframe logics and every cubic subframe of a product frame is in fact a product of some subframes of the components.

To prove the converse, we show that every frame $\mathfrak{F} = \mathfrak{F}_1 \times \cdots \times \mathfrak{F}_n$ with $\mathfrak{F}_i \models L_i$ is a p-morphic image of a cubic product frame, that is, a frame $\mathfrak{G} = \mathfrak{G}_1 \times \cdots \times \mathfrak{G}_n$ such that every $\mathfrak{G}_i$ has the same set of worlds and $\mathfrak{G}_i \models L_i$. Indeed, take a cardinal $\kappa \geq \max_{1 \leq i \leq n} |\mathfrak{F}_i|$ and let $\mathfrak{G}_i$ be the disjoint union of κ-many copies of $\mathfrak{F}_i$. Then $|\mathfrak{G}_i| = \kappa$ and $\mathfrak{F}_i$ is a p-morphic image of $\mathfrak{G}_i$, whenever $1 \leq i \leq n$. Therefore, by Proposition 3.10, $\mathfrak{F}$ is a p-morphic image of $\mathfrak{G}$. Since all the $\mathfrak{G}_i$ have the same cardinalities, we may assume that they are frames over the same set of worlds. ❑

Expanding and decreasing relativizations

First-order modal and intuitionistic logics as well as modal description logics motivate our third group of relativizations. Fix a subset N of $\{1, \dots, n\}$.

An n-frame $\mathfrak{G} = \langle W, S_1, \dots, S_n \rangle$ is called an *N-expanding* (or *N-decreasing*) *relativized product frame* if there are frames $\mathfrak{F}_1 = \langle U_1, R_1 \rangle, \dots, \mathfrak{F}_n = \langle U_n, R_n \rangle$ such that

- $\mathfrak{G}$ is a subframe of $\mathfrak{F}_1 \times \cdots \times \mathfrak{F}_n$;
- for all $\langle w_1, \dots, w_n \rangle \in W$, $j \in N$, and $u \in U_j$, if $w_j R_j u$ (or $u R_j w_j$) then $\langle w_1, \dots, w_{j-1}, u, w_{j+1}, \dots, w_n \rangle \in W$.

If $N = \{1\}$ then we call $\mathfrak{G}$ an (n-ary) *expanding* (*decreasing*) *relativized product frame*. Examples of decreasing relativized product frames are the two-dimensional frames for the interval temporal logic **HS** from Section 3.9 (they are also $\{2\}$-expanding). In what follows we consider only expanding relativizations. The reader should have no problem in reformulating all notions and results for the case of decreasing ones.

Define EX_n to be the class of all n-ary expanding relativized product frames. In case $n = 2$, we omit the index and write EX.

It is easy to see that every expanding relativized product frame has left commutativity and Church–Rosser properties between coordinates 1 and i, for all $i = 2, \dots, n$:

$$\forall x \forall y \forall z \left(x R_i y \wedge y R_1 z \to \exists u \, (x R_1 u \wedge u R_i z) \right),$$
$$\forall x \forall y \forall z \left(x R_i y \wedge x R_1 z \to \exists u \, (y R_1 u \wedge z R_i u) \right)$$

(cf. Section 5.1). Therefore, the formulas $\boldsymbol{com}^{l}_{1i}$ and $\boldsymbol{chr}_{1i}$ are valid in expanding relativized product frames for all $i = 2, \dots, n$ (see Chapter 8).

Let us consider first the axiomatization problem for two-dimensional expanding relativizations. Given logics L_1 and L_2, define

$$[L_1, L_2]^{\mathsf{EX}} = (L_1 \otimes L_2) \oplus \boldsymbol{com}^{l}_{12} \oplus \boldsymbol{chr}_{12}.$$

Theorem 9.10. *Suppose L_1 and L_2 are Kripke complete unimodal logics such that $L_1 \in \{\mathbf{K}, \mathbf{T}, \mathbf{K4}, \mathbf{S4}, \mathbf{S5}\}$ and L_2 is Horn axiomatizable. Then*

$$(L_1 \times L_2)^{\mathsf{EX}} = [L_1, L_2]^{\mathsf{EX}}.$$

Proof. It is easy to see that if $\boldsymbol{com}^{l}_{12}$ and $\boldsymbol{chr}_{12}$ are valid in $\mathfrak{F} = \langle W, R_1, R_2 \rangle$ with symmetric R_1, then $\boldsymbol{com}_{12}$ is valid in $\mathfrak{F}$ as well. By Theorem 5.9, we then have

$$(\mathbf{S5} \times L_2)^{\mathsf{EX}} = \mathbf{S5} \times L_2 = [\mathbf{S5}, L_2] = [\mathbf{S5}, L_2]^{\mathsf{EX}}.$$

In the other cases we can prove, similarly to Lemma 5.2, that every countable rooted 2-frame validating $\boldsymbol{com}^{l}_{12}$ and $\boldsymbol{chr}_{12}$ is a p-morphic image of an expanding relativized product frame. Then, like in the proof of Lemma 5.8, we add the missing pairs to R_1 and R_2, if needed. By adding new pairs to R_1 we are not forced to extend the set of worlds, because $L_1 \in \{\mathbf{T}, \mathbf{K4}, \mathbf{S4}\}$. ❑

Question 9.11. What can we say about axiomatizations of higher-dimensional expanding relativized products?

As to decidability, expanding relativizations can be reduced to products almost in the same way as first-order modal logics with expanding domains were reduced to logics with constant domains in Section 3.6 (cf. also Proposition 3.32). Let φ be an $\mathcal{ML}_n$-formula and e a propositional variable which does not occur in φ. Define by induction on the construction of φ an $\mathcal{ML}_n$-formula φ^e as follows:

$$\begin{aligned} p^e &= p \quad (p \text{ a propositional variable}),\\ (\psi \wedge \chi)^e &= \psi^e \wedge \chi^e,\\ (\neg\psi)^e &= \neg\psi^e,\\ (\Box_1\psi)^e &= \Box_1\psi^e,\\ (\Box_i\psi)^e &= \Box_i(e \to \psi^e) \quad (i = 2, \dots, n). \end{aligned}$$

Theorem 9.12. *For all Kripke complete unimodal logics $L_1, \dots, L_n$ and all $\mathcal{ML}_n$-formulas φ, the following conditions are equivalent:*

- $\varphi \in (L_1 \times \dots \times L_n)^{\mathsf{EX}_n}$;
- $\Big(e \wedge \Box_1^{\leq md(\varphi)} \boldsymbol{M}_{(2,n)}^{\leq md(\varphi)}(e \to \Box_1 e)\Big) \to \varphi^e \in L_1 \times \dots \times L_n,$

where $\boldsymbol{M}_{(2,n)}^{\leq 0}\psi = \psi$ and $\boldsymbol{M}_{(2,n)}^{\leq k+1}\psi = \boldsymbol{M}_{(2,n)}^{\leq k}\psi \wedge \bigwedge_{i=2}^{n} \Box_i \boldsymbol{M}_{(2,n)}^{\leq k}$.

In particular, for $n = 2$,

$$\varphi \in (L_1 \times L_2)^{\mathsf{EX}} \quad \textit{iff} \quad \Big(e \wedge \Box_1^{\leq md(\varphi)}\Box_2^{\leq md(\varphi)}(e \to \Box_1 e)\Big) \to \varphi^e \in L_1 \times L_2.$$

Proof. Similar to the proof of Proposition 3.20. ❑

As a consequence of this theorem we obtain that expanding relativized products are decidable in all those cases when the corresponding products are decidable.

Question 9.13. Does the decidability of an expanding relativized product logic imply that the corresponding product logic is decidable as well?

We conjecture that in some higher-dimensional cases the answer may be affirmative in the sense that expanding and decreasing relativized products—like the corresponding products—are undecidable. In particular, it was shown in (Hodkinson *et al.* 2002) that the product of any Kripke complete modal logic between **K** and **S5** with the $\Box_F$-fragment of branching time temporal logic **CTL*** is undecidable. We believe that a similar proof can show the undecidability of the decreasing relativization of $\mathbf{K4} \times \mathbf{S5} \times L$, for any Kripke complete modal logic L between $\mathbf{K}$ and $\mathbf{S5}$.

Let us conclude this section by observing the (lack of) connections between expanding relativized products and finite variable fragments of first-order modal logics with expanding domains. To begin with, as we saw in Section 3.6, modal product logics of the form

$$L \times \overbrace{\mathbf{S5} \times \cdots \times \mathbf{S5}}^{n}$$

can be reduced to n-variable fragments of first-order modal logics $\mathbf{Q}L$ with constant domains (cf. Theorem 3.21). It is readily checked that if $n = 1$ then the translation † defined in Section 3.6 reduces $(L \times \mathbf{S5})^{\mathsf{EX}}$ to the one-variable fragment of the first-order modal logic $\mathbf{Q}^e L$ having models with expanding domains.

On the other hand, as far as we see, for $n \geq 3$ there is no such reduction of expanding relativized products of the form

$$(L \times \overbrace{\mathbf{S5} \times \cdots \times \mathbf{S5}}^{n})^{\mathsf{EX}_{n+1}} \tag{9.5}$$

to $\mathbf{Q}^e L$, since quantifiers $\forall x_i$ and $\forall x_j$ of the latter always commute, while there is no interaction between the boxes $\Box_i$ and $\Box_j$ of the former whenever $i \neq j$ and $i, j > 1$. An alternative approach can be to consider instead of (9.5) the two-dimensional expanding relativized product

$$(L \times \mathbf{S5}^n)^{\mathsf{EX}}.$$

Note that for $n \geq 3$ it is not known whether the $n + 1$-dimensional product logic $L \times \mathbf{S5} \times \cdots \times \mathbf{S5}$ and the two-dimensional product logic $L \times \mathbf{S5}^n$ are the same; see Section 3.3. Moreover, since we do not know what frames for $\mathbf{S5}^n$ look like when $n \geq 3$ (cf. Theorem 8.29), it is not clear how to turn a model for $(L \times \mathbf{S5}^n)^{\mathsf{EX}}$ into a model for $\mathbf{Q}^e L$.

For $n = 2$ we do have a characterization of (countable) $\mathbf{S5} \times \mathbf{S5}$-frames; see Lemma 5.8. Therefore, it is not hard to see that we have the required reduction: for every $\mathcal{ML}_3$-formula φ,

$$\varphi \in (L \times (\mathbf{S5} \times \mathbf{S5}))^{\mathsf{EX}} \quad \text{iff} \quad \varphi^\dagger \in \mathbf{Q}^e L.$$

9.2 Valuation restrictions

One may try to loosen the strong interaction between the components of product logics by imposing restrictions on possible valuations in models.

Examples of specifying 'acceptable' valuations we have already met in this book are the finite state assumption (**FSA**) and the finite change assumption (**FCA**) of Section 3.2.

We also face the problem of valuation restrictions if we try to extend the definition of products of frames to products of models. In order to keep the notation transparent, in what follows we confine ourselves to the two-dimensional case (however, all the definitions and results can be generalized to higher dimensions in a straightforward manner). Suppose that $\mathfrak{M}_1 = \langle \mathfrak{F}_1, \mathfrak{V}_1 \rangle$ and $\mathfrak{M}_2 = \langle \mathfrak{F}_2, \mathfrak{V}_2 \rangle$ are models based on frames $\mathfrak{F}_i = \langle W_i, R_i \rangle$, $i = 1, 2$. Recall that a model over the product frame $\mathfrak{F}_1 \times \mathfrak{F}_2$ is a pair $\mathfrak{M} = \langle \mathfrak{F}_1 \times \mathfrak{F}_2, \mathfrak{V} \rangle$, where $\mathfrak{V}$ is a function mapping propositional variables to subsets of $W_1 \times W_2$. Now we call a model $\mathfrak{M}$ over $\mathfrak{F}_1 \times \mathfrak{F}_2$ an *i-flat product of $\mathfrak{M}_1$ and $\mathfrak{M}_2$* ($i = 1, 2$) if, for all propositional variables p and all worlds $u_1 \in W_1$, $u_2 \in W_2$,

$$\langle u_1, u_2 \rangle \in \mathfrak{V}(p) \quad \text{iff} \quad u_i \in \mathfrak{V}_i(p).$$

$\mathfrak{M}$ is called a *flat model* if it is an i-flat product model, for some $i = 1, 2$, and $\mathfrak{V}$ is called a *flat valuation*. Flat valuations are discussed for many-dimensional temporal logics in (Gabbay and Guenthner 1982, Gabbay *et al.* 1994) and for temporal arrow logics in (Marx and Venema 1997).

A more general way of classifying valuation restrictions (or defining products of models) is as follows. Take the first-order language with two unary predicate symbols V_1, V_2 and two binary predicate symbols, and let $\Phi(x_1, x_2)$ be a formula of this language. Then a model

$$\mathfrak{M}^\Phi = \langle \mathfrak{F}_1 \times \mathfrak{F}_2, \mathfrak{V}^\Phi \rangle$$

is said to be a *Φ-flat product of $\mathfrak{M}_1$ and $\mathfrak{M}_2$* if, for all propositional variables p and all $u_1 \in W_1$, $u_2 \in W_2$,

$$\langle u_1, u_2 \rangle \in \mathfrak{V}^\Phi(p) \quad \text{iff} \quad I_p \models \Phi[u_1, u_2],$$

where I_p is the first-order structure

$$I_p = \langle W_1 \cup W_2, \mathfrak{V}_1(p), \mathfrak{V}_2(p), R_1, R_2 \rangle.$$

For example, a 1-flat product model defined above is Φ-flat with $\Phi = V_1(x_1)$. If Φ is a Boolean combination of $V_1(x_1)$ and $V_2(x_2)$ then we say that $\mathfrak{M}^\Phi$ is a *Boolean-flat* model (see (Hasimoto 2002) for an example). Rabinovich (2003) considers flat products of Kripke models in a wider perspective by showing that they are special cases of the generalized product construction of Feferman and Vaught (1959).

Satisfiability in Boolean-flat models can be reduced to satisfiability in the component models, as the following 'flat product decomposition theorem' of Gabbay and Shehtman (1999) shows:

Theorem 9.14. *Let $\mathfrak{M}^\Phi$ be a Boolean-flat product of models $\mathfrak{M}_1$ and $\mathfrak{M}_2$. Then for every $\mathcal{ML}_2$-formula φ, there are a finite set I_φ and unimodal formulas φ_i^1 (with $\Box_1$) and φ_i^2 (with $\Box_2$), $i \in I_\varphi$, such that, for all worlds $\langle u_1, u_2\rangle$ in $\mathfrak{M}^\Phi$,*

$$(\mathfrak{M}^\Phi, \langle u_1, u_2\rangle) \models \varphi \quad \textit{iff} \quad \exists i \in I_\varphi\big((\mathfrak{M}_1, u_1) \models \varphi_i^1 \text{ and } (\mathfrak{M}_2, u_2) \models \varphi_i^2\big).$$

Corollary 9.15. *Let L_1 and L_2 be unimodal logics having the fmp. Assume that an $\mathcal{ML}_2$-formula φ is satisfied in a Boolean-flat product $\mathfrak{M}^\Phi$ of models $\mathfrak{M}_1$ and $\mathfrak{M}_2$, where $\mathfrak{M}_i \models L_i$, for $i = 1, 2$. Then φ is also satisfied in a Boolean-flat product $\mathfrak{N}^\Phi$ of finite models $\mathfrak{N}_1$ and $\mathfrak{N}_2$ such that $\mathfrak{N}_i \models L_i$, for $i = 1, 2$.*

Proof of Theorem 9.14. First note that, since Φ is a Boolean combination of $V_1(x_1)$ and $V_2(x_2)$, we may assume that

$$\Phi(x_1, x_2) = \bigvee_{i \in I} (\Phi_i^1(x_1) \wedge \Phi_i^2(x_2)),$$

where, for each $i \in I$, Φ_i^j is a (possibly empty) conjunction of $V_j(x_j)$ and $\neg V_j(x_j)$ ($j = 1, 2$).

We prove the theorem by induction on the construction of φ. First assume that $\varphi = p$, for some propositional variable p. Then let $I_\varphi = I$ and, for each $i \in I$, take

$$p_i^1 = \begin{cases} p & \text{if } \Phi_i^1 = V_1(x_1), \\ \neg p & \text{if } \Phi_i^1 = \neg V_1(x_1), \\ \bot & \text{if } \Phi_i^1 = V_1(x_1) \wedge \neg V_1(x_1), \\ \top & \text{if } \Phi_i^1 \text{ is empty}, \end{cases}$$

$$p_i^2 = \begin{cases} p & \text{if } \Phi_i^2 = V_2(x_2), \\ \neg p & \text{if } \Phi_i^2 = \neg V_2(x_2), \\ \bot & \text{if } \Phi_i^2 = V_2(x_2) \wedge \neg V_2(x_2), \\ \top & \text{if } \Phi_i^2 \text{ is empty}. \end{cases}$$

The cases when $\varphi = \psi \vee \chi$ or $\varphi = \neg\psi$ are straightforward. Suppose now that $\varphi = \Diamond_1 \psi$. Then

$$\begin{aligned}
&(\mathfrak{M}^\Phi, \langle u_1, u_2\rangle) \models \Diamond_1\psi \quad \text{iff} \\
&\exists u_1' \ \big(u_1 R_1 u_1' \ \& \ (\mathfrak{M}^\Phi, \langle u_1', u_2\rangle) \models \psi\big) \quad \text{iff} \quad \text{(by the induction hypothesis)} \\
&\exists u_1' \ \exists i \in I_\psi \ \big(u_1 R_1 u_1' \ \& \ (\mathfrak{M}_1, u_1') \models \psi_i^1 \ \& \ (\mathfrak{M}_2, u_2) \models \psi_i^2\big) \quad \text{iff} \\
&\exists i \in I_\psi \ \big((\mathfrak{M}_1, u_1) \models \Diamond_1 \psi_i^1 \ \& \ (\mathfrak{M}_2, u_2) \models \psi_i^2\big).
\end{aligned}$$

Now the statement follows by taking $I_\varphi = I_\psi$, $\varphi_i^1 = \Diamond_1\psi_i^1$, and $\varphi_i^2 = \psi_i^2$ $(i \in I_\varphi)$. The case of $\varphi = \Diamond_2\psi$ is similar. ❑

Question 9.16. Find other types of Φ-flat product models for which the corresponding variant of Theorem 9.14 holds. Does it hold for arbitrary first-order formulas Φ?

Chapter 10

Intuitionistic modal logics

In Section 3.11, we introduced the intuitionistic modal logic **FS** as the set of unimodal formulas φ whose standard translations $\varphi^\star$ belong to intuitionistic first-order logic **QInt**. In other words, **FS** can be regarded as a 'solution' to the equation $\frac{\mathbf{QCl}}{\mathbf{K}} = \frac{\mathbf{QInt}}{x}$, i.e., as an intuitionistic analog of classical **K**.

MIPC is an intuitionistic analog of classical **S5**. It was defined as the set of unimodal formulas φ whose translations $\varphi^\dagger$ into the one-variable fragment of first-order logic belong to **QInt** and thereby can be regarded as a 'solution' to the equation $\frac{\mathbf{QCl}}{\mathbf{S5}} = \frac{\mathbf{QInt}}{x}$.

In this chapter we provide axiomatizations of these two logics and show that both of them enjoy the finite model property relative to a certain class of so-called **FS**-frames. Remember that they do not have the finite model property with respect to their standard semantics (see Proposition 3.46). The proofs will use axiomatization results for products of classical modal logics and clearly show the two-dimensional character of **FS** and **MIPC**. To 'warm up,' we begin by investigating the simpler intuitionistic modal logic $\mathbf{IntK}_\Box$ having just one necessity operator and no possibility operator at all. This logic turns out to be embedded into the fusion $\mathbf{S4} \otimes \mathbf{K}$.

10.1 Intuitionistic modal logics with $\Box$

The language we consider in this section is $\mathcal{L}_\Box$ (the language of propositional logic extended with a single box operator) and the basic logic we are interested in is $\mathbf{IntK}_\Box$ which is obtained from **Int** by adding the axioms

$$\Box(p \wedge q) \leftrightarrow \Box p \wedge \Box q \quad \text{and} \quad \Box\top$$

and taking the closure under modus ponens, substitution and the regularity rule for $\Box$:

$$\frac{\varphi \to \psi}{\Box\varphi \to \Box\psi}.$$

Our first aim is to develop a semantical machinery for both this logic and all intuitionistic modal logics (im-logics, for short) containing it, i.e., for all subsets L of $\mathcal{L}_\Box$ containing $\mathbf{IntK}_\Box$ and closed under the regularity rule, modus ponens and substitution.

As a technical tool we first introduce an algebraic semantics. An $\mathbf{IntK}_\Box$-*algebra* is a structure of the form

$$\mathfrak{A} = \left\langle A, \to^{\mathfrak{A}}, \wedge^{\mathfrak{A}}, \vee^{\mathfrak{A}}, \Box^{\mathfrak{A}}, 0^{\mathfrak{A}}, 1^{\mathfrak{A}} \right\rangle \tag{10.1}$$

such that $\langle A, \to^{\mathfrak{A}}, \wedge^{\mathfrak{A}}, \vee^{\mathfrak{A}}, 0^{\mathfrak{A}}, 1^{\mathfrak{A}} \rangle$ is a Heyting (= pseudo-Boolean) algebra (see Section 2.7) with unit element $1^{\mathfrak{A}}$ and, for all $a, b \in A$,

$$\Box^{\mathfrak{A}} 1^{\mathfrak{A}} = 1^{\mathfrak{A}}, \quad \Box^{\mathfrak{A}}(a \wedge^{\mathfrak{A}} b) = \Box^{\mathfrak{A}} a \wedge^{\mathfrak{A}} \Box^{\mathfrak{A}} b.$$

Similar to classical modal logics (see Section 1.5), every im-logic containing $\mathbf{IntK}_\Box$ corresponds to a variety of $\mathbf{IntK}_\Box$-algebras. More precisely, $\mathcal{L}_\Box$-formulas are interpreted in $\mathbf{IntK}_\Box$-algebras $\mathfrak{A}$ by means of *valuations* $\mathfrak{V}$ which map $\mathcal{L}_\Box$ into A in such a way that, for all $\varphi, \psi \in \mathcal{L}_\Box$, we have

$$\begin{aligned}
\mathfrak{V}(\varphi \wedge \psi) &= \mathfrak{V}(\varphi) \wedge^{\mathfrak{A}} \mathfrak{V}(\psi),\\
\mathfrak{V}(\varphi \to \psi) &= \mathfrak{V}(\varphi) \to^{\mathfrak{A}} \mathfrak{V}(\psi),\\
\mathfrak{V}(\varphi \vee \psi) &= \mathfrak{V}(\varphi) \vee^{\mathfrak{A}} \mathfrak{V}(\psi),\\
\mathfrak{V}(\bot) &= 0^{\mathfrak{A}},\\
\mathfrak{V}(\Box\varphi) &= \Box^{\mathfrak{A}} \mathfrak{V}(\varphi).
\end{aligned}$$

A formula φ is *true* in the *algebraic model* $\langle \mathfrak{A}, \mathfrak{V} \rangle$ if $\mathfrak{V}(\varphi) = 1^{\mathfrak{A}}$. We say that φ is *valid* in $\mathfrak{A}$ and write $\mathfrak{A} \models \varphi$ if φ is true in all models based on $\mathfrak{A}$. An im-logic L containing $\mathbf{IntK}_\Box$ is said to be *characterized* (or *determined*) by a class $\mathcal{C}$ of $\mathbf{IntK}_\Box$-algebras when $\varphi \in L$ iff φ is valid in every algebra in $\mathcal{C}$.

Theorem 10.1. *Every im-logic containing* $\mathbf{IntK}_\Box$ *is determined by a class of* $\mathbf{IntK}_\Box$*-algebras. Conversely, the set of all formulas valid in a class of* $\mathbf{IntK}_\Box$*-algebras is an im-logic containing* $\mathbf{IntK}_\Box$.

Proof. The proof is based on the standard Lindenbaum construction (see also the proof of Theorem 4.5). Let L be an im-logic containing $\mathbf{IntK}_\Box$. Define an equivalence relation $\sim$ on the set $\mathcal{L}_\Box$ by taking

$$\varphi \sim \psi \quad \text{iff} \quad (\varphi \to \psi) \wedge (\psi \to \varphi) \in L$$

and denote by $[\varphi]$ the $\sim$-equivalence class generated by φ. Using the fact that L is closed under the regularity rule, it is easily seen that $\sim$ is a congruence relation on $\mathcal{L}_\Box$, i.e., for all $\mathcal{L}_\Box$-formulas φ_1, φ_2, ψ_1, ψ_2, if $\varphi_1 \sim \varphi_2$ and $\psi_1 \sim \psi_2$ then $\varphi_1 \odot \varphi_2 \sim \psi_1 \odot \psi_2$, where $\odot$ is any binary connective of $\mathcal{L}_\Box$, and $\varphi_1 \sim \varphi_2$ implies $\Box\varphi_1 \sim \Box\varphi_2$. Therefore, we can define an algebra[1] $\mathfrak{A} = \left\langle A, \to^{\mathfrak{A}}, \wedge^{\mathfrak{A}}, \vee^{\mathfrak{A}}, \Box^{\mathfrak{A}}, 0^{\mathfrak{A}}, 1^{\mathfrak{A}} \right\rangle$, where $A = \{[\varphi] \mid \varphi \in \mathcal{L}_\Box\}$ and

$$\begin{aligned} [\varphi] \wedge^{\mathfrak{A}} [\psi] &= [\varphi \wedge \psi], \\ [\varphi] \to^{\mathfrak{A}} [\psi] &= [\varphi \to \psi], \\ [\varphi] \vee^{\mathfrak{A}} [\psi] &= [\varphi \vee \psi], \\ 0^{\mathfrak{A}} &= [\bot], \\ \Box^{\mathfrak{A}}[\varphi] &= [\Box\varphi]. \end{aligned}$$

It is not difficult to show that $\mathfrak{A}$ is an $\mathbf{IntK}_\Box$-algebra validating L (i.e., $\mathfrak{A} \models \varphi$ for all $\varphi \in L$). Moreover, $\mathfrak{A} \not\models \psi$ whenever $\psi \notin L$. To prove this, it is enough to consider the valuation $\mathfrak{V}$ defined by taking $\mathfrak{V}(p) = [p]$ for all propositional variables p. In this case $\mathfrak{V}(\varphi) \neq 1^{\mathfrak{A}}$ because $\mathfrak{V}(\varphi) = [\varphi]$ and $\varphi \leftrightarrow \top \notin L$.

This proves the former claim of the theorem; the latter one is left to the reader as an easy exercise. ❑

Now we apply this completeness theorem for algebraic semantics to establish completeness of $\mathbf{IntK}_\Box$ with respect to the intended Kripke semantics defined in Section 3.11. Kripke completeness of various natural extensions of $\mathbf{IntK}_\Box$ can be proved in a similar way; see e.g., (Božić and Došen 1984, Sotirov 1984, Wolter and Zakharyaschev 1999a).

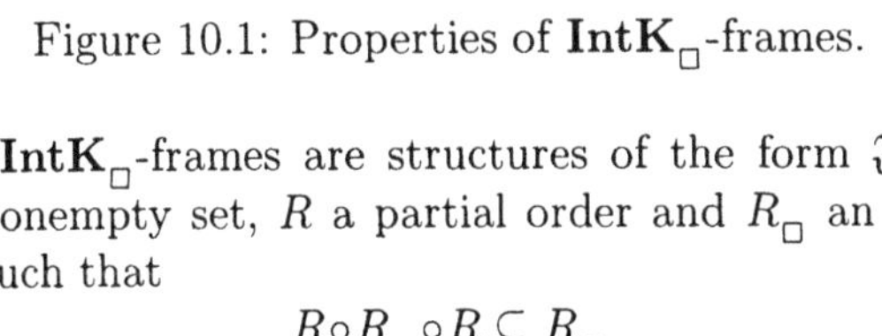

Figure 10.1: Properties of $\mathbf{IntK}_\Box$-frames.

Recall that $\mathbf{IntK}_\Box$-frames are structures of the form $\mathfrak{F} = \langle W, R, R_\Box \rangle$, where W is a nonempty set, R a partial order and $R_\Box$ an arbitrary binary relation on W such that

$$R \circ R_\Box \circ R \subseteq R_\Box$$

or, equivalently,

$$R \circ R_\Box = R_\Box \circ R = R_\Box$$

[1] This algebra is often called the *Lindenbaum algebra* for L.

(see Fig. 10.1). A *valuation* in $\mathfrak{F}$ is a map $\mathfrak{V}$ from the set of propositional variables into the set $Up\mathfrak{F}$ of R-closed subsets of W. Given a model $\mathfrak{M} = \langle \mathfrak{F}, \mathfrak{V} \rangle$, the truth-relation $(\mathfrak{M}, x) \models \varphi$ is defined in such a way that $\to$ is interpreted by R (as in intuitionistic logic) and $\Box$ by $R_\Box$ (as in classical modal logic); for details consult Section 3.11. A formula φ is valid in $\mathfrak{F}$ if $(\mathfrak{M}, x) \models \varphi$ for every $x \in W$ and every model $\mathfrak{M}$ based on $\mathfrak{F}$. It is easily checked that every formula from $\mathbf{IntK}_\Box$ is valid in every $\mathbf{IntK}_\Box$-frame.

Similarly to the classical modal case (see Section 1.5), every $\mathbf{IntK}_\Box$-frame $\mathfrak{F} = \langle W, R, R_\Box \rangle$ gives rise to its dual $\mathbf{IntK}_\Box$-algebra

$$\mathfrak{F}^+ = \left\langle Up\,\mathfrak{F}, \to, \cap, \cup, \emptyset, W, \Box \right\rangle,$$

where, for all $X, Y \in Up\,\mathfrak{F}$,

$$\begin{aligned} X \to Y &= \{x \in W \mid \forall y \in W\ (xRy \land y \in X \ \to\ y \in Y)\}, \\ \Box X &= \{x \in W \mid \forall y \in W\ (xR_\Box y \ \to\ y \in X)\}. \end{aligned}$$

Moreover, $\mathfrak{F} \models \varphi$ is obviously equivalent to $\mathfrak{F}^+ \models \varphi$, for every $\mathcal{L}_\Box$-formula φ.

Conversely, with every $\mathbf{IntK}_\Box$-algebra $\mathfrak{A}$ of the form (10.1) we can associate an $\mathbf{IntK}_\Box$-frame $\kappa\mathfrak{A} = \langle W, R, R_\Box \rangle$ by taking W to be the set of all prime filters in $\mathfrak{A}$ and, for $x, y \in W$,

$$\begin{aligned} xRy &\quad \text{iff} \quad x \subseteq y, \\ xR_\Box y &\quad \text{iff} \quad \forall a \in A\ (\Box^{\mathfrak{A}} a \in x \to a \in y). \end{aligned}$$

We remind the reader that a *prime filter* x in $\mathfrak{A}$ is a subset of the power-set of A such that, for all $a, b \in A$,

- $0^{\mathfrak{A}} \notin x$ and $1^{\mathfrak{A}} \in x$;
- $b \in x$ whenever $a \in x$ and $a \leq b$ (here and in what follows $\leq$ is the lattice partial order on A defined by $a \leq b$ iff $a \land^{\mathfrak{A}} b = a$);
- $a \land^{\mathfrak{A}} b \in x$ whenever $a, b \in x$;
- $a \in x$ or $b \in x$ whenever $a \lor^{\mathfrak{A}} b \in x$.

The following two lemmas are required to prove the completeness theorem. The first one is a standard lemma on the existence of prime filters with certain properties; see, e.g., (Rasiowa and Sikorski 1963).

Lemma 10.2. *Suppose that* $\mathfrak{A} = \langle A, \to^{\mathfrak{A}}, \land^{\mathfrak{A}}, \lor^{\mathfrak{A}}, 0^{\mathfrak{A}}, 1^{\mathfrak{A}} \rangle$ *is a Heyting algebra and* B, C *are nonempty subsets of* A *such that* (i) $b_1 \land^{\mathfrak{A}} \cdots \land^{\mathfrak{A}} b_n \not\leq c$, *for any* $b_1, \ldots, b_n \in B$, $c \in C$, *and* (ii) *for all* $c_1, c_2 \in C$, *there is* $c \in C$ *for which* $c_1 \lor^{\mathfrak{A}} c_2 \leq c$. *Then there exists a prime filter* ∇ *in* $\mathfrak{A}$ *such that* $B \subseteq \nabla$ *and* $C \cap \nabla = \emptyset$.

The second lemma connects $\mathbf{IntK}_\Box$-algebras $\mathfrak{A}$ with their $\mathbf{IntK}_\Box$-frames $\kappa\mathfrak{A}$.

Lemma 10.3. *Let* $\mathfrak{A} = \left\langle A, \to^{\mathfrak{A}}, \wedge^{\mathfrak{A}}, \vee^{\mathfrak{A}}, \Box^{\mathfrak{A}}, 0^{\mathfrak{A}}, 1^{\mathfrak{A}} \right\rangle$ *be an* $\mathbf{IntK}_\Box$*-algebra and* $\kappa\mathfrak{A} = \langle W, R, R_\Box \rangle$. *Then the map* $h : A \to Up\,\kappa\mathfrak{A}$, *defined by taking*

$$h(a) = \{x \in W \mid a \in x\},$$

for each $a \in A$, *is an injective homomorphism from* $\mathfrak{A}$ *to* $(\kappa\mathfrak{A})^+$.

Proof. We show only that h is injective and leave it to the reader to check that h is a homomorphism. Suppose $a \neq b$. Without loss of generality we may assume that $a \not\leq b$. Then, by Lemma 10.2, we can find a prime filter $x \in W$ such that $a \in x$ and $b \notin x$. Hence $h(a) \neq h(b)$. ❑

The completeness theorem follows now almost immediately:

Theorem 10.4. $\mathbf{IntK}_\Box$ *is determined by the class of* $\mathbf{IntK}_\Box$*-frames.*

Proof. We know already that every formula from $\mathbf{IntK}_\Box$ is valid in every $\mathbf{IntK}_\Box$-frame. Conversely, suppose that $\varphi \notin \mathbf{IntK}_\Box$. Then there exists an $\mathbf{IntK}_\Box$-algebra $\mathfrak{A}$ such that $\mathfrak{A} \not\models \varphi$. By Lemma 10.3, $\mathfrak{A}$ is isomorphic to a subalgebra of $(\kappa\mathfrak{A})^+$. Hence $(\kappa\mathfrak{A})^+ \not\models \varphi$, and so $\kappa\mathfrak{A} \not\models \varphi$. ❑

Recall that one of the main reasons for introducing modal logics, in particular **S4**, was the desire to find a classical interpretation of intuitionistic logic. This was done via the Gödel translation T of **Int** into **S4**; see Section 2.7. Now we show that this translation can be extended to an embedding of $\mathbf{IntK}_\Box$ into the fusion $\mathbf{S4} \otimes \mathbf{K}$. (Actually, it can be lifted to an embedding of all im-logics containing $\mathbf{IntK}_\Box$ into normal modal logics containing $\mathbf{S4} \otimes \mathbf{K}$; see (Wolter and Zakharyaschev 1997, 1999a).)

Let us assume that $\mathbf{S4} \otimes \mathbf{K}$ is formulated in the language $\mathcal{ML}_2$ with two necessity operators $\Box_I$ and $\Box_M$. Define inductively a translation T^* from $\mathcal{L}_\Box$ into $\mathcal{ML}_2$ by taking:

$$\begin{aligned}
\mathsf{T}^*(p) &= \Box_I p, \ p \text{ a variable},\\
\mathsf{T}^*(\bot) &= \Box_I \bot,\\
\mathsf{T}^*(\varphi \to \psi) &= \Box_I(\mathsf{T}^*(\varphi) \to \mathsf{T}^*(\psi)),\\
\mathsf{T}^*(\varphi \wedge \psi) &= \Box_I(\mathsf{T}^*(\varphi) \wedge \mathsf{T}^*(\psi)),\\
\mathsf{T}^*(\varphi \vee \psi) &= \Box_I(\mathsf{T}^*(\varphi) \vee \mathsf{T}^*(\psi)),\\
\mathsf{T}^*(\Box\varphi) &= \Box_I \Box_M \mathsf{T}^*(\varphi).
\end{aligned}$$

Theorem 10.5. *For every* $\mathcal{L}_\Box$*-formula* φ,

$$\varphi \in \mathbf{IntK}_\Box \quad \textit{iff} \quad \mathsf{T}^*(\varphi) \in \mathbf{S4} \otimes \mathbf{K}.$$

Proof. Suppose $\varphi \notin \mathbf{IntK}_\Box$. Then there is a model $\mathfrak{M} = \langle \mathfrak{F}, \mathfrak{V} \rangle$ based on an $\mathbf{IntK}_\Box$-frame $\mathfrak{F} = \langle W, R, R_\Box \rangle$ such that $(\mathfrak{M}, x) \not\models \varphi$ for some x in $\mathfrak{F}$. Construct a 2-frame $\sigma\mathfrak{F} = \langle W, R_I, R_M \rangle$ simply by taking $R_I = R$, $R_M = R_\Box$. It should be clear that $\sigma\mathfrak{F}$ is a frame for $\mathbf{S4} \otimes \mathbf{K}$. Define a valuation $\sigma\mathfrak{V}$ in $\sigma\mathfrak{F}$ by putting $\sigma\mathfrak{V}(p) = \mathfrak{V}(p)$ for all propositional variables p. Let $\sigma\mathfrak{M} = \langle \sigma\mathfrak{F}, \sigma\mathfrak{V} \rangle$. Then one can easily show by induction that

$$(\mathfrak{M}, y) \models \psi \quad \text{iff} \quad (\sigma\mathfrak{M}, y) \models \mathsf{T}^*(\psi),$$

for every $\mathcal{L}_\Box$-formula ψ and every y in $\mathfrak{F}$. Hence $\sigma\mathfrak{F}$ refutes $\mathsf{T}^*(\varphi)$, and so $\mathsf{T}^*(\varphi) \notin \mathbf{S4} \otimes \mathbf{K}$.

Conversely, suppose that $\mathsf{T}^*(\varphi) \notin \mathbf{S4} \otimes \mathbf{K}$. Let $\mathfrak{F} = \langle W, R_I, R_M \rangle$ be a 2-frame validating $\mathbf{S4} \otimes \mathbf{K}$ and refuting $\mathsf{T}^*(\varphi)$. We construct an $\mathbf{IntK}_\Box$-frame $\rho\mathfrak{F}$ refuting φ in three steps.

First, define a 2-frame $\mathfrak{F}^* = \langle W, R_I, R_M^* \rangle$ by taking, for all $x, y \in W$,

$$x R_M^* y \quad \text{iff} \quad x(R_I \circ R_M \circ R_I)y.$$

Since R_I is a quasi-order, we clearly have $R_M \subseteq R_M^*$. Moreover, the following is easily checked:

(i) The equality

$$R_I \circ R_M^* = R_M^* \circ R_I = R_M^*$$

holds in $\mathfrak{F}^*$ or, which is equivalent, $\mathfrak{F}^*$ validates the formula

$$\boldsymbol{mix} = (\Box_I \Box_M p \leftrightarrow \Box_M p) \wedge (\Box_M \Box_I p \leftrightarrow \Box_M p).$$

(ii) For every $\mathcal{L}_\Box$-formula ψ,

$$\mathfrak{F}^* \models \mathsf{T}^*(\psi) \quad \text{iff} \quad \mathfrak{F} \models \mathsf{T}^*(\psi).$$

Suppose now that a 2-frame $\mathfrak{G} = \langle V, S_I, S_M \rangle$ for $\mathbf{S4} \otimes \mathbf{K}$ validates $\boldsymbol{mix}$. Define an equivalence relation $\sim$ on V by taking $x \sim y$ iff x and y belong to the same S_I-cluster in $\mathfrak{G}$ (i.e., xS_Iy and yS_Ix) and let $[x] = x/_\sim$, for any $x \in V$. Put

$$[x]\,[S_I]\,[y] \quad \text{iff} \quad xS_Iy,$$
$$[x]\,[S_M]\,[y] \quad \text{iff} \quad xS_My.$$

(Since S_I is transitive and since, by $\boldsymbol{mix}$, xS_My iff zS_My, for every x and z belonging to the same S_I-cluster, the definition of $[S_I]$ and $[S_M]$ does not depend on the choice of representatives in the classes $[x]$ and $[y]$.) The structure $[\mathfrak{G}] = \langle [V], [S_I], [S_M] \rangle$, where $[V] = \{[y] \mid y \in V\}$, is called the *skeleton* of $\mathfrak{G}$. It is easy to see that if $\mathfrak{G}$ validates $\boldsymbol{mix}$ and S_I is a partial order then $\mathfrak{G} \simeq [\mathfrak{G}^*]$. The following is easily checked:

- the map $x \mapsto [x]$ is a p-morphism from $\mathfrak{G}$ onto $[\mathfrak{G}]$;
- $[S_I]$ is a partial order on $[V]$ and
$$[S_I] \circ [S_M] = [S_M] \circ [S_I] = [S_M];$$
- $\mathfrak{G} \models \mathsf{T}^*(\psi)$ iff $[\mathfrak{G}] \models \mathsf{T}^*(\psi)$, for every $\mathcal{L}_\Box$-formula ψ.

Now, given our original 2-frame $\mathfrak{F} = \langle W, R_I, R_M \rangle$, we first form the frame $[\mathfrak{F}^*] = \langle [W], [R_I], [R_M^*] \rangle$ and then define $\rho\mathfrak{F} = \langle [W], R, R_\Box \rangle$ by taking $R = [R_I]$ and $R_\Box = [R_M^*]$. It should be clear that $\rho\mathfrak{F}$ is an $\mathbf{IntK}_\Box$-frame. By induction on the construction of ψ one can readily show that

$$\mathfrak{F} \models \psi \quad \text{iff} \quad \rho\mathfrak{F} \models \mathsf{T}^*(\psi),$$

for every $\mathcal{L}_\Box$-formula ψ. It follows that φ is refuted in $\rho\mathfrak{F}$. ❑

Since the fusion $\mathbf{S4} \otimes \mathbf{K}$ is PSPACE-complete and has the fmp (see Theorem 4.19), as an immediate consequence of Theorem 10.5 and its proof we obtain the following result (the fmp was first proved in (Sotirov 1984)):

Corollary 10.6. $\mathbf{IntK}_\Box$ *is* PSPACE-*complete and has the fmp.*

10.2 Intuitionistic modal logics with □ and ◇

Recall from Section 3.11 that an im-logic in the language $\mathcal{L}_{\Box\Diamond}$ with both □ and ◇ is a set of $\mathcal{L}_{\Box\Diamond}$-formulas containing $\mathbf{IntK}_{\Box\Diamond}$ and closed under modus ponens, substitution and the regularity rules for both □ and ◇. In this section we concentrate on two such logics, **FS** and **MIPC**, which were introduced in Section 3.11 as

$$\mathbf{FS} = \{\varphi \in \mathcal{L}_{\Box\Diamond} \mid \varphi^\star \in \mathbf{QInt}\},$$
$$\mathbf{MIPC} = \{\varphi \in \mathcal{L}_{\Box\Diamond} \mid \varphi^\dagger \in \mathbf{QInt}\},$$

where $\varphi^\star$ is the standard translation of φ into first-order logic **QCl** and $\varphi^\dagger$ is the translation of φ into the one-variable fragment of **QCl** (see Section 1.3).

Our first aim is to axiomatize **FS** and **MIPC**, and then to prove their decidability by means of embedding them into relativized products $(\mathbf{S4}\times\mathbf{K})^{\mathsf{EX}}$ and $(\mathbf{S4} \times \mathbf{S5})^{\mathsf{EX}}$, respectively.

Theorem 10.7.

$$\begin{aligned}\mathbf{FS} &= \mathbf{IntK}_{\Box\Diamond} \oplus \Diamond(p \to q) \to (\Box p \to \Diamond q) \oplus \\ &\qquad (\Diamond p \to \Box q) \to \Box(p \to q), \\ \mathbf{MIPC} &= \mathbf{FS} \oplus \Box p \to p \oplus \Box p \to \Box\Box p \oplus \Diamond p \to \Box\Diamond p \oplus \\ &\qquad p \to \Diamond p \oplus \Diamond\Diamond p \to \Diamond p \oplus \Diamond\Box p \to \Box p.\end{aligned}$$

Proof. Let us denote the logics in the right-hand sides of these equalities by $\mathbf{FS}'$ and $\mathbf{MIPC}'$, respectively. Thus, we have to prove that $\mathbf{FS} = \mathbf{FS}'$ and $\mathbf{MIPC} = \mathbf{MIPC}'$. This will be done in four steps, each of which is of independent interest on its own:

Step 1. First we provide an algebraic and a Kripke-type semantics for im-logics in the language $\mathcal{L}_{\Box\Diamond}$. In particular, we obtain completeness results for $\mathbf{FS}'$ and $\mathbf{MIPC}'$.

Step 2. Then we observe that $\mathbf{FS} \supseteq \mathbf{FS}'$ and $\mathbf{MIPC} \supseteq \mathbf{MIPC}'$.

Step 3. Next, we extend the translation T^* from the previous section to a translation from $\mathcal{L}_{\Box\Diamond}$ into the bimodal language $\mathcal{ML}_2$ with the boxes $\Box_I$ and $\Box_M$ by taking

$$\mathsf{T}^*(\Diamond\varphi) = \Box_I \Diamond_M \mathsf{T}^*(\varphi).$$

We show that, for every $\mathcal{L}_{\Box\Diamond}$-formula φ,

$$\varphi \in \mathbf{FS} \quad\Longrightarrow\quad \mathsf{T}^*(\varphi) \in (\mathbf{S4} \times \mathbf{K})^{\mathsf{EX}} \tag{10.2}$$

and

$$\mathsf{T}^*(\varphi) \in [\mathbf{S4}, \mathbf{K}]^{\mathsf{EX}} \quad\Longrightarrow\quad \varphi \in \mathbf{FS}'. \tag{10.3}$$

Similarly, we show that, for every $\mathcal{L}_{\Box\Diamond}$-formula φ,

$$\varphi \in \mathbf{MIPC} \quad\Longrightarrow\quad \mathsf{T}^*(\varphi) \in (\mathbf{S4} \times \mathbf{S5})^{\mathsf{EX}} \tag{10.4}$$

and

$$\mathsf{T}^*(\varphi) \in [\mathbf{S4}, \mathbf{S5}]^{\mathsf{EX}} \quad\Longrightarrow\quad \varphi \in \mathbf{MIPC}'. \tag{10.5}$$

Step 4. Finally, we apply Theorem 9.10, according to which

$$(\mathbf{S4} \times \mathbf{K})^{\mathsf{EX}} = [\mathbf{S4}, \mathbf{K}]^{\mathsf{EX}} \quad \text{and} \quad (\mathbf{S4} \times \mathbf{S5})^{\mathsf{EX}} = [\mathbf{S4}, \mathbf{S5}]^{\mathsf{EX}},$$

and obtain for all $\mathcal{L}_{\Box\Diamond}$-formulas φ the following equivalences:

$$\begin{aligned}\varphi \in \mathbf{FS} &\iff \mathsf{T}^*(\varphi) \in (\mathbf{S4} \times \mathbf{K})^{\mathsf{EX}} \iff \varphi \in \mathbf{FS}', \\ \varphi \in \mathbf{MIPC} &\iff \mathsf{T}^*(\varphi) \in (\mathbf{S4} \times \mathbf{S5})^{\mathsf{EX}} \iff \varphi \in \mathbf{MIPC}'.\end{aligned}$$

This will prove Theorem 10.7. Moreover, we shall clearly have:

Theorem 10.8. *For every $\mathcal{L}_{\Box\Diamond}$-formula φ,*

$$\varphi \in \mathbf{FS} \quad \textit{iff} \quad \mathsf{T}^*(\varphi) \in (\mathbf{S4} \times \mathbf{K})^{\mathsf{EX}}$$

and

$$\varphi \in \mathbf{MIPC} \quad \textit{iff} \quad \mathsf{T}^*(\varphi) \in (\mathbf{S4} \times \mathbf{S5})^{\mathsf{EX}}.$$

As the products $\mathbf{S4} \times \mathbf{K}$ and $\mathbf{S4} \times \mathbf{S5}$ are decidable by Theorems 6.20 and 5.28, and $(\mathbf{S4} \times \mathbf{K})^{\mathsf{EX}}$ and $(\mathbf{S4} \times \mathbf{S5})^{\mathsf{EX}}$ are reducible, respectively, to $\mathbf{S4} \times \mathbf{K}$ and $\mathbf{S4} \times \mathbf{S5}$ by Theorem 9.12, we also obtain the following results of Bull (1965), Ono (1977), Simpson (1994) and Grefe (1998):

Theorem 10.9. *Both* **MIPC** *and* **FS** *are decidable.*

We begin the realization of this plan by providing an algebraic semantics for im-logics under consideration. In fact, it can be obtained by generalizing the algebraic semantics from the previous section in a straightforward way. An $\mathbf{IntK}_{\Box\Diamond}$-*algebra* is a structure of the form

$$\mathfrak{A} = \left\langle A, \to^{\mathfrak{A}}, \wedge^{\mathfrak{A}}, \vee^{\mathfrak{A}}, \Box^{\mathfrak{A}}, \Diamond^{\mathfrak{A}}, 0^{\mathfrak{A}}, 1^{\mathfrak{A}} \right\rangle$$

such that $\left\langle A, \to^{\mathfrak{A}}, \wedge^{\mathfrak{A}}, \vee^{\mathfrak{A}}, \Box^{\mathfrak{A}}, 0^{\mathfrak{A}}, 1^{\mathfrak{A}} \right\rangle$ is an $\mathbf{IntK}_{\Box}$-algebra and, for all elements $a, b \in A$,

$$\neg \Diamond^{\mathfrak{A}} 0^{\mathfrak{A}} = 1^{\mathfrak{A}}, \quad \Diamond^{\mathfrak{A}}(a \vee^{\mathfrak{A}} b) = \Diamond^{\mathfrak{A}} a \vee^{\mathfrak{A}} \Diamond^{\mathfrak{A}} b.$$

$\mathcal{L}_{\Box\Diamond}$-formulas are interpreted in $\mathbf{IntK}_{\Box\Diamond}$-algebras $\mathfrak{A}$ by means of *valuations* $\mathfrak{V}$ which map $\mathcal{L}_{\Box\Diamond}$ into A in such a way that the restriction of $\mathfrak{V}$ to $\mathcal{L}_{\Box}$ is a valuation in the sense of the previous section into the reduct of $\mathfrak{A}$ without $\Diamond^{\mathfrak{A}}$ and, for every $\mathcal{L}_{\Box\Diamond}$-formula ψ,

$$\mathfrak{V}(\Diamond\psi) \quad = \quad \Diamond^{\mathfrak{A}}\mathfrak{V}(\psi).$$

As before, a formula φ is said to be *true* in the model $\langle \mathfrak{A}, \mathfrak{V} \rangle$ if $\mathfrak{V}(\varphi) = 1^{\mathfrak{A}}$; φ is *valid* in $\mathfrak{A}$ ($\mathfrak{A} \models \varphi$, in symbols) if φ is true in all models based on $\mathfrak{A}$.

Given a class $\mathcal{C}$ of $\mathbf{IntK}_{\Box\Diamond}$-algebras and an im-logic L containing $\mathbf{IntK}_{\Box\Diamond}$, we say that L is *characterized* (or *determined*) by $\mathcal{C}$ when $\varphi \in L$ iff φ is valid in every algebra in $\mathcal{C}$.

Theorem 10.10. *Every im-logic containing* $\mathbf{IntK}_{\Box\Diamond}$ *is determined by a class of* $\mathbf{IntK}_{\Box\Diamond}$*-algebras. Conversely, the set of all formulas valid in a class of* $\mathbf{IntK}_{\Box\Diamond}$*-algebras is an im-logic containing* $\mathbf{IntK}_{\Box\Diamond}$.

Proof. Similar to the proof of Theorem 10.1. ❑

To obtain completeness results for $\mathbf{IntK}_{\Box\Diamond}$, $\mathbf{FS}'$ and $\mathbf{MIPC}'$ with respect to certain classes of so-called $\mathbf{FS}$-frames (which are generalizations of the standard $\mathbf{FS}$-frames), we require the *Stone–Jónsson–Tarski representations* of $\mathbf{IntK}_{\Box\Diamond}$-algebras. But first we remind the reader what such representations of Heyting algebras look like.

A *general intuitionistic frame* is a structure of the form $\mathfrak{F} = \langle W, R, P\rangle$, where $\langle W, R\rangle$ is an intuitionistic (Kripke) frame and P is a collection of sets in $Up\,\mathfrak{F}$ containing $\emptyset$ and closed under $\cap$, $\cup$ and the operation

$$X \to Y = \{x \in W \mid \forall y \in W\ (xRy \wedge y \in X \to y \in Y)\}.$$

If P contains all the upward closed subsets of W then we identify $\mathfrak{F}$ with $\langle W, R\rangle$ and call it, as before, an intuitionistic (Kripke) frame. Given a Heyting algebra $\mathfrak{A} = \langle A, \to^{\mathfrak{A}}, \wedge^{\mathfrak{A}}, \vee^{\mathfrak{A}}, 0^{\mathfrak{A}}, 1^{\mathfrak{A}}\rangle$, we define its *dual* $\mathfrak{A}_+$ to be the structure $\langle W, R, P\rangle$, where

- W is the set of prime filters in $\mathfrak{A}$,
- xRy iff $x \subseteq y$, for all $x, y \in W$,
- $P = \{X_a \mid a \in A\}$, where $X_a = \{x \in W \mid a \in x\}$.

(For more details on duality between Heyting algebras and general intuitionistic frames consult (Chagrov and Zakharyaschev 1997).)

Now, given an $\mathbf{IntK}_{\Box\Diamond}$-algebra $\mathfrak{A} = \left\langle A, \to^{\mathfrak{A}}, \wedge^{\mathfrak{A}}, \vee^{\mathfrak{A}}, \Box^{\mathfrak{A}}, \Diamond^{\mathfrak{A}}, 0^{\mathfrak{A}}, 1^{\mathfrak{A}}\right\rangle$, we define its *dual* $\mathfrak{A}_+$ as the structure $\langle W, R, R_\Box, R_\Diamond, P\rangle$, where $\langle W, R, P\rangle$ is the dual of the Heyting algebra underlying $\mathfrak{A}$ and, for all $x, y \in W$,

$$xR_\Box y \quad \text{iff} \quad \forall a \in A\ (\Box a \in x \to a \in y),$$
$$xR_\Diamond y \quad \text{iff} \quad \forall a \in A\ (a \in y \to \Diamond a \in x).$$

It follows immediately from the definition that

$$R \circ R_\Box \circ R \subseteq R_\Box, \tag{10.6}$$
$$R \circ R_\Diamond^{-1} \circ R \subseteq R_\Diamond^{-1}. \tag{10.7}$$

Observe that condition (10.6) was already introduced to characterize $\mathbf{IntK}_\Box$-frames. Structures of the form $\mathfrak{F} = \langle W, R, R_\Box, R_\Diamond, P\rangle$, where $\langle W, R, P\rangle$ is a general intuitionistic frame, $R_\Box$, $R_\Diamond$ are binary relations on W satisfying (10.6) and (10.7), and P is closed under the operations $\Box$ and $\Diamond$ defined by

$$\Box X = \{x \in W \mid \forall y \in X\ (xR_\Box y \to y \in X)\},$$
$$\Diamond X = \{x \in W \mid \exists y \in X\ xR_\Diamond y\},$$

will be called *general* $\mathbf{IntK}_{\Box\Diamond}$*-frames*. The *dual* of a general $\mathbf{IntK}_{\Box\Diamond}$-frame $\mathfrak{F}$ is then the algebra $\mathfrak{F}^+ = \left\langle P, \to, \cap, \cup, \Box, \Diamond, \emptyset, W \right\rangle$. It is not hard to check that $\mathfrak{F}^+$ is an $\mathbf{IntK}_{\Box\Diamond}$-algebra and that $\mathfrak{A}$ is isomorphic to $(\mathfrak{A}_+)^+$, for every $\mathbf{IntK}_{\Box\Diamond}$-algebra $\mathfrak{A}$. Say that a general $\mathbf{IntK}_{\Box\Diamond}$-frame $\mathfrak{F}$ is *descriptive* if $\mathfrak{F}$ is isomorphic to $(\mathfrak{F}^+)_+$.

A *valuation* $\mathfrak{V}$ in a general frame $\mathfrak{F} = \langle W, R, R_\Box, R_\Diamond, P\rangle$ associates with every propositional variable p a set $\mathfrak{V}(p) \in P$. Given a model $\mathfrak{M} = \langle \mathfrak{F}, \mathfrak{V}\rangle$, the truth-relation $(\mathfrak{M}, x) \models \varphi$ is defined by extending the truth-relation from the previous section in a straightforward way:

$$(\mathfrak{M}, x) \models \Diamond\varphi \quad \text{iff} \quad \exists y \in W\ (xR_\Diamond y\ \&\ y \models \varphi).$$

A formula φ is *valid* in $\mathfrak{F}$ if $(\mathfrak{M}, x) \models \varphi$, for every model $\mathfrak{M}$ based on $\mathfrak{F}$ and every x in $\mathfrak{F}$.

Since the general frames of the form $\mathfrak{A}_+$ are clearly descriptive, we have:

Proposition 10.11. *Every im-logic containing* $\mathbf{IntK}_{\Box\Diamond}$ *is determined by a suitable class of descriptive* $\mathbf{IntK}_{\Box\Diamond}$*-frames.* $\mathbf{IntK}_{\Box\Diamond}$ *is determined by the class of all descriptive* $\mathbf{IntK}_{\Box\Diamond}$*-frames.*

The following internal characterization of descriptive $\mathbf{IntK}_{\Box\Diamond}$-frames is obtained by a straightforward combination of the corresponding characterizations of descriptive modal and intuitionistic frames; for details consult (Goldblatt 1993, Chagrov and Zakharyaschev 1997).

Proposition 10.12. *A general* $\mathbf{IntK}_{\Box\Diamond}$*-frame* $\mathfrak{F} = \langle W, R, R_\Box, R_\Diamond, P\rangle$ *is descriptive iff* $\mathfrak{F}$ *is tight*$_R$*, tight*$_{R_\Box}$ *and tight*$_{R_\Diamond}$*, i.e.,*

$$\begin{aligned} xRy &\quad \textit{iff} \quad \forall X \in P\ (x \in X \to y \in X),\\ xR_\Box y &\quad \textit{iff} \quad \forall X \in P\ (x \in \Box X \to y \in X),\\ xR_\Diamond y &\quad \textit{iff} \quad \forall X \in P\ (y \in X \to x \in \Diamond X),\end{aligned}$$

and compact, i.e., for all $\mathcal{X} \subseteq P$ *and* $\mathcal{Y} \subseteq \{W - X \mid X \in P\}$*, if* $\mathcal{X} \cup \mathcal{Y}$ *has the finite intersection property*[2] *then* $\bigcap(\mathcal{X} \cup \mathcal{Y}) \neq \emptyset$*.*

A general $\mathbf{IntK}_{\Box\Diamond}$-frame $\mathfrak{F} = \langle W, R, R_\Box, R_\Diamond, P\rangle$ is called a *full* (or *Kripke*) $\mathbf{IntK}_{\Box\Diamond}$-frame if $\langle W, R, P\rangle$ is an intuitionistic Kripke frame. The underlying full frame of a general $\mathbf{IntK}_{\Box\Diamond}$-frame $\mathfrak{F}$ is denoted by $\kappa\mathfrak{F}$. An im-logic L containing $\mathbf{IntK}_{\Box\Diamond}$ is said to be *d-persistent* if $\kappa\mathfrak{F} \models L$ whenever $\mathfrak{F}$ is a descriptive frame validating L. All d-persistent im-logics are clearly determined by full $\mathbf{IntK}_{\Box\Diamond}$-frames.

We are about to show that $\mathbf{FS}'$ and $\mathbf{MIPC}'$ are d-persistent.

[2]A collection $\mathcal{Z}$ of sets is said to have the *finite intersection property* if the intersection of any finite number of sets in $\mathcal{Z}$ is nonempty.

Proposition 10.13. **FS**$'$ *is d-persistent. Hence it is determined by a class of full* $\mathbf{IntK}_{\Box\Diamond}$*-frames.*

Proof. It suffices to show that any full $\mathbf{IntK}_{\Box\Diamond}$-frame satisfying the conditions

$$\forall x \forall y \ (xR_{\Diamond}y \to \exists z \ (yRz \wedge xR_{\Box}z \wedge xR_{\Diamond}z)), \tag{10.8}$$

$$\forall x \forall y \ (xR_{\Box}y \to \exists z \ (xRz \wedge zR_{\Box}y \wedge zR_{\Diamond}y)) \tag{10.9}$$

(see Fig. 10.2) validates **FS**$'$, and that (10.8) and (10.9) hold in any descriptive frame validating **FS**$'$. To prove the former claim, suppose that a full $\mathbf{IntK}_{\Box\Diamond}$-

Figure 10.2: Properties (10.8) and (10.9).

frame $\mathfrak{F} = \langle W, R, R_{\Box}, R_{\Diamond} \rangle$ satisfies (10.8), but $\Diamond(p \to q) \to (\Box p \to \Diamond q)$ is refuted in $\mathfrak{F}$ under some valuation. Then $x \models \Diamond(p \to q)$, $x \models \Box p$, $x \not\models \Diamond q$, for some x in $\mathfrak{F}$, and so there is y such that $xR_{\Diamond}y$ and $y \models p \to q$. By (10.8), we have yRz, $xR_{\Box}z$ and $xR_{\Diamond}z$ for some point z. But then $z \models p \to q$ (since the truth-set of any formula is R-closed), $z \models p$ and $z \not\models q$, which is impossible. The second axiom of **FS**$'$ is considered analogously with the help of (10.9) and (10.6).

Suppose now that $\mathfrak{F} = \langle W, R, R_{\Box}, R_{\Diamond}, P \rangle$ is a descriptive frame validating **FS**$'$ and show that it satisfies (10.9). Without loss of generality we may assume that $\mathfrak{F} \simeq \mathfrak{A}_+$ for some $\mathbf{IntK}_{\Box\Diamond}$-algebra $\mathfrak{A}$ validating **FS**$'$. Thus, points in $\mathfrak{F}$ are prime filters in $\mathfrak{A}$. Let $x, y \in W$ and $xR_{\Box}y$. Put

$$B = x \cup \{\Diamond b \mid b \in y\}, \quad C = \{\Box c \mid c \notin y\}$$

and show that B and C satisfy (i) and (ii) in Lemma 10.2. Suppose

$$a \wedge \Diamond b_1 \wedge \cdots \wedge \Diamond b_n \leq \Box c$$

for some $a \in x$ (x is closed under $\wedge$), $b_1, \ldots, b_n \in y$ and $c \notin y$. Then

$$a \wedge \Diamond b_1 \wedge \cdots \wedge \Diamond b_n \to \Box c = \top$$

in $\mathfrak{A}$, from which, by the second axiom of **FS**$'$, we obtain

$$a \to \Box(b_1 \wedge \cdots \wedge b_n \to c) = \top.$$

It follows that $\Box(b \to c) \in x$ for some $b \in y$ and $c \notin y$. Since $xR_{\Box}y$, we then have $b \to c \in y$ and $c \in y$, which is a contradiction. Therefore, (i) holds. To show (ii), suppose $c_1, c_2 \notin y$. Since y is prime, $c_1 \vee c_2 \notin y$ and so $\Box(c_1 \vee c_2) \in C$ and $\Box c_1 \vee \Box c_2 \leq \Box(c_1 \vee c_2)$.

By Lemma 10.2, there is a prime filter $z \in W$ such that $B \subseteq z$ and $C \cap z = \emptyset$. This means that xRz, $zR_{\Box}y$ and $zR_{\Diamond}y$, as required by (10.9).

In the same way, using Lemma 10.2 and the first axiom of $\mathbf{FS}'$, one can show that $\mathfrak{F}$ satisfies (10.8). We leave this to the reader. ❑

Using the same sort of technique it is not hard to prove the following proposition, in which $\Box^n$ and $\Diamond^n$ are strings of n boxes and diamonds, respectively.

Proposition 10.14. *For all $k, l, m, n \geq 0$, the logic*

$$\mathbf{L}(k,l,m,n) = \mathbf{IntK}_{\Box\Diamond} \oplus \Diamond^k\Box^l p \to \Box^m\Diamond^n p$$

is d-persistent, with every descriptive $\mathbf{IntK}_{\Box\Diamond}$-frame $\mathfrak{F} = \langle W, R, R_{\Box}, R_{\Diamond}, P\rangle$ for $\mathbf{L}(k,l,m,n)$ satisfying the condition

$$\forall x \forall y \forall z \left(xR^k_{\Diamond}y \wedge xR^m_{\Box}z \to \exists u \ (yR^l_{\Box}u \wedge zR^n_{\Diamond}u)\right).$$

As $\mathbf{MIPC}'$ is axiomatized by adding axioms of the form $\mathbf{L}(k,l,m,n)$ to $\mathbf{FS}'$, we also obtain:

Corollary 10.15. $\mathbf{MIPC}'$ *is d-persistent. Hence it is determined by a class of full $\mathbf{IntK}_{\Box\Diamond}$-frames.*

We now introduce a more compact representation of $\mathbf{IntK}_{\Box\Diamond}$-frames for $\mathbf{FS}'$, so-called **FS**-frames. An **FS**-*frame* is a triple of the form $\mathfrak{F} = \langle W, R, S\rangle$, where R is a partial order on W, and R, S satisfy the *Church–Rosser property*

$$\forall x \forall y \forall z \left(xRy \wedge xSz \to \exists u \ (ySu \wedge zRu)\right) \tag{10.10}$$

and *left-commutativity*

$$\forall x \forall y \forall z \left(xSy \wedge yRz \to \exists u \ (xRu \wedge uSz)\right) \tag{10.11}$$

(cf. Sections 5.1 and 9.1). A valuation $\mathfrak{V}$ in $\mathfrak{F}$ associates with every variable p an R-closed subset $\mathfrak{V}(p)$ of W. The truth-relation $\models$ is defined in the standard way for the intuitionistic connectives; the truth-conditions for $\Box$ and $\Diamond$ look as follows:

$$w \models \Box\varphi \quad \text{iff} \quad \forall v \in W \ (wRv \to \forall u \ (vSu \to u \models \varphi)),$$

$$w \models \Diamond\varphi \quad \text{iff} \quad \exists v \in W \ (wSv \wedge v \models \varphi).$$

It is readily checked that all **FS**-frames validate **FS**$'$ and that an **FS**-frame $\langle W, R, S\rangle$ validates **MIPC**$'$ iff S is an equivalence relation on W.

The following lemma establishes a close connection between $\mathbf{IntK}_{\Box\Diamond}$-frames for **FS**$'$ and **FS**-frames.

Lemma 10.16. *For each descriptive* $\mathbf{IntK}_{\Box\Diamond}$*-frame* $\mathfrak{F} = \langle W, R, R_\Box, R_\Diamond, P\rangle$, *the following conditions are equivalent:*

(i) $\mathfrak{F} \models \mathbf{FS}'$,

(ii) $\mathfrak{F}' = \langle W, R, R_\Box \cap R_\Diamond\rangle$ *satisfies* (10.10) *and* (10.11), *i.e., is an* **SF**-*frame.*

Moreover, $\mathfrak{F}$ *and* $\mathfrak{F}'$ *validate precisely the same* $\mathcal{L}_{\Box\Diamond}$*-formulas.*

Proof. Exercise (hint: use (10.8) and (10.9)). ❑

As a consequence we obtain the following completeness results and thereby complete *Step 1.*

Theorem 10.17. **FS**$'$ *is characterized by the class of* **FS**-*frames.* **MIPC**$'$ *is characterized by the class of* **FS**-*frames* $\mathfrak{F} = \langle W, R, S\rangle$ *in which* S *is an equivalence relation.*

Step 2. We have to show that $\mathbf{FS} \supseteq \mathbf{FS}'$ and $\mathbf{MIPC} \supseteq \mathbf{MIPC}'$. This can be easily done by proving that all the axioms of **FS**$'$ and **MIPC**$'$ belong to **FS** and **MIPC**, respectively, and that **FS** and **MIPC** are closed under the inference rules of $\mathbf{Int}_{\Box\Diamond}$.

Another way of proving the claim is to show that every standard **FS**-frame can be transformed into an **FS**-frame. Indeed, assume that we have a standard **FS**-frame $\mathfrak{F} = \langle W, \lhd, \mathfrak{d}\rangle$ with $\mathfrak{d}(w) = \langle \Delta^w, S^w\rangle$. First, we make the sets Δ^w, $w \in W$, disjoint by subscribing each element $x \in \Delta^w$ with w. The set of worlds V of the **FS**-frame $\langle V, R, S\rangle$ under construction will consist of all x_w, where $w \in W$ and $x \in \Delta_w$. Now define relations S and R on V by taking

$$x_u S y_v \quad \text{iff} \quad u = v \text{ and } x S^u y,$$
$$x_u R y_v \quad \text{iff} \quad u \lhd v \text{ and } x = y.$$

(Thus, the relation S is the disjoint union of the relations S^w for all $w \in W$.) It should be clear that $\langle V, R, S\rangle$ is an **FS**-frame validating the same formulas as $\mathfrak{F}$.

Step 3. We start with the proof of (10.2). Suppose $\mathsf{T}^*(\varphi) \notin (\mathbf{S4} \times \mathbf{K})^{\mathsf{EX}}$. Then we can find a product $\mathfrak{F}_1 \times \mathfrak{F}_2$ of frames $\mathfrak{F}_1 = \langle U_1, R_1\rangle$ with transitive and reflexive R_1 and $\mathfrak{F}_2 = \langle U_2, R_2\rangle$, a subframe $\mathfrak{G} = \langle V, S_1, S_2\rangle$ of $\mathfrak{F}_1 \times \mathfrak{F}_2$ such that $\langle u_1', u_2\rangle \in V$ whenever $\langle u_1, u_2\rangle \in V$ and $u_1 R_1 u_1'$, a model $\mathfrak{M} = \langle \mathfrak{G}, \mathfrak{V}\rangle$, and a point $\langle w_1, w_2\rangle \in V$ for which $(\mathfrak{M}, \langle w_1, w_2\rangle) \not\models \mathsf{T}^*(\varphi)$. In fact, we may assume that R_1 is a partial order (if this is not the case, we can take the skeleton of $\langle U_1, R_1\rangle$ as in the previous section). Moreover, we may also assume that for every $u_1 \in U_1$ there exists a $u_2 \in U_2$ such that $\langle u_1, u_2\rangle \in V$.

Define a standard **FS**-frame $\mathfrak{G}' = \langle W, \lhd, \mathfrak{d}\rangle$ by taking $W = U_1$, $\lhd = R_1$ and $\mathfrak{d}(w) = \langle \Xi^w, S^w\rangle$, where $\Xi^w = \{u \mid \langle w, u\rangle \in V\}$ and $S^w = R_2 \cap (\Delta^w \times \Delta^w)$. Let $\mathfrak{M}' = \langle \mathfrak{G}', \mathfrak{V}'\rangle$, where

$$\mathfrak{V}'(w, p) = \{u \in \Xi^w \mid (\mathfrak{M}, \langle w, u\rangle) \models \Box p\}$$

for all $w \in W$ and all propositional variables p. One can show by induction that for every $\mathcal{L}_{\Box\Diamond}$-formula ψ and every $\langle u_1, u_2\rangle \in V$,

$$(\mathfrak{M}', \langle u_1, u_2\rangle) \models \chi \quad \text{iff} \quad (\mathfrak{M}, \langle u_1, u_2\rangle) \models \mathsf{T}^*(\psi).$$

It follows that $\mathfrak{G}' \not\models \varphi$, and so $\varphi \notin$ **FS**, which proves (10.2).

To show (10.3), suppose $\varphi \notin$ **FS**$'$. By Corollary 10.17, we have a model $\mathfrak{M} = \langle \mathfrak{F}, \mathfrak{V}\rangle$ based on an **FS**-frame $\mathfrak{F} = \langle W, R, S\rangle$ and refuting φ. Note that $\mathfrak{F}$ is clearly a frame for $[\mathbf{S4}, \mathbf{K}]^{\mathsf{EX}}$. Moreover, it is easily proved by induction that, for every $\mathcal{L}_{\Box\Diamond}$-formula ψ,

$$(\mathfrak{M}, w) \models' \psi \quad \text{iff} \quad (\mathfrak{M}, w) \models'' \mathsf{T}^*(\psi),$$

where $\models'$ is the truth-relation in **FS**-models and $\models''$ the standard truth-relation for classical bimodal logic. It follows that $\mathsf{T}^*(\varphi) \notin [\mathbf{S4}, \mathbf{K}]^{\mathsf{EX}}$.

The proof of (10.4) and (10.5) is similar and left to the reader. This completes Step 3 and thereby the proof of Theorem 10.7 as well. ❑

10.3 The finite model property

Unfortunately, the embeddings of **FS** and **MIPC**, constructed in the previous section, do not provide us with any information as to whether these logics have the fmp with respect to **FS**-frames. In this section we show how to establish the fmp of **FS** by means of an elaborated version of the filtration method. The proof is due to Grefe (1998); a somewhat different proof can be found in (Simpson 1994). But before that we illustrate the difference between standard **FS**-frames and (nonstandard) **FS**-frames by a simple example.

Example 10.18. Remember that according to the proof of Proposition 3.46 the formula

$$\varphi = \Box\neg\neg p \to \neg\neg\Box p$$

does not belong to **MIPC**, but is valid in all standard **FS**-frames. On the other hand, Fig. 10.3 shows a three-point **FS**-frame for **MIPC** refuting φ.

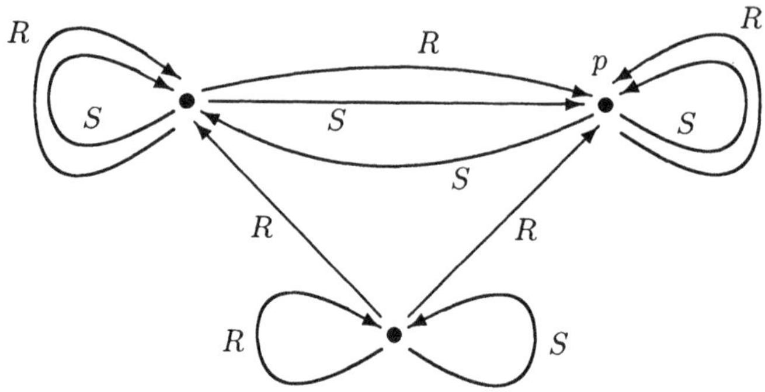

Figure 10.3: An **FS**-frame for **MIPC** refuting φ.

Theorem 10.19. **FS** *has the finite model property with respect to* **FS**-*frames.*

Proof. Suppose $\varphi \notin \mathbf{FS}$. Then there exists a descriptive $\mathbf{IntK}_{\Box\Diamond}$-frame $\mathfrak{G} = \langle W, R, R_\Box, R_\Diamond, P\rangle$ and a valuation $\mathfrak{V}$ in it such that φ is refuted in $\langle \mathfrak{G}, \mathfrak{V}\rangle$. Let $S = R_\Box \cap R_\Diamond$. As we know from the previous section, the triple $\langle W, R, S\rangle$ is an **FS**-frame. Our aim is to find a countermodel for φ based on a finite **FS**-frame $\mathfrak{F} = \langle W', R', S'\rangle$.

To this end, we will construct a sequence of frames $\mathfrak{F}_h = \langle W_h, R_h, S_h\rangle$, for $h < \omega$, such that the sets W_h and both relations, regarded as sets, grow with increasing h. It will turn out that there is $h_0 < \omega$ such that for all $h \geq h_0$, $\mathfrak{F}_h = \mathfrak{F}_{h_0}$; the frame $\mathfrak{F}_{h_0}$ will be the frame $\mathfrak{F}$ we are looking for. We refer to the elements of R_h and S_h as *intuitionistic* and *modal arrows*, respectively. With every point t in one of the sets W_h, we associate a set $\Sigma(t)$ of subformulas of φ and a point $\bar{t}$ from $\mathfrak{G}$. $\Sigma(t)$ can be regarded as the set of formulas whose truth-value at the point t is relevant for the construction of a countermodel for φ. Every point t of the model to be constructed can be thought of as a point $\bar{t}$ selected from $\mathfrak{G}$. However, we do not identify t and $\bar{t}$, because the same point in $\mathfrak{G}$ might be selected several times, producing distinct points in the model to be constructed. For every h, the structure $\langle W_h, R_h\rangle$ will be a forest, i.e., a disjoint union of trees.

To select points in $\mathfrak{G}$, we use the following fundamental property of descriptive frames observed by Fine (1974b) (for a proof see (Chagrov and Zakharyaschev 1997)). Say that a point x in $\mathfrak{G}$ is *maximal relative to* a formula ψ, if $x \not\models \psi$ and for all $y \neq x$ such that xRy, we have $y \not\models \psi$.

Lemma 10.20. *If $x \not\models \psi$ then there is a point $y \in W$ such that xRy and y is maximal relative to ψ.*

We are now in a position to define the whole construction in full detail. We start with the frame $\mathfrak{F}_0 = \langle\{t_0\}, \{(t_0, t_0)\}, \emptyset\rangle$ and $\Sigma(t_0) = sub\,\varphi$, where

$\bar{t_0}$ is any point x in $\mathfrak{G}$ such that $x \not\models \varphi$. $\mathfrak{F}_{h+1}$ is constructed from $\mathfrak{F}_h$ in the following three steps.

Step A: where possible, we apply recursively the following rules $(A_\Diamond)$ and $(A_\Box)$:

$(A_\Diamond)$ Suppose that $t \in W_h$ or t was constructed previously in Step A of the construction of $\mathfrak{F}_{h+1}$, and $\psi \in \Sigma(t)$ is a formula of the form $\Diamond\delta$ such that $\bar{t} \models \Diamond\delta$, but for no point $s \in W_h$ do we have $tS_h s$ and $\bar{s} \models \delta$.

Then choose a point x in $\mathfrak{G}$ such that $\bar{t}Sx$ and $x \models \delta$. Add a new point s to W_h, set

$$\Sigma(s) = \Sigma^-(t) = \bigcup\{sub\,\psi \mid \Box\psi \in \Sigma(t) \text{ or } \Diamond\psi \in \Sigma(t)\},$$

$\bar{s} = x$, and, finally, add the arrow (t, s) to S_h.

$(A_\Box)$ Suppose that $t \in W_h$ or t was constructed previously in Step A of the construction of $\mathfrak{F}_{h+1}$, and $\psi \in \Sigma(t)$ is a formula of the form $\Box\beta$ such that $\bar{t}$ is maximal relative to ψ. Suppose further that for no point $s \in W_h$ do we have $tS_h s$ and $\bar{s} \not\models \beta$.

Then choose a point x in $\mathfrak{G}$ such that $\bar{t}Sx$ and $x \not\models \beta$. Add a new point s to W_h, set $\bar{s} = x$, $\Sigma(s) = \Sigma^-(t)$ and, finally, add the arrow (t, s) to S_h.

Step B: where possible, we apply the following rule (B).

(B) Suppose that $t \in W_h$ and $\psi \in \Sigma(t)$ is a formula either of the form $\alpha \to \sigma$ or of the form $\Box\beta$, such that $\bar{t} \not\models \psi$, but $\bar{t}$ is not maximal relative to ψ.

Then choose a point x in $\mathfrak{G}$ such that $\bar{t}Rx$ and x is maximal relative to ψ. Add a new point s to W_h, set $\bar{s} = x$, $\Sigma(s) = \Sigma(t)$ and add the arrow (t, s) to the relation R_h.

Step C: where possible, apply recursively the following rules $(C1)$, $(C2)$:

$(C1)$ Suppose that $t, s \in W_h$, $tS_h s$ and that t' was constructed previously in Steps B or C of the construction of $\mathfrak{F}_{h+1}$ such that $tR_h t'$. Suppose further that there is no point $s' \in W_h$ such that $t'S_h s'$ and $sR_h s'$.

Then choose a point x such that $\bar{t'}Sx$ and $\bar{s}Rx$. If $\bar{s} = x$ then add (t', s) to S_h. Otherwise, add a new point s' to W_h, set $\bar{s'} = x$, $\Sigma(s') = \Sigma(s)$ and add the arrows (t', s') and (s, s') to S_h and R_h, respectively.

($C2$) Suppose that $t, s \in W_h$, sS_ht and t' was constructed previously in Steps B or C of the construction of $\mathfrak{F}_{h+1}$ so that tR_ht'. Suppose further that there is no point $s' \in W_h$ such that $s'S_ht'$ and sR_hs'.

Then choose a point x such that $xS\bar{t'}$ and $\bar{s}Rx$. If $\bar{s} = x$ then add (s, t') to S_h. Otherwise, add a new point s' to W_h, set $\bar{s'} = x$, $\Sigma(s') = \Sigma(s)$ and add the arrows (s', t') and (s, s') to S_h and R_h, respectively.

End of the construction: after closing the structure under these rules, we replace R_h by its reflexive and transitive closure, and denote the result by $\mathfrak{F}_{h+1} = \langle W_{h+1}, R_{h+1}, S_{h+1}\rangle$. Finally, we set

$$\mathfrak{F} = \left\langle \bigcup_{h<\omega} W_h, \bigcup_{h<\omega} R_h, \bigcup_{h<\omega} S_h \right\rangle.$$

Observe that all the choices we have to make during the construction are possible in the sense that there really is at least one point with the desired properties. This is immediate from the definition of a model (Step A), Lemma 10.20 (Step B) and the fact that **FS** is d-persistent and thus the descriptive frame $\mathfrak{G}$ satisfies (10.10) and (10.11) (Step C).

Lemma 10.21. *$\mathfrak{F}_h$ satisfies* (10.10) *and* (10.11), *and so is an* **FS***-frame.*

Proof. A straightforward induction on h is left to the reader as an exercise. ❑

Lemma 10.22. *For every h, the frame $\langle W_h, R_h\rangle$ is a forest.*

Proof. The relation R_h is transitive and reflexive by definition. So we just have to show that it contains no infinite descending chains and that no point has two distinct immediate R_h-predecessors. But both claims follow immediately from the fact that none of the above rules allows the introduction of an intuitionistic arrow that leads to an already constructed point. ❑

We will refer to the trees in the forest $\langle W_h, R_h\rangle$ as *R_h-trees.* The following observations are readily checked.

Claim 10.23. *If s and t belong to the same R_h-tree, then $\Sigma(s) = \Sigma(t)$. If sR_ht then $\bar{s}R\bar{t}$.*

The points introduced in Step A will be referred to as *original m-points.* These points, together with the very first point t_0, are obviously the roots of the R_h-trees. The points introduced in Step B are called *original i-points.* Every point s' which is constructed in Step C is called an *immediate copy.* More precisely, in the case of rule ($C1$), it is a copy of its immediate S_h-predecessor t', and in the case of rule ($C2$) it is a copy of its immediate

S_h-successor t'. Let $\approx$ be the least equivalence relation containing all pairs (s,t) such that one of the points is an immediate copy of the other. Let us declare the starting point to be an original as well. Then we easily find: $s \approx t$ iff s and t are iterated copies of the same original, and the following holds:

Claim 10.24. *Every equivalence class of $\approx$ contains exactly one original.*

Lemma 10.25.

(a) *Every chain in $\mathfrak{F}_h$ contains at most $\ell(\varphi)$ original i-points.*

(b) *The number of original m-points which are S_h-successors of the points of the same chain in $\mathfrak{F}_h$ is not greater than $\ell(\varphi)$.*

(c) *Every point in $\mathfrak{F}_h$ has at most $\ell(\varphi)$ R_h-incomparable S_h-successors. No point has R_h-incomparable S_h-predecessors.*

Proof. (a) Let the original s be introduced by applying rule (B) to t with respect to the formula ψ. Then $\bar{s}$ is maximal relative to ψ. Thus (B) cannot be applied to any successor of s with respect to ψ. Hence, for a fixed element of $\Sigma(t) \subseteq sub\varphi$, we have at most one original i-point per chain.

(b) Let the original m-point s be introduced by applying either rule $(A_\Diamond)$ or $(A_\Box)$ to t with respect to ψ. We show that the same rule cannot be applied to a proper successor of t with regard to the same formula. Let $(A_\Diamond)$ be applied to t with regard to $\Diamond\delta$. Then, t has an S_h-successor s such that $\bar{s} \models \delta$. Now, let t' be a proper R_h-successor of t and assume that $t' \in W_{h+1} - W_h$. Since $\mathfrak{F}_{h+1}$ satisfies (10.10) and (10.11), there is a point s' such that $t' S_{h+1} s'$ and $s R_{h+1} s'$, whence $\bar{s'} \models \delta$. Thus, rule $(A_\Diamond)$ cannot be applied to t with regard to $\Diamond\delta$. Let $(A_\Box)$ be applied to t with respect to $\Box\beta$. Then $\bar{t}$ is maximal relative to $\Box\beta$. Since from tR_ht' we have $\bar{t}R\bar{t'}$ by Claim 10.23, we get $\bar{t'} \models \Box\beta$, whence $(A_\Box)$ is not applicable with respect to $\Box\beta$.

(c) For the first claim, observe that the number of incomparable S_h-successors of a point t does not exceed the number of original m-points that are S_h-successors of points $t'R_ht$. Indeed, suppose otherwise. Then there is a point t which is R_h-minimal with respect to having two incomparable modal successors s_1 and s_2 that are R_h-successors of the same original m-point s'. From the minimality of t, it follows that s_1 and s_2 have the same immediate R_h-predecessor, and we may assume that this is s'. Suppose that s_1 was introduced earlier than s_2. Then s_2 must have been created by rule $(C1)$. But s' is a S_h-successor of a point t' which is either t itself or its immediate R_h-predecessor. Thus, we have tS_hs_1 and $s'R_hs_1$, so it is impossible to apply rule $(C1)$ to the points t, t' and s' in order to create s_2, contrary to the assumption that s_2 actually exists. The second claim is proved analogously: we just have to exchange the roles of $(C1)$ and $(C2)$ and to reverse all modal arrows involved. ❑

Lemma 10.26. *There is no sequence* $s_0 S_h s_1 S_h \ldots S_h s_{md(\varphi)+1}$.

Proof. If $tS_h s$, then the maximal modal depth of a formula in $\Sigma(s)$ is exactly one less than the corresponding value for $\Sigma(t)$. The claim follows from the fact that the maximal modal depth of a formula in $\Sigma(t_0) = sub\varphi$ is $md(\varphi)$. ❑

Let us define ℓ_h to be the set of R_h-leaves in $\mathfrak{F}_h$.

Lemma 10.27. *The frame* $\langle \ell_h, S_h \rangle$ *is a forest of intransitive trees. These intransitive trees are* $\ell(\varphi)$*-ary and of depth* $\leq md(\varphi)$. *Moreover,* ℓ_h *is finite.*

Proof. By putting together Lemmas 10.25 (c) and 10.26, we obtain the first two statements. Each intransitive S_h-tree contains at most $K = \sum_{i=0}^{md(\varphi)} [\ell(\varphi)]^i$ nodes. So it remains to show that ℓ_h is finite. Clearly, ℓ_0 is finite. So, let us assume that ℓ_k is finite. In Step A of the construction of $\mathfrak{F}_{k+1}$, to every leaf $t \in \ell_k$ an intransitive S_h-tree with at most K nodes is appended. Thus, when entering Step B, there are still finitely many leaves. In Step B, rule (B) is applied at most once to every leaf t with respect to some fixed subformula ψ of φ. Now consider Step C. Obviously, the points of a fixed equivalence class of $\approx$ all belong to the same S_h-tree. In particular, there are at most K iterated copies made of the same original i-point. It follows that $\mathfrak{F}_{k+1}$ is finite. ❑

For every point $t \in W_{h+1} - W_h$, let us denote by $[t]$ the S_h-tree that t belongs to in ℓ_{h+1}.

Say that a set of points in a **FS**-frame $\langle W, R, S \rangle$ is a *chain* (or an *antichain*) if any two distinct elements in it are comparable (or, respectively, incomparable) with regard to R.

Lemma 10.28. *Let* $\langle s_i \rangle_{i<\lambda}$, $\lambda \leq \omega$, *be a chain in* $\mathfrak{F}$. *Then there is no antichain with* $K + 1$ *elements in* $\bigcup_{i<\lambda} [s_i]$.

Proof. Suppose otherwise. Then there is a natural number $n \leq \lambda$ such that the antichain is contained in $\bigcup_{i<n} [s_i]$. From this we can conclude that every element of the antichain has an R_h-successor in $[s_n]$. Since $\langle W, R_h \rangle$ is a forest, all these R_h-successors are distinct. But this implies $\|[s_n]\| \geq K + 1$, contrary to Lemma 10.27. ❑

Lemma 10.29. $\mathfrak{F}$ *is finite.*

Proof. Suppose otherwise. Since $\mathfrak{F}_h$ has finitely many leaves for every $h < \omega$, there is an infinite ascending chain $\langle s_i \rangle_{i<\omega}$ such that $s_i < s_{i+1}$ for all $i < \omega$. By Lemma 10.28, $U = \bigcup_{i<\omega} [s_i]$ contains no antichain of size $K + 1$. On the other hand, U contains an original from every set $[s_i]$, $i < \omega$. Thus, there are infinitely many comparable originals, contrary to Lemma 10.25. ❑

Now, having constructed the desired finite frame, we have to show that φ is refuted in $\mathfrak{F}$. Consider the model $\mathfrak{M} = \langle \mathfrak{F}, \mathfrak{V} \rangle$, where $\mathfrak{V}$ is defined by taking $t \in \mathfrak{V}(p)$ iff $p \in \Sigma(t)$ and $\bar{t} \models p$.

Lemma 10.30. $(\mathfrak{M}, t) \models \psi$ *iff* $\bar{t} \models \psi$ *for all* t *in* $\mathfrak{F}$ *and* $\psi \in \Sigma(t)$. *In particular,* $(\mathfrak{M}, t_0) \not\models \varphi$.

Proof. The proof proceeds by induction and is straightforward. We show the least trivial, though still simple, cases for $\rightarrow$ and $\Diamond$.

Let $\psi = \alpha \rightarrow \sigma \in \Sigma(t)$ and $t \in W_{h+1} - W_h$. Assume $\bar{t} \not\models \psi$. If t is maximal relative to ψ, then $\bar{t} \models \alpha$ and $\bar{t} \not\models \sigma$. By the induction hypothesis, $(\mathfrak{M}, t) \models \alpha$ and $(\mathfrak{M}, t) \not\models \sigma$. Hence $(\mathfrak{M}, t) \not\models \psi$. So, take t not to be maximal relative to ψ. Then in $\mathfrak{F}_{h+2}$, t is given an R_{h+2}-successor s such that $\bar{s} \models \alpha$ and $\bar{s} \not\models \sigma$. By the induction hypothesis, we get $(\mathfrak{M}, s) \not\models \psi$, and hence $(\mathfrak{M}, t) \not\models \psi$. Conversely, suppose that $(\mathfrak{M}, t) \not\models \psi$. Then there is s with $tR_h s$ for some h such that $(\mathfrak{M}, s) \models \alpha$ and $(\mathfrak{M}, s) \not\models \sigma$. By the induction hypothesis, we have $\bar{s} \models \alpha$ and $\bar{s} \not\models \sigma$. Since $\bar{t}R\bar{s}$, we get $\bar{t} \not\models \psi$, as required.

Let $\psi = \Diamond\delta$, $t \in W_{h+1} - W_h$. Assume that $\bar{t} \models \psi$. Then there is $s \in W_{h+1}$ such that $tS_{h+1}s$ and $\bar{s} \models \delta$. By the induction hypothesis, $(\mathfrak{M}, s) \models \delta$ and hence $(\mathfrak{M}, t) \models \Diamond\delta$. Conversely, let $(\mathfrak{M}, t) \models \Diamond\delta$. Then there is an s such that $tS_h s$, for some h, and $(\mathfrak{M}, s) \models \delta$. Hence $\bar{s} \models \delta$ by the induction hypothesis and, since $\bar{t}S\bar{s}$, we eventually have $\bar{t} \models \psi$. ❑

Thus, our construction really provides us with a finite countermodel for any formula which is not in **FS**. ❑

In a similar but much simpler way one can prove the following theorem the algebraic version of which is due to Bull (1965):

Theorem 10.31. **MIPC** *has the fmp with respect to* **FS**-*frames.*

Unfortunately, the following problem is still open:

Question 10.32. What is the computational complexity of the decision problem for **FS** and **MIPC**?

While for **FS** no elementary upper bound of its computational complexity is known (observe that the size of the model constructed in the proof above is not bounded by any elementary recursive function), it is not difficult to see that any $\varphi \notin$ **MIPC** can be satisfied in an **FS**-frame validating **MIPC** and containing at most $2^{2^{p(\ell(\varphi))}}$ points, for some polynomial p. So the satisfiability problem for **MIPC** is decidable in N2EXPTIME. It is not known whether this upper bound is optimal.

Question 10.33. Is the 'transitive analog' of **FS** decidable? Does it have the fmp with respect to **FS**-frames?

For more information about intuitionistic modal logics (in particular, their connections with classical modal logics) see (Wolter and Zakharyaschev 1997, 1999a).

Part III

First-order modal logics

This part seems to need some introductory words. Indeed, how can the 'monster' logics mentioned in its title be discussed in a book the main concern of which is decidability, axiomatizability, and complexity?

First-order modal and temporal logics contain classical predicate logic; so they cannot be decidable. Moreover, as we saw in Section 8.4, even the *two-variable* fragment of many first-order modal logics is undecidable. Further, as we shall see later on in this part, the two-variable *monadic* fragment of some natural first-order temporal logics is not even recursively enumerable. The picture is completely different from what we have in classical predicate logic, where the early undecidability results of Turing and Church stimulated research and led to a rich and profound theory concerned with classifying fragments of first-order logic according to their decidability. Here are only three (out of dozens) examples of decidable fragments of classical first-order logic:

- the monadic fragment containing only unary predicate symbols (Löwenheim 1915);
- the fragment with only two individual variables (Scott 1962, Mortimer 1975);[1]
- the guarded fragment containing formulas of the form

  $$\exists \overline{y}\ (G(\overline{x},\overline{y}) \land \varphi(\overline{x},\overline{y})),$$

  where the guard $G(\overline{x},\overline{y})$ is atomic[2] (Andréka *et al.* 1998).

The current state of the art in this field is presented in the monograph (Börger *et al.* 1997).

As none of the results above holds for first-order modal and temporal logics, the question arises as to whether these logics contain anything at all which can be nontrivial, decidable and axiomatizable? It turns out that they do.

The main aim of this part is to define and investigate a new kind of sublanguage of the first-order modal and temporal languages which, on the one hand, is considerably more expressive than the propositional language, and yet, on the other hand, gives rise to decidable fragments of first-order modal and temporal logics. Roughly speaking, these fragments are obtained by:

(1) restricting the pure classical (nonmodal) part of the language to a decidable fragment of first-order logic, and

[1] The fragment with binary predicates and three variables is undecidable (Surányi 1943).
[2] For a precise definition see Section 11.2.

(2) restricting the modal or temporal part of the language to the *monodic* formulas whose subformulas beginning with a modal/temporal operator have at most one free variable.

Condition (1) allows the use of classical decidability results to select a suitable first-order part of the language, while (2) leaves enough room for nontrivial interactions between quantifiers and modal/temporal operators. Thus, we can talk about objects in the intended domain using the full power of the selected fragment of first-order logic; however, modal or temporal operators may be used to describe the behavior of only one object.

Besides proving the decidability of various monodic fragments, our concern in this part is

- to provide Hilbert-style axiomatizations for full monodic fragments of first-order temporal logic $\mathsf{QLog}_{\mathcal{U}}(\mathbb{N})$ and standard first-order epistemic logics,

- to determine the computational complexity of the most important decidable monodic fragments of $\mathsf{QLog}_{\mathcal{U}}(\mathbb{N})$, and

- to investigate the possibility of adding equality to decidable monodic fragments.

Some of these results will be used in Part IV to prove the decidability and determine the computational complexity of certain modal description and spatio-temporal logics.

Chapter 11

Fragments of first-order temporal logics

11.1 Undecidable fragments

First-order temporal logics have become notorious for their bad computational behavior since the unpublished results of Scott and Lindström in the 1960s and a series of incompleteness theorems (Abadi 1987, Andréka *et al.* 1979, Gabbay *et al.* 1994, Garson 1984, Merz 1992, Szałas 1986, Szałas and Holenderski 1988) which show that many of the first-order temporal logics most useful in computer science are not even recursively enumerable. In this section we prove two such theorems in order to indicate some limits outside which one cannot hope to find decidable fragments of first-order temporal logics.

We remind the reader that given a class $\mathcal{C}$ of strict linear orders, we denote by $\mathsf{QLog}_{\mathcal{SU}}(\mathcal{C})$ the *temporal logic of* $\mathcal{C}$, i.e., the set of $\mathcal{QTL}$-formulas (see Section 3.7) that are true in all models based on frames in $\mathcal{C}$:

$$\mathsf{QLog}_{\mathcal{SU}}(\mathcal{C}) = \{\varphi \in \mathcal{QTL} \mid (\mathfrak{M}, w) \models^{\mathfrak{a}} \varphi \text{ for all } \mathfrak{M} = \langle \mathfrak{F}, D, I\rangle \text{ with } \mathfrak{F} \in \mathcal{C}, \text{ all } w \text{ in } \mathfrak{F}, \text{ and all assignments } \mathfrak{a} \text{ in } D\}.$$

$\mathsf{QLog}^{fin}_{\mathcal{SU}}(\mathcal{C})$ stands for the set of those $\mathcal{QTL}$-formulas that are true in all models based on linear orders in $\mathcal{C}$ and having *finite domains*. Instead of $\mathsf{QLog}_{\mathcal{SU}}(\{\langle\mathbb{N}, <\rangle\})$, $\mathsf{QLog}^{fin}_{\mathcal{SU}}(\{\langle\mathbb{N}, <\rangle\})$ we write $\mathsf{QLog}_{\mathcal{SU}}(\mathbb{N})$ and $\mathsf{QLog}^{fin}_{\mathcal{SU}}(\mathbb{N})$, respectively; similar notation is used for $\langle\mathbb{Z}, <\rangle$, $\langle\mathbb{Q}, <\rangle$, and $\langle\mathbb{R}, <\rangle$. We will also be considering here the sublanguage $\mathcal{QTL}_{\mathcal{U}}$ of $\mathcal{QTL}$ without the temporal operator $\mathcal{S}$, and the corresponding logics

$$\mathsf{QLog}_{\mathcal{U}}(\mathcal{C}) = \mathsf{QLog}_{\mathcal{SU}}(\mathcal{C}) \cap \mathcal{QTL}_{\mathcal{U}},$$
$$\mathsf{QLog}^{fin}_{\mathcal{U}}(\mathcal{C}) = \mathsf{QLog}^{fin}_{\mathcal{SU}}(\mathcal{C}) \cap \mathcal{QTL}_{\mathcal{U}}.$$

For $\ell < \omega$, let $\mathcal{QTL}^\ell$ be the ℓ-variable fragment of $\mathcal{QTL}$ (i.e., every formula in $\mathcal{QTL}^\ell$ contains at most ℓ distinct individual variables). And by $\mathcal{QTL}^{mo}$ we denote the *monadic* fragment of $\mathcal{QTL}$ (i.e., the set of formulas which contain only unary predicates and propositional variables).

Theorem 11.1. *Let $\mathcal{C}$ be either $\{\langle \mathbb{N}, <\rangle\}$ or $\{\langle \mathbb{Z}, <\rangle\}$. Then the set*

$$\mathcal{QTL}^2 \cap \mathcal{QTL}^{mo} \cap \mathsf{QLog}_{\mathcal{U}}(\mathcal{C})$$

is not recursively enumerable. In fact, already those formulas in

$$\mathcal{QTL}^2 \cap \mathcal{QTL}^{mo} \cap \mathsf{QLog}_{\mathcal{U}}(\mathcal{C})$$

that contain only the temporal operators $\Diamond_F$, $\Box_F$ and $\bigcirc$ are not recursively enumerable.

Proof. We show this by reducing the following recurrent tiling problem to the satisfiability problem for the monadic $\mathcal{QTL}^2$-formulas without the operator $\mathcal{S}$ in $\mathcal{C}$:

- Given a finite set T of tile types and a $t_0 \in T$, can T tile $\mathbb{N} \times \mathbb{N}$ in such a way that t_0 appears infinitely often in the first row?

Harel (1986) proved that this problem is Σ_1^1-complete (see also Section 7.3).

Given a set T of tile types, we associate with each $t \in T$ a unary predicate P_t. We also require two unary predicates, Q_1 and Q_2, which will be used in the formula

$$R(x, y) = \Diamond_F(Q_1(x) \wedge Q_2(y)).$$

Now define a first-order temporal formula φ_T in $\mathcal{QTL}^2 \cap \mathcal{QTL}^{mo}$ as the conjunction of the following formulas:

$$\exists x \Box_F \Diamond_F P_{t_0}(x),$$

$$\forall x \exists y R(x, y),$$

$$\forall x \forall y \big(R(x, y) \to \Box_F R(x, y)\big),$$

$$\Box_F^+ \forall x \Big(\bigvee_{t \in T} P_t(x) \wedge \bigwedge_{\substack{t, t' \in T \\ t \neq t'}} (P_t(x) \to \neg P_{t'}(x))\Big),$$

$$\Box_F^+ \forall x \forall y \bigwedge_{t \in T} \Big(P_t(x) \wedge R(x, y) \to \bigvee_{\substack{t' \in T \\ up(t) = down(t')}} P_{t'}(y)\Big),$$

$$\Box_F^+ \forall x \bigwedge_{t \in T} \Big(P_t(x) \to \bigcirc \bigvee_{\substack{t' \in T \\ right(t) = left(t')}} P_{t'}(x)\Big).$$

Let us show that φ_T is satisfiable in a model based on the frame in $\mathcal{C}$ iff T tiles $\mathbb{N} \times \mathbb{N}$ with t_0 appearing infinitely often in the first row.

Suppose first that $\boldsymbol{\tau} : \mathbb{N} \times \mathbb{N} \to T$ defines a tiling with t_0 appearing infinitely often in the first row. Put $D = \mathbb{N}$,

$$P_t^{I(n)} = \{m \in D \mid \boldsymbol{\tau}(n, m) = t\},$$

for $n \in \mathbb{N}$, and select for every $i \in \mathbb{N}$ an infinite set $M_i \subseteq \mathbb{N}$ such that $M_i \cap M_{i'} = \emptyset$ whenever $i \neq i'$. Now put, for $i \in D$ and $n \in \mathbb{N}$,

$$i \in Q_1^{I(n)} \text{ and } i + 1 \in Q_2^{I(n)} \quad \text{iff} \quad n \in M_i.$$

Also specify that $0 \notin Q_2^{I(n)}$. It should be clear that φ_T is satisfied in $\langle \langle \mathbb{N}, < \rangle, D, I \rangle$. It follows that φ_T is satisfiable in $\mathcal{C}$.

Conversely, suppose that φ_T is satisfied in a first-order temporal model $\mathfrak{M} = \langle \mathfrak{F}, D, I \rangle$, for $\mathfrak{F} \in \mathcal{C}$. Then $\mathfrak{F} = \langle W, < \rangle$ contains an infinite ascending chain, say $0, 1, 2, \ldots$ such that $(\mathfrak{M}, 0) \models \varphi_T$ and $i + 1$ is the immediate successor of i in $\mathfrak{F}$. By the first conjunct of φ_T, we find an $a_0 \in D$ for which the set $\{n \in \mathbb{N} \mid (\mathfrak{M}, n) \models P_{t_0}[a_0]\}$ is infinite. Let

$$R^{I(n)} = \{\langle a, b \rangle \in D^2 \mid (\mathfrak{M}, n) \models \Diamond_F (Q_1 \wedge Q_2)[a, b]\}.$$

By the second conjunct, we have an R-ascending chain $a_0 R^{I(0)} a_1 R^{I(0)} a_2 \ldots$ of elements in D. And by the third conjunct, $a_0 R^{I(n)} a_1 R^{I(n)} a_2 \ldots$, for every $n \in \mathbb{N}$. Now define a function $\boldsymbol{\tau}$ by taking, for all $i, j \in \mathbb{N}$,

$$\boldsymbol{\tau}(i, j) = t \quad \text{iff} \quad (\mathfrak{M}, i) \models P_t[a_j].$$

It is straightforward to check that $\boldsymbol{\tau}$ is a recurrent tiling of $\mathbb{N} \times \mathbb{N}$. ❑

It follows, in particular, that $\mathsf{QLog}_{\mathcal{U}}(\mathbb{N})$ and $\mathsf{QLog}_{\mathcal{U}}(\mathbb{Z})$ are not recursively axiomatizable, cf. (Gabbay *et al.* 1994).

Note that although the two-variable fragment of classical first-order logic has the finite model property (that is, each satisfiable formula is satisfied in a model with a finite domain; see (Mortimer 1975, Börger *et al.* 1997)), this is not the case for first-order temporal logics over many flows of time, even if we consider formulas with only one individual variable and unary predicates:

Theorem 11.2. (i) *Let $1 \leq \ell < \omega$ and let $\mathcal{C}$ be a class of strict linear orders at least one of which is infinite. Then*

$$\mathcal{QTL}^{\ell} \cap \mathcal{QTL}^{mo} \cap \mathsf{QLog}_{\mathcal{SU}}(\mathcal{C}) \neq \mathcal{QTL}^{\ell} \cap \mathcal{QTL}^{mo} \cap \mathsf{QLog}^{fin}_{\mathcal{SU}}(\mathcal{C}).$$

(ii) *Let $1 \leq \ell < \omega$ and let $\mathcal{C}$ be a class of strict linear orders at least one of which contains an infinite ascending chain. Then*

$$\mathcal{QTL}^{\ell} \cap \mathcal{QTL}^{mo} \cap \mathsf{QLog}_{\mathcal{U}}(\mathcal{C}) \neq \mathcal{QTL}^{\ell} \cap \mathcal{QTL}^{mo} \cap \mathsf{QLog}^{fin}_{\mathcal{U}}(\mathcal{C}).$$

Proof. (i) Consider the following formula

$$\psi = \Box_F^+ \Box_P^+ \exists x \big(Q(x) \wedge \neg \Diamond_P Q(x)\big),$$

and let φ be the conjunction of ψ and the formula

$$p_0 \wedge \big(\Box_F^+(p_0 \to \Diamond_F p_0) \vee \Box_P^+(p_0 \to \Diamond_P p_0)\big).$$

It is readily checked that φ is satisfiable in a model based on an infinite flow of time, and that if ψ is satisfied in a model with a set W of time points and domain D, then $|D| \geq |W|$. On the other hand, φ cannot be satisfied over finite flows of time.

(ii) is proved similarly using the formulas

$$\Box_F^+ \exists x \big(Q(x) \wedge \neg \Diamond_F Q(x)\big) \quad \text{and} \quad p_0 \wedge \Box_F^+(p_0 \to \Diamond_F p_0).$$

Details are left to the reader. ❑

The negative result of Theorem 11.1 also holds for first-order temporal logics determined by models with finite domains:

Theorem 11.3. *Let $\mathcal{C}$ be one of the following classes of temporal frames: $\{\langle \mathbb{N}, < \rangle\}$, $\{\langle \mathbb{Z}, < \rangle\}$, the class of all strict linear orders. Then*

$$\mathcal{QTL}^2 \cap \mathcal{QTL}^{mo} \cap \mathsf{QLog}_{\mathcal{U}}^{fin}(\mathcal{C})$$

is not recursively enumerable.

Proof. We are going to reduce the undecidable halting problem for Turing machines (see Section 5.4) to the satisfiability problem for the monadic two-variable $\mathcal{QTL}_{\mathcal{U}}$-formulas in models with finite domains. Given a Turing machine $\boldsymbol{A}$, we will construct a monadic $\mathcal{QTL}_{\mathcal{U}}$-formula $\varphi_{\boldsymbol{A}}$ having two variables which is satisfiable in a model with a finite domain D (based on a frame in $\mathcal{C}$) iff $\boldsymbol{A}$ comes to a stop having started from the configuration $\langle \pounds, \langle s_0, b\rangle, b, b, \dots \rangle$. This will mean that the set $\mathcal{QTL}^2 \cap \mathcal{QTL}^{mo} \cap \mathsf{QLog}_{\mathcal{U}}^{fin}(\mathcal{C})$ is undecidable. On the other hand, its complement (in the set of monadic $\mathcal{QTL}$-formulas) is recursively enumerable. For it is not hard to see that satisfiability of monadic and indeed arbitrary $\mathcal{QTL}$-formulas in models based on frames in $\mathcal{C}$ and having domains of $\leq n$ elements, for a fixed n, can be reduced to satisfiability of propositional temporal formulas in $\mathcal{C}$, which is known to be decidable (see e.g. Gabbay *et al.* 1994).

So, let us define the required formula $\varphi_{\boldsymbol{A}}$. Roughly, the idea is to represent configurations of $\boldsymbol{A}$ by elements $x \in D$ using the behavior of x over time. We use the notation introduced in Section 5.4. First, the formula

$$\Box_F^+ \bigcirc \top \tag{11.1}$$

ensures that every moment of time (starting from the one satisfying this formula) has an immediate successor. Next, with every $\alpha \in A'$ we associate a unary predicate P_α. The formulas

$$\forall x\big(P_{\pounds}(x) \wedge \bigwedge_{\alpha\in A',\,\alpha\neq\pounds} \neg P_\alpha(x)\big), \tag{11.2}$$

$$\Box_F \forall x \bigvee_{\alpha\in A',\,\alpha\neq\pounds} (P_\alpha(x) \wedge \bigwedge_{\beta\in A',\,\beta\neq\alpha} \neg P_\beta(x)), \tag{11.3}$$

mean that 'now' all objects in D are in $P_{\pounds}$ and not in P_α for any other $\alpha \in A'$, while later each of them belongs to precisely one of the sets P_α, for $\alpha \in A'$, $\alpha \neq \pounds$.

To mark the object representing the active cell of a given configuration and its immediate predecessor and successor, we use three unary predicates, S, L, and R, defined by the formulas:

$$\Box_F^+ \forall x\big(S(x) \leftrightarrow \bigvee_{\langle s,a\rangle\in S\times A} P_{\langle s,a\rangle}(x)\big), \tag{11.4}$$

$$\Box_F^+ \forall x\big((L(x) \leftrightarrow \bigcirc S(x)) \wedge (S(x) \leftrightarrow \bigcirc R(x))\big), \tag{11.5}$$

$$\Box_F^+ \forall x(S(x) \rightarrow \neg\Diamond_F S(x)). \tag{11.6}$$

The transition from one configuration to another is simulated by means of the formula:

$$\begin{aligned}\chi(x,y) \;=\; & \bigvee_{\delta(\alpha,\beta,\gamma)=\langle\alpha',\beta',\gamma'\rangle} \Big[\Diamond_F^+\Big(L(x) \wedge P_\alpha(x) \wedge \bigcirc\big(P_\beta(x) \wedge \bigcirc P_\gamma(x)\big)\Big) \wedge \\ & \Box_F^+\Big((L(x) \rightarrow P_{\alpha'}(y)) \wedge (S(x) \rightarrow P_{\beta'}(y)) \wedge (R(x) \rightarrow P_{\gamma'}(y)) \\ & \wedge \bigwedge_{\alpha\in A'} (\neg L(x) \wedge \neg S(x) \wedge \neg R(x) \wedge P_\alpha(x) \rightarrow P_\alpha(y))\Big)\Big].\end{aligned}$$

We have to ensure that each configuration save the start one on the empty tape has a predecessor:

$$\forall y\Big(\neg(P_{\pounds}(y) \wedge \bigcirc(P_{\langle s_0,b\rangle}(y) \wedge \Box_F P_b(y)) \rightarrow \exists x\, \chi(x,y)\Big), \tag{11.7}$$

and that there exists a domain point representing a halt configuration:

$$\exists x \Diamond_F \bigvee_{a\in A} P_{\langle s_1,a\rangle}(x). \tag{11.8}$$

Finally, the following two formulas define a unary predicate C (clock); its intended meaning is to ensure that there are no loops in the 'time line' along which the Turing machine 'runs:'

$$\forall x(\Diamond_F^+ C(x) \wedge \Box_F^+(C(x) \to \neg\Diamond_F C(x))), \tag{11.9}$$

$$\forall x \forall y(\chi(x, y) \to \Diamond_F^+(C(x) \wedge \bigcirc C(y))). \tag{11.10}$$

Let $\varphi_{\boldsymbol{A}}$ be the conjunction of (11.1)–(11.10). It is not hard to check that $\varphi_{\boldsymbol{A}}$ is satisfied in a model with a finite domain (based on a frame in $\mathcal{C}$) iff $\boldsymbol{A}$ comes to a stop having started from the empty tape.

Indeed, the '$\Leftarrow$'-part of the proof should be clear. For the converse, suppose that $\varphi_{\boldsymbol{A}}$ is satisfied in a world w of a model based on some strict linear order $\langle W, < \rangle$ and having a finite domain D. By (11.8), there are $h \in D$ and $v > w$ in W such that $v \models P_{\langle s_1, a\rangle}[h]$ for some $a \in A$. We shall see that v is just 'finitely many steps' from w, and so h represents a halt configuration. First, observe that, by (11.9) and (11.10), we cannot have objects $c_0, \dots, c_n \in D$ such that

$$c_0 = c_n \quad \text{and} \quad w \models \chi[c_0, c_1] \wedge \dots \wedge \chi[c_{n-1}, c_n].$$

Now let $c_0, \dots, c_n$ be a maximal chain in D for which $w \models \chi[c_i, c_{i+1}]$ $(i < n)$ and $c_n = h$. Such a chain exists since D is finite. So there is no $c \in D$ with $w \models \chi[c, c_0]$. In view of (11.7), this can only mean that c_0 represents the start configuration on the empty tape. Thus, by (11.2)–(11.6) and the definition of χ, the sequence $c_0, \dots, c_n$ represents a halting computation of $\boldsymbol{A}$ starting from the empty tape. ❑

Theorem 11.4. *$\mathsf{QLog}_{\mathcal{SU}}(\mathbb{N})$ is polynomially reducible to $\mathsf{QLog}_{\mathcal{SU}}(\mathbb{R})$, and $\mathsf{QLog}^{fin}_{\mathcal{SU}}(\mathbb{N})$ is polynomially reducible to $\mathsf{QLog}^{fin}_{\mathcal{SU}}(\mathbb{R})$.*

Proof. Given a $\mathcal{QTL}$-formula φ, introduce a new propositional variable p and define the variable-free $\mathcal{QTL}$-formula

$$\nu = \diamondsuit \neg \Diamond_P p \wedge \boxdot(\Diamond_F p \wedge \neg p \mathcal{S} \top \wedge \neg p \mathcal{U} \top)$$

(recall from Section 2.1 that $\diamondsuit\psi$ abbreviates $\psi \vee \Diamond_F \psi \vee \Diamond_P \psi$ and $\boxdot\psi$ abbreviates $\neg\diamondsuit\neg\psi$). So ν states that p is bounded below, unbounded above, and that there is no accumulation point of p. Clearly, the models of ν with flow of time $\langle \mathbb{R}, < \rangle$ are precisely those in which the interpretation of p is isomorphic to $\langle \mathbb{N}, < \rangle$. Now define the relativization φ^p of the temporal connectives in φ to p, by induction in the usual way: $\alpha^p = \alpha$ for atomic α, $(\neg\psi)^p = \neg\psi^p$, $(\psi_1 \wedge \psi_2)^p = \psi_1^p \wedge \psi_2^p$, $(\forall x \psi)^p = \forall x \psi^p$, and $(\psi_1 \mathcal{U} \psi_2)^p = (p \to \psi_1^p) \mathcal{U} (p \wedge \psi_2^p)$, plus a similar clause for $\mathcal{S}$. Then it is easily seen that φ has a first-order temporal model with flow of time $\langle \mathbb{N}, < \rangle$ (and finite domains) iff $\nu \wedge \varphi^p$ has a model with flow of time $\langle \mathbb{R}, < \rangle$ (and finite domains). ❑

Since the formula $\nu \wedge \varphi^p$ above belongs to $\mathcal{QTL}^2 \cap \mathcal{QTL}^{mo}$ whenever φ does, from Theorems 11.1 and 11.3 we obtain (see also (Gabbay *et al.* 1994)):

Theorem 11.5. *The fragments*

$$\mathcal{QTL}^2 \cap \mathcal{QTL}^{mo} \cap \mathsf{QLog}_{\mathcal{SU}}(\mathbb{R}) \quad and \quad \mathcal{QTL}^2 \cap \mathcal{QTL}^{mo} \cap \mathsf{QLog}^{fin}_{\mathcal{SU}}(\mathbb{R})$$

are not recursively enumerable.

It is not clear whether the $\mathcal{S}$-free fragment $\mathsf{QLog}_{\mathcal{U}}(\mathbb{N})$ of $\mathsf{QLog}_{\mathcal{SU}}(\mathbb{N})$ is polynomially reducible to $\mathsf{QLog}_{\mathcal{U}}(\mathbb{R})$. We conjecture, however, that the fragment

$$\mathcal{QTL}^2 \cap \mathcal{QTL}^{mo} \cap \mathsf{QLog}_{\mathcal{U}}(\mathbb{R})$$

is not recursively enumerable either.

11.2 Monodic formulas, decidable fragments

Note that both undecidability proofs of Section 11.1 use temporal formulas of the form $\varphi\mathcal{U}\psi$ ($\Box_F\varphi$, to be more precise) with *two free variables*. We now consider the 'monodic' fragment of $\mathcal{QTL}$ without formulas of that sort.[1]

Denote by $\mathcal{QTL}_{\boxed{1}}$ the set of all $\mathcal{QTL}$-formulas φ such that any subformula of φ of the form $\psi_1\mathcal{U}\psi_2$ or $\psi_1\mathcal{S}\psi_2$ has at most one free variable. Such formulas will be called *monodic*. In other words, monodic formulas allow quantification into temporal contexts only with one free variable.

Here are some examples of monodic formulas:

- all nontemporal first-order formulas, i.e., $\mathcal{QL} \subseteq \mathcal{QTL}_{\boxed{1}}$;
- all $\mathcal{QTL}$-formulas which contain at most one individual variable (i.e., $\mathcal{QTL}^1 \subseteq \mathcal{QTL}_{\boxed{1}}$);
- $\exists x \Diamond_F \varphi(x) \leftrightarrow \Diamond_F \exists x \varphi(x)$ (the Barcan formula);
- $\Box_F^+ \exists x \left(\bigcirc P(x) \land \neg(\top \mathcal{S} P(x))\right)$ ('at every moment, someone starts to get old');
- $\Box_F^+ \Box_P^+ \exists x \left(Q(x) \land \neg \Diamond_P Q(x)\right)$ ('every day has its dog');
- $\forall x \Box_F^+ \left(Sub(x) \to \Box_F \neg Sub(x)\right)$ (this is a constraint for temporal databases from (Chomicki and Niwinski 1995): 'an order can be submitted only once');
- $\Diamond_P \exists y\, Works(x,y) \land \neg \exists y\, Works(x,y) \land \Diamond_F \exists y\, Works(x,y)$ (this is a query to a temporal database from (Chomicki and Toman 1998): 'list all persons who have been unemployed between jobs').

[1] *Monody* is a composition with only one melodic line.

The following formula—one more query from (Chomicki and Toman 1998)—is not monodic:

- $\Box_P^+\Box_F^+\neg\exists y\,\big(\mathit{Works}(x,y)\wedge\bigcirc\,\mathit{Works}(x,y)\wedge\bigcirc\bigcirc\,\mathit{Works}(x,y)\big)$ ('find all job-hoppers—people who never spent more than two years in one place').

Now imagine that we need to find out whether a $\mathcal{QTL}$-formula φ is satisfiable. Following the motto 'divide and conquer,' we separate the temporal and the pure first-order parts of $\mathcal{QTL}$, focusing attention mainly on the former and pretending that we have a friend who knows how to deal with the latter. As we will see, this approach really works if our formula is in $\mathcal{QTL}_{\boxed{1}}$. (Note that we may confine ourselves to dealing with $\mathcal{QTL}_{\boxed{1}}$-sentences only, because an arbitrary monodic formula $\varphi(y_1,\dots,y_m)$ is satisfied in a first-order temporal model iff the monodic sentence $\exists y_1\dots\exists y_m\varphi(y_1,\dots,y_m)$ is satisfied in the same model.)

For every formula $\psi(x)=\varphi_1\mathcal{U}\varphi_2$ or $\psi(x)=\varphi_1\mathcal{S}\varphi_2$ with one free variable x, we reserve a unary predicate symbol $R_\psi(x)$, and for every sentence $\psi=\varphi_1\mathcal{U}\varphi_2$ or $\psi=\varphi_1\mathcal{S}\varphi_2$ we fix a propositional variable p_ψ. $R_\psi(x)$ and p_ψ are called the *surrogates* of $\psi(x)$ and ψ, respectively. For clarity of presentation, we assume that these surrogates are *not* in the signature of $\mathcal{QL}$, and define $\overline{\mathcal{QL}}$ as the first-order language obtained by extending the signature of $\mathcal{QL}$ with countably infinitely many fresh propositional variables and unary predicates. Given a $\mathcal{QTL}_{\boxed{1}}$-formula φ, we denote by $\overline{\varphi}$ the formula that results from φ by replacing all its subformulas of the form $\psi_1\mathcal{U}\psi_2$ and $\psi_1\mathcal{S}\psi_2$, which are not within the scope of another occurrence of $\mathcal{U}$ or $\mathcal{S}$, by their surrogates. Thus, $\overline{\varphi}$ contains no occurrences of temporal operators at all—i.e., it is a $\overline{\mathcal{QL}}$-formula. Observe that, for all $\mathcal{QTL}_{\boxed{1}}$-formulas φ and ψ, we have

$$\overline{\varphi\wedge\psi}=\overline{\varphi}\wedge\overline{\psi},\quad \overline{\neg\varphi}=\neg\overline{\varphi}\quad\text{and}\quad \overline{\forall x\,\varphi}=\forall x\,\overline{\varphi}. \tag{11.11}$$

For a $\mathcal{QTL}$-formula φ, denote by $\mathit{sub}_n\,\varphi$ the closure under negation of the set of all subformulas of φ containing $\leq n$ free variables; $\mathit{sub}\,\varphi$ denotes the set of all subformulas in φ, and $\mathit{con}\,\varphi$ the set of all constants in φ. Without loss of generality, we may identify ψ and $\neg\neg\psi$; so $\mathit{sub}_n\,\varphi$ is finite. Let x be a variable not occurring in φ. Put

$$\mathit{sub}_x\,\varphi=\{\psi\{x/y\}\mid\psi(y)\in\mathit{sub}_1\,\varphi\}.$$

Given a $\mathcal{QTL}_{\boxed{1}}$-sentence φ, by a *type* for φ we mean any Boolean-saturated subset $\boldsymbol{t}$ of

$$\{\overline{\psi}\mid\psi\in\mathit{sub}_x\,\varphi\},$$

that is,

- $\overline{\psi}\wedge\overline{\chi}\in\boldsymbol{t}$ iff $\overline{\psi}\in\boldsymbol{t}$ and $\overline{\chi}\in\boldsymbol{t}$, for every $\psi\wedge\chi\in\mathit{sub}_x\,\varphi$;

- $\neg\overline{\psi} \in \boldsymbol{t}$ iff $\overline{\psi} \notin \boldsymbol{t}$, for every $\psi \in sub_x\,\varphi$.

We say that two types $\boldsymbol{t}$ and $\boldsymbol{t}'$ *agree on* $sub_0\,\varphi$ if

$$\boldsymbol{t} \cap \{\overline{\psi} \mid \psi \in sub_0\,\varphi\} = \boldsymbol{t}' \cap \{\overline{\psi} \mid \psi \in sub_0\,\varphi\}$$

(i.e., $\boldsymbol{t}$ and $\boldsymbol{t}'$ contain the same sentences). Given a type $\boldsymbol{t}$ for φ and a constant $c \in con\,\varphi$, the pair $\langle c, \boldsymbol{t}\rangle$ will be called an *indexed type* for φ (indexed by c).

To a certain extent, every state $I(w)$ in a first-order temporal model can be characterized—modulo φ—by the set of types that are 'realized' in this state under some assignment to x and the set of types that hold on constants in $con\,\varphi$. This motivates the following definition. A pair $\mathfrak{C} = \langle T_{\mathfrak{C}}, T_{\mathfrak{C}}^{con}\rangle$ is called a *state candidate* for φ if $T_{\mathfrak{C}}$ is a (nonempty) set of types for φ that agree on $sub_0\,\varphi$ and

$$T_{\mathfrak{C}}^{con} \subseteq con\,\varphi \times T_{\mathfrak{C}}$$

is a set of indexed types such that for each $c \in con\,\varphi$ there is a unique $\boldsymbol{t} \in T_{\mathfrak{C}}$ with $\langle c, \boldsymbol{t}\rangle \in T_{\mathfrak{C}}^{con}$. Indexed types $\langle c, \boldsymbol{t}\rangle$ in $T_{\mathfrak{C}}^{con}$ will also be denoted by $\boldsymbol{t}_{\mathfrak{C}}^{c}$.

Not all state candidates can represent states in first-order temporal models. To single out those that can, we require one more definition. Consider a $\overline{\mathcal{QL}}$-structure

$$I = \langle D, P_0^I, \ldots, c_0^I, \ldots\rangle \tag{11.12}$$

and suppose that $a \in D$. The set

$$\boldsymbol{t}^I(a) = \{\overline{\psi} \mid \psi \in sub_x\,\varphi,\ I \models \overline{\psi}[a]\}$$

is clearly a type for φ. Say that I *realizes* a state candidate $\mathfrak{C} = \langle T_{\mathfrak{C}}, T_{\mathfrak{C}}^{con}\rangle$ if the following conditions hold:

- $T_{\mathfrak{C}} = \{\boldsymbol{t}^I(a) \mid a \in D\}$,
- $T_{\mathfrak{C}}^{con} = \{\langle c, \boldsymbol{t}^I(c^I)\rangle \mid c \in con\,\varphi\}$.

A state candidate said to be (*finitely*) *realizable* if there is a (finite) $\overline{\mathcal{QL}}$-structure realizing it.

Given a state candidate $\mathfrak{C} = \langle T_{\mathfrak{C}}, T_{\mathfrak{C}}^{con}\rangle$, consider the $\overline{\mathcal{QL}}$-sentence:

$$\mathsf{real}_{\mathfrak{C}} = \bigwedge_{\boldsymbol{t}\in T_{\mathfrak{C}}} \exists x \bigwedge_{\psi\in\boldsymbol{t}} \psi(x) \wedge \bigwedge_{c\in con\,\varphi}\ \bigwedge_{\psi\in\boldsymbol{t}_{\mathfrak{C}}^{c}} \psi\{c/x\} \wedge \forall x \bigvee_{\boldsymbol{t}\in T_{\mathfrak{C}}} \bigwedge_{\psi\in\boldsymbol{t}} \psi(x). \tag{11.13}$$

Note that the number of different types for φ is bounded by

$$\flat(\varphi) = 2^{|sub_x\,\varphi|}.$$

The number $\sharp(\varphi)$ of distinct realizable state candidates for φ is bounded by

$$\sharp(\varphi) \leq 2^{\flat(\varphi)} \cdot \flat(\varphi)^{|con\varphi|}.$$

It follows immediately from the definitions that we have:

Lemma 11.6. *A state candidate $\mathfrak{C}$ for φ is (finitely) realizable iff* $\mathsf{real}_{\mathfrak{C}}$ *is satisfied in some (respectively, finite)* $\overline{\mathcal{QL}}$*-structure.*

We are now in a position to formulate the general decidability results of Hodkinson *et al.* (2000):

Theorem 11.7. *Let $\mathcal{QTL}' \subseteq \mathcal{QTL}_{\boxed{1}}$ and suppose that there is an algorithm which is capable of deciding, for any $\mathcal{QTL}'$-sentence φ, whether an arbitrarily given state candidate for φ is realizable. Let $\mathcal{C}$ be one of the following classes of flows of time:*

(1) $\{\langle \mathbb{N}, < \rangle\}$,

(2) $\{\langle \mathbb{Z}, < \rangle\}$,

(3) $\{\langle \mathbb{Q}, < \rangle\}$,

(4) *the class of all finite strict linear orders,*

(5) *any first-order definable[2] class of strict linear orders—for example, the class of all strict linear orders.*

Then the satisfiability problem for $\mathcal{QTL}'$-sentences in models based on flows of time from $\mathcal{C}$, and so the decision problem for the fragment $\mathsf{QLog}_{\mathcal{SU}}(\mathcal{C}) \cap \mathcal{QTL}'$, *are decidable.*

We prove this theorem in a more general form in Section 11.3 (see Theorem 11.21). However, the following problem is still open:

Question 11.8. Let $\mathcal{QTL}' \subseteq \mathcal{QTL}_{\boxed{1}}$ and suppose that there is an algorithm which is capable of deciding, for any $\mathcal{QTL}'$-sentence φ, whether an arbitrarily given state candidate for φ is realizable. Does it follow that

$$\mathsf{QLog}_{\mathcal{SU}}(\mathbb{R}) \cap \mathcal{QTL}'$$

is decidable?

Similar results hold for satisfiability in models with finite domains:

Theorem 11.9. *Let $\mathcal{QTL}' \subseteq \mathcal{QTL}_{\boxed{1}}$ and suppose that there is an algorithm which is capable of deciding, for any $\mathcal{QTL}'$-sentence φ, whether an arbitrarily given state candidate for φ is finitely realizable. Let $\mathcal{C}$ be one of the following classes of flows of time:*

(1) $\{\langle \mathbb{R}, < \rangle\}$,

(2) $\{\langle \mathbb{N}, < \rangle\}$,

[2]We mean definability in the language with equality and a binary predicate symbol $<$.

(3) $\{\langle \mathbb{Z}, < \rangle\}$,

(4) $\{\langle \mathbb{Q}, < \rangle\}$,

(5) *the class of all finite strict linear orders,*

(6) *any first-order definable class of strict linear orders.*

Then $\mathsf{QLog}^{fin}_{\mathcal{SU}}(\mathcal{C}) \cap \mathcal{QTL}'$ *is decidable.*

This theorem will be proved in Sections 11.5 and 11.6.

The same kind of decidability results can be obtained for fragments of the two-sorted first-order language $\mathcal{TS}$ introduced in Section 3.7. Recall that the set $\mathcal{TS}_{1t}$ consists of all $\mathcal{TS}$-formulas without subformulas of the form $\forall x \psi$ such that ψ has more than one free temporal variable. Similarly, define $\mathcal{TS}_{1x}$ as the set of all $\mathcal{TS}$-formulas without subformulas of the form $\forall t \psi$ such that ψ contains more than one free domain variable. Let $\mathcal{TS}_1 = \mathcal{TS}_{1t} \cap \mathcal{TS}_{1x}$.

Theorem 11.10. *Let* $\mathcal{C}$ *be any class of Dedekind-complete*[3] *flows of time (for example, the class* $\{\langle \mathbb{N}, < \rangle, \langle \mathbb{Z}, < \rangle, \langle \mathbb{R}, < \rangle\} \cup \{\mathfrak{F} \mid \mathfrak{F}$ *a finite linear order*$\}$*). Then* $\mathcal{QTL}_{\boxdot}$ *is expressively complete for* $\mathcal{TS}_1$ *over* $\mathcal{C}$.

Proof. It is enough to observe that if $\psi \in \mathcal{TS}_1$ then $\widehat{\psi} \in \mathcal{QTL}_{\boxdot}$ (see the proof of Theorem 3.28). ❑

For a class $\mathcal{C}$ of flows of time, denote by $\mathsf{TSLog}(\mathcal{C})$ the set of all $\mathcal{TS}$-sentences that are true in all first-order temporal models based on flows in $\mathcal{C}$, and by $\mathsf{TSLog}^{fin}(\mathcal{C})$ the set of $\mathcal{TS}$-sentences true in all models based on frames in $\mathcal{C}$ and having finite domains. Given a set $\mathcal{QTL}' \subseteq \mathcal{QTL}_{\boxdot}$, let

$$\mathcal{TS}' = \{\varphi \in \mathcal{TS}_1 \mid \widehat{\varphi} \in \mathcal{QTL}'\},$$

where $\widehat{\varphi}$ is as defined in the proof of Theorem 3.28. Since $\widehat{\varphi}$ is constructed effectively from φ (see Kamp 1968), as an immediate consequence of Lemma 3.27 and Theorem 3.28 we obtain the following:

Theorem 11.11. *Suppose that every* $\mathfrak{F} \in \mathcal{C}$ *is Dedekind-complete, and that* $\mathcal{QTL}' \subseteq \mathcal{QTL}_{\boxdot}$. *If the fragment* $\mathsf{QLog}_{\mathcal{SU}}(\mathcal{C}) \cap \mathcal{QTL}'$ *is decidable, then the fragment* $\mathsf{TSLog}(\mathcal{C}) \cap \mathcal{TS}'$ *is decidable. If the fragment* $\mathsf{QLog}^{fin}_{\mathcal{SU}}(\mathcal{C}) \cap \mathcal{QTL}'$ *is decidable, then the fragment* $\mathsf{TSLog}^{fin}(\mathcal{C}) \cap \mathcal{TS}'$ *is decidable.*

Now we apply the conditional decidability criteria of Theorems 11.7, 11.9 and 11.11 to single out a number of decidable fragments of various first-order temporal logics.

[3] See Section 7.3.

Two-variable fragment

Denote by $\mathcal{QTL}^2_{\boxed{1}}$ the language that contains all monodic $\mathcal{QTL}$-formulas with at most two variables, that is, $\mathcal{QTL}^2_{\boxed{1}} = \mathcal{QTL}^2 \cap \mathcal{QTL}_{\boxed{1}}$. Let $\mathcal{TS}^2_1$ be the sublanguage of $\mathcal{TS}_1$ whose formulas contain at most two domain variables. Clearly, $\mathcal{TS}^2_1 = \{\varphi \in \mathcal{TS}_1 \mid \widehat{\varphi} \in \mathcal{QTL}^2_{\boxed{1}}\}$.

Theorem 11.12. *Let $\mathcal{C}$ be any of the following classes of flows of time: $\{\langle \mathbb{N}, < \rangle\}$, $\{\langle \mathbb{Z}, < \rangle\}$, $\{\langle \mathbb{Q}, < \rangle\}$, the class of all finite strict linear orders, any first-order definable class of strict linear orders. Then the fragment*

$$\mathsf{QLog}_{\mathcal{SU}}(\mathcal{C}) \cap \mathcal{QTL}^2_{\boxed{1}}$$

is decidable. If $\mathcal{C}$ is one of the listed classes and all frames in $\mathcal{C}$ are Dedekind-complete then $\mathsf{TSLog}(\mathcal{C}) \cap \mathcal{TS}^2_1$ is decidable.

Proof. The $\overline{\mathcal{QL}}$-sentence $\mathsf{real}_{\mathfrak{C}}$ corresponding to a state candidate $\mathfrak{C}$ for a sentence $\varphi \in \mathcal{QTL}^2_{\boxed{1}}$ (see (11.13)) contains at most two individual variables. As is well known (see Scott 1962), the satisfiability problem for such formulas is decidable. All that remains is to use Theorems 11.7 and 11.11. ❑

Theorem 11.13. *Let $\mathcal{H}$ be any of the following classes of flows of time: $\{\langle \mathbb{R}, < \rangle\}$, $\{\langle \mathbb{N}, < \rangle\}$, $\{\langle \mathbb{Z}, < \rangle\}$, $\{\langle \mathbb{Q}, < \rangle\}$, the class of all finite strict linear orders, any first-order definable class of strict linear orders. Then the fragment*

$$\mathsf{QLog}^{fin}_{\mathcal{SU}}(\mathcal{H}) \cap \mathcal{QTL}^2_{\boxed{1}}$$

is decidable. If $\mathcal{H}$ is one of the listed classes and all frames in $\mathcal{H}$ are Dedekind-complete then $\mathsf{TSLog}^{fin}(\mathcal{H}) \cap \mathcal{TS}^2_1$ is decidable.

Proof. As the two-variable fragment of first-order logic has the finite model property (see Mortimer 1975), finite satisfiability for two-variable $\overline{\mathcal{QL}}$-sentences is decidable. Frow now on the proof is the same as the previous one, but this time we use Theorem 11.9 in place of Theorem 11.7. ❑

As $\mathcal{QTL}^2_{\boxed{1}}$ contains the set $\mathcal{QTL}^1$ of $\mathcal{QTL}$-formulas with at most one individual variable, $\mathcal{TS}^2_1$ contains the set $\mathcal{TS}^1_1$ of $\mathcal{TS}_1$-formulas with at most one domain variable, and $\mathcal{TS}^1_1 = \{\varphi \in \mathcal{TS}_1 \mid \widehat{\varphi} \in \mathcal{QTL}^1\}$, we also have:

Corollary 11.14. *Let $\mathcal{C}$ be as in Theorem 11.12, and $\mathcal{H}$ as in Theorem 11.13. Then the fragments $\mathsf{QLog}_{\mathcal{SU}}(\mathcal{C}) \cap \mathcal{QTL}^1$, $\mathsf{TSLog}(\mathcal{C}) \cap \mathcal{TS}^1_1$, $\mathsf{QLog}^{fin}_{\mathcal{SU}}(\mathcal{H}) \cap \mathcal{QTL}^1$ and $\mathsf{TSLog}^{fin}(\mathcal{H}) \cap \mathcal{TS}^1_1$ are decidable.*

We remind the reader that in many cases the one-variable constant-free fragment of $\mathsf{QLog}_{\mathcal{SU}}(\mathcal{C})$ is 'equivalent' to the product logic $\mathsf{Log}_{\mathcal{SU}}(\mathcal{C}) \times \mathbf{S5}$; see Theorems 3.29, 6.29, 6.30, and 6.31.

Monadic fragment

One more interesting fragment of $\mathcal{QTL}$ is the set $\mathcal{QTL}^{mo}$ of monadic temporal formulas. The corresponding two-sorted fragment $\mathcal{TS}^{mo}$ consists of those $\mathcal{TS}$-formulas involving only predicate symbols of the sort 'temporal × domain' or 'temporal.' As was shown in Section 11.1, both $\mathcal{QTL}^2 \cap \mathcal{QTL}^{mo} \cap \mathsf{QLog}_{\mathcal{SU}}(\mathbb{N})$ and $\mathcal{QTL}^2 \cap \mathcal{QTL}^{mo} \cap \mathsf{QLog}^{fin}_{\mathcal{SU}}(\mathbb{N})$ are undecidable. However, this is not the case for the languages

$$\mathcal{QTL}^{mo}_{\boxed{1}} = \mathcal{QTL}_{\boxed{1}} \cap \mathcal{QTL}^{mo} \quad \text{and} \quad \mathcal{TS}^{mo}_1 = \mathcal{TS}_1 \cap \mathcal{TS}^{mo}.$$

For then the sentence $\mathsf{real}_{\mathfrak{C}}$ corresponding to a state candidate $\mathfrak{C}$ for φ in $\mathcal{QTL}^{mo}_{\boxed{1}}$ is a monadic $\overline{\mathcal{QL}}$-sentence, and as is well-known (see Löwenheim 1915), the monadic fragment of first-order logic is decidable and has the finite model property. This yields:

Theorem 11.15. *Let $\mathcal{C}$ be as in Theorem 11.12, and $\mathcal{H}$ as in Theorem 11.13. Then $\mathsf{QLog}_{\mathcal{SU}}(\mathcal{C}) \cap \mathcal{QTL}^{mo}_{\boxed{1}}$, $\mathsf{TSLog}(\mathcal{C}) \cap \mathcal{TS}^{mo}_1$, $\mathsf{QLog}^{fin}_{\mathcal{SU}}(\mathcal{H}) \cap \mathcal{QTL}^{mo}_{\boxed{1}}$ and $\mathsf{TSLog}^{fin}(\mathcal{H}) \cap \mathcal{TS}^{mo}_1$ are decidable.*

Fluted fragment

The monodic fragment can be naturally combined also with the *fluted fragment* of classical first-order logic, which was shown to be decidable and to have the finite model property in (Purdy 1996a, Purdy 1996b).

Let $X_m = \langle x_0, \dots, x_{m-1} \rangle$ be the ordered list of the first m individual variables. For any $i < \omega$, an *atomic temporal fluted formula over X_i* is an atom of the form $P(x_k, x_{k+1}, \dots, x_{i-1})$ for some $k \le i-1$. *Temporal fluted formulas* are now defined inductively as follows:

- any atomic temporal fluted formula over X_i is a temporal fluted formula over X_i;
- any Boolean combination of temporal fluted formulas over X_i is a temporal fluted formula over X_i;
- if φ and ψ are temporal fluted formulas over X_i then $\varphi \mathcal{U} \psi$ and $\varphi \mathcal{S} \psi$ are temporal fluted formulas over X_i;
- if φ is a temporal fluted formula over X_{i+1}, then both $\exists x_{i+1} \varphi$ and $\forall x_{i+1} \varphi$ are temporal fluted formulas over X_i.

Denote by $\mathcal{TFLU}$ the set of all temporal fluted formulas in $\mathcal{QTL}$, and let $\mathcal{FLU} = \{\overline{\varphi} \mid \varphi \in \mathcal{TFLU}\}$.

Theorem 11.16. *Let $\mathcal{C}$ be any of the following classes of flows of time: $\{\langle \mathbb{N}, < \rangle\}$, $\{\langle \mathbb{Z}, < \rangle\}$, $\{\langle \mathbb{Q}, < \rangle\}$, the class of all finite strict linear orders, any first-order definable class of strict linear orders. Then*

$$\mathsf{QLog}_{\mathcal{SU}}(\mathcal{C}) \cap \mathcal{QTL}_{\boxed{1}} \cap \mathcal{TFLU}$$

is decidable. If $\mathcal{H}$ is any of the listed classes or $\mathcal{H} = \{\langle R, < \rangle\}$ then

$$\mathsf{QLog}^{fin}_{\mathcal{SU}}(\mathcal{H}) \cap \mathcal{QTL}_{\boxed{1}} \cap \mathcal{TFLU}$$

is decidable.

Proof. By Theorems 11.7, 11.9 and Lemma 11.6 it is enough to show that, given a monodic $\mathcal{TFLU}$-sentence φ, the $\overline{\mathcal{QL}}$-sentence $\mathsf{real}_{\mathfrak{C}}$ corresponding to a state candidate $\mathfrak{C}$ for φ belongs to $\mathcal{FLU}$. But this is almost obvious. Indeed, by only renaming the variables in formulas from the types for φ, we can rewrite them as fluted $\overline{\mathcal{QL}}$-formulas over X_1 with at most one free variable x_0. The resulting $\mathsf{real}_{\mathfrak{C}}$ clearly belongs to $\mathcal{FLU}$. ❑

Guarded fragments

Let us consider now the following natural generalization of the first-order guarded formulas of Andréka *et al.* (1998).

Denote by $\mathcal{TGF}$ the smallest set of $\mathcal{QTL}$-formulas such that

- every atomic formula is in $\mathcal{TGF}$;
- if φ and ψ are in $\mathcal{TGF}$, then so are $\varphi \wedge \psi$, $\neg\varphi$, $\varphi \mathcal{S} \psi$, and $\varphi \mathcal{U} \psi$;
- if G is an atomic formula (called the *guard*), $\varphi \in \mathcal{TGF}$, every free variable of φ occurs in G, and $\overline{y}$ is a tuple of variables occurring in G, then $\exists \overline{y}\,(G \wedge \varphi)$ is in $\mathcal{TGF}$.

The set $\mathcal{TGF}$ is called the *guarded fragment* of the first-order temporal language (or the *temporal guarded fragment*, for short). We write $\mathcal{GF}$ for the guarded fragment $\{\overline{\varphi} \mid \varphi \in \mathcal{TGF}\}$ of the first-order language $\overline{\mathcal{QL}}$.

Note that unlike $\mathcal{GF}$, which is known to be decidable (see Andréka *et al.* 1998), the temporal guarded fragment interpreted over the flows of time $\langle \mathbb{N}, < \rangle$ and $\langle \mathbb{Z}, < \rangle$ turns out to be not recursively enumerable.

Theorem 11.17. *Let $\mathcal{C}$ be either $\{\langle \mathbb{N}, < \rangle\}$ or $\{\langle \mathbb{Z}, < \rangle\}$. Then*

$$\mathsf{QLog}_{\mathcal{U}}(\mathcal{C}) \cap \mathcal{QTL}^2 \cap \mathcal{TGF}$$

is not recursively enumerable.

Proof. The proof is similar to that of Theorem 11.1. We simply write down the required formula φ_T for a given set T of tile types and a special tile type $t_0 \in T$.

Let R be a binary predicate symbol and P_t $(t \in T)$, Q unary ones. Define φ_T to be the conjunction of the following formulas:

$$\exists x(Q(x) \wedge \Box_F \Diamond_F P_{t_0}(x)),$$
$$\forall x(Q(x) \to \exists y(R(x,y) \wedge Q(y))),$$
$$\Box_F^+ \forall x(Q(x) \to \bigcirc Q(x)),$$
$$\forall x \forall y(R(x,y) \to \Box_F R(x,y)),$$
$$\Box_F^+ \forall x\Big(Q(x) \to \bigvee_{t \in T} P_t(x) \wedge \bigwedge_{\substack{t,t' \in T \\ t \neq t'}} (P_t(x) \to \neg P_{t'}(x))\Big),$$
$$\Box_F^+ \bigwedge_{t \in T} \forall x(P_t(x) \to \forall y(R(x,y) \to \bigvee_{\substack{t' \in T \\ up(t) = down(t')}} P_{t'}(y)),$$
$$\Box_F^+ \bigwedge_{t \in T} \forall x(P_t(x) \to \bigcirc \bigvee_{\substack{t' \in T \\ right(t) = left(t')}} P_{t'}(x)).$$

Clearly, φ_T belongs to $\mathcal{QTL}^2 \cap \mathcal{TGF}$ and does not contain $\mathcal{S}$. It is readily seen that φ_T is satisfiable in a model based on the frame in $\mathcal{C}$ iff T tiles $\mathbb{N} \times \mathbb{N}$ with t_0 appearing infinitely often in the first row. ❑

However, if we restrict it to monodic formulas, the temporal guarded fragment becomes decidable. Moreover, we can allow for more complex formulas as guards. A $\mathcal{QL}$-formula γ is said to be a *packing guard* if γ is a conjunction of atomic and existentially-quantified atomic formulas such that for any two distinct free variables x, y of γ, there is a conjunct of γ in which x, y both occur free. Now, a natural temporal generalization of the first-order *packed fragment*[4] of Marx (2001) can be defined as follows.

Let $\mathcal{TPF}$ be the smallest set of $\mathcal{QTL}$-formulas such that

- every atomic formula is in $\mathcal{TPF}$;
- if φ and ψ are in $\mathcal{TPF}$, then so are $\varphi \wedge \psi$, $\neg\varphi$, $\varphi\mathcal{S}\psi$, and $\varphi\mathcal{U}\psi$;
- if γ is a packing guard, $\varphi \in \mathcal{TPF}$, every free variable of φ is free in γ, and $\overline{y}$ is a tuple of variables free in γ, then $\exists \overline{y}\,(\gamma \wedge \varphi)$ is in $\mathcal{TPF}$.

Clearly, $\mathcal{TPF}$ contains the temporal guarded fragment $\mathcal{TGF}$. We write $\mathcal{PF}$ for the packed fragment $\{\overline{\varphi} \mid \varphi \in \mathcal{TPF}\}$ of the first-order language $\overline{\mathcal{QL}}$. Let $\mathcal{TPF}_{\boxed{1}} = \mathcal{TPF} \cap \mathcal{QTL}_{\boxed{1}}$.

[4] A syntactic variant of the packed fragment was considered by Grädel (1999a) under the name *clique guarded fragment*.

Theorem 11.18. *Let $\mathcal{C}$ be any of the following classes of flows of time: $\{\langle \mathbb{N}, < \rangle\}$, $\{\langle \mathbb{Z}, < \rangle\}$, $\{\langle \mathbb{Q}, < \rangle\}$, the class of all finite strict linear orders, any first-order definable class of strict linear orders. Then*

$$\mathsf{QLog}_{\mathcal{SU}}(\mathcal{C}) \cap \mathcal{TPF}_{\boxed{1}}$$

is decidable. If $\mathcal{H}$ is any of the listed classes or $\mathcal{H} = \{\langle \mathbb{R}, < \rangle\}$ then

$$\mathsf{QLog}^{fin}_{\mathcal{SU}}(\mathcal{H}) \cap \mathcal{TPF}_{\boxed{1}}$$

is decidable.

Proof. Since $\mathcal{PF}$ is known to be decidable (Grädel 1999a, Marx 2001) and to have the finite model property (Hodkinson 2002a, Hodkinson and Otto 2003), by Theorems 11.7, 11.9 and Lemma 11.6 it would suffice to show that, given a $\mathcal{TPF}_{\boxed{1}}$-sentence φ, the $\overline{\mathcal{QL}}$-sentence $\mathsf{real}_{\mathfrak{C}}$ corresponding to a state candidate $\mathfrak{C}$ for φ belongs to $\mathcal{PF}$. Although this is not the case, we can transform $\mathsf{real}_{\mathfrak{C}}$ into a $\mathcal{PF}$-sentence as follows. Let P be a new unary predicate symbol. For any $\overline{\mathcal{QL}}$-formula ψ, define the *relativization* ψ^P *of* ψ *to* P by taking, for atomic ψ, $\psi^P = \psi$, $(\neg\psi)^P = \neg\psi^P$, $(\psi_1 \wedge \psi_2)^P = \psi_1^P \wedge \psi_2^P$, $(\exists x \psi)^P = \exists x (P(x) \wedge \psi^P)$. Observe that if $\psi \in \mathcal{PF}$ then ψ^P is logically equivalent to a $\mathcal{PF}$-formula. The atomic and the Boolean cases are trivial; and for the case of 'packing guarded quantification' $\big((\exists y_1 \dots \exists y_n(\gamma \wedge \psi)\big)^P$ is logically equivalent to the $\mathcal{PF}$-formula

$$\exists y_1 \dots \exists y_n \big(\bigwedge_{1 \le i \le n} P(y_i) \wedge \gamma \wedge \psi^P \big).$$

Now, $\mathsf{real}^P_{\mathfrak{C}}$ is (equivalent to)

$$\bigwedge_{t \in T} \exists x \big(P(x) \wedge \bigwedge_{\psi \in t} \psi^P(x)\big) \wedge$$
$$\bigwedge_{\langle c, t \rangle \in T^{con}} \bigwedge_{\psi \in t} \psi^P(c) \;\wedge\; \forall x \big(P(x) \to \bigvee_{t \in T} \bigwedge_{\psi \in t} \psi^P(x)\big)$$

which is in $\mathcal{PF}$. Finally, by classical model theory, $\mathsf{real}_{\mathfrak{C}}$ has a (finite) model iff $\mathsf{real}^P_{\mathfrak{C}}$ has a (respectively, finite) model. Thus, it is decidable whether $\mathsf{real}_{\mathfrak{C}}$ has a (finite) model, as required. ❑

We can also define the packed fragment of the two-sorted language $\mathcal{TS}$. We call a $\mathcal{TS}$-formula γ a $\mathcal{TSPF}$*-guard* if γ is a conjunction of atomic and existentially 'domain-quantified' atomic formulas such that for any two distinct free domain variables x, y of γ, there is a conjunct of γ in which x, y both occur free. Now let $\mathcal{TSPF}$ be the smallest set of $\mathcal{TS}$-formulas such that every atomic formula is in $\mathcal{TSPF}$, $\mathcal{TSPF}$ is closed under the Boolean connectives and temporal quantification $\forall t$, and if γ is a $\mathcal{TSPF}$-guard, $\varphi \in \mathcal{TSPF}$, every free variable of φ is free in γ, and $\overline{y}$ is a tuple of domain variables free in γ, then $\exists \overline{y}(\gamma \wedge \varphi)$ is in $\mathcal{TSPF}$. Let $\mathcal{TSPF}_1 = \mathcal{TSPF} \cap \mathcal{TS}_1$.

Theorem 11.19. *Let $\mathcal{C}$ be either $\{\langle \mathbb{N}, < \rangle\}$, or $\{\langle \mathbb{Z}, < \rangle\}$, or the class of all finite strict linear orders. Then $\mathsf{TSLog}(\mathcal{C}) \cap \mathcal{TSPF}_1$ is decidable. If $\mathcal{H}$ is any of the listed classes or $\mathcal{H} = \{\langle \mathbb{R}, < \rangle\}$ then $\mathsf{TSLog}^{fin}(\mathcal{H}) \cap \mathcal{TSPF}_1$ is decidable.*

Proof. It is obvious from the proof of Theorem 3.28 that

$$\mathcal{TSPF}_1 = \{\varphi \in \mathcal{TS}_1 \mid \widehat{\varphi} \in \mathcal{TPF}_{\boxed{1}}\}.$$

So the theorem follows from Theorems 11.11 and 11.18. ❑

Remark 11.20. It maybe of interest to note that, over the flow of time $\langle \mathbb{N}, < \rangle$, one can extend the monodic fragment by allowing applications of the next-time operator $\bigcirc$ to arbitrary formulas (i.e., with any number of free variables), provided that no occurrence of such a $\bigcirc$ is in the scope of $\mathcal{U}$. In fact, as was shown by A. Degtyarev, M. Fisher and B. Konev (2003a), the satisfiability problem for this extended fragment can be polynomially reduced to satisfiability for the original monodic fragment. By using the resolution-based approach of (Degtyarev *et al.* 2003b), they also prove the decidability of some 'prefix' fragments with additional requirements on subformulas starting with temporal operators, e.g., the temporalized Gödel and Maslov classes.

The computational properties of monodic first-order branching temporal logics have been recently investigated in (Hodkinson *et al.* 2002, Bauer *et al.* 2002).

11.3 Embedding into monadic second-order theories

In this section we first formulate and prove our most general decidability criterion (Theorem 11.21) concerning monodic fragments of first-order temporal logics, and then show how this criterion yields Theorem 11.7 as a special case. An application of Theorem 11.21 to proving decidability of temporal epistemic logics can be found in Section 13.1.

We generalize Theorem 11.7 by considering satisfiability in first-order temporal models whose first-order 'bits' are not arbitrary, but taken from a specific class of $\mathcal{QL}$-structures, say, those in which the interpretation of a binary relation S is the transitive closure of the interpretation of another binary relation R. To be more precise, given a class $\mathcal{K}$ of $\mathcal{QL}$-structures, we call a first-order temporal model $\mathfrak{M} = \langle \mathfrak{F}, D, I \rangle$ a *$\mathcal{K}$-model* if for every time point w

in $\mathfrak{F}$, the $\mathcal{QL}$-structure $I(w)$ belongs to $\mathcal{K}$. Now, given a $\mathcal{QTL}_{\boxed{1}}$-sentence φ, a state candidate $\mathfrak{C}$ for φ is said to be *$\mathcal{K}$-realizable* if there is a $\overline{\mathcal{QL}}$-structure realizing $\mathfrak{C}$ and such that its *$\mathcal{QL}$-reduct* belongs to $\mathcal{K}$ (the $\mathcal{QL}$-reduct of a $\overline{\mathcal{QL}}$-structure is obtained by omitting the interpretations of the surrogates).

Theorem 11.21. *Let $\mathcal{QTL}' \subseteq \mathcal{QTL}_{\boxed{1}}$, and let $\mathcal{K}$ be a class of $\mathcal{QL}$-structures such that the following two conditions hold:*

(a) *there is an algorithm which is capable of deciding, for every $\mathcal{QTL}'$-sentence φ, whether an arbitrarily given state candidate for φ is $\mathcal{K}$-realizable;*

(b) *for every $\mathcal{QTL}'$-sentence φ, there is an infinite cardinal κ_φ such that for every cardinal $\kappa \geq \kappa_\varphi$ and every $\mathcal{K}$-realizable state candidate $\mathfrak{C}$ for φ, there is a $\overline{\mathcal{QL}}$-structure I realizing $\mathfrak{C}$ and such that the $\mathcal{QL}$-reduct of I is in $\mathcal{K}$ and the sets*

$$I_{\boldsymbol{t}} = \{a \in D^I \mid I \models \bigwedge_{\psi \in \boldsymbol{t}} \psi[a]\}$$

are of cardinality κ, for all types $\boldsymbol{t} \in T_{\mathfrak{C}}$.

Then the satisfiability problem for $\mathcal{QTL}'$-sentences in first-order temporal $\mathcal{K}$-models that are based on a flow of time from $\mathcal{C}$ is decidable, whenever $\mathcal{C}$ is one of the following classes:

(1) $\{\langle \mathbb{N}, < \rangle\}$,

(2) $\{\langle \mathbb{Z}, < \rangle\}$,

(3) $\{\langle \mathbb{Q}, < \rangle\}$,

(4) *the class of all finite strict linear orders,*

(5) *any first-order definable class of strict linear orders.*

Proof. Fix some $\mathcal{QTL}' \subseteq \mathcal{QTL}_{\boxed{1}}$ and a class $\mathcal{K}$ of $\mathcal{QL}$-structures as above. First we show that, modulo a given $\mathcal{QTL}'$-sentence φ, every first-order temporal $\mathcal{K}$-model can be represented as a structure which we again call a *quasimodel*. Roughly, quasimodels in this case consist of $\mathcal{K}$-realizable state candidates for φ linked together by special functions (runs) coding the flow of time and tracing the evolution of the domain elements. Having represented first-order temporal models as quasimodels, we translate the statement that a quasimodel satisfying φ exists into monadic second-order logic and use decidability results of Theorem 1.28.

Let us begin with the definition of a quasimodel. A *$\mathcal{K}$-basic structure* for φ is a pair $\langle \mathfrak{F}, \boldsymbol{q} \rangle$ such that $\mathfrak{F} = \langle W, < \rangle$ is a strict linear order and $\boldsymbol{q}$ is a map associating with each $w \in W$ a $\mathcal{K}$-realizable state candidate

$$\boldsymbol{q}(w) = \langle T_w, T_w^{con} \rangle$$

for φ. Such a map $\boldsymbol{q}$ is called a *state function* over $\mathfrak{F}$. By a *run through* $\langle \mathfrak{F}, \boldsymbol{q} \rangle$ we mean a function r from W into the set $\bigcup_{w \in W} T_w$ such that $r(w) \in T_w$, for all $w \in W$. A run r is called *coherent* and *saturated* if

- for every $\psi_1 \mathcal{U} \psi_2 \in sub_x \varphi$ and every $w \in W$, we have $\overline{\psi_1 \mathcal{U} \psi_2} \in r(w)$ iff there is $v > w$ such that $\overline{\psi_2} \in r(v)$ and $\overline{\psi_1} \in r(u)$ for all $u \in (w, v)$, and
- for every $\psi_1 \mathcal{S} \psi_2 \in sub_x \varphi$ and every $w \in W$, we have $\overline{\psi_1 \mathcal{S} \psi_2} \in r(w)$ iff there is $v < w$ such that $\overline{\psi_2} \in r(v)$ and $\overline{\psi_1} \in r(u)$ for all $u \in (v, w)$.

Finally, we say that a triple $\mathfrak{Q} = \langle \mathfrak{F}, \boldsymbol{q}, \mathfrak{R} \rangle$ is a *(first-order temporal) $\mathcal{K}$-quasimodel for φ (based on $\mathfrak{F}$)* if $\langle \mathfrak{F}, \boldsymbol{q} \rangle$ is a $\mathcal{K}$-basic structure for φ such that

(tqm1) there is a $w \in W$ such that $\overline{\varphi} \in \boldsymbol{t}$, for some (or, equivalently, all) $\boldsymbol{t} \in T_w$,

and $\mathfrak{R}$ is a set of coherent and saturated runs through $\langle \mathfrak{F}, \boldsymbol{q} \rangle$ satisfying the following conditions:

(tqm2) for every $c \in con\,\varphi$, the function r_c defined by $r_c(w) = \boldsymbol{t}$, for $\langle c, \boldsymbol{t} \rangle \in T_w^{con}$, $w \in W$, is a run in $\mathfrak{R}$, and

(tqm3) for every $w \in W$ and every $\boldsymbol{t} \in T_w$, there exists a run $r \in \mathfrak{R}$ such that $r(w) = \boldsymbol{t}$.

In this case the state candidates $\boldsymbol{q}(w)$ are called *quasistates* of $\mathfrak{Q}$.

Note that, for any two sets $\mathfrak{R}_1$ and $\mathfrak{R}_2$ of coherent and saturated runs through $\langle \mathfrak{F}, \boldsymbol{q} \rangle$, if $\mathfrak{R}_1 \subseteq \mathfrak{R}_2$ and $\langle \mathfrak{F}, \boldsymbol{q}, \mathfrak{R}_1 \rangle$ is a $\mathcal{K}$-quasimodel for φ then $\langle \mathfrak{F}, \boldsymbol{q}, \mathfrak{R}_2 \rangle$ is a $\mathcal{K}$-quasimodel for φ as well. Consequently, we may always assume that a $\mathcal{K}$-quasimodel for φ is of the form $\langle \mathfrak{F}, \boldsymbol{q}, \mathfrak{R}_{\boldsymbol{q}} \rangle$, where $\mathfrak{R}_{\boldsymbol{q}}$ denotes the set of *all* coherent and saturated runs through $\langle \mathfrak{F}, \boldsymbol{q} \rangle$.

Lemma 11.22. *Let $\mathcal{QTL}'$ and $\mathcal{K}$ satisfy condition* (b) *of Theorem 11.21, and let $\mathfrak{F} = \langle W, < \rangle$ be a strict linear order in $\mathcal{C}$. Then a $\mathcal{QTL}'$-sentence φ is satisfiable in a first-order temporal $\mathcal{K}$-model based on $\mathfrak{F}$ iff there is a $\mathcal{K}$-quasimodel for φ based on $\mathfrak{F}$.*

Proof. Suppose that our formula φ is satisfied in a first-order temporal $\mathcal{K}$-model $\mathfrak{M} = \langle \mathfrak{F}, D, I \rangle$. For every $a \in D$, $w \in W$, let

$$\boldsymbol{t}_{a,w} = \{\overline{\psi} \mid \psi \in sub_x\,\varphi,\ (\mathfrak{M}, w) \models \psi[a]\}.$$

For all $w \in W$, define $\boldsymbol{q}(w) = \langle T_w, T_w^{con}\rangle$ by taking

$$\begin{aligned} T_w &= \{\boldsymbol{t}_{a,w} \mid a \in D\}, \\ T_w^{con} &= \{\langle c, \boldsymbol{t}_{c^{I(w)},w}\rangle \mid c \in con\,\varphi\}. \end{aligned}$$

It is easy to see that for every $a \in D$, the function $r(w) = \boldsymbol{t}_{a,w}$ ($w \in W$) is a coherent and saturated run through $\langle \mathfrak{F}, \boldsymbol{q}\rangle$. Let $\mathfrak{R}$ be the set of all such runs. Then $\langle \mathfrak{F}, \boldsymbol{q}, \mathfrak{R}\rangle$ is clearly a $\mathcal{K}$-quasimodel for φ.

Conversely, suppose that $\mathfrak{Q} = \langle \mathfrak{F}, \boldsymbol{q}, \mathfrak{R}_{\boldsymbol{q}}\rangle$ is a $\mathcal{K}$-quasimodel for φ. We intend to build a first-order temporal $\mathcal{K}$-model satisfying φ by using the $\mathcal{QL}$-reducts of the $\overline{\mathcal{QL}}$-structures which realize the state candidates $\boldsymbol{q}(w)$, $w \in W$. The problem is that they do not necessarily have the same domains. Here we make use of condition (b).

Take an infinite cardinal κ exceeding both κ_φ and the cardinality of the set $\mathfrak{R}_{\boldsymbol{q}}$, and put

$$D = \{\langle r, \xi\rangle \mid r \in \mathfrak{R}_{\boldsymbol{q}},\ \xi < \kappa\}.$$

Fix some $w \in W$. Then, by **(tqm3)**, for any type $\boldsymbol{t} \in T_w$,

$$\big|\{\langle r, \xi\rangle \in D \mid r(w) = \boldsymbol{t}\}\big| = \kappa. \tag{11.14}$$

By condition (b), there exists a $\overline{\mathcal{QL}}$-structure $I(w)$ with domain $D(w)$ such that $I(w)$ realizes the state candidate $\boldsymbol{q}(w)$, the $\mathcal{QL}$-reduct $I'(w)$ of $I(w)$ is in $\mathcal{K}$, and for every $\boldsymbol{t} \in T_w$ there are κ many elements in $D(w)$ realizing $\boldsymbol{t}$. Hence, by (11.14), we can identify each $D(w)$ with D in a 'type-preserving' way, that is, we may assume that, for all $w \in W$, $\boldsymbol{t} \in T_w$,

$$\{a \in D(w) \mid I(w) \models \bigwedge_{\psi \in \boldsymbol{t}} \psi[a]\} = \{\langle r, \xi\rangle \in D \mid r(w) = \boldsymbol{t}\},$$

and $c^{I(w)} = \langle r_c, 0\rangle$, for every $c \in con\,\varphi$ (where r_c is defined in **(tqm2)**). In other words, for all $w \in W$, $r \in \mathfrak{R}_{\boldsymbol{q}}$ and $\xi < \kappa$, we have

$$r(w) = \{\overline{\psi} \mid \psi \in sub_x\varphi,\ I(w) \models \overline{\psi}[\langle r, \xi\rangle]\}. \tag{11.15}$$

Let $\mathfrak{M} = \langle \mathfrak{F}, D, I'\rangle$. We show by induction on ψ that for all $\psi \in sub\,\varphi$, $w \in W$, and assignments $\mathfrak{a}$ in D,

$$I(w) \models^{\mathfrak{a}} \overline{\psi} \quad \text{iff} \quad (\mathfrak{M}, w) \models^{\mathfrak{a}} \psi. \tag{11.16}$$

The basis of induction—i.e., the case when $\psi = P_i(x_1, \dots, x_\ell)$—is clear; for then $\psi = \overline{\psi}$. The induction steps for $\psi = \psi_1 \wedge \psi_2$, $\psi = \neg\psi_1$, and $\psi = \forall x \psi_1$ follow by the induction hypothesis from (11.11).

Let $\psi = \chi_1 \mathcal{U} \chi_2$. By renaming the free variable in ψ, we may assume that $\psi \in sub_x \varphi$. Suppose that $\mathfrak{a}(x) = \langle r, \xi \rangle$. By (11.15) and the induction hypothesis, we have

$$\begin{aligned} I(w) &\models^{\mathfrak{a}} \overline{\chi_1 \mathcal{U} \chi_2} \\ &\text{iff} \quad \overline{\chi_1 \mathcal{U} \chi_2} \in r(w) \\ &\text{iff} \quad \exists v > w \, \big(\overline{\chi_2} \in r(v) \text{ and } \forall u \in (w,v)\ \overline{\chi_1} \in r(u)\big) \\ &\text{iff} \quad \exists v > w \, \big(I(v) \models^{\mathfrak{a}} \overline{\chi_2} \text{ and } \forall u \in (w,v)\ I(u) \models^{\mathfrak{a}} \overline{\chi_1}\big) \\ &\text{iff} \quad \exists v > w \, \big((\mathfrak{M}, v) \models^{\mathfrak{a}} \chi_2 \text{ and } \forall u \in (w,v)\ (\mathfrak{M}, u) \models^{\mathfrak{a}} \chi_1\big) \\ &\text{iff} \quad (\mathfrak{M}, w) \models^{\mathfrak{a}} \chi_1 \mathcal{U} \chi_2. \end{aligned}$$

The formula $\psi = \chi_1 \mathcal{S} \chi_2$ is considered analogously.

Since, by **(tqm1)** and **(tqm3)**, $\overline{\varphi} \in r(w)$ for some $w \in W$ and $r \in \mathfrak{R}_q$, in view of (11.15) we have $I(w) \models \overline{\varphi}$, and so (11.16) gives $(\mathfrak{M}, w) \models \varphi$, as required. ❑

We can now deduce Theorem 11.21 by translating into monadic second-order logic the statement that there exists a $\mathcal{K}$-quasimodel for φ. We will use some auxiliary formulas. Let $\Sigma_{\mathcal{K}}$ be the set of all $\mathcal{K}$-realizable state candidates for φ. Introduce a unary predicate variable $P_{\mathfrak{C}}$ for each $\mathfrak{C} = \langle T_{\mathfrak{C}}, T_{\mathfrak{C}}^{con} \rangle$ in $\Sigma_{\mathcal{K}}$ and a unary predicate variable R_ψ for each $\psi \in sub_x \varphi$.

Given a type $\boldsymbol{t}$ for φ, let

$$\chi_{\boldsymbol{t}}(\overline{R}(x)) = \bigwedge_{\overline{\psi} \in \boldsymbol{t}} R_\psi(x) \wedge \bigwedge_{\substack{\overline{\psi} \notin \boldsymbol{t} \\ \psi \in sub_x \varphi}} \neg R_\psi(x),$$

saying that the type $\boldsymbol{t}$ at x is defined with the help of

$$\overline{R}(x) = \langle R_\psi(x) \mid \psi \in sub_x \varphi \rangle .$$

Let $\mathsf{run}(\overline{P}, \overline{R})$ denote the conjunction of the three formulas

$$\forall x \bigwedge_{\mathfrak{C} \in \Sigma_{\mathcal{K}}} \big(P_{\mathfrak{C}}(x) \to \bigvee_{\boldsymbol{t} \in T_{\mathfrak{C}}} \chi_{\boldsymbol{t}}(\overline{R}(x))\big),$$

$$\forall x \bigwedge_{\psi_1 \mathcal{U} \psi_2 \in sub_x \varphi} \big[R_{\psi_1 \mathcal{U} \psi_2}(x) \leftrightarrow \exists y \big(x < y \wedge R_{\psi_2}(y) \wedge \forall z (x < z < y \to R_{\psi_1}(z))\big)\big],$$

$$\forall x \bigwedge_{\psi_1 \mathcal{S} \psi_2 \in sub_x \varphi} \big[R_{\psi_1 \mathcal{S} \psi_2}(x) \leftrightarrow \exists y \big(y < x \wedge R_{\psi_2}(y) \wedge \forall z (y < z < x \to R_{\psi_1}(z))\big)\big]$$

which says that $\overline{R}$ defines a coherent and saturated run through a sequence of $\mathcal{K}$-realizable state candidates defined with the help of $\overline{P} = \langle P_{\mathfrak{C}} \mid \mathfrak{C} \in \Sigma_{\mathcal{K}} \rangle$.

We now define the monadic second-order sentence $\mathsf{qm}_{\mathcal{K},\varphi}$ by taking

$$
\begin{aligned}
\mathsf{qm}_{\mathcal{K},\varphi} = \exists_{\mathfrak{C}\in\Sigma_{\mathcal{K}}} P_{\mathfrak{C}} \Big(& \forall x \bigvee_{\mathfrak{C}\in\Sigma_{\mathcal{K}}} \Big[P_{\mathfrak{C}}(x) \wedge \bigwedge_{\substack{\mathfrak{C}'\in\Sigma_{\mathcal{K}}\\ \mathfrak{C}\neq\mathfrak{C}'}} \neg P_{\mathfrak{C}'}(x) \Big] \\
& \wedge \bigvee_{\substack{\mathfrak{C}\in\Sigma_{\mathcal{K}}\\ \overline{\varphi}\in\bigcup T_{\mathfrak{C}}}} \exists x\, P_{\mathfrak{C}}(x) \\
& \wedge \bigwedge_{c\in con\,\varphi} \exists_{\psi\in sub_x\,\varphi} R_{\psi} \Big[\mathsf{run}(\overline{P},\overline{R}) \wedge \forall x \bigwedge_{\mathfrak{C}\in\Sigma_{\mathcal{K}}} \big(P_{\mathfrak{C}}(x) \to \chi_{\boldsymbol{t}^c_{\mathfrak{C}}}(\overline{R}(x)) \big) \Big] \\
& \wedge \forall x \bigwedge_{\mathfrak{C}\in\Sigma_{\mathcal{K}}} \Big[P_{\mathfrak{C}}(x) \to \bigwedge_{\boldsymbol{t}\in T_{\mathfrak{C}}} \exists_{\psi\in sub_x\,\varphi} R_{\psi} \big(\mathsf{run}(\overline{P},\overline{R}) \wedge \chi_{\boldsymbol{t}}(\overline{R}(x)) \big) \Big] \Big).
\end{aligned}
$$

Evaluated in a flow of time $\mathfrak{F} = \langle W, < \rangle$, the first line of $\mathsf{qm}_{\mathcal{K},\varphi}$ says that the sets $P_{\mathfrak{C}} \subseteq W$ ($\mathfrak{C} \in \Sigma_{\mathcal{K}}$) form a partition of W. By defining the map $\boldsymbol{q} : W \to \Sigma_{\mathcal{K}}$ as

$$\boldsymbol{q}(w) = \mathfrak{C} \quad \text{iff} \quad w \in P_{\mathfrak{C}}$$

we obtain a quasimodel $\langle \mathfrak{F}, \boldsymbol{q}, \mathfrak{R}_{\boldsymbol{q}} \rangle$ for φ: the second, third and fourth lines of $\mathsf{qm}_{\mathcal{K},\varphi}$ state the conditions **(tqm1)**, **(tqm2)** and **(tqm3)**, respectively. Therefore, it is easy to see that we obtain the following:

Lemma 11.23. *For any strict linear order $\mathfrak{F}$, $\mathfrak{F} \models \mathsf{qm}_{\mathcal{K},\varphi}$ iff there exists a $\mathcal{K}$-quasimodel for φ based on $\mathfrak{F}$.*

Clearly, if $\Sigma_{\mathcal{K}}$ can be constructed from φ by an algorithm (and by condition (a), it can be), then so can $\mathsf{qm}_{\mathcal{K},\varphi}$. Hence we can now apply known facts on decidable theories of monadic second-order logic to obtain the decidability results of Theorem 11.21. The first four statements of the theorem follow from Theorem 1.28.

To prove statement (5), take a class $\mathcal{C}$ of strict linear orders definable in the first-order language with equality and a binary predicate symbol $<$. By considering the translation $\varphi^{\ddagger}$ of φ into the two-sorted first-order language $\mathcal{TS}$ (see Section 3.7) and applying the downward Löwenheim–Skolem–Tarski theorem, it can be seen that φ has a model with flow of time in $\mathcal{C}$ iff it has a model with a countable flow of time in $\mathcal{C}$. By Lemmas 11.22 and 11.23, this holds iff $\mathsf{qm}_{\mathcal{K},\varphi}$ is true in some countable strict linear order in $\mathcal{C}$.

For every formula ψ of monadic second-order logic and a monadic predicate variable P not occurring in ψ, define the relativization ψ^P of ψ to P inductively by taking $\psi^P = \psi$ for atomic ψ, $(\neg\psi)^P = \neg\psi^P$, $(\psi_1 \wedge \psi_2)^P = \psi_1^P \wedge \psi_2^P$, $(\forall x\psi)^P = \forall x(P(x) \to \psi^P)$, and $(\forall Q\psi)^P = \forall Q\psi^P$. Obviously, for any sentence ψ and any strict linear order $\mathfrak{F}$, we have $\mathfrak{F} \models \exists P(\exists x P(x) \wedge \psi^P)$ iff $\mathfrak{F}' \models \psi$ for some (nonempty) suborder $\mathfrak{F}'$ of $\mathfrak{F}$—the intended interpretation of P is the domain of $\mathfrak{F}'$.

Now any countable strict linear order is a suborder of $\langle \mathbb{Q}, < \rangle$. Let λ be the first-order sentence defining $\mathcal{C}$. Then $\mathsf{qm}_{\mathcal{K},\varphi}$ (assumed not to involve P) is satisfiable in some countable $\mathfrak{F} \in \mathcal{C}$ iff

$$\langle \mathbb{Q}, < \rangle \models \exists P \left(\exists x P(x) \wedge (\lambda \wedge \mathsf{qm}_{\mathcal{K},\varphi})^P \right).$$

By Theorem 1.28, this last statement is decidable.

This completes the proof of Theorem 11.21. ❑

We can prove now Theorem 11.7. Suppose $\mathcal{QTL}'$ is a sublanguage of $\mathcal{QTL}_{\boxdot}$ such that there is an algorithm which is capable of deciding, for any $\mathcal{QTL}'$-sentence φ, whether an arbitrarily given state candidate for φ is realizable. We claim that conditions (a) and (b) of Theorem 11.21 are satisfied if we take $\mathcal{K}$ to be the class of all $\mathcal{QL}$-structures. Indeed, (a) holds by assumption. The following claim shows that condition (b) also holds (where $\kappa_\varphi = \aleph_0$, for all $\mathcal{QTL}'$-sentences φ):

Claim 11.24. *For every $\mathcal{QTL}'$-sentence φ, every infinite cardinal κ, and every realizable state candidate $\mathfrak{C}$ for φ, there is a $\overline{\mathcal{QL}}$-structure I realizing $\mathfrak{C}$ and such that, for every type $\boldsymbol{t} \in T_{\mathfrak{C}}$, the set $I_{\boldsymbol{t}}$ is of cardinality κ.*

Proof. Let φ be a $\mathcal{QTL}'$-sentence and $\mathfrak{C}$ a realizable state candidate for φ. Since the language $\overline{\mathcal{QL}}$ is countable, by the downward Löwenheim–Skolem–Tarski theorem there is a countable $\overline{\mathcal{QL}}$-structure $J = \langle D^J, P_0^J, \dots, c_0^J, \dots \rangle$ realizing $\mathfrak{C}$. Take an infinite cardinal κ. If there is some $\boldsymbol{t} \in T_{\mathfrak{C}}$ such that $|J_{\boldsymbol{t}}| < \kappa$ then we 'blow up' J by making κ copies of each element of its domain as follows. Define a new $\overline{\mathcal{QL}}$-structure I by taking

$$\begin{aligned} D^I &= \{\langle a, \xi \rangle \mid a \in D^J,\ \xi < \kappa\}, \\ c^I &= \{\langle c^J, 0 \rangle\}, \\ P^I &= \{\langle \langle a_1, \xi_1 \rangle, \dots, \langle a_n, \xi_n \rangle \rangle \mid \langle a_1, \dots, a_n \rangle \in P^J,\ \xi_1, \dots, \xi_n < \kappa\}, \end{aligned}$$

for each constant symbol c and each n-ary predicate symbol P. Given an assignment $\mathfrak{a}$ in I, we can define an assignment $\mathfrak{a}^-$ in J by putting, for each variable y, $\mathfrak{a}^-(y) = a$ iff $\mathfrak{a}(y) = \langle a, \xi \rangle$ for some $\xi < \kappa$. Then a straightforward induction shows that for all $\overline{\mathcal{QL}}$-formulas ψ and assignments $\mathfrak{a}$ in I,

$$I \models^{\mathfrak{a}} \psi \quad \text{iff} \quad J \models^{\mathfrak{a}^-} \psi.$$

Here we use the fact that our language does not contain equality. Hence, I also realizes $\mathfrak{C}$ and the sets

$$I_{\boldsymbol{t}} = \{\langle a, \xi \rangle \mid a \in J_{\boldsymbol{t}},\ \xi < \kappa\}$$

are of cardinality κ, for all $\boldsymbol{t} \in T_{\mathfrak{C}}$. ❑

11.4 Elementary decision procedures for fragments of $\mathsf{QLog}_{\mathcal{SU}}(\mathbb{N})$ and their complexity

The translation into monadic second-order logic given in the preceding section reduces the satisfiability problem for monodic sentences to decidable problems of high computational complexity—for example, the complexity of the monadic second-order theory of $\langle \mathbb{N}, < \rangle$ is itself nonelementary (see (Rabin 1977) and references therein). In this section we demonstrate another way of proving decidability of monodic fragments of first-order temporal logics over $\langle \mathbb{N}, < \rangle$, which is more direct, makes plain the structure of the models, and does yield an elementary decision procedure, provided of course that determining the realizability of state candidates is elementary. Assuming an algorithm for the latter with a better complexity bound, we show how to obtain better bounds for satisfiability of monodic sentences. A summary of the obtained complexity results can be found in Table 11.1 on page 504.

First we show that, as far as decidability and computational complexity are concerned, it is sufficient to deal with the $\mathcal{S}$-free part $\mathsf{QLog}_{\mathcal{U}}(\mathbb{N})$ of $\mathsf{QLog}_{\mathcal{SU}}(\mathbb{N})$:

Proposition 11.25. $\mathsf{QLog}_{\mathcal{SU}}(\mathbb{N})$ *is polynomially reducible to* $\mathsf{QLog}_{\mathcal{U}}(\mathbb{N})$.

Proof. We simply 'lift' the proof of Proposition 2.8 to the first-order case in a straightforward way. The only difference is that, given a $\mathcal{QTL}$-formula φ and its subformula of the form $\psi_1 \mathcal{S} \psi_2$ with $\mathcal{S}$-free ψ_1, ψ_2 and with free variables $\bar{x} = x_1, \dots, x_m$, we introduce a fresh m-ary predicate symbol $P_{\psi_1 \mathcal{S} \psi_2}$ and construct the formula

$$\varphi'' = \varphi' \wedge \forall \bar{x}\, \neg P_{\psi_1 \mathcal{S} \psi_2}(\bar{x}) \wedge \Box_F^+ \Big(\forall \bar{x}\, \big(\bigcirc P_{\psi_1 \mathcal{S} \psi_2}(\bar{x}) \leftrightarrow \big(\psi_2 \vee (\psi_1 \wedge P_{\psi_1 \mathcal{S} \psi_2}(\bar{x}))\big)\big)\Big),$$

where φ' is the result of replacing every occurrence of $(\psi_1 \mathcal{S} \psi_2)(\bar{x})$ in φ by $P_{\psi_1 \mathcal{S} \psi_2}(\bar{x})$. Then φ'' is satisfiable at time point 0 iff φ is satisfiable at 0. By iterating this process sufficiently many times we end up with an $\mathcal{S}$-free formula $\bar{\varphi}$ as required (for details, see the proof of Proposition 2.8). ❑

In view of this proposition, we will be considering here only fragments of $\mathsf{QLog}_{\mathcal{U}}(\mathbb{N})$. All the results presented in this section will hold true for the corresponding fragments of $\mathsf{QLog}_{\mathcal{SU}}(\mathbb{N})$ as well. (For one can easily check that the translation $\varphi \mapsto \bar{\varphi}$ defined in the proof of Proposition 11.25 does not carry us outside the fragments of $\mathcal{QTL}$ we consider below. The only case when it does is the temporal guarded fragment. However, in this case we can add equality to the language (see the proof of Theorem 11.86), and then make use of the harmless guard $x_1 = x_1$ in the translation above as $\bar{x}$ contains at most one variable.)

Let $\mathcal{QTL}_{\mathcal{U}\boxdot}$ denote the monodic fragment of the $\mathcal{S}$-free sublanguage $\mathcal{QTL}_{\mathcal{U}}$ of $\mathcal{QTL}$, i.e.,

$$\mathcal{QTL}_{\mathcal{U}\boxdot} = \mathcal{QTL}_{\boxdot} \cap \mathcal{QTL}_{\mathcal{U}}.$$

Roughly, the idea in the elementary decidability proofs is to show that every quasimodel for a given $\mathcal{QTL}_{\mathcal{U}\boxdot}$-sentence φ can be converted into another quasimodel for φ which has a 'periodical' state function, with the period being of some appropriately bounded length.

Fix a $\mathcal{QTL}_{\mathcal{U}\boxdot}$-sentence φ. A pair $\langle \boldsymbol{t}, \boldsymbol{t}' \rangle$ of types for φ is called *suitable* if for every $\psi_1 \mathcal{U} \psi_2 \in sub_x \varphi$,

$$\overline{\psi_1 \mathcal{U} \psi_2} \in \boldsymbol{t} \quad \text{iff} \quad \text{either } \overline{\psi_2} \in \boldsymbol{t}' \text{ or } \overline{\psi_1} \in \boldsymbol{t}' \text{ and } \overline{\psi_1 \mathcal{U} \psi_2} \in \boldsymbol{t}'.$$

Suppose that $\langle \boldsymbol{t}_0, \dots, \boldsymbol{t}_n \rangle$ is a finite sequence of types for φ and $\overline{\psi_1 \mathcal{U} \psi_2} \in \boldsymbol{t}_0$. We say that $\langle \boldsymbol{t}_0, \dots, \boldsymbol{t}_n \rangle$ *realizes* $\overline{\psi_1 \mathcal{U} \psi_2}$, if

- there is l with $0 < l \leq n$ such that $\overline{\psi_2} \in \boldsymbol{t}_l$ and $\overline{\psi_1} \in \boldsymbol{t}_k$ for all $k \in (0, l)$,
- the pair $\langle \boldsymbol{t}_i, \boldsymbol{t}_{i+1} \rangle$ is suitable for every $i < n$.

Recall that $\flat(\varphi)$ and $\sharp(\varphi)$ bound the number of distinct types for φ and the number of distinct realizable state candidates for φ, respectively.

Theorem 11.26. *A $\mathcal{QTL}_{\mathcal{U}\boxdot}$-sentence φ is satisfiable in a first-order temporal model based on $\langle \mathbb{N}, < \rangle$ iff there are natural numbers l_1, l_2 and a sequence*

$$\langle \langle T_0, T_0^{con} \rangle, \dots, \langle T_{l_1+l_2-1}, T_{l_1+l_2-1}^{con} \rangle \rangle$$

of realizable state candidates for φ such that

$$l_1 \leq \sharp(\varphi), \qquad 0 < l_2 \leq |sub_x \varphi| \cdot \flat^2(\varphi) \cdot \sharp(\varphi) + \sharp(\varphi)$$

and the following conditions hold:

(1) $\overline{\varphi} \in \boldsymbol{t}$ *for some* $\boldsymbol{t} \in T_0$;

(2) *for every $i < l_1 + l_2 - 1$ and every $\boldsymbol{t} \in T_i$ there is a $\boldsymbol{t}' \in T_{i+1}$ such that the pair $\langle \boldsymbol{t}, \boldsymbol{t}' \rangle$ is suitable, and for every $\boldsymbol{t} \in T_{l_1+l_2-1}$ there is a $\boldsymbol{t}' \in T_{l_1}$ such that the pair $\langle \boldsymbol{t}, \boldsymbol{t}' \rangle$ is suitable;*

(3) *for every i with $0 < i < l_1 + l_2$ and every $\boldsymbol{t} \in T_i$ there is a $\boldsymbol{t}' \in T_{i-1}$ such that the pair $\langle \boldsymbol{t}', \boldsymbol{t} \rangle$ is suitable, and for every $\boldsymbol{t} \in T_{l_1}$ there is also a $\boldsymbol{t}'' \in T_{l_1+l_2-1}$ such that the pair $\langle \boldsymbol{t}'', \boldsymbol{t} \rangle$ is suitable;*

(4) *for every type $\boldsymbol{t} \in T_{l_1}$ there are types $\boldsymbol{t}_1, \dots, \boldsymbol{t}_{l_2}$ such that $\boldsymbol{t}_i \in T_{l_1+i}$ for $1 \leq i < l_2$, $\boldsymbol{t}_{l_2} \in T_{l_1}$ and all formulas of the form $\overline{\psi_1 \mathcal{U} \psi_2}$ in $\boldsymbol{t}$ are realized by the sequence $\langle \boldsymbol{t}, \boldsymbol{t}_1, \dots, \boldsymbol{t}_{l_2} \rangle$;*

(5) *for every* $c \in con\,\varphi$, *the pairs* $\langle t_i^c, t_{i+1}^c \rangle$ *are suitable whenever* $i < l_1$ *and all formulas of the form* $\overline{\psi_1 \mathcal{U} \psi_2}$ *in* $t_{l_1}^c$ *are realized by the sequence* $\langle t_{l_1}^c, \dots, t_{l_1+l_2-1}^c, t_{l_1}^c \rangle$.

Proof. We slightly modify the definition of *quasimodels* introduced in Section 11.3 (page 483). First, as now $\mathcal{K}$ is the class of all $\mathcal{QL}$-structures, we will not mention it at all. Second, since all quasimodels for φ in this section are based on $\langle \mathbb{N}, < \rangle$, instead of triples we use pairs of the form $\langle \boldsymbol{q}, \mathfrak{R} \rangle$ (with $\boldsymbol{q}(n) = \langle T_n, T_n^{con} \rangle$, $n \in \mathbb{N}$) to denote them. We also assume the following modification of condition **(tqm1)**:

(tqm1)$_0$ $\quad \overline{\varphi} \in \boldsymbol{t}$ for some (or, equivalently, all) $\boldsymbol{t} \in T_0$.

Further, we identify a state function $\boldsymbol{q}$ over $\langle \mathbb{N}, < \rangle$ with the infinite sequence of realizable state candidates

$$\boldsymbol{q} = \langle \boldsymbol{q}(0), \boldsymbol{q}(1), \dots, \boldsymbol{q}(n), \dots \rangle$$

and a run r with the infinite sequence of types

$$r = \langle r(0), r(1), \dots, r(n), \dots \rangle .$$

Thus, an infinite subsequence $\langle \boldsymbol{q}(i_0), \boldsymbol{q}(i_1), \dots \rangle$ of a state function $\boldsymbol{q}$ for φ will also be understood as a state function $\boldsymbol{q}'$ for φ over $\langle \mathbb{N}, < \rangle$ defined by $\boldsymbol{q}'(n) = \boldsymbol{q}(i_n)$, $n \in \mathbb{N}$.

We will use the following notation regarding sequences. Given a sequence $s = \langle s(0), s(1), \dots \rangle$ and $n \geq 0$, we denote by $s^{\leq n}$ and $s^{>n}$ the head $\langle s(0), \dots, s(n) \rangle$ and the tail $\langle s(n+1), s(n+2), \dots \rangle$ of s, respectively; $s_1 * s_2$ denotes the concatenation of sequences s_1 and s_2; $|s|$ denotes the length of s, and

$$s^* = s * s * s * \dots .$$

Now, the '$\Leftarrow$'-direction of the theorem is easy. Given numbers l_1, l_2 and a sequence of realizable state candidates as above, we are going to construct a quasimodel for φ. Then, by Lemma 11.22, we are done.

Define two sequences $\boldsymbol{q}_1$ and $\boldsymbol{q}_2$ of realizable state candidates by taking

$$\boldsymbol{q}_1 = \langle \langle T_0, T_0^{con} \rangle, \dots, \langle T_{l_1-1}, T_{l_1-1}^{con} \rangle \rangle,$$
$$\boldsymbol{q}_2 = \langle \langle T_{l_1}, T_{l_1}^{con} \rangle, \dots, \langle T_{l_1+l_2-1}, T_{l_1+l_2-1}^{con} \rangle \rangle,$$

and let

$$\boldsymbol{q} = \boldsymbol{q}_1 * \boldsymbol{q}_2^*.$$

Since $l_2 > 0$, $\boldsymbol{q}$ is clearly a state function over $\langle \mathbb{N}, < \rangle$ and **(tqm1)**$_0$ holds by condition (1). Let $\boldsymbol{q}(n) = \langle Q_n, Q_n^{con} \rangle$. First, we observe that by condition (5) the sequence

$$\langle \boldsymbol{t}_0^c, \dots, \boldsymbol{t}_{l_1-1}^c \rangle * \langle \boldsymbol{t}_{l_1}^c, \dots, \boldsymbol{t}_{l_1+l_2-1}^c \rangle^*$$

is a coherent and saturated run through $\boldsymbol{q}$, for every $c \in con\,\varphi$. Hence we have **(tqm2)**.

To show **(tqm3)**, we have to construct a coherent and saturated run coming through an arbitrarily given type $\boldsymbol{t}_n \in Q_n$, for every $n \in \mathbb{N}$. Let $k \in \mathbb{N}$ be such that $m = l_1 + k \cdot l_2 \geq n$. We first use conditions (2) and (3) to construct a sequence $\langle \boldsymbol{t}_0, \dots, \boldsymbol{t}_n, \dots, \boldsymbol{t}_m \rangle$ such that $\boldsymbol{t}_i \in Q_i$ for all $i \leq m$ and the pairs $\langle \boldsymbol{t}_i, \boldsymbol{t}_{i+1} \rangle$ are suitable for all $i < m$. After that, in accordance with condition (4), we continue this sequence to $\langle \boldsymbol{t}_0, \dots, \boldsymbol{t}_m, \dots, \boldsymbol{t}_{m+l_2} \rangle$ in order to realize all formulas of the form $\overline{\psi_1 \mathcal{U} \psi_2}$ in $\boldsymbol{t}_m$ (and thus in $\boldsymbol{t}_n$). Then we again use (4) to continue it to $\langle \boldsymbol{t}_0, \dots, \boldsymbol{t}_{m+l_2}, \dots, \boldsymbol{t}_{m+2l_2} \rangle$, realizing all formulas of the form $\overline{\psi_1 \mathcal{U} \psi_2}$ in $\boldsymbol{t}_{m+l_2}$. And so forth. The resulting sequence is clearly a coherent and saturated run through $\boldsymbol{q}$. By collecting all runs constructed this way into a set $\mathfrak{R}$ we obtain a quasimodel $\langle \boldsymbol{q}, \mathfrak{R} \rangle$ for φ.

For the '$\Rightarrow$'-direction, we need a series of lemmas. First we show that it is always possible to delete the interval between two identical quasistates:

Lemma 11.27. *Let $\langle \boldsymbol{q}, \mathfrak{R} \rangle$ be a quasimodel for φ such that $\boldsymbol{q}(n) = \boldsymbol{q}(m)$ for some $n < m$. Then $\langle \boldsymbol{q}^{\leq n} * \boldsymbol{q}^{>m}, \mathfrak{R}^{\leq n} * \mathfrak{R}^{>m} \rangle$ is also a quasimodel for φ, where*

$$\mathfrak{R}^{\leq n} * \mathfrak{R}^{>m} = \{ r_1^{\leq n} * r_2^{>m} \mid r_1, r_2 \in \mathfrak{R},\ r_1(n) = r_2(m) \}.$$

Proof. Observe that if $r_1, r_2 \in \mathfrak{R}$ and $r_1(n) = r_2(m)$, then $r_1^{\leq n} * r_2^{>m}$ is a coherent and saturated run through $\boldsymbol{q}^{\leq n} * \boldsymbol{q}^{>m}$. Indeed, suppose first that $\overline{\psi_1 \mathcal{U} \psi_2} \in r_1(k)$ for some $k \leq n$. Then, since r_1 is saturated, there is $l > k$ such that $\overline{\psi_2} \in r_1(l)$ and $\overline{\psi_1} \in r_1(l')$ for all $l' \in (k, l)$. If $l \leq n$ then we are done. If $l > n$ then, by the coherency of r_1, we have $\overline{\psi_1 \mathcal{U} \psi_2} \in r_1(n) = r_2(m)$, and so are done again, since r_2 is saturated. For the other direction, the only interesting case is when $k \leq n$ and there is an $l \geq m$ such that $\overline{\psi_2} \in r_2(l)$, $\overline{\psi_1} \in r_1(l')$ for all $l' \in (k, n)$, and $\overline{\psi_1} \in r_2(l')$ for all $l' \in (m, l)$. Since r_2 is coherent, we have $\overline{\psi_1 \mathcal{U} \psi_2} \in r_2(m) = r_1(n)$. Finally, since r_1 is saturated and coherent, we obtain that $\overline{\psi_1 \mathcal{U} \psi_2} \in r_1(k)$.

The pair $\langle \boldsymbol{q}^{\leq n} * \boldsymbol{q}^{>m}, \mathfrak{R}^{\leq n} * \mathfrak{R}^{>m} \rangle$ satisfies $\textbf{(tqm1)}_0$ because $\textbf{(tqm1)}_0$ holds for $\langle \boldsymbol{q}, \mathfrak{R} \rangle$. By condition **(tqm3)** for $\langle \boldsymbol{q}, \mathfrak{R} \rangle$, we have that for every $r_1 \in \mathfrak{R}$ there is $r_2 \in \mathfrak{R}$ such that $r_1(n) = r_2(m)$, and vice versa (by swapping n and m). It follows that $\langle \boldsymbol{q}^{\leq n} * \boldsymbol{q}^{>m}, \mathfrak{R}^{\leq n} * \mathfrak{R}^{>m} \rangle$ also satisfies **(tqm3)**.

Finally, observe that for every $c \in con\,\varphi$, we have $r_c^{\leq n} * r_c^{>m} \in \mathfrak{R}^{\leq n} * \mathfrak{R}^{>m}$, and hence **(tqm2)** holds. ❑

If $\boldsymbol{q}'$ is a subsequence of $\boldsymbol{q}$, and both $\langle \boldsymbol{q}, \mathfrak{R} \rangle$ and $\langle \boldsymbol{q}', \mathfrak{R}' \rangle$ are quasimodels for φ, then we call $\langle \boldsymbol{q}', \mathfrak{R}' \rangle$ a *subquasimodel* of $\langle \boldsymbol{q}, \mathfrak{R} \rangle$. For instance, $\langle \boldsymbol{q}^{\leq n} * \boldsymbol{q}^{>m}, \mathfrak{R}^{\leq n} * \mathfrak{R}^{>m} \rangle$ above is a subquasimodel of $\langle \boldsymbol{q}, \mathfrak{R} \rangle$.

Lemma 11.28. *Every quasimodel $\langle \boldsymbol{q}, \mathfrak{R}\rangle$ for φ contains a subquasimodel of the form $\langle \boldsymbol{q}_1 * \boldsymbol{q}_2, \mathfrak{R}'\rangle$ such that $|\boldsymbol{q}_1| \leq \sharp(\varphi)$ and each quasistate in the sequence $\boldsymbol{q}_2$ occurs in it infinitely many times.*

Proof. If each $\boldsymbol{q}(n)$, for $n \in \mathbb{N}$, occurs infinitely often in $\boldsymbol{q}$ then let $\boldsymbol{q}_1$ be empty, $\boldsymbol{q}_2 = \boldsymbol{q}$ and $\mathfrak{R}' = \mathfrak{R}$. Otherwise, we take n to be the maximal number such that $\boldsymbol{q}(n) \neq \boldsymbol{q}(m)$, for all $m > n$. Then put $\boldsymbol{q}_2 = \boldsymbol{q}^{>n}$ and apply Lemma 11.27 to the quasimodel $\langle \boldsymbol{q}, \mathfrak{R}\rangle$ deleting from the head $\boldsymbol{q}^{\leq n}$ of $\boldsymbol{q}$ all repeating quasistates, which yields us a subquasimodel $\langle \boldsymbol{q}_1 * \boldsymbol{q}_2, \mathfrak{R}'\rangle$ satisfying the required properties. ❑

Lemma 11.29. *Let $\langle \boldsymbol{q}_1 * \boldsymbol{q}_2, \mathfrak{R}'\rangle$ be a quasimodel for φ (with quasistates of the form $\langle T_i, T_i^{con}\rangle$ for $i \in \mathbb{N}$) such that $|\boldsymbol{q}_1| \leq \sharp(\varphi)$ and each quasistate in $\boldsymbol{q}_2$ occurs in it infinitely often. Then there is a subquasimodel of the form $\left\langle \boldsymbol{q}_1 * \boldsymbol{q}_0 * \boldsymbol{q}_2^{>l}, \mathfrak{R}''\right\rangle$, for some $l \geq 0$, such that*

(i) *$|\boldsymbol{q}_0| \leq |sub_x\, \varphi| \cdot \sharp(\varphi) \cdot \flat^2(\varphi) + \sharp(\varphi)$;*

(ii) *for every type $\boldsymbol{t} \in T_{|\boldsymbol{q}_1|}$ there is a run $r \in \mathfrak{R}''$ such that $r(|\boldsymbol{q}_1|) = \boldsymbol{t}$ and all formulas $\overline{\psi_1 \mathcal{U} \psi_2} \in r(|\boldsymbol{q}_1|)$ are realized by the sequence*

$$\langle r(|\boldsymbol{q}_1|), r(|\boldsymbol{q}_1| + 1), \dots, r(|\boldsymbol{q}_1| + |\boldsymbol{q}_0|)\rangle$$

of types;

(iii) *for every $c \in con\, \varphi$ the sequence*

$$\langle r_c(|\boldsymbol{q}_1|), r_c(|\boldsymbol{q}_1| + 1), \dots, r_c(|\boldsymbol{q}_1| + |\boldsymbol{q}_0|)\rangle$$

realizes all formulas $\overline{\psi_1 \mathcal{U} \psi_2} \in r_c(|\boldsymbol{q}_1|)$;

(iv) *$\boldsymbol{q}_0(0) = \boldsymbol{q}_2^{>l}(0)$.*

Proof. Observe that for any coherent and saturated run r and $i \in \mathbb{N}$, the pair $\langle r(i), r(i+1)\rangle$ of types is always suitable.

Now let $\mathfrak{Q} = \langle \boldsymbol{q}_1 * \boldsymbol{q}_2, \mathfrak{R}'\rangle$ and $n = |\boldsymbol{q}_1|$. Suppose that $\boldsymbol{t} \in T_n$, $\overline{\psi_1 \mathcal{U} \psi_2} \in \boldsymbol{t}$ and $r(n) = \boldsymbol{t}$, for $r \in \mathfrak{R}'$. Take the minimal $m > 0$ such that $\overline{\psi_2} \in r(n+m)$ and $\overline{\psi_1} \in r(n+k)$ for all $k \in (0, m)$. Assume now that we have i, j such that $0 < i < j < m$, $r(n+i) = r(n+j)$ and $\boldsymbol{q}_2(i) = \boldsymbol{q}_2(j)$. In view of Lemma 11.27, there is a subquasimodel $\left\langle \boldsymbol{q}_1 * \boldsymbol{q}_2^{\leq i} * \boldsymbol{q}_2^{>j}, \mathfrak{S}\right\rangle$ of $\mathfrak{Q}$, and $r^{\leq n+i} * r^{>n+j}$ is a run in $\mathfrak{S}$ coming through $\boldsymbol{t}$. It follows that we can construct a subquasimodel $\left\langle \boldsymbol{q}_1 * \boldsymbol{q}_2^{\leq 0} * \boldsymbol{q}_3, \mathfrak{R}_1\right\rangle$ of $\mathfrak{Q}$ and a run r_1 in $\mathfrak{R}_1$ such that $r_1(n) = \boldsymbol{t}$ and the sequence $\langle r_1(n), \dots, r_1(n+m_1)\rangle$ realizes $\overline{\psi_1 \mathcal{U} \psi_2}$ for some $m_1 \leq \flat(\varphi) \cdot \sharp(\varphi)$.

Then we consider another formula of the form $\overline{\psi_1' \mathcal{U} \psi_2'} \in \boldsymbol{t}$ and assume that $\langle r_1(n), \dots, r_1(n+m')\rangle$ realizes it for some $m' > m_1$. Using Lemma 11.27 once again (and deleting repeating quasistates between $\boldsymbol{q}_3(m_1)$ and $\boldsymbol{q}_3(m')$)

we select a subquasimodel $\left\langle \boldsymbol{q}_1 * \boldsymbol{q}_2^{\leq 0} * \boldsymbol{q}_3^{\leq m_1} * \boldsymbol{q}_4, \mathfrak{R}_2 \right\rangle$ of $\mathfrak{Q}$ and a run r_2 in $\mathfrak{R}_2$ such that $r_2(n) = \boldsymbol{t}$ and $\langle r_2(n), \dots, r_2(n+m_2)\rangle$ realizes both $\overline{\psi_1 \mathcal{U} \psi_2}$ and $\overline{\psi_1' \mathcal{U} \psi_2'}$ for some $m_2 \leq 2 \cdot \flat(\varphi) \cdot \sharp(\varphi)$.

Having analyzed all distinct formulas of the form $\overline{\psi_1 \mathcal{U} \psi_2}$ in $\boldsymbol{t}$, we obtain a subquasimodel $\left\langle \boldsymbol{q}_1 * \boldsymbol{q}_2^{\leq 0} * \boldsymbol{q}' * \boldsymbol{q}^{>k}, \mathfrak{R}_{\boldsymbol{t}} \right\rangle$ of $\mathfrak{Q}$ and a run $r_{\boldsymbol{t}} \in \mathfrak{R}_{\boldsymbol{t}}$ such that $r_{\boldsymbol{t}}(n) = \boldsymbol{t}$ and $\langle r_{\boldsymbol{t}}(n), \dots, r_{\boldsymbol{t}}(n+m_{\boldsymbol{t}})\rangle$ realizes all such 'Until-formulas' for some $m_{\boldsymbol{t}} \leq |sub_x\,\varphi| \cdot \flat(\varphi) \cdot \sharp(\varphi)$.

After that we consider in the same manner another type $\boldsymbol{t}' \in T_n$. However, this time we can delete quasistates only after $\boldsymbol{q}'(m_{\boldsymbol{t}})$, and so to realize in some run through $\boldsymbol{t}'$ a formula $\overline{\psi_1 \mathcal{U} \psi_2} \in \boldsymbol{t}'$, we need again $\leq \flat(\varphi) \cdot \sharp(\varphi)$ new steps. Since $|T_n| \leq \flat(\varphi)$, at most $|sub_x\,\varphi| \cdot \flat^2(\varphi) \cdot \sharp(\varphi)$ quasistates are required to satisfy (ii). Observe that if we 'cut' the runs $r_c \in \mathfrak{R}'$ corresponding to $c \in con\,\varphi$ this way, then we obtain runs satisfying (iii).

Finally, not more than $\sharp(\varphi)$ quasistates may be needed to comply with (iv). So we end up with a subquasimodel $\langle \boldsymbol{q}_1 * \boldsymbol{q}_0 * \boldsymbol{q}_2^{>l}, \mathfrak{R}''\rangle$ of $\mathfrak{Q}$ satisfying (i)–(iv). ❑

We are now in a position to prove the '⇒'-direction of Theorem 11.26. Suppose that φ is satisfiable in a first-order temporal model based on $\langle \mathbb{N}, <\rangle$. Then by Lemma 11.22, there is a quasimodel $\langle \boldsymbol{q}, \mathfrak{R}\rangle$ for φ with $\boldsymbol{q}(n) = \langle T_n, T_n^{con}\rangle$ for $n \in \mathbb{N}$. Clearly, we may assume that $\overline{\varphi} \in \boldsymbol{t}$ for some $\boldsymbol{t} \in T_0$. By applying Lemmas 11.28 and 11.29 we obtain a quasimodel for φ of the form $\langle \boldsymbol{q}_1 * \boldsymbol{q}_0 * \boldsymbol{q}_2^{>l}, \mathfrak{R}''\rangle$ described in Lemma 11.29. It remains to observe that the numbers $l_1 = |\boldsymbol{q}_1|$, $l_2 = |\boldsymbol{q}_0|$ and the sequence $\boldsymbol{q}_1 * \boldsymbol{q}_0$ of realizable state candidates for φ satisfy conditions (1)–(5) of Theorem 11.26. ❑

Given two numbers l_1, l_2 and a finite sequence $\boldsymbol{q}$ of state candidates for φ, we can effectively check whether they satisfy conditions (1)–(5) of Theorem 11.26. The only missing thing to make the criterion of Theorem 11.26 effective is therefore an algorithm for detecting whether a given state candidate for φ is realizable. If such an algorithm is *elementary* then the resulting satisfiability checking algorithm for monodic formulas is elementary as well. In particular, we have the following:

Theorem 11.30. *Let $\mathcal{QTL}'$ be a sublanguage of $\mathcal{QTL}_{\mathcal{U}\boxdot}$, and suppose that there is an algorithm which, given a state candidate $\mathfrak{C}$ for a $\mathcal{QTL}'$-sentence φ, can recognize whether $\mathfrak{C}$ is realizable using space $\leq 2^{p(\ell(\varphi))}$ for some polynomial function p. Then the fragment* $\mathsf{QLog}_{\mathcal{U}}(\mathbb{N}) \cap \mathcal{QTL}'$ *is decidable in* EXPSPACE.

Proof. We present a nondeterministic EXPSPACE satisfiability checking algorithm for $\mathcal{QTL}'$-sentences. It is based on the criterion of Theorem 11.26. First we guess two numbers $l_1 \leq \sharp(\varphi)$ and $l_2 \leq |sub_x\,\varphi| \cdot \flat^2(\varphi) \cdot \sharp(\varphi) + \sharp(\varphi)$.

As neither of the numbers exceeds $2^{2^{d\cdot\ell(\varphi)}}$, for some constant $d > 0$, we can write them in binary using space exponential in $\ell(\varphi)$.

Then we guess a state candidate $\boldsymbol{q}(0) = \langle T_0, T_0^{con}\rangle$ for φ such that some type in T_0 contains $\overline{\varphi}$. The algorithm provided by the formulation of the theorem can check whether $\boldsymbol{q}(0)$ is realizable using space exponential in $\ell(\varphi)$. Suppose that $\boldsymbol{q}(0)$ is realizable. Then we guess another state candidate $\boldsymbol{q}(1) = \langle T_1, T_1^{con}\rangle$ for φ, check whether it is realizable and whether the pair $\langle \boldsymbol{q}(0), \boldsymbol{q}(1)\rangle$ satisfies the following conditions, for $i = 0$:

(a) for every $\boldsymbol{t} \in T_i$ there is $\boldsymbol{t}' \in T_{i+1}$ such that the pair $\langle \boldsymbol{t}, \boldsymbol{t}'\rangle$ is suitable;

(b) for every $\boldsymbol{t}' \in T_{i+1}$ there is $\boldsymbol{t} \in T_i$ such that the pair $\langle \boldsymbol{t}, \boldsymbol{t}'\rangle$ is suitable;

(c) all pairs of the form $\langle \boldsymbol{t}_i^c, \boldsymbol{t}_{i+1}^c\rangle$, for $c \in con\,\varphi$ are suitable.

Clearly, this can be done using space exponential in $\ell(\varphi)$. After that we remove $\boldsymbol{q}(0)$, guess a state candidate $\boldsymbol{q}(2)$ and check conditions (a)–(c) for $i = 1$. We proceed in this way till we reach $\boldsymbol{q}(l_1)$. If this state candidate is realizable, we keep it in memory together with the set U containing all pairs of the form $\langle \boldsymbol{t}, \Xi\rangle$, where $\boldsymbol{t}$ is a type in T_{l_1} and Ξ is the set of all formulas of the form $\overline{\psi\mathcal{U}\chi}$ in $\boldsymbol{t}$. Then we guess a state candidate $\boldsymbol{q}(l_1+1)$, check whether it is realizable and whether conditions (a)–(c) hold for $i = l_1$. Besides, for every pair $p = \langle \boldsymbol{t}, \Xi\rangle \in U$ we guess a type $\boldsymbol{t}_0(p) \in T_{l_1+1}$ such that the pair $\langle \boldsymbol{t}, \boldsymbol{t}_0(p)\rangle$ is suitable and $\boldsymbol{t}_0(p) = \boldsymbol{t}_{l_1+1}^c$ whenever $\boldsymbol{t} = \boldsymbol{t}_{l_1}^c$ for some $c \in con\,\varphi$. Now we update U by replacing each pair $p = \langle \boldsymbol{t}, \Xi\rangle$ in U by $\langle \boldsymbol{t}_0(p), \Xi'\rangle$, where

$$\Xi' = \Xi - \{\overline{\psi\mathcal{U}\chi} \mid \overline{\chi} \in \boldsymbol{t}_0(p)\}.$$

(Note that U may contain more than one pair starting with the same $\boldsymbol{t}_0(p)$ and that not all $\boldsymbol{t} \in T_{l_1+1}$ are in the range of $\boldsymbol{t}_0$.) The updated U, $\boldsymbol{q}(l_1)$ and $\boldsymbol{q}(l_1+1)$ are stored in memory. The next step is similar to the previous one: we guess $\boldsymbol{q}(l_1+2)$, check whether it is realizable and whether conditions (a)–(c) hold for $i = l_1+1$. We also guess for every $p = \langle \boldsymbol{t}, \Xi\rangle \in U$ a $\boldsymbol{t}_1(p) \in T_{l_1+2}$ such that the pair $\langle \boldsymbol{t}, \boldsymbol{t}_1(p)\rangle$ is suitable and $\boldsymbol{t}_1(p) = \boldsymbol{t}_{l_1+2}^c$ whenever $\boldsymbol{t} = \boldsymbol{t}_{l_1+1}^c$ for some $c \in con\,\varphi$. We update U as before and store in memory only U, $\boldsymbol{q}(l_1)$ and $\boldsymbol{q}(l_1+2)$. We proceed in this way till we reach $\boldsymbol{q}(l_1+l_2-1)$. Then we check whether $\boldsymbol{q}(l_1+l_2-1)$ and $\boldsymbol{q}(l_1+l_2) = \boldsymbol{q}(l_1)$ satisfy (a)–(c) for $i = l_1+l_2-1$ and update U. If these conditions hold and, for every $\langle \boldsymbol{t}, \Xi\rangle \in U$, we have $\Xi = \emptyset$ then the algorithm returns: 'φ is satisfiable.'

It should be clear that this nondeterministic algorithm is sound and complete, and that it uses space exponential in $\ell(\varphi)$. And by Savitch's (1970) theorem, there is a deterministic algorithm checking satisfiability of an arbitrary $\mathcal{QTL}'$-sentence φ and requiring space exponential in $\ell(\varphi)$. ❑

As a consequence of this result and Theorems 3.30 and 5.43, we obtain:[5]

[5] This and the next two theorems are joint results with R. Kontchakov.

Theorem 11.31. *The fragments* $\mathsf{QLog}_{\mathcal{U}}(\mathbb{N}) \cap \mathcal{QTL}^{mo}_{\boxdot}$, $\mathsf{QLog}_{\mathcal{U}}(\mathbb{N}) \cap \mathcal{QTL}^1$ *and* $\mathsf{QLog}_{\mathcal{U}}(\mathbb{N}) \cap \mathcal{QTL}^2_{\boxdot}$ *are* EXPSPACE-*complete.*

Proof. The lower bounds follow from Theorems 3.30 and 5.43. (They also follow from Theorem 11.33 below.)

Let us establish the matching upper bounds, using the criterion provided by Theorem 11.30. Suppose first that $\mathfrak{C}$ is a state candidate for a $\mathcal{QTL}^{mo}_{\boxdot}$-sentence φ. As we know from Lemma 11.6, $\mathfrak{C}$ is realizable iff the $\overline{\mathcal{QL}}$-sentence $\mathsf{real}_{\mathfrak{C}}$ is satisfiable. And by Proposition 6.2.1 of (Börger *et al.* 1997), $\mathsf{real}_{\mathfrak{C}}$ is satisfiable iff it is satisfied in a model of cardinality $\leq 2^{c \cdot \ell(\varphi)}$. Thus, we have to check only models of exponential size, which can be done in nondeterministic exponential time, and so in EXPSPACE as well.

Now consider a state candidate $\mathfrak{C}$ for a $\mathcal{QTL}^2_{\boxdot}$-sentence φ. For each formula $\psi \in sub_x\varphi$ take a fresh unary predicate $Q_\psi(x)$. Let $\mathsf{real}^{\dagger}_{\mathfrak{C}}$ be the sentence

$$\bigwedge_{t \in T_{\mathfrak{C}}} \forall y \exists x \bigwedge_{\psi \in t} Q_\psi(x) \wedge \bigwedge_{c \in con\,\varphi} \bigwedge_{\psi \in t^c_{\mathfrak{C}}} Q_\psi(c) \wedge \forall y \forall x \bigvee_{t \in T_{\mathfrak{C}}} \bigwedge_{\psi \in t} Q_\psi(x) \wedge \bigwedge_{\psi \in sub_x\varphi} \forall x \big(Q_\psi(x) \leftrightarrow \psi\big),$$

for some (dummy) variable y. Clearly, $\mathsf{real}_{\mathfrak{C}}$ is satisfiable iff $\mathsf{real}^{\dagger}_{\mathfrak{C}}$ is satisfiable. Following the proof of Lemma 8.1.2 from (Börger *et al.* 1997), transform the last conjunct of $\mathsf{real}^{\dagger}_{\mathfrak{C}}$ into its *Scott's normal form*

$$\sigma = \forall x \forall y \alpha \wedge \bigwedge_{i=1}^{n} \forall x \exists y \beta_i,$$

where α and the β_i are quantifier-free. The length of σ can be bounded by a polynomial of the length of φ. Replace that conjunct with σ and denote the resultant formula by $\mathsf{real}^{*}_{\mathfrak{C}}$. According to the construction of σ, $\mathsf{real}^{\dagger}_{\mathfrak{C}}$ is satisfiable iff $\mathsf{real}^{*}_{\mathfrak{C}}$ is satisfiable. Now we can apply Proposition 8.1.4 from (Börger *et al.* 1997) to $\mathsf{real}^{*}_{\mathfrak{C}}$, which says that if $\mathsf{real}^{*}_{\mathfrak{C}}$ is satisfiable then it is satisfiable in a model of size $\ell(\mathsf{real}^{*}_{\mathfrak{C}}) \cdot 2^{p(r)}$ for some polynomial function p, where r is the number of predicate symbols in $\mathsf{real}^{*}_{\mathfrak{C}}$. The remaining part of the proof is the same as in the previous case. ❑

Next, we will use the results of (Grädel 1999b) on the complexity of the so-called *loosely guarded fragment*[6] of first-order logic to prove the following theorem:

Theorem 11.32. *The fragment* $\mathsf{QLog}_{\mathcal{U}}(\mathbb{N}) \cap \mathcal{TPF}_{\boxdot}$ *is* 2EXPTIME-*complete.*

[6]The loosely guarded fragment was introduced in (Andréka *et al.* 1998).

Proof. The lower bound follows from Theorem 4.4 of (Grädel 1999b) stating that the guarded fragment $\mathcal{GF}$ of first-order logic is 2EXPTIME-hard. Let us establish the matching upper bound.

It follows from the proof of Theorem 11.30 that it suffices to find an algorithm which, given a state candidate $\mathfrak{C}$ for a $\mathcal{TPF}_{\boxed{\text{I}}}$-sentence φ, is capable of checking whether $\mathfrak{C}$ is realizable in deterministic double exponential time of the length $\ell(\varphi)$ of φ. As we know from the proof of Theorem 11.18, $\mathfrak{C}$ is realizable iff the $\mathcal{PF}$-formula $\mathsf{real}^P_{\mathfrak{C}}$ is satisfiable. Note that $\ell(\mathsf{real}^P_{\mathfrak{C}})$ is an exponential function of $\ell(\varphi)$, and the number of variables and the number and arities of predicate symbols in $\mathsf{real}^P_{\mathfrak{C}}$ are bounded by a polynomial function of $\ell(\varphi)$. Now we apply to $\mathsf{real}^P_{\mathfrak{C}}$ two transformations. First, we turn $\mathsf{real}^P_{\mathfrak{C}}$ to a loosely guarded formula $\mathsf{real}'_{\mathfrak{C}}$ (of an extended signature) as was done in the proof of Theorem 3.3 of (Hodkinson 2002a), the length of which is still bounded by an exponential function of $\ell(\varphi)$ and the number of variables and the number and arities of predicate symbols in $\mathsf{real}'_{\mathfrak{C}}$ are still bounded by a polynomial function of $\ell(\varphi)$. Then transform $\mathsf{real}'_{\mathfrak{C}}$ to the normal form of Lemma 3.1 of (Grädel 1999b). Both transformations preserve satisfiability. The length of the resulting formula, $\mathsf{real}^*_{\mathfrak{C}}$, is still bounded by an exponential function of $\ell(\varphi)$, and the number of variables and the number and arities of predicate symbols in $\mathsf{real}^*_{\mathfrak{C}}$ are still bounded by a polynomial function of $\ell(\varphi)$. It remains to use the proof of Theorem 4.3 of (Grädel 1999b), according to which one can check whether $\mathsf{real}^*_{\mathfrak{C}}$ is satisfiable in deterministic double exponential time of $\ell(\varphi)$. ❑

We have seen above that even the one-variable fragment of $\mathsf{QLog}_{\mathcal{U}}(\mathbb{N})$ is considerably more complex (namely, EXPSPACE-complete) than its propositional fragment **PTL** (which is PSPACE-complete). But where precisely is the borderline between PSPACE and EXPSPACE? To answer this question we will now define two rather similar languages, located between propositional $\mathcal{ML}_{\mathcal{U}}$ and one-variable $\mathcal{QTL}^1_{\mathcal{U}}$, and show that one of them is PSPACE-complete, while the other one is EXPSPACE-complete.

Denote by $\mathcal{QTL}^1_{\circled{1}}$ the fragment of $\mathcal{QTL}^1$ in which only the next-time operator $\bigcirc$ can be applied to open formulas, while $\mathcal{U}$ and $\mathcal{S}$ are applied to sentences only (thus, we regard $\bigcirc$ as a primitive operator).

Theorem 11.33. *The fragment* $\mathsf{QLog}_{\mathcal{U}}(\mathbb{N}) \cap \mathcal{QTL}^1_{\circled{1}}$ *is* EXPSPACE-*hard.*

Proof. We will appropriately modify the reduction given in the proof of Theorem 5.43.

To this end, first consider $\mathcal{U}$, $\Box_F$ and $\bigcirc$ as primitive temporal operators of $\mathcal{QTL}$ ($\Diamond_F$ is regarded as an abbreviation). Let the sublanguage $\mathcal{QTL}^{1+}$ of $\mathcal{QTL}^1$ consist of those $\mathcal{QTL}^1$-formulas φ for which the following hold:

- $\mathcal{U}$ is applied only to subsentences of φ;

- for every subformula of φ of the form $\Box_F\psi(x)$ (with free variable x), if $\Box_F\psi(x)$ is under the scope of $\neg$ in φ, then ψ is a $\mathcal{QTL}^1_{\circledcirc}$-formula and φ contains a conjunct $\Diamond_F\forall x\Box_F\psi$.

Claim 11.34. *The fragment* $\mathsf{QLog}_{\mathcal{U}}(\mathbb{N})\cap\mathcal{QTL}^{1+}$ *is polynomially reducible to* $\mathsf{QLog}_{\mathcal{U}}(\mathbb{N})\cap\mathcal{QTL}^1_{\circledcirc}$.

Proof. Given a $\mathcal{QTL}^{1+}$-formula φ, denote by φ° the result of replacing every subformula of the form $\Box_F\psi(x)$ (with free x) in φ with a fresh unary predicate $P_{\Box\psi}(x)$. Let

$$\mathcal{R}_{\circ}(\varphi)=\{P_{\Box\psi}(x)\leftrightarrow\big(\bigcirc\psi^{\circ}(x)\wedge\bigcirc P_{\Box\psi}(x)\big)\mid\Box_F\psi(x)\in sub\,\varphi\}.\qquad(11.17)$$

We will show that, for every $\mathcal{QTL}^{1+}$-formula φ, φ is satisfiable in a first-order temporal model over $\langle\mathbb{N},<\rangle$ iff

$$\Big(\Box_F^+\forall x\bigwedge\mathcal{R}_{\circ}(\varphi)\Big)\wedge\varphi^{\circ}\qquad(11.18)$$

is satisfiable in a first-order temporal model over $\langle\mathbb{N},<\rangle$.

Suppose first that

$$(\mathfrak{M},0)\models^{\mathfrak{a}}\Big(\Box_F^+\forall x\bigwedge\mathcal{R}_{\circ}(\varphi)\Big)\wedge\varphi^{\circ}$$

for some model $\mathfrak{M}$ and assignment $\mathfrak{a}$. We claim that for every subformula α of φ, every assignment $\mathfrak{b}$ and every $n\in\mathbb{N}$,

$$\text{if}\quad(\mathfrak{M},n)\models^{\mathfrak{b}}\alpha^{\circ}\qquad\text{then}\qquad(\mathfrak{M},n)\models^{\mathfrak{b}}\alpha.$$

The proof is by induction on the construction of α. Clearly, if α is a $\mathcal{QTL}^1_{\circledcirc}$-formula then $\alpha^{\circ}=\alpha$. Since $\neg\bigcirc\neg\vartheta\leftrightarrow\bigcirc\vartheta$ is a valid formula, we may assume that φ is in a kind of 'normal form:' $\neg$ is always 'pushed down to' $\Box_F$. So the only nontrivial cases are $\alpha=\Box_F\psi(x)$ and $\alpha=\neg\Box_F\psi(x)$.

First let $\alpha=\Box_F\psi(x)$. Suppose $(\mathfrak{M},n)\models^{\mathfrak{b}}P_{\Box\psi}$. Then $(\mathfrak{M},n+1)\models^{\mathfrak{b}}\psi^{\circ}$, $(\mathfrak{M},n+1)\models^{\mathfrak{b}}P_{\Box\psi}$, and so, by the induction hypothesis, $(\mathfrak{M},n+1)\models^{\mathfrak{b}}\psi$. By iterating this we obtain that $(\mathfrak{M},k)\models^{\mathfrak{b}}\psi$ for all $k>n$, from which $(\mathfrak{M},n)\models^{\mathfrak{b}}\Box_F\psi$.

Next, let $\alpha=\neg\Box_F\psi(x)$. Suppose $(\mathfrak{M},n)\models^{\mathfrak{b}}\Box_F\psi$. We will show that $(\mathfrak{M},n)\models^{\mathfrak{b}}P_{\Box\psi}$ follows. Indeed, we have, for all $m>n$, $(\mathfrak{M},m)\models^{\mathfrak{b}}\psi$, and so (since ψ is a $\mathcal{QTL}^1_{\circledcirc}$-formula) $(\mathfrak{M},m)\models^{\mathfrak{b}}\psi^{\circ}$, for all $m>n$. Therefore,

$$\text{for all } m\geq n,\ (\mathfrak{M},m)\models^{\mathfrak{b}}\bigcirc\psi^{\circ}.\qquad(11.19)$$

On the other hand, since $\Diamond_F \forall x \Box_F \psi$ is a conjunct of φ, $(\Diamond_F \forall x \Box_F \psi)^\circ = \Diamond_F \forall x P_{\Box\psi}$ is a conjunct of φ°. So there is a $k \in \mathbb{N}$ such that $(\mathfrak{M}, k) \models^{\mathfrak{b}} P_{\Box\psi}$. Now by (11.17) and (11.19), we obtain $(\mathfrak{M}, n) \models^{\mathfrak{b}} P_{\Box\psi}$, as required.

Thus, we have $(\mathfrak{M}, 0) \models^{\mathfrak{a}} \varphi$.

Conversely, suppose that $(\mathfrak{M}, 0) \models^{\mathfrak{a}} \varphi$ for some model $\mathfrak{M} = \langle \langle \mathbb{N}, < \rangle, D, I \rangle$ and assignment $\mathfrak{a}$ in D. For each $n \in \mathbb{N}$, extend $I(n)$ to $I^+(n)$ by taking, for all $\Box_F \psi(x) \in sub\, \varphi$,

$$P_{\Box\psi}^{I^+(n)} = \{a \in D \mid (\mathfrak{M}, n) \models \Box_F \psi[a]\},$$

and let $\mathfrak{M}^+ = \langle \langle \mathbb{N}, < \rangle, D, I^+ \rangle$. We leave it to the reader to show that for all $n \in \mathbb{N}$, all assignments $\mathfrak{a}$ in D, and all $\psi \in sub\, \varphi$,

$$(\mathfrak{M}^+, n) \models \forall x \bigwedge \mathcal{R}_\circ(\varphi), \qquad \text{and}$$
$$(\mathfrak{M}, n) \models^{\mathfrak{a}} \psi \qquad \text{iff} \qquad (\mathfrak{M}^+, n) \models^{\mathfrak{a}} \psi^\circ.$$

Therefore, (11.18) is satisfiable in $\mathfrak{M}^+$. ❑

Now recall the formula $\varphi_{n,T}$ from the proof of Theorem 5.43. It is shown there that $\varphi_{n,T}$ is **PTL** $\times$ **S5**-satisfiable iff there is an $m \in \mathbb{N}$ such that T tiles the $m \times 2^n$-corridor as required. Consider the first-order temporal translation $\varphi^\dagger_{n,T}$ (see Section 3.7) of $\varphi_{n,T}$. By Theorem 3.30,

$$\varphi_{n,T} \text{ is } \mathbf{PTL} \times \mathbf{S5}\text{-satisfiable} \quad \text{iff} \quad \varphi^\dagger_{n,T} \text{ is satisfiable over } \langle \mathbb{N}, < \rangle. \quad (11.20)$$

A close inspection shows that $\varphi^\dagger_{n,T}$ 'almost' belongs to $\mathcal{QTL}^{1+}$. Its only 'problematic' part is the subformula $\mathsf{right}^\dagger = \mathsf{right}(x)$ which contains an occurrence of $\mathcal{U}$ applied to the open formula $\mathsf{tile}^\dagger = \mathsf{tile}(x)$. Now replace $\mathsf{right}(x)$ in $\varphi^\dagger_{n,T}$ with $\mathsf{equ}(x) \wedge R(x)$, where $\mathsf{equ}(x) = \mathsf{equ}^\dagger$ and R is a fresh unary predicate symbol, and add the conjuncts

$$\Box_F^+ \forall x \Big(R(x) \leftrightarrow \big(\bigcirc \mathsf{tile}(x) \vee (\bigcirc \neg \mathsf{equ}(x) \wedge \bigcirc R(x)) \big) \Big), \quad (11.21)$$

$$\Box_F^+ \forall x \big(R(x) \rightarrow \Diamond_F \mathsf{tile}(x) \big) \quad (11.22)$$

to $\varphi^\dagger_{n,T}$. Denote the resulting formula by $\psi_{n,T}$. By Theorem 3.30 and the proof of Claim 6.25, we obtain that

$$\varphi^\dagger_{n,T} \text{ is satisfiable over } \langle \mathbb{N}, < \rangle \quad \text{iff} \quad \psi_{n,T} \text{ is satisfiable over } \langle \mathbb{N}, < \rangle. \quad (11.23)$$

Moreover, we may assume that the models satisfying $\varphi^\dagger_{n,T}$ and $\psi_{n,T}$ have the same domains (so if one of them is finite, then the other is finite as well). Let $\chi_{n,T}$ be obtained from $\psi_{n,T}$ by omitting the conjunct (11.22). We claim that

$$\psi_{n,T} \text{ is satisfiable over } \langle \mathbb{N}, < \rangle \quad \text{iff} \quad \chi_{n,T} \text{ is satisfiable over } \langle \mathbb{N}, < \rangle, \quad (11.24)$$

and that again we may assume the models satisfying $\psi_{n,T}$ and $\chi_{n,T}$ to have the same domains (so if one of them is finite then the other is finite as well). Indeed, suppose that $\chi_{n,T}$ is satisfied at time point 0 in some first-order temporal model $\mathfrak{M}$ over $\langle \mathbb{N}, < \rangle$ with domain D. Assume that

$$(\mathfrak{M}, n) \not\models \Diamond_F \mathsf{tile}[a]$$

for some $n \in \mathbb{N}$ and $a \in D$. As is shown in the proof of Theorem 5.43, we may assume that D is finite and so there is some $k > n$ such that

$$(\mathfrak{M}, k) \models \mathsf{equ}[a].$$

Let N be the smallest k with this property. But then, by (11.21), we obtain $(\mathfrak{M}, N-1) \not\models R[a]$ from which, using (11.21) $N - n$ times, $(\mathfrak{M}, n) \not\models R[a]$, as required.

Finally, we claim that $\chi_{n,T}$ belongs to $\mathcal{QTL}^{1+}$. Indeed, the only subformula of $\chi_{n,T}$ of the form $\Box_F \psi(x)$ (with free x) that is under the scope of $\neg$ in $\chi_{n,T}$ is the occurrence of $\Box_F \neg \mathsf{mark}(x)$ in (11.21) (it is a conjunct of $\mathsf{tile}(x)$). But $\neg\mathsf{mark}(x)$ is a $\mathcal{QTL}^1_{①}$-formula, and $\Diamond_F \forall x \Box_F \neg \mathsf{mark}(x)$ can be considered as a conjunct of $\chi_{n,T}$ by (5.63).

So the theorem follows from (11.20), (11.23), (11.24), Claim 11.34, and Theorem 5.43. ❑

Thus, as soon as we allow for applications of only $\bigcirc$ to formulas with one free variable, exponential space is required for satisfiability checking. Let us consider now the case when *none* of the temporal operators can be applied to open formulas.

Denote by $\mathcal{QTL}_{\boxed{0}}$ the sublanguage of $\mathcal{QTL}_{\boxed{1}}$ in which all temporal operators are applied only to sentences, cf. (Finger and Gabbay 1992). The $\mathcal{S}$-free fragment of $\mathcal{QTL}_{\boxed{0}}$ is denoted by $\mathcal{QTL}_{\mathcal{U}\boxed{0}}$. Then the following result holds:

Theorem 11.35. *Let $\mathcal{QTL}'$ be a sublanguage of $\mathcal{QTL}_{\mathcal{U}\boxed{0}}$, φ a $\mathcal{QTL}'$-sentence, and suppose that the problem 'given a set $\Sigma \subseteq sub_0\varphi$, decide whether the formula $\bigwedge_{\psi\in\Sigma}\overline{\psi}$ is satisfiable' belongs to a complexity class $\mathcal{C} \supseteq$ PSPACE. Then the satisfiability problem for the fragment $\mathsf{QLog}_{\mathcal{U}}(\mathbb{N}) \cap \mathcal{QTL}'$ is in $\mathcal{C}$.*

Proof. The proof—a modification of the proof of Theorem 11.30—is left to the reader as an exercise. (We note only that state candidates for a $\mathcal{QTL}_{\mathcal{U}\boxed{0}}$-sentence φ are of the form $\{\overline{\psi} \mid \psi \in sub_0\varphi\}$ and we do not need runs in the quasimodels.) ❑

This theorem means that the complexity of the satisfiability problem for a fragment of $\mathsf{QLog}_{\mathcal{U}}(\mathbb{N}) \cap \mathcal{QTL}_{\boxed{0}}$ is the maximum of the complexity of the satisfiability problem for **PTL** (i.e., PSPACE) and that of the pure first-order part of the fragment. For example, denote by $\mathcal{QTL}^1_{\boxcircle}$, $\mathcal{QTL}^2_{\boxcircle}$ and $\mathcal{QTL}^{mo}_{\boxcircle}$ the one-variable, two-variable, and monadic fragments of $\mathcal{QTL}_{\boxcircle}$, respectively.

Then we have the following:

Theorem 11.36. (i) *The satisfiability problem for* $\mathsf{QLog}_{\mathcal{U}}(\mathbb{N}) \cap \mathcal{QTL}^1_{\boxcircle}$ *is* PSPACE-*complete.*

(ii) *The satisfiability problems for the fragments* $\mathsf{QLog}_{\mathcal{U}}(\mathbb{N}) \cap \mathcal{QTL}^2_{\boxcircle}$ *and* $\mathsf{QLog}_{\mathcal{U}}(\mathbb{N}) \cap \mathcal{QTL}^{mo}_{\boxcircle}$ *are* NEXPTIME-*complete.*

Proof. Follows from Theorem 11.35 and the complexity results for the corresponding fragments of first-order logic, which can be found, e.g., in (Börger *et al.* 1997). ❑

These results will be used for establishing the complexity of some temporalized description logics (Section 14.3) and spatio-temporal logics (Section 16.3).

In this section we have discussed the complexity of some monodic fragments of first-order temporal logics over $\langle \mathbb{N}, < \rangle$. The following questions remain open:

Question 11.37. What is the computational complexity of the decision problem for the other decidable monodic fragments, mentioned in Section 11.2, over $\langle \mathbb{N}, < \rangle$? What is the complexity of decidable monodic fragments over other flows of time?

Note that as a consequence of Theorems 3.29 and 6.63 we obtain the following result of (Hodkinson *et al.* 2003):

Theorem 11.38. *Let $\mathcal{C}$ be any class of strict linear orders at least one of which contains an infinite ascending chain. Then* $\mathsf{QLog}_{\mathcal{U}}(\mathcal{C}) \cap \mathcal{QTL}^{mo}_{\boxbar}$ *and* $\mathsf{QLog}_{\mathcal{U}}(\mathcal{C}) \cap \mathcal{QTL}^1$ *are* EXPSPACE-*hard.*

By Theorems 3.29, 6.30, 6.31 and 6.61,[7] we also have:

Theorem 11.39. *The fragment* $\mathsf{QLog}_{\mathcal{SU}}(\mathcal{C}) \cap \mathcal{QTL}^1$ *is in* 2EXPTIME *whenever* $\mathcal{C} = \{\langle \mathbb{Q}, < \rangle\}$ *or $\mathcal{C}$ is the class of all strict linear orders.*

Remark 11.40. Tableau decision algorithms for a number of decidable fragments of the logic $\mathsf{QLog}_{\mathcal{U}}(\mathbb{N}) \cap \mathcal{QTL}_{\boxbar}$ have been constructed in (Kontchakov *et al.* 2003).

---

[7] Theorem 6.61 is formulated only for the case when $\Box_F$ and $\Box_P$ are the only temporal operators, but it is not hard to generalize it for the case of $\mathcal{S}$ and $\mathcal{U}$ as well.

11.5 Satisfiability in models over $\langle \mathbb{N}, < \rangle$ with finite domains

Our aims in this section are to prove Theorem 11.9 (2) and to determine the complexity of the satisfiability problem in models with finite domains based on the flow of time $\langle \mathbb{N}, < \rangle$ for a number of monodic fragments of $\mathcal{QTL}$. As we will actually see, for all these fragments, the complexity does not depend on whether domains are (arbitrarily) finite or infinite (see Table 11.1).

Suppose that we are given a $\mathcal{QTL}_{\boxed{1}}$-sentence φ and a strict linear order $\mathfrak{F} = \langle W, < \rangle$. We call a quasimodel $\langle \mathfrak{F}, \boldsymbol{q}, \mathfrak{R} \rangle$ for φ *finitary* if $\boldsymbol{q}(w)$ is a finitely realizable state candidate for every $w \in W$ and $\mathfrak{R}$ is finite. Now, the finitary analog of Lemma 11.22 (with $\mathcal{K}$ being the class of all $\mathcal{QL}$-structures) is the following:

Lemma 11.41. *A $\mathcal{QTL}_{\boxed{1}}$-sentence φ is satisfiable in a first-order temporal model based on $\mathfrak{F}$ and having finite domains iff there is a finitary quasimodel for φ based on $\mathfrak{F}$.*

Proof. First, suppose that φ is satisfied in a first-order temporal model $\mathfrak{M} = \langle \mathfrak{F}, D, I \rangle$ with finite D. It is easy to see that the quasimodel $\langle \mathfrak{F}, \boldsymbol{q}, \mathfrak{R} \rangle$ defined in the proof of Lemma 11.22 is finitary.

Conversely, suppose that $\mathfrak{Q} = \langle \mathfrak{F}, \boldsymbol{q}, \mathfrak{R} \rangle$ is a finitary quasimodel for φ. We require the following claim which is a 'finite version' of Claim 11.24:

Claim 11.42. *There is a natural number m_φ such that, for every finitely realizable state candidate $\mathfrak{C} = \langle T, T^{con} \rangle$ and every sequence $\langle n_{\boldsymbol{t}} \mid \boldsymbol{t} \in T \rangle$ of numbers with $n_{\boldsymbol{t}} > m_\varphi$, $\boldsymbol{t} \in T$, there is a $\mathcal{QL}$-structure I realizing $\mathfrak{C}$ and such that $|I_{\boldsymbol{t}}| = n_{\boldsymbol{t}}$, for every $\boldsymbol{t} \in T$.*

Proof. Suppose that $\mathfrak{C}_0, \dots, \mathfrak{C}_k$ are all the distinct finitely realizable state candidates for φ (hence $k < \sharp(\varphi)$) and that for each $j \leq k$, I^j is a finite $\mathcal{QL}$-structure realizing $\mathfrak{C}_j = \langle T_j, T_j^{con} \rangle$. Then, using the 'blow up' technique of the proof of Claim 11.24, it is not hard to see that

$$m_\varphi = \max\{ |I^j_{\boldsymbol{t}}| \mid \boldsymbol{t} \in T_j,\ j \leq k \}$$

does the job. Again we use here the fact that our language does not contain equality. ❑

Now let m_φ be the number supplied by Claim 11.42. Put

$$D = \{ \langle r, \xi \rangle \mid r \in \mathfrak{R},\ \xi < m_\varphi \}.$$

Fix some $w \in W$. Then for any type $\boldsymbol{t} \in T_w$,

$$\big|\{ \langle r, \xi \rangle \in D \mid r(w) = \boldsymbol{t} \}\big| = m_\varphi \cdot \big|\{ r \in \mathfrak{R} \mid r(w) = \boldsymbol{t} \}\big| = m_\varphi \cdot k_{\boldsymbol{t}}. \quad (11.25)$$

By Claim 11.42, there exists a $\overline{\mathcal{QL}}$-structure $I(w)$ with domain $D(w)$ such that $I(w)$ realizes the state candidate $\boldsymbol{q}(w)$ and for every $\boldsymbol{t} \in T_w$ there are $m_\varphi \cdot k_{\boldsymbol{t}}$ elements in $D(w)$ realizing $\boldsymbol{t}$. Hence, by (11.25), we can identify each $D(w)$ with D in a 'type-preserving' way, that is, we may assume that, for all $w \in W$, $r \in \mathfrak{R}$,

$$r(w) = \{\overline{\psi} \mid \psi \in sub_x\varphi,\ I(w) \models \overline{\psi}[\langle r, \xi\rangle]\},$$

and that $c^{I(w)} = \langle r_c, 0\rangle$, for every $c \in con\,\varphi$. Let $\mathfrak{M} = \langle \mathfrak{F}, D, I'\rangle$ be the first-order temporal model, where I' is the $\mathcal{QL}$-reduct of I. In precisely the same way as in the proof of Lemma 11.22 one can show that φ is satisfied in $\mathfrak{M}$. ❑

It is worth noting that models with finite domains are closely connected to models satisfying the finite state assumption (which was introduced in Section 3.2 for topological temporal models).

Say that a first-order temporal model $\mathfrak{M} = \langle \mathfrak{F}, D, I\rangle$, where $\mathfrak{F} = \langle W, <\rangle$, satisfies the *finite state assumption* (**FSA**) if the set $\{I(w) \mid w \in W\}$ is finite. A quasimodel $\langle \mathfrak{F}, \boldsymbol{q}, \mathfrak{R}\rangle$ satisfies **FSA** if the set $\mathfrak{R}$ of its runs is finite. (Note that every finitary quasimodel satisfies **FSA**, but there are quasimodels with **FSA** that are not finitary: their realizable state candidates are not necessarily finitely realizable.)

The following lemma shows that first-order temporal models with **FSA** correspond to quasimodels with **FSA**:

Lemma 11.43. *A $\mathcal{QTL}_{\boxdot}$-sentence φ is satisfiable in a first-order temporal model with* **FSA** *based on a flow of time $\mathfrak{F}$ iff there is a quasimodel for φ with* **FSA** *based on $\mathfrak{F}$.*

Proof. The implication ($\Rightarrow$) was shown in the proof of Lemma 11.22, because the set of runs $\mathfrak{R}$ constructed in that proof is finite whenever the first-order temporal model satisfies **FSA**. (This is because every run r is determined by its values on a *finite* subset of its domain.)

Actually, the converse implication ($\Leftarrow$) also follows from that proof. Indeed, suppose that we are given a quasimodel $\langle \mathfrak{F}, \boldsymbol{q}, \mathfrak{R}\rangle$ satisfying **FSA**. Then instead of $\mathfrak{R}_{\boldsymbol{q}}$ we take the finite set $\mathfrak{R}$ of runs and choose the $\overline{\mathcal{QL}}$-structures $I(w)$, $w \in W$, in such a way that $I(w) = I(w')$ whenever $\boldsymbol{q}(w) = \boldsymbol{q}(w')$ and $r(w) = r(w')$ for all $r \in \mathfrak{R}$. ❑

As a consequence we obtain the following:

Theorem 11.44. *Suppose that $\mathcal{QTL}' \subseteq \mathcal{QTL}_{\boxdot}$ and that every realizable state candidate for an arbitrary $\mathcal{QTL}'$-sentence is finitely realizable. Then a $\mathcal{QTL}'$-sentence φ is satisfiable in a first-order temporal model based on a flow of time $\mathfrak{F}$ and having finite domains iff φ is satisfiable in a first-order temporal model with* **FSA** *based on $\mathfrak{F}$.*

Proof. The implication ($\Rightarrow$) is trivial. Suppose now that φ is satisfied in a first-order temporal model with **FSA** based on a flow of time $\mathfrak{F}$. Then, by Lemma 11.43, there is a quasimodel $\langle \mathfrak{F}, \boldsymbol{q}, \mathfrak{R} \rangle$ for φ with **FSA**. We know that all $\boldsymbol{q}(w)$ are realizable state candidates for φ. Hence they are finitely realizable and $\langle \mathfrak{F}, \boldsymbol{q}, \mathfrak{R} \rangle$ is actually a finitary quasimodel for φ. So, by Lemma 11.41, our sentence φ is satisfied in a first-order temporal model with finite domains based on $\mathfrak{F}$. ❑

We will use this theorem in Section 16.3.

Now we show how to obtain an *elementary* decision algorithm for any monodic fragment of $\mathsf{QLog}^{fin}_{\mathcal{SU}}(\mathbb{N})$ for which the finite realizability of state candidates can be decided by an elementary procedure. The complexity results we are going to prove are presented in Table 11.1.

To begin with, we have the following 'finite domain' analog of Proposition 11.25:

Proposition 11.45. $\mathsf{QLog}^{fin}_{\mathcal{SU}}(\mathbb{N})$ *is polynomially reducible to* $\mathsf{QLog}^{fin}_{\mathcal{U}}(\mathbb{N})$.

Proof. Similar to the proof of Proposition 11.25. ❑

Thus, similar to Section 11.4, we may confine ourselves to considering the temporal language without $\mathcal{S}$. As explained in Section 11.4, all the results below will hold true for the corresponding fragments of $\mathsf{QLog}^{fin}_{\mathcal{SU}}(\mathbb{N})$ as well.

The following theorem is an appropriate modification of the criterion in Theorem 11.26:

Theorem 11.46. *A* $\mathcal{QTL}_{\mathcal{U}\boxplus}$*-sentence* φ *is satisfiable in a first-order temporal model based on* $\langle \mathbb{N}, < \rangle$ *and having finite domains iff there are natural numbers* l_1, l_2 *and a sequence*

$$\langle \langle T_0, T_0^{con} \rangle, \dots, \langle T_{l_1+l_2-1}, T^{con}_{l_1+l_2-1} \rangle \rangle$$

of finitely realizable state candidates for φ *such that*

$$l_1 \leq \sharp(\varphi), \qquad 0 < l_2 \leq |sub_x\varphi| \cdot \flat(\varphi)^2 \cdot \sharp(\varphi) \cdot 2^{\flat(\varphi)^2} + \sharp(\varphi) \cdot 2^{\flat(\varphi)^2},$$

and conditions (1)–(3), (5) *of Theorem 11.26 and the following condition* $(4)^f$ *hold:*

$(4)^f$ *for every type* $\boldsymbol{t} \in T_{l_1}$ *there are types* $\boldsymbol{t}_1, \dots, \boldsymbol{t}_{l_2-1}$ *such that* $\boldsymbol{t}_i \in T_{l_1+i}$ *for* $1 \leq i < l_2$ *and all formulas of the form* $\psi_1 \mathcal{U} \psi_2$ *in* $\boldsymbol{t}$ *are realized by the sequence* $\langle \boldsymbol{t}, \boldsymbol{t}_1, \dots, \boldsymbol{t}_{l_2-1}, \boldsymbol{t} \rangle$.

	arbitrary domains	finite domains	models with **FSA**
$\mathcal{ML}_{\mathcal{SU}}$	PSPACE-complete (Thms. 2.7, 2.9)	PSPACE-complete (Thms. 2.7, 2.9)	PSPACE-complete (Thms. 2.7, 2.9)
$\mathcal{QTL}^1_{\boxed{0}}$	PSPACE-complete (Thm. 11.36)	PSPACE-complete (Thm. 11.55)	PSPACE-complete (Thms. 11.44, 11.55)
$\mathcal{QTL}^2_{\boxed{0}}$	NEXPTIME-complete (Thm. 11.36)	NEXPTIME-complete (Thm. 11.55)	NEXPTIME-complete (Thms. 11.44, 11.55)
$\mathcal{QTL}^{mo}_{\boxed{0}}$	NEXPTIME-complete (Thm. 11.36)	NEXPTIME-complete (Thm. 11.55)	NEXPTIME-complete (Thms. 11.44, 11.55)
$\mathcal{QTL}^1_{\textcircled{1}}$	EXPSPACE-complete (Thms. 11.31, 11.33)	EXPSPACE-complete (Thms. 11.52, 11.53)	EXPSPACE-complete (Thms. 11.44, 11.52, 11.53)
$\mathcal{QTL}^1$	EXPSPACE-complete (Thm. 11.31)	EXPSPACE-complete (Thm. 11.53)	EXPSPACE-complete (Thms. 11.44, 11.53)
$\mathcal{QTL}^2_{\boxed{1}}$	EXPSPACE-complete (Thm. 11.31)	EXPSPACE-complete (Thm. 11.53)	EXPSPACE-complete (Thms. 11.44, 11.53)
$\mathcal{QTL}^{mo}_{\boxed{1}}$	EXPSPACE-complete (Thm. 11.31)	EXPSPACE-complete (Thm. 11.53)	EXPSPACE-complete (Thms. 11.44, 11.53)
$\mathcal{TPF}_{\boxed{1}}$	2EXPTIME-complete (Thm. 11.32)	2EXPTIME-complete (Thm. 11.54)	2EXPTIME-complete (Thms. 11.44, 11.54)
$\mathcal{QTL}_{\boxed{1}}$	r.e. (Thm. 11.71)	not r.e. (Trakhtenbrot 1950)	not r.e.?
$\mathcal{QTL}^{=}_{\boxed{1}}$	not r.e. (Thm. 11.80)	not r.e. (Trakhtenbrot 1950)	not r.e.?
$\mathcal{QTL}^2 \cap \mathcal{QTL}^{mo}$	not r.e. (Thm. 11.1)	not r.e. (Thm. 11.3)	not r.e.?

Table 11.1: Complexity of first-order temporal logics over $\langle \mathbb{N}, < \rangle$ (with and without $\mathcal{S}$).

Proof. For the '$\Leftarrow$'-direction, given numbers l_1, l_2 and a sequence of finitely realizable state candidates as above, we have to construct a finitary quasimodel for φ. Then, by Lemma 11.41, we are done.

Define the state function $\boldsymbol{q} = \boldsymbol{q}_1 * \boldsymbol{q}_2^*$ as in the proof of Theorem 11.26. A finite set $\mathfrak{R}$ of runs through $\boldsymbol{q}$ will be defined using condition $(4)^f$ which guarantees that formulas of the form $\overline{\psi_1 \mathcal{U} \psi_2} \in \boldsymbol{t}_{l_1}$ are realized in 'loops' $\langle \boldsymbol{t}_{l_1}, \dots, \boldsymbol{t}_{l_1+l_2} = \boldsymbol{t}_{l_1} \rangle$. Say that a sequence $\langle \boldsymbol{t}_0, \dots, \boldsymbol{t}_k \rangle$ $(k > 0)$ of types is *suitable* if every pair of adjacent types in the sequence is suitable. We call a sequence $\langle \boldsymbol{t}_0, \dots, \boldsymbol{t}_k \rangle$ *root saturated* if the sequence $\langle \boldsymbol{t}_0, \dots, \boldsymbol{t}_k, \boldsymbol{t}_0 \rangle$ is suitable and realizes all formulas of the form $\overline{\psi_1 \mathcal{U} \psi_2} \in \boldsymbol{t}_0$.

Now let $\mathfrak{R}$ consist of all infinite words of the form

$$s_1 * (s_2 * s_3)^* \quad \text{and} \quad s_1 * (s_3 * s_2)^*,$$

where

- $s_1 = \langle \boldsymbol{t}_0, \dots, \boldsymbol{t}_{l_1+l_2-1} \rangle$ is a suitable sequence such that $\boldsymbol{t}_i \in T_i$, for all $i < l_1 + l_2$;
- $s_2 = \langle \boldsymbol{t}'_0, \dots, \boldsymbol{t}'_{l_2-1} \rangle$ is a suitable sequence such that $\boldsymbol{t}'_j \in T_{l_1+j}$, for all $j < l_2$;
- $s_3 = \langle \boldsymbol{t}''_0, \dots, \boldsymbol{t}''_{l_2-1} \rangle$ is a root saturated sequence such that $\boldsymbol{t}''_j \in T_{l_1+j}$, for all $j < l_2$;
- the pairs $\langle \boldsymbol{t}_{l_1+l_2-1}, \boldsymbol{t}'_0 \rangle$ and $\langle \boldsymbol{t}'_{l_2-1}, \boldsymbol{t}''_0 \rangle$ are suitable and
- $\boldsymbol{t}'_0 = \boldsymbol{t}''_0$.

It is readily checked that every such word is a coherent and saturated run through $\boldsymbol{q}$. Conditions (1)–(3), $(4)^f$ and (5) guarantee that $(\mathbf{tqm1})_0$–$(\mathbf{tqm3})$ hold, and hence $\langle \boldsymbol{q}, \mathfrak{R} \rangle$ is a quasimodel for φ. Needless to say $\mathfrak{R}$ is finite.

For the '$\Rightarrow$'-direction, we again need a series of lemmas which are stronger than the corresponding Lemmas 11.27–11.29. Suppose that $\langle \boldsymbol{q}, \mathfrak{R} \rangle$ is a quasimodel for φ. Define an equivalence relation $\sim_{\mathfrak{R}}$ on $\mathbb{N}$ by taking

$$i \sim_{\mathfrak{R}} j \quad \text{iff} \quad \boldsymbol{q}(i) = \boldsymbol{q}(j) \ \text{ and } \ \forall r \in \mathfrak{R}\ r(i) = r(j),$$

and denote by $[n]_{\mathfrak{R}}$ the $\sim_{\mathfrak{R}}$-equivalence class of n.

Besides, for each $n \in \mathbb{N}$, we define one more equivalence relation $\sim^n_{\mathfrak{R}}$ on $\mathbb{N}$ by taking $i \sim^n_{\mathfrak{R}} j$ iff $\boldsymbol{q}(i) = \boldsymbol{q}(j)$ and

- for every $r \in \mathfrak{R}$ there is $r' \in \mathfrak{R}$ such that $r(n) = r'(n)$ and $r(i) = r'(j)$,
- for every $r \in \mathfrak{R}$ there is $r'' \in \mathfrak{R}$ such that $r(n) = r''(n)$ and $r(j) = r''(i)$.

Lemma 11.47. *For every $n \in \mathbb{N}$, the number of pairwise distinct $\sim_{\mathfrak{R}}^{n}$-equivalence classes does not exceed*

$$\natural(\varphi) = \sharp(\varphi) \cdot 2^{\flat(\varphi)^2}.$$

Proof. Let $\langle \boldsymbol{t}_0, \dots, \boldsymbol{t}_{n_\varphi - 1} \rangle$ be an enumeration of all types for φ, $n_\varphi \leq \flat(\varphi)$. Fix some $n \in \mathbb{N}$ and define a function $\sigma_i(k,l)$, for $i \in \mathbb{N}$, $k, l < n_\varphi$, by taking

$$\sigma_i(k,l) = \begin{cases} 1, & \text{if } \exists r \in \mathfrak{R}\ (r(n) = \boldsymbol{t}_k \text{ and } r(i) = \boldsymbol{t}_l), \\ 0, & \text{otherwise.} \end{cases}$$

We then have $i \sim_{\mathfrak{R}}^{n} j$ iff $\boldsymbol{q}(i) = \boldsymbol{q}(j)$ and $\sigma_i(k,l) = \sigma_j(k,l)$, for all $k, l < n_\varphi$. It remains to observe that the number of functions from $\{0, \dots, n_\varphi - 1\}^2$ into $\{0,1\}$ is $2^{n_\varphi^2} \leq 2^{\flat(\varphi)^2}$. ❑

We again need to show that one can delete the interval between two identical quasistates. However, in the finitary case a somewhat subtler deleting technique than the one in Lemma 11.27 is required:

Lemma 11.48. *Let $\langle \boldsymbol{q}, \mathfrak{R} \rangle$ be a quasimodel for φ and $i \sim_{\mathfrak{R}}^{n} j$ for $n < i < j$. Then $\langle \boldsymbol{q}^{\leq i} * \boldsymbol{q}^{>j}, \mathfrak{S} \rangle$ is also a quasimodel for φ, where*

$$\mathfrak{S} = \mathfrak{R}^{\leq i} *_n \mathfrak{R}^{>j} = \{r_1^{\leq i} * r_2^{>j} \mid r_1, r_2 \in \mathfrak{R},\ r_1(i) = r_2(j),\ r_1(n) = r_2(n)\}.$$

Moreover, for all $n' > j$, if $n \sim_{\mathfrak{R}} n'$ then $n \sim_{\mathfrak{S}} n' - (j - i)$.

Proof. Follows immediately from the definition of $i \sim_{\mathfrak{R}}^{n} j$. ❑

The following finite analog of Lemma 11.28 is proved with the help of Lemma 11.48 in a way similarly to how Lemma 11.28 is proved by using Lemma 11.27:

Lemma 11.49. *Every quasimodel $\langle \boldsymbol{q}, \mathfrak{R} \rangle$ for φ with finite $\mathfrak{R}$ contains a subquasimodel $\langle \boldsymbol{q}_1 * \boldsymbol{q}_2, \mathfrak{R}' \rangle$ with finite $\mathfrak{R}'$ such that $|\boldsymbol{q}_1| \leq \natural(\varphi)$ and $[n]_{\mathfrak{R}'}$ is infinite, for every $n \geq |\boldsymbol{q}_1|$.*

The finite analog of Lemma 11.29 is the following:

Lemma 11.50. *Let $\langle \boldsymbol{q}_1 * \boldsymbol{q}_2, \mathfrak{R}' \rangle$ be a quasimodel for φ (with quasistates of the form $\langle T_i, T_i^{con} \rangle$ for $i \in \mathbb{N}$) such that $|\boldsymbol{q}_1| \leq \natural(\varphi)$, $\mathfrak{R}$ is finite, and $[m]_{\mathfrak{R}}$ is infinite for all $m > |\boldsymbol{q}_1|$. Then there is a subquasimodel of the form $\langle \boldsymbol{q}_1 * \boldsymbol{q}_0 * \boldsymbol{q}_2^{\geq l}, \mathfrak{R}'' \rangle$, for some $l \geq 0$, such that $\mathfrak{R}''$ is finite and*

(i) *$|\boldsymbol{q}_0| \leq |sub_x\varphi| \cdot \flat(\varphi)^2 \cdot \natural(\varphi) + \natural(\varphi)$;*

(ii) *for every type $\boldsymbol{t} \in T_{|\boldsymbol{q}_1|}$ there is a run $r \in \mathfrak{R}''$ such that $r(|\boldsymbol{q}_1|) = \boldsymbol{t}$ and all formulas $\overline{\psi_1 \mathcal{U} \psi_2} \in r(|\boldsymbol{q}_1|)$ are realized by the sequence*

$$\langle r(|\boldsymbol{q}_1|), r(|\boldsymbol{q}_1| + 1), \dots, r(|\boldsymbol{q}_1| + |\boldsymbol{q}_0|) \rangle\,;$$

(iii) *for every* $c \in con\,\varphi$ *the sequence*

$$\langle r_c(|\boldsymbol{q}_1|), r_c(|\boldsymbol{q}_1|+1), \dots, r_c(|\boldsymbol{q}_1|+|\boldsymbol{q}_0|)\rangle$$

realizes all formulas $\overline{\psi_1 \mathcal{U} \psi_2} \in r_c(|\boldsymbol{q}_1|)$;

(iv) $|\boldsymbol{q}_1| \sim_{\mathfrak{R}''} |\boldsymbol{q}_1 * \boldsymbol{q}_0|$.

Proof. Let $\mathfrak{Q} = \langle \boldsymbol{q}_1 * \boldsymbol{q}_2, \mathfrak{R}' \rangle$ and $n = |\boldsymbol{q}_1|$. Suppose $\boldsymbol{t} \in T_n$, $\overline{\psi_1 \mathcal{U} \psi_2} \in \boldsymbol{t}$ and $r \in \mathfrak{R}'$ with $r(n) = \boldsymbol{t}$. Then there exists $m > 0$ such that $\overline{\psi_2} \in r(n+m)$ and $\overline{\psi_1} \in r(n+k)$ for all $k \in (0, m)$. Assume now that $0 < i < j < m$, $r(n+i) = r(n+j)$ and $n+i \sim^n_{\mathfrak{R}'} n+j$. In view of Lemma 11.48, there is a subquasimodel $\left\langle \boldsymbol{q}_1 * \boldsymbol{q}_2^{\leq i} * \boldsymbol{q}_2^{>j}, \mathfrak{S} \right\rangle$ of $\mathfrak{Q}$ with $\mathfrak{S}$ being finite, $r^{\leq n+i} * r^{>n+j}$ is a run in $\mathfrak{S}$ through $\boldsymbol{t}$, and for all $n' > n+j$ we have $n \sim_{\mathfrak{S}} n' - (j-i)$ whenever $n \sim_{\mathfrak{R}'} n'$. Thus we can construct a subquasimodel $\left\langle \boldsymbol{q}_1 * \boldsymbol{q}_2^{\leq 0} * \boldsymbol{q}_3, \mathfrak{R}_1 \right\rangle$ of $\mathfrak{Q}$ with $\mathfrak{R}_1$ being finite, and a run $r_1 \in \mathfrak{R}_1$ such that $r_1(n) = \boldsymbol{t}$, the sequence $\langle r_1(n), \dots, r_1(n+m_1)\rangle$ realizes $\overline{\psi_1 \mathcal{U} \psi_2}$ for some $m_1 \leq \flat(\varphi) \cdot \natural(\varphi)$ and, for all $n' > n + m_1$ we have $n \sim_{\mathfrak{R}_1} n' - (j-i)$ whenever $n \sim_{\mathfrak{R}'} n'$. In particular, $[n]_{\mathfrak{R}_1}$ is infinite.

After that we consider another formula of the form $\overline{\psi_1' \mathcal{U} \psi_2'} \in \boldsymbol{t}$ and assume that $\langle r_1(n), \dots, r_1(n+m')\rangle$ realizes it for some $m' > m_1$. Using Lemma 11.48 once again (and deleting repeating quasistates between $\boldsymbol{q}_3(m_1)$ and $\boldsymbol{q}_3(m')$) we select a subquasimodel $\left\langle \boldsymbol{q}_1 * \boldsymbol{q}_2^{\leq 0} * \boldsymbol{q}_3^{\leq m_1} * \boldsymbol{q}_4, \mathfrak{R}_2 \right\rangle$ of $\mathfrak{Q}$ with $\mathfrak{R}_2$ being finite, and a run r_2 in $\mathfrak{R}_2$ such that $r_2(n) = \boldsymbol{t}$, $\langle r_2(n), \dots, r_2(n+m_2)\rangle$ realizes both $\overline{\psi_1 \mathcal{U} \psi_2}$ and $\overline{\psi_1' \mathcal{U} \psi_2'}$ for some $m_2 \leq 2 \cdot \flat(\varphi) \cdot \natural(\varphi)$ and $[n]_{\mathfrak{R}_2}$ is infinite.

Having analyzed all distinct formulas of the form $\overline{\psi_1 \mathcal{U} \psi_2}$ in $\boldsymbol{t}$ we obtain a subquasimodel $\left\langle \boldsymbol{q}_1 * \boldsymbol{q}_2^{\leq 0} * \boldsymbol{q}' * \boldsymbol{q}^{>k}, \mathfrak{R}_{\boldsymbol{t}} \right\rangle$ of $\mathfrak{Q}$ with finite $\mathfrak{R}_{\boldsymbol{t}}$, and a run $r_{\boldsymbol{t}} \in \mathfrak{R}_{\boldsymbol{t}}$ such that $r_{\boldsymbol{t}}(n) = \boldsymbol{t}$ and $\langle r_{\boldsymbol{t}}(n), \dots, r_{\boldsymbol{t}}(n+m_{\boldsymbol{t}})\rangle$ realizes all such 'Until-formulas' for some $m_{\boldsymbol{t}} \leq |sub_x \varphi| \cdot \flat(\varphi) \cdot \natural(\varphi)$. The equivalence class $[n]_{\mathfrak{R}_{\boldsymbol{t}}}$ is still infinite.

Then we consider in the same manner another type $\boldsymbol{t}' \in T_n$. However, this time we can delete quasistates only after $\boldsymbol{q}'(m_{\boldsymbol{t}})$. And so forth. Observe that if we 'cut' the runs $r_c \in \mathfrak{R}'$ corresponding to $c \in con\,\varphi$ this way, then we obtain runs satisfying (iii). Finally, not more than $\natural(\varphi)$ quasistates may be needed to comply with (iv). So we end up with a subquasimodel $\langle \boldsymbol{q}_1 * \boldsymbol{q}_0 * \boldsymbol{q}_2^{>l}, \mathfrak{R}'' \rangle$ of $\mathfrak{Q}$ satisfying (i)–(iv). ❑

We can now complete the proof of the '⇒'-direction of Theorem 11.46 as follows. Suppose that φ is satisfiable in a first-order temporal model based on $\langle \mathbb{N}, < \rangle$ and having finite domains. Then by Lemma 11.41, there is a finitary quasimodel $\langle \boldsymbol{q}, \mathfrak{R} \rangle$ for φ with $\boldsymbol{q}(n) = \langle T_n, T_n^{con} \rangle$ for $n \in \mathbb{N}$. Clearly, we may assume that $\overline{\varphi} \in \boldsymbol{t}$ for some $\boldsymbol{t} \in T_0$. By applying Lemmas 11.49 and 11.50 we obtain a finitary quasimodel for φ of the form $\langle \boldsymbol{q}_1 * \boldsymbol{q}_0 * \boldsymbol{q}_2^{\geq l}, \mathfrak{R}'' \rangle$ as described in Lemma 11.50. It remains to observe that the numbers $l_1 = |\boldsymbol{q}_1|$, $l_2 = |\boldsymbol{q}_0|$ and the sequence $\boldsymbol{q}_1 * \boldsymbol{q}_0$ of finitely realizable state candidates for φ satisfy conditions (1)–(3), $(4)^f$ and (5) of Theorem 11.46. ❑

We are now in a position to show that over the flow of time $\langle \mathbb{N}, < \rangle$ the complexity of the satisfiability problem in models with finite domains coincides with that of arbitrary domains. First, we have the following analog of Theorem 11.30:

Theorem 11.51. *Let $\mathcal{QTL}'$ be a sublanguage of $\mathcal{QTL}_{\mathcal{U}\boxed{1}}$, and suppose that there is an algorithm which, given a state candidate $\mathfrak{C}$ for a $\mathcal{QTL}'$-sentence φ, can recognize whether $\mathfrak{C}$ is finitely realizable using space $\leq 2^{p(\ell(\varphi))}$ for some polynomial function p. Then the fragment $\mathsf{QLog}_{\mathcal{U}}^{fin}(\mathbb{N}) \cap \mathcal{QTL}'$ is decidable in* EXPSPACE.

Proof. A straightforward modification of the proof of Theorem 11.30 is left to the reader. (Use Theorem 11.46 instead of Theorem 11.26.) ❑

As far as the lower bound is concerned, we have the following analogue of Theorem 11.33 which can actually be established using the very same proof, since the formula $\chi_{n,T}$ constructed in it is satisfiable in a first-order model over $\langle \mathbb{N}, < \rangle$ iff it is satisfiable in such a model with finite domains.

Theorem 11.52. *The fragment $\mathsf{QLog}_{\mathcal{U}}^{fin}(\mathbb{N}) \cap \mathcal{QTL}^1_{\textcircled{1}}$ is* EXPSPACE-*hard.*

Now, since finite realizability of state candidates coincides with realizability in arbitrary models for all the fragments considered in this section (Börger *et al.* 1997, Hodkinson 2002a), we have the following 'finite domain' versions of Theorems 11.31 and 11.32:

Theorem 11.53. *The fragments $\mathsf{QLog}_{\mathcal{U}}^{fin}(\mathbb{N}) \cap \mathcal{QTL}^{mo}_{\boxed{1}}$, $\mathsf{QLog}_{\mathcal{U}}^{fin}(\mathbb{N}) \cap \mathcal{QTL}^1$ and $\mathsf{QLog}_{\mathcal{U}}^{fin}(\mathbb{N}) \cap \mathcal{QTL}^2_{\boxed{1}}$ are* EXPSPACE-*complete.*

Theorem 11.54. $\mathsf{QLog}_{\mathcal{U}}^{fin}(\mathbb{N}) \cap \mathcal{TPF}_{\boxed{1}}$ *is* 2EXPTIME-*complete.*

Finally, it should be clear that for fragments of $\mathcal{QTL}_{\mathcal{U}\boxed{0}}$ the complexity of the satisfiability problem does not depend on whether we take models with finite or arbitrary domains—as long as any satisfiable formula of the first-order part of the fragment is satisfiable in a finite model. So, we obtain the following analog of Theorem 11.36:

Theorem 11.55. (i) *The satisfiability problem for $\mathsf{QLog}_{\mathcal{U}}^{fin}(\mathbb{N}) \cap \mathcal{QTL}^1_{\boxed{0}}$ is* PSPACE-*complete.*

(ii) *The satisfiability problem for the fragments $\mathsf{QLog}_{\mathcal{U}}^{fin}(\mathbb{N}) \cap \mathcal{QTL}^2_{\boxed{0}}$ and $\mathsf{QLog}_{\mathcal{U}}^{fin}(\mathbb{N}) \cap \mathcal{QTL}^{mo}_{\boxed{0}}$ is* NEXPTIME-*complete.*

In this section we have discussed the complexity of some monodic fragments of first-order temporal logics over $\langle \mathbb{N}, < \rangle$.

Question 11.56. What is the computational complexity of the decision problem for the other decidable monodic fragments, mentioned in Section 11.2, in models with finite domains over $\langle \mathbb{N}, < \rangle$? What is the complexity of such fragments over other flows of time?

Note that as a consequence of the proof of Theorem 3.29 and Theorem 6.63 we obtain the following result of (Hodkinson *et al.* 2003):

Theorem 11.57. *Let $\mathcal{C}$ be any class of strict linear orders at least one of which contains an infinite ascending chain. Then $\mathsf{QLog}^{fin}_{\mathcal{U}}(\mathcal{C}) \cap \mathcal{QTL}^{mo}_{\boxed{1}}$ and $\mathsf{QLog}^{fin}_{\mathcal{U}}(\mathcal{C}) \cap \mathcal{QTL}^1$ are* EXPSPACE*-hard.*

11.6 Satisfiability in models over $\langle \mathbb{R}, < \rangle$ with finite domains

Now we prove all statements in Theorem 11.9 by presenting a third method, due to Hodkinson *et al.* (2000), of reducing decidability of monodic fragments to classical decision problems. We will consider only case (1), that is, the following statement:

(∗) Let $\mathcal{QTL}' \subseteq \mathcal{QTL}_{\boxed{1}}$ and suppose that there is an algorithm which is capable of deciding, for any $\mathcal{QTL}'$-sentence φ, whether an arbitrarily given state candidate for φ is finitely realizable. Then the fragment $\mathsf{QLog}^{fin}_{\mathcal{SU}}(\mathbb{R}) \cap \mathcal{QTL}'$ is decidable.

Before starting the rather involved proof, let us first see how the other cases in Theorem 11.9, that is, the satisfiability problems for models with finite domains over the other listed flows of time reduce to the case of $\langle \mathbb{R}, < \rangle$. Consider first $\langle \mathbb{N}, < \rangle$ as the flow of time. Given a $\mathcal{QTL}'$-sentence φ, recall the $\mathcal{QTL}$-sentence $\nu \wedge \varphi^p$ defined in the proof of Theorem 11.4 and observe that it is monodic. It is not hard to see that types (and thus state candidates) for φ and φ^p are 'isomorphic' (they differ only in the names of their surrogate variables). So by the condition on $\mathcal{QTL}'$, it is decidable whether an arbitrarily given state candidate for φ^p is finitely realizable. Further, since $\overline{\nu}$ is a Boolean combination of propositional variables, it is also decidable whether an arbitrarily given state candidate for $\nu \wedge \varphi^p$ is finitely realizable. Thus, the decidability of $\mathsf{QLog}^{fin}_{\mathcal{SU}}(\mathbb{N}) \cap \mathcal{QTL}'$ follows from Theorem 11.4. The cases of $\langle \mathbb{Z}, < \rangle$ and the class of all finite linear orders can be proved by similar reductions.

Let σ be a first-order sentence (in the language with equality and a binary predicate symbol $<$) defining a class $\mathcal{C}$ of strict linear orders, and let P be a fresh unary predicate symbol. Consider the first-order formula

$$\sigma'(t) = \exists x P(x) \wedge \sigma^P \wedge \big(P(t) \to P(t)\big).$$

By Theorem 2.5 (see also Gabbay *et al.* 1994, p.356), there is an $\mathcal{ML}_{\mathcal{SU}}$-formula α such that

$$\langle \mathbb{R}, < \rangle \models \forall t \; (\sigma' \leftrightarrow \alpha^\star) \tag{11.26}$$

and α contains a propositional variable p with $p^\star = P(t)$ (here $\cdot^\star$ denotes the standard translation of $\mathcal{ML}_{\mathcal{SU}}$-formulas). Given a $\mathcal{QTL}'$-sentence φ, we then have the following equivalences:

φ is satisfiable in a model with finite domains and a flow of time in $\mathcal{C}$

iff φ is satisfiable in a model with finite domains and a countable flow of time in $\mathcal{C}$ (by considering the translation $\cdot^\ddagger$ into the two-sorted first-order language $\mathcal{TS}$ (see Section 3.7) and applying the downward Löwenheim–Skolem–Tarski theorem)

iff φ is satisfiable in a model with finite domains over $\langle \mathbb{R}, < \rangle$ (since every countable strict linear order is a suborder of $\langle \mathbb{R}, < \rangle$)

iff $\varphi^p \wedge \alpha$ is satisfiable in a model with finite domains over $\langle \mathbb{R}, < \rangle$ (by (11.26)).

Now one can complete the proof as in the case of $\langle \mathbb{N}, < \rangle$ above. The case of $\langle \mathbb{Q}, < \rangle$ follows because φ has a first-order temporal model with dense flow of time without endpoints (a first-order definable property) iff it has a model over $\langle \mathbb{Q}, < \rangle$. The details are left to the interested reader.

The rest of this section is devoted to the proof of $(*)$. The method is model-theoretic, based on that of (Burgess and Gurevich 1985, Gurevich 1977, Läuchli and Leonard 1966); see also (Gabbay *et al.* 1994, Chapter 6.9). Very roughly, the idea of the proof is as follows. By Lemma 11.22, we need only decide whether there is a finitary quasimodel for a given sentence $\varphi \in \mathcal{QTL}'$ based on flow of time $\langle \mathbb{R}, < \rangle$. Such a quasimodel has a finite set of runs through $\langle \mathbb{R}, < \rangle$, a 'snapshot' of the runs at any moment of time giving a finitely realizable state candidate. Thus, each finitely realizable state candidate gives an instantaneous description of the runs in the quasimodel. We will show how to describe the runs over longer intervals of $\mathbb{R}$, ranging from one-point intervals as above, to the whole of $\mathbb{R}$. We may decide whether each possible description of the runs is satisfiable: for one-point intervals using the algorithm provided by the assumption in $(*)$, and for more complex ones by decomposing them into simpler parts for which we can already decide satisfiability (cf. Lemma 11.58 (2, 4) below). We will then show that a description of the runs on the whole of $\mathbb{R}$ can always be built up in finitely many steps from instantaneous descriptions (finitely realizable state candidates)—cf. Lemma 11.58 (3). Combining these ideas serves to prove (∗); formally, (∗) follows from Lemma 11.58.

3-theories. We begin our proof with the definitions needed to describe runs over intervals of $\mathbb{R}$. Let $\mathcal{L}_\varphi$ denote the sublanguage of the first-order language $\mathcal{QL}$ with the signature $\{<, R_\psi \mid \psi \in sub_x\, \varphi\}$, where the R_ψ are unary predicate symbols. An $\mathcal{L}_\varphi$-*order* is an $\mathcal{L}_\varphi$-structure

$$I = \langle W, <, R_\psi^I \rangle_{\psi \in sub_x \varphi},$$

where $\langle W, < \rangle$ is a linear order and the R_ψ^I are subsets of W.

A 3-*theory* (in $\mathcal{L}_\varphi$) is a set σ of $\mathcal{L}_\varphi$-sentences of the form

$$\textit{3-th}(I) = \{\theta \mid \theta \text{ an } \mathcal{L}_\varphi\text{-sentence of quantifier depth at most 3, } I \models \theta\},$$

for some $\mathcal{L}_\varphi$-order I.

Up to logical equivalence, there are finitely many 3-theories. Note that by definition, any 3-theory has a model. Let T_φ be the set of all types for φ; recall that T_φ is finite, with $|T_\varphi| \le \flat(\varphi)$. If $\langle W, < \rangle$ is a linear order and $r : W \to T_\varphi$, define the $\mathcal{L}_\varphi$-order

$$I_r = \langle W, <, \{w \in W \mid \overline{\psi} \in r(w)\} \rangle_{\psi \in sub_x \varphi}.$$

That is, $I_r \models R_\psi(w)$ iff $\overline{\psi} \in r(w)$, for $w \in W$ and $\psi \in sub_x\, \varphi$. We let *3-th*(r) denote the 3-theory *3-th*(I_r).

Now let σ be a 3-theory. We say that

- σ *has a left endpoint*, if $\sigma \models \exists x \forall y \neg(y < x)$,
- that σ *has a right endpoint*, if $\sigma \models \exists x \forall y \neg(x < y)$,
- and that σ *is degenerate*, if $\sigma \models \forall x \forall y \neg(x < y)$.

Let $\langle U, < \rangle$ be a linear order and

$$I_u = \langle W_u, <_u, R_\psi^u \rangle_{\psi \in sub_x \varphi} \qquad (u \in U)$$

be $\mathcal{L}_\varphi$-orders. We write

$$I = \sum_{u \in U} I_u = \Big\langle \sum_{u \in U} W_u, <^+, R_\psi^I \Big\rangle_{\psi \in sub_x \varphi}$$

for the $\mathcal{L}_\varphi$-order where $\sum_{u \in U} W_u = \bigcup_{u \in U} W_u \times \{u\}$, ordered lexicographically by $\langle w, u \rangle <^+ \langle w', u' \rangle$ iff either $u < u'$ or $u = u'$ and $w <_u w'$, and with

$$\langle w, u \rangle \in R_\psi^I \quad \text{iff} \quad w \in R_\psi^u,$$

for $\langle w, u\rangle \in W$ and $\psi \in sub_x \varphi$. We also write the underlying linear order $<^+$ of I as $\sum_{u\in U} <_u$. When $U = \{0,1\}$ with $0 < 1$, we write simply $I_0 + I_1 = \langle W_0 + W_1, <_0 + <_1, \dots\rangle$.

A well-known Feferman–Vaught argument (see, e.g., Theorem A.6.2 of (Hodges 1993)) shows that if I_u and J_u, for $u \in U$, are $\mathcal{L}_\varphi$-orders and $3\text{-}th(I_u) = 3\text{-}th(J_u)$ for all u, then $3\text{-}th(\sum_{u\in U} I_u) = 3\text{-}th(\sum_{u\in U} J_u)$. Hence, we may use the following notation. Let $\langle U, <\rangle$ be a linear order and for each $u \in U$ let σ_u be a 3-theory. We write $\sum_{u\in U} \sigma_u$ for the unique 3-theory σ such that $\sigma = 3\text{-}th(\sum_{u\in U} I_u)$ for any $\mathcal{L}_\varphi$-orders I_u with $I_u \models \sigma_u$ $(u \in U)$. As with $\mathcal{L}_\varphi$-orders, we write $\sigma_0 + \sigma_1$ when $U = \{0,1\}$ with $0 < 1$.

Characters. Given a state function $\boldsymbol{q}$ for φ over a linear order $\langle W, <\rangle$, a coherent and saturated run r through $\boldsymbol{q}$ is completely described by the $\mathcal{L}_\varphi$-order I_r. The 3-theory $3\text{-}th(r)$ does not completely determine r, but it does carry a great deal of information about r. For example, for an arbitrary run through $\boldsymbol{q}$, $3\text{-}th(r)$ determines whether r is coherent and saturated, and whether $\overline{\varphi} \in r(w)$ for some $w \in W$. Moreover, 3-theories are finite syntactic objects and can be used in algorithms. So we will use them to represent quasimodels.

We aim to decide satisfiability of φ by deciding whether a finitary quasimodel for φ exists. Such a quasimodel has a set of coherent and saturated runs, and it can be described by a set of 3-theories—simply the 3-theories of its runs. The quasimodel also contains distinguished runs associated with constants, so we will also distinguish certain of the descriptive 3-theories. This leads us to the following definition.

A *character* is a pair $\langle S, S^{con}\rangle$, where S is a set of 3-theories and S^{con} is a function from $con\,\varphi$ to S. Clearly, there are only finitely many characters. A character $\langle S, S^{con}\rangle$ is said to have a *left* (*right*) *endpoint* if every $\sigma \in S$ has a left (right) endpoint. A character $\langle S, S^{con}\rangle$ is said to be *degenerate* if

- each $\sigma \in S$ is degenerate,
- for each $\sigma \in S$, the set $\boldsymbol{t}_\sigma = \{\overline{\psi} \mid \psi \in sub_x \varphi,\ \ \sigma \models \exists x R_\psi(x)\}$ is a type for φ,
- $\langle \{\boldsymbol{t}_\sigma \mid \sigma \in S\}, \{\langle c, \boldsymbol{t}_{S^{con}(c)}\rangle \mid c \in con\,\varphi\}\rangle$ is a finitely realizable state candidate for φ.

Suppose that $\langle\langle \mathbb{R}, <\rangle, \boldsymbol{q}, \mathfrak{R}\rangle$ is a finitary quasimodel for φ. Then the values of the state function $\boldsymbol{q}$ are finitely realizable state candidates, and $\mathfrak{R}$ is a finite set of coherent and saturated runs through $\boldsymbol{q}$. We may 'restrict' such a quasimodel to any suborder $\langle W, <\rangle$ of $\langle \mathbb{R}, <\rangle$, by restricting $\boldsymbol{q}$ and the runs in $\mathfrak{R}$ to W. In general, such a restriction need not be a quasimodel, since its runs are not necessarily coherent and saturated (we will call it a 'pre-quasimodel'), but it still has a character associated with it in the same way as for a 'full' quasimodel, by taking the 3-theories of the restrictions of the runs to W. The smallest possibility is when W consists of a single point of $\mathbb{R}$—the restriction of the quasimodel to W is then essentially a finitely realizable state candidate, and the associated character is degenerate.

We aim to try to build a quasimodel for φ from smaller pre-quasimodels which are restrictions of it. These smaller pre-quasimodels are in turn built from even smaller ones, and so on, leading eventually to one-point restrictions. We will calculate the character of each successively larger pre-quasimodel from the characters of the next smaller ones, starting from degenerate characters describing the one-point restrictions, and stopping when the character tells us that we have a genuine quasimodel. The allowed operations in building a pre-quasimodel from smaller ones are, roughly speaking: concatenating two pre-quasimodels; iterating a fixed pre-quasimodel ω times, forwards or backwards; and merging finitely many pre-quasimodels together in a densely ordered 'shuffle'. We note that these operations can in general be effected in more than one way, so are nondeterministic, and that certain preconditions borrowed from (Burgess and Gurevich 1985) have to be met in order to ensure that the final quasimodel is based on $\langle \mathbb{R}, < \rangle$.

Since we are representing pre-quasimodels by their characters, we need to calculate the character of a pre-quasimodel resulting from smaller ones by these operations. The following definition will allow us to do this. The building operations cited above are represented by clauses **(b1)**–**(b4)** in the definition. We should note that there can be more than one pre-quasimodel with a given character, and given that the building operations are also nondeterministic, the character of the resulting pre-quasimodel is not uniquely determined by the characters of the smaller ones. Therefore, we define only a *relation* '$\equiv$' between the 'input' and 'output' characters, not a function.

We will need the notion of a *condensation of* $\langle \mathbb{R}, < \rangle$: namely, a linear order $\langle I, <_I \rangle$ where I is the set of equivalence classes of some equivalence relation on $\mathbb{R}$ whose equivalence classes are convex, the ordering $<_I$ on I being induced from the ordering $<$ on $\mathbb{R}$ in the obvious way. For more information on condensations see, e.g., (Rosenstein 1982).

Now let $\langle I, <_I \rangle$ be a linear order, and $\chi = \langle S, S^{con} \rangle$ and $\chi_i = \langle S_i, S_i^{con} \rangle$ $(i \in I)$ be characters. We write

$$\chi \approx \sum_{i \in I} \chi_i$$

if

(a1) for each $c \in con\,\varphi$, $S^{con}(c) = \sum_{i \in I} S_i^{con}(c)$,

(a2) for each $\sigma \in S$ there are $\sigma_i \in S_i$ $(i \in I)$ such that $\sigma = \sum_{i \in I} \sigma_i$, and

(a3) for all $i \in I$ and $\sigma_i \in S_i$, there are $\sigma_j \in S_j$ ($j \in I - \{i\}$) such that $\sum_{j \in I} \sigma_j \in S$.

We write

$$\chi \equiv \sum_{i \in I} \chi_i$$

if one of the following holds:

(b1) $\langle I, <_I \rangle$ is a 2-element order, say $I = \{0, 1\}$ with $0 <_I 1$, either χ_0 has a right endpoint or χ_1 a left endpoint (not both), and $\chi \approx \chi_0 + \chi_1$,

(b2) $\langle I, <_I \rangle = \langle \mathbb{N}, < \rangle$, $\chi_i = \chi_0$ for all $i \in \mathbb{N}$, χ_0 has either a left or a right endpoint (not both), condition **(a1)** above holds, and

$$S = \Big\{ \sum_{i \in I} \sigma_i \mid \sigma_i \in S_0,\ \sigma_i = \sigma_0 \text{ for all } i \in I \Big\}.$$

(b3) As for **(b2)** but with $\langle I, <_I \rangle = \langle \mathbb{N}, > \rangle$.

(b4) $\langle I, <_I \rangle$ is a dense condensation of $\langle \mathbb{R}, < \rangle$ without endpoints, conditions **(a1)** and **(a2)** above hold, and for all $i \in I$ (so that i is a convex subset of $\mathbb{R}$):

- i and χ_i have a left and a right endpoint,
- i is a singleton subset of $\mathbb{R}$ iff $\sigma \models \forall x \forall y \neg (x < y)$ for all $\sigma \in S_i$,
- for each $\sigma \in S_i$ there are $\sigma_j \in S_j$ ($j \in I$) with $\sum_{j \in I} \sigma_j \in S$, $\langle \chi_j, \sigma_j \rangle = \langle \chi_i, \sigma \rangle$ for some $j \in I$, and for each $j \in I$, the set $\{k \in I \mid \langle \chi_k, \sigma_k \rangle = \langle \chi_j, \sigma_j \rangle\}$ is dense in $\langle I, <_I \rangle$.

We will see later that the conditions for $\chi \equiv \sum_{i \in I} \chi_i$ are decidable.

Legal and perfect characters. We now define those characters that are reachable from degenerate ones by finitely many applications of **(b1)**–**(b4)** above. Let Λ denote the smallest set of characters containing all degenerate characters and such that if $\langle I, <_I \rangle$ is a linear order, $\chi_i \in \Lambda$ for $i \in I$, and $\chi \equiv \sum_{i \in I} \chi_i$, then $\chi \in \Lambda$. A character χ is said to be *legal* if $\chi \in \Lambda$.

We also define those characters that may be descriptions of quasimodels. A character $\chi = \langle S, S^{con} \rangle$ is said to be *perfect* if for every $\sigma \in S$,

- $\sigma \models \forall x \big(R_{\psi_1 \mathcal{U} \psi_2}(x) \leftrightarrow \exists y (x < y \wedge R_{\psi_2}(y) \wedge \forall z (x < z < y \to R_{\psi_1}(z))) \big)$ for every $\psi_1 \mathcal{U} \psi_2 \in sub_x \varphi$,
- $\sigma \models \forall x \big(R_{\psi_1 \mathcal{S} \psi_2}(x) \leftrightarrow \exists y (y < x \wedge R_{\psi_2}(y) \wedge \forall z (y < z < x \to R_{\psi_1}(z))) \big)$ for every $\psi_1 \mathcal{S} \psi_2 \in sub_x \varphi$,

- $\sigma \models \forall x \exists y \exists z \, (y < x < z)$, and
- $\sigma \models \exists x R_{\varphi}(x)$, for some $\sigma \in S$.

By an *interval of* $\langle \mathbb{R}, < \rangle$ we mean a linear order whose domain is a nonempty convex subset of $\mathbb{R}$, the ordering on it being induced from $\langle \mathbb{R}, < \rangle$. We will often abuse notation by identifying the subset of $\mathbb{R}$ with the linear order. Note that up to isomorphism there are just five intervals of $\mathbb{R}$, represented by $[0,1]$, $[0,1)$, $(0,1]$, $(0,1)$, and $\{0\}$. Here and below, we use standard notation for intervals: $[x,y) = \{z \in \mathbb{R} \mid x \leq z < y\}$ if $x \leq y$, etc.

Characters describe runs over some interval of a potential finitary quasimodel. We now make this precise. We call a triple $\mathfrak{Q} = \langle \mathfrak{F}, \boldsymbol{q}, \mathfrak{R} \rangle$ a *pre-quasimodel* if $\mathfrak{F}$ is a linear order isomorphic to an interval of $\langle \mathbb{R}, < \rangle$, $\langle \mathfrak{F}, \boldsymbol{q} \rangle$ is a basic structure for φ and $\mathfrak{R}$ is a set of (not necessarily coherent and saturated) runs through $\langle \mathfrak{F}, \boldsymbol{q} \rangle$ satisfying **(tqm2)** and **(tqm3)**. Clearly any quasimodel for φ based on $\langle \mathbb{R}, < \rangle$ is a pre-quasimodel.

A pre-quasimodel $\mathfrak{Q} = \langle \mathfrak{F}, \boldsymbol{q}, \mathfrak{R} \rangle$ is a *model of a character* $\chi = \langle S, S^{con} \rangle$ ($\mathfrak{Q} \models \chi$, in symbols) if

- $\{\textit{3-th}(r) \mid r \in \mathfrak{R}\} = S$, and
- $\textit{3-th}(r_c) = S^{con}(c)$ for each $c \in con\,\varphi$.

Our main lemma is the following:

Lemma 11.58.

(1) *If χ is a perfect character, $\mathfrak{Q} = \langle \mathfrak{F}, \boldsymbol{q}, \mathfrak{R} \rangle$ is a pre-quasimodel, and $\mathfrak{Q} \models \chi$, then $\mathfrak{Q}$ is a finitary quasimodel for φ based on $\mathfrak{F}$ and $\mathfrak{F}$ is isomorphic to $\langle \mathbb{R}, < \rangle$.*

(2) *If χ is a legal character, then there exists a pre-quasimodel $\mathfrak{Q}$ with $\mathfrak{Q} \models \chi$.*

(3) *If $\mathfrak{Q} = \langle \langle \mathbb{R}, < \rangle, \boldsymbol{q}, \mathfrak{R} \rangle$ is a finitary quasimodel for φ, then there is a perfect legal character χ with $\mathfrak{Q} \models \chi$.*

(4) *Given an algorithm which is capable of deciding, for any $\mathcal{QTL}'$-sentence φ, whether a given state candidate for φ is finitely realizable, it is decidable whether there exists a perfect legal character. The decision algorithm is uniform in φ.*

Having this lemma, we can prove the statement $(*)$ (and so Theorem 11.9) as follows. Lemma 11.41 and parts (1)–(3) of Lemma 11.58 show that φ is satisfiable in a first-order temporal model over flow of time $\langle \mathbb{R}, < \rangle$ having finite domains iff there exists a perfect legal character. Finally, by part (4) of Lemma 11.58, given $\mathcal{QTL}' \subseteq \mathcal{QTL}_{\boxdot}$ and an algorithm that decides for any sentence $\varphi \in \mathcal{QTL}'$ whether a given state candidate for φ is finitely realizable, it is decidable whether such a character exists.

Proof of Lemma 11.58 (1). This is straightforward. Let $\chi = \langle S, S^{con}\rangle$ be a perfect character, $\mathfrak{Q} = \langle \mathfrak{F}, \boldsymbol{q}, \mathfrak{R}\rangle$ a pre-quasimodel with $\mathfrak{F} = \langle W, <_W\rangle$, and let $\mathfrak{Q} \models \chi$. Then by the definitions, $\mathfrak{R}$ is finite, and for every $r \in \mathfrak{R}$ we have $3\text{-}th(r) \in S$, and so r is coherent and saturated. Let $\sigma \in S$ be such that $\sigma \models \exists x R_\varphi(x)$, and let $r \in \mathfrak{R}$ such that $3\text{-}th(r) = \sigma$. Then clearly, $\overline{\varphi} \in r(w)$ for some $w \in W$. So $\mathfrak{Q}$ is a finitary quasimodel for φ based on $\mathfrak{F}$. Since $\sigma \models \forall x \exists y \exists z (y < x < z)$, $\mathfrak{F}$ is isomorphic to an interval of $\langle \mathbb{R}, <\rangle$. And $\mathfrak{F}$ has no endpoints, so it must be isomorphic to $\langle \mathbb{R}, <\rangle$. ❑

Proof of Lemma 11.58 (2). By definition of Λ, it suffices to prove that

- if χ is a degenerate character, then there is a pre-quasimodel $\mathfrak{Q}$ with $\mathfrak{Q} \models \chi$, and
- if $\langle I, <_I\rangle$ is a linear order, χ_i $(i \in I)$ are characters having pre-quasimodels, and $\chi \equiv \sum_{i\in I} \chi_i$, then $\mathfrak{Q} \models \chi$ for some pre-quasimodel $\mathfrak{Q}$.

So first let $\chi = \langle S, S^{con}\rangle$ be a degenerate character. Then by definition, $\boldsymbol{t}_\sigma = \{\overline{\psi} \mid \psi \in sub_x\,\varphi,\ \sigma \models \exists x R_\psi(x)\}$ is a type for φ, for every $\sigma \in S$. Let $\mathfrak{F}$ be a one-point ordering with domain $\{w\}$, define $\boldsymbol{q}(w)$ to be the finitely realizable state candidate $\langle \{\boldsymbol{t}_\sigma \mid \sigma \in S\}, \{\langle c, \boldsymbol{t}_{S^{con}(c)}\rangle \mid c \in con\,\varphi\}\rangle$ for φ, and for each $\sigma \in S$ put $r_\sigma(w) = \boldsymbol{t}_\sigma$. Observe that $3\text{-}th(r_\sigma) = \sigma$. For, by definition of I_{r_σ}, for every $\psi \in sub_x\,\varphi$ we have $I_{r_\sigma} \models R_\psi(w)$ iff $\overline{\psi} \in r_\sigma(w)$ iff $\sigma \models \exists x R_\psi(x)$. As $\sigma \models \forall x \forall y \neg(x < y)$, we see that if $J \models \sigma$ then J must be isomorphic to I_{r_σ}. Since such a J exists, we have $I_{r_\sigma} \models \sigma$. Hence, $3\text{-}th(r_\sigma) = 3\text{-}th(I_{r_\sigma}) = \sigma$. As $\mathfrak{F}$ is isomorphic to a (one-point) interval of $\langle \mathbb{R}, <\rangle$, we clearly have that $\mathfrak{Q} = \langle \mathfrak{F}, \boldsymbol{q}, \{r_\sigma \mid \sigma \in S\}\rangle$ is a pre-quasimodel such that $\mathfrak{Q} \models \chi$.

For the inductive step, let $\langle I, <_I\rangle$ be a linear order, $\chi = \langle S, S^{con}\rangle$, $\chi_i = \langle S_i, S_i^{con}\rangle$, $\mathfrak{Q}_i = \langle \langle W_i, <_i\rangle, \boldsymbol{q}_i, \mathfrak{R}_i\rangle$ $(i \in I)$ characters and pre-quasimodels with $\mathfrak{Q}_i \models \chi_i$ (for all $i \in I$), and suppose that $\chi \equiv \sum_{i\in I} \chi_i$. We will define a pre-quasimodel $\mathfrak{Q} = \langle \langle W, <_W\rangle, \boldsymbol{q}, \mathfrak{R}\rangle$ and show that $\mathfrak{Q} \models \chi$.

Let $W = \sum_{i\in I} W_i$ and $<_W = \sum_{i\in I} <_i$. We show first that

$$\langle W, <_W\rangle \text{ is isomorphic to an interval of } \langle \mathbb{R}, <\rangle. \tag{11.27}$$

If I is the 2-element order $0 < 1$ on $\{0, 1\}$, then our assumptions show that either $\langle W_0, <_0\rangle$ has a right endpoint or $\langle W_1, <_1\rangle$ a left endpoint, and not both, so that (11.27) is clear. (For example, if $\langle W_0, <_0\rangle$ is isomorphic to $\langle [0,1], <\rangle$ and $\langle W_1, <_1\rangle$ isomorphic to $\langle (1,2), <\rangle$ then $\langle W, <_W\rangle$ is isomorphic to the interval $\langle [0,2), <\rangle$ of $\langle \mathbb{R}, <\rangle$.) If $\langle I, <_I\rangle = \langle \mathbb{N}, <\rangle$, then each $\langle W_i, <_i\rangle$ has a left (say) endpoint, so again, $\langle W, <_W\rangle$ is isomorphic to an interval of $\langle \mathbb{R}, <\rangle$; the case $\langle I, <_I\rangle = \langle \mathbb{N}, >\rangle$ is similar. Finally, suppose that $\langle I, <_I\rangle$ is a dense condensation of $\langle \mathbb{R}, <\rangle$ without endpoints whose elements have left and right endpoints. Then by definition of $\equiv$, $\langle W_i, <_i\rangle$ is isomorphic to the interval i

for each $i \in I$, so (11.27) follows. All cases **(b1)**–**(b4)** in the definition of $\equiv$ are now covered, so we are done.

Next, for any functions g_i defined on W_i ($i \in I$), we write $\sum_{i \in I} g_i$ for the function g on W defined by $g(w, i) = g_i(w)$.

Claim 11.59. *If $r_i : W_i \to T_\varphi$ ($i \in I$), then $3\text{-}th(\sum_{i \in I} r_i) = \sum_{i \in I} 3\text{-}th(r_i)$.*

Proof. Write r for $\sum_{i \in I} r_i$. By definition,

$$3\text{-}th(r) = 3\text{-}th(I_r) \quad \text{and} \quad 3\text{-}th(r_i) = 3\text{-}th(I_{r_i}),$$

for each $i \in I$. Clearly, $I_r = \sum_{i \in I} I_{r_i}$. So $\sum_{i \in I} 3\text{-}th(r_i)$ by definition equals $3\text{-}th(r)$. ❑

Define a state function $\boldsymbol{q} = \sum_{i \in I} \boldsymbol{q}_i$ on W, and write $\boldsymbol{q}(w) = \langle T_w, T_w^{con} \rangle$ for $w \in W$. The definition of $\mathfrak{R}$ will divide into cases according to the parts of the definition of $\equiv$, but in all cases we will arrange that each $r \in \mathfrak{R}$ has the form $\sum_{i \in I} r_i$ for some $r_i \in \mathfrak{R}_i$ ($i \in I$), and that $r_c^{\boldsymbol{q}} = \sum_{i \in I} r_c^{\boldsymbol{q}_i}$ is in $\mathfrak{R}$ for each $c \in con\,\varphi$. Given this much, we can already check that

$$r(w) \in T_w, \qquad \text{for all } r \in \mathfrak{R},\ w \in W, \tag{11.28}$$

$$3\text{-}th(r_{\boldsymbol{q}}^f) = S^{con}(c). \tag{11.29}$$

For (11.28), let $\langle w, i \rangle \in W$ and $r = \sum_{i \in I} r_i \in \mathfrak{R}$. Then

$$\boldsymbol{q}_i(w) = \boldsymbol{q}(w, i) = \left\langle T_{\langle w,i \rangle}, T_{\langle w,i \rangle}^{con} \right\rangle.$$

So as $\mathfrak{Q}_i$ is a pre-quasimodel, $r(w, i) = (\sum_{i \in I} r_i)(w, i) = r_i(w) \in T_{\langle w,i \rangle}$, as required. For (11.29), as $\mathfrak{Q}_i \models \chi_i$ for each i, we have $3\text{-}th(r_c^{\boldsymbol{q}_i}) = S_i^{con}(c)$. By the definitions and Claim 11.59, we obtain

$$r_c^{\boldsymbol{q}} = \sum_{i \in I} r_c^{\boldsymbol{q}_i}, \text{ and} \tag{11.30}$$

$$\begin{aligned} 3\text{-}th(r_c^{\boldsymbol{q}}) &= 3\text{-}th\Big(\sum_{i \in I} r_c^{\boldsymbol{q}_i}\Big) = \sum_{i \in I} 3\text{-}th(r_c^{\boldsymbol{q}_i}) = \\ &= \sum_{i \in I} S_i^{con}(c) = S^{con}(c) \in S. \end{aligned} \tag{11.31}$$

Now we go through the cases **(b1)**–**(b4)** above in defining $\mathfrak{R}$ and checking that $\mathfrak{Q} = \langle \langle W, <_W \rangle, \boldsymbol{q}, \mathfrak{R} \rangle$ is a pre-quasimodel and $\mathfrak{Q} \models \chi$.

Case **(b1)**: $\langle I, <_I \rangle$ is a 2-element order on $\{0, 1\}$ with $0 <_I 1$. We define

$$\mathfrak{R} = \{r_0 + r_1 \mid r_0 \in \mathfrak{R}_0\ r_1 \in \mathfrak{R}_1,\ 3\text{-}th(r_0 + r_1) \in S\}.$$

This $\mathfrak{R}$ is clearly finite, since $\mathfrak{R}_0$ and $\mathfrak{R}_1$ are finite. By (11.30) and (11.31), we have $r_c^{\boldsymbol{q}} \in \mathfrak{R}$.

Let $w \in W$ and $\boldsymbol{t} \in T_w$; we seek $r \in \mathfrak{R}$ with $r(w) = \boldsymbol{t}$. Let $w = \langle w', i\rangle$ for $w' \in W_i$, $i \in I$. As $\mathfrak{Q}_i$ is a pre-quasimodel, there is $r_i \in \mathfrak{R}_i$ with $r_i(w') = \boldsymbol{t}$. As $\mathfrak{Q}_i \models \chi_i$, we have $\mathit{3\text{-}th}(r_i) = \sigma_i \in S_i$. As $\chi \approx \chi_0 + \chi_1$, there is $\sigma_{1-i} \in S_{1-i}$ with $\sigma_0 + \sigma_1 \in S$, and as $\mathfrak{Q}_{1-i} \models \chi_{1-i}$, there is $r_{1-i} \in \mathfrak{R}_{1-i}$ with $\mathit{3\text{-}th}(r_{1-i}) = \sigma_{1-i}$. Then by Claim 11.59, $r = r_0 + r_1$ satisfies

$$\mathit{3\text{-}th}(r) = \mathit{3\text{-}th}(r_0 + r_1) = \mathit{3\text{-}th}(r_0) + \mathit{3\text{-}th}(r_1) = \sigma_0 + \sigma_1 \in S,$$

so clearly $r \in \mathfrak{R}$ and $r(w) = r_i(w') = \boldsymbol{t}$.

To prove $S = \{\mathit{3\text{-}th}(r) \mid r \in \mathfrak{R}\}$, we only need check that if $\sigma \in S$ then there is $r \in \mathfrak{R}$ with $\sigma = \mathit{3\text{-}th}(r)$. By **(a2)** above, there are $\sigma_i \in S_i$ $(i = 0, 1)$ with $\sigma = \sigma_0 + \sigma_1$, and since $\mathfrak{Q}_i \models \chi_i$, there are $r_i \in \mathfrak{R}_i$ with $\mathit{3\text{-}th}(r_i) = \sigma_i$, for each i. We may take $r = r_0 + r_1$.

Case **(b2)**: $\langle I, <_I\rangle = \langle \mathbb{N}, <\rangle$. We may assume that $\mathfrak{Q}_i = \mathfrak{Q}_0$ for all $i \in I$, since $\chi_i = \chi_0$. We define

$$\mathfrak{R} = \{r \mid \mathit{3\text{-}th}(r) \in S,\ r = \textstyle\sum_{i\in I} r_i \text{ for some } r_i \in \mathfrak{R}_i\ (i \in I), \\ \text{and } r_i = r_0 \text{ for all } i \in I\}.$$

Clearly $|\mathfrak{R}| \leq |\mathfrak{R}_0|$, so $\mathfrak{R}$ is finite. If $c \in con\,\varphi$ then $r_c^{\boldsymbol{q}_i} = r_c^{\boldsymbol{q}_0}$ for all $i \in I$, since $\mathfrak{Q}_i = \mathfrak{Q}_0$. It now follows from (11.30) and (11.31) that $r_c^{\boldsymbol{q}} \in \mathfrak{R}$.

We let $w \in W$ and $\boldsymbol{t} \in T_w$ and find $r \in \mathfrak{R}$ with $r(w) = \boldsymbol{t}$. Suppose that $w = \langle w', n\rangle$, for $w' \in W_n$, $n \in \mathbb{N}$. As $\mathfrak{Q}_n$ is a pre-quasimodel, we may pick $r_n \in \mathfrak{R}_n$ with $r_n(w') = \boldsymbol{t}$. As $\mathfrak{Q}_n \models \chi_n$, we have $\mathit{3\text{-}th}(r_n) \in S_n$. Define $r_i = r_n$ for all $i \in I$. Then by definition of $\equiv$, we have

$$\mathit{3\text{-}th}(\sum_{i\in I} r_i) = \sum_{i\in I} \mathit{3\text{-}th}(r_i) \in S,$$

so $r = \sum_{i\in I} r_i \in \mathfrak{R}$ and $r(w) = r_n(w') = \boldsymbol{t}$.

By definition of $\equiv$, each $\sigma \in S$ has the form $\sum_{i\in I} \sigma_i$ for $\sigma_i \in S_i$ $(i \in I)$ with all σ_i equal to σ_0. By $\mathfrak{Q}_0 \models \chi_0$, there is $r_0 \in \mathfrak{R}_0$ with $\mathit{3\text{-}th}(r_0) = \sigma_0$. Let $r_i = r_0$, for each i, and $r = \sum_{i\in I} r_i$. Then $\mathit{3\text{-}th}(r) = \sigma$, so $r \in \mathfrak{R}$. Hence, $S \subseteq \{\mathit{3\text{-}th}(r) \mid r \in \mathfrak{R}\}$. The converse inclusion is clear by definition of $\mathfrak{R}$.

Case **(b3)**: $\langle I, <_I\rangle = \langle \mathbb{N}, >\rangle$. This is similar to the preceding case.

Case **(b4)**: $\langle I, <_I\rangle$ is a dense condensation of $\langle \mathbb{R}, <\rangle$. This is the most involved case. Again, we may as well suppose that if $\chi_i = \chi_j$ then $\mathfrak{Q}_i = \mathfrak{Q}_j$, for $i, j \in I$. The definition of $\mathfrak{R}$ has two parts. First, observe that by condition **(a2)**, for each $\sigma \in S$ there are $\sigma_i \in S_i$ $(i \in I)$ such that $\sigma = \sum_{i\in I} \sigma_i$. For each i, pick $r_i \in \mathfrak{R}_i$ with $\mathit{3\text{-}th}(r_i) = \sigma_i$, and let $r_\sigma = \sum_{i\in I} r_i$. Next, noting that it follows from the definition of $\equiv$ that for each character χ, the set

$$I_\chi = \{i \in I \mid \chi_i = \chi\}$$

is either empty or dense in $\langle I, <_I \rangle$, choose an equivalence relation $\sim$ on I with the following properties:

$$\forall i, j \in I \ (i \sim j \implies \chi_i = \chi_j), \text{ and} \tag{11.32}$$
$$\forall i \in I \ (I_{\chi_i} \text{ is partitioned by } \sim \text{ into } |S_i| \text{ equivalence classes, each dense in } I). \tag{11.33}$$

If $r_i \in \mathfrak{R}_i$ for $i \in I$, the sequence $\langle r_i \mid i \in I \rangle$ is said to be *simple* if $i \sim j$ implies $r_i = r_j$, for all $i, j \in I$. Note that there are only finitely many simple sequences. We let

$$\mathfrak{R} = \{r_\sigma \mid \sigma \in S\} \cup \Big\{ \textstyle\sum_{i \in I} r_i \mid \langle r_i \mid i \in I \rangle \text{ a simple sequence, } \textit{3-th}\Big(\sum_{i \in I} r_i\Big) \in S \Big\}.$$

Observe that if $c \in con\ \varphi$ then by (11.32) and (11.33), $\langle r_c^{q_i} \mid i \in I \rangle$ is simple, so by (11.30) and (11.31), $r_c^{q} \in \mathfrak{R}$.

Since by Claim 11.59 $\textit{3-th}(r_\sigma) = \sigma \in S$, we have $S = \{\textit{3-th}(r) \mid r \in \mathfrak{R}\}$.

Let $\langle w, j \rangle \in W$, and $\boldsymbol{t} \in T_w$. We require $r \in \mathfrak{R}$ with $r(w, j) = \boldsymbol{t}$. As $\mathfrak{Q}_j$ is a pre-quasimodel, we may pick $r_j \in \mathfrak{R}_j$ with $r_j(w) = \boldsymbol{t}$. By **(b4)**, there are $\sigma_i \in S_i$ for $i \in I$ such that $\sum_{i \in I} \sigma_i \in S$, $\langle \chi_i, \sigma_i \rangle = \langle \chi_j, \textit{3-th}(r_j) \rangle$ for some $i \in I$, and $\{k \in I \mid \langle \chi_k, \sigma_k \rangle = \langle \chi_i, \sigma_i \rangle\}$ is dense in $\langle I, <_I \rangle$ for each $i \in I$. We may therefore choose a new equivalence relation $\sim'$ on I satisfying the conditions (11.32) and (11.33) such that if $i \sim' i'$ then $\sigma_i = \sigma_{i'}$. So, writing $i/{\sim'}$ for the $\sim'$-class of i (and similarly for $\sim$), we may define $\sigma_{i/\sim'}$ to be σ_i, for $i \in I$. Let $I_\chi/{\sim}$ denote the set of $\sim$-classes contained in I_χ, and define $I_\chi/{\sim'}$ similarly. By (11.32) and (11.33), $|I_\chi/{\sim}| = |I_\chi/{\sim'}|$ for every χ, and we know that $\textit{3-th}(r_j) = \sigma_e$ for some $e \in I_{\chi_j}/{\sim'}$. Since $j \in I_{\chi_j}$, we may pick a bijection $\theta : I/{\sim} \to I/{\sim'}$ such that

- $\theta(I_\chi/{\sim}) = I_\chi/{\sim'}$, for all characters χ,
- $\theta(j/{\sim}) = e$, so that $\sigma_{\theta(j/\sim)} = \textit{3-th}(r_j)$.

Now pick $r_i \in \mathfrak{R}_i$ for each $i \in I - \{j\}$ in such a way that for all $i \in I$, $\textit{3-th}(r_i) = \sigma_{\theta(i/\sim)} \in S_i$ and for all $i, k \in I$, $i \sim k$ imply $r_i = r_k$. Thus, the sequence $\langle r_i \mid i \in I \rangle$ is simple. For every $i \in I$, the set

$$\{k \in I \mid \langle \chi_k, \textit{3-th}(r_k) \rangle = \langle \chi_i, \textit{3-th}(r_i) \rangle\}$$

contains $i/{\sim}$, so by (11.32) and (11.33) it is dense in I. We saw that an analogous property holds for $(\sigma_i)_{i \in I}$. A Feferman–Vaught argument (cf. Theorem A.6.2 of (Hodges 1993)) now shows that $\sum_{i \in I} \textit{3-th}(r_i) = \sum_{i \in I} \sigma_i \in S$. Hence, $r = \sum_{i \in I} r_i \in \mathfrak{R}$, and $r(w, j) = r_j(w) = \boldsymbol{t}$. ❑

Remark 11.60. The last paragraph of the above argument seems to fail in the arbitrary-domain case— then there is no obvious analog for the last, density condition of **(b4)**. This does not necessarily mean that the finite-domain case is 'easier', as opposed to 'different.' We conjecture that the argument of the first half of (Burgess and Gurevich 1985) may apply to arbitrary domains.

Proof of Lemma 11.58 (3). The argument is very similar to one in (Burgess and Gurevich 1985). Let $\mathfrak{Q} = \langle \langle \mathbb{R}, < \rangle, \boldsymbol{q}, \mathfrak{R} \rangle$ be a finitary quasimodel for φ. For any interval $\langle E, < \rangle$ of $\langle \mathbb{R}, < \rangle$, we put

$$\mathfrak{Q}{\restriction}E = \langle \langle E, < \rangle, \boldsymbol{q}{\restriction}E, \{r{\restriction}E \mid r \in \mathfrak{R}\} \rangle .$$

Note that $\mathfrak{Q}{\restriction}E$ is a pre-quasimodel. We write χ_E for the character

$$\chi_E = \langle \{\mathit{3\text{-}th}(r{\restriction}E) \mid r \in \mathfrak{R}\}, \ \{\langle c, \mathit{3\text{-}th}(r_c^{\boldsymbol{q}}{\restriction}E)\rangle \mid c \in \mathit{con}\,\varphi\} \rangle .$$

It is clear that $\mathfrak{Q}{\restriction}E \models \chi_E$ for all E, and that $\chi_{\mathbb{R}}$ is perfect. We are going to show that $\chi_{\mathbb{R}}$ is legal.

Claim 11.61. *Let $\langle I, <_I \rangle$ be a linear order and let $\langle E_i, < \rangle$ be an interval of $\langle \mathbb{R}, < \rangle$ for each $i \in I$ such that for $E = \bigcup_{i \in I} E_i$, $\langle E, < \rangle$ is also an interval of $\langle \mathbb{R}, < \rangle$, and $x < y$ whenever $i < j$ in I, $x \in E_i$, and $y \in E_j$. Then*

(i) *$\mathit{3\text{-}th}(r{\restriction}E) = \sum_{i \in I} \mathit{3\text{-}th}(r{\restriction}E_i)$ for each $r \in \mathfrak{R}$, and*

(ii) *$\chi_E \approx \sum_{i \in I} \chi_{E_i}$.*

Proof. Let $r \in \mathfrak{R}$. Then by definition,

$$\mathit{3\text{-}th}(r{\restriction}E) = \mathit{3\text{-}th}(I_{r{\restriction}E}) \quad \text{and} \quad \mathit{3\text{-}th}(r{\restriction}E_i) = \mathit{3\text{-}th}(I_{r{\restriction}E_i}),$$

for each i. Clearly, $I_{r{\restriction}E} = \sum_{i \in I} I_{r{\restriction}E_i}$. So $\sum_{i \in I} \mathit{3\text{-}th}(r{\restriction}E_i)$ by definition equals $\mathit{3\text{-}th}(r{\restriction}E)$.

We now check that $\chi_E \approx \sum_{i \in I} \chi_{E_i}$. Let $\chi_E = \langle S, S^{con} \rangle$, and $\chi_{E_i} = \langle S_i, S_i^{con} \rangle$ for $i \in I$. If $c \in \mathit{con}\,\varphi$ then by definition, $S^{con}(c) = \mathit{3\text{-}th}(r_c^{\boldsymbol{q}}{\restriction}E)$ and $S_i^{con}(c) = 3th(r_c^{\boldsymbol{q}}{\restriction}E_i)$ for $i \in I$. Of course, $r_c^{\boldsymbol{q}} \in \mathfrak{R}$. By (i) we conclude that $S^{con}(c) = \sum_{i \in I} S_i^{con}(c)$.

Conditions **(a2)** and **(a3)** follow easily from the fact that

$$S = \{\mathit{3\text{-}th}(r{\restriction}E) \mid r \in \mathfrak{R}\} = \Big\{ \sum_{i \in I} \mathit{3\text{-}th}(r{\restriction}E_i) \mid r \in \mathfrak{R} \Big\}, \tag{11.34}$$

which completes the proof of the claim. ❑

We say that an interval $\langle E, < \rangle$ of $\langle \mathbb{R}, < \rangle$ is *good* if the character χ_E is legal.

Claim 11.62. *Any one-point interval of $\langle \mathbb{R}, < \rangle$ is good.*

Proof. Let $E = \{e\}$. We claim that $\chi_E = \langle S, S^{con} \rangle$, say, is degenerate. Each $\sigma \in S$ has the form *3-th*$(r \upharpoonright E)$ for some $r \in \mathfrak{R}$. Then $I_{r \upharpoonright E} \models \forall x \forall y \neg (x < y)$, so $\sigma =$ *3-th*$(I_{r \upharpoonright E})$ is degenerate. Further,

$$\boldsymbol{t}_\sigma = \{\overline{\psi} \mid \psi \in sub_x \varphi, \sigma \models \exists x R_\psi(x)\} = \{\overline{\psi} \mid \psi \in sub_x \varphi, I_{r \upharpoonright E} \models R_\psi(e)\} = r(e)$$

is a type for φ. As $\mathfrak{Q}$ is a finitary quasimodel for φ,

$$\langle \{\boldsymbol{t}_\sigma \mid \sigma \in S\}, \{\langle c, \boldsymbol{t}_{S^{con}(c)} \rangle \mid c \in con\, \varphi\} \rangle = \langle \{r(e) \mid r \in \mathfrak{R}\}, \{\langle c, r_c^{\boldsymbol{q}}(e) \rangle \mid c \in con\, \varphi\} \rangle = \boldsymbol{q}(e)$$

is a finitely realizable state candidate for φ. ❑

Claim 11.63. *Assume the conditions of Claim 11.61, that $\langle I, <_I \rangle$ is a 2-element linear order with $I = \{0, 1\}$ and $0 <_I 1$, and that $\langle E_0, < \rangle$ and $\langle E_1, < \rangle$ are good. Then $\langle E, < \rangle$ is good too.*

Proof. It suffices to prove that $\chi_E \equiv \chi_{E_0} + \chi_{E_1}$. As $\langle E, < \rangle$ is an interval of $\langle \mathbb{R}, < \rangle$, either $\langle E_0, < \rangle$ has a right endpoint or $\langle E_1, < \rangle$ a left endpoint. Assume the former; the other case is similar. If $r \in \mathfrak{R}$ then, by definition, we have *3-th*$(r \upharpoonright E_0) =$ *3-th*$(I_{r \upharpoonright E_0})$. So as $I_{r \upharpoonright E_0} \models \exists x \forall y \neg (x < y)$, we also have *3-th*$(r \upharpoonright E_0) \models \exists x \forall y \neg (x < y)$. Hence, χ_{E_0} has a left endpoint.

By Claim 11.61(ii), we have $\chi_E \approx \chi_{E_0} + \chi_{E_1}$, and so we can conclude that $\chi_E \equiv \chi_{E_0} + \chi_{E_1}$. ❑

Claim 11.64. *Assume the conditions of Claim 11.61, that*

$$\langle I, <_I \rangle \in \{\langle \mathbb{N}, < \rangle, \langle \mathbb{N}, > \rangle, \langle \mathbb{Z}, < \rangle\},$$

and that every $\langle E_i, < \rangle$ $(i \in I)$ is good. Then $\langle E, < \rangle$ is good.

Proof. We only consider the case $\langle I, <_I \rangle = \langle \mathbb{N}, < \rangle$; the case $\langle \mathbb{N}, > \rangle$ is similar, and $\langle \mathbb{Z}, < \rangle$ is handled using $\langle \mathbb{N}, > \rangle$, $\langle \mathbb{N}, < \rangle$, and Claim 11.63. For $i < j$ in $\mathbb{N}$, let $E_{ij} = \bigcup_{i \le k < j} E_k$. By Claim 11.63 and induction on $j - i$, E_{ij} is good. There are only finitely many characters, so by Ramsey's theorem (Ramsey 1930), there is an infinite $X \subseteq \mathbb{N}$ such that $\chi_{E_{ij}}$ is constant for all $i < j$ in X. Let $x \in X$ be minimal. As $\langle E_{0,x}, < \rangle$ is good, by Claim 11.63 it suffices to prove that $\left\langle \bigcup_{i \ge x} E_i, < \right\rangle$ is good. Therefore, by renaming, we may assume that $\chi_{E_{ij}}$ is constant for all $i < j$ in $\mathbb{N}$. As $\mathfrak{R}$ is finite, we may further assume (by Ramsey's theorem) that for each $r \in \mathfrak{R}$, *3-th*$(r \upharpoonright E_{ij})$ is the same for all $i < j$ in $\mathbb{N}$.

We will show that $\chi_E \equiv \sum_{i \in I} \chi_{E_i}$. We know that $\chi_{E_i} = \chi_{E_0}$ for all $i \in I$. Since E_0, E_1 are disjoint convex subsets of $\mathbb{R}$ whose union is convex, either

$\langle E_0, <\rangle$ has a right endpoint or $\langle E_1, <\rangle$ a left endpoint—and not both. It follows as in Claim 11.61 that χ_{E_0} has either a left or right endpoint. Let $\chi_E = \langle S, S^{con}\rangle$ and $\chi_{E_i} = \langle S_i, S_i^{con}\rangle$ for $i \in I$, as usual. Then by Claim 11.61,

$$S^{con}(c) = \textit{3-th}(r_c^q \restriction E) = \sum_{i\in I} \textit{3-th}(r_c^q \restriction E_i) = \sum_{i\in I} S_i^{con}(c),$$

for each $c \in con\,\varphi$. We also have

$$S = \Big\{\sum_{i\in I} \sigma_i \mid \sigma_i \in S_0,\ \sigma_i = \sigma_0 \text{ for all } i \in I\Big\}$$

because of (11.34). And, by the above, $r\restriction E_i = r\restriction E_0$ for each $r \in \mathfrak{R}$, $i \in I$. Now **(b2)** gives $\chi_E \equiv \sum_{i\in I} \chi_{E_i}$. Since the χ_{E_i} are assumed legal, so is χ_E, and we conclude that E is good. ❑

We define a binary relation $\sim$ on $\mathbb{R}$ by $x \sim y$ iff $x = y$, or $x < y$ and every convex subset contained in $[x, y]$ is good, or $y < x$ and every convex subset contained in $[y, x]$ is good.

Claim 11.65. *The relation $\sim$ is an equivalence relation on $\mathbb{R}$, and any $\sim$-class is itself an interval of $\mathbb{R}$.*

Proof. Only transitivity needs a proof. Assume that $x \sim y \sim z$ in $\mathbb{R}$; we check that $x \sim z$. There are various cases, depending on the order-type of x, y, z. If $x < z < y$, it is clear. Assume that $x < y < z$, let E be a convex subset of $[x, z]$, $E_0 = E \cap [x, y)$, and $E_1 = E \cap [y, z]$. If either E_0 or E_1 is empty, then certainly $\langle E, <\rangle$ is good. Otherwise, we are in the situation of Claim 11.63, so again $\langle E, <\rangle$ is good. The other cases are similar. Hence, $x \sim z$, as required.

It is clear by definition that any $\sim$-class is convex. ❑

Claim 11.66. *Any subinterval $\langle E, <\rangle$ of any $\sim$-class is good.*

Proof. There are four cases, depending on the endpoints of E. If $E = [x, y]$ for some $x < y$ in $\mathbb{R}$, then $x \sim y$ and the result is trivial. Assume that $\langle E, <\rangle$ has a left-hand endpoint x_0 but no right-hand endpoint. Choose an increasing sequence $x_0 < x_1 < \cdots$ in E, of order type $\langle \mathbb{N}, <\rangle$ and unbounded in E, and let $E_i = [x_i, x_{i+1})$. Since $x_i \sim x_{i+1}$, $\langle E_i, <\rangle$ is good. Now we are in the situation of Claim 11.64, and we conclude that $\langle E, <\rangle$ is good. The other two cases, when $\langle E, <\rangle$ has no left-hand endpoint, can be covered using the cases $\langle \mathbb{N}, >\rangle$ and $\langle \mathbb{Z}, <\rangle$ of Claim 11.64. ❑

Claim 11.67. *Each $\sim$-class is a closed interval of $\mathbb{R}$.*

Proof. Let E be a $\sim$-class, and suppose that E has a least upper bound $b \in \mathbb{R}$. We show that $b \in E$. Take $e \in E$, and any interval $\langle D, < \rangle$ of $\langle \mathbb{R}, < \rangle$ with $D \subseteq [e, b]$. Claim 11.66 shows that $\langle D \cap E, < \rangle$ is good. If $D \subseteq E$, we are done. Otherwise, $D = (D \cap E) \cup \{b\}$, and Claims 11.62, 11.63, and 11.66 show that $\langle D, < \rangle$ is good. So $b \sim e$ and $b \in E$. Similarly, E contains any greatest lower bound for it. So it is closed. ❑

We aim to show that $\mathbb{R}$ is a single $\sim$-class. To this end, assume not: so the condensation $\langle C, <_C \rangle$ where $C = \mathbb{R}/\sim$ has at least two elements. Because $\langle \mathbb{R}, < \rangle$ is dense, Claim 11.67 now shows that $\langle C, <_C \rangle$ is a dense ordering. Let $N = |\mathfrak{R}|$, enumerate $\mathfrak{R}$ as $\langle r^n \mid n < N \rangle$, and choose an open interval $\langle I, <_I \rangle$ of $\langle C, <_C \rangle$ such that the finite set

$$\{\langle \chi_E, \textit{3-th}(r^0 \restriction E), \ldots, \textit{3-th}(r^{N-1} \restriction E) \rangle \mid E \in I\}$$

has least possible cardinality. It follows that for each open interval $J \subseteq I$ and each sequence $\xi = \langle \chi, \sigma_0, \ldots, \sigma_{N-1} \rangle$ of a character and N 3-theories, the set

$$\{E \in J \mid \langle \chi_E, \textit{3-th}(r^0 \restriction E), \ldots, \textit{3-th}(r^{N-1} \restriction E) \rangle = \xi\}$$

is empty or dense in $\langle J, <_I \rangle$.

It can now be seen that $\chi_{\bigcup J} \equiv \sum_{E \in J} \chi_E$ by dint of **(b4)**. Certainly, $\langle J, <_I \rangle$ is isomorphic to a dense condensation of $\langle \mathbb{R}, < \rangle$ without endpoints. By Claim 11.61 (2), conditions **(a1)** and **(a2)** hold. By Claim 11.67, each $E \in J$ has a right and a left endpoint, and since if $r \in \mathfrak{R}$ and $E \in J$ then $I_{r \restriction E} \models \textit{3-th}(r \restriction E)$ and the underlying order of $I_{r \restriction E}$ is $\langle E, < \rangle$, χ_E has left and right endpoints too. Similarly, $|E| = 1$ iff $\textit{3-th}(r \restriction E) \models \forall x \forall y \neg(x < y)$ for all $r \in \mathfrak{R}$. The last part of **(b4)** holds because for any $r \in \mathfrak{R}$ and $E \in J$, the set

$$\{E' \in J \mid \langle \chi_{E'}, \textit{3-th}(r \restriction E') \rangle = \langle \chi_E, \textit{3-th}(r \restriction E) \rangle\}$$

is dense in $\langle J, <_I \rangle$.

So $\bigcup J$ is good. By Claim 11.66, each $E \in J$ is good, and Claim 11.63 now shows that if $\langle J, <_I \rangle$ is any subinterval of $\langle I, <_I \rangle$ then $\bigcup J$ is good.

Take $x < y$ in $\bigcup I$ with $x \not\sim y$. So there is an interval $X \subseteq [x, y]$ that is not good. Let $\overline{X} = \{E \in I \mid E \subseteq X\}$. Then $\langle \overline{X}, <_I \rangle$ is a subinterval of $\langle I, <_I \rangle$, so $\bigcup \overline{X}$ is good. Let $X_< = \{z \in X \mid z < v \text{ for all } v \in \bigcup \overline{X}\}$, and define $X_>$ similarly. By Claim 11.66, $X_<$ and $X_>$ are good. We have $X = X_< + \bigcup \overline{X} + X_>$, so by Claim 11.63, X is itself good, a contradiction.

Hence indeed, $\mathbb{R}$ is a single $\sim$-class, so is good—$\chi_{\mathbb{R}}$ is legal. This completes the proof of Lemma 11.58 (3). ❑

Proof of Lemma 11.58 (4). Assume that we have an oracle telling whether a given state candidate for φ is finitely realizable. We show how to use it to decide whether there exists a legal perfect character. The decision procedure is uniform in φ. Our method is to reduce the problem to the satisfiability of certain existential monadic second-order sentences in $\langle \mathbb{R}, < \rangle$. By Theorem 2.9 (d) of (Burgess and Gurevich 1985), such problems are decidable. This reduction is quite quick to present, avoiding several semantic subtleties, but since (Burgess and Gurevich 1985) uses much the same methods as here, it is a very convoluted way of obtaining decidability. It is easy but tedious to give a more direct algorithm.

Recall that up to logical equivalence there are finitely many 3-theories. Indeed, we may easily construct from φ a finite set τ_φ of $\mathcal{L}_\varphi$-sentences of quantifier depth at most 3, closed under single negations and containing every such sentence up to logical equivalence, and in particular containing the sentences $\exists x \forall y \neg(y < x)$, $\exists x \forall y \neg(x < y)$, $\forall x \forall y \neg(x < y)$, and $\exists x R_\psi(x)$ for $\psi \in sub_x\,\varphi$, and their negations. Any 3-theory can be taken to be a certain subset of τ_φ, and a character a pair $\langle S, S^{con} \rangle$ where $S \subseteq 2^{\tau_\varphi}$ and S^{con} is a map from $con\,\varphi$ to S.

Note that not every such object is a 3-theory (or character). Nonetheless, we have:

Claim 11.68. *Given $\sigma \subseteq \tau_\varphi$ and $\chi = \langle S, S^{con} \rangle$ where $S \subseteq 2^{\tau_\varphi}$ and S^{con} a function from $con\,\varphi$ to S, it is decidable whether σ is a 3-theory and χ is a character.*

Proof. $\sigma \subseteq \tau_\varphi$ is a 3-theory iff it contains every sentence in τ_φ or its negation, and the sentence $\exists_{\psi \in sub_x \varphi} R_\psi \bigwedge \sigma$ is true in some linear order. Hence, by the decidability of the universal monadic second-order theory of linear order (Gurevich 1964, Burgess and Gurevich 1985), it is decidable whether σ is a 3-theory or not. Therefore, whether χ is a character is also decidable. ❑

By this result, it suffices to show that it is decidable (using the oracle) whether a given character is legal or perfect. We can decide by inspection whether a character is perfect. For legality, there are two parts.

Claim 11.69. *Given $S \subseteq 2^{\tau_\varphi}$ and $S^{con} : con\,\varphi \to S$, it is decidable (using the oracle) whether $\chi = \langle S, S^{con} \rangle$ is a degenerate character.*

Proof. We simply check that χ is a character and that each $\sigma \in S$ contains $\forall x \forall y \neg(x < y)$. Then we check by inspection that for each $\sigma \in S$, the set $\boldsymbol{t}_\sigma = \{\overline{\psi} \mid \psi \in sub_x\,\varphi,\ \exists x R_\psi(x) \in \sigma\}$ is a type for φ. Finally, we check with the oracle that $\langle \{\boldsymbol{t}_\sigma \mid \sigma \in S\}, \{\langle c, \boldsymbol{t}_{S^{con}(c)} \rangle \mid c \in con\,\varphi\} \rangle$ is a finitely realizable state candidate for φ. Our χ is a degenerate character iff all these checks succeed. ❑

Claim 11.70. *Let Σ be a set of characters and χ be a character. It is decidable whether there exist a linear order $\langle I, <_I \rangle$ and characters $\chi_i \in \Sigma$, $i \in I$, such that $\chi \equiv \sum_{i \in I} \chi_i$.*

Proof. We can certainly decide whether a character has a left or right endpoint.

For the remainder, we need some notation. If α is a $\mathcal{QL}$-formula with x and perhaps other variables free, and θ is a $\mathcal{QL}$-formula, we define the *relativization* θ^α of θ *to* α in the usual way, by first renaming variables of θ so that they do not occur in α, and then setting $\theta^\alpha = \theta$ for atomic θ, $(\theta \wedge \theta')^\alpha = \theta^\alpha \wedge \theta'^\alpha$, $(\neg\theta)^\alpha = \neg\theta^\alpha$, and $(\exists y \theta)^\alpha = \exists y(\alpha\{y/x\} \wedge \theta^\alpha)$. We will always use the variable x for relativization, and θ will always be a sentence, so that it is harmless to rename its variables.

We note that any 3-theory σ is satisfiable in a countable $\mathcal{L}_\varphi$-order, and that any countable linear order embeds in $\langle \mathbb{R}, < \rangle$. Hence, if P is a new unary predicate symbol, $\sigma^{P(x)}$ is true in some expansion of $\langle \mathbb{R}, < \rangle$ interpreting the symbols of $\mathcal{L}_\varphi \cup \{P\}$.

Now we go through the cases **(b1)**–**(b4)** once more:

Case **(b1)**. Introduce new unary predicate symbols P_0, P_1. For 3-theories $\sigma, \sigma_0, \sigma_1$, we have $\sigma = \sigma_0 + \sigma_1$ iff the following sentence is true in some expansion of $\langle \mathbb{R}, < \rangle$:

$$(\bigwedge \sigma_0)^{P_0(x)} \wedge (\bigwedge \sigma_1)^{P_1(x)} \wedge (\bigwedge \sigma)^{P_0 \vee P_1(x)} \wedge \bigwedge_{i<2} \exists x P_i(x) \wedge \\ \forall x \forall y (P_0(x) \wedge P_1(y) \to x < y).$$

By the result of (Burgess and Gurevich 1985) already mentioned, this is decidable. The definition of $\chi \equiv \chi_0 + \chi_1$ is a Boolean combination of such conditions, and is therefore decidable. So we can decide whether $\chi \equiv \chi_0 + \chi_1$ for some $\chi_0, \chi_1 \in \Sigma$, by considering all of the finitely many possibilities for χ_0, χ_1.

Case **(b2)**. Let P, Q be new unary predicate symbols and let ν be the conjunction of the sentences

$$\begin{aligned}
&\forall x \neg(P(x) \wedge Q(x)), \\
&\exists x\big(Q(x) \wedge \forall y < x(\neg P(y) \wedge \neg Q(y))\big), \\
&\forall x \exists y > x Q(y), \\
&\forall x \exists y < x \forall z (y < z < x \to \neg Q(z)), \\
&\forall x \exists y > x \forall z (x < z < y \to \neg Q(z)), \\
&\forall x \exists y > x P(x).
\end{aligned}$$

An expansion of $\langle \mathbb{R}, < \rangle$ is a model of ν iff the interpretations of P and Q are disjoint and unbounded above in $\mathbb{R}$, Q has order type $\langle \mathbb{N}, < \rangle$, and there is no P-point before the first Q-point. Let

$$\alpha(x, y) = P(x) \wedge \forall z \big((x \le z \le y \vee y \le z \le x) \to \neg Q(z)\big).$$

Let σ, σ_i $(i \in I)$ be 3-theories with $\sigma_i = \sigma_0$ for all i. Then $\sigma = \sum_{i \in I} \sigma_i$ iff

$$\nu \wedge \sigma^{P(x)} \wedge \forall y(P(y) \rightarrow (\sigma_0)^{\alpha(x,y)})$$

is true in some expansion of $\langle \mathbb{R}, < \rangle$ (relativizing on x as said before). This statement is decidable, so given characters χ, $\chi_0 = \chi_1 = \cdots$, we can check effectively whether $S^{con}(c) = \sum_{i \in I} S_i^{con}(c)$ for all $c \in con\,\varphi$ and whether $S = \{\sum_{i \in I} \sigma_i \mid \sigma_i \in S_i,\ \sigma_i = \sigma_0 \text{ for all } i\}$. Thus, whether $\chi \equiv \sum_{i \in I} \chi_i$ for some $\chi_0 = \chi_1 = \cdots$ is in Σ is decidable.

Case **(b3)** is analogous to **(b2)**.

Case **(b4)**. We will need to make 'copies' $\mathcal{L}_s$ of $\mathcal{L}_\varphi$, for various objects s, by renaming the symbols R_ψ of the signature of $\mathcal{L}_\varphi$. We assume that if $s \neq s'$ then the intersection of the signatures of $\mathcal{L}_s$ and $\mathcal{L}_{s'}$ consists of just the symbol $<$. If $\mathcal{L}_s$ is such a copy, and θ is an $\mathcal{L}_\varphi$-sentence, we write $\theta_{\mathcal{L}_s}$ for the result of replacing the predicate symbols R_ψ of $\mathcal{L}_\varphi$ in θ by the corresponding ones of $\mathcal{L}_s$.

For a unary predicate symbol P, we let

$$\alpha(x, y, P) = \forall z\big((x \leq z \leq y \vee y \leq z \leq x) \rightarrow P(z)\big).$$

Let $\{\chi_0, \ldots, \chi_{n-1}\}$ be a set of characters, with $n \geq 2$, and let $\chi = \langle S, S^{con} \rangle$ be another character. Write $\chi_i = \langle S_i, S_i^{con} \rangle$, as usual. Introduce new unary predicate symbols X_i $(i < n)$, and consider the following sentences:

$$\forall x \bigvee_{i<n} (X_i(x) \wedge \bigwedge_{j \neq i} \neg X_j(x)),$$

$$\bigwedge_{i<n} \forall x \exists y \exists z \big(y < x < z \wedge X_i(y) \wedge X_i(z)\big),$$

$$\forall x \forall y \bigwedge_{i \neq j} \big(x < y \wedge X_i(x) \wedge X_j(y) \rightarrow \bigwedge_{k<n} \exists z(x < z < y \wedge X_k(z))\big).$$

These three sentences say that the condensation given by

$$x \sim y \quad \text{iff} \quad \bigvee_{i<n} \alpha(x, y, X_i)$$

is dense without endpoints, and indeed that the classes included in any X_i occur densely.

Now for each $c \in con\,\varphi$, take a copy $\mathcal{L}_c$ of $\mathcal{L}_\varphi$ and add the sentences

$$(\bigwedge S^{con}(c))_{\mathcal{L}_c} \quad \text{and} \quad \forall y\big(X_i(y) \rightarrow (\bigwedge S_i^{con}(c))_{\mathcal{L}_c}^{\alpha(x,y,X_i)}\big),$$

for each $i < n$.

Then for each $\sigma \in S$, take a copy $\mathcal{L}_\sigma$ of $\mathcal{L}_\varphi$, and add the sentences

$$(\bigwedge \sigma)_{\mathcal{L}_\sigma} \quad \text{and} \quad \forall y \bigwedge_{i<n} \big(X_i(y) \to \bigvee_{\sigma_i \in S_i} (\bigwedge \sigma_i)_{\mathcal{L}_\sigma}^{\alpha(x,y,X_i)}\big).$$

Finally, for each $\pi = \langle j, \sigma \rangle$ where $j < n$ and $\sigma \in S_j$, introduce new unary predicate symbols $Q_{\pi,i,\sigma'}$ for $i < n$ and $\sigma' \in S_i$, and add the sentences:

- 'the $Q_{\pi,i,\sigma'}$ are pairwise disjoint',
- $\bigwedge_{i<n} \forall x \big(X_i(x) \leftrightarrow \bigvee_{\sigma' \in S_i} Q_{\pi,i,\sigma'}(x)\big)$,
- $\big(\exists x Q_\eta(x)\big) \to \forall x \forall y (x < y \wedge Q_{\eta'}(x) \wedge Q_{\eta''}(y) \to \exists z (x < z < y \wedge Q_\eta(z))\big)$, for any three triples η, η', η'' of the form $\langle \pi, i, \sigma' \rangle$ for fixed π as above and with $\eta' \neq \eta''$,
- $\forall y \big(Q_{\pi,i,\sigma'}(y) \to (\bigwedge \sigma')_{\mathcal{L}_\pi}^{\alpha(x,y,Q_{\pi,i,\sigma'})}\big)$, for each i, σ',
- $\bigvee_{\sigma \in S} (\bigwedge \sigma)_{\mathcal{L}_\pi}$.

It is not so hard to check that the conjunction of these sentences is true in some expansion of $\langle \mathbb{R}, < \rangle$ iff $\chi \equiv \sum_{i \in I} \chi_i$, where $\langle I, <_I \rangle$ is a condensation of $\langle \mathbb{R}, < \rangle$, $\{\chi_i \mid i \in I\} = \{\chi_0, \dots, \chi_{n-1}\}$, and the provisions of **(b4)** are met. Hence, as before, it is decidable whether $\chi \equiv \sum_{i \in I} \chi_i$ via **(b4)** for some $\chi_i \in \Sigma$. ❑

Now we decide whether a character λ is legal as follows. Build the set Λ_0 of all degenerate characters, using Claims 11.68 and 11.69. Given Λ_n, check for each character $\chi \notin \Lambda_n$ whether $\chi \equiv \sum_{i \in I} \chi_i$ for some linear order $\langle I, < \rangle$ and some $\chi_i \in \Lambda_n$, using Claim 11.70. If so, put χ in Λ_{n+1}. Increment n, and repeat. Terminate when $\Lambda_{n+1} = \Lambda_n$, and check whether $\lambda \in \Lambda_n$. This determines whether λ is legal, and completes the proof of Lemma 11.58 and Theorem 11.9. ❑

11.7 Axiomatizing monodic fragments

The full monodic fragments of first-order temporal logics are certainly undecidable: they contain full classical predicate logic. However, unlike, say, $\mathsf{QLog}_{\mathcal{U}}(\mathbb{N})$ which is not recursively enumerable (cf. Theorem 11.1), the monodic fragments may be finitely axiomatizable. To present an example of such an axiomatization is the aim of this section.

We are going to axiomatize the monodic fragment of first-order temporal logic over the flow of time $\langle \mathbb{N}, < \rangle$. To simplify presentation, we consider only the 'future' sublanguage $\mathcal{QTL}_{\mathcal{U}}$ of the language $\mathcal{QTL}$ having temporal operators $\bigcirc$, $\Box_F$, and $\mathcal{U}$.

Define an axiomatic system $\mathcal{MON}$ by putting together the axiomatic systems for classical first-order logic **QCl** (Section 1.3) and propositional temporal logic **PTL** (Section 2.1), and adding the Barcan formula for $\bigcirc$. More precisely, let $\mathcal{MON}$ be the calculus with the following axiom schemata and inference rules (all instances of which are restricted to monodic formulas only):

Axiom schemata (ranging over monodic $\mathcal{QTL}_{\mathcal{U}}$-formulas):

the axiom schemata of classical first-order logic **QCl**,

$$\Box_F(\varphi \to \psi) \to (\Box_F\varphi \to \Box_F\psi), \tag{11.35}$$

$$\bigcirc(\varphi \to \psi) \to (\bigcirc\varphi \to \bigcirc\psi), \tag{11.36}$$

$$\bigcirc\neg\varphi \leftrightarrow \neg\bigcirc\varphi, \tag{11.37}$$

$$\Box_F\varphi \leftrightarrow \bigcirc\varphi \land \bigcirc\Box_F\varphi, \tag{11.38}$$

$$\Box_F(\varphi \to \bigcirc\varphi) \to \bigcirc(\varphi \to \Box_F\varphi), \tag{11.39}$$

$$\varphi\mathcal{U}\psi \to \Diamond_F\psi, \tag{11.40}$$

$$\varphi\mathcal{U}\psi \leftrightarrow \bigcirc\psi \lor \bigcirc(\varphi \land \varphi\mathcal{U}\psi), \tag{11.41}$$

$$\bigcirc\forall x\varphi \leftrightarrow \forall x\bigcirc\varphi. \tag{11.42}$$

Inference rules (ranging over monodic $\mathcal{QTL}_{\mathcal{U}}$-formulas):

the rules of **QCl**,

given φ, derive $\Box_F\varphi$. (11.43)

A monodic $\mathcal{QTL}_{\mathcal{U}}$-formula φ is *$\mathcal{MON}$-derivable* (in symbols: $\vdash_{\mathcal{MON}} \varphi$) if there is a sequence of monodic $\mathcal{QTL}_{\mathcal{U}}$-formulas ending with φ and such that each member of the sequence is either a substitution instance of an axiom schema, or obtained from some earlier members of the sequence by applying one of the inference rules.

In the remainder of this section we prove the following result of Wolter and Zakharyaschev (2002):

Theorem 11.71. *For every monodic $\mathcal{QTL}_{\mathcal{U}}$-formula φ, we have*

$$\vdash_{\mathcal{MON}} \varphi \quad \textit{iff} \quad \varphi \in \mathsf{QLog}_{\mathcal{U}}(\mathbb{N}).$$

Proof. It is easy to check the *soundness* part ($\Rightarrow$) of the theorem.

To prove the *completeness* part ($\Leftarrow$), we have to show that if $\not\vdash_{\mathcal{MON}} \varphi$ then there is a first-order temporal model based on $\langle\mathbb{N}, <\rangle$ in which φ is not true. To put it another way, we can show that if $\not\vdash_{\mathcal{MON}} \neg\varphi$—i.e., φ is *consistent with* $\mathcal{MON}$—then φ is satisfiable in a first-order temporal model based on $\langle\mathbb{N}, <\rangle$.

(Without loss of generality we may assume that φ is a sentence. Indeed, if a monodic $\mathcal{QTL}_{\mathcal{U}}$-formula $\varphi(x_1, \dots, x_n)$ is in $\mathsf{QLog}_{\mathcal{U}}(\mathbb{N})$, then so is the monodic sentence $\forall x_1 \dots \forall x_n \varphi(x_1, \dots, x_n)$. So if we succeed in proving that $\vdash_{\mathcal{MON}} \forall x_1 \dots \forall x_n \varphi(x_1, \dots, x_n)$, then we will also have $\vdash_{\mathcal{MON}} \varphi(x_1, \dots, x_n)$, because $\mathcal{MON}$ contains the axiom schemata of classical first-order logic.)

Thus, we need some means of constructing models. As in Sections 11.3 and 11.4, we will be using for this purpose some kind of quasimodels, appropriately modified for the needs of this proof. First, note the following formula and rule can be derived in $\mathcal{MON}$ using (11.36), (11.38) and (11.43):

$$\bigcirc(\varphi \wedge \psi) \leftrightarrow (\bigcirc\varphi \wedge \bigcirc\psi), \tag{11.44}$$

$$\text{given } \varphi, \text{ derive } \bigcirc\varphi. \tag{11.45}$$

Fix a monodic $\mathcal{QTL}_{\mathcal{U}}$-sentence φ. Recall that $sub\,\varphi$ and $con\,\varphi$ denote the sets of all subformulas and constant symbols in φ, respectively. Let

$$sub_{\bigcirc\neg}\,\varphi = sub\,\varphi \cup \{\neg\psi \mid \psi \in sub\,\varphi\} \cup \{\bigcirc\psi \mid \psi \in sub\,\varphi\} \cup \{\bigcirc\neg\psi \mid \psi \in sub\,\varphi\}.$$

Denote by $sub_n\,\varphi$ the subset of $sub_{\bigcirc\neg}\,\varphi$ containing formulas with $\leq n$ free variables. Without loss of generality we may assume that $sub_n\,\varphi$ is closed under negation, at least modulo equivalences $\neg\neg\psi \leftrightarrow \psi$ and (11.37). Let x be a variable not occurring in φ. Put

$$sub_x\,\varphi = \{\psi\{x/y\} \mid \psi(y) \in sub_1\,\varphi\}.$$

Now by a *type* for φ we mean any Boolean-saturated subset $\boldsymbol{t}$ of $sub_x\,\varphi$. As before, we say that two types $\boldsymbol{t}$ and $\boldsymbol{t}'$ *agree on* $sub_0\,\varphi$ if $\boldsymbol{t} \cap sub_0\,\varphi = \boldsymbol{t}' \cap sub_0\,\varphi$. Given a type $\boldsymbol{t}$ for φ and a constant $c \in con\,\varphi$, the pair $\langle c, \boldsymbol{t}\rangle$ will be called an *indexed type* for φ (indexed by c).

A pair $\mathfrak{C} = \langle T_{\mathfrak{C}}, T_{\mathfrak{C}}^{con}\rangle$ is called a *state candidate* for φ if $T_{\mathfrak{C}}$ is a (nonempty) set of types for φ that agree on $sub_0\,\varphi$, and

$$T_{\mathfrak{C}}^{con} \subseteq con\,\varphi \times T_{\mathfrak{C}}$$

is a set of indexed types such that for each $c \in con\,\varphi$ there is a unique $\boldsymbol{t} \in T_{\mathfrak{C}}$ with $\langle c, \boldsymbol{t}\rangle \in T_{\mathfrak{C}}^{con}$. As before, indexed types $\langle c, \boldsymbol{t}\rangle$ in $T_{\mathfrak{C}}^{con}$ will also be denoted by $\boldsymbol{t}_{\mathfrak{C}}^{c}$.

In what follows we will often identify a type $\boldsymbol{t}$ with the $\mathcal{QTL}_{\mathcal{U}}$-formula $\bigwedge_{\psi \in \boldsymbol{t}} \psi$. Given a state candidate $\mathfrak{C}$, we put

$$\mathsf{real}_{\mathfrak{C}} = \bigwedge_{\boldsymbol{t} \in T_{\mathfrak{C}}} \exists x\, \boldsymbol{t}(x) \wedge \bigwedge_{c \in con\,\varphi} \boldsymbol{t}_{\mathfrak{C}}^{c}\{c/x\} \wedge \forall x \bigvee_{\boldsymbol{t} \in T_{\mathfrak{C}}} \boldsymbol{t}(x).$$

Say that a state candidate $\mathfrak{C}$ is *consistent* if the sentence $\mathsf{real}_{\mathfrak{C}}$ is consistent with $\mathcal{MON}$. A pair $\langle \boldsymbol{t}_1, \boldsymbol{t}_2 \rangle$ of types for φ is called *suitable* if the formula $\boldsymbol{t}_1 \wedge \bigcirc \boldsymbol{t}_2$ is consistent with $\mathcal{MON}$. A pair of state candidates $\langle \mathfrak{C}_1, \mathfrak{C}_2 \rangle$ is *suitable* if $\mathsf{real}_{\mathfrak{C}_1} \wedge \bigcirc \mathsf{real}_{\mathfrak{C}_2}$ is consistent with $\mathcal{MON}$. Note that if the pair $\langle \boldsymbol{t}_1, \boldsymbol{t}_2 \rangle$ is suitable then, by (11.37) and (11.45), both $\boldsymbol{t}_1$ and $\boldsymbol{t}_2$ are consistent. The same applies to suitable pairs of state candidates.

Let $\boldsymbol{q} = \langle \mathfrak{C}_n = \langle T_n, T_n^{con} \rangle \mid n \in \mathbb{N} \rangle$ be a sequence of state candidates for φ. A *run* through $\boldsymbol{q}$ is a map r associating with every $n \in \mathbb{N}$ a type $r(n)$ in T_n. We call such an r *coherent and saturated* if the following hold:

- the pairs $\langle r(n), r(n+1) \rangle$ are suitable for all $n \in \mathbb{N}$,
- $\varphi \mathcal{U} \psi \in r(n)$ iff there exists $m > n$ such that $\psi \in r(m)$ and $\varphi \in r(k)$ for all $k \in (n, m)$.

A $\mathcal{MON}$-*quasimodel* for φ is a pair $\mathfrak{Q} = \langle \boldsymbol{q}, \mathfrak{R} \rangle$, where $\boldsymbol{q} = \langle \mathfrak{C}_n \mid n \in \mathbb{N} \rangle$ is a sequence of state candidates for φ such that

(mqm1) $\varphi \in \boldsymbol{t}$ for some $n \in \mathbb{N}$ and $\boldsymbol{t} \in T_n$,

(mqm2) the pairs $\langle \mathfrak{C}_n, \mathfrak{C}_{n+1} \rangle$ are suitable for all $n \in \mathbb{N}$,

and $\mathfrak{R}$ is a set of coherent and saturated runs through $\boldsymbol{q}$ satisfying the following conditions:

(mqm3) for every $c \in con\,\varphi$, the function r_c defined by $r_c(n) = \boldsymbol{t}$, for $\langle c, \boldsymbol{t} \rangle \in T_n^{con}$, $n \in \mathbb{N}$ is a run in $\mathfrak{R}$,

(mqm4) for every $n \in \mathbb{N}$ and every type $\boldsymbol{t}$ in T_n there exists a run r in $\mathfrak{R}$ such that $r(n) = \boldsymbol{t}$.

Lemma 11.72. *Given a monodic $\mathcal{QTL}_{\mathcal{U}}$-sentence φ, if there is a $\mathcal{MON}$-quasimodel for φ, then φ is satisfiable in a first-order temporal model based on $\langle \mathbb{N}, < \rangle$.*

Proof. The proof is almost the same as the corresponding part of the proof of Lemma 11.22. The only difference is that now we are not given that the state candidates in $\mathfrak{Q}$ are realizable. We know, however, that every state candidate $\mathfrak{C}_n$ is consistent. By treating subformulas of $\mathsf{real}_{\mathfrak{C}_n}$ of the form $\bigcirc\psi$, $\Box_F\psi$, and $\chi\mathcal{U}\psi$ that are not in the scope of another temporal operator as unary predicate symbols or propositional variables, we obtain that the resulting sentence $\overline{\mathsf{real}_{\mathfrak{C}_n}}$ is consistent with the axiomatic system of classical first-order logic (since $\mathcal{MON}$ contains the axiom schemata and rules of the latter). So, by Gödel's completeness theorem, there is a first-order structure 'realizing' $\mathfrak{C}_n$. The remaining part is precisely the same as that of the proof mentioned above; see also Claim 11.24. ❑

Thus, to prove Theorem 11.71, it suffices to show the following:

Lemma 11.73. *Suppose a monodic $\mathcal{QTL}_{\mathcal{U}}$-sentence φ is consistent with $\mathcal{MON}$. Then there is a $\mathcal{MON}$-quasimodel for φ.*

Proof. We require a series of claims.

Claim 11.74. (i) *Let $\langle \boldsymbol{t}_1, \boldsymbol{t}_2 \rangle$ be a suitable pair of types for φ. Then*

- *for every $\bigcirc\psi \in sub_x\,\varphi$, $\bigcirc\psi \in \boldsymbol{t}_1$ implies $\psi \in \boldsymbol{t}_2$,*
- *for every $\chi\mathcal{U}\psi \in sub_x\,\varphi$, $\chi\mathcal{U}\psi \in \boldsymbol{t}_1$ implies that either $\psi \in \boldsymbol{t}_2$ or $\chi \in \boldsymbol{t}_2$ and $\chi\mathcal{U}\psi \in \boldsymbol{t}_2$.*

(ii) *Let $\langle \mathfrak{C}_1, \mathfrak{C}_2 \rangle$ be a suitable pair of state candidates, $\mathfrak{C}_1 = \langle T_1, T_1^{con} \rangle$ and $\mathfrak{C}_2 = \langle T_2, T_2^{con} \rangle$. Then*

- *for every $\boldsymbol{t}_1 \in T_1$, there exists a $\boldsymbol{t}_2 \in T_2$ such that the pair $\langle \boldsymbol{t}_1, \boldsymbol{t}_2 \rangle$ is suitable,*
- *for every $\boldsymbol{t}_2 \in T_2$, there exists a $\boldsymbol{t}_1 \in T_1$ such that $\langle \boldsymbol{t}_1, \boldsymbol{t}_2 \rangle$ is suitable, and*
- *for every $c \in con\,\varphi$, the pair $\left\langle \boldsymbol{t}^c_{\mathfrak{C}_1}, \boldsymbol{t}^c_{\mathfrak{C}_2} \right\rangle$ is suitable.*

Proof. (i) Suppose $\bigcirc\psi \in \boldsymbol{t}_1$, but $\psi \notin \boldsymbol{t}_2$. Then $\neg\psi \in \boldsymbol{t}_2$. Since $t_1 \wedge \bigcirc t_2$ is consistent with $\mathcal{MON}$, by (11.44) the formula $\bigcirc\psi \wedge \bigcirc\neg\psi$ is also consistent with $\mathcal{MON}$, which is impossible, again by (11.44).

Suppose now that $\chi\mathcal{U}\psi \in \boldsymbol{t}_1$. In view of (11.41), (11.44) and consistency of $\boldsymbol{t}_1$, we then have either $\bigcirc\psi \in \boldsymbol{t}_1$ or $\bigcirc\chi$, $\bigcirc(\chi\mathcal{U}\psi) \in \boldsymbol{t}_1$. And as we have just shown, either $\psi \in \boldsymbol{t}_2$ or χ, $\chi\mathcal{U}\psi \in \boldsymbol{t}_2$ follow.

(ii) Assume that there is $\boldsymbol{t}_1 \in T_1$ such that none of the pairs $\langle \boldsymbol{t}_1, \boldsymbol{t}_2 \rangle$, for $\boldsymbol{t}_2 \in T_2$, is suitable. It follows that

$$\vdash_{\mathcal{MON}} \boldsymbol{t}_1 \to \bigcirc \bigwedge_{\boldsymbol{t}_2 \in T_2} \neg \boldsymbol{t}_2,$$

from which

$$\vdash_{\mathcal{MON}} \exists x \boldsymbol{t}_1 \to \exists x \bigcirc \bigwedge_{\boldsymbol{t}_2 \in T_2} \neg \boldsymbol{t}_2,$$

and so, by (11.37) and (11.42),

$$\vdash_{\mathcal{MON}} \neg(\exists x \boldsymbol{t}_1 \wedge \bigcirc \forall x \bigvee_{\boldsymbol{t}_2 \in T_2} \boldsymbol{t}_2),$$

contrary to $\langle \mathfrak{C}_1, \mathfrak{C}_2 \rangle$ being suitable.

Now suppose that there is $t_2 \in T_2$ such that none of the pairs $\langle t_1, t_2\rangle$, for $t_1 \in T_1$, is suitable. Then

$$\vdash_{\mathcal{MON}} \exists x \bigcirc t_2 \to \exists x \neg \bigvee_{t_1 \in T_1} t_1,$$

which is equivalent to

$$\vdash_{\mathcal{MON}} \neg(\forall x \bigvee_{t_1 \in T_1} t_1 \wedge \exists x \bigcirc t_2),$$

contrary to $\langle \mathfrak{C}_1, \mathfrak{C}_2\rangle$ being suitable.

Finally, assume that $c \in con\,\varphi$. Then $t^c_{\mathfrak{C}_1} \wedge \bigcirc t^c_{\mathfrak{C}_2}$ is consistent with $\mathcal{MON}$, and so the pair $\langle t^c_{\mathfrak{C}_1}, t^c_{\mathfrak{C}_2}\rangle$ is suitable. ❑

A *pointed state candidate* for φ is a pair $\mathfrak{P} = \langle \mathfrak{C}, t\rangle$, where $\mathfrak{C} = \langle T, T^{con}\rangle$ is a state candidate for φ and t a type in T.

Say that $\mathfrak{P} = \langle \mathfrak{C}, t\rangle$ is *consistent* if the formula $\mathsf{real}_{\mathfrak{C}} \wedge t$ is consistent with $\mathcal{MON}$. A pair $\mathfrak{P}_1 = \langle \mathfrak{C}_1, t_1\rangle$, $\mathfrak{P}_2 = \langle \mathfrak{C}_2, t_2\rangle$ of pointed state candidates for φ is called *suitable* (in symbols, $\mathfrak{P}_1 \prec \mathfrak{P}_2$) if the formula $\mathsf{real}_{\mathfrak{C}_1} \wedge t_1 \wedge \bigcirc(\mathsf{real}_{\mathfrak{C}_2} \wedge t_2)$ is consistent with $\mathcal{MON}$. Given a $c \in con\,\varphi$, a pair $\mathfrak{P}_1 = \langle \mathfrak{C}_1, t_1\rangle$, $\mathfrak{P}_2 = \langle \mathfrak{C}_2, t_2\rangle$ of pointed state candidates for φ is called *suitable for c* (in symbols, $\mathfrak{P}_1 \prec_c \mathfrak{P}_2$) if $\mathfrak{P}_1 \prec \mathfrak{P}_2$, $\langle c, t_1\rangle \in T_1^{con}$ and $\langle c, t_2\rangle \in T_2^{con}$, where $\mathfrak{C}_i = \langle T_i, T_i^{con}\rangle$, $i = 1, 2$.

Claim 11.75. (i) *There is a consistent state candidate $\mathfrak{C} = \langle T, T^{con}\rangle$ for φ such that $\varphi \in t$ for all $t \in T$.*

(ii) *For every consistent pointed state candidate $\mathfrak{P}_1 = \langle \mathfrak{C}_1, t_1\rangle$ for φ, there is a pointed state candidate $\mathfrak{P}_2 = \langle \mathfrak{C}_2, t_2\rangle$ such that $\mathfrak{P}_1 \prec \mathfrak{P}_2$.*

(iii) *Let $c \in con\,\varphi$. For every consistent pointed state candidate for φ of the form $\mathfrak{P}_1 = \langle \mathfrak{C}_1, t^c_{\mathfrak{C}_1}\rangle$ and every state candidate $\mathfrak{C}_2$ such that the pair $\langle \mathfrak{C}_1, \mathfrak{C}_2\rangle$ is suitable, we have $\mathfrak{P}_1 \prec_c \mathfrak{P}_2$ for $\mathfrak{P}_2 = \langle \mathfrak{C}_2, t^c_{\mathfrak{C}_2}\rangle$.*

Proof. Let π_φ be the disjunction of formulas $\mathsf{real}_{\mathfrak{C}} \wedge t$, for all pointed state candidates $\langle \mathfrak{C}, t\rangle$ for φ. By treating subformulas of π_φ of the form $\bigcirc\psi$, $\Box_F\psi$, and $\chi\mathcal{U}\psi$ that are not in the scope of another temporal operator as unary predicate symbols or propositional variables, we obtain that the resulting sentence $\overline{\pi_\varphi}$ is clearly true in all (classical) first-order structures. Since $\mathcal{MON}$ contains the axiom schemata and rules of classical first-order logic, by Gödel's completeness theorem we obtain

$$\vdash_{\mathcal{MON}} \pi_\varphi. \tag{11.46}$$

(i) Since φ is consistent with $\mathcal{MON}$, from (11.46) we obtain that $\pi_\varphi \wedge \varphi$ is consistent with $\mathcal{MON}$. Then there is a disjunct $\mathsf{real}_{\mathfrak{C}} \wedge t$ of π_φ such that

$\mathsf{real}_{\mathfrak{C}} \wedge \boldsymbol{t} \wedge \varphi$ is also consistent with $\mathcal{MON}$, so $\varphi \in \boldsymbol{t}$. Since φ is a sentence, $\varphi \in \boldsymbol{t}'$ follows for all types $\boldsymbol{t}'$ of $\mathfrak{C}$.

(ii) By (11.45) and (11.46), we have $\vdash_{\mathcal{MON}} \bigcirc\pi_\varphi$. Hence, $\mathsf{real}_{\mathfrak{C}_1} \wedge \boldsymbol{t}_1 \wedge \bigcirc\pi_\varphi$ is consistent with $\mathcal{MON}$, and so, by (11.37) and (11.44), there must be a state candidate $\mathfrak{C}_2 = \langle T_2, T_2^{con}\rangle$ and a type $\boldsymbol{t}_2 \in T_2$ such that the formula $\mathsf{real}_{\mathfrak{C}_1} \wedge \boldsymbol{t}_1 \wedge \bigcirc(\mathsf{real}_{\mathfrak{C}_2} \wedge \boldsymbol{t}_2)$ is consistent with $\mathcal{MON}$.

(iii) Suppose that $c \in con\,\varphi$, $\mathfrak{P}_1 = \langle \mathfrak{C}_1, \boldsymbol{t}^c_{\mathfrak{C}_1}\rangle$ and that the pair $\langle \mathfrak{C}_1, \mathfrak{C}_2\rangle$ is suitable. Let $\mathfrak{P}_2 = \langle \mathfrak{C}_2, \boldsymbol{t}^c_{\mathfrak{C}_2}\rangle$. Then $\mathfrak{P}_1 \prec_c \mathfrak{P}_2$ must hold, for otherwise we would have

$$\vdash_{\mathcal{MON}} \mathsf{real}_{\mathfrak{C}_1} \wedge \boldsymbol{t}^c_{\mathfrak{C}_1} \to \neg\bigcirc(\mathsf{real}_{\mathfrak{C}_2} \wedge \boldsymbol{t}^c_{\mathfrak{C}_2}),$$

and so

$$\vdash_{\mathcal{MON}} \mathsf{real}_{\mathfrak{C}_1} \wedge \boldsymbol{t}^c_{\mathfrak{C}_1}\{c/x\} \to \neg\bigcirc(\mathsf{real}_{\mathfrak{C}_2} \wedge \boldsymbol{t}^c_{\mathfrak{C}_2}\{c/x\}),$$

i.e., $\vdash_{\mathcal{MON}} \mathsf{real}_{\mathfrak{C}_1} \to \neg\bigcirc\mathsf{real}_{\mathfrak{C}_2}$, which is a contradiction. ❑

Suppose $\mathfrak{P}_0 = \langle \mathfrak{C}_0, \boldsymbol{t}_0\rangle$ is a consistent pointed state candidate for φ and $\chi\mathcal{U}\psi \in \boldsymbol{t}_0$. Suppose also that $\langle \mathfrak{P}_0, \dots, \mathfrak{P}_n\rangle$, for some $n \in \mathbb{N}$, is a sequence of pointed state candidates $\mathfrak{P}_i = \langle \mathfrak{C}_i, \boldsymbol{t}_i\rangle$ such that

$$\mathfrak{P}_0 \prec \mathfrak{P}_1 \prec \dots \prec \mathfrak{P}_n,$$

and there exists $0 < k \leq n$ for which $\psi \in \boldsymbol{t}_k$ and $\chi \in \boldsymbol{t}_i$ for all $0 < i < k$. Then we say that this sequence *realizes* $\chi\mathcal{U}\psi$ in $\boldsymbol{t}_0$. If for some $c \in con\,\varphi$

$$\mathfrak{P}_0 \prec_c \mathfrak{P}_1 \prec_c \dots \prec_c \mathfrak{P}_n,$$

then we say that the sequence $\langle \mathfrak{P}_0, \dots, \mathfrak{P}_n\rangle$ *c-realizes* $\chi\mathcal{U}\psi$ in $\boldsymbol{t}_0$.

Claim 11.76. *For every consistent pointed state candidate* $\mathfrak{P}_0 = \langle \mathfrak{C}_0, \boldsymbol{t}_0\rangle$ *and every formula* $\chi\mathcal{U}\psi \in \boldsymbol{t}_0$, *there is a sequence* $\langle \mathfrak{P}_0, \dots, \mathfrak{P}_n\rangle$ *realizing* $\chi\mathcal{U}\psi$ *in* $\boldsymbol{t}_0$. *Moreover, if* $\boldsymbol{t}_0 = \boldsymbol{t}^c_{\mathfrak{C}_0}$ *then we can find a sequence* $\langle \mathfrak{P}_0, \dots, \mathfrak{P}_n\rangle$ *which c-realizes* $\chi\mathcal{U}\psi$ *in* $\boldsymbol{t}_0$.

Proof. Suppose otherwise. Let $\mathcal{K}$ be the set of all pointed state candidates $\mathfrak{P}$ such that there exist $n \geq 1$ and pointed state candidates $\mathfrak{P}_1, \dots, \mathfrak{P}_n$ with

$$\mathfrak{P}_0 \prec \mathfrak{P}_1 \prec \dots \prec \mathfrak{P}_n \quad \text{and} \quad \mathfrak{P}_n = \mathfrak{P}.$$

Consider the (nonempty, by Claim 11.75 (i)) disjunction

$$\vartheta = \bigvee_{\langle \mathfrak{C}, \boldsymbol{t}\rangle \in \mathcal{K}} (\mathsf{real}_{\mathfrak{C}} \wedge \boldsymbol{t}).$$

Note first that

$$\vdash_{\mathcal{MON}} \vartheta \to \neg\psi. \tag{11.47}$$

Indeed, otherwise the formula $\vartheta \wedge \psi$ is consistent with $\mathcal{MON}$, and so

$$\bigvee_{\langle \mathfrak{C}, \boldsymbol{t}\rangle \in \mathcal{K}} (\mathsf{real}_{\mathfrak{C}} \wedge \boldsymbol{t} \wedge \psi)$$

is consistent with $\mathcal{MON}$ as well. Hence there is $\mathfrak{P} = \langle \mathfrak{C}, \boldsymbol{t}\rangle$ in $\mathcal{K}$ such that $\mathsf{real}_{\mathfrak{C}} \wedge \boldsymbol{t} \wedge \psi$ is consistent with $\mathcal{MON}$, which means, in particular, that ψ is in $\boldsymbol{t}$ (for otherwise $\neg\psi \in \boldsymbol{t}$ and $\mathsf{real}_{\mathfrak{C}} \wedge \boldsymbol{t} \wedge \psi$ cannot be consistent). Thus we have a sequence

$$\mathfrak{P}_0 \prec \mathfrak{P}_1 \prec \cdots \prec \mathfrak{P}_n$$

such that $n \geq 1$ and $\psi \in \boldsymbol{t}_n$ for $\mathfrak{P}_i = \langle \mathfrak{C}_i, \boldsymbol{t}_i\rangle$. As all pairs $\langle \boldsymbol{t}_i, \boldsymbol{t}_{i+1}\rangle$, $i < n$, are suitable, it follows from Claim 11.74 (i) that the sequence $\langle \mathfrak{P}_0, \dots, \mathfrak{P}_n\rangle$ realizes $\chi \mathcal{U} \psi$ in $\boldsymbol{t}_0$, contrary to our assumption. Thus, we have (11.47).

Let us show now that

$$\vdash_{\mathcal{MON}} \vartheta \rightarrow \bigcirc\vartheta. \tag{11.48}$$

If this is not the case then the formula $\vartheta \wedge \bigcirc\neg\vartheta$ is consistent with $\mathcal{MON}$, and so there is $\mathfrak{P} = \langle \mathfrak{C}, \boldsymbol{t}\rangle$ in $\mathcal{K}$ such that $\mathsf{real}_{\mathfrak{C}} \wedge \boldsymbol{t} \wedge \bigcirc\neg\vartheta$ is consistent with $\mathcal{MON}$. By Claim 11.75 (i), we have a pointed state candidate $\mathfrak{P}' = \langle \mathfrak{C}', \boldsymbol{t}'\rangle$ for which $\mathfrak{P} \prec \mathfrak{P}'$. But then $\mathfrak{P}' \in \mathcal{K}$ and $\mathsf{real}_{\mathfrak{C}} \wedge \boldsymbol{t} \wedge \bigcirc(\mathsf{real}_{\mathfrak{C}'} \wedge \boldsymbol{t}')$ is consistent with $\mathcal{MON}$, contrary to consistency of $\mathsf{real}_{\mathfrak{C}} \wedge \boldsymbol{t} \wedge \bigcirc \bigwedge_{\langle \mathfrak{C}', \boldsymbol{t}'\rangle \in \mathcal{K}} \neg(\mathsf{real}_{\mathfrak{C}'} \wedge \boldsymbol{t}')$. Thus, we have (11.48).

Now, from (11.47) we obtain, by (11.43) and (11.35), that

$$\vdash_{\mathcal{MON}} \Box_F\vartheta \rightarrow \Box_F\neg\psi. \tag{11.49}$$

Further, from (11.48) we have:

$$\begin{array}{ll}
\vdash_{\mathcal{MON}} \Box_F(\vartheta \rightarrow \bigcirc\vartheta) & \text{(by (11.43))},\\
\vdash_{\mathcal{MON}} \bigcirc(\vartheta \rightarrow \Box_F\vartheta) & \text{(by (11.39))},\\
\vdash_{\mathcal{MON}} \bigcirc\vartheta \rightarrow \bigcirc\Box_F\vartheta & \text{(by (11.36))},\\
\vdash_{\mathcal{MON}} \bigcirc\vartheta \rightarrow \Box_F\vartheta & \text{(by (11.38))},\\
\vdash_{\mathcal{MON}} \vartheta \rightarrow (\neg\psi \wedge \Box_F\neg\psi) & \text{(by (11.47), (11.48) and (11.49))}.
\end{array}$$

Now take any $\mathfrak{P}_1 = \langle \mathfrak{C}_1, \boldsymbol{t}_1\rangle$ from $\mathcal{K}$ with $\mathfrak{P}_0 \prec \mathfrak{P}_1$. As $\mathsf{real}_{\mathfrak{C}_1} \wedge \boldsymbol{t}_1$ is a disjunct of ϑ, we then have

$$\vdash_{\mathcal{MON}} (\mathsf{real}_{\mathfrak{C}_1} \wedge \boldsymbol{t}_1) \rightarrow (\neg\psi \wedge \Box_F\neg\psi),$$

from which, by (11.45), (11.36), (11.44), and (11.38),

$$\vdash_{\mathcal{MON}} \bigcirc(\mathsf{real}_{\mathfrak{C}_1} \wedge \boldsymbol{t}_1) \rightarrow \Box_F\neg\psi,$$

and so

$$\vdash_{\mathcal{MON}} \left(\mathsf{real}_{\mathfrak{C}_0} \wedge \boldsymbol{t}_0 \wedge \bigcirc(\mathsf{real}_{\mathfrak{C}_1} \wedge \boldsymbol{t}_1)\right) \to \Box_F \neg\psi. \tag{11.50}$$

On the other hand, by $\chi\mathcal{U}\psi \in \boldsymbol{t}_0$ and (11.40), we have

$$\vdash_{\mathcal{MON}} \left(\mathsf{real}_{\mathfrak{C}_0} \wedge \boldsymbol{t}_0 \wedge \bigcirc(\mathsf{real}_{\mathfrak{C}_1} \wedge \boldsymbol{t}_1)\right) \to \Diamond_F \psi,$$

contrary to (11.50) and $\mathfrak{P}_0 \prec \mathfrak{P}_1$.

The existence of a c-realizing sequence, for each $c \in con\,\varphi$, is proved analogously. ❑

Now we can complete the proof of Lemma 11.73 as follows. In view of Claim 11.75 (i), there is a consistent pointed state candidate $\langle \mathfrak{C}_0, \boldsymbol{t}_0\rangle$ such that $\varphi \in \boldsymbol{t}_0$. $\mathfrak{C}_0$ will be the starting state candidate in the underlying sequence

$$\boldsymbol{q} = \langle \mathfrak{C}_i = \langle T_i, T_i^{con}\rangle \mid i \in \mathbb{N}\rangle$$

of the quasimodel $\mathfrak{Q} = \langle \boldsymbol{q}, \mathfrak{R}\rangle$ to be constructed.

Take some $\boldsymbol{t} \in T_0$ and $\chi\mathcal{U}\psi \in \boldsymbol{t}$. The pointed state candidate $\langle \mathfrak{C}_0, \boldsymbol{t}\rangle$ is clearly consistent. So, by Claim 11.76, there is a sequence

$$\langle \langle \mathfrak{C}_0, \boldsymbol{t}\rangle, \langle \mathfrak{C}_1, \boldsymbol{t}_1\rangle, \ldots, \langle \mathfrak{C}_k, \boldsymbol{t}_k\rangle\rangle \tag{11.51}$$

of pointed state candidates realizing $\chi\mathcal{U}\psi$ in $\boldsymbol{t}$. Next we take another formula $\chi'\mathcal{U}\psi' \in \boldsymbol{t}$, if any, which is not realized in this sequence. In this case, by Claim 11.74 (i), we have χ', $\chi'\mathcal{U}\psi' \in \boldsymbol{t}_k$. Using Claim 11.76 once again, we extend (11.51) to

$$\langle \langle \mathfrak{C}_0, \boldsymbol{t}\rangle, \langle \mathfrak{C}_1, \boldsymbol{t}_1\rangle, \ldots, \langle \mathfrak{C}_k, \boldsymbol{t}_k\rangle, \ldots, \langle \mathfrak{C}_l, \boldsymbol{t}_l\rangle\rangle \tag{11.52}$$

realizing $\chi'\mathcal{U}\psi'$ in $\boldsymbol{t}$. Following this way, we can construct a sequence extending (11.52) and realizing all formulas of the form $\chi\mathcal{U}\psi$ in $\boldsymbol{t}$. Let (11.52) be such a sequence.

Now take another type $\boldsymbol{t}' \in T_0$. By Claim 11.74 (ii), there are types $\boldsymbol{t}'_i \in T_i$, $i \leq l$, such that $\langle \mathfrak{C}_0, \boldsymbol{t}'\rangle \prec \langle \mathfrak{C}_1, \boldsymbol{t}'_1\rangle \prec \cdots \prec \langle \mathfrak{C}_l, \boldsymbol{t}'_l\rangle$. In precisely the same manner as before we extend the sequence

$$\langle \langle \mathfrak{C}_0, \boldsymbol{t}'\rangle, \langle \mathfrak{C}_1, \boldsymbol{t}'_1\rangle, \ldots, \langle \mathfrak{C}_l, \boldsymbol{t}'_l\rangle\rangle$$

to a sequence realizing all formulas of the form $\chi\mathcal{U}\psi$ in $\boldsymbol{t}'$. After that we consider yet another type $\boldsymbol{t}'' \in T_0$, and so forth. When all types are exhausted, we shall have a sequence $\langle \mathfrak{C}_0, \ldots, \mathfrak{C}_n\rangle$ of state candidates. (If no type in $\mathfrak{C}_0$ contains formulas of the form $\chi\mathcal{U}\psi$, we take a state candidate $\mathfrak{C}_1$ such that the pair $\langle \mathfrak{C}_0, \mathfrak{C}_1\rangle$ is suitable and put $\mathfrak{C}_n = \mathfrak{C}_1$.)

We have not taken care of the constants yet. So suppose $c \in con\,\varphi$ and $\chi\mathcal{U}\psi \in t^c_{\mathfrak{C}_0}$. By Claim 11.75 (ii), we have

$$\langle \mathfrak{C}_0, t^c_{\mathfrak{C}_0}\rangle \prec_c \langle \mathfrak{C}_1, t^c_{\mathfrak{C}_1}\rangle \prec_c \cdots \prec_c \langle \mathfrak{C}_n, t^c_{\mathfrak{C}_n}\rangle .$$

If $\chi\mathcal{U}\psi$ is not c-realized by the sequence

$$\langle\langle \mathfrak{C}_0, t^c_{\mathfrak{C}_0}\rangle, \langle \mathfrak{C}_1, t^c_{\mathfrak{C}_1}\rangle, \dots, \langle \mathfrak{C}_n, t^c_{\mathfrak{C}_n}\rangle\rangle \tag{11.53}$$

in $t^c_{\mathfrak{C}_0}$, then χ, $\chi\mathcal{U}\psi \in t^c_{\mathfrak{C}_n}$. By Claim 11.75 (ii), we can extend (11.53) to

$$\left\langle\langle \mathfrak{C}_0, t^c_{\mathfrak{C}_0}\rangle, \langle \mathfrak{C}_1, t^c_{\mathfrak{C}_1}\rangle, \dots, \langle \mathfrak{C}_n, t^c_{\mathfrak{C}_n}\rangle, \dots, \left\langle \mathfrak{C}_{n'}, t^c_{\mathfrak{C}_{n'}}\right\rangle\right\rangle \tag{11.54}$$

c-realizing $\chi\mathcal{U}\psi$ in $t^c_{\mathfrak{C}_0}$. Next we take another $d \in con\,\varphi$ and $\chi'\mathcal{U}\psi' \in t^d_{\mathfrak{C}_0}$ which is not d-realized by the sequence

$$\left\langle\langle \mathfrak{C}_0, t^d_{\mathfrak{C}_0}\rangle, \langle \mathfrak{C}_1, t^d_{\mathfrak{C}_1}\rangle, \dots, \langle \mathfrak{C}_n, t^d_{\mathfrak{C}_n}\rangle, \dots, \left\langle \mathfrak{C}_{n'}, t^d_{\mathfrak{C}_{n'}}\right\rangle\right\rangle . \tag{11.55}$$

We extend (11.55) so that $\chi'\mathcal{U}\psi'$ is d-realized by the new sequence. After that we consider yet another $e \in con\,\varphi$ and $\chi''\mathcal{U}\psi'' \in t^e_{\mathfrak{C}_0}$, and so forth. When all constants $c \in con\,\varphi$ and all $\chi\mathcal{U}\psi \in t^c_{\mathfrak{C}_0}$ are exhausted, we have a sequence $\langle \mathfrak{C}_0, \dots, \mathfrak{C}_m\rangle$ of state candidates for φ.

Then we consider the types and indexed types from $\mathfrak{C}_m$ and construct a sequence $\langle \mathfrak{C}_m, \dots, \mathfrak{C}_{m'}\rangle$ as if $\mathfrak{C}_m$ were $\mathfrak{C}_0$. After that we take care of $\mathfrak{C}_{m'}$, and so on. Let $\boldsymbol{q} = \langle \mathfrak{C}_i \mid i \in \mathbb{N}\rangle$ be the resulting infinite sequence of state candidates.

It is readily seen (using Claim 11.74) that $\mathfrak{Q} = \langle \boldsymbol{q}, \mathfrak{R}_{\boldsymbol{q}}\rangle$ is a quasimodel for φ, where $\mathfrak{R}_{\boldsymbol{q}}$ is the set of all coherent and saturated runs through $\boldsymbol{q}$. ❑

This completes the proof of Theorem 11.71. ❑

Question 11.77. Give axiomatizations of the monodic fragments of other first-order temporal logics considered above.

It is of interest to note that the very same proof provides an axiomatization of the one-variable constant-free fragment of $\mathsf{QLog}_{\mathcal{U}}(\mathbb{N})$—in other words, the propositional product logic **PTL** $\times$ **S5**. Indeed, define the axiomatic system $\mathcal{MON}^1$ by taking the axiom schemata and inference rules as above, but now ranging over one-variable constant-free $\mathcal{QTL}_{\mathcal{U}}$-formulas only. Observe that

- all the types for a given one-variable constant-free $\mathcal{QTL}_{\mathcal{U}}$-sentence φ (and so the sentences $\mathsf{real}_{\mathfrak{C}}$ for any state candidate $\mathfrak{C}$) contain only constant-free one-variable formulas as well, and

- the restriction of the axiomatic system for classical first-order logic (given in Section 1.3) to one-variable constant-free formulas axiomatizes the one-variable constant-free fragment of **QCL** (see, e.g., Henkin *et al.* 1971).

So from the above proof we obtain that, for every one-variable constant-free $\mathcal{QTL}_{\mathcal{U}}$-formula φ,

$$\vdash_{\mathcal{MON}^1} \varphi \quad \text{iff} \quad \varphi \in \mathsf{QLog}_{\mathcal{U}}(\mathbb{N}). \tag{11.56}$$

Using this result we can show now that the product logic $\mathbf{PTL} \times \mathbf{S5}$ is finitely axiomatizable (in fact, a kind of product-matching). Indeed, define the logic $[\mathbf{PTL}, \mathbf{S5}]$ by putting together the axioms and rules of **PTL** (see Theorem 2.6) and **S5**, plus the commutativity axiom for the $\bigcirc$ of **PTL** and the $\Box$ of **S5**. Then we have:

Theorem 11.78. $\mathbf{PTL} \times \mathbf{S5} = [\mathbf{PTL}, \mathbf{S5}]$.

Proof. The inclusion $[\mathbf{PTL}, \mathbf{S5}] \subseteq \mathbf{PTL} \times \mathbf{S5}$ is clear. To prove the converse, take some $\mathcal{ML}_{\mathcal{U}} \otimes \mathcal{ML}$-formula φ such that $\varphi \in \mathbf{PTL} \times \mathbf{S5}$. Consider the translation $\varphi^{\dagger}$ of φ defined in Section 3.7 which is a one-variable constant-free $\mathcal{QTL}_{\mathcal{U}}$-formula. Then, by Theorem 3.29, we have $\varphi^{\dagger} \in \mathsf{QLog}_{\mathcal{U}}(\mathbb{N})$, and so $\vdash_{\mathcal{MON}^1} \varphi^{\dagger}$, by (11.56). We claim that $\varphi \in [\mathbf{PTL}, \mathbf{S5}]$ follows. To show this, observe first that each one-variable constant-free $\mathcal{QTL}_{\mathcal{U}}$-formula ψ actually coincides with $\chi^{\dagger}$, for some $\mathcal{ML}_{\mathcal{U}} \otimes \mathcal{ML}$-formula χ. Now consider a $\mathcal{MON}^1$-proof of $\varphi^{\dagger}$ and replace each formula ψ in it with its '†-inverse', say $\psi^{\bullet}$. The resulting sequence is 'almost' a $[\mathbf{PTL}, \mathbf{S5}]$-proof of φ. Indeed, the translations of axiom schemata (11.35)–(11.42) and rule (11.43) are clearly $[\mathbf{PTL}, \mathbf{S5}]$-valid. Let us discuss briefly what to do with translations of the axiom schemata and rules of **QCl**. The axiom schema

- '$\forall x \psi \to \psi\{\tau/x\}$, where τ is free for x in ψ'

translates to an instance of the **S5**-axiom $\Box p \to p$ (since the only term now is x). The rule

- 'given $\psi \to \chi$, derive $\psi \to \forall x \chi$, whenever x is not free in ψ'

translates to 'given $\psi^{\bullet} \to \chi^{\bullet}$ and either $\psi^{\bullet} \leftrightarrow \Box\psi^{\bullet}$ or $\psi^{\bullet} \leftrightarrow \Diamond\psi^{\bullet}$, derive $\psi^{\bullet} \to \Box\chi^{\bullet}$.' It is not hard to show, using the axioms and rules of **S5**, that this is a valid inference in $[\mathbf{PTL}, \mathbf{S5}]$. The axiom schema and rule involving the existential quantifier are treated analogously. ❑

Remark 11.79. A resolution type semi-decision procedure for the full monodic fragment of $\mathsf{QLog}_{\mathcal{U}}(\mathbb{N})$ has been developed in (Degtyarev and Fisher 2001, Degtyarev *et al.* 2003b).

11.8 Monodicity and equality

So far we have considered first-order languages *without equality and function symbols*. A natural question is whether our decidability and axiomatizability results concerning monodic fragments can be generalized to the language with these ingredients. It should be clear that function symbols easily destroy the nice properties of monodic fragments: in the proof of Theorem 11.1 we can replace $Q_2(y)$ and $P_t(y)$ with $Q_2(f(x))$ and $P_t(f(x))$, respectively, thus obtaining a monodic monadic one-variable formula ψ_T, associated with a set T of tile types, such that ψ_T is satisfiable in a first-order temporal model iff T recurrently tiles $\mathbb{N} \times \mathbb{N}$.

In this section we investigate the possibility of adding equality to the first-order temporal language $\mathcal{QTL}_{\boxed{\mathrm{U}}}$. Let $\mathcal{QTL}^{=}_{\boxed{\mathrm{U}}}$ denote the resulting language. First we prove the following result of Wolter and Zakharyaschev (2002), which is in contrast with Theorem 11.71:

Theorem 11.80. *The set of $\mathcal{QTL}^{=}_{\boxed{\mathrm{U}}}$-formulas that are valid in all first-order temporal models based on $\langle \mathbb{N}, < \rangle$ is not recursively enumerable, and so not recursively axiomatizable.*

Proof. Let us fix a unary predicate symbol P and denote by χ the conjunction of the following formulas:

$$\exists x P(x) \land \forall x \forall y (P(x) \land P(y) \to x = y), \tag{11.57}$$

$$\Box_F^+ \forall x (P(x) \to \bigcirc P(x)), \tag{11.58}$$

$$\Box_F^+ \forall x \forall y (\bigcirc P(x) \land \bigcirc P(y) \land \neg P(x) \land \neg P(y) \to x = y), \tag{11.59}$$

$$\Diamond_F \forall x (P(x) \leftrightarrow \Diamond_F P(x)). \tag{11.60}$$

The reader can readily check that the following lemma holds:

Lemma 11.81. *For every first-order temporal model $\mathfrak{M} = \langle \langle \mathbb{N}, < \rangle, D, I \rangle$, we have $(\mathfrak{M}, 0) \models \chi$ iff the following conditions are satisfied:*

- $\left|P^{I(0)}\right| = 1$;
- *for all $n \in \mathbb{N}$, $P^{I(n)} \subseteq P^{I(n+1)}$ and $\left|P^{I(n+1)} - P^{I(n)}\right| \leq 1$;*
- *there is an $m \in \mathbb{N}$ such that for all $k \geq m$, $P^{I(m)} = P^{I(k)}$.*

(In other words, there is a unique element $a_0 \in D$ for which $P(a_0)$ holds true at moment 0; $P(a_0)$ remains true always in the future. At moment 1 there may be only two elements $a_0, a_1 \in D$ for which P is true, at moment 2 only three such elements, etc. We eventually reach a moment m starting from which P is stable.)

Suppose now that we are given an arbitrary $\mathcal{QL}$-sentence ψ which does not contain occurrences of P. Let Q be a unary predicate symbol not occurring in ψ either. Put

$$\chi' = \forall x(Q(x) \leftrightarrow \Diamond P(x)),$$

and denote by ψ^Q the relativization of ψ to Q (i.e., $\varphi^Q = \varphi$ for atomic φ, $(\neg\varphi)^Q = \neg\varphi^Q$, $(\varphi_1 \wedge \varphi_2)^Q = \varphi_1^Q \wedge \varphi_2^Q$, and $(\forall x\, \varphi)^Q = \forall x\, (Q(x) \rightarrow \varphi^Q))$. Clearly, all the formulas χ, χ', and ψ^Q are $\mathcal{QTL}^{=}_{\square}$-formulas.

Lemma 11.82. *The following conditions are equivalent:*

- ψ *is valid in all finite* $\mathcal{QL}$*-structures;*
- $\chi \wedge \chi' \rightarrow \psi^Q$ *is valid in all first-order temporal models based on* $\langle \mathbb{N}, < \rangle$.

Proof. Suppose $\chi \wedge \chi' \rightarrow \psi^Q$ is refuted in $\mathfrak{M} = \langle \langle \mathbb{N}, < \rangle, D, I \rangle$. Without loss of generality we may assume that $(\mathfrak{M}, 0) \models \chi \wedge \chi'$ and $(\mathfrak{M}, 0) \not\models \psi^Q$. By Lemma 11.81, $Q^{I(0)}$ is finite. Let J be the $\mathcal{QL}$-structure with domain $Q^{I(0)}$ and n-ary predicates R^J, $n \geq 0$, defined by taking, for every n-tuple $\langle a_1, \dots, a_n \rangle$ of elements in $Q^{I(0)}$,

$$\langle a_1, \dots, a_n \rangle \in R^J \quad \text{iff} \quad \langle a_1, \dots, a_n \rangle \in R^{I(0)}.$$

It is easily checked by induction that for every assignment $\mathfrak{a}$ in $Q^{I(0)}$ and every $\mathcal{QL}$-formula ϑ, we have $(\mathfrak{M}, 0) \models^{\mathfrak{a}} \vartheta^Q$ iff $J \models^{\mathfrak{a}} \vartheta$. It follows that the finite $\mathcal{QL}$-structure J refutes ψ.

Conversely, suppose that $J = \langle D, \dots, R^J, \dots \rangle$ is a finite $\mathcal{QL}$-structure refuting ψ and having domain $D = \{a_0, \dots, a_n\}$. Define a first-order temporal model $\mathfrak{M} = \langle \langle \mathbb{N}, < \rangle, D, I \rangle$ by taking $R^{I(0)} = R^J$ for the predicate symbols R in ψ, $Q^{I(0)} = D$, and for every $i \in \mathbb{N}$,

$$P^{I(i)} = \begin{cases} \{a_0, \dots, a_i\}, & \text{if } i \leq n, \\ D, & \text{if } i > n. \end{cases}$$

Clearly, we have $(\mathfrak{M}, 0) \not\models \psi^Q$. On the other hand, $(\mathfrak{M}, 0) \models \chi \wedge \chi'$ holds by Lemma 11.81. ❑

Now recall that by Trakhtenbrot's (1950) theorem (see also Börger *et al.* 1997) the set of $\mathcal{QL}$-formulas that are valid in all finite first-order structures is not recursively enumerable. As a consequence we obtain our theorem. ❑

We can formulate a general decidability criterion, similar to Theorem 11.21, for fragments of $\mathcal{QTL}^{=}_{\square}$ as well. To this end, for every $\mathcal{QTL}^{=}_{\square}$-sentence φ, define

$$C_x\varphi = sub_x\,\varphi \cup \{x = c, x \neq c \mid c \in con\,\varphi\}.$$

By a *type* for φ we mean this time any Boolean saturated[8] subset $\boldsymbol{t}$ of the set

$$\{\overline{\psi} \mid \psi \in C_x\varphi\}$$

(see Sections 11.2 and 11.3). A type $\boldsymbol{t}$ is said to be a *constant type* if $(x = c) \in \boldsymbol{t}$ for some $c \in con\,\varphi$. A *state candidate* for φ is a set T of types. Given a class $\mathcal{K}$ of $\mathcal{QL}$-structures, a state candidate T is called $\mathcal{K}$-*realizable* if there is a $\overline{\mathcal{QL}}$-structure I (with domain D^I) such that its $\mathcal{QL}$-reduct belongs to $\mathcal{K}$ and

$$T = \{\boldsymbol{t}^I(a) \mid a \in D^I\},$$

where $\boldsymbol{t}^I(a) = \{\overline{\psi} \mid \psi \in C_x\varphi,\ I \models \overline{\psi}[a]\}$. If $\mathcal{K}$ is the class of all (finite) $\mathcal{QL}$-structures, then we simply say that such a state candidate is (*finitely*) *realizable*. Recall that for $\overline{\mathcal{QL}}$-structure I and type $\boldsymbol{t}$ for φ,

$$I_{\boldsymbol{t}} = \{a \in D^I \mid I \models \bigwedge_{\psi \in \boldsymbol{t}} \psi[a]\}.$$

The following general decidability criterion is an analog of Theorem 11.21:

Theorem 11.83. *Let $\mathcal{QTL}' \subseteq \mathcal{QTL}^{=}_{\boxed{\text{U}}}$, and let $\mathcal{K}$ be a class of $\mathcal{QL}$-structures such that the following two conditions hold:*

(a) *there is an algorithm which is capable of deciding, for every $\mathcal{QTL}'$-sentence φ, whether an arbitrarily given state candidate for φ is $\mathcal{K}$-realizable;*

(b) *for every $\mathcal{QTL}'$-sentence φ, there is an infinite cardinal κ_φ such that for every cardinal $\kappa \geq \kappa_\varphi$ and every $\mathcal{K}$-realizable state candidate T for φ, there is a $\overline{\mathcal{QL}}$-structure I realizing T and such that the $\mathcal{QL}$-reduct of I is in $\mathcal{K}$ and the sets $I_{\boldsymbol{t}}$ are of cardinality κ, for all nonconstant types $\boldsymbol{t} \in T$.*

Then the satisfiability problem for $\mathcal{QTL}'$-sentences in first-order temporal $\mathcal{K}$-models that are based on a flow of time from $\mathcal{C}$ is decidable, whenever $\mathcal{C}$ is one of the classes from the following list: $\{\langle \mathbb{N}, <\rangle\}$, $\{\langle \mathbb{Z}, <\rangle\}$, $\{\langle \mathbb{Q}, <\rangle\}$, the class of all finite strict linear orders, any first-order definable class of strict linear orders.

Proof. We modify the proof of Theorem 11.21. Fix some $\mathcal{QTL}' \subseteq \mathcal{QTL}^{=}_{\boxed{\text{U}}}$ and a class $\mathcal{K}$ of $\mathcal{QL}$-structures. We again define quasimodels. Suppose that $\mathfrak{F} = \langle W, <\rangle$ is a strict linear order in $\mathcal{C}$. A $\mathcal{K}$-*state function* over $\mathfrak{F}$ is a map $\boldsymbol{q}$ associating with each $w \in W$ a $\mathcal{K}$-realizable state candidate for φ. By a *run*

[8] One might also require types to be closed under 'equational reasoning,' but we do not need this in the proofs.

through $\boldsymbol{q}$ we mean a function r from W into the set $\bigcup_{w \in W} \boldsymbol{q}(w)$ such that $r(w) \in \boldsymbol{q}(w)$, for all $w \in W$ and

$$\forall c \in con\,\varphi\ \forall w, w' \in W\ \big((x = c) \in r(w) \text{ iff } (x = c) \in r(w')\big). \qquad (11.61)$$

Coherent and saturated runs are defined as in the proof of Theorem 11.21. A *$\mathcal{K}$-quasimodel* for φ based on $\mathfrak{F}$ is a triple $\mathfrak{Q} = \langle \mathfrak{F}, \boldsymbol{q}, \mathfrak{R} \rangle$, where $\boldsymbol{q}$ is a $\mathcal{K}$-state function and $\mathfrak{R}$ is a set of coherent and saturated runs through $\boldsymbol{q}$ satisfying **(tqm1)** and **(tqm3)** with $T_w = \boldsymbol{q}(w)$, $w \in W$ (see page 483).

We have the following analog of Lemma 11.22:

Lemma 11.84. *Let $\mathcal{QTL}'$ and $\mathcal{K}$ satisfy* (b) *of Theorem 11.83, and let $\mathfrak{F}$ be a strict linear order in $\mathcal{C}$. Then a $\mathcal{QTL}'$-sentence φ is satisfiable in a first-order temporal $\mathcal{K}$-model based on $\mathfrak{F}$ iff there is a $\mathcal{K}$-quasimodel for φ based on $\mathfrak{F}$.*

Proof. The '$\Rightarrow$'-direction of this lemma is proved in precisely the same way as the corresponding part of Lemma 11.22.

For the '$\Leftarrow$'-direction, suppose that $\mathfrak{F} = \langle W, < \rangle$ is in $\mathcal{C}$ and $\langle \mathfrak{F}, \boldsymbol{q}, \mathfrak{R} \rangle$ is a $\mathcal{K}$-quasimodel for φ. Take an infinite cardinal κ exceeding both κ_φ supplied by condition (b) and the cardinality of the set $\mathfrak{R}$ of runs, and put

$$D = \{\langle r, \xi \rangle \mid r \in \mathfrak{R},\ r(w) \text{ is not a constant type, for all } w \in W,\ \xi < \kappa\} \cup \{\langle r, 0 \rangle \mid r \in \mathfrak{R},\ r(w) \text{ is a constant type, for some } w \in W\}.$$

Fix some $w \in W$. For each type $\boldsymbol{t}$ in $\boldsymbol{q}(w)$, let

$$\lambda_{\boldsymbol{t}}(w) = |\{\langle r, \xi \rangle \in D \mid r(w) = \boldsymbol{t}\}|.$$

We claim that

- $\lambda_{\boldsymbol{t}}(w) = 1$ if $\boldsymbol{t}$ is a constant type, and
- $\lambda_{\boldsymbol{t}}(w) = \kappa$ otherwise.

The second claim clearly holds. For the first, if $\boldsymbol{t}$ is a constant type, then $\langle r, 0 \rangle \in D$ for some $r \in \mathfrak{R}$, so $\lambda_{\boldsymbol{t}}(w) \geq 1$. Suppose that $r, r' \in \mathfrak{R}$ satisfy $r(w) = r'(w) = \boldsymbol{t}$. We show that $r = r'$ must hold, that is, for all $u \in W$, we have $r(u) = r'(u)$. Choose a $c \in con\,\varphi$ with $x = c \in \boldsymbol{t}$. So $x = c \in r(w) \cap r'(w)$. Since r and r' are runs, by (11.61) we have $x = c \in r(u) \cap r'(u)$ for all $u \in W$. Pick a first-order structure J realizing $\boldsymbol{q}(u)$, and let a, a' be elements of its domain such that $\boldsymbol{t}^J(a) = r(u)$ and $\boldsymbol{t}^J(a') = r'(u)$. Then $a = c^J$ and $a' = c^J$, so $a = a'$, which implies $r(u) = r'(u)$. As u was arbitrary, $r = r'$ as claimed.

By condition (b), for each $w \in W$ there exists a $\overline{\mathcal{QL}}$-structure $I(w)$ with domain $D(w)$ such that $I(w)$ realizes the state candidate $\boldsymbol{q}(w)$, the $\mathcal{QL}$-reduct $I'(w)$ of $I(w)$ is in $\mathcal{K}$, and for every $\boldsymbol{t} \in \boldsymbol{q}(w)$ there are $\lambda_{\boldsymbol{t}}(w)$ elements in $D(w)$ realizing $\boldsymbol{t}$. So we can identify $D(w)$ and D in a 'type preserving and constant respecting' way and complete the proof as for Lemma 11.22. ❑

We can now deduce Theorem 11.83 by translating into monadic second-order logic the statement that there exists a $\mathcal{K}$-quasimodel for φ, and using Theorem 1.28, as it was done in the proof of Theorem 11.21. ❑

Although, as we saw, the full monodic fragment with equality is not recursively enumerable, one might still hope that the criterion of Theorem 11.83 applies to the monodic fragments listed in Section 11.2, and these fragments remain *decidable* with equality added to the language. The next result of Degtyarev *et al.* (2002) shows that this is not the case, at least for the *monodic monadic two-variable fragment* over the flow of time $\langle \mathbb{N}, < \rangle$ (cf. Theorems 11.12 and 11.15).

Theorem 11.85. *The set of monadic two-variable $\mathcal{QTL}^{=}_{\boxed{\scriptstyle\cup}}$-formulas that are valid in all first-order temporal models based on $\langle \mathbb{N}, < \rangle$ is not recursively enumerable.*

The proof goes via encoding of the behavior of Minsky machines (see Minsky 1961). We leave it to the reader as an exercise.

Better news is the following analog of Theorem 11.18 for the *monodic temporal packed fragment with equality*, due to Hodkinson (2002b). Define the fragment $\mathcal{TPF}^{=}$ of $\mathcal{QTL}^{=}$ the same way as $\mathcal{TPF}$ was defined in Section 11.2, but now also allowing equations as atomic formulas (in the guards as well), and let

$$\begin{aligned} \mathcal{PF}^{=} &= \{\overline{\varphi} \mid \varphi \in \mathcal{TPF}^{=}\} \\ \mathcal{TPF}^{=}_{\boxed{\scriptstyle\cup}} &= \mathcal{TPF}^{=} \cap \mathcal{QTL}^{=}_{\boxed{\scriptstyle\cup}}. \end{aligned}$$

Theorem 11.86. *Let $\mathcal{C}$ be any of the following classes of flows of time: $\{\langle \mathbb{N}, < \rangle\}$, $\{\langle \mathbb{Z}, < \rangle\}$, $\{\langle \mathbb{Q}, < \rangle\}$, the class of all finite strict linear orders, any first-order definable class of strict linear orders. Then it is decidable whether a $\mathcal{TPF}^{=}_{\boxed{\scriptstyle\cup}}$-sentence is satisfiable in a first-order temporal model based on a flow of time in $\mathcal{C}$.*

If $\mathcal{H}$ is any of the listed classes or $\mathcal{H} = \{\langle \mathbb{R}, < \rangle\}$ then it is decidable whether a $\mathcal{TPF}^{=}_{\boxed{\scriptstyle\cup}}$-sentence is satisfiable in a first-order temporal model based on a flow of time in $\mathcal{H}$ and having finite domains.

Proof. Let us consider first satisfiability in first-order temporal models with arbitrary domains. We show that conditions (a) and (b) of Theorem 11.83 hold, for $\mathcal{QTL}' = \mathcal{TPF}^{=}_{\boxed{\scriptstyle\cup}}$ and $\mathcal{K}$ being the class of all $\mathcal{QL}$-structures. Indeed, (a) has already been shown in the proof of Theorem 11.18, since the packed fragment $\mathcal{PF}$ of first-order logic is decidable even with equality added. In order to establish (b), we prove the following analog of Claims 11.24 and 11.42:

Claim 11.87. *For every $\mathcal{TPF}^{=}_{\square}$-sentence φ, there is a (finite) cardinal κ_φ such that, for every (finitely) realizable state candidate T for φ, and every sequence*

$$\langle \lambda_{\boldsymbol{t}} \mid \boldsymbol{t} \in T, \boldsymbol{t} \text{ is not a constant type} \rangle$$

of (finite) cardinals with $\lambda_{\boldsymbol{t}} \geq \kappa_\varphi$, there is a $\overline{\mathcal{QL}}$-structure I realizing T and such that, for every type $\boldsymbol{t} \in T$,

$$|I_{\boldsymbol{t}}| = \begin{cases} \lambda_{\boldsymbol{t}}, & \text{if } \boldsymbol{t} \text{ is not a constant type,} \\ 1, & \text{otherwise.} \end{cases}$$

Proof. Suppose that $T_0, \dots, T_k$ are all the distinct (finitely) realizable state candidates for φ and that for each $j \leq k$, I^j is a (finite) $\overline{\mathcal{QL}}$-structure realizing T_j. Put

$$\kappa_\varphi = \max\{|I^j_{\boldsymbol{t}}| \mid \boldsymbol{t} \in T_j, j \leq k\}.$$

Suppose that T is a (finitely) realizable state candidate for φ and for each non-constant type $\boldsymbol{t}$ in T, we are given a (finite) cardinal $\lambda_{\boldsymbol{t}} \geq \kappa_\varphi$. By the definition of κ_φ, we can choose a (finite) $\overline{\mathcal{QL}}$-structure $J = \langle D^J, P_0^J, \dots, c_0^J, \dots \rangle$ realizing T and such that all the sets $J_{\boldsymbol{t}}$, $\boldsymbol{t} \in T$, have cardinality $\leq \lambda_{\boldsymbol{t}}$.

Now the idea again is to make copies of the elements of D^J, but this time carefully, only of those that are not named by constants. For each type $\boldsymbol{t} \in T$, choose an arbitrary element $a_{\boldsymbol{t}} \in J_{\boldsymbol{t}}$ and define the cardinal $\lambda_{a_{\boldsymbol{t}}}$ such that

$$|J_{\boldsymbol{t}} - \{a_{\boldsymbol{t}}\}| + \lambda_{a_{\boldsymbol{t}}} = \lambda_{\boldsymbol{t}},$$

and for all other elements $b \in J_{\boldsymbol{t}}$, let $\lambda_b = 1$. Note that $\lambda_{c^J} = 1$ for each $c \in con\,\varphi$. Now define a new domain D^I by taking

$$D^I = \{\langle a, \xi \rangle \mid a \in D^J, \xi < \lambda_a\}.$$

A subset S of D^I is said to be *thin* if whenever $\langle a, \xi_1 \rangle, \langle a, \xi_2 \rangle \in S$ then $\xi_1 = \xi_2$. We define a $\overline{\mathcal{QL}}$-structure I with domain D^I by taking, for each constant symbol c and each n-ary predicate symbol P,

$$\begin{aligned} c^I &= \{\langle c^J, 0 \rangle\}, \\ P^I &= \{\langle \langle a_1, \xi_1 \rangle, \dots, \langle a_n, \xi_n \rangle \rangle \mid \langle a_1, \dots, a_n \rangle \in P^J, \xi_i < \lambda_{a_i}, \\ &\qquad \{\langle a_1, \xi_1 \rangle, \dots, \langle a_n, \xi_n \rangle\} \text{ is thin}\}. \end{aligned}$$

We call an assignment $\mathfrak{a}$ in I *thin* if its range is a thin set.

Given an assignment $\mathfrak{a}$ in I, we can define an assignment $\mathfrak{a}^-$ in J by putting, for each variable y, $\mathfrak{a}^-(y) = a$ iff $\mathfrak{a}(y) = \langle a, \xi \rangle$ for some $\xi < \lambda_a$. Observe that for all atomic $\overline{\mathcal{QL}}^{=}$-formulas α and assignments $\mathfrak{a}$ in I,

$$\text{if } I \models^{\mathfrak{a}} \alpha \text{ then } J \models^{\mathfrak{a}^-} \alpha. \tag{11.62}$$

Conversely, given an assignment $\mathfrak{b}$ in J and a thin assignment $\mathfrak{a}$ in I such that $\mathfrak{b}$ agrees with $\mathfrak{a}^-$ except perhaps on some $\overline{y}$, we can 'lift' $\mathfrak{b}$ to an assignment $\mathfrak{b}^{\mathfrak{a}}$ in I by taking

$$\mathfrak{b}^{\mathfrak{a}}(v) = \begin{cases} \mathfrak{a}(z), & \text{if } \mathfrak{b}(v) = \mathfrak{a}^-(z) \text{ for some variable } z, \\ \langle \mathfrak{b}(v), 0 \rangle, & \text{otherwise.} \end{cases}$$

Then it is easily checked that $\mathfrak{b}^{\mathfrak{a}}$ is well-defined, thin, $(\mathfrak{b}^{\mathfrak{a}})^- = \mathfrak{b}$, and $\mathfrak{b}^{\mathfrak{a}}$ agrees with $\mathfrak{a}$ except perhaps on $\overline{y}$.

Now we claim that for all $\mathcal{PF}^=$-formulas ψ and thin assignments $\mathfrak{a}$ in I,

$$I \models^{\mathfrak{a}} \psi \quad \text{iff} \quad J \models^{\mathfrak{a}^-} \psi. \tag{11.63}$$

We prove this by induction on ψ. For ψ atomic, observe that (since $\lambda_{c^J} = 1$ for all $c \in con\,\varphi$) the union of the range of $\mathfrak{a}$ and the set $\{c^I \mid c \in con\,\varphi\}$ is still thin. The Boolean cases are easy and we leave them to the reader. So consider a $\mathcal{PF}^=$-formula θ of the form $\exists \overline{y}(\gamma \wedge \psi)$, where γ is a packing guard, and ψ is a $\mathcal{PF}^=$-formula for which (11.63) is assumed inductively.

Assume first that $J \models^{\mathfrak{a}^-} \theta$. Then there is an assignment $\mathfrak{b}$ in J agreeing with $\mathfrak{a}^-$ except perhaps on $\overline{y}$ such that $J \models^{\mathfrak{b}} \gamma \wedge \psi$. So by the induction hypothesis, we have $I \models^{\mathfrak{b}^{\mathfrak{a}}} \psi$. To see that $I \models^{\mathfrak{b}^{\mathfrak{a}}} \gamma$ holds as well, take any conjunct $\exists \overline{z} \alpha$ of γ ($\overline{z}$ can be empty). Since $J \models^{\mathfrak{b}} \gamma$, there is an assignment $\mathfrak{d}$ in J agreeing with $\mathfrak{b}$ except perhaps on $\overline{z}$ such that $J \models^{\mathfrak{d}} \alpha$. By the induction hypothesis for atoms, we have $I \models^{\mathfrak{d}^{\mathfrak{b}^{\mathfrak{a}}}} \alpha$, and so $I \models^{\mathfrak{b}^{\mathfrak{a}}} \exists \overline{z} \alpha$. Hence $I \models^{\mathfrak{b}^{\mathfrak{a}}} \gamma \wedge \psi$, which certainly implies $I \models^{\mathfrak{b}^{\mathfrak{a}}} \theta$. Since $\mathfrak{a}$ and $\mathfrak{b}^{\mathfrak{a}}$ agree on the free variables of θ, we obtain $I \models^{\mathfrak{a}} \theta$.

For the converse, assume that $I \models^{\mathfrak{a}} \theta$. Then there is some assignment $\mathfrak{b}$ in I agreeing with $\mathfrak{a}$ except perhaps on $\overline{y}$ such that $I \models^{\mathfrak{b}} \gamma \wedge \psi$. Then for any conjunct $\exists \overline{z} \alpha$ of γ ($\overline{z}$ can be empty), there is an assignment $\mathfrak{d}$ in I agreeing with $\mathfrak{b}$ except perhaps on $\overline{z}$ such that $I \models^{\mathfrak{d}} \alpha$. By (11.62), we have $J \models^{\mathfrak{d}^-} \alpha$, and so $J \models^{\mathfrak{b}^-} \exists \overline{z} \alpha$. Hence, we have $J \models^{\mathfrak{b}^-} \gamma$.

We need to show that $J \models^{\mathfrak{b}^-} \psi$ also holds. Since, by assumption, all variables in $\overline{y}$ and all free variables in ψ occur free in γ, we may assume that $\overline{y}$ is a nonempty tuple of free variables of γ, and that for some y in $\overline{y}$ we have $\mathfrak{b}(v) = \mathfrak{b}(y)$ for every variable v that does not occur free in γ.

We claim that $\mathfrak{b}$ is thin. For, let v and w be distinct variables such that $\mathfrak{b}(v) = \langle a, \xi_1 \rangle$ and $\mathfrak{b}(w) = \langle a, \xi_2 \rangle$. We need to show that $\xi_1 = \xi_2$. By the assumption just made, we can suppose that v, w occur free in γ. So there is a conjunct $\exists \overline{z} \alpha$ of γ in which both v, w occur free. If α is an equality then we have $\mathfrak{b}(v) = \mathfrak{b}(w)$, and so $\xi_1 = \xi_2$. Suppose that α is of the form $P(x_1, \dots, x_n)$. We have $I \models^{\mathfrak{b}} \exists \overline{z} P(x_1, \dots, x_n)$, so there is an assignment $\mathfrak{d}$ in I agreeing with $\mathfrak{b}$ except perhaps on $\overline{z}$ (in particular, $\mathfrak{d}(v) = \mathfrak{b}(v)$ and $\mathfrak{d}(w) = \mathfrak{b}(w)$) such that $I \models^{\mathfrak{d}} P(x_1, \dots, x_n)$. Then, by the definition of P^I, the set $\{\mathfrak{d}(x_1), \dots, \mathfrak{d}(x_n)\}$

must be thin. Since this set contains $\langle a, \xi_1 \rangle$ and $\langle a, \xi_2 \rangle$, it follows again that $\xi_1 = \xi_2$.

So by the induction hypothesis, $J \models^{\mathfrak{b}^-} \psi$ holds. Since $\mathfrak{a}^-$ and $\mathfrak{b}^-$ agree except perhaps on $\overline{y}$, we have $J \models^{\mathfrak{a}^-} \theta$, proving (11.63).

Now by (11.63) we have, for all $\langle a, \xi \rangle \in D^I$,

$$\boldsymbol{t}^I(\langle a, \xi \rangle) = \boldsymbol{t}^J(a),$$

since types consist of $\mathcal{PF}^=$-formulas with at most one free variable, and the set $\{\langle a, \xi \rangle\}$ is thin. So for all types $\boldsymbol{t}$ in the state candidate T, we have

$$|I_t| = \left|\{\langle a_t, \xi \rangle \mid \xi < \lambda_{a_t}\} \cup \{\langle b, 0 \rangle \mid b \in J_t - \{a_t\}\}\right| = \lambda_{a_t} + \left|J_t - \{a_t\}\right| = \lambda_t,$$

as required. ❑

For satisfiability of $\mathcal{TPF}^=_{\square}$-sentences in first-order temporal models with finite domains, we proceed as follows. Given a $\mathcal{TPF}^=_{\square}$-sentence φ and a strict linear order $\mathfrak{F} = \langle W, < \rangle$ in $\mathcal{H}$, we call a quasimodel $\langle \mathfrak{F}, \boldsymbol{q}, \mathfrak{R} \rangle$ for φ *finitary*, if all the state candidates $\boldsymbol{q}(w)$, $w \in W$, are finitely realizable, and $\mathfrak{R}$ is finite. Now one can repeat the proof of Theorem 11.9 given in Sections 11.5 and 11.6 for the case $\mathcal{QTL}' = \mathcal{TPF}^=_{\square}$, using Claim 11.87 in place of Claim 11.42. ❑

Question 11.88. Do the other decidable monodic fragments of first-order temporal logics mentioned in Section 11.2 remain decidable after adding equality to the language?

Chapter 12

Fragments of first-order dynamic and epistemic logics

Now we extend some of the results of the previous chapter to the monodic fragments of the dynamic first-order logics **QDL** and **CQDL**, and the epistemic first-order logics $\mathbf{Q}L$, for $L \in \{\mathbf{K}_n^C, \mathbf{T}_n^C, \mathbf{K4}_n^C, \mathbf{S4}_n^C, \mathbf{KD45}_n^C, \mathbf{S5}_n^C\}$ (these logics were introduced in Section 3.6). In Section 12.1 we formulate decidability criteria and single out a number of decidable monodic fragments. In Section 12.2 we give Hilbert-style axiomatizations for the full monodic fragments of the logics above. These results are due to (Sturm *et al.* 2002).

12.1 Decision problems

Non-recursively enumerable fragments

To begin with, we show that, similarly to the temporal case, the expressive first-order modal logics mentioned above and even the restrictions to the two-variable, the monadic or the guarded fragments of some of them are undecidable; in fact, they are not even recursively enumerable. Define the *guarded fragments* $\mathcal{DGF}$ and $\mathcal{EGF}_n$ of $\mathcal{QDL}$ and $\mathcal{QML}_n^C$ in the same way as we defined the temporal guarded fragment $\mathcal{TGF}$ of $\mathcal{QTL}$ in Section 11.2.

Theorem 12.1. (i) *The two-variable monadic fragment of* **QDL** *and the two-variable fragment of* $\mathcal{DGF} \cap$ **QDL** *are not recursively enumerable.*

(ii) *Let* $L \in \{\mathbf{K}_1^C, \mathbf{T}_2^C, \mathbf{K4}_2^C, \mathbf{S4}_2^C, \mathbf{KD45}_2^C\}$. *Then the two-variable monadic fragment of* $\mathbf{Q}L$ *and the two-variable fragment of* $\mathcal{EGF} \cap \mathbf{Q}L$ *are not recursively enumerable.*

Proof. By 'lifting' the reductions given in the proofs of Theorems 2.36, 2.38 and 2.39 to the first-order case, one can reduce the decision problems for the two-variable, the monadic and the guarded fragments of $\mathsf{QLog}_{\mathcal{U}}(\mathbb{N})$ to the decision problems for the fragments mentioned in the formulation of the theorem (we leave the more or less obvious details to the reader). It remains to use Theorems 11.1 and 11.17. ❑

However, the following problem is still open:

Question 12.2. Is the two-variable fragment of $\mathbf{QS5}_2^C$ recursively enumerable?

Note that this fragment is undecidable because, by Theorem 8.35, already the two-variable fragment of any logic between **QK** and **QS5** is undecidable. The reader may find a proof of the following partial result in (Wolter 2000a):

Theorem 12.3. *The three-variable monadic fragment of* $\mathbf{QS5}_2^C$ *is not recursively enumerable.*

For undecidable and decidable fragments of first-order provability logic see (Japaridze and de Jongh 1998) and references therein.

Decidable monodic fragments

Similarly to the temporal case, several *monodic fragments* of the logics under consideration—i.e., fragments in which the modal operators can be applied to formulas with at most one free variable—turn out to be decidable. We denote the monodic fragments of the languages $\mathcal{CQDL}$ and $\mathcal{QML}_n^C$ by $\mathcal{CQDL}_{\boxed{1}}$ and $\mathcal{QML}_{n\boxed{1}}^C$. The corresponding monodic fragments with equality are denoted by $\mathcal{CQDL}_{\boxed{1}}^{=}$ and $\mathcal{QML}_{n\boxed{1}}^{C=}$.

As in the temporal case, for every $\mathcal{QML}_{n\boxed{1}}^{C=}$-formula $\psi(y)$ of the form $\Box_i\chi(y)$ or $\mathsf{C}_M\chi(y)$ with one free variable y, we reserve a unary predicate $P_\psi(y)$ that does not occur in φ. Likewise, for every $\mathcal{QML}_{n\boxed{1}}^{C=}$-sentence $\psi = \Box_i\chi$ or $\psi = \mathsf{C}_M\chi$, we fix a propositional variable p_ψ not occurring in φ. $P_\psi(y)$ and p_ψ are called the *surrogates* for $\psi(y)$ and ψ, respectively. Now, given a $\mathcal{QML}_{n\boxed{1}}^{C=}$-formula φ, we denote by $\overline{\varphi}$ the formula that results from φ by replacing all subformulas of the form $\Box_i\psi(y)$, $\Box_i\psi$, $\mathsf{C}_M\psi(y)$, and $\mathsf{C}_M\psi$ which are not within the scope of another epistemic operator ($\Box_j$ or C_M) with their surrogates. Observe that the formula $\overline{\varphi}$ contains no occurrences of epistemic operators at all—i.e., it is a $\mathcal{QL}$-formula (see Section 11.2). We can define $\overline{\varphi}$ for a $\mathcal{CQDL}_{\boxed{1}}^{=}$-formula φ a similar way, by replacing all subformulas of the form $[\alpha]\psi$, α an action term, which are not within the scope of another $[\beta]$ operator with their surrogates.

Types and state candidates for monodic formulas, and their $\mathcal{K}$-realizability, are defined as it was done for the temporal case in Section 11.2 (and in Section 11.8, for monodic formulas with equality).

Now decidable monodic fragments of first-order dynamic and epistemic logics can be singled out using a criterion similar to Theorem 11.83 in the temporal case.

Theorem 12.4. *Let $\mathcal{L}'$ be a sublanguage of either $\mathcal{CQDL}^{=}_{\boxed{1}}$ or $\mathcal{QML}^{C=}_{n\boxed{1}}$, and let $\mathcal{K}$ be a class of $\mathcal{QL}$-structures such that the following two conditions hold:*

(a) *there is an algorithm which is capable of deciding, for any $\mathcal{L}'$-sentence φ, whether an arbitrarily given state candidate for φ is $\mathcal{K}$-realizable;*

(b) *for every $\mathcal{L}'$-sentence φ, there is an infinite cardinal κ_φ such that for every cardinal $\kappa \geq \kappa_\varphi$ and every $\mathcal{K}$-realizable state candidate T for φ, there is a $\overline{\mathcal{QL}}$-structure I realizing T and such that the $\mathcal{QL}$-reduct of I is in $\mathcal{K}$ and the sets $I_{\boldsymbol{t}}$ are of cardinality κ, for all nonconstant types $\boldsymbol{t} \in T$.*

Then the satisfiability problem for $\mathcal{L}'$-sentences in, respectively,

- *first-order dynamic $\mathcal{K}$-models for* $\mathbf{CQDL}^{=}$,
- *first-order modal $\mathcal{K}$-models for $\mathbf{Q}L^{=}$, where $L \in \{\mathbf{K}^C_n, \mathbf{T}^C_n, \mathbf{K4}^C_n, \mathbf{S4}^C_n, \mathbf{KD45}^C_n, \mathbf{S5}^C_n\}$*

is decidable.

Proof. By 'lifting' the reductions given in the proof of Theorem 2.39 to the first-order case, one can reduce the satisfiability problems for all the logics listed in the theorem to the satisfiability problem of the corresponding monodic $\mathbf{CQDL}^{=}$-fragment, so it suffices to prove the theorem for this case. Quasimodels for monodic $\mathcal{CQDL}^{=}$-sentences can be defined in a way similar to the first-order temporal case (see Sections 11.2 and 11.8). Then one can prove the analog of the 'quasimodel' Lemma 11.84. Finally, we can obtain decidability with the help of an analog of the block-technique used in the proof of decidability of $\mathbf{CPDL} \times \mathbf{S5}$ (Theorem 6.49). Details are left to the reader. ❑

Now the following analog of Theorem 11.7 can be obtained as a corollary:

Theorem 12.5. (i) *Suppose $\mathcal{L}' \subseteq \mathcal{CQDL}_{\boxed{1}}$ and there is an algorithm which is capable of deciding, for any $\mathcal{L}'$-sentence φ, whether an arbitrarily given state candidate for φ is realizable. Then $\mathcal{L}' \cap$ $\mathbf{CQDL}$ is decidable.*

(ii) *Suppose* $\mathcal{L}' \subseteq \mathcal{QML}^C_{n\boxed{1}}$, *and* $L \in \{\mathbf{K}^C_n, \mathbf{T}^C_n, \mathbf{K4}^C_n, \mathbf{S4}^C_n, \mathbf{KD45}^C_n, \mathbf{S5}^C_n\}$. *If there is an algorithm which is capable of deciding, for any* $\mathcal{L}'$*-sentence* φ, *whether an arbitrarily given state candidate for* φ *is realizable, then* $\mathcal{L}' \cap \mathbf{QL}$ *is decidable.*

Let the *packed fragments* of $\mathcal{CQDL}^=$ and $\mathcal{QML}^{C=}_n$ be defined similarly to the temporal packed fragment $\mathcal{TPF}^=$ of $\mathcal{QTL}^=$ in Section 11.8. As a consequence of the theorems above we obtain, in particular, the following decidability results:

Theorem 12.6. *Suppose* $L \in \{\mathbf{QK}^C_n, \mathbf{QT}^C_n, \mathbf{QK4}^C_n, \mathbf{QS4}^C_n, \mathbf{QKD45}^C_n, \mathbf{QS5}^C_n, \mathbf{CQDL}\}$. *Then*

- *the monadic monodic fragment of* L,
- *the two-variable monodic fragment of* L,
- *the monodic packed fragment of* L *with equality*

are decidable.

We remind the reader that the one-variable constant-free fragment of $\mathbf{Q}L$, for all the dynamic and epistemic logics L considered above, is 'equivalent' to the propositional product logic $L \times \mathbf{S5}$; see Theorem 3.21.

For some other results on decidable first-order epistemic logics see (Japaridze 2000).

12.2 Axiomatizing monodic fragments

In contrast to Theorem 12.1 and similar to the temporal case, the monodic fragments of first-order epistemic and dynamic logics turn out to be axiomatizable. The Hilbert-style axiomatizations below can be obtained by putting together the axiom schemata and inference rules of classical first-order logic **QCl** (Section 1.3), those of the corresponding propositional epistemic or dynamic logic (Sections 2.3 and 2.4), the corresponding Barcan axioms, and by restricting the range of the schemata and rules to monodic formulas. More precisely, let $\mathcal{MONK}^C_n$ be the axiomatic system with the following axiom schemata and inference rules:

Axiom schemata (ranging over monodic $\mathcal{QML}^C_n$-formulas):

the axiom schemata of classical first-order logic **QCl**,

$$\Box_i(\varphi \to \psi) \to (\Box_i\varphi \to \Box_i\psi), \quad \text{for } 1 \le i \le n, \tag{12.1}$$

$$\mathsf{C}_M\varphi \leftrightarrow (\varphi \wedge \mathsf{E}_M\mathsf{C}_M\varphi), \quad \text{for } M \subseteq \{1,\dots,n\},\ |M| \ge 1, \tag{12.2}$$

$$\Box_i\forall x\psi \leftrightarrow \forall x\Box_i\psi, \quad \text{for } 1 \le i \le n. \tag{12.3}$$

Inference rules (ranging over monodic $\mathcal{QML}_n^C$-formulas):

the rules of **QCl**,

given φ, derive $\Box_i\varphi$, for $1 \le i \le n$, (12.4)

given $\varphi \to \psi \wedge \mathsf{E}_M\varphi$, derive $\varphi \to \mathsf{C}_M\psi$,

for $M \subseteq \{1, \dots, n\}$, $|M| \ge 1$. (12.5)

Let $\mathcal{MONT}_n^C$ be the axiomatic system obtained by adding to $\mathcal{MONK}_n^C$ the schema (A_T), $\mathcal{MONK4}_n^C$ the system obtained from $\mathcal{MONK}_n^C$ by adding the schema (A_4), $\mathcal{MONS4}_n^C$ the axiomatic system obtained from $\mathcal{MONT}_n^C$ by adding (A_4), let $\mathcal{MONKD45}_n^C$ be $\mathcal{MONK}_n^C$ extended by (A_D), (A_4), and (A_5), and let $\mathcal{MONS5}_n^C$ be $\mathcal{MONKD45}_n^C$ plus (A_T), where

(A_D) $\Box_i\varphi \to \Diamond_i\varphi$, for $1 \le i \le n$;

(A_T) $\Box_i\varphi \to \varphi$, for $1 \le i \le n$;

(A_4) $\Box_i\varphi \to \Box_i\Box_i\varphi$, for $1 \le i \le n$;

(A_5) $\neg\Box_i\varphi \to \Box_i\neg\Box_i\varphi$, for $1 \le i \le n$,

all ranging over monodic $\mathcal{QML}_n^C$-formulas.

Let $\mathcal{MONC}$ be one of the axiomatic systems

$$\mathcal{MONK}_n^C,\ \mathcal{MONT}_n^C,\ \mathcal{MONK4}_n^C,\ \mathcal{MONS4}_n^C,\ \mathcal{MONKD45}_n^C,\ \mathcal{MONS5}_n^C.$$

A monodic $\mathcal{QML}_n^C$-formula φ is *$\mathcal{MONC}$-derivable* (in symbols: $\vdash_{\mathcal{MONC}} \varphi$) if there is a sequence of monodic $\mathcal{QML}_n^C$-formulas ending with φ and such that each member of the sequence is either a substitution instance of an axiom schema of $\mathcal{MONC}$, or obtained from some earlier members of the sequence by applying one of the inference rules of $\mathcal{MONC}$.

Remark 12.7. Using the axiomatizations of propositional epistemic logics formulated in Remark 2.18, one can also give the corresponding alternative axiomatizations for the monodic fragment of first-order epistemic logics.

Theorem 12.8. *Let $\mathcal{MONC}$ be one of the axiomatic systems $\mathcal{MONK}_n^C$, $\mathcal{MONT}_n^C$, $\mathcal{MONK4}_n^C$, $\mathcal{MONS4}_n^C$, $\mathcal{MONKD45}_n^C$, $\mathcal{MONS5}_n^C$, and let L be the corresponding logic from the list* $\mathbf{K}_n^C$, $\mathbf{T}_n^C$, $\mathbf{K4}_n^C$, $\mathbf{S4}_n^C$, $\mathbf{KD45}_n^C$, $\mathbf{S5}_n^C$. *Then for every monodic $\mathcal{QML}_n^C$-formula φ, we have*

$$\vdash_{\mathcal{MONC}} \varphi \quad \textit{iff} \quad \varphi \in \mathbf{Q}L.$$

Proof. The *soundness* part ($\Rightarrow$) is easy; we leave it to the reader as an exercise and concentrate on the *completeness* part ($\Leftarrow$). It suffices to show that every $\mathcal{MONC}$-consistent monodic $\mathcal{QML}_n^C$-formula φ (i.e., a formula φ such that $\not\vdash_{\mathcal{MONC}} \neg\varphi$) is satisfiable in a first-order Kripke model based on a frame for L. Fix such a φ. As before, without loss of generality we may assume that φ is a sentence. Indeed, if a monodic $\mathcal{QML}_n^C$-formula $\varphi(x_1, \dots, x_n)$ is in $\mathbf{Q}L$ then so is the monodic sentence $\forall x_1 \dots \forall x_n \varphi(x_1, \dots, x_n)$. So if we succeed to prove that $\vdash_{\mathcal{MONC}} \forall x_1 \dots \forall x_n \varphi(x_1, \dots, x_n)$, then we will also have $\vdash_{\mathcal{MONC}} \varphi(x_1, \dots, x_n)$, since $\mathcal{MONC}$ contains the axiom schemata of classical first-order logic. To simplify notation, we will also assume that φ contains occurrences of the operators $\mathsf{C} = \mathsf{C}_{\{1,\dots,n\}}$ and $\mathsf{E} = \mathsf{E}_{\{1,\dots,n\}}$ only. (Recall that E is expressible via the $\Box_i$.)

Similarly to the proof of Theorem 11.71, we will be constructing models using a kind of 'syntactical quasimodels.'

Given a set Γ of monodic $\mathcal{QML}_n^C$-formulas, we denote by $con\,\Gamma$ and $sub\,\Gamma$ the sets of all constants and all subformulas of formulas in Γ, respectively, and denote by $sub_C\,\Gamma$ the following set:

$$sub_C\,\Gamma = sub\,\Gamma \cup \{\mathsf{E}\mathsf{C}\,\psi \mid \mathsf{C}\,\psi \in sub\,\Gamma\} \cup \{\Box_i \mathsf{C}\,\psi \mid \mathsf{C}\,\psi \in sub\,\Gamma,\ i = 1, \dots, n\}.$$

Further, let

$$sub_C^{\neg}\,\Gamma = sub_C\,\Gamma \cup \{\neg\psi \mid \psi \in sub_C\,\Gamma\},$$

and let $sub_n\,\Gamma$ be the subset of $sub_C^{\neg}\,\Gamma$ containing only formulas with $\leq n$ free variables. Note that modulo equivalence $\vartheta \leftrightarrow \neg\neg\vartheta$ we may assume that $sub_n\,\Gamma$ is closed under $\neg$. If Γ is a singleton set, say $\Gamma = \{\psi\}$, we write $con\,\psi$ instead of $con\,\{\psi\}$, $sub_n\,\psi$ instead of $sub_n\,\{\psi\}$, etc. As before, we do not distinguish between a finite set Γ of formulas and the conjunction $\bigwedge \Gamma$ of formulas in it. Let x be a variable not occurring in Γ. Put

$$sub_x\,\Gamma \;=\; \{\psi\{x/y\} \mid \psi(y) \in sub_1\,\Gamma\} \;\cup\; \{\Box_i\bot, \neg\Box_i\bot \mid i \leq n\} \;\cup\; \{\top, \bot\}.$$

All formulas in $sub_x\,\Gamma$ have at most one free variable, and that variable is x. If Γ is finite, then $sub_x\,\Gamma$ is finite as well.

As before, by a *type* for φ we mean a Boolean-saturated subset $\boldsymbol{t}$ of $sub_x\,\varphi$. We say that two types $\boldsymbol{t}$ and $\boldsymbol{t}'$ *agree on* $sub_0\,\varphi$ if $\boldsymbol{t} \cap sub_0\,\varphi = \boldsymbol{t}' \cap sub_0\,\varphi$. Given a type $\boldsymbol{t}$ for φ and a constant $c \in con\,\varphi$, the pair $\langle c, \boldsymbol{t}\rangle$ is called an *indexed type* for φ (indexed by c).

A pair $\mathfrak{C} = \langle T_{\mathfrak{C}}, T_{\mathfrak{C}}^{con}\rangle$ is called a *state candidate* for φ if $T_{\mathfrak{C}}$ is a (nonempty) set of types for φ that agree on $sub_0\,\varphi$, and

$$T_{\mathfrak{C}}^{con} \subseteq con\,\varphi \times T_{\mathfrak{C}}$$

is a set of indexed types such that for each $c \in con\,\varphi$ there is a unique $\boldsymbol{t} \in T_{\mathfrak{C}}$ with $\langle c, \boldsymbol{t}\rangle \in T_{\mathfrak{C}}^{con}$. As before, indexed types $\langle c, \boldsymbol{t}\rangle$ in $T_{\mathfrak{C}}^{con}$ will also be denoted by $\boldsymbol{t}_{\mathfrak{C}}^{c}$.

Given a state candidate $\mathfrak{C}$, we put

$$\mathsf{real}_{\mathfrak{C}} \;=\; \bigwedge_{\boldsymbol{t} \in T_{\mathfrak{C}}} \exists x\, \boldsymbol{t}(x) \;\wedge\; \bigwedge_{c \in con\,\varphi} \boldsymbol{t}_{\mathfrak{C}}^{c}\{c/x\} \;\wedge\; \forall x \bigvee_{\boldsymbol{t} \in T_{\mathfrak{C}}} \boldsymbol{t}(x).$$

Say that a state candidate $\mathfrak{C}$ is *$\mathcal{MONC}$-consistent* if the sentence $\mathsf{real}_{\mathfrak{C}}$ is consistent with $\mathcal{MONC}$. A pair $\langle \boldsymbol{t}_1, \boldsymbol{t}_2 \rangle$ of types for φ is called *i-suitable for $\mathcal{MONC}$*, $i = 1, \dots, n$, if the formula $\boldsymbol{t}_1 \wedge \Diamond_i \boldsymbol{t}_2$ is consistent with $\mathcal{MONC}$. A pair of state candidates $\langle \mathfrak{C}_1, \mathfrak{C}_2 \rangle$ is *i-suitable for $\mathcal{MONC}$*, $i = 1, \dots, n$, if $\mathsf{real}_{\mathfrak{C}_1} \wedge \Diamond_i \mathsf{real}_{\mathfrak{C}_2}$ is consistent with $\mathcal{MONC}$. Note that if the pair $\langle \boldsymbol{t}_1, \boldsymbol{t}_2 \rangle$ is i-suitable for some $i = 1, \dots, n$, then the $\mathcal{MONC}$-consistency of both $\boldsymbol{t}_1$ and $\boldsymbol{t}_2$ follows by (12.4). The same applies to suitable pairs of state candidates.

A *basic $\mathcal{MONC}$-structure* for φ is a pair $\langle \mathfrak{F}, \boldsymbol{q} \rangle$ such that

$$\mathfrak{F} = \langle W, <_1, \dots, <_n \rangle$$

is an intransitive tree, and $\boldsymbol{q}$ is a map associating with every $w \in W$ an $\mathcal{MONC}$-consistent state candidate $\boldsymbol{q}(w) = \langle T_w, T_w^{con} \rangle$ for φ.

A *run* through $\langle \mathfrak{F}, \boldsymbol{q} \rangle$ is a map r associating with every $w \in W$ a type $r(w)$ in T_w. We call such an r *coherent* if

- the pair $\langle r(w_1), r(w_2) \rangle$ is i-suitable for $\mathcal{MONC}$, whenever $w_1 <_i w_2$;

and *saturated* if the following hold:

- if $\neg \Box_i \psi \in r(w)$ then there is $w' \in W$ such that $w <_i w'$ and $\psi \notin r(w')$;
- if $\neg \mathsf{C}\, \psi \in r(w)$ then there is $w' \in W$ such that $w (\bigcup_{1 \le i \le n} <_i)^* w'$ and $\psi \notin r(w')$.

A *$\mathcal{MONC}$-quasimodel* for φ is a triple $\mathfrak{Q} = \langle \mathfrak{F}, \boldsymbol{q}, \mathfrak{R} \rangle$, where $\langle \mathfrak{F}, \boldsymbol{q} \rangle$ is a basic $\mathcal{MONC}$-structure for φ such that

(meqm1) $\varphi \in \boldsymbol{t}$ for some $w \in W$ and $\boldsymbol{t} \in T_w$,

(meqm2) the pair $\langle \boldsymbol{q}(w_1), \boldsymbol{q}(w_2) \rangle$ is i-suitable for $\mathcal{MONC}$, whenever $w_1 <_i w_2$, $i = 1, \dots, n$,

and $\mathfrak{R}$ is a set of coherent and saturated runs through $\langle \mathfrak{F}, \boldsymbol{q} \rangle$ satisfying the following conditions:

(meqm3) for every $c \in con\,\varphi$, the function r_c defined by $r_c(w) = \boldsymbol{t}$, for $\langle c, \boldsymbol{t} \rangle \in T_w^{con}$, $w \in W$, is a run in $\mathfrak{R}$,

(meqm4) for every $w \in W$ and every type $\boldsymbol{t}$ in T_w there exists a run r in $\mathfrak{R}$ such that $r(w) = \boldsymbol{t}$.

Note that, for any two sets $\mathfrak{R}_1$ and $\mathfrak{R}_2$ of coherent and saturated runs through $\langle \mathfrak{F}, \boldsymbol{q} \rangle$, if $\mathfrak{R}_1 \subseteq \mathfrak{R}_2$ and $\langle \mathfrak{F}, \boldsymbol{q}, \mathfrak{R}_1 \rangle$ is a $\mathcal{MONC}$-quasimodel for φ then $\langle \mathfrak{F}, \boldsymbol{q}, \mathfrak{R}_2 \rangle$ is a $\mathcal{MONC}$-quasimodel for φ as well. Consequently, we may always assume that a $\mathcal{MONC}$-quasimodel for φ is of the form $\langle \mathfrak{F}, \boldsymbol{q}, \mathfrak{R}_{\boldsymbol{q}} \rangle$, where $\mathfrak{R}_{\boldsymbol{q}}$ denotes the set of all coherent and saturated runs through $\langle \mathfrak{F}, \boldsymbol{q} \rangle$.

Lemma 12.9. *Suppose $\mathcal{MONC}$ is one of the axiomatic systems $\mathcal{MONK}^C_n$, $\mathcal{MONT}^C_n$, $\mathcal{MONK4}^C_n$, $\mathcal{MONS4}^C_n$, $\mathcal{MONKD45}^C_n$, $\mathcal{MONS5}^C_n$, and let L be the corresponding logic from the list $\mathbf{K}^C_n$, $\mathbf{T}^C_n$, $\mathbf{K4}^C_n$, $\mathbf{S4}^C_n$, $\mathbf{KD45}^C_n$, $\mathbf{S5}^C_n$. Then for every monodic $\mathcal{QML}^C_n$-sentence φ, if there is a $\mathcal{MONC}$-quasimodel for φ, then φ is satisfiable in a first-order Kripke model based on a frame for L.*

Proof. The proof is similar to the corresponding part of the proof of Lemma 11.22. Suppose that $\mathfrak{Q} = \langle \mathfrak{F}, \boldsymbol{q}, \mathfrak{R}_{\boldsymbol{q}} \rangle$ is a $\mathcal{MONC}$-quasimodel for φ, where $\mathfrak{F} = \langle W, <_1, \dots, <_n \rangle$ and $\boldsymbol{q}(w) = \langle T_w, T_w^{con} \rangle$, for $w \in W$. We know that the state candidate $\boldsymbol{q}(w)$ is $\mathcal{MONC}$-consistent, for every $w \in W$. As before, by turning subformulas of $\mathsf{real}_{\boldsymbol{q}(w)}$ of the form $\Box_i \psi$ and $\mathsf{C}\,\psi$ that are not in the scope of another epistemic operator to unary predicate symbols or propositional variables, we obtain a sentence $\overline{\mathsf{real}_{\boldsymbol{q}(w)}}$ of the first-order language we used to denote by $\overline{\mathcal{QL}}$ (cf. Section 11.2). The $\overline{\mathcal{QL}}$-sentence $\overline{\mathsf{real}_{\boldsymbol{q}(w)}}$ is consistent with the axiomatic system of classical first-order logic (since $\mathcal{MONC}$ contains the axiom schemata and rules of the latter). So, by Gödel's completeness theorem, for every $w \in W$, there is a $\overline{\mathcal{QL}}$-structure $J(w)$ such that $J(w) \models \overline{\mathsf{real}_{\boldsymbol{q}(w)}}$. We intend to build a first-order Kripke model satisfying φ by using the $\mathcal{QL}$-reducts of these $\overline{\mathcal{QL}}$-structures. The problem again is that they do not necessarily have the same domains.

To overcome this, take an infinite cardinal κ exceeding the cardinality of the set $\mathfrak{R}_{\boldsymbol{q}}$, and put

$$D = \{\langle r, \xi \rangle \mid r \in \mathfrak{R}_{\boldsymbol{q}},\ \xi < \kappa\}.$$

Fix some $w \in W$. Then for any type $\boldsymbol{t} \in T_w$,

$$\big|\{\langle r, \xi \rangle \in D \mid r(w) = \boldsymbol{t}\}\big| = \kappa. \tag{12.6}$$

A proof similar to that of Claim 11.24 shows that one can 'blow up' each $J(w)$ to obtain a $\overline{\mathcal{QL}}$-structure $I(w)$ with domain $D(w)$ such that $I(w) \models \overline{\mathsf{real}_{\boldsymbol{q}(w)}}$ also holds, and for every $\boldsymbol{t} \in T_w$ there are κ many elements in $D(w)$ 'realizing' $\overline{\boldsymbol{t}}$:

$$\big|\{a \in D(w) \mid I(w) \models \overline{\boldsymbol{t}}[a]\}\big| = \kappa.$$

(Here we use the fact that our language does not contain equality.) It is not difficult to see that, using (12.6), we can identify each $D(w)$ with D in a 'type-preserving' way, that is, we may assume that, for all $w \in W$, $\boldsymbol{t} \in T_w$, $\langle r, \xi \rangle \in D$,

$$I(w) \models \overline{\boldsymbol{t}}[\langle r, \xi \rangle] \quad \text{iff} \quad r(w) = \boldsymbol{t},$$

and $c^{I(w)} = \langle r_c, 0 \rangle$, for every $c \in con\,\varphi$. In other words, for all $w \in W$, $r \in \mathfrak{R}_{\boldsymbol{q}}$ and $\xi < \kappa$, we have

$$r(w) = \{\psi \in sub_x \varphi \mid I(w) \models \overline{\psi}[\langle r, \xi \rangle]\}. \tag{12.7}$$

Let us now define the frame

$$\mathfrak{H} = \langle W, R_1, \dots, R_n \rangle$$

on which the first-order Kripke model we are constructing is based. The definition of the accessibility relations R_i ($i = 1, \dots, n$) depends on the choice of $\mathcal{MONC}$. In particular, if $\mathcal{MONC} = \mathcal{MON}K_n^C$ then each $R_i = <_i$, that is, $\mathfrak{H} = \mathfrak{F}$. In the other cases, we define R_i to be

- $<_i \cup \{\langle w, w \rangle \mid w \in W\}$, if $\mathcal{MONC} = \mathcal{MON}T_n^C$;
- the transitive closure $<_i^+$ of $<_i$, if $\mathcal{MONC} = \mathcal{MON}K4_n^C$;
- the reflexive and transitive closure of $<_i$, if $\mathcal{MONC} = \mathcal{MON}S4_n^C$;
- $<_i^+ \cup \{\langle w, w' \rangle \mid \exists v \in W(v <_i^+ w,\ v <_i^+ w',\text{ and } \neg\exists u\ u <_i v)\}$, whenever $\mathcal{MONC} = \mathcal{MON}KD45_n^C$;
- the reflexive, symmetric and transitive closure of $<_i$, if $\mathcal{MONC} = \mathcal{MON}S5_n^C$.

The reader can easily prove that in each case, $\mathfrak{H}$ is a frame for the corresponding epistemic logic (for instance, the frame for $\mathcal{MON}KD45_n^C$ is serial because by (A_D) the formula $\neg\Box_i\bot$ belongs to all types in all state candidates). We claim that, for each choice of $\mathcal{MONC}$, we have the following:

Claim 12.10. *For all $r \in \mathfrak{R}_q$, $w \in W$, $\Box_i\psi \in sub_x\varphi$, $\mathsf{C}\psi \in sub_x\varphi$,*

$$\begin{array}{lll}\Box_i\psi \in r(w) & \textit{iff} & \forall w'(wR_iw' \longrightarrow \psi \in r(w'));\\ \mathsf{C}\psi \in r(w) & \textit{iff} & \forall w'(w(\bigcup_{1\le j\le n} R_j)^*w' \longrightarrow \psi \in r(w')).\end{array}$$

Proof. The $(\Leftarrow)$ directions of both statements follow from the saturation conditions on runs and from the fact that $<_i \subseteq R_i$, for each choice of $\mathcal{MONC}$.

Let us prove the $(\Rightarrow)$ directions. Fix some $r \in \mathfrak{R}_q$ and $w \in W$. First we prove the statement for $\Box_i$, so suppose that $\Box_i\psi \in r(w)$.

(i) Let $\mathcal{MONC} = \mathcal{MON}K_n^C$. Then $R_i = <_i$. Take any w' with $w <_i w'$, and suppose $\psi \notin r(w')$. Then we have $\vdash_{\mathcal{MONC}} r(w') \to \neg\psi$, from which, by contraposition, (12.4) and (12.1), we obtain

$$\vdash_{\mathcal{MONC}} \Box_i\psi \to \Box_i\neg r(w'). \tag{12.8}$$

On the other hand, as $\Box_i\psi \in r(w)$, we have $\vdash_{\mathcal{MONC}} r(w) \to \Box_i\psi$. Now, by (12.8) we obtain $\vdash_{\mathcal{MONC}} r(w) \to \Box_i\neg r(w')$, contrary to the $\mathcal{MONC}$-consistency of $r(w) \wedge \Diamond_i r(w')$, which follows from the coherency of r.

(ii) Let $\mathcal{MONC} = \mathcal{MON}T_n^C$. If wR_iw' then either $w <_i w'$ or $w = w'$. In the former case, it follows from (i) that $\psi \in r(w')$. In the latter, $\psi \in r(w)$ follows from (A_T).

(iii) Let $\mathcal{MONC} = \mathcal{MON}K4_n^C$. Take any w' with wR_iw'. Then there are worlds $v_0, \dots, v_{m+1}$ such that $v_0 = w$, $v_{m+1} = w'$, and $v_j <_i v_{j+1}$ for every $j \leq m$. By induction on j one can show that $\Box_i\psi \in r(v_j)$, for all $j \leq m$. (Indeed, for $j = 0$ this holds by assumption. Let $\Box_i\psi \notin r(v_j)$ for some $j \geq 1$. Then we have $\vdash_{\mathcal{MONC}} \Box_i\psi \to \neg r(v_j)$, and so $\vdash_{\mathcal{MONC}} \Box_i\Box_i\psi \to \Box_i\neg r(v_j)$, by (12.4) and (12.1). On the other hand, by the induction hypothesis, we have $\Box_i\psi \in r(v_{j-1})$, and so $\vdash_{\mathcal{MONC}} r(v_{j-1}) \to \Box_i\psi$. Finally, by (A_4), we obtain $\vdash_{\mathcal{MONC}} r(v_{j-1}) \to \Box_i\neg r(v_j)$, contrary to $\langle r(v_{j-1}), r(v_j)\rangle$ being i-suitable for $\mathcal{MON}K4_n^C$.) In particular, we have $\Box_i\psi \in r(v_m)$. Since $v_m <_i w'$, we have $\psi \in r(w')$ by (i).

(iv) Let $\mathcal{MONC} = \mathcal{MON}S4_n^C$. In this case, a combination of the arguments given in (ii) and (iii) works.

(v) Let $\mathcal{MONC} = \mathcal{MON}S5_n^C$. Take any w' with wR_iw'. Then there are worlds $v_0, \dots, v_{m+1}$ such that $v_0 = w$, $v_{m+1} = w'$ and, for every $j \leq m$, either $v_j <_i v_{j+1}$ or $v_j = v_{j+1}$, $v_{j+1} <_i v_j$. By induction on j one can show that $\Box_i\psi \in r(v_j)$, for all $j \leq m+1$. (Indeed, for $j = 0$ this holds by assumption. Now suppose $\Box_i \notin r(v_j)$ for some $j \geq 1$. By the argument in (iii), this can only happen if $v_j <_i v_{j-1}$. Then we have $\vdash_{\mathcal{MONC}} r(v_j) \to \neg\Box_i\psi$, so $\vdash_{\mathcal{MONC}} r(v_j) \to \Box_i\neg\Box_i\psi$, by (A_5). On the other hand, by the induction hypothesis, we have $\Box_i\psi \in r(v_{j-1})$, and so $\vdash_{\mathcal{MONC}} r(v_{j-1}) \to \Box_i\psi$. This implies $\vdash_{\mathcal{MONC}} \Box_i\neg\Box_i\psi \to \Box_i\neg r(v_{j-1})$, by contraposition, (12.4) and (12.1). Finally, we obtain $\vdash_{\mathcal{MONC}} r(v_j) \to \Box_i\neg r(v_{j-1})$, which is a contradiction.) In particular, we have $\Box_i\psi \in r(w')$, and so $\psi \in r(w')$ by (A_T).

(vi) Let $\mathcal{MONC} = \mathcal{MON}KD45_n^C$. By the definition of R_i, either $w <_i^+ w'$ or there exists a v such that $v <_i^+ w$, $v <_i^+ w'$ and $\neg\exists u\; u <_i v$. In the former case, we have $\psi \in r(w)$ by (iii). In the latter case, there are worlds $v_0, \dots, v_{m+1}$ such that $v_0 = v$, $v_{m+1} = w$ and $v_j <_i v_{j+1}$ for every $j \leq m$. As the arguments in (iii)–(v) show, we have $\Box_i\psi \in r(v_j)$, for all $j \leq m$, from which $\psi \in r(w')$.

Now, to prove the claim for C, we suppose that $\mathsf{C}\psi \in r(w)$. Take any w' with $w(\bigcup_{1\leq i\leq n} R_i)^*w'$. Then there are worlds $v_0, \dots, v_{m+1}$ such that $v_0 = w$, $v_{m+1} = w'$ and, for every $j \leq m$, either $v_jR_iv_{j+1}$ for some $i = 1, \dots, n$, or $v_j = v_{j+1}$. By induction on j one can show that $\mathsf{C}\psi \in r(v_j)$, for all $j \leq m+1$. (Indeed, for $j = 0$ this holds by assumption. Now suppose that $\mathsf{C}\psi \in r(v_j)$ and $v_jR_iv_{j+1}$. Since both $\mathsf{EC}\psi$ and $\Box_i\mathsf{C}\psi$ are in $sub_x\,\varphi$, in view of (12.2) we then have $\mathsf{EC}\psi \in r(v_j)$, and so $\Box_i\mathsf{C}\psi \in r(v_j)$. As we have shown above,

$\mathsf{C}\psi \in r(v_{j+1})$ follows, for all choices of $\mathcal{MONC}$.) So we have $\mathsf{C}\psi \in r(w')$. Using (12.2) again, we obtain $\psi \in r(w')$, as required. ❑

Now we can complete the proof of Lemma 12.9 as follows. For each $w \in W$, let $I'(w)$ be the $\mathcal{QL}$-reduct of $I(w)$. Consider the first-order Kripke model $\mathfrak{M} = \langle \mathfrak{H}, D, I' \rangle$. We show by induction on ψ that for all $\psi \in sub\,\varphi$, $w \in W$, and all assignments $\mathfrak{a}$ in D,

$$I(w) \models^{\mathfrak{a}} \overline{\psi} \quad \text{iff} \quad (\mathfrak{M}, w) \models^{\mathfrak{a}} \psi. \tag{12.9}$$

The basis of induction, i.e., the case when $\psi = P_i(\tau_1, \dots, \tau_m)$, is clear; for then $\psi = \overline{\psi}$. The induction step for $\psi = \psi_1 \wedge \psi_2$, $\psi = \neg\psi_1$, and $\psi = \forall y \psi_1$ follows by the induction hypothesis from the equations

$$\overline{\psi_1 \wedge \psi_2} = \overline{\psi_1} \wedge \overline{\psi_2}, \quad \overline{\neg\psi_1} = \neg\overline{\psi_1}, \quad \overline{\forall y \psi_1} = \forall y \overline{\psi_1}.$$

Let $\psi = \Box_i \chi$. By renaming the free variable in ψ, we may assume that $\psi \in sub_x\,\varphi$. Suppose that $\mathfrak{a}(x) = \langle r, \xi \rangle$. By (12.7), Claim 12.10, and the induction hypothesis, we have

$$\begin{array}{lll} I(w) \models^{\mathfrak{a}} \overline{\Box_i \chi} & \text{iff} & \Box_i \chi \in r(w) \\ & \text{iff} & \forall w' \left(w R_i w' \longrightarrow \chi \in r(w') \right) \\ & \text{iff} & \forall w' \left(w R_i w' \longrightarrow I(w') \models^{\mathfrak{a}} \overline{\chi} \right) \\ & \text{iff} & \forall w' \left(w R_i w' \longrightarrow (\mathfrak{M}, w') \models^{\mathfrak{a}} \chi \right) \\ & \text{iff} & (\mathfrak{M}, w) \models^{\mathfrak{a}} \Box_i \chi. \end{array}$$

The formula $\psi = \mathsf{C}\chi$ is considered analogously.

Since, by **(meqm1)** and **(meqm4)**, $\varphi \in r(w)$ for some $w \in W$ and $r \in \mathfrak{R}_q$, by (12.7) we have $I(w) \models \overline{\varphi}$, and so (12.9) gives $(\mathfrak{M}, w) \models \varphi$, as required. ❑

Thus, to prove Theorem 12.8, it suffices to show the following:

Lemma 12.11. *Suppose that a monodic $\mathcal{QML}_n^C$-sentence φ is consistent with $\mathcal{MONC}$. Then there is a $\mathcal{MONC}$-quasimodel for φ.*

Proof. We require a series of claims.

Claim 12.12. *Let $\langle \mathfrak{C}_1, \mathfrak{C}_2 \rangle$ be a pair of state candidates that is i-suitable for $\mathcal{MONC}$, $\mathfrak{C}_1 = \langle T_1, T_1^{con} \rangle$ and $\mathfrak{C}_2 = \langle T_2, T_2^{con} \rangle$. Then*

(i) *for every $\mathbf{t}_1 \in T_1$, there is a $\mathbf{t}_2 \in T_2$ such that $\langle \mathbf{t}_1, \mathbf{t}_2 \rangle$ is i-suitable for $\mathcal{MONC}$;*

(ii) *for every $\mathbf{t}_2 \in T_2$, there is a $\mathbf{t}_1 \in T_1$ such that $\langle \mathbf{t}_1, \mathbf{t}_2 \rangle$ is i-suitable for $\mathcal{MONC}$;*

(iii) *for every $c \in con\,\varphi$, the pair $\langle \mathbf{t}^c_{\mathfrak{C}_1}, \mathbf{t}^c_{\mathfrak{C}_2} \rangle$ is i-suitable for $\mathcal{MONC}$.*

Proof. We first show that for every monodic $\mathcal{QML}_n^C$-formula of the form $\Box_i\psi$ we have

$$\vdash_{\mathcal{MONC}} \exists x\Box_i\psi \to \Box_i\exists x\psi. \tag{12.10}$$

Indeed, we have $\vdash_{\mathcal{MONC}} \psi \to \exists x\psi$, and so, by (12.4), (12.1) and contraposition, $\vdash_{\mathcal{MONC}} \neg\Box_i\exists x\psi \to \neg\Box_i\psi$. It follows from classical first-order logic that $\vdash_{\mathcal{MONC}} \neg\Box_i\exists x\psi \to \forall x\neg\Box_i\psi$, from which we obtain (12.10) by the definition of $\exists$ and contraposition.

(i) Suppose now that $\boldsymbol{t}_1 \in T_1$, but there is no $\boldsymbol{t}_2 \in T_2$ for which $\langle \boldsymbol{t}_1, \boldsymbol{t}_2\rangle$ is i-suitable for $\mathcal{MONC}$. This means that $\vdash_{\mathcal{MONC}} \boldsymbol{t}_1 \to \Box_i\neg\boldsymbol{t}_2$ for each $\boldsymbol{t}_2 \in T_2$, and so

$$\vdash_{\mathcal{MONC}} \boldsymbol{t}_1 \to \Box_i\neg \bigvee_{\boldsymbol{t}_2\in T_2} \boldsymbol{t}_2.$$

Then we have

$$\vdash_{\mathcal{MONC}} \exists x\, \boldsymbol{t}_1 \to \exists x\Box_i\neg \bigvee_{\boldsymbol{t}_2\in T_2} \boldsymbol{t}_2,$$

from which, by (12.10),

$$\vdash_{\mathcal{MONC}} \exists x\, \boldsymbol{t}_1 \to \Box_i\exists x\neg \bigvee_{\boldsymbol{t}_2\in T_2} \boldsymbol{t}_2.$$

Since

$$\vdash_{\mathcal{MONC}} \exists x\neg \bigvee_{\boldsymbol{t}_2\in T_2} \boldsymbol{t}_2 \to \neg\mathsf{real}_{\mathfrak{C}_2} \quad \text{and} \quad \vdash_{\mathcal{MONC}} \mathsf{real}_{\mathfrak{C}_1} \to \exists x\, \boldsymbol{t}_1,$$

we finally obtain $\vdash_{\mathcal{MONC}} \mathsf{real}_{\mathfrak{C}_1} \to \Box_i\neg\mathsf{real}_{\mathfrak{C}_2}$, contrary to $\langle \mathfrak{C}_1, \mathfrak{C}_2\rangle$ being i-suitable for $\mathcal{MONC}$.

(ii) Now suppose that $\boldsymbol{t}_2 \in T_2$, but there is no $\boldsymbol{t}_1 \in T_1$ for which $\langle \boldsymbol{t}_1, \boldsymbol{t}_2\rangle$ is i-suitable for $\mathcal{MONC}$. This means that

$$\vdash_{\mathcal{MONC}} \bigvee_{\boldsymbol{t}_1\in T_1} \boldsymbol{t}_1 \to \Box_i\neg\boldsymbol{t}_2.$$

Hence

$$\vdash_{\mathcal{MONC}} \forall x \bigvee_{\boldsymbol{t}_1\in T_1} \boldsymbol{t}_1 \to \forall x\Box_i\neg\boldsymbol{t}_2$$

and, by (12.3),

$$\vdash_{\mathcal{MONC}} \forall x \bigvee_{\boldsymbol{t}_1\in T_1} \boldsymbol{t}_1 \to \Box_i\forall x\neg\boldsymbol{t}_2,$$

contrary to $\langle \mathfrak{C}_1, \mathfrak{C}_2\rangle$ being i-suitable for $\mathcal{MONC}$.

(iii) Finally, assume that $c \in con\,\varphi$. Then $\boldsymbol{t}^c_{\mathfrak{C}_1} \wedge \Diamond_i\boldsymbol{t}^c_{\mathfrak{C}_2}$ is consistent with $\mathcal{MON}$, and so the pair $\langle \boldsymbol{t}^c_{\mathfrak{C}_1}, \boldsymbol{t}^c_{\mathfrak{C}_2}\rangle$ is suitable. ❑

A *pointed state candidate* for φ is a pair $\mathfrak{P} = \langle \mathfrak{C}, \boldsymbol{t} \rangle$, where $\mathfrak{C} = \langle T, T^{con} \rangle$ is a state candidate for φ and $\boldsymbol{t}$ a type in T, called the *point* of $\mathfrak{P}$.

Say that a pointed state candidate $\mathfrak{P} = \langle \mathfrak{C}, \boldsymbol{t} \rangle$ is $\mathcal{MONC}$*-consistent* if the formula

$$\mathsf{point}_{\mathfrak{P}} = \mathsf{real}_{\mathfrak{C}} \wedge \boldsymbol{t}$$

is consistent with $\mathcal{MONC}$. A pair $\mathfrak{P}_1 = \langle \mathfrak{C}_1, \boldsymbol{t}_1 \rangle$, $\mathfrak{P}_2 = \langle \mathfrak{C}_2, \boldsymbol{t}_2 \rangle$ of pointed world candidates is i*-suitable for* $\mathcal{MONC}$, $i = 1, \dots, n$, if the formula

$$\mathsf{point}_{\mathfrak{P}_1} \wedge \Diamond_i \mathsf{point}_{\mathfrak{P}_2}$$

is consistent with $\mathcal{MONC}$. In this case we write $\mathfrak{P}_1 \prec_i \mathfrak{P}_2$. Given a $c \in con\,\varphi$, a pair $\mathfrak{P}_1 = \langle \mathfrak{C}_1, \boldsymbol{t}_1 \rangle$, $\mathfrak{P}_2 = \langle \mathfrak{C}_2, \boldsymbol{t}_2 \rangle$ of pointed state candidates is called i*-suitable for* $\mathcal{MONC}$ *relative to* c (in symbols, $\mathfrak{P}_1 \prec_i^c \mathfrak{P}_2$) if $\mathfrak{P}_1 \prec_i \mathfrak{P}_2$, $\langle c, \boldsymbol{t}_1 \rangle \in T_1^{con}$ and $\langle c, \boldsymbol{t}_2 \rangle \in T_2^{con}$, where $\mathfrak{C}_1 = \langle T_1, T_1^{con} \rangle$ and $\mathfrak{C}_2 = \langle T_2, T_2^{con} \rangle$.

Claim 12.13. (i) *There is a* $\mathcal{MONC}$*-consistent state candidate* $\mathfrak{C} = \langle T, T^{con} \rangle$ *for* φ *such that* $\varphi \in \boldsymbol{t}$ *for all* $\boldsymbol{t} \in T$.

Let $\mathfrak{P} = \langle \mathfrak{C}, \boldsymbol{t} \rangle$ *be a* $\mathcal{MONC}$*-consistent pointed state candidate for* φ, *where* $\mathfrak{C} = \langle T, T^{con} \rangle$. *Then the following hold:*

(ii) *If* $\neg \Box_i \psi \in \boldsymbol{t}$, *then there exists* $\mathfrak{P}' = \langle \mathfrak{C}', \boldsymbol{t}' \rangle$ *such that* $\mathfrak{P} \prec_i \mathfrak{P}'$ *and* $\neg \psi \in \boldsymbol{t}'$.

(iii) *Suppose* $c \in con\,\varphi$. *If* $\neg \Box_i \psi \in \boldsymbol{t}_{\mathfrak{C}}^c$, *then there exists* $\mathfrak{P}' = \langle \mathfrak{C}', \boldsymbol{t}' \rangle$ *such that* $\mathfrak{P} \prec_i^c \mathfrak{P}'$ *and* $\neg \psi \in \boldsymbol{t}'$.

(iv) *If* $\neg \mathsf{C} \psi \in \boldsymbol{t}$, *then there is a sequence* $\langle \mathfrak{P}_0, \dots, \mathfrak{P}_k \rangle$, $k < \omega$, *of pointed state candidates* $\mathfrak{P}_j = \langle \mathfrak{C}_j, \boldsymbol{t}_j \rangle$ *such that*

$$\mathfrak{P} = \mathfrak{P}_0 \prec_{i_1} \mathfrak{P}_1 \prec_{i_2} \cdots \prec_{i_k} \mathfrak{P}_k,$$

for some $1 \le i_1, \dots, i_k \le n$, *and* $\neg \psi \in \boldsymbol{t}_k$.

(v) *Suppose* $c \in con\,\varphi$. *If* $\neg \mathsf{C} \psi \in \boldsymbol{t}_{\mathfrak{C}}^c$, *then there is a sequence* $\langle \mathfrak{P}_0, \dots, \mathfrak{P}_k \rangle$, $k < \omega$, *of pointed state candidates* $\mathfrak{P}_j = \langle \mathfrak{C}_j, \boldsymbol{t}_j \rangle$ *such that*

$$\mathfrak{P} = \mathfrak{P}_0 \prec_{i_1}^c \mathfrak{P}_1 \prec_{i_2}^c \cdots \prec_{i_k}^c \mathfrak{P}_k,$$

for some $1 \le i_1, \dots, i_k \le n$, *and* $\neg \psi \in \boldsymbol{t}_k$.

Proof. Let π_φ be the disjunction of formulas $\mathsf{point}_{\mathfrak{P}}$ for all pointed state candidates $\mathfrak{P}$ for φ. By treating subformulas of π_φ of the form $\Box_i \psi$ and $\mathsf{C}\,\psi$ that are not in the scope of another epistemic operator as unary predicate symbols or propositional variables, we obtain that the resulting sentence $\overline{\pi_\varphi}$ is clearly true in all (classical) first-order structures. Since $\mathcal{MONC}$ contains the axiom schemata and rules of classical first-order logic, by Gödel's completeness theorem we obtain

$$\vdash_{\mathcal{MONC}} \pi_\varphi. \tag{12.11}$$

(i) Since φ is consistent with $\mathcal{MONC}$, from (12.11) we obtain that $\pi_\varphi \wedge \varphi$ is consistent with $\mathcal{MONC}$. Then there is a disjunct $\mathsf{real}_{\mathfrak{C}} \wedge \boldsymbol{t}$ of π_φ such that $\mathsf{real}_{\mathfrak{C}} \wedge \boldsymbol{t} \wedge \varphi$ is also consistent with $\mathcal{MONC}$, from which $\varphi \in \boldsymbol{t}$. Since φ is a sentence, we have $\varphi \in \boldsymbol{t}'$ for all types $\boldsymbol{t}'$ of $\mathfrak{C}$.

(ii) We claim that

$$\mathsf{point}_{\mathfrak{P}} \wedge \Diamond_i(\pi_\varphi \wedge \neg\psi) \text{ is consistent with } \mathcal{MONC}. \tag{12.12}$$

Suppose otherwise. Clearly, $\mathsf{point}_{\mathfrak{P}} \wedge \neg\Box_i\psi$ is consistent with $\mathcal{MONC}$. So we have $\vdash_{\mathcal{MONC}} \neg(\Diamond_i(\pi_\varphi \wedge \neg\psi))$, that is, $\vdash_{\mathcal{MONC}} \Box_i(\pi_\varphi \to \psi)$. By (12.4) and (12.11), $\vdash_{\mathcal{MONC}} \Box_i\pi_\varphi$, and so, by (12.1), we obtain $\vdash_{\mathcal{MONC}} \Box_i\psi$, contrary to the $\mathcal{MONC}$-consistency of $\mathsf{point}_{\mathfrak{P}} \wedge \neg\Box_i\psi$. Thus we have (12.12). By (12.1), it follows that there is a pointed state candidate $\mathfrak{P}'$ with point $\boldsymbol{t}'$ such that $\mathsf{point}_{\mathfrak{P}} \wedge \Diamond_i(\mathsf{point}_{\mathfrak{P}'} \wedge \neg\psi)$ is consistent with $\mathcal{MONC}$. Therefore, $\neg\psi \in \boldsymbol{t}'$.

(iii) is proved analogously to (ii).

(iv) Suppose that such a sequence does not exist. Let $\mathcal{T}$ be the minimal set of pointed state candidates for φ such that

- $\mathfrak{P} \in \mathcal{T}$,
- if $\mathfrak{D}_1 \in \mathcal{T}$ and $\mathfrak{D}_1 \prec_i \mathfrak{D}_2$ for some i, then $\mathfrak{D}_2 \in \mathcal{T}$.

First, we claim that

$$\vdash_{\mathcal{MONC}} \vartheta \to \psi. \tag{12.13}$$

Indeed, otherwise the formula $\vartheta \wedge \neg\psi$ is consistent with $\mathcal{MONC}$, and so

$$\bigvee_{\mathfrak{D} \in \mathcal{T}} (\mathsf{point}_{\mathfrak{D}} \wedge \neg\psi)$$

is consistent with $\mathcal{MONC}$ as well. Hence there is $\mathfrak{D}$ in $\mathcal{T}$ such that $\mathsf{point}_{\mathfrak{D}} \wedge \neg\psi$ is consistent with $\mathcal{MONC}$, which means, in particular, that $\neg\psi$ is in $\boldsymbol{t}'$ for the point $\boldsymbol{t}'$ of $\mathfrak{D}$. Thus we have a sequence

$$\mathfrak{P} = \mathfrak{P}_0 \prec_{i_1} \mathfrak{P}_1 \prec_{i_2} \cdots \prec_{i_k} \mathfrak{P}_k = \mathfrak{D}$$

such that $\neg\psi \in \boldsymbol{t}'$, contrary to our assumption. Thus, we have (12.13).

Let us now show that

$$\vdash_{\mathcal{MONC}} \vartheta \to \Box_i\vartheta, \quad \text{for all } i = 1, \dots, n. \tag{12.14}$$

If this is not the case then the formula $\vartheta \wedge \neg\Box_i\vartheta$ is consistent with $\mathcal{MONC}$ for some i, and so there is $\mathfrak{D}$ in $\mathcal{T}$ such that $\mathsf{point}_{\mathfrak{D}} \wedge \Diamond_i\neg\vartheta$ is consistent with $\mathcal{MONC}$. By (ii) above, there is a pointed state candidate $\mathfrak{P}' = \langle \mathfrak{C}', \boldsymbol{t}' \rangle$ for which $\mathfrak{D} \prec_i \mathfrak{P}'$. But then $\mathfrak{P}' \in \mathcal{T}$ and $\mathsf{point}_{\mathfrak{D}} \wedge \Diamond_i\mathsf{point}_{\mathfrak{P}'}$ is consistent with

$\mathcal{MONC}$, contrary to the consistency of $\mathsf{point}_{\mathfrak{D}} \wedge \Diamond_i \bigwedge_{\mathfrak{D}' \in T} \neg \mathsf{point}_{\mathfrak{D}'}$. Thus, we have (12.14), and so $\vdash_{\mathcal{MONC}} \vartheta \to \mathsf{E}\,\vartheta$.

Together with (12.13) this yields $\vdash_{\mathcal{MONC}} \vartheta \to \psi \wedge \mathsf{E}\,\vartheta$. By (12.5), we obtain $\vdash_{\mathcal{MONC}} \vartheta \to \mathsf{C}\,\psi$, and so $\vdash_{\mathcal{MONC}} \mathsf{point}_{\mathfrak{P}} \to \mathsf{C}\,\psi$, since $\mathfrak{P} \in \mathcal{T}$. But $\mathfrak{P}$ is a $\mathcal{MONC}$-consistent pointed state candidate and $\neg \mathsf{C}\,\psi \in \boldsymbol{t}$ for its point $\boldsymbol{t}$, which is a contradiction.

(v) is proved analogously to (iv). ❑

We are now in a position to complete the proof of Lemma 12.11. By Claim 12.13 (i), there is a $\mathcal{MONC}$-consistent state candidate $\mathfrak{C}^* = \langle T^*, T^{con*} \rangle$ for φ such that $\varphi \in \boldsymbol{t}$ for all $\boldsymbol{t} \in T^*$. We are going to construct a basic $\mathcal{MONC}$-structure underlying the required quasimodel as the limit of a sequence

$$\langle \mathfrak{F}_m, \boldsymbol{q}_m \rangle = \langle \langle W_m, <_1^m, \dots, <_n^m \rangle, \boldsymbol{q}_m \rangle,$$

of basic $\mathcal{MONC}$-structures, $m < \omega$.

Let $W_0 = \{w^*\}$ and $\boldsymbol{q}_0(w^*) = \mathfrak{C}^*$. Suppose now that $\langle \mathfrak{F}_m, \boldsymbol{q}_m \rangle$ has already been defined. For every $w \in W_m - W_{m-1}$ we then construct a number of new points 'saturating' $\boldsymbol{q}_m(w)$ (where $W_{-1} = \emptyset$).

Let $\mathfrak{C} = \boldsymbol{q}_m(w)$, $\mathfrak{C} = \langle T, T^{con} \rangle$, and let $\boldsymbol{t} \in T$. We then do the following.

(a1) For every $\chi = \neg \Box_i \psi$ in $\boldsymbol{t}$ we take *two* points a_χ and b_χ, add them to W_m, put $w <_i^{m+1} a_\chi$, $w <_i^{m+1} b_\chi$, and $\boldsymbol{q}_{m+1}(a_\chi) = \boldsymbol{q}_{m+1}(b_\chi) = \mathfrak{C}'$, for some $\mathfrak{P}' = \langle \mathfrak{C}', \boldsymbol{t}' \rangle$ such that $\langle \mathfrak{C}, \boldsymbol{t} \rangle \prec_i \mathfrak{P}'$ and $\psi \notin \boldsymbol{t}'$. That such a $\mathfrak{P}'$ exists is guaranteed by Claim 12.13 (ii).

(a2) Suppose $con\,\varphi \neq \emptyset$. Then we also do the following for all $c \in con\,\varphi$: for every $\chi = \neg \Box_i \psi$ in $\boldsymbol{t}_{\mathfrak{C}}^c$ we take a point a_χ, add it to W_m, put $w <_i^{m+1} a_\chi$, and $\boldsymbol{q}_{m+1}(a_\chi) = \mathfrak{C}'$, for some $\mathfrak{P}' = \langle \mathfrak{C}', \boldsymbol{t}' \rangle$ such that $\langle \mathfrak{C}, \boldsymbol{t}_{\mathfrak{C}}^c \rangle \prec_i^c \mathfrak{P}'$ and $\psi \notin \boldsymbol{t}'$. That such a $\mathfrak{P}'$ exists is guaranteed by Claim 12.13 (iii).

(b1) For every $\chi = \neg \mathsf{C}\,\psi$ in $\boldsymbol{t}$ we take *two* sequences $a_\chi^1, \dots, a_\chi^k$ and $b_\chi^1, \dots, b_\chi^k$ and put

$$w <_{i_1}^{m+1} a_\chi^1 <_{i_2}^{m+1} \cdots <_{i_k}^{m+1} a_\chi^k, \qquad w <_{i_1}^{m+1} b_\chi^1 <_{i_2}^{m+1} \cdots <_{i_k}^{m+1} b_\chi^k,$$

and

$$\boldsymbol{q}_{m+1}(a_\chi^j) = \boldsymbol{q}_{m+1}(b_\chi^j) = \mathfrak{C}^j, \text{ for all } 1 \leq j \leq k,$$

where the $\langle \mathfrak{C}^j, \boldsymbol{t}^j \rangle$ form a sequence of pointed state candidates such that

$$\langle \mathfrak{C}, \boldsymbol{t} \rangle \prec_{i_1} \langle \mathfrak{C}^1, \boldsymbol{t}^1 \rangle \prec_{i_2} \cdots \prec_{i_k} \langle \mathfrak{C}^k, \boldsymbol{t}^k \rangle$$

and $\neg \psi \in \boldsymbol{t}^k$. Claim 12.13 (iv) ensures the existence of such a sequence.

(b2) Suppose $con\,\varphi \neq \emptyset$. Then we also do the following for all $c \in con\,\varphi$: for every $\chi = \neg \mathsf{C} \psi$ in $\boldsymbol{t}_{\mathfrak{C}}^c$ we take a sequence $a_\chi^1, \dots, a_\chi^k$ and put

$$w <_{i_1}^{m+1} a_\chi^1 <_{i_2}^{m+1} \cdots <_{i_k}^{m+1} a_\chi^k$$

and

$$q_{m+1}(a^j_\chi) = \mathfrak{C}^j, \text{ for all } 1 \le j \le k,$$

where the $\langle \mathfrak{C}^j, t^j \rangle$ form a sequence of pointed state candidates such that

$$\langle \mathfrak{C}, t^c_{\mathfrak{C}} \rangle \prec^c_{i_1} \langle \mathfrak{C}^1, t^1 \rangle \prec^c_{i_2} \cdots \prec^c_{i_k} \langle \mathfrak{C}^k, t^k \rangle$$

and $\neg\psi \in t^k$. Claim 12.13 (v) ensures the existence of such a sequence.

In the same manner we consider all the other types in T and all the other worlds $v \in W_m - W_{m-1}$. W_{m+1} is then defined as the (disjoint) union of W_m and the new points constructed by performing steps (a1)–(b2). The relations $<^{m+1}_i$ and the function q_{m+1} coincide with, respectively, $<^m_i$ and q_m on W_m and are defined by (a1)–(b2) for the new points. This gives us $\langle \mathfrak{F}_{m+1}, q_{m+1} \rangle$.

Finally, we put $\langle \mathfrak{F}, q \rangle = \langle \langle W, <^m_1, \dots, <^m_n \rangle, q \rangle$, where

$$W = \bigcup_{m<\omega} W_m, \quad <_i = \bigcup_{m<\omega} <^m_i, \quad q = \bigcup_{m<\omega} q_m.$$

Let $\mathfrak{R}_q$ be the set of all coherent and saturated runs through $\langle \mathfrak{F}, q \rangle$. Let us prove that $\mathfrak{Q} = \langle \mathfrak{F}, q, \mathfrak{R}_q \rangle$ is a $\mathcal{MONC}$-quasimodel for φ.

First, conditions **(meqm1)** and **(meqm2)** hold by the definition of $\mathfrak{Q}$. For **(meqm3)**, it is enough to show that the r_c are coherent and saturated: the coherency condition follows from Claim 12.12 (iii), and the two saturation conditions from the construction described under (a2) and (b2), respectively.

It remains to show that **(meqm4)** holds, that is, for every $w \in W$ and every type t in $q(w)$, there exists a run $r \in \mathfrak{R}_q$ such that $r(w) = t$. Using Claim 12.12 (ii), we find a sequence

$$w^* = w_0 <_{i_1} w_1 <_{i_2} \cdots <_{i_k} w_k = w$$

and types t_j in $q(w_j)$, $j \le k$, such that $t_k = t$ and $t_j \wedge \Diamond_{i_{j+1}} t_{j+1}$ is consistent with $\mathcal{MONC}$ for all $j < k$.

Define r on the worlds w_j in $V = \{w_0, \dots, w_k\}$ by taking $r(w_j) = t_j$. Now by induction we extend r to the sets $V \cup W_i$. Suppose that we have defined r on $V \cup W_n$. Then for every $v \in W_n - W_{n-1}$ with $r(v) = t$ we do the following:

- If $\neg\Box_i\psi \in t$, then we take $v' \in W_{n+1} - (W_n \cup V)$ and t' in $q(v')$ such that $v <_i v'$, $t \wedge \Diamond_i t'$ is consistent with $\mathcal{MONC}$ and $\psi \notin t'$. This can be done because, according to (a1), we always took *two* saturating worlds. Put $r(v') = t'$.

- If $\neg\mathsf{C}\,\psi \in t$, then we take a sequence $v_1, \dots, v_k$ from $W_{n+1} - (W_n \cup V)$ and types t_j in $q(v_j)$, $1 \le j \le k$, such that

  $$v <_{i_0} v_1 <_{i_1} \cdots <_{i_{k-1}} v_{i_k}$$

  and

- $\langle \boldsymbol{t}, \boldsymbol{t}_1 \rangle$ is i_0-suitable for $\mathcal{MONC}$,
- $\langle \boldsymbol{t}_j, \boldsymbol{t}_{j+1} \rangle$ is i_j-suitable for $\mathcal{MONC}$, $1 \leq j < k$,
- $\psi \notin \boldsymbol{t}_k$.

Again this can be done because, according to (b1), we always took two saturating sequences. Put $r(v_j) = \boldsymbol{t}_j$, for all $1 \leq j \leq k$.

To define r on the remaining $v \in W_{n+1}$, we do inductively the following. Suppose that r is not yet defined on v, but defined on the unique v' such that $v' <_i v$. Then, using Claim 12.12 (i), we take a $\boldsymbol{t}$ in $\boldsymbol{q}(v)$ such that $\langle r(v'), \boldsymbol{t} \rangle$ is i-suitable for $\mathcal{MONC}$ and put $r(v) = \boldsymbol{t}$.

It is now straightforward to see that the constructed r is a coherent and saturated run. ❑

This completes the proof of Theorem 12.8. ❑

One can also prove a similar theorem for the monodic fragment of the first-order dynamic logic **CQDL**:

Theorem 12.14. *A monodic $\mathcal{CQDL}$-formula*[1] φ *belongs to* **CQDL** *iff* φ *is derivable in the axiomatic system defined by the following axiom schemata and inference rules:*

Axiom schemata (*ranging over monodic $\mathcal{CQDL}$-formulas*):

- *the axiom schemata of* **QCl**,
- $[\alpha](\varphi \to \psi) \to ([\alpha]\varphi \to [\alpha]\psi)$,
- $[\alpha ; \beta]\varphi \leftrightarrow [\alpha][\beta]\varphi$,
- $[\alpha \cup \beta]\varphi \leftrightarrow [\alpha]\varphi \wedge [\beta]\varphi$,
- $[\alpha^*]\varphi \leftrightarrow \varphi \wedge [\alpha][\alpha^*]\varphi$,
- $[\alpha^*](\varphi \to [\alpha]\varphi) \to (\varphi \to [\alpha^*]\varphi)$,
- $[\alpha]\forall x\psi \leftrightarrow \forall x[\alpha]\psi$.

[1]Recall that $\mathcal{CQDL}$-formulas do not contain tests.

Inference rules (*ranging over monodic $\mathcal{CQDL}$-formulas*):

- *the rules of* **QCl**,
- *given* φ, *derive* $[\alpha]\varphi$, *for all action terms* α.

Proof. The proof is similar to the proof of Theorem 12.8 (observe also the similarities with the axiomatic systems in Remarks 12.7 and 2.18). We leave the details to the reader. ❑

Note that the above proofs provide axiomatizations for the one-variable fragments of the logics under consideration, and so we can obtain alternative proofs of Theorems 6.54 and 6.55 on the axiomatization of the corresponding products with **S5** (see the proof of Theorem 11.78 for details).

Part IV

Applications to knowledge representation and reasoning

Chapter 13

Temporal epistemic logics

In Section 3.4 we introduced combinations of temporal and epistemic logics intended for reasoning about multi-agent systems. For any epistemic logic L from the list $\mathbf{K}_n$, $\mathbf{T}_n$, $\mathbf{K4}_n$, $\mathbf{S4}_n$, $\mathbf{KD45}_n$, $\mathbf{S5}_n$ and any class $\mathcal{C}$ of strict linear orders, we considered the class $\mathcal{TE}_{L,\mathcal{C}}$ of all temporal epistemic structures of the form

$$\langle T \times \mathcal{R}, <, R_1, \ldots, R_n \rangle$$

such that $\langle T, < \rangle \in \mathcal{C}$ and $\langle T \times \mathcal{R}, R_1, \ldots, R_n \rangle \models L$. Theorem 3.19 showed that if $\mathcal{C}$ consists of only one flow of time $\mathfrak{F}$, then the temporal epistemic logic $\mathsf{ELog}_{\mathcal{SU}}(\mathcal{TE}_{L,\mathfrak{F}})$ determined by this class coincides with the fusion of L (or L^C, if we consider epistemic logics with the common knowledge operators) and the propositional temporal logic $\mathsf{Log}_{\mathcal{SU}}(\mathfrak{F})$.

Different features of agents—that they know the time, do not learn, or do not forget—were reflected by imposing various constraints on the temporal epistemic structures. The results obtained and techniques introduced in Part III *cannot* be directly applied to all logics determined by classes of temporal epistemic structures corresponding to possible combinations of these constraints. However, besides the simplest case of fusions considered above, at least two nonempty sets of constraints can be treated using the methodology developed so far:

(1) For *synchronous* systems, that is, for classes of temporal epistemic structures modeling agents who *know the time*, one can show that the resulting logics can easily be embedded into decidable monodic fragments of first-order temporal logics. Moreover, for various important flows of time (like $\langle \mathbb{Z}, < \rangle$, $\langle \mathbb{Q}, < \rangle$, and $\langle \mathbb{R}, < \rangle$), the resulting logics do not reflect any interaction between time and knowledge, i.e., we again obtain the fusions of the corresponding temporal and epistemic logics.

(2) Temporal epistemic structures modeling agents who *know the time, do not forget and do not learn* can be regarded as product frames (see page 139) and therefore we can apply the results and techniques introduced for dealing with products of modal logics.

In the next section we consider case (1), and then, in Section 13.2, turn to case (2). For complexity results including 'intermediate logics' which are not covered by (1) and (2) we refer the reader to Table 13.1 which lists the results[1] of (Halpern and Vardi 1989). For these 'intermediate' constraints, so far only logics based on $\mathbf{S5}_n$ or $\mathbf{S5}_n^C$ and the flow of time $\langle \mathbb{N}, < \rangle$ have been considered.

	S5	$\mathbf{S5}_n$, $n \geq 2$	$\mathbf{S5}_n^C$, $n \geq 2$
no constraints	PSPACE-complete	PSPACE-complete	EXPTIME-complete
sync	PSPACE-complete	PSPACE-complete	EXPTIME-complete
nf	2EXPTIME-complete	not in ELEM	Σ_1^1-complete
nl	EXPSPACE-complete	not in ELEM	Σ_1^1-complete
sync, nf	2EXPTIME-complete	not in ELEM	Σ_1^1-complete
sync, nl	EXPSPACE-complete	not in ELEM	Σ_1^1-complete
nl, nf	EXPSPACE-complete	not in ELEM	Σ_1^1-complete
sync, nl, nf	EXPSPACE-complete	not in ELEM	Σ_1^1-complete

Table 13.1: The results of Halpern and Vardi (1989) on the complexity of the satisfiability problem for some temporal epistemic logics based on the flow of time $\langle \mathbb{N}, < \rangle$, with the sole temporal operator $\mathcal{U}$, interpreted in models combining the constraints of synchronicity (sync), not forgetting (nf), and not learning (nl).

[1]Spaan (1993) proved that the Σ_1^1-completeness results of Table 13.1 hold for the language with the sole temporal operator $\Box_F$ as well.

13.1 Synchronous systems

Let us recall from Section 3.4 that synchronous systems, that is, multi-agent systems with agents who know the time, are modeled by temporal epistemic structures $\langle T \times \mathcal{R}, <, R_1, \dots, R_n \rangle$, where, for all $t, t' \in T$, $f, f' \in \mathcal{R}$, and $i \leq n$,

$$\langle t, f \rangle R_i \langle t', f' \rangle \quad \text{implies} \quad t = t'.$$

In this section, we consider temporal epistemic logics determined by these kinds of structures. It turns out that the interaction between the temporal and epistemic operators interpreted in synchronous structures is rather limited. In some important cases there is no interaction at all, i.e., we obtain fusions of the temporal and epistemic components. Therefore, it should not come as a surprise that the resulting logics are almost always decidable, no matter whether we consider languages with or without common knowledge operators.

Given an epistemic logic L and a class $\mathcal{C}$ of strict linear orders, let

$$\mathcal{SYNC}_{L,\mathcal{C}} = \mathcal{TE}_{L,\mathcal{C}} \cap \mathcal{SYNC},$$

where $\mathcal{SYNC}$ denotes the class of all synchronous temporal epistemic structures. Observe that for every structure $\langle T \times \mathcal{R}, <, R_1, \dots, R_n \rangle$ in $\mathcal{SYNC}$, the n-frame $\langle T \times \mathcal{R}, R_1, \dots, R_n \rangle$ is in fact the disjoint union of the n-frames $\langle \{t\} \times \mathcal{R}, R_1^t, \dots, R_n^t \rangle$ for $t \in T$, where each R_i^t is the restriction of R_i to $\{t\} \times \mathcal{R}$. This observation provides a key to the reduction of temporal epistemic logics to first-order temporal logics presented below.

Recall from Section 1.3 the standard translation $\cdot^\star$ of the unimodal language $\mathcal{ML}$ into the sublanguage of $\mathcal{QL}$ having a binary predicate symbol and countably many unary predicate symbols. Now consider the sublanguage of $\mathcal{QL}$ with countably many unary predicate symbols $P_0, P_1, \dots$, binary predicate symbols $R_1, \dots, R_n$, plus a binary predicate symbol R_M for each nonempty set $M \subseteq \{1, \dots, n\}$. The following natural generalization of $\cdot^\star$ translates formulas of $\mathcal{ML}_n^C$ into this first-order language:

$$\begin{aligned}
p_i^\star &= P_i(x)\\
(\neg\varphi)^\star &= \neg\varphi^\star\\
(\varphi \wedge \psi)^\star &= \varphi^\star \wedge \psi^\star\\
(\Box_i \psi)^\star &= \forall y \, (x R_i y \to \psi^\star\{y/x\})\\
(\mathsf{C}_M \psi)^\star &= \forall y \, (x R_M y \to \psi^\star\{y/x\}).
\end{aligned}$$

Here, as before, x is a fixed individual variable and y is a fresh variable not occurring in $\psi^\star$. Now we extend this standard translation to a translation of the temporal epistemic language $\mathcal{ML}_{\mathcal{SU}} \otimes \mathcal{ML}_n^C$ into the first-order temporal language $\mathcal{QTL}$ by taking

$$(\psi_1 \mathcal{U} \psi_2)^\star = \psi_1^\star \mathcal{U} \psi_2^\star$$
$$(\psi_1 \mathcal{S} \psi_2)^\star = \psi_1^\star \mathcal{S} \psi_2^\star.$$

Observe that x is the only variable that can occur free in $\varphi^\star$ and that $\varphi^\star$ is always a *monodic* $\mathcal{QTL}$-formula.

For each $L \in \{\mathbf{K}_n, \mathbf{T}_n, \mathbf{K4}_n, \mathbf{S4}_n, \mathbf{KD45}_n, \mathbf{S5}_n\}$, define classes $\mathcal{K}_L$ and $\mathcal{K}_{L^C}$ of $\mathcal{QL}$-structures by taking, respectively,

$$\mathcal{K}_L = \Big\{ I = \langle D^I, R_1^I, \dots, R_n^I, P_0^I, \dots \rangle \mid \langle D^I, R_1^I, \dots, R_n^I \rangle \models L \Big\},$$

and

$$\begin{aligned}\mathcal{K}_{L^C} = \Big\{ I = \Big\langle D^I, R_1^I, \dots, R_n^I, R_{\{1\}}^I, \dots, R_{\{1,\dots,n\}}^I, P_0^I, \dots \Big\rangle \mid \\ \Big\langle D^I, R_1^I, \dots, R_n^I \Big\rangle \models L \text{ and } R_M^I \text{ is the reflexive and transitive} \\ \text{closure of } \bigcup_{i \in M} R_i^I, \text{ for all nonempty } M \subseteq \{1, \dots, n\} \Big\}.\end{aligned}$$

Now every model $\mathfrak{M} = \langle \mathfrak{G}, \mathfrak{V} \rangle$ based on a temporal epistemic structure $\mathfrak{G} = \langle T \times \mathcal{R}, <, R_1, \dots, R_n \rangle$ in $\mathcal{SYNC}_{L, \langle T, < \rangle}$ can be turned into a first-order temporal $\mathcal{K}_{L^C}$-model $\langle \langle T, < \rangle, \mathcal{R}, I_{\mathfrak{M}} \rangle$, where, for every $t \in T$,

- $R_i^{I_{\mathfrak{M}}(t)} = \{ \langle f, f' \rangle \mid \langle t, f \rangle \, R_i \, \langle t, f' \rangle \}$, for $i = 1, \dots, n$,
- $P_j^{I_{\mathfrak{M}}(t)} = \{ f \mid f(t) \in \mathfrak{V}(p_j) \}$, for $j < \omega$.

It is easily seen that, for every $\mathcal{ML}_{\mathcal{SU}} \otimes \mathcal{ML}_n^C$-formula φ and every $\langle t, f \rangle$ in $T \times \mathcal{R}$, we have

$$(\mathfrak{M}, \langle t, f \rangle) \models \varphi \quad \text{iff} \quad (\langle \langle T, < \rangle, \mathcal{R}, I_{\mathfrak{M}} \rangle, t) \models \varphi^\star[t].$$

Conversely, every first-order temporal $\mathcal{K}_{L^C}$-model $\mathfrak{N} = \langle \langle T, < \rangle, D, I \rangle$ can be turned into a temporal epistemic model $\mathfrak{M}_{\mathfrak{N}}$ as follows. Define a set S of states as $S = T \times D$. For every $a \in D$, define a function f_a from T to S by taking

$$f_a(t) = \langle t, a \rangle,$$

and let

$$D^+ = \{ f_a \mid a \in D \}.$$

Now define a temporal epistemic structure

$$\mathfrak{G}_{\mathfrak{N}} = \big\langle T \times D^+, <, R_1, \dots, R_n \big\rangle$$

by taking, for each $i = 1, \ldots, n$,

$$R_i = \{\langle\langle t, f_a\rangle, \langle t', f_{a'}\rangle\rangle \mid t = t' \text{ and } aR_i^{I(t)}a'\}.$$

Then define the model $\mathfrak{M}_{\mathfrak{N}} = \langle \mathfrak{G}_{\mathfrak{N}}, \mathfrak{V}_{\mathfrak{N}}\rangle$ by letting, for each propositional variable p_i,

$$\mathfrak{V}_{\mathfrak{N}}(p_i) = \{\langle t, a\rangle \in S \mid a \in P_i^{I(t)}\}.$$

Again we have, for every $\mathcal{ML}_{\mathcal{SU}} \otimes \mathcal{ML}_n^C$-formula φ, every t in T, and every a in D,

$$(\langle\langle T, <\rangle, D, I\rangle, t) \models \varphi^\star[a] \quad \text{iff} \quad (\mathfrak{M}_{\mathfrak{N}}, \langle t, f_a\rangle) \models \varphi.$$

As a consequence we obtain the following:

Theorem 13.1. *Suppose that $L \in \{\mathbf{K}_n, \mathbf{T}_n, \mathbf{K4}_n, \mathbf{S4}_n, \mathbf{KD45}_n, \mathbf{S5}_n\}$ and that $\mathcal{C}$ is a class of strict linear orders. Then*

(i) *for every $\mathcal{ML}_{\mathcal{SU}} \otimes \mathcal{ML}_n$-formula φ, $\varphi \in \mathsf{ELog}_{\mathcal{SU}}(\mathcal{SYNC}_{L,\mathcal{C}})$ iff $\neg\varphi^\star$ is not satisfiable in any first-order temporal $\mathcal{K}_L$-model based on a flow of time in $\mathcal{C}$;*

(ii) *for every $\mathcal{ML}_{\mathcal{SU}} \otimes \mathcal{ML}_n^C$-formula φ, $\varphi \in \mathsf{ELog}^C_{\mathcal{SU}}(\mathcal{SYNC}_{L,\mathcal{C}})$ iff $\neg\varphi^\star$ is not satisfiable in any first-order temporal $\mathcal{K}_{L^C}$-model based on a flow of time in $\mathcal{C}$.*

We can now apply the criterion of Theorem 11.21 to obtain the following result:

Theorem 13.2. *Suppose that $L \in \{\mathbf{K}_n, \mathbf{T}_n, \mathbf{K4}_n, \mathbf{S4}_n, \mathbf{KD45}_n, \mathbf{S5}_n\}$, and let $\mathcal{C}$ be one of the following classes of flows of time: $\{\langle\mathbb{N}, <\rangle\}$, $\{\langle\mathbb{Z}, <\rangle\}$, $\{\langle\mathbb{Q}, <\rangle\}$, the class of all finite strict linear orders, any first-order definable class of strict linear orders (for example, the class of all linear orders). Then the temporal epistemic logics*

$$\mathsf{ELog}_{\mathcal{SU}}(\mathcal{SYNC}_{L,\mathcal{C}}) \quad \textit{and} \quad \mathsf{ELog}^C_{\mathcal{SU}}(\mathcal{SYNC}_{L,\mathcal{C}})$$

are decidable.

Proof. We only consider the language with common knowledge operators. Fix a logic $L' \in \{\mathbf{K}, \mathbf{T}, \mathbf{K4}, \mathbf{S4}, \mathbf{KD45}, \mathbf{S5}\}$, and let

$$L = \overbrace{L' \otimes \cdots \otimes L'}^{n}.$$

Let $\mathcal{QTL}' = \{\varphi^\star \mid \varphi \text{ is an } \mathcal{ML}_{\mathcal{SU}} \otimes \mathcal{ML}_n^C\text{-formula}\}$. Then $\mathcal{QTL}' \subseteq \mathcal{QTL}_{\boxdot}$. We show that $\mathcal{QTL}'$ and $\mathcal{K} = \mathcal{K}_{L^C}$ satisfy conditions (a) and (b) of Theorem 11.21.

To this end, recall first that, for every $\mathcal{QTL}_{\boxed{\mathrm{I}}}$-formula ψ, we denote by $\overline{\psi}$ the $\overline{\mathcal{QL}}$-formula that results from ψ by replacing all its subformulas of the form $\chi_1 \mathcal{U} \chi_2$ and $\chi_1 \mathcal{S} \chi_2$, which are not within the scope of another occurrence of $\mathcal{U}$ or $\mathcal{S}$, by their surrogates. Now observe that, given an $\mathcal{ML}_{\mathcal{SU}} \otimes \mathcal{ML}_n^C$-formula φ, we can obtain the $\overline{\mathcal{QL}}$-formula $\overline{\varphi^\star}$ in a different way. First, we turn φ into an $\mathcal{ML}_n^C$-formula $\tilde{\varphi}$ by replacing each of its subformulas of the form $\chi_1 \mathcal{U} \chi_2$ and $\chi_1 \mathcal{S} \chi_2$, that is not within the scope of another occurrence of $\mathcal{U}$ or $\mathcal{S}$, by a fresh propositional variable (its *surrogate*). Then, by applying the standard translation *, we turn $\tilde{\varphi}$ into a $\overline{\mathcal{QL}}$-formula $\tilde{\varphi}^\star$, see Fig. 13.1. It should be clear that we have $\overline{\varphi^\star} = \tilde{\varphi}^\star$.

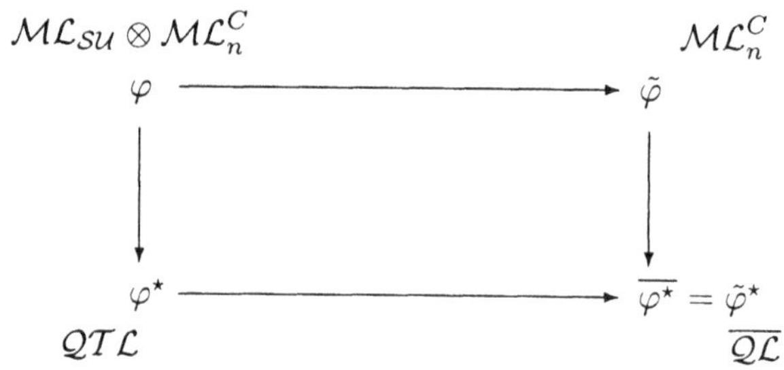

Figure 13.1: Translations from $\mathcal{ML}_{\mathcal{SU}} \otimes \mathcal{ML}_n^C$ to $\overline{\mathcal{QL}}$.

Now fix an $\mathcal{ML}_{\mathcal{SU}} \otimes \mathcal{ML}_n^C$-formula φ. Recall that a *type* for $\varphi^\star$ is any Boolean-saturated subset $\boldsymbol{t}$ of $\{\overline{\psi} \mid \psi \in sub_x\, \varphi^\star\}$. For every such type $\boldsymbol{t}$, define a set $\widehat{\boldsymbol{t}}$ of $\mathcal{ML}_n^C$-formulas by taking

$$\widehat{\boldsymbol{t}} = \{\tilde{\psi} \mid \psi \text{ is an } \mathcal{ML}_{\mathcal{SU}} \otimes \mathcal{ML}_n^C\text{-formula and } \overline{\psi^\star} \in \boldsymbol{t}\}.$$

It is not hard to see (since for every $\psi \in sub_x\, \varphi^\star$, there is an $\mathcal{ML}_{\mathcal{SU}} \otimes \mathcal{ML}_n^C$-formula χ such that $\psi = \chi^\star$) that

$$\boldsymbol{t} = \{\psi^\star \mid \psi \in \widehat{\boldsymbol{t}}\}. \tag{13.1}$$

As $\varphi^\star$ does not contain any constants, a *state candidate* for $\varphi^\star$ is just a set of types for $\varphi^\star$. Given such a state candidate T, define

$$\widehat{T} = \{\widehat{\boldsymbol{t}} \mid \boldsymbol{t} \in T\}.$$

Say that $\widehat{T}$ is L^C-*realizable* if there exist an n-frame $\mathfrak{F} = \langle W, R_1, \ldots, R_n\rangle$ for L and a model $\mathfrak{M} = \langle \mathfrak{F}, \mathfrak{V}\rangle$ such that the following hold:

- for every $\boldsymbol{t} \in \widehat{T}$, there exists $w \in W$ with $(\mathfrak{M}, w) \models \bigwedge_{\psi \in \boldsymbol{t}} \psi$;

- for every $w \in W$, there exists $\boldsymbol{t} \in \widehat{T}$ such that $(\mathfrak{M}, w) \models \bigwedge_{\psi \in \boldsymbol{t}} \psi$.

It should be clear that a state candidate T is $\mathcal{K}_{I,C}$-realizable iff $\widehat{T}$ is L^C-realizable. Hence it suffices to prove that the realizability problem for sets of 'types' of the form $\widehat{T}$ is decidable. Of course, this would follow from the decidability of the extension of L^C with the universal modality. Since we have not proved this decidability result, here we provide a different argument which implicitly uses the fact that in many respects the common knowledge operator $\mathsf{C}_{\{1,\dots,n+1\}}$ of $\mathcal{ML}^C_{n+1}$ behaves similarly to the universal modality added to the language $\mathcal{ML}^C_n$. Observe that $\widehat{T}$ is L^C-realizable iff the $\mathcal{ML}^C_{n+1}$-formula

$$\mathsf{C}_{\{1,\dots,n+1\}} \bigvee_{\boldsymbol{t} \in \widehat{T}} \bigwedge_{\psi \in \boldsymbol{t}} \psi \wedge \bigwedge_{\boldsymbol{t} \in \widehat{T}} \neg \mathsf{C}_{\{1,\dots,n+1\}} \neg \bigwedge_{\psi \in \boldsymbol{t}} \psi$$

is satisfiable in a frame for $(L \otimes L')^C$. Since, by Theorem 2.17, $(L \otimes L')^C$ is decidable, we have proved that condition (a) of Theorem 11.21 holds.

To prove (b), observe that again by Theorem 2.17, $(L \otimes L')^C$ has the fmp. So any L^C-realizable set $\widehat{T}$ as above is in fact 'realizable' in a finite model $\mathfrak{M}$ based on a frame for L^C. Now take $\kappa_\varphi = \aleph_0$, and let κ be an infinite cardinal. Take the disjoint union $\mathfrak{M}'$ of κ isomorphic copies of $\mathfrak{M}$. Then $\mathfrak{M}'$ still realizes $\widehat{T}$ and, moreover,

$$\left| \{ w \mid (\mathfrak{M}', w) \models \bigwedge_{\psi \in \widehat{\boldsymbol{t}}} \psi \} \right| = \kappa,$$

for every $\widehat{\boldsymbol{t}} \in \widehat{T}$. Let $I(\mathfrak{M}')$ denote the $\overline{\mathcal{QL}}$-structure corresponding to $\mathfrak{M}'$ (see Section 1.3). By (13.1), we also have

$$\left| \{ w \mid I(\mathfrak{M}') \models \bigwedge_{\psi \in \boldsymbol{t}} \psi[w] \} \right| = \kappa,$$

for every $\boldsymbol{t} \in T$, as required in (b). ❑

The result above does not cover the flow of time $\langle \mathbb{R}, < \rangle$ simply because we do not know of any significant decidability result for monodic fragments of first-order temporal logics based on $\langle \mathbb{R}, < \rangle$ and arbitrary (possibly infinite) domains. However, it turns out that the interaction between the temporal and epistemic operators in synchronous structures is much weaker than the interaction between temporal operators and quantifiers in monodic first-order temporal logic.

Call a flow of time $\mathfrak{F} = \langle T, < \rangle$ *homogeneous* if, for any $t, t' \in T$, there exists an isomorphism f from $\mathfrak{F}$ onto $\mathfrak{F}$ such that $f(t) = t'$. For example, $\langle \mathbb{Z}, < \rangle$, $\langle \mathbb{Q}, < \rangle$, and $\langle \mathbb{R}, < \rangle$ are clearly homogeneous, while $\langle \mathbb{N}, < \rangle$ is not.

Theorem 13.3. *Let L be one of the epistemic logics* $\mathbf{K}_n$, $\mathbf{T}_n$, $\mathbf{K4}_n$, $\mathbf{S4}_n$, $\mathbf{KD45}_n$, $\mathbf{S5}_n$, *and let* $\mathfrak{F} = \langle T, < \rangle$ *be a homogeneous flow of time. Then*

(i) *the temporal epistemic logic* $\mathsf{ELog}_{\mathcal{SU}}(\mathcal{SYNC}_{L,\mathfrak{F}})$ *coincides with the fusion of the temporal logic* $\mathsf{Log}_{\mathcal{SU}}(\mathfrak{F})$ *and* L;

(ii) *for* $\mathfrak{F} \in \{\langle \mathbb{Z}, < \rangle, \langle \mathbb{Q}, < \rangle, \langle \mathbb{R}, < \rangle\}$, $\mathsf{ELog}_{\mathcal{SU}}(\mathcal{SYNC}_{L,\mathfrak{F}})$ *is decidable;*

(iii) *the same results hold for* $\mathsf{ELog}^C_{\mathcal{SU}}(\mathcal{SYNC}_{L,\mathfrak{F}})$.

Proof. Let $\mathfrak{F} = \langle T, < \rangle$ be a homogeneous flow of time. The inclusion

$$\mathsf{ELog}_{\mathcal{SU}}(\mathcal{SYNC}_{L,\mathfrak{F}}) \supseteq \mathsf{Log}_{\mathcal{SU}}(\mathfrak{F}) \otimes L$$

(as well as its version with common knowledge operators) is clear.

For the converse inclusion, we consider only the case $L = \mathbf{K}$ and show that $\mathsf{ELog}^C_{\mathcal{SU}}(\mathcal{SYNC}_{L,\mathfrak{F}}) \subseteq \mathsf{Log}_{\mathcal{SU}}(\mathfrak{F}) \otimes \mathbf{K}^C$. The remaining cases are similar and left to the reader. It follows from the proof of Theorem 4.1 (and can also be proved by means of an unraveling argument) that $\mathsf{Log}_{\mathcal{SU}}(\mathfrak{F}) \otimes \mathbf{K}^C$ is determined by the class of frames $\langle W, S, R \rangle$ (which may be called $\mathfrak{F}$-*cactuses*) satisfying the following conditions:

- $\langle W, S \rangle$ is the disjoint union of a family $\mathfrak{F}_i = \langle T_i, <_i \rangle$, $i \in I$, of isomorphic copies of $\mathfrak{F} = \langle T, < \rangle$;
- $\langle W, R \rangle$ is the disjoint union of intransitive trees;
- the frame $\langle I, \lhd \rangle$, where $\lhd$ is defined by taking

  $$i \lhd j \quad \text{iff} \quad \exists t \in T_i \, \exists t' \in T_j \; tRt',$$

  is an intransitive tree; moreover, if we have $t, t' \in T_i$, $s, s' \in T_j$, tRs and $t'Rs'$, then $t = t'$ and $s = s'$.

So it is enough to show that any $\mathcal{ML}_{\mathcal{SU}} \otimes \mathcal{ML}^C$-formula φ satisfiable in such an $\mathfrak{F}$-cactus is satisfiable in a temporal epistemic structure from $\mathcal{SYNC}_{\mathbf{K},\mathfrak{F}}$. Suppose a formula φ is satisfied in a model $\mathfrak{M} = \langle \mathfrak{G}, \mathfrak{V} \rangle$, where $\mathfrak{G} = \langle W, S, R \rangle$ is an $\mathfrak{F}$-cactus. For every i in I, we define an isomorphism $f_i : \mathfrak{F} \to \mathfrak{F}_i$. With the root i_0 of $\langle I, \lhd \rangle$ we associate an arbitrary isomorphism f_{i_0} from $\mathfrak{F}$ onto $\mathfrak{F}_{i_0}$. Suppose, inductively, that f_j is defined for the $\lhd$-predecessor j of i. Take the (uniquely determined) $t \in T_i$ and $t' \in T_j$ such that $t'Rt$ and let f_i be an isomorphism from $\mathfrak{F}$ onto $\mathfrak{F}_i$ such that $f_i^{-1}(t) = f_j^{-1}(t')$. Let

$$I^+ = \{f_i \mid i \in I\}.$$

We can regard W as a set of states, and members of I^+ as functions from T to W. So we can define a temporal epistemic structure $\mathfrak{G}' = \langle T \times I^+, <, R' \rangle$ by taking

$$\langle t, f_i \rangle \, R' \, \langle t', f_{i'} \rangle \quad \text{iff} \quad t = t' \text{ and } f_i(t) R f_{i'}(t),$$

and a model based on $\mathfrak{G}'$ by taking $\mathfrak{M}' = \langle \mathfrak{G}', \mathfrak{V} \rangle$. An easy induction shows that, for all $i \in I$, $t \in T$, and formulas ψ,

$$(\mathfrak{M}, f_i(t)) \models \psi \quad \text{iff} \quad (\mathfrak{M}', \langle t, f_i \rangle) \models \psi.$$

Note that the induction steps for the temporal operators follow from the equivalence $t < t'$ iff $f_i(t) <_i f_i(t')$. ❑

Observe that this result cannot be extended to nonhomogeneous flows of time. For example, $\Box_P \bot \rightarrow \Box\Box_P \bot$ is valid in all synchronous systems based on $\langle \mathbb{N}, < \rangle$ but does not belong to the fusion $\mathsf{Log}_{\mathcal{SU}}(\mathbb{N}) \otimes \mathbf{S5}$.

13.2 Agents who know the time and neither forget nor learn

As we saw in Section 3.4, if our agents know the time, do not forget and do not learn simultaneously, then the corresponding temporal epistemic structure

$$\langle T \times \mathcal{R}, <, R_1, \dots, R_n \rangle$$

is isomorphic to the product of the frames $\langle T, < \rangle$ and $\langle \mathcal{R}, S_1, \dots, S_n \rangle$, where

$$f S_i f' \quad \text{iff} \quad \exists t, t' \in T \ \langle t, f \rangle R_i \langle t', f' \rangle \quad \text{iff} \quad \forall t \in T \ \langle t, f \rangle R_i \langle t, f' \rangle .$$

This observation enables us to use the machinery developed in Parts II and III to analyze the computational behavior of the logics modeled by such structures.

Denote by $\mathcal{KN}$ the class of all temporal epistemic structures of this form. Given an epistemic logic L and a class $\mathcal{C}$ of flows of time, let

$$\mathcal{KN}_{L,\mathcal{C}} = \mathcal{TE}_{L,\mathcal{C}} \cap \mathcal{KN}$$

and let $\mathsf{ELog}_{\mathcal{SU}}(\mathcal{KN}_{L,\mathcal{C}})$ denote the temporal epistemic logic formulated in the language $\mathcal{ML}_{\mathcal{SU}} \otimes \mathcal{ML}_n$ and determined by the class $\mathcal{KN}_{L,\mathcal{C}}$. Similarly, $\mathsf{ELog}^{C}_{\mathcal{SU}}(\mathcal{KN}_{L,\mathcal{C}})$ is the corresponding temporal epistemic logic in the language $\mathcal{ML}_{\mathcal{SU}} \otimes \mathcal{ML}^{C}_n$. We also let

- $\mathsf{ELog}(\mathcal{KN}_{L,\mathcal{C}}) = \mathsf{ELog}_{\mathcal{SU}}(\mathcal{KN}_{L,\mathcal{C}}) \cap (\mathcal{ML} \otimes \mathcal{ML}_n)$, where $\mathcal{ML}$ is the unimodal language with the sole temporal operator $\Box_F$, and
- $\mathsf{ELog}_{FP}(\mathcal{KN}_{L,\mathcal{C}}) = \mathsf{ELog}_{\mathcal{SU}}(\mathcal{KN}_{L,\mathcal{C}}) \cap (\mathcal{ML}_2 \otimes \mathcal{ML}_n)$, where $\mathcal{ML}_2$ is the bimodal language with the temporal operators $\Box_F$ and $\Box_P$.

If the language contains the common knowledge operators, then the logics are denoted by $\mathsf{ELog}^C(\mathcal{KN}_{L,\mathcal{C}})$ and $\mathsf{ELog}^C_{FP}(\mathcal{KN}_{L,\mathcal{C}})$, respectively.

The plan of this section is as follows. First we consider the temporal epistemic logics defined above and containing no common knowledge operators. It turns out that we have two different cases. If the epistemic component is **K4** or **S4** (or their multimodal versions $\mathbf{K4}_n$ or $\mathbf{S4}_n$) then the logics are undecidable (at least for the flow of time $\langle \mathbb{N}, < \rangle$). All the other epistemic logics ($\mathbf{K}_n$, $\mathbf{T}_n$, $\mathbf{KD45}_n$, $\mathbf{S5}_n$) give rise to decidable combinations (at least for important flows of time like $\langle \mathbb{N}, < \rangle$, $\langle \mathbb{Q}, < \rangle$, and the class of all strict linear orders). On the other hand, for almost all interesting flows of time, temporal epistemic logics *with* the common knowledge operator modeling agents who know the time, do not forget and do not learn are undecidable.

Without common knowledge

Some of the decidability and complexity results follow immediately from those obtained in Sections 6.4–6.6, since our temporal epistemic logics coincide with product logics:

Theorem 13.4. *Let* $L \in \{\mathbf{K}_n, \mathbf{T}_n, \mathbf{K4}_n, \mathbf{S4}_n, \mathbf{KD45}_n, \mathbf{S5}_n\}$ *and let* $\mathcal{C}$ *be a class of strict linear orders. Then*

$$\mathsf{ELog}(\mathcal{KN}_{L,\mathcal{C}}) = \mathsf{Log}(\mathcal{C} \times \mathsf{Fr}\, L).$$

Proof. Fix an $\mathcal{ML}\otimes\mathcal{ML}_n$-formula φ. Suppose first that $\varphi \notin \mathsf{Log}(\mathcal{C} \times \mathsf{Fr}\, L)$. Then φ is refuted in some model $\mathfrak{M} = \langle \mathfrak{G}, \mathfrak{V} \rangle$ based on the product of some $\mathfrak{G} = \langle T, < \rangle$ in $\mathcal{C}$ and a frame $\mathfrak{F} = \langle W, S_1, \dots, S_n \rangle$ for L. We can turn $\mathfrak{M}$ into a temporal epistemic model as follows. Let us regard $T \times W$ as a set of states. For every $w \in W$, define a function f_w from T to $T \times W$ by taking, for every $t \in T$, $f_w(t) = \langle t, w \rangle$, and let

$$\mathcal{R} = \{f_w \mid w \in W\}.$$

Now define a temporal epistemic structure $\mathfrak{H} = \langle T \times \mathcal{R}, <, R_1, \dots, R_n \rangle$ by taking, for each $i = 1, \dots, n$,

$$R_i = \{\langle \langle t, f_w \rangle, \langle t', f_{w'} \rangle \rangle \mid t = t' \text{ and } wS_iw'\}.$$

It is straightforward to see that $\langle T \times \mathcal{R}, R_1, \dots, R_n \rangle$ is isomorphic to a disjoint union of isomorphic copies of $\mathfrak{F}$ (cf. the proof of Proposition 3.8), and so $\mathfrak{H} \in \mathcal{KN}_{L,\mathcal{C}}$. Finally, φ is clearly refuted in the temporal epistemic model $\langle \mathfrak{H}, \mathfrak{V} \rangle$.

Conversely, if $\varphi \notin \mathsf{ELog}(\mathcal{KN}_{L,\mathcal{C}})$, then φ is refuted in a model $\mathfrak{M} = \langle \mathfrak{H}, \mathfrak{V} \rangle$ based on a temporal epistemic structure

$$\mathfrak{H} = \langle T \times \mathcal{R}, <, R_1, \dots, R_n \rangle = \langle T, < \rangle \times \langle \mathcal{R}, S_1, \dots, S_n \rangle,$$

where $\langle T, <\rangle$ is in $\mathcal{C}$, $\mathcal{R}$ is a set of functions from T to some set of states, and

$$\langle T \times \mathcal{R}, R_1, \ldots, R_n\rangle$$

is a frame for L. Since $\langle T \times \mathcal{R}, R_1, \ldots, R_n\rangle$ is the disjoint union of isomorphic copies of $\mathfrak{F} = \langle \mathcal{R}, S_1, \ldots, S_n\rangle$, we obtain that $\mathfrak{F}$ is a frame for L as well, and so $\mathfrak{H}$ is a frame in $\mathcal{C} \times \mathsf{Fr}\, L$. Define a valuation $\mathfrak{W}$ in $\mathfrak{H}$ by taking, for every propositional variable p,

$$\mathfrak{W}(p) = \{\langle t, f\rangle \mid f(t) \in \mathfrak{V}(p)\}.$$

It should be clear that φ is refuted in the model $\langle \mathfrak{H}, \mathfrak{W}\rangle$. ❑

In particular, we have:

Theorem 13.5. *Let* $L \in \{\mathbf{K}_n, \mathbf{T}_n, \mathbf{K4}_n, \mathbf{S4}_n, \mathbf{KD45}_n, \mathbf{S5}_n\}$. *Then the following equalities hold:*

(i) *if* $\mathbb{F} \in \{\mathbb{N}, \mathbb{Q}\}$, *then*
$\mathsf{ELog}(\mathcal{KN}_{L,\{\langle \mathbb{F},<\rangle\}}) = \mathsf{Log}\{\langle \mathbb{F}, <\rangle\} \times L$,
$\mathsf{ELog}_{\mathcal{FP}}(\mathcal{KN}_{L,\{\langle \mathbb{F},<\rangle\}}) = \mathsf{Log}_{\mathcal{FP}}(\mathbb{F}) \times L$,
$\mathsf{ELog}_{\mathcal{SU}}(\mathcal{KN}_{L,\{\langle \mathbb{F},<\rangle\}}) = \mathsf{Log}_{\mathcal{SU}}(\mathbb{F}) \times L$;

(ii) *if* $\mathcal{C}$ *is the class of all strict linear orders, then*
$\mathsf{ELog}(\mathcal{KN}_{L,\mathcal{C}}) = \mathbf{K4.3} \times L$,
$\mathsf{ELog}_{\mathcal{FP}}(\mathcal{KN}_{L,\mathcal{C}}) = \mathbf{Lin} \times L$,
$\mathsf{ELog}_{\mathcal{SU}}(\mathcal{KN}_{L,\mathcal{C}}) = \mathbf{Lin}_{\mathcal{SU}} \times L$.

Proof. To prove $\mathsf{ELog}(\mathcal{KN}_{L,\{\langle \mathbb{N},<\rangle\}}) = \mathsf{Log}\{\langle \mathbb{N}, <\rangle\} \times L$, observe that by Theorem 6.29, we have

$$\mathsf{Log}(\{\langle \mathbb{N}, <\rangle\} \times \mathsf{Fr}\, L) = \mathsf{Log}\{\langle \mathbb{N}, <\rangle\} \times L.$$

The proofs of the other equalities in (i) are similar. For (ii), one has to use Theorems 6.30 and 6.31, respectively, in place of Theorem 6.29. (Note that these theorems are not stated for the logics when both $\mathcal{S}$ and $\mathcal{U}$ are present, but the reader should have no difficulty in proving them for these cases.) ❑

Thus we can use results of Chapter 6 to obtain decidability and complexity results on some of the temporal epistemic logics above. Instead of doing this, here we prove a general decidability theorem, comparable to Theorem 11.7, by combining the kind of quasimodels introduced in Chapter 6 with the embeddings into monadic second-order theories of strict linear orders used in Chapter 11.

Theorem 13.6. *Suppose that* $L \in \{\mathbf{K}_n, \mathbf{T}_n, \mathbf{KD45}_n, \mathbf{S5}_n\}$ *and* $\mathcal{C}$ *is one of the following classes of strict linear orders:*

(1) $\{\langle \mathbb{N}, < \rangle\}$,

(2) $\{\langle \mathbb{Z}, < \rangle\}$,

(3) $\{\langle \mathbb{Q}, < \rangle\}$,

(4) *the class of all finite strict linear orders,*

(5) *any first-order definable class of strict linear orders—for example, the class of all strict linear orders.*

Then $\mathsf{ELog}_{\mathcal{SU}}(\mathcal{KN}_{L,\mathcal{C}})$ *is decidable.*

Proof. We prove this theorem first for $L = \mathbf{K}$. Let us begin with a straightforward modification of the notion of a quasimodel used in the proof of decidability of $\mathbf{K} \times \mathbf{K}$ (Theorem 6.1). Fix an $\mathcal{ML}_{\mathcal{SU}} \otimes \mathcal{ML}$-formula φ.

By a *type* for φ we mean any Boolean-saturated subset of $sub\,\varphi$. A *quasistate* for φ is a pair $\boldsymbol{q} = \langle \langle T, < \rangle, \boldsymbol{t} \rangle$, where $\langle T, < \rangle$ is a finite intransitive tree of depth $\leq md(\varphi)$ and $\boldsymbol{t}$ is a labeling function associating with each $x \in T$ a type $\boldsymbol{t}(x)$ for φ such that conditions **(qm1)** and **(qm1′)** from the proof of Theorem 6.1 hold. Two quasistates $\langle \langle T, < \rangle, \boldsymbol{t} \rangle$ and $\langle \langle T', <' \rangle, \boldsymbol{t}' \rangle$ are called *isomorphic* if there is an isomorphism f between the trees $\langle T, < \rangle$ and $\langle T', <' \rangle$ such that $\boldsymbol{t}(x) = \boldsymbol{t}'(f(x))$, for all $x \in T$. In what follows we assume that non-isomorphic quasistates are disjoint and that isomorphic quasistates actually coincide.

Now fix a flow of time $\mathfrak{F} = \langle W, < \rangle$ from $\mathcal{C}$. A *basic structure* of depth m for φ is a pair $\langle \mathfrak{F}, \boldsymbol{q} \rangle$, where $\boldsymbol{q}$ is a function associating with each $w \in W$ a quasistate

$$\boldsymbol{q}(w) = \langle \langle T_w, <_w \rangle, \boldsymbol{t}_w \rangle$$

for φ such that the depth of each $\langle T_w, <_w \rangle$ is m.

Let $\langle \mathfrak{F}, \boldsymbol{q} \rangle$ be a basic structure for φ of depth m and let $k \leq m$. A *k-run through* $\langle \mathfrak{F}, \boldsymbol{q} \rangle$ is a function r giving for each $w \in W$ a point $r(w) \in T_w$ of co-depth k. Given a set $\mathfrak{R}$ of runs, we denote by $\mathfrak{R}_k$ the set of all k-runs from $\mathfrak{R}$.

A run r is called *coherent and saturated* if the following holds:

- for every $\psi_1 \mathcal{U} \psi_2 \in sub\,\varphi$ and every $w \in W$, we have $\psi_1 \mathcal{U} \psi_2 \in \boldsymbol{t}_w(r(w))$ iff there is $v > w$ such that $\psi_2 \in \boldsymbol{t}_v(r(v))$ and $\psi_1 \in \boldsymbol{t}_u(r(u))$ for all $u \in (w, v)$, and

- for every $\psi_1 \mathcal{S} \psi_2 \in sub\,\varphi$ and every $w \in W$, we have $\psi_1 \mathcal{S} \psi_2 \in \boldsymbol{t}_w(r(w))$ iff there is $v < w$ such that $\psi_2 \in \boldsymbol{t}_v(r(v))$ and $\psi_1 \in \boldsymbol{t}_u(r(u))$ for all $u \in (v, w)$.

We say that a quadruple $\mathfrak{Q} = \langle \mathfrak{F}, \boldsymbol{q}, \mathfrak{R}, \lhd \rangle$ is a *quasimodel for* φ *based on* $\mathfrak{F}$ if $\langle \mathfrak{F}, \boldsymbol{q} \rangle$ is a basic structure for φ of depth $m \leq md(\varphi)$, $\mathfrak{R}$ is a set of coherent and saturated runs through $\langle \mathfrak{F}, \boldsymbol{q} \rangle$ and $\lhd$ is a binary relation on $\mathfrak{R}$ satisfying the following conditions:

(eqm2) $\exists w_0 \in W \; \varphi \in \boldsymbol{t}_{w_0}(r_0(w_0))$, where $r_0 \in \mathfrak{R}_0$;

(eqm3) for all $r, r' \in \mathfrak{R}$, if $r \lhd r'$ then $r(w) <_w r'(w)$ for all $w \in W$;

(eqm4) for all $k < m$, $r \in \mathfrak{R}_k$, $w \in W$ and $x \in T_w$, if $r(w) <_w x$ then there is $r' \in \mathfrak{R}_{k+1}$ such that $r'(w) = x$ and $r \lhd r'$.

The following can be proved in the same way as Lemma 6.2:

Lemma 13.7. *An* $\mathcal{ML}_{\mathcal{SU}} \otimes \mathcal{ML}$*-formula* φ *is satisfiable in a product frame* $\mathfrak{F} \times \mathfrak{G}$ *iff there is a quasimodel for* φ *based on* $\mathfrak{F}$.

As was shown in the proof of Theorem 13.5, an $\mathcal{ML}_{\mathcal{SU}} \otimes \mathcal{ML}$-formula φ is satisfiable in a product frame $\mathfrak{F} \times \mathfrak{G}$ iff φ is satisfiable in a temporal epistemic structure $\mathfrak{F} \times \mathfrak{G}'$ from $\mathcal{KN}$, where $\mathfrak{G}'$ is isomorphic to $\mathfrak{G}$.

We can now deduce the decidability of $\mathsf{ELog}_{\mathcal{SU}}(\mathcal{KN}_{\mathbf{K},\mathcal{C}})$ by translating into monadic second-order logic the statement that there exists a quasimodel for φ based on some $\mathfrak{F} \in \mathcal{C}$. We require a number of auxiliary formulas. Fix some $m \leq md(\varphi)$. Denote by Σ_m the set of all quasistates for φ of depth m. Given a quasistate $\boldsymbol{q} = \langle \langle T_{\boldsymbol{q}}, <_{\boldsymbol{q}} \rangle, \boldsymbol{t}_{\boldsymbol{q}} \rangle$ from Σ_m and a point a in $T_{\boldsymbol{q}}$, we denote the co-depth of a by $cd_{\boldsymbol{q}}(a)$.

Introduce a unary predicate variable $P_{\boldsymbol{q}}$ for each $\boldsymbol{q} \in \Sigma_m$ and a unary predicate variable R^k_ψ for each $\psi \in sub\,\varphi$ and each $k \leq m$. Given a type $\boldsymbol{t}$ for φ and $k \leq m$, let

$$\chi_{\boldsymbol{t}}(\overline{R^k}(x)) = \bigwedge_{\psi \in \boldsymbol{t}} R^k_\psi(x) \wedge \bigwedge_{\substack{\psi \notin \boldsymbol{t} \\ \psi \in sub\,\varphi}} \neg R^k_\psi(x),$$

saying that the type $\boldsymbol{t}$ at point x of co-depth k is defined with the help of

$$\overline{R^k}(x) = \left\langle R^k_\psi(x) \mid \psi \in sub\,\varphi \right\rangle.$$

For each $k \leq m$, let $\mathsf{run}_0(\overline{P}, \overline{R^k})$ denote the conjunction of the three formulas

$$\forall x \bigwedge_{\boldsymbol{q} \in \Sigma_m} \Big(P_{\boldsymbol{q}}(x) \to \bigvee_{\substack{a \in T_q \\ cd_q(a) = k}} \chi_{\boldsymbol{t}_q(a)}(\overline{R^k}(x))\Big),$$

$$\forall x \bigwedge_{\psi_1 \mathcal{U} \psi_2 \in sub\,\varphi} \left[R^k_{\psi_1 \mathcal{U} \psi_2}(x) \leftrightarrow \exists y \big(x < y \wedge R^k_{\psi_2}(y) \wedge \forall z (x < z < y \to R^k_{\psi_1}(z))\big)\right],$$

$$\forall x \bigwedge_{\psi_1 \mathcal{S} \psi_2 \in sub\,\varphi} \left[R^k_{\psi_1 \mathcal{S} \psi_2}(x) \leftrightarrow \exists y \big(y < x \wedge R^k_{\psi_2}(y) \wedge \forall z (y < z < x \to R^k_{\psi_1}(z))\big)\right]$$

—this is intended to say that $\overline{R^k}$ defines a coherent and saturated k-run through a sequence of quasistates defined with the help of $\overline{P} = \langle P_{\boldsymbol{q}} \mid \boldsymbol{q} \in \Sigma_m \rangle$.

However, we have to refine this definition in order to ensure that condition **(eqm4)** holds. To this end, we define, by 'backwards' induction on k, another formula $\mathsf{run}(\overline{P}, \overline{R^k})$ as follows. If $k = m$ (that is, we are at the 'leaf-level') then take $\mathsf{run}(\overline{P}, \overline{R^m}) = \mathsf{run}_0(\overline{P}, \overline{R^m})$.

Suppose, inductively, that for $k \leq m$ we have already defined $\mathsf{run}(\overline{P}, \overline{R^k})$. Then let $\mathsf{run}(\overline{P}, \overline{R^{k-1}})$ be the following formula:

$$\begin{aligned}
&\mathsf{run}_0(\overline{P}, \overline{R^{k-1}}) \wedge \\
&\forall x \bigwedge_{\boldsymbol{q}\in\Sigma_m} \bigwedge_{\substack{a\in T_q \\ cd_q(a)=k-1}} \Big[P_{\boldsymbol{q}}(x) \wedge \chi_{\boldsymbol{t}_q(a)}(\overline{R^{k-1}}(x)) \rightarrow \\
&\quad \bigwedge_{\substack{b\in T_q \\ a<_q b}} \exists_{\psi\in sub\,\varphi} R^k_\psi \Big(\mathsf{run}(\overline{P}, \overline{R^k}) \wedge \chi_{\boldsymbol{t}_q(b)}(\overline{R^k}(x)) \wedge \\
&\quad \forall z \bigwedge_{\boldsymbol{s}\in\Sigma_m} \bigwedge_{\substack{c\in T_s \\ cd_s(c)=k-1}} (P_{\boldsymbol{s}}(z) \wedge \chi_{\boldsymbol{t}_s(c)}(\overline{R^{k-1}}(z)) \rightarrow \bigvee_{\substack{d\in T_s \\ c<_s d}} \chi_{\boldsymbol{t}_s(d)}(\overline{R^k}(z)))\Big)\Big].
\end{aligned}$$

Finally, we define a monadic second-order sentence qm^m_φ by taking

$$\begin{aligned}
\mathsf{qm}^m_\varphi = \exists_{\boldsymbol{q}\in\Sigma_m} P_{\boldsymbol{q}} \Big[&\forall x \bigvee_{\boldsymbol{q}\in\Sigma_m} (P_{\boldsymbol{q}}(x) \wedge \bigwedge_{\substack{\boldsymbol{q}'\in\Sigma_m \\ \boldsymbol{q}\neq\boldsymbol{q}'}} \neg P_{\boldsymbol{q}'}(x)) \wedge \\
&\bigvee_{\substack{\boldsymbol{s}\in\Sigma_m,\, a\in T_s \\ cd_s(a)=0 \\ \varphi\in\boldsymbol{t}_s(a)}} \exists x \Big(P_{\boldsymbol{s}}(x) \wedge \exists_{\psi\in sub\,\varphi} R^0_\psi \left(\mathsf{run}(\overline{P}, \overline{R^0}) \wedge \chi_{\boldsymbol{t}_s(a)}(\overline{R^0}(x)) \right) \Big) \Big].
\end{aligned}$$

Evaluated in a flow of time $\mathfrak{F} = \langle W, < \rangle$, the first line of qm^m_φ says that the sets $P_{\boldsymbol{q}} \subseteq W$ ($\boldsymbol{q} \in \Sigma_m$) form a partition of W. By defining the map $\boldsymbol{q} : W \rightarrow \Sigma_m$ as

$$\boldsymbol{q}(w) = \boldsymbol{q} \quad \text{iff} \quad w \in P_{\boldsymbol{q}}$$

and a relation $\lhd$ on the runs by taking $r \lhd r'$ iff r is defined by $\overline{R^{k-1}}$ and r' is defined by $\overline{R^k}$ for some $k \leq m$, we obtain a quasimodel $\mathfrak{Q} = \langle \mathfrak{F}, \boldsymbol{q}, \mathfrak{R}, \lhd \rangle$ for φ: the second line of qm^m_φ states condition **(eqm2)**; conditions **(eqm3)** and **(eqm4)** are satisfied by the definitions of $\lhd$ and the formulas $\mathsf{run}(\overline{P}, \overline{R^k})$, respectively. Therefore, it is easy to see that we obtain the following:

Lemma 13.8. *For any strict linear order $\mathfrak{F}$, $\mathfrak{F} \models \mathsf{qm}^m_\varphi$ for some $m \leq md(\varphi)$ iff there exists a quasimodel for φ based on $\mathfrak{F}$.*

Clearly, Σ_m can be constructed from φ by an algorithm. So we can now apply Theorem 1.28 stating the decidability of certain theories of monadic second-order logic to obtain the first four statements of our theorem for the case $L = \mathbf{K}$. To prove statement (5), the reader should repeat the corresponding part of the proof of Theorem 11.21.

Straightforward modifications of the above proof give the statements when L is multimodal $\mathbf{K}_n$ or $\mathbf{T}_n$. For $L = \mathbf{KD45}_n$ and $L = \mathbf{S5}_n$, the reader should have no difficulty in repeating the proof above by appropriately modifying quasimodels similarly to what was done in the proofs of Theorems 6.49 and 6.68. ❑

In Table 13.2 we summarized the upper bounds for the computational complexity of temporal epistemic logics. All the decidability (but not the complexity) results of Table 13.2 follow from Theorem 13.6.

$\mathsf{ELog}_{\mathcal{SU}}(\mathcal{KN}_{L,\mathcal{C}})$		$\mathcal{C}$: $\{\langle \mathbb{N}, < \rangle\}$	$\mathcal{C}$: $\{\langle \mathbb{Q}, < \rangle\}$, all strict linear orders
L	$\mathbf{K}_n$, $\mathbf{T}_n$ $(n \geq 1)$	decidable	decidable
L	$\mathbf{S5}$, $\mathbf{KD45}$	in EXPSPACE (Thms. 3.30, 11.31, Prop. 11.25)	in 2EXPTIME (Thms. 6.61[2], 13.5)
L	$\mathbf{S5}_n$, $\mathbf{KD45}_n$ $(n \geq 2)$	decidable	decidable

Table 13.2: Upper bounds for the complexity of temporal epistemic logics with $\mathcal{S}$ and $\mathcal{U}$, but without common knowledge operators.

As concerns lower bounds, by Theorems 6.63 and 13.4 we have:

Theorem 13.9. *Let $\mathcal{C}$ be a class of strict linear orders such that at least one flow of time in $\mathcal{C}$ contains an infinite ascending chain. Then $\mathsf{ELog}(\mathcal{KN}_{\mathbf{S5},\mathcal{C}})$ is* EXPSPACE-*hard.*

As a consequence of Theorems 7.24 and 13.5 we obtain:

Theorem 13.10. *The logics $\mathsf{ELog}(\mathcal{KN}_{\mathbf{K4},\{\langle \mathbb{N},<\rangle\}})$ and $\mathsf{ELog}(\mathcal{KN}_{\mathbf{S4},\{\langle \mathbb{N},<\rangle\}})$ are undecidable.*

[2]Theorem 6.61 is formulated only for the case when $\Box_F$ and $\Box_P$ are the only temporal operators, but it is not hard to generalize it for the case of $\mathcal{S}$ and $\mathcal{U}$ as well.

With common knowledge

Similarly to Theorem 13.4, we again see that our temporal epistemic logics coincide with the logics of the corresponding product frames:

Theorem 13.11. *Let* $L \in \{\mathbf{K}_n, \mathbf{T}_n, \mathbf{K4}_n, \mathbf{S4}_n, \mathbf{KD45}_n, \mathbf{S5}_n\}$ *and let* $\mathcal{C}$ *be a class of strict linear orders. Then*

$$\mathsf{ELog}^C(\mathcal{KN}_{L,\mathcal{C}}) = \mathsf{Log}(\mathcal{C} \times \mathsf{Fr}\, L^C).$$

Thus, by Theorem 7.19, we obtain that the addition of the common knowledge operators to temporal epistemic logics modeling agents who know time, do not forget and do not learn almost always results in undecidable or even not recursively enumerable formalisms:

Theorem 13.12. *Let* $\mathcal{C}$ *be a class of linear orders such that at least one flow of time in* $\mathcal{C}$ *contains an infinite ascending chain of distinct points. Then* $\mathsf{ELog}^C(\mathcal{KN}_{L,\mathcal{C}})$ *is not recursively enumerable, whenever* $L \in \{\mathbf{K}, \mathbf{T}_2, \mathbf{K4}_2, \mathbf{S4}_2, \mathbf{KD45}_2\}$. $\mathsf{ELog}^C(\mathcal{KN}_{\mathbf{S5}_2,\mathcal{C}})$ *is undecidable.*

Proof. For $\mathsf{ELog}^C(\mathcal{KN}_{\mathbf{K},\mathcal{C}})$ the statement follows from Theorems 13.11 and 7.19.

The proof of Theorem 6.23 shows that $\mathsf{Log}(\mathcal{C} \times \mathsf{Fr}\, \mathbf{K}_1^C)$ is polynomially reducible to $\mathsf{Log}(\mathcal{C} \times \mathsf{Fr}\, L^C)$, for any $L \in \{\mathbf{T}_2, \mathbf{K4}_2, \mathbf{S4}_2, \mathbf{KD45}_2\}$. Therefore, the statements for $L \neq \mathbf{S5}_2$ follow from Theorems 13.11 and 7.19.

The proof of Theorem 6.71 (3) shows that $\mathsf{Log}(\mathcal{C} \times \mathsf{Fr}\, \mathbf{K}_u)$ is polynomially reducible to $\mathsf{Log}(\mathcal{C} \times \mathsf{Fr}\, \mathbf{S5}_2^C)$. But $\mathsf{Log}(\mathcal{C} \times \mathsf{Fr}\, \mathbf{K}_u)$ is undecidable, by Theorem 7.19, so the undecidability of $\mathsf{ELog}^C(\mathcal{KN}_{\mathbf{S5}_2,\mathcal{C}})$ follows from Theorem 13.11. ❑

Chapter 14

Modal description logics

In this chapter we investigate the decision problem for description logics with temporal, epistemic, dynamic, and standard modal operators. In most cases we obtain decidability and complexity results by means of reductions to products of modal logics or suitable fragments of first-order modal logics and using results of Chapters 6 and 11. We consider the decidability and complexity of three different reasoning tasks for 'modalized' description languages with modal component L.

Section 14.1 investigates the *concept satisfiability problem relative to empty knowledge base* for concepts *without modalized roles* for modal extensions of $\mathcal{ALC}$. This reasoning problem is important for knowledge representation systems because the global concept satisfiability problem relative to a knowledge base consisting of a simple and acyclic TBox is reducible to that problem by 'unfolding' the knowledge base (see below for definitions).

As was proved in Section 3.8, for a given Kripke complete modal logic L, the satisfiability problem mentioned above is equivalent to the satisfiability problem for $L \times \mathbf{K}_m$—at least when we consider the language without local role names. So numerous decidability results can be obtained as direct consequences of our investigation of $L \times \mathbf{K}_m$ in Chapter 6. We also see that, although decidable in many cases, this reasoning problem can be nonelementary, since satisfiability for $L \times \mathbf{K}_m$ is nonelementary if $L \in \{\mathbf{PTL}, \mathbf{K}_1^C, \mathbf{T}_2^C, \mathbf{K4}_2^C, \mathbf{S4}_2^C, \mathbf{KD45}_2^C, \mathbf{PDL}, \mathbf{CPDL}\}$, cf. Chapter 6. What happens if we add local role names to the language? Fortunately, it turns out that we can do this 'for free.' More precisely, we show that the concept satisfiability problem for $L_{\mathcal{ALC}}$ with local role names is polynomially reducible to the same problem for $L_{\mathcal{ALC}}$ without local role names. Finally, we address the question of whether these decidability results can be extended to more expressive description logics, like $\mathcal{ALC}$ with primitive transitive roles or $\mathcal{CQ}$. The answer is 'no,' and this will be proved again by a reduction to undecidability results for products of modal logics.

Section 14.2 investigates the full *formula satisfiability problem* (that is, the satisfiability problem for formulas having both local and global role names, as well as modalized roles). Or, equivalently, it investigates the local concept satisfiability problem (with possibly nonempty knowledge base) in the full modal description language. Of course, this reasoning problem is much harder than the previous one. In fact, it turns out that only for very few logics—like $\mathbf{K}_{\mathcal{ALC}}$ and $\mathbf{S5}_{\mathcal{ALC}}$—is the problem decidable. This is the only part of this chapter where useful reductions to products or first-order modal logics are not available. Products are useless here, because we do not have anything like modalized accessibility relations. Our results on first-order modal logics are not helpful either, because the translation of a modalized role is not monodic. Decidability results will be obtained by employing the method of quasimodels once again, namely by generalizing the proof of the decidability of $\mathbf{K}_n \times \mathbf{K}_m$.

Next, we consider reasoning tasks which can be analyzed by means of embeddings into monodic fragments of first-order modal logics. Section 14.3 is concerned with the *formula satisfiability problem* for formulas *without modalized roles and global role names*. Such formulas can be regarded (via the embedding of Section 3.8) as members of the monodic fragment of the corresponding first-order modal logic. Thus, if the description logic part of the modal description logic is contained in a decidable fragment of first-order logic without equality—say, its two-variable or guarded fragment—(as is indeed the case for $\mathcal{ALC}$, see Section 3.8), then the decidability of the satisfiability problem, as well as upper bounds for its computational complexity, are immediate consequences of results obtained for monodic fragments of first-order modal logics. On the other hand, lower bounds for the computational complexity of this reasoning problem can be quickly derived from the polynomial reduction of $L \times \mathbf{S5}$-satisfiability to satisfiability of $\mathcal{ML}_{\mathcal{ALC}}$-formulas without any roles at all; see Theorem 3.35 and our results on the complexity of $L \times \mathbf{S5}$ in Sections 5.5 and 6.5. The results on monodic fragments of first-order temporal logics are not directly applicable when the description logic component contains number restrictions or transitive closure operators, which are present, for example, in $\mathcal{CQO}$. In Section 14.3 we show, however, that even for such strong description logics the criteria provided by Theorem 11.83 (for temporal logics) and Theorem 12.4 (for **CPDL** and epistemic logics) can be applied to obtain decidability results.

Finally, in Section 14.4 we consider various reasoning tasks for modal description logics interpreted in models with *finite* domains.

The syntax and semantics of basic modal description logics were introduced in Section 3.8, so a few remarks on the definition of expressive modal description logics like $\mathbf{CPDL}_{\mathcal{ALC}}$, $\mathbf{PTL}_{\mathcal{ALC}}$, or $(\mathbf{S5}_n^C)_{\mathcal{ALC}}$ should be enough. In what follows we omit the test-operator '?' from $\mathcal{CPDL}$. Without '?' the language $\mathcal{CPDL}$ can be regarded as an ordinary modal language with infinitely many modal operators $[\alpha]$, where α is composed from atomic actions $\alpha_0, \alpha_1, \ldots$ using ;, $\cup$, and $\cdot^*$, and interpreted by relations T_α as defined in Section 2.4. Now $\mathcal{CPDL}_{\mathcal{ALC}}$ is defined in the same way as $\mathcal{ML}_{\mathcal{ALC}}$ and can express, for example, that

$$\mathsf{Ordered\_object} \sqsubseteq [\mathsf{construct}; \mathsf{send}]\, \mathsf{Delivered\_object}$$

(an ordered object is a delivered one after the actions 'construct' and 'send' have been performed). By $\mathbf{CPDL}_{\mathcal{ALC}}$ we denote the set of all $\mathcal{CPDL}_{\mathcal{ALC}}$-formulas that are valid in all models. We omit '?' only to simplify the definitions of syntax and semantics. (Recall that test-free **CPDL** is a Kripke complete multimodal logic, see Remark 2.23.) All our results can be extended to full $\mathcal{CPDL}_{\mathcal{ALC}}$ (with appropriately extended semantics); for details see (Wolter 2000a).

Similarly, epistemic logics with common knowledge operators, say $\mathbf{S5}_n^C$, can be regarded as standard modal logics. We denote the resulting epistemic description logic by $(\mathbf{S5}_n^C)_{\mathcal{ALC}}$.

14.1 Concept satisfiability

In this section we are concerned with the following reasoning task. Suppose L is some Kripke complete modal logic. Then the problem is to decide, given an $\mathcal{ML}_{\mathcal{ALC}}$-concept C without modalized roles, whether C is satisfiable in a model for $L_{\mathcal{ALC}}$.

A decision procedure for this problem can be used to provide the following standard reasoning service in description logic systems. As in Section 2.5, we call a set Σ of $\mathcal{ML}_{\mathcal{ALC}}$-formulas a *simple and acyclic TBox* if Σ consists of definitions $A = C$, where A is a concept name and C is an $\mathcal{ML}_{\mathcal{ALC}}$-concept without modalized roles, such that every concept name is defined at most once in Σ and no defined concept name is used in its own definition, explicitly or implicitly. (The first two 'modalized' equations of the 'car salesman knowledge base' and the definition of 'mortal' in Section 3.8 are typical (toy) examples of simple and acyclic knowledge bases.)

Now, the *global concept satisfiability problem for $L_{\mathcal{ALC}}$ relative to simple and acyclic TBoxes* is formulated as follows: given a simple and acyclic TBox Σ and a concept C, decide whether there exists a model $\mathfrak{M} = \langle \mathfrak{F}, I \rangle$ such that $\mathfrak{F}$ is a frame for L, $(\mathfrak{M}, w) \models A = D$ for every world w in $\mathfrak{F}$ and every definition $A = D$ in Σ, and $C^{I(v)} \neq \emptyset$ for some v in $\mathfrak{F}$. (This reasoning service is important because quite often knowledge bases, used in applications, are acyclic and simply introduce abbreviations for complex concepts.) It is not difficult to see that $\mathfrak{M}$ meets the conditions above if and only if the concept, obtained from C by replacing recursively every defined concept with its definition, is satisfied in $\mathfrak{M}$. So the reasoning task above is polynomially reducible to the concept satisfiability problem relative to empty knowledge base (see Table 14.1). In this section we concentrate on the latter problem.

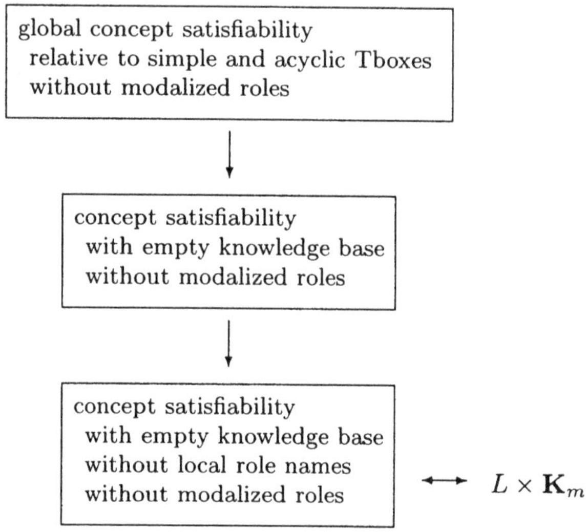

Table 14.1: Reductions between some reasoning tasks for $L_{\mathcal{ALC}}$.

We are going to prove the following decidability results:

Theorem 14.1. *The satisfiability problem for concepts without modalized roles relative to empty knowledge base is decidable for the following logics:*

(1) *the dynamic description logic* $\mathbf{CPDL}_{\mathcal{ALC}}$,

(2) *the epistemic description logics with common knowledge operators* $L^C_{\mathcal{ALC}}$, *where* $L \in \{\mathbf{K}_n, \mathbf{T}_n, \mathbf{K4}_n, \mathbf{S4}_n, \mathbf{KD45}_n, \mathbf{S5}_n\}$,

(3) *the temporal description logics* $\mathbf{PTL}_{\mathcal{ALC}}$, $\mathbf{Lin}_{\mathcal{ALC}}$, *and* $\mathsf{Log}_{FP}(\mathbb{Q})_{\mathcal{ALC}}$,

(4) $\mathbf{K4.3}_{\mathcal{ALC}}$, $\mathsf{Log}\{\langle\mathbb{N}, <\rangle\}_{\mathcal{ALC}}$, *and* $\mathsf{Log}\{\langle\mathbb{Q}, <\rangle\}_{\mathcal{ALC}}$.

These satisfiability problems are not in ELEM *for* $\mathbf{PDL}_{\mathcal{ALC}}$, $\mathbf{PTL}_{\mathcal{ALC}}$ *and* $L^C_{\mathcal{ALC}}$, *where* $L \in \{\mathbf{K}_1, \mathbf{T}_2, \mathbf{K4}_2, \mathbf{S4}_2, \mathbf{KD45}_2\}$.

If we consider the language without local role names and modalized roles, then these decidability results follow already from Theorems 6.10 and 6.40 (stating the decidability of $\mathbf{CPDL} \times \mathbf{K}_m$, $\mathbf{Lin} \times \mathbf{K}_m$ and $\mathsf{Log}_{FP}(\mathbb{Q}) \times \mathbf{K}_m$), and the reductions in Tables 3.1 and 6.1. Further, the nonelementary lower bound is a consequence of the reductions in Table 3.1 and Theorems 6.15, 6.26 and 6.37.

Thus, we only have to prove the following (cf. Table 14.1):

Theorem 14.2. *Let L be a Kripke complete multimodal logic. Then the concept satisfiability problem for $L_{\mathcal{ALC}}$-concepts without modalized roles relative to empty knowledge base is polynomially reducible to the same problem for concepts without local role names and modalized roles.*

Proof. We assume for simplicity that L is a unimodal logic formulated in the language $\mathcal{ML}$. In what follows, we call a model $\langle \mathfrak{F}, I\rangle$ an *$L_{\mathcal{ALC}}$-model* if $\mathfrak{F}$ is a frame for L.

First we show that any $L_{\mathcal{ALC}}$-satisfiable concept C (with both local and global role names, but without modalized roles) is satisfied in an $L_{\mathcal{ALC}}$-model $\langle \mathfrak{F}, I\rangle$ with a set of worlds W and the domain Δ of I such that, for every $x \in \Delta$ and every (global or local) role name T,

$$\left|\{y \in \Delta \mid \exists w \in W \; yT^{I(w)}x\}\right| \leq 1. \tag{14.1}$$

Indeed, suppose that C is satisfied in a model $\langle \mathfrak{F}, I\rangle$. Suppose also that $x_0 \in C^{I(w)}$, $\mathfrak{F} = \langle W, \lhd\rangle$ and

$$I(w) = \left\langle \Delta, C_0^{I(w)}, \dots, R_0^{I(w)}, \dots, S_0^{I(w)}, \dots \right\rangle.$$

(Throughout the proof, we omit interpretations of object names from models, since object names do not occur in concepts and we are dealing with empty knowledge bases.) For each local role name S_i, let

$$R[S_i]^I = \bigcup_{w \in W} S_i^{I(w)},$$

and suppose that $Q_{2i} = R_i^{I(w)}$ and $Q_{2i+1} = R[S_i]^I$, for $i < \omega$. Using the unraveling technique we construct a model $\langle \mathfrak{F}, J\rangle$ by taking, for $w \in W$,

$$J(w) = \left\langle \Delta', C_0^{J(w)}, \dots, R_0^{J(w)}, \dots, S_0^{J(w)}, \dots \right\rangle,$$

where

$$\Delta' = \{\langle x_0, Q_{i_1}, x_1, \dots, Q_{i_m}, x_m\rangle \mid m < \omega, \; \forall j \, (1 \leq j \leq m \to x_{j-1} Q_{i_j} x_j)\},$$

$R_i^{J(w)}$ is defined by taking

$$\langle x_0, Q_{i_1}, \dots, Q_{i_m}, x_m\rangle \, R_i^{J(w)} x \quad \text{iff} \quad \exists y \; x = \left\langle x_0, Q_{i_1}, \dots, Q_{i_m}, x_m, R_i^{I(w)}, y\right\rangle,$$

$S_i^{J(w)}$ is defined by

$$\langle x_0, Q_{i_1}, \ldots, Q_{i_m}, x_m \rangle \, S_i^{J(w)} x \quad \text{iff}$$
$$\exists y \; \big(x = \big\langle x_0, Q_{i_1}, \ldots, Q_{i_m}, x_m, R[S_i]^I, y \big\rangle \text{ and } x_m S_i^{I(w)} y\big),$$

and $C_i^{J(w)}$ is defined by taking

$$\langle x_0, Q_{i_1}, \ldots, Q_{i_m}, x_m \rangle \in C_i^{J(w)} \quad \text{iff} \quad x_m \in C_i^{I(w)}.$$

Clearly, $\langle \mathfrak{F}, J \rangle$ satisfies (14.1). By induction on the construction of a concept D, one can readily prove that

$$\langle x_0, Q_{i_1}, \ldots, Q_{i_m}, x_m \rangle \in D^{J(w)} \quad \text{iff} \quad x_m \in D^{I(w)}.$$

It follows that $\langle x_0 \rangle \in C^{J(w)}$, as required.

Next, for any local role name S take a new concept name reach_S and a new global role name R_S. Define a translation $\cdot^{\bowtie}$ from the set of $\mathcal{ML}_{\mathcal{ALC}}$-concepts without modalized roles into the set of $\mathcal{ML}_{\mathcal{ALC}}$-concepts without modalized roles and local role names by taking:

$$\begin{aligned}
(C_i)^{\bowtie} &= C_i, \; C_i \text{ a concept name,} \\
(C \sqcap D)^{\bowtie} &= C^{\bowtie} \sqcap D^{\bowtie}, \\
(\neg C)^{\bowtie} &= \neg C^{\bowtie}, \\
(\Box C)^{\bowtie} &= \Box C^{\bowtie}, \\
(\exists R.C)^{\bowtie} &= \exists R.C^{\bowtie}, \; R \text{ a global role name,} \\
(\exists S.C)^{\bowtie} &= \exists R_S.(\mathsf{reach}_S \sqcap C^{\bowtie}), \; S \text{ a local role name.}
\end{aligned}$$

We claim that for every $\mathcal{ML}_{\mathcal{ALC}}$-concept C without modalized roles,

$$C \text{ is satisfied in an } L_{\mathcal{ALC}}\text{-model} \quad \text{iff} \quad C^{\bowtie} \text{ is satisfied in an } L_{\mathcal{ALC}}\text{-model.}$$

($\Rightarrow$) Suppose that C is satisfied in an $L_{\mathcal{ALC}}$-model $\mathfrak{M} = \langle \mathfrak{F}, I \rangle$ for which (14.1) holds. Let $x_0 \in C^{I(w)}$, $\mathfrak{F} = \langle W, \lhd \rangle$ and

$$I(w) = \Big\langle \Delta, C_0^{I(w)}, \ldots, R_0^{I(w)}, \ldots, S_0^{I(w)}, \ldots \Big\rangle.$$

Define a model $\langle \mathfrak{F}, J \rangle$ with

$$J(w) = \Big\langle \Delta, C_0^{J(w)} \ldots, \mathsf{reach}_{S_0}^{J(w)}, \ldots, R_0^{J(w)}, \ldots, R_{S_0}^{J(w)}, \ldots \Big\rangle$$

by taking:

- $C_i^{J(w)} = C_i^{I(w)}$, for any concept name C_i different from reach_{S_j};

- $\mathsf{reach}_{S_i}^{J(w)} = \{x \in \Delta \mid \exists y \in \Delta\; y S_i^{I(w)} x\}$;
- $R_{S_i}^{J,w} = \bigcup_{v \in W} S_i^{I(v)}$, for any local role name S_i;
- $R_i^{J(w)} = R_i^{I(w)}$, for any global role name R_i.

By induction on the construction of a concept D one can now prove that $D^{I(w)} = (D^{\bowtie})^{J(w)}$, for all $w \in W$. We only show the induction step for $C = \exists S_i.D$. Suppose that $x \in C^{I(w)}$. Then there exists y such that $x S_i^{I(w)} y$ and $y \in D^{I(w)}$. By the induction hypothesis, we have $x R_{S_i}^{J(w)} y$, $y \in \mathsf{reach}_{S_i}^{J(w)}$, and $y \in (D^{\bowtie})^{J(w)}$. Hence $y \in (\exists R_{S_i}.(\mathsf{reach}_{S_i} \sqcap (D^{\bowtie})))^{J(w)}$. Conversely, suppose that $x \in (C^{\bowtie})^{J(w)}$. Then we find $y \in \mathsf{reach}_{S_i}^{J(w)}$ with $x R_{S_i}^{J(w)} y$ and $y \in (D^{\bowtie})^{J(w)}$. By the definition of $R_{S_i}^{J(w)}$, there exists $v \in W$ such that $x S_i^{I(v)} y$ and, by the definition of $\mathsf{reach}_{S_i}^{J(w)}$, we find x' with $x' S_i^{I(w)} y$. By (14.1), $x = x'$ and so $x S_i^{I(w)} y$. By the induction hypothesis, $y \in D^{I(w)}$ from which we obtain $x \in C^{I(w)}$.

($\Leftarrow$) Suppose that $C^{\bowtie}$ is satisfied in a $L_{\mathcal{ALC}}$-model $\langle \mathfrak{F}, I \rangle$ of the form

$$I(w) = \left\langle \Delta, C_0^{I(w)}, \dots, \mathsf{reach}_{S_0}^{I(w)}, \dots, R_0^{I(w)}, \dots, R_{S_0}^{I(w)}, \dots \right\rangle.$$

Define

$$J(w) = \left\langle \Delta, C_0^{J(w)}, \dots, R_0^{J(w)}, \dots, S_0^{J(w)}, \dots \right\rangle$$

by leaving the interpretations of concepts C_i and global role names R_i unchanged and putting for any local role name S,

$$x S^{J(w)} y \quad \text{iff} \quad x R_S^{I(w)} y \text{ and } y \in \mathsf{reach}_S^{I(w)}.$$

Again, by an easy induction on the construction of D one can show that $D^{J(w)} = (D^{\bowtie})^{I(w)}$, for all concepts D and all $w \in W$. ❑

Note, however, that the following problems are still open:

Question 14.3. What is the computational complexity of the satisfiability problem for concepts without modalized roles for $(\mathbf{S5}_n^C)_{\mathcal{ALC}}$, $\mathbf{Lin}_{\mathcal{ALC}}$, $\mathsf{Log}_{FP}(\mathbb{Q})_{\mathcal{ALC}}$, $\mathbf{K4.3}_{\mathcal{ALC}}$, $\mathsf{Log}\{\langle \mathbb{N}, < \rangle\}_{\mathcal{ALC}}$ and $\mathsf{Log}\{\langle \mathbb{Q}, < \rangle\}_{\mathcal{ALC}}$?

Thus we see that, although of high computational complexity, the satisfiability problem for concepts without modalized roles relative to empty knowledge base is decidable for modalized $\mathcal{ALC}$ with rather expressive modal components. Unfortunately, this result cannot be extended to modal description logics with expressive description components. Here we prove a 'negative' result for the description logics $\mathcal{ALC}_{R^+}$ (alias $\mathcal{S}$) and $\mathcal{CQ}$ introduced in Section 2.5.

Theorem 14.4. *The satisfiability problem for concepts without modalized roles relative to empty knowledge base is undecidable for the following logics:*

(1) *the dynamic description logic* $\mathbf{PDL}_{\mathcal{ALC}_{R^+}}$*,*

(2) *the epistemic description logics* $L^C_{\mathcal{ALC}_{R^+}}$ *with common knowledge operators, where* $L \in \{\mathbf{K}_1, \mathbf{T}_2, \mathbf{K4}_2, \mathbf{S4}_2, \mathbf{KD45}_2\}$*,*

(3) *the temporal description logic* $\mathbf{PTL}_{\mathcal{ALC}_{R^+}}$*.*

Proof. Let $L \in \{\mathbf{PTL}, \mathbf{PDL}, \mathbf{K}_1^C, \mathbf{T}_2^C, \mathbf{K4}_2^C, \mathbf{S4}_2^C, \mathbf{KD45}_2^C\}$. To begin with, recall that the translation $\cdot^\dagger$ (given in Section 3.8) polynomially reduces the satisfiability problem for $L \times \mathbf{K}_m$ to the satisfiability problem for $L_{\mathcal{ALC}}$-concepts without local role names and modalized roles relative to empty knowledge base (see Table 3.1). Now, set in the translation $\cdot^\dagger$ for the **K4**-operator $\Diamond$ of $L \times \mathbf{K4}$:

$$(\Diamond\varphi)^\dagger = \exists R.\varphi^\dagger,$$

where R is a transitive role of $\mathcal{ALC}_{R^+}$. Then we obtain a polynomial reduction of the satisfiability problem for $L \times \mathbf{K4}$ to the concept satisfiability problem with empty knowledge base for $L_{\mathcal{ALC}_{R^+}}$. By Theorem 7.25, $L \times \mathbf{K4}$ is undecidable, for all the listed logics L. ❑

Note that $\mathbf{S5}_2$ does not occur in the list of epistemic logics above:

Question 14.5. Is the concept satisfiability problem with empty knowledge base for $(\mathbf{S5}_2^C)_{\mathcal{ALC}_{R^+}}$ decidable?

Theorem 14.6. *The satisfiability problem for concepts without modalized roles relative to empty knowledge base is undecidable for the following logics:*

(1) *the dynamic description logic* $\mathbf{PDL}_{\mathcal{CQ}}$*,*

(2) *the epistemic description logics with common knowledge operators* $L^C_{\mathcal{CQ}}$*, where* $L \in \{\mathbf{K}_1, \mathbf{T}_2, \mathbf{K4}_2, \mathbf{S4}_2, \mathbf{KD45}_2, \mathbf{S5}_2\}$*,*

(3) $\mathsf{Log}(\mathcal{C})_{\mathcal{CQ}}$*, where* $\mathcal{C}$ *is any class of strict linear orders such that at least one of them contains an infinite ascending chain of distinct points.*

Proof. For any L listed in the theorem, $L \times \mathbf{K}_1^C$ is polynomially reducible to the satisfiability problem for $L_{\mathcal{CQ}}$-concepts without local and modalized roles relative to the empty knowledge base by extending in a straightforward manner the reduction of $L \times \mathbf{K}_m$ to $L_{\mathcal{ALC}}$, given in Section 3.8; cf. Table 3.1 (details are left to the reader as an exercise). Now the theorem follows from Theorems 7.19 and 7.20. ❑

14.2 General formula satisfiability

In this section we consider the formula satisfiability problem for modalized description logics in which modal operators can be applied to concepts, global and local roles, and formulas. In other words, we deal with the full modal description languages introduced in Section 3.8. The price we have to pay for this expressive power is high—only very few logics turn out to be decidable.

We start with the following 'negative' result:

Theorem 14.7. *The satisfiability problem for formulas without modalized roles and local role names is undecidable for the following logics:*

(1) *the dynamic description logic* $\mathbf{PDL}_{\mathcal{ALC}}$,

(2) *the epistemic description logics* $L^{C}_{\mathcal{ALC}}$ *with common knowledge operators, where* $L \in \{\mathbf{K}_1, \mathbf{T}_2, \mathbf{K4}_2, \mathbf{S4}_2, \mathbf{KD45}_2, \mathbf{S5}_2\}$,

(3) $\mathsf{Log}(\mathcal{C})_{\mathcal{ALC}}$, *where* $\mathcal{C}$ *is any class of strict linear orders such that at least one of them contains an infinite ascending chain of distinct points.*

Proof. First, we know from Theorems 7.19 and 7.20 that $L \times \mathbf{K}_u$ is undecidable, for any modal logic L listed in the theorem. And, by Theorem 3.36, $L \times \mathbf{K}_u$ is polynomially reducible to the formula satisfiability problem for $L_{\mathcal{ALC}}$ without modalized roles and local role names. ❑

On the other hand, the following positive result is shown in (Wolter and Zakharyaschev 1999b):

Theorem 14.8. *Let* $L \in \{\mathbf{K}_n, \mathbf{T}_n, \mathbf{KD45}_n, \mathbf{S5}_n\}$. *Then the formula satisfiability problem for* $L_{\mathcal{ALC}}$ *is decidable.*

Proof. To simplify presentation, we begin by considering the modal description logic $\mathbf{K}_{\mathcal{ALC}}$ with only one modal operator. It is straightforward (and left to the reader) to generalize the proof to the other logics mentioned in the theorem—some hints will be given at the end of the proof.

The proof is a generalization of the proof of Theorem 6.1 (stating the decidability of $\mathbf{K}_n \times \mathbf{K}_m$), and is organized as follows. First, we represent $\mathbf{K}_{\mathcal{ALC}}$-models in the form of *quasimodels* and then show that these quasimodels can be constructed like mosaics from a finite number of relatively small finite pattern pieces (which again are called *blocks*).

A number of notions—types, quasistates, basic structures, runs, quasimodels, blocks, etc.—which were used in the proof of Theorem 6.1 will be used here as well. While their role in the present decidability proof is quite similar to the role they played before, the definitions do not coincide. As before, the use of the same name for different objects in different proofs turns out to be rather helpful, since this clarifies the similarities (and the differences) between these proofs.

Let us fix an arbitrary $\mathcal{ML}_{\mathcal{ALC}}$-formula φ and try to define a suitable notion of $\mathbf{K}_{\mathcal{ALC}}$-quasimodel for φ, following the pattern from the proof of Theorem 6.1. We again require a number of auxiliary definitions. Let $ob\,\varphi$, $con\,\varphi$, and $rol\,\varphi$ be the sets of all object names, concepts, and roles in φ, respectively, and let $sub\,\varphi$ denote the set of all subformulas of φ.

A *concept type* for φ is a subset $\boldsymbol{c}$ of $con\,\varphi$ such that

- $C \sqcap D \in \boldsymbol{c}$ iff $C, D \in \boldsymbol{c}$, for every $C \sqcap D \in con\,\varphi$;
- $\neg C \in \boldsymbol{c}$ iff $C \notin \boldsymbol{c}$, for every $\neg C \in con\,\varphi$.

A *named concept type* is a pair $\boldsymbol{c}_a = \langle \boldsymbol{c}, a \rangle$ in which $\boldsymbol{c}$ is a concept type and $a \in ob\,\varphi$. A *formula type* for φ is a subset $\boldsymbol{f}$ of $sub\,\varphi$ such that

- $\psi \wedge \chi \in \boldsymbol{f}$ iff $\psi, \chi \in \boldsymbol{f}$, for every $\psi \wedge \chi \in sub\,\varphi$;
- $\neg\psi \in \boldsymbol{f}$ iff $\psi \notin \boldsymbol{f}$, for every $\neg\psi \in sub\,\varphi$.

A *named formula type* is the pair $\boldsymbol{f}_a = \langle \boldsymbol{f}, a \rangle$ in which $\boldsymbol{f}$ is a formula type and $a \in ob\,\varphi$. Finally, by a *type* for φ we will mean the pair $\boldsymbol{t} = \langle \boldsymbol{c}, \boldsymbol{f} \rangle$, where $\boldsymbol{c}$ is a concept type and $\boldsymbol{f}$ a formula type for φ; $\boldsymbol{t}_a = \langle \boldsymbol{c}_a, \boldsymbol{f}_a \rangle$ is a *named type* for φ.

To simplify notation we will write $C \in \boldsymbol{t}$ and $\psi \in \boldsymbol{t}$ whenever $\boldsymbol{t} = \langle \boldsymbol{c}, \boldsymbol{f} \rangle$, $C \in \boldsymbol{c}$ and $\psi \in \boldsymbol{f}$ (in the case of named types, $C \in \boldsymbol{t}_a$ and $\psi \in \boldsymbol{t}_a$ mean that $\boldsymbol{t}_a = \langle \boldsymbol{c}_a, \boldsymbol{f}_a \rangle$, $\boldsymbol{c}_a = \langle \boldsymbol{c}, a \rangle$, $\boldsymbol{f}_a = \langle \boldsymbol{f}, a \rangle$, and $C \in \boldsymbol{c}$, $\psi \in \boldsymbol{f}$). Two types $\boldsymbol{t}_1 = \langle \boldsymbol{c}_1, \boldsymbol{f}_1 \rangle$ and $\boldsymbol{t}_2 = \langle \boldsymbol{c}_2, \boldsymbol{f}_2 \rangle$ are said to be *formula-equivalent* if $\boldsymbol{f}_1 = \boldsymbol{f}_2$.

A *quasistate candidate* for φ is a pair $\langle \langle T, < \rangle, \boldsymbol{t} \rangle$, where $\langle T, < \rangle$ is a finite intransitive tree of depth $\leq md(\varphi)$ and $\boldsymbol{t}$ a labeling function associating with each $x \in T$ a type $\boldsymbol{t}(x)$ for φ. (So we can think of a quasistate candidate as a tree of types.) Two quasistate candidates $\langle \langle T, < \rangle, \boldsymbol{t} \rangle$ and $\langle \langle T', <' \rangle, \boldsymbol{t}' \rangle$ are called *isomorphic* if there is an isomorphism f between the trees $\langle T, < \rangle$ and $\langle T', <' \rangle$ such that $\boldsymbol{t}(x) = \boldsymbol{t}'(f(x))$, for all $x \in T$.

A quasistate candidate $\langle \langle T, < \rangle, \boldsymbol{t} \rangle$ is called a *quasistate* for φ if the following conditions hold:

(dlqm1) For all $x \in T$, $\Diamond C \in con\,\varphi$, and $\Diamond\psi \in sub\,\varphi$,

$$\Diamond C \in \boldsymbol{t}(x) \quad \text{iff} \quad \exists y \in T\ \big(x < y \wedge C \in \boldsymbol{t}(y)\big),$$
$$\Diamond \psi \in \boldsymbol{t}(x) \quad \text{iff} \quad \exists y \in T\ \big(x < y \wedge \psi \in \boldsymbol{t}(y)\big).$$

(dlqm1′) For all $x, x_1, x_2 \in T$ such that $x < x_1$, $x < x_2$ and $x_1 \neq x_2$, the structures $\langle \langle T^{x_1}, <^{x_1} \rangle, \boldsymbol{t}^{x_1} \rangle$ and $\langle \langle T^{x_2}, <^{x_2} \rangle, \boldsymbol{t}^{x_2} \rangle$ are not isomorphic, where $\langle T^{x_i}, <^{x_i}\rangle$ is the subtree of $\langle T, <\rangle$ generated by x_i, and $\boldsymbol{t}^{x_i}$ is the restriction of $\boldsymbol{t}$ to T^{x_i}, $i = 1, 2$.

Quasistates are intended to represent the 'behavior' of a single object in models (modulo φ).

As the number of different types for φ does not exceed $2^{|con\,\varphi|} \cdot 2^{|sub\,\varphi|}$, the number of pairwise nonisomorphic quasistates for φ of depth 0 is at most $2^{|con\,\varphi|} \cdot 2^{|sub\,\varphi|}$ as well. Now define inductively

$$n_0(\varphi) = 2^{|con\,\varphi|} \cdot 2^{|sub\,\varphi|}, \qquad n_{k+1}(\varphi) = 2^{|con\,\varphi|} \cdot 2^{|sub\,\varphi|} \cdot 2^{n_k(\varphi)}.$$

Clearly, $n_k(\varphi)$ is an upper bound for the number of nonisomorphic quasistates for φ of depth k, and so

$$b(\varphi) = \sum_{k=0}^{md(\varphi)} n_k(\varphi)$$

is an upper bound for the number of different quasistates for φ. The number of points in any quasistate for φ is bounded by

$$n_0(\varphi) + \sum_{k=1}^{md(\varphi)} \prod_{j=1}^{k} n_{md(\varphi)-j}(\varphi) \leq b(\varphi)^{md(\varphi)} = p(\varphi).$$

In what follows we assume that nonisomorphic quasistates are disjoint and that isomorphic quasistates actually coincide.

A *basic structure* of depth m for φ is a pair $\langle \Delta, \boldsymbol{q}\rangle$ such that Δ is a nonempty set and $\boldsymbol{q}$ a function associating with each $w \in \Delta$ a quasistate

$$\boldsymbol{q}(w) = \langle \langle T_w, <_w\rangle, \boldsymbol{t}_w\rangle$$

for φ such that the depth of each $\langle T_w, <_w\rangle$ is m and, for every $a \in \Delta \cap ob\,\varphi$, the set $\{\boldsymbol{t}_a(x) \mid x \in T_a\}$ consists of only named types of the form $\boldsymbol{t}_a$.

Let $\langle \Delta, \boldsymbol{q}\rangle$ be a basic structure for φ of depth m and let $k \leq m$. A *k-run through* $\langle \Delta, \boldsymbol{q}\rangle$ is pair of the form

$$\boldsymbol{r} = \langle r, \{R_r \mid R \in rol\,\varphi\}\rangle$$

in which R_r is a binary relation on Δ, for each $R \in rol\,\varphi$, and r is a function giving, for each $w \in \Delta$, a point $r(w) \in T_w$ of co-depth k such that all the types $\boldsymbol{t}_w(r(w))$, $w \in \Delta$, are formula-equivalent to each other. Given a set $\mathfrak{R}$ of runs, we denote by $\mathfrak{R}_k$ the set of all k-runs from $\mathfrak{R}$.

A run $\boldsymbol{r}$ is called *coherent* if the following conditions hold, for all $w \in \Delta$:

- for every $C = D$ in $sub\,\varphi$, $(C = D) \in \boldsymbol{t}_w(r(w))$ iff for all $v \in \Delta$, we have $(C \in \boldsymbol{t}_v(r(v)) \leftrightarrow D \in \boldsymbol{t}_v(r(v)))$;

- for every $a : C$ in $sub\,\varphi$, $(a : C) \in \boldsymbol{t}_w(r(w))$ iff $C \in \boldsymbol{t}_a(r(a))$, provided that $a \in \Delta$;
- for every aRb in $sub\,\varphi$, we have $(aRb) \in \boldsymbol{t}_w(r(w))$ iff $aR_r b$, provided that $a, b \in \Delta$;
- for every $\exists R.C$ in $con\,\varphi$, if there exists a $v \in \Delta$ such that $wR_r v$ and $C \in \boldsymbol{t}_v(r(v))$ then $\exists R.C \in \boldsymbol{t}_w(r(w))$.

A run $\boldsymbol{r}$ is called *w-saturated* for $w \in \Delta$ if

- for every $\exists R.C$ in $con\,\varphi$, $\exists R.C \in \boldsymbol{t}_w(r(w))$ implies that there is a $v \in \Delta$ such that $wR_r v$ and $C \in \boldsymbol{t}_v(r(v))$.

A run is *saturated* if it is w-saturated for all $w \in \Delta$.

Finally, we say that a quadruple $\mathfrak{Q} = \langle \Delta, \boldsymbol{q}, \mathfrak{R}, \lhd \rangle$ is a *$\mathbf{K}_{\mathcal{ALC}}$-quasimodel for φ (based on Δ)* if $\langle \Delta, \boldsymbol{q} \rangle$ is a basic structure for φ of depth $m \leq md(\varphi)$ such that $\Delta \supseteq ob\,\varphi$, $\mathfrak{R}$ is a set of coherent and saturated runs through $\langle \Delta, \boldsymbol{q} \rangle$, and $\lhd$ is a binary relation on $\mathfrak{R}$ satisfying the following conditions:

(dlqm2) there is an $\boldsymbol{r} \in \mathfrak{R}$ such that $\varphi \in \boldsymbol{t}_w(r(w))$, for some (or, equivalently, all) $w \in \Delta$;

(dlqm3) for all $\boldsymbol{r}, \boldsymbol{r}' \in \mathfrak{R}$, if $\boldsymbol{r} \lhd \boldsymbol{r}'$ then $r(w) <_w r'(w)$ for all $w \in \Delta$;

(dlqm4) $\mathfrak{R}_0 \neq \emptyset$, and for all $k < m$, $\boldsymbol{r} \in \mathfrak{R}_k$, $w \in \Delta$ and $x \in T_w$, if $r(w) <_w x$ then there is $\boldsymbol{r}' \in \mathfrak{R}_{k+1}$ such that $r'(w) = x$ and $\boldsymbol{r} \lhd \boldsymbol{r}'$;

(dlqm5) for all $w, v \in \Delta$, $\Diamond R \in rol\,\varphi$, $k \leq m$, and $\boldsymbol{r} \in \mathfrak{R}_k$, we have $w(\Diamond R)_r v$ iff there is $\boldsymbol{r}' \in \mathfrak{R}_{k+1}$ such that $\boldsymbol{r} \lhd \boldsymbol{r}'$ and $wR_{r'}v$;
for all $w, v \in \Delta$, $\Box R \in rol\,\varphi$, $k \leq m$, and $\boldsymbol{r} \in \mathfrak{R}_k$, we have $w(\Box R)_r v$ iff for all $\boldsymbol{r}' \in \mathfrak{R}_{k+1}$, $\boldsymbol{r} \lhd \boldsymbol{r}'$ imply $wR_{r'}v$;

(dlqm6) for all $\boldsymbol{r}, \boldsymbol{r}' \in \mathfrak{R}$, we have $R_r = R_{r'}$, whenever R is a global role name in $rol\,\varphi$.

The notion of quasimodel has been defined, and now we have to prove the 'quasimodel lemma' (cf. Lemma 6.2):

Lemma 14.9. *An $\mathcal{ML}_{\mathcal{ALC}}$-formula φ is satisfied in a $\mathbf{K}_{\mathcal{ALC}}$-model iff there is a $\mathbf{K}_{\mathcal{ALC}}$-quasimodel for φ.*

Proof. ($\Leftarrow$) Suppose that $\langle \Delta, \boldsymbol{q}, \mathfrak{R}, \lhd \rangle$ is a quasimodel for φ. Construct a $\mathbf{K}_{\mathcal{ALC}}$-model $\mathfrak{M} = \langle \mathfrak{F}, I \rangle$ based on the frame $\mathfrak{F} = \langle \mathfrak{R}, \lhd \rangle$ by taking, for all $\boldsymbol{r} \in \mathfrak{R}$,

$$I(\boldsymbol{r}) = \left\langle \Delta, R_0^{I(\boldsymbol{r})}, \dots, C_0^{I(\boldsymbol{r})}, \dots, a_0^{I(\boldsymbol{r})}, \dots \right\rangle,$$

where,

- $wR_i^{I(\boldsymbol{r})}v$ iff $w(R_i)_r v$,
- $C_i^{I(\boldsymbol{r})} = \{w \in \Delta \mid C_i \in \boldsymbol{t}_w(r(w))\}$,
- $a_i^{I(\boldsymbol{r})} = a_i$,

whenever $R_i \in rol\,\varphi$, $C_i \in con\,\varphi$, and $a_i \in ob\,\varphi$, and arbitrary otherwise. By a straightforward induction on the construction of concepts, roles and formulas one can check (using conditions **(dlqm3)**–**(dlqm5)**) that for all $C \in con\,\varphi$, $R \in rol\,\varphi$, $\psi \in sub\,\varphi$, $w, v \in \Delta$, and $\boldsymbol{r} \in \mathfrak{R}$, we have:

$$\begin{array}{lll} wR^{I(\boldsymbol{r})}v & \text{iff} & wR_r v, \\ w \in C^{I(\boldsymbol{r})} & \text{iff} & C \in \boldsymbol{t}_w(r(w)), \\ (\mathfrak{M}, \boldsymbol{r}) \models \psi & \text{iff} & \psi \in \boldsymbol{t}_w(r(w)) \ \text{ for some (or, equivalently, all) } w \in \Delta. \end{array}$$

Therefore, by **(dlqm2)**, φ is satisfied in $\mathfrak{M}$.

($\Rightarrow$) Suppose now that φ is satisfied in a $\mathbf{K}_{\mathcal{ALC}}$-model $\mathfrak{M} = \langle \mathfrak{F}, I \rangle$ with domain $\Delta \supseteq ob\,\varphi$. An argument similar to the one proving Proposition 1.8 shows that we may assume $\mathfrak{F} = \langle W, < \rangle$ to be an intransitive tree of depth $m \leq md(\varphi)$ and

$$(\mathfrak{M}, x_0) \models \varphi$$

for the root x_0 of $\mathfrak{F}$. For every pair $w \in \Delta$, $x \in W$, let

$$\begin{aligned} \boldsymbol{c}(w, x) &= \{C \in con\,\varphi \mid w \in C^{I(x)}\}, \\ \boldsymbol{f}(x) &= \{\psi \in sub\,\varphi \mid (\mathfrak{M}, x) \models \psi\}, \\ \boldsymbol{t}(w, x) &= \langle \boldsymbol{c}(w, x), \boldsymbol{f}(x) \rangle . \end{aligned}$$

Clearly, $\boldsymbol{t}(w, x)$ is a type for φ; $\boldsymbol{t}(a, x)$ is regarded to be a type named by a, for $a \in ob\,\varphi$. Now we have to construct a quasistate $\langle \langle T_w, <_w \rangle, \boldsymbol{t}_w \rangle$ for each $w \in \Delta$. The obvious choice of $T_w = W$, $<_w = <$ and $\boldsymbol{t}_w(x) = \boldsymbol{t}(w, x)$ does not work, because W can be infinite. So let us make it finite in such a way that the resulting structure still satisfies **(dlqm1)** and also complies with the smallness condition **(dlqm1′)**. Fix a $w \in \Delta$ and define a binary relation $\sim_w$ on W as follows. If $x, y \in W$ are of depth 0 (i.e., they are leaves of $\mathfrak{F}$) then

$$x \sim_w y \quad \text{iff} \quad \boldsymbol{t}(w, x) = \boldsymbol{t}(w, y).$$

For $x, y \in W$ of depth k ($0 < k \leq md(\varphi)$), let

$$\begin{array}{lll} x \sim_w y & \text{iff} & \boldsymbol{t}(w, x) = \boldsymbol{t}(w, y) \\ & & \wedge\, \forall z \in W \,\big(x < z \ \rightarrow\ \exists z' \in W \,(y < z' \ \wedge\ z \sim_w z')\big) \\ & & \wedge\, \forall z \in W \,\big(y < z \ \rightarrow\ \exists z' \in W \,(x < z' \ \wedge\ z \sim_w z')\big). \end{array}$$

Clearly $\sim_w$ is an equivalence relation on W. Denote by $[x]_w$ the $\sim_w$-equivalence class of x and put

$$\begin{array}{rcl} W_w & = & \{[x]_w \mid x \in W\}, \\ [x]_w S_w [y]_w & \text{iff} & \exists y' \in [y]_w \; x < y', \\ \boldsymbol{l}_w([x]_w) & = & \boldsymbol{t}(w, x). \end{array}$$

Then, by the definition of $\sim_w$, S_w is well-defined and the structure

$$\langle \langle W_w, S_w \rangle, \boldsymbol{l}_w \rangle$$

clearly satisfies **(dlqm1′)**. Note that the map $f_w : x \mapsto [x]_w$ is a p-morphism from $\langle W, < \rangle$ onto $\langle W_w, S_w \rangle$, and so it also satisfies **(dlqm1)**. However, $\langle W_w, S_w \rangle$ is not necessarily a tree. The tree $\langle T_w, <_w \rangle$ we need can be obtained by unraveling $\langle W_w, S_w \rangle$:

$$\begin{array}{l} T_w = \{\langle [x_0]_w, \dots, [x_k]_w \rangle \mid k \leq m, \; [x_0]_w S_w [x_1]_w S_w \dots S_w [x_k]_w\}, \\ u <_w v \quad \text{iff} \quad u = \langle [x_0]_w, \dots, [x_k]_w \rangle, \; v = \langle [x_0]_w, \dots, [x_k]_w, [x_{k+1}]_w \rangle \\ \qquad\qquad\qquad\quad \text{and} \quad [x_k]_w S_w [x_{k+1}]_w. \end{array}$$

Let

$$\boldsymbol{t}_w(\langle [x_0]_w, \dots, [x_k]_w \rangle) = \boldsymbol{l}_w([x_k]_w) = \boldsymbol{t}(w, x_k).$$

It is not hard to see that, for any $w \in \Delta$,

$$\boldsymbol{q}(w) = \langle \langle T_w, <_w \rangle, \boldsymbol{t}_w \rangle$$

is a quasistate for φ. It remains to define appropriate runs through the basic structure $\langle \Delta, \boldsymbol{q} \rangle$. To this end, for each $k \leq m$ and each sequence $\bar{x} = \langle x_0, \dots, x_k \rangle$ of points in W such that $x_0 < \dots < x_k$, take the map

$$r_{\bar{x}} : w \mapsto \langle [x_0]_w, \dots, [x_k]_w \rangle,$$

and, for each $R \in rol\, \varphi$, define a binary relation $R_{r_{\bar{x}}}$ on Δ by taking

$$w R_{r_{\bar{x}}} v \quad \text{iff} \quad w R^{I(x_k)} v. \tag{14.2}$$

It is easy to check that

$$\boldsymbol{r}_{\bar{x}} = \langle r_{\bar{x}}, \{R_{r_{\bar{x}}} \mid R \in rol\, \varphi\} \rangle$$

is a coherent and saturated k-run. Let $\mathfrak{R}$ be the set of all such runs. Then **(dlqm6)** holds by definition. For $\boldsymbol{r}_{\bar{x}}, \boldsymbol{r}_{\bar{y}} \in \mathfrak{R}$, let $\boldsymbol{r}_{\bar{x}} \lhd \boldsymbol{r}_{\bar{y}}$ iff $\bar{x} = \langle x_0, \dots, x_k \rangle$, $\bar{y} = \langle x_0, \dots, x_k, x_{k+1} \rangle$ for some points $x_0 < \dots < x_k < x_{k+1}$ in W. Then **(dlqm3)** holds by the definition of $<_w$. Condition **(dlqm2)** holds since we have $\varphi \in \boldsymbol{t}_w(r_0(w))$ for the run $\boldsymbol{r}_0 \in \mathfrak{R}_0$ with $r_0(w) = \langle [x_0]_w \rangle$, for all $w \in \Delta$. For **(dlqm4)**, let $\boldsymbol{r} \in \mathfrak{R}_k$, $v \in \Delta$ and $z \in T_v$ be such that $r(v) <_v z$. We have to show that there is $\boldsymbol{r}' \in \mathfrak{R}_{k+1}$ such that $\boldsymbol{r} \lhd \boldsymbol{r}'$ and $r'(v) = z$. Since $r(v) <_v z$, we have $r(v) = \langle [x_0]_v, \dots, [x_k]_v \rangle$ and $z = \langle [x_0]_v, \dots, [x_k]_v, [x_{k+1}]_v \rangle$, for some $x_1, \dots, x_k, x_{k+1}$ such that $x_0 < x_1 < \dots < x_k$ and $[x_k]_v S_v [x_{k+1}]_v$. By the definition of S_v, there is $y \in [x_{k+1}]_v$ such that $x_k < y$. Take the map

$$r' : w \mapsto \langle [x_0]_w, \dots, [x_k]_w, [y]_w \rangle ,$$

and, for each $R \in rol\,\varphi$, define a binary relation $R_{r'}$ on Δ by taking

$$wR_{r'}v \quad \text{iff} \quad wR^{I(y)}v.$$

Then the pair $\boldsymbol{r}' = \langle r', \{R_{r'} \mid R \in rol\,\varphi\} \rangle$ is in $\mathfrak{R}$.

It remains to prove **(dlqm5)**. We check only the condition for $\Box$; the $\Diamond$ case is treated analogously. Suppose that $w(\Box R)_r v$ and $\boldsymbol{r} \lhd \boldsymbol{r}'$. Then $\boldsymbol{r} = \boldsymbol{r}_{\bar{x}}$, $\boldsymbol{r}' = \boldsymbol{r}_{\bar{y}}$, for some sequences $\bar{x} = \langle x_0, \dots, x_k \rangle$, $\bar{y} = \langle x_0, \dots, x_k, x_{k+1} \rangle$ of points from W such that $x_0 < \dots < x_k < x_{k+1}$, and $w(\Box R)^{I(x_k)} v$ holds by (14.2). It follows by definition that we have $wR^{I(y)}v$ for all $y > x_k$, and so in particular, $wR^{I(x_{k+1})}v$. Using (14.2) again, we obtain $wR_{r'}v$. Conversely, suppose that $wR_{r'}v$ holds whenever $\boldsymbol{r} \lhd \boldsymbol{r}'$, for some $\boldsymbol{r} = \boldsymbol{r}_{\bar{x}}$, $\bar{x} = \langle x_0, \dots, x_k \rangle$. We need to show that $w(\Box R)_r v$, that is, $wR^{I(y)}v$ hold for all $y > x_k$ in W. For every $y > x_k$, take the run $\boldsymbol{r}_{\bar{y}}$ corresponding to the sequence $\bar{y} = \langle x_0, \dots, x_k, y \rangle$. By the definition of $\lhd$ we have $\boldsymbol{r} \lhd \boldsymbol{r}_{\bar{y}}$, from which $wR_{r_{\bar{y}}}v$. Therefore, by (14.2), $wR^{I(y)}v$, as required.

Thus, $\langle \Delta, \boldsymbol{q}, \mathfrak{R}, \lhd \rangle$ is a quasimodel for φ. ❑

Suppose that $\mathfrak{Q} = \langle \Delta, \boldsymbol{q}, \mathfrak{R}, \lhd \rangle$ is a quasimodel for φ, $w \in \Delta$, $R \in rol\,\varphi$, and $R = \mathsf{M}R_i$ for some (possibly empty) string M of $\Diamond$ and $\Box$, and role name R_i. Consider the tree $\mathfrak{F}_w = \langle T_w, <_w \rangle$ as a usual Kripke frame. If we have $(\mathfrak{F}_w, r(w)) \models \mathsf{M}\bot$, for $\boldsymbol{r} \in \mathfrak{R}$, then let us say that R is $\boldsymbol{r}$-*universal*. This name is explained by the fact that if R is $\boldsymbol{r}$-universal then $R_r = \Delta \times \Delta$, which can be easily established by induction on the length of the string M.

We also say that objects $v, v' \in \Delta$ are *twins* (*in* $\mathfrak{Q}$) *relative to* $w \in \Delta$ if

- $\boldsymbol{q}(v) = \boldsymbol{q}(v')$;
- for all $\boldsymbol{r} \in \mathfrak{R}$, $r(v) = r(v')$; and
- for all $\boldsymbol{r} \in \mathfrak{R}$ and $R \in rol\,\varphi$, we have $wR_r v$ iff $wR_r v'$.

Lemma 14.10. *For every satisfiable $\mathcal{ML}_{\mathcal{ALC}}$-formula φ, there is a quasimodel $\mathfrak{Q}^* = \langle \Delta^*, \boldsymbol{q}^*, \mathfrak{R}^*, \lhd^* \rangle$ for φ such that the following conditions hold:*

(i) *for any distinct $w, v \in \Delta^*$, the object v has infinitely many twins in $\mathfrak{Q}^*$ relative to w;*

(ii) *the relation*

$$\{\langle v, v'\rangle \mid v, v' \notin ob\,\varphi \textit{ and } vR_r v' \textit{ for some } \boldsymbol{r} \in \mathfrak{R}^*,$$
$$R \in rol\,\varphi \textit{ is not } \boldsymbol{r}\textit{-universal}\}$$

is a disjoint union of intransitive tree orders on the set $\Delta - ob\,\varphi$.

Proof. Suppose that there is a quasimodel $\mathfrak{Q} = \langle \Delta, \boldsymbol{q}, \mathfrak{R}, \lhd\rangle$ for φ. For each $w \in \Delta$ we take an infinite set X_w containing w so that $X_w \cap X_{w'} = \emptyset$ whenever $w \neq w'$. For every $v \in X_w$ let $\boldsymbol{q}'(v)$ be an isomorphic copy of $\boldsymbol{q}(w)$, and let $\Delta' = \bigcup\{X_w \mid w \in \Delta\}$. Thus we have got a basic structure $\langle \Delta', \boldsymbol{q}'\rangle$. Now we extend every run $\boldsymbol{r} \in \mathfrak{R}$ to a run $\boldsymbol{r}'$ through $\langle \Delta', \boldsymbol{q}'\rangle$ simply by taking $r'(v) = r(w)$ for all $v \in X_w$, and $w' R_{r'} v'$ iff $wR_r v$, for all $w' \in X_w$, $v' \in X_v$. The resulting set of runs is denoted by $\mathfrak{R}'$; we put $\boldsymbol{r}_1' \lhd' \boldsymbol{r}_2'$ iff $\boldsymbol{r}_1 \lhd \boldsymbol{r}_2$, for all $\boldsymbol{r}_1, \boldsymbol{r}_2 \in \mathfrak{R}$. It is readily seen that $\mathfrak{Q}' = \langle \Delta', \boldsymbol{q}', \mathfrak{R}', \lhd'\rangle$ is a quasimodel for φ satisfying condition (i).

To satisfy (ii), we apply the unraveling technique to $\mathfrak{Q}'$. Denote by Δ^* the set of all finite tuples $\langle w_1, \dots, w_n\rangle$ of objects in Δ' such that $w_i \notin ob\,\varphi$ for $i \neq 1$, and let $\boldsymbol{q}^*(\langle w_1, \dots, w_n\rangle) = \boldsymbol{q}'(w_n)$, which yields us a basic structure $\langle \Delta^*, \boldsymbol{q}^*\rangle$. Given a run $\boldsymbol{r} \in \mathfrak{R}'$, we construct $\boldsymbol{r}^*$ by taking

$$r^*(\langle w_1, \dots, w_n\rangle) = r(w_n)$$

and, for every $R \in rol\,\varphi$, $\langle w_1, \dots, w_n\rangle\, R_{r^*}\, \langle v_1, \dots, v_m\rangle$ iff either R is $\boldsymbol{r}$-universal or $\langle w_1, \dots, w_n\rangle = \langle v_1, \dots, v_{m-1}\rangle$ and $w_n R_r v_m$. It is not hard to check that $\boldsymbol{r}^*$ is a run through $\langle \Delta^*, \boldsymbol{q}^*\rangle$. Finally, we put $\boldsymbol{r}_1^* \lhd^* \boldsymbol{r}_2^*$ iff $\boldsymbol{r}_1 \lhd' \boldsymbol{r}_2$, for all $\boldsymbol{r}_1, \boldsymbol{r}_2 \in \mathfrak{R}'$. The structure $\mathfrak{Q}^* = \langle \Delta^*, \boldsymbol{q}^*, \mathfrak{R}^*, \lhd^*\rangle$ is then a quasimodel for φ satisfying both conditions (i) and (ii). ❑

Our next task is to provide an algorithm for deciding whether there exists a $\mathbf{K}_{\mathcal{ALC}}$-quasimodel for φ. In fact, we will show that instead of finding such a quasimodel, it is enough to find a finite set of finite 'pattern blocks' out of which a quasimodel for φ can be constructed, with the size of the set and the size of blocks in it being effectively computable.

A *w-block for* φ is a quadruple $\mathfrak{B} = \langle \Delta, \boldsymbol{q}, \mathfrak{R}, \lhd\rangle$ such that

- Δ is a finite set disjoint from $ob\,\varphi$, $w \in \Delta$, and $\langle \Delta, \boldsymbol{q}\rangle$ is a basic structure for φ of depth m, for some $m \leq md(\varphi)$;
- $\mathfrak{R}$ is a set of coherent and w-saturated runs through $\langle \Delta, \boldsymbol{q}\rangle$ satisfying **(dlqm6)**;
- $\lhd$ is a binary relation on $\mathfrak{R}$ satisfying **(dlqm3)**–**(dlqm5)**; and

- the relation

$$\{\langle v, v'\rangle \mid vR_r v' \text{ for some } \boldsymbol{r} \in \mathfrak{R},\ R \in rol\,\varphi \text{ is not } \boldsymbol{r}\text{-universal}\}$$

is an intransitive tree order on Δ with root w.

A *kernel block* over $ob\,\varphi \neq \emptyset$ is a structure of the form $\mathfrak{B}_o = \langle ob\,\varphi, \boldsymbol{q}_o, \mathfrak{R}_o, \triangleleft_o\rangle$ in which

- $\langle ob\,\varphi, \boldsymbol{q}_o\rangle$ is a basic structure for φ of depth $m \leq md(\varphi)$ (recall that for every $a \in ob\,\varphi$, the set $\{\boldsymbol{t}_a(x) \mid x \in T_a\}$ consists of only types named by a, whenever $\boldsymbol{q}_o(a) = \langle\langle T_a, <_a\rangle, \boldsymbol{t}_a\rangle$);
- $\mathfrak{R}_o$ is a set of coherent runs through $\langle ob\,\varphi, \boldsymbol{q}_o\rangle$ satisfying **(dlqm6)**; and
- $\triangleleft_o$ is a binary relation on $\mathfrak{R}_o$ satisfying **(dlqm3)**–**(dlqm5)**.

Kernel and w-blocks defined above are also called *blocks for* φ. A nonempty set $\mathcal{S}$ of blocks for φ is called *satisfying* if

- all blocks in $\mathcal{S}$ are of the same depth m, for some $m \leq md(\varphi)$;
- $\mathcal{S}$ contains a single kernel block for φ whenever $ob\,\varphi \neq \emptyset$;
- every block in $\mathcal{S}$ satisfies **(dlqm2)**;
- for every block $\mathfrak{B} = \langle\Delta, \boldsymbol{q}, \mathfrak{R}, \triangleleft\rangle$ in $\mathcal{S}$ and every $w \in \Delta$, there is precisely one w-block $\mathfrak{B}' = \langle\Delta', \boldsymbol{q}', \mathfrak{R}', \triangleleft'\rangle$ in $\mathcal{S}$ such that $\boldsymbol{q}(w) = \boldsymbol{q}'(w)$;
- for every $a \in ob\,\varphi$, there is precisely one w-block $\mathfrak{B} = \langle\Delta, \boldsymbol{q}, \mathfrak{R}, \triangleleft\rangle$ in $\mathcal{S}$ such that $\boldsymbol{q}_o(a) = \langle\langle T_a, <_a\rangle, \boldsymbol{t}_a\rangle$ is 'isomorphic' to $\boldsymbol{q}(w) = \langle\langle T_w, <_w\rangle, \boldsymbol{t}_w\rangle$ in the following sense: There is an isomorphism f between the trees $\langle T_a, <_a\rangle$ and $\langle T_w, <_w\rangle$ such that, for all $x \in T_a$, if $\boldsymbol{t}_a(x) = \langle \boldsymbol{c}_a, \boldsymbol{f}_a\rangle$ then $\boldsymbol{t}_w(f(x)) = \langle \boldsymbol{c}, \boldsymbol{f}\rangle$.

Lemma 14.11. *There is a* $\mathbf{K}_{\mathcal{ALC}}$*-quasimodel for* φ *iff there is a satisfying set of blocks for* φ *such that the domain of each nonkernel block contains at most*

$$1 + (md(\varphi) + 1) \cdot p(\varphi) \cdot |con\,\varphi|$$

objects.

Proof. ($\Leftarrow$) First we show how a quasimodel for φ can be constructed from a satisfying set $\mathcal{S}$ of blocks for φ. To begin with, we call a quadruple $\langle\Delta, \boldsymbol{q}, \mathfrak{R}, \triangleleft\rangle$ a *weak quasimodel for* φ if the following conditions hold:

(wdlq1) Δ is a finite set containing $ob\,\varphi$ and $\langle\Delta, \boldsymbol{q}\rangle$ is a basic structure for φ;

(wdlq2) $\mathfrak{R}$ is a set of runs through $\langle \Delta, \boldsymbol{q} \rangle$ and $\lhd$ is a binary relation on $\mathfrak{R}$ satisfying **(dlqm2)**–**(dlqm6)**;

(wdlq3) for all $w, v \in \Delta$ such that $wR_r v$ for some $\boldsymbol{r} \in \mathfrak{R}$, there exists a block $\mathfrak{B}^{wv} = \langle \Delta^{wv}, \boldsymbol{q}^{wv}, \mathfrak{R}^{wv}, \lhd^{wv} \rangle$ in $\mathcal{S}$ such that

- $\Delta^{wv} \subseteq \Delta$ and $w, v \in \Delta^{wv}$,
- for all $u \in \Delta^{wv}$, $\boldsymbol{q}(u) = \boldsymbol{q}^{wv}(u)$,
- for all $\boldsymbol{r} \in \mathfrak{R}$, the restriction $\boldsymbol{r}^{wv}$ of $\boldsymbol{r}$ to Δ^{wv} is a run in $\mathfrak{R}^{wv}$.

We construct by induction a sequence $\mathfrak{Q}_n = \langle \Delta_n, \boldsymbol{q}_n, \mathfrak{R}_n, \lhd_n \rangle$, $n < \omega$, $n > 0$, of weak quasimodels that 'converges' to a quasimodel for φ. If $ob\,\varphi \neq \emptyset$ then let $\mathfrak{Q}_1$ be the kernel block $\mathfrak{B}_o$ in $\mathcal{S}$. Otherwise, let $\mathfrak{Q}_1$ be any nonkernel block from $\mathcal{S}$. Clearly, in both cases $\mathfrak{Q}_1$ is a weak quasimodel for φ.

Now let $n \geq 1$, and suppose that we have already constructed $\mathfrak{Q}_k$, for $0 < k \leq n$. Let $\Delta_0 = \emptyset$. For each $w \in \Delta_n - \Delta_{n-1}$, select a w-block $\mathfrak{B}^w = \langle \Delta^w, \boldsymbol{q}^w, \mathfrak{R}^w, \lhd^w \rangle$ from $\mathcal{S}$ such that $\boldsymbol{q}_n(w) = \boldsymbol{q}^w(w)$. (The existence of such a block follows from **(wdlq3)**.) We may assume that all of the selected blocks are pairwise disjoint and $\Delta^w \cap \Delta_n = \{w\}$. Define $\langle \Delta_{n+1}, \boldsymbol{q}_{n+1} \rangle$ by taking

$$\Delta_{n+1} = \Delta_n \cup \bigcup \{\Delta^w \mid w \in \Delta_n - \Delta_{n-1}\},$$
$$\boldsymbol{q}_{n+1}(v) = \begin{cases} \boldsymbol{q}^w(v), & \text{if } v \in \Delta^w,\ w \in \Delta_n - \Delta_{n-1}, \\ \boldsymbol{q}_n(v), & \text{if } v \in \Delta_n. \end{cases}$$

In other words, we 'glue together' the basic structures $\langle \Delta_n, \boldsymbol{q}_n \rangle$ and $\langle \Delta^w, \boldsymbol{q}^w \rangle$ at object w.

Next we define $\mathfrak{R}_{n+1}$ and $\lhd_{n+1}$. Suppose that we have $\boldsymbol{r} \in \mathfrak{R}_n$ and a sequence $\bar{\boldsymbol{s}} = \langle \boldsymbol{s}^w \in \mathfrak{R}^w \mid w \in \Delta_n - \Delta_{n-1} \rangle$ such that $r(w) = s^w(w)$, for all $w \in \Delta_n - \Delta_{n-1}$. Define the *extension* $\boldsymbol{r} \cup \bar{\boldsymbol{s}}$ of $\boldsymbol{r}$ by taking $\boldsymbol{r} \cup \bar{\boldsymbol{s}} = \langle r \cup \bar{s}, R_{r \cup \bar{s}} \rangle$ where, for all $u, v \in \Delta_{n+1}$,

$$r \cup \bar{s}(v) = \begin{cases} s^w(v), & \text{if } v \in \Delta^w,\ w \in \Delta_n - \Delta_{n-1}, \\ r(v), & \text{if } v \in \Delta_n, \end{cases}$$

$R_{r \cup \bar{s}} = \Delta_{n+1} \times \Delta_{n+1}$ if R is $\boldsymbol{r}$- or $\boldsymbol{s}^w$-universal for some $w \in \Delta_n - \Delta_{n-1}$, and

$$uR_{r \cup \bar{s}} v \quad \text{iff} \quad \begin{cases} uR_r v, & \text{if } u, v \in \Delta_n, \\ uR_{s^w} v, & \text{if } u, v \in \Delta^w,\ w \in \Delta_n - \Delta_{n-1}. \end{cases}$$

Let $\mathfrak{R}_{n+1}$ be the set of all such extensions and let

$$(\boldsymbol{r}_1 \cup \bar{\boldsymbol{s}}_1) \lhd_{n+1} (\boldsymbol{r}_2 \cup \bar{\boldsymbol{s}}_2) \quad \text{iff} \quad \boldsymbol{r}_1 \lhd_n \boldsymbol{r}_2 \text{ and } \boldsymbol{s}_1^w \lhd^w \boldsymbol{s}_2^w, \text{ for all } w \in \Delta_n - \Delta_{n-1}.$$

It can be readily checked that $\mathfrak{R}_{n+1}$ and $\lhd_{n+1}$ satisfy **(dlqm3)**–**(dlqm6)**, and so $\mathfrak{Q}_{n+1} = \langle \Delta_{n+1}, \boldsymbol{q}_{n+1}, \mathfrak{R}_{n+1}, \lhd_{n+1} \rangle$ is a weak quasimodel.

The 'limit quasimodel' is defined as follows. First, let

$$\Delta = \bigcup_{n<\omega} \Delta_n, \qquad \boldsymbol{q} = \bigcup_{0<n<\omega} \boldsymbol{q}_n.$$

Next, for each sequence of runs $\langle \boldsymbol{r}_n \in \mathfrak{R}_n \mid 0 < n < \omega \rangle$ such that $\boldsymbol{r}_{n+1}$ is an extension of $\boldsymbol{r}_n$ take $\boldsymbol{r} = \bigcup_{0<n<\omega} \boldsymbol{r}_n$. Let $\mathfrak{R}$ be the set of all such runs. For $\boldsymbol{r}, \boldsymbol{r}' \in \mathfrak{R}$, define

$$\boldsymbol{r} \lhd \boldsymbol{r}' \quad \text{iff} \quad \boldsymbol{r}_n \lhd_n \boldsymbol{r}'_n \text{ for all } 0 < n < \omega,$$

where $\boldsymbol{r}' = \bigcup_{0<n<\omega} \boldsymbol{r}'_n$.

It is not hard to see, using **(wdlq1)**–**(wdlq3)**, that all the runs in $\mathfrak{R}$ are coherent and saturated, and $\langle \Delta, \boldsymbol{q}, \mathfrak{R}, \lhd \rangle$ is a quasimodel for φ (see the proof of Lemma 6.3 for some details).

($\Rightarrow$) Now we have to show how to extract a set of 'small' blocks from a given quasimodel $\mathfrak{Q} = \langle \Delta, \boldsymbol{q}, \mathfrak{R}, \lhd \rangle$ for φ of depth $m \leq md(\varphi)$. Note that we may assume that our quasimodel $\mathfrak{Q}$ meets conditions (i) and (ii) of Lemma 14.10.

To begin with, it is readily seen that if $ob\,\varphi \neq \emptyset$ then $\mathfrak{B}_o = \langle ob\,\varphi, \boldsymbol{q}_o, \mathfrak{R}_o, \lhd_o \rangle$ is a kernel block for φ, where $\boldsymbol{q}_o$ is the restriction of $\boldsymbol{q}$ to $ob\,\varphi$, each run $\boldsymbol{r}_o$ in $\mathfrak{R}_o$ is the restriction of some run $\boldsymbol{r} \in \mathfrak{R}$ to $ob\,\varphi$, and for all $\boldsymbol{r}_o, \boldsymbol{r}'_o \in \mathfrak{R}_o$, $\boldsymbol{r}_o \lhd_o \boldsymbol{r}'_o$ iff $\boldsymbol{r} \lhd \boldsymbol{r}'$.

To construct a satisfying set $\mathcal{S}$ of blocks, we will associate with each w in $\Delta - ob\,\varphi$ a w-block $\mathfrak{B}^w = \langle \Delta^w, \boldsymbol{q}^w, \mathfrak{R}^w, \lhd^w \rangle$ such that $\boldsymbol{q}^w(w) = \boldsymbol{q}(w)$, and put

$$\mathcal{S} = \{\mathfrak{B}^w \mid w \in \Delta - ob\,\varphi\} \cup \{\mathfrak{B}_o\}.$$

The resulting $\mathcal{S}$ will clearly be a satisfying set of blocks for φ.

So, fix a $w \in \Delta - ob\,\varphi$. First we define inductively sets of k-runs $\mathfrak{S}_k \subseteq \mathfrak{R}_k$, for $k \leq m$:

- Let $\mathfrak{S}_0 = \{\boldsymbol{r}_0\}$ for the unique run $\boldsymbol{r}_0$ in $\mathfrak{R}_0$.
- Given $\mathfrak{S}_k$, we construct $\mathfrak{S}_{k+1}$ as follows. For every run $\boldsymbol{r} \in \mathfrak{S}_k$ and every $x \in T_w$ with $r(w) <_w x$, select an $\boldsymbol{r}' \in \mathfrak{R}_{k+1}$ such that $\boldsymbol{r} \lhd \boldsymbol{r}'$ and $r'(w) = x$, and put it into $\mathfrak{S}_{k+1}$. (Such a run $\boldsymbol{r}'$ exists by **(dlqm4)**.)

Finally, let $\mathfrak{S} = \bigcup_{k \leq m} \mathfrak{S}_k$. Clearly, $|\mathfrak{S}| \leq p(\varphi)$.

For every $\boldsymbol{r} \in \mathfrak{S}$ and every $\exists R.C \in \boldsymbol{t}_w(r(w))$, we then let

$$Sat(\boldsymbol{r}, \exists R.C) = \{v \in \Delta \mid wR_r v,\ C \in \boldsymbol{t}_v(r(v))\}.$$

As $\boldsymbol{r}$ is saturated, $Sat(\boldsymbol{r}, \exists R.C) \neq \emptyset$. We select an $m+1$-element subset

$$\Delta^w(\boldsymbol{r}, \exists R.C) = \{v_1, \dots, v_{m+1}\}$$

of $Sat(\boldsymbol{r}, \exists R.C)$ such that $v_1, \dots, v_{m+1}$ are twins relative to w. We may assume that the obtained sets $\Delta^w(\boldsymbol{r}, \exists R.C)$ are pairwise disjoint.

Now we define

- $\Delta^w = \{w\} \cup \bigcup\{\Delta^w(\boldsymbol{r}, \exists R.C) \mid \boldsymbol{r} \in \mathfrak{S},\ \exists R.C \in \boldsymbol{t}_w(r(w))\}$, and
- for all $v \in \Delta^w$, $\boldsymbol{q}^w(v) = \boldsymbol{q}(v)$.

Then $\langle \Delta^w, \boldsymbol{q}^w \rangle$ is a basic structure for φ, and the cardinality of Δ^w is clearly bounded by $1 + (md(\varphi) + 1) \cdot p(\varphi) \cdot |con\,\varphi|$.

According to Lemma 14.10, we may assume that for every run $\boldsymbol{r} \in \mathfrak{R}$ and every $R \in rol\,\varphi$, we have:

- if $(\langle T_w, <_w \rangle, r(w)) \not\models \mathsf{M}\bot$ then, for all $u, v \in \Delta^w$, $uR_r v$ implies $u = w$ and $v \neq w$;
- if $(\langle T_w, <_w \rangle, r(w)) \models \mathsf{M}\bot$ then $R_r = \Delta \times \Delta$.

Let $v \in \Delta^w$, $v \neq w$, and suppose that the pairs $\boldsymbol{r} = \langle r, \{R_r \mid R \in rol\,\varphi\} \rangle$ and $\boldsymbol{r}' = \langle r', \{R_{r'} \mid R \in rol\,\varphi\} \rangle$ are such that the domains of the functions r and r' contain Δ^w, $r(w) = r'(w)$, and R_r, $R_{r'}$ are binary relations on Δ, for all $R \in rol\,\varphi$. We define the pair $\boldsymbol{r} +_v \boldsymbol{r}' = \langle r +_v r', \{R_{r+_v r'} \mid R \in rol\,\varphi\} \rangle$ as follows. For all $z \in \Delta^w$,

$$(r +_v r')(z) = \begin{cases} r(z), & \text{if } z = v, \\ r'(z), & \text{if } z \neq v, \end{cases}$$

and, for each $R \in rol\,\varphi$,

- $R_{r+_v r'} = \Delta^w \times \Delta^w$, whenever R is $\boldsymbol{r}$-universal (and so $\boldsymbol{r}'$-universal as well),
- $wR_{r+_v r'}u$ iff $u = v$ and $wR_r v$, or $u \neq v$ and $wR_{r'}u$, otherwise.

Using this 'addition' function, we now define sets $\mathfrak{R}_k^w$ of k-runs through $\langle \Delta^w, \boldsymbol{q}^w \rangle$, for every $k \leq m$. Let $\mathfrak{R}_0^w$ consist of the restriction of $\boldsymbol{r}_0$ to Δ^w. For $k > 0$, we put all the restrictions of runs from $\mathfrak{S}_k$ into $\mathfrak{R}_k^w$ and also add there

$$\boldsymbol{r}_1 +_{v_1} (\boldsymbol{r}_2 +_{v_2} (\dots (\boldsymbol{r}_l +_{v_l} \boldsymbol{r}) \dots)),$$

where $1 \leq l \leq k$, $\boldsymbol{r} \in \mathfrak{S}_k$, $\boldsymbol{r}_1, \dots, \boldsymbol{r}_l \in \mathfrak{R}_k$ are such that $r(w) = r_i(w)$, for $1 \leq i \leq l$, and $v_1, \dots, v_l$ are pairwise distinct points in Δ^w different from w.

Obviously, every run $\boldsymbol{s} \in \mathfrak{R}^w$ is coherent. We show that it is w-saturated. This is clear if $\boldsymbol{s}$ is the restriction of some run in $\mathfrak{S}$. Otherwise, $\boldsymbol{s}$ is of the form

$$\boldsymbol{r}_1 +_{v_1} (\boldsymbol{r}_2 +_{v_2} (\dots (\boldsymbol{r}_k +_{v_k} \boldsymbol{r}) \dots))$$

for some $k \leq m$. So, we modified the w-saturated run $\boldsymbol{r}$ at $\leq m$ places. Take some concept $\exists R.C \in \boldsymbol{t}_w(s(w))$. Since we selected for Δ^w $m+1$ twins for each point in $Sat(\boldsymbol{r}, \exists R.C)$, there is still at least one v left to 'saturate s with respect to $\exists R.C$,' that is, such that $\exists R.C \in \boldsymbol{t}_v(s(v))$.

Finally, let

$$s = r_1 +_{v_1} (r_2 +_{v_2} (\dots (r_l +_{v_l} r) \dots)), \tag{14.3}$$

$$s' = r'_1 +_{v'_1} (r'_2 +_{v'_2} (\dots (r'_n +_{v'_n} r')) \dots) \tag{14.4}$$

be two runs in $\mathfrak{R}^w$. (If either s or s' is the restriction of a run in $\mathfrak{S}$, then we consider l or n as 0, respectively.) We let $s \lhd s'$ iff the following hold:

- $s \in \mathfrak{R}^w_k$ and $s' \in \mathfrak{R}^w_{k+1}$, for some $k < m$,
- $r \lhd r'$,
- $l \leq n$ and $v_i = v'_i$, for all $1 \leq i \leq l$,
- for all $z \in \Delta^w$, $r_i(z) <_z r'_i(z)$ whenever $1 \leq i \leq l$, and $r(z) <_z r'_i(z)$ whenever $l + 1 \leq i \leq n$.

Then **(dlqm3)** holds by definition. We show that **(dlqm4)** also holds. Suppose that s is of the form (6.2), $z \in \Delta^w$, $x \in T_z$ and $s(z) <_z x$. We need a run s' in $\mathfrak{R}^w$ such that $s \lhd^w s'$ and $s'(z) = x$.

Case 1: $z = v_j$ for some $1 \leq j \leq l$. Then $s(z) = r_j(z) = v_j$ for some $r_i \in \mathfrak{R}$. As the original quasimodel $\mathfrak{Q}$ satisfies **(dlqm4)**, we have a run $r'_j \in \mathfrak{R}$ such that $r_j \lhd r'_j$ and $r'_j(z) = x$. Similarly, for all $i \neq j$, $1 \leq i \leq l$, take a run r'_i from $\mathfrak{R}$ such that $r_i \lhd r'_i$ and $r'_i(w) = r'_j(w)$. Finally, take a run r' from $\mathfrak{S}$ such that $r \lhd r'$ and $r'(w) = r'_j(w)$. (Such a run exists by the definition of $\mathfrak{S}$.) Then

$$s' = r'_1 +_{v_1} (r'_2 +_{v_2} (\dots (r'_l +_{v_l} r') \dots))$$

is a run in $\mathfrak{R}^w$, as required.

Case 2: $z \neq v_j$ for any $1 \leq j \leq l$. Then $s(z) = r(z)$. Select a run r'_{l+1} from $\mathfrak{R}$ such that $r \lhd r'_{l+1}$ and $r'_{l+1}(z) = x$. For each i, $1 \leq i \leq l$, take a run r'_i from $\mathfrak{R}$ such that $r_i \lhd r'_i$ and $r'_i(w) = r'_{l+1}(w)$. Finally, take a run r' from $\mathfrak{S}$ such that $r \lhd r'$ and $r'(w) = r'_{l+1}(w)$. Then

$$s' = r'_1 +_{v_1} (r'_2 +_{v_2} (\dots (r'_{l+1} +_z r') \dots))$$

is a run in $\mathfrak{R}^w$, as required.

We check only the first condition of **(dlqm5)**; the second is similar and left to the reader. First, suppose that $s \in \mathfrak{R}^w_k$ is of the form (14.3) and $w(\Diamond R)_s z$, for some $z \in \Delta^w$. Let us assume first that $\Diamond R$ is not s-universal.

Case 1: $z = v_j$ for some $1 \leq j \leq l$. By **(dlqm5)** in the quasimodel $\mathfrak{Q}$, we have a run $r'_j \in \mathfrak{R}_{k+1}$ such that $r_j \lhd r'_j$ and $w R_{r'_j} v_j$. For all $i \neq j$, $1 \leq i \leq l$, we select (by **(dlqm4)** in $\mathfrak{Q}$) a run $\boldsymbol{r}'_i \in \mathfrak{R}_{k+1}$ such that $\boldsymbol{r}_i \lhd \boldsymbol{r}'_i$ and $r'_i(w) = r'_j(w)$. Finally, take a run $\boldsymbol{r}'$ from $\mathfrak{S}$ such that $\boldsymbol{r} \lhd \boldsymbol{r}'$ and $r'(w) = r'_j(w)$. Then

$$\boldsymbol{s}' = \boldsymbol{r}'_1 +_{v_1} (\boldsymbol{r}'_2 +_{v_2} (\dots (\boldsymbol{r}'_l +_{v_l} \boldsymbol{r}') \dots))$$

is in $\mathfrak{R}^w_{k+1}$, $\boldsymbol{s} \lhd^w \boldsymbol{s}'$ and $wR_{\boldsymbol{s}'}z$.

Case 2: $z \neq v_j$ for any $1 \leq j \leq l$. By **(dlqm5)** in $\mathfrak{Q}$, we have a run $\boldsymbol{r}'_{l+1}$ in $\mathfrak{R}_{k+1}$ such that $\boldsymbol{r} \lhd \boldsymbol{r}'_{l+1}$ and $wR_{\boldsymbol{r}'_{l+1}}z$. For each i, $1 \leq i \leq l$, take a run $\boldsymbol{r}'_i$ from $\mathfrak{R}_{k+1}$ such that $\boldsymbol{r}_i \lhd \boldsymbol{r}'_i$ and $r'_i(w) = r'_{l+1}(w)$. Finally, take a run $\boldsymbol{r}'$ from $\mathfrak{S}$ such that $\boldsymbol{r} \lhd \boldsymbol{r}'$ and $r'(w) = r'_{l+1}(w)$. Then

$$\boldsymbol{s}' = \boldsymbol{r}'_1 +_{v_1} (\boldsymbol{r}'_2 +_{v_2} (\dots (\boldsymbol{r}'_{l+1} +_{z} \boldsymbol{r}') \dots))$$

is in $\mathfrak{R}^w_{k+1}$, $\boldsymbol{s} \lhd^w \boldsymbol{s}'$, and $wR_{\boldsymbol{s}'}z$.

Assume now that $\Diamond R$ is $\boldsymbol{s}$-universal and $u(\Diamond R)_{\boldsymbol{s}}v$, for $u, v \in \Delta^w$, with $\boldsymbol{s}$ being of the form (14.3). Then $\Diamond R$ is $\boldsymbol{r}_i$- and $\boldsymbol{r}$-universal too. Suppose $R = \mathsf{M}R_j$, R_j a role name. Then there exists $x \in T_w$ such that $r(w) <_w x$ and $(\langle \mathfrak{T}_w, <_w \rangle, x) \models \mathsf{M}\bot$. Take an $\boldsymbol{r}' \in \mathfrak{R}_{k+1}$ such that $\boldsymbol{r} \lhd \boldsymbol{r}'$, $r'(w) = r(w)$, and also take $\boldsymbol{r}'_i \in \mathfrak{R}_{k+1}$ such that $\boldsymbol{r}_i \lhd \boldsymbol{r}'_i$ and $r'_i(w) = r_i(w)$, for all $1 \leq i \leq l$. Then

$$\boldsymbol{s}' = \boldsymbol{r}'_1 +_{v_1} (\boldsymbol{r}'_2 +_{v_2} (\dots (\boldsymbol{r}'_l +_{v_l} \boldsymbol{r}') \dots))$$

is in $\mathfrak{R}^w_{k+1}$, $\boldsymbol{s} \lhd^w \boldsymbol{s}'$, and R is $\boldsymbol{s}'$-universal. Thus we have proved the ($\Rightarrow$)-part of **(dlqm5)**.

To show the converse, suppose that $\boldsymbol{s}$ and $\boldsymbol{s}'$ are of the form (14.3) and (14.4), respectively, $\boldsymbol{s} \lhd^w \boldsymbol{s}'$ and $wR_{\boldsymbol{s}'}z$, for some $\Diamond R \in rol\,\varphi$ that is not $\boldsymbol{s}$-universal. If $z = v_i$, for some $1 \leq i \leq l$, then we have $\boldsymbol{r}_i \lhd \boldsymbol{r}'_i$, and so, in view of **(dlqm 5)** in $\mathfrak{Q}$, $w(\Diamond R)_{r_i}z$, from which $w(\Diamond R)_{\boldsymbol{s}}z$ follows. Let $z = v_j$ for some $l + 1 \leq j \leq k$. Then $\boldsymbol{r} \lhd \boldsymbol{r}_j$, $w(\Diamond R)_{\boldsymbol{r}}z$, and so $w(\Diamond R)_{\boldsymbol{s}}z$. Finally, if $z \neq v_i$ for any i with $1 \leq i \leq k$, then $\boldsymbol{r} \lhd \boldsymbol{r}'$, and we again have $w(\Diamond R)_{\boldsymbol{r}}z$, from which $w(\Diamond R)_{\boldsymbol{s}}z$ follows. The case of an $\boldsymbol{s}$-universal $\Diamond R$ is trivial.

Finally, it is straightforward to see that **(dlqm6)** holds. Thus, the structure $\langle \Delta^w, \boldsymbol{q}^w, \mathfrak{R}^w, \lhd^w \rangle$ is indeed a w-block. ❑

Since one can effectively check, given a $\mathcal{ML}_{\mathcal{ALC}}$-formula φ, whether there exists a satisfying set for φ, as an immediate consequence of Lemmas 14.9 and 14.11 we obtain a proof of Theorem 14.8 for $\mathbf{K}_{\mathcal{ALC}}$. It should be noted that the obvious 'brute-force' algorithm is nonelementary. The complexity of the satisfiability problem for $\mathcal{ML}_{\mathcal{ALC}}$-formulas remains open. Recall, however, that even for $\mathbf{K} \times \mathbf{K}$—which is embedded in our logic—no elementary algorithm is known either (see Question 6.5).

It is not hard to adopt the developed technique in order to prove Theorem 14.8 for the other modal description logics listed in its formulation. In particular, for **S5** one can modify the proof above in the following way (cf. also the proofs of Theorems 5.22 and 6.44).

First of all, *quasistates* for a given formula φ are now simply sets $\boldsymbol{T}$ of types for φ such that, for all $\boldsymbol{t} \in \boldsymbol{T}$, $\Diamond C \in con\,\varphi$, and $\Diamond\psi \in sub\,\varphi$,

$$\Diamond C \in \boldsymbol{t} \quad \text{iff} \quad \exists \boldsymbol{t}' \in \boldsymbol{T}\ C \in \boldsymbol{t}',$$

$$\Diamond \psi \in \boldsymbol{t} \quad \text{iff} \quad \exists \boldsymbol{t}' \in \boldsymbol{T}\ \psi \in \boldsymbol{t}'.$$

Besides, no ordering of the runs is needed. Thus, an $\mathbf{S5}_{\mathcal{ALC}}$-*quasimodel* for φ is a triple $\langle \Delta, \boldsymbol{q}, \mathfrak{R} \rangle$, where $\Delta \supseteq ob\,\varphi$, $\boldsymbol{q}$ is a function associating with each $w \in \Delta$ a quasistate $\boldsymbol{q}(w) = \boldsymbol{T}_w$ and $\mathfrak{R}$ is a set of runs through $\langle \Delta, \boldsymbol{q} \rangle$ such that

- for all $w \in \Delta$, $\boldsymbol{t} \in \boldsymbol{T}_w$ there is a run $\boldsymbol{r} \in \mathfrak{R}$ such that $r(w) = \boldsymbol{t}$;
- for all $w, v \in \Delta$, $\Diamond R \in rol\,\varphi$, and $\boldsymbol{r} \in \mathfrak{R}$, we have $w(\Diamond R)_r v$ iff there is $\boldsymbol{r}' \in \mathfrak{R}$ such that $wR_{r'}v$;
- for all $w, v \in \Delta$, $\Box R \in rol\,\varphi$, and $\boldsymbol{r} \in \mathfrak{R}$, we have $w(\Box R)_r v$ iff $wR_{r'}v$ hold for all $\boldsymbol{r}' \in \mathfrak{R}$.

The remaining part of the proof is similar to the proof given above. It may be worth noting that now in the construction of the block Δ^w, it is enough to take only two twins v_1, v_2 relative to w. ❑

14.3 Restricted formula satisfiability

In this section we consider the satisfiability problem for formulas without modalized roles and global role names. It turns out that in this case we can prove decidability for description logics which are considerably more expressive than $\mathcal{ALC}$, say, $\mathcal{CQO}$ defined in Section 2.5.

The modal description languages $\mathcal{ML}_{\mathcal{CQO}}$, $(\mathcal{ML}_{\mathcal{SU}})_{\mathcal{CQO}}$ and $\mathcal{CPDL}_{\mathcal{CQO}}$ are obtained from $\mathcal{ML}_{\mathcal{ALC}}$, $(\mathcal{ML}_{\mathcal{SU}})_{\mathcal{ALC}}$ and $\mathcal{CPDL}_{\mathcal{ALC}}$, respectively, by allowing the use of arbitrary $\mathcal{CQO}$-concepts; the modal operators are applicable only to concepts and formulas. Nominals in $\mathcal{CQO}$ are interpreted as rigid designators, since in every world w, we have $\{a\}^{I(w)} = \{a^{I(w)}\}$ for all object names a.

Theorem 14.12. *The satisfiability problem for formulas without modalized roles and global role names is decidable for the logics $L_{\mathcal{CQO}}$, where L is one of the following dynamic, epistemic and temporal logics:*

(1) **CPDL**,

(2) $\mathbf{K}_n^C$, $\mathbf{T}_n^C$, $\mathbf{K4}_n^C$, $\mathbf{S4}_n^C$, $\mathbf{KD45}_n^C$, $\mathbf{S5}_n^C$,

(3) $\mathsf{Log}_{\mathcal{SU}}(\mathcal{C})$, *where* $\mathcal{C}$ *is one of the following classes:* $\{\langle\mathbb{N},<\rangle\}$, $\{\langle\mathbb{Z},<\rangle\}$, $\{\langle\mathbb{Q},<\rangle\}$, *the class of finite strict linear orders, any first-order definable class of strict linear orders.*

Proof. The result is proved using Theorem 11.83 (for temporal logics) and Theorem 12.4 (for **CPDL** and epistemic logics). We will confine ourselves to considering the temporal case and leave the remaining dynamic and epistemic logics to the reader as an exercise. So let $L = \mathsf{Log}_{\mathcal{SU}}(\mathcal{C})$ for any of the listed classes $\mathcal{C}$.

Recall that in Section 3.8 we introduced a translation $\cdot^T$ from $\mathcal{ML}_{\mathcal{ALC}}$ into first-order modal logic. As modalized roles are available in $\mathcal{ML}_{\mathcal{ALC}}$, the map $\cdot^T$ is not an embedding into the monodic fragment of $\mathcal{QTL}$. However, since we consider here the language without global role names and modalized roles, $\cdot^T$ can be extended to a translation from $(\mathcal{ML}_{\mathcal{SU}})_{\mathcal{CQO}}$ without such roles into the monodic fragment $\mathcal{QTL}^{=}_{\square}$ of first-order temporal logic with equality. First, define inductively for all roles R and S (we consider the new clauses only)

$$\begin{aligned}(R \sqcup S)^T &= R^T(x,y) \vee S^T(x,y),\\ (R \circ S)^T &= \exists z\ (R^T(x,z) \wedge S^T(z,y)),\\ (R^*)^T &= \overline{R}(x,y),\end{aligned}$$

where $\overline{R}$ is a fresh binary predicate symbol. Now, for every basic role B, any concepts C and D, and every object name a, we let:

$$(\exists_{\geq n} B.C)^T = \exists x_1 \dots \exists x_n (\bigwedge_{1\leq i\leq n} B^T(x,x_i) \wedge \bigwedge_{1\leq i\leq n} C^T\{x_i/x\} \wedge \bigwedge_{1\leq i<j\leq n} x_i \neq x_j),$$

$$\begin{aligned}(C\mathcal{U}D)^T &= C^T\mathcal{U}D^T,\\ (C\mathcal{S}D)^T &= C^T\mathcal{S}D^T,\\ (\{a\})^T &= (x = a).\end{aligned}$$

Finally, for all formulas φ and ψ we set:

$$\begin{aligned}(\varphi\mathcal{U}\psi)^T &= \varphi^T\mathcal{U}\psi^T,\\ (\varphi\mathcal{S}\psi)^T &= \varphi^T\mathcal{S}\psi^T.\end{aligned}$$

Denote by $\mathcal{K}$ the class of all first-order structures

$$I = \left\langle D^I, R_0^I, \dots, C_0^I, \dots, \overline{R}^I, \dots, a_0^I, \dots \right\rangle,$$

where $\overline{R}^I$ is the transitive and reflexive closure of the relation

$$\{\langle a,b\rangle \in D^I \times D^I \mid I \models R^T[a,b]\},$$

for each $\mathcal{CQO}$-role R. Let

$$\mathcal{QTL}' = \{\varphi^T \mid \varphi \text{ is an } (\mathcal{ML}_{\mathcal{SU}})_{\mathcal{CQO}}\text{-formula}\}.$$

Clearly, $\mathcal{QTL}'$ is a set of sentences from $\mathcal{QTL}^{=}_{\boxed{1}}$. We claim that conditions (a) and (b) of Theorem 11.83 hold for $\mathcal{QTL}'$ and $\mathcal{K}$.

To show (a), recall first that, having concepts of the form $\{a\}$, there is no need to define $a : C$ and aRb as atomic formulas: they are equivalent to $\{a\} \to C = \top$ and $\{a\} \to \exists R.\{b\} = \top$, respectively. Therefore, we may assume that all atomic $(\mathcal{ML}_{\mathcal{SU}})_{\mathcal{CQO}}$-formulas are of the form $C = \top$, and so their translations are of the form $\forall x C^T(x)$.

Further, we remind the reader that, for every $\mathcal{QTL}^{=}_{\boxed{1}}$-formula ψ, we denote by $\overline{\psi}$ the $\overline{\mathcal{QL}}$-formula that results from ψ by replacing all of its subformulas of the form $\chi_1 \mathcal{U} \chi_2$ and $\chi_1 \mathcal{S} \chi_2$, which are not within the scope of another occurrence of $\mathcal{U}$ or $\mathcal{S}$, with their surrogates. Now, observe that, given an $(\mathcal{ML}_{\mathcal{SU}})_{\mathcal{CQO}}$-concept C (or $(\mathcal{ML}_{\mathcal{SU}})_{\mathcal{CQO}}$-formula φ) without modalized roles, we can obtain the $\overline{\mathcal{QL}}$-formula $\overline{C^T}$ (or $\overline{\varphi^T}$) in a different way. First, we turn C into a $\mathcal{CQO}$-concept $\tilde{C}$ (or φ into a $\mathcal{CQO}$-formula $\tilde{\varphi}$) by replacing each of its subconcepts of the form $D_1 \mathcal{U} D_2$ and $D_1 \mathcal{S} D_2$ (or subformulas of the form $\chi_1 \mathcal{U} \chi_2$ and $\chi_1 \mathcal{S} \chi_2$) that is not within the scope of another occurrence of $\mathcal{U}$ or $\mathcal{S}$, by a fresh concept name (or, in case of formulas, by an atom $C = \top$ with a fresh concept name C). Then, by applying the translation $\cdot^T$, we turn $\tilde{C}$ into a $\overline{\mathcal{QL}}$-formula $\tilde{C}^T$ (or $\tilde{\varphi}$ into a $\overline{\mathcal{QL}}$-sentence $\tilde{\varphi}^T$). It should be clear that we have $\overline{C^T} = \tilde{C}^T$ (and $\overline{\varphi^T} = \tilde{\varphi}^T$).

Now fix an $\mathcal{QTL}'$-sentence φ^T. Recall that a *type* for φ^T is any Boolean-saturated subset $\boldsymbol{t}$ of $\{\overline{\psi} \mid \psi \in sub_x\, \varphi^T \cup \{x = a, x \neq a \mid a \in ob\, \varphi\}\}$. For every such type $\boldsymbol{t}$, define a set $\widehat{\boldsymbol{t}}$ consisting of $\mathcal{CQO}$-concepts and $\mathcal{CQO}$-formulas by taking

$$\begin{aligned}\widehat{\boldsymbol{t}} = \; & \{\tilde{C} \mid C \text{ is an } (\mathcal{ML}_{\mathcal{SU}})_{\mathcal{CQO}}\text{-concept and } \overline{C^T} \in \boldsymbol{t}\} \cup \\ & \{\tilde{\psi} \mid \psi \text{ is an } (\mathcal{ML}_{\mathcal{SU}})_{\mathcal{CQO}}\text{-formula and } \overline{\psi^T} \in \boldsymbol{t}\}.\end{aligned}$$

It is not hard to see (since for every $\psi \in sub_x\, \varphi^T \cup \{x = a, x \neq a \mid a \in ob\, \varphi\}$, either there is an $(\mathcal{ML}_{\mathcal{SU}})_{\mathcal{CQO}}$-concept C such that $\psi = C^T$ or there is an $(\mathcal{ML}_{\mathcal{SU}})_{\mathcal{CQO}}$-formula χ such that $\psi = \chi^T$) that

$$\boldsymbol{t} = \{C^T \mid C \in \widehat{\boldsymbol{t}}\} \cup \{\psi^T \mid \psi \in \widehat{\boldsymbol{t}}\}. \tag{14.5}$$

Recall that a *state candidate* for φ^T is a set of types for φ^T. Given such a state candidate S, define

$$\widehat{S} = \{\widehat{\boldsymbol{t}} \mid \boldsymbol{t} \in S\}.$$

Then it is not hard to see that S is $\mathcal{K}$-realizable iff the $\mathcal{CQO}$-formula

$$\varphi_S = \bigwedge_{\widehat{t}\in\widehat{S}} \Big[\bigwedge_{\psi\in\widehat{t}} \psi \wedge \neg((\bigsqcap_{C\in\widehat{t}} C) = \bot) \Big] \wedge$$
$$\bigvee_{\widehat{t}\in\widehat{S}} \bigwedge_{\psi\in\widehat{t}} \psi \wedge \Big((\bigsqcup_{\widehat{t}\in\widehat{S}} \bigsqcap_{C\in\widehat{t}} C) = \top \Big) \tag{14.6}$$

is satisfied in a $\mathcal{CQO}$-model. So, the decidability of the problem of whether a given state candidate for φ^T is $\mathcal{K}$-realizable follows from the decidability of the formula satisfiability problem for $\mathcal{CQO}$.

To prove (b), it suffices to show that there exists a cardinal κ_0 such that, for any $\kappa \geq \kappa_0$ and any satisfiable $\mathcal{CQO}$-formula φ_S of the form (14.6), there exists a $\mathcal{CQO}$-model $I = \langle \Delta, R_0^I, \dots, C_0^I, \dots, a_0^I, \dots \rangle$ satisfying φ_S and such that the set

$$\{w \in \Delta \mid w \in C^I \text{ for all } C \in \widehat{t}\}$$

is of cardinality κ for any $\widehat{t} \in \widehat{S}$, whenever no nominal $\{a\}$ occurs in $\widehat{t}$.

To prove this, define κ_0 to be the smallest infinite cardinal such that any satisfiable φ_S is satisfiable in a model of cardinality $\leq \kappa_0$. (Note that we could actually choose $\kappa_0 = \aleph_0$ by a Löwenheim–Skolem–Tarski argument.) Assume now that a given φ_S is satisfiable. Take a model

$$I = \langle \Delta, R_0^I, \dots, C_0^I, \dots, a_0^I, \dots \rangle$$

of cardinality $\leq \kappa_0$ satisfying φ_S and take a $\kappa \geq \kappa_0$. Let $N = \{a^I \mid a \in ob\,\varphi_S\}$ and

$$J = \langle \Delta', R_0^J, \dots, C_0^J, \dots, a_0^J, \dots \rangle,$$

where

- $\Delta' = N \cup \{\langle v, \xi\rangle \mid v \in \Delta - N,\ \xi < \kappa\}$,
- $C_i^J = (C_i^I \cap N) \cup \{\langle v, \xi\rangle \mid v \in (\Delta - N) \cap C_i^I,\ \xi < \kappa\}$,
- $a^J = a^I$, for all $a \in ob\,\varphi_S$,
- $\langle v, \xi\rangle\, R_i^J\, \langle w, \xi\rangle$ iff $vR^I w$,
- $a^J R_i^J b^J$ iff $a^I R_i^I b^I$,
- $\langle v, \xi\rangle\, R^J a^J$ iff $vR^I a^I$,
- $a^J R_i^J \langle v, \xi\rangle$ iff $a^I R_i^I v$ and $\xi = 0$.

It is not difficult to show that J is as required (we leave this to the reader). Our theorem follows now from Theorem 11.83. ❑

What is the computational complexity of the algorithmic problems considered in the theorem above? Actually, not too much is known. To begin with, here is a 'negative' result:

Theorem 14.13. *If* $L \in \{\mathbf{PDL}, \mathbf{PTL}, \mathbf{K4.3}, \mathsf{Log}\{\langle \mathbb{N}, < \rangle\}, \mathsf{Log}\{\langle \mathbb{Q}, < \rangle\}, \mathsf{Log}_{FP}(\mathbb{Q}), \mathbf{K}_1^C, \mathbf{T}_2^C, \mathbf{K4}_2^C, \mathbf{S4}_2^C, \mathbf{KD45}_2^C\}$ *then the* $L_{\mathcal{ALC}}$*-satisfiability problem for formulas without modalized roles and global role names is* EXPSPACE-*hard.*

Proof. Follows from Theorems 6.64 and 6.66, and the reduction of Theorem 3.35. ❑

By Theorems 3.35 and 5.42, we also have:

Theorem 14.14. *Let* L *be any Kripke complete multimodal logic between* $\mathbf{K}_n$ *and* $\mathbf{S5}_n$. *Then the* $L_{\mathcal{ALC}}$*-satisfiability problem for formulas (without any roles at all) is* NEXPTIME-*hard.*

Note that for $(\mathbf{K}_n)_{\mathcal{ALC}}$ a matching upper bound is given in Theorem 15.15 below. The following theorem establishes one more matching upper bound:

Theorem 14.15. *The* $\mathbf{PTL}_{\mathcal{ALC}}$*-satisfiability problem for formulas without global role names and modalized roles is* EXPSPACE-*complete.*

Proof. EXPSPACE-hardness was shown in Theorem 14.13. To obtain the matching upper bound, we first extend the translation $\cdot^T$ from the language of $\mathcal{ML}_{\mathcal{ALC}}$ defined in Section 3.8 to a translation from $(\mathcal{ML}_{\mathcal{U}})_{\mathcal{ALC}}$ to $\mathcal{QTL}_{\mathcal{U}}$ by taking

$$(C\mathcal{U}D)^T = C^T\mathcal{U}D^T.$$

Let $\mathcal{QTL}' = \{\varphi^T \mid \varphi \text{ an } (\mathcal{ML}_{\mathcal{U}})_{\mathcal{ALC}}\text{-formula}\}$. Now, to prove our theorem it suffices to observe that $\mathcal{QTL}' \subseteq \mathcal{QTL}^2_{\boxed{1}}$ and that, by Theorem 11.31, $\mathsf{QLog}_{\mathcal{U}}(\mathbb{N}) \cap \mathcal{QTL}^2_{\boxed{1}}$ is EXPSPACE-complete. A more direct proof can be given as follows: it actually suffices to show that the language $\mathcal{QTL}'$ satisfies the conditions of Theorem 11.30. But this is fairly clear if we observe that state candidates for a $\mathcal{QTL}'$-sentence φ^T correspond to model candidates for φ defined in the proof of Theorem 2.27. This proof actually provides an algorithm which, given a state candidate $\mathfrak{C}$ for φ^T, can recognize whether $\mathfrak{C}$ is realizable using space $\leq 2^{p(\ell(\varphi))}$ for some polynomial function p. ❑

It is worth noting that if we consider the fragment of $\mathbf{PTL}_{\mathcal{ALC}}$ in which the temporal operators can be applied only to formulas then $\mathcal{QTL}' \subseteq \mathcal{QTL}_{\mathcal{U}\boxed{0}}$. Therefore, by using Theorem 11.35 instead of Theorem 11.30, from Theorem 2.27 we obtain an EXPTIME upper bound for the satisfiability problem.

14.4 Satisfiability in models with finite domains

Some applications of description logics may require satisfiability-checking in models with *finite domains.* For example, if the description logic is used for conceptual data modeling, then finite domains seem more appropriate than infinite ones; see, e.g., (Calvanese 1996, Calvanese *et al.* 1998).

Note first that, as was shown in Section 2.5, for most (nonmodal) description logics there is no difference between satisfiability in finite and arbitrary models. (However, there are description logics which do not enjoy the fmp. Typical examples are logics with number restrictions and inverse roles; see, e.g., (Calvanese 1996).) On the other hand, many modalized description logics can distinguish—often already on the concept level (with empty knowledge base)—between models with finite domains and models with infinite domains. We explain this observation by connecting finite models for modal description logics with finite product models. For simplicity, we will confine ourselves to considering logics based on $\mathcal{ALC}$.

Theorem 14.16. *Suppose L is a Kripke complete multimodal logic with the fmp. Assume also that Ξ is the set of $\mathcal{ML}_{\mathcal{ALC}}$-concepts containing no modalized roles. Then the following are equivalent:*

(i) *a concept in Ξ is $L_{\mathcal{ALC}}$-satisfiable (relative to empty knowledge base) iff it is satisfiable in an $L_{\mathcal{ALC}}$-model with finite domains;*

(ii) *a concept in Ξ is $L_{\mathcal{ALC}}$-satisfiable (relative to empty knowledge base) iff it is satisfiable in a finite $L_{\mathcal{ALC}}$-model;*

(iii) *the product logic $L \times \mathbf{K}_n$ has the product fmp, for every $n \geq 1$.*

Proof. In view of (the multimodal generalization of) Proposition 5.35, for concepts without modalized roles and local role names the equivalence follows from the fact that the reduction on page 174 associates $L_{\mathcal{ALC}}$-models having finite domains with product models in which one component is finite, and finite $L_{\mathcal{ALC}}$-models with finite product models.

As to local role names, their reduction to global ones provided in the proof of Theorem 14.2 is easily modified in such a way that it preserves finiteness of the $\mathcal{ALC}$-domain: when unraveling the $\mathcal{ALC}$-part of a model for a concept C, we take only those points from the resulting tree that are reachable by a path of length $\leq rd(C)$ from the root. ❑

Since $\mathbf{K}_m$ has the fmp (see Theorem 1.16) and $\mathbf{K}_m \times \mathbf{K}_n$ has the product fmp (see Theorem 6.4), we obtain:

Proposition 14.17. *Every $(\mathbf{K}_m)_{\mathcal{ALC}}$-satisfiable concept without modalized roles is satisfiable in a finite $(\mathbf{K}_m)_{\mathcal{ALC}}$-model (and so in a $(\mathbf{K}_m)_{\mathcal{ALC}}$-model with finite domains).*

A similar statement holds for $\mathbf{T}_m$ in place of $\mathbf{K}_m$. On the other hand, by Theorems 1.16, 2.2, 2.17, 2.22, Remark 2.11 and Theorems 5.32, 6.13, 6.21, we have:

Proposition 14.18. *For all modal description logics below, one can find concepts without modalized roles which are satisfiable (relative to empty knowledge base) in models with infinite domains but not in models with finite domains:*

(1) *the dynamic description logic* $\mathbf{CPDL}_{\mathcal{ALC}}$,

(2) *the epistemic description logics* $L_{\mathcal{ALC}}$, *where* $L \in \{\mathbf{K4}_n, \mathbf{S4}_n\}$,

(3) *the epistemic description logics with common knowledge operators* $L^C_{\mathcal{ALC}}$, *where* $L \in \{\mathbf{K}_n, \mathbf{T}_n, \mathbf{K4}_n, \mathbf{S4}_n, \mathbf{KD45}_n, \mathbf{S5}_n\}$,

(4) *the temporal description logics* $\mathbf{PTL}_{\mathcal{ALC}}$, $\mathbf{Lin}_{\mathcal{ALC}}$, *and* $(\mathsf{Log}_{FP}(\mathbb{Q}))_{\mathcal{ALC}}$,

(5) $\mathbf{K4.3}_{\mathcal{ALC}}$, $\mathsf{Log}\{\langle \mathbb{N}, <\rangle\}_{\mathcal{ALC}}$, *and* $\mathsf{Log}\{\langle \mathbb{Q}, <\rangle\}_{\mathcal{ALC}}$.

However, the following questions remain open:

Question 14.19. Is the satisfiability problem for concepts without modalized roles relative to empty knowledge base decidable in models with finite domains for any of the logics listed in Proposition 14.18?

As concerns formula satisfiability, if we allow global role names, then the following holds:

Proposition 14.20. *Suppose L is a Kripke complete multimodal logic with the fmp. If $L \times \mathbf{K}_u$ does not have the product fmp, then the sets of formulas with global role names (but without modalized roles and local role names) satisfiable in arbitrary $L_{\mathcal{ALC}}$-models and in those with finite domains are different.*

Proof. By (the multimodal generalization of) Proposition 5.35, if $L \times \mathbf{K}_u$ does not have the product fmp, then it is not determined by frames of the form $\mathfrak{F}_1 \times \mathfrak{F}_2$, where $\mathfrak{F}_1 \in \mathsf{Fr}L$ and $\mathfrak{F}_2$ is a finite frame for $\mathbf{K}_u$. Take any formula φ of the language of $L \times \mathbf{K}_u$ which is $L \times \mathbf{K}_u$ satisfiable but not in a product model for $L \times \mathbf{K}_u$ with finite second component. It is not hard to see (by repeating the proof of Theorem 3.36 for the finite domain case) that there is a formula of the language of $L_{\mathcal{ALC}}$ such that it is satisfiable in an $L_{\mathcal{ALC}}$-model but not in an $L_{\mathcal{ALC}}$-model with finite domains. ❑

Since almost all logics L considered in this book have an infinite frame $\langle W, R, \dots\rangle$ with a point $x \in W$ such that xRy, for all $y \in W$, $y \neq x$, we can use Theorem 5.34, according to which, for every such L, $L \times \mathbf{K}_u$ does not have the product fmp. So, if such a logic L has the fmp (like $\mathbf{K}$ or $\mathbf{S5}$), then the set of formulas with global role names satisfiable in $L_{\mathcal{ALC}}$-models with finite domains is properly contained in the set of formulas satisfiable in arbitrary $L_{\mathcal{ALC}}$-models.

Question 14.21. Is there an 'interesting' modal logic L such that the formula satisfiability problem in $L_{\mathcal{ALC}}$-models with finite domains is decidable?

Let us consider now the satisfiability problem in models with finite domains for formulas having neither modalized roles nor global role names. To begin with, we note that $(\mathbf{K}_m)_{\mathcal{ALC}}$ does not 'feel' the difference between finite and infinite domains:

Proposition 14.22. *If a formula without modalized roles and global role names is $(\mathbf{K}_m)_{\mathcal{ALC}}$-satisfiable, then it is satisfiable in a finite $(\mathbf{K}_m)_{\mathcal{ALC}}$-model (and so in a model with finite domains).*

Proof. The tableau algorithm of Section 15.2 constructs a finite model for any satisfiable formula without modalized roles and global role names. ❑

Mostly, however, the set of formulas satisfiable in models with finite domains is properly contained in the set of formulas satisfiable in models with arbitrarily large domains:

Proposition 14.23. *Suppose L is a Kripke complete multimodal logic with the fmp. If $L \times \mathbf{S5}$ does not have the product fmp, then the sets of formulas (containing neither modalized roles nor global role names) satisfiable in arbitrary $L_{\mathcal{ALC}}$-models and only in those with finite domains are different.*

Proof. Assume for simplicity that L is a unimodal logic. By Proposition 5.35, if $L \times \mathbf{S5}$ does not have the product fmp, then it is not determined by frames of the form $\mathfrak{F}_1 \times \mathfrak{F}_2$, where $\mathfrak{F}_1 \in \mathsf{Fr}L$ and $\mathfrak{F}_2$ is a finite frame for $\mathbf{S5}$. Take any $\mathcal{ML}_2$-formula φ which is $L \times \mathbf{S5}$ satisfiable but not in a product model for $L \times \mathbf{S5}$ with finite second component. It is not hard to see (by repeating the proof of Theorem 3.35 for the finite domain case) that the $\mathcal{ML}_{\mathcal{ALC}}$-formula $\neg(\varphi^\sharp = \bot) \wedge \Box_1^{\leq md(\varphi)} \chi$ defined in the proof of Theorem 3.35 is satisfiable in an $L_{\mathcal{ALC}}$-model but not in an $L_{\mathcal{ALC}}$-model with finite domains. ❑

So, by Theorems 1.16, 2.2, 2.17, 2.22 and Theorems 5.32, 5.33, 6.51, we have that, for all dynamic and epistemic logics L mentioned in Theorem 14.12, for $L = \mathsf{Log}_{\mathcal{SU}}(\mathcal{C})$, $\mathcal{C} \in \{\{\langle \mathbb{N}, < \rangle\}, \{\langle \mathbb{Z}, < \rangle\}, \{\langle \mathbb{Q}, < \rangle\}\}$, and for $L = \mathbf{GL.3}$, the sets of formulas (without modalized roles and global role names) satisfiable in arbitrary $L_{\mathcal{ALC}}$-models and only in those with finite domains are different.

It is of interest to note that if we have neither global role names nor modalized roles, then satisfiability in models with finite domains is decidable at least for some temporal description logics:

Theorem 14.24. *The satisfiability problem in models with finite domains for formulas containing neither modalized roles nor global role names is decidable for the logics* $L_{\mathcal{ALC}}$, *where* $L = \mathsf{Log}_{\mathcal{SU}}(\mathcal{C})$ *and* $\mathcal{C}$ *is one of the following classes:* $\{\langle\mathbb{N},<\rangle\}$, $\{\langle\mathbb{Z},<\rangle\}$, $\{\langle\mathbb{Q},<\rangle\}$, $\{\langle\mathbb{R},<\rangle\}$, *any first-order definable class of strict linear orders.*

Proof. This can be proved in the same way as Theorem 14.12. Just apply Theorem 11.9 instead of Theorem 11.83 and use the fact that any satisfiable $\mathcal{ALC}$-formula is satisfiable in a finite model (see Proposition 2.28). ❑

For $\mathbf{PTL}_{\mathcal{ALC}}$ we also have the following complexity result:

Theorem 14.25. *The satisfiability problem for formulas (containing neither global role names nor modalized roles) in* $\mathbf{PTL}_{\mathcal{ALC}}$*-models with finite domains is* EXPSPACE*-complete.*

Proof. As was shown in the proof of Theorem 5.43, the logic determined by products of **PTL**-frames and finite **S5**-frames is EXPSPACE-hard. Now, the EXPSPACE-hardness of the satisfiability problem for our formulas in $\mathbf{PTL}_{\mathcal{ALC}}$-models with finite domains follows from the fact that the reduction of Theorem 3.35 associates this kind of 'half finite' product frames with models having finite domains.

The proof of the upper bound is similar to the proof of Theorem 14.15. In this case use Theorems 2.27 and 11.51, together with Proposition 2.28. ❑

This result is of particular interest when temporal description logics are used for reasoning about conceptual schemas, where it is natural to assume domains to be finite. It shows that various results presented in (Artale *et al.* 2002) can be lifted to finite domain models.

The following question remains open:

Question 14.26. Is the satisfiability problem for formulas (containing neither global role names nor modalized roles) in $L_{\mathcal{ALC}}$-models with finite domains decidable whenever L is one of the logics mentioned in Theorem 14.12 (1), (2)?

Chapter 15

Tableau decision algorithms for modal description logics

The proofs of decidability presented so far are based on a semantical approach. They do not provide us with any 'practical' decision procedures that could be implemented in reasoning systems which are reasonably fast on reasonably large sets of problems.

The aim of this chapter is to show how potentially 'implementable' sound and complete *tableau algorithms* (D'Agostino *et al.* 1999) deciding the satisfiability problem for various modal description logics can be designed. Tableau-based algorithms have been shown to be 'practical' for standard description logics of rather high complexity such as $\mathcal{ALC}$ with number restrictions and transitive roles; see, e.g., (Haarslev and Möller 1999, Horrocks 1998, Horrocks *et al.* 2000a). Here we explore how tableaux can be lifted to modal description logics.

To make this chapter self-contained, we start with a tableau decision algorithm for 'pure' $\mathcal{ALC}$. Then, in Section 15.2, we show in detail how to extend it to a tableau system for the modal description logic $\mathbf{K}_{\mathcal{ALC}}$ (without global and modalized roles) interpreted in models with constant domains. The tableau procedure will be shown to run in NEXPTIME, which matches the lower bound of Theorem 14.14. Section 15.3 provides tableaux for two extensions of $\mathbf{K}_{\mathcal{ALC}}$. First, by adding two rules we obtain a tableau procedure for $\mathbf{K}_{\mathcal{ALCU}}$, the extension of $\mathbf{K}_{\mathcal{ALC}}$ with the universal role U. And second, we modify the resulting tableaux to obtain an algorithm checking not only formula satisfiability but also the more complex global concept satisfiability (see Section 3.8).

15.1 Tableaux for $\mathcal{ALC}$

To decide whether a given $\mathcal{ALC}$-formula ϑ is satisfiable, a tableau algorithm tries to construct a model for ϑ by repeatedly applying so-called *completion rules* to an appropriate data structure. Usually, in modal logic these data structures are just sets of formulas, cf. (Gore 1999). In the case of $\mathcal{ALC}$, which contains assertions of the form $a : C$ and aRb, we require also *variables*. The data structures are then *constraint systems*, where each *constraint* can be

- an $\mathcal{ALC}$-formula,
- an expression of the form $x : C$, where x is either a variable or an object name and C is a concept,
- or an expression of the form xRy, where x and y are variables or object names and R is a role

(see, e.g., Hollunder and Nutt 1990, Baader and Hanschke 1991, Baader and Laux 1995).

For now, it is convenient to think of variables as representing domain objects: the expression $x : C$ says that concept C applies to the object represented by x, while xRy says that the object represented by x stands in relation R to y.

A tableau algorithm checking satisfiability of ϑ starts with a constraint system containing only ϑ (and $v : \top$, for some technical reasons). The completion rules are applied until (i) an 'obviously' contradictory constraint system is obtained or (ii) a contradiction-free (or clash-free) and *complete* constraint system is found, complete in the sense that no further rule is applicable to it. By an 'obvious contradiction' we mean that the constraint system contains, for example, both φ and $\neg\varphi$ for some formula φ. To illustrate what completion rules look like, we sketch some standard rules which can be found in most tableau algorithms for description logics: see, e.g., (Hollunder and Nutt 1990, Baader and Hanschke 1991, Baader and Laux 1995, Horrocks *et al.* 1999).

1. If a constraint system S contains the formula $C = \top$ and a variable or an object name x, then we add $x : C$ to S.
2. If a constraint system S contains $x : C \sqcup D$, then we add to S either $x : C$ or $x : D$.
3. If a constraint system S contains $x : \exists R.C$, then we add to S two constraints $v : C$ and xRv, where v is a fresh variable that was not used in S before.

The second rule above is nondeterministic: its application yields more than one possible outcome. In the presence of nondeterministic rules, a tableau algorithm terminates successfully if the completion rules can be applied in such a way that the result is a complete and clash-free constraint system. The tableau algorithm is *sound* if, whenever it terminates successfully on input ϑ, then ϑ is satisfiable. The tableau algorithm is *complete* if, whenever it does not terminate successfully on input ϑ, then ϑ is not satisfiable. Finally, the tableau algorithm *terminates* if, on any input ϑ after finitely many applications of completion rules to ϑ, it terminates in the sense that no completion rule is applicable any more. Of course, a sound, complete and terminating tableau algorithm provides a decision procedure for the satisfiability problem for formulas. We now describe such a tableau calculus for $\mathcal{ALC}$ in full detail.

Say that an $\mathcal{ALC}$-formula φ is *equivalent* to an $\mathcal{ALC}$-formula ψ if $\{\varphi\} \models \psi$ and $\{\psi\} \models \varphi$. Similarly, an $\mathcal{ALC}$-concept C is *equivalent* to a concept D if $C^I = D^I$ for all $\mathcal{ALC}$-models I. The formula $C = D$ is clearly equivalent to $(\neg C \sqcup D) \sqcap (\neg D \sqcup C) = \top$. So without loss of generality we can assume that in every atomic formula of the form $C = D$ the concept D is $\top$. Furthermore, we generally assume formulas and concepts to be in negation normal form which is defined as follows.

A concept C is said to be in *negation normal form* (NNF, for short) if negation occurs in C only in front of concept names. A formula φ is in *negation normal form* if negation occurs in φ only in front of concept names and atomic formulas of the form $C = D$ or aRb.

Each concept C can be transformed into an equivalent concept in NNF by pushing negation inwards with the help of De Morgan's laws and the duality between $\exists$ and $\forall$. The NNF of $\neg C$ will be denoted by $\sim C$. Similarly, each formula can be transformed into an equivalent one in NNF by employing De Morgan's laws and the fact that $\neg(a : C)$ is equivalent to $a : \neg C$.

Let us now define formally what we mean by a constraint system for a given $\mathcal{ALC}$-formula ϑ. As before, we denote by

- $ob\,\vartheta$ the set of all object names occurring in ϑ;
- $con\,\vartheta$ the set of all concepts occurring in ϑ;
- $sub\,\vartheta$ the set of all subformulas of ϑ;
- $rol\,\vartheta$ the set of roles occurring in ϑ.

The *fragment induced by* ϑ is defined as the set

$$Fg\,\vartheta = ob\,\vartheta \cup sub\,\vartheta \cup con\,\vartheta \cup rol\,\vartheta \cup \{\sim C \mid C \in con\,\vartheta\} \cup \{\top\}.$$

Fix a countably infinite set V of *(individual) variables*. The variables in V and the object names in $ob\,\vartheta$ will be called *terms* for ϑ. We will assume that we have a well-ordering $\ll$ on the set of terms. Throughout this chapter, we denote variables by v and u, and terms by x and y.

An $\mathcal{ALC}$-*constraint* for ϑ is either a formula in $sub\,\vartheta$, an expression xRx', where $R \in rol\,\vartheta$ and x, x' are terms for ϑ, or an atom of the form $x : C$, where C is a concept in $Fg\,\vartheta$ and x a term for ϑ. A *constraint system* for ϑ is a finite set S of constraints for ϑ. A variable v is called *fresh* for S if v does not occur in S.

To ensure termination of repeated applications of the completion rules, we use the so-called 'blocking' technique (see e.g., (Baader and Laux 1995) and references therein). Say that a variable v in a constraint system S is *blocked by a variable* v' in S if $v' \ll v$ and

$$\{C \mid (v : C) \in S\} \subseteq \{C \mid (v' : C) \in S\}.$$

Note that only variables, rather than object names, may block terms. Also, only variables can be blocked.

A constraint system S is said to be *clash-free* if it contains no formulas $\neg\top$ and $x : \neg\top$, and neither a pair of the form $x : C_i$, $x : \neg C_i$, nor a pair of the form xRy, $\neg(xRy)$ occurs in it. Otherwise we say that S contains a *clash*. A constraint system S is *complete* if no completion rule from Fig. 15.1 is applicable to S.

To decide whether a given formula ϑ in negation normal form is satisfiable, we form the initial constraint system $S_\vartheta = \{\vartheta, v : \top\}$, where v is $\ll$-minimal. After that we repeatedly apply the $\mathcal{ALC}$-completion rules from Fig. 15.1 in such a way that the $\mathcal{ALC}$-generating rules are applied only if no other rule is applicable. This strategy prevents the introduction of a large number of variables to which the same concepts apply. It is, however, not required for termination or correctness. The tableau algorithm is shown in Fig. 15.2 in a pseudocode notation.

We prove now that this tableau algorithm is sound, complete and terminates.

Theorem 15.1 (soundness). *Suppose that S is a complete clash-free constraint system for ϑ. Then ϑ is satisfiable.*

Proof. Given a complete and clash-free S, we construct a model

$$I = \left\langle \Delta, R_0^I, \ldots, C_0^I, \ldots, a_0^I, \ldots \right\rangle,$$

where

- Δ is the set of terms occurring in S;
- $x \in C_i^I$ iff $(x : C) \in S$, for all $x \in \Delta$;
- $a^I = a$, for all $a \in ob\,\vartheta$;
- xR^Iy iff $xRy \in S$ or $zRy \in S$ for some z which blocks x in S.

$\mathcal{ALC}$-rules on formulas

$\mathsf{R}_\wedge$ If $(\varphi \wedge \psi) \in S$ and $\{\varphi, \psi\} \not\subseteq S$,
then set $S := S \cup \{\varphi, \psi\}$.

$\mathsf{R}_\vee$ If $(\varphi \vee \psi) \in S$ and $\{\varphi, \psi\} \cap S = \emptyset$,
then set $S := S \cup \{\theta\}$, where $\theta = \varphi$ or $\theta = \psi$.

$\mathcal{ALC}$-nongenerating rules on concepts

$\mathsf{R}_\sqcap$ If $(x : C \sqcap D) \in S$ for a term x and $\{x : C, x : D\} \not\subseteq S$,
then set $S := S \cup \{x : C, x : D\}$.

$\mathsf{R}_\sqcup$ If $(x : C \sqcup D) \in S$ and $\{x : C, x : D\} \cap S = \emptyset$,
then set $S := S \cup \{x : E\}$, where $E = C$ or $E = D$.

$\mathsf{R}_=$ If $(C = \top) \in S$, a term x occurs in S, but $(x : C) \notin S$,
then set $S := S \cup \{x : C\}$.

$\mathsf{R}_\forall$ If $\{x : \forall R.C, xRy\} \subseteq S$ but $(y : C) \notin S$,
then set $S := S \cup \{y : C\}$.

$\mathcal{ALC}$-generating rules

$\mathsf{R}_\neq$ If $\neg(C = \top) \in S$ and there is no term x in S such that
$(x : {\sim}C) \in S$,
then choose the $\ll$-minimal fresh variable v for S and set
$S := S \cup \{v : {\sim}C\}$.

$\mathsf{R}_\exists$ If $(x : \exists R.C) \in S$, x is not blocked in S and there is no term y
in S such that $\{xRy, y : C)\} \subseteq S$,
then choose the $\ll$-minimal fresh variable v for S and set
$S := S \cup \{v : C, xRv\}$.

Figure 15.1: Completion rules for $\mathcal{ALC}$.

Claim 15.2. *For all concepts $C \in con\,\vartheta$ and all $x \in \Delta$, if $(x : C) \in S$ then $x \in C^I$.*

Proof. The proof is by induction on the construction of concepts. Recall that all concepts we deal with are in NNF. For atomic concepts the claim follows from the definition. Suppose now that $C = \neg C_i$, for an atomic concept C_i, and $(x : C) \in S$. Then $(x : C_i) \notin S$, since S is clash-free, and so $x \notin C_i^I$ by definition. Hence $x \in (\neg C_i)^I$.

Suppose $C = D \sqcap E$ and $(x : C) \in S$. Since S is closed under $\mathsf{R}_\sqcap$, we then have $(x : D) \in S$ and $(x : E) \in S$. By the induction hypothesis, $x \in D^I$ and $x \in E^I$, and so $x \in (D \sqcap E)^I$.

define procedure *sat*(*S*)
  if *S* contains a clash **then**
    return *unsatisfiable*
  if a nongenerating rule *r* is applicable to *S*
    then apply *r* to *S*
    return *sat*(*S*)
  if a rule $r \in \{\mathsf{R}_{\neq}, \mathsf{R}_{\exists}\}$ is applicable to *S*
    then apply *r* to *S*
    return *sat*(*S*)
  return *satisfiable*

Figure 15.2: The satisfiability-checking algorithm for $\mathcal{ALC}$.

Suppose $C = D \sqcup E$ and $(x : C) \in S$. Since S is closed under $\mathsf{R}_{\sqcup}$, we have $(x : D) \in S$ or $(x : E) \in S$, and so, by the induction hypothesis, $x \in D^I$ or $x \in E^I$, from which $x \in (D \sqcup E)^I$.

Suppose $C = \forall R.D$ and $(x : C) \in S$. Let $xR^I y$. Then, by definition, either $xRy \in S$ or there is z which blocks x in S and such that $zRy \in S$. Since S is closed under $\mathsf{R}_{\forall}$, we have $(y : D) \in S$. Hence, by the induction hypothesis, $y \in D^I$. This holds for all y with $xR^I y$, and so $x \in (\forall R.D)^I$.

Suppose $C = \exists R.D$ and $(x : C) \in S$. Assume first that x is not blocked in S. Then, since S is closed under $\mathsf{R}_{\exists}$, we find y such that $xRy \in S$ and $(y : D) \in S$. Hence $xR^I y$ and, by the induction hypothesis, $y \in D^I$ so that $x \in (\exists R.D)^I$. Assume now that x is blocked by a variable y in S. As $\ll$ is a well-ordering, we can find a $\ll$-minimal y which blocks x in S. It follows that y is not blocked in S by another variable and that

$$\{E \mid (x : E) \in S\} \subseteq \{E \mid (y : E) \in S\}.$$

As shown above, we then have a variable z such that $yRz \in S$ (and so $yR^I z$) and $z \in D^I$. But then $xR^I z$, and so $x \in (\exists R.D)^I$. ❑

Claim 15.3. *For all formulas $\varphi \in sub\,\vartheta$, if $\varphi \in S$ then $I \models \varphi$.*

Proof. The proof is again by induction on the construction of φ.

Case 1: $(a : C) \in S$. Then, by Claim 15.2, $a^I \in C^I$ and so $I \models a : C$.

Case 2: $aRb \in S$. Then, by definition, $a^I R^I b^I$ and so $I \models aRb$.

Case 3: $\neg(aRb) \in S$. Then, since S is clash-free, $a^I R^I b^I$ does not hold, and so $I \models \neg aRb$.

Case 4: $(C = \top) \in S$. Let $x \in \Delta$. Then, since S is closed under $\mathsf{R}_{=}$, $(x : C) \in S$, and so, by Claim 15.2, $x \in C^I$. Hence $C^I = \Delta$.

Case 5: $\neg(C = \top) \in S$. Since S is closed under $\mathsf{R}_{\neq}$, we find y such that $(y : \sim C) \in S$. Hence, by Claim 15.2, $y \in (\neg C)^I$.

Case 6: $\psi_1 \wedge \psi_2 \in S$. Since S is closed under $\mathsf{R}_\wedge$, we then have $\psi_1 \in S$ and $\psi_2 \in S$. By the induction hypothesis, $I \models \psi_1$ and $I \models \psi_2$, from which $I \vdash \psi_1 \wedge \psi_2$.

Case 7: $\psi_1 \vee \psi_2 \in S$. Since S is closed under $\mathsf{R}_\vee$, $\psi_1 \in S$ or $\psi_2 \in S$. By the induction hypothesis, $I \models \psi_1$ or $I \models \psi_2$, and so $I \models \psi_1 \vee \psi_2$. ❑

To complete the proof of soundness, it remains to observe that $I \models \vartheta$ follows from $\vartheta \in S$. ❑

Theorem 15.4 (termination). *The number of iterated rule applications to S_ϑ does not exceed $2^{p(|Fg\,\vartheta|)}$, for some polynomial function p.*

Proof. Note first that the only rules introducing new variables are the generating rules $\mathsf{R}_{\neq}$ and $\mathsf{R}_\exists$. $\mathsf{R}_{\neq}$ can introduce at most $|Fg\,\vartheta|$ variables. Because of the priority of nongenerating rules over generating ones, if $\mathsf{R}_\exists$ is applied to $v : \exists R.C$, then v will never be blocked by another variable. Now, there are at most $2^{|Fg\,\vartheta|}$ unblocked variables, and so the number of terms to which $\mathsf{R}_\exists$ can be applied is bounded by $|ob\,\vartheta| + 2^{|Fg\,\vartheta|}$. Therefore, the number of terms does not exceed

$$|Fg\,\vartheta| + (|ob\,\vartheta| + 2^{|Fg\,\vartheta|}) \cdot (|Fg\,\vartheta| + 1).$$

The upper bound we need follows now from the observation that every rule introduces a new member of $Fg\,\vartheta$ or an expression of the form $x : C$, for $C \in Fg\,\vartheta$. ❑

Theorem 15.5 (completeness). *Suppose ϑ is satisfiable. Then there exists a complete clash-free constraint system containing S_ϑ.*

Proof. Take a model $I = \langle \Delta, R_0^I, \ldots, C_0^I, \ldots, a_0^I, \ldots \rangle$ which satisfies ϑ. We use I as a 'guide' for applications of the nondeterministic rules to construct a complete and clash-free constraint system. Say that a constraint system S for ϑ is *compatible* with I if (i) $I \models \varphi$ whenever φ is a formula and $\varphi \in S$, and (ii) there exists a map π from the set of terms in S into Δ such that

- $\pi(a) = a^I$ for all $a \in ob\,\vartheta$;
- $\pi(x) \in C^I$ whenever $(x : C) \in S$;
- $\pi(x) R^I \pi(y)$ whenever $xRy \in S$.

We claim that if a constraint system S is compatible with I and a rule R is applicable to S, then it can be applied in such a way that the result is compatible with I as well.

Indeed, suppose that S is a constraint system for ϑ compatible with I and that π is a function satisfying the conditions listed under (ii). Consider all possible cases for rule applications:

(a) Let the $\mathsf{R}_\wedge$ rule be applicable to a formula $\varphi \wedge \psi$ in S. Since S is I-compatible, $I \models \varphi \wedge \psi$ and so $I \models \varphi$ and $I \models \psi$. The application of $\mathsf{R}_\wedge$ to S adds φ and ψ to S, so (i) holds after the application. The very same function π satisfies (ii).

(b) Suppose that the $\mathsf{R}_\vee$ rule is applicable to a formula $\varphi \vee \psi$ in S. Then, as we know, $I \models \varphi \vee \psi$, and so either $I \models \varphi$ or $I \models \psi$. By applying the $\mathsf{R}_\vee$ rule to S accordingly, we clearly obtain an S' for which (i) holds; π remains unchanged.

(c) Suppose that the $\mathsf{R}_\sqcap$ rule is applicable to a constraint $x : C \sqcap D$ in S. Then $\pi(x) \in (C \sqcap D)^I$, and so $\pi(x) \in C^I$ and $\pi(x) \in D^I$. The application of the $\mathsf{R}_\sqcap$ rule adds $x : C$ and $x : D$ to S. Clearly, (i) still holds. The function π is as required for (ii).

(d) Suppose the $\mathsf{R}_\sqcup$ rule is applicable to a constraint $x : C \sqcup D$ in S. Then $\pi(x) \in (C \sqcup D)^I$ and so $\pi(x) \in C^I$ or $\pi(x) \in D^I$. By applying the $\mathsf{R}_\sqcup$ rule to S accordingly, we see that (i) still holds and the function π is still as required.

(e) Suppose that the $\mathsf{R}_=$ rule is applicable to a formula $C = \top$ and a term x in S. Then $I \models C = \top$. Hence $\pi(x) \in C^I$ and so π is still as required after the application of $\mathsf{R}_=$ to S.

(f) The application of $\mathsf{R}_\forall$ is treated in the same manner.

(g) Suppose $\mathsf{R}_{\neq}$ is applicable to $\neg(C = \top)$ in S. Then $I \models \neg(C = \top)$ and there is $d \in \Delta$ such that $d \notin C^I$. We introduce the $\ll$-minimal fresh variable x and define π' as an extension of π to x by taking $\pi'(x) = d$. The function π' is then as required for the resulting constraint system.

(h) Let $\mathsf{R}_\exists$ be applicable to $x : \exists R.C$ in S. We have $\pi(x) \in (\exists R.C)^I$. Let $d \in \Delta$ with $\pi(x) R^I d$ and $d \in C^I$. Now we proceed as in the $\mathsf{R}_{\neq}$-case.

By Theorem 15.4, after finitely many rule applications we obtain a complete constraint system S which is compatible with I. Obviously, S is clash-free. ❑

It can be shown that the exponential upper bound in Theorem 15.4 cannot be improved. Thus, the tableau procedure checking satisfiability of $\mathcal{ALC}$-formulas presented above runs in NEXPTIME and does *not* have the optimal worst case behavior: according to Theorem 2.27, this satisfiability problem is EXPTIME-complete. For a discussion and comparison of different approaches to satisfiability checking in modal and description logic see (Baader and Tobies 2001).

15.2 Tableaux for $\mathbf{K}_{\mathcal{ALC}}$ with constant domains

In this section we construct a tableau-based decision procedure for $\mathbf{K}_{\mathcal{ALC}}$ containing neither global nor modalized roles and interpreted in models with constant domains. The algorithm runs in NEXPTIME and thus matches the lower bound established in Theorem 14.14. The result is due to Lutz *et al.* (2002).

We know from the preceding section what tableaux for $\mathcal{ALC}$ look like. Having recalled from Section 2.5 the connection between modal and description logics, we can easily construct a tableau system for $\mathbf{K}$. So at first sight it should not be a problem to design a tableau algorithm for $\mathbf{K}_{\mathcal{ALC}}$. Indeed, given an $\mathcal{ML}_{\mathcal{ALC}}$-formula ϑ, we can first apply to it the tableau rules for $\mathcal{ALC}$, thus constructing an $\mathcal{ALC}$-model for the nonmodal part of ϑ in the initial world w_0. Then we apply the rules of a tableau system for $\mathbf{K}$ to the modalized concepts and formulas in this model and thereby introduce a number of new worlds w_i 'populated' by *the same objects* as w_0. After that we use the $\mathcal{ALC}$-rules in the w_i and possibly *extend* their domains. And so forth. However, this straightforward approach, first proposed and investigated in (Baader and Laux 1995), works perfectly well only if $\mathcal{ML}_{\mathcal{ALC}}$ is interpreted in models with *expanding domains.* In the case of models with constant domains, after expanding the domain of w_i with a new object a, we have to add a to the domain of w_0 which, in turn, may force us to expand this domain—and so the domains of the w_i as well—with some new objects, and so on. As we shall see a little later, the resulting algorithm does not terminate.

The main technical contribution of this section is that it shows how the quasimodel technique can be used to solve this problem and to design a machinery for constructing tableaux with constant domains. The fundamental idea is that the tableau algorithm constructs not a model itself but its representation in the form of a quasimodel, the worlds in which are 'populated' by (partial) *types* of objects rather than real objects.

The section is organized in the following way. First we discuss in more detail some difficulties in designing tableau procedures for modalized description logics under the constant domain assumption and give an overview of the tableau algorithm developed later in this section. In the next subsection we define constraint systems for $\mathcal{ML}_{\mathcal{ALC}}$-formulas and then show how to encode $\mathbf{K}_{\mathcal{ALC}}$-models in the form of quasimodels. Finally, the tableau decision algorithm is presented and analyzed.

All $\mathcal{ML}_{\mathcal{ALC}}$-formulas we deal with in this chapter contain neither global nor modalized roles.

Tableau algorithms and constant domains

To begin with, we generalize some basic definitions of the previous section from $\mathcal{ALC}$ to $\mathcal{ML}_{\mathcal{ALC}}$. Say that an $\mathcal{ML}_{\mathcal{ALC}}$-formula φ is *equivalent* to an $\mathcal{ML}_{\mathcal{ALC}}$-formula ψ when $(\mathfrak{M}, w) \models \varphi$ iff $(\mathfrak{M}, w) \models \psi$, for every model $\mathfrak{M} = \langle \mathfrak{F}, I \rangle$ and every world w in it. Similarly, an $\mathcal{ML}_{\mathcal{ALC}}$-concept C is *equivalent* to an $\mathcal{ML}_{\mathcal{ALC}}$-concept D if $C^{I(w)} = D^{I(w)}$ for all models $\mathfrak{M}$ and their worlds w. Without loss of generality we may assume that in every atomic formula of the form $C = D$ the concept D is $\top$. The negation normal form (NNF) is defined in precisely the same manner as for $\mathcal{ALC}$. Again, we will assume that all formulas and concepts are in NNF.

The completion rules for $\mathcal{ALC}$ operate on constraint systems. In the case of $\mathcal{ML}_{\mathcal{ALC}}$, a more complex data structure is required: we need *completion trees* whose edges represent the accessibility relation and whose nodes are labeled with constraint systems representing $\mathcal{ALC}$ models. The tableau algorithm starts with a completion tree consisting of a single node labeled with a constraint system containing only the input formula ϑ (and some additional constraints). Again the completion rules are applied until a clash is obtained or a complete clash-free completion tree is found. Besides the rules introduced in the previous section, we now obviously need rules of the following kind:

- If the label S of a node g in a completion tree $\boldsymbol{T}$ contains the constraint $x : \Diamond C$, then we add to $\boldsymbol{T}$ a new node g' as a successor of g and label it with the constraint system containing $x : C$ and $x : D$, for every $x : \Box D$ in S.

But then we are facing the problem of keeping the domain of the model under construction the same in every world. To illustrate this problem, let us consider the following example from (Baader and Laux 1995). Suppose that we have a completion tree $\boldsymbol{T}$ with one node g labeled with the constraint system

$$\mathcal{L}(g) = \{v : \top,\ (\Diamond \exists R.C) = \top\}.$$

An application of the rule $\mathsf{R}_=$ from Fig. 15.1 yields an additional constraint $v : \Diamond \exists R.C$. By applying the rule above, we construct a new node g' in $\boldsymbol{T}$ with the label $\mathcal{L}(g') = \{v : \exists R.C\}$, which is then extended to

$$\mathcal{L}(g') = \{v : \exists R.C,\ vRv',\ v' : C\}$$

by an application of $\mathsf{R}_\exists$ from Fig. 15.1. Since we assume constant domains and since the variables represent domain objects, the presence of v' in $\mathcal{L}(g')$ forces us to add v' to $\mathcal{L}(g)$. This can be done by extending $\mathcal{L}(g)$ with the constraint $v' : \top$, which triggers the rule $\mathsf{R}_=$ again: now it adds $v' : \Diamond \exists R.C$ to $\mathcal{L}(g)$. This constraint and the rule above give a new node g'' with label $\mathcal{L}(g'') = \{v' : \exists R.C\}$ which is then extended by the rule $\mathsf{R}_\exists$ with $v'Rv''$ and

$v'' : C$. Thus we obtain a new variable v'' which has to be added to both $\mathcal{L}(g)$ and $\mathcal{L}(g')$. As these steps are to be repeated infinitely many times, the algorithm does not terminate.

What can we do to prevent the introduction of more and more variables? The key idea is that similar to types in the quasimodels introduced in Section 5.2 the variables in tableaux can represent *partial types* of domain objects rather than domain objects themselves. Dealing only with types, we construct not a model satisfying the input formula, but its representation in the form of a quasimodel. We illustrate this idea by the following example. Suppose that $\boldsymbol{T}$ is a completion tree consisting of a single node g labeled with

$$\mathcal{L}(g) = \{\Box D = \top,\ v : \Diamond\exists R.C,\ v : \Box\neg C,\ v : \Box D\},$$

An application of the rule above generates a successor g' of g with the label $\{v : \exists R.C,\ v : \neg C,\ v : D\}$, which is then extended by $\mathsf{R}_\exists$ to

$$\mathcal{L}(g') = \{v : \exists R.C,\ vRv',\ v : \neg C,\ v : D,\ v' : C\}.$$

The constructed completion tree represents models with the set $W = \{w, w'\}$ of worlds such that

- $\lhd = \{\langle w, w'\rangle\}$,
- in the interpretation $I(w)$, there are domain objects 'of type v,' and
- in the interpretation $I(w')$, there are domain objects of type v and of type v'.

Let us now see how the algorithm copes with constant domains. Fix a model described by the completion tree and let d be an object in $I(w')$ of type v'. As we make the constant domain assumption, d is also an element of the domain of $I(w)$. However, in $I(w)$ this element cannot be of type v because otherwise d would satisfy $\neg C$ in $I(w')$ which is impossible, since it also satisfies C. A straightforward approach to attack this problem would be to introduce a new type into $\mathcal{L}(g)$ (thus overruling blocking). But then again we would face the problem of termination. Lutz *et al.* (2002) take a different way: the solution is to generate a set of *minimal partial types* in *each* constraint system $\mathcal{L}(g)$ so that every domain object in the corresponding $\mathcal{ALC}$-interpretation $I(w)$ is of exactly one of the types in the set. To this end we distinguish between two kinds of variables. A variable may be *marked* in a constraint system, which indicates that it represents a minimal (partial) type, or it may be *unmarked*, which means that the variable represents an 'ordinary' type. We illustrate the difference between marked and unmarked variables as well as the role of minimal partial types by reconsidering the example above.

According to the minimal type strategy, we have to introduce into $\mathcal{L}(g)$ a marked variable v_m together with the constraint $v_m : \top$ *before* generating

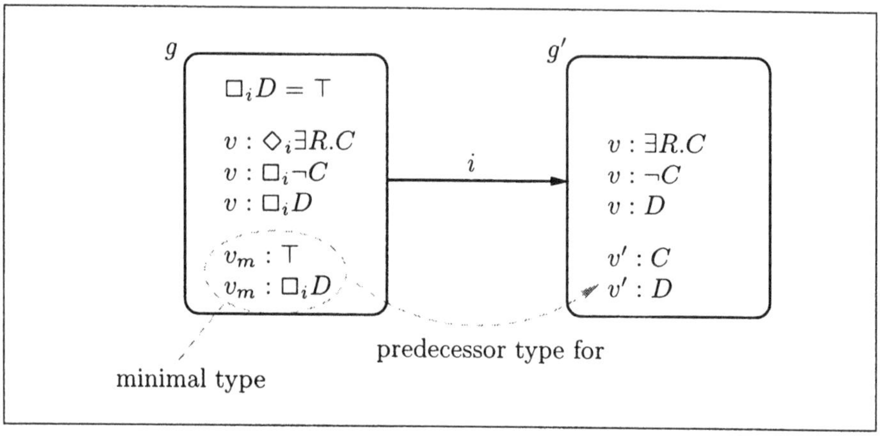

Figure 15.3: The fully expanded completion tree.

the node g'. In nearly all completion rules marked variables are treated like unmarked ones. An application of the first rule adds $v_m : \Box D$ to $\mathcal{L}(g)$. This constraint means that *every* domain object in the $\mathcal{ALC}$-interpretation $I(w)$ is in $(\Box D)^{I(w)}$. After that we construct the node g' and the variable v' as above. In models described by the resulting completion tree, domain objects may be of types v and v' in $I(w')$ and of types v and v_m in $I(w)$. Again, we face the problem of finding a 'predecessor type' for v', i.e., a type for objects in $I(w)$ which are of type v' in $I(w')$. According to the minimal type strategy, we must choose this predecessor among the marked variables in $\mathcal{L}(g)$; in our case this can only be v_m. However, since the constraint $v_m : \Box_i D$ is in $\mathcal{L}(g)$ and v_m was chosen as the predecessor type for v', we must add $v' : D$ to $\mathcal{L}(g')$. Figure 15.3 shows the resulting completion tree. Note that using the minimal type strategy, there is no need to reconsider constraint systems that have already been treated, which helps to avoid the termination problem.

To conclude this subsection, we give a brief overview of how the set of minimal types is generated. Consider a completion tree consisting of a node g labeled with

$$\mathcal{L}(g) = \{A = \top,\ B \sqcup C = \top,\ v : C\}.$$

Again we start by introducing a single marked variable v_m together with the constraint $v_m : \top$. Applications of the rule $\mathsf{R}_=$ from Fig. 15.1 above add both $v_m : A$ and $v_m : B \sqcup C$. According to the rule for $\sqcup$, we must now decide where to put v_m: to B or to C. However, it may be the case that neither of these two choices is the correct one: that all domain objects in interpretations corresponding to $\mathcal{L}(g)$ satisfy $B \sqcup C$ does not imply that all of them satisfy B or that all of them satisfy C. So for marked variables, disjunction must be treated in a special way. Namely, first we introduce a new marked variable

v_m' which is a 'copy' of v_m, i.e., we have $v_m' : A$ and $v_m' : B \sqcup C$ in $\mathcal{L}(g)$. And then we add constraints $v_m : B$ and $v_m' : C$ saying that each object is either of type v_m, and so belongs to B, or of type v_m', and so belongs to C. To be more precise, we need a nondeterministic rule. In one case, we explore both disjuncts as has just been described; in the two additional cases, we explore only one of the disjuncts (which is necessary to deal with disjuncts that lead to a contradiction). Similar modifications are required for all nondeterministic rules dealing with marked variables.

Constraint systems

Given a $\mathcal{ML}_{\mathcal{ALC}}$-formula ϑ, define the sets $ob\,\vartheta$, $con\,\vartheta$, $sub\,\vartheta$, $rol\,\vartheta$ and $Fg\,\vartheta$ in precisely the same manner as in the previous section. Terms are again variables or object names.

A *constraint* for ϑ is either a formula in $sub\,\vartheta$ or an atom of the form xRy or $x : C$, where $R \in rol\,\vartheta$, $C \in con\,\vartheta$ and x, y are terms for ϑ. A *constraint system* for ϑ is a finite set S of constraints for ϑ such that

1. each variable occurring in S is either *marked* or *unmarked*;

2. $a : \top$ is in S for every $a \in ob\,\vartheta$;

3. S contains at least one atom of the form $x : C$.

We assume again that the set of variables is well-ordered by $\ll$ and use the same notion of *blocking* as before. The *completion rules* of the tableau algorithm are divided into two classes:

- *local rules* operate exclusively on constraint systems, while
- *global rules* operate on completion trees; they involve more than one constraint system.

The local rules are the rules in Fig. 15.1 of the previous section, where the formulas and constants now range over $\mathcal{ML}_{\mathcal{ALC}}$ and

- the $\mathcal{ALC}$-rules on formulas are now called *local rules on formulas*,
- the $\mathcal{ALC}$-generating rules are now called *local generating rules*; they introduce *unmarked* variables v,
- the $\mathcal{ALC}$-nongenerating rules are now called *local nongenerating rules on concepts* and the rule $\mathsf{R}_\sqcup$ is replaced with the two rules shown in Fig. 15.4, where the operation '+' is defined as follows:

Let S be a constraint system and Φ a set of concepts. Then

$\mathsf{R}_\sqcup$	If $(x : C \sqcup D) \in S$ for unmarked x and $\{x : C, x : D\} \cap S = \emptyset$, then set $S := S \cup \{x : E\}$, where $E = C$ or $E = D$.
$\mathsf{R}_{\sqcup'}$	If $(v : C \sqcup D) \in S$ for a marked v and $\{v : C, v : D\} \cap S = \emptyset$, then either (i) set $S := S \cup \{v : E\}$, where $E = C$ or $E = D$, or (ii) set $S := (S \cup \{v : C\}) + (\{D\} \cup \{E \mid (v : E) \in S\})$.

Figure 15.4: Local rules for $\sqcup$.

- $S + \Phi$ is S if S contains a marked variable v for which
$$\Phi = \{E \mid (v : E) \in S\};$$
- $S + \Phi$ is $S \cup \{(v : E) \mid E \in \Phi\}$ otherwise, where v is $\ll$-minimal fresh for S and marked in $S + \Phi$.

Note that the rule $\mathsf{R}_\sqcup$ is as before: it takes care of unmarked variables. The need for the rule $\mathsf{R}_{\sqcup'}$ operating with marked variables was explained above.

Quasimodels

In this subsection we show how $\mathbf{K}_{\mathcal{ALC}}$-models can be represented in the form of quasimodels. As in Sections 11.7 and 12.2, here we characterize quasimodels syntactically. Quasimodels of this sort were first introduced in (Sturm and Wolter 2002).

Let ϑ be a $\mathcal{ML}_{\mathcal{ALC}}$-formula. A *quasistate* for ϑ is a complete clash-free constraint system for ϑ all variables in which are unmarked.

A frame $\mathfrak{F} = \langle W, \lhd \rangle$ whose worlds are (labeled with) quasistates for ϑ will be called a *ϑ-frame*. More precisely, a ϑ-frame is a triple $\mathfrak{F} = \langle W, \lhd, \sigma \rangle$, where $W \neq \emptyset$, $\lhd \subseteq W \times W$ and σ is a map from W into the set of quasistates for ϑ.

Let $\mathfrak{F} = \langle W, \lhd, \sigma \rangle$ be a ϑ-frame. A *run* r in $\mathfrak{F}$ is a function associating with every $w \in W$ a term $r(w)$ occurring in the quasistate $\sigma(w)$ in such a way that

- if $(r(w) : \Diamond C) \in \sigma(w)$ then there exists a $w' \in W$ such that $w \lhd w'$ and $(r(w') : C) \in \sigma(w')$;
- if $(r(w) : \Box C) \in \sigma(w)$ and $w \lhd w'$, then $(r(w') : C) \in \sigma(w')$.

A ϑ-frame $\mathfrak{F} = \langle W, \lhd, \sigma \rangle$ is called a *quasimodel* for ϑ if the following conditions hold:

1. for every object name $a \in ob\,\vartheta$, the function r_a defined by $r_a(w) = a$, for $w \in W$, is a run in $\mathfrak{F}$;

2. for every $w \in W$ and every variable v in $\sigma(w)$, there exists a run r in $\mathfrak{F}$ such that $r(w) = v$;

3. for every $w \in W$ and every $\Diamond\varphi \in \sigma(w)$, there exists a $w' \in W$ such that $w \lhd w'$ and $\varphi \in \sigma(w')$;

4. for every $w \in W$ and every $\Box\varphi \in \sigma(w)$, whenever $w \lhd w'$ then $\varphi \in \sigma(w')$.

We say that ϑ is *quasisatisfiable* if there is a quasimodel $\mathfrak{F} = \langle W, \lhd, \sigma\rangle$ for ϑ such that $\vartheta \in \sigma(w)$ for some $w \in W$.

Theorem 15.6. *An $\mathcal{ML}_{\mathcal{ALC}}$-formula ϑ in NNF is satisfiable iff it is quasisatisfiable.*

Proof. ($\Rightarrow$) Suppose ϑ is satisfiable. Then there is a model $\mathfrak{M} = \langle\langle W, \lhd\rangle, I\rangle$ such that $(\mathfrak{M}, w_\vartheta) \models \vartheta$ for some $w_\vartheta \in W$. Let Δ be the domain of $\mathfrak{M}$. For all $w \in W$ and $d \in \Delta$ we then put

$$\tau^{I(w)}(d) = \{C \in Fg\,\vartheta \mid d \in C^{I(w)}\}.$$

Let

$$T_w = \{\tau^{I(w)}(d) \mid d \in \Delta\}.$$

For each $t = \tau^{I(w)}(d)$, take an individual variable v_t and define a constraint system $\sigma(w)$ as the union of the following sets:

$$\{\varphi \in sub\,\vartheta \mid (\mathfrak{M}, w) \models \varphi\},$$
$$\{a : C \mid a \in ob\,\vartheta,\ C \in Fg\,\vartheta,\ a^{I(w)} \in C^{I(w)}\},$$
$$\{v_t : C \mid C \in t\} \text{ for } t \in T_w,$$
$$\{v_{t_1} R v_{t_2} \mid \exists d_1, d_2\ (t_1 = \tau^{I(w)}(d_1)\ \&\ t_2 = \tau^{I(w)}(d_2)\ \&\ d_1 R^{I(w)} d_2)\},$$
$$\{aRv_t \mid a \in ob\,\vartheta \text{ and } \exists d\ (t = \tau^{I(w)}(d)\ \&\ a^{I(w)} R^{I(w)} d)\}.$$

All variables are unmarked in $\sigma(w)$. We show that $\mathfrak{F} = \langle W, \lhd, \sigma\rangle$ is a quasimodel for ϑ.

It should be clear that the $\sigma(w)$ are quasistates for ϑ and that $\mathfrak{F}$ satisfies conditions (1), (3) and (4) in the definition of quasimodels. Let us check (2). Suppose that $w \in W$ and v_t is a variable from $\sigma(w)$. Take a $d \in \Delta$ such that $\tau^{I(w)}(d) = t$ and define a function r with domain W by putting $r(u) = v_{\tau^{I(u)}(d)}$, for each $u \in W$. It is easy to see that r is a run in $\mathfrak{F}$ coming through v_t. That ϑ is quasisatisfiable follows from $\vartheta \in \sigma(w_\vartheta)$.

($\Leftarrow$) Suppose that ϑ is quasisatisfied in a world $w_\vartheta \in W$ of a quasimodel $\langle W, \lhd, \sigma\rangle$, i.e., $\vartheta \in \sigma(w_\vartheta)$.

Define $\mathfrak{M} = \langle\langle W, \lhd\rangle, I\rangle$ with $I(w) = \left\langle \Delta, R_0^{I(w)}, \dots, C_0^{I(w)}, \dots, a_0^{I(w)}, \dots \right\rangle$ as follows:

- Δ is the set of all runs in $\langle W, \lhd, \sigma \rangle$;
- $a^{I(w)} = r_a$, for all $a \in ob\,\vartheta$;
- $C_i^{I(w)} = \{r \in \Delta \mid (r(w) : C_i) \in \sigma(w)\}$, for all concept names C_i in $Fg\,\vartheta$;
- for every pair $r_1, r_2 \in \Delta$ and every role name R, we have $r_1 R^{I(w)} r_2$ iff $r_1(w) R r_2(w) \in \sigma(w)$ or $z R r_2(w) \in \sigma(w)$ for some z which blocks $r_1(w)$ in $\sigma(w)$.

We are about to show that ϑ is satisfied in $\mathfrak{M}$.

Claim 15.7. *For all $w \in W$, $C \in Fg\,\vartheta$, and $r \in \Delta$, if $(r(w) : C) \in \sigma(w)$ then $r \in C^{I(w)}$.*

Proof. The proof is by induction on the construction of C. All steps save $C = \Diamond D$ and $C = \Box D$ can be proved in the same manner as in the proof of Claim 15.2.

Suppose $C = \Diamond D$ and $(r(w) : \Diamond D) \in \sigma(w)$. By the first clause in the definition of runs, there is $w' \in W$ such that $w \lhd w'$ and $(r(w') : D) \in \sigma(w')$. So, by the induction hypothesis, $r \in D^{I(w')}$, from which $r \in (\Diamond D)^{I(w)}$.

Suppose $C = \Box D$. By the second clause in the definition of runs, we then have $(r(w') : D) \in \sigma(w')$, and so $r \in D^{I(w')}$, for all $w' \in W$ such that $w \lhd w'$. It follows that $r \in (\Box D)^{I(w)}$. ❑

Claim 15.8. *For every $w \in W$ and every $\varphi \in sub\,\vartheta$, if $\varphi \in \sigma(w)$ then $(\mathfrak{M}, w) \models \varphi$.*

Proof. This claim is also proved by induction. Let $\varphi \in \sigma(w)$ be atomic. Consider three cases. First, suppose $\varphi = (a : C)$. By the first clause in the definition of quasimodels, we have $(r_a(w) : C) \in \sigma(w)$. Hence, by Claim 15.7, $r_a \in C^{I(w)}$. Recall that $a^{I(w)}$ was defined as r_a. So $(\mathfrak{M}, w) \models a : C$. Second, assume $\varphi = (C = \top)$. Let $r \in \Delta$. As $\sigma(w)$ is closed under $\mathsf{R}_=$, we then have $(r(w) : C) \in \sigma(w)$. It follows from Claim 15.7 that $r \in C^{I(w)}$. Finally, for $\varphi = aRb$ the claim follows immediately from the definition of $R^{I(w)}$.

Next, let $\varphi = \neg\psi$ for atomic ψ. Since ϑ is in NNF, ψ has the form $(C = \top)$ or aRb. We consider only the former case. As $\sigma(w)$ is closed under $\mathsf{R}_{\neq}$, we have $(x : {\sim}C) \in \sigma(w)$ for some x. By the second clause in the definition of quasimodels, there exists a run r such that $r(w) = x$. Moreover, it follows from Claim 15.7 that $r \in ({\sim}C)^{I(w)}$. So there is a $d \in \Delta$ such that $d \in ({\sim}C)^{I(w)}$, from which $(\mathfrak{M}, w) \models \neg(C = \top)$.

The induction step is straightforward (it is based on (3) and (4) in the definition of quasimodels; see also the proof of Theorem 15.1). ❑

It follows from Claim 15.8 that $(\mathfrak{M}, w_\vartheta) \models \vartheta$. ❑

The algorithm

We are now in a position to define completion trees, the global completion rules and the tableau algorithm itself. Fix a countably infinite set N of *nodes.*

A *completion tree* for a $\mathbf{K}_{\mathcal{ALC}}$-formula ϑ is a tree $\boldsymbol{T}$ whose nodes $g \in N$ are labeled with constraint systems $\mathcal{L}(g)$ for ϑ. If there is an edge (g, g') in $\boldsymbol{T}$, then we say that g' is a *successor* of g in $\boldsymbol{T}$.

The global completion rules operate on completion trees. To introduce the rules we require the following definitions. Given a constraint system S, define an equivalence relation $\sim_S$ on the set of variables (not *terms*) occurring in S by taking

$$v \sim_S v' \quad \text{iff} \quad \{C \mid (v : \Box C) \in S\} = \{C \mid (v' : \Box C) \in S\}.$$

Denote by $[v]_S$ the equivalence class (with respect to $\sim_S$) generated by a variable v, by $\min(X)$ the $\ll$-minimal member of a set X of variables, and put

$$S_{\sim} = \bigcup_{v \text{ occurs in } S} \{\min([v]_S)\}.$$

The global completion rules intended for constructing a completion tree for a formula ϑ are shown in Figs 15.5 and 15.6. Note that there we have two versions of the $\mathsf{R}_{\downarrow}$ rule: one for unmarked variables and one for marked ones. This can be explained analogously to the double $\mathsf{R}_{\sqcup}$ rule above. Indeed, the two versions of the rule are needed, since $\mathsf{R}_{\downarrow}$ is nondeterministic. As for $R_{\sqcup}$, it is not sufficient to explore each nondeterministic choice separately, but, additionally, we must explore all possible combinations of nondeterministic choices simultaneously. The interested reader may check, for example, that the satisfiable formula

$$(\top = \Box\Box C \sqcup \Box\Box D) \wedge (a : \Diamond\Diamond(\exists R.C \sqcap \exists R.\neg C))$$

would be judged unsatisfiable if the $R_{\downarrow}$ rule is used for marked variables instead of the $R_{\downarrow'}$ rule.

We say that a completion tree $\boldsymbol{T}$ contains a *clash* if there exists a node g in $\boldsymbol{T}$ such that $\mathcal{L}(g)$ contains a clash; otherwise $\boldsymbol{T}$ is called *clash-free.* $\boldsymbol{T}$ is said to be *complete* if no completion rule is applicable to $\boldsymbol{T}$.

To decide whether a given formula ϑ in NNF is satisfiable, we form the initial completion tree $\boldsymbol{T}_{\vartheta}$ consisting of a single node g_0 labeled with the initial constraint system

$$S_{\vartheta} = \{\vartheta\} \cup \{a : \top \mid a \in ob\,\vartheta\} \cup \{v : \top\},$$

where v is $\ll$-minimal and marked. After that we repeatedly apply both local and global completion rules with the following priority: the rules $\mathsf{R}_{\Diamond_f}$ and

$\mathsf{R}_{\Diamond f}$ If $\Diamond\varphi \in \mathcal{L}(g)$ and $\varphi \notin \mathcal{L}(g')$, for all successors g' of g, then construct a new successor g' of g and set $\mathcal{L}(g')$ to the union of the following sets:

$$\begin{array}{ll} \{\varphi\} & \{\psi \mid \Box\psi \in \mathcal{L}(g)\} \\ \{a : \top \mid a \in ob\,\vartheta\} & \{v : \top\} \\ \{a : C \mid (a : \Box C) \in \mathcal{L}(g)\} & \bigcup_{u \in (\mathcal{L}(g))_\sim} \{u : C \mid (u : \Box C) \in \mathcal{L}(g)\} \end{array}$$

where v is the only marked variable in $\mathcal{L}(g')$ and $v \notin (\mathcal{L}(g))_\sim$.

$\mathsf{R}_{\Diamond c}$ If $(x : \Diamond C) \in \mathcal{L}(g)$ and for all successors g' of g and terms y,

$$\{C\} \cup \{E \mid (x : \Box E) \in \mathcal{L}(g)\} \not\subseteq \{E \mid (y : E) \in \mathcal{L}(g')\}$$

then construct a new successor g' of g and set $\mathcal{L}(g')$ to the union of the following sets:

$$\begin{array}{ll} \{v' : C\} & \{\psi \mid \Box\psi \in \mathcal{L}(g)\} \\ \{v' : D \mid (x : \Box D) \in \mathcal{L}(g)\} & \\ \{a : \top \mid a \in ob\,\vartheta\} & \{v : \top\} \\ \{a : C \mid (a : \Box C) \in \mathcal{L}(g)\} & \bigcup_{u \in (\mathcal{L}(g))_\sim} \{u : C \mid (u : \Box C) \in \mathcal{L}(g)\} \end{array}$$

where v is the only marked variable in $\mathcal{L}(g')$, $v \neq v'$, and $v, v' \notin (\mathcal{L}(g))_\sim$.

Figure 15.5: Global generating rules for $\mathbf{K}_{\mathcal{ALC}}$.

$\mathsf{R}_{\Diamond c}$ are applied only if no other rule is applicable, and the local generating rules are applied only if no rule different from $\mathsf{R}_{\Diamond f}$ and $\mathsf{R}_{\Diamond c}$ is applicable; see Fig. 15.7.

Theorem 15.9 (soundness). *If there is a complete clash-free completion tree for a $\mathbf{K}_{\mathcal{ALC}}$-formula ϑ, then ϑ is satisfiable.*

Proof. Let $\boldsymbol{T}$ be a complete clash-free completion tree for ϑ. By Theorem 15.6, it is sufficient to show that ϑ is quasisatisfiable. Define a structure $\mathfrak{F} = \langle W, \lhd, \sigma\rangle$ by taking

- W to be the set of nodes in $\boldsymbol{T}$,
- $w \lhd w'$ iff w' is a successor of w in $\boldsymbol{T}$,
- $\sigma(w) = \mathsf{unmark}(\mathcal{L}(w))$,

where $\mathsf{unmark}(\mathcal{L}(w))$ is the constraint system obtained by 'unmarking' the marked variables in $\mathcal{L}(w)$. Obviously, $\mathfrak{F}$ is a ϑ-frame. We now show that $\mathfrak{F}$ has the following properties:

$\mathsf{R}_{\downarrow}$ If g' is a successor of g, v an unmarked variable in $\mathcal{L}(g')$, and for no term x in $\mathcal{L}(g)$ do we have

$$\{C \mid (x : \Box C) \in \mathcal{L}(g)\} \subseteq \{\, C \mid (v : C) \in \mathcal{L}(g')\},$$

then nondeterministically choose a marked variable v' in $\mathcal{L}(g)$ and set $\mathcal{L}(g') := \mathcal{L}(g') \cup \{v : C \mid (v' : \Box C) \in \mathcal{L}(g)\}$.

$\mathsf{R}_{\downarrow'}$ If g' is a successor of g, v a marked variable in $\mathcal{L}(g')$, for no term x in $\mathcal{L}(g)$ do we have

$$\{C \mid (x : \Box C) \in \mathcal{L}(g)\} \subseteq \{\, C \mid (v : C) \in \mathcal{L}(g')\},$$

and X is the set of marked variables occurring in $\mathcal{L}(g)$
then nondeterministically choose
a nonempty subset $Y = \{v_1, \dots, v_k\}$ of X
and set $S_1 := \mathcal{L}(g') \cup \{v : D \mid (v_1 : \Box D) \in \mathcal{L}(g)\}$,
$S_j := S_{j-1} + (\{D \mid (v_j : \Box D) \in \mathcal{L}(g)\} \cup \{E \mid (v : E) \in \mathcal{L}(g')\})$,
for all $1 < j \leq k$, and $\mathcal{L}(g') := S_k$.

Figure 15.6: Global nongenerating rules for $\mathbf{K}_{\mathcal{ALC}}$.

(i) if $(\Diamond\varphi) \in \sigma(w)$ for some $w \in W$, then there exists a $w' \in W$ such that $w \lhd w'$ and $\varphi \in \sigma(w')$;

(ii) if $(a : \Diamond C) \in \sigma(w)$ for some $w \in W$ and $a \in ob\,\vartheta$, then there exists a $w' \in W$ such that $w \lhd w'$ and $(a : C) \in \sigma(w')$;

(iii) if $(v : \Diamond C) \in \sigma(w)$ for some $w \in W$ and $\Psi = \{E \mid (v : \Box E) \in \sigma(w)\}$, then there exist a world $w' \in W$ and a term x such that $w \lhd w'$ and

$$\Psi \cup \{C\} \subseteq \{E \mid (x : E) \in \sigma(w')\};$$

(iv) if $w \lhd w'$ then:

(a) $(\Box\varphi) \in \sigma(w)$ implies $\varphi \in \sigma(w')$,

(b) $\{E \mid (a : \Box E) \in \sigma(w)\} \subseteq \{E \mid (a : E) \in \sigma(w')\}$ for all $a \in ob\,\vartheta$,

(c) for each variable v in $\sigma(w)$, there exists a term x in $\sigma(w')$ such that $\{E \mid (v : \Box E) \in \sigma(w)\} \subseteq \{E \mid (x : E) \in \sigma(w')\}$,

(d) for each variable v in $\sigma(w')$, there exists a term x in $\sigma(w)$ such that $\{E \mid (x : \Box E) \in \sigma(w)\} \subseteq \{E \mid (v : E) \in \sigma(w')\}$.

Conditions (i)–(iii) are satisfied simply because the rules $\mathsf{R}_{\Diamond_f}$ and $\mathsf{R}_{\Diamond_c}$ are not applicable to $\boldsymbol{T}$ in view of its completeness. Let $\Box\varphi \in \mathcal{L}(w)$ and $w \lhd w'$. Then

define procedure $sat(\boldsymbol{T})$
  if $\boldsymbol{T}$ contains a clash **then**
    return *unsatisfiable*
  if a rule $r \notin \{\mathsf{R}_\exists, \mathsf{R}_{\neq}, \mathsf{R}_{\Diamond f}, \mathsf{R}_{\Diamond c}\}$ is applicable to $\boldsymbol{T}$
    then apply r to $\boldsymbol{T}$
    return $sat(\boldsymbol{T})$
  if a rule $r \in \{\mathsf{R}_\exists, \mathsf{R}_{\neq}\}$ is applicable to $\boldsymbol{T}$
    then apply r to $\boldsymbol{T}$
    return $sat(\boldsymbol{T})$
  if a rule $r \in \{\mathsf{R}_{\Diamond f}, \mathsf{R}_{\Diamond c}\}$ is applicable to $\boldsymbol{T}$
    then apply r to $\boldsymbol{T}$
    return $sat(\boldsymbol{T})$
  return *satisfiable*

Figure 15.7: The satisfiability-checking algorithm for $\mathbf{K}_{\mathcal{ALC}}$.

w' has been generated by an application of a global generating rule (either $\mathsf{R}_{\Diamond f}$ or $\mathsf{R}_{\Diamond c}$). As these rules are applied only when no other rule is applicable, $\Box\varphi$ was already in $\mathcal{L}(w)$ by the moment of the application of that rule, and so $\varphi \in \mathcal{L}(w')$. This proves (iv.a). Conditions (iv.b) and (iv.c) are proved analogously, and (iv.d) follows from the fact that the rules $\mathsf{R}_\downarrow$ and $\mathsf{R}_{\downarrow'}$ are not applicable to $\boldsymbol{T}$.

Conditions (i)–(iv) do not mean that $\mathfrak{F}$ is a quasimodel. However, it is not difficult to modify $\mathfrak{F}$ in such a way that the resulting structure is a quasimodel. Namely, we can just introduce sufficiently many 'copies' of worlds and convert the ϑ-frame $\mathfrak{F}$ into a structure $\mathfrak{F}' = \langle W', \lhd', \sigma' \rangle$ which additionally has the following property:

(v) if $(v : \Diamond C) \in \sigma'(w)$, for some $w \in W'$, and $k < \omega$, then there exist pairwise distinct $w_1, \ldots, w_k \in W'$ such that $w \lhd' w_j$ for $1 \le j \le k$, and there are terms $x_1, \ldots, x_k$ for which

$$\{E \mid (v : \Box E) \in \sigma'(w)\} \cup \{C\} \subseteq \{E \mid (x_j : E) \in \sigma'(w_j)\}.$$

We now show that $\mathfrak{F}'$ is a quasimodel for ϑ. Conditions 1, 3 and 4 in the definition of quasimodels follow immediately from (ii), (i) and (iv). Let us prove condition 2 claiming that, for every variable v in every $\sigma'(w_0)$, $w_0 \in W'$, there is a run r coming through v. We construct r by induction. To begin with, we put $r(w_0) = v$. Now two cases are possible.

Case $\downarrow$: Suppose that $r(w')$ has already been defined and $w \lhd' w'$ with undefined $r(w)$. By (iv.d), there is a term x such that

$$\{E \mid (x : \Box E) \in \sigma'(w)\} \subseteq \{E \mid (r(w') : E) \in \sigma'(w')\}.$$

Then we put $r(w) = x$. We proceed with Case $\downarrow$ till (in finitely many steps) we reach the root of $\mathfrak{F}'$. After that we switch to

Case $\uparrow$: Suppose that $r(w)$ has already been defined, but there is $w' \rhd w$ with undefined $r(w')$. Let $\Diamond C_1, \dots, \Diamond C_l$ be all distinct concepts in $Fg\,\vartheta$ of the form $\Diamond C$ such that $(r(w) : \Diamond C_j) \in \sigma(w)$, $1 \le j \le l$. For every such $\Diamond C$ we choose by (v) a world w_j with undefined $r(w_j)$ and a term x_j such that

$$\{E \mid (v : \Box E) \in \sigma'(w)\} \cup \{C_j\} \subseteq \{E \mid (x_j : E) \in \sigma'(w_j)\}$$

and $w_j \neq w_i$ whenever $j \neq i$. Put $r(w_j) = x_j$. If we still have a world $w' \rhd w$ with undefined $r(w')$, then we use condition (iv.c), according to which there is a term x such that

$$\{E \mid (r(w) : \Box E) \in \sigma'(w)\} \subseteq \{E \mid (x : E) \in \sigma'(w')\}.$$

Then we set $r(w') = x$.

It should be clear from the definition that r is a run in $\mathfrak{F}'$ coming through v in $\sigma'(w)$. Thus $\mathfrak{F}'$ is a quasimodel for ϑ. That ϑ is satisfied in $\mathfrak{F}'$ follows from $\vartheta \in \sigma'(w_\vartheta)$. ❑

Theorem 15.10 (termination). *Having started on the initial completion tree $\boldsymbol{T}_\vartheta$, the (nondeterministic) completion algorithm terminates after at most $2^{p(|Fg\,\vartheta|}$ steps, where p is a polynomial function.*

Proof. Recall that the *depth* of a tree is the number of edges in its longest branch; the *outdegree* of the tree is the maximal number of immediate successors of nodes in it.

Claim 15.11. *Let g be a node in $\boldsymbol{T}$. Then the number of constraints of the form $x : C$ in $\mathcal{L}(g)$ does not exceed $2^{p_1(|Fg\,\vartheta|)}$, where p_1 is a polynomial function.*

Proof. We determine an upper bound for the number of distinct terms per node label. By the definition of completion trees and constraint systems, all object names occurring in node labels are from $ob\,\vartheta$. So the number of distinct object names in a label does not exceed $|Fg\,\vartheta|$.

At the moment of its generation, the node g (its label, to be more precise) contains not more than $2^{|Fg\,\vartheta|}$ distinct unmarked variables and a single marked one. Consider now the rules that can introduce new variables in $\mathcal{L}(g)$. First, the marked variables. They are introduced by the $\mathsf{R}_{\sqcup'}$ and $\mathsf{R}_{\downarrow'}$ rules. Define a tree T whose nodes are the marked variables in $\mathcal{L}(g)$ and whose edges are labeled with either $\mathsf{R}_{\sqcup'}$ or $\mathsf{R}_{\downarrow'}$ as follows:

- the root node is the initial marked variable in $\mathcal{L}(g)$;

- if a completion rule $r \in \{\mathsf{R}_{\sqcup'}, \mathsf{R}_{\downarrow'}\}$ is applied to a marked variable v generating new marked variables $v_1, \ldots, v_k$, then v_i is a successor of v in T and the edge between v and v_i is labeled with r for $1 \leq i \leq k$.

Using the definition of $\mathsf{R}_{\sqcup'}$, $\mathsf{R}_{\downarrow'}$ and $Fg\,\vartheta$, it is not hard to see that the depth of T is bounded by $|Fg\,\vartheta|$. Moreover, each node has at most $|Fg\,\vartheta| + 2^{|Fg\,\vartheta|}$ successors: at most $|Fg\,\vartheta|$ outgoing edges labeled with $\mathsf{R}_{\sqcup'}$ and at most $2^{|Fg\,\vartheta|}$ outgoing edges labeled with $\mathsf{R}_{\downarrow'}$. Hence, the number of nodes in the tree is bounded by $2^{4|Fg\,\vartheta|^2}$, which is therefore the maximum number of marked variables in $\mathcal{L}(g)$. Now, for the unmarked variables: $\mathsf{R}_{\neq}$ can add to $\mathcal{L}(g)$ at most $|Fg\,\vartheta|$ new variables. There are at most $2^{|Fg\,\vartheta|}$ unblocked variables and so the number of terms to which $\mathsf{R}_{\exists}$ can be applied is at most $|ob\,\vartheta| + 2^{|Fg\,\vartheta|}$. So the number of marked variables does not exceed

$$|Fg\,\vartheta| + (|ob\,\vartheta| + 2^{|Fg\,\vartheta|}) \cdot (|Fg\,\vartheta| + 1).$$

Claim 15.11 follows immediately. ❑

Claim 15.12. *The depth of $\boldsymbol{T}$ is bounded by $|Fg\,\vartheta|$ and the outdegree of $\boldsymbol{T}$ does not exceed $2^{d|Fg\,\vartheta|}$, where d is a constant.*

Proof. If g' is a successor of g in $\boldsymbol{T}$, then clearly

$$\max\{md(C) \mid (x : C) \in \mathcal{L}(g)\} > \max\{md(C) \mid (x : C) \in \mathcal{L}(g')\}$$

and

$$\max\{md(\varphi) \mid \varphi \in \mathcal{L}(g)\} > \max\{md(\varphi) \mid \varphi \in \mathcal{L}(g')\},$$

where $md(C)$ and $md(\varphi)$ denote the modal depth of a concept C and a formula φ defined in Section 3.8. So the depth of $\boldsymbol{T}$ is at most $md(\vartheta)$. Now we compute the outdegree. Let g be a node in $\boldsymbol{T}$. Each successor of g in $\boldsymbol{T}$ is generated by an application of the $\mathsf{R}_{\Diamond_f}$ rule to some formula $\Diamond\varphi$ or by an application of the $\mathsf{R}_{\Diamond_c}$ rule to some constraint $x : \Diamond C$ in $\mathcal{L}(g)$. The number of applications of the $\mathsf{R}_{\Diamond_f}$ rule is obviously bounded by the number of distinct formulas in $\mathcal{L}(g)$, i.e., by $|Fg\,\vartheta|$. Moreover, by the definition of the $\mathsf{R}_{\Diamond_c}$ rule, the number of applications of this rule is bounded by $2^{|Fg\,\vartheta|}$ (i.e., the number of distinct subsets of concepts in $Fg\,\vartheta$). ❑

We are now ready to prove the theorem. By Claim 15.12, there is a constant e such that the number of nodes in each completion tree constructed by the algorithm is at most $2^{|Fg\,\vartheta|^e}$. As every global generating rule adds a new node, the number of applications of such rules is bounded by the same number. Now let us compute the number of applications of local rules on formulas. Each local rule on formulas introduces a new formula to a node label. Hence there may be at most $|Fg\,\vartheta|$ applications of rules of this type per node. So the total number of applications of local rules on formulas is bounded by $2^{|Fg\,\vartheta|^e} \cdot |Fg\,\vartheta|$.

Finally, each of the local nongenerating rules on concepts, local generating rules and global nongenerating rules adds a new constraint of the form $x : C$ to a constraint system $\mathcal{L}(g)$. By Claim 15.11, the number of such constraints per node is $\leq 2^{p_1(|Fg\,\vartheta|)}$ for some polynomial function p_1. Thus, the number of applications of these rules per node is $\leq 2^{p_1(|Fg\,\vartheta|)}$. The total number of such rule applications is then bounded by $2^{(|Fg\,\vartheta|^e} \cdot 2^{p_1(|Fg\,\vartheta|)}$. ❑

Theorem 15.13 (completeness). *If a $\mathbf{K}_{\mathcal{ALC}}$-formula ϑ is satisfiable then, having started from $\boldsymbol{T}_\vartheta$, the satisfiability-checking algorithm for $\mathbf{K}_{\mathcal{ALC}}$ constructs a complete clash-free completion tree for ϑ.*

Proof. Consider a model $\mathfrak{M} = \langle\langle W, \lhd\rangle, I\rangle$ and a world $w_\vartheta \in W$ such that $w_\vartheta \models \vartheta$. We use $\mathfrak{M}$ as a 'guide' for applications of the nondeterministic rules to construct a complete clash-free completion tree for ϑ.

Say that a completion tree $\boldsymbol{T}$ for ϑ is $\mathfrak{M}$-*compatible* if the following holds:

1. there is a map π from the set of nodes in $\boldsymbol{T}$ to W such that
   - if g' is a successor of g in $\boldsymbol{T}$, then $\pi(g) \lhd \pi(g')$ and
   - if $\varphi \in \mathcal{L}(g)$ then $\pi(g) \models \varphi$, for every $\varphi \in sub\,\vartheta$;
2. for each node g in $\boldsymbol{T}$, there is a total surjective function τ_g from Δ to the set of marked variables in $\mathcal{L}(g)$ such that if $(v : C) \in \mathcal{L}(g)$ and $\tau_g(d) = v$ then $d \in C^{I(\pi(g))}$, and
3. for each node g in $\boldsymbol{T}$, there is a total function π_g from the set of unmarked terms in $\mathcal{L}(g)$ to Δ such that if $(x : C) \in \mathcal{L}(g)$ then $\pi_g(x) \in C^{I(\pi(g))}$.

Claim 15.14. *If a completion tree $\boldsymbol{T}$ for ϑ is $\mathfrak{M}$-compatible and $\boldsymbol{T}'$ is the result of an application of a rule* R *to $\boldsymbol{T}$, then $\boldsymbol{T}'$ is $\mathfrak{M}$-compatible as well.*

Proof. Let $\boldsymbol{T}$ be an $\mathfrak{M}$-compatible completion tree, g a node in $\boldsymbol{T}$ and let π, τ_g and π_g be the functions supplied by the definition of $\mathfrak{M}$-compatibility. Consider all possible cases for R.

Suppose that the $\mathsf{R}_\wedge$ rule is applicable to a formula $\varphi \wedge \psi$ in $\mathcal{L}(g)$. Since $\boldsymbol{T}$ is $\mathfrak{M}$-compatible, $\pi(g) \models \varphi \wedge \psi$. The application of $\mathsf{R}_\wedge$ to $\mathcal{L}(g)$ adds φ and ψ to $\mathcal{L}(g)$. Then the very same functions π, τ_g and π_g ensure that the resulting completion tree $\boldsymbol{T}'$ is $\mathfrak{M}$-compatible.

Suppose that the $\mathsf{R}_\vee$ rule is applicable to a formula $\varphi \vee \psi$ in $\mathcal{L}(g)$. Then, as we know, $\pi(g) \models \varphi \vee \psi$, and so either φ or ψ is in $\sigma(\pi(g))$. By applying the $\mathsf{R}_\vee$ rule to $\mathcal{L}(g)$ accordingly, we clearly obtain an $\mathfrak{M}$-compatible completion tree.

Suppose that the $R_\sqcap$ rule is applicable to a constraint $x : C \sqcap D$ in $\mathcal{L}(g)$. Let $d \in \Delta$ be such that either $\pi_g(x) = d$ (x is unmarked in $\mathcal{L}(g)$) or $\tau_g(d) = x$ (x is marked in $\mathcal{L}(g)$). In both cases we have $d \in (C \sqcap D)^{I(\pi(g))}$. So $d \in C^{I(\pi(g))}$ and $d \in C^{I(\pi(g))}$. An application of the $R_\sqcap$ rule adds $x : C$ and $x : D$ to $\mathcal{L}(g)$. Hence, the functions π, τ_g and π_g are as required for the resulting completion tree $\boldsymbol{T}'$.

Suppose that the $R_\sqcup$ rule is applicable to a constraint $x : C \sqcup D$ in $\mathcal{L}(g)$. Then x is unmarked in $\mathcal{L}(g)$. Clearly, either $\pi_g(x) \in C^{I(\pi(g))}$ or $\pi_g(x) \in D^{I(\pi(g))}$. By applying the $R_\sqcup$ rule to $\mathcal{L}(g)$ accordingly, we see that the functions π, π_g and τ_g are as required.

Suppose that the $R_{\sqcup'}$ rule is applicable to $v : C \sqcup D$ in $\mathcal{L}(g)$. Then v is marked in $\mathcal{L}(g)$. Let Y be the set of d in Δ for which $\tau_g(d) = v$. Since τ_g is surjective, Y is nonempty. Clearly, $d \in C^{I(\pi(g))}$ or $d \in D^{I(\pi(g))}$ for any $d \in Y$. Put

$$Y_C = \{d \in Y \mid d \in C^{I(\pi(g))}\},$$
$$Y_D = Y - Y_C.$$

An application of $R_{\sqcup'}$ adds either (i) $v : C$ or (ii) $v : D$ to $\mathcal{L}(g)$, or (iii) it creates 'a marked copy' v' of v, for which, additionally, $(v' : D) \in \mathcal{L}(g)$ holds, and then adds $v : C$ to $\mathcal{L}(g)$. If $Y_C = \emptyset$, apply the rule in such a way that $v : D$ is added. If $Y_D = \emptyset$, apply the rule so that $v : C$ is added. Otherwise we apply the rule in the third possible way. In the first two cases, π, π_g and τ_g are as required for the resulting completion tree $\boldsymbol{T}'$. In the third case, define

$$\tau'_g(d) = \begin{cases} v', & \text{if } d \in Y_D \\ \tau_g(d), & \text{otherwise} \end{cases}$$

and $\tau'_h = \tau_h$ for all $h \neq g$. The functions π, π_g and τ'_g ensure that $\boldsymbol{T}'$ is $\mathfrak{M}$-compatible.

Suppose that the $R_=$ rule is applicable to a formula $C = \top$ and a term x in $\mathcal{L}(g)$. Then $\pi(g) \models C = \top$ and we find $d \in \Delta$ such that either $\pi_g(x) = d$ (x is unmarked in $\mathcal{L}(g)$) or $\tau_g(d) = x$ (x is marked in $\mathcal{L}(g)$). We have $d \in C^{I(\pi(g))}$. Hence, after an application of $R_=$ (which adds $x : C$ to $\mathcal{L}(g)$), the functions π, τ_g and π_g will be as required for the resulting completion tree $\boldsymbol{T}'$.

Suppose that $R_{\neq}$ is applicable to $\neg(C = \top)$ in $\mathcal{L}(g)$. Then there is $d \in \Delta$ with $d \in (\sim C)^{I((\pi(g))}$. By applying $R_{\neq}$ to $\mathcal{L}(g)$, we introduce a new (unmarked) variable x. Define π'_g as the extension of π_g to x with $\pi'_g(x) = d$, and put $\pi'_h = \pi_h$ for all $h \neq g$. The functions π, τ_g and π'_g are then as required for the resulting completion tree $\boldsymbol{T}'$.

Let $R_\exists$ be applicable to $x : \exists R.C$ in $\mathcal{L}(g)$. Let $d \in \Delta$ be such that either $\pi_g(x) = d$ (x is unmarked in $\mathcal{L}(g)$) or $\tau_g(d) = x$ (x is marked in $\mathcal{L}(g)$). In both cases we have $d \in (\exists R.C)^{I(\pi(g))}$ and can proceed as in the $R_{\neq}$ case (note that the newly generated variable is unmarked in any case).

Now we come to the global rules and suppose that $\mathsf{R}_{\Diamond f}$ is applicable to $\Diamond\varphi$ in $\mathcal{L}(g)$. Let $\pi(g) = w$. Then $\Diamond\varphi \in \sigma(w)$. The rule application generates a successor g' of g. We find $w' \in W$ such that $w \lhd w'$ and $w' \models \varphi$. Set $\pi(g') = w'$. It remains to define $\pi_{g'}$ and $\tau_{g'}$. The terms occurring in $\mathcal{L}(g')$ are the object names in $ob\,\vartheta$, one marked variable v, and a set of unmarked variables $v_1, \dots, v_k$. Put

1. $\pi_{g'}(a) = a^I$ for every $a \in ob\,\vartheta$,
2. $\pi_{g'}(v_j) \in \{d \mid \{E \mid (v_j : E) \in \mathcal{L}(g')\} \subseteq \{E \mid d \in E^{I(w')}\}\}$ for $1 \le j \le k$,
3. $\tau_{g'}(d) = v$ for every $d \in \Delta$.

The function $\pi_{g'}$ is well-defined for all unmarked variables $v_1, \dots, v_k$ in $\mathcal{L}(g)$. Indeed, fix a $j \in \{1, \dots, k\}$. By the definition of the $\mathsf{R}_{\Diamond f}$ rule, there is a variable v such that

$$\{E \mid (v : \Box E) \in \mathcal{L}(g)\} = \{E \mid (v_j : E) \in \mathcal{L}(g')\}.$$

By the definition of $\mathfrak{M}$-compatibility, it follows that there is a term $d \in \Delta$ such that

$$\{E \mid (v_j : E) \in \mathcal{L}(g')\} \subseteq \{E \mid d \in E^{I(w)}\}.$$

It is easy to see that the defined functions π, τ_g and π_g are as required. The case of $\mathsf{R}_{\Diamond c}$ is considered analogously.

Suppose that $\mathsf{R}_{\downarrow}$ is applicable to a variable v in a $\mathcal{L}(g')$ and $\pi(g') = w'$. Then v is unmarked in $\mathcal{L}(g')$ and there is a node g such that g' is a successor of g in $\boldsymbol{T}$. Let $\pi_{g'}(v) = d$ and $\pi(g) = w$. By the definition of $\mathfrak{M}$-compatibility, we have $w \lhd w'$ The rule application nondeterministically chooses a marked variable v' in $\mathcal{L}(g)$ and augments $\mathcal{L}(g')$ with $\{v : D \mid (v' : \Box D) \in \mathcal{L}(g)\}$. Take $v' = \tau_g(d)$. The functions π, τ_g and π_g are as required for the resulting completion tree $\boldsymbol{T}'$.

Finally, assume that the $\mathsf{R}_{\downarrow'}$ rule is applicable to a marked variable v_1 in $\mathcal{L}(g')$ and that $\pi(g') = w'$. Then there is a node g such that g' is a successor of g. Let X be the set of $d \in \Delta$ for which $\tau_{g'}(d) = v_1$ and let $\pi(g) = w$. As $\tau_{g'}$ is surjective, $X \neq \emptyset$. The rule application chooses a nonempty subset Y of the marked variables in $\mathcal{L}(g)$. Let

$$Y = \{v' \mid \exists d \in X\ \tau_g(d) = v'\}.$$

Since τ_g is total, Y is nonempty. Let $v_1', \dots, v_k'$ be all its elements. The application of $\mathsf{R}_{\downarrow'}$ does the following:

- it generates $k-1$ 'marked copies' $v_2, \dots, v_k$ of v_1 and then
- augments $\mathcal{L}(g')$ with $\{v_j : D \mid (v_j' : \Box D) \in \mathcal{L}(g)\}$ for $1 \le j \le k$.

Put

$$\tau'_g(d) = \begin{cases} v_j, & \text{if } d \in X \text{ and } \tau_g(d) = v'_j \\ \tau_g(d), & \text{otherwise} \end{cases}$$

and $\tau'_h = \tau_h$ if $h \neq g$. It is obvious that the functions π and π_g are as required for the resulting completion tree $\boldsymbol{T}'$ (note that $\pi_g(v_j)$ is undefined for $1 \leq j \leq k$). We show that τ'_g is also as required. Assume that the rule application added a constraint $(v_j : D)$ to $\mathcal{L}(g')$ and fix a term $d \in \Delta$ such that $\tau_{g'}(d) = v_j$. Then $(v'_j : \Box D) \in \mathcal{L}(g)$. By the definition of X and $\tau_{g'}$, we have $\tau_g(d) = v'_j$, which yields $d \in (\Box D)^{I(w)}$. Then $d \in D^{I(w')}$. ❑

Now, returning to the proof of the completeness theorem, we show that it follows from the claim above. Let $\boldsymbol{T}_\vartheta$ be the initial completion tree for ϑ, g the node in $\boldsymbol{T}_\vartheta$, and v the marked variable in $\mathcal{L}(g)$. Set $\pi(g) = w_\vartheta$ (recall that we have $\vartheta \in w_\vartheta$), $\tau_g(d) = v$ for all $d \in \Delta$, and $\pi_g(a) = a^I(w)$ for all $a \in ob\,\vartheta$. It is readily checked that these functions ensure that $\boldsymbol{T}_\vartheta$ is $\mathfrak{M}$-compatible.

By the claim above, the completion rules can be applied in such a way that the resulting completion trees are $\mathfrak{M}$-compatible. According to Theorem 15.10, we then eventually construct a complete $\mathfrak{M}$-compatible completion tree $\boldsymbol{T}$. Obviously, $\boldsymbol{T}$ is clash-free. ❑

Theorem 15.10 states that the nondeterministic tableau algorithm terminates after exponentially many steps (in the length of the input formula). Together with the soundness and completeness theorems this provides us with a NEXPTIME satisfiability-checking procedure for $\mathbf{K}_{\mathcal{ALC}}$-formulas, which matches the lower bound of Theorem 14.14. Thus we have:

Theorem 15.15. *The satisfiability problem for $\mathcal{ML}_{\mathcal{ALC}}$-formulas without global role names and modalized roles is* NEXPTIME-*complete.*

15.3 Adding expressive power to $\mathbf{K}_{\mathcal{ALC}}$

It is not difficult to extend the tableau procedure for $\mathbf{K}_{\mathcal{ALC}}$ (without global and modalized roles) introduced above in both the modal and the description logic dimensions. For example, Sturm and Wolter (2002) and Lutz *et al.* (2001) present tableau systems for the temporal description logic $\mathbf{PTL}_{\mathcal{ALC}}$. In this section we show tableaux for the two following extensions. First, we add to $\mathcal{ALC}$ the universal role U, thus obtaining $\mathcal{ALCU}$, and extend the tableaux of the preceding section to $\mathbf{K}_{\mathcal{ALCU}}$ (again without global and modalized roles). Second, we construct a tableau algorithm deciding the global concept satisfiability problem for $\mathbf{K}_{\mathcal{ALCU}}$ without global and modalized roles, which is more difficult than the formula satisfiability problem (see Section 3.8).

$\mathsf{R}_{\forall U}$	If $\{x : \forall U.C\} \subseteq S$, y occurs in S, but $(y : C) \notin S$, then set $S := S \cup \{y : C\}$.
$\mathsf{R}_{\exists U}$	If $(x : \exists U.C) \in S$, and there is no term y in S such that $y : C \in S$, then choose the $\ll$-minimal fresh variable v for S and set $S := S \cup \{v : C\}$.

Figure 15.8: Additional local rules for $\mathbf{K}_{\mathcal{ALCU}}$.

define procedure $sat(\boldsymbol{T})$
  if $\boldsymbol{T}$ contains a clash **then**
    return *unsatisfiable*
  if a rule $r \notin \{\mathsf{R}_\exists, \mathsf{R}_{\neq}, \mathsf{R}_{\Diamond f}, \mathsf{R}_{\Diamond c}, \mathsf{R}_{\exists U}\}$ is applicable to $\boldsymbol{T}$
    then apply r to $\boldsymbol{T}$
    return $sat(\boldsymbol{T})$
  if a rule $r \in \{\mathsf{R}_\exists, \mathsf{R}_{\neq}, \mathsf{R}_{\exists U}\}$ is applicable to $\boldsymbol{T}$
    then apply r to $\boldsymbol{T}$
    return $sat(\boldsymbol{T})$
  if a rule $r \in \{\mathsf{R}_{\Diamond f}, \mathsf{R}_{\Diamond c}\}$ is applicable to $\boldsymbol{T}$
    then apply r to $\boldsymbol{T}$
    return $sat(\boldsymbol{T})$
  return *satisfiable*

Figure 15.9: The satisfiability-checking algorithm for $\mathbf{K}_{\mathcal{ALCU}}$.

Adding the universal role

Denote by $\mathcal{ALCU}$ the description logic obtained by adding to $\mathcal{ALC}$ the *universal role* U. Models of $\mathcal{ALCU}$ are $\mathcal{ALC}$-models

$$I = \langle \Delta, U^I, S_0^I, \dots, C_0^I, \dots, a_0^I, \dots \rangle$$

in which $U^I = \Delta \times \Delta$.

A tableau procedure for $\mathbf{K}_{\mathcal{ALCU}}$ is obtained by extending the tableau system for $\mathbf{K}_{\mathcal{ALC}}$ with the two local rules shown in Fig. 15.8. The rule $\mathsf{R}_{\forall U}$ is nongenerating and has the same priority as the 'old' local nongenerating rules. The rule $\mathsf{R}_{\exists U}$ is local generating and has the same priority as the local generating rules $\mathsf{R}_\exists$ and $\mathsf{R}_{\neq}$. The tableau algorithm for $\mathbf{K}_{\mathcal{ALCU}}$ is presented in Fig. 15.9.

Soundness, termination and completeness can be proved in precisely the same manner as in the previous section. The algorithm runs in NEXPTIME

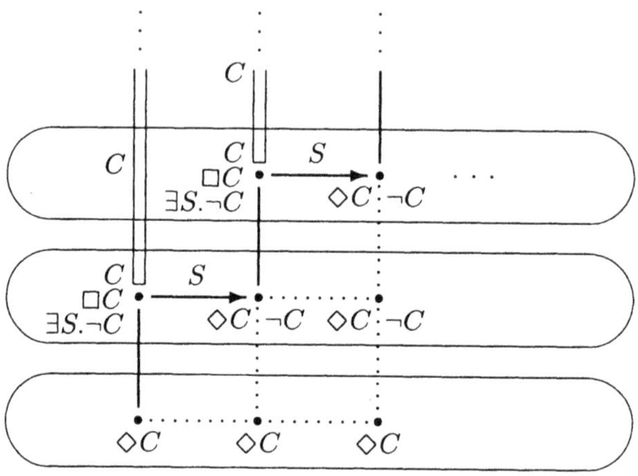

Figure 15.10: Satisfying $\top$ relative to Σ.

as well. Thus, in view of Theorem 15.15, we obtain

Theorem 15.16. *The satisfiability problem for $\mathcal{ML}_{\mathcal{ALCU}}$-formulas containing neither global role names nor modalized roles is* NEXPTIME*-complete.*

Global concept satisfiability

Let us fix a finite set Σ of $\mathcal{ML}_{\mathcal{ALCU}}$-formulas and an $\mathcal{ML}_{\mathcal{ALCU}}$-concept F. Remember that F is called *globally satisfiable relative to* Σ if there exists a model $\mathfrak{M} = \langle \mathfrak{F}, I \rangle$ such that $F^{I(v)} \neq \emptyset$, for some world v in $\mathfrak{F}$, and $w \models \varphi$ for all worlds w and all $\varphi \in \Sigma$.

The global concept satisfiability problem for $\mathbf{K}_{\mathcal{ALCU}}$ is obviously more difficult than formula satisfiability. For example, $\top$ is satisfiable relative to

$$\Sigma = \{C \sqsubseteq \Box C, \quad \exists S.\neg C = \top, \quad \Diamond C = \top\}$$

only in infinite models (see Fig. 15.10), whereas every satisfiable $\mathcal{ML}_{\mathcal{ALCU}}$-formula is satisfied in a finite model.

As was shown in the proof of Theorem 15.6, every finite quasimodel can be transformed into a finite model. So finite quasimodels are not enough to characterize global concept satisfiability. The tableau algorithm we present below constructs what one might call finite 'quasi-quasimodels' from which (possibly infinite) quasimodels can be obtained by a sort of unraveling (see Section 1.4). The related new ingredient of the tableaux is that—to ensure termination—we have to apply a blocking strategy also in the modal dimension.

To implement the global blocking strategy, we now assume that the set N of nodes of completion trees is well-ordered by some relation $\ll_m$. Roughly, a node g *blocks* another node g' if $g \ll_m g'$ and $\mathcal{L}(g)$ coincides with $\mathcal{L}(g')$ modulo renaming of variables. More precisely, let S and S' be constraint systems. S' is called a *variant* of S if there is a bijective function π from the variables occurring in S onto the variables occurring in S' which respects markedness (i.e., unmarked variables are mapped to unmarked variables and marked variables to marked variables) and S' is obtained from S by replacing each variable v in S with $\pi(v)$. A node g is said to be *blocked by a node* g' in $\boldsymbol{T}$ if $g' \ll_m g$ and $\mathcal{L}(g')$ is a variant of $\mathcal{L}(g)$.

Denote by $Fg(\Sigma, F)$ the union of all $Fg\,\vartheta$ for $\vartheta \in \Sigma \cup \{\neg(F = \bot)\}$. To decide whether F is globally satisfiable relative to Σ, we form the initial tree $\boldsymbol{T}_{\Sigma,F}$ consisting of the $\ll_m$-minimal node g_0 labeled with the initial constraint system

$$S_{\Sigma,F} = \Sigma \cup \{\neg(F = \bot)\} \cup \{a : \top \mid a \in ob\,\Sigma\} \cup \{v : \top\},$$

where v is the $\ll$-minimal variable and marked. The tableau algorithm applied to $S_{\Sigma,F}$ is obtained by replacing the rules $\mathsf{R}_{\Diamond f}$ and $\mathsf{R}_{\Diamond c}$ in Fig. 15.9 with $\mathsf{R}^{\Sigma}_{\Diamond f}$ and $\mathsf{R}^{\Sigma}_{\Diamond c}$ in Fig. 15.11, respectively (we use the notation of the previous section).

Note that a new node is added as a successor of a node g only if g is not blocked. Another difference from the previously introduced rules is that each formula from Σ is automatically put into each $\mathcal{L}(g)$ because all of them has to be true in every world of the model constructed using the completion tree.

The completeness and termination of this algorithm are proved similarly to Theorems 15.10 and 15.13; we leave this to the reader as an exercise. Note only that the proof of Claim 15.11 still goes through. As far as Claim 15.12 is concerned, it is no longer true that the depth of $\boldsymbol{T}$ is bounded by $|Fg(\Sigma, F)|$. However, the blocking strategy ensures that there does not exist a branch in $\boldsymbol{T}$ containing distinct nodes g and g' such that $\mathcal{L}(g)$ is a variant of $\mathcal{L}(g')$. Hence, the depth of $\boldsymbol{T}$ does not exceed

$$2^{|Fg(\Sigma,F)| + 2^{p_1(|Fg(\Sigma,F)|)}},$$

where p_1 is the polynomial function from Claim 15.11. The remaining steps are again the same as in the proof of Theorem 15.10.

Theorem 15.17 (soundness). *If there is a clash-free completion of* $\boldsymbol{T}_{\Sigma,F}$, *then* F *is globally satisfiable relative to* Σ.

$\mathsf{R}^{\Sigma}_{\Diamond f}$ If $\Diamond\varphi \in \mathcal{L}(g)$, g is not blocked in $\boldsymbol{T}$, and $\varphi \notin \mathcal{L}(g')$, for all successors g' of g, then take the $\ll_m$-minimal fresh g' from N as a new successor of g and set $\mathcal{L}(g')$ to the union of the following sets:

$$\begin{array}{ll} \{\varphi\} \cup \Sigma & \{\psi \mid \Box\psi \in \mathcal{L}(g)\} \\ \{a : \top \mid a \in ob\,\vartheta\} & \{v : \top\} \\ \{a : C \mid (a : \Box C) \in \mathcal{L}(g)\} & \bigcup\limits_{u \in (\mathcal{L}(g))_\sim} \{u : C \mid (u : \Box C) \in \mathcal{L}(g)\} \end{array}$$

where v is the only marked variable in $\mathcal{L}(g')$ and $v \notin (\mathcal{L}(g))_\sim$.

$\mathsf{R}^{\Sigma}_{\Diamond c}$ If $(x : \Diamond C) \in \mathcal{L}(g)$, g is not blocked in $\boldsymbol{T}$, and for all successors g' of g and all terms y,

$$\{C\} \cup \{E \mid (x : \Box E) \in \mathcal{L}(g)\} \not\subseteq \{E \mid (y : E) \in \mathcal{L}(g')\}$$

then take the $\ll_m$-minimal fresh g' from N as a new successor of g and set $\mathcal{L}(g')$ to the union of the following sets:

$$\begin{array}{ll} \{v' : C\} \cup \Sigma & \{\psi \mid \Box\psi \in \mathcal{L}(g)\} \\ \{v' : D \mid (x : \Box D) \in \mathcal{L}(g)\} & \\ \{a : \top \mid a \in ob\,\vartheta\} & \{v : \top\} \\ \{a : C \mid (a : \Box C) \in \mathcal{L}(g)\} & \bigcup\limits_{u \in (\mathcal{L}(g))_\sim} \{u : C \mid (u : \Box C) \in \mathcal{L}(g)\} \end{array}$$

where v is the only marked variable in $\mathcal{L}(g')$, $v \neq v'$ and $v, v' \notin (\mathcal{L}(g))_\sim$.

Figure 15.11: Σ-global generating rules with blocking.

Proof. Let $\boldsymbol{T}$ be a clash-free completion of $\boldsymbol{T}_{\Sigma,F}$ with root g_0. By the definition of the Σ-global generating rules with blocking, $\Sigma \subseteq \mathcal{L}(g)$ for all nodes g in $\boldsymbol{T}$. Besides, $\neg(F = \bot) \in \mathcal{L}(g_0)$. Now, the proof of Theorem 15.9 shows that it is sufficient to build a structure $\mathfrak{F} = \langle W, \lhd, \sigma\rangle$ which is

- a $(\bigwedge \Sigma \wedge \neg(F = \bot))$-frame and
- satisfies conditions (i)–(iv) listed in the proof of Theorem 15.9.

Define $\mathfrak{F}$ by taking

- W to be the set of finite sequences $\langle g_0, g_1, \ldots, g_n\rangle$ of nodes in $\boldsymbol{T}$ such that (a) g_{i+1} is a successor of g_i in $\boldsymbol{T}$ or (b) g_i is blocked by a node with successor g_{i+1} in $\boldsymbol{T}$;

- $w_1 \lhd w_2$ iff $w_1, w_2 \in W$ and $w_2 = w_1 * g$ for some node g (where $*$ is the operation of concatenation);

- $\sigma(w) = \mathsf{unmark}(\mathcal{L}(g))$, where g is the final node of w and $\mathsf{unmark}(\mathcal{L}(g))$ is the constraint system obtained by 'unmarking' the marked variables in $\mathcal{L}(g)$.

Obviously, $\mathfrak{F}$ is a $(\bigwedge \Sigma \wedge \neg(F = \bot))$-frame. So it remains to show that $\mathfrak{F}$ satisfies (i)–(iv). Conditions (i)–(iii) are satisfied because the rules $\mathsf{R}^{\Sigma}_{\Diamond f}$ and $\mathsf{R}^{\Sigma}_{\Diamond c}$ are not applicable to $\boldsymbol{T}$ in view of its completeness. Let $\Box\varphi \in \sigma(w)$ and $w \lhd w'$. Then, for $w = \langle g_0, \dots, g_n \rangle$ and $w' = w * g$, the node g has been generated by an application of a global generating rule (either $\mathsf{R}^{\Sigma}_{\Diamond f}$ or $\mathsf{R}^{\Sigma}_{\Diamond c}$) to g_n or g_n is blocked by a node g'_n such that g has been generated by an application of a global generating rule to g'_n. As these rules are applied only when no other rule is applicable, $\Box\varphi$ was already in $\mathcal{L}(g_n)$ (respectively $\mathcal{L}(g'_n)$) by the moment of the application of that rule, and so $\varphi \in \mathcal{L}(g)$. This proves (iv.a). Conditions (iv.b) and (iv.c) are proved analogously, and (iv.d) follows from the fact that the rules $\mathsf{R}_{\downarrow}$ and $\mathsf{R}_{\downarrow'}$ are not applicable to $\boldsymbol{T}$. ❑

Chapter 16

Spatio-temporal logics

In this chapter, we analyze the algorithmic properties of the spatio-temporal languages

$$\mathcal{ST}_0 \subseteq \mathcal{ST}_1 \subseteq \mathcal{ST}_2$$

introduced in Section 3.2 and interpreted in various kinds of topological temporal models. These languages are combinations of the fragment $\mathcal{BRCC}$-8 of the region connection calculus with the standard point-based temporal language $\mathcal{ML}_{\mathcal{SU}}$. According to Theorem 3.5, all these languages can be embedded into the propositional spatio-temporal language

$$\mathcal{PST} = \mathcal{ML}_{\mathcal{SU}} \otimes \mathcal{ML}^u.$$

We obtain the following results. First, in Section 16.1, we connect satisfiability of $\mathcal{PST}$-formulas in topological $\mathcal{PST}$-models with satisfiability in Kripke models based on products of linear orders and $\mathbf{S4}_u$-frames. We use this connection to show that the satisfiability problem for the full propositional spatio-temporal language $\mathcal{PST}$ in topological $\mathcal{PST}$-models over discrete flows of time like $\langle \mathbb{N}, < \rangle$ is undecidable, no matter whether we adopt the finite state assumption **FSA** or not.

In Section 16.2, we embed the smaller spatio-temporal languages $\mathcal{ST}_i$ ($i = 0, 1, 2$) into the one-variable fragment of the first-order temporal language $\mathcal{QTL}$, and prove that the satisfiability problem for $\mathcal{ST}_i$-formulas in tt-models over various flows of time (like $\langle \mathbb{N}, < \rangle$, $\langle \mathbb{Q}, < \rangle$, the class of all strict linear orders) is decidable (in the case of $\mathcal{ST}_2$ we assume that models satisfy **FSA**).

In Section 16.3, we analyze the computational complexity of the satisfiability problem for $\mathcal{ST}_i$-formulas ($i = 0, 1, 2$) in tt-models over the flow $\langle \mathbb{N}, < \rangle$. For $\mathcal{ST}_2$-formulas (interpreted in tt-models satisfying **FSA**) and for $\mathcal{ST}_1$-formulas we show EXPSPACE-completeness, while $\mathcal{ST}_0$-formula satisfiability is shown to be PSPACE-complete. We also consider the fragment

$\mathcal{ST}_1^-$ of $\mathcal{ST}_1$, which is based on $\mathcal{RCC}$-8 rather than $\mathcal{BRCC}$-8, and prove that the satisfiability problem for $\mathcal{ST}_1^-$-formulas in tt-models over $\langle \mathbb{N}, < \rangle$ is PSPACE-complete—i.e., considerably less complex than the corresponding $\mathcal{ST}_1$-formula satisfiability problem.

And finally, in Section 16.4, we show that $\mathcal{ST}_2$-formulas can distinguish between tt-models based on arbitrary and Euclidean topological spaces. On the other hand, we prove that over countable discrete flows of time $\mathcal{ST}_1^-$-formula satisfiability in tt-models based on arbitrary topological spaces is equivalent to satisfiability in tt-models based on Euclidean spaces $\langle \mathbb{R}^n, \mathbb{I} \rangle$, $n \geq 1$.

16.1 Modal formalisms for spatio-temporal reasoning

To begin with, we remind the reader (see Section 3.2) that the propositional spatio-temporal language $\mathcal{PST}$ contains the temporal operators $\mathcal{S}$ and $\mathcal{U}$ of $\mathcal{ML}_{\mathcal{SU}}$ as well as the modal operators of $\mathcal{ML}^u$ which are denoted by $\mathbf{I}$ (interior), $\mathbf{C}$ (closure), $\boxtimes$ and $\diamondsuit$ (universal box and diamond). $\mathcal{PST}$-formulas are interpreted in *topological $\mathcal{PST}$-models* which are triples of the form $\mathfrak{M} = \langle \mathfrak{F}, \mathfrak{T}, \mathfrak{U} \rangle$, where $\mathfrak{F} = \langle W, < \rangle$ is a strict linear order, $\mathfrak{T} = \langle U, \mathbb{I} \rangle$ a topological space, and $\mathfrak{U}$ is a valuation associating with every propositional variable p and every $w \in W$ a set $\mathfrak{U}(p, w) \subseteq U$. $\mathfrak{U}$ is extended to arbitrary $\mathcal{PST}$-formulas in a standard way. For example,

- $\mathfrak{U}(\psi \wedge \chi, w) = \mathfrak{U}(\psi, w) \cap \mathfrak{U}(\chi, w)$;

- $\mathfrak{U}(\boxtimes\psi, w) = U$ if $\mathfrak{U}(\psi, w) = U$, and $\mathfrak{U}(\boxtimes\psi, w) = \emptyset$ otherwise;

- $\mathfrak{U}(\mathbf{I}\psi, w) = \mathbb{I}\mathfrak{U}(\psi, w)$;

- $\mathfrak{U}(\psi\mathcal{U}\chi, w) = \bigcup_{v>w} \big(\mathfrak{U}(\chi, v) \cap \bigcap_{u\in(w,v)} \mathfrak{U}(\psi, u)\big)$.

We say that a topological $\mathcal{PST}$-model $\langle \mathfrak{F}, \mathfrak{T}, \mathfrak{U} \rangle$ *satisfies* **FSA** if, for every propositional variable p, the set

$$\{\mathfrak{U}(p, w) \mid w \in W\}$$

is finite. It is easy to show by induction that actually in topological $\mathcal{PST}$-models satisfying **FSA**, for *every* $\mathcal{PST}$-formula ψ, the set $\{\mathfrak{U}(\psi, w) \mid w \in W\}$ is finite.

As we know from Section 3.2, $\mathcal{PST}$-formulas can also be interpreted in usual Kripke models based on the product of a strict linear order $\mathfrak{F}$ and a rooted $\mathbf{S4}_u$-frame[1] $\mathfrak{G}$. Since the $\mathbf{S4}_u$-frame $\mathfrak{G}$ gives rise to a topological space $\mathfrak{T}_{\mathfrak{G}}$ (see Section 2.6), every Kripke model based on $\mathfrak{F} \times \mathfrak{G}$ can be transformed into a topological $\mathcal{PST}$-model of the form $\langle \mathfrak{F}, \mathfrak{T}_{\mathfrak{G}}, \mathfrak{U} \rangle$. As Proposition 3.6 and the $\mathcal{PST}$-formula

$$\Diamond_F \mathbf{C}p \leftrightarrow \mathbf{C}\Diamond_F p \tag{16.1}$$

show, the set of $\mathcal{PST}$-formulas satisfiable in such product models is properly contained in the set of $\mathcal{PST}$-formulas satisfiable in topological $\mathcal{PST}$-models.

Our aim now is twofold. First, we want to identify classes of formulas which do not distinguish between topological $\mathcal{PST}$-models and product Kripke models. And second, we want to show that $\mathcal{PST}$-formulas do not distinguish between topological $\mathcal{PST}$-models and product Kripke models, if these models satisfy **FSA**.

Say that a Kripke model $\mathfrak{M} = \langle \mathfrak{F} \times \mathfrak{G}, \mathfrak{V} \rangle$ *satisfies* **FSA** if for every propositional variable p, the set

$$\big\{\{x \mid \langle w, x \rangle \in \mathfrak{V}(p)\} \mid w \text{ in } \mathfrak{F}\big\}$$

is finite. Again, it is easy to show by induction that if $\mathfrak{M}$ satisfies **FSA** then, for *every* $\mathcal{PST}$-formula ψ, the set $\big\{\{x \mid (\mathfrak{M}, \langle w, x \rangle) \models \psi\} \mid w \text{ in } \mathfrak{F}\big\}$ is finite.

Denote by $\mathcal{PST}_{\bigcirc}$ the sublanguage of $\mathcal{PST}$ in which $\bigcirc$ is the only temporal operator. A $\mathcal{PST}$-formula of the form $\boxdot\psi$ or $\diamondplus\psi$, where ψ is a $\mathcal{PST}_{\bigcirc}$-formula, will be called a *basic u-formula*. And by a *u-formula* we mean a $\mathcal{PST}$-formula constructed from basic u-formulas using arbitrary connectives of $\mathcal{PST}$. It is to be noted that the formula (16.1) is not a u-formula. On the other hand, the translation $\varphi^{\bowtie}$ of an $\mathcal{ST}_1$-formula φ defined in Section 3.2 is a u-formula.

Now we have the following:

Lemma 16.1. (i) *If a $\mathcal{PST}_{\bigcirc}$-formula or a u-formula φ is satisfied in a topological $\mathcal{PST}$-model based on a flow of time $\mathfrak{F}$, then φ is satisfied in a Kripke model based on the product of $\mathfrak{F}$ and a rooted $\mathbf{S4}_u$-frame $\mathfrak{G}$.*

(ii) *If a $\mathcal{PST}$-formula φ is satisfied in a topological $\mathcal{PST}$-model satisfying* **FSA** *and based on a flow of time $\mathfrak{F}$, then φ is satisfied in a Kripke model satisfying* **FSA** *and based on the product of $\mathfrak{F}$ and a rooted $\mathbf{S4}_u$-frame $\mathfrak{G}$.*

Moreover, in both cases we can choose the $\mathbf{S4}_u$-frame $\mathfrak{G} = \langle V, R_{\mathbf{I}}, R_{\forall} \rangle$ and the Kripke model $\mathfrak{M}$ based on $\mathfrak{F} \times \mathfrak{G}$ in such a way that, for all w in $\mathfrak{F}$, x in $\mathfrak{G}$ and ψ, the set

$$A_{w,x,\psi} = \{y \in V \mid xR_{\mathbf{I}}y \text{ and } (\mathfrak{M}, \langle w, y \rangle) \models \psi\}$$

contains an $R_{\mathbf{I}}$-maximal point.[2]

[1] We remind the reader that a *rooted* $\mathbf{S4}_u$-frame is a frame $\mathfrak{G} = \langle V, R_{\mathbf{I}}, R_{\forall} \rangle$, where $\langle V, R_{\mathbf{I}} \rangle$ is a (not necessarily rooted) quasi-order and $R_{\forall}$ is the universal relation on V.

[2] A point z is said to be *$R_{\mathbf{I}}$-maximal in* $A \subseteq V$ if, for every $z' \in A$, we have $z'R_{\mathbf{I}}z$ whenever $zR_{\mathbf{I}}z'$.

Proof. The proof is based on the Stone–Jónsson–Tarski representation of topological Boolean algebras (in particular, topological spaces) in the form of general frames (see Goldblatt 1976 or Chagrov and Zakharyaschev 1997). All the necessary definitions are given below.

(i) Suppose φ is satisfied in a topological $\mathcal{PST}$-model $\mathfrak{N} = \langle \mathfrak{F}, \mathfrak{T}, \mathfrak{U} \rangle$ based on a flow of time $\mathfrak{F} = \langle W, < \rangle$ and a topological space $\mathfrak{T} = \langle U, \mathbb{I} \rangle$. An *ultrafilter* $\boldsymbol{x}$ *over* U is a subset of the powerset 2^U of U such that

- $U \in \boldsymbol{x}$;
- if $A \in \boldsymbol{x}$ and $A \subseteq B$, then $B \in \boldsymbol{x}$;
- $A \cap B \in \boldsymbol{x}$ iff $A, B \in \boldsymbol{x}$, for all $A, B \subseteq U$;
- $A \cup B \in \boldsymbol{x}$ iff either $A \in \boldsymbol{x}$ or $B \in \boldsymbol{x}$, for all $A, B \subseteq U$;
- $A \in \boldsymbol{x}$ iff $U - A \notin \boldsymbol{x}$, for every $A \subseteq U$.

Denote by V the set of all ultrafilters over U. This set is not empty: as is well known, for every set $\boldsymbol{y} \subseteq 2^U$ with the *finite intersection property* (i.e., such that $A_1 \cap \cdots \cap A_k \neq \emptyset$ for any $A_1, \ldots, A_k \in \boldsymbol{y}$ and $k < \omega$) there exists an ultrafilter $\boldsymbol{x} \supseteq \boldsymbol{y}$. (For instance, the set $\{A \subseteq U \mid u \in A\}$ is an ultrafilter, for every $u \in U$.) So if a set $A \subseteq U$ is in *every* ultrafilter over U, then A must be U itself. For any two ultrafilters $\boldsymbol{x}_1, \boldsymbol{x}_2 \in V$, put

$$\boldsymbol{x}_1 R_{\mathbf{I}} \boldsymbol{x}_2 \quad \text{iff} \quad \forall A \subseteq U \ (\mathbb{I}A \in \boldsymbol{x}_1 \rightarrow A \in \boldsymbol{x}_2).$$

It is easy to see that $R_{\mathbf{I}}$ is a quasi-order on V. Let $R_\forall$ be the universal relation on V. Define a Kripke model $\mathfrak{M} = \langle \mathfrak{F} \times \mathfrak{G}, \mathfrak{V} \rangle$ by taking $\mathfrak{G} = \langle V, R_{\mathbf{I}}, R_\forall \rangle$ and

$$\mathfrak{V}(p) = \{\langle w, \boldsymbol{x} \rangle \in W \times V \mid \mathfrak{U}(p, w) \in \boldsymbol{x}\}.$$

We show by induction on the construction of ψ that, for all $\mathcal{PST}_{\mathcal{O}}$- and u-formulas ψ, for all $w \in W$ and $\boldsymbol{x} \in V$,

$$(\mathfrak{M}, \langle w, \boldsymbol{x} \rangle) \models \psi \quad \text{iff} \quad \mathfrak{U}(\psi, w) \in \boldsymbol{x}. \tag{16.2}$$

For propositional variables (16.2) follows from the definition of $\mathfrak{M}$. Let us prove it for $\mathcal{PST}_{\mathcal{O}}$-formulas.

Case $\psi = \psi_1 \wedge \psi_2$. We have: $(\mathfrak{M}, \langle w, \boldsymbol{x} \rangle) \models \psi$ iff $(\mathfrak{M}, \langle w, \boldsymbol{x} \rangle) \models \psi_1$ and $\langle w, \boldsymbol{x} \rangle \models \psi_2$ iff (by IH) $\mathfrak{U}(\psi_1, w) \in \boldsymbol{x}$ and $\mathfrak{U}(\psi_2, w) \in \boldsymbol{x}$ iff (by the definition of ultrafilters) $\mathfrak{U}(\psi_1, w) \cap \mathfrak{U}(\psi_2, w) \in \boldsymbol{x}$ iff $\mathfrak{U}(\psi_1 \wedge \psi_2, w) \in \boldsymbol{x}$.

Case $\psi = \neg \psi'$. In this case, $(\mathfrak{M}, \langle w, \boldsymbol{x} \rangle) \models \psi$ iff $(\mathfrak{M}, \langle w, \boldsymbol{x} \rangle) \not\models \psi'$ iff (by IH) $\mathfrak{U}(\psi', w) \notin \boldsymbol{x}$ iff $U - \mathfrak{U}(\psi', w) \in \boldsymbol{x}$ iff $\mathfrak{U}(\psi, w) \in \boldsymbol{x}$.

Case $\psi = \mathbf{I}\psi'$. Suppose that $(\mathfrak{M}, \langle w, \boldsymbol{x} \rangle) \models \mathbf{I}\psi'$, but $\mathfrak{U}(\mathbf{I}\psi', w) \notin \boldsymbol{x}$. Then $\mathbb{I}\mathfrak{U}(\psi', w) \notin \boldsymbol{x}$ which means that $\mathbb{C}(U - \mathfrak{U}(\psi', w)) \in \boldsymbol{x}$. Observe that the set

$$\boldsymbol{y}_0 = \{U - \mathfrak{U}(\psi', w)\} \cup \{A \subseteq U \mid \mathbb{I}A \in \boldsymbol{x}\}$$

has the finite intersection property. Indeed, otherwise we would have sets $A_1, \dots, A_k \subseteq U$ such that $\mathbb{I}A_i \in \boldsymbol{x}$, for $1 \leq i \leq k$, and

$$(U - \mathfrak{U}(\psi', w)) \cap A_1 \cap \dots \cap A_k = \emptyset.$$

But then, by (2.21) on page 84,

$$\begin{aligned} \emptyset &= \mathbb{C}(U - \mathfrak{U}(\psi', w)) \cap \mathbb{I}(A_1 \cap \dots \cap A_k) \\ &= \mathbb{C}(U - \mathfrak{U}(\psi', w)) \cap \mathbb{I}A_1 \cap \dots \cap \mathbb{I}A_k \in \boldsymbol{x}, \end{aligned}$$

which is impossible, since in this case $U \notin \boldsymbol{x}$. Take an ultrafilter $\boldsymbol{y} \supseteq \boldsymbol{y}_0$. Then $\boldsymbol{x}R_{\mathbf{I}}\boldsymbol{y}$, and hence $(\mathfrak{M}, \langle w, \boldsymbol{y}\rangle) \models \psi'$, i.e., by IH, $\mathfrak{U}(\psi', w) \in \boldsymbol{y}$, contrary to $(U - \mathfrak{U}(\psi', w)) \in \boldsymbol{y}_0 \subseteq \boldsymbol{y}$.

Conversely, suppose $(\mathfrak{M}, \langle w, \boldsymbol{x}\rangle) \not\models \mathbf{I}\psi'$. Then we can find $\boldsymbol{y}$ such that $\boldsymbol{x}R_{\mathbf{I}}\boldsymbol{y}$ and $(\mathfrak{M}, \langle w, \boldsymbol{y}\rangle) \not\models \psi'$. By IH, $\mathfrak{U}(\psi', w) \notin \boldsymbol{y}$, and so, by the definition of $R_{\mathbf{I}}$, $\mathbb{I}\mathfrak{U}(\psi', w) \notin \boldsymbol{x}$, which means that $\mathfrak{U}(\psi, w) \notin \boldsymbol{x}$.

Case $\psi = \boxtimes\psi'$. Suppose that $(\mathfrak{M}, \langle w, \boldsymbol{x}\rangle) \models \boxtimes\psi'$. Then $(\mathfrak{M}, \langle w, \boldsymbol{y}\rangle) \models \psi'$ for all $\boldsymbol{y} \in V$, and so, by IH, $\mathfrak{U}(\psi', w) \in \boldsymbol{y}$ for all $\boldsymbol{y} \in V$. But then

$$\mathfrak{U}(\boxtimes\psi', w) = \mathfrak{U}(\psi', w) = U \in \boldsymbol{x}.$$

Conversely, if $\mathfrak{U}(\boxtimes\psi', w) \in \boldsymbol{x}$ then $\mathfrak{U}(\boxtimes\psi', w) \neq \emptyset$, and so $\mathfrak{U}(\boxtimes\psi', w) = U$. It follows that $\mathfrak{U}(\psi', w) = U$, i.e., $\mathfrak{U}(\psi', w) \in \boldsymbol{y}$ for all $\boldsymbol{y} \in V$, from which, by IH, $(\mathfrak{M}, \langle w, \boldsymbol{x}\rangle) \models \boxtimes\psi'$.

Case $\psi = \bigcirc\psi'$. We have $(\mathfrak{M}, \langle w, \boldsymbol{x}\rangle) \models \bigcirc\psi'$ iff there exists an immediate successor w' of w and $(\mathfrak{M}, \langle w', \boldsymbol{x}\rangle) \models \psi'$ iff, by IH, there is an immediate successor w' of w and $\mathfrak{U}(\psi', w') \in \boldsymbol{x}$. It remains to recall that

$$\mathfrak{U}(\bigcirc\psi', w) = \begin{cases} \mathfrak{U}(\psi', w') & \text{if } w' \text{ is an immediate successor of } w, \\ \emptyset & \text{if } w \text{ has no immediate successor.} \end{cases}$$

So we have proved (16.2) for every $\mathcal{PST}_{\bigcirc}$-formula ψ. In order to show (16.2) for every u-formula, first observe the following properties of u-formulas:

Claim 16.2. *For every u-formula* ψ,

- *for all* $\mathcal{PST}$*-models* $\langle \mathfrak{F}, \langle U, \mathbb{I}\rangle, \mathfrak{U}\rangle$ *and points* w *in* $\mathfrak{F}$, *either* $\mathfrak{U}(\psi, w) = \emptyset$ *or* $\mathfrak{U}(\psi, w) = U$;

- *if* $(\mathfrak{M}, \langle w, x\rangle) \models \psi$ *for some point* $\langle w, x\rangle$ *in a model* $\mathfrak{M}$ *based on a product frame* $\mathfrak{F} \times \mathfrak{G}$, *then* $(\mathfrak{M}, \langle w, y\rangle) \models \psi$ *holds for every* y *in* $\mathfrak{G}$.

The claim can be proved by a straightforward induction on the construction of ψ. Using this claim and (16.2) for $\mathcal{PST}_{\bigcirc}$-formulas, we obtain (16.2) for every u-formula as well.

It follows immediately that φ is satisfied in $\mathfrak{M}$. Indeed, take $w \in W$ such that $\mathfrak{U}(\varphi, w) \neq \emptyset$, and let $\boldsymbol{x}$ be an ultrafilter containing $\mathfrak{U}(\varphi, w)$. Then $(\mathfrak{M}, \langle w, \boldsymbol{x} \rangle) \models \varphi$. Thus, we have proved (i).

(ii) Suppose that a $\mathcal{PST}$-formula φ is satisfied in a topological $\mathcal{PST}$-model $\mathfrak{N} = \langle \mathfrak{F}, \mathfrak{T}, \mathfrak{U} \rangle$ with **FSA** based on a flow of time $\mathfrak{F} = \langle W, < \rangle$ and a topological space $\mathfrak{T} = \langle U, \mathbb{I} \rangle$. The construction of the Kripke model $\mathfrak{M}$ is the same as in (i). Observe that this time $\mathfrak{M}$ satisfies **FSA**. We show by induction that, for every subformula ψ of φ, we have

$$(\mathfrak{M}, \langle w, \boldsymbol{x} \rangle) \models \psi \quad \text{iff} \quad \mathfrak{U}(\psi, w) \in \boldsymbol{x}.$$

The proof is almost the same as above. This time, however, instead of $\bigcirc$ we need induction steps for $\mathcal{U}$ and $\mathcal{S}$.

Case $\psi = \psi_1 \mathcal{U} \psi_2$. Assume that $(\mathfrak{M}, \langle w, \boldsymbol{x} \rangle) \models \psi_1 \mathcal{U} \psi_2$. Then there is $v > w$ such that $(\mathfrak{M}, \langle v, \boldsymbol{x} \rangle) \models \psi_2$ and $(\mathfrak{M}, \langle u, \boldsymbol{x} \rangle) \models \psi_1$ for all u in the interval (w, v). By IH, $\mathfrak{U}(\psi_2, v) \in \boldsymbol{x}$ and $\mathfrak{U}(\psi_1, u) \in \boldsymbol{x}$ for all $u \in (w, v)$. Since

$$\mathfrak{U}(\psi_1 \mathcal{U} \psi_2, w) \supseteq \mathfrak{U}(\psi_2, v) \cap \bigcap_{u \in (w,v)} \mathfrak{U}(\psi_1, u),$$

we shall have $\mathfrak{U}(\psi_1 \mathcal{U} \psi_2, w) \in \boldsymbol{x}$ if we show that

$$\Big(\mathfrak{U}(\psi_2, v) \cap \bigcap_{u \in (w,v)} \mathfrak{U}(\psi_1, u)\Big) \in \boldsymbol{x}. \tag{16.3}$$

In view of **FSA**, we can find time points $u_1, \dots, u_l \in (w, v)$ such that

$$\mathfrak{U}(\psi_1, u_1) \cap \dots \cap \mathfrak{U}(\psi_1, u_l) = \bigcap_{u \in (w,v)} \mathfrak{U}(\psi_1, u),$$

which yields (16.3) because ultrafilters are closed under *finite* intersections.

Conversely, let $\mathfrak{U}(\psi_1 \mathcal{U} \psi_2, w) \in \boldsymbol{x}$. By **FSA**, there are time points $v_1, \dots v_l$ such that

$$\mathfrak{U}(\psi_1 \mathcal{U} \psi_2, w) = \bigcup_{1 \leq i \leq l} \Big(\mathfrak{U}(\psi_2, v_i) \cap \bigcap_{u \in (w, v_i)} \mathfrak{U}(\psi_1, u)\Big).$$

And since $\boldsymbol{x}$ is an ultrafilter, we have

$$\mathfrak{U}(\psi_2, v_i) \cap \bigcap_{u \in (w, v_i)} \mathfrak{U}(\psi_1, u) \in \boldsymbol{x},$$

for some i, $1 \leq i \leq l$. So, by IH, $(\mathfrak{M}, \langle v_i, \boldsymbol{x} \rangle) \models \psi_2$ and $(\mathfrak{M}, \langle u, \boldsymbol{x} \rangle) \models \psi_1$ for all $u \in (w, v_i)$. Hence $(\mathfrak{M}, \langle w, \boldsymbol{x} \rangle) \models \psi_1 \mathcal{U} \psi_2$.

Case $\psi = \psi_1 \mathcal{S} \psi_2$ is considered analogously.

The existence of R_I-maximal points in sets of the form $A_{w,\boldsymbol{x},\psi}$ (where $w \in W$, $\boldsymbol{x} \in V$, and ψ is a $\mathcal{PST}$-formula) follows from a result of Fine (1974b) (see Theorem 10.36 in Chagrov and Zakharyaschev 1997). Here is a sketch of the proof. Consider the family

$$\mathcal{X} = \{X \subseteq A_{w,\boldsymbol{x},\psi} \mid R_\mathrm{I} \cap (X \times X) \text{ is a linear order with smallest element } \boldsymbol{x}\}.$$

Let C be a $\subseteq$-maximal set in $\mathcal{X}$ (i.e., for every $C' \in \mathcal{X}$, $C \subseteq C'$ implies $C' = C$); its existence can be readily proved with the help of Zorn's lemma. Now take the set

$$\boldsymbol{y}_0 = \{A \subseteq U \mid \exists \boldsymbol{z} \in C\, \forall \boldsymbol{z}' \in C\, (\boldsymbol{z} R_\mathrm{I} \boldsymbol{z}' \rightarrow A \in \boldsymbol{z}')\}.$$

This set is not empty, since $\mathfrak{U}(\psi, w) \in \boldsymbol{y}_0$, and clearly $\boldsymbol{y}_0$ has the finite intersection property. Hence we can find an ultrafilter $\boldsymbol{y}$ containing $\boldsymbol{y}_0$. Then it is easy to see, using the definition of R_I, that

$$\forall \boldsymbol{z} \in C\ \boldsymbol{z} R_\mathrm{I} \boldsymbol{y}. \tag{16.4}$$

We claim that $\boldsymbol{y}$ is R_I-maximal in $A_{w,\boldsymbol{x},\psi}$. Indeed, take some $\boldsymbol{y}' \in A_{w,\boldsymbol{x},\psi}$ such that $\boldsymbol{y} R_\mathrm{I} \boldsymbol{y}'$. If $\boldsymbol{y}' \in C$ then $\boldsymbol{y}' R_\mathrm{I} \boldsymbol{y}$ holds by (16.4). If $\boldsymbol{y}' \notin C$ then, by the $\subseteq$-maximality of C in $\mathcal{X}$, $C \cup \{\boldsymbol{y}'\}$ is not linearly ordered by R_I. Since by (16.4) and $\boldsymbol{y} R_\mathrm{I} \boldsymbol{y}'$, we have $\boldsymbol{z} R_\mathrm{I} \boldsymbol{y}'$ for all $\boldsymbol{z} \in C$, there exists a $\boldsymbol{z}' \in C$ such that $\boldsymbol{y}' R_\mathrm{I} \boldsymbol{z}'$, and so, again by (16.4), $\boldsymbol{y}' R_\mathrm{I} \boldsymbol{y}$ as required. ❑

We now use Lemma 16.1 to show that the full language $\mathcal{PST}$ is 'too expressive,' at least when interpreted in topological temporal models over discrete flows of time.

Theorem 16.3. *Suppose that* $\mathcal{C}$ *is one of the following classes of flows of time:* $\{\langle \mathbb{N}, < \rangle\}$, $\{\langle \mathbb{Z}, < \rangle\}$, *or the class of all finite strict linear orders. Then*

(i) *the satisfiability problem for* $\mathcal{PST}$*-formulas in topological* $\mathcal{PST}$*-models over the flows of time in* $\mathcal{C}$ *is undecidable;*

(ii) *the satisfiability problem for* $\mathcal{PST}$*-formulas in topological* $\mathcal{PST}$*-models with* **FSA** *over the flows of time in* $\mathcal{C}$ *is undecidable.*

Proof. We consider only the case of $\mathcal{C} = \{\langle \mathbb{N}, < \rangle\}$; the other cases are similar and left to the reader. The proof is a slight modification of the proof of Theorem 7.24 (note that (i) does not follow from Theorem 7.24 because satisfiability in topological $\mathcal{PST}$-models is not equivalent to $\mathsf{Log}\{\langle \mathbb{N}, < \rangle\} \times \mathbf{S4}_u$-satisfiability).

Given a finite alphabet A and a set $P = \{\langle v_1, w_1 \rangle, \dots, \langle v_k, w_k \rangle\}$ of pairs of words over A, we construct the formula $\psi_{A,P}$ as in the proof of Theorem 7.24

using **I**, **C**, $\Box_F$ and $\Diamond_F$ instead of $\boxbar$, $\diamondbar$, $\boxminus$ and $\diamondminus$, respectively. Let r be a fresh variable. Consider the $\mathcal{PST}$-formula

$$\psi'_{A,P} = r \wedge \Diamond_F \neg r \wedge \Box_F(\neg r \to \Box_F^+ \boxtimes \neg r) \wedge \Box_F^+(r \to \boxtimes r) \wedge \psi^r_{A,P},$$

where $\psi^r_{A,P}$ is obtained from $\psi_{A,P}$ by relativizing all of its temporal operators with respect to r, i.e., by replacing recursively $\Diamond_F \chi$ with $\Diamond_F(r \wedge \chi)$, and $\Box_F \chi$ with $\Box_F(r \to \chi)$.

We show that the following statements are equivalent:

(1) $\psi'_{A,P}$ is satisfied in a topological $\mathcal{PST}$-model over $\langle \mathbb{N}, < \rangle$;

(2) $\psi'_{A,P}$ is satisfied in a topological $\mathcal{PST}$-model with **FSA** over $\langle \mathbb{N}, < \rangle$;

(3) $\psi'_{A,P}$ is $\mathsf{Log}\{\langle \mathbb{N}, < \rangle\} \times \mathbf{S4}_u$-satisfiable;

(4) $\psi_{A,P}$ is $\mathsf{Log}\{\langle \mathbb{N}, < \rangle\} \times \mathbf{S4}$-satisfiable;

(5) there exist a natural number $N \geq 1$ and a sequence $i_1, \dots, i_N$ of indices such that $v_{i_1} * \cdots * v_{i_N} = w_{i_1} * \cdots * w_{i_N}$.

Since (5) is undecidable, this is enough to prove our theorem.

The implication (1) $\Rightarrow$ (2) is obvious, (2) $\Rightarrow$ (3) follows from Lemma 16.1 (ii), and (3) $\Rightarrow$ (1) follows from Theorem 6.29 and Proposition 3.6. The implication (3) $\Rightarrow$ (4) is again obvious, and (4) $\Rightarrow$ (5) was shown in the proof of Theorem 7.24. Finally, (5) $\Rightarrow$ (3) can be proved by appropriately modifying the valuation $\mathfrak{V}$ in the product frame $\langle \mathbb{N}, < \rangle \times \langle \mathbb{N}, \leq \rangle$ that was given in the proof of Theorem 7.24. ❑

16.2 Embedding spatio-temporal logics in first-order temporal logic

As we saw in Section 3.2 (Theorem 3.5), the spatio-temporal languages $\mathcal{ST}_i$ ($i = 0, 1, 2$) are embeddable into the propositional modal language $\mathcal{PST}$. However, because of the undecidability results for $\mathcal{PST}$ above, this fact alone does not shed any light on the algorithmic behavior of $\mathcal{ST}_i$-formulas in tt-models. In this section we first provide a fine-tuned analysis of the modal models required to satisfy the modal translations of $\mathcal{ST}_i$-formulas and show that actually rather simple ones are enough to do the job. Then we use the obtained results to embed $\mathcal{ST}_2$ into the one-variable fragment of the first-order temporal language $\mathcal{QTL}$.

The modal translations of $\mathcal{ST}_2$-formulas form a rather special fragment of the modal language $\mathcal{PST}$. Renz (1998) showed that an $\mathcal{RCC}$-8 formula φ is satisfied in a topological space iff its translation $\varphi^{\bowtie}$ is satisfied in a Kripke model based on some special $\mathbf{S4}_u$-frame that we call a *quasisaw*. Recall from Section 2.6 that a quasisaw is a 2-frame $\mathfrak{F} = \langle W, R, R_U \rangle$ such that R_U is the universal relation on W and $\langle W, R \rangle$ is a partial order of depth ≤ 1 and width ≤ 2 (that is, no R-chain has more than two distinct points, and no point has more than two distinct proper successors). It turns out that Renz's result can be generalized to $\mathcal{ST}_1$- and $\mathcal{ST}_2$-formulas:

Theorem 16.4. (i) *An $\mathcal{ST}_1$-formula φ is satisfied in a tt-model based on a flow of time $\mathfrak{F}$ iff $\varphi^{\bowtie}$ is satisfied in a Kripke model based on the product of $\mathfrak{F}$ and a quasisaw $\mathfrak{G}$.*

(ii) *An $\mathcal{ST}_2$-formula φ is satisfied in a tt-model satisfying* **FSA** *and based on a flow of time $\mathfrak{F}$ iff $\varphi^{\bowtie}$ is satisfied in a Kripke model satisfying* **FSA** *and based on the product of $\mathfrak{F}$ and a quasisaw $\mathfrak{G}$.*

Proof. The proof proceeds via a series of lemmas. To begin with, we define a set of $\mathcal{PST}$-formulas which contains the modal translations of all $\mathcal{ST}_2$-formulas.

Say that a $\mathcal{PST}$-formula is a **CI**-*term* if it can be obtained by prefixing **CI** to every subformula of an $\mathcal{ML}_{\mathcal{SU}}$-formula χ. If χ contains occurrences of only the temporal operator $\bigcirc$, then the corresponding **CI**-term is called a $\mathbf{CI}_{\bigcirc}$-*term*. By definition, the translation $t^{\bowtie}$ of every region term t of $\mathcal{ST}_2$ is a **CI**-term, while the translation $t^{\bowtie}$ of every region term t of $\mathcal{ST}_1$ is a $\mathbf{CI}_{\bigcirc}$-term.

A **CI**-*formula* ($\mathbf{CI}_{\bigcirc}$-*formula*) is a formula constructed, using $\mathcal{U}$, $\mathcal{S}$, and the Booleans, from formulas of the form $\diamondplus\psi$, where each ψ has one of the forms

$$\psi_1 \wedge \psi_2, \quad \neg\psi_1 \wedge \psi_2, \quad \psi_1 \wedge \neg\mathbf{I}\psi_2 \quad \text{or} \quad \mathbf{I}\psi_1 \wedge \mathbf{I}\psi_2,$$

with ψ_1 and ψ_2 being **CI**-terms (respectively, $\mathbf{CI}_{\bigcirc}$-terms). Note that every $\mathbf{CI}_{\bigcirc}$-formula is a u-formula.

By replacing every $\boxplus$ with $\neg\diamondplus\neg$ in the modal translation of atomic $\mathcal{ST}_2$-formulas given in Section 3.2, we obtain:

$$
\begin{aligned}
(\mathsf{DC}(t_1,t_2))^{\bowtie} &= \neg\diamondplus(t_1^{\bowtie} \wedge t_2^{\bowtie}),\\
(\mathsf{EQ}(t_1,t_2))^{\bowtie} &= \neg\diamondplus(t_1^{\bowtie} \wedge \neg t_2^{\bowtie}) \wedge \neg\diamondplus(\neg t_1^{\bowtie} \wedge t_2^{\bowtie}),\\
(\mathsf{PO}(t_1,t_2))^{\bowtie} &= \diamondplus(\mathbf{I}t_1^{\bowtie} \wedge \mathbf{I}t_2^{\bowtie}) \wedge \diamondplus(t_1^{\bowtie} \wedge \neg t_2^{\bowtie}) \wedge \diamondplus(\neg t_1^{\bowtie} \wedge t_2^{\bowtie}),\\
(\mathsf{EC}(t_1,t_2))^{\bowtie} &= \diamondplus(t_1^{\bowtie} \wedge t_2^{\bowtie}) \wedge \neg\diamondplus(\mathbf{I}t_1^{\bowtie} \wedge \mathbf{I}t_2^{\bowtie}),\\
(\mathsf{TPP}(t_1,t_2))^{\bowtie} &= \neg\diamondplus(t_1^{\bowtie} \wedge \neg t_2^{\bowtie}) \wedge \diamondplus(t_1^{\bowtie} \wedge \neg\mathbf{I}t_2^{\bowtie}) \wedge \diamondplus(\neg t_1^{\bowtie} \wedge t_2^{\bowtie}),\\
(\mathsf{NTPP}(t_1,t_2))^{\bowtie} &= \neg\diamondplus(t_1^{\bowtie} \wedge \neg\mathbf{I}t_2^{\bowtie}) \wedge \diamondplus(\neg t_1^{\bowtie} \wedge t_2^{\bowtie}).
\end{aligned}
$$

It follows immediately from the definition that

(sp1) the modal translation of every $\mathcal{ST}_1$-formula is equivalent (in topological $\mathcal{PST}$-models) to a $\mathbf{CI}_{\bigcirc}$-formula;

(sp2) the modal translation of every $\mathcal{ST}_2$-formula is equivalent (in topological $\mathcal{PST}$-models) to a $\mathbf{CI}$-formula.

Lemma 16.5. *Suppose that a* $\mathbf{CI}$*-formula* φ *is satisfied in a Kripke model* $\mathfrak{M}$ *based on the product of a strict linear order* $\mathfrak{F} = \langle W, < \rangle$ *and a rooted* $\mathbf{S4}_u$*-frame* $\mathfrak{G} = \langle V, R_{\mathbf{I}}, R_\forall \rangle$*. Suppose also that for any* $w \in W$*,* $x \in V$*, and any* $\mathbf{CI}$*-formula* ψ *the set*

$$A_{w,x,\psi} = \{y \in V \mid xR_{\mathbf{I}}y \text{ and } (\mathfrak{M}, \langle w, y\rangle) \models \psi\}$$

contains an $R_{\mathbf{I}}$*-maximal point. Then* φ *is satisfied in a Kripke model* $\mathfrak{M}'$ *based on the product of* $\mathfrak{F}$ *and a rooted* $\mathbf{S4}_u$*-frame* $\mathfrak{G}' = \langle V', R'_{\mathbf{I}}, R'_\forall \rangle$ *such that* $\langle V', R'_{\mathbf{I}} \rangle$ *is a partial order of depth* ≤ 1*. If* $\mathfrak{M}$ *satisfies* **FSA***, then* $\mathfrak{M}'$ *satisfies* **FSA** *as well.*

Proof. Suppose that $\mathfrak{M} = \langle \mathfrak{F} \times \mathfrak{G}, \mathfrak{V} \rangle$. Define $\mathfrak{G}' = \langle V', R'_{\mathbf{I}}, R'_\forall \rangle$ by taking $V' = V$, $R'_\forall = R_\forall = V \times V$, and $R'_{\mathbf{I}}$ to be the reflexive closure of $R_{\mathbf{I}} \cap (V_1 \times V_0)$, where

$$V_0 = \{x \in V \mid \forall y\, (xR_{\mathbf{I}}y \to yR_{\mathbf{I}}x)\}, \qquad V_1 = V - V_0.$$

In other words, $\mathfrak{G}'$ keeps the same set of worlds as $\mathfrak{G}$, but only those $R_{\mathbf{I}}$-arrows from the latter that lead to points in final clusters ($R_{\mathbf{I}}$-arrows within these clusters are also omitted). By the condition of the lemma, for every $x \in V$ there exists a $y \in V_0$ such that $xR_{\mathbf{I}}y$ (take $\psi = \top$). Finally, we put $\mathfrak{V}' = \mathfrak{V}$ and $\mathfrak{M}' = \langle \mathfrak{F} \times \mathfrak{G}', \mathfrak{V}' \rangle$. Clearly, if $\mathfrak{M}$ satisfies **FSA**, then $\mathfrak{M}'$ satisfies **FSA** as well.

First we show that for every $\mathbf{CI}$-term ψ, every $w \in W$, and every $x \in V$,

$$(\mathfrak{M}', \langle w, x\rangle) \models \psi \quad \text{iff} \quad (\mathfrak{M}, \langle w, x\rangle) \models \psi. \tag{16.5}$$

The proof is by induction on the construction of ψ. If ψ is a propositional variable then (16.5) follows from the definition of $\mathfrak{M}'$. Since the truth values of $\mathcal{ML}_{\mathcal{SU}}$-formulas do not depend on the $\mathbf{S4}_u$-component of the underlying product frame, we have (16.5) for every $\mathcal{ML}_{\mathcal{SU}}$-formula ψ as well.

Now it is not hard to see that, for every $\mathcal{ML}_{\mathcal{SU}}$-formula ψ, we have:

$$\begin{aligned}
(\mathfrak{M}, \langle w, x\rangle) \models \mathbf{CI}\psi \quad & \text{iff} \quad \exists y\, \big(xR_{\mathbf{I}}y \text{ and } \forall z\, (yR_{\mathbf{I}}z \to (\mathfrak{M}, \langle w, z\rangle) \models \psi)\big)\\
& \text{iff} \quad \exists y \in V_0\, \big(xR'_{\mathbf{I}}y \text{ and } (\mathfrak{M}, \langle w, y\rangle) \models \psi\big)\\
& \text{iff} \quad \exists y \in V_0\, \big(xR'_{\mathbf{I}}y \text{ and } (\mathfrak{M}', \langle w, y\rangle) \models \psi\big)\\
& \text{iff} \quad \exists y \in V_0\, \big(xR'_{\mathbf{I}}y \text{ and } (\mathfrak{M}', \langle w, y\rangle) \models \mathbf{I}\psi\big)\\
& \text{iff} \quad (\mathfrak{M}', \langle w, x\rangle) \models \mathbf{CI}\psi.
\end{aligned}$$

Next, we extend (16.5) to formulas of the form $\mathbf{I}\chi$, where χ is a **CI**-term. If $(\mathfrak{M}, \langle w, x\rangle) \models \mathbf{I}\chi$ then $(\mathfrak{M}, \langle w, y\rangle) \models \chi$ whenever $xR_{\mathbf{I}}y$, and so, by $R'_{\mathbf{I}} \subseteq R_{\mathbf{I}}$, we have $(\mathfrak{M}', \langle w, x\rangle) \models \mathbf{I}\chi$. Conversely, suppose $(\mathfrak{M}', \langle w, x\rangle) \models \mathbf{I}\chi$. Take any y with $xR_{\mathbf{I}}y$ and any $z \in V_0$ with $yR_{\mathbf{I}}z$. We claim that $(\mathfrak{M}, \langle w, z\rangle) \models \chi$. Indeed, if $x \in V_1$ then this follows by IH from $xR'_{\mathbf{I}}z$. If $x \in V_0$ then $zR_{\mathbf{I}}x$. Since we have $(\mathfrak{M}', \langle w, x\rangle) \models \chi$, by IH and $\chi = \mathbf{CI}\psi$, we obtain $(\mathfrak{M}, \langle w, z\rangle) \models \chi$. Now $(\mathfrak{M}, \langle w, y\rangle) \models \chi$ follows by $yR_{\mathbf{I}}z$ and $\chi = \mathbf{CI}\psi$. Since y was arbitrary with $xR_{\mathbf{I}}y$, we have $(\mathfrak{M}, \langle w, x\rangle) \models \mathbf{I}\chi$.

Finally, we can easily extend (16.5) to arbitrary **CI**-formulas simply because they are constructed from terms of the form $\mathbf{I}\chi$ and χ, where χ is a **CI**-term, by means of the Boolean operators, temporal operators, and $\diamondsuit$, and because none of these operators depends on the structure of the underlying partial order. ❑

Lemma 16.6. *If a* **CI**-*formula* φ *is satisfied in a Kripke model* $\mathfrak{M}$ *based on the product of a strict linear order* $\mathfrak{F}$ *and a rooted* $\mathbf{S4}_u$-*frame* $\mathfrak{G} = \langle V, R_{\mathbf{I}}, R_{\forall}\rangle$ *such that* $\langle V, R_{\mathbf{I}}\rangle$ *is a partial order of depth* ≤ 1, *then* φ *is satisfied in a Kripke model* $\mathfrak{M}'$ *based on the product of* $\mathfrak{F}$ *and a quasisaw* $\mathfrak{G}'$. *If* $\mathfrak{M}$ *satisfies* **FSA**, *then* $\mathfrak{M}'$ *satisfies* **FSA** *as well.*

Proof. Suppose that φ is satisfied in a Kripke model $\mathfrak{M} = \langle \mathfrak{F} \times \mathfrak{G}, \mathfrak{V}\rangle$ such that $\mathfrak{G} = \langle V, R_{\mathbf{I}}, R_{\forall}\rangle$ and $\langle V, R_{\mathbf{I}}\rangle$ is a partial order of depth ≤ 1. It is not hard to see that then V is the disjoint union of two sets, say, V_0 and V_1, such that $R_{\mathbf{I}}$ is the reflexive closure of a subset of $V_1 \times V_0$. The points in V_i are said to be of *depth* i, for $i = 0, 1$.

Every **CI**-formula φ is composed (using the temporal operators and the Booleans) from formulas $\Sigma_\varphi = \{\diamondsuit\chi_1, \ldots, \diamondsuit\chi_n\}$, where the χ_i are of the form

$$\psi_1 \wedge \psi_2, \quad \psi_1 \wedge \neg\psi_2, \quad \mathbf{I}\psi_1 \wedge \mathbf{I}\psi_2, \quad \psi_1 \wedge \neg\mathbf{I}\psi_2,$$

with ψ_1, ψ_2 being **CI**-terms. We write $(\mathfrak{M}, w) \models \diamondsuit\psi$ if there is $x \in V$ such that $(\mathfrak{M}, \langle w, x\rangle) \models \diamondsuit\psi$ (or, equivalently, if $(\mathfrak{M}, \langle w, x\rangle) \models \diamondsuit\psi$ for all $x \in V$).

For every $\diamondsuit\psi \in \Sigma_\varphi$ and every $w \in W$ with $(\mathfrak{M}, w) \models \diamondsuit\psi$, we fix a point $x_{\psi,w} \in V$ such that $(\mathfrak{M}, \langle w, x_{\psi,w}\rangle) \models \psi$. We may assume that the $x_{\psi,w}$ are pairwise distinct and that all x of depth 1 are of the form $x_{\psi,w}$ for some $w \in W$ and $\diamondsuit\psi \in \Sigma_\varphi$. Moreover, we may assume that no point $y \in V$ has more than one proper $R_{\mathbf{I}}$-predecessor.

Let us construct a new model $\mathfrak{M}' = \langle \mathfrak{F} \times \mathfrak{G}', \mathfrak{V}'\rangle$ as follows. First, for every $w \in W$ and every $\diamondsuit\psi \in \Sigma_\varphi$ such that there is a point x of depth 0 with $(\mathfrak{M}, \langle w, x\rangle) \models \psi$, we remove the point $x_{\psi,w}$ from V whenever it is of depth 1. Denote the resulting set of points by V'.

Next, given a point $x = x_{\psi,w} \in V'$, we delete some $R_{\mathbf{I}}$-arrows coming from x depending on the form of ψ. There are four possible cases.

Case 1. $\psi = \mathbf{CI}\psi_1 \wedge \mathbf{CI}\psi_2$. Then we select points $x_1, x_2 \in V'$ of depth 0 such that $(\mathfrak{M}, \langle w, x_i \rangle) \models \mathbf{CI}\psi_i$ and $xR_{\mathbf{I}}x_i$, for $i = 1, 2$, and remove all $R_{\mathbf{I}}$-arrows leading from x to points different from x_1, x_2.

Case 2. $\psi = \mathbf{CI}\psi_1 \wedge \neg\mathbf{CI}\psi_2$. Then we select $x_1, x_2 \in V'$ of depth 0 such that $(\mathfrak{M}, \langle w, x_1 \rangle) \models \mathbf{CI}\psi_1$, $(\mathfrak{M}, \langle w, x_2 \rangle) \models \neg\mathbf{CI}\psi_2$, $xR_{\mathbf{I}}x_i$, and remove all $R_{\mathbf{I}}$-arrows leading from x to points different from x_1, x_2.

Case 3. $\psi = \mathbf{ICI}\psi_1 \wedge \mathbf{ICI}\psi_2$. Then we select $x_1, x_2 \in V'$ of depth 0 such that $(\mathfrak{M}, \langle w, x_i \rangle) \models \mathbf{ICI}\psi_i$, $xR_{\mathbf{I}}x_i$, and remove all $R_{\mathbf{I}}$-arrows leading from x to points different from x_1, x_2.

Case 4. $\psi = \mathbf{CI}\psi_1 \wedge \neg\mathbf{ICI}\psi_2$. Then we select $x_1, x_2 \in V'$ of depth 0 such that $(\mathfrak{M}, \langle w, x_1 \rangle) \models \mathbf{CI}\psi_1$, $(\mathfrak{M}, \langle w, x_2 \rangle) \models \neg\mathbf{ICI}\psi_2$, $xR_{\mathbf{I}}x_i$, and remove all $R_{\mathbf{I}}$-arrows leading from x to points different from x_1, x_2.

Denote by $R'_{\mathbf{I}}$ the resulting relation and put $R'_{\forall} = V' \times V'$ and $\mathfrak{G}' = \langle V', R'_{\mathbf{I}}, R'_{\forall} \rangle$. It should be clear that $\mathfrak{G}'$ is a quasisaw. Finally, we define $\mathfrak{V}'$ by taking, for every propositional variable p, every $w \in W$, and every $x \in V'$,

$$\langle w, x \rangle \in \mathfrak{V}'(p) \quad \text{iff} \quad \text{there is } y \in V' \text{ of depth 0 such that } xR'_{\mathbf{I}}y \text{ and } \langle w, y \rangle \in \mathfrak{V}(p).$$

Clearly, if $\mathfrak{M}$ satisfies **FSA**, then $\mathfrak{M}'$ satisfies **FSA** as well.

To show that φ is satisfied in $\mathfrak{M}'$, we first prove that, for all $w \in W$ and all $\diamondsuit\psi \in \Sigma_\varphi$,

$$(\mathfrak{M}', w) \models \diamondsuit\psi \quad \text{iff} \quad (\mathfrak{M}, w) \models \diamondsuit\psi.$$

It is readily proved by induction that $(\mathfrak{M}', \langle w, x \rangle) \models \chi$ iff $(\mathfrak{M}, \langle w, x \rangle) \models \chi$ for all points x of depth 0, all $w \in W$ and all **CI**-terms χ. Then, by the construction, we also have that, for all $\diamondsuit\psi \in \Sigma_\varphi$ and all $w \in W$, $(\mathfrak{M}, w) \models \diamondsuit\psi$ implies $(\mathfrak{M}', w) \models \diamondsuit\psi$. So it remains to show that $(\mathfrak{M}, w) \models \neg\diamondsuit\psi$ implies $(\mathfrak{M}', w) \models \neg\diamondsuit\psi$, for all $\diamondsuit\psi \in \Sigma_\varphi$ and all $w \in W$. Consider all four possible cases for ψ.

Case 1. $\psi = \mathbf{CI}\psi_1 \wedge \mathbf{CI}\psi_2$. Suppose that we have $(\mathfrak{M}', \langle w, x \rangle) \models \psi$ and $(\mathfrak{M}, w) \models \neg\diamondsuit\psi$. Then there are $x_1, x_2 \in V'$ of depth 0 such that $xR'_{\mathbf{I}}x_i$ and $(\mathfrak{M}', \langle w, x_i \rangle) \models \mathbf{CI}\psi_i$. But then $(\mathfrak{M}, \langle w, x \rangle) \models \psi$, which is a contradiction.

Case 2. $\psi = \mathbf{CI}\psi_1 \wedge \neg\mathbf{CI}\psi_2$. Suppose that we have $(\mathfrak{M}', \langle w, x \rangle) \models \psi$ but $(\mathfrak{M}, w) \models \neg\diamondsuit\psi$. Then $(\mathfrak{M}', \langle w, y \rangle) \models \mathbf{CI}\psi_1$ for some y of depth 0 with $xR'_{\mathbf{I}}y$. But then $(\mathfrak{M}', \langle w, y \rangle) \models \neg\psi_2$, so $(\mathfrak{M}, \langle w, y \rangle) \models \psi$, which is a contradiction.

Case 3. $\psi = \mathbf{ICI}\psi_1 \wedge \mathbf{ICI}\psi_2$. Suppose we have $(\mathfrak{M}', \langle w, x \rangle) \models \psi$ but $(\mathfrak{M}, w) \models \neg\diamondsuit\psi$. Then for every $R'_{\mathbf{I}}$-successor y of x of depth 0, we have $(\mathfrak{M}', \langle w, y \rangle) \models \psi$, and so $(\mathfrak{M}, \langle w, y \rangle) \models \psi$, which is again a contradiction.

Case 4. $\psi = \mathbf{CI}\psi_1 \wedge \neg\mathbf{ICI}\psi_2$. Suppose that we have $(\mathfrak{M}', \langle w, x \rangle) \models \psi$ and $(\mathfrak{M}, w) \models \neg\diamondsuit\psi$. Then $(\mathfrak{M}', \langle w, y \rangle) \models \mathbf{CI}\psi_1$ for some y of depth 0 with $xR'_{\mathbf{I}}y$. Thus $(\mathfrak{M}, \langle w, y \rangle) \models \mathbf{CI}\psi_1$, and so $(\mathfrak{M}, \langle w, x \rangle) \models \mathbf{ICI}\psi_2$, which means that $(\mathfrak{M}, \langle w, z\rangle) \models \mathbf{CI}\psi_2$ whenever $xR_{\mathbf{I}}z$. So $(\mathfrak{M}', \langle w, z\rangle) \models \mathbf{CI}\psi_2$ whenever $xR'_{\mathbf{I}}z$, contrary to $(\mathfrak{M}', \langle w, x\rangle) \models \neg\mathbf{ICI}\psi_2$.

Now, by a straightforward induction we can easily show that, for all $w \in W$ and all **CI**-formulas ψ built from Σ_φ using the temporal operators and the Booleans, we have

$$(\mathfrak{M}', w) \models \psi \quad \text{iff} \quad (\mathfrak{M}, w) \models \psi.$$

It follows that φ is satisfied in $\mathfrak{M}'$. ❑

We are now in a position to complete the proof of Theorem 16.4.

(i) Suppose that an $\mathcal{ST}_1$-formula φ is satisfied in a tt-model based on a flow of time $\mathfrak{F}$. By **(sp1)**, $\varphi^{\bowtie}$ is (equivalent to) a $\mathbf{CI}_{\bigcirc}$-formula satisfied in a topological $\mathcal{PST}$-model based on $\mathfrak{F}$. Since every $\mathbf{CI}_{\bigcirc}$-formula is a u-formula, it follows from Lemmas 16.1 (i), 16.5, and 16.6 that $\varphi^{\bowtie}$ is satisfied in a Kripke model based on the product of $\mathfrak{F}$ and a quasisaw $\mathfrak{G}$. The converse implication follows from Proposition 3.6.

(ii) is proved in the same way using **(sp2)** and Lemmas 16.1 (ii), 16.5, 16.6. ❑

Now we define a translation from $\mathcal{ST}_2$-formulas into the one-variable fragment $\mathcal{QTL}^1$ of first-order temporal logic. This translation in a sense extends the translation $\cdot^{\odot}$ from $\mathcal{BRCC}$-8 into **S5** (which was introduced in Section 2.6), so—with a slight abuse of notation—we also denote it by $\cdot^{\odot}$. First, we 'extend' the translations $\cdot^b$, $\cdot^l$, and $\cdot^r$ of $\mathcal{ML}^u$ into $\mathcal{ML}$ to translations (denoted also by $\cdot^b$, $\cdot^l$, and $\cdot^r$) from $\mathcal{PST}$ into $\mathcal{QTL}^1$. Fix an individual variable x. For every propositional variable p, we reserve three different unary predicate symbols B_p, L_p, R_p, and set

$$p^b = B_p(x), \quad p^l = L_p(x), \quad p^r = R_p(x).$$

Then set inductively

$$\begin{aligned}
(\psi \wedge \chi)^i &= \psi^i \wedge \chi^i, \text{ for } i \in \{b, l, r\},\\
(\neg\psi)^i &= \neg\psi^i, \text{ for } i \in \{b, l, r\},\\
(\mathbf{I}\psi)^b &= \psi^b \wedge \psi^l \wedge \psi^r,\\
(\mathbf{I}\psi)^i &= \psi^i, \text{ for } i \in \{l, r\},\\
(\mathbf{C}\psi)^b &= \psi^b \vee \psi^l \vee \psi^r,\\
(\mathbf{C}\psi)^i &= \psi^i, \text{ for } i \in \{l, r\},\\
(\diamondsuit\psi)^i &= \exists x\, (\psi^b \vee \psi^l \vee \psi^r), \text{ for } i \in \{b, r, l\},\\
(\boxdot\psi)^i &= \forall x\, (\psi^b \wedge \psi^l \wedge \psi^r), \text{ for } i \in \{b, r, l\},
\end{aligned}$$

$$(\psi_1 \mathcal{U} \psi_2)^i = \psi_1^i \mathcal{U} \psi_2^i, \text{ for } i \in \{b, l, r\},$$
$$(\psi_1 \mathcal{S} \psi_2)^i = \psi_1^i \mathcal{S} \psi_2^i, \text{ for } i \in \{b, l, r\}.$$

Finally, we define the *$\mathcal{QTL}^1$-translation* $\varphi^{\odot}$ of an $\mathcal{ST}_2$-formula φ as $(\varphi^{\bowtie})^b$. The length of $\varphi^{\odot}$ is polynomial in the length of φ. We now have the following analog of Theorem 2.34:

Theorem 16.7. *Suppose $\mathfrak{F}$ is a flow of time and φ an $\mathcal{ST}_2$-formula. Then the following conditions are equivalent:*

- *$\varphi^{\bowtie}$ is satisfiable in a Kripke model (with* **FSA***) based on the product of $\mathfrak{F}$ and a quasisaw;*
- *$\varphi^{\odot}$ is satisfiable in a first-order temporal model (with* **FSA***) based on $\mathfrak{F}$.*

Proof. The proof is a straightforward modification of the proof of Theorem 2.34 and left to the reader. ❑

Now we obtain the following:

Theorem 16.8. (i) *The map $\cdot^{\odot}$ is a polynomial translation of $\mathcal{ST}_1$-formulas into the one-variable fragment $\mathcal{QTL}^1$ of $\mathcal{QTL}$ such that, for any flow of time $\mathfrak{F}$, an $\mathcal{ST}_1$-formula φ is satisfiable in a tt-model based on $\mathfrak{F}$ iff $\varphi^{\odot}$ is satisfiable in a first-order temporal model based on $\mathfrak{F}$.*

(ii) *The map $\cdot^{\odot}$ is a polynomial translation of $\mathcal{ST}_2$-formulas into the one-variable fragment $\mathcal{QTL}^1$ of $\mathcal{QTL}$ such that, for any flow of time $\mathfrak{F}$, an $\mathcal{ST}_2$-formula φ is satisfiable in a tt-model with* **FSA** *based on $\mathfrak{F}$ iff $\varphi^{\odot}$ is satisfiable in a first-order temporal model based on $\mathfrak{F}$ and having finite domains.*

Proof. (i) Follows from Theorems 16.4 (i) and 16.7.

(ii) Follows from Theorems 16.4 (ii), 16.7 and Theorem 11.44. ❑

The decidability results for first-order temporal logics obtained Section 11.2 now yield the decidability of the satisfiability problem for $\mathcal{ST}_i$-formulas:

Theorem 16.9. *Suppose $\mathcal{C}$ is any of the following classes of flows of time: $\{\langle \mathbb{N}, <\rangle\}$, $\{\langle \mathbb{Z}, <\rangle\}$, $\{\langle \mathbb{Q}, <\rangle\}$, the class of all finite strict linear orders, any first-order definable class of strict linear orders. Then the satisfiability problem for $\mathcal{ST}_1$-formulas in tt-models based on flows of time in $\mathcal{C}$ is decidable.*

Proof. Follows from Theorem 16.8 (i) and Corollary 11.14. ❑

Theorem 16.10. *Let $\mathcal{H}$ be any of the following classes of flows of time: $\{\langle \mathbb{N}, <\rangle\}$, $\{\langle \mathbb{Z}, <\rangle\}$, $\{\langle \mathbb{Q}, <\rangle\}$, $\{\langle \mathbb{R}, <\rangle\}$, the class of all finite strict linear orders, any first-order definable class of strict linear orders. Then the satisfiability problem for $\mathcal{ST}_2$-formulas in tt-models with* **FSA** *based on flows of time in $\mathcal{H}$ is decidable.*

Proof. Follows from Theorem 16.8 (ii) and Corollary 11.14. ❑

Remark 16.11. The (properly optimized) tableau- and resolution-type procedures for the one-variable fragment of $\mathsf{QLog}_{\mathcal{U}}(\mathbb{N})$ mentioned in Remarks 11.40 and 11.79 may provide 'practical' satisfiability-checking procedures for the spatio-temporal logics $\mathcal{ST}_i$ over the flow of time $\langle \mathbb{N}, < \rangle$.

16.3 Complexity of spatio-temporal logics

Theorems 16.8, 11.31 and 11.53 provide us with EXPSPACE satisfiability checking algorithms for $\mathcal{ST}_1$- and $\mathcal{ST}_2$-formulas in tt-models based on the flow of time $\langle \mathbb{N}, < \rangle$ (and satisfying **FSA** in the latter case). To obtain the matching lower bounds, we encode

- the constant-free fragment of $\mathcal{QTL}^1$ in $\mathcal{ST}_2$,
- the constant-free fragment of $\mathcal{QTL}^1_{①}$ in $\mathcal{ST}_1$,

and then use Theorems 11.52 and 11.33, respectively (constants were not involved in the proofs of these theorems). We will also see that our translation encodes

- the constant-free fragment of $\mathcal{QTL}^1_{\boxdot}$ in $\mathcal{ST}_0$,

from which we can conclude, by Theorem 11.36, that the satisfiability problem for $\mathcal{ST}_0$-formulas in tt-models based on the flow of time $\langle \mathbb{N}, < \rangle$ is PSPACE-complete.

Besides, we will single out a PSPACE-complete fragment sitting between $\mathcal{ST}_0$ and $\mathcal{ST}_1$ by forbidding applications of the Boolean operators to region terms in $\mathcal{ST}_1$-formulas.

The results of this section were obtained in (Gabelaia *et al.* 2003).

Complexity of the satisfiability problem for $\mathcal{ST}_i$-formulas over $\langle \mathbb{N}, < \rangle$

In order to define an embedding of the constant-free fragment of $\mathcal{QTL}^1$ into $\mathcal{ST}_2$,we first require some simple facts about $\mathcal{QTL}^1$-formulas. A *basic Q-formula* is a $\mathcal{QTL}^1$-formula of the form $\forall x\, \vartheta(x)$, where $\vartheta(x)$ is quantifier-free and contains neither constant symbols nor propositional variables. Say that a $\mathcal{QTL}^1$-sentence φ is in *Q-normal form* if it is built from basic Q-formulas using the Booleans and temporal operators. In other words, sentences in Q-normal form contain neither nested quantifiers nor constants, and use only unary predicate symbols. Then we have the following:

Lemma 16.12. *For every constant-free $\mathcal{QTL}^1$-sentence φ, one can effectively construct a $\mathcal{QTL}^1$-sentence $\widehat{\varphi}$ in Q-normal form such that φ is satisfiable in a first-order temporal model based on a flow of time $\mathfrak{F}$ (and having finite domains) iff $\widehat{\varphi}$ is satisfiable in a first-order temporal model based on $\mathfrak{F}$ (and having finite domains). Moreover, the length of $\widehat{\varphi}$ is linear in the length of φ.*

Proof. The proof is similar to that of Theorem 3.35. Without loss of generality we may assume that φ contains no occurrences of $\exists$. To transform φ into its Q-normal form, we first introduce a fresh unary predicate symbol P_i for every propositional variable p_i in φ and replace each occurrence of p_i with $\forall x\, P_i(x)$. Denote the resulting formula by φ_0. For every subformula ψ of φ_0 define a formula $\psi^\sharp$ by taking inductively

$$
\begin{aligned}
(P(x))^\sharp &= P(x),\\
(\psi_1 \wedge \psi_2)^\sharp &= \psi_1^\sharp \wedge \psi_2^\sharp,\\
(\neg\psi)^\sharp &= \neg\psi^\sharp,\\
(\psi_1 \mathcal{U} \psi_2)^\sharp &= \psi_1^\sharp \mathcal{U} \psi_2^\sharp,\\
(\psi_1 \mathcal{S} \psi_2)^\sharp &= \psi_1^\sharp \mathcal{S} \psi_2^\sharp,\\
(\forall x \psi)^\sharp &= P_{\forall x\psi}(x),
\end{aligned}
$$

where $P_{\forall x\psi}$ is a fresh unary predicate symbol.

Now, denote by χ the formula

$$
\bigwedge_{\forall x\psi \in sub\,\varphi_0} (\forall x P_{\forall x\psi}(x) \vee \forall x \neg P_{\forall x\psi}(x)) \wedge \bigwedge_{\forall x\psi \in sub\,\varphi_0} (\forall x P_{\forall x\psi}(x) \leftrightarrow \forall x \psi^\sharp).
$$

One can readily show by induction that

$$
\widehat{\varphi} = \neg\forall x \neg\varphi_0^\sharp \wedge \Box_F \Box_P \chi
$$

is satisfiable in a first-order temporal model based on $\mathfrak{F}$ (and having finite domains) iff φ is satisfiable in a first-order temporal model based on $\mathfrak{F}$ (and having finite domains). Moreover, $\widehat{\varphi}$ is in Q-normal form. ❑

We are now in a position to define a translation $\cdot^\heartsuit$ from $\mathcal{QTL}^1$-sentences in Q-normal form into $\mathcal{ST}_2$-formulas. Given such a sentence φ, denote by $\varphi^\heartsuit$ the result of replacing all occurrences of basic Q-formulas $\forall x\, \vartheta(x)$ in φ with $\mathsf{EQ}(\vartheta^\heartsuit, U \sqcup \neg U)$, where U is a region variable and the translation $\vartheta^\heartsuit$ of quantifier-free formulas $\vartheta(x)$ is defined by taking:

$$
\begin{aligned}
(P_i(x))^\heartsuit &= X_i, \quad (P_i \text{ a unary predicate symbol, } X_i \text{ a region variable}),\\
(\neg\psi)^\heartsuit &= \neg\psi^\heartsuit,
\end{aligned}
$$

$$\begin{aligned}
(\psi_1 \wedge \psi_2)^\heartsuit &= \psi_1^\heartsuit \sqcap \psi_2^\heartsuit,\\
(\psi_1 \vee \psi_2)^\heartsuit &= \psi_1^\heartsuit \sqcup \psi_2^\heartsuit,\\
(\bigcirc\psi)^\heartsuit &= \bigcirc\psi^\heartsuit,\\
(\psi_1 \mathcal{U} \psi_2)^\heartsuit &= \psi_1^\heartsuit \mathcal{U} \psi_2^\heartsuit,\\
(\psi_1 \mathcal{S} \psi_2)^\heartsuit &= \psi_1^\heartsuit \mathcal{S} \psi_2^\heartsuit.
\end{aligned}$$

Lemma 16.13. *A $\mathcal{QTL}^1$-sentence φ in Q-normal form is satisfiable in a first-order temporal model based on a flow of time $\mathfrak{F}$ (and having finite domains) iff $(\varphi^\heartsuit)^\bowtie$ is satisfiable in a Kripke model based on the product of $\mathfrak{F}$ and a (finite) quasisaw.*

Proof. ($\Rightarrow$) Suppose that φ is in Q-normal form and $\mathfrak{M} = \langle \mathfrak{F}, D, I \rangle$ is a first-order temporal model, where $\mathfrak{F} = \langle W, < \rangle$ and, for all $w \in W$,

$$I(w) = \left\langle D, P_0^{I(w)}, \dots \right\rangle.$$

Assume also that $(\mathfrak{M}, w) \models \varphi$, for some $w \in W$, and construct a Kripke model $\mathfrak{M}' = \langle \mathfrak{F} \times \mathfrak{G}, \mathfrak{V} \rangle$ by taking the quasisaw $\mathfrak{G} = \langle D, R_{\mathbf{I}}, R_\forall \rangle$, where

$$R_{\mathbf{I}} = \{\langle a, a \rangle \mid a \in D\}, \quad R_\forall = D \times D$$

and

$$\mathfrak{V}(p_i) = \{\langle w, a \rangle \mid a \in P_i^{I(w)}\}.$$

Note that the topological space $\mathfrak{T}_\mathfrak{G} = \langle D, \mathbb{I}_\mathfrak{G} \rangle$ induced by $\mathfrak{G}$ is *discrete*, i.e., for all $X \subseteq D$,

$$\mathbb{I}_\mathfrak{G} X = \mathbb{C}_\mathfrak{G} X = X.$$

It follows by induction that, for every constant-free and quantifier-free $\mathcal{QTL}^1$-formula ϑ, all $w \in W$ and all $a \in D$, we have

$$(\mathfrak{M}, w) \models \vartheta[a] \quad \text{iff} \quad (\mathfrak{M}', \langle w, a \rangle) \models (\vartheta^\heartsuit)^\bowtie.$$

Therefore, for every basic Q-formula $\forall x\, \vartheta(x)$, all $w \in W$ and all (or, equivalently, some) $a \in D$,

$$(\mathfrak{M}, w) \models \forall x\, \vartheta(x) \quad \text{iff} \quad (\mathfrak{M}', \langle w, a \rangle) \models \left(\mathsf{EQ}(\vartheta^\heartsuit, U \sqcup \neg U)\right)^\bowtie.$$

It follows by induction that $(\varphi^\heartsuit)^\bowtie$ is satisfied in $\mathfrak{M}'$.

($\Leftarrow$) Suppose that $(\varphi^\heartsuit)^\bowtie$ is satisfied in a Kripke model $\mathfrak{M} = \langle \mathfrak{F} \times \mathfrak{G}, \mathfrak{V} \rangle$ based on the product of $\mathfrak{F}$ and a quasisaw $\mathfrak{G} = \langle V, R_{\mathbf{I}}, R_\forall \rangle$. Denote by $V_0 \subseteq V$ the set of points of depth 0 in $\langle V, R_{\mathbf{I}} \rangle$ and define a first-order temporal model $\mathfrak{M}' = \langle \mathfrak{F}, V_0, I \rangle$ by taking

$$I(w) = \left\langle V_0, P_0^{I(w)}, \dots \right\rangle$$

and

$$P_i^{I(w)} = \{a \in V_0 \mid (\mathfrak{M}, \langle w, a \rangle) \models p_i\}.$$

Clearly, for every $X \subseteq V$, we have

$$\mathbb{I}_{\mathfrak{G}} X \cap V_0 = \mathbb{C}_{\mathfrak{G}} X \cap V_0 = X \cap V_0,$$

where $\mathfrak{T}_{\mathfrak{G}} = \langle V, \mathbb{I}_{\mathfrak{G}} \rangle$ is the topological space induced by $\mathfrak{G}$. So we obtain by induction that for every constant-free and quantifier-free $\mathcal{QTL}_1$-formula ϑ, all $w \in W$ and all $a \in V_0$

$$(\mathfrak{M}', w) \models \vartheta[a] \quad \text{iff} \quad (\mathfrak{M}, \langle w, a \rangle) \models (\vartheta^{\heartsuit})^{\bowtie}.$$

A regular closed set $X \subseteq V$ in $\mathfrak{T}_{\mathfrak{G}}$ coincides with V if and only if it contains V_0. So, for all basic Q-formulas $\forall x\, \vartheta(x)$, all $w \in W$ and all (or, equivalently, some) $a \in V_0$,

$$(\mathfrak{M}', w) \models \forall x\, \vartheta(x) \quad \text{iff} \quad (\mathfrak{M}, \langle w, a \rangle) \models \big(\mathsf{EQ}(\vartheta^{\heartsuit}, U \sqcup \neg U)\big)^{\bowtie}.$$

It follows by induction that φ is satisfied in $\mathfrak{M}'$. ❑

Now it is easy to define the fragments of $\mathcal{QTL}^1$ corresponding—in the sense of Lemma 16.13—to $\mathcal{ST}_0$ and $\mathcal{ST}_1$. First, observe:

Lemma 16.14. *For the translation $\varphi \mapsto \widehat{\varphi}$ of Lemma 16.12 the following hold true, for all constant-free $\mathcal{QTL}^1$-formulas φ:*

- *$\widehat{\varphi}$ is a $\mathcal{QTL}^1_{①}$-formula whenever φ is a $\mathcal{QTL}^1_{①}$-formula.*
- *$\widehat{\varphi}$ is a $\mathcal{QTL}^1_{\boxed{0}}$-formula whenever φ is a $\mathcal{QTL}^1_{\boxed{0}}$-formula.*

So, $\widehat{\varphi}^{\heartsuit}$ is an $\mathcal{ST}_1$-formula whenever φ is a constant-free $\mathcal{QTL}^1_{①}$-sentence, and it is an $\mathcal{ST}_0$-formula whenever φ is a constant-free $\mathcal{QTL}^1_{\boxed{0}}$-sentence. We then obtain:

Theorem 16.15. (i) *A constant-free $\mathcal{QTL}^1_{\boxed{0}}$-sentence φ is satisfied in a first-order temporal model based on a flow of time $\mathfrak{F}$ (and having finite domains) iff the $\mathcal{ST}_0$-formula $\widehat{\varphi}^{\heartsuit}$ is satisfied in a tt-model based on $\mathfrak{F}$ (and satisfying* **FSA**).

(ii) *A constant-free $\mathcal{QTL}^1_{①}$-sentence φ is satisfied in a first-order temporal model based on a flow of time $\mathfrak{F}$ (and having finite domains) iff the $\mathcal{ST}_1$-formula $\widehat{\varphi}^{\heartsuit}$ is satisfied in a tt-model based on $\mathfrak{F}$ (and satisfying* **FSA**).

(iii) *A constant-free $\mathcal{QTL}^1$-sentence φ is satisfied in a first-order temporal model based on a flow of time $\mathfrak{F}$ and having finite domains iff the $\mathcal{ST}_2$-formula $\widehat{\varphi}^{\heartsuit}$ is satisfied in a tt-model based on $\mathfrak{F}$ and satisfying* **FSA**.

Proof. This follows from Lemmas 16.12, 16.13, 16.14 and Theorem 16.4. ❑

Finally, we can derive the following tight complexity results:

Theorem 16.16. (i) *The satisfiability problem for $\mathcal{ST}_0$-formulas in tt-models (satisfying* **FSA**) *based on* $\langle \mathbb{N}, < \rangle$ *is* PSPACE-*complete. This holds true for* $\mathcal{S}$*-free* $\mathcal{ST}_0$*-formulas as well.*

(ii) *The satisfiability problem for* $\mathcal{ST}_1$*-formulas in tt-models (satisfying* **FSA**) *based on* $\langle \mathbb{N}, < \rangle$ *is* EXPSPACE-*complete. This holds true for* $\mathcal{S}$*-free* $\mathcal{ST}_1$*-formulas as well.*

(iii) *The satisfiability problem for* $\mathcal{ST}_2$*-formulas in tt-models with* **FSA** *based on* $\langle \mathbb{N}, < \rangle$ *is* EXPSPACE-*complete. This holds true for* $\mathcal{S}$*-free* $\mathcal{ST}_2$*-formulas as well.*

Proof. For the upper bound in (i), observe that $\varphi^{\odot}$ is a $\mathcal{QTL}^1_{\square}$-formula whenever φ is an $\mathcal{ST}_0$-formula, and apply Theorems 11.36 and 16.8. The upper bounds in (ii) and (iii) follow from Theorems 11.31, 11.53 and 16.8.

The PSPACE lower bound in (i) follows from Theorems 16.15 (i) and 11.36, because the PSPACE lower bound proof of the latter goes through for the constant-free fragment of $\mathcal{QTL}^1_{\square}$. The EXPSPACE lower bound in (ii) follows from Theorems 16.15 (ii) and 11.33, and from the observation that the EXPSPACE lower bound proof of the latter goes through for the constant-free fragment of $\mathcal{QTL}^1_{\bigcirc}$. The EXPSPACE lower bound in (iii) is shown in the same way using Theorems 16.15 (iii) and 11.53. As explained in Section 11.4, everything will hold true for $\mathcal{S}$-free formulas as well. ❑

A PSPACE-complete fragment between $\mathcal{ST}_0$ and $\mathcal{ST}_1$

As we saw above, the satisfiability problem for $\mathcal{ST}_1$-formulas in tt-models over $\langle \mathbb{N}, < \rangle$ is EXPSPACE-complete. It turns out, however, that if we do not allow applications of the Boolean operators to region terms then satisfiability becomes PSPACE-complete. Let us denote the resulting spatio-temporal language by

$$\mathcal{ST}_1^-$$

(its region terms are of the form $\bigcirc^n X$, $n \geq 0$, where X is a region variable). Our aim is to show that the satisfiability problem for $\mathcal{ST}_1^-$-formulas in tt-models over $\langle \mathbb{N}, < \rangle$ is PSPACE-complete. To this end, we first prove that $\mathcal{RCC}$-8 has a kind of 'completion property' (Theorem 16.17). It turns out that, because of this property, we can check the satisfiability of $\mathcal{ST}_1^-$-formulas by a simple, almost modular, combination of the satisfiability checking algorithm of Sistla and Clarke (1985) for **PTL** and any algorithm checking satisfiability of $\mathcal{RCC}$-8 formulas. This approach to determine the computational complexity of combinations of **PTL** with constraint systems like Allen's interval algebra $\mathcal{A}\ell\ell$-13 (see Section 2.2) and the orientation logic of Ligozat (1998) has been introduced by Balbiani and Condotta (2002) and further developed for constraint systems without the 'completion property' by Demri and D'Souza (2002).

To simplify presentation, throughout the remaining part of this section we assume that at each moment of time region variables are interpreted as *nonempty* regular closed sets. This means, in particular, that the eight $\mathcal{RCC}$-8 relations are jointly exhaustive and pairwise disjoint; in other words, in any model and at any moment of time precisely one of the eight relations holds true between the interpretations of two $\mathcal{ST}_1^-$-region terms, while the other seven do not hold. The results presented in this section are easily generalized to the framework in which empty regions are permitted; we refer the reader to the discussion at the end of this section.

To begin with, we remind the reader that a *fork* is a frame $\mathfrak{f} = \langle W_\mathfrak{f}, R_\mathfrak{f} \rangle$ such that $W_\mathfrak{f} = \{b_\mathfrak{f}, l_\mathfrak{f}, r_\mathfrak{f}\}$ and $R_\mathfrak{f}$ is the reflexive closure of $\{\langle b_\mathfrak{f}, l_\mathfrak{f}\rangle, \langle b_\mathfrak{f}, r_\mathfrak{f}\rangle\}$. A *saw* is a disjoint union of forks (in which the universal modality is interpreted by the universal relation).

A *fork model* is a Kripke model $\mathfrak{m} = \langle \mathfrak{f}, \mathfrak{v} \rangle$, where $\mathfrak{f}$ is a fork and, for every variable p, we have

$$b_\mathfrak{f} \in \mathfrak{v}(p) \quad \text{iff} \quad l_\mathfrak{f} \in \mathfrak{v}(p) \text{ or } r_\mathfrak{f} \in \mathfrak{v}(p).$$

A *saw model* is a disjoint union of fork models. Note that every saw model validates $p \leftrightarrow \mathbf{CI}p$, for every propositional variable p. By Theorem 2.33 (or 16.4), an $\mathcal{RCC}$-8 formula φ is satisfiable iff $\varphi^{\bowtie}$ is satisfiable in a saw model.

Given a set $\mathcal{V}$ of propositional variables, say that fork models $\mathfrak{m}_1 = \langle \mathfrak{f}_1, \mathfrak{v}_1 \rangle$ and $\mathfrak{m}_2 = \langle \mathfrak{f}_2, \mathfrak{v}_2 \rangle$ are *$\mathcal{V}$-equivalent* when $x_{\mathfrak{f}_1} \in \mathfrak{v}_1(p)$ iff $x_{\mathfrak{f}_2} \in \mathfrak{v}_2(p)$, for every $p \in \mathcal{V}$ and every $x \in \{l, r\}$. If $\mathcal{V}$ is of cardinality n, then there exist precisely 4^n pairwise non-$\mathcal{V}$-equivalent fork models. Denote by $\mathsf{Fork}_\mathcal{V}$ the set of fork models containing one member of each $\mathcal{V}$-equivalence class (over the variables in $\mathcal{V}$).

To simplify (and slightly abuse) notation, from now on we denote by X the propositional variable associated by the translation $\cdot^{\bowtie}$ with a region variable X. As we are going to consider only saws models, without loss of generality we may assume that $X^{\bowtie} = X$ (not $\mathbf{CI}X$ as in the original definition). For instance,

$$(\mathsf{DC}(X,Y))^{\bowtie} = \boxed{\forall}(\neg X \vee \neg Y),$$

$$(\mathsf{EQ}(X,Y))^{\bowtie} = \boxed{\forall}(X \to Y) \wedge \boxed{\forall}(Y \to X).$$

(cf. Section 2.6 where $\cdot^{\bowtie}$ was introduced in a somewhat different but equivalent form). Define the *universal part* $\cdot^{\forall}$ of the translation $\cdot^{\bowtie}$ by taking

$$(\mathsf{DC}(X,Y))^{\forall} = \boxed{\forall}(\neg X \vee \neg Y),$$

$$(\mathsf{EQ}(X,Y))^{\forall} = \boxed{\forall}(X \to Y) \wedge \boxed{\forall}(Y \to X),$$

$$(\mathsf{PO}(X,Y))^{\forall} = \top,$$

$$\begin{aligned}
(\mathsf{EC}(X,Y))^{\forall} &= \boxtimes(\neg\mathbf{I}X \vee \neg\mathbf{I}Y),\\
(\mathsf{TPP}(X,Y))^{\forall} &= \boxtimes(X \to Y),\\
(\mathsf{TPPi}(X,Y))^{\forall} &= \boxtimes(Y \to X),\\
(\mathsf{NTPP}(X,Y))^{\forall} &= \boxtimes(X \to \mathbf{I}Y),\\
(\mathsf{NTPPi}(X,Y))^{\forall} &= \boxtimes(Y \to \mathbf{I}X).
\end{aligned}$$

Fix some set $\mathcal{V}$ of region variables $X_1, \dots, X_n$. For every pair $\langle X_i, X_j \rangle$, where $1 \leq i < j \leq n$, we also fix a unique $\mathcal{RCC}$-8 relation R_{ij} and let

$$\Phi = \{\mathsf{R}_{ij}(X_i, X_j) \mid 1 \leq i < j \leq n\}. \tag{16.6}$$

Say that Φ is *satisfied* in a saw model $\mathfrak{M}$ and write $\mathfrak{M} \models \Phi$ if

$$\mathfrak{M} \models \bigwedge_{\mathsf{R}(X,Y)\in\Phi} (\mathsf{R}(X,Y))^{\bowtie}.$$

We are now going to introduce some special models satisfying Φ (if Φ is satisfiable at all). First we let

$$\mathsf{Fork}_{\Phi} = \left\{ \mathfrak{m} \in \mathsf{Fork}_{\mathcal{V}} \;\middle|\; \mathfrak{m} \models \bigwedge_{\mathsf{R}(X,Y)\in\Phi} (\mathsf{R}(X,Y))^{\forall} \right\}$$

and then take the disjoint union of countably infinitely many isomorphic copies of each member of Fork_{Φ}. The resulting saw model $\mathfrak{M}$ will be called the Φ-*exhaustive model*. It should be clear that this model satisfies Φ whenever Φ is satisfiable.

Theorem 16.17. *Let Y be a fresh region variable not occurring in $\mathcal{V}$ and let*

$$\Psi = \Phi \cup \{\mathsf{R}_i(X_i, Y) \mid 1 \leq i \leq n\},$$

for some $\mathcal{RCC}$-8-relations R_i. Suppose Ψ is satisfiable and let $\mathfrak{M} = \langle \mathfrak{G}, \mathfrak{V} \rangle$ be the Φ-exhaustive model. Then there exists a valuation $\mathfrak{V}'$ in $\mathfrak{G}$ which coincides with $\mathfrak{V}$ on $\mathcal{V}$ and such that the saw model $\langle \mathfrak{G}, \mathfrak{V}' \rangle$ is Ψ-exhaustive.

Proof. It is sufficient to show that, for every fork model $\mathfrak{m} \in \mathsf{Fork}_{\Phi}$, there exists a $\mathcal{V}$-equivalent fork model $\mathfrak{m}'$ which is $\mathcal{V} \cup \{Y\}$-equivalent to a member of Fork_{Ψ} (we will simply say that $\mathfrak{m}'$ is a member of Fork_{Ψ}).

So suppose that $\mathfrak{m} = \langle \mathfrak{f}, \mathfrak{v} \rangle$ is a member of Fork_{Φ}, where $\mathfrak{f} = \langle W_{\mathfrak{f}}, R_{\mathfrak{f}} \rangle$, $W_{\mathfrak{f}} = \{b_{\mathfrak{f}}, l_{\mathfrak{f}}, r_{\mathfrak{f}}\}$ and $R_{\mathfrak{f}}$ is the reflexive closure of $\{\langle b_{\mathfrak{f}}, l_{\mathfrak{f}} \rangle, \langle b_{\mathfrak{f}}, r_{\mathfrak{f}} \rangle\}$. The set $\mathsf{Fork}_{\mathcal{V}\cup\{Y\}}$ contains four fork models which are $\mathcal{V}$-equivalent to $\mathfrak{m}$, namely, $\mathfrak{m}_{00}, \mathfrak{m}_{01}, \mathfrak{m}_{10}, \mathfrak{m}_{11}$, where

- $\mathfrak{m}_{LR} = \langle \mathfrak{f}, \mathfrak{v}_{LR} \rangle$ and $\mathfrak{v}_{LR}(X_i) = \mathfrak{v}(X_i)$, for every $X_i \in \mathcal{V}$ and $L, R \in \{0, 1\}$,

- $\mathfrak{v}_{00}(Y) = \emptyset$, $\mathfrak{v}_{01}(Y) = \{r_\mathfrak{f}, b_\mathfrak{f}\}$, $\mathfrak{v}_{10}(Y) = \{l_\mathfrak{f}, b_\mathfrak{f}\}$ and $\mathfrak{v}_{11}(Y) = \{l_\mathfrak{f}, r_\mathfrak{f}, b_\mathfrak{f}\}$.

We will show that Fork_Ψ contains at least one of $\mathfrak{m}_{00}, \mathfrak{m}_{01}, \mathfrak{m}_{10}, \mathfrak{m}_{11}$.

First observe that if $\mathsf{EQ}(X_i, Y) \in \Psi$, for some $X_i \in \mathcal{V}$, then Fork_Ψ contains the fork model $\mathfrak{m}_{LR}$ with $\mathfrak{v}_{LR}(X_i) = \mathfrak{v}_{LR}(Y)$. Therefore, we may assume that for no X_i does $\mathsf{EQ}(X_i, Y) \in \Psi$ hold. Second, all the four fork models satisfy $(\mathsf{PO}(X_i, Y))^\forall$ and so the formulas $\mathsf{PO}(X_i, Y)$ in Ψ need no consideration.

Now consider the following sets of regions:

$$\mathcal{N} = \{X_i \mid \mathsf{NTPP}(X_i, Y) \in \Psi \text{ and } (\mathfrak{m}, b_\mathfrak{f}) \models X_i\},$$
$$\mathcal{T}_L = \{X_i \mid \mathsf{TPP}(X_i, Y) \in \Psi \text{ and } (\mathfrak{m}, l_\mathfrak{f}) \models X_i\},$$
$$\mathcal{T}_R = \{X_i \mid \mathsf{TPP}(X_i, Y) \in \Psi \text{ and } (\mathfrak{m}, r_\mathfrak{f}) \models X_i\}.$$

Fours cases are possible.

Case 1: $\mathcal{N} \cup \mathcal{T}_L \cup \mathcal{T}_R = \emptyset$ (i.e., $b_\mathfrak{f}$ does not belong to any region in $\mathcal{V}$ which is a proper part of Y). Then $\mathfrak{m}_{00}$ is contained in Fork_Ψ, since $\mathfrak{m}_{00} \models (\mathsf{R}(X_i, Y))^\forall$ for every $\mathsf{R}(X_i, Y) \in \Psi$. Indeed, for every X_i we have:

$$\begin{aligned} \mathfrak{m}_{00} &\models \boxtimes(\neg X_i \vee \neg Y) & \text{if} \quad & \mathsf{DC}(X_i, Y) \in \Psi, \\ \mathfrak{m}_{00} &\models \boxtimes(\neg \mathbf{I}X_i \vee \neg \mathbf{I}Y) & \text{if} \quad & \mathsf{EC}(X_i, Y) \in \Psi, \\ \mathfrak{m}_{00} &\models \boxtimes(Y \to X_i) & \text{if} \quad & \mathsf{TPPi}(X_i, Y) \in \Psi, \\ \mathfrak{m}_{00} &\models \boxtimes(Y \to \mathbf{I}X_i) & \text{if} \quad & \mathsf{NTPPi}(X_i, Y) \in \Psi. \end{aligned}$$

If $\mathsf{TPP}(X_i, Y) \in \Psi$ or $\mathsf{NTPP}(X_i, Y) \in \Psi$, then by assumption $(\mathfrak{m}_{00}, b_\mathfrak{f}) \models \neg X_i$. Therefore,

$$\begin{aligned} \mathfrak{m}_{00} &\models \boxtimes(X_i \to Y) & \text{if} \quad & \mathsf{TPP}(X_i, Y) \in \Psi, \\ \mathfrak{m}_{00} &\models \boxtimes(X_i \to \mathbf{I}Y) & \text{if} \quad & \mathsf{NTTP}(X_i, Y) \in \Psi. \end{aligned}$$

Case 2: $\mathcal{N} \cup \mathcal{T}_L = \emptyset$ and $\mathcal{T}_R \neq \emptyset$ (i.e., $b_\mathfrak{f}$ is on the border of some region from $\mathcal{V}$ which is a tangential proper part of Y, and $b_\mathfrak{f}$ does not belong to any region in $\mathcal{V}$ which is a non-tangential proper part of Y). Then $\mathfrak{m}_{01}$ is in Fork_Ψ, since $\mathfrak{m}_{01} \models (\mathsf{R}(X_i, Y))^\forall$ for every $\mathsf{R}(X_i, Y) \in \Psi$. Indeed, for every X_i we have

$$\begin{aligned} \mathfrak{m}_{01} &\models \boxtimes(X_i \to Y) & \text{if} \quad & \mathsf{TPP}(X_i, Y) \in \Psi, \\ \mathfrak{m}_{01} &\models \boxtimes(X_i \to \mathbf{I}Y) & \text{if} \quad & \mathsf{NTTP}(X_i, Y) \in \Psi, \end{aligned}$$

since in both cases $(\mathfrak{m}_{01}, l_\mathfrak{f}) \models \neg X_i$ (by assumption) and $(\mathfrak{m}_{01}, r_\mathfrak{f}) \models Y$. Further, let $Z \in \mathcal{T}_R$, i.e., $\mathsf{TPP}(Z, Y) \in \Psi$, $(\mathfrak{m}, l_\mathfrak{f}) \models \neg Z$ and $(\mathfrak{m}, r_\mathfrak{f}) \models Z$. Consider four remaining cases for X_i.

Let $\mathsf{DC}(X_i, Y) \in \Psi$. We have to show that $\mathfrak{m}_{01} \models \boxdot(\neg X_i \vee \neg Y)$. Suppose otherwise. Then $(\mathfrak{m}_{01}, b_\mathfrak{f}) \models X_i$ and $(\mathfrak{m}, b_\mathfrak{f}) \models X_i$. On the other hand, as both $\mathsf{DC}(X_i, Y)$ and $\mathsf{TPP}(Z, Y)$ are in Ψ (which is satisfiable), we obtain that $\mathsf{DC}(Z, X_i) \in \Phi$, contrary to $(\mathfrak{m}, b_\mathfrak{f}) \models X_i \wedge Z$.

Let $\mathsf{EC}(X_i, Y) \in \Psi$. Suppose $\mathfrak{m}_{01} \not\models \boxdot(\neg \mathbf{I}X_i \vee \neg \mathbf{I}Y)$. Then $(\mathfrak{m}_{01}, r_\mathfrak{f}) \models X_i$ and $(\mathfrak{m}, r_\mathfrak{f}) \models X_i$. On the other hand, $\mathsf{EC}(X_i, Y) \in \Psi$ and $\mathsf{TPP}(Z, Y) \in \Psi$ imply that we have either $\mathsf{DC}(Z, X_i) \in \Phi$ or $\mathsf{EC}(Z, X_i) \in \Phi$, contrary to $(\mathfrak{m}, r_\mathfrak{f}) \models X_i \wedge Z$.

Let $\mathsf{TPPi}(X_i, Y) \in \Psi$. Suppose $\mathfrak{m}_{01} \not\models \boxdot(Y \to X_i)$. Then $(\mathfrak{m}_{01}, r_\mathfrak{f}) \models \neg X_i$ and $(\mathfrak{m}, r_\mathfrak{f}) \models \neg X_i$. On the other hand, $\mathsf{TPPi}(X_i, Y) \in \Psi$ and $\mathsf{TPP}(Z, Y) \in \Psi$ imply that we have either $\mathsf{TPP}(Z, X_i) \in \Phi$ or $\mathsf{NTPP}(Z, X_i) \in \Phi$, contrary to $(\mathfrak{m}, r_\mathfrak{f}) \models \neg X_i \wedge Z$.

Let $\mathsf{NTPPi}(X_i, Y) \in \Psi$. Suppose that $\mathfrak{m}_{01} \not\models \boxdot(Y \to \mathbf{I}X_i)$. Then we have $(\mathfrak{m}_{01}, r_\mathfrak{f}) \models \neg X_i$ and $(\mathfrak{m}, r_\mathfrak{f}) \models \neg X_i$. But $\mathsf{NTPPi}(X_i, Y) \in \Psi$ and $\mathsf{TPP}(Z, Y) \in \Psi$ imply $\mathsf{NTPP}(Z, X_i) \in \Phi$, contrary to $(\mathfrak{m}, r_\mathfrak{f}) \models \neg X_i \wedge Z$.

Case 3: $\mathcal{N} \cup \mathcal{T}_R = \emptyset$ and $\mathcal{T}_L \neq \emptyset$. Then $\mathfrak{m}_{10}$ is in Fork_Ψ. This case is a mirror image of Case 2.

Case 4: $\mathcal{N} \cup \mathcal{T}_L \neq \emptyset$ and $\mathcal{N} \cup \mathcal{T}_R \neq \emptyset$. Then $\mathfrak{m}_{11}$ is in Fork_Ψ, since $\mathfrak{m}_{11} \models (\mathsf{R}(X_i, Y))^\forall$ for every $\mathsf{R}(X_i, Y) \in \Psi$. Indeed, for every X_i we have:

$$\mathfrak{m}_{11} \models \boxdot(X_i \to Y) \quad \text{if} \quad \mathsf{TPP}(X_i, Y) \in \Psi,$$

$$\mathfrak{m}_{11} \models \boxdot(X_i \to \mathbf{I}Y) \quad \text{if} \quad \mathsf{NTTP}(X_i, Y) \in \Psi.$$

Now consider four remaining cases for X_i.

Let $\mathsf{DC}(X_i, Y) \in \Psi$ and suppose that $\mathfrak{m}_{11} \not\models \boxdot(\neg X_i \vee \neg Y)$. Then we have $(\mathfrak{m}_{11}, b_\mathfrak{f}) \models X_i$ and $(\mathfrak{m}, b_\mathfrak{f}) \models X_i$. On the other hand, there is $Z \in \mathcal{V}$ such that $(\mathfrak{m}, b_\mathfrak{f}) \models Z$ and either $\mathsf{TPP}(Z, Y) \in \Psi$ or $\mathsf{NTPP}(Z, Y) \in \Psi$, which together with $\mathsf{DC}(X_i, Y) \in \Psi$ imply $\mathsf{DC}(Z, X_i) \in \Phi$, contrary to $(\mathfrak{m}, b_\mathfrak{f}) \models X_i \wedge Z$.

Let $\mathsf{EC}(X_i, Y) \in \Psi$ and suppose $\mathfrak{m}_{11} \not\models \boxdot(\neg \mathbf{I}X_i \vee \neg \mathbf{I}Y)$. Then we have $(\mathfrak{m}_{11}, b_\mathfrak{f}) \models \mathbf{I}X_i$ and $(\mathfrak{m}, b_\mathfrak{f}) \models \mathbf{I}X_i$. On the other hand, there is a $Z \in \mathcal{V}$ such that $(\mathfrak{m}, b_\mathfrak{f}) \models Z$ and either $\mathsf{TPP}(Z, Y) \in \Psi$ or $\mathsf{NTPP}(Z, Y) \in \Psi$, which together with $\mathsf{EC}(X_i, Y) \in \Psi$ imply either $\mathsf{DC}(Z, X_i) \in \Phi$ or $\mathsf{EC}(Z, X_i) \in \Phi$, contrary to $(\mathfrak{m}, b_\mathfrak{f}) \models \mathbf{I}X_i \wedge Z$.

Let $\mathsf{TPPi}(X_i, Y) \in \Psi$. Suppose that $\mathfrak{m}_{11} \not\models \boxdot(Y \to X_i)$. Then either $(\mathfrak{m}_{11}, l_\mathfrak{f}) \models \neg X_i$ or $(\mathfrak{m}_{11}, r_\mathfrak{f}) \models \neg X_i$. Therefore, $(\mathfrak{m}_{11}, b_\mathfrak{f}) \models \neg \mathbf{I}X_i$ and $(\mathfrak{m}, b_\mathfrak{f}) \models \neg \mathbf{I}X_i$. Then two cases are possible:

(1) There is a $Z \in \mathcal{V}$ such that $(\mathfrak{m}, b_\mathfrak{f}) \models Z$ and $\mathsf{NTPP}(Z, Y) \in \Psi$. Then we have $\mathsf{NTTP}(Z, X_i) \in \Phi$, contrary to $(\mathfrak{m}, b_\mathfrak{f}) \models \neg \mathbf{I}X_i \wedge Z$.

(2) There are $Z_l, Z_r \in \mathcal{V}$ with $(\mathfrak{m}, l_\mathfrak{f}) \models Z_l$, $(\mathfrak{m}, r_\mathfrak{f}) \models Z_r$ and both $\mathsf{TPP}(Z_l, Y)$ and $\mathsf{TPP}(Z_r, Y)$ are in Ψ. Then either $\mathsf{TPP}(Z_l, X_i) \in \Phi$ or $\mathsf{NTTP}(Z_l, X_i) \in \Phi$, and either $\mathsf{TPP}(Z_r, X_i) \in \Phi$ or $\mathsf{NTTP}(Z_r, X_i) \in \Phi$.

In all these four cases we get a contradiction with $(\mathfrak{m}, l_\mathfrak{f}) \models \neg X_i \wedge Z_l$ or $(\mathfrak{m}, r_\mathfrak{f}) \models \neg X_i \wedge Z_r$.

Let $\mathsf{NTPPi}(X_i, Y) \in \Psi$. Suppose that $\mathfrak{m}_{11} \not\models \boxdot(Y \to \mathbf{I}X_i)$. Then either $(\mathfrak{m}_{11}, l_\mathfrak{f}) \models \neg X_i$ or $(\mathfrak{m}_{11}, r_\mathfrak{f}) \models \neg X_i$. So, $(\mathfrak{m}_{11}, b_\mathfrak{f}) \models \neg \mathbf{I}X_i$ and $(\mathfrak{m}, b_\mathfrak{f}) \models \neg \mathbf{I}X_i$. On the other hand, there exists $Z \in \mathcal{V}$ such that $(\mathfrak{m}, b_\mathfrak{f}) \models Z$ and either $\mathsf{TPP}(Z, Y) \in \Psi$ or $\mathsf{NTPP}(Z, Y) \in \Psi$, which together with $\mathsf{NTPPi}(X_i, Y) \in \Psi$ imply $\mathsf{NTTP}(Z, X_i) \in \Phi$, contrary to $(\mathfrak{m}, b_\mathfrak{f}) \models \neg \mathbf{I}X_i \wedge Z$. ❑

As a consequence of the proof of Theorem 16.17 and the uniqueness of exhaustive models we obtain the following:

Corollary 16.18. *Suppose that $\mathcal{V}' \subseteq \mathcal{V}$ and*

$$\Phi' = \{\mathsf{R}(X_i, X_j) \mid \mathsf{R}(X_i, X_j) \in \Phi,\ X_i, X_j \in \mathcal{V}'\}.$$

Then by restricting the valuation of the Φ-exhaustive model to $\mathcal{V}'$ we obtain a Φ'-exhaustive model.

We are now in a position to prove:

Theorem 16.19. *The satisfiability problem for $\mathcal{ST}_1^-$-formulas in tt-models over the flow of time $\langle \mathbb{N}, < \rangle$ is decidable in* PSPACE (*and so is* PSPACE-*complete*).

Proof. The proof of Proposition 11.25 shows that it is sufficient to prove this result for $\mathcal{ST}_1^-$-formulas without $\mathcal{S}$. Suppose that we have an $\mathcal{ST}_1^-$-formula φ without $\mathcal{S}$. Note first that without loss of generality we may assume that every region term occurring in φ is of the from X or $\bigcirc X$, where X is a region variable (if this is not the case, then for each $\bigcirc^n X$, $n > 1$, in φ we introduce fresh region variables $X_1, \dots, X_n$, replace $\bigcirc^n X$ with X_n and add the formulas

$$\Box_F^+ \mathsf{EQ}(X_1, \bigcirc X) \quad \text{and} \quad \Box_F^+ \mathsf{EQ}(X_{i+1}, \bigcirc X_i),$$

for $i = 1, \dots, n-1$, as conjuncts to the resulting formula).

Let $\mathcal{V}$ be the set of region variables occurring in φ and let

$$\mathcal{V}^\circ = \mathcal{V} \cup \{\bigcirc X \mid X \in \mathcal{V}\}.$$

Replace every occurrence of an $\mathcal{RCC}$-8 relation $\mathsf{R}(t_1, t_2)$ in φ (remember that $t_1, t_2 \in \mathcal{V}^\circ$) with a propositional variable $\mathsf{R}^{t_1 t_2}$ and add to the result the following conjunct

$$\Box_F^+ \bigwedge_{\substack{X,Y \in \mathcal{V} \\ \mathsf{R} \in \mathcal{R}}} (\mathsf{R}^{\bigcirc X \bigcirc Y} \leftrightarrow \bigcirc \mathsf{R}^{XY}), \tag{16.7}$$

where $\mathcal{R} = \{\mathsf{EQ}, \mathsf{EC}, \mathsf{DC}, \mathsf{PO}, \mathsf{TPP}, \mathsf{TPPi}, \mathsf{NTPP}, \mathsf{NTPPi}\}$. Denote the resulting $\mathcal{ML}_\mathcal{U}$-formula by $\widetilde{\varphi}$. It should be clear that the length of $\widetilde{\varphi}$ is a polynomial function in the length of φ.

We claim that φ is satisfiable in a tt-model over $\langle \mathbb{N}, < \rangle$ iff

(i) there exists a Kripke model $\mathfrak{N} = \langle\langle\mathbb{N}, <\rangle, \mathfrak{U}\rangle$ satisfying $\widetilde{\varphi}$ and

(ii) for every $n \in \mathbb{N}$, the set

$$\Phi_n = \{\mathsf{R}(t_1, t_2) \mid (\mathfrak{N}, n) \models \mathsf{R}^{t_1 t_2},\ t_1, t_2 \in \mathcal{V}^\circ\}$$

of $\mathcal{RCC}$-8 relations is satisfiable if we regard all region terms $t \in \mathcal{V}^\circ$ as region variables.

The implication ($\Rightarrow$) is obvious. To show ($\Leftarrow$), given a Kripke model $\mathfrak{N} = \langle\langle\mathbb{N}, <\rangle, \mathfrak{U}\rangle$ satisfying $\widetilde{\varphi}$ and condition (ii) above, we construct inductively a model $\mathfrak{M} = \langle\langle\mathbb{N}, <\rangle \times \mathfrak{G}, \mathfrak{V}\rangle$ such that $\mathfrak{G}$ is a saw and $\mathfrak{M}$ satisfies $\varphi^{\bowtie}$. Then, by Theorem 16.4 (i), φ is satisfiable in a tt-model over $\langle\mathbb{N}, <\rangle$.

To begin with, we take the Φ_0-exhaustive model $\mathfrak{M}_0 = \langle\mathfrak{G}, \mathfrak{V}_0\rangle$. It exists because Φ_0 is satisfiable. Set

$$\langle 0, x\rangle \in \mathfrak{V}(X) \quad \text{iff} \quad x \in \mathfrak{V}_0(X),$$

for all points x in $\mathfrak{G}$, and for all region variables $X \in \mathcal{V}$.

Consider now the model $\mathfrak{M}_1' = \langle\mathfrak{G}, \mathfrak{V}_1'\rangle$, where $\mathfrak{V}_1'(X) = \mathfrak{V}_0(\bigcirc X)$, $X \in \mathcal{V}$. By Corollary 16.18, this model can be regarded as the Φ_1'-exhaustive model for

$$\Phi_1' = \{\mathsf{R}(X, Y) \mid \mathsf{R}(\bigcirc X, \bigcirc Y) \in \Phi_0\}$$

over the variables in $\mathcal{V}$. By the second conjunct of (16.7), $\Phi_1' \subseteq \Phi_1$ and by Theorem 16.17, there is a valuation $\mathfrak{V}_1$ coinciding with $\mathfrak{V}_1'$ on $\mathcal{V}$ and such that $\mathfrak{M}_1 = \langle\mathfrak{G}, \mathfrak{V}_1\rangle$ is a Φ_1-exhaustive model over $\mathcal{V}^\circ$. Set

$$\langle 1, x\rangle \in \mathfrak{V}(X) \quad \text{iff} \quad x \in \mathfrak{V}_1(X),$$

for all x in $\mathfrak{G}$, and for all $X \in \mathcal{V}$.

Then we consider the model $\mathfrak{M}_2' = \langle\mathfrak{G}, \mathfrak{V}_2'\rangle$, where $\mathfrak{V}_2'(X) = \mathfrak{V}_1(\bigcirc X)$, $X \in \mathcal{V}$, and use it in precisely the same way as above to define when $\langle 2, x\rangle$ belongs to $\mathfrak{V}(X)$. And so forth.

Now by induction on the construction of φ we show that $\mathfrak{M}$ satisfies $\varphi^{\bowtie}$. The basis of the induction follows from the fact that, for every $n \in \mathbb{N}$ and all $t_1, t_2 \in \mathcal{V}^\circ$, we have

$$(\mathfrak{N}, n) \models \mathsf{R}^{t_1 t_2} \quad \text{iff} \quad (\mathfrak{M}, \langle n, x\rangle) \models (\mathsf{R}(t_1, t_2))^{\bowtie},$$

for all (some) x. The induction steps are trivial and left to the reader.

We are now in a position to formulate a PSPACE satisfiability checking algorithm for $\mathcal{ST}_1^-$-formulas. Given such a formula φ, we construct $\widetilde{\varphi}$. Take the well-known PSPACE satisfiability checking algorithm for **PTL** of Sistla and Clarke (1985) or the proof of Theorem 19.8.1 from (Gabbay *et al.* 1994). (Actually, this algorithm is a simple variant of the algorithm presented in the proof of Theorem 11.30 above.) To comply with condition (ii), at each step of the algorithm which guesses a set of subformulas of $\widetilde{\varphi}$ that are true at a certain time point n, we should now check whether the corresponding Φ_n is satisfiable in a topological space. According to Theorem 2.35, this can be done by a nondeterministic polynomial time algorithm. ❑

As mentioned above, in Theorems 16.17, 16.19, Corollary 16.18, and their proofs we made the assumption that region variables are interpreted as non-empty sets. In this case the set Φ of (16.6) is equivalent to the set

$$\Phi \cup \bigcup \{\neg \mathsf{S}_{ij}(X_i, X_j) \mid \mathsf{S}_{ij} \neq \mathsf{R}_{ij}, 1 \leq i < j \leq n\},$$

where S_{ij} are $\mathcal{RCC}$-8 relations. In the proofs above, the negated $\mathcal{RCC}$-8 relations are covered implicitly because the $\mathcal{RCC}$-8 relations are pairwise disjoint and jointly exhaustive. This is no longer the case for empty regions. However, the proofs can be easily modified to cover the empty regions *by taking care of negated relations explicitly*. For example, in (16.6) we should include for any pair $\langle X_i, X_j \rangle$, $1 \leq i, j \leq n$, and any $\mathcal{RCC}$-8 relation R either $\mathsf{R}(X_i, X_j)$ or $\neg\mathsf{R}(X_i, X_j)$. Note that $\mathsf{DC}(X_i, X_i)$ implies that X_i is empty, while $\neg\mathsf{DC}(X_i, X_i)$ means that X_i is nonempty. Now, with this modification of Φ and corresponding modifications of the definitions of Ψ, Φ', and Φ_n, one can easily obtain a proof for the general case.

Figure 16.1 summarizes the obtained complexity results. Here the spatio-temporal logics $\mathcal{ST}_i^-$, for $i = 0, 1, 2$, are the $\mathcal{ST}_i$ with the restriction that only the corresponding temporal operators (but not the Booleans) and region variables can be used to construct region terms.

16.4 Spatio-temporal models based on Euclidean spaces

Although the region connection calculus $\mathcal{RCC}$ (see Section 2.6) was formulated as a first-order theory that can be interpreted in arbitrary topological spaces, of course the intended models for various applications are one-, two-, or three-dimensional Euclidean spaces, i.e., $\langle \mathbb{R}^n, \mathbb{I} \rangle$ for $n = 1, 2, 3$ with the standard interior operator.[3] Renz (1998) showed that for pure $\mathcal{RCC}$-8 formulas satisfiability in arbitrary topological spaces coincides with satisfiability in $\langle \mathbb{R}, \mathbb{I} \rangle$, and so in $\langle \mathbb{R}^n, \mathbb{I} \rangle$ for any $n > 0$; $\langle \mathbb{R}^3, \mathbb{I} \rangle$ is enough to realize any set of satisfiable $\mathcal{RCC}$-8 formulas using only *connected* regions.

Let us observe first that this result of (Renz 1998) cannot be generalized to $\mathcal{RCC}$-8 extended with the operation $\sqcup$ intended to form unions of regions.

[3]Cohn (1997) notes, however, that in some applications discrete or even finite topological spaces may be preferable.

	in tt-models	in tt-models with **FSA**
$\mathcal{PST}$	undecidable (Thm. 16.3 (i))	undecidable (Thm. 16.3 (ii))
$\mathcal{ST}_2$	?	EXPSPACE-complete (Thm. 16.16 (iii))
$\mathcal{ST}_1$	EXPSPACE-complete (Thm. 16.16 (ii))	EXPSPACE-complete (Thm. 16.16 (ii))
$\mathcal{ST}_0$	PSPACE-complete (Thm. 16.16 (i))	PSPACE-complete (Thm. 16.16 (i))
$\mathcal{ST}_2^-$	?	in EXPSPACE (Thm. 16.16 (iii)) ?
$\mathcal{ST}_1^-$	PSPACE-complete (Thm. 16.19)	in EXPSPACE (Thm. 16.16 (ii)) ?
$\mathcal{ST}_0^-$	PSPACE-complete (Thms. 2.7, 16.16 (i))	PSPACE-complete (Thms. 2.7, 16.16 (i))

Table 16.1: Complexity of spatio-temporal logics over $\langle \mathbb{N}, < \rangle$.

Recall that a topological space is called *connected* if it cannot be represented as a union of two disjoint nonempty open sets.

Proposition 16.20. *There exists a satisfiable $\mathcal{BRCC}$-8 formula φ which is not satisfiable in any connected topological space. In particular, φ is not satisfiable in $\langle \mathbb{R}^n, \mathbb{I} \rangle$, for any $n \geq 1$.*

Proof. Take the conjunction φ of the following predicates:

$$\mathsf{EQ}(X_1 \sqcup X_2, Y), \quad \mathsf{NTPP}(X_1, Y), \quad \mathsf{NTPP}(X_2, Y), \quad \mathsf{NTPP}(Y, Z).$$

Clearly, φ is satisfied in the topological space consisting of three points and having the interior operator $\mathbb{I}$ such that $\mathbb{I}X = X$, for every set X.

Now suppose that $\mathfrak{T} \models^{\mathfrak{a}} \varphi$ for some topological space $\mathfrak{T} = \langle U, \mathbb{I} \rangle$. Then the region $\mathfrak{a}(X_1 \sqcup X_2)$ is closed and included in $\mathbb{I}\mathfrak{a}(Y)$. On the other hand, it coincides with $\mathfrak{a}(Y)$. Hence $\mathfrak{a}(Y)$ is both closed and open. However, $\mathfrak{a}(Y)$ is not the whole space because it is a proper part of $\mathfrak{a}(Z)$. So U is the union of the disjoint nonempty open sets $\mathfrak{a}(Y)$ and $U - \mathfrak{a}(Y)$. ❑

Thus, if we want to generalize Renz's result to spatio-temporal logics, we should base them on $\mathcal{RCC}$-8 rather than $\mathcal{BRCC}$-8. Denote the corresponding reducts of the $\mathcal{ST}_i$ by $\mathcal{ST}_i^-$, $i = 0, 1, 2$. (Remember that $\mathcal{ST}_1^-$ was already defined in Section 16.3.) However, in general even this restriction is not enough. Since the operation of forming unions of regions is implicitly available in the language $\mathcal{ST}_2^-$ in the form of $\Diamond_F$, we obtain the following:

Proposition 16.21. *There is an $\mathcal{ST}_2^-$-formula satisfiable in some tt-model with* **FSA**, *but not in a model based on a connected topological space, in particular not in $\langle \mathbb{R}^n, \mathbb{I} \rangle$, for any $n \geq 1$.*

Proof. Let φ be the conjunction of the predicates:

$$\mathsf{EQ}(\Diamond_F X, Y), \quad \mathsf{NTPP}(\bigcirc X, Y), \quad \mathsf{NTPP}(\bigcirc \Diamond_F X, Y), \quad \mathsf{NTPP}(Y, Z).$$

Suppose that $(\mathfrak{M}, w) \models \varphi$ for some tt-model $\mathfrak{M} = \langle \mathfrak{F}, \mathfrak{T}, \mathfrak{a} \rangle$ with **FSA**, where $\mathfrak{F} = \langle W, < \rangle$ is a discrete flow of time, $\mathfrak{T} = \langle U, \mathbb{I} \rangle$ a topological space and $w \in W$. Let w' be the immediate successor of w. Then we have:

$$\begin{aligned}
\mathfrak{a}(\Diamond_F X, w) &= \mathbb{C}\mathbb{I} \bigcup_{v > w} \mathfrak{a}(X, v) = \mathbb{C}\mathbb{I}\big(\mathfrak{a}(X, w') \cup \bigcup_{v > w'} \mathfrak{a}(X, v)\big) \\
&\overset{\mathbf{FSA}}{=} \mathbb{C}\mathbb{I}\mathfrak{a}(X, w') \cup \mathbb{C}\mathbb{I} \bigcup_{v > w'} \mathfrak{a}(X, v) \\
&= \mathfrak{a}(X, w') \cup \mathfrak{a}(\Diamond_F X, w') = \mathfrak{a}(\bigcirc X, w) \cup \mathfrak{a}(\bigcirc \Diamond_F X, w).
\end{aligned}$$

The remaining part of the proof is the same as that of Proposition 16.20. ❑

Fortunately, this is not the case for $\mathcal{ST}_1^-$.

Theorem 16.22. *The following conditions are equivalent for every $\mathcal{ST}_1^-$-formula φ, every countable discrete flow of time $\mathfrak{F}$, and every $n \geq 1$:*

- *φ is satisfiable in a tt-model based on $\mathfrak{F}$;*
- *φ is satisfiable in a tt-model based on $\mathfrak{F}$ and the Euclidean space $\langle \mathbb{R}^n, \mathbb{I} \rangle$.*

Proof. Fix some $n \geq 1$, a countable discrete flow of time $\mathfrak{F} = \langle W, < \rangle$ and an $\mathcal{ST}_1^-$-formula φ which is satisfied in a tt-model $\mathfrak{M} = \langle \mathfrak{F}, \mathfrak{T}, \mathfrak{a} \rangle$.

Suppose $\{X_1, \dots, X_m\}$ are the region variables occurring in φ. For every $w \in W$, take m fresh region variables X_i^w, $1 \leq i \leq m$, and define a set Γ of $\mathcal{RCC}$-8 formulas by taking

$$\Gamma = \{\mathsf{R}(X_i^{w_1}, X_j^{w_2}) \mid w_1, w_2 \in W,\ 1 \leq i, j \leq m, \text{ and } \mathsf{R} \text{ is an } \mathcal{RCC}\text{-8 relation or its negation for which } \mathsf{R}(\mathfrak{a}(X_i, w_1), \mathfrak{a}(X_j, w_2)) \text{ holds in } \mathfrak{T}\}.$$

We claim that if the set Γ is satisfied in a topological space $\mathfrak{T}'$ under an assignment $\mathfrak{b}$ (i.e., $\mathfrak{T}' \models^{\mathfrak{b}} \psi$ for all $\psi \in \Gamma$), then φ is satisfied in a tt-model $\mathfrak{M}' = \langle \mathfrak{F}, \mathfrak{T}', \mathfrak{a}' \rangle$. Indeed, for every $w \in W$ and every $i = 1, \dots, m$, set $\mathfrak{a}'(X_i, w) = \mathfrak{b}(X_i^w)$. We show by induction that for every subformula χ of φ and every $w \in W$,

$$(\mathfrak{M}, w) \models \chi \quad \text{iff} \quad (\mathfrak{M}', w) \models \chi.$$

Given $w \in W$ and $k < \omega$, define w^k by taking $w^0 = w$ and w^{k+1} to be the immediate successor of w^k. Let $\chi = \mathsf{R}(\bigcirc^{n_i} X_i, \bigcirc^{n_j} X_j)$. Suppose first that $(\mathfrak{M}, w) \models \chi$. Then $\mathsf{R}(\mathfrak{a}(X_i, w^{n_i}), \mathfrak{a}(X_j, w^{n_j}))$ holds and $\mathsf{R}(X_i^{w^{n_i}}, X_j^{w^{n_j}})$ belongs to Γ. Thus, $\mathfrak{T}' \models^{\mathfrak{b}} \mathsf{R}(X_i^{w^{n_i}}, X_j^{w^{n_j}})$, and so $\mathsf{R}(\mathfrak{a}'(X_i, w^{n_i}), \mathfrak{a}'(X_j, w^{n_j}))$ holds, from which $(\mathfrak{M}', w) \models \chi$.

Conversely, suppose that $(\mathfrak{M}', w) \models \chi$. Then $\mathfrak{T}' \models^{\mathfrak{b}} \mathsf{R}(X_i^{w^{n_i}}, X_j^{w^{n_j}})$, which implies (since $\mathfrak{T}' \models^{\mathfrak{b}} \psi$ for all $\psi \in \Gamma$) that $\mathsf{R}(X_i^{w^{n_i}}, X_j^{w^{n_j}})$ belongs to Γ. Hence $(\mathfrak{M}, w) \models \chi$. The induction steps for the Booleans, $\mathcal{U}$ and $\mathcal{S}$ are straightforward and left to the reader.

So it is enough to prove that Γ is satisfiable in $\langle \mathbb{R}^n, \mathbb{I} \rangle$. To this end, we first show that

$$\Gamma \text{ is satisfiable in } \langle \mathbb{R}, \mathbb{I} \rangle. \tag{16.8}$$

Indeed, Γ is clearly satisfiable in $\mathfrak{T}$ (simply put $\mathfrak{b}(X_i^w) = \mathfrak{a}(X_i, w)$). Now an inspection of the proof of Theorem 16.4 shows that Theorem 2.33 can be generalized from $\mathcal{BRCC}$-8 formulas to *arbitrary sets* of $\mathcal{BRCC}$-8 formulas: such a set Σ is satisfiable in a topological space iff the set

$$\Sigma^{\bowtie} = \{\psi^{\bowtie} \mid \psi \in \Sigma\}$$

of $\mathcal{ML}^u$-formulas is *satisfiable* in a quasisaw $\mathfrak{G}$ (in the sense that there is a model $\mathfrak{M}$ based on $\mathfrak{G}$ and such that $(\mathfrak{M}, x) \models \psi^{\bowtie}$ for all $\psi \in \Sigma$ and all—or, equivalently, some—points x in $\mathfrak{G}$). So $\Gamma^{\bowtie}$ is satisfiable in a quasisaw. Since $\mathfrak{F}$ is countable, $\Gamma^{\bowtie}$ is countable as well. Now, by using the standard first-order translation of $\mathcal{ML}^u$-formulas and then applying the downward Löwenheim–Skolem–Tarski theorem, we can assume that this quasisaw is the disjoint union of countably many forks.

As before, with a slight abuse of notation, we identify each region variable X in Γ with its translation $X^{\bowtie}$. Moreover, as we are going to work in a quasisaw, without loss of generality we may assume that $X^{\bowtie} = X$ is a propositional variable (and not $\mathbf{CI}X$ as in the original definition), that is, we may assume that the model satisfying $\Gamma^{\bowtie}$ is a saw model (see Section 16.3).

So suppose that $\Gamma^{\bowtie}$ is satisfiable in a saw model $\mathfrak{M} = \langle \mathfrak{G}, \mathfrak{V} \rangle$ such that $\mathfrak{G}$ is the disjoint union of countably many forks $\mathfrak{f}_k$, $k < \omega$, where $\mathfrak{f}_k = \langle W_k, R_k \rangle$, $W_k = \{b_k, l_k, r_k\}$ and R_k is the reflexive closure of $\{\langle b_k, l_k \rangle, \langle b_k, r_k \rangle\}$.

Denote by $\mathcal{X}_{bl}^k$ the set of all region variables X such that

$$\mathfrak{V}(X) \cap W_k = \{b_k, l_k\}.$$

Analogously, $\mathcal{X}_{br}^k$ and $\mathcal{X}_{blr}^k$ are the sets of all region variables X such that

$$\mathfrak{V}(X) \cap W_k = \{b_k, r_k\},$$
$$\mathfrak{V}(X) \cap W_k = \{b_k, l_k, r_k\},$$

respectively, and let $\mathcal{X}^k = \mathcal{X}_{bl}^k \cup \mathcal{X}_{br}^k \cup \mathcal{X}_{blr}^k$.

For each $k < \omega$, we then choose three maps

$$f_{bl}^k : \mathcal{X}_{bl}^k \to (0, 0.2),$$
$$f_{br}^k : \mathcal{X}_{br}^k \to (0, 0.2),$$
$$f_{blr}^k : \mathcal{X}_{blr}^k \to (0.3, 0.4)$$

in such a way that, for every $\bar{e} \in \{bl, br, blr\}$ and all $X, Y \in \mathcal{X}_{\bar{e}}^k$,

$$f_{\bar{e}}^k(X) \le f_{\bar{e}}^k(Y) \quad \text{if} \quad \mathfrak{V}(X) \subseteq \mathfrak{V}(Y), \tag{16.9}$$

and

$$f_{\bar{e}}^k(X) \ne f_{\bar{e}}^k(Y) \quad \text{if} \quad \mathfrak{V}(X) \ne \mathfrak{V}(Y). \tag{16.10}$$

Clearly, such maps exist. Then we put, for every X in Γ,

$$\mathfrak{b}^k(X) = \begin{cases} [k - f_{bl}^k(X), k], & \text{if } X \in \mathcal{X}_{bl}^k, \\ [k, k + f_{br}^k(X)], & \text{if } X \in \mathcal{X}_{br}^k, \\ [k - f_{blr}^k(X), k + f_{blr}^k(X)], & \text{if } X \in \mathcal{X}_{blr}^k, \\ \emptyset, & \text{if } X \notin \mathcal{X}^k, \end{cases}$$

see Fig. 16.1. Finally, let $\mathfrak{b}(X) = \bigcup_{k<\omega} \mathfrak{b}^k(X)$, for every region variable X.

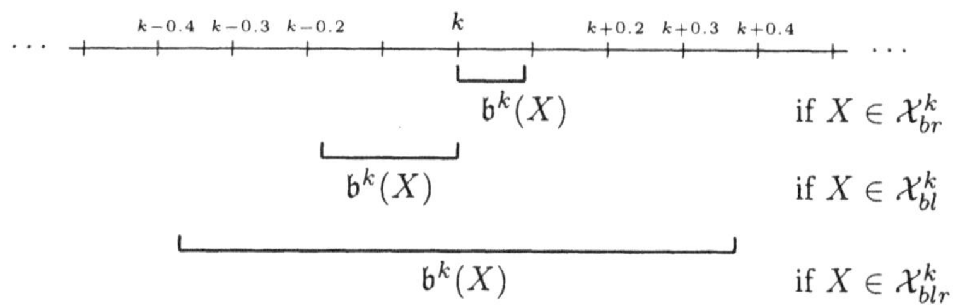

Figure 16.1: The assignment $\mathfrak{b}^k$.

It is a matter of routine to show now that $\langle \mathbb{R}, \mathbb{I}\rangle \models^{\mathfrak{b}} \psi$ for all $\psi \in \Gamma$. We will consider here only two cases.

Suppose $\mathsf{EC}(X, Y) \in \Gamma$. Then

$$\mathfrak{M} \models \diamondsuit(X \wedge Y) \wedge \neg\diamondsuit(\mathbf{I}X \wedge \mathbf{I}Y). \tag{16.11}$$

By the first conjunct in (16.11), there is a $k < \omega$ such that $W_k \cap \mathfrak{V}(X) \cap \mathfrak{V}(Y)$ is not empty, and so $X, Y \in \mathcal{X}^k$. It follows that $\mathfrak{b}^k(X) \cap \mathfrak{b}^k(Y) \neq \emptyset$ (see Fig. 16.1). On the other hand, by the second conjunct in (16.11), we have that $\mathfrak{V}(X) \cap \mathfrak{V}(Y)$ is disjoint from $\bigcup_{i<\omega}\{l_i, r_i\}$. So if $W_k \cap \mathfrak{V}(X) \cap \mathfrak{V}(Y) \neq \emptyset$ then either $X \in \mathcal{X}^k_{bl}$ and $Y \in \mathcal{X}^k_{br}$, or $Y \in \mathcal{X}^k_{bl}$ and $X \in \mathcal{X}^k_{br}$. In both cases, $\mathfrak{b}^k(X)$ and $\mathfrak{b}^k(Y)$ are externally connected.

Suppose $\mathsf{NTPP}(X, Y) \in \Gamma$. Then

$$\mathfrak{M} \models \boxdot(\neg X \vee \mathbf{I}Y) \wedge \diamondsuit(\neg X \wedge Y). \tag{16.12}$$

First, observe that, by the first conjunct in (16.12) and the reflexivity of $\mathfrak{G}$, we have $\mathfrak{V}(X) \subseteq \mathfrak{V}(Y)$ and, by the second conjunct in (16.12), $\mathfrak{V}(Y) - \mathfrak{V}(X) \neq \emptyset$. So, by (16.9) and (16.10), there is a $k < \omega$ such that $\mathfrak{b}^k(Y) - \mathfrak{b}^k(X) \neq \emptyset$. It suffices to show that $\mathfrak{b}(X)$ is included in the interior of $\mathfrak{b}(Y)$. Suppose $\mathfrak{b}^k(X) \neq \emptyset$, for some $k < \omega$, that is, $X \in \mathcal{X}^k$. Then, by the first conjunct in (16.12), we have $Y \in \mathcal{X}^k_{blr}$. There are two cases:

Case 1: $X \in \mathcal{X}^k_{bl} \cup \mathcal{X}^k_{br}$. Then $\mathfrak{b}^k(X)$ is included in the interior of $\mathfrak{b}^k(Y)$ by the definition of $\mathfrak{b}^k$ (see Fig. 16.1).

Case 2: $X \in \mathcal{X}^k_{blr}$. Then $f^k_{blr}(X) < f^k_{blr}(Y)$, by (16.9) and (16.10). So $\mathfrak{b}^k(X)$ is included in the interior of $\mathfrak{b}^k(Y)$ (see Fig. 16.1).

The cases of the other $\mathcal{RCC}$-8 predicates and their negations are similar and left to the reader. So we have shown (16.8).

It is not hard to see that if Γ satisfiable in $\langle \mathbb{R}, \mathbb{I}\rangle$, then it is also satisfiable in $\langle \mathbb{R}^n, \mathbb{I}\rangle$ for any $n > 1$ (if $\mathfrak{b}$ is an assignment in $\langle \mathbb{R}, \mathbb{I}\rangle$ satisfying Γ, then put $\mathfrak{b}'(X) = \mathfrak{b}(X) \times \mathbb{R}^{n-1}$). This completes the proof of Theorem 16.22. ❑

As a consequence of Theorems 16.9 and 16.22 we finally obtain:

Theorem 16.23. *Let $\mathcal{C}$ be any of the following classes of flows of time: $\{\langle \mathbb{N}, <\rangle\}$, $\{\langle \mathbb{Z}, <\rangle\}$, the class of all finite strict linear orders, any discrete first-order definable class of strict linear orders. Then satisfiability of $\mathcal{ST}_1^-$-formulas in tt-models based on a flow of time in $\mathcal{C}$ and on the topological space $\langle \mathbb{R}^n, \mathbb{I}\rangle$, for any $n \geq 1$, is decidable.*

Moreover, as a consequence of Theorems 16.19 and 16.22 we obtain:

Theorem 16.24. *The satisfiability of $\mathcal{ST}_1^-$-formulas in tt-models based on the flow of time $\langle \mathbb{N}, <\rangle$ and on the topological space $\langle \mathbb{R}^n, \mathbb{I}\rangle$, for any $n \geq 1$, is* PSPACE*-complete.*

Epilogue

We have considered so many different languages and logics in this book that, although most of the presented results are collected and systematized in numerous tables and diagrams, a brief summary from a 'bird's-eye view' of what has been done and what lies ahead may still be helpful.

Our main objects of investigation have been *modal-like* languages interpreted in *many-dimensional* structures. Three important observations motivate our interest in these formalisms:

(1) Various kinds of languages stemming (directly or indirectly) from Modal Logic[1] have been recognized as major representation and reasoning tools in many fields of computer science, artificial intelligence, philosophy, computational linguistics, the foundations of mathematics, etc. To a large extent, this success of 'modal' logics is due to their being a reasonable compromise between expressivity and effectiveness. In fact, one can often implement a reasoning procedure for a standard modal logic in a rather straightforward way, even without bothering about its computational behavior.

(2) On the other hand, realistic applications of modal logics in computer science, artificial intelligence, philosophy and other disciplines usually require a number of *interacting* modal operators. It is not sufficient to model time, space, belief, terminology, action, etc., independently of each other. What we actually need is semantically well-founded *composed* logics, which leads us to models of many (at least two) dimensions.

(3) While standard 'one-dimensional' modal logics were celebrated for their robust computational behavior (Vardi 1997, Grädel 2001), many-dimensional ones often exhibit rather nasty computational properties, and

[1] Unfortunately, the name of the discipline—Modal Logic—no longer reflects the diversity of logical systems sheltered under its roof, implying that all of them belong to Philosophical Logic. However, no better name has been suggested so far (shall we announce a competition?), and the only consolation is that many fields of traditional modal logic have become full-fledged research areas with their own names: temporal logic, dynamic logic, description logic, etc.

> the toolkit of standard modal logic is no longer directly applicable. Moreover, straightforward naïve constructions of many-dimensional formalisms from one-dimensional ones will almost certainly result in computationally useless 'monsters.'

In this monograph we develop a mathematical framework which could guide (nonmathematician) users when constructing their many-dimensional formalisms. We provide two fundamental and interrelated kinds of mathematical abstractions for such a framework: *products of propositional modal logics* and *(decidable fragments of) first-order temporal, epistemic, etc. logics*. We developed a mathematical machinery for dealing with these logics in Parts II and III and then, in Part IV, we applied it to three case studies from the field of knowledge representation and reasoning, in particular, with the aim of illustrating the subtle trade-off between expressivity and computational properties of multidimensional formalisms. In all three cases—temporal epistemic logics, modalized description logics, and spatio-temporal logics—we suggested hierarchies of more and more expressive languages and showed how the computational complexity of the resulting logical systems increases. Although the formalisms we discussed cannot be directly transformed into 'real systems for industrial application,' we believe that they provide sufficient guidance for the designer of such systems.

Now we very briefly sum up the major technical results and open problems of this book.

Products of modal logics

The investigation of products of modal logics constitutes the mathematical core of our approach. As traditional methods of handling modal logics (say, canonical models and filtration) are very rarely applicable to products, we introduced the new fundamental method of quasimodels for obtaining positive results (decidability, axiomatizability) and used various (sometimes rather sophisticated) 'encoding techniques' for establishing lower bounds of the computational complexity (in particular, undecidability or nonrecursive enumerability). Numerous results for applied many-dimensional logics were obtained by reductions to products of modal logics or by generalizing methodologies first explored in the context of products. Three main conclusions can be drawn from our investigation:

1. While 'standard' one-dimensional modal logics are usually finitely axiomatizable and decidable in PSPACE or EXPTIME, already two-dimensional products of such (or even 'simpler') logics are often not finitely axiomatizable and at least NEXPTIME-hard.

2. There is a huge gap between two- and higher-dimensional products: while in the former case we obtain an unexpected diversity of decidable and undecidable, finitely axiomatizable and nonrecursively enumerable logics, in the latter case almost all logics are undecidable and nonfinitely axiomatizable.

3. When in trying to improve the bad computational behavior of product logics, we move from interpretations in full product frames to interpretations in their substructures, the resulting logics may unexpectedly lose the genuine many-dimensional character and degenerate to fusions of modal logics (with all their nice computational properties, but without interacting modal operators).

Let us now consider in more detail the three main problems we focused on in this book.

Decidability. As mentioned above, two-dimensional product logics exhibit a rather diverging computational behavior. The first 'rule of thumb' can be formulated as follows:

☞ *The product of two modal logics is decidable if one of them is 'similar' to* $\mathbf{K}_m$ *or* $\mathbf{S5}_m$*, while the other one can be 'very expressive,' say, dynamic logic* **CPDL**, *epistemic logic* $\mathbf{S5}_n^C$*, temporal logic* **PTL**, *etc. The formation of two-dimensional products of other kinds of logics may be 'dangerous:' there is a good chance that they may be undecidable.*

The term 'similar' in this rule excludes all logics which are 'nonlocal' in the sense that one can quantify over 'sufficiently many points with sufficiently rich structure,' e.g., **K4**, $\mathbf{K}_u$, or **S4.3** (for instance, $\mathsf{Log}\{\langle\mathbb{N},<\rangle\} \times \mathbf{K4}$ is undecidable). We still do not have enough information to refine the 'dangerous' part of this rule. The following approximation may serve as the second 'rule of thumb' reflecting our current knowledge:

☞ *Products of modal logics determined by linear orders are usually undecidable.*

The picture becomes absolutely unclear if both components of the product are sort of 'weak standard' modal logics[2] different from $\mathbf{K}_m$ and $\mathbf{S5}_m$. In fact, the main challenging open question is whether products of *transitive* logics like $\mathbf{K4} \times \mathbf{K4}$ or $\mathbf{K4} \times \mathbf{K4.3}$ are decidable. In this connection, we should admit that a deep and systematic investigation into the structure of abstract (i.e., not necessarily product) frames for these logics is still missing.

As mentioned above, higher-dimensional products are often very complex:

[2]We mean logics similar to those in Fig. 1.1.

☞ *For $n \geq 3$, n-dimensional products of modal logics are usually undecidable.*

Complexity. Two-dimensional products of standard modal logics are at least NEXPTIME-hard, but often they are even more complex (for example, **K4.3** $\times$ **S5** is EXPSPACE-hard). The main distinction here is between products of expressive one-dimensional modal logics with (a) logics similar to **S5**, which are usually in ELEM, and (b) products of those expressive logics with logics similar to $\mathbf{K}_m$ (for $m \geq 1$) or $\mathbf{S5}_n$ (for $n \geq 2$), which are usually not in ELEM. If both modal logics are weak standard logics, however, numerous complexity questions remain open. Perhaps, the most challenging problem is whether **K** $\times$ **K** belongs to ELEM or not.

It is important to remember that the formation of products is not monotonic: logics L_1 and L_2 may be less complex than L'_1 and L'_2, and yet $L_1 \times L_2$ is more complex than $L'_1 \times L'_2$. For example, **S4.3** is NP-complete and therefore less complex than, say, **K**; however, **S4.3** $\times$ **S4.3** is undecidable, while **K** $\times$ **K** is decidable. It is this fact that makes it rather difficult to describe a general picture of the computational complexity of product logics.

Axiomatizability. Here again the landscape is rather diverse. Our first 'rule of thumb' says:

☞ *The product of any number of modal logics whose classes of frames are definable by recursive sets of first-order formulas is recursively enumerable. Moreover, the product of two Horn definable logics is product-matching: it can be axiomatized by putting together the axioms of the components and the natural interaction axioms. So if the two components are finitely axiomatizable and Horn definable, then their product is finitely axiomatizable as well. Two-dimensional products of other kinds of logics may not be such.*

The term 'may not be' in the last sentence refers to the fact that products of transitive logics with either **K4.3** or **Grz** are usually not product-matching. However, we do not know any example of a pair of finitely axiomatizable logics such that their product is recursively enumerable but nonfinitely axiomatizable. It can happen that the product of two finitely axiomatizable logics is not even recursively enumerable: such product logics are, for example, $\mathsf{Log}\{\langle \mathbb{N}, < \rangle\} \times \mathsf{Log}\{\langle \mathbb{N}, < \rangle\}$ and **Grz.3** $\times$ **Grz.3**.

In some cases the main obstacle in proving an axiomatization result for products is that we do not know that the logic, obtained by putting together the axioms for the components and the natural interaction axioms, is Kripke complete. It would be interesting to find an example when it is not.

In higher dimensions good news is rare:

☞ *For $n \geq 3$, n-dimensional products of modal logics are usually nonfinitely axiomatizable.*

Fragments of first-order modal logics

First-order modal logics have become notorious for their extremely bad computational behavior since the 1960s: their two-variable or monadic fragments are usually undecidable; such fragments of expressive modal logics (like quantified **PTL** or epistemic modal logics with the common knowledge operator) are not even recursively enumerable. Given the 'negative' results on three- (or more) dimensional products of propositional modal logics, this fact should not be too surprising, however, because the $n + 1$-dimensional product logic $L \times \mathbf{S5} \times \cdots \times \mathbf{S5}$ is embeddable into the n-variable fragment of the first-order extension of L, for any Kripke complete modal logic L.

In Part III we introduced the first general methodology for constructing relatively expressive, but still reasonably well-behaved fragments of first-order modal logics. Roughly, the idea behind the construction is that we have to be 'closer' to two-dimensional products than to three-dimensional ones.[3] It turned out that this can be achieved by restricting applications of modal operators to formulas with at most one free variable; the resulting formulas were called *monodic*. The two main results on monodic fragments can be formulated as follows:

☞ *The monodic fragments of first-order extensions of very expressive one-dimensional modal logics (like* **PTL** *or epistemic modal logics with common knowledge operators) are usually recursively axiomatizable.*

☞ *If the pure first-order (nonmodal) part of a subfragment of the monodic fragment of a very expressive first-order modal logic is decidable, then usually the subfragment itself is decidable as well. This applies, for example, to the two-variable monodic fragment, the monadic monodic fragment, and various guarded monodic fragments.*

These results require a few of comments: (a) they apply only to languages without function symbols and equality (although equality can be used in the guarded fragments); (b) they apply to models with arbitrary constant domains and finite constant domains; (c) the qualification 'usually' should be taken seriously: for example, it is one of the main open problems in the field whether the latter result holds for the first-order extension of the temporal logic over the real line $\langle \mathbb{R}, < \rangle$. (It holds true under finite domains!)

[3]In this connection it may be of interest to refer to a result of Hodkinson *et al.* (2002) which shows that no '$2\frac{1}{2}$D' product of any (1D) modal logic located between **K** and **S5** with the ($1\frac{1}{2}$D) computational tree logic **CTL*** is decidable.

We have already mentioned that the properties of monodic fragments can be partly explained by the fact that they are closer to two-dimensional products than to three-dimensional ones. In fact, the monodic fragments of the first-order extension of a modal logic L are in a sense similar to $L \times \mathbf{S5}$ (which, in turn, is equivalent to the one-variable fragment of the first-order extension of L), so that the quasimodel technique we used to investigate the monodic fragments closely resembles the quasimodel technique for products with **S5**. Therefore, again, it should not come as a surprise that decidable subfragments of monodic fragments are mostly in ELEM, in contrast to the nonelementarity of products with $\mathbf{K}_n$ (for $n \geq 1$) and $\mathbf{S5}_n$ (for $n > 1$).

Bibliography

Abadi 1987. M. Abadi. The power of temporal proofs. In *Proceedings of the 2nd Annual IEEE Symposium on Logic in Computer Science (LICS'87)*, pages 176–186. IEEE Computer Society, 1987.

Abiteboul *et al.* 1996. S. Abiteboul, L. Herr, and J. van den Bussche. Temporal versus first-order logic in query temporal databases. In *ACM Symposium on Principles of Database Systems*, pages 49–57, Montreal, Canada, 1996.

Aiello and van Benthem 2000. M. Aiello and J. van Benthem. Logical patterns in space. In *Proceedings of the First CSLI Workshop on Visual Reasoning*. CSLI Publications, Stanford, 2000.

Aiello 2001. M. Aiello. A spatial similarity measure based on games: theory and practice. *Logic Journal of the IGPL*, 10:1–22, 2001.

Allen and Hayes 1985. J. Allen and P. Hayes. A common sense theory of time. In *Proceedings of the 9th International Joint Conference on Artificial Intelligence (IJCAI'85)*, pages 528–531. Morgan Kaufmann, 1985.

Allen and Koomen 1983. J. Allen and J. Koomen. Planning using a temporal world model. In *Proceedings of the 8th International Joint Conference on Artificial Intelligence (IJCAI'83)*, pages 741–747. Morgan Kaufmann, 1983.

Allen *et al.* 1991. J. Allen, H. Kautz, R. Pelavin, and J. Tenenberg. *Reasoning About Plans*. Morgan Kaufmann, 1991.

Allen 1983. J. Allen. Maintaining knowledge about temporal intervals. *Communications of the ACM*, 26:832–843, 1983.

Allen 1984. J. Allen. Towards a general theory of action and time. *Artificial Intelligence*, 26:123–154, 1984.

Anderson and Belnap 1975. A. Anderson and N. Belnap. *Entailment. The Logic of Relevance and Necessity. I.* Princeton University Press, 1975.

Andréka and Mikulás 1994. H. Andréka and Sz. Mikulás. Lambek calculus and its relational semantics: completeness and incompleteness. *Journal of Logic, Language, and Information*, 3:1–37, 1994.

Andréka and Németi 1994. H. Andréka and I. Németi. Simple proof for decidability of the universal theory of cylindric set algebras of dimension 2. Handout, "Algebraic Logic and the Methodology of Applying It," TEMPUS Summer School, Budapest, Hungary, 1994.

Andréka *et al.* 1979. H. Andréka, I. Németi, and I. Sain. Completeness problems in verification of programs and program schemes. In Jirí Becvár, editor, *Mathematical Foundations of Computer Science 1979*, volume 74 of *Lecture Notes in Computer Science*, pages 208–218. Springer, 1979.

Andréka *et al.* 1994. H. Andréka, A. Kurucz, I. Németi, I. Sain, and A. Simon. Exactly which logics touched by the dynamic trend are decidable? In *Proceedings of the 9th Amsterdam Colloquium*, pages 67–85, 1994.

Andréka *et al.* 1997. H. Andréka, S. Givant, and I. Németi. Decision problems for equational theories of relation algebras. *Memoirs of the AMS*, 126(604), 1997.

Andréka *et al.* 1998. H. Andréka, I. Németi, and J. van Benthem. Modal languages and bounded fragments of predicate logic. *Journal of Philosophical Logic*, 27:217–274, 1998.

Andréka *et al.* 2001. H. Andréka, I. Németi, and I. Sain. Algebraic logic. In D. Gabbay and F. Guenthner, editors, *Handbook of Philosophical Logic*, volume II, pages 133–247. Kluwer Academic Publishers, 2nd edition, 2001.

Andréka 1997. H. Andréka. Complexity of equations valid in algebras of relations. Part I: strong non-finitizability. *Annals of Pure and Applied Logic*, 89:149–209, 1997.

Areces *et al.* 2000. C. Areces, P. Blackburn, and M. Marx. The computational complexity of hybrid temporal logics. *Logic Journal of the IGPL*, 8:653–679, 2000.

Armando 2002. A. Armando, editor. *Frontiers of Combining Systems IV*, volume 2309 of *Lecture Notes in Artificial Intelligence*. Springer, 2002.

Artale and Franconi 1998. A. Artale and E. Franconi. A temporal description logic for reasoning about actions and plans. *Journal of Artificial Intelligence Research*, 9:463–506, 1998.

Artale and Franconi 2003. A. Artale and E. Franconi. Temporal description logics. In *Handbook of Time and Temporal Reasoning in Artificial Intelligence.* MIT Press, 2003. (To appear).

Artale *et al.* 2002. A. Artale, E. Franconi, F. Wolter, and M. Zakharyaschev. A temporal description logic for reasoning about conceptual schemas and queries. In S. Flesca, S. Greco, N. Leone, and G. Ianni, editors, *Proceedings of JELIA'02*, volume 2424 of *Lecture Notes in Computer Science*, pages 98–110. Springer, 2002.

Artemov 2001. S. Artemov. Explicit provability and constructive semantics. *Bulletin of Symbolic Logic*, 7:1–36, 2001.

Aumann 1976. R. Aumann. Agreeing to disagree. *The Annals of Statistics*, 4:1236–1239, 1976.

Baader and Hanschke 1991. F. Baader and P. Hanschke. A scheme for integrating concrete domains into concept languages. In J. Mylopoulos and R. Reiter, editors, *Proceedings of the 12th International Joint Conference on Artificial Intelligence (IJCAI'91)*, pages 452–457. Morgan Kaufmann, 1991.

Baader and Laux 1995. F. Baader and A. Laux. Terminological logics with modal operators. In *Proceedings of the 14th International Joint Conference on Artificial Intelligence (IJCAI'95)*, pages 808–814. Morgan Kaufmann, 1995.

Baader and Ohlbach 1995. F. Baader and H.J. Ohlbach. A multi-dimensional terminological knowledge representation language. *Journal of Applied Non-Classical Logics*, 5:153–197, 1995.

Baader and Schulz 1996. F. Baader and K. Schulz, editors. *Frontiers of Combining Systems.* Kluwer Academic Publishers, 1996.

Baader and Tinelli 1997. F. Baader and C. Tinelli. A new approach to combining decision procedures for the word problem, and its connection to the Nelson–Oppen combination method. In W. McCune, editor, *Proceedings of CADE-14*, volume 1249 of *Lecture Notes in Artificial Intelligence*, pages 19–33. Springer, 1997.

Baader and Tinelli 2002. F. Baader and C. Tinelli. Deciding the word problem in the union of equational theories. *Information and Computation*, 178:346–390, 2002.

Baader and Tobies 2001. F. Baader and S. Tobies. The inverse method implements the automata approach for modal satisfiability. In F. Baader and

U. Sattler, editors, *Automated Reasoning, IJCAR 2001*, volume 2083 of *Lecture Notes in Artificial Intelligence*, pages 92–106. Springer, 2001.

Baader *et al.* 2002. F. Baader, C. Lutz, H. Sturm, and F. Wolter. Fusions of description logics and abstract description systems. *Journal of Artificial Intelligence Research*, 16:1–58, 2002.

Baader *et al.* 2003. F. Baader, D. Calvanese, D. McGuinness, D. Nardi, and P. Patel-Schneider, editors. *The Description Logic Handbook: Theory, Implementation, and Applications*. Cambridge University Press, 2003.

Baader 1990. F. Baader. Terminological cycles in KL-ONE-based knowledge representation languages. In *Proceedings of the 8th National Conference on Artificial Intelligence*, pages 621–626. AAAI Press/MIT Press, 1990.

Bacharach 1994. M. Bacharach. The epistemic structure of a theory of games. *Theory and Decision*, 37:7–48, 1994.

Balbiani and Condotta 2002. P. Balbiani and J.-F. Condotta. Computational complexity of propositional linear temporal logics based on qualitative spatial or temporal reasoning. In A. Armando, editor, *Proceedings of Frontiers of Combining Systems (FroCoS 2002)*, volume 2309 of *Lecture Notes in Computer Science*, pages 162–176. Springer, 2002.

Barwise 1977. J. Barwise, editor. *Handbook of Mathematical Logic*. Elsevier, North-Holland, 1977.

Bauer *et al.* 2002. S. Bauer, I. Hodkinson, F. Wolter, and M. Zakharyaschev. On non-local propositional and local one variable CTL*. In *Proceedings of TIME 2002*, pages 2–9. IEEE Computer Society, 2002.

Bennett *et al.* 1997. B. Bennett, A. Cohn, and A. Isli. A logical approach to incorporating qualitative spatial reasoning into GIS. In S. Hirtle and A. Frank, editors, *Proceedings of the International Conference on Spatial Information Theory (COSIT)*, volume 1329 of *Lecture Notes in Computer Science*, pages 503–504. Springer, 1997.

Bennett 1977. M. Bennett. A guide to the logic of tense and aspect in English. *Logique et Analyse*, 20:137–163, 1977.

Bennett 1994. B. Bennett. Spatial reasoning with propositional logic. In *Proceedings of the 4th International Conference on Knowledge Representation and Reasoning*, pages 51–62. Morgan Kaufmann, 1994.

Bennett 1996. B. Bennett. Modal logics for qualitative spatial reasoning. *Logic Journal of the IGPL*, 4:23–45, 1996.

Bennett 1998. B. Bennett. Determining consistency of topological relations. *Constraints*, 3:213–225, 1998.

Berger 1966. R. Berger. The undecidability of the domino problem. *Memoirs of the AMS*, 66, 1966.

Bergstra and Klop 1984. J. Bergstra and J. Klop. The algebra of recursively defined processes and the algebra of regular processes. In J. Paredaens, editor, *Proceedings of the 11th ICALP*, volume 172 of *Lecture Notes in Computer Science*, pages 82–95. Springer, 1984.

Berman 1979. F. Berman. A completeness technique for D-axiomatizable semantics. In *Proceedings of the 11th Annual ACM Symposium on Theory of Computing (STOC)*, pages 160–166. ACM, 1979.

Bessonov 1977. A. Bessonov. New operations in intuitionistic calculus. *Mathematical Notes*, 22:503–506, 1977.

Beth 1956. E.W. Beth. Semantic construction of intuitionistic logic. *Mededelingen der Koninklijke Nederlandse Akademie van Wetenschappen, Afd. Letterkunde*, 19:357–388, 1956.

Bezhanishvili 1998. G. Bezhanishvili. Varieties of monadic algebras. Part I. *Studia Logica*, 61:367–402, 1998.

Blackburn *et al.* 2001. P. Blackburn, M. de Rijke, and Y. Venema. *Modal Logic.* Cambridge University Press, 2001.

Blackburn 1992. P. Blackburn. Fine grained theories of time. In M. Aurnague, A. Borillo, M. Borillo, and M. Bras, editors, *Proceedings of the 4th European Workshop on Semantics of Time, Space, and Movement and Spatio-Temporal Reasoning*, pages 299–320, Château de Bonas, France, 1992. Groupe "Langue, Raisonnement, Calcul", Toulouse.

Blackburn 1993. P. Blackburn. Nominal tense logic. *Notre Dame Journal of Formal Logic*, 34:56–83, 1993.

Blok and Pigozzi 1989. W. Blok and D. Pigozzi. Algebraizable Logics. *Memoirs of the AMS*, 77(396), 1989.

Blok 1978. W. Blok. On the degree of incompleteness in modal logics and the covering relation in the lattice of modal logics. Technical Report 78-07, Department of Mathematics, University of Amsterdam, 1978.

Boolos 1993. G. Boolos. *The Logic of Provability.* Cambridge University Press, 1993.

Börger *et al.* 1997. E. Börger, E. Grädel, and Yu. Gurevich. *The Classical Decision Problem.* Perspectives in Mathematical Logic. Springer, 1997.

Božić and Došen 1984. M. Božić and K. Došen. Models for normal intuitionistic logics. *Studia Logica*, 43:217–245, 1984.

Brachman and Levesque 1985. R. Brachman and H. Levesque. *Readings in Knowledge Representation.* Morgan Kaufmann, 1985.

Brachman and Schmolze 1985. R. Brachman and J. Schmolze. An overview of the KL-ONE knowledge representation system. *Cognitive Science*, 9:171–216, 1985.

Brouwer 1907. L.E.J. Brouwer. *Over de Grondslagen der Wiskunde.* PhD thesis, Amsterdam, 1907. Translation: "On the foundation of mathematics" in A. Heyting, editor, *Brouwer, Collected Works I*, pages 11–101. Elsevier, North-Holland, 1975.

Brouwer 1908. L.E.J. Brouwer. De onbetrouwbaarheid der logische principes. *Tijdschrift voor Wijsbegeerte*, 2:152–158, 1908. Translation: "The unreliability of the logical principles," ibid, pages 107–111.

Büchi 1962. J.R. Büchi. On a decision method in restricted second order arithmetic. In *Logic, Methodology and Philosophy of Science: Proceedings of the 1960 International Congress*, pages 1–11. Stanford University Press, 1962.

Bull 1965. R.A. Bull. A modal extension of intuitionistic logic. *Notre Dame Journal of Formal Logic*, 6:142–146, 1965.

Bull 1966. R.A. Bull. That all normal extensions of **S4.3** have the finite model property. *Zeitschrift für Mathematische Logik und Grundlagen der Mathematik*, 12:341–344, 1966.

Bull 1968. R.A. Bull. An algebraic study of tense logic with linear time. *Journal of Symbolic Logic*, 33:27–38, 1968.

Burgess and Gurevich 1985. J. Burgess and Yu. Gurevich. The decision problem for linear temporal logic. *Notre Dame Journal of Formal Logic*, 26:115–128, 1985.

Burris and Sankappanavar 1981. S. Burris and H. Sankappanavar. *A Course in Universal Algebra.* Springer, 1981.

Calvanese *et al.* 1998. D. Calvanese, M. Lenzerini, and D. Nardi. Description logics for conceptual data modeling. In J. Chomicki and G. Saake, editors, *Logics for Databases and Information Systems*, pages 229–263. Kluwer Academic Publishers, 1998.

Calvanese *et al.* 2001. D. Calvanese, G. De Giacomo, M. Lenzerini, and D. Nardi. Reasoning in expressive description logics. In A. Robinson and A. Voronkov, editors, *Handbook of Automated Reasoning*, pages 1581–1634. Elsevier, North-Holland, 2001.

Calvanese 1996. D. Calvanese. Finite model reasoning in description logics. In *Proceedings of the 5th International Conference on Principles of Knowledge Representation and Reasoning (KR'96)*, pages 292–303. Morgan Kaufmann, 1996.

Carnap 1942. R. Carnap. *Introduction to Semantics.* Harvard University Press, Cambridge, 1942.

Carnap 1947. R. Carnap. *Meaning and Necessity. A Study in Semantics and Modal Logic.* The University of Chicago Press, 1947.

Casati and Varzi 1999. R. Casati and A. Varzi. *Parts and Places. The Structures of Spatial Representations.* MIT Press, 1999.

Chagrov and Zakharyaschev 1997. A. Chagrov and M. Zakharyaschev. *Modal Logic*, volume 35 of *Oxford Logic Guides.* Clarendon Press, Oxford, 1997.

Chomicki and Niwinski 1995. J. Chomicki and D. Niwinski. On the feasibility of checking temporal integrity constraints. *Journal of Computer and Systems Sciences*, 51:523–535, 1995.

Chomicki and Toman 1998. J. Chomicki and D. Toman. Temporal logic in information systems. In J. Chomicki and G. Saake, editors, *Logics for Databases and Information Systems*, pages 31–70. Kluwer Academic Publishers, 1998.

Chomicki *et al.* 2001. J. Chomicki, D. Toman, and M. Böhlen. Querying ATSQL databases with temporal logic. *ACM Transactions on Database Systems*, 26:145–178, 2001.

Chomicki 1994. J. Chomicki. Temporal query languages: a survey. In D. Gabbay and H.J. Ohlbach, editors, *Temporal Logic, First International Conference (ICTL'94)*, volume 827 of *Lecture Notes in Computer Science*, pages 506–534. Springer, 1994.

Clarke and Emerson 1981. E. Clarke and E. Emerson. Design and synthesis of synchronisation skeletons using branching time temporal logic. In D. Kozen, editor, *Logic of Programs*, volume 131 of *Lecture Notes in Computer Science*, pages 52–71. Springer, 1981.

Clarke and Emerson 1982. E. Clarke and E. Emerson. Using branching time temporal logic to synthesize synchronisation skeletons. *Science of Computer Programming*, 2:241–266, 1982.

Clarke *et al.* 2000. E. Clarke, O. Grumberg, and D. Peled. *Model Checking*. MIT Press, 2000.

Cohn 1997. A. Cohn. Qualitative spatial representation and reasoning techniques. In G. Brewka, C. Habel, and B. Nebel, editors, *KI-97: Advances in Artificial Intelligence*, volume 1303 of *Lecture Notes in Computer Science*, pages 1–30. Springer, 1997.

Craig 1953. W. Craig. On axiomatizability within a system. *Journal of Symbolic Logic*, 18:30–32, 1953.

Craig 1957. W. Craig. Three uses of the Herbrandt–Gentzen theorem in relating model theory and proof theory. *Journal of Symbolic Logic*, 22:269–285, 1957.

Craig 1974. W. Craig. *Logic in Algebraic Form*. North-Holland, 1974.

D'Agostino and Hollenberg 1998. G. D'Agostino and M. Hollenberg. Uniform interpolation, automata and the modal mu-calculus. In M. Kracht, M. de Rijke, H. Wansing, and M. Zakharyaschev, editors, *Advances in Modal Logic*, volume 1, pages 73–84. CSLI Publications, Stanford, 1998.

D'Agostino *et al.* 1999. M. D'Agostino, D. Gabbay, R. Hähnle, and J. Posegga, editors. *Handbook of Tableau Methods*. Kluwer Academic Publishers, 1999.

Daigneault and Monk 1963. A. Daigneault and J.D. Monk. Representation theory for polyadic algebras. *Fundamenta Mathematicae*, 52:151–176, 1963.

De Giacomo and Lenzerini 1994. G. De Giacomo and M. Lenzerini. Boosting the correspondence between description logics and propositional dynamic logics. In *Proceedings of the 12th National Conference on Artificial Intelligence (AAAI'94)*, pages 205–212. AAAI Press/MIT Press, 1994.

De Giacomo and Lenzerini 1995. G. De Giacomo and M. Lenzerini. **PDL**-based framework for reasoning about actions. In *Proceedings of the 4th Congress of the Italian Association for Artificial Intelligence*, pages 103–114. Springer, 1995.

De Giacomo and Lenzerini 1996. G. De Giacomo and M. Lenzerini. TBox and ABox reasoning in expressive description logics. In *Proceedings of the 5th Conference on Principles of Knowledge Representation and Reasoning (KR'96)*, pages 316–327. Morgan Kaufmann, 1996.

De Giacomo 1995. G. De Giacomo. *Decidability of Class-Based Knowledge Representation Formalisms.* PhD thesis, Università di Roma, 1995.

De Morgan 1860. A. De Morgan. On the syllogism: IV, and on the logic of relations. Read before the Cambridge Philosophical Society on April 23, 1860. *Transactions of the Cambridge Philosophical Society*, 10:331–358, 1964.

de Rijke and Blackburn 1996. M. de Rijke and P. Blackburn, editors. *Notre Dame Journal of Formal Logic*, volume 37(2), 1996. Special Issue on Combining Logics.

de Rijke and Gabbay 2000. M. de Rijke and D. Gabbay, editors. *Frontiers of Combining Systems II.* Research Studies Press, England, 2000.

Degtyarev and Fisher 2001. A. Degtyarev and M. Fisher. Towards first-order temporal resolution. In F. Baader, G. Brewka, and T. Eiter, editors, *Advances in Artificial Intelligence (KI 2001)*, volume 2174 of *Lecture Notes in Artificial Intelligence*, pages 18–32. Springer, 2001.

Degtyarev *et al.* 2002. A. Degtyarev, M. Fisher, and A. Lisitsa. Equality and monodic first-order temporal logic. *Studia Logica*, 72:147–156, 2002.

Degtyarev *et al.* 2003a. A. Degtyarev, M. Fisher, and B. Konev. Exploring the monodic fragment of first-order temporal logic using clausal temporal resolution. Technical report, 2003. Submitted. Available as Technical report ULCS-03-012 from `http://www.csc.liv.ac.uk/research/`.

Degtyarev *et al.* 2003b. A. Degtyarev, M. Fisher, and B. Konev. Monodic temporal resolution. In *Proceedings of CADE-19*, Lecture Notes in Artificial Intelligence. Springer, 2003.

Demri and D'Souza 2002. S. Demri and D. D'Souza. An automata-theoretic approach to constraint LTL. In *Proceedings of the 22nd Conference on Foundations of Software Technology and Theoretical Computer Science*, volume 2556 of *Lecture Notes in Computer Science*, pages 121–132. Springer, 2002.

Domenjoud *et al.* 1994. E. Domenjoud, F. Klay, and C. Ringeissen. Combination techniques for non-disjoint equational theories. In A. Bundy, editor, *Proceedings of CADE-12*, volume 814 of *Lecture Notes in Artificial Intelligence*, pages 267–281, 1994.

Donini and Masacci 2000. F. Donini and F. Masacci. Exptime tableaux for $\mathcal{ALC}$. *Artificial Intelligence*, 124:87–138, 2000.

Donini *et al.* 1995. F. Donini, M. Lenzerini, D. Nardi, and W. Nutt. The complexity of concept languages. Technical Report RR-95-07, Deutsches Forschungszentrum für Künstliche Intelligenz (DFKI), 1995.

Donini *et al.* 1996. F. Donini, M. Lenzerini, D. Nardi, and A. Schaerf. Reasoning in description logics. In G. Brewka, editor, *Principles of Knowledge Representation*, pages 191–236. CSLI Publications, Stanford, 1996.

Dowty 1979. D. Dowty. *Word Meaning and Montague Grammar.* Reidel, Dordrecht, 1979.

Dummett and Lemmon 1959. M. Dummett and E. Lemmon. Modal logics between **S4** and **S5**. *Zeitschrift für Mathematische Logik und Grundlagen der Mathematik*, 5:250–264, 1959.

Dummett 1977. M. Dummett. *Elements of Intuitionism.* Clarendon Press, Oxford, 1977.

Egenhofer and Franzosa 1991. M. Egenhofer and R. Franzosa. Point-set topological spatial relations. *International Journal of Geographical Information Systems*, 5:161–174, 1991.

Egenhofer and Mark 1995. M. Egenhofer and D. Mark. Naive geography. In A. Frank and W. Kuhn, editors, *Spatial Information Theory: a Theoretical Basis for GIS*, volume 988 of *Lecture Notes in Computer Science*, pages 1–16. Springer, Berlin, 1995.

Emerson and Halpern 1985. E. Emerson and J. Halpern. Decision procedures and expressiveness in the temporal logic of branching time. *Journal of Computer and System Sciences*, 30:1–24, 1985.

Emerson 1990. E. Emerson. Temporal and modal logic. In J. van Leeuwen, editor, *Handbook of Theoretical Computer Science*, pages 996–1076, 1990.

Enderton 2001. H. Enderton. *A Mathematical Introduction to Logic.* Academic Press, New York, 2nd edition, 2001.

Ershov *et al.* 1965. Yu. Ershov, I. Lavrov, A. Taimanov, and M. Taitslin. Elementary theories. *Russian Mathematical Surveys*, 20:35–105, 1965.

Ewald 1986. W. Ewald. Intuitionistic tense and modal logic. *Journal of Symbolic Logic*, 51:166–179, 1986.

Fagin *et al.* 1995. R. Fagin, J. Halpern, Y. Moses, and M. Vardi. *Reasoning about Knowledge.* MIT Press, 1995.

Feferman and Vaught 1959. S. Feferman and R. Vaught. The first-order properties of products of algebraic systems. *Fundamenta Mathematicae*, 47:57–103, 1959.

Fine and Schurz 1996. K. Fine and G. Schurz. Transfer theorems for stratified modal logics. In J. Copeland, editor, *Logic and Reality, Essays in Pure and Applied Logic. In memory of Arthur Prior*, pages 169–213. Oxford University Press, 1996.

Fine 1971. K. Fine. The logics containing **S4.3**. *Zeitschrift für Mathematische Logik und Grundlagen der Mathematik*, 17:371–376, 1971.

Fine 1972. K. Fine. In so many possible worlds. *Notre Dame Journal of Formal Logic*, 13:516–520, 1972.

Fine 1974a. K. Fine. An incomplete logic containing **S4**. *Theoria*, 40:23–29, 1974.

Fine 1974b. K. Fine. Logics containing **K4**, part I. *Journal of Symbolic Logic*, 39:229–237, 1974.

Fine 1975a. K. Fine. Normal forms in modal logic. *Notre Dame Journal of Formal Logic*, 16:31–42, 1975.

Fine 1975b. K. Fine. Some connections between elementary and modal logic. In S. Kanger, editor, *Proceedings of the Third Scandinavian Logic Symposium*, pages 15–31. North-Holland, 1975.

Fine 1985. K. Fine. Logics containing **K4**, part II. *Journal of Symbolic Logic*, 50:619–651, 1985.

Finger and Gabbay 1992. M. Finger and D. Gabbay. Adding a temporal dimension to a logic system. *Journal of Logic, Language and Information*, 2:203–233, 1992.

Finger and Reynolds 1999. M. Finger and M. Reynolds. Imperative history: two-dimensional executable temporal logic. In U. Reyle and H.J. Ohlbach, editors, *Logic, Language and Reasoning: Essays in Honour of Dov Gabbay*, pages 81–107. Kluwer Academic Publishers, 1999.

Fischer and Immerman 1987. M. Fischer and N. Immerman. Interpreting logics of knowledge in propositional dynamic logic with converse. *Information Processing Letters*, 25:175–182, 1987.

Fischer and Ladner 1977. M. Fischer and R. Ladner. Propositional modal logic of programs. In *Proceedings of the 9th ACM Symposium on Theory of Computing*, pages 286–294. ACM, 1977.

Fischer and Ladner 1979. M. Fischer and R. Ladner. Propositional dynamic logic of regular programs. *Journal of Computer and System Sciences*, 18:194–211, 1979.

Fischer Servi 1980. G. Fischer Servi. Semantics for a class of intuitionistic modal calculi. In M. L. Dalla Chiara, editor, *Italian Studies in the Philosophy of Science*, pages 59–72. Reidel, Dordrecht, 1980.

Fischer Servi 1984. G. Fischer Servi. Axiomatizations for some intuitionistic modal logics. *Rendiconti di Matematica di Torino*, 42:179–194, 1984.

Fitting and Mendelson 1998. M. Fitting and R. Mendelson. *First-Order Modal Logic*. Kluwer Academic Publishers, Dordrecht, 1998.

Fitting 1983. M. Fitting. *Proof Methods for Modal and Intuitionistic Logics*. Reidel, Dordrecht, 1983.

Floyd 1967. R. Floyd. Assigning meanings to programs. In *Proceedings of the AMS Symposium on Applied Mathematics*, pages 361–390, 1967.

Gabbay and Guenthner 1982. D. Gabbay and F. Guenthner. A note on systems of n-dimensional tense logics. In T. Pauli, editor, *Essays dedicated to L. Aqvist*, pages 63–71. Uppsala, 1982.

Gabbay and Hodkinson 1990. D. Gabbay and I. Hodkinson. An axiomatization of the temporal logic with until and since over the real numbers. *Journal of Logic and Computation*, 1:229–260, 1990.

Gabbay and Pirri 1997. D. Gabbay and F. Pirri, editors. *Studia Logica*, volume 59 (1,2), 1997. Special Issue on Combining Logics.

Gabbay and Shehtman 1993. D. Gabbay and V. Shehtman. Undecidability of modal and intermediate first-order logics with two individual variables. *Journal of Symbolic Logic*, 58:800–823, 1993.

Gabbay and Shehtman 1998. D. Gabbay and V. Shehtman. Products of modal logics. Part I. *Logic Journal of the IGPL*, 6:73–146, 1998.

Gabbay and Shehtman 1999. D. Gabbay and V. Shehtman. Flow products of modal logics. Manuscript, 1999.

Gabbay and Shehtman 2000. D. Gabbay and V. Shehtman. Products of modal logics. Part II. *Logic Journal of the IGPL*, 2:165–210, 2000.

Gabbay *et al.* 1980. D. Gabbay, A. Pnueli, S. Shelah, and J. Stavi. On the temporal analysis of fairness. In *Proceedings of the 7th ACM Symposium on Principles of Programming Languages*, volume 1384 of *Lecture Notes in Computer Science*, pages 163–173. Springer, 1980.

Gabbay *et al.* 1994. D. Gabbay, I. Hodkinson, and M. Reynolds. *Temporal Logic: Mathematical Foundations and Computational Aspects, Volume 1.* Oxford University Press, 1994.

Gabbay *et al.* 2000. D. Gabbay, M. Reynolds, and M. Finger. *Temporal Logic: Mathematical Foundations and Computational Aspects, Volume 2.* Oxford University Press, 2000.

Gabbay *et al.* 2003. D. Gabbay, V. Shehtman, and D. Skvortsov. Quantification in Non-Classical Logic. (In preparation), 2003.

Gabbay 1977a. D. Gabbay. Axiomatisations of logics of programs. Manuscript, Bar-Ilan University, Ramat-Gan, Israel, 1977.

Gabbay 1977b. D. Gabbay. On some new intuitionistic propositional connectives. 1. *Studia Logica*, 36:127–139, 1977.

Gabbay 1981a. D. Gabbay. An irreflexivity lemma with application to axiomatizations of conditions on linear frames. In U. Mönnich, editor, *Aspects of Philosophical Logic*, pages 67–89. Reidel, Dordrecht, 1981.

Gabbay 1981b. D. Gabbay. *Semantical Investigations in Heyting's Intuitionistic Logic.* Reidel, Dordrecht, 1981.

Gabbay 1996. D. Gabbay. Fibred semantics and the weaving of logics, part 1: Modal and intuitionistic logics. *Journal of Symbolic Logic*, 61:1057–1120, 1996.

Gabbay 1999. D. Gabbay. *Fibring Logics*, volume 38 of *Oxford Logic Guides.* Clarendon Press, Oxford, 1999.

Gabelaia *et al.* 2003. D. Gabelaia, R. Kontchakov, A. Kurucz, F. Wolter, and M. Zakharyaschev. Computational complexity of spatio-temporal logics. (Submitted), 2003.

Galton 2000. A. Galton. *Qualitative Spatial Change.* Oxford University Press, 2000.

Garey and Johnson 1979. M. Garey and D. Johnson. *Computers and Intractability. A Guide to the Theory of NP-Completeness.* Freemann, San Francisco, 1979.

Garson 1984. J. Garson. Quantification in modal logic. In D. Gabbay and F. Guenthner, editors, *Handbook of Philosophical Logic*, volume 2, pages 249–307. Kluwer Academic Publishers, 1984.

Ghilardi and Zawadowski 1995. S. Ghilardi and M. Zawadowski. Undefinability of propositional quantifiers in modal system **S4**. *Studia Logica*, 55:259–271, 1995.

Ghilardi 1995. S. Ghilardi. An algebraic theory of normal forms. *Annals of Pure and Applied Logic*, 71:189–245, 1995.

Gibbons 1992. R. Gibbons. *Game Theory for Applied Economics*. Princeton University Press, 1992.

Glivenko 1929. V. Glivenko. Sur quelques points de la logique de L. Brouwer. *Bulletin de la Classe des Sciences de l'Académie Royale de Belgique*, 15:183–188, 1929.

Gödel 1933. K. Gödel. Eine Interpretation des intuitionistischen Aussagenkalküls. *Ergebnisse eines mathematischen Kolloquiums*, 4:39–40, 1933.

Gödel 1958. K. Gödel. Über eine bisher noch nicht benützte Erweiterung des finiten Standpunktes. *Dialectica*, 12:280–287, 1958. Translation: *Journal of Philosophical Logic*, 9:133–142, 1980.

Goldblatt 1976. R. Goldblatt. Metamathematics of modal logic, Part I. *Reports on Mathematical Logic*, 6:41–78, 1976.

Goldblatt 1982. R. Goldblatt. *Axiomatizing the Logic of Computer Programming*, volume 130 of *Lecture Notes in Computer Science*. Springer, 1982.

Goldblatt 1987. R. Goldblatt. *Logics of Time and Computation*. Number 7 in CSLI Lecture Notes. CSLI Publications, Stanford, 1987.

Goldblatt 1989. R. Goldblatt. Varieties of complex algebras. *Annals of Pure and Applied Logic*, 38:173–241, 1989.

Goldblatt 1993. R. Goldblatt. *Mathematics of Modality*. Number 43 in CSLI Lecture Notes. CSLI Publications, Stanford, 1993.

Goranko and Passy 1992. V. Goranko and S. Passy. Using the universal modality: gains and questions. *Journal of Logic and Computation*, 2:5–30, 1992.

Gore 1999. R. Gore. Tableau methods for modal and temporal logics. In M. D'Agostino, D. Gabbay, R. Hähnle, and J. Posegga, editors, *Handbook of Tableau Methods*, pages 297–396. Kluwer Academic Publishers, 1999.

Görnemann 1991. S. Görnemann. A logic stronger than intuitionism. *Journal of Symbolic Logic*, 36:249–261, 1991.

Gotts 1996a. N. Gotts. An axiomatic approach to topology for spatial information systems. Technical Report 96.25, School of Computer Studies, University of Leeds, 1996.

Gotts 1996b. N. Gotts. Using the RCC formalism to describe the topology of spherical regions. Technical Report 96.24, School of Computer Studies, University of Leeds, 1996.

Gräber *et al.* 1995. A. Gräber, H. Bürckert, and A. Laux. Terminological reasoning with knowledge and belief. In A. Laux and H. Wansing, editors, *Knowledge and Belief in Philosophy and Artificial Intelligence*, pages 29–61. Akademie Verlag, 1995.

Grädel *et al.* 1997. E. Grädel, P. Kolaitis, and M. Vardi. On the decision problem for two-variable first order logic. *Bulletin of Symbolic Logic*, 3:53–69, 1997.

Grädel 1999a. E. Grädel. Decision procedures for guarded logics. In *Proceedings of CADE-16*, volume 1632 of *Lecture Notes in Computer Science*, pages 31–51. Springer, 1999.

Grädel 1999b. E. Grädel. On the restraining power of guards. *Journal of Symbolic Logic*, 64:1719–1742, 1999.

Grädel 2001. E. Grädel. Why are modal logics so robustly decidable? In G. Paun, G. Rozenberg, and A. Salomaa, editors, *Current Trends in Theoretical Computer Science. Entering the 21st Century*, pages 393–408. World Scientific, 2001.

Grefe 1998. C. Grefe. Fischer Servi's intuitionistic modal logic has the finite model property. In M. Kracht, M. de Rijke, H. Wansing, and M. Zakharyaschev, editors, *Advances in Modal Logic*, volume 1, pages 85–98. CSLI Publications, Stanford, 1998.

Grzegorczyk 1951. A. Grzegorczyk. Undecidability of some topological theories. *Fundamenta Mathematicae*, 38:137–152, 1951.

Grzegorczyk 1964. A. Grzegorczyk. A philosophically plausible formal interpretation of intuitionistic logic. *Indagationes Mathematicae*, 26:596–601, 1964.

Gurevich and Shelah 1982. Yu. Gurevich and S. Shelah. Monadic theory of order and topology in *ZFC*. *Annals of Mathematical Logic*, 23:179–198, 1982.

Gurevich 1964. Yu. Gurevich. Elementary properties of ordered Abelian groups. *Algebra and Logic*, 3:5–39, 1964.

Gurevich 1977. Yu. Gurevich. Intuitionistic logic with strong negation. *Studia Logica*, 36:49–59, 1977.

Gurevich 1990. Yu. Gurevich. On the classical decision problem. *The Bulletin of EATCS*, 42:140–150, 1990.

Gurevich 1995. Yu. Gurevich. The value, if any, of decidability. *The Bulletin of EATCS*, 55:129–135, 1995.

Haarslev and Möller 1999. V. Haarslev and R. Möller. An empirical evaluation of optimization strategies for ABox reasoning in expressive description logics. In P. Lambrix, A. Borgida, M. Lenzerini, R. Möller, and P. Patel-Schneider, editors, *Proceedings of the International Workshop on Description Logics (DL'99)*, number 22 in CEUR-WS, pages 115–119, 1999.

Haarslev *et al.* 1999. V. Haarslev, C. Lutz, and R. Möller. A description logic with concrete domains and role-forming predicates. *Journal of Logic and Computation*, 9:351–384, 1999.

Halmos 1957. P. Halmos. Algebraic logic, IV. *Transactions of the AMS*, 86:1–27, 1957.

Halmos 1962. P. Halmos. *Algebraic Logic.* Chelsea Publishing Company, New York, 1962.

Halpern and Moses 1992. J. Halpern and Y. Moses. A guide to completeness and complexity for modal logics of knowledge and belief. *Artificial Intelligence*, 54:319–379, 1992.

Halpern and Shoham 1986. J. Halpern and Y. Shoham. A propositional modal logic of time intervals. In *Proceedings of the 1st Annual IEEE Symposium on Logic in Computer Science, (LICS'86)*, pages 279–292. IEEE Computer Society, 1986.

Halpern and Shoham 1991. J. Halpern and Y. Shoham. A propositional modal logic of time intervals. *Journal of the ACM*, 38:935–962, 1991.

Halpern and Vardi 1989. J. Halpern and M. Vardi. The complexity of reasoning about knowledge and time I: lower bounds. *Journal of Computer and System Sciences*, 38:195–237, 1989.

Halpern and Vardi 1991. J. Halpern and M. Vardi. Model checking vs. theorem proving: a manifesto. In *Artificial Intelligence and Mathematical Theory of Computation (Papers in Honor of John McCarthy)*, pages 151–176. Academic Press, San Diego, California, 1991.

Harel *et al.* 1983. D. Harel, A. Pnueli, and J. Stavi. Propositional dynamic logic of nonregular programs. *Journal of Computer and System Sciences*, 26:222–243, 1983.

Harel *et al.* 2000. D. Harel, D. Kozen, and J. Tiuryn. *Dynamic Logic*. MIT Press, 2000.

Harel 1984. D. Harel. Dynamic logic. In D. Gabbay and F. Guenthner, editors, *Handbook of Philosophical Logic*, volume 2, pages 497–604. Reidel, Dordrecht, 1984.

Harel 1986. D. Harel. Effective transformations on infinite trees, with applications to high undecidability, dominoes, and fairness. *Journal of the ACM*, 33:224–248, 1986.

Hasimoto 2002. Y. Hasimoto. Normal products of modal logics. In F. Wolter, H. Wansing, M. de Rijke, and M. Zakharyaschev, editors, *Advances in Modal Logic*, volume 3, pages 241–255. World Scientific, 2002.

Hemaspaandra 1996. E. Hemaspaandra. The price of universality. *Notre Dame Journal of Formal Logic*, 37:174–203, 1996.

Henkin *et al.* 1971. L. Henkin, J.D. Monk, and A. Tarski. *Cylindric Algebras, Part I.* Studies in Logic. Elsevier, North-Holland, 1971.

Henkin *et al.* 1981. L. Henkin, J.D. Monk, A. Tarski, H. Andréka, and I. Németi. *Cylindric Set Algebras*, volume 883 of *Lecture Notes in Mathematics*. Springer, 1981.

Henkin *et al.* 1985. H. Henkin, J.D. Monk, and A. Tarski. *Cylindric Algebras, Part II.* Studies in Logic. Elsevier, North-Holland, 1985.

Heyting 1930. A. Heyting. Die formalen Regeln der intuitionistischen Logik. *Sitzungsberichte der Preussischen Akademie der Wissenschaften*, pages 42–56, 1930.

Hintikka 1957. J. Hintikka. Quantifiers in deontic logic. *Societas Scientiarum Fennica, Commentationes humanarum litterarum*, 23:1–23, 1957.

Hintikka 1961. J. Hintikka. Modality and quantification. *Theoria*, 27:119–128, 1961.

Hintikka 1962. J. Hintikka. *Knowledge and Belief. An Introduction into the Logic of the Two Notions.* Cornell University Press, Ithaca, 1962.

Hintikka 1963. J. Hintikka. The modes of modality. *Acta Philosophica Fennica*, 16:65–82, 1963.

Hirsch and Hodkinson 1997. R. Hirsch and I. Hodkinson. Step by step — building representations in algebraic logic. *Journal of Symbolic Logic*, 62:225–279, 1997.

Hirsch and Hodkinson 2001. R. Hirsch and I. Hodkinson. Representability is not decidable for finite relation algebras. *Transactions of the AMS*, 353:1403–1425, 2001.

Hirsch and Hodkinson 2002. R. Hirsch and I. Hodkinson. *Relation Algebras by Games*. Studies in Logic. Elsevier, North Holland, 2002.

Hirsch *et al.* 2002. R. Hirsch, I. Hodkinson, and A. Kurucz. On modal logics between $\mathbf{K} \times \mathbf{K} \times \mathbf{K}$ and $\mathbf{S5} \times \mathbf{S5} \times \mathbf{S5}$. *Journal of Symbolic Logic*, 67:221–234, 2002.

Hoare 1967. C.A.R. Hoare. An axiomatic basis for computer programming. *Communications of the ACM*, 12:516–580, 1967.

Hoare 1985. C.A.R. Hoare. *Communicating Sequential Processes*. Prentice-Hall International, 1985.

Hodges 1993. W. Hodges. *Model Theory*. Cambridge University Press, 1993.

Hodkinson and Otto 2003. I. Hodkinson and M. Otto. Finite conformal hypergraph covers and Gaifman cliques in finite structures. *Bulletin of Symbolic Logic*, 2003. (In print).

Hodkinson *et al.* 2000. I. Hodkinson, F. Wolter, and M. Zakharyaschev. Decidable fragments of first-order temporal logics. *Annals of Pure and Applied Logic*, 106:85–134, 2000.

Hodkinson *et al.* 2002. I. Hodkinson, F. Wolter, and M. Zakharyaschev. Decidable and undecidable fragments of first-order branching temporal logics. In *Proceedings of the 17th Annual IEEE Symposium on Logic in Computer Science (LICS 2002)*, pages 393–402. IEEE Computer Society, 2002.

Hodkinson *et al.* 2003. I. Hodkinson, R. Kontchakov, A. Kurucz, F. Wolter, and M. Zakharyaschev. On the computational complexity of decidable fragments of first-order linear temporal logics. In *Proceedings of TIME-ICTL*. IEEE Computer Society, 2003.

Hodkinson 2002a. I. Hodkinson. Loosely guarded fragment of first-order logic has the finite model property. *Studia Logica*, 70:205–240, 2002.

Hodkinson 2002b. I. Hodkinson. Monodic packed fragment with equality is decidable. *Studia Logica*, 72:185–197, 2002.

Hollunder and Nutt 1990. B. Hollunder and W. Nutt. Subsumption algorithms for concept languages. Technical Report DFKI Research Report RR-90-04, Deutsches Forschungszentrum für Künstliche Intelligenz, Kaiserslautern, 1990.

Hopcroft and Ullman 1979. J.E. Hopcroft and J.D. Ullman. *Introduction to Automata Theory, Languages, and Computation*. Addison-Wesley, 1979.

Hopcroft *et al.* 2001. J.E. Hopcroft, R. Motwani, and J.D. Ullman. *Introduction to Automata Theory, Languages, and Computation*. Addison-Wesley, 2001.

Horrocks *et al.* 1999. I. Horrocks, U. Sattler, and S. Tobies. Practical reasoning for expressive description logics. In H. Ganzinger, D. McAllester, and A. Voronkov, editors, *Proceedings of the 6th International Conference on Logic for Programming and Automated Reasoning (LPAR'99)*, volume 1705 of *Lecture Notes in Artificial Intelligence*, pages 161–180. Springer, 1999.

Horrocks *et al.* 2000a. I. Horrocks, P. Patel-Schneider, and R. Sebastiani. An analysis of empirical testing for modal decision procedures. *Logic Journal of the IGPL*, 8:293–323, 2000.

Horrocks *et al.* 2000b. I. Horrocks, U. Sattler, and S. Tobies. Practical reasoning for very expressive description logics. *Logic Journal of the IGPL*, 8:239–264, 2000.

Horrocks 1998. I. Horrocks. Using an expressive description logic: FACT or fiction? In A. Cohn, L. Schubert, and S. Shapiro, editors, *Proceedings of the 6th International Conference on Principles of Knowledge Representation and Reasoning (KR'98)*, pages 636–647, 1998.

Hughes and Cresswell 1996. G.E. Hughes and M.J. Cresswell. *A New Introduction to Modal Logic*. Methuen, London, 1996.

Humberstone 1979. I.L. Humberstone. Interval semantics for tense logics. *Journal of Philosophical Logic*, 8:171–196, 1979.

Hustadt and Schmidt 2000. U. Hustadt and R. Schmidt. MSPASS: modal reasoning by translation and first-order resolution. In D. Dyckhoff, editor, *Proceedings of TABLEAUX 2000*, volume 1847 of *Lecture Notes in Artificial Intelligence*, pages 67–71. Springer, 2000.

Janiczak 1953. A. Janiczak. Undecidability of some simple formalized theories. *Fundamenta Mathematicae*, 40:131–139, 1953.

Japaridze and de Jongh 1998. G. Japaridze and D. de Jongh. The logic of provability. In S. Buss, editor, *Handbook of Proof Theory*, pages 475–545. North-Holland, 1998.

Japaridze 2000. G. Japaridze. A decidable first-order epistemic logic. *Proceedings of the Georgian Academy of Sciences*, (1–2):81–95, 2000.

Johnson 1969. J.S. Johnson. Nonfinitizability of classes of representable polyadic algebras. *Journal of Symbolic Logic*, 34:344–352, 1969.

Johnson 1990. D. Johnson. A catalog of complexity classes. In J. van Leeuwen, editor, *Handbook of Theoretical Computer Science*. Elsevier, North-Holland, 1990.

Jónsson and Tarski 1951. B. Jónsson and A. Tarski. Boolean algebras with operators. I. *American Journal of Mathematics*, 73:891–939, 1951.

Kamp 1968. H. Kamp. *Tense Logic and the Theory of Linear Order*. PhD Thesis, University of California, Los Angeles, 1968.

Kamp 1971. H. Kamp. Formal properties of 'now'. *Theoria*, 37:237–273, 1971.

Kamp 1979. H. Kamp. Instants, events and temporal discourse. In R. Bäuerle, C. Schwarze, and A. von Stechow, editors, *Semantics from Different Points of View*, pages 376–417. Springer, 1979.

Kaneko and Nagashima 1997. M. Kaneko and T. Nagashima. Game logic and its applications 2. *Studia Logica*, 58:273–303, 1997.

Kanger 1957a. S. Kanger. The Morning Star paradox. *Theoria*, 23:1–11, 1957.

Kanger 1957b. S. Kanger. A note on quantification and modalities. *Theoria*, 23:131–134, 1957.

Kirchner and Ringeissen 2000. H. Kirchner and C. Ringeissen, editors. *Frontiers of Combining Systems III*, volume 1794 of *Lecture Notes in Artificial Intelligence*. Springer, 2000.

Klarlund *et al.* 2002. N. Klarlund, A. Møller, and M. Schwartzbach. MONA implementation secrets. *International Journal of Foundations of Computer Science*, 13:571–586, 2002.

Kleene 1945. S. Kleene. On the interpretation of intuitionistic number theory. *Journal of Symbolic Logic*, 10:109–124, 1945.

Kolmogorov 1925. A.N. Kolmogorov. On the principle tertium non datur. *Mathematics of the USSR, Sbornik*, 32:646–667, 1925. Translation in: *From Frege to Gödel: A Source Book in Mathematical Logic 1879–1931* (J. van Heijenoord ed.), Harvard University Press, Cambridge 1967.

Kontchakov *et al.* 2003. R. Kontchakov, C. Lutz, F. Wolter, and M. Zakharyaschev. Temporalising tableaux. *Studia Logica*, 2003. (In print).

Koppelberg 1988. S. Koppelberg. General theory of Boolean algebras. In D.J. Monk, editor, *Handbook of Boolean Algebras*, volume 1. Elsevier, North-Holland, 1988.

Kowalski and Sergot 1985. R. Kowalski and M. Sergot. A logic-based calculus of events. Technical report, Department of Computing, Imperial College London, 1985.

Kozen and Parikh 1981. D. Kozen and R. Parikh. An elementary proof of the completeness of **PDL**. *Theoretical Computer Science*, 14:113–118, 1981.

Kozen and Tiuryn 1990. D. Kozen and J. Tiuryn. Logics of programs. In J. van Leeuwen, editor, *Handbook of Theoretical Computer Science*, volume B, pages 789–840. Elsevier, North-Holland, 1990.

Kozen 1981. D. Kozen. On induction vs. *-continuity. In D. Kozen, editor, *Proceedings of the Workshop on Logics of Programs*, volume 131 of *Lecture Notes in Computer Science*, pages 167–176. Springer, 1981.

Kozen 1990. D. Kozen. On Kleene algebras and closed semirings. In Rovan, editor, *Proceedings of the Symposium on Mathematical Foundations of Computer Science*, volume 452 of *Lecture Notes in Computer Science*, pages 26–47. Springer, 1990.

Kracht and Wolter 1991. M. Kracht and F. Wolter. Properties of independently axiomatizable bimodal logics. *Journal of Symbolic Logic*, 56:1469–1485, 1991.

Kracht and Wolter 1999. M. Kracht and F. Wolter. Normal monomodal logics can simulate all others. *Journal of Symbolic Logic*, 64:99–138, 1999.

Kracht 1999. M. Kracht. *Tools and Techniques in Modal Logic*. Studies in Logic. Elsevier, North-Holland, 1999.

Kreisel 1962. G. Kreisel. Foundations of intuitionistic logic. In E. Nagel, P. Suppes, and A. Tarski, editors, *Logic, Methodology and Philosophy of Science*, pages 198–210. Stanford, 1962.

Kripke 1959. S.A. Kripke. A completeness theorem in modal logic. *Journal of Symbolic Logic*, 24:1–14, 1959.

Kripke 1963a. S.A. Kripke. Semantical analysis of modal logic, Part I. *Zeitschrift für Mathematische Logik und Grundlagen der Mathematik*, 9:67–96, 1963.

Kripke 1963b. S.A. Kripke. Semantical considerations on modal logic. *Acta Philosophica Fennica*, 16:83–94, 1963.

Kripke 1965. S.A. Kripke. Semantical analysis of intuitionistic logic. I. In J.N. Crossley and M.A.E. Dummett, editors, *Formal Systems and Recursive Functions. Proceedings of the 8th Logic Colloquium*, pages 92–130. North-Holland, 1965.

Kuhn 1980. S.T. Kuhn. Quantifiers as modal operators. *Studia Logica*, 39:145–158, 1980.

Kurtonina 1995. N. Kurtonina. *Frames and Labels: a Modal Analysis of Categorial Deduction*. PhD thesis, University of Utrecht and ILLC, University of Amsterdam, 1995.

Kurucz and Zakharyaschev 2003. A. Kurucz and M. Zakharyaschev. A note on relativised products of modal logics. In P. Balbiani, N.-Y. Suzuki, F. Wolter, and M. Zakharyaschev, editors, *Advances in Modal Logic*, volume 4, 2003. (In print).

Kurucz *et al.* 1995. A. Kurucz, I. Németi, I. Sain, and A. Simon. Decidable and undecidable logics with a binary modality. *Journal of Logic, Language, and Information*, 4:191–206, 1995.

Kurucz *et al.* 2002. A. Kurucz, F. Wolter, and M. Zakharyaschev, editors. *Studia Logica*, volume 72, 2002. Special Issue on Many-Dimensional Logical Systems.

Kurucz 1997. A. Kurucz. *Decision Problems in Algebraic Logic*. PhD thesis, Hungarian Academy of Sciences, 1997.

Kurucz 2000a. A. Kurucz. Arrow logic and infinite counting. *Studia Logica*, 65:199–222, 2000.

Kurucz 2000b. A. Kurucz. On axiomatising products of Kripke frames. *Journal of Symbolic Logic*, 65:923–945, 2000.

Kurucz 2002. A. Kurucz. $\mathbf{S5} \times \mathbf{S5} \times \mathbf{S5}$ lacks the finite model property. In F. Wolter, H. Wansing, M. de Rijke, and M. Zakharyaschev, editors, *Advances in Modal Logic*, volume 3, pages 321–328. World Scientific, 2002.

Kutz *et al.* 2002. O. Kutz, F. Wolter, and M. Zakharyaschev. Connecting abstract description systems. In D. Fensel, F. Giunchiglia, D. McGuinness, and A. Williams, editors, *Proceedings of the 8th International Conference on Principles of Knowledge Representation and Reasoning (KR'02)*, pages 215–226. Morgan Kaufmann, 2002.

Kuznetsov and Muravitskij 1986. A. Kuznetsov and A. Muravitskij. On superintuitionistic logics as fragments of proof logic extensions. *Studia Logica*, 45:77–99, 1986.

Kuznetsov 1985. A. Kuznetsov. Proof-intuitionistic propositional calculus. *Doklady Academii Nauk SSSR*, 283:27–30, 1985. (In Russian).

Ladner and Reif 1986. R. Ladner and J. Reif. The logic of distributed protocols. In J. Halpern, editor, *Proceedings of "Theoretical Aspects of Reasoning about Knowledge"*, pages 207–221. Morgan Kaufmann, 1986.

Ladner 1977. R. Ladner. The computational complexity of provability in systems of modal logic. *SIAM Journal on Computing*, 6:467–480, 1977.

Lambek 1958. J. Lambek. The mathematics of sentence structure. *American Mathematical Monthly*, 65:154–170, 1958.

Lamport 1985. L. Lamport. Interprocessor communication: final report. Technical report, SRI International, Menlo Park, California, 1985.

Läuchli and Leonard 1966. H. Läuchli and J. Leonard. On the elementary theory of linear order. *Fundamenta Informaticae*, 59:109–116, 1966.

Laux and Wansing 1995. A. Laux and H. Wansing, editors. *Knowledge and Belief in Philosophy and Artificial Intelligence.* Akademie Verlag, 1995.

Laux 1994. A. Laux. Beliefs in multi-agent worlds: a terminological approach. In A. Cohn, editor, *Proceedings of the 11th European Conference on Artificial Intelligence (ECAI'94)*, pages 299–303. John Wiley and Sons, 1994.

Lehmann 1984. D.J. Lehmann. Knowledge, common knowledge, and related puzzles. In *Proceedings of the 3rd ACM Symposium on Principles of Distributed Computing*, pages 62–67, 1984.

Lenzen 1978. W. Lenzen. *Recent Work in Epistemic Logic.* Elsevier, North-Holland, 1978.

Levin 1973. L. Levin. Universal search problems. *Problems of Information Transmission*, 9:265–266, 1973.

Lewis and Langford 1932. C. Lewis and C. Langford. *Symbolic Logic.* Appleton-Century-Crofts, New York, 1932.

Lewis and Papadimitriou 1981. H. Lewis and C. Papadimitriou. *Elements of the Theory of Computation.* Prentice Hall, 1981.

Lewis 1918. C. Lewis. *A Survey of Symbolic Logic.* University of California Press, Berkeley, 1918.

Lewis 1978. H. Lewis. Complexity of solvable cases of the decision problem for the predicate calculus. In *Proceedings of the 19th Symposium on the Foundations of Computer Science*, pages 35–47, 1978.

Ligozat 1998. G. Ligozat. Reasoning about cardinal directions. *Journal of Visual Languages and Computing*, 9:23–44, 1998.

Löwenheim 1915. L. Löwenheim. Über Möglichkeiten im Relativkalkül. *Mathematische Annalen*, 76:447–470, 1915.

Lutz *et al.* 2001. C. Lutz, H. Sturm, F. Wolter, and M. Zakharyaschev. Tableaux for temporal description logic with constant domains. In R. Goré, A. Leitsch, and T. Nipkow, editors, *Automated Reasoning, Proceedings of the First Joint Conference, IJCAR 2001*, volume 2083 of *Lecture Notes in Artificial Intelligence*, pages 121–136. Springer, 2001.

Lutz *et al.* 2002. C. Lutz, H. Sturm, F. Wolter, and M. Zakharyaschev. A tableau decision algorithm for modalized $\mathcal{ALC}$ with constant domains. *Studia Logica*, 72:199–232, 2002.

Lutz 1999. C. Lutz. Complexity of terminological reasoning revisited. In *Proceedings of the 6th International Conference on Logic for Programming and Automated Reasoning (LPAR'99)*, volume 1705 of *Lecture Notes in Artificial Intelligence*, pages 181–200. Springer, 1999.

Lyndon 1950. R. Lyndon. The representation of relational algebras. *Annals of Mathematics*, 51:707–729, 1950.

Maddux 1978. R. Maddux. *Topics in Relation Algebras.* PhD thesis, University of California, Berkeley, 1978.

Maddux 1980. R. Maddux. The equational theory of CA_3 is undecidable. *Journal of Symbolic Logic*, 45:311–315, 1980.

Maddux 1982. R. Maddux. Some varieties containing relation algebras. *Transactions of the AMS*, 272:501–526, 1982.

Maddux 1991. R. Maddux. Introductory course on relation algebras, finite-dimensional cylindric algebras, and their interconnections. In H. Andréka, J.D. Monk, and I. Németi, editors, *Algebraic Logic*, Colloquia Mathematica Societatis J. Bolyai, pages 361–392. Elsevier, North-Holland, 1991.

Makinson 1971. D. Makinson. Some embedding theorems for modal logic. *Notre Dame Journal of Formal Logic*, 12:252–254, 1971.

Maksimova 1979. L. Maksimova. Interpolation theorems in modal logic and amalgamable varieties of topological Boolean algebras. *Algebra and Logic*, 18:348–370, 1979.

Manna and Pnueli 1992. Z. Manna and A. Pnueli. *The Temporal Logic of Reactive and Concurrent Systems*. Springer, 1992.

Manna and Pnueli 1995. Z. Manna and A. Pnueli. *Temporal Verification of Reactive Systems: Safety*. Springer, 1995.

Marx and Mikulás 1999. M. Marx and Sz. Mikulás. Decidability of cylindric set algebras of dimension two and first-order logic with two variables. *Journal of Symbolic Logic*, 64:1563–1572, 1999.

Marx and Mikulás 2001. M. Marx and Sz. Mikulás. Products, or how to create modal logics of high complexity. *Logic Journal of the IGPL*, 9:77–88, 2001.

Marx and Mikulás 2002. M. Marx and Sz. Mikulás. An elementary construction for a non-elementary procedure. *Studia Logica*, 72:253–263, 2002.

Marx and Reynolds 1999. M. Marx and M. Reynolds. Undecidability of compass logic. *Journal of Logic and Computation*, 9:897–914, 1999.

Marx and Venema 1997. M. Marx and Y. Venema. *Multi-Dimensional Modal Logic*. Kluwer Academic Publishers, 1997.

Marx *et al.* 1996. M. Marx, L. Pólos, and M. Masuch, editors. *Arrow Logic and Multi-Modal Logic*. CSLI Publications, Stanford, 1996.

Marx 1995. M. Marx. *Algebraic Relativization and Arrow Logic*. PhD thesis, University of Amsterdam, 1995.

Marx 1999. M. Marx. Complexity of products of modal logics. *Journal of Logic and Computation*, 9:197–214, 1999.

Marx 2001. M. Marx. Tolerance logic. *Journal of Logic, Language and Information*, 10:353–374, 2001.

Marx 2002. M. Marx. Computing with cylindric modal logics and arrow logics, lower bounds. *Studia Logica*, 72:233–252, 2002.

McCarthy *et al.* 1979. J. McCarthy, M. Sato, T. Hayashi, and S. Igarishi. On the model theory of knowledge. Technical Report STAN-CS-78-657, Stanford University, 1979.

McColl 1906. H. McColl. *Symbolic Logic and its Applications.* Ballentyne, Hanson & Co., Edinburgh/London, 1906.

McKinsey and Tarski 1944. J.C.C. McKinsey and A. Tarski. The algebra of topology. *Annals of Mathematics*, 45:141–191, 1944.

McKinsey and Tarski 1946. J.C.C. McKinsey and A. Tarski. On closed elements in closure algebras. *Annals of Mathematics*, 47:122–162, 1946.

McKinsey and Tarski 1948. J.C.C. McKinsey and A. Tarski. Some theorems about the sentential calculi of Lewis and Heyting. *Journal of Symbolic Logic*, 13:1–15, 1948.

McKinsey 1941. J.C.C. McKinsey. A solution of the decision problem for the Lewis systems **S2** and **S4**, with an application to topology. *Journal of Symbolic Logic*, 6:117–134, 1941.

Medvedev 1962. Yu. Medvedev. Finite problems. *Soviet Mathematics Doklady*, 3:227–230, 1962.

Merz 1992. S. Merz. Decidability and incompleteness results for first-order temporal logics of linear time. *Journal of Applied Non-classical Logic*, 2:139–156, 1992.

Meyer and van der Hoek 1995. J. Meyer and W. van der Hoek. *Epistemic Logic for AI and Computer Science.* Cambridge University Press, 1995.

Meyer 1974. A. Meyer. The inherent complexity of theories of ordered sets. In *Proceedings of the International Congress of Mathematicians*, pages 477–482, 1974.

Meyer 1988. J. Meyer. A different approach to deontic logic: deontic logic viewed as a variant of dynamic logic. *Notre Dame Journal of Formal Logic*, 29:109–136, 1988.

Mikulás and Marx 2000. Sz. Mikulás and M. Marx. Relativized products of modal logics. In *Proceedings of Advances in Modal Logic (AiML 2000)*, pages 201–208, Leipzig, Germany, 2000.

Mikulás 1995. Sz. Mikulás. *Taming Logics.* PhD thesis, University of Amsterdam, 1995.

Mikulás 2000. Sz. Mikulás. Personal communication, 2000.

Milgrom 1981. P. Milgrom. An axiomatic characterization of common knowledge. *Econometrica*, 49:219–222, 1981.

Milner 1980. R. Milner. *A Calculus of Communicating Systems*, volume 92 of *Lecture Notes in Computer Science*. Springer, 1980.

Minsky 1961. M. Minsky. Recursive unsolvability of Post's problem of "tag" and other topics in the theory of Turing machines. *Annals of Mathematics*, 74:437–455, 1961.

Minsky 1975. M. Minsky. A framework for representing knowledge. In P. Winston, editor, *The Psychology of Computer Vision*. McGraw-Hill, 1975.

Monk 1961. J.D. Monk. *Studies in Cylindric Algebra*. PhD thesis, University of California, Berkeley, 1961.

Monk 1964. J.D. Monk. On representable relation algebras. *Michigan Mathematical Journal*, 11:207–210, 1964.

Monk 1969. J.D. Monk. Nonfinitizability of classes of representable cylindric algebras. *Journal of Symbolic Logic*, 34:331–343, 1969.

Monk 1988. J.D. Monk, editor. *Handbook of Boolean Algebras, Volume I*. Elsevier, North-Holland, 1988.

Moortgat 1996. M. Moortgat. Categorial type logics. In J. van Benthem and A. ter Meulen, editors, *Handbook of Logic and Language*, pages 93–177. Elsevier, North-Holland, 1996.

Mortimer 1975. M. Mortimer. On languages with two variables. *Zeitschrift für Mathematische Logik und Grundlagen der Mathematik*, 21:135–140, 1975.

Nash 1991. J. Nash. Noncooperative games. *Annals of Mathematics*, 54:286–295, 1991.

Nebel 1990. B. Nebel. Terminological reasoning is inherently intractable. *Artificial Intelligence*, 43:235–249, 1990.

Nebel 1991. B. Nebel. Terminological cycles: semantics and computational properties. In J. Sowa, editor, *Principles of Semantic Networks*, pages 331–361. Morgan Kaufmann, 1991.

Németi 1984. I. Németi. Neither CA_α nor Gs_α is generated by its finite members if $\alpha \geq 3$. Manuscript, 1984.

Németi 1987. I. Németi. Decidability of relation algebras with weakened associativity. *Proceedings of the AMS*, 100:340–344, 1987.

Németi 1991. I. Németi. Algebraization of quantifier logics: an overview. *Studia Logica*, 3,4:485–569, 1991.

Németi 1995. I. Németi. Decidable versions of first-order predicate logic and cylindric relativized set algebras. In L. Csirmaz, D. Gabbay, and M. de Rijke, editors, *Logic Colloquium'92*, pages 177–241. CSLI Publications, Stanford, 1995.

Nishimura 1979. I. Nishimura. Sequential method in propositional dynamic logic. *Acta Informatica*, 12:377–400, 1979.

Nishimura 1980. H. Nishimura. Interval logics with applications to study of tense and aspect in English. *Publications of the Research Institute for Mathematical Sciences, Kyoto University*, 16:417–459, 1980.

Ono and Nakamura 1980. H. Ono and A. Nakamura. On the size of refutation Kripke models for some linear modal and tense logics. *Studia Logica*, 39:325–333, 1980.

Ono and Suzuki 1988. H. Ono and N.-Y. Suzuki. Relations between intuitionistic modal logics and intermediate predicate logics. *Reports on Mathematical Logic*, 22:65–87, 1988.

Ono 1977. H. Ono. On some intuitionistic modal logics. *Publications of the Research Institute for Mathematical Sciences, Kyoto University*, 13:55–67, 1977.

Orlov 1928. I. Orlov. The calculus of compatibility of propositions. *Mathematics of the USSR, Sbornik*, 35:263–286, 1928. (In Russian).

Papadimitriou 1994. Ch. Papadimitriou. *Computational Complexity*. Addison-Wesley, 1994.

Parikh and Ramanujam 1985. R. Parikh and R. Ramanujam. Distributed processing and the logic of knowledge. In R. Parikh, editor, *Proceedings of the Workshop on Logics of Programs*, volume 193 of *Lecture Notes in Computer Science*, pages 256–268. Springer, 1985.

Parikh 1978. R. Parikh. The completeness of propositional dynamic logic. In *Proceedings of the 7th Symposium on Mathematical Foundations of Computer Science*, volume 64 of *Lecture Notes in Computer Science*, pages 403–415. Springer, 1978.

Peirce 1870. C. Peirce. Description of a notation for the logic of relatives, resulting from an amplification of the conceptions of Boole's calculus of logic. *Memoirs of the American Academy of Sciences*, 9:317–378, 1870.

Pigozzi 1974. D. Pigozzi. The join of equational theories. *Colloquium Mathematicum*, 30:15–25, 1974.

Pinter 1973. C. Pinter. A simple algebra of first order logic. *Notre Dame Journal of Formal Logic*, 1:361–366, 1973.

Pinter 1975. C. Pinter. Algebraic logic with generalized quantifiers. *Notre Dame Journal of Formal Logic*, 16:511–516, 1975.

Pitts 1992. A. Pitts. On an interpretation of second order quantification in first order intuitionistic propositional logic. *Journal of Symbolic Logic*, 57:33–52, 1992.

Plotkin and Stirling 1986. G. Plotkin and C. Stirling. A framework for intuitionistic modal logic. In J. Halpern, editor, *Proceedings of the 1st Conference on Theoretical Aspects of Reasoning about Knowledge (TARK'86)*, pages 399–406. Morgan Kaufmann, 1986.

Pnueli 1986. A. Pnueli. Applications of temporal logic to the specification and verification of reactive systems: a survey of current trends. In J. de Bakker, W.-P. de Roever, and G. Rozenberg, editors, *Current Trends in Concurrency*, volume 224 of *Lecture Notes in Computer Science*, pages 510–584. Springer, 1986.

Ponse *et al.* 1996. A. Ponse, M. de Rijke, and Y. Venema, editors. *Modal Logic and Process Algebra*. CSLI Publications, Stanford, 1996.

Post 1946. E. Post. A variant of a recursively unsolvable problem. *Bulletin of the AMS*, 52:264–268, 1946.

Pratt 1978. V. Pratt. A practical decision method for propositional dynamic logic. In *Proceedings of the 10th ACM Symposium on the Theory of Computation*, pages 326–337, 1978.

Pratt 1979. V. Pratt. Models of program logics. In *Proceedings of the 20th IEEE Symposium on Foundations of Computer Science*, pages 115–122, 1979.

Pratt 1990. V. Pratt. Action logic and pure induction. In J. van Eijck, editor, *Proceedings of JELIA'90*, volume 478 of *Lecture Notes in Computer Science*, pages 97–120. Springer, 1990.

Pratt 1991. V. Pratt. Dynamic algebras: examples, constructions, applications. *Studia Logica*, 50:571–605, 1991.

Prendinger and Schurz 1996. H. Prendinger and G. Schurz. Reasoning about action and change. *Journal of Logic, Language and Information*, 5:209–245, 1996.

Prior 1957. A. Prior. *Time and Modality.* Clarendon Press, Oxford, 1957.

Purdy 1996a. W. Purdy. Decidability of fluted logic with identity. *Notre Dame Journal of Formal Logic*, 37:84–104, 1996.

Purdy 1996b. W. Purdy. Fluted formulas and the limits of decidability. *Journal of Symbolic Logic*, 61:608–620, 1996.

Quillian 1967. M. Quillian. Word concepts: A theory and simulation of some basic capabilities. *Behavioral Science*, 12:410–430, 1967. Republished in [Brachman and Levesque 1985].

Quillian 1968. M. Quillian. Semantic memory. In M. Minsky, editor, *Semantic Information Processing*, pages 216–270. MIT Press, 1968.

Quine 1971. W.V. Quine. Algebraic logic and predicate functors. In R. Rudner and I. Scheffer, editors, *Logic and Art: Essays in Honor of Nelson Goodman.* Bobbs-Merrill, 1971. Reprinted with amendments in *The Ways of Paradox and Other Essays*, 2nd edition, Harvard University Press, Cambridge, 1976.

Rabin 1969. M.O. Rabin. Decidability of second order theories and automata on infinite trees. *Transactions of the AMS*, 141:1–35, 1969.

Rabin 1977. M.O. Rabin. Decidable theories. In J. Barwise, editor, *Handbook of Mathematical Logic*, pages 595–630. Elsevier, North-Holland, 1977.

Rabinovich 2003. A. Rabinovich. On compositionality and its limitations. *Journal of Logic and Computation*, 2003. (In print).

Ramsey 1930. F. Ramsey. On a problem of formal logic. *Proceedings of London Mathematical Society*, 30:264–286, 1930.

Randell *et al.* 1992. D. Randell, Z. Cui, and A. Cohn. A spatial logic based on regions and connection. In *Proceedings of the 3rd International Conference on Knowledge Representation and Reasoning (KR'92)*, pages 165–176. Morgan Kaufmann, 1992.

Raphael 1968. B. Raphael. SIR: Semantic information retrieval. In M. Minsky, editor, *Semantic Information Processing*, pages 33–134. MIT Press, 1968.

Rasiowa and Sikorski 1963. H. Rasiowa and R. Sikorski. *The Mathematics of Metamathematics.* Polish Scientific Publishers, 1963.

Reif and Sistla 1985. J. Reif and A. Sistla. A multiprocess network logic with temporal and spatial modalities. *Journal of Computer and System Sciences*, 30:41–53, 1985.

Renz and Nebel 1998. J. Renz and B. Nebel. Spatial reasoning with topological information. In C. Freksa, C. Habel, and K. Wender, editors, *Spatial Cognition—An Interdisciplinary Approach to Representation and Processing of Spatial Knowledge*, volume 1404 of *Lecture Notes in Computer Science*, pages 351–372. Springer, 1998.

Renz and Nebel 1999. J. Renz and B. Nebel. On the complexity of qualitative spatial reasoning. *Artificial Intelligence*, 108:69–123, 1999.

Renz 1998. J. Renz. A canonical model of the region connection calculus. In A. Cohn, L. Schubert, and S. Shapiro, editors, *Proceedings of the 6th International Conference on Knowledge Representation and Reasoning (KR'98)*, pages 330–341. Morgan Kaufmann, 1998.

Reynolds and Zakharyaschev 2001. M. Reynolds and M. Zakharyaschev. On the products of linear modal logics. *Journal of Logic and Computation*, 11:909–931, 2001.

Reynolds 1992. M. Reynolds. An axiomatisation for Until and Since over the reals without the IRR rule. *Studia Logica*, 51:165–194, 1992.

Reynolds 1996. M. Reynolds. Axiomatizing first-order temporal logic: Until and Since over linear time. *Studia Logica*, 57:279–302, 1996.

Reynolds 1997. M. Reynolds. A decidable temporal logic of parallelism. *Notre Dame Journal of Formal Logic*, 38:419–436, 1997.

Reynolds 1999. M. Reynolds. The complexity of the temporal logic over the reals, 1999. (Submitted).

Reynolds 2003. M. Reynolds. The complexity of the temporal logic with 'until' over general linear time. *Journal of Computer and System Science*, 66:393–426, 2003.

Robertson 1974. E. Robertson. Structure of complexity in weak monadic second order theories of the natural numbers. In *Proceedings of the 6th Symposium on Theory of Computing*, pages 161–171, 1974.

Robinson 1971. R. Robinson. Undecidability and nonperiodicity for tilings of the plane. *Inventiones Mathematicae*, 12:177–209, 1971.

Roorda 1991. D. Roorda. *Resource Logics, a Proof-Theoretic Investigation.* PhD thesis, ILLC, University of Amsterdam, 1991.

Rosenstein 1982. J. Rosenstein. *Linear Orderings.* Academic Press, New York, 1982.

Sági 2002. G. Sági. A note on algebras of substitutions. *Studia Logica*, 72:265–284, 2002.

Sahlqvist 1975. H. Sahlqvist. Completeness and correspondence in the first and second order semantics for modal logic. In S. Kanger, editor, *Proceedings of the Third Scandinavian Logic Symposium*, pages 110–143. North-Holland, 1975.

Sain and Thompson 1991. I. Sain and R. Thompson. Strictly finite schema axiomatization of quasipolyadic algebras. In H. Andréka, J.D. Monk, and I. Németi, editors, *Algebraic Logic*, pages 539–571. Elsevier, North-Holland, 1991.

Sato 1977. M. Sato. A study of Kripke-style methods of some modal logics by Gentzen's sequential method. *Publications Research Institute for Mathematical Sciences, Kyoto University*, 13(2), 1977.

Sattler 1996. U. Sattler. A concept language extended with different kinds of transitive roles. In G. Görz and S. Hölldobler, editors, *20. Deutsche Jahrestagung für Künstliche Intelligenz*, volume 1137 of *Lecture Notes in Artificial Intelligence*, pages 333–345. Springer, 1996.

Savitch 1970. W. Savitch. Relationship between nondeterministic and deterministic tape classes. *Journal of Computer and System Sciences*, 4:177–192, 1970.

Schild 1991. K. Schild. A correspondence theory for terminological logics: preliminary report. In *Proceedings of the 12th International Joint Conference on Artificial Intelligence (IJCAI'91)*, pages 466–471. Morgan Kaufmann, 1991.

Schild 1993. K. Schild. Combining terminological logics with tense logic. In *Proceedings of the 6th Portuguese Conference on Artificial Intelligence*, pages 105–120, Porto, 1993.

Schmidt and Tishkovsky 2003. R. Schmidt and D. Tishkovsky. Logics with commuting action and informational modalities. In P. Balbiani, N.-Y. Suzuki, F. Wolter, and M. Zakharyaschev, editors, *Advances in Modal Logic*, volume 4, 2003. (In print).

Schmidt-Schauß and Smolka 1991. M. Schmidt-Schauß and G. Smolka. Attributive concept descriptions with complements. *Artificial Intelligence*, 48:1–26, 1991.

Schmiedel 1990. A. Schmiedel. A temporal terminological logic. In *Proceedings of the 9th National Conference of the American Association for Artificial Intelligence*, pages 640–645, Boston, 1990.

Schröder 1895. E. Schröder. *Vorlesungen über die Algebra der Logik (exacte Logik), Vol. 3, "Algebra und Logik der Relative", Part I.* Leipzig, 1895.

Schütte 1968. K. Schütte. *Vollständige Systeme modaler und intuitionistischer Logik.* Springer, 1968.

Scott 1962. D. Scott. A decision method for validity of sentences in two variables. *Journal of Symbolic Logic*, 27:477, 1962.

Segerberg 1970. K. Segerberg. Modal logics with linear alternative relations. *Theoria*, 36:301–322, 1970.

Segerberg 1971. K. Segerberg. An essay in classical modal logic. *Philosophical Studies*, 13, 1971.

Segerberg 1973. K. Segerberg. Two-dimensional modal logic. *Journal of Philosophical Logic*, 2:77–96, 1973.

Segerberg 1977. K. Segerberg. A completeness theorem in the modal logic of programs. *Notices of the AMS*, 24(6):A–522, 1977.

Segerberg 1980. K. Segerberg. Applying modal logic. *Studia Logic*, 39:275–295, 1980.

Segerberg 1989. K. Segerberg. Von Wright's tense logic. In P. Schilpp and L. Hahn, editors, *The Philosophy of Georg Henrik von Wright*, pages 603–635. La Salle, IL: Open Court, 1989.

Shehtman and Skvortsov 1990. V. Shehtman and D. Skvortsov. Semantics of non-classical first order predicate logics. In P. Petkov, editor, *Proceedings of Summer School and Conference on Mathematical Logic, Chaika, Bulgaria*, pages 105–116. Plenum Press, 1990.

Shehtman 1978. V. Shehtman. Two-dimensional modal logics. *Mathematical Notices of the USSR Academy of Sciences*, 23:417–424, 1978. (Translated from Russian).

Shelah 1975. S. Shelah. The monadic theory of order. *Annals of Mathematics*, 102:379–419, 1975.

Shoenfield 1967. J.R. Shoenfield. *Mathematical Logic.* Addison-Wesley, 1967.

Sikorski 1969. R. Sikorski. *Boolean Algebras.* Springer, 3rd edition, 1969.

Simpson 1994. A. Simpson. *The Proof Theory and Semantics of Intuitionistic Modal Logic.* PhD thesis, University of Edinburgh, 1994.

Sistla and Clarke 1985. A. Sistla and E. Clarke. The complexity of propositional linear temporal logics. *Journal of the Association for Computing Machinery*, 32:733–749, 1985.

Sistla and Zuck 1987. A. Sistla and L. Zuck. On the eventuality operator in temporal logic. In *Proceedings of the 2nd Annual IEEE Symposium on Logic in Computer Science (LICS'87)*, pages 153–166. IEEE Computer Society, 1987.

Skvortsov 1979. D. Skvortsov. On some propositional logics connected with the concept of Yu.T. Medvedev's types of information. *Semiotics and Information Science*, 13:142–149, 1979. (In Russian).

Solovay 1976. R. Solovay. Provability interpretations of modal logic. *Israel Journal of Mathematics*, 25:287–304, 1976.

Sotirov 1984. V. Sotirov. Modal theories with intuitionistic logic. In *Proceedings of the Conference on Mathematical Logic, Sofia, 1980*, pages 139–171. Bulgarian Academy of Sciences, 1984.

Spaan 1993. E. Spaan. *Complexity of Modal Logics.* PhD thesis, Department of Mathematics and Computer Science, University of Amsterdam, 1993.

Spielmann 2000. M. Spielmann. Verification of relational transducers for electronic commerce. In *Proceedings of the 19th ACM SIGMOD-SIGACT-SIGART Symposium on Principles of Database Systems (PODS 2000)*, pages 92–103. ACM, 2000.

Stebletsova 2000a. V. Stebletsova. *Algebras, Relations and Geometries.* PhD thesis, Universiteit Utrecht, 2000.

Stebletsova 2000b. V. Stebletsova. Weakly associative relation algebras with polyadic composition operations. *Studia Logica*, 66:297–323, 2000.

Stirling 1987. C. Stirling. Modal logics for communicating systems. *Theoretical Computer Science*, 49:311–347, 1987.

Stock 1997. O. Stock, editor. *Spatial and Temporal Reasoning.* Kluwer Academic Publishers, 1997.

Stockmeyer 1974. L. Stockmeyer. *The Complexity of Decision Problems in Automata Theory and Logic.* PhD thesis, MIT, 1974.

Stone 1937. M. Stone. Application of the theory of Boolean rings to general topology. *Transactions of the AMS*, 41:321–364, 1937.

Sturm and Wolter 2002. H. Sturm and F. Wolter. A tableau calculus for temporal description logic: the expanding domain case. *Journal of Logic and Computation*, 12:809–838, 2002.

Sturm *et al.* 2002. H. Sturm, F. Wolter, and M. Zakharyaschev. Common knowledge and quantification. *Economic Theory*, 19:157–186, 2002.

Surányi 1943. J. Surányi. Zur Reduktion des entscheidungsproblems des logischen Funktioskalküls. *Mathematikai és Fizikai Lapok*, 50:51–74, 1943.

Szałas and Holenderski 1988. A. Szałas and L. Holenderski. Incompleteness of first-order temporal logic with until. *Theoretical Computer Science*, 57:317–325, 1988.

Szałas 1986. A. Szałas. Concerning the semantic consequence relation in first-order temporal logic. *Theoretical Computer Science*, 47:329–334, 1986.

Tarski and Givant 1987. A. Tarski and S. Givant. *A Formalization of Set Theory without Variables*, volume 41 of *AMS Colloquium Publications*. Providence RI, 1987.

Tarski 1938. A. Tarski. Der Aussagenkalkül und die Topologie. *Fundamenta Mathematicae*, 31:103–134, 1938.

Tarski 1941. A. Tarski. On the calculus of relations. *Journal of Symbolic Logic*, 6:73–89, 1941.

Tarski 1948. A. Tarski. A decision method for elementary algebra and geometry. Manuscript. Berkeley, 1948.

Thomason 1974a. S. Thomason. An incompleteness theorem in modal logic. *Theoria*, 40:30–34, 1974.

Thomason 1974b. S. Thomason. Reduction of tense logic to modal logic I. *Journal of Symbolic Logic*, 39:549–551, 1974.

Thomason 1975. S. Thomason. Reduction of tense logic to modal logic II. *Theoria*, 41:154–169, 1975.

Thomason 1980. S. Thomason. Independent propositional modal logics. *Studia Logica*, 39:143–144, 1980.

Thomason 1984. R. Thomason. Combinations of tense and modality. In D. Gabbay and F. Guenthner, editors, *Handbook of Philosophical Logic*, volume 2, pages 135–165. Reidel, Dordrecht, 1984.

Trakhtenbrot 1950. B. Trakhtenbrot. The impossibility of an algorithm for the decision problem for finite models. *Doklady Akademii Nauk USSR*, 70:569–572, 1950.

Tsao Chen 1938. T. Tsao Chen. Algebraic postulates and a geometric interpretation of the Lewis calculus of strict implication. *Bulletin of the AMS*, 44:737–744, 1938.

Vakarelov 1996a. D. Vakarelov. Many-dimensional arrow logics. *Journal of Applied Non-Classical Logics*, 6:303–345, 1996.

Vakarelov 1996b. D. Vakarelov. Many-dimensional arrow structures. In M. Marx, L. Pólos, and M. Masuch, editors, *Arrow Logic and Multi-modal Logics*. CSLI Publications, Stanford, 1996.

Vakarelov 1997. D. Vakarelov. Modal logics of arrows. In M. de Rijke, editor, *Advances in Intensional Logic*, pages 137–171. Kluwer Academic Publishers, 1997.

van Benthem 1983. J. van Benthem. *Modal Logic and Classical Logic*. Bibliopolis, Napoli, 1983.

van Benthem 1984. J. van Benthem. Correspondence theory. In D. Gabbay and F. Guenthner, editors, *Handbook of Philosophical Logic*, volume 2, pages 167–247. Reidel, Dordrecht, 1984.

van Benthem 1991. J. van Benthem. *Language in Action. Categories, Lambdas and Dynamic Logic*. Elsevier, North-Holland, 1991.

van Benthem 1994. J. van Benthem. A note on dynamic arrow logic. In J. van Eijck and A. Visser, editors, *Logic and Information Flow*, Foundations of Computing, pages 15–29. MIT Press, 1994.

van Benthem 1995. J. van Benthem. Temporal logic. In D. Gabbay, C. Hogger, and J. Robinson, editors, *Handbook of Logic in Artificial Intelligence and Logic Programming*, volume 4, pages 241–350. Oxford Scientific Publishers, 1995.

van Benthem 1996. J. van Benthem. *Exploring Logical Dynamics*. CSLI Publications, Stanford, 1996.

van Dalen 1986. D. van Dalen. Intuitionistic Logic. In D. Gabbay and F. Guenthner, editors, *Handbook of Philosophical Logic*, volume 3, pages 225–339. Reidel, Dordrecht, 1986.

van der Hoek 1992. W. van der Hoek. *Modalities for Reasoning about Knowledge and Quantities.* PhD thesis, University of Amsterdam, 1992.

van Eijck and Visser 1994. J. van Eijck and A. Visser, editors. *Logic and Information Flow.* Foundations of Computing. MIT Press, 1994.

van Emde Boas 1997. P. van Emde Boas. The convenience of tilings. In A. Sorbi, editor, *Complexity, Logic and Recursion Theory*, volume 187 of *Lecture Notes in Pure and Applied Mathematics*, pages 331–363. Marcel Dekker Inc., 1997.

Vardi and Wolper 1986. M. Vardi and P. Wolper. Automata-theoretic techniques for modal logics of programs. *Journal of Computer and System Sciences*, 32:183–221, 1986.

Vardi 1985. M. Vardi. The taming of converse: Reasoning about two way computations. In R. Parikh, editor, *Logics of Programs*, volume 193 of *Lecture Notes in Computer Science*, pages 413–424. Springer, 1985.

Vardi 1997. M. Vardi. Why is modal logic so robustly decidable? In *DIMACS Series in Discrete Mathematics and Theoretical Computer Science*, volume 31, pages 149–184. AMS, 1997.

Venema and Marx 1999. Y. Venema and M. Marx. A modal logic of relations. In E. Orlowska, editor, *Logic at Work*, pages 124–167. Springer, 1999.

Venema 1991. Y. Venema. *Many-Dimensional Modal Logics.* PhD thesis, Universiteit van Amsterdam, 1991.

Vilain and Kautz 1986. M.B. Vilain and H. Kautz. Constraint propagation algorithms for temporal reasoning. In *Proceedings of the 5th AAAI Conference, Philadelphia*, pages 377–382, 1986.

Visser 1996. A. Visser. Uniform interpolation and layered bisimulation. In P. Hajek, editor, *Gödel'96*, pages 139–164. Springer, 1996.

Wajsberg 1933. M. Wajsberg. Ein erweiterter Klassenkalkül. *Monatsh Math. Phys.*, 40:113–126, 1933.

Wansing 1996. H. Wansing. *Proof Theory of Modal Logic.* Kluwer Academic Publishers, 1996.

Wijesekera 1990. D. Wijesekera. Constructive modal logic I. *Annals of Pure and Applied Logic*, 50:271–301, 1990.

Williamson 1992. T. Williamson. On intuitionistic modal epistemic logic. *Journal of Philosophical Logic*, 21:63–89, 1992.

Wolter and Zakharyaschev 1997. F. Wolter and M. Zakharyaschev. On the relation between intuitionistic and classical modal logics. *Algebra and Logic*, 36:121–155, 1997.

Wolter and Zakharyaschev 1998. F. Wolter and M. Zakharyaschev. Satisfiability problem in description logics with modal operators. In A. Cohn, L. Schubert, and S. Shapiro, editors, *Proceedings of the 6th Conference on Principles of Knowledge Representation and Reasoning (KR'98)*, pages 512–523. Morgan Kaufmann, 1998.

Wolter and Zakharyaschev 1999a. F. Wolter and M. Zakharyaschev. Intuitionistic modal logics as fragments of classical bimodal logics. In E. Orlowska, editor, *Logic at Work*, pages 168–186. Springer, 1999.

Wolter and Zakharyaschev 1999b. F. Wolter and M. Zakharyaschev. Modal description logics: modalizing roles. *Fundamenta Informaticae*, 39:411–438, 1999.

Wolter and Zakharyaschev 2000a. F. Wolter and M. Zakharyaschev. Spatial reasoning in RCC-8 with Boolean region terms. In W. Horn, editor, *Proceedings of the 14th European Conference on Artificial Intelligence (ECAI 2000)*, pages 244–248. IOS Press, 2000.

Wolter and Zakharyaschev 2000b. F. Wolter and M. Zakharyaschev. Spatio-temporal representation and reasoning based on RCC-8. In A. Cohn, F. Giunchiglia, and B. Seltman, editors, *Proceedings of the 7th Conference on Principles of Knowledge Representation and Reasoning (KR 2000)*, pages 3–14. Morgan Kaufmann, 2000.

Wolter and Zakharyaschev 2000c. F. Wolter and M. Zakharyaschev. Temporalizing description logics. In D. Gabbay and M. de Rijke, editors, *Frontiers of Combining Systems II*, pages 379–401. Studies Press/Wiley, 2000.

Wolter and Zakharyaschev 2001a. F. Wolter and M. Zakharyaschev. Decidable fragments of first-order modal logics. *Journal of Symbolic Logic*, 66:1415–1438, 2001.

Wolter and Zakharyaschev 2001b. F. Wolter and M. Zakharyaschev. Dynamic description logic. In M. Zakharyaschev, K. Segerberg, M. de Rijke, and H. Wansing, editors, *Advances in Modal Logic*, volume 2, pages 431–445. CSLI Publications, Stanford, 2001.

Wolter and Zakharyaschev 2002. F. Wolter and M. Zakharyaschev. Axiomatizing the monodic fragment of first-order temporal logic. *Annals of Pure and Applied Logic*, 118:133–145, 2002.

Wolter 1996. F. Wolter. Tense logics without tense operators. *Mathematical Logic Quarterly*, 42:145–171, 1996.

Wolter 1997. F. Wolter. The structure of lattices of subframe logics. *Annals of Pure and Applied Logic*, 86:47–100, 1997.

Wolter 1998. F. Wolter. Fusions of modal logics revisited. In M. Kracht, M. de Rijke, H. Wansing, and M. Zakharyaschev, editors, *Advances in Modal Logic*, volume 1, pages 361–379. CSLI Publications, Stanford, 1998.

Wolter 2000a. F. Wolter. First-order common knowledge logics. *Studia Logica*, 65:249–271, 2000.

Wolter 2000b. F. Wolter. The product of converse **PDL** and polymodal **K**. *Journal of Logic and Computation*, 10:223–251, 2000.

Yashin 1994. A. Yashin. The Smetanich logic T^{ϕ} and two definitions of a new intuitionistic connective. *Mathematical Notes*, 56:135–142, 1994. (In Russian).

Zakharyaschev and Alekseev 1995. M. Zakharyaschev and A. Alekseev. All finitely axiomatizable normal extensions of **K4.3** are decidable. *Mathematical Logic Quarterly*, 41:15–23, 1995.

Zakharyaschev *et al.* 2001. M. Zakharyaschev, F. Wolter, and A. Chagrov. Advanced modal logic. In D. Gabbay and F. Guenthner, editors, *Handbook of Philosophical Logic*, volume 3, pages 83–266. Kluwer Academic Publishers, 2nd edition, 2001.

Zanardo 1990. A. Zanardo. Axiomatization of 'Peircean' branching-time logic. *Studia Logica*, 49:183–195, 1990.

Zanardo 1996. A. Zanardo. Branching-time logic with quantification over branches: the point of view of modal logic. *Journal of Symbolic Logic*, 61:1–39, 1996.

Zeman 1973. J.J. Zeman. *Modal Logic. The Lewis-Modal Systems.* Clarendon Press, Oxford, 1973.

List of tables

2.1 Reasoning tasks in $\mathcal{ALC}$. 72
2.2 'Model level' reductions between modal, epistemic, dynamic, and temporal logics. 96

3.1 Reasoning tasks in $L_{\mathcal{ALC}}$. 176

6.1 Reductions between decidable product logics. 299
6.2 Products of unimodal logics with **K**. 339
6.3 Products of unimodal logics with **S5**. 340
6.4 Complexity of decidable products of multimodal logics with **K** and $\mathbf{S5}_n$. 341

7.1 Reductions between undecidable product logics. 370
7.2 Products of transitive unimodal logics. 375

8.1 Some higher-dimensional product logics. 378

11.1 Complexity of first-order temporal logics over $\langle \mathbb{N}, < \rangle$. 504

13.1 The results of Halpern and Vardi (1989) on the complexity of the satisfiability problem for some temporal epistemic logics. . 568
13.2 Upper bounds for the complexity of temporal epistemic logics with $\mathcal{S}$ and $\mathcal{U}$, but without common knowledge operators. . . . 581

14.1 Reductions between some reasoning tasks for $L_{\mathcal{ALC}}$. 586

16.1 Complexity of spatio-temporal logics over $\langle \mathbb{N}, < \rangle$. 673

List of languages and logics

$\mathcal{AL}$, 180
$\mathbf{AL}_{min}$, 182
$\mathbf{AL}_{NA}$, 184
$\mathbf{AL}_{RA}$, 184
$\mathbf{AL}_{sq}$, 183
$\mathbf{AL}_{WA}$, 184
$\mathcal{ALC}$, 67
$\mathcal{ALC}_{R^+}$, 78
$\mathcal{ALCH}_{R^+}$, 78
$\mathcal{ALCU}$, 641
$\mathcal{A}\ell\ell$-13, 50
Alt, 12, 13
$\mathbf{Alt}^n$, 378, 410
$\mathcal{ARCC}$-8, 125

$\mathcal{BRCC}$-8, 86

Cl, 5
$\mathcal{CPDL}$, 64
  $\mathcal{PDL}$, 62
CPDL, 64, 108
$\mathbf{CPDL}^{-?}$, 65
$\mathbf{CPDL} \times L$, 298
  $\mathbf{CPDL} \times \mathbf{K}$, 296
  $\mathbf{CPDL} \times \mathbf{K}_n$, 285
  $\mathbf{CPDL} \times \mathbf{K4}$, 374
  $\mathbf{CPDL} \times \mathbf{KD45}$, 320
  $\mathbf{CPDL} \times \mathbf{KD45}_n$, 334
  $\mathbf{CPDL} \times \mathbf{S4}$, 374
  $\mathbf{CPDL} \times \mathbf{S5}$, 320, 323, 334
  $\mathbf{CPDL} \times \mathbf{S5}_n$, 334
$\mathcal{CPDL} \otimes \mathcal{ML}$, 285

$\mathbf{CPDL} \cdot \mathbf{S5}$, 321
$\mathcal{CPDL}_{\mathcal{ALC}}$, 585
$\mathcal{CPDL}_{\mathcal{CQO}}$, 605
$\mathcal{CQ}$, 76
$\mathcal{CQDL}$, 153
  $\mathcal{CQDL}_{\boxed{1}}$, 548
  $\mathcal{CQDL}_{\boxed{1}}^{=}$, 548
  $\mathcal{DGF}$, 547
  $\mathcal{QDL}$, 153
$\mathcal{CQO}$, 77

D, 9, 12, 36
$\mathbf{D} \times L$, 230
$\mathbf{D}_n$, 37
$\mathbf{D}_n \times L$, 244
$\mathbf{D}_u$, 38
DAlt, 12, 13
$\mathbf{DAlt}^n$, 410

$\mathsf{ELog}(\mathcal{KN}_{L,\mathcal{C}})$, 575
$\mathsf{ELog}^{\mathcal{C}}(\mathcal{KN}_{L,\mathcal{C}})$, 576, 582
$\mathsf{ELog}_{FP}(\mathcal{KN}_{L,\mathcal{C}})$, 575
$\mathsf{ELog}_{FP}^{\mathcal{C}}(\mathcal{KN}_{L,\mathcal{C}})$, 576
$\mathsf{ELog}_{\mathcal{SU}}(\mathcal{K})$, 137
  $\mathsf{ELog}_{\mathcal{SU}}(\mathcal{KN}_{L,\mathcal{C}})$, 575, 581
  $\mathsf{ELog}_{\mathcal{SU}}(\mathcal{TE}_{L,\mathfrak{F}})$, 567
$\mathsf{ELog}_{\mathcal{SU}}^{\mathcal{C}}(\mathcal{K})$, 137
  $\mathsf{ELog}_{\mathcal{SU}}^{\mathcal{C}}(\mathcal{KN}_{L,\mathcal{C}})$, 575

FS, 188, 445

GL, 8, 12, 35, 36

$\mathbf{GL}_u$, 38
GL.3, 12, 13, 26, 34
GL.3 $\times L$, 249
  GL.3 $\times$ **GL.3**, 360, 364
  GL.3 $\times$ **Grz.3**, 360, 364
Grz, 13
Grz $\times L$, 233
Grz.3, 13, 26
Grz.3 $\times L$, 233, 249
  Grz.3 $\times$ **Grz.3**, 360, 364

HS, 53, 353
  $\mathbf{HS}_{\mathcal{C}}$, 353
  $\mathbf{HS}_{\mathcal{C}}^{fw}$, 353

$\mathbf{I}^{\triangle}$, 189
Int, 92
$\mathbf{IntK}_{\Box}$, 189
$\mathbf{IntK}_{\Box\Diamond}$, 190
$\mathbf{IntK}_{\Diamond}$, 189

K, 7, 19, 35, 36
K $\times L$, 230, 267, 337
  K $\times$ **Alt**, 339
  K $\times$ **CPDL**, 296
  K $\times$ **K**, 227, 339
  K $\times$ $\mathbf{K}_1^C$, 341
  K $\times$ **K4**, 339
  K $\times$ $\mathbf{K4}_2^C$, 341
  K $\times$ **K4.3**, 339
  K $\times$ $\mathbf{KD45}_2^C$, 341
  K $\times$ **Lin**, 341
  K $\times$ $\mathsf{Log}\{\langle\mathbb{N},<\rangle\}$, 339
  K $\times$ $\mathsf{Log}\{\langle\mathbb{Q},<\rangle\}$, 339
  K $\times$ **PDL**, 296, 341
  K $\times$ **PTL**, 341
  K $\times$ **S5**, 325, 339
  K $\times$ $\mathbf{S5}_2^C$, 341
$\mathbf{K}_{\mathcal{ALCCU}}$, 640
$\mathbf{K}_n$, 20, 35, 37
$\mathbf{K}_n \times L$, 175, 244, 284, 285
  $\mathbf{K}_n \times \mathbf{K}_m$, 274, 284
$\mathbf{K}_n^C$, 60, 96, 103, 106, 108
$\mathbf{K}_n^C \times L$, 298, 337, 367
  $\mathbf{K}_n^C$ $\times$ **CPDL**, 369
  $\mathbf{K}_n^C \times \mathbf{K}_m^C$, 369
  $\mathbf{K}_n^C$ $\times$ **K4**, 374
  $\mathbf{K}_n^C \times \mathbf{K4}_m^C$, 369
  $\mathbf{K}_n^C$ $\times$ **PDL**, 369
  $\mathbf{K}_n^C$ $\times$ **S4**, 374
  $\mathbf{K}_n^C \times \mathbf{S4}_m^C$, 369
  $\mathbf{K}_n^C$ $\times$ **S5**, 323, 334
  $\mathbf{K}_n^C \times \mathbf{S5}_m$, 336
  $\mathbf{K}_n^C \times \mathbf{S5}_m^C$, 369
  $\mathbf{K}_n^C \times \mathbf{T}_m^C$, 369
$\mathbf{K}^n$, 378, 397
$\mathbf{K}_u$, 38, 103
$\mathbf{K}_u \times L$, 178, 251, 337, 367
  $\mathbf{K}_u$ $\times$ **CPDL**, 369
  $\mathbf{K}_u \times \mathbf{K}_n^C$, 369
  $\mathbf{K}_u \times \mathbf{K}_u$, 254
  $\mathbf{K}_u \times \mathbf{K4}_n^C$, 369
  $\mathbf{K}_u$ $\times$ **PDL**, 369
  $\mathbf{K}_u \times \mathbf{S4}_n^C$, 369
  $\mathbf{K}_u \times \mathbf{S5}_n^C$, 369
  $\mathbf{K}_u \times \mathbf{T}_n^C$, 369
K4, 8, 12, 25, 35, 36
K4 $\times L$, 230, 233
  K4 $\times$ **K4**, 375
$\mathbf{K4}_n$, 21, 35
$\mathbf{K4}_n \times L$, 244
$\mathbf{K4}_n^C$, 60, 96, 108
$\mathbf{K4}_n^C \times L$, 298
  $\mathbf{K4}_n^C$ $\times$ **K4**, 374
  $\mathbf{K4}_n^C$ $\times$ **S4**, 374
  $\mathbf{K4}_n^C$ $\times$ **S5**, 323, 334
  $\mathbf{K4}_n^C \times \mathbf{S5}_m$, 336
$\mathbf{K4}^n$, 378, 397
$\mathbf{K4}_u$, 38
K4.3, 12, 13, 26, 34, 36
K4.3 $\times L$, 232, 233, 304
  K4.3 $\times$ **K**, 319
  K4.3 $\times$ $\mathbf{K}_n$, 314
  K4.3 $\times$ **K4**, 375
  K4.3 $\times$ **K4.3**, 344, 375

$\mathbf{K4.3} \times \mathsf{Log}\{\langle \mathbb{N}, < \rangle\}$, 344
$\mathbf{K4.3} \times \mathsf{Log}\{\langle \mathbb{Q}, < \rangle\}$, 344
$\mathbf{K4.3} \times \mathbf{S4.3}$, 344
$\mathbf{K4.3} \times \mathbf{S5}$, 325, 333
$\mathbf{K4.3} \times \mathbf{S5}_n$, 336
$\mathbf{K4.3}_u$, 38
$\mathbf{K45}$, 427
$\mathbf{K5}$, 426
$\mathbf{KD45}$, 12, 13, 34, 36
$\mathbf{KD45} \times L$, 230
$\mathbf{KD45}_n$, 21, 35
$\mathbf{KD45}_n \times L$, 244
$\mathbf{KD45}_n^C$, 60, 96, 108
$\mathbf{KD45}_n^C \times L$, 298
$\mathbf{KD45}_n^C \times \mathbf{K4}$, 374
$\mathbf{KD45}_n^C \times \mathbf{S4}$, 374
$\mathbf{KD45}_n^C \times \mathbf{S5}$, 323, 334
$\mathbf{KD45}_n^C \times \mathbf{S5}_m$, 336
$\mathbf{KD45}_u$, 38

$\mathcal{L}$, 5
$[L_1, L_2]$, 223
$[\mathbf{K}, \mathbf{K}]$, 223, 227
$[L_1, L_2]^{\mathsf{EX}}$, 432
$L_1 \otimes L_2$, 111, 198, 205, 209, 216
$L_1 \otimes \cdots \otimes L_n$, 112, 420
$L_1 \times L_2$, 126, 222
$L_1 \times \cdots \times L_n$, 129, 130, 377
$(L_1 \times \cdots \times L_n)^{\mathcal{K}}$, 419
$(L_1 \times L_2)^{\mathsf{EX}}$, 432
$(L_1 \times \cdots \times L_n)^{\mathsf{EX}_n}$, 433
$(L_1 \times \cdots \times L_n)^{\mathsf{LC}_n}$, 430
$(L_1 \times \cdots \times L_n)^{\mathsf{SF}_n}$, 420
$L_{\mathcal{ALC}}$, 174
$\mathbf{CPDL}_{\mathcal{ALC}}$, 584, 586
$\mathbf{K}_{\mathcal{ALC}}$, 623
$(\mathbf{K}_n)_{\mathcal{ALC}}$, 591
$\mathbf{K4.3}_{\mathcal{ALC}}$, 586, 591, 609
$(\mathbf{KD45}_n)_{\mathcal{ALC}}$, 591
$\mathbf{Lin}_{\mathcal{ALC}}$, 586
$\mathbf{PDL}_{\mathcal{ALC}}$, 586, 591, 609
$\mathbf{PTL}_{\mathcal{ALC}}$, 584, 586, 591, 609
$(\mathbf{S5}_n)_{\mathcal{ALC}}$, 591
$(\mathbf{T}_n)_{\mathcal{ALC}}$, 591
$L^C_{\mathcal{ALC}}$, 585
$(\mathbf{K}_n^C)_{\mathcal{ALC}}$, 586, 591, 609
$(\mathbf{K4}_n^C)_{\mathcal{ALC}}$, 586, 591, 609
$(\mathbf{KD45}_n^C)_{\mathcal{ALC}}$, 586, 591, 609
$(\mathbf{S4}_n^C)_{\mathcal{ALC}}$, 586, 591, 609
$(\mathbf{S5}_n^C)_{\mathcal{ALC}}$, 584, 586, 591
$(\mathbf{T}_n^C)_{\mathcal{ALC}}$, 586, 591, 609
$L_{\mathcal{ALC}_{R+}}$
$(\mathbf{K}_n^C)_{\mathcal{ALC}_{R+}}$, 590
$(\mathbf{K4}_n^C)_{\mathcal{ALC}_{R+}}$, 590
$(\mathbf{KD45}_n^C)_{\mathcal{ALC}_{R+}}$, 590
$\mathbf{PDL}_{\mathcal{ALC}_{R+}}$, 590
$\mathbf{PTL}_{\mathcal{ALC}_{R+}}$, 590
$(\mathbf{S4}_n^C)_{\mathcal{ALC}_{R+}}$, 590
$(\mathbf{T}_n^C)_{\mathcal{ALC}_{R+}}$, 590
$L_{\mathcal{CQ}}$
$(\mathbf{K}_n^C)_{\mathcal{CQ}}$, 590
$(\mathbf{K4}_n^C)_{\mathcal{CQ}}$, 590
$\mathbf{K4.3}_{\mathcal{CQ}}$, 590
$(\mathbf{KD45}_n^C)_{\mathcal{CQ}}$, 590
$\mathbf{PDL}_{\mathcal{CQ}}$, 590
$\mathbf{PTL}_{\mathcal{CQ}}$, 590
$(\mathbf{S4}_n^C)_{\mathcal{CQ}}$, 590
$(\mathbf{S5}_n^C)_{\mathcal{CQ}}$, 590
$(\mathbf{T}_n^C)_{\mathcal{CQ}}$, 590
$L_{\mathcal{CQO}}$
$\mathbf{CPDL}_{\mathcal{CQO}}$, 605
$(\mathbf{K}_n^C)_{\mathcal{CQO}}$, 605
$(\mathbf{K4}_n^C)_{\mathcal{CQO}}$, 605
$\mathbf{K4.3}_{\mathcal{CQO}}$, 606
$(\mathbf{KD45}_n^C)_{\mathcal{CQO}}$, 605
$\mathbf{PTL}_{\mathcal{CQO}}$, 606
$(\mathbf{S4}_n^C)_{\mathcal{CQO}}$, 605
$(\mathbf{S5}_n^C)_{\mathcal{CQO}}$, 605
$(\mathbf{T}_n^C)_{\mathcal{CQO}}$, 605
L_u, 38
$\mathbf{Lin}$, 43
$\mathbf{Lin} \times L$, 304
$\mathbf{Lin} \times \mathbf{K}$, 319
$\mathbf{Lin} \times \mathbf{K}_n$, 314
$\mathbf{Lin} \times \mathbf{S5}$, 325, 333

$\mathbf{Lin} \times \mathbf{S5}_n$, 336
$\mathbf{Lin}_{\mathcal{SU}}$, 46
$\mathsf{Log}\,\mathcal{C}$, 10, 21, 29
$\mathsf{Log}\{\langle \mathbb{N}, < \rangle\}$, 42
$\mathsf{Log}\{\langle \mathbb{N}, < \rangle \times \langle \mathbb{N}, < \rangle\}$, 363
$\mathsf{Log}\{\langle \mathbb{N}, < \rangle \times \langle \mathbb{Q}, < \rangle\}$, 363
$\mathsf{Log}\{\langle \mathbb{N}, < \rangle \times \langle \mathbb{R}, < \rangle\}$, 363
$\mathsf{Log}\{\langle \mathbb{N}, \leq \rangle\}$, 42
$\mathsf{Log}\{\langle \mathbb{N}, > \rangle \times \langle \mathbb{N}, > \rangle\}$, 366
$\mathsf{Log}\{\langle \mathbb{N}, \geq \rangle \times \langle \mathbb{N}, \geq \rangle\}$, 366
$\mathsf{Log}\{\langle \mathbb{Q}, < \rangle\}$, 42
$\mathsf{Log}\{\langle \mathbb{Q}, \leq \rangle\}$, 42
$\mathsf{Log}\{\langle \mathbb{R}, < \rangle\}$, 42
$\mathsf{Log}\{\langle \mathbb{Z}, < \rangle\}$, 42
$\mathsf{Log}\{\langle \mathbb{N}, < \rangle\} \times L$
$\mathsf{Log}\{\langle \mathbb{N}, < \rangle\} \times \mathbf{K4}$, 371
$\mathsf{Log}\{\langle \mathbb{N}, < \rangle\} \times \mathsf{Log}\{\langle \mathbb{N}, < \rangle\}$, 375
$\mathsf{Log}\{\langle \mathbb{N}, < \rangle\} \times \mathbf{S4}$, 371, 375
$\mathsf{Log}\{\langle \mathbb{N}, < \rangle\} \times \mathbf{S5}$, 333
$\mathsf{Log}\{\langle \mathbb{Q}, < \rangle\} \times L$
$\mathsf{Log}\{\langle \mathbb{Q}, < \rangle\} \times \mathbf{K}$, 319
$\mathsf{Log}\{\langle \mathbb{Q}, < \rangle\} \times \mathbf{K}_n$, 314
$\mathsf{Log}\{\langle \mathbb{Q}, < \rangle\} \times \mathbf{S5}$, 325, 333
$\mathsf{Log}\{\langle \mathbb{Q}, < \rangle\} \times \mathbf{S5}_n$, 336
$\mathsf{Log}_{FP}\mathcal{C}$, 43
$\mathsf{Log}_{FP}(\mathbb{N})$, 43
$\mathsf{Log}_{FP}(\mathbb{Q})$, 43
$\mathsf{Log}_{FP}(\mathbb{R})$, 43
$\mathsf{Log}_{FP}(\mathbb{Z})$, 43
$\mathsf{Log}_{FP}(\mathbb{Q}) \times \mathbf{S5}_n$, 336
$\mathsf{Log}_{\mathcal{SU}}(\mathcal{C})$, 46
$\mathsf{Log}_{\mathcal{SU}}(\mathbb{N})$, 46
$\mathsf{Log}_{\mathcal{SU}}(\mathbb{Q})$, 46
$\mathsf{Log}_{\mathcal{SU}}(\mathbb{R})$, 46
$\mathsf{Log}_{\mathcal{SU}}(\mathbb{Z})$, 46

MIPC, 188, 445
$\mathcal{ML}_{\mathcal{ALCU}}$, 641
$\mathcal{ML}_{\mathcal{CQO}}$, 605
$\mathcal{ML}_n^C$, 59
$\mathcal{L}_{\mathsf{M}}$, 189
$\mathcal{ML}$, 4
$\mathcal{ML}_2$, 42
$\mathcal{ML}_n$, 20
$\mathcal{ML}_n^u$, 37
$\mathcal{ML}^u$, 87
$\mathcal{ML}_{\mathcal{SU}}$, 44, 504
$\mathcal{ML}_{\mathcal{U}}$, 46
$\mathcal{ML}_{\mathcal{SU}} \otimes \mathcal{ML}_n^C$, 137
$\mathcal{ML}_{\mathcal{SU}} \otimes \mathcal{ML}_n$, 136
$(\mathcal{ML}_{\mathcal{SU}})_{\mathcal{CQO}}$, 605
$\mathcal{MSOL}$, 39

PDL, 63, 108
$\mathbf{PDL} \times L$
$\mathbf{PDL} \times \mathbf{K}$, 296
$\mathbf{PDL} \times \mathbf{K4}$, 374
$\mathbf{PDL} \times \mathbf{S4}$, 374
$\mathbf{PDL} \times \mathbf{S5}$, 323, 334
$\mathbf{PDL} \times \mathbf{S5}_n$, 336
PTL, 46, 106, 109
$\mathbf{PTL} \times L$, 301, 304
$\mathbf{PTL} \times \mathbf{K}_n$, 306
$\mathbf{PTL} \times \mathbf{K4}$, 374
$\mathbf{PTL} \times \mathbf{PTL}$, 258
$\mathbf{PTL} \times \mathbf{S4}$, 374
$\mathbf{PTL} \times \mathbf{S5}$, 164, 268, 325, 537
$\mathbf{PTL} \times \mathbf{S5}_n$, 336
$\mathbf{PTL}_{\Box\bigcirc}$, 47
$\mathbf{PTL}_{\Box\bigcirc} \times \mathbf{K4}$, 258
$\mathbf{PTL}_{\Box\bigcirc} \times \mathbf{PTL}_{\Box\bigcirc}$, 256
$\mathcal{PST}$, 122, 648, 673
$\mathcal{PST}_{\bigcirc}$, 649

QCl, 17, 143
$\mathbf{QCl}^=$, 17
QInt, 153
QIntCD, 156
$\mathcal{QL}$, 15
$\mathcal{FLU}$, 477
$\mathcal{GF}$, 478
$\mathcal{L}_\varphi$, 511
$\mathcal{PF}$, 479
$\mathcal{QL}_t$, 45
$r\mathcal{QL}^n$, 141
$\mathcal{QL}^=$, 16

$\mathcal{PF}^=$, 542
$\mathcal{QL}_2^{3,=}$, 186, 381
$\overline{\mathcal{QL}}$, 472
$\boldsymbol{QL}$, 145, 398
CQDL, 153
QDL, 153, 547
QK, 145, 398
$\mathbf{Q}^v\mathbf{K}_l$, 146
$\mathbf{Q}_v\mathbf{K}_l$, 146
$\mathbf{QK}_n^C$, 150, 153, 548
QK4, 145, 398
$\mathbf{QK4}_n^C$, 150, 153, 548
$\mathbf{QKD45}_n^C$, 150, 153, 548
QS4, 145, 398
$\mathbf{Q}^e\mathbf{S4}$, 156
$\mathbf{QS4}_n^C$, 150, 153, 548
QS5, 398
$\mathbf{QS5}_n^C$, 150, 153, 548
QT, 145
$\mathbf{QT}_n^C$, 150, 153, 548
$\mathsf{QLog}_{\mathcal{SU}}(\mathcal{C})$, 158, 465
$\mathsf{QLog}_{\mathcal{SU}}(\mathcal{LO})$, 159
$\mathsf{QLog}_{\mathcal{SU}}(\mathbb{N})$, 465
$\mathsf{QLog}_{\mathcal{SU}}(\mathbb{Q})$, 465
$\mathsf{QLog}_{\mathcal{SU}}(\mathbb{R})$, 465
$\mathsf{QLog}_{\mathcal{SU}}^{fin}(\mathcal{C})$, 465
$\mathsf{QLog}_{\mathcal{SU}}^{fin}(\mathbb{N})$, 465
$\mathsf{QLog}_{\mathcal{SU}}^{fin}(\mathbb{Q})$, 465
$\mathsf{QLog}_{\mathcal{SU}}^{fin}(\mathbb{R})$, 465
$\mathsf{QLog}_{\mathcal{U}}(\mathcal{C})$, 158, 466
$\mathsf{QLog}_{\mathcal{U}}(\mathbb{N})$, 164
$\mathsf{QLog}_{\mathcal{U}}^{fin}(\mathcal{C})$, 466
$\mathcal{QML}_n^C$, 150
$\mathcal{EGF}_n$, 547
$\mathcal{QML}_l$, 143
$\mathcal{QML}_{n\boxed{1}}^C$, 548
$\mathcal{QML}_{n\boxed{1}}^{C=}$, 548
$\mathcal{QTL}$, 157
$\mathcal{QTL}_{\boxed{0}}$, 499
$\mathcal{QTL}_{\boxed{0}}^1$, 500, 504
$\mathcal{QTL}_{\boxed{0}}^2$, 500, 504
$\mathcal{QTL}_{\boxed{0}}^{mo}$, 500, 504
$\mathcal{QTL}^1$, 466, 476, 504
$\mathcal{QTL}^{1+}$, 496
$\mathcal{QTL}_{\textcircled{1}}^1$, 496, 504
$\mathcal{QTL}_{\boxed{1}}$, 471, 504
$\mathcal{QTL}_{\boxed{1}}^2$, 476, 504
$\mathcal{QTL}_{\boxed{1}}^{mo}$, 477, 504
$\mathcal{QTL}^2$, 466, 476
$\mathcal{QTL}^{mo}$, 466, 477
$\mathcal{QTL}^n$, 466
$\mathcal{QTL}_{\mathcal{U}}$, 158, 465
$\mathcal{QTL}_{\mathcal{U}\boxed{0}}$, 499
$\mathcal{QTL}_{\mathcal{U}\boxed{1}}$, 489
$\mathcal{TFLU}$, 477
$\mathcal{TGF}$, 478
$\mathcal{TPF}$, 479
$\mathcal{TPF}_{\boxed{1}}$, 479, 504
$\mathcal{TPF}_{\boxed{1}}^=$, 542
$\mathcal{QTL}^=$
$\mathcal{QTL}_{\boxed{1}}^=$, 504, 538
$\mathcal{TPF}^=$, 542

$\mathcal{RCC}$, 80
$\mathcal{RCC}$-8, 81

S4, 8, 12, 25, 35, 36
S4 $\times L$, 230, 233
S4 $\times$ **S4**, 375
$\mathbf{S4}_n$, 21, 35
$\mathbf{S4}_n \times L$, 244
$\mathbf{S4}_n^C$, 60, 96, 108
$\mathbf{S4}_n^C \times L$, 298
$\mathbf{S4}_n^C \times \mathbf{K4}$, 374
$\mathbf{S4}_n^C \times \mathbf{S4}$, 374
$\mathbf{S4}_n^C \times \mathbf{S5}_m$, 336
$\mathbf{S4}^n$, 397
$\mathbf{S4}_u$, 38, 87
S4.3, 13, 26, 34, 36
S4.3 $\times L$, 233
$\mathbf{S4.3}_u$, 38
S5, 8, 12, 19, 34, 36, 90
S5 $\times L$, 177, 230, 267, 320, 323, 333, 334
S5 $\times$ **Alt**, 340

$\mathbf{S5} \times \mathbf{CPDL}$, 334
$\mathbf{S5} \times \mathbf{K}_1^C$, 341
$\mathbf{S5} \times \mathbf{K4}$, 340
$\mathbf{S5} \times \mathbf{K4}_2^C$, 341
$\mathbf{S5} \times \mathbf{K4.3}$, 340
$\mathbf{S5} \times \mathbf{KD45}_2^C$, 341
$\mathbf{S5} \times \mathbf{Lin}$, 341
$\mathbf{S5} \times \mathsf{Log}\{\langle \mathbb{N}, < \rangle\}$, 340
$\mathbf{S5} \times \mathsf{Log}\{\langle \mathbb{Q}, < \rangle\}$, 340
$\mathbf{S5} \times \mathbf{PDL}$, 334, 341
$\mathbf{S5} \times \mathbf{PTL}$, 341, 537
$\mathbf{S5} \times \mathbf{S5}$, 239, 244, 262, 340
$\mathbf{S5} \times \mathbf{S5}_2^C$, 341
$\mathbf{S5}_n$, 21, 35
$\mathbf{S5}_n \times L$, 244, 336
$\mathbf{S5}_n \times \mathbf{K}_1^C$, 341
$\mathbf{S5}_n \times \mathbf{K4}_2^C$, 341
$\mathbf{S5}_n \times \mathbf{KD45}_2^C$, 341
$\mathbf{S5}_n \times \mathbf{Lin}$, 341
$\mathbf{S5}_n \times \mathbf{PDL}$, 341
$\mathbf{S5}_n \times \mathbf{PTL}$, 341
$\mathbf{S5}_n \times \mathbf{S5}_2^C$, 341
$\mathbf{S5}_n^C$, 60, 103, 108
$\mathbf{S5}_n^C \times L$, 298
$\mathbf{S5}_n^C \times \mathbf{S5}$, 323
$\mathbf{S5}_n^C \times \mathbf{S5}_m$, 336
$[\mathbf{S5}, \mathbf{S5}, \dots, \mathbf{S5}]$, 379, 381, 387
$\mathbf{S5}^n$, 143, 378, 381, 387, 397
$\mathbf{S5}_u$, 38
$\mathcal{ST}_i$, 654
$\mathcal{ST}_0$, 115, 673
$\mathcal{ST}_1$, 117, 673
$\mathcal{ST}_2$, 118, 673
$\mathcal{ST}_0^-$, 672, 673
$\mathcal{ST}_1^-$, 665, 673
$\mathcal{ST}_2^-$, 672, 673
$\mathcal{ST}_i^-$, 674

$\mathbf{T}$, 8, 12, 35, 36
$\mathbf{T} \times L$, 230
$\mathbf{T}_n$, 21, 35, 37
$\mathbf{T}_n \times L$, 244
$\mathbf{T}_n^C$, 60, 96, 108
$\mathbf{T}_n^C \times L$, 298
$\mathbf{T}_n^C \times \mathbf{K4}$, 374
$\mathbf{T}_n^C \times \mathbf{S4}$, 374
$\mathbf{T}_n^C \times \mathbf{S5}$, 323, 334
$\mathbf{T}_n^C \times \mathbf{S5}_m$, 336
$\mathbf{T}_u$, 38
Triv, 13
$\mathcal{TS}$, 159
$\mathcal{TS}_1$, 475
$\mathcal{TS}_1^1$, 476
$\mathcal{TS}_1^2$, 476
$\mathcal{TS}_{1t}$, 162, 475
$\mathcal{TS}_{1x}$, 475
$\mathcal{TS}^{mo}$, 477
$\mathcal{TS}_1^{mo}$, 477
$\mathcal{TSPF}$, 480
$\mathcal{TSPF}_1$, 481
$\mathsf{TSLog}(\mathcal{C})$, 475
$\mathsf{TSLog}^{fin}(\mathcal{C})$, 475

Verum, 13

Symbol index

CALCULI:

$\mathcal{MON}$, 528
$\mathcal{MON}^1$, 536
$\mathcal{MONC}$, 551
$\mathcal{MON}K_n^C$, 550
$\mathcal{MON}K4_n^C$, 551
$\mathcal{MON}KD45_n^C$, 551
$\mathcal{MON}S4_n^C$, 551
$\mathcal{MON}S5_n^C$, 551
$\mathcal{MON}T_n^C$, 551

CLASSES OF STRUCTURES:

$\mathsf{Alg}L$, 29
$\mathsf{Atg}L$, 201
$\mathcal{C}_1 \otimes \mathcal{C}_2$, 112
$\mathcal{C}_1 \times \mathcal{C}_2$, 126
C_n, 429
Df_n, 379
EX, 432
EX_n, 432
$\mathsf{Fr}L$, 11
$\mathsf{Fr}^r L$, 127
$\mathcal{KN}$, 575
$\mathcal{KN}_{L,\mathcal{C}}$, 575
LC_n, 429
$\mathcal{LO}$, 159
RDf_n, 379
SF_n, 419
$\mathcal{SYNC}$, 569
$\mathcal{SYNC}_{L,\mathcal{C}}$, 569
$\mathcal{TE}_{L,\mathcal{C}}$, 137, 567
$\mathcal{TE}_{L,\mathfrak{F}}$, 137

CONSEQUENCE RELATIONS:

$\vdash_L$, 35
$\vdash_L^*$, 35
$\vdash_{L_1}^* \times \vdash_{L_2}^*$, 129
$\vdash_{\mathcal{MON}}$, 528
$\vdash_{\mathcal{MONC}}$, 551

FORMULAS:

$\boldsymbol{cd}$, 156
$\boldsymbol{chr}$, 222
$\boldsymbol{chr}_{ij}$, 378
$\boldsymbol{com}$, 223
$\boldsymbol{com}_{ij}$, 378
$\boldsymbol{com}^{\boldsymbol{l}}$, 222
$\boldsymbol{com}_{ij}^{\boldsymbol{l}}$, 378
$\boldsymbol{com}^{\boldsymbol{r}}$, 222
$\boldsymbol{com}_{ij}^{\boldsymbol{r}}$, 378
$\boldsymbol{cub}^{ijk}$, 380
$\boldsymbol{cub}_{\ell}$, 397
$\overline{\varphi}$, 472
$\varphi^{\mathbf{1}}$, 198
$\varphi^{\mathbf{2}}$, 198
$\varphi_{\mathfrak{A}}$, 401
$\boldsymbol{mix}$, 444
$\mathsf{real}_{\mathfrak{C}}$, 473, 530, 553

FRAMES:

$\mathfrak{F}_1 \times \mathfrak{F}_2$, 126
$\mathfrak{F}_1 \times \cdots \times \mathfrak{F}_n$, 377
$\mathfrak{F}_{\mathfrak{A},n}$, 401

$\mathfrak{I}(\mathfrak{F})$, 179
$\kappa\mathfrak{F}$, 449

LOGICAL CONNECTIVES:
$\bot$, 4
$\top$, 4
$\neg$, 4
$\wedge$, 4
$\vee$, 4
$\to$, 4
$\leftrightarrow$, 4
$\Box$, 4
$\Diamond$, 4
$\forall$, 15
$\exists$, 15
$\Box^+$, 13
$\Diamond^+$, 13
$\Box_i$, 20
$\Diamond_i$, 20
$\Box^{\leq k}$, 36
$\Box^k$, 36
$\boldsymbol{M}^{\leq k}_{(n)}$, 36
$\boldsymbol{M}^{k}_{(n)}$, 36
$\boxtimes$, 37
$\diamondtimes$, 37
$\Box_F$, 42
$\Diamond_F$, 42
$\Box_P$, 42
$\Diamond_P$, 42
$\bigcirc$, 44
$\Diamond_f$, 53
$\Diamond_f^{-1}$, 53
$\Diamond_s$, 53
$\Diamond_s^{-1}$, 53
$\mathcal{S}$, 44
$\mathcal{U}$, 44
C_M, 59
E_M, 59
$[\alpha]$, 62
$\langle\alpha\rangle$, 63
$\boxminus$, 221
$\diamondminus$, 221
$\boxbar$, 221
$\diamondbar$, 221
$\ominus$, 256
$\obar$, 256
$\boxtimes$, 251
$\blacksquare$, 346
$\blacklozenge$, 346
$\blacksquare$, 346
$\blacklozenge$, 346
$a : C$, 67
$\sqcap$, 67
$\sqcup$, 67
$\sqsubseteq$, 68
$\exists R.C$, 67
$\exists_{\geq n} B.C$, 76

LOGICS:
$L \oplus \Delta$, 7, 20
$[L_1, L_2]$, 223
$[L_1, \dots, L_n]$, 378
$L_1 \otimes L_2$, 111, 198
$L_1 \otimes \dots \otimes L_n$, 112
$L_1 \times L_2$, 126, 222
$L_1 \times \dots \times L_n$, 129, 130, 377
$(L_1 \times \dots \times L_n)^{\mathsf{EX}_n}$, 433
$(L_1 \times \dots \times L_n)^{\mathcal{K}}$, 419
$(L_1 \times \dots \times L_n)^{\mathsf{LC}_n}$, 430
$(L_1 \times \dots \times L_n)^{\mathsf{SF}_n}$, 420
L_u, 38

NUMBERS:
$\flat(\varphi)$, 473
$\natural(\varphi)$, 506
$\sharp(\varphi)$, 473
$a(\varphi)$, 210
$cd(x)$, 24
$d(x)$, 739
$d^1(\varphi)$, 210
$d^2(\varphi)$, 210
$\ell(\varphi)$, 31
$md(C)$, 169
$md(\varphi)$, 20, 147, 169
$md(R)$, 169

$rd(C)$, 170
$vmd(\varphi)$, 306

SETS OF FORMULAS:
$3\text{-}th(I)$, 511
$3\text{-}th(r)$, 511
$C_x\varphi$, 540
$\boldsymbol{f}$, 592
$flc(\varphi)$, 287
$sub\,\varphi$, 5, 72, 472, 592
$sub^{\bigcirc}\varphi$, 106
$sub_{\bigcirc\neg}\varphi$, 529
$sub_n\,\varphi$, 472, 529
$sub_x\,\varphi$, 472, 529
$sub\,\Gamma$, 552
$sub_C\,\Gamma$, 552
$sub_C^{\neg}\,\Gamma$, 552
$sub_n\,\Gamma$, 552
$sub_x\,\Gamma$, 552
$\boldsymbol{t}$, 239, 472
$\Theta^{\mathbf{1}}(\varphi)$, 198
$\Theta^{\mathbf{2}}(\varphi)$, 198

MISCELLANOUS:
$\oplus$, 7, 20
$\otimes$, 111, 112
$\times$, 126
£, 255
2^W, 28
$\mathfrak{a}$, 6, 144, 158, 161
$\boldsymbol{A}$, 254
α^*, 62
$\alpha\cup\beta$, 62
$\alpha\,;\beta$, 62
$\mathbf{c}$, 592
$\mathbb{C}$, 83
co$\mathcal{C}$, 33
$con\,\varphi$, 72, 472, 592
$con\,\Gamma$, 552
$down(t)$, 252
$\varphi?$, 62
$\mathfrak{F}^+$, 29
$Fg\,\vartheta$, 617
$G(\mathfrak{G})$, 224
$\mathbb{I}$, 82
Λ, 514
$left(t)$, 252
$\mathfrak{M}^{\Phi}$, 435
$ob\,\varphi$, 72, 592
$\boldsymbol{q}$, 483
R^*, 11
$right(t)$, 252
$rol\,\varphi$, 592
s^*, 490
$s_1 * s_2$, 490
Σ_1^1, 40
$\boldsymbol{\tau}$, 252
(t_1, t_2), 44
$\mathfrak{T}_{\mathfrak{F}}$, 87
$Up\mathfrak{F}$, 94
$up(t)$, 252
$var\varphi$, 216

Subject index

ABox, 68
abstract fmp, 236
accessibility relation, 9
action term, 62
acyclic TBox, 75, 585
agents who do know learn, 138
agents who do not forget, 139
agents who know time, 138
algebra for $\mathbf{IntK}_{\Box}$, 440
algebra for $\mathbf{IntK}_{\Box\Diamond}$, 447
algebra for a logic, 29
algebraic model for $\mathcal{ML}_n$, 28
alternation depth, 210
alternativeness relation, 9
antisymmetric relation, 13
arbitrarily relativized product, 419
arrow frame, 181
pair, 183
square, 183
arrow logic, 182
arrow model, 181
ascending ω-type chain, 249, 344
assignment, 16, 144, 158, 161
in first-order structure, 16
in first-order Kripke model, 144
in first-order temporal model, 158, 161
atom, 4, 201
atomic action, 62
atomic Boolean algebra, 201
atomic formula, 4
atomic program, 62
atomic relation algebra, 401
atomless Boolean algebra, 201
axiom, 6
axiom schema, 18
axiomatization problem, xi

balloon, 49
Barcan formula, 147
converse, 147
basic Q-formula, 661
basic role, 76
basic structure, 237
for $\mathbf{CPDL} \times \mathbf{K}$, 288
for $\mathsf{ELog}_{\mathcal{SU}}(\mathcal{KN}_{L,C})$, 578
for $\mathbf{K} \times \mathbf{K}$, 275
for $\mathbf{K} \times \mathbf{S5}_2$, 335
for $\mathbf{K}_{\mathcal{ALC}}$, 593
for $\mathbf{K4.3} \times \mathbf{S5}$, 326
for $\mathcal{MONC}$, 553
for $\mathcal{QTL}_{\boxed{\mathrm{I}}}$, 483
for $\mathbf{S5} \times \mathbf{S5}$, 239
basic u-formula, 649
binary generated model, 384
Birkhoff variety theorem, 31
block
for $\mathbf{CPDL} \times \mathbf{K}$, 290
for $\mathbf{CPDL} \times \mathbf{S5}$, 320
for $\mathbf{K} \times \mathbf{K}$, 278
for $\mathbf{K}_{\mathcal{ALC}}$, 598
for $\mathbf{K4.3} \times \mathbf{K}$, 314
for $\mathbf{K4.3} \times \mathbf{S5}$, 326
for $\mathsf{Log}\{\langle\mathbb{Q},<\rangle\} \times \mathbf{K}$, 318
blocking, 618, 643
Boolean algebra, 27
atomic, 201
atomless, 201

Boolean logical connective, 4
Boolean region term, 86
Boolean truth-table, 5
Boolean-flat product of models, 435
Boolean-saturation, 239
bound variable, 16
bounded tiling problem, 262
bulldozing, 24

c.i.a.-algebra, 201
calculus, 6
  Hilbert-style, 6
  $\mathcal{MON}$, 528
  $\mathcal{MON}^1$, 536
  $\mathcal{MON}K_n^C$, 550
  $\mathcal{MON}K4_n^C$, 551
  $\mathcal{MON}KD45_n^C$, 551
  $\mathcal{MON}S4_n^C$, 551
  $\mathcal{MON}S5_n^C$, 551
  $\mathcal{MON}T_n^C$, 551
categorial grammar, 187
chain
  ascending ω-type, 249, 344
  infinite ascending, 249
  of clusters, 345
  strictly ascending, 12
character, 512
  legal, 514
  perfect, 514
Church–Rosser property, 222
clash, 618, 631
classical first-order logic, 17
  with equality, 17
classical propositional logic, 5
classical semantics, 5
clique guarded fragment, 479
closed interval, 53
closure operator, 83
cluster, 24
  degenerate, 25
  proper, 25
  simple, 25
co-depth of point in frame, 24
coherent run, 239, 275, 288, 483, 530, 553, 578
common knowledge, 59
common knowledge logic, 137
common knowledge operator, 59
commutativity, 222
  left, 222
  right, 222
commutator, 378
compass relations, 79, 179
completeness, 10
completion rule, 616, 627
completion tree, 624, 631
  clash-free, 631
  complete, 631
complexity class, 32
  2EXPTIME, 32
  ELEM, 32
  EXPSPACE, 32
  EXPTIME, 32
  N2EXPTIME, 33
  NEXPTIME, 32
  NP, 32
  P, 32
  PSPACE, 32
complexity problem, xi
complexity theory, 32
composition, 61
composition of relations, 63
computation of Turing machine, 256
concept, 67
  defined, 75
  satisfiable, 69
concept name, 67
concept satisfiability
  for $\mathcal{ALC}$, 69
  for $L_{\mathcal{ALC}}$, 174
  global, 174
concept type, 72, 592
  named, 592
condensation, 513
connected partial order, 26
connected topological space, 83
consistency, 69
consistent logic, 13
constant domain assumption, 145, 167
constraint, 616, 618
  for $\mathcal{ML}_{\mathcal{ALC}}$, 627
constraint system, 616, 618

clash-free, 618
complete, 616, 618
for $\mathcal{ML}_{ACC}$, 627
converse operator, 64
cubic frame, 429
cubic universal product frame, 130
cyclic TBox, 75
cylindric modal logic, 429

decidable logic, 31
decision problem, xi, 31, 33
decreasing relativized product frame, 432
Dedekind completeness, 53, 361
deduction theorem, 35
degenerate cluster, 25
dense flow of time, 41
dense relation, 41
deontic logic, 8
depth of a partial order, 89
depth of point in frame, 24
derivation, 6
descriptive $\mathbf{IntK}_{\Box\Diamond}$-frame, 449
deterministic algorithm, 32
diagonal constant, 427
diagonal element, 428
diagonal-free cylindric algebra, 379
direct product, 30
discrete flow of time, 41
discrete relation, 41
disjoint union, 26
disjunction property, 95
domain, 16
domain sort, 160
domain term, 160
dovetailing, 199
d-persistent im-logic, 449
dual of Heyting algebra, 448
dual of $\mathbf{IntK}_{\Box\Diamond}$-algebra, 448
dynamic algebra, 63

efmp, 34
epistemic logic, 8, 56
equality, 16
equivalence relation, 11
Euclidean relation, 13
Euclidean space, 83

existential quantifier, 15
expanding relativized product frame, 432
expressive completeness, 45, 162
extra axiom, 7

fibring semantics, 199
filtration, 244, 395
finitary quasimodel, 501
finite change assumption (**FCA**), 120
finite depth property, 394
finite intersection property, 449, 650
finite model property (fmp), 34
finite state assumption (**FSA**)
in first-order temporal model, 502
in product Kripke model, 649
in quasimodel, 502
in topological $\mathcal{PST}$-model, 122, 648
in tt-model, 120
finitely axiomatizable logic, 7
finitely realizable state candidate, 473
first-order definable class of structures, 17
first-order dynamic logic, 153
first-order epistemic logic, 150
first-order intuitionistic logic, 153
first-order Kripke model, 144
intuitionistic, 154
with constant domains, 145
with decreasing domains, 146
with expanding domains, 146, 167
with varying domains, 145, 167
first-order language, 15
first-order modal language, 143
first-order structure, 16
first-order temporal language, 157
two-sorted, 159
first-order temporal logic
of class of flows of time, 158
first-order temporal model, 158
$\mathcal{K}$-model, 481
Fischer–Ladner closure, 287
flat model, 435
flat product of models, 435
flat valuation, 435

flow of time, 41
  dense, 41
  discrete, 41
  homogeneous, 573
fluted fragment, 477
  temporal, 477
fmp, 34
  abstract, 236
  effective, 34
  exponential, 34
  polynomial, 34
  product, 132, 236
fork, 666
fork model, 666
  $\mathcal{V}$-equivalent, 666
formula, 4
  atomic, 4
  classical first-order, 16
  classically valid, 17
  derivable from a set of formulas, 18
  Γ-satisfiable, 10
  Horn, 228
  monadic, 19
  positive, 228
  refuted in Kripke model, 10
  satisfiable in frame, 10
  satisfied in algebraic model, 29
  satisfied in Kripke model, 10
  spatial, 81
  true in algebraic model, 29
  true in first-order structure, 17
  true in Kripke model, 10
  true under assignment, 16
  valid in algebra, 29
  valid in frame, 10
  variable free, 228
formula satisfiability, 174
formula type, 72, 592
  named, 592
frame, 9
  antisymmetric, 13
  connected, 26
  cubic, 429
  cubic universal product, 130
  dense, 41
  discrete, 41
  Euclidean, 13
  for **HS**, 53
  for set of formulas, 10
  functional, 13
  general intuitionistic, 448
  intransitive, 23
  irreflexive, 12
  quasi-ordered, 12
  rooted, 23
  serial, 11
  symmetric, 11
  transitive, 11
  universal, 11
  universal product, 130
  validating formula, 10
  weakly connected, 13, 344
frame formula, 232
free variable, 16
FS-frame, 451
full diagonal-free cylindric set algebra, 379
full $\mathbf{IntK}_{\Box\Diamond}$-frame, 449
functional relation, 13
fusion of frames, 112
fusion of logics, 111

game, 224
general $\mathbf{IntK}_{\Box\Diamond}$-frame, 449
general intuitionistic frame, 448
generalized substitution, 428
generated subframe, 25
generated submodel, 26
global concept satisfiability, 174, 585, 642
  relative to simple, acyclic TBox, 585
global consequence, 35
  determined by countable frames, 36
  determined by finite frames, 36
global instance checking, 174
global Kripke completeness, 36
global role name, 166
global rule, 627
global subsumption, 174

Gödel translation, 95
graded modality, 76
guarded fragment, 478
clique, 479
dynamic, 547
epistemic, 547
loosely, 495
packed, 479
temporal, 478
temporal packed, 479

Halpern–Shoham interval logic, 53
halting problem, 256
Heyting algebra, 93
Hilbert-style calculus, 6
classical first-order, 17
homogeneous flow of time, 573
homomorphic image, 30
homomorphism, 23, 30
Horn axiomatizable logic, 228
Horn formula, 228

im-logic, 189
inconsistent logic, 13
indexed type, 473, 529, 552
individual constant, 15
individual variable, 15
inference rule, 6
inference system, 6
infinite ascending chain, 249
instance checking, 69, 174
global, 174
interior operator, 82
intermediate logic, 93
interpolant, 216
uniform, 216
interpolation
uniform, 216
interpolation property, 216
interval topological model, 125
interval variable, 50
intransitive frame, 23
intransitive tree, 23
intuitionistic Kripke model, 94
first-order, 154
intuitionistic modal logic, 189
intuitionistic propositional logic, 92
intuitionistic provability logic, 189
irreflexive relation, 12
iteration, 61

Kamp's theorem, 45
kernel block, 599
$\mathcal{K}$-model, 481
knowledge base, 68
Kripke completeness, 11
global, 36
Kripke frame, 9
Kripke model, 9, 21
based on a frame, 10
k-run, 275
Kuratowski axioms, 82

Lambek calculus, 187
language
first-order, 15
first-order modal, 143
first-order temporal, 157
monadic second-order, 39
propositional modal, 4
propositional n-modal, 20
propositional spatio-temporal, 122
least upper bound, 361
left commutativity, 222
legal character, 514
length of formula, 31
Lindenbaum algebra, 441
linear order, 26
strict, 26
local consequence, 35
local n-cube, 429
local role name, 166
local rule, 627
logic, 6
characterized by class of frames, 11
classical first-order, 17
classical propositional, 5
common knowledge, 137
consistent, 13
cylindric modal, 429
decidable, 31
deontic, 8

determined by class of algebras, 29
determined by class of frames, 11
epistemic, 8, 56
finitely axiomatizable, 7
first-order dynamic, 153
first-order epistemic, 150
first-order intuitionistic, 153
first-order temporal, 158
Halpern–Shoham interval, 53
Horn axiomatizable, 228
inconsistent, 13
intermediate, 93
intuitionistic modal, 189
intuitionistic propositional, 92
intuitionistic provability, 189
Kripke complete, 11
modal, 7
monadic second-order, 38
multimodal, 20
of class of frames, 10
of class of n-frames, 21
product, 129, 222
propositional dynamic, 61
propositional temporal, 46
provability, 8
superintuitionistic, 93
tabular, 34
temporal epistemic, 137
undecidable, 31
weakly transitive, 393
logic of proofs, 8
logical connective, 4
logical consequence, 35
global, 35
local, 35
of knowledge base, 69
logical constant, 4
logical omniscience, 56
loosely guarded fragment, 495
Löb axiom, 8

Makinson's theorem, 15
marked variable, 625
maximal point (relative to formula), 454
method of quasimodels, 236
minimal deontic logic, 9
minimal modal logic, 7
minimal partial type, 625
modal depth
of $\mathcal{ALC}$-formula, 169
of concept, 169
of $\mathcal{CPDL} \otimes \mathcal{ML}$-formula, 288
of $\mathcal{ML}_n$-formula, 20
of $\mathcal{QML}_n$-formula, 147
of role, 169
modal logic, 7
finitely axiomatizable, 20
of relations, 429
recursively axiomatizable, 20
modal operator, 4
modality *de dicto*, 143
modality *de re*, 143
modalized role, 166
model
binary generated, 384
first-order Kripke, 144
first-order temporal, 158
flat, 435
for $\mathcal{ALC}$, 68
for $\mathcal{ML}_{\mathcal{ALC}}$, 166
interval topological, 125
intuitionistic Kripke, 94
topological, 87
topological $\mathcal{PST}$-, 122, 648
model candidate, 72
modus ponens, 6
moment of time, 41
monadic formula, 19
monadic fragment of $\mathcal{QTL}$, 466
monadic second-order language, 39
monadic second-order logic, 38
monadic second-order theory, 39
monodic formulas, 471
monodic fragment, 471, 548
MP, 6
multimodal logic, 20

named concept type, 592
named formula type, 592
named type, 592

Nash equilibrium, 151
n-cube, 429
necessitation, 7
necessity, 3
negation normal form, 617
negative introspection, 56
network, 224, 421
next-time operator, 44
n-frame, 21
n-modal algebra, 28
n-modal logic, 20
NNF, 617
Noetherian relation, 12
nominal, 77
nondeterministic algorithm, 32
nondeterministic choice, 61

object name, 67
1-depth, 210
open interval, 44

packed fragment, 479
  of $\mathcal{CQDL}^{=}$, 550
  of $\mathcal{QML}_n^{C=}$, 550
  temporal, 479
packing guard, 479
partial order, 13
  connected, 26
  strict, 12
partial type, 625
  minimal, 625
partition of Boolean algebra, 203
path in frame, 24
PCP, 258
perfect character, 514
perfect recall, 139
p-morphic image, 23
p-morphism, 22
point, 9
pointed state candidate, 532, 559
polynomial reduction, 33
positive formula, 228
positive introspection, 56
possibility, 3
possible world semantics, 9
Post's correspondence problem, 258
post-condition, 61
precedence relation, 41
precondition, 61
predecessor, 9
predicate symbol, 15
predicate variable, 39
prime filter, 442
prisoner's dilemma, 150
problem
  $\mathcal{C}$-complete, 33
  $\mathcal{C}$-hard, 33
  Σ_1^1-complete, 40
  Σ_1^1-hard, 40
product finite model property, 132, 236
product fmp, 132, 236
product frame, 126
  decreasing relativized, 432
  expanding relativized, 432
product logic, 129, 222
product of consequence relations, 129
product of frames
  higher-dimensional, 129, 377
  two-dimensional, 126, 222
product of logics
  higher-dimensional, 129, 377
  relativized, 419
  two-dimensional, 126, 222
product-matching logics, 223
proof interpretation, 92
proper cluster, 25
propositional dynamic logic, 61
propositional modal language, 4
propositional n-modal language, 20
propositional spatio-temporal language, 122
propositional temporal logic, 46
propositional variable, 4
provability logic, 8
pseudo-Boolean algebra, 93

$\mathcal{QL}$-reduct, 482
$\mathcal{QL}$-structure, 16
Q-normal form, 661
qualified number restriction, 76
quantifier
  existential, 15
  universal, 15

quasi-order, 11, 12
quasimodel, 238
 finitary, 501
 for $\mathcal{ALC}$, 73
 for **CPDL** × **K**, 289
 for **CPDL** × **S5**, 320
 for $\mathsf{ELog}_{\mathcal{SU}}(\mathcal{KN}_{L,C})$, 579
 for **K** × **K**, 276
 for **K** × $\mathbf{S5}_2$, 336
 for $\mathbf{K}_{\mathcal{ALC}}$, 594
 for $\mathbf{K}_{\mathcal{ALC}}$ tableaux, 628
 for **K4.3** × **K**, 314
 for **K4.3** × **S5**, 326
 for $\mathsf{Log}\{\langle\mathbb{Q},<\rangle\}$ × **K**, 318
 for $\mathcal{MON}$, 530
 for $\mathcal{MONC}$, 553
 for $\mathcal{QTL}_{\boxdot}$, 483
 for $\mathcal{QTL}^{\equiv}_{\boxdot}$, 541
 for **S5** × **S5**, 240
 for $\mathbf{S5}_{\mathcal{ALC}}$, 605
quasisaw, 89, 655
quasistate, 237
 for **CPDL** × **K**, 287
 for **CPDL** × **S5**, 320
 for $\mathsf{ELog}_{\mathcal{SU}}(\mathcal{KN}_{L,C})$, 578
 for **K** × **K**, 274
 for **K** × $\mathbf{S5}_2$, 335
 for $\mathbf{K}_{\mathcal{ALC}}$, 592
 for $\mathbf{K}_{\mathcal{ALC}}$ tableaux, 628
 for **K4.3** × **S5**, 326
 for $\mathcal{QTL}_{\boxdot}$, 483
 for **S5** × **S5**, 239
 for $\mathbf{S5}_{\mathcal{ALC}}$, 605
quasistate candidate
 for **K** × **K**, 274
 for **K** × $\mathbf{S5}_2$, 334
 for $\mathbf{K}_{\mathcal{ALC}}$, 592

realizable state candidate, 473
recurrent Turing machine, 256
reflexive and transitive closure, 11
reflexive relation, 11
region O-term, 117
region connection calculus $\mathcal{RCC}$, 80
region variable, 81
regular closed set, 83
regularity rule, 189
relation algebra, 400
 atomic, 401
 representable, 401
 simple, 401
relational semantics, 9
relativization of algebra, 203
relativized product logic, 419
 arbitrarily, 419
 cubic, 429
 decreasing, 432
 expanding, 432
 locally cubic, 429
representable cylindric algebra, 428
representable diagonal-free cylindric algebra, 379
representable relation algebra, 401
right commutativity, 222
rigid designator, 144, 167
RN, 7
role, 76
 basic, 76
 modalized, 166
 universal, 641
role depth, 170
role inclusion axiom, 78
role name, 67
 global, 166
 local, 166
 transitive, 78
root of frame, 23
rooted frame, 23
run, 238
 for **CPDL** × **K**, 288
 for $\mathsf{ELog}_{\mathcal{SU}}(\mathcal{KN}_{L,C})$, 578
 for **K** × **K**, 275
 for **K** × $\mathbf{S5}_2$, 336
 for $\mathbf{K}_{\mathcal{ALC}}$, 593
 for $\mathbf{K}_{\mathcal{ALC}}$ tableaux, 628
 for **K4.3** × **S5**, 326
 for $\mathcal{MON}$, 530
 for $\mathcal{MONC}$, 553
 for $\mathcal{QTL}_{\boxdot}$, 483
 for $\mathcal{QTL}^{\equiv}_{\boxdot}$, 541
 for **S5** × **S5**, 239

satisfiability problem, 33

satisfying set of blocks, 279
for **CPDL** × **K**, 291
for **CPDL** × **S5**, 320
for $\mathbf{K}_{\mathcal{ALC}}$, 599
for **K4.3** × **K**, 315
for **K4.3** × **S5**, 326
saturated run, 239, 276, 289, 483, 530, 553, 578
Savitch's theorem, 32
saw, 666
saw model, 666
Φ-exhaustive, 667
Scott's normal form, 495
semantic network, 66
sentence, 16
sequencing, 61
serial relation, 11
signature, 15
simple cluster, 25
simple relation algebra, 401
simple TBox, 75, 585
since operator, 44
skeleton subframe, 218
soundness, 10
spatial formula, 81
square, 346
p-, 346, 357
standard **FS**-frame, 191
standard **FS**-model, 191
standard translation, 18, 45
state, 9, 63
state candidate
finitely realizable, 473
for $\mathcal{MON}$, 529
for $\mathcal{MONC}$, 552
for $\mathcal{QTL}_{\boxed{1}}$, 473
for $\mathcal{QTL}^{=}_{\boxed{1}}$, 540
$\mathcal{K}$-realizable, 482, 540
pointed, 532, 559
realizable, 473
state function
for $\mathcal{QTL}_{\boxed{1}}$, 483
for $\mathcal{QTL}^{=}_{\boxed{1}}$, 540
Stone–Jónsson–Tarski representation, 448
Stone representation theorem, 28
strict linear order, 26
strict partial order, 12
strictly ascending chain, 12
subalgebra, 30
subformula, 5
subframe, 25
generated, 25
subframe logic, 419
substitution, 6
Subst, 6
subsumption, 69, 174
global, 174
successor, 9
suitable pair of state candidates
for $\mathcal{MON}$, 530
for $\mathcal{MONC}$, 553
suitable pair of types, 489
for $\mathcal{MON}$, 530
for $\mathcal{MONC}$, 553
superintuitionistic logic, 93
surrogate, 198, 472, 548
symmetric relation, 11
synchronous system, 138
system with perfect recall, 139

tableau algorithm
complete, 617
sound, 617
terminating, 617
tabular logic, 34
TBox, 68, 176, 585
acyclic, 75, 585
cyclic, 75
simple, 75, 585
temporal epistemic logic, 137
temporal epistemic structure, 136
temporal fluted fragment, 477
temporal guarded fragment, 478
temporal interval, 50
temporal region term, 118
temporal sort, 160
term, 16
free for variable, 16
test, 61
theory of first-order structures, 17
3-theory, 511

tile type, 252
tiling, 252
tiling problem, 252
    2^n-corridor, 267
        infinite version, 328
    for $k \times k$, 262
    for $\mathbb{N} \times \mathbb{N}$, 252
topological interpretation of **Int**, 93
topological model, 87
topological $\mathcal{PST}$-model, 122, 648
topological space, 82
    connected, 83
    discrete, 663
    Euclidean, 83
topological temporal model, 115
transition relation, 63
transitive relation, 11
tree, 23
    intransitive, 23
truth-functional operator, 6
truth-relation, 10
    classical first-order, 16
    first-order intuitionistic, 154
    first-order modal, 144
    first-order temporal, 158
    for $\mathcal{ALC}$, 69
    for arrow logic, 182
    for **FS**, 451
    for interval logic, 50
    for intuitionistic modal logic, 190
    for $\mathcal{ML}_{\mathcal{ALC}}$, 167
    for **PDL**, 63
    for $\mathcal{RCC}$-8, 83
    for $\mathcal{ST}_i$, 116
    for standard **FS**-model, 191
    for $\mathcal{TS}$, 161
    intuitionistic, 94
    monadic second-order, 39
    propositional modal, 10
    propositional n-modal, 21
    propositional temporal, 44
    temporal epistemic, 137
truth-set, 10
tt-model, 115
Turing machine, 254
    recurrent, 256
twin, 242, 281, 293, 597
2-depth, 210
type, 237
    for $\mathcal{CPDL} \otimes \mathcal{ML}$, 287, 320
    for $\mathcal{ML}_2$, 239, 274
    for $\mathcal{ML}_{\mathcal{ALC}}$, 592
    for $\mathcal{ML}_{\mathcal{SU}} \otimes \mathcal{ML}$, 578
    for $\mathcal{MON}$, 529
    for $\mathcal{MONC}$, 552
    for $\mathcal{QTL}_{\boxed{1}}$, 472
    for $\mathcal{QTL}^{\equiv}_{\boxed{1}}$, 540
    indexed, 473, 529, 552
    minimal partial, 625
    named, 592

u-formula, 649
    basic, 649
ultrafilter, 650
undecidable logic, 31
uniform interpolant, 216
uniform interpolation, 216
universal Horn sentence, 228
universal modality, 37
universal product frame, 130
    cubic, 130
universal quantifier, 15
universal relation, 11
universal role, 641
universe of algebra, 27
unmarked variable, 625
unraveling, 23
until operator, 44
upward closed set, 26
upward closure, 26

validity problem, 33
valuation, 9, 21
    in modal algebra, 28
value of formula in modal algebra, 28
variable
    bound, 16
    free, 16
    individual, 15
    marked, 625
    unmarked, 625
variable free formula, 228

weakly connected relation, 13, 344
weakly transitive logic, 393
width of a partial order, 89
world, 9

www.ingramcontent.com/pod-product-compliance
Lightning Source LLC
Chambersburg PA
CBHW071307180926
51613CB00001B/148
9780444508263